Plant Physiology
Fourth Edition

Plant Physiology
Fourth Edition

Lincoln Taiz
University of California, Santa Cruz

Eduardo Zeiger
University of California, Los Angeles

Sinauer Associates, Inc., Publishers
Sunderland, Massachusetts

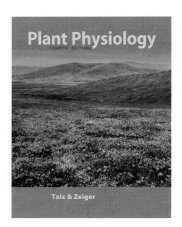

Front Cover
Poppy (*Eschscholzia californica*), goldfield (*Lasthenia californica*), and lupine (*Lupinus* sp.). Photograph © Mel Weinstein.

Back Cover
Clockwise from top left: Scarlet barrel cactus, (*Notocactus haselbergii*), orchid cactus (*Epiphyllum* sp.), primrose (*Primula* sp.), Siberian squill (*Scilla siberica*), fringed polygala (*Polygala paucifolia*), water lily (*Nymphaea* sp.), paper whites (*Narcissus* sp.), butterfly weed (*Asclepsias tuberosa*), yellow iris (*Iris pseudacorus*), passionflower (*Passiflora caerulea*). Photographs by David McIntyre.

Plant Physiology, Fourth Edition
Copyright © 2006 by Sinauer Associates, Inc.
All rights reserved.
This book may not be reprinted in whole or in part without permission from the publisher. For information or to order, address:

Sinauer Associates, Inc.
23 Plumtree Road/PO Box 407
Sunderland, MA 01375 U.S.A.
FAX: 413-549-1118
Email: publish@sinauer.com
www.sinauer.com

Library of Congress Cataloging-in-Publication Data
Taiz, Lincoln.
    Plant physiology / Lincoln Taiz, Eduardo Zeiger.—4th ed.
      p. cm.
    Includes bibliographical references and index.
    ISBN 0-87893-856-7 (hardcover)
Plant physiology. I. Zeiger, Eduardo. II. Title.
QK711.2.T35 2006
571.2—dc22

2006010133

4 3 2 1

Preface

In the preface to the First Edition of *Plant Physiology*, published in 1991, we remarked that during the previous decade "the biological sciences had experienced a period of unprecedented progress, and nowhere is the excitement of this new era more apparent than in the field of plant physiology." The Fourth Edition marks another milestone in the tradition that began 15 years ago. And once again we find ourselves amazed by the stunning progress in plant biology that has occurred since the last edition. Staying abreast of all the new developments has been challenging enough. Doing so without sacrificing either clarity or the appropriate pedagogical level has presented an even more daunting task. From the outset, we realized that to achieve both of these goals in as broad and diverse a field as plant physiology required a group effort. This insight has proved to be correct, and whatever success the book has achieved owes much to the original concept of a multi-author approach.

For the Fourth Edition we are blessed once again with an absolutely outstanding group of contributing authors, each one a leader in his or her respective subfield. Some of them, Bob Blankenship, Dan Cosgrove, and Susan Dunford, have been with us since the First Edition. Rick Amasino, Arnold Bloom, Ray Bressan, John Browse, Bob Buchanan, Paul Hasegawa, Joe Kieber, and Ricardo Wolusiuk, joined us for the Second Edition, and Ruth Finkelstein, Michele Holbrook, Ian Møller, Angus Murphy, and Allan Rasmusson joined us for the Third Edition. We welcome Sarah Assmann, Tom Brutnell, Joanne Chory, James Ehleringer, Jürgen Engelberth, Sigal Savaldi-Goldstein, Valerie Sponsel, and Bruce Veit to our team for the Fourth Edition. In addition, our textbook has greatly benefited from the contributing authors who are no longer with us, but whose scientific insights and pedagogical acumen continue to permeate the book. As in previous editions, we have divided responsibilities for the oversight of the chapters and their integration into a cohesive whole. L.T was in charge of chapters 1, 2, 13–17 and 19–25, while E.Z. was in charge of chapters 3–12, 18 and 26.

Perhaps the most obvious innovation in the Fourth Edition is the new chapter on brassinosteroids. A sufficient body of scientific work has now accumulated on various aspects of brassinosteroid biosynthesis, physiology, and signal transduction to justify incorporation into a chapter of its own. Other changes include a complete rewrite of Chapter 16, *Growth and Development*, which reflects the rapidly evolving progress in this field. Updated material incorporated into other chapters includes recent, important progress in the light reactions and biochemistry of photosynthesis, respiration, ion transport, and water relations. In the hormone chapters, new information on transport mechanisms, signaling pathways, and regulatory networks has been added. One of the most exciting new findings presented in the Fourth Edition is the tentative identification of the long sought-after photoperiodic floral stimulus as a specific macromolecule.

Users of the textbook will continue to benefit from the free-access, fully dedicated website, www.plantphys.net, that has become a landmark in the field of plant physiology in its own right. The website has been extensively revised and updated, making it possible to provide our readers with a wealth of information without lengthening the textbook.

As in past editions, we wish to express our deep appreciation to the outstanding team of professionals assembled by our publisher, Sinauer Associates. We are indebted to our developmental editor, James Funston, for helping us achieve high pedagogic standards; to our production editor, Kathaleen Emerson, for her superb managing skills and her nearly infinite patience; and to Elizabeth Morales for her beautiful and illuminating illustrations. We also thank Maggie Brown, our copy editor, David McIntyre, our photography specialist, Jason Dirks, our Webmaster, Susan McGlew and Marie Scavotto, our marketing specialists, and Chris Small, Joan Gemme, and Sydney Carroll for their expertise.

We offer our most special acknowledgment and gratitude to our publisher, Andy Sinauer, for his commitment to excellence and the highest professional standards. Our thanks also to our reviewers, to those sharp-eyed readers who have taken the time to contact us regarding errors, and to all the instructors and students who have made this textbook a genuine collective endeavor.

Last, but not least, we wish to thank Lee Taiz and Yael Zeiger-Fischmann for their continued enthusiasm, patience, and support.

Lincoln Taiz
Eduardo Zeiger
April 2006

Media & Supplements to Accompany *Plant Physiology*, Fourth Edition

For the Student

Companion Website (www.plantphys.net)

Available free of charge, this website supplements the coverage provided in the textbook with additional and more advanced material on selected topics of interest and current research. The site includes the following:

- **Web Topics:** Additional coverage of selected topics across all chapters
- **Web Essays:** Articles on cutting-edge research, written by the researchers themselves
- **Study Questions:** A set of short answer-style questions for each chapter
- **Suggested Readings:** Chapter-specific recommended readings for further study or research

References to specific Web Topics and Essays are included throughout the printed text, as well as at the end of each chapter. Also included on the website are two textbook chapters that are provided online only: Chapter 2: *Energy and Enzymes* and Chapter 14: *Gene Expression and Signal Transduction*. These chapters are provided as PDF files, and may either be read online or printed.

For the Instructor
(Available to qualified adopters)

Instructor's Resource Library (ISBN 0-87893-858-3)

The *Plant Physiology*, Fourth Edition IRL includes the complete collection of visual resources from the textbook for use in preparing lectures and other course materials. The textbook figures have all been sized, formatted, and color-adjusted for optimal image quality and legibility when projected. The IRL includes:

- All textbook art, photos, and tables in JPEG format, both high-resolution and low-resolution versions
- All textbook art, photos, and tables in PowerPoint® format

The Authors

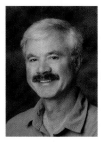
Lincoln Taiz is a Professor of Biology at the University of California at Santa Cruz. He received his Ph.D. in Botany from the University of California at Berkeley in 1971. Dr. Taiz worked for many years on the structure and function of vacuolar H^+-ATPases. He has also worked on plant metal tolerance and the role of flavonoids and amino-peptidases in auxin transport. His current research is on UV-B receptors and their roles in phototropism and stomatal opening.

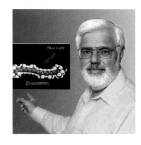
Eduardo Zeiger is a Professor of Biology at the University of California at Los Angeles. He received a Ph.D. in Plant Genetics at the University of California at Davis in 1970. His research interests include stomatal function, the sensory transduction of blue light responses, and the study of stomatal acclimations associated with increases in crop yields.

Principal Contributors

Richard Amasino is a Professor in the Department of Biochemistry at the University of Wisconsin-Madison. He received a Ph.D. in Biology from Indiana University in 1982 in the laboratory of Carlos Miller, where his interests in the induction of flowering were kindled. One of his research interests continues to be the mechanisms by which plants regulate the timing of flower initiation. (Chapter 25)

Sarah M. Assmann is a Professor in the Biology Department at the Pennsylvania State University. She received a Ph.D. in the Biological Sciences at Stanford University. Dr. Assmann's present research focuses on cellular signaling cascades in the regulation of guard cell function, and the roles of heterotrimeric G-proteins in plant growth, plasticity, and environmental response. (Chapter 6)

Robert E. Blankenship is a Professor of Biology and Chemistry at Washingon University in St. Louis. He received his Ph. D. in Chemistry from the University of California at Berkeley in 1975. His professional interests include mechanisms of energy and electron transfer in photosynthetic organisms, and the origin and early evolution of photosynthesis. (Chapter 7)

Arnold J. Bloom is a Professor in the Department of Vegetable Crops at the University of California at Davis. He received a Ph.D. in Biological Sciences at Stanford University in 1979. His research focuses on plant-nitrogen relationships, especially the differences in plant responses to ammonium and nitrate as nitrogen sources. (Chapters 5 & 12)

Ray A. Bressan is a Professor of Plant Physiology at Purdue University. He received a Ph.D. in Plant Physiology from Colorado State University in 1976. Dr. Bressan has studied the basis of salinity and drought tolerance for several years. His recent interests have also turned toward the way plants defend themselves against insects and fungal disease. (Chapter 26)

John Browse is a Professor in the Institute of Biological Chemistry at Washington State University. He received his Ph.D. from the University of Aukland, New Zealand, in 1977. Dr. Browse's research interests include the biochemistry of lipid metabolism and the responses of plants to low temperatures. (Chapter 11)

Thomas Brutnell is an Associate Scientist at the Boyce Thompson Institute for Plant Research at Cornell University. He received his Ph.D. in Biology from Yale University in 1995. His research interests focus on C_4 photosynthesis in maize and dissecting the role of phytochromes in maize and other agronomically important grasses. (Chapter 17)

Bob B. Buchanan is a Professor of Plant and Microbial Biology at the University of California at Berkeley. After working on photosynthesis, Dr. Buchanan turned his attention to seed germination, where his findings have given new insight into germination and led to promising technologies. (Chapter 8)

Authors and Contributors

Joanne Chory is an Investigator with the Howard Hughes Medical Institute and Professor at The Salk Institute for Biological Studies. She is also Adjunct Professor of Biology at the University of California, San Diego. She received a Ph.D. in microbiology from the University of Illinois at Urbana-Champaign. Dr. Chory's research focuses on the mechanisms by which plants respond to changes in their light environment. Her group's genetic studies have led to the identification of the plant steroid receptor and several components in the steroid signaling pathway. (Chapter 24)

Daniel J. Cosgrove is a Professor of Biology at the Pennsylvania State University at University Park. His Ph.D. in Biological Sciences was earned at Stanford University. Dr. Cosgrove's research interest is focused on plant growth, specifically the biochemical and molecular mechanisms governing cell enlargement and cell wall expansion. His research team discovered the cell wall loosening proteins called expansins and is currently studying the structure, function, and evolution of this gene family. (Chapter 15)

Susan Dunford is an Associate Professor of Biological Sciences at the University of Cincinnati. She received her Ph.D. from the University of Dayton in 1973 with a specialization in plant and cell physiology. Dr. Dunford's research interests include long-distance transport systems in plants, especially translocation in the phloem, and plant water relations. (Chapter 10)

James Ehleringer is at the University of Utah where he is a Distinguished Professor of Biology and serves as Director of the Stable Isotope Ratio Facility for Environmental Research (SIRFER). His research focuses on understanding terrestrial ecosystem processes through stable isotope analyses, gas exchange and biosphere-atmosphere interactions, water relations, and stable isotopes applied to homeland security issues. He serves as editor-in-chief for *Oecologia* and chairs the Biosphere-Atmosphere Stable Isotope Network (BASIN). (Chapter 9)

Jurgen Engelberth is an Assistant Professor of Plant Biochemistry at the University of Texas at San Antonio. He received his Ph.D. in Plant Physiology at the Ruhr-University Bochum, Germany in 1995 and did postdoctoral work at the Max Planck Institute for Chemical Ecology, at USDA, ARS, CMAVE in Gainesville, and at Penn State University. His research focuses on signaling involved in plant-insect and plant-plant interaction. (Chapter 13)

Ruth Finkelstein is a Professor in the Department of Molecular, Cellular and Developmental Biology at the University of California at Santa Barbara. She received her Ph.D., also in Molecular, Cellular and Developmental Biology, from Indiana University in 1986. Her research interests include mechanisms of abscisic acid response, and their interactions with other hormonal, environmental, and nutrient signaling pathways. (Chapter 23)

Paul M. Hasegawa is a Professor of Plant Physiology at Purdue University. He earned a Ph.D. in Plant Physiology at the University of California at Riverside. His research has focused on plant morphogenesis and the genetic transformation of plants. He has used his expertise in these areas to study many aspects of stress tolerance in plants, especially ion homeostasis. (Chapter 26)

N. Michele Holbrook is a Professor in the Department of Organismic and Evolutionary Biology at Harvard University. She received her Ph.D. from Stanford University in 1995. Dr. Holbrook's research group focuses on water relations and water transport through the xylem. (Chapters 3 & 4)

Joseph Kieber is an Associate Professor in the Biology Department at the University of North Carolina at Chapel Hill. He earned his Ph.D. in Biology from the Massachusetts Institute of Technology in 1990. Dr. Kieber's research interests include the role of hormones in plant development, with a focus on the signaling pathways for ethylene and ctyokinin, as well as circuitry regulating ethylene biosynthesis. (Chapters 21 & 22)

Robert D. Locy is a Professor of Biological Science at Auburn University in Auburn, Alabama. He received his Ph. D. in Plant Biochemistry from the Purdue University in 1974. His professional interests include biochemical and molecular mechanisms of plant tolerance to abiotic stress, and undergraduate plant science education. (Chapter 26)

Ian Max Møller is a Professor of plant biochemistry at Risø National Laboratory in Denmark. He received his Ph.D. in plant biochemistry from Imperial College, London, U.K. and has worked for a number of years at Lund University, Sweden. Professor Møller has investigated plant respiration throughout his career and his current interests include the respiratory NAD(P)H dehydrogenases, formation of reactive oxygen species and the functional proteomics of plant mitochondria. (Chapter 11)

Angus Murphy is an Assistant Professor in the Department of Horticulture and Landscape Architecture at the Purdue University. He earned his Ph.D. in Biology from the University of California, Santa Cruz in 1996. Dr. Murphy studies the regulation of auxin transport and the mechanisms by which transport proteins are asymmetrically distributed in plant cells. (Chapter 19)

Allan G. Rasmusson is an Associate Professor at Lund University in Sweden. He received his Ph.D. in plant physiology at the same university in 1994. Dr. Rasmusson's current research centers on expressional regulation of respiratory chain enzymes, especially NAD(P)H dehydrogenases, and their physiological significance. (Chapter 11)

Sigal Savaldi-Goldstein is a research associate in Joanne Chory's lab at the Salk Institute. She received her B.Sc. from the Faculty of Life Sciences in Tel Aviv University and her Ph.D. from the Dept. of Plant Sciences at the Weizmann Institute, Israel. Dr. Savaldi-Goldstein's postdoctoral research is specializing in the plant hormone field. Her main interest is to understand the mechanisms by which brassinosteroids coordinate plant growth. (Chapter 24)

Valerie Sponsel is an Associate Professor in the Biology Department at the University of Texas at San Antonio. She received her Ph.D. from the University of Wales, U.K. in 1972 and a D.Sc from the University of Bristol, U.K. in 1984. Her research has focused on the biosynthesis and catabolism of gibberellins, and more recently, on the interaction of auxin with gibberellin biosynthesis and signaling pathways. (Chapter 20)

Bruce Veit is a senior scientist at AgResearch in Palmerston North, New Zealand. He received his Ph.D. in Genetics from University of Washington, Seattle in 1986 before undertaking postdoctoral research at the Plant Gene Expression Center in Albany, California. Dr. Veit's current research interests focus on mechanisms that influence the determination of cell fate. (Chapter 16)

Ricardo A. Wolosiuk is a Professor in the Instituto de Investigaciones Bioquímicas at the University of Buenos Aires. He received his Ph.D. in Chemistry from the same university in 1974. Dr. Wolosiuk's research interests concern the modulation of chloroplast metabolism and the structure and function of plant proteins. (Chapter 8)

Reviewers

Heidi Appel
The Pennsylvania State University

Ivan Baxter
Purdue University

Philip Becraft
Iowa State University

Gerald Berkowitz
University of Connecticut, Storrs

Wade Berry
University of California, Los Angeles

Nick Carpita
Purdue University

Kent Chapman
University of North Texas

Steven Clouse
North Carolina State University

George Coupland
Max Planck Institute for Plant Breeding

Nigel Crawford
University of California, San Diego

Wolf B. Frommer
Stanford University

Larry Griffing
Texas A&M University

Peter Hedden
Rothamsted Research

Steve Huber
University of Illinois

Ross Koning
Eastern Connecticut State University

Steve Long
University of Illinois

Richard Malkin
University of California, Berkeley

Patrick Masson
University of Wisconsin, Madison

David Mok
Oregon State University

Keith Mott
Utah State University

Ana Rus
Purdue University

David Salt
Purdue University

Eric Schaller
Dartmouth College

Julian I. Schroeder
University of California, San Diego

Harry Smith
University of Nottingham

John Sperry
University of Utah

Tai-Ping Sun
Duke University

Gary Tallman
Willamette University

Robert Turgeon
Cornell University

Contents in Brief

   1 *Plant Cells*   1
   2 *Available at www.plantphys.net*   *Energy and Enzymes*   33

UNIT I *Transport and Translocation of Water and Solutes*   35
   3 *Water and Plant Cells*   37
   4 *Water Balance of Plants*   53
   5 *Mineral Nutrition*   73
   6 *Solute Transport*   95

UNIT II *Biochemistry and Metabolism*   123
   7 *Photosynthesis: The Light Reactions*   125
   8 *Photosynthesis: Carbon Reactions*   159
   9 *Photosynthesis: Physiological and Ecological Considerations*   197
 10 *Translocation in the Phloem*   221
 11 *Respiration and Lipid Metabolism*   253
 12 *Assimilation of Mineral Nutrients*   289
 13 *Secondary Metabolites and Plant Defense*   315

UNIT III *Growth and Development*   345
 14 *Available at www.plantphys.net*   *Gene Expression and Signal Transduction*   347
 15 *Cell Walls: Structure, Biogenesis, and Expansion*   349
 16 *Growth and Development*   377
 17 *Phytochrome and Light Control of Plant Development*   417
 18 *Blue-Light Responses: Stomatal Movements and Morphogenesis*   445
 19 *Auxin: The Growth Hormone*   467
 20 *Gibberellins: Regulators of Plant Height and Seed Germination*   509
 21 *Cytokinins: Regulators of Cell Division*   543
 22 *Ethylene: The Gaseous Hormone*   571
 23 *Abscisic Acid: A Seed Maturation and Antistress Signal*   593
 24 *Brassinosteroids*   617
 25 *The Control of Flowering*   635
 26 *Stress Physiology*   671

Table of Contents

Preface v
Authors and Contributors vii

1 Plant Cells 1

Plant Life: Unifying Principles 1

Overview of Plant Structure 2
Plant cells are surrounded by rigid cell walls 2
New cells are produced by dividing tissues called meristems 2
Three major tissue systems make up the plant body 5

The Plant Cell 5
Biological membranes are phospholipid bilayers that contain proteins 6
The nucleus contains most of the genetic material of the cell 8
Protein synthesis involves transcription and translation 11
The endoplasmic reticulum is a network of internal membranes 11
Secretion of proteins from cells begins with the rough ER 11
Golgi stacks produce and distribute secretory products 14
Proteins and polysaccharides destined for secretion are processed in the Golgi apparatus 15
Two models for intra-Golgi transport have been proposed 15
Specific coat proteins facilitate vesicle budding 15
Vacuoles play multiple roles in plant cells 17
Mitochondria and chloroplasts are sites of energy conversion 17
Mitochondria and chloroplasts are semiautonomous organelles 20
Different plastid types are interconvertible 20
Microbodies play specialized metabolic roles in leaves and seeds 21
Oleosomes are lipid-storing organelles 21

The Cytoskeleton 23
Plant cells contain microtubules, microfilaments, and intermediate filaments 23
Microtubules and microfilaments can assemble and disassemble 24
Microtubules function in mitosis and cytokinesis 25
Motor proteins mediate cytoplasmic streaming and organelle movements 26

Cell Cycle Regulation 27
Each phase of the cell cycle has a specific set of biochemical and cellular activities 27
The cell cycle is regulated by cyclin-dependent kinases 27

Plasmodesmata 29
There are two types of plasmodesmata: primary and secondary 29
Plasmodesmata have a complex internal structure 29
Macromolecular traffic through plasmodesmata is important for developmental signaling 29

Summary 30

2 (Available at www.plantphys.net) Energy and Enzymes 33

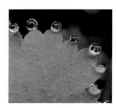

UNIT I

Transport and Translocation of Water and Solutes 35

3 Water and Plant Cells 37

Water in Plant Life 37

The Structure and Properties of Water 38

The polarity of water molecules gives rise to hydrogen bonds 38

The polarity of water makes it an excellent solvent 39

The thermal properties of water result from hydrogen bonding 39

The cohesive and adhesive properties of water are due to hydrogen bonding 39

Water has a high tensile strength 40

Water Transport Processes 41

Diffusion is the movement of molecules by random thermal agitation 41

Diffusion is rapid over short distances but extremely slow over long distances 41

Pressure-driven bulk flow drives long-distance water transport 42

Osmosis is driven by a water potential gradient 42

The chemical potential of water represents the free-energy status of water 43

Three major factors contribute to cell water potential 43

Water enters the cell along a water potential gradient 45

Water can also leave the cell in response to a water potential gradient 45

Small changes in plant cell volume cause large changes in turgor pressure 47

Water transport rates depend on driving force and hydraulic conductivity 48

Aquaporins facilitate the movement of water across cell membranes 49

The water potential concept helps us evaluate the water status of a plant 49

The components of water potential vary with growth conditions and location within the plant 50

Summary 51

4 Water Balance of Plants 53

Water in the Soil 54

A negative hydrostatic pressure in soil water lowers soil water potential 54

Water moves through the soil by bulk flow 55

Water Absorption by Roots 56

Water moves in the root via the apoplast, symplast, and transmembrane pathways 56

Solute accumulation in the xylem can generate "root pressure" 58

Water Transport through the Xylem 59

The xylem consists of two types of tracheary elements 59

Water movement through the xylem requires less pressure than movement through living cells 59

What pressure difference is needed to lift water 100 meters to a treetop? 61

The cohesion–tension theory explains water transport in the xylem 61

Xylem transport of water in trees faces physical challenges 63

Plants minimize the consequences of xylem cavitation 63

Water Movement from the Leaf to the Atmosphere 64

The driving force for water loss is the difference in water vapor concentration 65

Water loss is also regulated by the pathway resistances 65

Stomatal control couples leaf transpiration to leaf photosynthesis 66

The cell walls of guard cells have specialized features 67

An increase in guard cell turgor pressure opens the stomata 68

The transpiration ratio measures the relationship between water loss and carbon gain 69

Overview: The Soil–Plant–Atmosphere Continuum 69

Summary 70

5 Mineral Nutrition 73

Essential Nutrients, Deficiencies, and Plant Disorders 74
 Special techniques are used in nutritional studies 76
 Nutrient solutions can sustain rapid plant growth 77
 Mineral deficiencies disrupt plant metabolism and function 77
 Analysis of plant tissues reveals mineral deficiencies 82

Treating Nutritional Deficiencies 83
 Crop yields can be improved by addition of fertilizers 83
 Some mineral nutrients can be absorbed by leaves 84

Soil, Roots, and Microbes 84
 Negatively charged soil particles affect the adsorption of mineral nutrients 84
 Soil pH affects nutrient availability, soil microbes, and root growth 86
 Excess minerals in the soil limit plant growth 86
 Plants develop extensive root systems 86
 Root systems differ in form but are based on common structures 87
 Different areas of the root absorb different mineral ions 88
 Mycorrhizal fungi facilitate nutrient uptake by roots 89
 Nutrients move from the mycorrhizal fungi to the root cells 91

Summary 91

6 Solute Transport 95

Passive and Active Transport 96

Transport of Ions across a Membrane Barrier 98
 Different diffusion rates for cations and anions produce diffusion potentials 98
 How does membrane potential relate to ion distribution? 98
 The Nernst equation distinguishes between active and passive transport 100
 Proton transport is a major determinant of the membrane potential 101

Membrane Transport Processes 101
 Channel transporters enhance diffusion across membranes 103
 Carriers bind and transport specific substances 105
 Primary active transport requires energy 105
 Secondary active transport uses stored energy 106
 Kinetic analyses can elucidate transport mechanisms 106

Membrane Transport Proteins 108
 The genes for many transporters have been identified 108
 Transporters exist for diverse nitrogen-containing compounds 110
 Cation transporters are diverse 110
 Some anion transporters have been identified 112
 Metals are transported by ZIP proteins 112
 Aquaporins may have novel functions 113
 The plasma membrane H^+-ATPase has several functional domains 113
 The tonoplast H^+-ATPase drives solute accumulation into vacuoles 114
 H^+-pyrophosphatases also pump protons at the tonoplast 116

Ion Transport in Roots 116
 Solutes move through both apoplast and symplast 117
 Ions cross both symplast and apoplast 117
 Xylem parenchyma cells participate in xylem loading 118

Summary 119

UNIT II
Biochemistry and Metabolism 123

7 Photosynthesis: The Light Reactions 125

Photosynthesis in Higher Plants 126

General Concepts 126
- Light has characteristics of both a particle and a wave 126
- When molecules absorb or emit light, they change their electronic state 127
- Photosynthetic pigments absorb the light that powers photosynthesis 128

Key Experiments in Understanding Photosynthesis 130
- Action spectra relate light absorption to photosynthetic activity 130
- Photosynthesis takes place in complexes containing light-harvesting antennas and photochemical reaction centers 130
- The chemical reaction of photosynthesis is driven by light 132
- Light drives the reduction of NADP and the formation of ATP 132
- Oxygen-evolving organisms have two photosystems that operate in series 133

Organization of the Photosynthetic Apparatus 134
- The chloroplast is the site of photosynthesis 134
- Thylakoids contain integral membrane proteins 135
- Photosystems I and II are spatially separated in the thylakoid membrane 137
- Anoxygenic photosynthetic bacteria have a single reaction center 137

Organization of Light-Absorbing Antenna Systems 137
- The antenna funnels energy to the reaction center 138
- Many antenna complexes have a common structural motif 138

Mechanisms of Electron Transport 139
- Electrons from chlorophyll travel through the carriers organized in the "Z scheme" 139
- Energy is captured when an excited chlorophyll reduces an electron acceptor molecule 140
- The reaction center chlorophylls of the two photosystems absorb at different wavelengths 142
- The photosystem II reaction center is a multisubunit pigment–protein complex 142
- Water is oxidized to oxygen by photosystem II 142
- Pheophytin and two quinones accept electrons from photosystem II 143
- Electron flow through the cytochrome $b_6 f$ complex also transports protons 144
- Plastoquinone and plastocyanin carry electrons between photosystems II and I 146
- The photosystem I reaction center reduces $NADP^+$ 146
- Cyclic electron flow generates ATP but no NADPH 147
- Some herbicides block photosynthetic electron flow 147

Proton Transport and ATP Synthesis in the Chloroplast 148

Repair and Regulation of the Photosynthetic Machinery 151
- Carotenoids serve as photoprotective agents 151
- Some xanthophylls also participate in energy dissipation 152
- The photosystem II reaction center is easily damaged 152
- Photosystem I is protected from active oxygen species 153
- Thylakoid stacking permits energy partitioning between the photosystems 153

Genetics, Assembly, and Evolution of Photosynthetic Systems 153
- Chloroplast, cyanobacterial, and nuclear genomes have been sequenced 153
- Chloroplast genes exhibit non-Mendelian patterns of inheritance 153
- Many chloroplast proteins are imported from the cytoplasm 154
- The biosynthesis and breakdown of chlorophyll are complex pathways 154
- Complex photosynthetic organisms have evolved from simpler forms 154

Summary 156

8 Photosynthesis: Carbon Reactions 159

The Calvin Cycle 160
- The Calvin cycle has three stages: carboxylation, reduction, and regeneration 160
- The carboxylation of ribulose-1,5-bisphosphate is catalyzed by the enzyme rubisco 161
- Operation of the Calvin cycle requires the regeneration of ribulose-1,5-bisphosphate 163
- The Calvin cycle regenerates its own biochemical components 164
- The Calvin cycle uses energy very efficiently 164

Regulation of the Calvin Cycle 165
- Light regulates the Calvin cycle 165
- The activity of rubisco increases in the light 166
- The ferredoxin–thioredoxin system regulates the Calvin cycle 166
- Light-dependent ion movements regulate Calvin cycle enzymes 168

The C_2 Oxidative Photosynthetic Carbon Cycle 168
- Photosynthetic CO_2 fixation and photorespiratory oxygenation are competing reactions 168
- Photorespiration depends on the photosynthetic electron transport system 171
- The biological function of photorespiration is under investigation 171

CO_2-Concentrating Mechanisms 172

I. CO_2 and HCO_3^- Pumps 173

II. The C_4 Carbon Cycle 173
- Malate and aspartate are carboxylation products of the C_4 cycle 173
- Two different types of cells participate in the C_4 cycle 174
- The C_4 cycle concentrates CO_2 in the chloroplasts of bundle sheath cells 176
- The C_4 cycle also concentrates CO_2 in single cells 178
- The C_4 cycle has higher energy demand than the Calvin cycle 179
- Light regulates the activity of key C_4 enzymes 179
- In hot, dry climates, the C_4 cycle reduces photorespiration and water loss 180

III. Crassulacean Acid Metabolism (CAM) 180
- The stomata of CAM plants open at night and close during the day 180
- Some CAM plants change the pattern of CO_2 uptake in response to environmental conditions 180

Starch and Sucrose 182
- Chloroplast starch is synthesized during the day and degraded at night 183
- Starch is synthesized in the chloroplast 183
- Starch degradation requires phosphorylation of amylopectin 183
- Triose phosphates synthesized in the chloroplast build up the pool of hexose phosphates in the cytosol 186
- Fructose-6-phosphate can be converted to fructose-1,6-bisphosphate by two different enzymes 190
- Fructose-2,6-bisphosphate is an important regulatory compound 190
- The hexose phosphate pool is regulated by fructose-2,6-bisphosphate 190
- Sucrose is continuously synthesized in the cytosol 191

Summary 192

9 Photosynthesis: Physiological and Ecological Considerations 197

Light, Leaves, and Photosynthesis 198

Units in the Measurement of Light 199
- Leaf anatomy maximizes light absorption 200
- Plants compete for sunlight 201
- Leaf angle and leaf movement can control light absorption 202
- Plants acclimate and adapt to sun and shade 203

Photosynthetic Responses to Light by the Intact Leaf 203
- Light-response curves reveal photosynthetic properties 204
- Leaves must dissipate excess light energy 206
- Absorption of too much light can lead to photoinhibition 208

Photosynthetic Responses to Temperature 209
- Leaves must dissipate vast quantities of heat 209
- Photosynthesis is temperature sensitive 210

Photosynthetic Responses to Carbon Dioxide 211
- Atmospheric CO_2 concentration keeps rising 211
- CO_2 diffusion to the chloroplast is essential to photosynthesis 212
- Patterns of light absorption generate gradients of CO_2 fixation 213
- CO_2 imposes limitations on photosynthesis 214

Crassulacean Acid Metabolism 216
- Carbon isotope ratio variations reveal different photosynthetic pathways 216
- How do we measure the carbon isotopes of plants? 216
- Why are there carbon isotope ratio variations in plants? 217

Summary 218

10 Translocation in the Phloem 221

Pathways of Translocation 222
　Sugar is translocated in phloem sieve elements 222
　Mature sieve elements are living cells specialized for translocation 223
　Large pores in cell walls are the prominent feature of sieve elements 224
　Damaged sieve elements are sealed off 224
　Companion cells aid the highly specialized sieve elements 225

Patterns of Translocation: Source to Sink 227
　Source-to-sink pathways follow anatomic and developmental patterns 227

Materials Translocated in the Phloem 228
　Phloem sap can be collected and analyzed 229
　Sugars are translocated in nonreducing form 229

Rates of Movement 231

The Pressure-Flow Model for Phloem Transport 231
　A pressure gradient drives translocation in the pressure-flow model 231
　The predictions of mass flow have been confirmed 232
　Sieve plate pores are open channels 233
　There is no bidirectional transport in single sieve elements 233
　The energy requirement for transport through the phloem pathway is small 233
　Pressure gradients are sufficient to drive a mass flow of phloem sap 233
　Significant questions about the pressure-flow model still exist 235

Phloem Loading 235
　Phloem loading can occur via the apoplast or symplast 236
　Sucrose uptake in the apoplastic pathway requires metabolic energy 237
　Phloem loading in the apoplastic pathway involves a sucrose–H$^+$ symporter 237
　Phloem loading is symplastic in plants with intermediary cells 239
　The polymer-trapping model explains symplastic loading 239
　The type of phloem loading is correlated with plant family and with climate 239

Phloem Unloading and Sink-to-Source Transition 241
　Phloem unloading and short-distance transport can occur via symplastic or apoplastic pathways 241
　Transport into sink tissues requires metabolic energy 242
　The transition of a leaf from sink to source is gradual 242

Photosynthate Distribution: Allocation and Partitioning 244
　Allocation includes storage, utilization, and transport 244
　Various sinks partition transport sugars 244
　Source leaves regulate allocation 245
　Sink tissues compete for available translocated photosynthate 246
　Sink strength depends on sink size and activity 246
　The source adjusts over the long term to changes in the source-to-sink ratio 246

The Transport of Signaling Molecules 247
　Turgor pressure and chemical signals coordinate source and sink activities 247
　Signal molecules in the phloem regulate growth and development 247

Summary 250

11 Respiration and Lipid Metabolism 253

Overview of Plant Respiration 253

Glycolysis: A Cytosolic and Plastidic Process 256
　Glycolysis converts carbohydrates into pyruvate, producing NADH and ATP 256
　Plants have alternative glycolytic reactions 257
　In the absence of O_2, fermentation regenerates the NAD$^+$ needed for glycolysis 259
　Fermentation does not liberate all the energy available in each sugar molecule 259
　Plant glycolysis is controlled by its products 260
　The pentose phosphate pathway produces NADPH and biosynthetic intermediates 260

The Citric Acid Cycle: A Mitochondrial Matrix Process 262
　Mitochondria are semiautonomous organelles 262
　Pyruvate enters the mitochondrion and is oxidized via the citric acid cycle 263
　The citric acid cycle of plants has unique features 265

Mitochondrial Electron Transport and ATP Synthesis 265
　The electron transport chain catalyzes a flow of electrons from NADH to O_2 266
　Some electron transport enzymes are unique to plant mitochondria 266

ATP synthesis in the mitochondrion is coupled to electron transport 268
Transporters exchange substrates and products 269
Aerobic respiration yields about 60 molecules of ATP per molecule of sucrose 271
Several subunits of respiratory complexes are encoded by the mitochondrial genome 271
Plants have several mechanisms that lower the ATP yield 272
Mitochondrial respiration is controlled by key metabolites 273
Respiration is tightly coupled to other pathways 274

Respiration in Intact Plants and Tissues 274
Plants respire roughly half of the daily photosynthetic yield 274
Respiration operates during photosynthesis 275
Different tissues and organs respire at different rates 276
Mitochondrial function is crucial during pollen development 276
Environmental factors alter respiration rates 277

Lipid Metabolism 278
Fats and oils store large amounts of energy 278
Triacylglycerols are stored in oil bodies 278
Polar glycerolipids are the main structural lipids in membranes 279
Fatty acid biosynthesis consists of cycles of two-carbon addition 279
Glycerolipids are synthesized in the plastids and the ER 282
Lipid composition influences membrane function 283
Membrane lipids are precursors of important signaling compounds 283
Storage lipids are converted into carbohydrates in germinating seeds 283

Summary 285

12 Assimilation of Mineral Nutrients 289

Nitrogen in the Environment 290
Nitrogen passes through several forms in a biogeochemical cycle 290
Unassimilated ammonium or nitrate may be dangerous 291

Nitrate Assimilation 292
Many factors regulate nitrate reductase 293
Nitrite reductase converts nitrite to ammonium 293
Both roots and shoots assimilate nitrate 294

Ammonium Assimilation 294
Converting ammonium to amino acids requires two enzymes 294
Ammonium can be assimilated via an alternative pathway 296
Transamination reactions transfer nitrogen 296
Asparagine and glutamine link carbon and nitrogen metabolism 296

Amino Acid Biosynthesis 296

Biological Nitrogen Fixation 296
Free-living and symbiotic bacteria fix nitrogen 297
Nitrogen fixation requires anaerobic conditions 297
Symbiotic nitrogen fixation occurs in specialized structures 299
Establishing symbiosis requires an exchange of signals 300
Nod factors produced by bacteria act as signals for symbiosis 300
Nodule formation involves phytohormones 301
The nitrogenase enzyme complex fixes N_2 301
Amides and ureides are the transported forms of nitrogen 303

Sulfur Assimilation 304
Sulfate is the absorbed form of sulfur in plants 304
Sulfate assimilation requires the reduction of sulfate to cysteine 305
Sulfate assimilation occurs mostly in leaves 305
Methionine is synthesized from cysteine 305

Phosphate Assimilation 306

Cation Assimilation 306
Cations form noncovalent bonds with carbon compounds 306
Roots modify the rhizosphere to acquire iron 306
Iron forms complexes with carbon and phosphate 308

Oxygen Assimilation 308

The Energetics of Nutrient Assimilation 310

Summary 311

13 Secondary Metabolites and Plant Defense 315

Cutin, Waxes, and Suberin 316
  Cutin, waxes, and suberin are made up of hydrophobic compounds 316
  Cutin, waxes, and suberin help reduce transpiration and pathogen invasion 317

Secondary Metabolites 317
  Secondary metabolites defend plants against herbivores and pathogens 318
  Secondary metabolites are divided into three major groups 318

Terpenes 318
  Terpenes are formed by the fusion of five-carbon isoprene units 318
  There are two pathways for terpene biosynthesis 318
  Isopentenyl diphosphate and its isomer combine to form larger terpenes 319
  Some terpenes have roles in growth and development 319
  Terpenes defend against herbivores in many plants 321

Phenolic Compounds 322
  Phenylalanine is an intermediate in the biosynthesis of most plant phenolics 322
  Some simple phenolics are activated by ultraviolet light 323
  The release of phenolics into the soil may limit the growth of other plants 324
  Lignin is a highly complex phenolic macromolecule 325
  There are four major groups of flavonoids 326
  Anthocyanins are colored flavonoids that attract animals 326
  Flavonoids may protect against damage by ultraviolet light 327
  Isoflavonoids have antimicrobial activity 328
  Tannins deter feeding by herbivores 328

Nitrogen-Containing Compounds 329
  Alkaloids have dramatic physiological effects on animals 329
  Cyanogenic glycosides release the poison hydrogen cyanide 332
  Glucosinolates release volatile toxins 332
  Nonprotein amino acids defend against herbivores 333

Induced Plant Defenses against Insect Herbivores 334
  Plants can recognize specific components of insect saliva 334
  Jasmonic acid is a plant hormone that activates many defense responses 335
  Some plant proteins inhibit herbivore digestion 336
  Herbivore damage induces systemic defenses 336
  Herbivore-induced volatiles have complex ecological functions 336

Plant Defense against Pathogens 338
  Some antimicrobial compounds are synthesized before pathogen attack 338
  Infection induces additional antipathogen defenses 338
  Some plants recognize specific substances released from pathogens 340
  Exposure to elicitors induces a signal transduction cascade 340
  A single encounter with a pathogen may increase resistance to future attacks 340

Summary 341

UNIT III
Growth and Development 345

14 (Available at www.plantphys.net) Gene Expression and Signal Transduction 347

15 Cell Walls: Structure, Biogenesis, and Expansion 349

The Structure and Synthesis of Plant Cell Walls 350
   Plant cell walls have varied architecture 350
   The primary cell wall is composed of cellulose microfibrils embedded in a polysaccharide matrix 351
   Cellulose microfibrils are synthesized at the plasma membrane 353
   Matrix polymers are synthesized in the Golgi and secreted via vesicles 357
   Hemicelluloses are matrix polysaccharides that bind to cellulose 357
   Pectins are gel-forming components of the matrix 357
   Structural proteins become cross-linked in the wall 361
   New primary walls are assembled during cytokinesis 362
   Secondary walls form in some cells after expansion ceases 363

Patterns of Cell Expansion 364
   Microfibril orientation influences growth directionality of cells with diffuse growth 364
   Cortical microtubules influence the orientation of newly deposited microfibrils 366

The Rate of Cell Elongation 368
   Stress relaxation of the cell wall drives water uptake and cell elongation 368
   The rate of cell expansion is governed by two growth equations 368
   Acid-induced growth is mediated by expansins 370
   Glucanases and other hydrolytic enzymes may modify the matrix 371
   Structural changes accompany the cessation of wall expansion 372

Wall Degradation and Plant Defense 372
   Enzymes mediate wall hydrolysis and degradation 372
   Oxidative bursts accompany pathogen attack 372
   Wall fragments can act as signaling molecules 373

Summary 373

16 Growth and Development 377

Overview of Plant Growth and Development 378
   Sporophytic development can be divided into three major stages 378
   Development can be analyzed at the molecular level 381

Embryogenesis: The Origins of Polarity 382
   The pattern of embryogenesis differs in dicots and monocots 382
   The axial polarity of the plant is established by the embryo 384
   Position-dependent signaling guides embryogenesis 384
   Auxin may function as a morphogen during embryogenesis 386
   Genes control apical–basal patterning 387
   Embryogenesis genes have diverse biochemical functions 388
   MONOPTEROS activity is inhibited by a repressor protein 388
   Gene expression patterns correlate with auxin 389
   GNOM gene determines the distribution of efflux proteins 389
   Radial patterning establishes fundamental tissue layers 389
   Two genes regulate protoderm differentiation 392
   Cytokinin stimulates cell divisions for vascular elements 392
   Two genes control the differentiation of cortical and endodermal tissues through intercellular communication 393
   Intercellular communication is central to plant development 394

Shoot Apical Meristem 396
   The shoot apical meristem forms at a position where auxin is low 396
   Forming an embryonic SAM requires many genes 398
   Shoot apical meristems vary in size and shape 398
   The shoot apical meristem contains distinct zones and layers 398
   Groups of relatively stable initial cells have been identified 398
   SAM function may require intercellular protein movement 400
   Protein turnover may spatially restrict gene activity 400
   Stem cell population is maintained by a transcriptional feedback loop 400

Root Apical Meristem 402
   High auxin levels stimulate the formation of the root apical meristem 402
   The root tip has four developmental zones 403
   Specific root initials produce different root tissues 404
   Root apical meristems contain several types of initials 404

Vegetative Organogenesis 405

Periclinal cell divisions initiate leaf primordia 406
Local auxin concentrations in the SAM control leaf initiation 406
Three developmental axes describe the leaf's planar form 407
Spatially regulated gene expression controls leaf pattern 407
MicroRNAs regulate the sidedness of the leaf 409

Branch roots and shoots have different origins 409

Senescence and Programmed Cell Death 410
Plants exhibit various types of senescence 410
Senescence involves ordered cellular and biochemical changes 411
Programmed cell death is a specialized type of senescence 412

Summary 412

17 Phytochrome and Light Control of Plant Development 417

The Photochemical and Biochemical Properties of Phytochrome 418
Phytochrome can interconvert between Pr and Pfr forms 419
Pfr is the physiologically active form of phytochrome 420

Characteristics of Phytochrome-Induced Responses 420
Phytochrome responses vary in lag time and escape time 420
Phytochrome responses can be distinguished by the amount of light required 421
Very low–fluence responses are nonphotoreversible 421
Low-fluence responses are photoreversible 422
High-irradiance responses are proportional to the irradiance and the duration 422

Structure and Function of Phytochrome Proteins 423
Phytochrome has several important functional domains 424
Phytochrome is a light-regulated protein kinase 425
Pfr is partitioned between the cytosol and nucleus 425
Phytochromes are encoded by a multigene family 425

Genetic Analysis of Phytochrome Function 427
Phytochrome A mediates responses to continuous far-red light 427
Phytochrome B mediates responses to continuous red or white light 428
Roles for phytochromes C, D, and E are emerging 428
Phy gene family interactions are complex 428

PHY gene functions have diversified during evolution 428

Phytochrome Signaling Pathways 430
Phytochrome regulates membrane potentials and ion fluxes 430
Phytochrome regulates gene expression 430
Phytochrome interacting factors (PIFs) act early in phy signaling 430
Phytochrome associates with protein kinases and phosphatases 431
Phytochrome-induced gene expression involves protein degradation 432

Circadian Rhythms 433
The circadian oscillator involves a transcriptional negative feedback loop 433

Ecological Functions 435
Phytochrome regulates the sleep movements of leaves 435
Phytochrome enables plant adaptation to light quality changes 437
Decreasing the R:FR ratio causes elongation in sun plants 437
Small seeds typically require a high R:FR ratio for germination 438
Phytochrome interactions are important early in germination 439
Reducing shade avoidance responses can improve crop yields 439
Phytochrome responses show ecotypic variation 440
Phytochrome action can be modulated 440

Summary 440

18 Blue-Light Responses: Stomatal Movements and Morphogenesis 445

The Photophysiology of Blue-Light Responses 446
Blue light stimulates asymmetric growth and bending 446
How do plants sense the direction of the light signal? 448

Blue light rapidly inhibits stem elongation 448
Blue light regulates gene expression 448
Blue light stimulates stomatal opening 449
Blue light activates a proton pump at the guard cell plasma membrane 451

Blue-light responses have characteristic kinetics and lag times   452
Blue light regulates osmotic relations of guard cells   453
Sucrose is an osmotically active solute in guard cells   453

Blue-Light Photoreceptors   455
Cryptochromes are involved in the inhibition of stem elongation   455
Phototropins mediate blue light–dependent phototropism and chloroplast movements   456
The carotenoid zeaxanthin mediates blue-light photoreception in guard cells   457
Green light reverses blue light-stimulated opening   461
The xanthophyll cycle confers plasticity to the stomatal responses to light   462

Summary   462

19 Auxin: The Growth Hormone   467

The Emergence of the Auxin Concept   468

Identification, Biosynthesis, and Metabolism of Auxin   468
The principal auxin in higher plants is indole-3-acetic acid   470
IAA is synthesized in meristems and young dividing tissues   471
Multiple pathways exist for the biosynthesis of IAA   471
IAA can also be synthesized from indole-3-glycerol phosphate   472
Seeds and storage organs contain large amounts of covalently bound auxin   474
IAA is degraded by multiple pathways   475
IAA partitions between the cytosol and the chloroplasts   475

Auxin Transport   476
Polar transport requires energy and is gravity independent   476
A chemiosmotic model has been proposed to explain polar transport   477
P-glycoproteins are also auxin transport proteins   480
Inhibitors of auxin transport block auxin influx and efflux   482
Auxin is also transported nonpolarly in the phloem   482
Auxin transport is regulated by multiple mechanisms   483
Polar auxin transport is required for development   484

Actions of Auxin: Cell Elongation   484
Auxins promote growth in stems and coleoptiles, while inhibiting growth in roots   484
The outer tissues of dicot stems are the targets of auxin action   485
The minimum lag time for auxin-induced growth is ten minutes   485
Auxin rapidly increases the extensibility of the cell wall   486
Auxin-induced proton extrusion increases cell extension   486
Auxin-induced proton extrusion may involve both activation and synthesis   487

Actions of Auxin: Phototropism and Gravitropism   488
Phototropism is mediated by the lateral redistribution of auxin   488
Gravitropism involves lateral redistribution of auxin   490
Dense plastids serve as gravity sensors   490
Gravity sensing may involve pH and calcium as second messengers   492
Auxin is redistributed laterally in the root cap   494

Developmental Effects of Auxin   496
Auxin regulates apical dominance   496
Auxin transport regulates floral bud development and phyllotaxy   498
Auxin promotes the formation of lateral and adventitious roots   498
Auxin induces vascular differentiation   499
Auxin delays the onset of leaf abscission   500
Auxin promotes fruit development   500
Synthetic auxins have a variety of commercial uses   500

Auxin Signal Transduction Pathways   501
A ubiquitin E_3 ligase subunit is an auxin receptor   501
Auxin-induced genes are negatively regulated by AUX/IAA proteins   502
Auxin binding to SCF^{TIR1} stimulates AUX/IAA destruction   502
Auxin-induced genes fall into two classes: early and late   502
Rapid auxin responses may involve a different receptor protein   503

Summary   504

20 Gibberellins: Regulators of Plant Height and Seed Germination   509

Gibberellins: Their Discovery and Chemical Structure   510
Gibberellins were discovered by studying a disease of rice   510
Gibberellic acid was first purified from *Gibberella* culture filtrates   510
All gibberellins are based on an *ent*-gibberellane skeleton   511

Effects of Gibberellins on Growth and Development   512
Gibberellins can stimulate stem growth   512
Gibberellins regulate the transition from juvenile to adult phases   512
Gibberellins influence floral initiation and sex determination   513
Gibberellins promote pollen development and tube growth   513
Gibberellins promote fruit set and parthenocarpy   514
Gibberellins promote seed development and germination   514
Commercial uses of gibberellins and GA biosynthesis inhibitors   514

Biosynthesis and Catabolism of Gibberellins   514
Gibberellins are synthesized via the terpenoid pathway   515
Some enzymes in the GA pathway are highly regulated   515
Gibberellin regulates its own metabolism   518
GA biosynthesis occurs at multiple cellular sites   518
Environmental conditions can influence GA biosynthesis   519
GA_1 and GA_4 have intrinsic bioactivity for stem growth   519
Plant height can be genetically engineered   521
Dwarf mutants often have other defects in addition to dwarfism   521

Gibberellin Signaling: Significance of Response Mutants   522
Mutations of negative regulators of GA may produce slender or dwarf phenotypes   522
Negative regulators with DELLA domains have agricultural importance   523
Gibberellins signal the degradation of transcriptional repressors   524
F-box proteins target DELLA domain proteins for degradation   524
A possible GA receptor has been identified in rice   526

Gibberellin Responses: The Cereal Aleurone Layer   527
GA is synthesized in the embryo   528
Aleurone cells may have two types of GA receptors   529
GA signaling requires several second messengers   529
Gibberellins enhance the transcription of α-amylase mRNA   531
GAMYB is a positive regulator of α-amylase transcription   532
DELLA domain proteins are rapidly degraded   532

Gibberellin Responses: Flowering in Long-Day Plants   533
There are multiple independent pathways to flowering   533
The long day and gibberellin pathways interact   533
GAMYB regulates flowering and male fertility   534
MicroRNAs regulate MYBs after transcription   535

Gibberellin Responses: Stem Growth   535
The shoot apical meristem interior lacks bioactive GA   535
Gibberellins stimulate cell elongation and cell division   535
GAs regulate the transcription of cell cycle kinases   537
Auxin promotes GA biosynthesis and signaling   537

Summary   538

21 Cytokinins: Regulators of Cell Division   543

Cell Division and Plant Development   544
Differentiated plant cells can resume division   544
Diffusible factors may control cell division   544
Plant tissues and organs can be cultured   544

The Discovery, Identification, and Properties of Cytokinins   545
Kinetin was discovered as a breakdown product of DNA   545
Zeatin was the first natural cytokinin discovered   545
Some synthetic compounds can mimic or antagonize cytokinin action   546
Cytokinins occur in both free and bound forms   547
The hormonally active cytokinin is the free base   547
Some plant pathogenic bacteria, fungi, insects, and nematodes secrete free cytokinins   547

Biosynthesis, Metabolism, and Transport of Cytokinins 548
  Crown gall cells have acquired a gene for cytokinin synthesis 548
  IPT catalyzes the first step in cytokinin biosynthesis 551
  Cytokinins from the root are transported to the shoot via the xylem 551
  A signal from the shoot regulates the transport of zeatin ribosides from the root 552
  Cytokinins are rapidly metabolized by plant tissues 552

The Biological Roles of Cytokinins 552
  Cytokinins regulate cell division in shoots and roots 553
  Cytokinins regulate specific components of the cell cycle 555
  The auxin:cytokinin ratio regulates morphogenesis in cultured tissues 556
  Cytokinins modify apical dominance and promote lateral bud growth 557
  Cytokinins induce bud formation in a moss 557
  Cytokinin overproduction has been implicated in genetic tumors 558
  Cytokinins delay leaf senescence 559
  Cytokinins promote movement of nutrients 559
  Cytokinins promote chloroplast development 560
  Cytokinins promote cell expansion in leaves and cotyledons 561
  Cytokinin-regulated processes are revealed in plants that overproduce cytokinins 561

Cellular and Molecular Modes of Cytokinin Action 563
  A cytokinin receptor related to bacterial two-component receptors has been identified 563
  Cytokinins increase expression of the type-A response regulator genes via activation of the type-B *ARR* genes 564
  Histidine phosphotransferases are also involved in cytokinin signaling 565

Summary 567

22 Ethylene: The Gaseous Hormone 571

Structure, Biosynthesis, and Measurement of Ethylene 572
  The properties of ethylene are deceptively simple 572
  Bacteria, fungi, and plant organs produce ethylene 572
  Regulated biosynthesis determines the physiological activity of ethylene 573
  Environmental stresses and auxins promote ethylene biosynthesis 574
  Ethylene biosynthesis can be stimulated by ACC synthase stabilization 575
  Ethylene biosynthesis and action can be blocked by inhibitors 575
  Ethylene can be measured by gas chromatography 576

Developmental and Physiological Effects of Ethylene 576
  Ethylene promotes the ripening of some fruits 576
  Leaf epinasty results when ACC from the root is transported to the shoot 577
  Ethylene induces lateral cell expansion 579
  The hooks of dark-grown seedlings are maintained by ethylene production 580
  Ethylene breaks seed and bud dormancy in some species 580
  Ethylene promotes the elongation growth of submerged aquatic species 580
  Ethylene induces the formation of roots and root hairs 581
  Ethylene induces flowering in the pineapple family 581
  Ethylene enhances the rate of leaf senescence 581
  Some defense responses are mediated by ethylene 582
  Ethylene regulates changes in the abscission layer that cause abscission 582
  Ethylene has important commercial uses 584

Ethylene Signal Transduction Pathways 584
  Ethylene receptors are related to bacterial two-component system histidine kinases 585
  High-affinity binding of ethylene to its receptor requires a copper cofactor 586
  Unbound ethylene receptors are negative regulators of the response pathway 586
  A serine/threonine protein kinase is also involved in ethylene signaling 586
  EIN2 encodes a transmembrane protein 587
  Ethylene regulates gene expression 588
  Genetic epistasis reveals the order of the ethylene signaling components 588

Summary 588

23 Abscisic Acid: A Seed Maturation and Antistress Signal 593

Occurrence, Chemical Structure, and Measurement of ABA 594
  The chemical structure of ABA determines its physiological activity 594
  ABA is assayed by biological, physical, and chemical methods 594

Biosynthesis, Metabolism, and Transport of ABA 594
  ABA is synthesized from a carotenoid intermediate 594
  ABA concentrations in tissues are highly variable 596
  ABA can be inactivated by oxidation or conjugation 597
  ABA is translocated in vascular tissue 597

Developmental and Physiological Effects of ABA 598
  ABA regulates seed maturation 598
  ABA inhibits precocious germination and vivipary 598
  ABA promotes seed storage reserve accumulation and desiccation tolerance 599
  The seed coat and the embryo can cause dormancy 599
  Environmental factors control the release from seed dormancy 600
  Seed dormancy is controlled by the ratio of ABA to GA 600
  ABA inhibits GA-induced enzyme production 601
  ABA closes stomata in response to water stress 601
  ABA promotes root growth and inhibits shoot growth at low water potentials 601
  ABA promotes leaf senescence independently of ethylene 602
  ABA accumulates in dormant buds 602

ABA Signal Transduction Pathways 603
  ABA regulates ion channels and the PM-ATPase in guard cells 603
  ABA may be perceived by both cell surface and intracellular receptors 604
  ABA signaling involves both calcium-dependent and calcium-independent pathways 606
  ABA-induced lipid metabolism generates second messengers 608
  ABA signaling involves protein kinases and phosphatases 609
  ABA regulates gene expression 610
  Other negative regulators also influence the ABA response 611

Summary 613

24 Brassinosteroids 617

Brassinosteroid Structure, Occurrence, and Genetic Analysis 618
  BR-deficient mutants are impaired in photomorphogenesis 619

Biosynthesis, Metabolism, and Transport of Brassinosteroids 621
  Brassinolide is synthesized from campesterol 621
  Catabolism and negative feedback contribute to BR homeostasis 623
  Brassinosteroids act locally near their sites of synthesis 624

Brassinosteroids: Effects on Growth and Development 625
  BRs promote both cell expansion and cell division in shoots 626
  BRs both promote and inhibit root growth 627
  BRs promote xylem differentiation during vascular development 628
  BRs are required for the growth of pollen tubes 628
  BRs promote seed germination 629

The Brassinosteroid Signaling Pathway 629
  BR-insensitive mutants identified the BR cell surface receptor 629
  Phosphorylation activates the BRI1 receptor 630
  BIN2 is a repressor of BR-induced gene expression 630
  BES1 and BZR1 regulate different subsets of genes 631

Prospective Uses of Brassinosteroids in Agriculture 632

Summary 632

25 The Control of Flowering   635

Floral Meristems and Floral Organ Development   636
   The shoot apical meristems in Arabidopsis change with development   636
   The four different types of floral organs are initiated as separate whorls   637
   Three types of genes regulate floral development   638
   Meristem identity genes regulate meristem function   638
   Homeotic mutations led to the identification of floral organ identity genes   638
   Three types of homeotic genes control floral organ identity   639
   The ABC model explains the determination of floral organ identity   640

Floral Evocation: Internal and External Cues   641

The Shoot Apex and Phase Changes   641
   Shoot apical meristems have three developmental phases   642
   Juvenile tissues are produced first and are located at the base of the shoot   643
   Phase changes can be influenced by nutrients, gibberellins, and other chemical signals   644
   Competence and determination are two stages in floral evocation   644

Circadian Rhythms: The Clock Within   646
   Circadian rhythms exhibit characteristic features   646
   Phase shifting adjusts circadian rhythms to different day–night cycles   646
   Phytochromes and cryptochromes entrain the clock   648

Photoperiodism: Monitoring Day Length   648
   Plants can be classified according to their photoperiodic responses   649
   The leaf is the site of perception of the photoperiodic signal   650
   The floral stimulus is transported in the phloem   650
   Plants monitor day length by measuring the length of the night   651
   Night breaks can cancel the effect of the dark period   652
   The circadian clock and photoperiodic timekeeping   652
   The coincidence model is based on oscillating light sensitivity   653
   The coincidence of *CONSTANS* expression and light promotes flowering in LDPs   653
   The coincidence of *Heading-date* 1 expression and light inhibits flowering in SDPs   655
   Phytochrome is the primary photoreceptor in photoperiodism   655
   A blue-light photoreceptor regulates flowering in some LDPs   656

Vernalization: Promoting Flowering with Cold   657
   Vernalization results in competence to flower at the shoot apical meristem   658
   Vernalization involves epigenetic changes in gene expression   658
   A variety of vernalization mechanisms may have evolved   659

Biochemical Signaling Involved in Flowering   660
   Grafting studies have provided evidence for a transmissible floral stimulus   660
   Indirect induction implies that the floral stimulus is self-propagating   661
   Evidence for antiflorigen has been found in some LDPs   662
   Florigen may be a macromolecule   663
   FLOWERING LOCUS T is a candidate for the photoperiodic floral stimulus   663
   Gibberellins and ethylene can induce flowering in some plants   664
   The transition to flowering involves multiple factors and pathways   665

Summary   667

26 Stress Physiology   671

Water Deficit and Drought Tolerance   672
   Drought resistance strategies can vary   672
   Decreased leaf area is an early response to water deficit   673
   Water deficit stimulates leaf abscission   674
   Water deficit enhances root growth   674
   Abscisic acid induces stomatal closure during water deficit   674
   Water deficit limits photosynthesis   676
   Osmotic adjustment of cells helps maintain water balance   676
   Water deficit increases resistance to water flow   677
   Water deficit increases leaf wax deposition   678
   Water deficit alters energy dissipation from leaves   678
   CAM plants are adapted to water stress   678
   Osmotic stress changes gene expression   679
   ABA-dependent and ABA-independent signaling pathways regulate stress tolerance   680

Heat Stress and Heat Shock 682
  High leaf temperature and minimal evaporative cooling lead to heat stress 682
  At high temperatures, photosynthesis is inhibited before respiration 683
  Plants adapted to cool temperatures acclimate poorly to high temperatures 683
  Temperature affects membrane stability 684
  Several adaptations protect leaves against excessive heating 684
  At higher temperatures, plants produce protective proteins 684
  A transcription factor mediates HSP accumulation 685
  HSPs mediate tolerance to high temperatures 685
  Several signaling pathways mediate thermotolerance responses 686

Chilling and Freezing 687
  Membrane properties change in response to chilling injury 687
  Ice crystal formation and protoplast dehydration kill cells 689
  Limitation of ice formation contributes to freezing tolerance 689
  Some woody plants can acclimate to very low temperatures 689
  Some bacteria living on leaf surfaces increase frost damage 690
  Acclimation to freezing involves ABA and protein synthesis 690
  Numerous genes are induced during cold acclimation 691
  A transcription factor regulates cold-induced gene expression 692

Salinity Stress 692
  Salt accumulation in irrigated soils impairs plant function 692
  Plants show great diversity for salt tolerance 693
  Salt stress causes multiple injury effects 693
  Plants use multiple strategies to reduce salt stress 694
  Ion exclusion and compartmentation reduce salinity stress 695
  Plant adaptations to toxic trace elements 696

Oxygen Deficiency 698
  Anaerobic microorganisms are active in water-saturated soils 698
  Roots are damaged in anoxic environments 698
  Damaged O_2-deficient roots injure shoots 699
  Submerged organs can acquire O_2 through specialized structures 700
  Most plant tissues cannot tolerate anaerobic conditions 701
  Synthesis of anaerobic stress proteins leads to acclimation to O_2 deficit 702

Summary 702

Glossary 707

Author Index 739

Subject Index 745

Plant Physiology
Fourth Edition

Chapter 1 | Plant Cells

THE TERM *CELL* IS DERIVED from the Latin *cella*, meaning storeroom or chamber. It was first used in biology in 1665 by the English scientist Robert Hooke to describe the individual units of the honeycomb-like structure he observed in cork under a compound microscope. The "cells" Hooke observed were actually the empty lumens of dead cells surrounded by cell walls, but the term is an apt one because cells are the basic building blocks that define plant structure.

This book will emphasize the physiological and biochemical functions of plants, but it is important to recognize that these functions depend on structures, whether the process is gas exchange in the leaf, water conduction in the xylem, photosynthesis in the chloroplast, or ion transport across the plasma membrane. At every level, structure and function represent different frames of reference of a biological unity.

This chapter provides an overview of the basic anatomy of plants, from the organ level down to the ultrastructure of cellular organelles. In subsequent chapters we will treat these structures in greater detail from the perspective of their physiological functions in the plant life cycle.

Plant Life: Unifying Principles

The spectacular diversity of plant size and form is familiar to everyone. Plants range in size from less than 1 cm tall to greater than 100 m. Plant morphology, or shape, is also surprisingly diverse. At first glance, the tiny plant duckweed (*Lemna*) seems to have little in common with a giant saguaro cactus or a redwood tree. Yet regardless of their specific adaptations, all plants carry out fundamentally similar processes and are based on the same architectural plan. We can summarize the major design elements of plants as follows:

- As Earth's primary producers, green plants are the ultimate solar collectors. They harvest the energy of sunlight by converting light energy to chemical energy, which they store in bonds formed when they synthesize carbohydrates from carbon dioxide and water.
- Other than certain reproductive cells, plants are nonmotile. As a substitute for motility, they have evolved the ability to grow toward essential resources, such as light, water, and mineral nutrients, throughout their life span.
- Terrestrial plants are structurally reinforced to support their mass as they grow toward sunlight against the pull of gravity.
- Terrestrial plants lose water continuously by evaporation and have evolved mechanisms for avoiding desiccation.
- Terrestrial plants have mechanisms for moving water and minerals from the soil to the sites of photosynthesis and growth, as well as mechanisms for moving the products of photosynthesis to nonphotosynthetic organs and tissues.

Overview of Plant Structure

Despite their apparent diversity, all seed plants (see **Web Topic 1.1**) have the same basic body plan (Figure 1.1). The vegetative body is composed of three organs: leaf, stem, and root. The primary function of a leaf is photosynthesis, that of the stem is support, and that of the root is anchorage and absorption of water and minerals. Leaves are attached to the stem at nodes, and the region of the stem between two nodes is termed the internode. The stem together with its leaves is commonly referred to as the shoot.

There are two categories of seed plants: gymnosperms (from the Greek for "naked seed") and angiosperms (based on the Greek for "vessel seed," or seeds contained in a vessel). **Gymnosperms** are the less advanced type; about 700 species are known. The largest group of gymnosperms is the conifers ("cone-bearers"), which include such commercially important forest trees as pine, fir, spruce, and redwood.

Angiosperms, the more advanced type of seed plant, first became abundant during the Cretaceous period, about 100 million years ago. Today, they dominate the landscape, easily outcompeting the gymnosperms. About 250,000 species are known, but many more remain to be characterized. The major innovation of the angiosperms is the flower; hence they are referred to as *flowering plants* (see **Web Topic 1.2**).

FIGURE 1.1 Schematic representation of the body of a typical dicot. Cross sections of (A) the leaf, (B) the stem, and (C) the root are also shown. Inserts show longitudinal sections of a shoot tip and a root tip from flax (*Linum usitatissimum*), showing the apical meristems. (Photos © J. Robert Waaland/Biological Photo Service.)

Plant cells are surrounded by rigid cell walls

A fundamental difference between plants and animals is that each plant cell is surrounded by a rigid **cell wall**. In animals, embryonic cells can migrate from one location to another, resulting in the development of tissues and organs containing cells that originated in different parts of the organism.

In plants, such cell migrations are prevented because each walled cell and its neighbor are cemented together by a **middle lamella**. As a consequence, plant development, unlike animal development, depends solely on patterns of cell division and cell enlargement.

Plant cells have two types of walls: primary and secondary (Figure 1.2). **Primary cell walls** are typically thin (less than 1 μm) and are characteristic of young, growing cells. **Secondary cell walls** are thicker and stronger than primary walls and are deposited when most cell enlargement has ended. Secondary cell walls owe their strength and toughness to **lignin**, a brittle, gluelike material (see Chapter 13). Circular gaps in the secondary wall give rise to **simple pits**, which always occur opposite simple pits in the neighboring secondary wall. Adjoining simple pits are called **pit-pairs**.

The evolution of lignified secondary cell walls provided plants with the structural reinforcement necessary to grow vertically above the soil and to colonize the land. Bryophytes, which lack lignified cell walls, are unable to grow more than a few centimeters above the ground.

New cells are produced by dividing tissues called meristems

Plant growth is concentrated in localized regions of cell division called **meristems**. Nearly all nuclear divisions (mitosis) and cell divisions (cytokinesis) occur in these meristematic regions. In a young plant, the most active meristems are called **apical meristems**; they are located at the tips of the stem and the root (see Figure 1.1). At the nodes, **axillary buds** contain the apical meristems for branch shoots. Lateral roots arise from the **pericycle**, an internal meristematic tissue (see Figure 1.1C). Proximal to (i.e., next to) and overlapping the meristematic regions are zones of cell elongation in which cells increase dramatically in length and width. Cells usually differentiate into specialized types after they elongate.

The phase of plant development that gives rise to new organs and to the basic plant form is called **primary growth**. Primary growth results from the activity of apical meristems, in which cell division is followed by progressive cell enlargement, typically elongation. After elongation in a given region is complete, **secondary growth** may

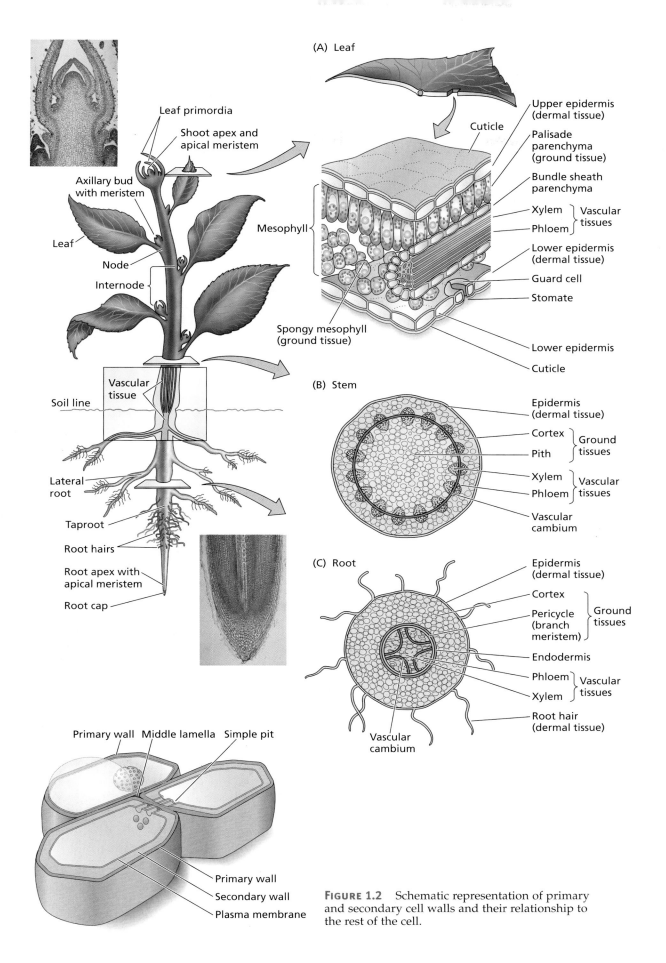

FIGURE 1.2 Schematic representation of primary and secondary cell walls and their relationship to the rest of the cell.

(A) Dermal tissue: epidermal cells

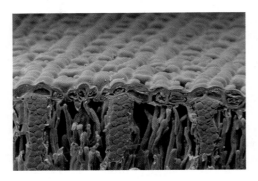

(B) Ground tissue: parenchyma cells

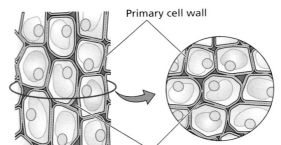

Primary cell wall

Middle lamella

(C) Ground tissue: collenchyma cells

Primary cell wall

Nucleus

(D) Ground tissue: sclerenchyma cells

Sclereids

Fibers

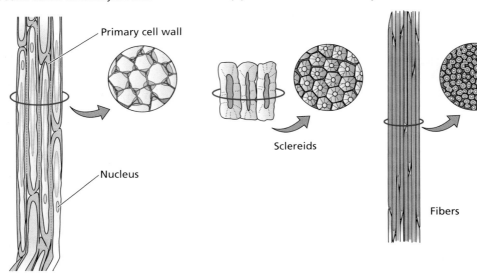

(E) Vascular tisssue: xylem and phloem

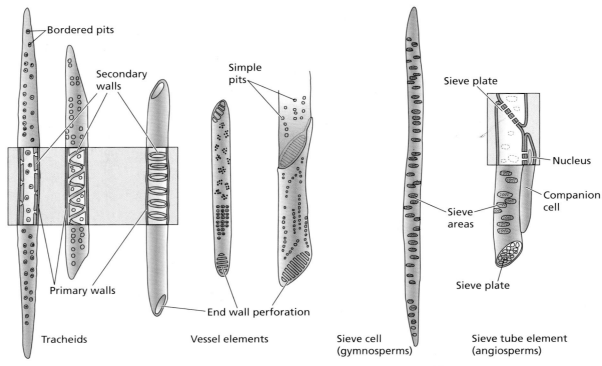

Bordered pits

Secondary walls

Simple pits

Sieve plate

Nucleus

Companion cell

Sieve areas

Primary walls

Sieve plate

End wall perforation

Tracheids

Vessel elements

Sieve cell (gymnosperms)

Sieve tube element (angiosperms)

Xylem

Phloem

◀ **FIGURE 1.3** (A) The outer epidermis (dermal tissue) of a leaf of *Welwischia mirabilis* (120×). Diagrammatic representations of three types of ground tissue: (B) parenchyma, (C) collenchyma, (D) sclerenchyma cells, and (E) conducting cells of the xylem and phloem. (A © Meckes/Ottawa/Photo Researchers, Inc.)

occur. Secondary growth involves two lateral meristems: the **vascular cambium** (plural *cambia*) and the **cork cambium**. The vascular cambium gives rise to secondary xylem (wood) and secondary phloem. The cork cambium produces the periderm, consisting mainly of cork cells.

Three major tissue systems make up the plant body

Three major tissue systems are found in all plant organs: dermal tissue, ground tissue, and vascular tissue. These tissues are illustrated and briefly characterized in Figure 1.3. For further details and characterizations of these plant tissues, see **Web Topic 1.3**.

The Plant Cell

Plants are multicellular organisms composed of millions of cells with specialized functions. At maturity, such specialized cells may differ greatly from one another in their structures. However, all plant cells have the same basic eukaryotic organization: They contain a nucleus, a cytoplasm, and subcellular organelles, and they are enclosed in a membrane that defines their boundaries (Figure 1.4). Certain structures, including the nucleus, can be lost during cell maturation, but all plant cells *begin* with a similar complement of organelles.

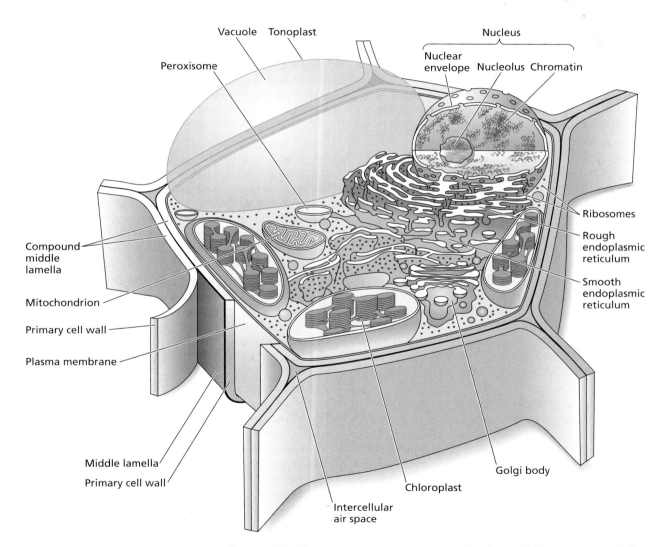

FIGURE 1.4 Diagrammatic representation of a plant cell. Various intracellular compartments are defined by their respective membranes, such as the tonoplast, the nuclear envelope, and the membranes of the other organelles. The two adjacent primary walls, along with the middle lamella, form a composite structure called the compound middle lamella.

An additional characteristic feature of plant cells is that they are surrounded by a cellulosic cell wall. The following sections provide an overview of the membranes and organelles of plant cells. The structure and function of the cell wall will be treated in detail in Chapter 15.

Biological membranes are phospholipid bilayers that contain proteins

All cells are enclosed in a membrane that serves as their outer boundary, separating the cytoplasm from the external environment. This **plasma membrane** (also called **plasmalemma**) allows the cell to take up and retain certain substances while excluding others. Various transport proteins embedded in the plasma membrane are responsible for this selective traffic of solutes across the membrane. The accumulation of ions or molecules in the cytosol through the action of transport proteins consumes metabolic energy. Membranes also delimit the boundaries of the specialized internal organelles of the cell and regulate the fluxes of ions and metabolites into and out of these compartments.

According to the **fluid-mosaic model**, all biological membranes have the same basic molecular organization. They consist of a double layer (*bilayer*) of either phospholipids or, in the case of chloroplasts, glycosylglycerides, in which proteins are embedded (Figure 1.5A and B). In most membranes, proteins make up about half of the membrane's mass. However, the composition of the lipid components and the properties of the proteins vary from membrane to membrane, conferring on each membrane its unique functional characteristics.

PHOSPHOLIPIDS Phospholipids are a class of lipids in which two fatty acids are covalently linked to glycerol, which is covalently linked to a phosphate group. Also attached to this phosphate group is a variable component, called the *head group*, such as serine, choline, glycerol, or inositol (Figure 1.5C). In contrast to the fatty acids, the head groups are highly polar; consequently, phospholipid molecules display both hydrophilic and hydrophobic properties (i.e., they are *amphipathic*). The nonpolar hydrocarbon chains of the fatty acids form a region that is exclusively hydrophobic—that is, that excludes water.

The membranes of specialized plant organelles called **plastids**, the group of membrane-bound organelles to which chloroplasts belong, are unique in that their lipid component consists almost entirely of **glycosylglycerides** rather than phospholipids. In glycosylglycerides, the polar head group consists of galactose, digalactose, or sulfated galactose, without a phosphate group (see **Web Topic 1.4**).

The fatty acid chains of phospholipids and glycosylglycerides are variable in length, but they usually consist of 14 to 24 carbons. One of the fatty acids is typically *saturated* (i.e., it contains no double bonds); the other fatty acid chain usually is *unsaturated* (i.e., it has one or more *cis* double bonds).

The presence of *cis* double bonds creates a kink in the chain that prevents tight packing of the phospholipids in the bilayer. As a result, the fluidity of the membrane is increased. The fluidity of the membrane, in turn, plays a critical role in many membrane functions. Membrane fluidity is also strongly influenced by temperature. Because plants generally cannot regulate their body temperatures, they are often faced with the problem of maintaining membrane fluidity under conditions of low temperature, which tends to decrease membrane fluidity. Thus, plant phospholipids have a high percentage of unsaturated fatty acids, such as *oleic acid* (one double bond), *linoleic acid* (two double bonds) and *α-linolenic acid* (three double bonds), which increase the fluidity of their membranes.

PROTEINS The proteins associated with the lipid bilayer are of three main types: integral, peripheral, and anchored. In addition, proteins and lipids can form transient aggregates in the membrane called *lipid rafts*.

Integral proteins are embedded in the lipid bilayer. Most integral proteins span the entire width of the phospholipid bilayer, so one part of the protein interacts with the outside of the cell, another part interacts with the hydrophobic core of the membrane, and a third part interacts with the interior of the cell, the cytosol. Proteins that serve as ion channels (see Chapter 6) are always integral membrane proteins, as are certain receptors that participate in signal transduction pathways (see Chapter 14 on the web site). Some receptor-like proteins on the outer surface of the plasma membrane recognize and bind tightly to cell wall constituents, effectively cross-linking the membrane to the cell wall.

Peripheral proteins are bound to the membrane surface by noncovalent bonds, such as ionic bonds or hydrogen bonds, and can be dissociated from the membrane with high-salt solutions or chaotropic agents, which break ionic and hydrogen bonds, respectively. Peripheral proteins serve a variety of functions in the cell. For example, some are involved in interactions between the plasma membrane and components of the cytoskeleton, such as microtubules and actin microfilaments, which are discussed later in this chapter.

Anchored proteins are bound to the membrane surface via lipid molecules, to which they are covalently attached. These lipids include fatty acids (myristic acid and palmitic acid), prenyl groups derived from the isoprenoid pathway (farnesyl and geranylgeranyl groups), and glycosylphosphatidylinositol (GPI)-anchored proteins (Figure 1.6) (Buchanan et al. 2000).

Lipid rafts were originally isolated from animal membranes as detergent-resistant aggregates of lipids and proteins. Lipid rafts have been proposed to represent transient, rigid microdomains of tightly packed fatty acid chains enriched in *sphingolipids* (lipids containing the amino alcohol sphingosine) and *sterols* (see Chapters 11 and 13). Proteins may be targeted to lipid rafts by covalently attached

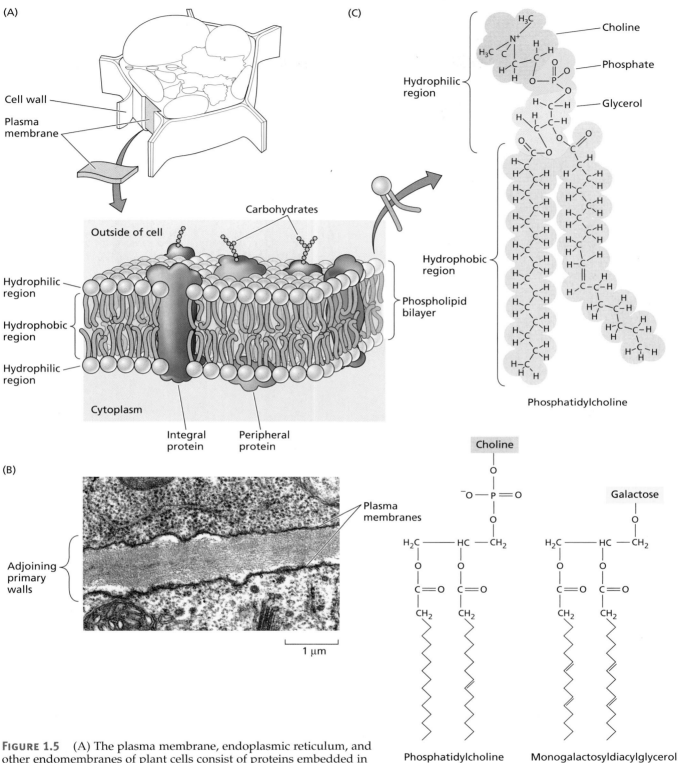

FIGURE 1.5 (A) The plasma membrane, endoplasmic reticulum, and other endomembranes of plant cells consist of proteins embedded in a phospholipid bilayer. (B) This transmission electron micrograph shows plasma membranes in cells from the meristematic region of a root tip of cress (*Lepidium sativum*). The overall thickness of the plasma membrane, viewed as two dense lines and an intervening space, is 8 nm. (C) Chemical structures and space-filling models of typical phospholipids: phosphatidylcholine and monogalactosyldiacylglycerol. (B from Gunning and Steer 1996.)

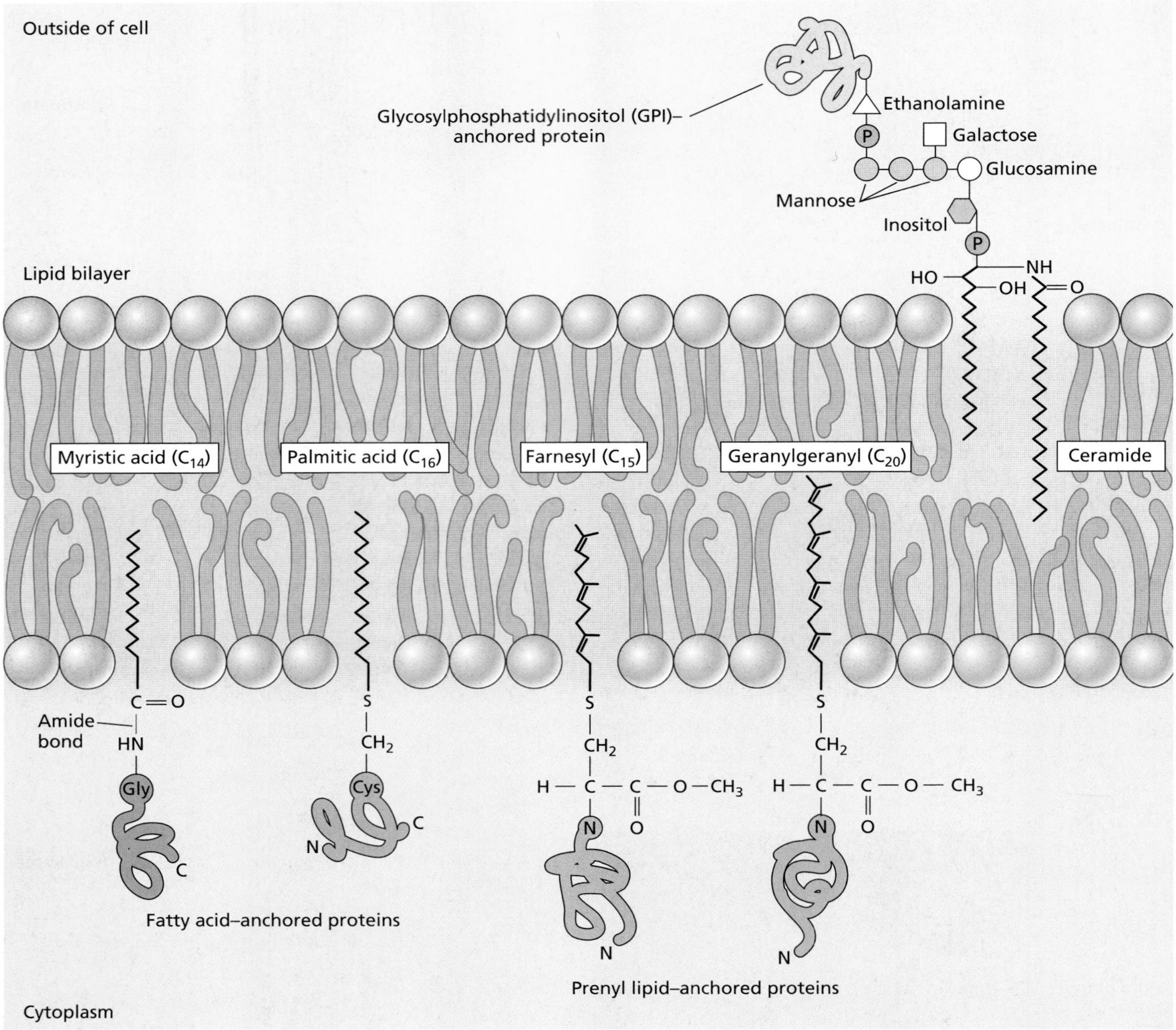

FIGURE 1.6 Different types of anchored membrane proteins that are attached to the membrane via fatty acids, prenyl groups, or phosphatidylinositol. (After Buchanan et al. 2000.)

lipids. By facilitating the interactions of specific membrane proteins involved in cell signaling, lipid rafts may play crucial roles in developmental pathways.

The nucleus contains most of the genetic material of the cell

The **nucleus** (plural *nuclei*) is the organelle that contains the genetic information primarily responsible for regulating the metabolism, growth, and differentiation of the cell. Collectively, these genes and their intervening sequences are referred to as the **nuclear genome**. The size of the nuclear genome in plants is highly variable, ranging from about 1.2 $\times 10^8$ base pairs for the dicot *Arabidopsis thaliana* to 1×10^{11} base pairs for the lily *Fritillaria assyriaca*. The remainder of the genetic information of the cell is contained in the two semiautonomous organelles—chloroplasts and mitochondria—which we will discuss a little later in this chapter.

The nucleus is surrounded by a double membrane called the **nuclear envelope** (Figure 1.7A). The space between the two membranes of the nuclear envelope is called the **perinuclear space**, and the two membranes of the nuclear envelope join at sites called **nuclear pores** (Figure 1.7B). There can be very few to many thousands of nuclear pores on an individual nuclear envelope.

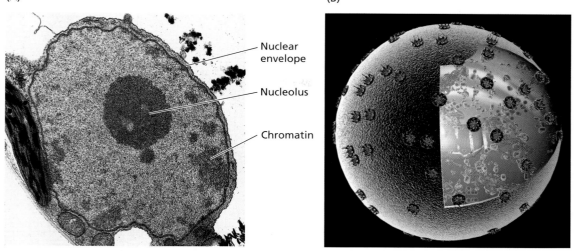

Figure 1.7 (A) Transmission electron micrograph of a plant cell, showing the nucleolus and the nuclear envelope. (B) Composite drawing of a nucleus with nuclear pores. The cutaway section shows the interior of the nucleus. (A courtesy of R. Evert; B courtesy of Samir S. Patel and Michael Rexach.)

The nuclear "pore" is actually an elaborate structure composed of more than a hundred different **nucleoporin** proteins arranged octagonally to form a 120 nm **nuclear pore complex** (**NPC**) (Figure 1.8). The nucleoporins lining the 40 nm channel of the NPC form a meshwork that acts as a supramolecular sieve (Denning et al. 2003). A specific amino acid sequence called the **nuclear localization signal** is required for a protein to gain entry into the nucleus. Three proteins required for nuclear import and export have been identified: Ran (a monomeric GTP-binding protein;

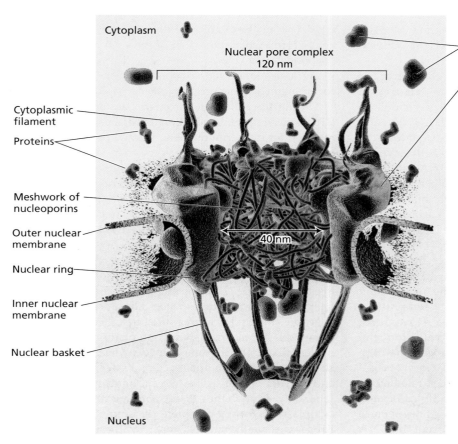

Figure 1.8 The nuclear pore complex acts as a supramolecular sieve that spans both membranes of the nuclear envelope. The proteins that make up the pore are called nucleoporins. Parallel rings, exhibiting octagonal symmetry, are situated near the inner and outer membranes of the nuclear envelope. Various nucleoporin proteins form the other structures, such as the nuclear ring, cytoplasmic filaments, and the nuclear basket. Small metabolites diffuse freely through the NPC, but the flow of large proteins and RNA is selective. According to a recent model, selectivity is derived from a meshwork of unstructured nucleoporins within the aqueous pore. To penetrate this meshwork into or out of the nucleus, macromolecules must be associated with specific carrier proteins. Ribosomes presumably exit the nucleus by physically moving through the meshwork. (Courtesy of Samir S. Patel and Michael Rexach.)

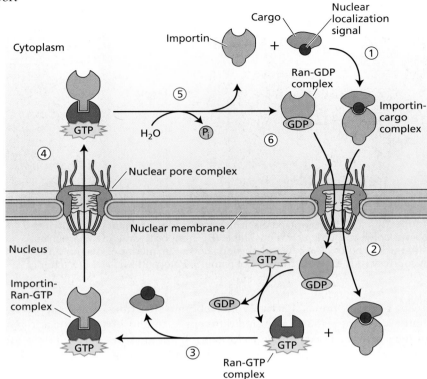

Figure 1.9 The mechanism of protein import into the nucleus. (After Lodish et al. 2004.)

1. Free heterodimers of α and β importin in the cytoplasm bind to the nuclear localization signal of the cargo protein, forming a trimolecular complex.

2. The importin–cargo complex interacts with specific sites on the meshwork of nucleoporins (see Figure 1.8) and diffuses into the nucleus.

3. In the nucleus, tight binding of importins to Ran-GTP liberates the cargo molecule from the complex. This dissociation makes the import irreversible.

4. The importin-Ran-GTP complex then exits the nucleus, returning the importins to the cytoplasm.

5. Hydrolysis of Ran-GTP to Ran-GDP releases importin enabling it to bind to another cargo molecule.

6. Ran-GDP re-enters the nucleus and is converted back to Ran-GTP.

see Chapter 14), and α and β importin. A model for the movement of proteins into and out of the nucleus is shown in Figure 1.9.

The nucleus is the site of storage and replication of the **chromosomes**, composed of DNA and its associated proteins. Collectively, this DNA–protein complex is known as **chromatin**. The linear length of all the DNA within any plant genome is usually millions of times greater than the diameter of the nucleus in which it is found. To solve the problem of packaging this chromosomal DNA within the nucleus, segments of the linear double helix of DNA are coiled twice around a solid cylinder of eight **histone** protein molecules, forming a **nucleosome**. Nucleosomes are arranged like beads on a string along the length of each chromosome.

During mitosis, the chromatin condenses, first by coiling tightly into a **30 nm chromatin fiber**, with six nucleosomes per turn, followed by further folding and packing processes that depend on interactions between proteins and nucleic acids (Figure 1.10). At interphase, two types of chromatin are visible: heterochromatin and euchromatin. About 10% of the DNA consists of **heterochromatin**, a highly compact and transcriptionally inactive form of chromatin. Most of the heterochromatin is concentrated along the periphery of the nuclear membrane and is associated with regions of the chromosome containing few genes, such as telomeres and centromeres. The rest of the DNA consists of **euchromatin**, the dispersed, transcriptionally active form. Only about 10% of the euchromatin is transcriptionally active at any given time. The remainder exists in an intermediate state of condensation, between heterochromatin and transcriptionally active euchromatin.

During the cell cycle chromatin undergoes dynamic structural changes. In addition to transient local changes required for transcription, heterochromatic regions can be converted to euchromatic regions, and vice-versa, by the addition or removal of functional groups on the histone proteins. Such global changes in the genome, called *chromatin remodeling*, can give rise to stable changes in gene expression. In general, stable changes in gene expression that occur without changes in the DNA sequence are referred to as *epigenetic regulation*.

Nuclei also contain a densely granular region, called the **nucleolus** (plural *nucleoli*), that is the site of ribosome synthesis (see Figure 1.7A). The nucleolus includes portions of one or more chromosomes where ribosomal RNA (rRNA) genes are clustered to form a structure called the **nucleolar organizer**. Typical cells have one or more nucleoli per nucleus. Each 80S ribosome is made of a large and a small subunit, and each subunit is a complex aggregate of rRNA and specific proteins. The two subunits exit the nucleus separately, through the nuclear pore, and then unite in the cytoplasm to form a complete ribosome (Figure 1.11A). **Ribosomes** are the sites of protein synthesis.

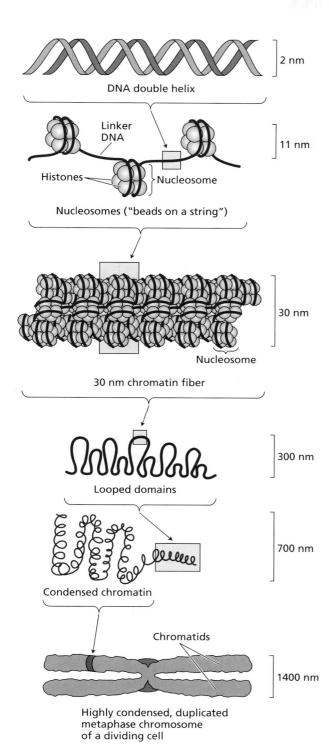

FIGURE 1.10 Packaging of DNA in a metaphase chromosome. The DNA is first aggregated into nucleosomes and then wound to form the 30 nm chromatin fibers. Further coiling leads to the condensed metaphase chromosome. (After Alberts et al. 2002.)

Protein synthesis involves transcription and translation

The complex process of protein synthesis starts with **transcription**—the synthesis of an RNA polymer bearing a base sequence that is complementary to a specific gene. The primary RNA transcript is processed to become messenger RNA (mRNA), which moves from the nucleus to the cytoplasm through the nuclear pore. In the cytoplasm, the mRNA attaches first to the small ribosomal subunit and then to the large subunit to initiate translation.

Translation is the process whereby a specific protein is synthesized from amino acids according to the sequence information encoded by the mRNA. The ribosome travels the entire length of the mRNA and serves as the site for the sequential bonding of amino acids as specified by the base sequence of the mRNA (Figure 1.11B).

The endoplasmic reticulum is a network of internal membranes

Cells have an elaborate network of internal membranes called the **endoplasmic reticulum (ER)**. The membranes of the ER are typical lipid bilayers with interspersed integral and peripheral proteins. These membranes behave in a highly dynamic way throughout the life of a cell, often forming flattened or tubular sacs known as **cisternae** (singular *cisterna*).

Ultrastructural studies have shown that the ER is continuous with the outer membrane of the nuclear envelope. There are two main types of ER—smooth and rough (Figure 1.12)—and the two types are interconnected. **Rough ER (RER)** differs from smooth ER in that it is covered with ribosomes that are actively engaged in protein synthesis; in addition, rough ER tends to be lamellar (a flat sheet composed of two unit membranes), while smooth ER tends to be tubular, although a gradation for each type can be observed in almost any cell.

The structural differences between the two forms of ER are accompanied by functional differences. **Smooth ER** functions as a major site of lipid synthesis and membrane assembly. Rough ER is the site of synthesis of membrane proteins and proteins to be secreted outside the cell or into the vacuoles.

In addition, two interconnected domains of ER tubules have been identified by their cellular location:

1. **Cortical ER** refers to ER tubules that line the plasma membrane.

2. **Perinuclear ER** consists of ER tubules that are continuous with the nuclear envelope.

These two ER domains are associated with distinct proteins and play different roles throughout the cell cycle, although their specific functions are poorly understood.

Secretion of proteins from cells begins with the rough ER

Proteins destined for secretion cross the RER membrane and enter the lumen of the ER. This is the first step in the secretion pathway that involves the Golgi body and vesicles that fuse with the plasma membrane.

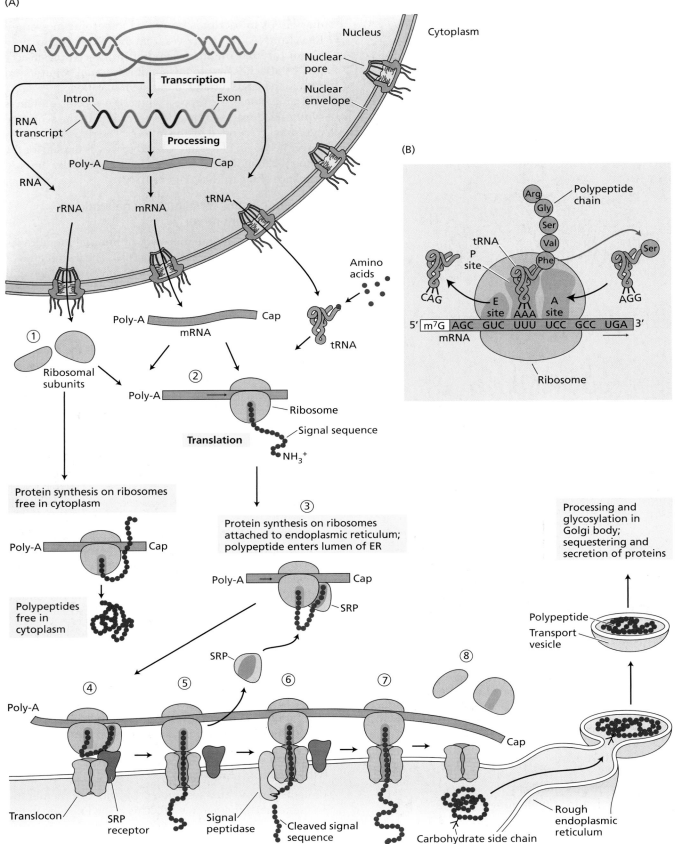

FIGURE 1.11 (A) Basic steps in gene expression, including transcription, processing, export to the cytoplasm, and translation. (1–2) Proteins may be synthesized on free or bound ribosomes. (3) Proteins destined for secretion are synthesized on the rough endoplasmic reticulum and contain a hydrophobic signal sequence. A signal recognition particle (SRP) binds the signal peptide and to the ribosome, interrupting translation. (4) SRP receptors associate with protein-transporting channels called translocons. The ribosome-SRP complex binds to the SRP receptor on the ER membrane, and the ribosome docks with the translocon. (5) The translocon pore opens, the SRP particle is released, and the elongating polypeptide enters the lumen of the endoplasmic reticulum. (6) Translation resumes. Upon entering the lumen of the ER, the signal sequence is cleaved off by a signal peptidase on the membrane. (7,8) After carbohydrate addition and chain folding, the newly synthesized polypeptide is shuttled to the Golgi apparatus via vesicles. (B) Amino acids are polymerized on the ribosome, with the help of tRNA, to form the elongating polypeptide chain.

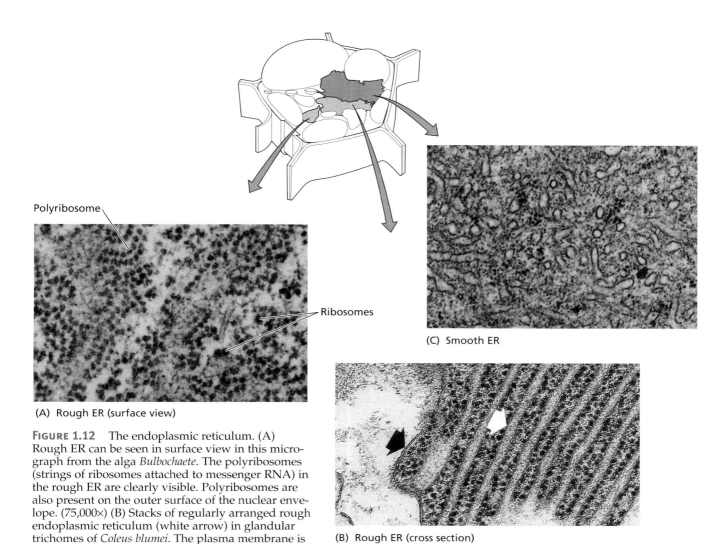

(A) Rough ER (surface view)

(C) Smooth ER

(B) Rough ER (cross section)

FIGURE 1.12 The endoplasmic reticulum. (A) Rough ER can be seen in surface view in this micrograph from the alga *Bulbochaete*. The polyribosomes (strings of ribosomes attached to messenger RNA) in the rough ER are clearly visible. Polyribosomes are also present on the outer surface of the nuclear envelope. (75,000×) (B) Stacks of regularly arranged rough endoplasmic reticulum (white arrow) in glandular trichomes of *Coleus blumei*. The plasma membrane is indicated by the black arrow, and the material outside the plasma membrane is the cell wall. (75,000×) (C) Smooth ER often forms a tubular network, as shown in this transmission electron micrograph from a young petal of *Primula kewensis*. (45,000×) (Micrographs from Gunning and Steer 1996.)

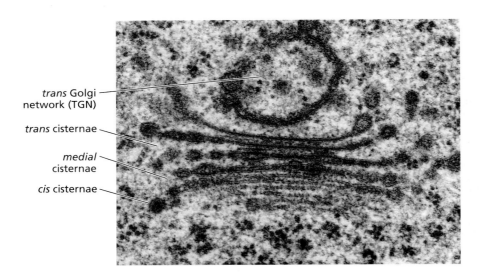

FIGURE 1.13 Electron micrograph of a Golgi apparatus in a tobacco (*Nicotiana tabacum*) root cap cell. The *cis*, *medial*, and *trans* cisternae are indicated. The *trans* Golgi network is associated with the *trans* cisterna. (60,000×) (From Gunning and Steer 1996.)

The mechanism of transport across the membrane is complex, involving the ribosomes, the mRNA that codes for the secretory protein, and a special receptor in the ER membrane (Figure 1.11B). All secretory proteins and most integral membrane proteins have been shown to have a hydrophobic leader sequence of 18 to 30 amino acid residues at the amino-terminal end of the chain called the **signal peptide** sequence. Early in translation, a **signal recognition particle** (**SRP**), made up of protein and RNA, binds both to this hydrophobic leader and to the ribosome, interrupting translation. The RER membrane contains **SRP receptors**, which can associate with protein-lined channels called **translocons**. The ribosome-SRP complex in the cytosol binds to the SRP receptor on the membrane, and the ribosome docks with the translocon. Docking opens the translocon pore, the SRP particle is released, translation resumes, and the elongating polypeptide enters the lumen of the ER.* Once inside the lumen of the ER, the signal sequence is cleaved off by a signal peptidase on the membrane.

In the vast majority of cases, a branched oligosaccharide chain made up of *N*-acetylglucosamine (GlcNac), mannose (Man), and glucose (Glc), having the stoichiometry GlcNac$_2$Man$_9$Glc$_3$, is attached to the free amino group of one or more specific asparagine residues of the secretory protein. This *N-linked glycan* is first assembled on a lipid molecule, **dolichol diphosphate** (see Chapter 13), which is embedded in the ER membrane. The completed 14-residue precursor is then transferred to the nascent polypeptide as it enters the lumen. The three terminal glucose residues and one specific mannose residue—which appear to serve as signals that the N-linked glycan is completed and ready to be attached to the nascent polypeptide—are then re-

moved by specific glucosidases, and the processed glycoprotein (i.e., a protein with covalently attached sugars) is ready for transport to the Golgi apparatus. The so-called **N-linked glycoproteins** are then transported to the Golgi apparatus via small vesicles. The vesicles move through the cytosol and fuse with cisternae on the *cis* face of the Golgi apparatus (Figure 1.13).

Golgi stacks produce and distribute secretory products

The Golgi apparatus typically appears in electron micrographs as a stack of three to ten flattened membrane sacs, or cisternae, and an irregular network of tubules and vesicles called the ***trans* Golgi network** (**TGN**) (Figure 1.13). As Figure 1.13 shows, the Golgi has distinct functional zones, each specialized by its enzyme complement for a different step in glycoprotein processing. The cisternae on the secreting side of the Golgi apparatus are called the *trans* face, and the cisternae on the forming side of the stack are called the *cis* face. The *medial* cisternae are between the *trans* and *cis* cisternae, and the *trans* Golgi network is located on the *trans* face. The entire structure is stabilized by the presence of **intercisternal elements**, protein cross-links that hold the cisternae together.

Plant cells contain up to several hundred separate Golgi stacks dispersed throughout the cytoplasm (Driouich et al. 1994). In contrast, those of animal cells tend to aggregate in one part of the cell and are interconnected via tubules. Plant Golgi also exhibit directed movement along *actin filaments*, cytoskeletal components discussed later in the chapter. The driving force for Golgi movement along the actin filaments is an ATP-dependent myosin motor, the same motor molecule that drives muscle contraction in animal cells. Each plant Golgi apparatus can thus be viewed as a "mobile factory" that produces, sorts, and distributes secretory products throughout the cell (Nebenführ and Staehelin 2001).

---

*In the case of integral membrane proteins, some parts of the polypeptide chain are translocated across the membrane, while others are not. The completed proteins are anchored to the membrane by one or more hydrophobic membrane-spanning domains.

Proteins and polysaccharides destined for secretion are processed in the Golgi apparatus

Glycoproteins destined for secretion reach the Golgi via vesicles that bud off from the RER and are enzymatically modified within the lumens of the Golgi cisternae. As the glycoproteins pass through the *cis*, *medial*, and *trans* Golgi cisternae, they are successively modified by the specific enzymes localized in these three compartments. Certain sugars, such as mannose, are removed from the oligosaccharide chains, and other sugars are added. In addition to these modifications, glycosylation of the —OH groups of hydroxyproline, serine, threonine, and tyrosine residues (**O-linked oligosaccharides**) also occurs in the Golgi. After being processed within the Golgi, the glycoproteins leave the organelle in other vesicles, usually deriving from the TGN. All of this processing appears to confer on each protein a tag or marker that specifies the ultimate destination of that protein inside or outside the cell. Oligosaccharides also help to stabilize proteins by providing protection against proteases. In addition, some signaling mechanisms involving cell surface receptors may involve recognition of sugar residues attached to proteins.

In plant cells, the Golgi apparatus plays an important role in both cell plate and cell wall formation (see Chapter 15). Noncellulosic cell wall polysaccharides (hemicellulose and pectin) are synthesized, and a variety of glycoproteins, including hydroxyproline-rich glycoproteins, are processed within the Golgi. Secretory vesicles derived from the Golgi carry the polysaccharides and glycoproteins to the plasma membrane, where the vesicles fuse with the plasma membrane and empty their contents into the region of the cell wall. Vesicles derived from the Golgi are also sorted and targeted to different vacuoles in the cell.

Two models for intra-Golgi transport have been proposed

Despite many years of study, the precise pathway of newly synthesized materials through the Golgi apparatus has not yet been determined. According to the **vesicular shuttle model**, the *cis*, *medial*, and *trans* cisternae are stable structures. Cargo molecules move from one cisterna to the next via small vesicles that bud off from the periphery of each cisterna. The vesicular shuttle model thus accounts for the stability of the specific enzymatic markers that distinguish *cis*, *medial*, and *trans* cisternae.

However, the vesicular shuttle model fails to explain the ability of some Golgi to synthesize and secrete large structures, such as the glycoprotein scales of algal cell walls and the procollagen aggregates of animal fibroblasts. Too bulky to enter small secretory vesicles, algal scales and procollagen filaments pass from the *cis* to the *trans* face of the Golgi stack without ever leaving their cisternae. The model proposed for this type of intra-Golgi transport is called the **cisternal maturation/progression model** (see Figure 1.14). In this model, the Golgi stack is not fixed, but is a dynamic structure in which cisternae progress through *cis*, *medial*, and *trans* faces, carrying their cargo with them. In the final phase, the cisterna itself becomes a large secretory vesicle that fuses with the plasma membrane. The movement of membrane cisternae or vesicles in the forward direction is referred to as **anterograde transport**.

Although the cisternal maturation/progression model solves the problem of the secretion of algal scales, it raises another question. If individual cisternae pass through the Golgi like floats in a parade, how are the distinct histochemical zones (*cis*, *medial*, and *trans*) maintained? This riddle was solved by the discovery of retrograde (reverse) transport of vesicles between Golgi cisternae (see Figure 1.14). **Retrograde vesicular transport** maintains the spatial distribution of enzymes and other functional proteins within the Golgi stack by acting as a countercurrent to cisternal progression. When a particular *cis* cisterna moves forward to the *medial* zone, vesicles budding off from the periphery shuttle its complement of enzymes and receptors by retrograde transport back to the newly formed *cis* cisterna. Simultaneously, vesicles arriving from the previous *medial* cisterna by retrograde transport fuse with the former *cis* cisterna, converting it into a functional *medial* cisterna (see Figure 1.14).

Most cell biologists favor the cisternal maturation/progression model over the vesicular shuttle model of intra-Golgi transport, although some combination of the two has not been ruled out.

Specific coat proteins facilitate vesicle budding

As in other eukaryotic organisms, the budding of vesicles that occurs along the secretory pathway is often assisted by **coat proteins**, and the type of coat protein employed is specific for the organelles involved and the direction of transport. For example, the budding of vesicles from the ER to the Golgi (anterograde transport) is assisted by **coat protein II** (**COPII**), while vesicles moving from the Golgi to the ER (retrograde transport) utilize a different coat protein, **COPI** (see Figure 1.14).

A third type of coat protein, **clathrin**, is required for the budding-off of vesicles from the plasma membrane and possibly from the *trans* Golgi. Clathrin-coated vesicles (Figure 1.15) participate in **endocytotic recycling**, the retrograde transport that brings soluble and membrane-bound proteins into the cell. After shedding their coats, these endocytotic vesicles fuse with the **prevacuolar compartment (PVC)***, where recycling of membrane receptors takes place. From the PVC, vesicles bud off and deliver their cargo to a lytic vacuole (see Figure 1.14).

The importance of endocytotic recycling can perhaps best be appreciated in the case of cells active in secretion, such as root cap cells. Root cap cells secrete copious

---

*The PVC is probably equivalent to the late endosome of mammalian cells. It has also been equated with multivesicular bodies, or MVBs.

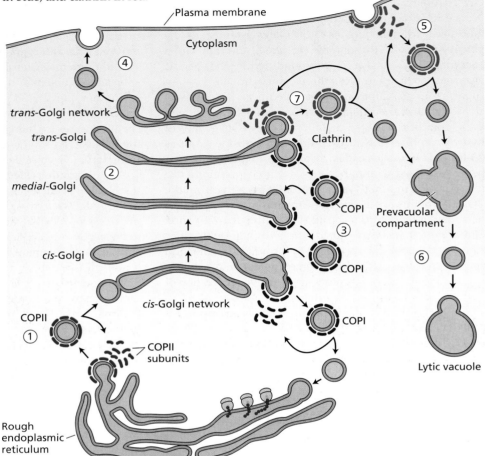

FIGURE 1.14 Vesicular traffic along the secretory and endocytotic pathways is mediated by three types of coat proteins. COPII is indicated in green, COPI in blue, and clathrin in red.

1. COPII-coated vesicles bud from the ER and are transported to the *cis* face of the Golgi apparatus.

2. Cisternae progress through the Golgi stack in the anterograde direction, carrying their cargo with them.

3. Retrograde movement of COPI-coated vesicles maintains the correct distribution of enzymes in the *cis-*, *medial-*, and *trans*-cisternae of the stack.

4. Uncoated vesicles bud from the *trans*-Golgi membrane and fuse with the plasma membrane.

5. Endocytotic clathrin-coated vesicles fuse with the prevacuolar compartment.

6. Uncoated vesicles bud off from the prevacuolar compartment and carry their cargo to a lytic vacuole.

7. Proteins destined for lytic vacuoles are secreted from the *trans* Golgi to the PVC via clathrin-coated vesicles, and then repackaged for delivery to the lytic vacuole.

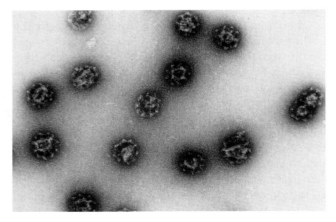

FIGURE 1.15 Preparation of clathrin-coated vesicles isolated from bean leaves. (102,000×) (Courtesy of D. G. Robinson.)

amounts of mucopolysaccharides ("slime"), which lubricate the root tip as it grows through the soil. The increase in membrane surface area caused by the fusion of large slime-filled vesicles with the plasma membrane would become excessive if it were not for the process of endocytosis, which constantly recycles plasma membrane back to the prevacuolar compartment. Proteins destined for **lytic vacuoles** (acidic vacuoles containing hydrolytic enzymes) are also transported via clathrin-coated vesicles from the *trans* Golgi network to the PVC before being repackaged and sent to the vacuole (see Figure 1.14).

Not all plant secretory vesicles have visible protein coats, however. Two examples of apparently smooth secretory vesicles are the previously mentioned slime-filled vesicles produced by root cap cells and the dense vesicles involved in the transport of storage proteins from the Golgi apparatus to specialized **protein-storing vacuoles** (PSVs). Although no coat proteins are visible on these vesicles, it is likely that analogous proteins too small to be seen by electron microscope are present.

Vacuoles play multiple roles in plant cells

Mature living plant cells contain large, water-filled **central vacuoles** that can occupy up to 95% of the total volume of the cell (see Figure 1.4). Each vacuole is surrounded by a vacuolar membrane, or tonoplast. Protein transporters embedded in the tonoplast regulate the fluxes of ions and organic molecules into and out of the lumen. Many cells also have cytoplasmic strands that run through the vacuole, but each transvacuolar strand is surrounded by the tonoplast.

In meristematic tissue, vacuoles are less prominent, though they are always present as small **provacuoles**. Provacuoles are produced by the *trans* Golgi network (see Figure 1.13). As the cell begins to mature, the provacuoles fuse to produce the large central vacuoles that are characteristic of most mature plant cells. In such cells, the cytoplasm is restricted to a thin layer surrounding the vacuole.

The central vacuole contains water and dissolved inorganic ions, organic acids, sugars, enzymes, and a variety of secondary metabolites (see Chapter 13), which often play roles in plant defense. Active solute accumulation provides the osmotic driving force for water uptake by the vacuole, which is required for plant cell enlargement. The turgor pressure generated by this water uptake provides the structural rigidity needed to keep herbaceous plants upright, since they lack the lignified support tissues of woody plants.

Like animal lysosomes, plant central vacuoles are large lytic compartments that contain hydrolytic enzymes, including proteases, ribonucleases, and glycosidases. Unlike animal lysosomes, however, plant vacuoles do not participate in the turnover of macromolecules throughout the life of the cell. Instead, their degradative enzymes leak out into the cytosol as the cell undergoes senescence or in response to cell damage, thereby helping to recycle valuable nutrients to the living portion of the plant.

Protein-storing vacuoles, called **protein bodies**, are abundant in seeds. During germination the storage proteins in the protein bodies are hydrolyzed to amino acids and exported to the cytosol for use in protein synthesis. The hydrolytic enzymes are stored in small **lytic vacuoles**, which fuse with the protein bodies to initiate the breakdown process (Figure 1.16). The acidic luminal pH of lytic vacuoles is maintained by the activity of the vacuolar H^+-ATPase, a proton pump that moves protons from the cytosol into the vacuole.

Similar small lytic vacuoles are found in senescing leaves (Otegui et al. 2005). Leaf lytic vacuoles are believed to participate in the breakdown of proteins and other cellular constituents that occurs during leaf senescence and in response to environmental stress. In this case, the lytic vacuoles interact with structures called autophagosomes.

Autophagosomes are double-membrane–bound organelles that engulf and sequester large portions of the cytoplasm, including entire organelles. Digestion of the engulfed cytoplasm is initiated when the autophagosome outer membrane fuses with the membrane of a lytic vacuole. As a result, the inner membrane of the autophagosome and its cytoplasmic contents are released into the lumen of the lytic compartment and degraded.

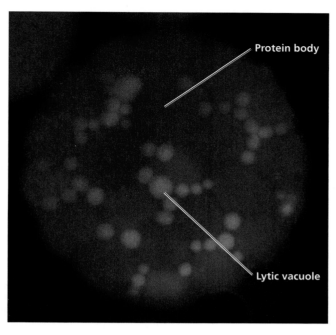

FIGURE 1.16 Light micrograph of a protoplast prepared from the aleurone layer of seeds. The fluorescent stain reveals two types of vacuoles: the larger protein bodies and the smaller lytic vacuoles. (Courtesy of P. Bethke and R. L. Jones.)

Mitochondria and chloroplasts are sites of energy conversion

A typical plant cell has two types of energy-producing organelles: mitochondria and chloroplasts. Both types are separated from the cytosol by a double membrane (an outer and an inner membrane). **Mitochondria** (singular *mitochondrion*) are the cellular sites of respiration, a process in which the energy released from sugar metabolism is used for the synthesis of ATP (adenosine triphosphate) from ADP (adenosine diphosphate) and inorganic phosphate (P_i) (see Chapter 11).

Mitochondria are highly dynamic structures that can undergo both fission and fusion. Mitochondrial fusion can result in long, tube-like structures that may branch to form mitochondrial networks. Regardless of shape, all mitochondria have a smooth outer membrane and a highly convoluted inner membrane (Figure 1.17). The infoldings of the inner membrane are called **cristae** (singular *crista*). The compartment enclosed by the inner membrane, the mitochondrial **matrix**, contains the enzymes of the pathway of intermediary metabolism called the Krebs cycle.

In contrast to the mitochondrial outer membrane and all other membranes in the cell, the inner membrane of a mitochondrion is almost 70% protein and contains some phos-

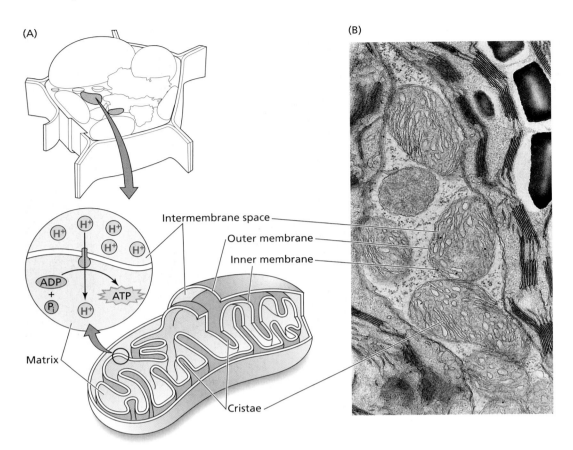

FIGURE 1.17 (A) Diagrammatic representation of a mitochondrion, including the location of the H⁺-ATPases involved in ATP synthesis on the inner membrane. (B) An electron micrograph of mitochondria from a leaf cell of Bermuda grass, *Cynodon dactylon*. (26,000×) (Micrograph by S. E. Frederick, courtesy of E. H. Newcomb.)

pholipids that are unique to the organelle (e.g., cardiolipin). The proteins in and on the inner membrane have special enzymatic and transport capacities.

The inner membrane is highly impermeable to the passage of H⁺; that is, it serves as a barrier to the movement of protons. This important feature allows the formation of electrochemical gradients. Dissipation of such gradients by the controlled movement of H⁺ ions through the transmembrane enzyme **ATP synthase** is coupled to the phosphorylation of ADP to produce ATP. ATP can then be released to other cellular sites where energy is needed to drive specific reactions.

Chloroplasts (Figure 1.18A) belong to another group of double-membrane–enclosed organelles called **plastids**. Chloroplast membranes are rich in glycosylglycerides (see Web Topic 1.4). Chloroplast membranes contain chlorophyll and its associated proteins and are the sites of photosynthesis. In addition to their inner and outer envelope membranes, chloroplasts possess a third system of membranes called **thylakoids**. A stack of thylakoids forms a **granum** (plural *grana*) (Figure 1.18B). Proteins and pigments (chlorophylls and carotenoids) that function in the photochemical events of photosynthesis are embedded in the thylakoid membrane. The fluid compartment surrounding the thylakoids, called the **stroma**, is analogous to the matrix of the mitochondrion. Adjacent grana are connected by unstacked membranes called **stroma lamellae** (singular *lamella*).

The different components of the photosynthetic apparatus are localized in different areas of the grana and the stroma lamellae. The ATP synthases of the chloroplast are located on the thylakoid membranes (Figure 1.18C). During photosynthesis, light-driven electron transfer reactions result in a proton gradient across the thylakoid membrane (Figure 1.18D). As in the mitochondria, ATP is synthesized when the proton gradient is dissipated via the ATP synthase.

Plastids that contain high concentrations of carotenoid pigments rather than chlorophyll are called **chromoplasts**. They are one of the causes of the yellow, orange, or red colors of many fruits and flowers, as well as of autumn leaves (Figure 1.19).

Nonpigmented plastids are called **leucoplasts**. The most important type of leucoplast is the **amyloplast**, a starch-storing plastid. Amyloplasts are abundant in storage tissues of the shoot and root, and in seeds. Specialized amy-

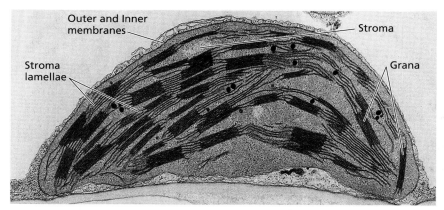
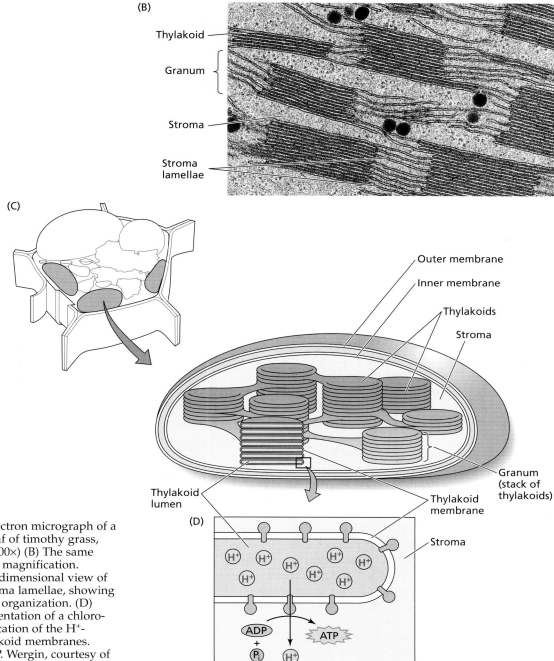

Figure 1.18 (A) Electron micrograph of a chloroplast from a leaf of timothy grass, *Phleum pratense*. (18,000×) (B) The same preparation at higher magnification. (52,000×) (C) A three-dimensional view of grana stacks and stroma lamellae, showing the complexity of the organization. (D) Diagrammatic representation of a chloroplast, showing the location of the H^+-ATPases on the thylakoid membranes. (Micrographs by W. P. Wergin, courtesy of E. H. Newcomb.)

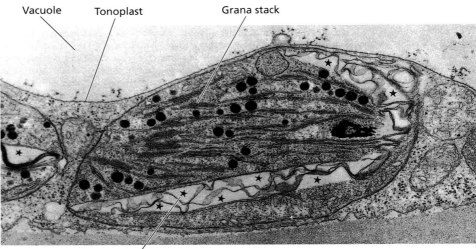

Figure 1.19 Electron micrograph of a chromoplast from tomato (*Lycopersicon esculentum*) fruit at an early stage in the transition from chloroplast to chromoplast. Small grana stacks are still visible. Stars indicate crystals of the carotenoid lycopene. (27,000×) (From Gunning and Steer 1996.)

loplasts in the root cap also serve as gravity sensors that direct root growth downward into the soil (see Chapter 19).

Mitochondria and chloroplasts are semiautonomous organelles

Both mitochondria and chloroplasts contain their own DNA and protein-synthesizing machinery (ribosomes, transfer RNAs, and other components) and are believed to have evolved from endosymbiotic bacteria (Margulis 1992). Both plastids and mitochondria divide by fission, and mitochondria can also undergo extensive fusion to form elongated structures or networks.

The DNA of these organelles is in the form of circular chromosomes, similar to those of bacteria and very different from the linear chromosomes in the nucleus. These DNA circles are localized in specific regions of the mitochondrial matrix or plastid stroma called **nucleoids**. DNA replication in both mitochondria and chloroplasts is independent of DNA replication in the nucleus. On the other hand, the numbers of these organelles within a given cell type remain approximately constant, suggesting that some aspects of organelle replication are under cellular regulation.

The mitochondrial genome of plants consists of about 200 kilobase pairs (200,000 base pairs), a size considerably larger than that of most animal mitochondria. The mitochondria of meristematic cells are typically polyploid; that is, they contain multiple copies of the circular chromosome. However, the number of copies per mitochondrion gradually decreases as cells mature, because the mitochondria continue to divide in the absence of DNA synthesis.

Most of the proteins encoded by the mitochondrial genome are prokaryotic-type 70S ribosomal proteins and components of the electron transfer system. The majority of mitochondrial proteins, including Krebs cycle enzymes, are encoded by nuclear genes and are imported from the cytosol.

The chloroplast genome is smaller than the mitochondrial genome, about 145 kilobase pairs (145,000 base pairs). Whereas mitochondria are polyploid only in the meristems, chloroplasts become polyploid during cell maturation. Thus the average amount of DNA per chloroplast in the plant is much greater than that of the mitochondria. The total amount of DNA from the mitochondria and plastids combined is about one-third of the nuclear genome (Gunning and Steer 1996).

Chloroplast DNA encodes rRNA; transfer RNA (tRNA); the large subunit of the enzyme that fixes CO_2, ribulose-1,5-bisphosphate carboxylase/oxygenase (rubisco); and several of the proteins that participate in photosynthesis. Nevertheless, the majority of chloroplast proteins, like those of mitochondria, are encoded by nuclear genes, synthesized in the cytosol, and transported to the organelle. Although mitochondria and chloroplasts have their own genomes and can divide independently of the cell, they are characterized as *semiautonomous organelles* because they depend on the nucleus for the majority of their proteins.

Different plastid types are interconvertible

Meristem cells contain proplastids, which have few or no internal membranes, no chlorophyll, and an incomplete complement of the enzymes necessary to carry out photosynthesis (Figure 1.20A). In angiosperms and some gymnosperms, chloroplast development from proplastids is triggered by light. Upon illumination, enzymes are formed inside the proplastid or imported from the cytosol, light-absorbing pigments are produced, and membranes proliferate rapidly, giving rise to stroma lamellae and grana stacks (Figure 1.20B).

Seeds usually germinate in the soil away from light, and chloroplasts develop only when the young shoot is exposed to light. If seeds are germinated in the dark, the proplastids differentiate into **etioplasts**, which contain

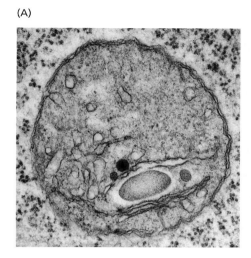
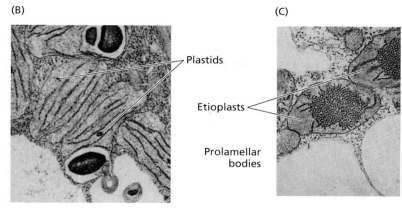

FIGURE 1.20 Electron micrographs illustrating several stages of plastid development. (A) A higher-magnification view of a proplastid from the root apical meristem of the broad bean (*Vicia faba*). The internal membrane system is rudimentary, and grana are absent. (47,000×) (B) A mesophyll cell of a young oat (*Avena sativa*) leaf at an early stage of differentiation in the light. The plastids are developing grana stacks. (C) A cell from a young oat leaf from a seedling grown in the dark. The plastids have developed as etioplasts, with elaborate semicrystalline lattices of membrane tubules called prolamellar bodies. When exposed to light, the etioplast can convert to a chloroplast by the disassembly of the prolamellar body and the formation of grana stacks. (7,200×) (From Gunning and Steer 1996.)

semicrystalline tubular arrays of membrane known as **prolamellar bodies** (Figure 1.20C). Instead of chlorophyll, the etioplast contains a pale yellow-green precursor pigment, protochlorophyllide.

Within minutes after exposure to light, the etioplast differentiates, converting the prolamellar body into thylakoids and stroma lamellae, and the protochlorophyllide into chlorophyll (for a discussion of chlorophyll biosynthesis, see **Web Topic 7.11**). The maintenance of chloroplast structure depends on the presence of light, and mature chloroplasts can revert to etioplasts during extended periods of darkness.

Chloroplasts can be converted to chromoplasts, as in autumn leaves and ripening fruit, and in some cases this process is reversible. And amyloplasts can be converted to chloroplasts, which explains why exposure of roots to light often results in greening of the roots.

Individual plastids may also be interconnected via stroma-filled tubules (**stromules**), forming a plastid network. The function of stromules is currently unknown, but presumably they facilitate communication among plastids and coordination of their activities.

Microbodies play specialized metabolic roles in leaves and seeds

Plant cells also contain **microbodies**, a class of spherical organelles surrounded by a single membrane and specialized for one of several metabolic functions. The two main types of microbodies are peroxisomes and glyoxysomes.

Peroxisomes are found in all eukaryotic organisms, and in plants they are present in photosynthetic cells (Figure 1.21). Peroxisomes function in the removal of hydrogens from organic substrates, consuming O_2 in the process, according to the following reaction:

$$RH_2 + O_2 \rightarrow R + H_2O_2$$

where R is the organic substrate. The potentially harmful peroxide produced in these reactions is broken down in peroxisomes by the enzyme catalase, according to the following reaction:

$$H_2O_2 \rightarrow H_2O + \tfrac{1}{2}O_2$$

Although some oxygen is regenerated during the catalase reaction, there is a net consumption of oxygen overall.

Another type of microbody, the **glyoxysome**, is present in oil-storing seeds. Glyoxysomes contain the *glyoxylate cycle* enzymes, which help convert stored fatty acids into sugars that can be translocated throughout the young plant to provide energy for growth (see Chapter 11). Because both types of microbodies carry out oxidative reactions, it has been suggested they may have evolved from primitive respiratory organelles that were superseded by mitochondria.

Oleosomes are lipid-storing organelles

In addition to starch and protein, many plants synthesize and store large quantities of triacylglycerol in the form of oil during seed development. These oils accumulate in organelles called oleosomes, also referred to as *lipid bodies* or *spherosomes* (Figure 1.22A).

FIGURE 1.21 Electron micrograph of a peroxisome from a mesophyll cell, showing a crystalline core. (27,000×) This peroxisome is seen in close association with two chloroplasts and a mitochondrion, probably reflecting the cooperative role of these three organelles in photorespiration. (Micrograph from Huang 1987.)

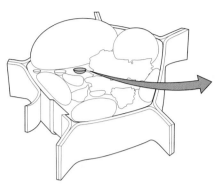

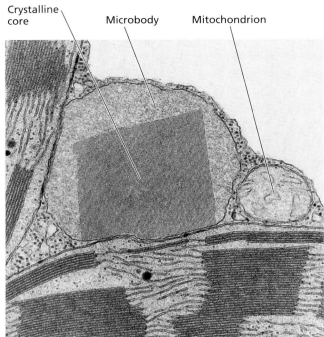

Oleosomes are unique among the organelles in that they are surrounded by a "half–unit membrane"—that is, a phospholipid monolayer—derived from the ER (Harwood 1997). The phospholipids in the half–unit membrane are oriented with their polar head groups toward the aqueous phase of the cytosol and their hydrophobic fatty acid tails facing the lumen, dissolved in the stored lipid. Oleosomes are thought to arise from the deposition of lipids within the bilayer itself (Figure 1.22B).

Proteins called **oleosins** are present in the half–unit membrane (see Figure 1.22B). One of the proposed func-

tions of the oleosins is to maintain each oleosome as a discrete organelle by preventing fusion with other oleosomes. Oleosins may also help other proteins bind to the organelle surface. As noted earlier, during seed germination the lipids in the oleosomes are broken down and converted to sucrose with the help of the glyoxysome. The first step in the process is the hydrolysis of the fatty acid chains from

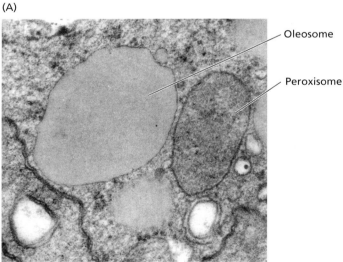

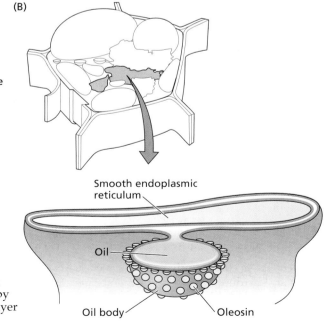

FIGURE 1.22 (A) Electron micrograph of an oleosome beside a peroxisome. (B) Diagram showing the formation of oleosomes by the synthesis and deposition of oil within the phospholipid bilayer of the ER. After budding off from the ER, the oleosome is surrounded by a phospholipid monolayer containing the protein oleosin. (A from Huang 1987; B after Buchanan et al. 2000.)

the glycerol backbone by the enzyme lipase. Lipase is tightly associated with the surface of the half–unit membrane and may be attached to the oleosins.

The Cytoskeleton

The cytosol is organized into a three-dimensional network of filamentous proteins called the **cytoskeleton**. This network provides the spatial organization for the organelles and serves as a scaffolding for the movements of organelles and other cytoskeletal components. It also plays fundamental roles in mitosis, meiosis, cytokinesis, wall deposition, the maintenance of cell shape, and cell differentiation.

Plant cells contain microtubules, microfilaments, and intermediate filaments

Three major types of cytoskeletal elements have been demonstrated in plant cells: microtubules, microfilaments, and intermediate filament–like structures. Each type is filamentous, having a fixed diameter and a variable length, up to many micrometers.

Microtubules and microfilaments are macromolecular assemblies of globular proteins. **Microtubules** are hollow cylinders with an outer diameter of 25 nm; they are composed of polymers of the protein **tubulin**. The tubulin monomer of microtubules is a heterodimer composed of two similar polypeptide chains (α- and β-**tubulin**), each having an apparent molecular mass of 55,000 daltons (Figure 1.23A). A single microtubule consists of hundreds of thousands of tubulin monomers arranged in 13 columns called *protofilaments*.

Microfilaments are solid, with a diameter of 7 nm; they are composed of a special form of the protein found in muscle: globular actin, or **G-actin**. Each actin molecule is

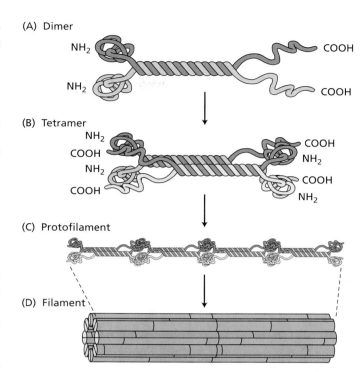

FIGURE 1.24 The current model for the assembly of intermediate filaments from protein monomers. (A) Coiled-coil dimer in parallel orientation (i.e., with amino and carboxyl termini at the same ends). (B) A tetramer of two dimers. Note that the dimers are arranged in an antiparallel fashion, and that one is slightly offset from the other. (C) Two tetramers. (D) Tetramers packed together to form the 10 nm intermediate filament. (After Alberts et al. 2002.)

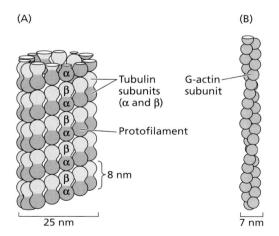

FIGURE 1.23 (A) Drawing of a microtubule in longitudinal view. Each microtubule is composed of 13 protofilaments. The organization of the α and β subunits is shown. (B) Diagrammatic representation of a microfilament, showing two strands of G-actin subunits.

composed of a single polypeptide with a molecular mass of approximately 42,000 daltons. A microfilament consists of two chains of polymerized actin subunits that intertwine in a helical fashion (Figure 1.23B).

Intermediate filaments (IFs) are a diverse group of tough, helically wound fibrous elements, 10 nm in diameter, that function in the structural support of membranes. Intermediate filaments are composed of linear polypeptide monomers of various types. In animal cells, for example, the **nuclear lamins** are composed of a specific polypeptide monomer, while the **keratins**, a type of intermediate filament found in the cytosol, are composed of a different polypeptide monomer. In animal cells, nuclear lamins and keratin IFs stabilize the nuclear envelope and plasma membrane, respectively.

Animal intermediate filaments are assembled from pairs of parallel monomers (i.e., aligned with their —NH$_2$ groups at the same ends) helically wound around each other in a **coiled coil**. Two coiled-coil dimers then align in an antiparallel fashion (i.e., with their —NH$_2$ groups at opposite ends) to form a staggered tetrameric unit. The tetrameric units then assemble into the final intermediate filament (Figure 1.24).

Immunocytochemical studies at first suggested that plant cells might contain nuclear lamins, but with the exception of a single protein from carrot, none were identified. However, a comprehensive database search of all the coiled-coil proteins in the Arabidopsis genome has now yielded several possible candidates for plant nuclear intermediate filament proteins (Rose 2004). Whether any of these proteins are functional equivalents of animal lamins in plant nuclei remains to be seen.

Microtubules and microfilaments can assemble and disassemble

In the cell, actin and tubulin subunits exist as pools of free proteins that are in dynamic equilibrium with the polymerized forms. Each of the monomers contains a bound nucleotide: ATP in the case of actin, guanosine triphosphate (GTP) in the case of tubulin. Both microtubules and microfilaments are polarized; that is, the two ends are different. In microfilaments, the polarity arises from the polarity of the actin monomer itself. In microtubules, the polarity arises from the polarity of the α- and β-tubulin heterodimer. The polarity is manifested by the different rates of growth of the two ends, with the more active end denoted the *plus* end and the less active end the *minus* end. In microtubules, the α-tubulin monomer is exposed on the minus end, while the β-tubulin is exposed on the plus end.

Microfilaments and microtubules have half-lives, usually counted in minutes, determined by accessory proteins called *actin-binding proteins* (*ABPs*) in microfilaments and *microtubule-associated proteins* (*MAPs*) in microtubules. A subset of these ABPs and MAPs stabilize microfilaments and microtubules, respectively, and prevent their depolymerization.

Polymerization of actin proceeds in three distinct stages. The rate-limiting step is the **nucleation phase**, in which several actin monomers aggregate to form an oligomeric structure. Once formed, these oligomeric structures act as stable "seeds" that bring about the **elongation phase** of assembly, which extend the length of the actin microfilament. Finally, the **steady-state phase** is reached, during which the length of the microfilament remains constant, although exchange with the pool of monomers continues to take place. The attachments between the actin subunits in the polymers are noncovalent, and no energy is required for assembly. However, after a monomer is incorporated into a protofilament, the bound ATP is hydrolyzed to ADP. The energy released is stored in the microfilament itself, making it more prone to dissociation.

The assembly of microtubules follows a similar pattern, involving nucleation, elongation and steady state phases. However, microtubule nucleation in vivo is not mediated by oligomerization of its constituent subunits, but by a small ring-shaped complex of a much less abundant type of tubulin called *γ-tubulin*. γ-Tubulin ring complexes (γ-TuRCs) are present at sites, called *microtubule organizing centers* (*MTOCs*), where microtubules are nucleated. The principal sites of MT nucleation include the cortical cytoplasm (outer layer of cytoplasm adjacent to the plasma membrane) in cells at interphase, and the spindle poles in dividing cells. The function of γ-TuRCs is to prime the polymerization of α- and β-tubulin heterodimers into short longitudinal **protofilaments**. Next, a group of 13 protofilaments associate laterally to form a flat sheet. Finally, the sheet composed of 13 protofilaments rolls into a cylindrical microtubule.

Each tubulin heterodimer contains two GTP molecules, one on the α-tubulin monomer and the other on the β-tubulin monomer. The GTP on the α-tubulin monomer is tightly bound and nonhydrolyzable, while the GTP on the β-tubulin site is readily exchangeable with the medium and is slowly hydrolyzed to GDP after the subunit assembles onto a microtubule. The hydrolysis of GTP to GDP on the β-tubulin subunit causes the dimer to bend slightly and, if not capped by recently added GTP-charged tubulin, the protofilaments come apart from each other, initiating a "catastrophic" depolymerization that is much more rapid than the rate of polymerization. Such depolymerization "catastrophes" can be "rescued" by the resulting increase in the local tubulin concentration, which (with the help of GTP) tends to favor polymerization (Figure 1.25).

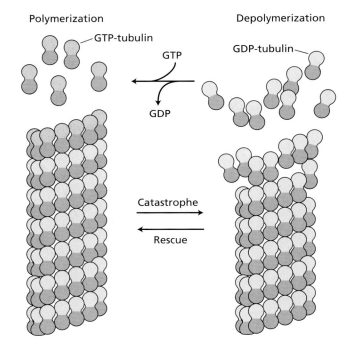

Figure 1.25 Model for the dynamic equilibrium between polymerization and depolymerization of a microtubule. Catastrophic depolymerization leads to an increase in the local tubulin concentration. Rescue occurs when the exchange of GTP for GDP promotes the tubulin polymerization reaction.

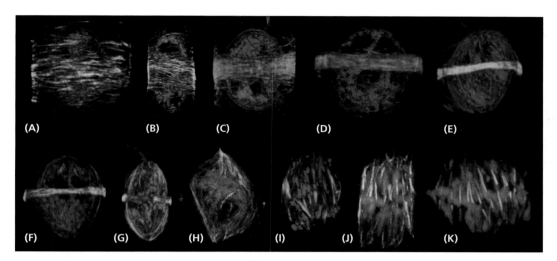

FIGURE 1.26 Fluorescence micrograph taken with a confocal microscope showing changes in microtubule arrangements at different stages in the cell cycle of wheat root meristem cells. Microtubules stain green and yellow; DNA is blue. (A–D) Cortical microtubules disappear and the preprophase band is formed around the nucleus at the site of the future cell plate. (E–H) The prophase spindle forms from foci of microtubules at the poles. (G, H) The preprophase band disappears in late prophase. (I–K) The nuclear membrane breaks down, and the two poles become more diffuse. The mitotic spindle forms in parallel arrays and the kinetochores bind to spindle microtubules. (From Gunning and Steer 1996.)

Neighboring microtubules in a cell do not undergo catastrophes at the same time. Microtubules are said to be *dynamically unstable*, and in any grouping of microtubules some are growing while others are rapidly shrinking. The frequency and extent of catastrophic depolymerization that characterizes dynamic instability can be controlled by specific MAPs that can stabilize or destabilize the linkages between the tubulin heterodimers in the wall of the microtubule.

Microtubules function in mitosis and cytokinesis

Mitosis is the process by which previously replicated chromosomes are aligned, separated, and distributed in an orderly fashion to daughter cells (Figure 1.26). Microtubules are an integral part of mitosis. Before mitosis begins, microtubules in the cortical (outer) cytoplasm depolymerize, breaking down into their constituent subunits. The subunits then repolymerize before the start of prophase to form the **preprophase band** (**PPB**), a ring of microtubules encircling the nucleus (see Figure 1.26C–F). The PPB appears in the region where the future cell wall will form after the completion of mitosis, and it is thought to be involved in regulating the plane of cell division.

During prophase, microtubules begin to assemble at two foci on opposite sides of the nucleus, forming the **prophase spindle** (Figure 1.27). Although not associated with any specific structure, these foci serve the same function as animal cell centrosomes in organizing and assembling microtubules.

In early metaphase the nuclear envelope breaks down, the PPB disassembles, and new microtubules polymerize to form the mitotic spindle. In animal cells the spindle microtubules radiate toward each other from two discrete foci at the poles (the centrosomes), resulting in an ellipsoidal, or football-shaped, array of microtubules. The mitotic spindle of plant cells, which lack centrosomes, is more boxlike in shape because the spindle microtubules arise from a diffuse zone consisting of multiple foci at opposite ends of the cell and extend toward the middle in nearly parallel arrays (see Figure 1.27).

Some of the microtubules of the spindle apparatus become attached to the chromosomes at their **kinetochores**, while others remain unattached. The kinetochores are located in the **centromeric** regions of the chromosomes. Some of the unattached microtubules overlap with microtubules from the opposite polar region in the spindle midzone.

Cytokinesis is the process whereby a cell is partitioned into two progeny cells. Cytokinesis usually begins late in mitosis. The precursor of the new wall, the **cell plate** that forms between incipient daughter cells, is rich in pectins (Figure 1.28). Cell plate formation in higher plants is a multistep process (see **Web Topic 1.5**). Vesicle aggregation in the **spindle midzone** is organized by the **phragmoplast**, a complex of microtubules and ER that forms during late anaphase or early telophase from dissociated spindle subunits. The phragmoplast facilitates the delivery of secretory vesicles containing cell wall material to the growing cell

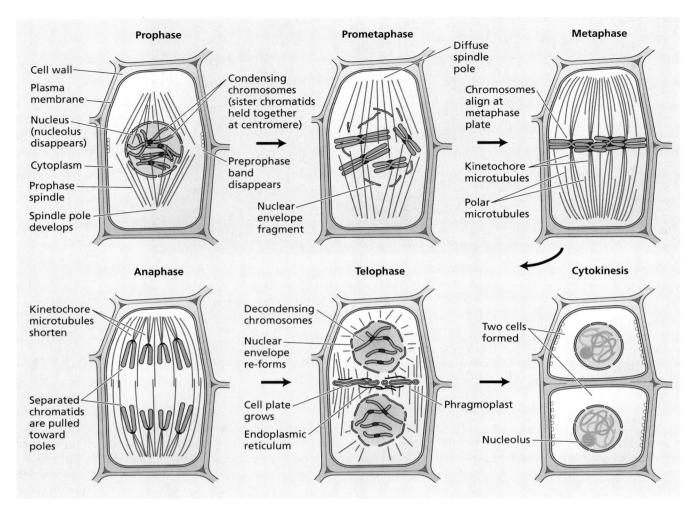

FIGURE 1.27 Diagram of mitosis in plants.

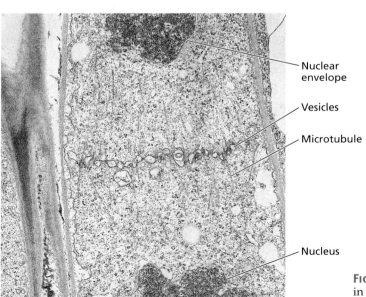

plate. Once thought to be peculiar to plant cells, the phragmoplast is now known to play a key role in cytokinesis in animals cells as well (Albertson et al. 2005).

Motor proteins mediate cytoplasmic streaming and organelle movements

Cytoplasmic streaming is the coordinated movement of particles and organelles through the cytosol in a helical path down one side of a cell and up the other side. Cytoplasmic streaming occurs in most plant cells and has been studied extensively in the giant cells of the green algae *Chara* and *Nitella*, in which speeds up to 75 μm s^{-1} have been measured.

FIGURE 1.28 Electron micrograph of a cell plate forming in a maple (*Acer*) seedling. (10,000×) (© E. H. Newcomb and B. A. Palevitz/Biological Photo Service.)

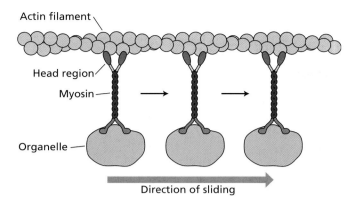

Figure 1.29 Myosin-mediated transport of organelles along actin microfilaments is the basis of cytoplasmic streaming. Myosin is a homodimer with two heads and two tails. The two heads, shown in red, have ATPase and motor activity, and the tails bind to the organelle. The blue shapes below the two heads indicate the regions that bind to calmodulin, a calcium-activated regulatory protein (see Chapter 14 on the web site).

The mechanism of cytoplasmic streaming involves bundles of actin microfilaments that are arranged parallel to the longitudinal direction of particle movement. The forces necessary for movement may be generated by an interaction of the microfilament protein actin with the protein myosin in a fashion comparable to that of the sliding protein interaction that occurs during muscle contraction in animals (Shimmen and Yokota 2004).

Myosins are proteins that have the ability to hydrolyze ATP to ADP and P_i when activated by binding to an actin microfilament. The energy released by ATP hydrolysis propels myosin molecules along the actin microfilament from the minus end to the plus end. Thus, myosins belong to the general class of **motor proteins** that drive cytoplasmic streaming and the movements of organelles within the cell (Figure 1.29). Examples of other motor proteins include the **kinesins** and **dyneins**, which drive movements of organelles and other cytoskeletal components along the surfaces of microtubules. The speed of movement along the microtubule system (~150 nm sec^{-1}) is much slower than that of streaming induced by the actin–myosin system.

Actin microfilaments also participate in the growth of the pollen tube. Upon germination, a pollen grain forms a tubular extension that grows down the style toward the embryo sac. As the tip of the pollen tube extends, new cell wall material is continually deposited to maintain the integrity of the wall. A network of microfilaments appears to guide vesicles containing wall precursors from their site of formation in the Golgi through the cytosol to the site of new wall formation at the tip. Fusion of these vesicles with the plasma membrane deposits wall precursors outside the cell, where they are assembled into wall material.

Microtubules, along with the motor proteins kinesin or dynein, are involved in chloroplast movements regulated by red and far-red light (see Chapter 17), while both microtubules and microfilaments mediate chloroplast movements in response to blue light (see Chapter 18).

Cell Cycle Regulation

The cell division cycle, or cell cycle, is the process by which cells reproduce themselves and their genetic material, the nuclear DNA. The four phases of the cell cycle are designated G_1, S, G_2, and M (Figure 1.30A).

Each phase of the cell cycle has a specific set of biochemical and cellular activities

Nuclear DNA is prepared for replication in G_1 by the assembly of a prereplication complex at the origins of replication along the chromatin. DNA is replicated during the S phase, and G_2 cells prepare for mitosis.

The whole architecture of the cell is altered as cells enter mitosis: The nuclear envelope breaks down, chromatin condenses to form recognizable chromosomes, the mitotic spindle forms, and the replicated chromosomes attach to the spindle fibers. The transition from metaphase to anaphase of mitosis marks a major transition point when the two chromatids of each replicated chromosome, which were held together at their kinetochores, are separated and the daughter chromosomes are pulled to opposite poles by spindle fibers.

At a key regulatory point early in G_1 of the cell cycle, the cell becomes committed to the initiation of DNA synthesis. In yeasts, this point is called START. Once a cell has passed START, it is irreversibly committed to initiating DNA synthesis and completing the cell cycle through mitosis and cytokinesis. After the cell has completed mitosis, it may initiate another complete cycle (G_1 through mitosis), or it may leave the cell cycle and differentiate. This choice is made at the critical G_1 point, before the cell begins to replicate its DNA.

DNA replication and mitosis are linked in mammalian cells. Often mammalian cells that have stopped dividing can be stimulated to reenter the cell cycle by a variety of hormones and growth factors. When they do so, they reenter the cell cycle at the critical point in early G_1. In contrast, plant cells can leave the cell division cycle either before or after replicating their DNA (i.e., during G_1 or G_2). As a consequence, whereas most animal cells are diploid (having two sets of chromosomes), plant cells frequently are tetraploid (having four sets of chromosomes), or even polyploid (having many sets of chromosomes), after going through additional cycles of nuclear DNA replication without mitosis.

The cell cycle is regulated by cyclin-dependent kinases

The biochemical reactions governing the cell cycle are evolutionarily highly conserved in eukaryotes, and plants have

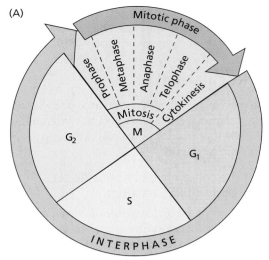
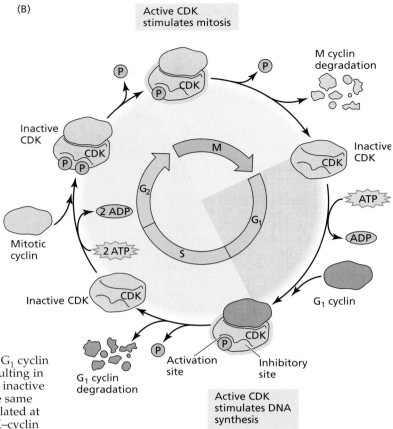

Figure 1.30 (A) Diagram of the cell cycle. (B) Diagram of the regulation of the cell cycle by cyclin-dependent protein kinase. During G_1, CDK is in its inactive form. CDK becomes activated by binding to G_1 cyclin and by being phosphorylated (P) at the activation site. The activated CDK–cyclin complex allows the transition to the S phase. At the end of the S phase, the G_1 cyclin is degraded and the CDK is dephosphorylated, resulting in an inactive CDK. The cell enters G_2. During G_2, the inactive CDK binds to the mitotic cyclin, or M cyclin. At the same time, the CDK–cyclin complex becomes phosphorylated at both its activation and its inhibitory sites. The CDK–cyclin complex is still inactive, because the inhibitory site is phosphorylated. The inactive complex becomes activated when the phosphate is removed from the inhibitory site by a protein phosphatase. The activated CDK then stimulates the transition from G_2 to mitosis. At the end of mitosis, the mitotic cyclin is degraded and the remaining phosphate at the activation site is removed by the phosphatase, and the cell enters G_1 again.

1. G_1/S cyclins: cyclin E, active late in G_1
2. S cyclins: cyclin A, active at the beginning of the S-phase
3. M cyclins: cyclin B, active just prior to the mitotic phase
4. G_1 cyclins: cyclin D, active early in G_1

retained the basic components of this mechanism (Renaudin et al. 1996). Progression through the cycle is regulated mainly at two control points: during the late G_1 phase and at the G_2/M boundary. The restriction point at late G_1 is especially significant, since cells must integrate extracellular signals and either commit to a further round of cell division or stop dividing and begin to differentiate.

The key enzymes that control the transitions between the different states of the cell cycle, and the entry of nondividing cells into the cell cycle, are the **cyclin-dependent protein kinases**, or **CDKs** (Figure 1.30B). Protein kinases are enzymes that phosphorylate proteins using ATP. Most multicellular eukaryotes use several protein kinases that are active in different phases of the cell cycle. All depend on regulatory subunits called **cyclins** for their activities. Four classes of cyclins have been identified in animals and yeast:

The critical restriction point late in G_1, which commits the cell to another round of cell division, is regulated primarily by the D-type cyclins. As we will see later in the book, plant hormones that promote cell division, including cytokinins (see Chapter 21) and brassinosteroids (see Chapter 24), appear to do so at least in part through an increase in CycD3, a plant D-type cyclin.

CDK activity can be regulated in various ways, but two of the most important mechanisms are (1) cyclin synthesis and destruction and (2) the phosphorylation and dephosphorylation of key amino acid residues within the CDK protein. In the first regulatory mechanism, CDKs are inactive unless they are associated with a cyclin, and most cyclins turn over rapidly. They are synthesized and then actively degraded (using ATP) at specific points in the cell cycle. Cyclins are degraded in the cytosol by a large proteolytic complex called the **proteasome** (see Chapter 14 on the web site). Before being degraded by the proteasome, the cyclins are marked for destruction by the attachment of a small protein called *ubiquitin*, a process that requires ATP. Ubiquitination is a general mechanism for tagging cellular proteins destined for turnover (see Chapter 14).

The second mechanism regulating CDK activity is phosphorylation and dephosphorylation. CDKs possess two tyrosine phosphorylation sites: One causes activation of the enzyme; the other causes inactivation. Specific kinases carry out both the stimulatory and the inhibitory phosphorylations. Similarly, protein phosphatases can remove phosphates from CDKs, either stimulating or inhibiting their activity depending on the position of the phosphate. The addition or removal of phosphate groups from CDKs is highly regulated and an important mechanism for the control of cell cycle progression (see Figure 1.30B). Cyclin inhibitors play an important role in regulating the cell cycle in animals, and probably in plants as well, although little is known about plant cyclin inhibitors.

Plasmodesmata

Plasmodesmata (singular *plasmodesma*) are tubular extensions of the plasma membrane, 40 to 50 nm in diameter, that traverse the cell wall and connect the cytoplasms of adjacent cells. Because most plant cells are interconnected in this way, their cytoplasms form a continuum referred to as the *symplast*. Intercellular transport of solutes through plasmodesmata is thus called *symplastic transport* (see Chapters 4 and 6).

There are two types of plasmodesmata: primary and secondary

Primary plasmodesmata form during cytokinesis when Golgi-derived vesicles containing cell wall precursors fuse to form the cell plate (the future middle lamella). Rather than forming a continuous uninterrupted sheet, the newly deposited cell plate is penetrated by numerous pores (Figure 1.31A and B) where remnants of the spindle apparatus, consisting of ER and microtubules, disrupt vesicle fusion. Further deposition of wall polymers increases the thickness of the two primary cell walls on either side of the middle lamella, generating linear membrane-lined channels (see Figure 1.31B). Development of primary plasmodesmata thus provides direct continuity and communication between cells that are clonally related (i.e., derived from the same mother cell).

Secondary plasmodesmata form between cells after their cell walls have been deposited. They arise either by evagination of the plasma membrane at the cell surface, or by branching from a primary plasmodesma (Lucas and Wolf 1993). In addition to increasing the communication between clonally related cells, secondary plasmodesmata allow symplastic continuity between cells that are not clonally related.

Plasmodesmata have a complex internal structure

Like nuclear pores, plasmodesmata have a complex internal structure that functions in regulating macromolecular traffic from cell to cell. Each plasmodesma contains a narrow tubule of ER called a *desmotubule* (see Figure 1.2). The desmotubule is continuous with the ER of the adjacent cells. Thus the symplast joins not only the cytosol of neighboring cells, but the contents of the ER lumens as well. However, it is not clear that the desmotubule actually represents a passage, since the membranes are tightly appressed. There is some evidence that lipophilic substances may diffuse from cell to cell via the desmotubule membrane, but the significance of desmotubular transport for intercellular signaling has not been established. Macromolecular trafficking via plasmodesmata occurs primarily in the **cytoplasmic sleeve**, a narrow space between the desmotubule and the plasma membrane (Roberts and Oparka 2003).

Within the cytoplasmic sleeve, both the desmotubule and the plasma membrane are studded with globular proteins arranged in helical rows (see Figure 1.31C). These globular subunits appear to be interconnected by spokelike extensions, dividing the sleeve into eight to ten spiraling microchannels (Ding et al. 1992). The size of these microchannels serves to restrict the sizes of the molecules that can pass through the pore. By following the movement of fluorescent dye molecules of different sizes through plasmodesmata connecting leaf epidermal cells, Robards and Lucas (1990) determined the limiting molecular mass for transport to be about 700 to 1000 daltons, equivalent to a molecular size of about 1.5 to 2.0 nm. This is the **size exclusion limit**, or **SEL**, of plasmodesmata.

If the SEL of plasmodesmata restricts traffic between cells to molecules 2.0 nm or less, how do particles as large as tobacco mosaic virus (a rod 18 nm wide and 300 nm long) pass through them? The answer is that the SELs of plasmodesmata are not fixed, but can be regulated. The mechanism for regulating the SEL is poorly understood, but the localization of both actin and myosin to the plasmodesmatal pore suggests that they may participate in the process (Roberts and Oparka 2003). Two possibilities have been suggested:

1. The actin and myosin may enable the pore to constrict, reducing the SEL of the pore.
2. The actin and myosin may themselves facilitate the movement of macromolecules and particles through the pore.

Macromolecular traffic through plasmodesmata is important for developmental signaling

Development in multicellular organisms is a complex process, and cell-to-cell communication is needed to coordinate the activities of the various cells and tissues. The signals involved in this process involve not only small molecules, such as hormones, but proteins and RNA as well. In Chapter 16, we will examine this type of cell-to-cell communication in greater detail. Although first discovered in plants, macromolecular trafficking between cells through tubular membrane channels has been shown to be a property of animal cells as well (Gallagher and Benfey 2005).

Summary

Despite their great diversity in form and size, all plants carry out similar physiological processes. As primary producers, plants convert solar energy to chemical energy. Being nonmotile, plants must grow toward light, and they

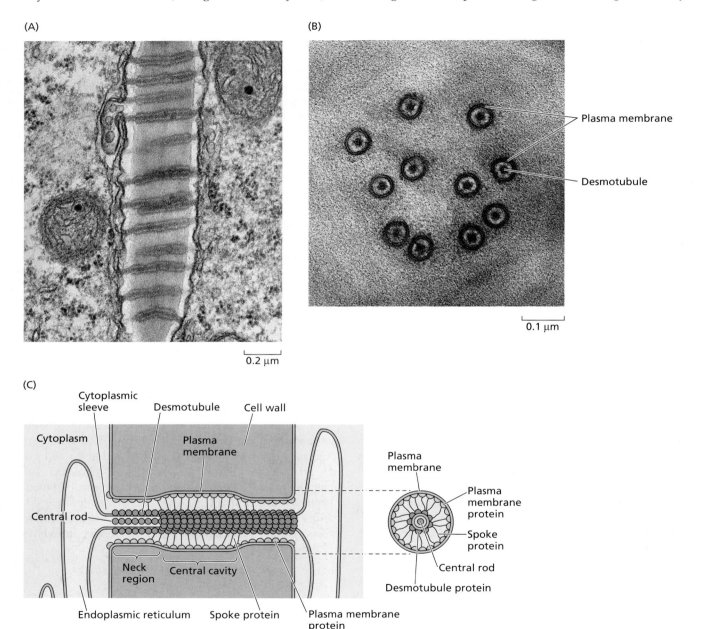

Figure 1.31 Plasmodesmata between cells. (A) Electron micrograph of a wall separating two adjacent cells, showing the plasmodesmata in longitudinal view. (B) Tangential section through a cell wall showing numerous plasmodesmata in cross-section. (C) Schematic view of a cell wall with a plasmodesma. The pore consists of a central cavity between two narrow neck regions. The desmotubule is continuous with the ER of the adjoining cells. Proteins line the outer surface of the desmotubule and the inner surface of the plasma membrane; the two surfaces are thought to be connected by filamentous proteins, which divide the cytoplasmic sleeve into microchannels. The size of the gaps between the proteins controls the molecular sieving properties of plasmodesmata. (A and B courtesy of Ray Evert; from Robinson-Beers and Evert 1991.)

must have efficient vascular systems for movement of water, mineral nutrients, and photosynthetic products throughout the plant body. Green land plants must also have mechanisms for avoiding desiccation.

The major vegetative organ systems of seed plants are the shoot and the root. The shoot consists of two types of organs: stems and leaves. Unlike animal development, plant growth is indeterminate because of the presence of permanent meristem tissue at the shoot and root apices, which gives rise to new tissues and organs during the entire vegetative phase of the life cycle. Lateral meristems (the vascular cambium and the cork cambium) produce growth in girth, or secondary growth.

Three major tissue systems are recognized: dermal, ground, and vascular. Each of these tissues contains a variety of cell types specialized for different functions.

Plants are eukaryotes and have the typical eukaryotic cell organization, consisting of nucleus and cytoplasm. The nuclear genome directs the growth and development of the organism. The cytoplasm is enclosed by a plasma membrane and contains numerous membrane-enclosed organelles, including plastids, mitochondria, microbodies, oleosomes, and a large central vacuole. Chloroplasts and mitochondria are semiautonomous organelles that contain their own DNA. Nevertheless, most of their proteins are encoded by nuclear DNA and are imported from the cytosol.

The cytoskeletal components—microtubules, microfilaments, and intermediate filaments—participate in a variety of processes involving intracellular movements, such as mitosis, cytoplasmic streaming, secretory vesicle transport, cell plate formation, and cellulose microfibril deposition. The process by which cells reproduce is called the cell cycle. The cell cycle consists of the G_1, S, G_2, and M phases. The transition from one phase to another is regulated by cyclin-dependent protein kinases. The activity of the CDKs is regulated by cyclins and by protein phosphorylation.

During cytokinesis, the phragmoplast gives rise to the cell plate in a multistep process that involves vesicle fusion. After cytokinesis, primary cell walls are deposited. The cytosol of adjacent cells is continuous through the cell walls because of the presence of membrane-lined channels called plasmodesmata, which facilitate and regulate cell-to-cell communication.

Web Material

Web Topics

1.1 The Plant Kingdom
The major groups of the plant kingdom are surveyed and described.

1.2 Flower Structure and the Angiosperm Life Cycle
The steps in the reproductive style of angiosperms are discussed and illustrated.

1.3 Plant Tissue Systems: Dermal, Ground, and Vascular
A more detailed treatment of plant anatomy is given.

1.4 The Structures of Chloroplast Glycosylglycerides
The chemical structures of the chloroplast lipids are illustrated.

1.5 The Multiple Steps in Construction of the Cell Plate Following Mitosis
Details of the production of the cell plate during cytokinesis in plants are described.

Chapter References

Alberts, B., Johnson, A., Lewis, J., Raff, M., Roberts, K., and Walter, P. (2002) *Molecular Biology of the Cell*, 4th ed. Garland, New York.

Albertson, R., Riggs, B., and Sullivan, W. (2005) Membrane traffic: A driving force in cytokinesis. *Trends Cell Biol.* 15: 92–101.

Buchanan, B. B., Gruissem, W., and Jones, R. L. (eds.) (2000) *Biochemistry and Molecular Biology of Plants*. Amer. Soc. Plant Physiologists, Rockville, MD.

Denning, D. P., Patel, S. S., Uversky, V., Fink, A. L., and Rexach, M. (2003) Disorder in the nuclear pore complex: The FG repeat regions of nucleoporins are natively unfolded. *Proc. Natl. Acad. Sci. USA* 100: 2450–2455.

Ding, B., Turgeon, R., and Parthasarathy, M. V. (1992) Substructure of freeze substituted plasmodesmata. *Protoplasma* 169: 28–41.

Driouich, A., Levy, S., Staehelin, L. A., and Faye, L. (1994) Structural and functional organization of the Golgi apparatus in plant cells. *Plant Physiol. Biochem.* 32: 731–749.

Frederick, S. E., Mangan, M. E., Carey, J. B., and Gruber, P. J. (1992) Intermediate filament antigens of 60 and 65 kDa in the nuclear matrix of plants: Their detection and localization. *Exp. Cell Res.* 199: 213–222.

Gallagher, K. L., and Benfey, P. N. (2005) Not just another hole in the wall: Understanding intercellular protein trafficking. *Genes Dev.* 19: 189–195.

Gunning, B. E. S., and Steer, M. W. (1996) *Plant Cell Biology: Structure and Function of Plant Cells*. Jones and Bartlett, Boston.

Harwood, J. L. (1997) Plant lipid metabolism. In *Plant Biochemistry*, P. M. Dey and J. B. Harborne, eds., Academic Press, San Diego, CA, pp. 237–272.

Huang, A. H. C. (1987) Lipases in *The Biochemistry of Plants: A Comprehensive Treatise*. In Vol. 9, *Lipids: Structure and Function*, P. K. Stumpf, ed. Academic Press, New York, pp. 91–119.

Lucas, W. J., and Wolf, S. (1993) Plasmodesmata: The intercellular organelles of green plants. *Trends Cell Biol.* 3: 308–315.

Lodish, H., Berk, A., Matsudaira, P., Kaiser, C. A., Krieger, M., Scott, M. P., Zipursky, S. L., and Darnell, J. (2004) *Molecular Cell Biology*, 5th ed. Freeman, New York.

Margulis, L. 1992. *Symbiosis in Cell Evolution: Microbial Communities in the Archean and Proterozoic Eons*. W. H. Freeman.

Nebenführ, A., and Staehelin, L. A. (2001) Mobile factories: Golgi dynamics in plant cells. *Trends Plant Sci.* 6: 160–167.

Otegui, M. S., Noh, Y.-S., Martinez, D. E., Vila Petroff, M. G., Staehelin, L. A., Amasino, R. M., and Guiamet, J. J. (2005) Senescence-associated vacuoles with intense proteolytic activity develop in leaves of *Arabidopsis* and soybean. *Plant J.* 41: 831–844.

Renaudin, J.-P., Doonan, J. H., Freeman, D., Hashimoto, J., Hirt, H., Inze, D., Jacobs, T., Kouchi, H., Rouze, P., Sauter, M., et al. (1996) Plant cyclins: A unified nomenclature for plant A-, B- and D-type cyclins based on sequence organization. *Plant Mol. Biol.* 32: 1003–1018.

Robards, A. W., and Lucas, W. J. (1990) Plasmodesmata. *Annu. Rev. Plant Physiol. Plant Mol. Biol.* 41: 369–420.

Roberts, A. G., and Oparka, K. J. (2003) Plasmodesmata and the control of symplastic transport. *Plant Cell Environ.* 26: 103–124.

Robinson-Beers, K., and Evert, R. F. (1991) Fine structure of plasmodesmata in mature leaves of sugar cane. *Planta* 184: 307–318.

Rose, A., Manikantan, S., Schraegle, S. J., Maloy, M. A., Stahlberg, E. A., and Meier, I. (2004) Genome-wide identification of *Arabidopsis* coiled-coil proteins and establishment of the ARABI-COIL database. *Plant Physiol.* 134: 927–939.

Shimmen, T., and Yokota, E. (2004) Cytoplasmic streaming in plants. *Curr. Opin. Cell Biol.* 16: 68–72.

Tilney, L. G., Cooke, T. J., Connelly, P. S., and Tilney, M. S. (1991) The structure of plasmodesmata as revealed by plasmolysis, detergent extraction, and protease digestion. *J. Cell Biol.* 112: 739–748.

White, R. G., Badelt, K., Overall, R. L., and Vesk, M. (1994) Actin associated with plasmodesmata. *Protoplasma* 180: 169–184.

Yang, C., Min, G. W., Tong, X. J., Luo, Z., Liu, Z. F., and Zhai, Z. H. (1995) The assembly of keratins from higher plant cells. *Protoplasma* 188: 128–132.

Web Chapter 2 | Energy and Enzymes

Content available at www.plantphys.net

Energy Flow through Living Systems

Energy and Work
The first law: The total energy is always conserved

The change in the internal energy of a system represents the maximum work it can do

Each type of energy is characterized by a capacity factor and a potential factor

The Direction of Spontaneous Processes
The second law: The total entropy always increases

A process is spontaneous if ΔS for the system and its surroundings is positive

Free Energy and Chemical Potential
ΔG is negative for a spontaneous process at constant temperature and pressure

The standard free-energy change, $\Delta G°$, is the change in free energy when the concentration of reactants and products is 1 M

The value of ΔG is a function of the displacement of the reaction from equilibrium

The enthalpy change measures the energy transferred as heat

Redox Reactions
The free-energy change of an oxidation–reduction reaction is expressed as the standard redox potential in electrochemical units

The Electrochemical Potential
Transport of and uncharged solute against its concentration gradient decreases the entropy of the system

The membrane potential is the work that must be done to move an ion from one side of the membrane to the other

The electrochemical-potential difference, $\Delta \tilde{\mu}$, includes both concentration and electric potentials

Energy flow is a universal feature of living systems. It is also at the core of atomic and molecular motions and reactions that characterize nonliving systems as well. Water molecules would not move through the soil, into the plant, up the xylem, into the leaf, and out into the atmosphere without the appropriate energy gradients to drive them. Sunlight, the visible portion of the electromagnetic spectrum, must be captured and converted to chemical energy to allow the synthesis of the complex organic molecules required for life. Indeed, there is not a single physical or chemical process in plants that does not involve energy transformations of some kind. Thus the study of energy in living systems—or bioenergetics—is central to an understanding of plant physiology.

This chapter reviews the basic concepts and principles of energy transformations, called *thermodynamics*, as they apply to plants and other organisms. The concept of free energy provides a framework for understanding the direction of spontaneous processes. The presence of semipermeable membranes in living cells allows the formation of energy gradients that drive a host of reactions necessary for life. This chapter provides an introduction to basic concepts of membrane transport, which are treated more extensively in later chapters on water relations (Chapters 3 and 4) and solute transport (Chapter 6).

Finally, life would be impossible without the presence of protein catalysts, called *enzymes*, which speed up reactions that would ordinarily occur much too slowly for living processes to be carried out. This chapter therefore ends with a brief overview of protein structure, the mechanisms of enzyme-catalyzed reactions, and the fundamental ways in which enzymes are regulated in living cells.

Enzymes: The Catalysts of Life
Proteins are chains of amino acids joined by peptide bonds
Protein structure is hierarchical
Enzymes are highly specific protein catalysts
Enzymes lower the free-energy barrier between substrates and products
A simple kinetic equation describes an enzyme-catalyzed reaction
Enzymes are subject to various kinds of inhibition
pH and temperature affect the rate of enzyme-catalyzed reactions
Cooperative systems increase the sensitivity to substrates and are usually allosteric
The kinetics of some membrane transport processes can be described by the Michaelis–Menten equation
Enzyme activity is often regulated

Summary

Unit I
Transport and Translocation of Water and Solutes

Unit I — TRANSPORT AND TRANSLOCATION OF WATER AND SOLUTES

3 Water and Plant Cells

4 Water Balance of the Plant

5 Mineral Nutrition

6 Solute Transport

*Guttation from a young leaf of lady's mantle, Alchemilla mollis.
Photograph by David McIntyre.*

Chapter 3 | Water and Plant Cells

WATER PLAYS A CRUCIAL ROLE in the life of the plant. Photosynthesis requires that plants draw carbon dioxide from the atmosphere, but doing so exposes them to water loss and the threat of dehydration. For every gram of organic matter made by the plant, approximately 500 g of water is absorbed by the roots, transported through the plant body, and lost to the atmosphere. Even slight imbalances in this flow of water can cause water deficits and severe malfunctioning of many cellular processes. Thus, balancing the uptake and loss of water represents an important challenge for land plants.

A major difference between plant and animal cells that affects virtually all aspects of their relation with water is the existence in plants of the cell wall. Cell walls allow plant cells to build up large internal hydrostatic pressures, called **turgor pressure**. Turgor pressure is essential for many physiological processes, including cell enlargement, stomatal opening, transport in the phloem, and various transport processes across membranes. Turgor pressure also contributes to the rigidity and mechanical stability of nonlignified plant tissues. In this chapter we will consider how water moves into and out of plant cells, emphasizing the molecular properties of water and the physical forces that influence water movement at the cell level.

Water in Plant Life

Of all the resources that plants need to grow and function, water is the most abundant and at the same time often the most limiting. The practice of crop irrigation reflects the fact that water is a key resource limiting agricultural productivity (Figure 3.1). Water availability likewise limits the productivity of natural ecosystems (Figure 3.2), leading to marked differences in vegetation type along precipitation gradients.

The reason that water is frequently a limiting resource for plants, but rarely so for animals, is that plants use water in huge amounts. This tremendous "thirst" is a direct consequence of the diffusional uptake of CO_2 for photosynthesis. Most (~97%) of the water absorbed by a plant's roots is carried through the plant and evaporates from leaf surfaces. Such water loss is called **transpiration**. In contrast, only a small amount of the water absorbed by roots actually remains in the plant to supply growth (~2%) or to be used in photosynthesis and other metabolic processes (~1%).

Water loss to the atmosphere appears to be an inevitable consequence of carrying out photosynthesis on land. The uptake of CO_2 is coupled to water loss because plants have not evolved a material that is differentially permeable to CO_2 compared to water. Compounding this problem is the fact that the concentration gradient for water loss from leaves is much larger than for CO_2 uptake. A typical exchange ratio for a healthy plant in well-watered soil is thus on the order of 500 water molecules lost for every CO_2 molecule gained. This unfavorable exchange has had a major influence on the evolution of plant form and function and explains why water plays such a key role in the physiology of plants.

We will begin our study of water by considering how its structure gives rise to some of its unique physical properties. We will then examine the physical basis for water movement, the concept of water potential, and the application of this concept to cell–water relations.

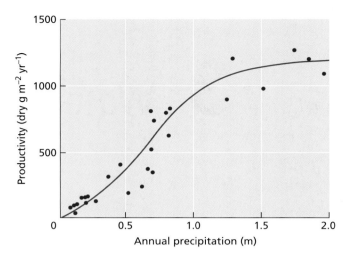

FIGURE 3.2 Productivity of various ecosystems as a function of annual precipitation. Productivity was estimated as net aboveground accumulation of organic matter through growth and reproduction. (After Whittaker 1970.)

The Structure and Properties of Water

Water has special properties that enable it to act as a solvent and to be readily transported through the body of the plant. These properties derive primarily from the polar structure of the water molecule. In this section we will examine how the formation of hydrogen bonds contributes to the high specific heat, surface tension, and tensile strength of water.

The polarity of water molecules gives rise to hydrogen bonds

The water molecule consists of an oxygen atom covalently bonded to two hydrogen atoms. The two O—H bonds form an angle of 105° (Figure 3.3). Because the oxygen atom is more **electronegative** than hydrogen, it tends to attract the electrons of the covalent bond. This attraction results in a partial negative charge at the oxygen end of the molecule and a partial positive charge at each hydrogen. These partial charges are equal, so the water molecule carries no *net* charge.

This separation of partial charges, together with the shape of the water molecule, makes water a *polar molecule*, and the opposite partial charges between neighboring water molecules tend to attract each other. The weak electrostatic attraction between water molecules, known as a **hydrogen bond**, is responsible for many of the unusual physical properties of water. Hydrogen bonds can also form between water and other molecules that contain electronegative atoms (O or N).

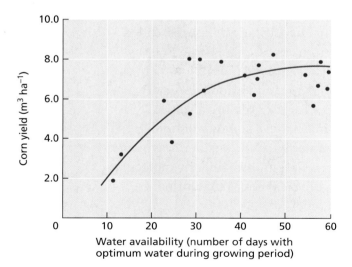

FIGURE 3.1 Corn yield as a function of water availability. The data plotted here were gathered at an Iowa farm over a four-year period. Water availability was assessed as the number of days without water stress during a nine-week growing period. (Data from Mitchell 1964.)

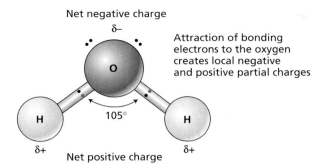

FIGURE 3.3 Diagram of the water molecule. The two intramolecular hydrogen–oxygen bonds form an angle of 105°. The opposite partial charges (δ^- and δ^+) on the water molecule lead to the formation of intermolecular hydrogen bonds with other water molecules. Oxygen has six electrons in the outer orbitals; each hydrogen has one.

The polarity of water makes it an excellent solvent

Water is an excellent solvent: It dissolves greater amounts of a wider variety of substances than do other related solvents. This versatility as a solvent is due in part to the small size of the water molecule and in part to its polar nature. The latter makes water a particularly good solvent for ionic substances and for molecules such as sugars and proteins that contain polar —OH or —NH$_2$ groups.

Hydrogen bonding between water molecules and ions, and between water and polar solutes, in solution effectively decreases the electrostatic interaction between the charged substances and thereby increases their solubility. Furthermore, the polar ends of water molecules can orient themselves next to charged or partially charged groups in macromolecules, forming **shells of hydration**. Hydrogen bonding between macromolecules and water reduces the interaction between the macromolecules and helps draw them into solution.

The thermal properties of water result from hydrogen bonding

The extensive hydrogen bonding between water molecules results in unusual thermal properties, such as high specific heat and high latent heat of vaporization. **Specific heat** is the heat energy required to raise the temperature of a substance by a specific amount.

When the temperature of water is raised, the molecules vibrate faster and with greater amplitude. To allow for this motion, energy must be added to the system to break the hydrogen bonds between water molecules. Thus, compared with other liquids, water requires a relatively large energy input to raise its temperature. This large energy input requirement is important for plants, because it helps buffer temperature fluctuations.

Latent heat of vaporization is the energy needed to separate molecules from the liquid phase and move them into the gas phase at constant temperature—a process that occurs during transpiration. For water at 25°C, the heat of vaporization is 44 kJ mol^{-1}—the highest value known for any liquid. Most of this energy is used to break hydrogen bonds between water molecules.

The high latent heat of vaporization of water enables plants to cool themselves by evaporating water from leaf surfaces, which are prone to heat up because of the radiant input from the sun. Transpiration is an important component of temperature regulation in plants.

The cohesive and adhesive properties of water are due to hydrogen bonding

Water molecules at an air–water interface are more strongly attracted to neighboring water molecules than to the gas phase in contact with the water surface. As a consequence of this unequal attraction, the lowest energy (i.e., most stable) configuration is one that minimizes the surface area of the air–water interface. To increase the area of an air–water interface, hydrogen bonds must be broken, which requires an input of energy. The energy required to increase the surface area of a gas–liquid interface is known as **surface tension**.

Surface tension has units of energy per area (J m^{-2}), but is generally expressed in the equivalent, but less intuitive, units of force per length (J m^{-2} = N m^{-1}). A Joule (J) is the SI unit of energy, with units of force × distance (N m); a Newton (N) is the SI unit of force, with units of mass × acceleration (1 kg m s^{-2}). Surface tension not only influences the shape of the air–water interface but also produces a net force if the air–water interface is curved (Figure 3.4). As we will see later, surface tension and adhesion (which is defined below) at the evaporative surfaces of leaves generate the physical forces that pull water through the plant's vascular system.

The extensive hydrogen bonding in water also gives rise to the property known as **cohesion**, the mutual attraction between molecules. A related property, called **adhesion**, is the attraction of water to a solid phase such as a cell wall or glass surface. The degree to which water is attracted to the solid phase versus to itself can be quantified by measuring the **contact angle** (Figure 3.5A). The contact angle describes the shape of the resulting air–water interface and thus the effect that the surface tension has on the pressure in the liquid.

Cohesion, adhesion, and surface tension give rise to a phenomenon known as **capillarity** (Figure 3.5B). Consider a vertically oriented glass capillary tube with wettable walls (contact angle < 90°). At equilibrium, the water level within the capillary tube will be higher than that of the water supply at its base. Water is drawn into the capillary tube due to (1) the attraction of water to the polar surface

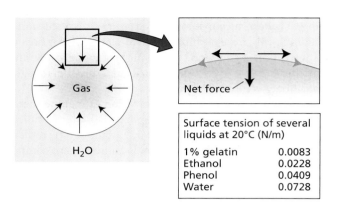

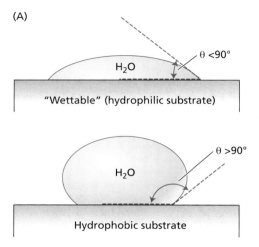

FIGURE 3.4 A gas bubble suspended within a liquid assumes a spherical shape such that its surface area is minimized. Because surface tension acts along the tangent to the gas–liquid interface, the resultant (net) force will be inward, leading to the compression of the bubble. The magnitude of the pressure (force/area) exerted by the interface is equal to $2T/r$, where T is the surface tension of the liquid (N/m) and r is the radius of the bubble (m). Water has an extremely high surface tension compared to other liquids at the same temperature.

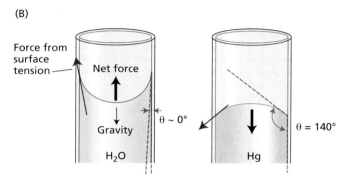

of the glass tube (adhesion) and (2) the surface tension of water. Together, adhesion and surface tension pull on the water molecules, causing them to move up the tube until this upward force is balanced by the weight of the water column. The narrower the tube, the higher the equilibrium water level. For calculations related to capillarity, see **Web Topic 3.1**.

Water has a high tensile strength

Cohesion gives water a high **tensile strength**, defined as the maximum force per unit area that a continuous column of water can withstand before breaking. We do not usually think of liquids as having tensile strength; however, such a property must exist for a water column to be pulled up a capillary tube.

We can demonstrate the tensile strength of water by placing it in a capped syringe (Figure 3.6). When we *push* on the plunger, the water is compressed and a positive **hydrostatic pressure** builds up. Pressure is measured in units called *pascals* (Pa) or, more conveniently, *megapascals* (MPa). One MPa equals approximately 9.9 atmospheres. Pressure is equivalent to a force per unit area (1 Pa = 1 N m^{-2}) and to an energy per unit volume (1 Pa = 1 J m^{-3}). Table 3.1 compares units of pressure.

If instead of pushing on the plunger we *pull* on it, a tension, or *negative hydrostatic pressure*, develops in the water to resist the pull. How hard must we pull on the plunger before the water molecules are torn away from each other and the water column breaks? Breaking the water column

FIGURE 3.5 (A) The shape of a droplet placed on a solid surface reflects the relative attraction of the liquid to the solid versus to itself. The contact angle (θ), defined as the angle from the solid surface through the liquid to the gas–liquid interface, is used to describe this interaction. "Wettable" surfaces have contact angles of less than 90°; a highly wettable (i.e., hydrophilic) surface (such as water on clean glass or primary cell walls) has a contact angle close to 0°. Water spreads out to form a thin film on highly wettable surfaces. In contrast, nonwettable (i.e., hydrophobic) surfaces have contact angles greater than 90°. Water "beads" up on such surfaces. (B) Capillarity can be demonstrated by observing what happens to liquids supplied to the bottom of a vertically oriented capillary tube. If the walls are highly wettable (e.g., water on clean glass has a contact angle ~0°), the net force is upward. The water column will thus rise until this upward force is balanced by the weight of the water column. In contrast, if the liquid does not "wet" the walls (e.g., Hg on clean glass has a contact angle of approximately 140°), the meniscus curve will curve downward, and the force resulting from surface tension will lower the level of the liquid in the tube.

requires sufficient energy to pull apart the hydrogen bonds that attract water molecules to one another.

Careful studies have demonstrated that water in small capillaries can resist tensions more negative than –30 MPa (the negative sign indicates tension, as opposed to compression). This value is only a fraction of the theoretical ten-

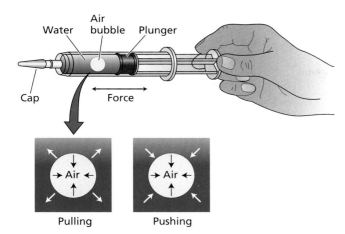

FIGURE 3.6 A sealed syringe can be used to create positive and negative pressures in fluids such as water. Pushing on the plunger cause the fluid to develop a positive, hydrostatic pressure (blue arrows) that acts in the same direction as the interfacial force resulting from surface tension (black arrows). Thus, a small air bubble trapped within the syringe will shrink as the pressure increases. Pulling on the plunger causes the fluid to develop a tension, or negative pressure. Air bubbles in the syringe will expand if the outward force on the bubble exerted by the fluid (white arrows) exceeds the inward force resulting from the surface tension of the gas–liquid interface (black arrows).

sile strength of water computed on the basis of the strength of hydrogen bonds. Nevertheless, it is quite substantial.

The presence of gas bubbles reduces the tensile strength of a water column. For example, in the syringe shown in Figure 3.6, microscopic bubbles may interfere with the ability of the water to resist the pull exerted by the plunger. If a tiny gas bubble forms in a column of water under tension, the gas bubble may expand until the tension in the liquid phase collapses (i.e., goes to zero), a phenomenon known as **cavitation**. As we will see in Chapter 4, cavitation can have a devastating effect on water transport through the xylem.

TABLE 3.1
Comparison of units of pressure

1 atmosphere = 14.7 pounds per square inch
= 760 mm Hg (at sea level, 45° latitude)
= 1.013 bar
= 0.1013 MPa
= 1.013 × 10^5 Pa

A car tire is typically inflated to about 0.2 MPa.

The water pressure in home plumbing is typically 0.2–0.3 MPa.

The water pressure under 15 feet (5 m) of water is about 0.05 MPa.

Water Transport Processes

When water moves from the soil through the plant to the atmosphere, it travels through a widely variable medium (e.g., cell wall, cytoplasm, membrane, air spaces), and the mechanisms of water transport vary with the type of medium.

We will now consider the two major processes in water transport: molecular diffusion and bulk flow.

Diffusion is the movement of molecules by random thermal agitation

Water molecules in a solution are not static; they are in continuous motion, colliding with one another and exchanging kinetic energy. The molecules intermingle as a result of their random thermal agitation. This random motion is called **diffusion**. As long as other forces are not acting on the molecules, diffusion causes the net movement of molecules from regions of high concentration to regions of low concentration—that is, down a concentration gradient (Figure 3.7).

In the 1880s the German scientist Adolf Fick discovered that the rate of diffusion is directly proportional to the concentration gradient ($\Delta c_s/\Delta x$)—that is, to the difference in concentration of substance s (Δc_s) between two points separated by the distance Δx. In symbols, we write this relation as Fick's first law:

$$J_s = -D_s \frac{\Delta c_s}{\Delta x} \tag{3.1}$$

The rate of transport, or the **flux density** (J_s), is the amount of substance s crossing a unit area per unit time (e.g., J_s may have units of moles per square meter per second [mol m^{-2} s^{-1}]). The **diffusion coefficient** (D_s) is a proportionality constant that measures how easily substance s moves through a particular medium. The diffusion coefficient is a characteristic of the substance (larger molecules have smaller diffusion coefficients) and depends on both the medium (diffusion in air is much faster than diffusion in a liquid, for example) and the temperature (substances diffuse faster at higher temperatures). The negative sign in the equation indicates that the flux moves down a concentration gradient.

Fick's first law says that a substance will diffuse faster when the concentration gradient becomes steeper (Δc_s is large) or when the diffusion coefficient is increased. Note that this equation accounts only for movement in response to a concentration gradient, and not for movement in response to other forces (e.g., pressure, electric fields, and so on).

Diffusion is rapid over short distances but extremely slow over long distances

From Fick's first law, one can derive an expression for the time it takes for a substance to diffuse a particular distance.

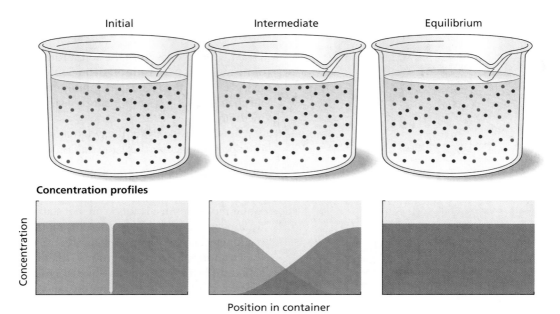

Figure 3.7 Thermal motion of molecules leads to diffusion—the gradual mixing of molecules and eventual dissipation of concentration differences. Initially, two materials containing different molecules are brought into contact. The materials may be gas, liquid, or solid. Diffusion is fastest in gases, slower in liquids, and slowest in solids. The initial separation of the molecules is depicted graphically in the upper panels, and the corresponding concentration profiles are shown in the lower panels as a function of position. With time, the mixing and randomization of the molecules diminishes net movement. At equilibrium the two types of molecules are randomly (evenly) distributed.

If the initial conditions are such that all the solute molecules are concentrated at the starting position (Figure 3.8A), then the concentration front moves away from the starting position, as shown for a later time point in Figure 3.8B. As the substance diffuses away from the starting point, the concentration gradient becomes less steep (Δc_s decreases), and thus net movement becomes slower.

The average time needed for a particle to diffuse a distance L is equal to L^2/D_s, where D_s is the diffusion coefficient, which depends on both the identity of the particle and the medium in which it is diffusing. In other words, the average time required for a substance to diffuse a given distance increases in proportion to the *square* of that distance. The diffusion coefficient for glucose in water is about 10^{-9} m^2 s^{-1}. Thus the average time required for a glucose molecule to diffuse across a cell with a diameter of 50 μm is 2.5 s. However, the average time needed for the same glucose molecule to diffuse a distance of 1 m in water is approximately 32 years. These values show that diffusion in solutions can be effective within cellular dimensions but is far too slow for mass transport over long distances. For additional calculations on diffusion times, see **Web Topic 3.2**.

Pressure-driven bulk flow drives long-distance water transport

A second process by which water moves is known as **bulk flow** or **mass flow**. Bulk flow is the concerted movement of groups of molecules en masse, most often in response to a pressure gradient. Among many common examples of bulk flow are water moving through a garden hose, a river flowing, and rain falling.

If we consider bulk flow through a tube, the rate of volume flow depends on the radius (r) of the tube, the viscosity (η) of the liquid, and the pressure gradient ($\Delta \Psi_p / \Delta x$) that drives the flow. Jean-Léonard-Marie Poiseuille (1797–1869) was a French physician and physiologist, and the relation just described is given by one form of Poiseuille's equation:

$$\text{Volume flow rate} = \left(\frac{\pi r^4}{8\eta}\right)\left(\frac{\Delta \Psi_p}{\Delta x}\right) \quad (3.2)$$

expressed in cubic meters per second (m^3 s^{-1}). This equation tells us that pressure-driven bulk flow is very sensitive to the radius of the tube. If the radius is doubled, the volume flow rate increases by a factor of 16 (2^4).

Pressure-driven bulk flow of water is responsible for long-distance transport of water in the xylem. It also accounts for much of the water flow through the soil and through the cell walls of plant tissues. In contrast to diffusion, pressure-driven bulk flow is independent of solute concentration gradients, as long as viscosity changes are negligible.

Osmosis is driven by a water potential gradient

Membranes of plant cells are **selectively permeable**; that is, they allow the movement of water and other small

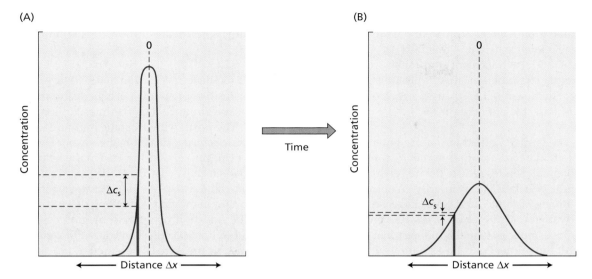

FIGURE 3.8 Graphical representation of the concentration gradient of a solute that is diffusing according to Fick's law. The solute molecules were initially located in the plane indicated on the x-axis. (A) The distribution of solute molecules shortly after placement at the plane of origin. Note how sharply the concentration drops off as the distance, x, from the origin increases. (B) The solute distribution at a later time point. The average distance of the diffusing molecules from the origin has increased, and the slope of the gradient has flattened out. (After Nobel 1999.)

uncharged substances across them more readily than the movement of larger solutes and charged substances (Stein 1986).

Like molecular diffusion and pressure-driven bulk flow, **osmosis** occurs spontaneously in response to a driving force. In simple diffusion, substances move down a concentration gradient; in pressure-driven bulk flow, substances move down a pressure gradient; in osmosis, both types of gradients influence transport (Finkelstein 1987). The direction and rate of water flow across a membrane are determined not solely by the concentration gradient of water or by the pressure gradient, but by the sum of these two driving forces.

We will soon see how osmosis drives the movement of water across membranes. First, however, let's discuss the concept of a composite or total driving force, representing the free-energy gradient of water.

The chemical potential of water represents the free-energy status of water

All living things, including plants, require a continuous input of free energy to maintain and repair their highly organized structures, as well as to grow and reproduce. Processes such as biochemical reactions, solute accumulation, and long-distance transport are all driven by an input of free energy into the plant. (For a detailed discussion of the thermodynamic concept of free energy, see Chapter 2 on the web site.)

The **chemical potential** of water is a quantitative expression of the free energy associated with water. In thermodynamics, free energy represents the potential for performing work. Note that chemical potential is a relative quantity: It is expressed as the difference between the potential of a substance in a given state and the potential of the same substance in a standard state. The unit of chemical potential is energy per mole of substance (J mol^{-1}).

For historical reasons, plant physiologists have most often used a related parameter called **water potential**, defined as the chemical potential of water divided by the partial molal volume of water (the volume of 1 mol of water): 18×10^{-6} m^3 mol^{-1}. Water potential is a measure of the free energy of water per unit volume (J m^{-3}). These units are equivalent to pressure units such as the pascal, which is the common measurement unit for water potential. Let's look more closely at the important concept of water potential.

Three major factors contribute to cell water potential

The major factors influencing the water potential in plants are *concentration*, *pressure*, and *gravity*. Water potential is symbolized by Ψ_w (the Greek letter psi), and the water potential of solutions may be dissected into individual components, usually written as the following sum:

$$\Psi_w = \Psi_s + \Psi_p + \Psi_g \qquad (3.3)$$

The terms Ψ_s and Ψ_p and Ψ_g denote the effects of solutes, pressure, and gravity, respectively, on the free energy of water. (Alternative conventions for components of water potential are discussed in **Web Topic 3.3**.) Energy levels must be defined in relation to a reference, analogous to how the contour lines on a map specify the distance above sea level. The reference state most often used to define water potential is pure water at ambient pressure and temperature. The reference height is generally set either at the base of the plant (for whole plant studies) or at the level of the tissue under examination (for studies of water movement at the cellular level). Let's consider each of the terms on the right-hand side of Equation 3.3.

SOLUTES The term Ψ_s, called the **solute potential** or the **osmotic potential**, represents the effect of dissolved solutes on water potential. Solutes reduce the free energy of water by diluting the water. This is primarily an entropy effect; that is, the mixing of solutes and water increases the disorder of the system and thereby lowers free energy. This means that the osmotic potential is independent of the specific nature of the solute. For dilute solutions of nondissociating substances such as sucrose, the osmotic potential may be estimated by the **van't Hoff equation**:

$$\Psi_s = -RTc_s \quad (3.4)$$

where R is the gas constant (8.32 J mol^{-1} K^{-1}), T is the absolute temperature (in degrees Kelvin, or K), and c_s is the solute concentration of the solution, expressed as **osmolality** (moles of total dissolved solutes per liter of water [mol L^{-1}]). The minus sign indicates that dissolved solutes reduce the water potential of a solution relative to the reference state of pure water.

Table 3.2 shows the values of RT at various temperatures and the Ψ_s values of solutions of different solute concentrations. For ionic solutes that dissociate into two or more particles, c_s must be multiplied by the number of dissociated particles to account for the increased number of dissolved particles.

Equation 3.4 is valid for "ideal" solutions. Real solutions frequently deviate from the ideal, especially at high concentrations—for example, greater than 0.1 mol L^{-1}. In our treatment of water potential, we will assume that we are dealing with ideal solutions (Friedman 1986; Nobel 1999).

PRESSURE The term Ψ_p is the **hydrostatic pressure** of the solution. Positive pressures raise the water potential; negative pressures reduce it. Sometimes Ψ_p is called *pressure potential*. The positive hydrostatic pressure within cells is the pressure referred to as *turgor pressure*. The value of Ψ_p can also be negative, as is the case in the xylem and in the walls between cells, where a *tension*, or *negative hydrostatic pressure*, can develop. As we will see, negative pressures are important in moving water long distances through the plant. The question of whether negative pressures can occur within living cells is considered in **Web Topic 3.4**.

Hydrostatic pressure is measured as the deviation from ambient pressure. Remember that water in the reference state is at ambient pressure, so by this definition $\Psi_p = 0$ MPa for water in the standard state. Thus the value of Ψ_p for pure water in an open beaker is 0 MPa, even though its absolute pressure is approximately 0.1 MPa (1 atmosphere).

GRAVITY Gravity causes water to move downward unless the force of gravity is opposed by an equal and opposite force. The term Ψ_g depends on the height (h) of the water above the reference-state water, the density of water (ρ_w), and the acceleration due to gravity (g). In symbols, we write the following:

$$\Psi_g = \rho_w g h \quad (3.5)$$

where $\rho_w g$ has a value of 0.01 MPa m^{-1}. Thus a vertical distance of 10 m translates into a 0.1 MPa change in water potential.

When dealing with water transport at the cell level, the gravitational component (Ψ_g) is generally omitted because it is negligible compared to the osmotic potential and the

TABLE 3.2
Values of RT and osmotic potential of solutions at various temperatures

Temperature (°C)	RT[a] (L MPa mol−1)	Osmotic potential (MPa) of solution with solute concentration in mol L⁻¹ water			Osmotic potential of seawater (MPa)
		0.01	0.10	1.00	
0	2.271	−0.0227	−0.227	−2.27	−2.6
20	2.436	−0.0244	−0.244	−2.44	−2.8
25	2.478	−0.0248	−0.248	−2.48	−2.8
30	2.519	−0.0252	−0.252	−2.52	−2.9

[a]R = 0.0083143 L MPa mol⁻¹ K⁻¹.

hydrostatic pressure. Thus, in these cases Equation 3.3 can be simplified as follows:

$$\Psi_w = \Psi_s + \Psi_p \qquad (3.6)$$

In discussions of water in dry soils and plant tissues with very low water contents such as seeds, one often finds reference to the matric potential, Ψ_m. Under these conditions, water exists as a very thin, perhaps one or two molecules deep, layer bound to solid surfaces by electrostatic interactions. These interactions are not easily separated into their effects on Ψ_s and Ψ_p, and are thus sometimes combined into a single term, Ψ_m. The matric potential is discussed further in **Web Topic 3.5**.

WATER POTENTIAL IN THE PLANT Cell growth, photosynthesis, and crop productivity are all strongly influenced by water potential and its components. Like the body temperature of humans, water potential is a good overall indicator of plant health. Plant scientists have thus expended considerable effort in devising accurate and reliable methods for evaluating the water status of plants. Some of the instruments that have been used to measure Ψ_w, Ψ_s, and Ψ_p are described in **Web Topic 3.6**.

Water enters the cell along a water potential gradient

In this section we will illustrate the osmotic behavior of plant cells with some numerical examples. First imagine an open beaker full of pure water at 20°C (Figure 3.9A). Because the water is open to the atmosphere, the hydrostatic pressure of the water is the same as atmospheric pressure ($\Psi_p = 0$ MPa). There are no solutes in the water, so $\Psi_s = 0$ MPa; therefore the water potential is 0 MPa ($\Psi_w = \Psi_s + \Psi_p$). Finally, because we focus here on transport processes that take place within the beaker, we define the reference height as equal to the level of the beaker and thus $\Psi_g = 0$ MPa.

Now imagine dissolving sucrose in the water to a concentration of 0.1 M (see Figure 3.9B). This addition lowers the osmotic potential (Ψ_s) to –0.244 MPa (see Table 3.2) and decreases the water potential (Ψ_w) to –0.244 MPa.

Next consider a "wilted" plant cell (i.e., a cell with no turgor pressure) that has a total internal solute concentration of 0.3 M (see Figure 3.9C). This solute concentration gives an osmotic potential (Ψ_s) of –0.732 MPa. Because the cell is flaccid, the internal pressure is the same as ambient pressure, so the hydrostatic pressure (Ψ_p) is 0 MPa and the water potential of the cell is –0.732 MPa.

What happens if this cell is placed in the beaker containing 0.1 M sucrose (see Figure 3.9C)? Because the water potential of the sucrose solution ($\Psi_w = -0.244$ MPa; see Figure 3.9B) is greater than the water potential of the cell ($\Psi_w = -0.732$ MPa), water will move from the sucrose solution to the cell (from high to low water potential).

Because plant cells are surrounded by relatively rigid cell walls, even a slight increase in cell volume causes a large increase in the hydrostatic pressure within the cell. As water enters the cell, the cell wall is stretched by the contents of the enlarging protoplast. The wall resists such stretching by pushing back on the cell. This phenomenon is analogous to inflating a basketball with air, except that air is compressible, whereas water is nearly incompressible.

As water moves into the cell, the hydrostatic pressure, or turgor pressure (Ψ_p), of the cell increases. Consequently, the cell water potential (Ψ_w) increases, and the difference between inside and outside water potentials ($\Delta\Psi_w$) is reduced. Eventually, cell Ψ_p increases enough to raise the cell Ψ_w to the same value as the Ψ_w of the sucrose solution. At this point, equilibrium is reached ($\Delta\Psi_w = 0$ MPa), and net water transport ceases.

Because the volume of the beaker is much larger than that of the cell, the tiny amount of water taken up by the cell does not significantly affect the solute concentration of the sucrose solution. Hence Ψ_s, Ψ_p, and Ψ_w of the sucrose solution are not altered. Therefore, at equilibrium, $\Psi_{w(cell)} = \Psi_{w(solution)} = -0.244$ MPa.

The exact calculation of cell Ψ_p and Ψ_s requires knowledge of the change in cell volume. However, if we assume that the cell has a very rigid cell wall, then the increase in cell volume will be small. Thus we can assume to a first approximation that $\Psi_{s(cell)}$ is unchanged during the equilibration process. We can obtain cell hydrostatic pressure by rearranging Equation 3.6 as follows: $\Psi_p = \Psi_w - \Psi_s = (-0.244) - (-0.732) = 0.488$ MPa.

Water can also leave the cell in response to a water potential gradient

Water can also leave the cell by osmosis. If, in the previous example, we remove our plant cell from the 0.1 M sucrose solution and place it in a 0.3 M sucrose solution (Figure 3.9D), $\Psi_{w(solution)}$ (–0.732 MPa) is more negative than $\Psi_{w(cell)}$ (–0.244 MPa), and water will move from the turgid cell to the solution.

As water leaves the cell, the cell volume decreases. As the cell volume decreases, cell Ψ_p and Ψ_w decrease also until $\Psi_{w(cell)} = \Psi_{w(solution)} = -0.732$ MPa. From the water potential equation (Equation 3.6) we can calculate that at equilibrium, $\Psi_p = 0$ MPa. As before, we assume that the change in cell volume is small, so we can ignore the change in Ψ_s due to the net outflow of water from the cell.

If we then slowly squeeze the turgid cell by pressing it between two plates (Figure 3.9E), we effectively raise the cell Ψ_p, consequently raising the cell Ψ_w and creating a $\Delta\Psi_w$ such that water now flows *out* of the cell. This is analogous to the industrial process of reverse osmosis in which externally applied pressure is used to separate water from dissolved solutes by forcing it across a semipermeable barrier. If we continue squeezing until half the cell water is removed and then hold the cell in this condition, the cell

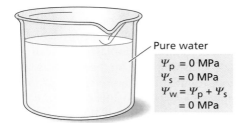
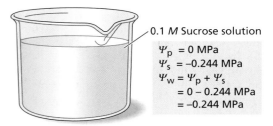
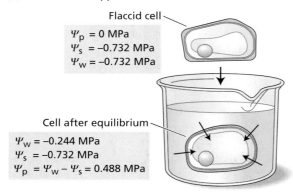
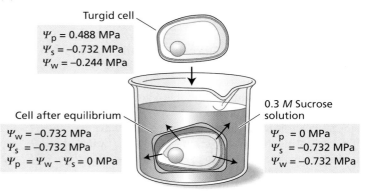
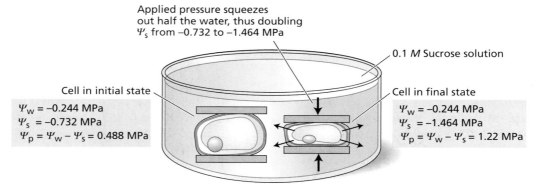

will reach a new equilibrium. As in the previous example, at equilibrium, $\Delta\Psi_w = 0$ MPa, and the amount of water added to the external solution is so small that it can be ignored. The cell will thus return to the Ψ_w value that it had before the squeezing procedure. However, the components of the cell Ψ_w will be quite different.

Because half of the water was squeezed out of the cell while the solutes remained inside the cell (the plasma membrane is selectively permeable), the cell solution is concentrated twofold, and thus Ψ_s is lower ($-0.732 \times 2 = -1.464$ MPa). Knowing the final values for Ψ_w and Ψ_s, we can calculate the turgor pressure, using Equation 3.6, as $\Psi_p = \Psi_w - \Psi_s = (-0.244) - (-1.464) = 1.22$ MPa. In our example we used an external force to change cell volume without a change in water potential. In nature, it is typically the water potential of the cell's environment that changes, and the cell gains or loses water until its Ψ_w matches that of its surroundings.

One point common to all these examples deserves emphasis: *Water flow across membranes is a passive process. That is, water moves in response to physical forces, toward regions of low water potential or low free energy.* There are no metabolic "pumps" (reactions driven by ATP hydrolysis) that can be used to drive water across a semipermeable membrane against its gradient in free energy.

The only situation in which water can be said to move across a semipermeable membrane against its water poten-

◀ **FIGURE 3.9** Five examples illustrating the concept of water potential and its components. (A) Pure water. (B) A solution containing 0.1 *M* sucrose. (C) A flaccid cell (in air) is dropped in the 0.1 *M* sucrose solution. Because the starting water potential of the cell is less than the water potential of the solution, the cell takes up water. After equilibration, the water potential of the cell rises to equal the water potential of the solution, and the result is a cell with a positive turgor pressure. (D) Increasing the concentration of sucrose in the solution makes the cell lose water. The increased sucrose concentration lowers the solution water potential, draws water out of the cell, and thereby reduces the cell's turgor pressure. In this case, the protoplast is able to pull away from the cell wall (i.e., the cell plasmolyzes), because sucrose molecules are able to pass through the relatively large pores of the cell walls. When this occurs, the difference in water potential between the cytoplasm and the solution is entirely across the plasma membrane, and thus the protoplast shrinks independently of the cell wall. In contrast, when a cell desiccates in air (e.g., the flaccid cell in C) plasmolysis does not occur, because the water held by capillary forces in the cell walls causes the plasma membrane to remain pressed against the cell wall even as the protoplast loses volume. Thus the cell (cytoplasm + wall) shrinks as a unit, resulting in the cell wall being mechanically deformed as the cell loses volume. (E) Another way to make the cell lose water is to press it slowly between two plates. In this case, half of the cell water is removed, so cell osmotic potential increases by a factor of 2.

tial gradient is when it is coupled to the movement of solutes. The transport of sugars, amino acids, or other small molecules by various membrane proteins can "drag" up to 260 water molecules across the membrane per molecule of solute transported (Loo et al. 1996). Such transport of water can occur even when the movement is against the usual water potential gradient (i.e., toward a larger water potential), because the loss of free energy by the solute more than compensates for the gain of free energy by the water. The net change in free energy remains negative. The amount of water transported in this way will generally be quite small compared to the passive movement of water down its water potential gradient. Thus, the net flow of water across a semipermeable membrane will typically be from a higher to a lower water potential.

In the absence of a semipermeable barrier, the movement of water and dissolved solutes occurs along gradients in pressure (i.e., from regions of higher to lower pressure). Although the diffusion of molecules from regions of higher to lower concentrations will still occur, without a semipermeable barrier the diffusion of water molecules in one direction will be countered by the diffusion of solutes in the other direction. Thus, the oft-repeated statement that water moves from higher to lower water potentials is true only when a semipermeable membrane is present.

Small changes in plant cell volume cause large changes in turgor pressure

Cell walls provide plant cells with a substantial degree of volume homeostasis relative to the large changes in water potential that they experience as the everyday consequence of the transpirational water losses associated with photosynthesis (see Chapter 4). Because plant cells have fairly rigid walls, a change in cell Ψ_w is generally accompanied by a large change in Ψ_p, with relatively little change in cell (protoplast) volume.

This phenomenon is illustrated in plots of Ψ_w, Ψ_p, and Ψ_s as a function of relative cell volume. In the example of a hypothetical cell shown in Figure 3.10, as Ψ_w decreases from 0 to about –2 MPa, the cell volume is reduced by only 5%. Most of this decrease is due to a reduction in Ψ_p (by about 1.2 MPa); Ψ_s decreases by about 0.3 MPa as a result of water loss by the cell and consequent increased concentration of cell solutes.

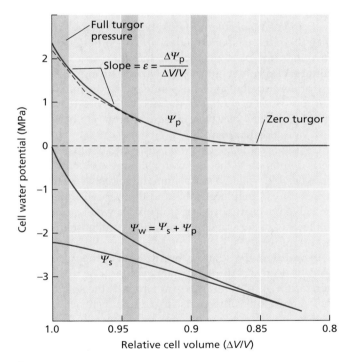

FIGURE 3.10 Hoffler diagrams showing the relation between cell water potential (Ψ_w) and its components (Ψ_p and Ψ_s), and relative cell volume ($\Delta V/V$). Note that turgor pressure (Ψ_p) decreases steeply with the initial 5% decrease in cell volume. In comparison, osmotic potential (Ψ_s) changes very little. As cell volume decreases below 0.9 in this example, the situation reverses: Most of the change in water potential is due to a drop in cell Ψ_s accompanied by relatively little change in turgor pressure. The slope of the curve that illustrates Ψ_p versus volume relationship is a measure of the cell's elastic modulus (ε) (a measurement of wall rigidity). Note also that ε is not constant but decreases as the cell loses turgor. (After Tyree and Jarvis 1982, based on a shoot of Sitka spruce [*Picea sitchensis*].)

Measurements of cell water potential and cell volume (see Figure 3.10) can be used to quantify how cell walls influence the water status of plant cells. Turgor pressure in most cells approaches zero as the relative cell volume decreases by 10 to 15%. However, for cells with very rigid cell walls (e.g., mesophyll cells in the leaves of many palm trees), the volume change associated with turgor loss can be much smaller. In cells with extremely elastic walls, such as the water-storing cells in the stems of many cacti, this volume change may be substantially larger.

The Ψ_p curve of Figure 3.10 provides a way to quantify the elasticity of the cell wall, symbolized by ε (the Greek letter epsilon): $\varepsilon = \Delta\Psi_p/\Delta$(relative volume). ε is the slope of the Ψ_p curve. ε is not constant, but decreases as turgor pressure is lowered, because nonlignified plant cell walls usually are rigid only when turgor pressure puts them under tension. Such cells act like a basketball: The wall is stiff (has high ε) when the ball is inflated but becomes soft and collapsible ($\varepsilon = 0$) when the ball loses pressure.

A comparison of the cell water relations within stems of cacti illustrates the important role of cell wall properties. Cacti are stem succulent plants, typically of arid regions. Their stems consist of an outer, photosynthetic layer that surrounds non-photosynthetic tissues that serve as a water storage reservoir (Figure 3.11). During drought, water is lost preferentially from these inner cells, despite the fact that the water potential of the two cell types remains in equilibrium (or "very close to equilibrium") (Nobel 1988). How does this happen?

Detailed studies of *Opuntia ficus-indica* (Goldstein et al. 1991) demonstrate that the water storage cells have much more flexible walls (lower ε) than the photosynthetic cells, in part due to their larger size and thinner cell walls. Thus for a given decrease in water potential, a water storage cell will lose a greater fraction of its water content than a photosynthetic cell.

In addition, the solute concentration of the water storage cells decreases during drought, in part due to the polymerization of soluble sugars into nonsoluble starch granules. A more typical response to drought is to accumulate solutes, in part to prevent water loss from cells. However, in the case of the cacti, the combination of more flexible cell walls and a decrease in solute concentration during drought allows water to be preferentially withdrawn from the water storage cells, thus helping to maintain the hydration of the photosynthetic tissues.

Water transport rates depend on driving force and hydraulic conductivity

So far, we have seen that water moves across a membrane in response to a water potential gradient. The direction of flow is determined by the direction of the Ψ_w gradient, and the rate of water movement is proportional to the magnitude of the driving gradient. However, for a cell that experiences a change in the water potential of its surroundings

FIGURE 3.11 Cross section of a cactus stem, showing an outer photosynthetic layer and inner, non-photosynthetic tissue that functions in water storage. During drought, water is lost preferentially from non-photosynthetic cells, thus serving to help maintain the water status of the photosynthetic tissue. (By David McIntyre.)

(e.g., see Figure 3.9), the movement of water across the cell membrane will decrease with time as the internal and external water potentials converge (Figure 3.12). The rate approaches zero in an exponential manner (see Dainty 1976), with a half-time, or $t_{1/2}$ (half-times conveniently characterize processes that change exponentially with time), given by the following equation:

$$t_{1/2} = \left(\frac{0.693}{(A)(Lp)}\right)\left(\frac{V}{\varepsilon - \Psi_s}\right) \qquad (3.7)$$

where V and A are, respectively, the volume and surface of the cell, and Lp is the **hydraulic conductivity** of the cell membrane. Hydraulic conductivity describes how readily water can move across a membrane, and has units of volume of water per unit area of membrane per unit time per unit driving force (i.e., $m^3\ m^{-2}\ s^{-1}\ MPa^{-1}$). For additional discussion on hydraulic conductivity, see **Web Topic 3.7**.

A short half-time means fast equilibration. Thus, cells with large surface-to-volume ratios, high membrane hydraulic conductivity, and stiff cell walls (large ε) will come rapidly into equilibrium with their surroundings. Cell half-times typically range from 1 to 10 s, although some are much shorter (Steudle 1989). These low half-times mean that single cells come to water potential equilibrium with their surroundings in less than 1 minute. For multicellular tissues, the half-times may be much larger.

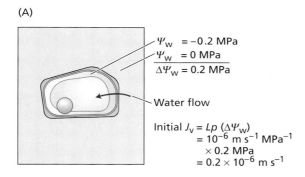
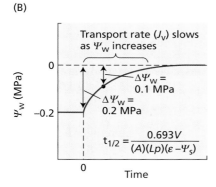

FIGURE 3.12 The rate of water transport into a cell depends on the water potential difference ($\Delta\Psi_w$) and the hydraulic conductivity of the cell membranes (Lp). (A) In this example, the initial water potential difference is 0.2 MPa and Lp is 10^{-6} m s^{-1} MPa^{-1}. These values give an initial transport rate (J_v) of 0.2×10^{-6} m s^{-1}. (B) As water is taken up by the cell, the water potential difference decreases with time, leading to a slowing in the rate of water uptake. This effect follows an exponentially decaying time course with a half-time ($t_{1/2}$) that depends on the following cell parameters: volume (V), surface area (A), conductivity (Lp), volumetric elastic modulus (ε), and cell osmotic potential (Ψ_s).

Aquaporins facilitate the movement of water across cell membranes

For many years there has been much uncertainty about how water moves across plant membranes. Specifically, it was unclear whether water movement into plant cells was limited to the diffusion of water molecules across the plasma membrane's lipid bilayer or also involved diffusion through protein-lined pores (Figure 3.13). Some studies suggested that diffusion directly across the lipid bilayer was not sufficient to account for observed rates of water movement across membranes, but the evidence in support of microscopic pores was not compelling.

This uncertainty was put to rest with the recent discovery of **aquaporins** (see Figure 3.13). Aquaporins are integral membrane proteins that form water-selective channels across the membrane. Because water diffuses much faster through such channels than through a lipid bilayer, aquaporins facilitate water movement into plant cells (Weig et

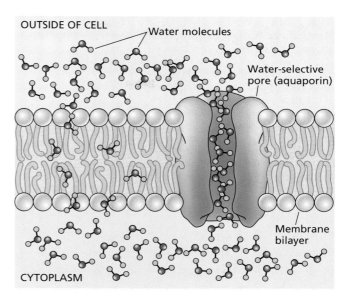

FIGURE 3.13 Water can cross plant membranes by diffusion of individual water molecules through the membrane bilayer, as shown on the left, and by the linear diffusion of water molecules through a water-selective pore formed by integral membrane proteins such as aquaporins.

al. 1997; Schäffner 1998; Tyerman et al. 1999). Note that although the presence of aquaporins may alter the *rate* of water movement across the membrane, they do not change the direction of transport or the driving force for water movement. However, aquaporins can be reversibly "gated" (i.e., transferred between an open and a closed state) in response to physiological parameters such as intercellular pH and Ca^{2+} levels (Tyerman et al. 2002). This opens the door to the possibility that plants actively regulate the permeability of their cell membranes to water.

The water potential concept helps us evaluate the water status of a plant

The concept of water potential has two principal uses: First, water potential governs transport across cell membranes, as we have described. Second, water potential is often used as a measure of the *water status* of a plant. Because of transpirational water loss to the atmosphere, plants are seldom fully hydrated. During periods of drought, they suffer from water deficits that lead to inhibition of plant growth and photosynthesis. Figure 3.14 lists some of the physiological changes that occur as plants experience increasingly drier conditions.

The sensitivity of any particular physiological process to water deficits is, to a large extent, a reflection of that plant's strategy for dealing with the range of water availability that it experiences in its environment. According to Figure 3.14, the process that is most affected by water deficit is cell expansion. However, cell expansion is not inherently more constrained by water availability than any of the other processes listed. By accumulating solutes, a cell could main-

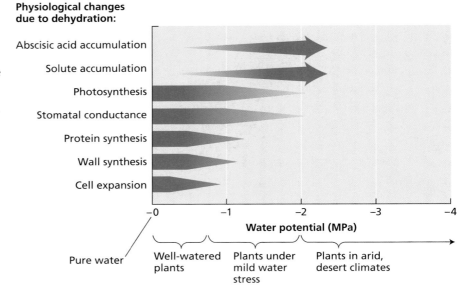

FIGURE 3.14 Water potential of plants under various growing conditions, and sensitivity of various physiological processes to water potential. The intensity of the bar color corresponds to the magnitude of the process. For example, cell expansion decreases as water potential falls (becomes more negative). Abscisic acid is a hormone that induces stomatal closure during water stress (see Chapter 23). (After Hsiao 1979.)

tain the positive turgor pressures needed for cell expansion despite decreases low water potential. In many plants reductions in water supply inhibit shoot growth and leaf expansion but *stimulate* root growth. A relative increase in roots relative to leaves is an appropriate response to reductions in water availability, and thus the sensitivity of shoot growth to decreases in water availability can be seen as an adaptation to drought rather than a physiological constraint.

However, what plants cannot do is to alter the availability of water in the soil. Thus, drought does impose some absolute limitations on physiological processes, although the actual water potentials at which such limitations occur vary between species. Figure 3.14 also shows representative values for Ψ_w at various stages of water stress. In leaves of well-watered plants, Ψ_w ranges from –0.2 to about –1.0 MPa. Leaves of plants in arid climates, however, can have much lower values, down to below –10 MPa under extreme conditions.

The ability to maintain physiological activity as water becomes less available typically incurs some costs. These costs can be in the form of solute accumulation to maintain turgor pressure, the need to invest more heavily in nonphotosynthetic organs such as roots to increase water uptake capacity, or in building structures capable of withstanding large negative pressures in the apoplast. Thus, physiological responses to water availability reflect a tradeoff between the benefits accrued by being able to carry out physiological processes (e.g., growth) over a wider range of environmental conditions versus the costs inherent in permitting such capability.

The components of water potential vary with growth conditions and location within the plant

Just as Ψ_w values depend on the growing conditions and the type of plant, so too, the values of Ψ_s can vary considerably. Within cells of well-watered garden plants (examples include lettuce, cucumber seedlings, and bean leaves), Ψ_s may be as high (i.e., less concentrated) as –0.5 MPa, although values of –0.8 to –1.2 MPa are more typical. The upper limit for cell Ψ_s is set probably by the minimum concentration of dissolved ions, metabolites, and proteins in the cytoplasm of living cells.

At the other extreme, plants under drought conditions sometimes attain a much more negative Ψ_s. For instance, water stress typically leads to an accumulation of solutes in the cytoplasm and vacuole, thus allowing the plant to maintain turgor pressure despite low water potentials.

Plant tissues that store high concentrations of sucrose or other sugars, such as sugar beet roots, sugarcane stems, or grape berries, also attain low values of Ψ_s. Values as low as –2.5 MPa are not unusual. Plants that grow in saline environments, called **halophytes**, typically have very low values of Ψ_s. A low Ψ_s lowers cell Ψ_w enough to extract water from salt water, without allowing excessive levels of salts to enter at the same time.

Although Ψ_s *within* cells may be quite negative, the apoplastic solution surrounding the cells—that is, in the cell walls and in the xylem—is generally quite dilute. The Ψ_s of the apoplast is typically –0.1 to 0 MPa, although in certain tissues (e.g., developing fruits) and habitats (e.g., halophytes) the concentration of solutes in the apoplast can be large. In general, water potentials in the xylem and cell walls are dominated by Ψ_p, which is typically less than zero. In contrast, values for Ψ_p within cells of well-watered plants may range from 0.1 to perhaps 1 MPa, depending on the value of Ψ_s inside the cell.

A positive turgor pressure (Ψ_p) is important for several reasons. First, growth of plant cells requires turgor pressure to stretch the cell walls. The loss of Ψ_p under water deficits can explain in part why cell growth is so sensitive to water stress, as well as why this sensitivity can be modified by varying the cell's osmotic potential (see Chapter 26). The second reason positive turgor is important is that turgor pressure increases the mechanical rigidity of cells and tissues. This function of cell turgor pressure is particularly important for young, nonlignified tissues, which cannot support themselves mechanically without a high internal pressure.

A plant **wilts** when the turgor pressure inside the cells of such tissues falls toward zero. Plasmolysis, however, does not occur, because the negative pressures in the cell wall mean that the plasma membrane remains pressed against the cell wall. Instead, dehydrating cells become mechanically deformed as the cell loses volume. Only if air can penetrate the cell wall will the protoplast be free to shrink independently of the cell wall. Web Topic 3.8 discusses plasmolysis, contrasting the situation when a cell is dehydrated osmotically (i.e., in solution) to when water is withdrawn from the cell due to lower (more negative) pressures in the apoplast.

Finally, some physiological processes appear to be influenced directly by turgor pressure. However, it is likely that the majority of metabolic processes are more sensitive to changes in the concentration of the cytoplasm than they are to changes in the internal hydrostatic pressure. The existence of stretch-activated signaling molecules in the plasma membrane suggests that plant cells may sense changes in their water status via changes in volume, rather than by directly measuring turgor pressure.

Summary

A consequence of conducting photosynthesis on land is that water is continually lost to the atmosphere and thus must be replenished from the environment, typically by uptake from the soil. Water moves passively along gradients in the components of free energy, and water may move by diffusion, by bulk flow, or by a combination of these fundamental transport mechanisms. Water diffuses because molecules are in constant thermal agitation, which tends to even out concentration differences. Water moves by bulk flow in response to a pressure difference. The movement of water across semipermeable barriers depends on a gradient in free energy of water across the membrane—a gradient commonly measured as a difference in water potential.

Solute concentration and hydrostatic pressure are the two major factors that affect water potential, although when large vertical distances are involved, gravity is also important. These components of the water potential may be summed as follows: $\Psi_w = \Psi_s + \Psi_p + \Psi_g$. Plant cells come into water potential equilibrium with their local environment by absorbing or losing water. Because cell walls are relatively strong, this change in cell volume typically results in a large change in cell Ψ_p, accompanied by minor changes in cell Ψ_s. The rate of water transport across a membrane depends on the water potential difference across the membrane and the hydraulic conductivity of the membrane.

In addition to its importance in transport, water potential is a useful measure of the water status of plants. As we will see in Chapter 4, diffusion, bulk flow, and osmosis are all involved in the movement of water from the soil through the plant to the atmosphere.

Web Material

Web Topics

3.1 Calculating Capillary Rise
Quantification of capillary rise allows us to assess its functional role in water movement of plants.

3.2 Calculating Half-Times of Diffusion
The assessment of the time needed for a molecule such as glucose to diffuse across cells, tissues, and organs shows that diffusion has physiological significance only over short distances.

3.3 Alternative Conventions for Components of Water Potential
Plant physiologists have developed several conventions to define water potential of plants. A comparison of key definitions in some of these convention systems provides us with a better understanding of the water relations literature.

3.4 Can Negative Turgor Pressures Exist in Living Cells?
It is assumed that Ψ_p is zero or greater in living cells; is this true for living cells with lignified walls?

3.5 The Matric Potential
Matric potential is used to quantify the chemical potential of water in soils, seeds, and cell walls.

3.6 Measuring Water Potential
Several methods are available to measure water potential in plant cells and tissues.

3.7 Understanding Hydraulic Conductivity
Hydraulic conductivity, a measurement of the membrane permeability to water, is one of the factors determining the velocity of water movements in plants.

3.8 Wilting and Plasmolysis
Plasmolysis is a major structural change resulting from major water loss by osmosis.

Chapter References

Dainty, J. (1976) Water relations of plant cells. In *Transport in Plants*, Vol. 2, Part A: *Cells* (Encyclopedia of Plant Physiology, New Series, Vol. 2.), U. Lüttge and M. G. Pitman, eds., Springer, Berlin, pp. 12–35.

Finkelstein, A. (1987) *Water Movement through Lipid Bilayers, Pores, and Plasma Membranes: Theory and Reality*. Wiley, New York.

Friedman, M. H. (1986) *Principles and Models of Biological Transport*. Springer Verlag, Berlin.

Goldstein, G., Andrade, J. L., and Nobel, P. S. (1991) Differences in water relations parameters for the chlorenchyma and parenchyma of *Opuntia ficus-indica* under wet versus dry conditions. *Aust. J. Plant Physiol.* 18: 95–107.

Hsiao, T. C. (1979) Plant responses to water deficits, efficiency, and drought resistance. *Agricult. Meteorol.* 14: 59–84.

Loo, D. D. F., Zeuthen, T., Chandy, G., and Wright, E. M. (1996) Cotransport of water by the Na^+/glucose cotransporter. *Proc. Natl. Acad. Sci. USA* 93: 13367–13370.

Nobel, P. S. (1999) *Physicochemical and Environmental Plant Physiology*, 2nd ed. Academic Press, San Diego, CA.

Nobel, P. S. (1988) *Environmental Biology of Agaves and Cacti*. Cambridge University Press, New York.

Schäffner, A. R. (1998) Aquaporin function, structure, and expression: Are there more surprises to surface in water relations? *Planta* 204: 131–139.

Stein, W. D. (1986) *Transport and Diffusion across Cell Membranes*. Academic Press, Orlando, FL.

Steudle, E. (1989) Water flow in plants and its coupling to other processes: An overview. *Methods Enzymol.* 174: 183–225.

Tyerman, S. D., Bohnert, H. J., Maurel, C., Steudle, E., and Smith, J. A. C. (1999) Plant aquaporins: Their molecular biology, biophysics and significance for plant–water relations. *J. Exp. Bot.* 50: 1055–1071.

Tyerman, S. D., Niemietz, C. M., and Bramley, H. (2002) Plant aquaporins: Multifunctional water and solute channels with expanding roles. *Plant Cell Environ.* 25: 173–194.

Tyree, M. T., and Jarvis, P. G. (1982) Water in tissues and cells. In *Physiological Plant Ecology*, Vol. 2: *Water Relations and Carbon Assimilation* (Encyclopedia of Plant Physiology, New Series, Vol. 12B), O. L. Lange, P. S. Nobel, C. B. Osmond, and H. Ziegler, eds., Springer, Berlin, pp. 35–77.

Mitchell, J. M., Jr. (1964) *Weather and Our Food Supply* (CAED Report 20). Center for Agricultural and Economic Development, Iowa State University of Science and Technology, Ames, IA.

Weig, A., Deswarte, C., and Chrispeels, M. J. (1997) The major intrinsic protein family of Arabidopsis has 23 members that form three distinct groups with functional aquaporins in each group. *Plant Physiol.* 114: 1347–1357.

Whittaker R. H. (1970) *Communities and Ecosystems*. Macmillan, New York.

Chapter 4 | Water Balance of Plants

LIFE IN EARTH'S ATMOSPHERE presents a formidable challenge to land plants. On one hand, the atmosphere is the source of carbon dioxide, which is needed for photosynthesis. Plants therefore need ready access to the atmosphere. On the other hand, the atmosphere is relatively dry. Because plants lack surfaces that can allow the internal diffusion of CO_2 while preventing water loss, CO_2 uptake exposes plants to the risk of dehydration. This problem is compounded because the concentration gradient for CO_2 uptake is much, much smaller than the concentration gradient that drives water loss. To meet the contradictory demands of maximizing carbon dioxide uptake while limiting water loss, plants have evolved adaptations to control water loss from leaves, and to replace the water lost to the atmosphere.

In this chapter we will examine the mechanisms and driving forces operating on water transport within the plant and between the plant and its environment. Transpirational water loss from the leaf is driven by a gradient in water vapor concentration. Long-distance transport in the xylem is driven by pressure gradients, as is water movement in the soil. Water transport through layers of living cells such as the root cortex is complex, but it responds to water potential gradients across the tissue.

Throughout this journey water transport is passive in the sense that metabolic energy is not used to "pump" water from the soil to the leaves. Instead, the spontaneous movement of water from regions of higher to lower free energy results in the transport of water from the soil to replenish water lost by transpiration. Despite its passive nature, water transport is finely regulated by the plant to minimize dehydration, largely by regulating transpiration. We will begin our examination of water transport by focusing on water in the soil.

Water in the Soil

The water content and the rate of water movement in soils depend to a large extent on soil type and soil structure. Table 4.1 shows that the physical characteristics of different soils can vary greatly. At one extreme is sand, in which the soil particles may be 1 mm or more in diameter. Sandy soils have a relatively low surface area per gram of soil and have large spaces or channels between particles.

At the other extreme is clay, in which particles are smaller than 2 μm in diameter. Clay soils have much greater surface areas and smaller channels between particles. With the aid of organic substances such as humus (decomposing organic matter), clay particles may aggregate into "crumbs" that help improve soil aeration and infiltration of water.

When a soil is heavily watered by rain or by irrigation (see **Web Topic 4.1**), the water percolates downward by gravity through the spaces between soil particles, partly displacing, and in some cases trapping, air in these channels. Water in the soil may exist as a film adhering to the surface of soil particles, or it may fill the entire channel between particles.

In sandy soils, the spaces between particles are so large that water tends to drain from them and remain only on the particle surfaces and at interstices between particles. In clay soils, the channels are small enough that water is retained against the forces due to gravity. The moisture-holding capacity of soils is called the **field capacity**. Field capacity is the water content of a soil after it has been saturated with water and excess water has been allowed to drain away. Clay soils or soils with a high humus content have a large field capacity. A few days after being saturated, they might retain 40% water by volume. In contrast, sandy soils typically retain 3% water by volume after saturation.

In the following sections we will examine how the physical structure of the soil influences soil water potential, how water moves in the soil, and how roots absorb the water needed by the plant.

TABLE 4.1
Physical characteristics of different soils

Soil	Particle diameter (μm)	Surface area per gram (m²)
Coarse sand	2000–200	<1–10
Fine sand	200–20	
Silt	20–2	10–100
Clay	<2	100–1000

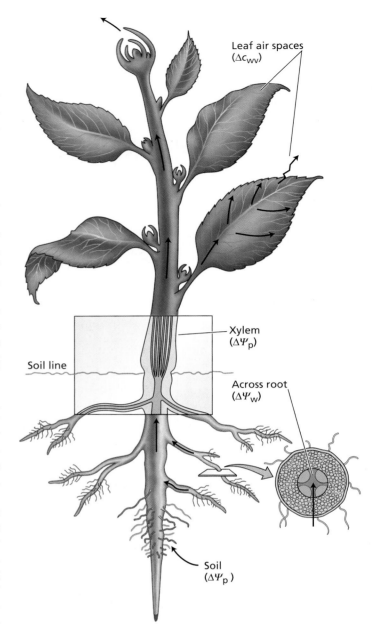

FIGURE 4.1 Main driving forces for water flow from the soil through the plant to the atmosphere: differences in water vapor concentration (Δc_{wv}), hydrostatic pressure ($\Delta \Psi_p$), and water potential ($\Delta \Psi_w$).

A negative hydrostatic pressure in soil water lowers soil water potential

Like the water potential of plant cells, the water potential of soils may be dissected into two components, the osmotic potential and the hydrostatic pressure. The **osmotic potential** (Ψ_s; see Chapter 3) of soil water is generally negligible, because solute concentrations are low; a typical value might be –0.02 MPa. For soils that contain a substantial concentration of salts, however, Ψ_s can be significant, perhaps –0.2 MPa or lower.

The second component of soil water potential is **hydrostatic pressure** (Ψ_p) (Figure 4.1). For wet soils, Ψ_p is very

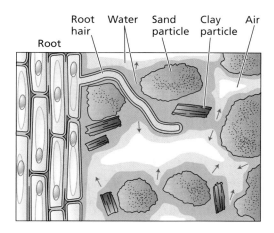

FIGURE 4.2 Root hairs make intimate contact with soil particles and greatly amplify the surface area that can be used for water absorption by the plant. The soil is a mixture of particles (sand, clay, silt, and organic material), water, dissolved solutes, and air. Water is adsorbed to the surface of the soil particles. As water is absorbed by the plant, the soil solution recedes into smaller pockets, channels, and crevices between the soil particles. At the air–water interfaces, this recession causes the surface of the soil solution to develop concave menisci (curved interfaces between air and water marked in the figure by arrows), and brings the solution into tension (negative pressure) by surface tension. As more water is removed from the soil, more curved menisci are formed, resulting in greater tensions (more negative pressures).

close to zero. As a soil dries out, Ψ_p decreases and can become quite negative. Where does the negative pressure in soil water come from?

Recall from our discussion of capillarity in Chapter 3 that water has a high surface tension that tends to minimize air–water interfaces. However, because of adhesive forces, water also tends to cling to the surfaces of soil particles (Figure 4.2).

As the water content of the soil decreases, the water recedes into the interstices between soil particles, forming air–water surfaces whose curvature represents the balance between the tendency to minimize the surface area of the air-water interface and the attraction of the water for the soil particles. Water under these curved surfaces develops a negative pressure that may be estimated by the following formula:

$$\Psi_p = \frac{-2T}{r} \qquad (4.1)$$

where T is the surface tension of water (7.28×10^{-8} MPa m) and r is the radius of curvature of the air–water interface. Note that this is the same capillarity equation discussed in **Web Topic 3.1** (see also Figure 3.5), where here the soil particles are assumed to be fully wettable (contact angle $\Theta = 0$; $\cos \Theta = 1$).

As a soil dries out, water is first removed from the largest spaces between soil particles. Thus, the value of Ψ_p in soil water can become quite negative during soil drying due to the increasing curvature of air–water surfaces in successively smaller diameter pores. For instance, a curvature $r = 1$ µm (about the size of the largest clay particles) corresponds to a Ψ_p value of –0.15 MPa. The value of Ψ_p may easily reach –1 to –2 MPa as the air–water interface recedes into the smaller cracks between clay particles.

Water moves through the soil by bulk flow

Water moves through soils predominantly by bulk flow driven by a pressure gradient. Because the pressure in soil water is due to the existence of curved air–water interfaces, water will flow from regions of higher soil water content, where the water-filled spaces are larger, to regions of lower soil water content, where the smaller size of the water-filled spaces is associated with more curved air–water interfaces. Diffusion of water vapor also accounts for some water movement, which can be important in dry soils.

As plants absorb water from the soil, they deplete the soil of water near the surface of the roots. This depletion reduces Ψ_p in the water near the root surface and establishes a pressure gradient with respect to neighboring regions of soil that have higher Ψ_p values. Because the water-filled pore spaces in the soil are interconnected, water moves to the root surface by bulk flow through these channels down the pressure gradient.

The rate of water flow in soils depends on two factors: the size of the pressure gradient through the soil, and the hydraulic conductivity of the soil. **Soil hydraulic conductivity** is a measure of the ease with which water moves through the soil, and it varies with the type of soil and water content. Sandy soils, with their large spaces between particles, have a large hydraulic conductivity, whereas clay soils, with the minute spaces between their particles, have an appreciably smaller hydraulic conductivity.

As the water content (and hence the water potential) of a soil decreases, the hydraulic conductivity decreases drastically (see **Web Topic 4.2**). This decrease in soil hydraulic conductivity is due primarily to the replacement of water in the soil spaces by air. When air moves into a soil channel previously filled with water, water movement through that channel is restricted to the periphery of the channel. As more of the soil spaces become filled with air, water can flow through fewer and narrower channels, and the hydraulic conductivity falls.

In very dry soils, the water potential (Ψ_w) may fall below what is called the **permanent wilting point**. At this point the water potential of the soil is so low that plants cannot regain turgor pressure even if all water loss through transpiration ceases. This means that the water potential of the soil (Ψ_w) is less than or equal to the osmotic potential (Ψ_s) of the plant. Because cell Ψ_s varies with plant species, the permanent wilting point is clearly not a unique property of the soil; it depends on the plant species as well.

Water Absorption by Roots

Intimate contact between the surface of the root and the soil is essential for effective water absorption by the root. This contact provides the surface area needed for water uptake and is maximized by the growth of the root and of root hairs into the soil. **Root hairs** are microscopic extensions of root epidermal cells that greatly increase the surface area of the root, thus providing greater capacity for absorption of ions and water from the soil. When 4-month-old rye (*Secale*) plants were examined, their root hairs were found to constitute more than 60% of the surface area of the roots (see Figure 5.6).

Water enters the root most readily near the root tip. More mature regions of the root often have an outer layer of protective tissue, called an *exodermis* or *hypodermis*, which contains hydrophobic materials in its walls and is relatively impermeable to water. Although it might at first seem counterintuitive that any portion of the root system should be impermeable to water, the older regions of the root must be sealed off if there is to be water uptake (and thus bulk flow of nutrients in the xylem) from the regions that are actively exploring new areas in the soil (Figure 4.3) (Zwieniecki et al. 2002). Breaks in the cortex associated with the outgrowth of secondary roots may allow some water to enter older regions, but these must also be sealed off to avoid "short-circuiting" the driving force for water uptake by root tips.

The intimate contact between the soil and the root surface is easily ruptured when the soil is disturbed. It is for this reason that newly transplanted seedlings and plants need to be protected from water loss for the first few days after transplantation. Thereafter, new root growth into the soil reestablishes soil–root contact, and the plant can better withstand water stress.

Let's consider how water moves within the root, and the factors that determine the rate of water uptake into the root.

Water moves in the root via the apoplast, symplast, and transmembrane pathways

In the soil, water is transported predominantly by bulk flow. However, when water comes in contact with the root surface, the nature of water transport becomes more complex. From the epidermis to the endodermis of the root, there are three pathways through which water can flow (Figure 4.4): the apoplast, symplast, and transmembrane pathways.

1. The apoplast is the continuous system of cell walls, intercellular air spaces, and lumen of cells that have lost their cytoplasm (e.g., xylem conduits and fibers). In the apoplast pathway, water moves through cell walls and any water-filled extracellular spaces (i.e., without crossing any membranes) as it travels across the root cortex.

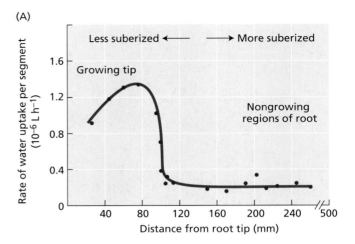

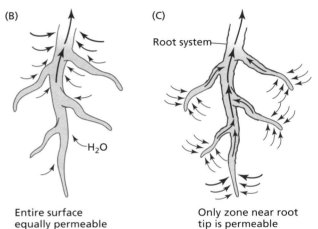

FIGURE 4.3 Rate of water uptake at various positions along a pumpkin root (A). Diagram of water uptake in which the entire root surface is equally permeable (B) or impermeable in older regions due to the deposition of suberin (C). When root surfaces are equally permeable, most of the water enters near the top of the root system, with more distal regions being hydraulically isolated as the suction in the xylem is relieved due to the inflow of water. Decreasing the permeability of older regions of the root allows xylem tensions to extend further into the root system, allowing water uptake from distal regions of the root system. (A after Kramer and Boyer 1995.)

2. The symplast consists of the entire network of cell cytoplasm interconnected by plasmodesmata. In the symplast pathway, water travels across the root cortex by passing from one cell to the next via the plasmodesmata (see Chapter 1). Because water movement in both the apoplast and symplast pathways does not have to cross any semipermeable membranes, the relevant driving force for mass flow is the gradient in hydrostatic pressure.

3. The transmembrane pathway is the route followed by water that sequentially enters a cell on one side, exits the cell on the other side, enters the next in the series, and so on. In this pathway, water crosses at least two

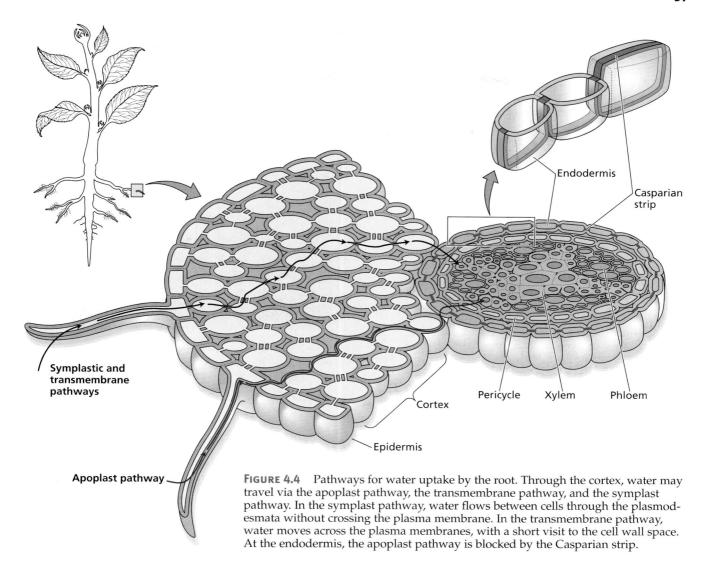

FIGURE 4.4 Pathways for water uptake by the root. Through the cortex, water may travel via the apoplast pathway, the transmembrane pathway, and the symplast pathway. In the symplast pathway, water flows between cells through the plasmodesmata without crossing the plasma membrane. In the transmembrane pathway, water moves across the plasma membranes, with a short visit to the cell wall space. At the endodermis, the apoplast pathway is blocked by the Casparian strip.

membranes for each cell in its path (the plasma membrane on entering and on exiting). Transport across the tonoplast may also be involved. In the transmembrane pathway, the presence of semipermeable membranes means that the relevant driving force is the total water potential gradient.

Although the relative importance of the apoplast, symplast, and transmembrane pathways has not yet been clearly established, experiments with the pressure probe technique (see **Web Topic 3.6**) indicate that the apoplast pathway is particularly important for water movement across the root cortex in young corn roots (Frensch et al. 1996; Steudle and Frensch 1996).

At the endodermis, water movement through the apoplast pathway is obstructed by the Casparian strip (see Figure 4.4). The **Casparian strip** is a band of radial cell walls in the endodermis that is impregnated with the waxlike, hydrophobic substance **suberin**. Suberin acts as a barrier to water and solute movement. The endodermis becomes suberized in the nongrowing part of the root, several millimeters behind the root tip, at about the same time that the first protoxylem elements mature (Esau 1953). The Casparian strip breaks the continuity of the apoplast pathway, forcing water and solutes to cross the endodermis by passing through the plasma membrane.

The requirement that water move symplastically across the endodermis also helps explain why the permeability of roots to water depends strongly on the presence of aquaporins. Down-regulating the expression of aquaporin genes markedly reduces the hydraulic conductivity of roots and can result in plants that wilt easily (Siefritz et al. 2002) or that produce larger root systems (Martre et al. 2002).

Water uptake decreases when roots are subjected to low temperature or anaerobic conditions, or treated with respiratory inhibitors. Submerged roots soon run out of oxygen, which is normally provided by diffusion through the air spaces in the soil (diffusion through gas is 10^4 times faster than diffusion through water). The anaerobic roots trans-

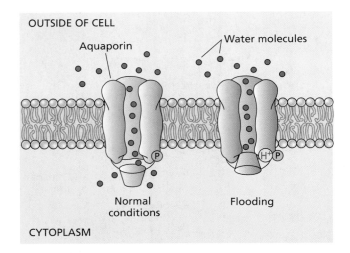

FIGURE 4.5 Stereo model of the spinach plasma membrane aquaporin (SoPIP2;1) in its open (blue) and closed (green) conformations. Closure in response to drought results from the dephosphorylation of two highly conserved serine residues, whereas closure during flooding results from the protonation of a conserved histidine. (After Törnroth-Horsefield et al. 2006).

FIGURE 4.6 Guttation in leaves from strawberry (*Fragaria grandiflora*). In the early morning, leaves secrete water droplets through the hydathodes, located at the margins of the leaves. Young flowers may also show guttation. (Courtesy of R. Aloni.)

port less water to the shoots, which consequently suffer net water loss and begin to wilt.

Until recently, there was no explanation for the connection between root respiration and water uptake, or the enigmatic wilting of flooded plants. We now know that the permeability of aquaporins can be regulated in response to intracellular pH (Tournaire-Roux et al. 2003). Decreased rates of respiration can lead to increases in intracellular pH. This increase in cytoplasmic pH alters the conductance of aquaporins involved in the movement of water across roots, resulting in roots that are markedly less permeable to water. The fact that aquaporins can be gated in response to pH and other signals (e.g., osmolarity), provides a mechanism by which roots can actively alter their permeability to water in response to their local environment (Figure 4.5). Whether the effects of flooding on root permeability represent an adaptive response, however, remains an open question.

Solute accumulation in the xylem can generate "root pressure"

Plants sometimes exhibit a phenomenon referred to as **root pressure**. For example, if the stem of a young seedling is cut off just above the soil, the stump will often exude sap from the cut xylem for many hours. If a manometer is sealed over the stump, positive pressures can be measured. These pressures can be as high as 0.05 to 0.5 MPa.

Roots generate positive hydrostatic pressure by absorbing ions from the dilute soil solution and concentrating them into the xylem. The buildup of solutes in the xylem sap leads to a decrease in the xylem osmotic potential (Ψ_s) and thus a decrease in the xylem water potential (Ψ_w). This lowering of the xylem Ψ_w provides a driving force for water absorption, which in turn leads to a positive hydrostatic pressure in the xylem. In effect, the whole root acts like an osmotic cell; the multicellular root tissue behaves as an osmotic membrane does, building up a positive hydrostatic pressure in the xylem in response to the accumulation of solutes.

Root pressure is most likely to occur when soil water potentials are high and transpiration rates are low. When transpiration rates are high, water is taken up so rapidly into the leaves and lost to the atmosphere that a positive pressure resulting from ion uptake never develops in the xylem.

Plants that develop root pressure frequently produce liquid droplets on the edges of their leaves, a phenomenon known as **guttation** (Figure 4.6). Positive xylem pressure causes exudation of xylem sap through specialized pores called *hydathodes* that are associated with vein endings at the leaf margin. The "dewdrops" that can be seen on the tips of grass leaves in the morning are actually guttation droplets exuded from such specialized pores. Guttation is most noticeable when transpiration is suppressed and the relative humidity is high, such as during the night. It is possible that root pressure reflects an unavoidable consequence of high rates of ion accumulation. However, the existence of positive pressures within the xylem at night can help to dissolve previously formed gas bubbles, and thus play a role in reversing the deleterious effects of cavitation described in the next section.

Water Transport through the Xylem

In most plants, the xylem constitutes the longest part of the pathway of water transport. In a plant 1 m tall, more than 99.5% of the water transport pathway through the plant is within the xylem, and in tall trees the xylem represents an even greater fraction of the pathway. Compared with the movement of water though layers of living cells, the xylem is a simple pathway of low resistivity. However, because the xylem constitutes the major portion of the water transport pathway, it plays an important role in constraining the movement of water from soil to leaves. In the following sections we will examine how the structure of the xylem contributes to the movement of water from the roots to the leaves, and how negative hydrostatic pressure generated by transpiration pulls water through the xylem.

The xylem consists of two types of tracheary elements

The conducting cells in the xylem have a specialized anatomy that enables them to transport large quantities of water with great efficiency. There are two important types of **tracheary elements** in the xylem: tracheids and vessel elements (Figure 4.7). Vessel elements are found in angiosperms, a small group of gymnosperms called the Gnetales, and some ferns. Tracheids are present in both angiosperms and gymnosperms, as well as in ferns and other groups of vascular plants.

The maturation of both tracheids and vessel elements involves the production of secondary cell walls and the subsequent "death" of the cell. This is a form of programmed cell death that results in the loss of the cytoplasm and all of its contents. Thus, functional water-conducting cells have no membranes and no organelles. What remains are the thick, lignified cell walls, which form hollow tubes through which water can flow with relatively little resistance.

Tracheids are elongated, spindle-shaped cells (see Figure 4.7A) that are arranged in overlapping vertical files (Figure 4.8). Water flows between tracheids by means of the numerous **pits** in their lateral walls (see Figure 4.7B). Pits are microscopic regions where the secondary wall is absent and the primary wall is thin and porous (see Figure 4.7C). Pits of one tracheid are typically located opposite pits of an adjoining tracheid, forming **pit pairs**. Pit pairs constitute a low-resistance path for water movement between tracheids. The porous layer between pit pairs, consisting of two primary walls and a middle lamella, is called the **pit membrane**.

Pit membranes in tracheids of conifers have a central thickening, called a **torus** (plural *tori*) (see Figure 4.7C). The torus acts like a valve to close the pit by lodging itself in the circular or oval wall thickenings bordering these pits. Such lodging of the torus is an effective way of preventing gas bubbles from spreading into neighboring tracheids (we will discuss this formation of bubbles, a process called cavitation, shortly). Pit membranes in all other plants, whether in tracheids or vessel elements, are structurally homogeneous, meaning they lack a torus. Because the water-filled pores in these pit membranes are very small, they also serve as an effective barrier against the movement of gas bubbles. Thus, pit membranes of both types play an important role in preventing the spread of gas bubbles, called emboli, within the xyem.

Vessel elements tend to be shorter and wider than tracheids and have perforated end walls that form a **perforation plate** at each end of the cell. Like tracheids, vessel elements have pits on their lateral walls (see Figure 4.7B). Unlike tracheids, the perforated end walls allow vessel members to be stacked end to end to form a larger conduit called a **vessel** (see Figure 4.8). Vessels are multicellular conduits that vary in length both within and between species. Maximum vessel lengths range from a few centimeters to many meters. Because of their open end walls, vessels provide an efficient pathway for water movement. The vessel members found at the extreme ends of a vessel lack perforations in their end walls and communicate with neighboring vessels via pit pairs.

Water movement through the xylem requires less pressure than movement through living cells

The xylem provides a pathway of low resistivity for water movement, thus reducing the pressure gradients needed to transport water from the soil to the leaves. Some numerical values will help us appreciate the extraordinary efficiency of the xylem. We will calculate the driving force required to move water through the xylem at a typical velocity and compare it with the driving force that would be needed to move water through a cell-to-cell pathway.

For the purposes of this comparison, we will use a value of 4 mm s^{-1} for the xylem transport velocity and 40 µm as the vessel radius. This is a high velocity for such a narrow vessel, so it will tend to exaggerate the pressure gradient required to support water flow in the xylem. Using a version of Poiseuille's equation (see Equation 3.2), we can calculate the pressure gradient needed to move water at a velocity of 4 mm s^{-1} through an *ideal* tube with a uniform inner radius of 40 µm. The calculation gives a value of 0.02 MPa m^{-1}. Elaboration of the assumptions, equations, and calculations can be found in **Web Topic 4.3**.

Of course, *real* xylem conduits have irregular inner wall surfaces, and water flow through perforation plates and pits adds additional resistance. Such deviations from an ideal tube will increase the frictional drag above that calculated from Poiseuille's equation. Measurements show that the actual resistance is greater by approximately a factor of 2 (Nobel 1999). Thus our estimate of 0.02 MPa m^{-1} is in the correct range for pressure gradients found in trees.

Let's now compare this value (0.02 MPa m^{-1}) with the driving force that would be necessary to move water at the same velocity from cell to cell, crossing the plasma mem-

FIGURE 4.7 Tracheary elements and their interconnections. (A) Structural comparison of tracheids and vessel elements, two classes of tracheary elements involved in xylem water transport. Tracheids are elongate, hollow, dead cells with highly lignified walls. The walls contain numerous pits—regions where secondary wall is absent but primary wall remains. The shape and pattern of wall pitting vary with species and organ type. Tracheids are present in all vascular plants. Vessels consist of a stack of two or more vessel elements. Like tracheids, vessel elements are dead cells and are connected to one another by perforation plates—regions of the wall where pores or holes have developed. Vessels are connected to other vessels and to tracheids through pits. Vessels are found in most angiosperms and are lacking in most gymnosperms. (B) Scanning electron micrograph of oak wood showing two vessel elements that make up a portion of a vessel. Large pits are visible on the side walls, and the end walls are open at the perforation plate. (420×) (C) Diagram of a coniferous bordered pit with the torus centered in the pit cavity (left) or lodged to one side of the cavity (right). When the pressure difference between two tracheids is small, the pit membrane will lie close to the center of the bordered pit, allowing water to flow through the porous margo region of the pit membrane; when the pressure difference between two tracheids is large, such as when one has cavitated and the other remains filled with water under tension, the pit membrane is displaced such that the torus becomes lodged against the overarching walls, thereby preventing the embolism from propagating between tracheids. (D) In contrast, the pit membranes of angiosperms and other nonconiferous vascular plants are relatively homogeneous in their structure. These pit membranes have very small pores, which prevents the spread of embolism but also imparts a significant hydraulic resistance compared with conifer pits. (B © G. Shih-R. Kessel/Visuals Unlimited; C after Zimmermann 1983.)

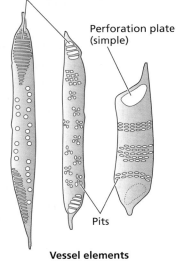
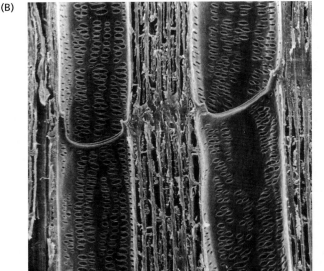
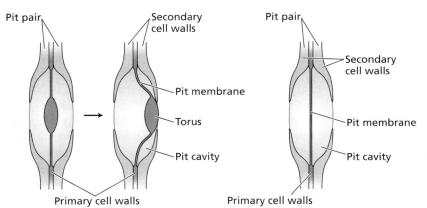

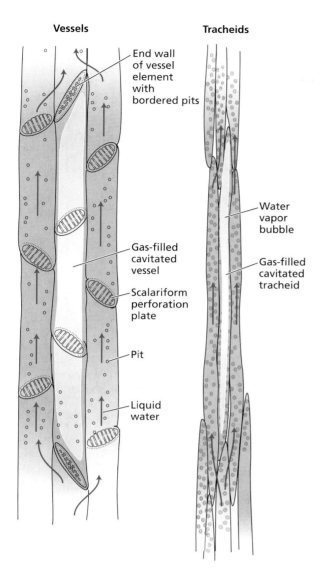

FIGURE 4.8 Vessels (left) and tracheids (right) form a series of parallel, interconnected pathways for water movement. Cavitation blocks water movement because of the formation of gas-filled (embolized) conduits. Because xylem conduits are interconnected through openings ("bordered pits") in their thick secondary walls, water can detour around the blocked vessel by moving through adjacent tracheary elements. The very small pores in the pit membranes help prevent embolisms from spreading between xylem conduits. Thus, in the diagram on the right the gas is contained within a single cavitated tracheid. In the diagram on the left, gas has filled the entire cavitated vessel, shown here as being made up of three vessel elements, each separated by scalariform perforation plates. In nature vessels can be very long (up to several meters in length) and thus made up of many vessel elements.

brane each time. Using Poiseuille's equation, as described in **Web Topic 4.3** the driving force needed to move water through a layer of cells at 4 mm s^{-1} is calculated to be 2×10^8 MPa m^{-1}. This is ten orders of magnitude greater than the driving force needed to move water through our 40-μm-radius xylem vessel. Our calculation clearly shows that water flow through the xylem is vastly more efficient than water flow across living cells. Nevertheless, because the xylem makes up most of the water transport pathway from soil to leaves, it makes a significant contribution to the total resistance to water flow through the plant.

What pressure difference is needed to lift water 100 meters to a treetop?

With the foregoing example in mind, let's see what pressure gradient is needed to move water up to the top of a very tall tree. The tallest trees in the world are the coast redwoods (*Sequoia sempervirens*) of North America and mountain ash (*Eucalyptus regnans*) of Australia. Individuals of both species can exceed 100 m. If we think of the stem of a tree as a long pipe, we can estimate the pressure difference that is needed to overcome the frictional drag of moving water from the soil to the top of the tree by multiplying our pressure gradient of 0.02 MPa m^{-1} by the height of the tree (0.02 MPa m^{-1} × 100 m = 2 MPa).

In addition to frictional resistance, we must consider gravity. The weight of a standing column of water 100 m tall creates a pressure of 1 MPa at the bottom of the water column (100 m × 0.01 MPa m^{-1}). This pressure gradient due to gravity must be added to that required to cause water movement through the xylem. Thus we calculate that a pressure difference of roughly 3 MPa, from the base to the top branches, is needed to carry water up the tallest trees.

The cohesion–tension theory explains water transport in the xylem

In theory, the pressure gradients needed to move water through the xylem could result from the generation of positive pressures at the base of the plant or negative pressures at the top of the plant. We mentioned previously that some roots can develop positive hydrostatic pressure in their xylem. However, root pressure is typically less than 0.1 MPa and disappears when the transpiration rate is high or when soils are dry, so it is clearly inadequate to move water up a tall tree. Furthermore, because root pressure is generated by the accumulation of ions in the xylem, relying on this for transporting water would require a mechanism for dealing with these solutes once the water evaporates from the leaves.

Instead, the water at the top of a tree develops a large tension (a negative hydrostatic pressure), and this tension *pulls* water through the xylem. This mechanism, first proposed toward the end of the nineteenth century, is called the *cohesion–tension theory of sap ascent* because it requires the cohesive properties of water to sustain large tensions in the xylem water columns. One can readily demonstrate xylem tensions by puncturing intact xylem through a drop of ink on the surface of a stem from a transpiring plant. When the tension in the xylem is relieved, the ink is drawn instantly into the xylem, resulting in visible streaks along the stem.

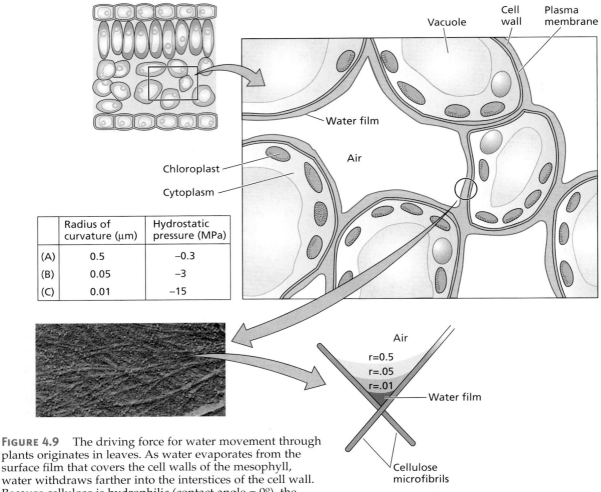

FIGURE 4.9 The driving force for water movement through plants originates in leaves. As water evaporates from the surface film that covers the cell walls of the mesophyll, water withdraws farther into the interstices of the cell wall. Because cellulose is hydrophilic (contact angle = 0°), the force resulting from surface tension causes a negative pressure in the liquid phase. As the radius of curvature of these air:water interfaces decreases, the hydrostatic pressure becomes more negative, as calculated from Equation 4.1. (Micrograph from Gunning and Steer 1996.)

What is the source of the negative pressure of water in leaves and how does it serve to pull water from the soil? The negative pressure that causes water to move up through the xylem develops at the surface of the cell walls in the leaf (Figure 4.9). The situation is analogous to that in the soil. The cell wall acts like a very fine capillary network or wick soaked with water. Water adheres to the cellulose microfibrils and other hydrophilic components of the wall. The mesophyll cells within the leaf are in direct contact with the atmosphere through an extensive system of intercellular air spaces. As water is lost to the air, the surface of the remaining water is drawn into the interstices of the cell wall (Figure 4.9), where it forms curved air–water interfaces. Because of the high surface tension of water, the curvature of these interfaces induces a tension, or negative pressure, in the water. As more water is removed from the wall, the curvature of these air–water interfaces increases and the pressure of the water becomes more negative (see Equation 4.1).

The cohesion–tension theory explains how the substantial movement of water through plants can occur without the direct expenditure of metabolic energy. Instead, the energy input that powers the movement water through plants comes from the sun, which by increasing the temperature of both the leaf and the surrounding air drives the evaporation of water.

The cohesion–tension theory has been a controversial subject for more than a century and continues to generate lively debate. The main controversy surrounds the question of whether water columns in the xylem can sustain the large tensions (negative pressures) necessary to pull water up tall trees. Most researchers believe that the cohesion–tension theory is sound (Steudle 2001). For details on the history of research on water transport in the xylem, including the controversy surrounding the cohesion-tension theory, see **Web Essays 4.1 and 4.2**.

Xylem transport of water in trees faces physical challenges

The large tensions that develop in the xylem of trees (see **Web Essay 4.3**) and other plants present significant physical challenges. First, the water under tension transmits an inward force to the walls of the xylem. If the cell walls were weak or pliant, they would collapse under the influence of this tension. The secondary wall thickenings and lignification of tracheids and vessels are adaptations that offset this tendency to collapse. Plants that experience large xylem tensions tend to have denser wood, reflecting the mechanical stresses imposed on the wood by water under tension (Hacke et al. 2001).

A second challenge is that water under such tensions is in a *physically metastable state*. When the hydrostatic pressure in liquid water becomes equal to its saturated vapor pressure, the water will undergo a phase change. In essence, it will boil. We are all familiar with the idea of vaporizing water by increasing its temperature (which raises its saturated vapor pressure). Less familiar, but still easily observed, is the fact that water can be made to boil at room temperature by placing it in a vacuum chamber (which lowers the hydrostatic pressure of the liquid phase by reducing the pressure of the atmosphere).

In our earlier example, we estimated that a pressure gradient of 3 MPa would be needed to supply water to leaves to the top of a very tall tree. If we assume that the soil surrounding this tree is fully hydrated and lacks significant concentrations of solutes (i.e., $\Psi_w = 0$), the cohesion–tension theory predicts that the hydrostatic pressure of water in the xylem at the top of the tree will be –3 MPa. This value is significantly below the saturated vapor pressure (~0.002 MPa at 20°C), raising the question of what maintains the water column in its liquid state? That is, why doesn't it boil?

Water in the xylem is described as being in a metastable state because of the persistence of liquid water despite the existence of a thermodynamically lower energy state (i.e., the vapor phase). This situation occurs because (1) the cohesion and adhesion of water makes the activation energy for the liquid-to-vapor phase change very high, and (2) the structure of the xylem minimizes the presence of nucleating sites that provide this activation energy.

The most important nucleating sites are gas bubbles of sufficient size that the inward force resulting from surface tension is less than the outward force due to the negative pressure in the liquid phase. When this occurs, the bubble will expand. Furthermore, once a bubble starts to expand the inward force due to surface tension decreases because the air–water interface has less curvature. Thus, a bubble that exceeds the critical size for expansion will expand until it fills the entire conduit.

The absence of gas bubbles of sufficient size to destabilize the water column when under tension is, in part, due to the filtering action of the roots as water is forced to flow symplasmically across the endodermis. However, as the tension in water increases, there is an increased tendency for air to be pulled through microscopic pores in the xylem cell walls. This phenomenon is called *air seeding*.

The most permeable regions of xylem walls are the pit membranes themselves, the sites where water flows between conduits. Normally, these prevent the spread of gas between conduits. However, when exposed to air on one side—due to injury, leaf abscission, or the existence of a neighboring gas-filled conduit they can serve as sites of air-entry. This will occur if the pressure difference across the pit membrane is sufficient to overcome either the capillary forces of the air-water interfaces within the cellulose microfibriller matrix of structurally homogeneous pit membranes (see Figure 4.7D), or to dislodge the torus of a coniferous pit membrane (see Figure 4.7C).

A second mode by which bubbles can form in xylem conduits is due to the very low solubility of gases in ice (Davis et al. 1999): Because water in the xylem contains dissolved gases, freezing of xylem conduits can lead to bubble formation.

This phenomenon of bubble expansion is known as *cavitation* and the resulting gas-filled void is referred to as an *embolism*. Its effect is similar to a vapor lock in the fuel line of an automobile or an embolism in a blood vessel. Cavitation breaks the continuity of the water column and prevents the transport of water under tension (Tyree and Sperry 1989).

Such breaks in the water columns in plants are not unusual. When plants are deprived of water, sound pulses can be detected (Jackson et al. 1999). The pulses or clicks are presumed to correspond to the formation and rapid expansion of air bubbles in the xylem, resulting in high-frequency acoustic shock waves through the rest of the plant. These breaks in xylem water continuity, if not repaired, would be disastrous to the plant. By blocking the main transport pathway of water, such embolisms increase flow resistance and can ultimately cause the dehydration and death of the leaves.

"Vulnerability curves" (see Figure 4.10) provide a way of quantifying a species' susceptibility to cavitation and its impact on flow through the xylem. A vulnerability curve plots the measured hydraulic conductivity (usually as a % of maximum) of a branch, stem or root segment versus the experimentally imposed level of xylem tension. In all species, xylem hydraulic conductivity decreases with increasing tensions (more negative pressures) until ultimately flow entirely ceases. However, the decrease in xylem hydraulic conductivity occurs at much less negative xylem pressures in species from wetter habitats, such as birch, than in species from more arid regions, such as sagebrush.

Plants minimize the consequences of xylem cavitation

The impact of xylem cavitation on the plant can be minimized by several means. Because the tracheary elements in the xylem are interconnected, one gas bubble might, in principle, expand to fill the whole network. In practice, gas bub-

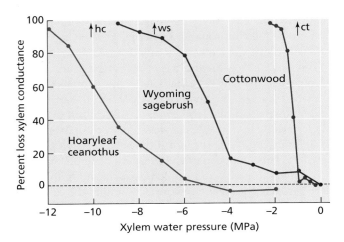

FIGURE 4.10 The percentage loss of hydraulic conductance in stem xylem versus xylem water pressure in three species of contrasting drought tolerance. Arrows on the upper pressure axis indicate minimum xylem pressures measured in the field for each species. (After Sperry 2000.)

bles do not spread far because the expanding gas bubble cannot easily pass through the small pores of the pit membranes. Because the capillaries in the xylem are interconnected, one gas bubble does not completely stop water flow. Instead, water can detour around the embolized conduit by traveling through neighboring, water-filled conduits (see Figure 4.8). Thus the finite length of the tracheid and vessel conduits of the xylem, while resulting in an increased resistance to water flow, also provides a way to restrict the impact of cavitation.

Gas bubbles can also be eliminated from the xylem. At night, when transpiration is low, xylem Ψ_p increases and the water vapor and gases may simply dissolve back into the solution of the xylem. Moreover, as we have seen, some plants develop positive pressures (root pressures) in the xylem. Such pressures shrink the gas bubble and cause the gases to dissolve. Recent studies indicate that cavitation may be repaired even when the water in the xylem is under tension (Holbrook et al. 2001; Salleo et al. 2004). A mechanism for such repair is not yet known and remains the subject of active research (see **Web Essay 4.4**).

Finally, many plants have secondary growth in which new xylem forms each year. The production of new xylem conduits allows plants to replace losses in water transport capacity due to cavitation.

Water Movement from the Leaf to the Atmosphere

On its way from the leaf to the atmosphere, water is pulled from the xylem into the cell walls of the mesophyll, where it evaporates into the air spaces of the leaf (Figure 4.11). The

FIGURE 4.11 Water pathway through the leaf. Water is pulled from the xylem into the cell walls of the mesophyll, where it evaporates into the air spaces within the leaf. Water vapor then diffuses through the leaf air space, through the stomatal pore, and across the boundary layer of still air found next to the leaf surface. CO_2 diffuses in the opposite direction along its concentration gradient (low inside, higher outside).

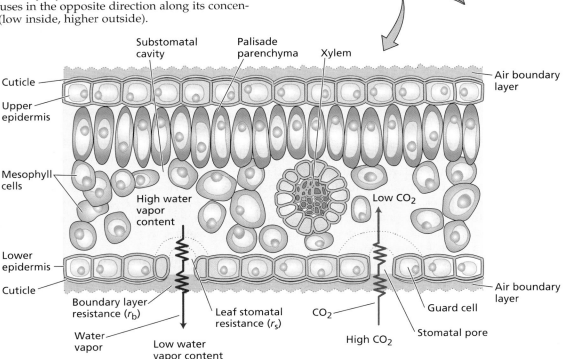

water vapor then exits the leaf through the stomatal pore. Water moves along this pathway predominantly by diffusion, so this water movement is controlled by the *concentration gradient* of water vapor.

The waxy cuticle that covers the leaf surface is a very effective barrier to water movement. It has been estimated that only about 5% of the water lost from leaves escapes through the cuticle. Almost all of the water lost from typical leaves is lost by diffusion of water vapor through the tiny pores of the stomatal apparatus, which are usually most abundant on the lower surface of the leaf.

We will now examine the driving force for leaf transpiration, the main resistances in the diffusion pathway from the leaf to the atmosphere, and the anatomical features of the leaf that regulate transpiration.

The driving force for water loss is the difference in water vapor concentration

Transpiration from the leaf depends on two major factors: (1) the **difference in water vapor concentration** between the leaf air spaces and the external air and (2) the **diffusional resistance** (r) of this pathway. The difference in water vapor concentration is expressed as $c_{wv(leaf)} - c_{wv(air)}$. The water vapor concentration of bulk air ($c_{wv(air)}$) can be readily measured, but that of the leaf ($c_{wv(leaf)}$) is more difficult to assess.

Whereas the volume of air space inside the leaf is small, the wet surface from which water evaporates is comparatively large. Air space volume is about 5% of the total leaf volume for pine needles, 10% for corn leaves, 30% for barley, and 40% for tobacco leaves. In contrast to the volume of the air space, the internal surface area from which water evaporates may be from 7 to 30 times the external leaf area. This high ratio of surface area to volume makes for rapid vapor equilibration inside the leaf. Thus we can assume that the air space in the leaf is close to water potential equilibrium with the cell wall surfaces from which liquid water is evaporating.

Within the range of water potentials experienced by transpiring leaves (generally < 2.0 MPa) the equilibrium water vapor concentration is within a few percentage points of the saturation water vapor concentration. This allows one to estimate the water vapor concentration within a leaf from its temperature, which is easy to measure. (**Web Topic 4.4** shows how we can calculate the water vapor concentration in the leaf air spaces and discusses other aspects of the water relations within a leaf.)

The concentration of water vapor, c_{wv}, changes at various points along the transpiration pathway. We see from Table 4.2 that c_{wv} decreases at each step of the pathway from the cell wall surface to the bulk air outside the leaf. The important points to remember are (1) that the driving force for water loss from the leaf is the *absolute* concentration difference (difference in c_{wv}, in mol m^{-3}), and (2) that this difference depends on leaf temperature, as shown in Figure 4.12.

Water loss is also regulated by the pathway resistances

The second important factor governing water loss from the leaf is the diffusional resistance of the transpiration pathway, which consists of two varying components (see Figure 4.11):

1. The resistance associated with diffusion through the stomatal pore, the **leaf stomatal resistance** (r_s).

2. The resistance due to the layer of unstirred air next to the leaf surface through which water vapor must diffuse to reach the turbulent air of the atmosphere. This second resistance, r_b, is called the leaf **boundary layer resistance.** We will discuss this type of resistance before considering stomatal resistance.

The thickness of the boundary layer is determined primarily by wind speed and leaf size. When the air surrounding the leaf is very still, the layer of unstirred air on the surface of the leaf may be so thick that it is the primary deterrent to water vapor loss from the leaf. Increases in stomatal apertures under such conditions have little effect on transpiration rate (Figure 4.13), although closing the stomata completely will still reduce transpiration.

When wind velocity is high, the moving air reduces the thickness of the boundary layer at the leaf surface, reducing the resistance of this layer. Under such conditions, the stomatal resistance will largely control water loss from the leaf.

Various anatomical and morphological aspects of the leaf can influence the thickness of the boundary layer. Hairs on the surface of leaves

TABLE 4.2
Representative values for relative humidity, absolute water vapor concentration, and water potential for four points in the pathway of water loss from a leaf

Location	Relative humidity	Water vapor Concentration (mol m^{-3})	Water vapor Potential (MPa)[a]
Inner air spaces (25°C)	0.99	1.27	−1.38
Just inside stomatal pore (25°C)	0.95	1.21	−7.04
Just outside stomatal pore (25°C)	0.47	0.60	−103.7
Bulk air (20°C)	0.50	0.50	−93.6

Source: Adapted from Nobel 1999.

Note: See Figure 4.11.

[a]Calculated using Equation 4.5.2 in Web Topic 4.4, with values for $RT/\bar{V}_w$ of 135 MPa at 20°C and 137.3 MPa at 25°C.

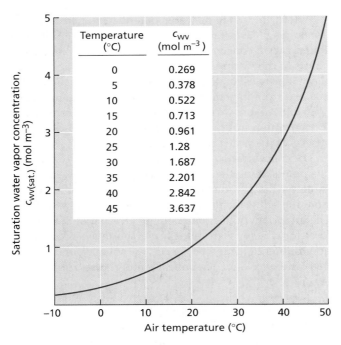

Figure 4.12 Concentration of water vapor in saturated air as a function of air temperature.

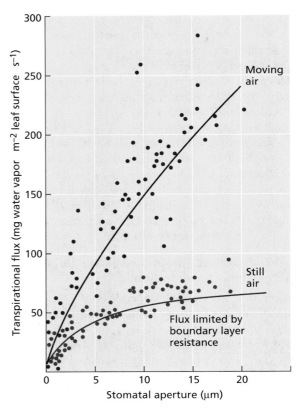

Figure 4.13 Dependence of transpiration flux on the stomatal aperture of zebra plant (*Zebrina pendula*) in still air and in moving air. The boundary layer is larger and more rate limiting in still air than in moving air. As a result, the stomatal aperture has less control over transpiration in still air. (After Bange 1953.)

can serve as microscopic windbreaks. Some plants have sunken stomata that provide a sheltered region outside the stomatal pore. The size and shape of leaves also influence the way the wind sweeps across the leaf surface. Although these and other factors may influence the boundary layer, they are not characteristics that can be altered on an hour-to-hour or even day-to-day basis. For short-term regulation, control of stomatal apertures by the guard cells plays a crucial role in the regulation of leaf transpiration.

Stomatal control couples leaf transpiration to leaf photosynthesis

Because the cuticle covering the leaf is nearly impermeable to water, most leaf transpiration results from the diffusion of water vapor through the stomatal pore (see Figure 4.11). The microscopic stomatal pores provide a *low-resistance pathway* for diffusional movement of gases across the epidermis and cuticle. That is, the stomatal pores lower the diffusional resistance for water loss from leaves. Changes in stomatal resistance are important for the regulation of water loss by the plant and for controlling the rate of carbon dioxide uptake necessary for sustained CO_2 fixation during photosynthesis.

All land plants are faced with competing demands of taking up CO_2 from the atmosphere while limiting water loss. The cuticle that covers exposed plant surfaces serves as an effective barrier to water loss and thus protects the plant from desiccation. However, plants cannot prevent outward diffusion of water through stomatal pores without simultaneously excluding CO_2 from the leaf. This problem is compounded because the concentration gradient for CO_2 uptake is much, much smaller than the concentration gradient that drives water loss.

When water is abundant, the functional solution to this dilemma is the *temporal* regulation of stomatal apertures—open during the day, closed at night. At night, when there is no photosynthesis and thus no demand for CO_2 inside the leaf, stomatal apertures are kept small, preventing unnecessary loss of water. On a sunny morning when the supply of water is abundant and the solar radiation incident on the leaf favors high photosynthetic activity, the demand for CO_2 inside the leaf is large, and the stomatal pores are wide open, decreasing the stomatal resistance to CO_2 diffusion. Water loss by transpiration is also substantial under these conditions, but since the water supply is plentiful, it is advantageous for the plant to trade water for the products of photosynthesis, which are essential for growth and reproduction.

On the other hand, when soil water is less abundant, the stomata will open less or even remain closed on a sunny morning. By keeping its stomata closed in dry conditions, the plant avoids dehydration. The values for $(c_{wv(leaf)} - c_{wv(air)})$ and for r_b are not readily amenable to biological control. However, stomatal resistance (r_s) can be regulated by opening and closing of the stomatal pore. This biological control is exerted by a pair of specialized epidermal cells, the **guard cells**, which surround the stomatal pore (Figure 4.14).

The cell walls of guard cells have specialized features

Guard cells can be found in leaves of all vascular plants, and they are also present in organs from more primitive plants, such as liverworts and mosses (Ziegler 1987). Guard cells show considerable morphological diversity, but we can distinguish two main types: One is typical of grasses and a

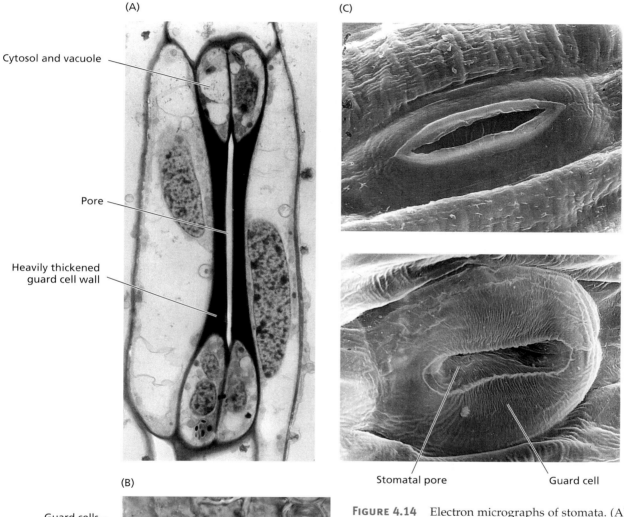

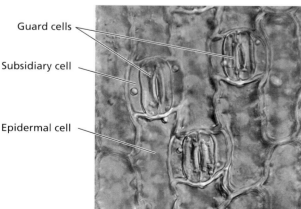

FIGURE 4.14 Electron micrographs of stomata. (A) A stoma from a grass. The bulbous ends of each guard cell show their cytosolic content and are joined by the heavily thickened walls. The stomatal pore separates the two midportions of the guard cells. (2560×) (B) Stomatal complexes of the sedge *Carex* viewed with differential interference contrast light microscopy. Each complex consists of two guard cells surrounding a pore and two flanking subsidiary cells. (550×) (C) Scanning electron micrographs of onion epidermis. The top panel shows the outside surface of the leaf, with a stomatal pore inserted in the cuticle. The bottom panel shows a pair of guard cells facing the stomatal cavity, toward the inside of the leaf. (1640×) (A from Palevitz 1981, B from Jarvis and Mansfield 1981, A and B courtesy of B. Palevitz; C from Zeiger and Hepler 1976 [top] and E. Zeiger and N. Burnstein [bottom].)

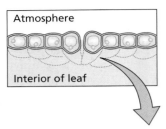

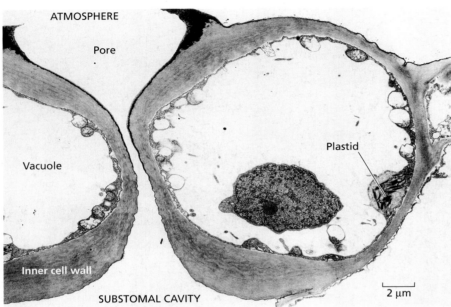

Figure 4.15 Electron micrograph showing a pair of guard cells from the dicot tobacco (*Nicotiana tabacum*). The section was made perpendicular to the main surface of the leaf. The pore faces the atmosphere; the bottom faces the substomatal cavity inside the leaf. Note the uneven thickening pattern of the walls, which determines the asymmetric deformation of the guard cells when their volume increases during stomatal opening. (Micrograph from Sack 1987, courtesy of F. Sack.)

few other monocots, such as palms; the other is found in all dicots, in many monocots, and in mosses, ferns, and gymnosperms.

In grasses (see Figure 4.14A), guard cells have a characteristic dumbbell shape, with bulbous ends. The pore proper is a long slit located between the two "handles" of the dumbbells. These guard cells are always flanked by a pair of differentiated epidermal cells called **subsidiary cells**, which help the guard cells control the stomatal pores (see Figure 4.14B). The guard cells, subsidiary cells, and pore are collectively called the **stomatal complex**.

In dicot plants and nongrass monocots, kidney-shaped guard cells have an elliptical contour with the pore at its center (see Figure 4.14C). Although subsidiary cells are not uncommon in species with kidney-shaped stomata, they are often absent, in which case the guard cells are surrounded by ordinary epidermal cells.

A distinctive feature of the guard cells is the specialized structure of their walls. Portions of these walls are substantially thickened (Figure 4.15) and may be up to 5 μm across, in contrast to the 1 to 2 μm typical of epidermal cells. In kidney-shaped guard cells, a differential thickening pattern results in very thick inner and outer (lateral) walls, a thin dorsal wall (the wall in contact with epidermal cells), and a somewhat thickened ventral (pore) wall. The portions of the wall that face the atmosphere extend into well-developed ledges, which form the pore proper.

The alignment of **cellulose microfibrils**, which reinforce all plant cell walls and are an important determinant of cell shape (see Chapter 15), plays an essential role in the opening and closing of the stomatal pore. In ordinary cells having a cylindrical shape, cellulose microfibrils are oriented transversely to the long axis of the cell. As a result, the cell expands in the direction of its long axis because the cellulose reinforcement offers the least resistance at right angles to its orientation.

In guard cells the microfibril organization is different. Kidney-shaped guard cells have cellulose microfibrils fanning out radially from the pore (Figure 4.16A). Thus the cell girth is reinforced like a steel-belted radial tire, and the guard cells curve outward during stomatal opening (Sharpe et al. 1987). In grasses, the dumbbell-shaped guard cells function like beams with inflatable ends. As the bulbous ends of the cells increase in volume and swell, the beams are separated from each other and the slit between them widens (Figure 4.16B).

An increase in guard cell turgor pressure opens the stomata

Guard cells function as multisensory hydraulic valves. Environmental factors such as light intensity and quality, temper-

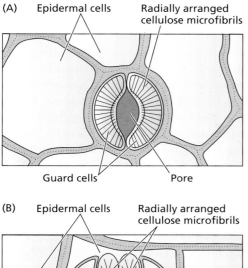

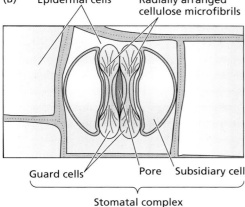

FIGURE 4.16 The radial alignment of the cellulose microfibrils in guard cells and epidermal cells of (A) a kidney-shaped stoma and (B) a grasslike stoma. (After Meidner and Mansfield 1968.)

ature, leaf water status, and intracellular CO_2 concentrations are sensed by guard cells, and these signals are integrated into well-defined stomatal responses. If leaves kept in the dark are illuminated, the light stimulus is perceived by the guard cells as an opening signal, triggering a series of responses that result in opening of the stomatal pore.

The early aspects of this process are ion uptake and other metabolic changes in the guard cells, which will be discussed in detail in Chapter 18. Here we will note the effect of decreases in osmotic potential (Ψ_s) resulting from ion uptake and from biosynthesis of organic molecules in the guard cells. Water relations in guard cells follow the same rules as in other cells. As Ψ_s decreases, the water potential decreases and water consequently moves into the guard cells. As water enters the cell, turgor pressure increases. Because of the elastic properties of their walls, guard cells can reversibly increase their volume by 40 to 100%, depending on the species. Because of the differential thickening of guard cell walls, such changes in cell volume lead to opening or closing of the stomatal pore.

The transpiration ratio measures the relationship between water loss and carbon gain

The effectiveness of plants in moderating water loss while allowing sufficient CO_2 uptake for photosynthesis can be assessed by a parameter called the **transpiration ratio**. This value is defined as the amount of water transpired by the plant, divided by the amount of carbon dioxide assimilated by photosynthesis.

For plants in which the first stable product of carbon fixation is a three-carbon compound (such plants are called C_3 plants; see Chapter 8), about 500 molecules of water are lost for every molecule of CO_2 fixed by photosynthesis, giving a transpiration ratio of 500. (Sometimes the reciprocal of the transpiration ratio, called the *water use efficiency*, is cited. Plants with a transpiration ratio of 500 have a water use efficiency of 1/500, or 0.002.)

The large ratio of H_2O efflux to CO_2 influx results from three factors:

1. The concentration gradient driving water loss is about 50 times larger than that driving the influx of CO_2. In large part, this difference is due to the low concentration of CO_2 in air (about 0.038%) and the relatively high concentration of water vapor within the leaf.

2. CO_2 diffuses about 1.6 times more slowly through air than water does (the CO_2 molecule is larger than H_2O and has a smaller diffusion coefficient).

3. CO_2 must cross the plasma membrane, the cytoplasm, and the chloroplast envelope before it is assimilated in the chloroplast. These membranes add to the resistance of the CO_2 diffusion pathway.

Some plants are adapted for life in particularly dry environments or seasons of the year. These plants, designated as C_4 and CAM plants, utilize variations in the usual photosynthetic pathway for fixation of carbon dioxide. Plants with C_4 photosynthesis (in which a four-carbon compound is the first stable product of photosynthesis; see Chapter 8) generally transpire less water per molecule of CO_2 fixed; a typical transpiration ratio for C_4 plants is about 250. Desert-adapted plants with crassulacean acid metabolism (CAM) photosynthesis, in which CO_2 is initially fixed into four-carbon organic acids at night, have even lower transpiration ratios; values of about 50 are not unusual.

Overview: The Soil–Plant–Atmosphere Continuum

We have seen that movement of water from the soil through the plant to the atmosphere involves different mechanisms of transport:

- In the soil and the xylem, water moves by bulk flow in response to a pressure gradient ($\Delta\Psi_p$).

- In the vapor phase, water moves primarily by diffusion, at least until it reaches the outside air, where convection (a form of bulk flow) becomes dominant.

- When water is transported across membranes, the driving force is the water potential difference across the membrane. Such osmotic flow occurs when cells absorb water and when roots transport water from the soil to the xylem.

However, the key element in the transport of water from the soil to the leaves is the generation of negative pressures within the xylem due to the capillary forces within the cell walls of transpiring leaves. At the other end of the plant, soil water is also held by capillary forces. This results in a "tug-of-war" on a rope of water by capillary forces at both ends. As a leaf loses water due to transpiration, water moves up the plant and out of the soil driven by physical forces, without the involvement of any metabolic pump.

This simple mechanism makes for tremendous efficiency energetically—which is critical when 500 molecules of water are being transported for every CO_2 molecule being taken up in exchange. Crucial elements that allow this transport system to function are a low-resistivity xylem flow path that is protected from cavitation, selective filtration of water across the root membranes, and a high-surface-area root system for mining the soil water.

Summary

Water is the essential medium of life. Land plants are faced with potentially lethal desiccation by water loss to the atmosphere. This problem is aggravated by the large surface area of leaves, their high radiant-energy gain, and their need to have an open pathway for CO_2 uptake. The absence of surfaces differentially permeable to CO_2 versus water vapor means that plants face a conflict between the need for water conservation and the need for CO_2 assimilation.

The need to resolve this vital conflict determines much of the structure of land plants: (1) an extensive root system to extract water from the soil; (2) a low-resistance pathway through the xylem vessel elements and tracheids to bring water to the leaves; (3) a hydrophobic cuticle covering the surfaces of the plant to reduce evaporation; (4) microscopic stomata on the leaf surface to allow gas exchange; and (5) guard cells to regulate the diameter (and diffusional resistance) of the stomatal aperture.

The result is an organism that transports water from the soil to the atmosphere purely in response to physical forces. No energy is expended directly by the plant to translocate water, although development and maintenance of the structures needed for efficient and controlled water transport require considerable energy input.

The mechanisms of water transport from the soil through the plant body to the atmosphere include diffusion, bulk flow, and osmosis. Each of these processes is coupled to different driving forces. However, ultimately capillarity is the key element that allows for the movement of water from the soil to the leaves.

Water in the plant can be considered a continuous hydraulic system, connecting the water in the soil with the water vapor in the atmosphere. Transpiration is regulated principally by the guard cells, which regulate the stomatal pore size to meet the photosynthetic demand for CO_2 uptake while minimizing water loss to the atmosphere. Water evaporation from the cell walls of the leaf mesophyll cells generates large negative pressures (or tensions) in the apoplastic water by capillarity. These negative pressures are transmitted to the xylem, and they pull water through the long xylem conduits.

Although aspects of the cohesion–tension theory of sap ascent are intermittently debated, an overwhelming body of evidence supports the idea that water transport in the xylem requires the existence of large negative pressures. When transpiration is high, such large negative pressures in the xylem water may cause cavitation. The resulting embolisms can block water transport and lead to severe water deficits in the leaf. Water deficits are commonplace in plants, necessitating a host of adaptive responses that modify the physiology and development of plants.

Web Material

Web Topics

4.1 **Irrigation**
An overview of different irrigation methods, and their impact on crop yield and soil salinity.

4.2 **Soil Hydraulic Conductivity and Water Potential**
Soil hydraulic conductivity determines the ease with which water moves through the soil, and it is closely related to soil water potential.

4.3 **Calculating Velocities of Water Movement in the Xylem and in Living Cells**
Calculations of velocities of water movement through the xylem, up a tree trunk, and across cell membranes in a tissue, and their implications for water transport mechanisms.

4.4 **Leaf Transpiration and Water Vapor Gradients**
Leaf transpiration and stomatal conductance affect leaf and air water vapor concentrations.

Web Essays

4.1 **A Brief History of the Study of Water Movement in the Xylem**
The history of our understanding of sap ascent in plants, especially in trees, is a beautiful example of how knowledge about plants is acquired.

4.2 The Cohesion–Tension Theory at Work
A detailed discussion of the cohesion–tension theory on sap ascent in plants and some alternative explanations.

4.3 How Water Climbs to the Top of a 112-Meter-Tall Tree
Measurements of photosynthesis and transpiration in 112-meter tall trees show that some of the conditions experienced by the top foliage compares to that of extreme deserts.

4.4 Cavitation and Refilling
A possible mechanism for cavitation repair is under active investigation.

Chapter References

Bange, G. G. J. (1953) On the quantitative explanation of stomatal transpiration. *Acta Botanica Neerlandica* 2: 255–296.

Davis, S. D., Sperry, J. S., and Hacke, U. G. (1999) The relationship between xylem conduit diameter and cavitation caused by freezing. *Am. J. Bot.* 86: 1367–1372.

Esau, K. (1953) *Plant Anatomy*. John Wiley & Sons, Inc. New York.

Frensch, J., Hsiao, T. C., and Steudle, E. (1996) Water and solute transport along developing maize roots. *Planta* 198: 348–355.

Gunning, B. S., and Steer, M. M. (1996) *Plant Cell Biology: Structure and Function*. Jones and Bartlett Publishers, Boston.

Hacke, U. G., Sperry, J. S., Pockman, W. T., Davis, S. D., and McCulloh, K. (2001) Trends in wood density and structure are linked to prevention of xylem implosion by negative pressure. *Oecologia* 126: 457–461.

Holbrook, N. M., Ahrens, E. T., Burns, M. J., and Zwieniecki, M. A. (2001) In vivo observation of cavitation and embolism repair using magnetic resonance imaging. *Plant Physiol.* 126: 27–31.

Jackson, G. E., Irvine, J., and Grace, J. (1999) Xylem acoustic emissions and water relations of *Calluna vulgaris* L. at two climatological regions of Britain. *Plant Ecol.* 140: 3–14.

Jarvis, P. G., and Mansfield, T. A. (1981) *Stomatal Physiology*. Cambridge University Press, Cambridge.

Kramer, P. J., and Boyer, J. S. (1995) *Water Relations of Plants and Soils*. Academic Press, San Diego, CA.

Martre, P., Morillon, R., Barrieu, F., North, G. B., Nobel, P. S., and Chrispeels, M. J. (2002) Plasma membrane aquaporins play a significant role during recovery from water deficit. *Plant Physiol.* 130: 2101–2110.

Meidner, H., and Mansfield, D. (1968) *Stomatal Physiology*. McGraw-Hill, London.

Nobel, P. S. (1999) *Physicochemical and Environmental Plant Physiology*, 2nd ed. Academic Press, San Diego, CA.

Palevitz, B. A. (1981) The structure and development of guard cells. In *Stomatal Physiology*, P. G. Jarvis and T. A. Mansfield, eds., Cambridge University Press, Cambridge, pp. 1–23.

Salleo, S., Lo Gullo, M. A., Trifilo, P., and Nardini, A. (2004) New evidence for a role of vessel-associated cells and phloem in the rapid xylem refilling of cavitated stems of *Laurus nobilis* L. *Plant, Cell and Environment* 27: 1065–1076.

Sack, F. D. (1987) The development and structure of stomata. In *Stomatal Function*, E. Zeiger, G. Farquhar, and I. Cowan, eds., Stanford University Press, Stanford, CA, pp. 59–90.

Sharpe, P. J. H., Wu, H.-I., and Spence, R. D. (1987) Stomatal mechanics. In *Stomatal Function*, E. Zeiger, G. Farquhar, and I. Cowan, eds., Stanford University Press, Stanford, CA, pp. 91–114.

Siefritz, F., Tyree, M. T., Lovisolo, C., Schubert, A., and Kaldenhoff, R. (2002) PIP1 plasma membrane aquaporins in tobacco: From cellular effects to function in plants. *Plant Cell* 14: 869–876.

Steudle, E. (2001) The cohesion-tension mechanism and the acquisition of water by plant roots. *Annu. Rev. Plant Physiol. Plant Mol. Biol.* 52: 847–875.

Steudle, E., and Frensch, J. (1996) Water transport in plants: Role of the apoplast. *Plant Soil* 187: 67–79.

Törnroth-Horsefield, S., Wang, Y., Hedfalk. K., Johanson, U., Karlsson, M., Tajkhorshid, E., Neutze, R., and Kjellbom, P. (2006). Structural mechanism of plant aquaporin gating. *Nature* 439: 688–694.

Tournaire-Roux, C., Sutka, M., Javot, H., Gout, E., Gerbeau, P., Luu, D. T., Bligny, R., and Maurel, C. (2003) Cytosolic pH regulates root water transport during anoxic stress through gating of aquaporins. *Nature* 425: 393–397.

Tyree, M. T., and Sperry, J. S. (1989) Vulnerability of xylem to cavitation and embolism. *Annu. Rev. Plant Physiol. Plant Mol. Biol.* 40: 19–38.

Zeiger, E., and Hepler, P. K. (1976) Production of guard cell protoplasts from onion and tobacco. *Plant Physiol.* 58: 492–498.

Ziegler, H. (1987) The evolution of stomata. In *Stomatal Function*, E. Zeiger, G. Farquhar, and I. Cowan, eds., Stanford University Press, Stanford, CA, pp. 29–58.

Zimmermann, M. H. (1983) *Xylem Structure and the Ascent of Sap*. Springer, Berlin.

Zwieniecki, M. A., Thompson, M. V., and Holbrook, N. M. (2002) Understanding the hydraulics of porous pipes: Tradeoffs between water uptake and root length utilization. *J. Plant Growth Reg.* 21: 315–323.

Chapter 5 | Mineral Nutrition

MINERAL NUTRIENTS ARE ELEMENTS such as nitrogen, phosphorus, and potassium acquired primarily in the form of inorganic ions from the soil. Although mineral nutrients continually cycle through all organisms, they enter the biosphere predominantly through the root systems of plants, so in a sense plants act as the "miners" of Earth's crust (Epstein 1999). The large surface area of roots and their ability to absorb inorganic ions at low concentrations from the soil solution make mineral absorption by plants a very effective process. After being absorbed by the roots, the mineral elements are translocated to the various parts of the plant, where they are utilized in numerous biological functions. Other organisms, such as mycorrhizal fungi and nitrogen-fixing bacteria, often participate with roots in the acquisition of nutrients.

The study of how plants obtain and use mineral nutrients is called **mineral nutrition**. This area of research is central to modern agriculture and environmental protection. High agricultural yields depend strongly on fertilization with mineral nutrients. In fact, yields of most crop plants increase linearly with the amount of fertilizer that they absorb (Loomis and Conner 1992). To meet increased demand for food, world consumption of the primary fertilizer mineral elements—nitrogen, phosphorus, and potassium—rose steadily from 112 million metric tons in 1980 to 143 million metric tons in 1990 and has remained constant during the last 15 years as fertilizer is used more judiciously in an attempt to balance rising costs.

Crop plants, however, typically use less than half of the fertilizer applied (Loomis and Conner 1992). The remaining minerals may leach into surface waters or groundwater, become attached to soil particles, or contribute to air pollution. As a consequence of fertilizer leaching, many water wells in the United States no longer meet federal standards for nitrate (NO_3^-) concentrations in drinking water (Nolan and Stoner 2000). Enhanced nitrogen availability through nitrate (NO_3^-) and ammonium (NH_4^+) dissolved in rainwater, a process known as atmospheric nitrogen deposition, is altering ecosystems throughout the U.S. (Aber et al. 2003; Fenn et al. 2003).

On a brighter note, plants are the traditional means for recycling animal wastes and are proving useful for removing deleterious minerals including heavy metals from toxic-waste dumps (Macek et al. 2000). Because of the complex nature of plant–soil–atmosphere relationships, studies in the area of mineral nutrition involve atmospheric chemists, soil scientists, hydrologists, microbiologists, and ecologists, as well as plant physiologists.

In this chapter we will discuss the nutritional needs of plants, the symptoms of specific nutritional deficiencies, and the use of fertilizers to ensure proper plant nutrition. Then we will examine how soil structure (the arrangement of solid, liquid, and gaseous components) and root morphology influence the transfer of inorganic nutrients from the environment into a plant. Finally, we will introduce the topic of mycorrhizal symbiotic associations. Chapters 6 and 12 address additional aspects of solute transport and nutrient assimilation, respectively.

Essential Nutrients, Deficiencies, and Plant Disorders

Only certain elements have been determined to be essential for plants. An **essential element** is defined as one that is an intrinsic component in the structure or metabolism of a plant or whose absence causes severe abnormalities in plant growth, development, or reproduction (Arnon and Stout 1939; Epstein and Bloom 2005). If plants are given these essential elements, as well as water and energy from sunlight, they can synthesize all the compounds they need for normal growth. Table 5.1 lists the elements that are considered to be essential for most, if not all, higher plants. The first three elements—hydrogen, carbon, and oxygen—are not considered mineral nutrients because they are obtained primarily from water or carbon dioxide.

TABLE 5.1
Adequate tissue levels of elements that may be required by plants

Element	Chemical symbol	Concentration in dry matter (% or ppm)[a]	Relative number of atoms with respect to molybdenum
Obtained from water or carbon dioxide			
Hydrogen	H	6	60,000,000
Carbon	C	45	40,000,000
Oxygen	O	45	30,000,000
Obtained from the soil			
Macronutrients			
Nitrogen	N	1.5	1,000,000
Potassium	K	1.0	250,000
Calcium	Ca	0.5	125,000
Magnesium	Mg	0.2	80,000
Phosphorus	P	0.2	60,000
Sulfur	S	0.1	30,000
Silicon	Si	0.1	30,000
Micronutrients			
Chlorine	Cl	100	3,000
Iron	Fe	100	2,000
Boron	B	20	2,000
Manganese	Mn	50	1,000
Sodium	Na	10	400
Zinc	Zn	20	300
Copper	Cu	6	100
Nickel	Ni	0.1	2
Molybdenum	Mo	0.1	1

Source: Epstein 1972, 1999.

[a]The values for the nonmineral elements (H, C, O) and the macronutrients are percentages. The values for micronutrients are expressed in parts per million.

Essential mineral elements are usually classified as macronutrients or micronutrients, according to their relative concentration in plant tissue. In some cases, the differences in tissue content of macronutrients and micronutrients are not as great as those indicated in Table 5.1. For example, some plant tissues, such as the leaf mesophyll, have almost as much iron or manganese as they do sulfur or magnesium. Many elements often are present in concentrations greater than the plant's minimum requirements.

Some researchers have argued that a classification into macronutrients and micronutrients is difficult to justify physiologically. Mengel and Kirkby (1987) have proposed that the essential elements be classified instead according to their biochemical role and physiological function. Table 5.2 shows such a classification, in which plant nutrients have been divided into four basic groups:

1. Nitrogen and sulfur constitute the first group of essential elements. Plants assimilate these nutrients via bio-

TABLE 5.2
Classification of plant mineral nutrients according to biochemical function

Mineral nutrient	Functions
Group 1	**Nutrients that are part of carbon compounds**
N	Constituent of amino acids, amides, proteins, nucleic acids, nucleotides, coenzymes, hexoamines, etc.
S	Component of cysteine, cystine, methionine, and proteins. Constituent of lipoic acid, coenzyme A, thiamine pyrophosphate, glutathione, biotin, adenosine-5′-phosphosulfate, and 3-phosphoadenosine.
Group 2	**Nutrients that are important in energy storage or structural integrity**
P	Component of sugar phosphates, nucleic acids, nucleotides, coenzymes, phospholipids, phytic acid, etc. Has a key role in reactions that involve ATP.
Si	Deposited as amorphous silica in cell walls. Contributes to cell wall mechanical properties, including rigidity and elasticity.
B	Complexes with mannitol, mannan, polymannuronic acid, and other constituents of cell walls. Involved in cell elongation and nucleic acid metabolism.
Group 3	**Nutrients that remain in ionic form**
K	Required as a cofactor for more than 40 enzymes. Principal cation in establishing cell turgor and maintaining cell electroneutrality.
Ca	Constituent of the middle lamella of cell walls. Required as a cofactor by some enzymes involved in the hydrolysis of ATP and phospholipids. Acts as a second messenger in metabolic regulation.
Mg	Required by many enzymes involved in phosphate transfer. Constituent of the chlorophyll molecule.
Cl	Required for the photosynthetic reactions involved in O_2 evolution.
Mn	Required for activity of some dehydrogenases, decarboxylases, kinases, oxidases, and peroxidases. Involved with other cation-activated enzymes and photosynthetic O_2 evolution.
Na	Involved with the regeneration of phosphoenolpyruvate in C_4 and CAM plants. Substitutes for potassium in some functions.
Group 4	**Nutrients that are involved in redox reactions**
Fe	Constituent of cytochromes and nonheme iron proteins involved in photosynthesis, N_2 fixation, and respiration.
Zn	Constituent of alcohol dehydrogenase, glutamic dehydrogenase, carbonic anhydrase, etc.
Cu	Component of ascorbic acid oxidase, tyrosinase, monoamine oxidase, uricase, cytochrome oxidase, phenolase, laccase, and plastocyanin.
Ni	Constituent of urease. In N_2-fixing bacteria, constituent of hydrogenases.
Mo	Constituent of nitrogenase, nitrate reductase, and xanthine dehydrogenase.

Source: After Evans and Sorger 1966 and Mengel and Kirkby 1987.

chemical reactions involving oxidation and reduction to form covalent bonds with carbon and create organic compounds.

2. The second group is important in energy storage reactions or in maintaining structural integrity. Elements in this group are often present in plant tissues as phosphate, borate, and silicate esters in which the elemental group is covalently bound to the hydroxyl group of an organic molecule (e.g., sugar–phosphate).

3. The third group is present in plant tissue as either free ions dissolved in the plant water or ions electrostatically bound to substances such as the pectic acids present in the plant cell wall. Of particular importance are their roles as enzyme cofactors and in the regulation of osmotic potentials.

4. The fourth group, comprising metals such as iron, has important roles in reactions involving electron transfer.

Naturally occurring elements such as aluminum, selenium, and cobalt are not listed in Table 5.1, but can also accumulate in plant tissues. For example, aluminum is not considered to be an essential element, but plants commonly contain from 0.1 to 500 ppm aluminum, and addition of low levels of aluminum to a nutrient solution may stimulate plant growth (Marschner 1995). Many species in the genera *Astragalus*, *Xylorhiza*, and *Stanleya* accumulate selenium, although plants have not been shown to have a specific requirement for this element.

Cobalt is part of cobalamin (vitamin B_{12} and its derivatives), a component of several enzymes in nitrogen-fixing microorganisms. Thus cobalt deficiency blocks the devel-

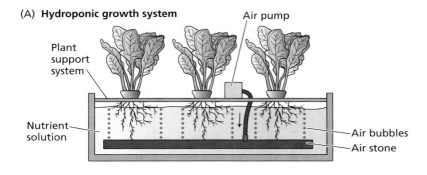

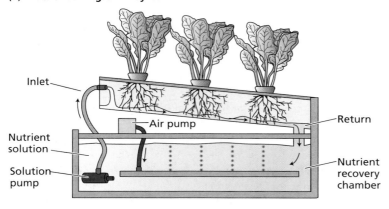

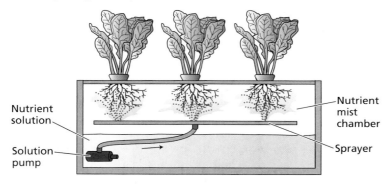

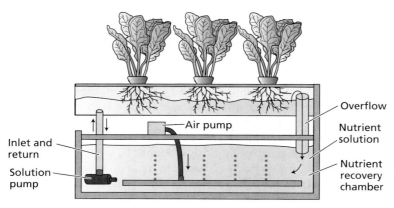

opment and function of nitrogen-fixing nodules. Nonetheless, plants that do not fix nitrogen, as well as nitrogen-fixing plants that are supplied with ammonium or nitrate, do not require cobalt. Crop plants normally contain only relatively small amounts of nonessential elements.

Special techniques are used in nutritional studies

To demonstrate that an element is essential requires that plants be grown under experimental conditions in which only the element under investigation is absent. Such conditions are extremely difficult to achieve with plants grown in a complex medium such as soil. In the nineteenth century, several researchers, including Nicolas-Théodore de Saussure, Julius von Sachs, Jean-Baptiste-Joseph-Dieudonné Boussingault, and Wilhelm Knop, approached this problem by growing plants with their roots immersed in a **nutrient solution** containing only inorganic salts. Their demonstration that plants could grow normally with no soil or organic matter proved unequivocally that plants can fulfill all their needs from only inorganic elements and sunlight.

The technique of growing plants with their roots immersed in nutrient solution without soil is called solution culture or **hydroponics** (Gericke 1937). Successful hydroponic culture (Figure 5.1A) requires a large volume of nutrient solution or frequent adjustment of the nutrient solution to prevent nutrient uptake by roots from producing radical changes in nutrient concentrations and pH of the medium. A sufficient supply of oxygen to the root system—also critical—may be achieved by vigorous bubbling of air through the medium.

FIGURE 5.1 Various types of solution culture systems. (A) In a standard hydroponic system, plants are suspended by the base of the stem over a tank containing nutrient solution. An air stone, a porous solid that generates an air stream of small bubbles, keeps the solution fully saturated with oxygen. (B) In the nutrient film technique, a solution pump drives nutrient solution from a main reservoir, along the bottom of a tilted tank, and down a return tube back to the reservoir. (C) In one type of aeroponics, a high-pressure pump sprays nutrient solution on the roots enclosed in a tank. (D) In an ebb-and-flow system, a pump periodically fills an upper chamber containing the plant roots with nutrient solution. When the pump is turned off, the solution drains back through the pump into a main reservoir. (From Epstein and Bloom 2005.)

Hydroponics is used in the commercial production of many greenhouse crops such as tomatoes (*Lycopersicon esculentum*). In one form of commercial hydroponic culture, plants are grown in a supporting material such as sand, gravel, vermiculite, or expanded clay (i.e., kitty litter). Nutrient solutions are then flushed through the supporting material, and old solutions are removed by leaching. In another form of hydroponic culture, plant roots lie on the surface of a trough, and nutrient solutions flow in a thin layer along the trough over the roots (Cooper 1979; Asher and Edwards 1983). This **nutrient film growth system** ensures that the roots receive an ample supply of oxygen (see Figure 5.1B).

Another alternative, which has sometimes been heralded as the medium of the future for scientific investigations, is to grow the plants **aeroponically** (Weathers and Zobel 1992). In this technique, plants are grown with their roots suspended in air while being sprayed continuously with a nutrient solution (see Figure 5.1C). This approach provides easy manipulation of the gaseous environment around the root, but it requires higher levels of nutrients than hydroponic culture does to sustain rapid plant growth. For this reason and other technical difficulties, the use of aeroponics is not widespread.

An ebb-and-flow system (see Figure 5.1D) is yet another approach to solution culture. In such systems, the nutrient solution periodically rises to immerse plant roots and then recedes, exposing the roots to a moist atmosphere. Like the aeroponics, ebb-and-flow systems require higher levels of nutrients than hydroponics or nutrient films.

Nutrient solutions can sustain rapid plant growth

Over the years, many formulations have been used for nutrient solutions. Early formulations developed by Knop in Germany included only KNO_3, $Ca(NO_3)_2$, KH_2PO_4, $MgSO_4$, and an iron salt. At the time, this nutrient solution was believed to contain all the minerals required by the plant, but these experiments were carried out with chemicals that were contaminated with other elements that are now known to be essential (such as boron or molybdenum). Table 5.3 shows a more modern formulation for a nutrient solution. This formulation is called a modified **Hoagland solution**, named after Dennis R. Hoagland, a researcher who was prominent in the development of modern mineral nutrition research in the U.S.

A modified Hoagland solution contains all of the known mineral elements needed for rapid plant growth. The concentrations of these elements are set at the highest possible levels without producing toxicity symptoms or salinity stress, and thus may be several orders of magnitude higher than those found in the soil around plant roots. For example, whereas phosphorus is present in the soil solution at concentrations normally less than 0.06 ppm, here it is offered at 62 ppm (Epstein and Bloom 2005). Such high initial levels permit plants to be grown in a medium for extended periods without replenishment of the nutrient, but may injure young plants. Therefore, many researchers dilute their nutrient solutions severalfold and replenish them frequently to minimize fluctuations of nutrient concentration in the medium and in plant tissue.

Another important property of the modified Hoagland formulation is that nitrogen is supplied as both ammonium (NH_4^+) and nitrate (NO_3^-). Supplying nitrogen in a balanced mixture of cations and anions tends to reduce the rapid rise in the pH of the medium that is commonly observed when the nitrogen is supplied solely as nitrate anion (Asher and Edwards 1983). Even when the pH of the medium is kept neutral, most plants grow better if they have access to both NH_4^+ and NO_3^-, because absorption and assimilation of the two nitrogen forms promotes cation–anion balances within the plant (Raven and Smith 1976; Bloom 1994).

A significant problem with nutrient solutions is maintaining the availability of iron. When supplied as an inorganic salt such as $FeSO_4$ or $Fe(NO_3)_2$, iron can precipitate out of solution as iron hydroxide. If phosphate salts are present, insoluble iron phosphate will also form. Precipitation of the iron out of solution makes it physically unavailable to the plant, unless iron salts are added at frequent intervals. Earlier researchers solved this problem by adding iron together with citric acid or tartaric acid. Compounds such as these are called **chelators** because they form soluble complexes with cations such as iron and calcium in which the cation is held by ionic forces, rather than by covalent bonds. Chelated cations thus remain physically available to a plant.

More modern nutrient solutions use the chemicals ethylenediaminetetraacetic acid (EDTA) or diethylenetriaminepentaacetic acid (DTPA, or pentetic acid) as chelating agents (Sievers and Bailar 1962). Figure 5.2 shows the structure of DTPA. The fate of the chelation complex during iron uptake by the root cells is not clear; iron may be released from the chelator when it is reduced from Fe^{3+} to Fe^{2+} at the root surface. The chelator may then diffuse back into the nutrient (or soil) solution and react with another Fe^{3+} ion or other metal ions.

After uptake into the root, iron is kept soluble by chelation with organic compounds present in plant cells. Citric acid may play a major role as such an organic iron chelator, and long-distance transport of iron in the xylem appears to involve an iron–citric acid complex.

Mineral deficiencies disrupt plant metabolism and function

An inadequate supply of an essential element results in a nutritional disorder manifested by characteristic deficiency symptoms. In hydroponic culture, withholding of an essential element can be readily correlated with a given set of symptoms for acute deficiencies. For example, a particular deficiency might elicit a specific pattern of leaf discol-

TABLE 5.3
Composition of a modified Hoagland nutrient solution for growing plants

Compound	Molecular weight	Concentration of stock solution	Concentration of stock solution	Volume of stock solution per liter of final solution	Element	Final concentration of element	
	g mol^{-1}	mM	g L^{-1}	mL		μM	ppm
Macronutrients							
KNO$_3$	101.10	1,000	101.10	6.0	N	16,000	224
Ca(NO$_3$)$_2$·4H$_2$O	236.16	1,000	236.16	4.0	K	6,000	235
NH$_4$H$_2$PO$_4$	115.08	1,000	115.08	2.0	Ca	4,000	160
MgSO$_4$·7H$_2$O	246.48	1,000	246.49	1.0	P	2,000	62
					S	1,000	32
					Mg	1,000	24
Micronutrients							
KCl	74.55	25	1.864		Cl	50	1.77
H$_3$BO$_3$	61.83	12.5	0.773		B	25	0.27
MnSO$_4$·H$_2$O	169.01	1.0	0.169	2.0	Mn	2.0	0.11
ZnSO$_4$·7H$_2$O	287.54	1.0	0.288		Zn	2.0	0.13
CuSO$_4$·5H$_2$O	249.68	0.25	0.062		Cu	0.5	0.03
H$_2$MoO$_4$ (85% MoO$_3$)	161.97	0.25	0.040		Mo	0.5	0.05
NaFeDTPA (10% Fe)	468.20	64	30.0	0.3–1.0	Fe	16.1–53.7	1.00–3.00
Optional[a]							
NiSO$_4$·6H$_2$O	262.86	0.25	0.066	2.0	Ni	0.5	0.03
Na$_2$SiO$_3$·9H$_2$O	284.20	1,000	284.20	1.0	Si	1,000	28

Source: After Epstein 1972.

Note: The macronutrients are added separately from stock solutions to prevent precipitation during preparation of the nutrient solution. A combined stock solution is made up containing all micronutrients except iron. Iron is added as sodium ferric diethylenetriaminepentaacetate (NaFeDTPA, trade name Ciba-Geigy Sequestrene 330 Fe; see Figure 5.2); some plants, such as maize, require the higher level of iron shown in the table.

[a]Nickel is usually present as a contaminant of the other chemicals, so it may not need to be added explicitly. Silicon, if included, should be added first and the pH adjusted with HCl to prevent precipitation of the other nutrients.

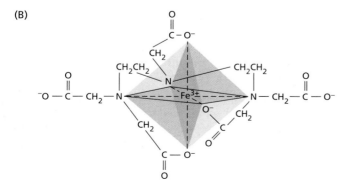

FIGURE 5.2 Chelator and chelated cation. Chemical structure of the chelator diethylenetriaminepentaacetic acid, (DTPA) by itself (A) and chelated to an Fe^{3+} ion (B). Iron binds to DTPA through interactions with three nitrogen atoms and the three ionized oxygen atoms of the carboxylate groups (Sievers and Bailar 1962). The resulting ring structure clamps the metallic ion and effectively neutralizes its reactivity in solution. During the uptake of iron at the root surface, Fe^{3+} appears to be reduced to Fe^{2+}, which is released from the DTPA–iron complex. The chelator can then bind to other available Fe^{3+} ions.

oration. Diagnosis of soil-grown plants can be more complex, for the following reasons:

- Deficiencies of several elements may occur simultaneously in different plant tissues.
- Deficiencies or excessive amounts of one element may induce deficiencies or excessive accumulations of another.
- Some virus-induced plant diseases may produce symptoms similar to those of nutrient deficiencies.

Nutrient deficiency symptoms in a plant are the expression of metabolic disorders resulting from the insufficient supply of an essential element. These disorders are related to the roles played by essential elements in normal plant metabolism and function. Table 5.2 lists some of the roles of essential elements.

Although each essential element participates in many different metabolic reactions, some general statements about the functions of essential elements in plant metabolism are possible. In general, the essential elements function in plant structure, metabolism, and cellular osmoregulation. More specific roles may be related to the ability of divalent cations such as calcium or magnesium to modify the permeability of plant membranes. In addition, research continues to reveal specific roles for these elements in plant metabolism; for example, calcium acts as a signal to regulate key enzymes in the cytosol (Hepler and Wayne 1985; Sanders et al. 1999). Thus, most essential elements have multiple roles in plant metabolism.

When relating acute deficiency symptoms to a particular essential element, an important clue is the extent to which an element can be recycled from older to younger leaves. Some elements, such as nitrogen, phosphorus, and potassium, can readily move from leaf to leaf; others, such as boron, iron, and calcium, are relatively immobile in most plant species (Table 5.4). If an essential element is mobile, deficiency symptoms tend to appear first in older leaves. Deficiency of an immobile essential element becomes evident first in younger leaves. Although the precise mechanisms of nutrient mobilization are not well understood, plant hormones such as cytokinins appear to be involved (see Chapter 21). In the discussion that follows, we will describe the specific deficiency symptoms and functional roles of the mineral essential elements as they are grouped in Table 5.2.

Group 1: Deficiencies in mineral nutrients that are part of carbon compounds. This first group consists of nitrogen and sulfur. Nitrogen availability in soils limits plant productivity in most natural and agricultural ecosystems. By contrast, soils generally contain sulfur in excess. Despite this difference, nitrogen and sulfur are similar in that their oxidation–reduction states range widely (see Chapter 12). Some of the most energy-intensive reactions in life convert the highly oxidized, inorganic forms such as nitrate and sulfate absorbed from the soil to the highly reduced forms found in organic compounds such as amino acids.

TABLE 5.4
Mineral elements classified on the basis of their mobility within a plant and their tendency to retranslocate during deficiencies

Mobile	Immobile
Nitrogen	Calcium
Potassium	Sulfur
Magnesium	Iron
Phosphorus	Boron
Chlorine	Copper
Sodium	
Zinc	
Molybdenum	

Note: Elements are listed in the order of their abundance in the plant.

NITROGEN Nitrogen is the mineral element that plants require in the greatest amounts. It serves as a constituent of many plant cell components, including amino acids, proteins, and nucleic acids. Therefore, nitrogen deficiency rapidly inhibits plant growth. If such a deficiency persists, most species show **chlorosis** (yellowing of the leaves), especially in the older leaves near the base of the plant (for pictures of nitrogen deficiency and the other mineral deficiencies described in this chapter, see **Web Topic 5.1**). Under severe nitrogen deficiency, these leaves become completely yellow (or tan) and fall off the plant. Younger leaves may not show these symptoms initially, because nitrogen can be mobilized from older leaves. Thus a nitrogen-deficient plant may have light green upper leaves and yellow or tan lower leaves.

When nitrogen deficiency develops slowly, plants may have markedly slender and often woody stems. This woodiness may be due to a buildup of excess carbohydrates that cannot be used in the synthesis of amino acids or other nitrogen compounds. Carbohydrates not used in nitrogen metabolism may also be used in anthocyanin synthesis, leading to accumulation of that pigment. This condition is revealed as a purple coloration in leaves, petioles, and stems of some nitrogen-deficient plants, such as tomato and certain varieties of corn (maize; *Zea mays*).

SULFUR Sulfur is found in two amino acids (cysteine and methionine) and is a constituent of several coenzymes and vitamins (acetyl coenzyme A, S-adenosylmethionine, biotin, Vitamin B_1, pantothenic acid) essential for metabolism. Many of the symptoms of sulfur deficiency are similar to those of nitrogen deficiency, including chlorosis, stunting of growth, and anthocyanin accumulation. This similarity is not surprising, since sulfur and nitrogen are both constituents of proteins. However, the chlorosis caused by sulfur deficiency generally arises initially in mature and young leaves, rather than in the old leaves as in nitrogen deficiency, because unlike nitrogen, sulfur is not easily remobilized to the younger leaves in most species. Nonetheless, in many plant species sulfur chlorosis may occur simultaneously in all leaves or even initially in the older leaves.

Group 2: Deficiencies in mineral nutrients that are important in energy storage or structural integrity. This group consists of phosphorus, silicon, and boron. Phosphorus and silicon are found at concentrations within plant tissue that warrant their classification as macronutrients, whereas boron is much less abundant and is considered a micronutrient. These elements are usually present in plants as ester linkages to a carbon molecule.

PHOSPHORUS Phosphorus (as phosphate, PO_4^{3-}) is an integral component of important compounds of plant cells, including the sugar–phosphate intermediates of respiration and photosynthesis, and the phospholipids that make up plant membranes. It is also a component of nucleotides used in plant energy metabolism (such as ATP) and in DNA and RNA. Characteristic symptoms of phosphorus deficiency include stunted growth in young plants and a dark green coloration of the leaves, which may be malformed and contain small spots of dead tissue called **necrotic spots** (for a picture, see **Web Topic 5.1**).

As in nitrogen deficiency, some species may produce excess anthocyanins, giving the leaves a slight purple coloration. In contrast to nitrogen deficiency, the purple coloration of phosphorus deficiency is not associated with chlorosis. In fact, the leaves may be a dark greenish purple. Additional symptoms of phosphorus deficiency include the production of slender (but not woody) stems and the death of older leaves. Maturation of the plant may also be delayed.

SILICON Only members of the family Equisetaceae—called *scouring rushes* because at one time their ash, rich in gritty silica, was used to scour pots—require silicon to complete their life cycle. Nonetheless, many other species accumulate substantial amounts of silicon within their tissues and show enhanced growth and fertility when supplied with adequate amounts of silicon (Epstein 1999).

Plants deficient in silicon are more susceptible to lodging (falling over) and fungal infection. Silicon is deposited primarily in the endoplasmic reticulum, cell walls, and intercellular spaces as hydrated, amorphous silica ($SiO_2 \cdot nH_2O$). It also forms complexes with polyphenols and thus serves as an alternative to lignin in the reinforcement of cell walls. In addition, silicon can lessen the toxicity of many metals, including aluminum and manganese.

BORON Although the precise function of boron in plant metabolism is unclear, evidence suggests that it plays roles in cell elongation, nucleic acid synthesis, hormone responses, and membrane function (Shelp 1993). Boron-deficient plants may exhibit a wide variety of symptoms, depending on the species and the age of the plant.

A characteristic symptom is black necrosis of the young leaves and terminal buds. The necrosis of the young leaves occurs primarily at the base of the leaf blade. Stems may be unusually stiff and brittle. Apical dominance may also be lost, causing the plant to become highly branched; however, the terminal apices of the branches soon become necrotic because of inhibition of cell division. Structures such as the fruit, fleshy roots, and tubers may exhibit necrosis or abnormalities related to the breakdown of internal tissues.

Group 3: Deficiencies in mineral nutrients that remain in ionic form. This group includes some of the most familiar mineral elements: the macronutrients potassium, calcium, and magnesium; and the micronutrients chlorine, manganese, and sodium. They may be found in solution in the cytosol or vacuoles, or they may be bound electrostatically or as ligands to larger, carbon-containing compounds.

POTASSIUM Potassium, present within plants as the cation K^+, plays an important role in regulation of the osmotic potential of plant cells (see Chapters 3 and 6). It also activates many enzymes involved in respiration and photosynthesis. The first observable symptom of potassium deficiency is mottled or marginal chlorosis, which then develops into necrosis primarily at the leaf tips, at the margins, and between veins. In many monocots, these necrotic lesions may initially form at the leaf tips and margins and then extend toward the leaf base.

Because potassium can be mobilized to the younger leaves, these symptoms appear initially on the more mature leaves toward the base of the plant. The leaves may also curl and crinkle. The stems of potassium-deficient plants may be slender and weak, with abnormally short internodal regions. In potassium-deficient corn, the roots may have an increased susceptibility to root-rotting fungi present in the soil, and this susceptibility, together with effects on the stem, results in an increased tendency for the plant to be easily bent to the ground (lodging).

CALCIUM Calcium ions (Ca^{2+}) are used in the synthesis of new cell walls, particularly the middle lamellae that separate newly divided cells. Calcium is also used in the mitotic spindle during cell division. It is required for the normal functioning of plant membranes and has been implicated as a second messenger for various plant responses to both environmental and hormonal signals (White and Broadley 2003). In its function as a second messenger, calcium may bind to **calmodulin**, a protein found in the cytosol of plant cells. The calmodulin–calcium complex then binds to several different types of proteins, including kinases, phosphatases, second messenger signaling proteins, and cytoskeletal proteins, and thereby regulates many cellular processes ranging from transcription control and cell survival to release of chemical signals.

Characteristic symptoms of calcium deficiency include necrosis of young meristematic regions, such as the tips of roots or young leaves, where cell division and wall formation are most rapid. Necrosis in slowly growing plants may be preceded by a general chlorosis and downward hooking of the young leaves. Young leaves may also appear deformed. The root system of a calcium-deficient plant may appear brownish, short, and highly branched. Severe stunting may result if the meristematic regions of the plant die prematurely.

MAGNESIUM In plant cells, magnesium ions (Mg^{2+}) have a specific role in the activation of enzymes involved in respiration, photosynthesis, and the synthesis of DNA and RNA. Magnesium is also a part of the ring structure of the chlorophyll molecule (see Figure 7.6A). A characteristic symptom of magnesium deficiency is chlorosis between the leaf veins, occurring first in the older leaves because of the mobility of this element. This pattern of chlorosis results because the chlorophyll in the vascular bundles remains unaffected for longer periods than the chlorophyll in the cells between the bundles does. If the deficiency is extensive, the leaves may become yellow or white. An additional symptom of magnesium deficiency may be premature leaf abscission.

CHLORINE The element chlorine is found in plants as the chloride ion (Cl^-). It is required for the water-splitting reaction of photosynthesis through which oxygen is produced (see Chapter 7) (Clarke and Eaton-Rye 2000). In addition, chlorine may be required for cell division in both leaves and roots (Harling et al. 1997). Plants deficient in chlorine develop wilting of the leaf tips followed by general leaf chlorosis and necrosis. The leaves may also exhibit reduced growth. Eventually, the leaves may take on a bronzelike color ("bronzing"). Roots of chlorine-deficient plants may appear stunted and thickened near the root tips.

Chloride ions are very soluble and generally available in soils because seawater is swept into the air by wind and is delivered to soil when it rains. Therefore, chlorine deficiency is only rarely observed in plants grown in native or agricultural habitats (Engel et al. 2001). Most plants absorb chlorine at levels much higher than those required for normal functioning.

MANGANESE Manganese ions (Mn^{2+}) activate several enzymes in plant cells. In particular, decarboxylases and dehydrogenases involved in the tricarboxylic acid (Krebs) cycle are specifically activated by manganese. The best-defined function of manganese is in the photosynthetic reaction through which oxygen (O_2) is produced from water (Marschner 1995). The major symptom of manganese deficiency is intervenous chlorosis associated with the development of small necrotic spots. This chlorosis may occur on younger or older leaves, depending on plant species and growth rate.

SODIUM Most species utilizing the C_4 and crassulacean acid metabolism (CAM) pathways of carbon fixation (see Chapter 8) require sodium ions (Na^+). In these plants, sodium appears vital for regenerating phosphoenolpyruvate, the substrate for the first carboxylation in the C_4 and CAM pathways (Johnstone et al. 1988). Under sodium deficiency, these plants exhibit chlorosis and necrosis, or even fail to form flowers. Many C_3 species also benefit from exposure to low levels of sodium ions. Sodium stimulates growth through enhanced cell expansion, and it can partly substitute for potassium as an osmotically active solute.

Group 4: Deficiencies in mineral nutrients that are involved in redox reactions. This group of five micronutrients includes the metals iron, zinc, copper, nickel, and molybdenum. All of these can undergo reversible oxidations and reductions (e.g., $Fe^{2+} \leftrightarrow Fe^{3+}$) and have important roles in electron transfer and energy transformation. They are usually found in association with larger molecules such as cytochromes, chlorophyll, and proteins (usually enzymes).

IRON Iron has an important role as a component of enzymes involved in the transfer of electrons (redox reactions), such as cytochromes. In this role, it is reversibly oxidized from Fe^{2+} to Fe^{3+} during electron transfer. As in magnesium deficiency, a characteristic symptom of iron deficiency is intervenous chlorosis. In contrast to magnesium deficiency symptoms, these symptoms appear initially on the younger leaves, because iron cannot be readily mobilized from older leaves. Under conditions of extreme or prolonged deficiency, the veins may also become chlorotic, causing the whole leaf to turn white.

The leaves become chlorotic because iron is required for the synthesis of some of the chlorophyll–protein complexes in the chloroplast. The low mobility of iron is probably due to its precipitation in the older leaves as insoluble oxides or phosphates, or to the formation of complexes with phytoferritin, an iron-binding protein found in the leaf and other plant parts (Oh et al. 1996). The precipitation of iron diminishes subsequent mobilization of the metal into the phloem for long-distance translocation.

ZINC Many enzymes require zinc ions (Zn^{2+}) for their activity, and zinc may be required for chlorophyll biosynthesis in some plants. Zinc deficiency is characterized by a reduction in internodal growth, and as a result plants display a rosette habit of growth in which the leaves form a circular cluster radiating at or close to the ground. The leaves may also be small and distorted, with leaf margins having a puckered appearance. These symptoms may result from loss of the capacity to produce sufficient amounts of the auxin indole-3-acetic acid (IAA). In some species (e.g., corn, sorghum, and beans), the older leaves may become intervenously chlorotic and then develop white necrotic spots. This chlorosis may be an expression of a zinc requirement for chlorophyll biosynthesis.

COPPER Like iron, copper is associated with enzymes involved in redox reactions being reversibly oxidized from Cu^+ to Cu^{2+}. An example of such an enzyme is plastocyanin, which is involved in electron transfer during the light reactions of photosynthesis (Haehnel 1984). The initial symptom of copper deficiency is the production of dark green leaves, which may contain necrotic spots. The necrotic spots appear first at the tips of the young leaves and then extend toward the leaf base along the margins. The leaves may also be twisted or malformed. Under extreme copper deficiency, leaves may abscise prematurely.

NICKEL Urease is the only known nickel-containing (Ni^{2+}) enzyme in higher plants, although nitrogen-fixing microorganisms require nickel (Ni^+ through Ni^{4+}) for the enzyme that reprocesses some of the hydrogen gas generated during fixation (hydrogen uptake hydrogenase) (see Chapter 12). Nickel-deficient plants accumulate urea in their leaves and, consequently, show leaf tip necrosis. Nickel deficiency in the field has been found in only one crop, pecan trees in the southeastern U.S. (Wood et al. 2003), because the amounts of nickel required are minuscule.

MOLYBDENUM Molybdenum ions (Mo^{4+} through Mo^{6+}) are components of several enzymes, including nitrate reductase and nitrogenase. Nitrate reductase catalyzes the reduction of nitrate to nitrite during its assimilation by the plant cell; nitrogenase converts nitrogen gas to ammonia in nitrogen-fixing microorganisms (see Chapter 12). The first indication of a molybdenum deficiency is general chlorosis between veins and necrosis of the older leaves. In some plants, such as cauliflower or broccoli, the leaves may not become necrotic but instead may appear twisted and subsequently die (whiptail disease). Flower formation may be prevented, or the flowers may abscise prematurely.

Because molybdenum is involved with both nitrate assimilation and nitrogen fixation, a molybdenum deficiency may bring about a nitrogen deficiency if the nitrogen source is primarily nitrate or if the plant depends on symbiotic nitrogen fixation. Although plants require only small amounts of molybdenum, some soils (for example, acid soils in Australia) supply inadequate levels. Small additions of molybdenum to such soils can greatly enhance crop or forage growth at negligible cost.

Analysis of plant tissues reveals mineral deficiencies

Requirements for mineral elements change during the growth and development of a plant. In crop plants, nutrient levels at certain stages of growth influence the yield of the economically important tissues (tuber, grain, and so on). To optimize yields, farmers use analyses of nutrient levels in soil and in plant tissue to determine fertilizer schedules.

Soil analysis is the chemical determination of the nutrient content in a soil sample from the root zone. As discussed later in the chapter, both the chemistry and the biology of soils are complex, and the results of soil analyses vary with sampling methods, storage conditions for the samples, and nutrient extraction techniques. Perhaps more important is that a particular soil analysis reflects the levels of nutrients *potentially* available to the plant roots from the soil, but soil analysis does not tell us how much of a particular mineral nutrient the plant actually needs or is able to absorb. This additional information is best determined by plant tissue analysis.

Proper use of **plant tissue analysis** requires an understanding of the relationship between plant growth (or yield) and the mineral concentration of plant tissue samples (Bouma 1983). Figure 5.3 identifies three zones (deficiency, adequate, and toxic) in the response of growth to increasing tissue concentrations of a nutrient. When the nutrient concentration in a tissue sample is low, growth is reduced. In this **deficiency zone** of the curve, an increase

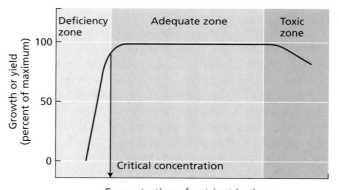

FIGURE 5.3 Relationship between yield (or growth) and the nutrient content of the plant tissue defines zones of deficiency, adequacy, and toxicity. Yield or growth may be expressed in terms of shoot dry weight or height. To yield data of this type, plants are grown under conditions in which the concentration of one essential nutrient is varied while all others are in adequate supply. The effect of varying the concentration of this nutrient during plant growth is reflected in the growth or yield. The critical concentration for that nutrient is the concentration below which yield or growth is reduced.

in nutrient availability is directly related to an increase in growth or yield. As the nutrient availability continues to increase, a point is reached at which further addition of nutrients is no longer related to increases in growth or yield but is reflected in increased tissue concentrations. This region of the curve is often called the **adequate zone**.

The transition between the deficiency and adequate zones of the curve reveals the **critical concentration** of the nutrient (see Figure 5.3), which may be defined as the minimum tissue content of the nutrient that is correlated with maximal growth or yield. As the nutrient concentration of the tissue increases beyond the adequate zone, growth or yield declines because of toxicity (this is the **toxic zone**).

To evaluate the relationship between growth and tissue nutrient concentration, researchers grow plants in soil or nutrient solution in which all the nutrients are present in adequate amounts except the nutrient under consideration. At the start of the experiment, the limiting nutrient is added in increasing concentrations to different sets of plants, and the concentrations of the nutrient in specific tissues are correlated with a particular measure of growth or yield. Several curves are established for each element, one for each tissue and tissue age.

Because agricultural soils are often limited in the elements nitrogen, phosphorus, and potassium, many farmers routinely take into account, at a minimum, growth or yield responses for these elements. If a nutrient deficiency is suspected, steps are taken to correct the deficiency before it reduces growth or yield. Plant analysis has proven useful in establishing fertilizer schedules that sustain yields and ensure the food quality of many crops.

Treating Nutritional Deficiencies

Many traditional and subsistence farming practices promote the recycling of mineral elements. Crop plants absorb the nutrients from the soil, humans and animals consume locally grown crops, and crop residues and manure from humans and animals return the nutrients to the soil. The main losses of nutrients from such agricultural systems ensue from leaching that carries dissolved ions, especially nitrate, away with drainage water. In acid soils, leaching may be decreased by the addition of lime—a mix of CaO, $CaCO_3$, and $Ca(OH)_2$—to make the soil more alkaline, because many mineral elements form less-soluble compounds when the pH is higher than 6 (Figure 5.4). This decrease in leaching, however, may be gained at the expense of decreased availability of some nutrients, especially iron.

In the high-production agricultural systems of industrial countries, the unidirectional removal of nutrients from the soil to the crop can become significant, because a large portion of crop biomass leaves the area of cultivation, and returning crop residues to the land where the crop was produced becomes difficult at best. Plants synthesize all their components from basic inorganic substances and sunlight,

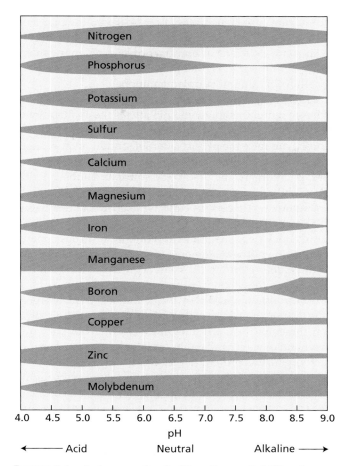

FIGURE 5.4 Influence of soil pH on the availability of nutrient elements in organic soils. The width of the shaded areas indicates the degree of nutrient availability to the plant root. All of these nutrients are available in the pH range of 5.5 to 6.5. (After Lucas and Davis 1961.)

so it is important to restore these lost nutrients to the soil through the addition of fertilizers containing one or more of these nutrients.

Crop yields can be improved by addition of fertilizers

Most chemical fertilizers contain inorganic salts of the macronutrients nitrogen, phosphorus, and potassium (see Table 5.1). Fertilizers that contain only one of these three nutrients are termed straight fertilizers. Some examples of straight fertilizers are superphosphate, ammonium nitrate, and muriate of potash (a source of potassium). Fertilizers that contain two or more mineral nutrients are called compound fertilizers or mixed fertilizers, and the numbers on the package label, such as 10-14-10, refer to the effective percentages of N, P as P_2O_5, and K as K_2O, respectively, in the fertilizer.

With long-term agricultural production, consumption of micronutrients can reach a point at which they, too, must be added to the soil as fertilizers. Adding micronutrients to the soil may also be necessary to correct a preexisting deficiency. For example, many acid, sandy soils in humid regions are deficient in boron, copper, zinc, manganese, molybdenum, or iron (Mengel and Kirkby 1987) and can benefit from nutrient supplementation.

Chemicals may also be applied to the soil to modify soil pH. As Figure 5.4 shows, soil pH affects the availability of all mineral nutrients. Addition of lime, as mentioned previously, can raise the pH of acidic soils; addition of elemental sulfur can lower the pH of alkaline soils. In the latter case, microorganisms absorb the sulfur and subsequently release sulfate and hydrogen ions that acidify the soil.

Organic fertilizers, in contrast to chemical fertilizers, originate from the residues of plant or animal life or from natural rock deposits. Plant and animal residues contain many of the nutrient elements in the form of organic compounds. Before crop plants can acquire the nutrient elements from these residues, the organic compounds must be broken down, usually by the action of soil microorganisms through a process called **mineralization**. Mineralization depends on many factors, including temperature, water and oxygen availability, and the type and number of microorganisms present in the soil.

As a consequence, the rate of mineralization is highly variable, and nutrients from organic residues become available to plants over periods that range from days to months to years. The slow rate of mineralization hinders efficient fertilizer use, so farms that rely solely on organic fertilizers may require the addition of substantially more nitrogen or phosphorus and suffer even higher nutrient losses than farms that use chemical fertilizers. Residues from organic fertilizers do improve the physical structure of most soils, enhancing water retention during drought and increasing drainage in wet weather.

Some mineral nutrients can be absorbed by leaves

In addition to nutrients being added to the soil as fertilizers, most plants can absorb mineral nutrients applied to their leaves as sprays, in a process known as **foliar application**. In some cases, this method can have agronomic advantages over the application of nutrients to the soil. Foliar application can reduce the lag time between application and uptake by the plant, which could be important during a phase of rapid growth. It can also circumvent the problem of restricted uptake of a nutrient from the soil. For example, foliar application of mineral nutrients such as iron, manganese, and copper may be more efficient than application through the soil, where they are adsorbed on soil particles and hence are less available to the root system.

Nutrient uptake by plant leaves is most effective when the nutrient solution remains on the leaf as a thin film (Mengel and Kirkby 1987). Production of a thin film often requires that the nutrient solutions be supplemented with surfactant chemicals, such as the detergent Tween 80, that reduce surface tension. Nutrient movement into the plant seems to involve diffusion through the cuticle and uptake by leaf cells. Although uptake through the stomatal pore could provide a pathway into the leaf, the architecture of the pore (see Figures 4.14 and 4.15) largely prevents liquid penetration (Ziegler 1987).

For foliar nutrient application to be successful, damage to the leaves must be minimized. If foliar sprays are applied on a hot day, when evaporation is high, salts may accumulate on the leaf surface and cause burning or scorching. Spraying on cool days or in the evening helps to alleviate this problem. Addition of lime to the spray diminishes the solubility of many nutrients and limits toxicity. Foliar application has proved economically successful mainly with tree crops and vines such as grapes, but it is also used with cereals. Nutrients applied to the leaves could save an orchard or vineyard when soil-applied nutrients would be too slow to correct a deficiency. In wheat (*Triticum aestivum*), nitrogen applied to the leaves during the later stages of growth enhances the protein content of seeds.

Soil, Roots, and Microbes

The soil is a complex physical, chemical, and biological substrate. It is a heterogeneous material containing solid, liquid, and gaseous phases (see Chapter 4). All of these phases interact with mineral elements. The inorganic particles of the solid phase provide a reservoir of potassium, calcium, magnesium, and iron. Also associated with this solid phase are organic compounds containing nitrogen, phosphorus, and sulfur, among other elements. The liquid phase of the soil constitutes the soil solution, which contains dissolved mineral ions and serves as the medium for ion movement to the root surface. Gases such as oxygen, carbon dioxide, and nitrogen are dissolved in the soil solution, but roots exchange gases with soils, predominantly through the air gaps between soil particles.

From a biological perspective, soil constitutes a diverse ecosystem in which plant roots and microorganisms compete strongly for mineral nutrients. Despite this competition, roots and microorganisms can form alliances for their mutual benefit (**symbioses**, singular *symbiosis*). In this section we will discuss the importance of soil properties, root structure, and mycorrhizal symbiotic relationships to plant mineral nutrition. Chapter 12 addresses symbiotic relationships with nitrogen-fixing bacteria.

Negatively charged soil particles affect the adsorption of mineral nutrients

Soil particles, both inorganic and organic, have predominantly negative charges on their surfaces. Many inorganic soil particles are crystal lattices that are tetrahedral arrangements of the cationic forms of aluminum and silicon (Al^{3+} and Si^{4+}) bound to oxygen atoms, thus forming aluminates and silicates. When cations of lesser charge

TABLE 5.5
Comparison of properties of three major types of silicate clays found in the soil

Property	Type of clay		
	Montmorillonite	Illite	Kaolinite
Size (μm)	0.01–1.0	0.1–2.0	0.1–5.0
Shape	Irregular flakes	Irregular flakes	Hexagonal crystals
Cohesion	High	Medium	Low
Water-swelling capacity	High	Medium	Low
Cation exchange capacity (milliequivalents 100 g^{-1})	80–100	15–40	3–15

Source: After Brady 1974.

replace Al^{3+} and Si^{4+} within the crystal lattice, these inorganic soil particles become negatively charged.

Organic soil particles originate from the products of the microbial decomposition of dead plants, animals, and microorganisms. The negative surface charges of organic particles result from the dissociation of hydrogen ions from the carboxylic acid and phenolic groups present in this component of the soil. Most of the world's soil particles, however, are inorganic.

Inorganic soils are categorized by particle size (see also Table 4.1):

- Gravel has particles larger than 2 mm.
- Coarse sand has particles between 0.2 and 2 mm.
- Fine sand has particles between 0.02 and 0.2 mm.
- Silt has particles between 0.002 and 0.02 mm.
- Clay has particles smaller than 0.002 mm.

The silicate-containing clay materials are further divided into three major groups—kaolinite, illite, and montmorillonite—based on differences in their structure and physical properties (Table 5.5). The kaolinite group is generally found in well-weathered soils; the montmorillonite and illite groups are found in less weathered soils.

Mineral cations such as ammonium (NH_4^+) and potassium (K^+) adsorb to the negative surface charges of inorganic and organic soil particles. This cation adsorption is an important factor in soil fertility. Mineral cations adsorbed on the surface of soil particles are not easily lost when the soil is leached by water, and they provide a nutrient reserve available to plant roots. Mineral nutrients adsorbed in this way can be replaced by other cations in a process known as **cation exchange** (Figure 5.5). The degree to which a soil can adsorb and exchange ions is termed its *cation exchange capacity* (*CEC*) and is highly dependent on the soil type. A soil with higher cation exchange capacity generally has a larger reserve of mineral nutrients.

Mineral anions such as nitrate (NO_3^-) and chloride (Cl^-) tend to be repelled by the negative charge on the surface of soil particles and remain dissolved in the soil solution. Thus the anion exchange capacity of most agricultural soils is small compared to the cation exchange capacity. Among anions, nitrate remains mobile in the soil solution, where it is susceptible to leaching by water moving through the soil.

Phosphate ions ($H_2PO_2^-$) may bind to soil particles containing aluminum or iron because the positively charged iron and aluminum ions (Fe^{2+}, Fe^{3+}, and Al^{3+}) have hydroxyl (OH^-) groups that exchange with phosphate. As a result, phosphate can be tightly bound, and its mobility and availability in soil can limit plant growth.

Sulfate (SO_4^{2-}) in the presence of calcium (Ca^{2+}) forms gypsum ($CaSO_4$). Gypsum is only slightly soluble, but it releases sufficient sulfate to support plant growth. Most nonacid soils contain substantial amounts of calcium; consequently, sulfate mobility in these soils is low, so sulfate is not highly susceptible to leaching.

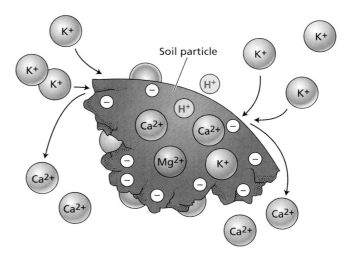

FIGURE 5.5 The principle of cation exchange on the surface of a soil particle. Cations are bound to the surface of soil particles, because the surface is negatively charged. Addition of a cation such as potassium (K^+) can displace another cation such as calcium (Ca^{2+}) from its binding on the surface of the soil particle and make it available for uptake by the root.

Soil pH affects nutrient availability, soil microbes, and root growth

Hydrogen ion concentration (pH) is an important property of soils because it affects the growth of plant roots and soil microorganisms. Root growth is generally favored in slightly acidic soils, at pH values between 5.5 and 6.5. Fungi generally predominate in acidic (pH below 7) soils; bacteria become more prevalent in alkaline (pH above 7) soils. Soil pH determines the availability of soil nutrients (see Figure 5.4). Acidity promotes the weathering of rocks that releases K^+, Mg^{2+}, Ca^{2+}, and Mn^{2+} and increases the solubility of carbonates, sulfates, and phosphates. Increasing the solubility of nutrients facilitates their availability to roots.

Major factors that lower the soil pH are the decomposition of organic matter and the amount of rainfall. Carbon dioxide is produced as a result of the decomposition of organic material and equilibrates with soil water in the following reaction:

$$CO_2 + H_2O \leftrightarrow H^+ + HCO_3^-$$

This reaction releases hydrogen ions (H^+), lowering the pH of the soil. Microbial decomposition of organic material also produces ammonia and hydrogen sulfide that can be oxidized in the soil to form the strong acids nitric acid (HNO_3) and sulfuric acid (H_2SO_4), respectively. Hydrogen ions also displace K^+, Mg^{2+}, Ca^{2+}, and Mn^{2+} from the cation exchange complex in a soil. Leaching then may remove these ions from the upper soil layers, leaving a more acid soil. By contrast, the weathering of rock in arid regions releases K^+, Mg^{2+}, Ca^{2+}, and Mn^{2+} to the soil, but because of the low rainfall, these ions do not leach from the upper soil layers, and the soil remains alkaline.

Excess minerals in the soil limit plant growth

When excess minerals are present in the soil, the soil is said to be saline, and plant growth may be restricted if these mineral ions reach levels that limit water availability or exceed the adequate zone for a particular nutrient (see Chapter 25). Sodium chloride and sodium sulfate are the most common salts in saline soils. Excess minerals in soils can be a major problem in arid and semiarid regions, because rainfall is insufficient to leach the mineral ions from the soil layers near the surface. Irrigated agriculture fosters soil salinization if the amount of water applied is insufficient to leach the salt below the rooting zone. Irrigation water can contain 100 to 1000 g of minerals per cubic meter. An average crop requires about 4000 m^3 of water per acre. Consequently, 400 to 4000 kg of minerals may be added to the soil per crop (Marschner 1995), and over a number of growing seasons, high levels of minerals may accumulate in the soil.

In saline soil, plants encounter **salt stress**. Whereas many plants are affected adversely by the presence of relatively low levels of salt, other plants can survive high levels (**salt-tolerant plants**) or even thrive (**halophytes**) under such conditions. The mechanisms by which plants tolerate salinity are complex (see Chapter 26), involving molecular synthesis, enzyme induction, and membrane transport. In some species, excess minerals are not taken up; in others, minerals are taken up but excreted from the plant by salt glands associated with the leaves. To prevent toxic buildup of mineral ions in the cytosol, many plants may sequester them in the vacuole (Stewart and Ahmad 1983). Efforts are under way to bestow salt tolerance on salt-sensitive crop species using both classic plant breeding and molecular biology (Blumwald 2003), as detailed in Chapter 26.

Another important problem with excess minerals is the accumulation of heavy metals in the soil, which can cause severe toxicity in plants as well as humans (see **Web Essay 5.1**). Heavy metals include zinc, copper, cobalt, nickel, mercury, lead, cadmium, silver, and chromium (Berry and Wallace 1981).

Plants develop extensive root systems

The ability of plants to obtain both water and mineral nutrients from the soil is related to their capacity to develop an extensive root system. In the late 1930s, H. J. Dittmer examined the root system of a single winter rye plant after 16 weeks of growth and estimated that the plant had 13×10^6 primary and lateral root axes, extending more than 500 km in length and providing 200 m^2 of surface area (Dittmer 1937). This plant also had more than 10^{10} root hairs, providing another 300 m^2 of surface area. In total, the surface area of roots from a single rye plant equaled that of a professional basketball court.

In the desert, the roots of mesquite (genus *Prosopis*) may extend down more than 50 m to reach groundwater. Annual crop plants have roots that usually grow between 0.1 and 2.0 m in depth and extend laterally to distances of 0.3 to 1.0 m. In orchards, the major root systems of trees planted 1 m apart reach a total length of 12 to 18 km per tree. The annual production of roots in natural ecosystems may easily surpass that of shoots, so in many respects, the aboveground portions of a plant represent only "the tip of an iceberg." Nonetheless, making observations on root systems is difficult and usually requires special techniques (see **Web Topic 5.2**).

Plant roots may grow continuously throughout the year. Their proliferation, however, depends on the availability of water and minerals in the immediate microenvironment surrounding the root, the so-called **rhizosphere**. If the rhizosphere is poor in nutrients or too dry, root growth is slow. As rhizosphere conditions improve, root growth increases. If fertilization and irrigation provide abundant nutrients and water, root growth may not keep pace with shoot growth. Plant growth under such conditions becomes carbohydrate-limited, and a relatively small root system meets the nutrient needs of the whole plant (Bloom et al. 1993). Indeed, crops under fertilization and irrigation allocate more resources to the shoot and reproductive

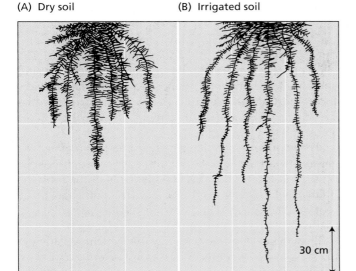

FIGURE 5.6 Fibrous root systems of wheat (a monocot). (A) The root system of a mature (3-month-old) wheat plant growing in dry soil. (B) The root system of a wheat plant growing in irrigated soil. It is apparent that the morphology of the root system is affected by the amount of water present in the soil. In a fibrous root system, the primary root axes are no longer distinguishable. (After Weaver 1926.)

structures than to roots, and this shift in allocation patterns often results in higher yields.

Root systems differ in form but are based on common structures

The *form* of the root system differs greatly among plant species. In monocots, root development starts with the emergence of three to six **primary** (or seminal) root axes from the germinating seed. With further growth, the plant extends new adventitious roots, called **nodal roots** or *brace roots*. Over time, the primary and nodal root axes grow and branch extensively to form a complex fibrous root system (Figure 5.6). In fibrous root systems, all the roots generally have the same diameter (except where environmental conditions or pathogenic interactions modify the root structure), so it is impossible to distinguish a main root axis.

In contrast to monocots, dicots develop root systems with a main single root axis, called a **taproot**, which may thicken as a result of secondary cambial activity. From this main root axis, lateral roots develop to form an extensively branched root system (Figure 5.7).

The development of the root system in both monocots and dicots depends on the activity of the root apical meristem and the production of lateral root meristems. Figure 5.8 shows a generalized diagram of the apical region of a plant root and identifies the three zones of activity: meristematic, elongation, and maturation.

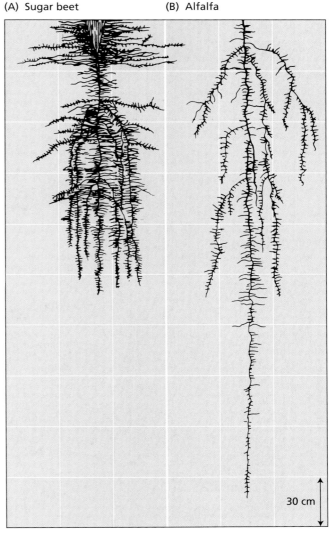

FIGURE 5.7 Taproot system of two adequately watered dicots: sugar beet (A) and alfalfa (B). The sugar beet root system is typical of 5 months of growth; the alfalfa root system is typical of 2 years of growth. In both dicots, the root system shows a major vertical root axis. In the case of sugar beet, the upper portion of the taproot system is thickened because of its function as storage tissue. (After Weaver 1926.)

In the **meristematic zone**, cells divide both in the direction of the root base to form cells that will differentiate into the tissues of the functional root and in the direction of the root apex to form the **root cap**. The root cap protects the delicate meristematic cells as the root expands into the soil. It also secretes a gelatinous material called *mucigel*, which commonly surrounds the root tip. The precise function of the mucigel is uncertain, but it may lubricate the penetration of the root through the soil, protect the root apex from desiccation, promote the transfer of nutrients to the root, or affect the interaction between roots and soil microorgan-

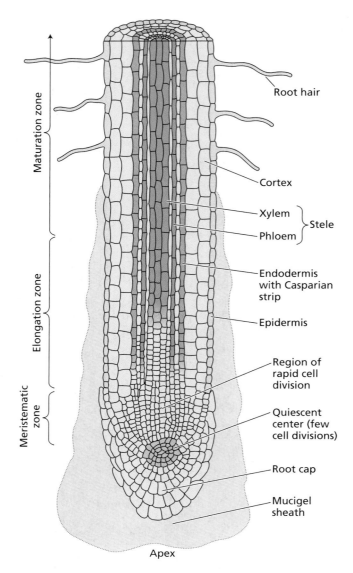

FIGURE 5.8 Diagrammatic longitudinal section of the apical region of the root. The meristematic cells are located near the tip of the root. These cells generate the root cap and the upper tissues of the root. In the elongation zone, cells differentiate to produce xylem, phloem, and cortex. Root hairs, formed in epidermal cells, first appear in the maturation zone.

The **elongation zone** begins approximately 0.7 to 1.5 mm from the apex (see Figure 5.8). In this zone, cells elongate rapidly and undergo a final round of divisions to produce a central ring of cells called the **endodermis**. The walls of this endodermal cell layer become thickened, and suberin (see Chapter 13) deposited on the radial walls forms the **Casparian strip**, a hydrophobic structure that prevents the apoplastic movement of water or solutes across the root (see Figure 4.4).

The endodermis divides the root into two regions: the **cortex** toward the outside and the **stele** toward the inside. The stele contains the vascular elements of the root: the **phloem**, which transports metabolites from the shoot to the root, and the **xylem**, which transports water and solutes to the shoot.

Phloem develops more rapidly than xylem, attesting to the fact that phloem function is critical near the root apex. Large quantities of carbohydrates must flow through the phloem to the growing apical zones in order to support cell division and elongation. Carbohydrates provide rapidly growing cells with an energy source and with the carbon skeletons required to synthesize organic compounds. Six-carbon sugars (hexoses) also function as osmotically active solutes in the root tissue. At the root apex, where the phloem is not yet developed, carbohydrate movement depends on symplastic diffusion and is relatively slow (Bret-Harte and Silk 1994). The low rates of cell division in the quiescent center may result from the fact that insufficient carbohydrates reach this centrally located region or that this area is kept in an oxidized state.

Root hairs, with their large surface area for absorption of water and solutes, first appear in the **maturation zone** (see Figure 5.8), and here the xylem develops the capacity to translocate substantial quantities of water and solutes to the shoot.

Different areas of the root absorb different mineral ions

The precise point of entry of minerals into the root system has been a topic of considerable interest. Some researchers have claimed that nutrients are absorbed only at the apical regions of the root axes or branches (Bar-Yosef et al. 1972); others claim that nutrients are absorbed over the entire root surface (Nye and Tinker 1977). Experimental evidence supports both possibilities, depending on the plant species and the nutrient being investigated:

- Root absorption of calcium in barley (*Hordeum vulgare*) appears to be restricted to the apical region.
- Iron may be taken up either at the apical region, as in barley (Clarkson 1985), or over the entire root surface, as in corn (Kashirad et al. 1973).
- Potassium, nitrate, ammonium, and phosphate can be absorbed freely at all locations of the root surface (Clarkson 1985), but in corn the elongation zone has

isms (Russell 1977). The root cap is central to the perception of gravity, the signal that directs the growth of roots downward. This process is termed the **gravitropic response** (see Chapter 19).

Cell division at the root apex proper is relatively slow; thus this region is called the **quiescent center**. After a few generations of slow cell divisions, root cells displaced from the apex by about 0.1 mm begin to divide more rapidly. Cell division again tapers off at about 0.4 mm from the apex, and the cells expand equally in all directions.

the maximum rates of potassium accumulation (Sharp et al. 1990) and nitrate absorption (Taylor and Bloom 1998).

- In corn and rice (*Oryza sativa*), the root apex absorbs ammonium more rapidly than the elongation zone does (Colmer and Bloom 1998).

- In several species, root hairs are the most active in phosphate absorption (Föhse et al. 1991).

The high rates of nutrient absorption in the apical root zones result from the strong demand for nutrients in these tissues and the relatively high nutrient availability in the soil surrounding them. For example, cell elongation depends on the accumulation of solutes such as potassium, chloride, and nitrate to increase the osmotic pressure within the cell (see Chapter 15). Ammonium is the preferred nitrogen source to support cell division in the meristem, because meristematic tissues are often carbohydrate-limited, and the assimilation of ammonium consumes less energy than that of nitrate (see Chapter 12). The root apex and root hairs grow into fresh soil, where nutrients have not yet been depleted.

Within the soil, nutrients can move to the root surface both by bulk flow and by diffusion (see Chapter 3). In bulk flow, nutrients are carried by water moving through the soil toward the root. The amount of nutrient provided to the root by bulk flow depends on the rate of water flow through the soil toward the plant, which depends on transpiration rates and on nutrient levels in the soil solution. When both the rate of water flow and the concentrations of nutrients in the soil solution are high, bulk flow can play an important role in nutrient supply.

In diffusion, mineral nutrients move from a region of higher concentration to a region of lower concentration. Nutrient uptake by the roots lowers the concentration of nutrients at the root surface, generating concentration gradients in the soil solution surrounding the root. Diffusion of nutrients down their concentration gradients and bulk flow resulting from transpiration can increase nutrient availability at the root surface.

When absorption of nutrients by the roots is high and the nutrient concentration in the soil is low, bulk flow can supply only a small fraction of the total nutrient requirement (Mengel and Kirkby 1987). Under these conditions, diffusion rates limit the movement of nutrients to the root surface. When diffusion is too slow to maintain high nutrient concentrations near the root, a **nutrient depletion zone** forms adjacent to the root surface (Figure 5.9). This zone extends from about 0.2 to 2.0 mm from the root surface, depending on the mobility of the nutrient in the soil.

The formation of a depletion zone tells us something important about mineral nutrition: Because roots deplete the mineral supply in the rhizosphere, their effectiveness in mining minerals from the soil is determined not only by the rate at which they can remove nutrients from the soil

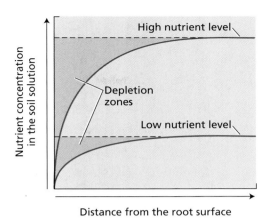

FIGURE 5.9 Formation of a nutrient depletion zone in the region of the soil adjacent to the plant root. A nutrient depletion zone forms when the rate of nutrient uptake by the cells of the root exceeds the rate of replacement of the nutrient by bulk flow and diffusion in the soil solution. This depletion causes a localized decrease in the nutrient concentration in the area adjacent to the root surface. (After Mengel and Kirkby 1987.)

solution, but by their continuous growth. *Without growth, roots would rapidly deplete the soil adjacent to their surface. Optimal nutrient acquisition therefore depends both on the capacity for nutrient uptake and on the ability of the root system to grow into fresh soil.*

Mycorrhizal fungi facilitate nutrient uptake by roots

Our discussion so far has centered on the direct acquisition of mineral elements by the root, but this process may be modified by the association of mycorrhizal fungi with the root system. The host plant provides associated **mycorrhizae** (singular *mycorrhiza*, from the Greek words for "fungus" and "root") with carbohydrates and in return receives nutrients or water from the mycorrhizae. Mycorrhizae are not unusual; in fact, they are widespread under natural conditions. Much of the world's vegetation appears to have roots associated with mycorrhizal fungi: 83% of dicots, 79% of monocots, and all gymnosperms regularly form mycorrhizal associations (Wilcox 1991).

In contrast, plants from the families Brassicaceae (e.g., cabbage [*Brassica oleracea*]), Chenopodiaceae (e.g., spinach [*Spinacea oleracea*]), and Proteaceae (e.g., macadamia nut [*Macadamia integrifolia*]), as well as aquatic plants, rarely if ever have mycorrhizae. Mycorrhizae are absent from roots in very dry, saline, or flooded soils, or where soil fertility is extreme, either high or low. In particular, plants grown by hydroponics and young, rapidly growing crop plants seldom have mycorrhizae.

Mycorrhizal fungi are composed of fine, tubular filaments called *hyphae* (singular *hypha*). The mass of hyphae that forms the body of the fungus is called the *mycelium* (plural *mycelia*). There are two major classes of mycorrhizal

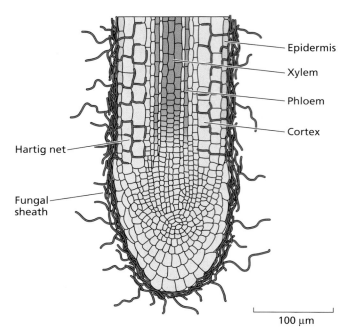

FIGURE 5.10 Root infected with ectotrophic mycorrhizal fungi. In the infected root, the fungal hyphae surround the root to produce a dense fungal sheath and penetrate the intercellular spaces of the cortex to form the Hartig net. The total mass of fungal hyphae may be comparable to the root mass itself. (After Rovira et al. 1983.)

fungi that are important in terms of mineral nutrient uptake: ectotrophic mycorrhizae and vesicular-arbuscular mycorrhizae (Brundrett 2004).

Ectotrophic mycorrhizal fungi typically form a thick sheath, or "mantle," of mycelium around the roots, and some of the mycelium penetrates between the cortical cells (Figure 5.10). The cortical cells themselves are not penetrated by the fungal hyphae, but instead are surrounded by a network of hyphae called the **Hartig net**. Often the amount of fungal mycelium is so extensive that its total mass is comparable to that of the roots themselves. The fungal mycelium also extends into the soil, away from this compact mantle, where it forms individual hyphae or strands containing fruiting bodies.

The capacity of the root system to absorb nutrients is improved by the presence of external fungal hyphae, because they are much finer than plant roots and can reach beyond the areas of nutrient-depleted soil near the roots (Clarkson 1985). Ectotrophic mycorrhizal fungi infect exclusively tree species, including gymnosperms and woody angiosperms.

Unlike the ectotrophic mycorrhizal fungi, **vesicular-arbuscular mycorrhizal fungi** do not produce a compact mantle of fungal mycelium around the root. Instead, the hyphae grow in a less dense arrangement, both within the root itself and extending outward from the root into the surrounding soil (Figure 5.11). After entering the root through either the epidermis or a root hair, the hyphae not only extend through the regions between cells but also penetrate individual cells of the cortex. Within the cells, the hyphae can form oval structures called **vesicles** and branched structures called **arbuscules**. The arbuscules appear to be sites of nutrient transfer between the fungus and the host plant.

Outside the root, the external mycelium can extend several centimeters away from the root and may contain spore-bearing structures. Unlike the ectotrophic mycorrhizae, vesicular-arbuscular mycorrhizae make up only a small mass of fungal material, which is unlikely to exceed 10% of the root weight. Vesicular-arbuscular mycorrhizae are found in association with the roots of most species of herbaceous angiosperms (Smith et al. 1997).

The association of vesicular-arbuscular mycorrhizae with plant roots facilitates the uptake of phosphorus and trace metals such as zinc and copper. By extending beyond the depletion zone for phosphorus around the root, the external mycelium improves phosphorus absorption. Calculations show that a root associated with mycorrhizal

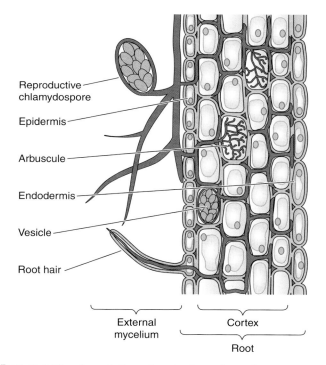

FIGURE 5.11 Association of vesicular–arbuscular mycorrhizal fungi with a section of a plant root. The fungal hyphae grow into the intercellular wall spaces of the cortex and penetrate individual cortical cells. As they extend into the cell, they do not break the plasma membrane or the tonoplast of the host cell. Instead, the hypha is surrounded by these membranes and forms structures known as arbuscules, which participate in nutrient ion exchange between the host plant and the fungus. (After Mauseth 1988.)

fungi can transport phosphate at a rate more than four times higher than that of a root not associated with mycorrhizae (Nye and Tinker 1977). The external mycelium of the ectotrophic mycorrhizae can also absorb phosphate and make it available to the plant. In addition, it has been suggested that ectotrophic mycorrhizae proliferate in the organic litter of the soil and hydrolyze organic phosphorus for transfer to the root (Smith et al. 1997).

Nutrients move from the mycorrhizal fungi to the root cells

Little is known about the mechanism by which the mineral nutrients absorbed by mycorrhizal fungi are transferred to the cells of plant roots. With ectotrophic mycorrhizae, inorganic phosphate may simply diffuse from the hyphae in the Hartig net and be absorbed by the root cortical cells. With vesicular-arbuscular mycorrhizae, the situation may be more complex. Nutrients may diffuse from intact arbuscules to root cortical cells. Alternatively, because some root arbuscules are continually degenerating while new ones are forming, degenerating arbuscules may release their internal contents to the host root cells.

A key factor in the extent of mycorrhizal association with the plant root is the nutritional status of the host plant. Moderate deficiency of a nutrient such as phosphorus tends to promote infection, whereas plants with abundant nutrients tend to suppress mycorrhizal infection.

Mycorrhizal association in well-fertilized soils may shift from a symbiotic relationship to a parasitic one in that the fungus still obtains carbohydrates from the host plant, but the host plant no longer benefits from improved nutrient uptake efficiency. Under such conditions, the host plant may treat mycorrhizal fungi as it does other pathogens (Brundrett 1991; Marschner 1995).

Summary

Plants are autotrophic organisms capable of using the energy from sunlight to synthesize all their components from carbon dioxide, water, and mineral elements. Studies of plant nutrition have shown that specific mineral elements are essential for plant life. These elements are classified as macronutrients or micronutrients, depending on the relative amounts found in plant tissue.

Certain visual symptoms are diagnostic for deficiencies in specific nutrients in higher plants. Nutritional disorders occur because nutrients have key roles within plant. They serve as components of organic compounds, in energy storage, in plant structures, as enzyme cofactors, and in electron transfer reactions. Mineral nutrition can be studied through the use of solution culture, which allow the characterization of specific nutrient requirements. Soil and plant tissue analysis can provide information on the nutritional status of the plant–soil system and can suggest corrective actions to avoid deficiencies or toxicities.

When crop plants are grown under modern high-production conditions, substantial amounts of nutrients, particularly nitrogen, phosphorus, or potassium, are removed from the soil. To prevent the development of deficiencies, nutrients can be added back to the soil in the form of fertilizers. Fertilizers that provide nutrients in inorganic forms are called chemical fertilizers; those that derive from plant or animal residues are considered organic fertilizers. In both cases, plants absorb the nutrients primarily as inorganic ions. Most fertilizers are applied to the soil, but some are sprayed on leaves.

The soil is a complex substrate—physically, chemically, and biologically. The size of soil particles and the cation exchange capacity of the soil determine the extent to which a soil provides a reservoir for water and nutrients. Soil pH also has a large influence on the availability of mineral elements to plants.

If mineral elements, especially sodium or heavy metals, are present in excess in the soil, plant growth may be adversely affected. Certain plants are able to tolerate excess mineral elements, and a few species—for example, halophytes in the case of sodium—may thrive under these extreme conditions.

To obtain nutrients from the soil, plants develop extensive root systems. Roots have a relatively simple structure with radial symmetry and few differentiated cell types. Roots continually deplete the nutrients from the immediate soil around them, and such a simple structure may permit rapid growth into fresh soil.

Plant roots often form associations with mycorrhizal fungi. The fine hyphae of mycorrhizae extend the reach of roots into the surrounding soil and facilitate the acquisition of mineral elements, particularly those like phosphorus that are relatively immobile in the soil. In return, plants provide carbohydrates to the mycorrhizae. Plants tend to suppress mycorrhizal associations under conditions of high nutrient availability.

Web Material

Web Topics

5.1 Symptoms of Deficiency in Essential Minerals
Deficiency symptoms are characteristic of each essential element and can be used as diagnostic for the deficiency. The color photographs in this topic illustrate deficiency symptoms for each essential element in a tomato.

5.2 Observing Roots below Ground
The study of roots growing under natural conditions requires means to observe roots below ground. State-of-the-art techniques are described in this topic.

> **Web Essay**
>
> **5.1 From Meals to Metals and Back**
> Heavy metal accumulation by plants is toxic. Understanding the molecular process involved is helping to develop better phytoremediation crops.

Chapter References

Aber, J. D., Goodale, C. L., Ollinger, S. V., Smith, M. L., Magill, A. H., Martin, M. E., Hallett, R. A., and Stoddard, J. L. (2003) Is nitrogen deposition altering the nitrogen status of northeastern forests? *BioScience* 53: 375–389.

Arnon, D. I., and Stout, P. R. (1939) The essentiality of certain elements in minute quantity for plants with special reference to copper. *Plant Physiol.* 14: 371–375.

Asher, C. J., and Edwards, D. G. (1983) Modern solution culture techniques. In *Inorganic Plant Nutrition* (Encyclopedia of Plant Physiology, New Series, Vol. 15B), A. Läuchli and R. L. Bieleski, eds., Springer, Berlin, pp. 94–119.

Bar-Yosef, B., Kafkafi, U., and Bresler, E. (1972) Uptake of phosphorus by plants growing under field conditions. I. Theoretical model and experimental determination of its parameters. *Soil Sci.* 36: 783–800.

Berry, W. L., and Wallace, A. (1981) Toxicity: The concept and relationship to the dose response curve. *J. Plant Nutr.* 3: 13–19.

Bloom, A. J. (1994) Crop acquisition of ammonium and nitrate. In *Physiology and Determination of Crop Yield*, K. J. Boote, J. M. Bennett, T. R. Sinclair, and G. M. Paulsen, eds., Soil Science Society of America, Crop Science Society of America, Madison, WI, pp. 303–309.

Bloom, A. J., Jackson, L. E., and Smart, D. R. (1993) Root growth as a function of ammonium and nitrate in the root zone. *Plant Cell Environ.* 16: 199–206.

Blumwald, E. (2003) Engineering salt tolerance in plants. *Biotechnol. Genet. Eng. Rev.* 20: 261–275.

Bouma, D. (1983) Diagnosis of mineral deficiencies using plant tests. In *Inorganic Plant Nutrition* (Encyclopedia of Plant Physiology, New Series, Vol. 15B), A. Läuchli and R. L. Bieleski, eds., Springer, Berlin, pp. 120–146.

Brady, N. C. (1974) *The Nature and Properties of Soils*, 8th ed. Macmillan, New York.

Bret-Harte, M. S., and Silk, W. K. (1994) Nonvascular, symplasmic diffusion of sucrose cannot satisfy the carbon demands of growth in the primary root tip of *Zea mays* L. *Plant Physiol.* 105: 19–33.

Brundrett, M. C. (1991) Mycorrhizas in natural ecosystems. *Adv. Ecol. Res.* 21: 171–313.

Brundrett, M. (2004) Diversity and classification of mycorrhizal associations. *Biol. Rev. Camb. Philos. Soc.* 79: 473–495.

Clarke, S. M., and Eaton-Rye, J. J. (2000) Amino acid deletions in loop C of the chlorophyll a-binding protein CP47 alter the chloride requirement and/or prevent the assembly of photosystem II. *Plant Mol. Biol.* 44: 591–601.

Clarkson, D. T. (1985) Factors affecting mineral nutrient acquisition by plants. *Annu. Rev. Plant Physiol.* 36: 77–116.

Colmer, T. D., and Bloom, A. J. (1998) A comparison of net NH_4^+ and NO_3^- fluxes along roots of rice and maize. *Plant Cell Environ.* 21: 240–246.

Cooper, A. (1979) *The ABC of NFT: Nutrient Film Technique: The World's First Method of Crop Production without a Solid Rooting Medium*. Grower Books, London.

Dittmer, H. J. (1937) A quantitative study of the roots and root hairs of a winter rye plant (*Secale cereale*). *Am. J. Bot.* 24: 417–420.

Engel, R. E., Bruebaker, L., and Emborg, T. J. (2001) A chloride deficient leaf spot of durum wheat. *Soil Sci. Soc. Am. J.* 65: 1448–1454.

Epstein, E. (1994) The anomaly of silicon in plant biology. *Proc. Natl. Acad. Sci. USA* 91: 11–17.

Epstein, E. (1999) Silicon. *Annu. Rev. Plant Physiol. Plant Mol. Biol.* 50: 641–664.

Epstein, E., and Bloom, A. J. (2005) *Mineral Nutrition of Plants: Principles and Perspectives*, 2nd ed. Sinauer Associates, Sunderland, MA.

Evans, H. J., and Sorger, G. J. (1966) Role of mineral elements with emphasis on the univalent cations. *Annu. Rev. Plant Physiol.* 17: 47–76.

Fenn, M. E., Baron, J. S., Allen, E. B., Rueth, H. M., Nydick, K. R., Geiser, L., Bowman, W. D., Sickman, J. O., Meixner, T., Johnson, D. W., et al. (2003) Ecological effects of nitrogen deposition in the western United States. *BioScience* 53: 404–420.

Föhse, D., Claassen, N., and Jungk, A. (1991) Phosphorus efficiency of plants. II. Significance of root radius, root hairs and cation-anion balance for phosphorus influx in seven plant species. *Plant Soil* 132: 261–272.

Gericke, W. F. (1937) Hydroponics—Crop production in liquid culture media. *Science* 85: 177–178.

Haehnel, W. (1984) Photosynthetic electron transport in higher plants. *Annu. Rev. Plant Physiol.* 35: 659–693.

Harling, H., Czaja, I., Schell, J., and Walden, R. (1997) A plant cation-chloride co-transporter promoting auxin-independent tobacco protoplast division. *EMBO J.* 16: 5855–5866.

Hasegawa, P. M., Bressan, R. A., Zhu, J.-K., and Bohnert, H. J. (2000) Plant cellular and molecular responses to high salinity. *Annu. Rev. Plant Physiol. Plant Mol. Biol.* 51: 463–499.

Hepler, P. K., and Wayne, R. O. (1985) Calcium and plant development. *Annu. Rev. Plant Physiol.* 36: 397–440.

Johnstone, M., Grof, C. P. L., and Brownell, P. F. (1988) The effect of sodium nutrition on the pool sizes of intermediates of the C_4 photosynthetic pathway. *Aust. J. Plant Physiol.* 15: 749–760.

Kashirad, A., Marschner, H., and Richter, C. H. (1973) Absorption and translocation of ^{59}Fe from various parts of the corn plant. *Z. Pflanzenernähr. Bodenk.* 134: 136–147.

Loomis, R. S., and Connor, D. J. (1992) *Crop Ecology: Productivity and Management in Agricultural Systems*. Cambridge University Press, Cambridge.

Lucas, R. E., and Davis, J. F. (1961) Relationships between pH values of organic soils and availabilities of 12 plant nutrients. *Soil Sci.* 92: 177–182.

Macek, T., Mackova, M., and Kas, J. (2000) Exploitation of plants for the removal of organics in environmental remediation. *Biotech. Adv.* 18: 23–34.

Marschner, H. (1995) *Mineral Nutrition of Higher Plants*, 2nd ed. Academic Press, London.

Mauseth, J. D. (1988) *Plant Anatomy*. Benjamin/Cummings, Menlo Park, CA.

Mengel, K., and Kirkby, E. A. (1987) *Principles of Plant Nutrition*. International Potash Institute, Worblaufen-Bern, Switzerland.

Nolan, B. T. and Stoner, J. D. (2000) Nutrients in groundwater of the center conterminous United States 1992–1995. *Environ. Sci. Tech.* 34: 1156–1165.

Nye, P. H., and Tinker, P. B. (1977) *Solute Movement in the Soil-Root System*. University of California Press, Berkeley.

Oh, S.-H., Cho, S.-W., Kwon, T.-H., and Yang, M.-S. (1996) Purification and characterization of phytoferritin. *J. Biochem. Mol. Biol.* 29: 540–544.

Raven, J. A., and Smith, F. A. (1976) Nitrogen assimilation and transport in vascular land plants in relation to intracellular pH regulation. *New Phytol.* 76: 415–431.

Rovira, A. D., Bowen, C. D., and Foster, R. C. (1983) The significance of rhizosphere microflora and mycorrhizas in plant nutrition. In *Inorganic Plant Nutrition* (Encyclopedia of Plant Physiology, New Series, Vol. 15B) A. Läuchli and R. L. Bieleskis, eds., Springer, Berlin, pp. 61–93.

Russell, R. S. (1977) *Plant Root Systems: Their Function and Interaction with the Soil*. McGraw-Hill, London.

Sanders, D., Brownlee, C., and Harper J. F. (1999) Communicating with calcium. *Plant Cell* 11: 691–706.

Sharp, R. E., Hsiao, T. C., and Silk, W. K. (1990) Growth of the maize primary root at low water potentials. 2. Role of growth and deposition of hexose and potassium in osmotic adjustment. *Plant Physiol.* 93: 1337–1346.

Shelp, B. J. (1993) Physiology and biochemistry of boron in plants. In *Boron and Its Role in Crop Production*, U. C. Gupta, ed., CRC Press, Boca Raton, FL, pp. 53–85.

Sievers, R. E., and Bailar, J. C., Jr. (1962) Some metal chelates of ethylenediaminetetraacetic acid, diethylenetriaminepentaacetic acid, and triethylenetriaminehexaacetic acid. *Inorganic Chem.* 1: 174–182.

Smith, S. E., Read, D. J., and Harley, J. L. (1997) *Mycorrhizal Symbiosis*. Academic Press, San Diego, CA.

Stewart, G. R., and Ahmad, I. (1983) Adaptation to salinity in angiosperm halophytes. In *Metals and Micronutrients: Uptake and Utilization by Plants*, D. A. Robb and W. S. Pierpoint, eds., Academic Press, New York, pp. 33–50.

Taylor, A. R., and Bloom, A. J. (1998) Ammonium, nitrate and proton fluxes along the maize root. *Plant Cell Environ.* 21: 1255–1263.

Weathers, P. J., and Zobel, R. W. (1992) Aeroponics for the culture of organisms, tissues, and cells. *Biotech. Adv.* 10: 93–115.

Weaver, J. E. (1926) *Root Development of Field Crops*. McGraw-Hill, New York.

White, P. J., and Broadley, M. R. (2003) Calcium in plants. *Ann. Bot.* 92: 487–511.

Wilcox, H. E. (1991) Mycorrhizae. In *Plant Roots: The Hidden Half*, Y. Waisel, A. Eshel, and U. Kafkafi, eds., Marcel Dekker, New York, pp. 731–765.

Wood, B. W., Reilly, C. C., and Nyczepir, A. P. (2003) Nickel corrects mouse-ear. *The Pecan Grower* 15: 3–5.

Ziegler, H. (1987) The evolution of stomata. In *Stomatal Function*, E. Zeiger, G. Farquhar, and I. Cowan, eds., Stanford University Press, Stanford, CA, pp. 29–57.

Chapter 6 | Solute Transport

THE INTERIOR OF A PLANT CELL IS SEPARATED from the plant cell wall and the environment by a plasma membrane that is only two lipid molecules thick. This thin layer separates a relatively constant internal environment from highly variable external surroundings. In addition to forming a hydrophobic barrier to diffusion, the membrane must facilitate and continuously regulate the inward and outward traffic of selected molecules and ions as the cell takes up nutrients, exports solutes, and regulates its turgor pressure. The same is true of the internal membranes that separate the various compartments within each cell.

The plasma membrane also relays information about the physical environment, about molecular signals from other cells, and about the presence of invading pathogens. Often these signal transduction processes are mediated by changes in ion fluxes across the membrane.

Molecular and ionic movement from one location to another is known as **transport**. Local transport of solutes into or within cells is regulated mainly by membranes. Larger-scale transport between plant and environment, or between plant organs, is also controlled by membrane transport at the cellular level. For example, the transport of sucrose from leaf to root through the phloem, referred to as *translocation*, is driven and regulated by membrane transport into the phloem cells of the leaf, and from the phloem to the storage cells of the root (see Chapter 10).

In this chapter we will consider first the physical and chemical principles that govern the movements of molecules in solution. Then we will show how these principles apply to membranes and to biological systems. We will also discuss the molecular mechanisms of transport in living cells and the great variety of membrane transport proteins that are responsible for the particular transport properties of plant cells. Finally, we will examine the pathway that ions take when they enter the root, as well as the mechanism of xylem loading, the process whereby ions are released into the vessel elements and tracheids of the stele.

Passive and Active Transport

According to Fick's first law (see Equation 3.1), the movement of molecules by diffusion always proceeds spontaneously, down a gradient of free energy or chemical potential (see Chapter 2 on the web site), until equilibrium is reached. The spontaneous "downhill" movement of molecules is termed **passive transport**. At equilibrium, no further net movements of solute can occur without the application of a driving force.

The movement of substances against or up a gradient of chemical potential is termed **active transport**. It is not spontaneous, and it requires that work be done on the system by the application of cellular energy. One way (but not the only way) of accomplishing this task is to couple transport to the hydrolysis of ATP.

Recall from Chapter 3 that we can calculate the driving force for diffusion, or, conversely, the energy input necessary to move substances against a gradient, by measuring the potential-energy gradient, which, for uncharged solutes, is often a simple function of the difference in concentration. Biological transport can be driven by four major forces: concentration, hydrostatic pressure, gravity, and electric fields. (However, recall from Chapter 3 that in biological systems, gravity seldom contributes substantially to the force that drives transport.)

The **chemical potential** for any solute is defined as the sum of the concentration, electric, and hydrostatic potentials (and the chemical potential under standard conditions). *The importance of the concept of chemical potential is that it sums all the forces that may act on a molecule to drive net transport* (Nobel 1991).

$$\tilde{\mu}_j = \mu_j^* + RT \ln C_j$$

Chemical potential for a given solute, j | Chemical potential of j under standard conditions | Concentration (activity) component

$$+ \; z_j FE \; + \; \bar{V}_j P$$

Electric-potential component | Hydrostatic-pressure component

(6.1)

Here $\tilde{\mu}_j$ is the chemical potential of the solute species j in joules per mole (J mol^{-1}), μ_j^* is its chemical potential under standard conditions (a correction factor that will cancel out in future equations and so can be ignored), R is the universal gas constant, T is the absolute temperature, and C_j is the concentration (more accurately the activity) of j.

The electrical term, $z_j FE$, applies only to ions; z is the electrostatic charge of the ion (+1 for monovalent cations, –1 for monovalent anions, +2 for divalent cations, and so on), F is Faraday's constant (equivalent to the electric charge on 1 mol of protons), and E is the overall electric potential of the solution (with respect to ground). The final term, $\bar{V}_j P$, expresses the contribution of the partial molal volume of j ($\bar{V}_j$) and pressure (P) to the chemical potential of j. (The partial molal volume of j is the change in volume per mole of substance j added to the system, for an infinitesimal addition.)

This final term, $\bar{V}_j P$, makes a much smaller contribution to $\tilde{\mu}_j$ than do the concentration and electrical terms, except in the very important case of osmotic water movements. As discussed in Chapter 3, the chemical potential of water (i.e., the water potential) depends on the concentration of dissolved solutes and the hydrostatic pressure on the system.

In general, diffusion (passive transport) always moves molecules from areas of higher chemical potential downhill to areas of lower chemical potential. Movement against a chemical-potential gradient is indicative of active transport (Figure 6.1).

If we take the diffusion of sucrose across a cell membrane as an example, we can accurately approximate the chemical potential of sucrose in any compartment by the concentration term alone (unless a solution is concentrated, causing hydrostatic pressure to build up within the plant cell). From Equation 6.1, the chemical potential of sucrose inside a cell can be described as follows (in the next three equations, the subscript s stands for sucrose, and the superscripts i and o stand for inside and outside, respectively):

$$\tilde{\mu}_s^i = \mu_s^* + RT \ln C_s^i$$

Chemical potential of sucrose solution inside the cell | Chemical potential of sucrose solution under standard conditions | Concentration component

(6.2)

The chemical potential of sucrose outside the cell is calculated as follows:

$$\tilde{\mu}_s^o = \mu_s^* + RT \ln C_s^o \qquad (6.3)$$

We can calculate the difference in the chemical potential of sucrose between the solutions inside and outside the cell,

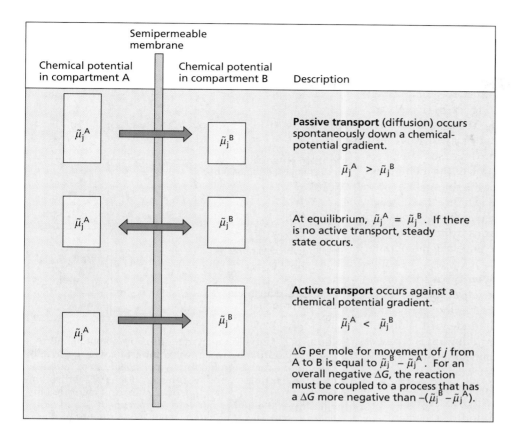

Figure 6.1 Relationship between the chemical potential, $\tilde{\mu}$, and the transport of molecules across a permeability barrier. The net movement of molecular species j between compartments A and B depends on the relative magnitude of the chemical potential of j in each compartment, represented here by the size of the boxes. Movement down a chemical gradient occurs spontaneously and is called passive transport; movement against or up a gradient requires energy and is called active transport.

$\Delta \tilde{\mu}_s$, regardless of the mechanism of transport. To get the signs right, remember that for inward transport, sucrose is being removed (−) from outside the cell and added (+) to the inside, so the change in free energy in joules per mole of sucrose transported will be as follows:

$$\Delta \tilde{\mu}_s = \tilde{\mu}_s^i - \tilde{\mu}_s^o \tag{6.4}$$

Substituting the terms from Equations 6.2 and 6.3 into Equation 6.4, we get the following:

$$\begin{aligned}\Delta \tilde{\mu}_s &= \left(\mu_s^* + RT \ln C_s^i\right) - \left(\mu_s^* + RT \ln C_s^o\right) \\ &= RT\left(\ln C_s^i - \ln C_s^o\right) \\ &= RT \ln \frac{C_s^i}{C_s^o}\end{aligned} \tag{6.5}$$

If this difference in chemical potential is negative, sucrose can diffuse inward spontaneously (provided the membrane had a finite permeability to sucrose; see the next section). In other words, the driving force ($\Delta \tilde{\mu}_s$) for solute diffusion is related to the magnitude of the concentration gradient (C_s^i/C_s^o).

If the solute carries an electric charge (as does, for example, the potassium ion), the electrical component of the chemical potential must also be considered. Suppose the membrane is permeable to K^+ and Cl^- rather than to sucrose. Because the ionic species (K^+ and Cl^-) diffuse independently, each has its own chemical potential. Thus for inward K^+ diffusion,

$$\Delta \tilde{\mu}_K = \tilde{\mu}_K^i - \tilde{\mu}_K^o \tag{6.6}$$

Substituting the appropriate terms from Equation 6.1 into Equation 6.6, we get

$$\Delta \tilde{\mu}_s = (RT \ln [K^+]^i + zFE^i) - (RT \ln [K^+]^o + zFE^o) \tag{6.7}$$

and because the electrostatic charge of K^+ is +1, $z = +1$ and

$$\Delta \tilde{\mu}_K = RT \ln \frac{[K^+]^i}{[K^+]^o} + F(E^i - E^o) \tag{6.8}$$

The magnitude and sign of this expression will indicate the driving force for K^+ diffusion across the membrane, and its direction. A similar expression can be written for Cl^- (but remember that for Cl^-, $z = -1$).

Equation 6.8 shows that ions, such as K^+, diffuse in response to both their concentration gradients ($[K^+]^i/[K^+]^o$) and any electric-potential difference between the two compartments ($E^i - E^o$). One very important implication of this equation is that ions can be driven passively against their concentration gradients if an appropriate voltage (electric field) is applied between the two compartments. Because of the importance of electric fields in the biological transport of any charged molecule, $\tilde{\mu}$ is often called the **electrochemical potential**, and $\Delta \tilde{\mu}$ is the difference in electrochemical potential between two compartments.

Transport of Ions across a Membrane Barrier

If the two KCl solutions in the previous example are separated by a biological membrane, diffusion is complicated by the fact that the ions must move through the membrane as well as across the open solutions. The extent to which a membrane permits the movement of a substance is called **membrane permeability**. As will be discussed later, permeability depends on the composition of the membrane, as well as on the chemical nature of the solute. In a loose sense, permeability can be expressed in terms of a diffusion coefficient for the solute through the membrane. However, permeability is influenced by several additional factors, such as the ability of a substance to enter the membrane, that are difficult to measure.

Despite its theoretical complexity, we can readily measure permeability by determining the rate at which a solute passes through a membrane under a specific set of conditions. Generally the membrane will hinder diffusion and thus reduce the speed with which equilibrium is reached. For a permeable solute, the permeability or resistance of the membrane itself, however, cannot alter the final equilibrium conditions. Equilibrium occurs when $\Delta \tilde{\mu}_j = 0$.

In the sections that follow we will discuss the factors that influence the passive distribution of ions across a membrane. These parameters can be used to predict the relationship between the electrical gradient and the concentration gradient of an ion.

Different diffusion rates for cations and anions produce diffusion potentials

When salts diffuse across a membrane, an electric membrane potential (voltage) can develop. Consider the two KCl solutions separated by a membrane in Figure 6.2. The K^+ and Cl^- ions will permeate the membrane independently as they diffuse down their respective gradients of electrochemical potential. And unless the membrane is very porous, its permeability for the two ions will differ.

As a consequence of these different permeabilities, K^+ and Cl^- initially diffuse across the membrane at different rates. The result is a slight separation of charge, which instantly creates an electric potential across the membrane. In biological systems, membranes are usually more permeable to K^+ than to Cl^-. Therefore, K^+ will diffuse out of the cell (see compartment A in Figure 6.2) faster than Cl^-, causing the cell to develop a negative electric charge with respect to the medium. A potential that develops as a result of diffusion is called a **diffusion potential**.

An important principle that must always be kept in mind when the movement of ions across membranes is considered is the principle of electrical neutrality. Bulk solutions always contain equal numbers of anions and cations. The existence of a membrane potential implies that the distribution of charges across the membrane is uneven;

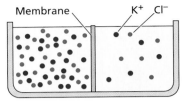
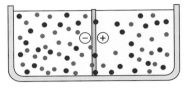
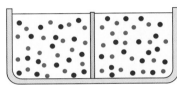

FIGURE 6.2 Development of a diffusion potential and a charge separation between two compartments separated by a membrane that is preferentially permeable to potassium. If the concentration of potassium chloride is higher in compartment A ($[KCl]_A > [KCl]_B$), potassium and chloride ions will diffuse at a higher rate into compartment B, and a diffusion potential will be established. When membranes are more permeable to potassium than to chloride, potassium ions will diffuse faster than chloride ions, and a charge separation (+ and –) will develop.

however, the actual number of unbalanced ions is negligible in chemical terms. For example, a membrane potential of –100 millivolts (mV), like that found across the plasma membranes of many plant cells, results from the presence of only one extra anion out of every 100,000 within the cell—a concentration difference of only 0.001%!

As Figure 6.2 shows, all of these extra anions are found immediately adjacent to the surface of the membrane; there is no charge imbalance throughout the bulk of the cell. In our example of KCl diffusion across a membrane, electrical neutrality is preserved because as K^+ moves ahead of Cl^- in the membrane, the resulting diffusion potential retards the movement of K^+ and speeds that of Cl^-. Ultimately, both ions diffuse at the same rate, but the diffusion potential persists and can be measured. As the system moves toward equilibrium and the concentration gradient collapses, the diffusion potential also collapses.

How does membrane potential relate to ion distribution?

Because the membrane is permeable to both K^+ and Cl^- ions, equilibrium in the preceding example will not be reached for either ion until the concentration gradients decrease to zero. However, if the membrane were permeable to only K⁺, diffusion of K⁺ would carry charges across the membrane until the membrane potential balanced the concentration gradient. Because a change in potential requires very few ions, this balance would be reached instantly. Potassium ions would then be at equilibrium, even though the change in the concentration gradient for K⁺ would be negligible.

When the distribution of any solute across a membrane reaches equilibrium, the passive flux, J (i.e., the amount of solute crossing a unit area of membrane per unit time), is the same in the two directions—outside to inside and inside to outside:

$$J_{o \to i} = J_{i \to o}$$

Fluxes are related to $\Delta \tilde{\mu}$ (for a discussion on fluxes and $\Delta \tilde{\mu}$, see Chapter 2 on the web site); thus at equilibrium, the electrochemical potentials will be the same:

$$\tilde{\mu}_j^o = \tilde{\mu}_j^i$$

and for any given ion (the ion is symbolized here by the subscript j):

$$\mu_j^* + RT \ln C_j^o + z_j F E^o = \mu_j^* + RT \ln C_j^i + z_j F E^i \quad (6.9)$$

By rearranging Equation 6.9, we can obtain the difference in electric potential between the two compartments at equilibrium ($E^i - E^o$):

$$E^i - E^o = \frac{RT}{z_j F} \left(\ln \frac{C_j^o}{C_j^i} \right)$$

This electric-potential difference is known as the **Nernst potential** (ΔE_j) for that ion:

$$\Delta E_j = E^i - E^o$$

and

$$\Delta E_j = \frac{RT}{z_j F} \left(\ln \frac{C_j^o}{C_j^i} \right) \quad (6.10)$$

or

$$\Delta E_j = \frac{2.3 RT}{z_j F} \left(\log \frac{C_j^o}{C_j^i} \right)$$

This relationship, known as the *Nernst equation*, states that at equilibrium the difference in concentration of an ion between two compartments is balanced by the voltage difference between the compartments. The Nernst equation can be further simplified for a univalent cation at 25°C:

$$\Delta E_j = 59 \text{mV} \log \frac{C_j^o}{C_j^i} \quad (6.11)$$

Note that a tenfold difference in concentration corresponds to a Nernst potential of 59 mV ($C_o/C_i = 10/1$; log 10 = 1). That is, a membrane potential of 59 mV would maintain a tenfold concentration gradient of an ion whose movement across the membrane is driven by passive diffusion. Similarly, if a tenfold concentration gradient of an ion existed across the membrane, passive diffusion of that ion down its concentration gradient (if it were allowed to come to equilibrium) would result in a difference of 59 mV across the membrane.

All living cells exhibit a membrane potential that is due to the asymmetric ion distribution between the inside and outside of the cell. We can determine these membrane potentials by inserting a microelectrode into the cell and measuring the voltage difference between the inside of the cell and the external bathing medium (Figure 6.3).

The Nernst equation can be used at any time to determine whether a given ion is at equilibrium across a membrane. However, a distinction must be made between equilibrium and steady state. Steady state is the condition in which influx and efflux of a given solute are equal and therefore the ion concentrations are constant with respect

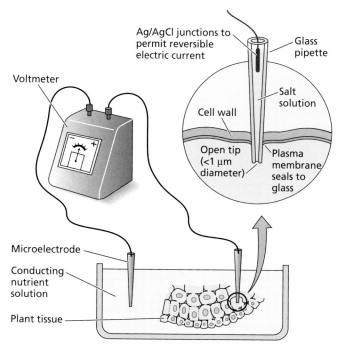

FIGURE 6.3 Diagram of a pair of microelectrodes used to measure membrane potentials across cell membranes. One of the glass micropipette electrodes is inserted into the cell compartment under study (usually the vacuole or the cytoplasm), while the other is kept in an electrolytic solution that serves as a reference. The microelectrodes are connected to a voltmeter, which records the electric-potential difference between the cell compartment and the solution. Typical membrane potentials across plant cell membranes range from −60 to −240 mV. The insert shows how electrical contact with the interior of the cell is made through the open tip of the glass micropipette, which contains an electrically conducting salt solution.

to time. Steady state is not necessarily the same as equilibrium (see Figure 6.1); in steady state, the existence of active transport across the membrane prevents many diffusive fluxes from ever reaching equilibrium.

The Nernst equation distinguishes between active and passive transport

Table 6.1 shows how the experimentally measured ion concentrations at steady state for pea root cells compare with predicted values calculated from the Nernst equation (Higinbotham et al. 1967). In this example, the external concentration of each ion in the solution bathing the tissue, and the measured membrane potential, were substituted into the Nernst equation, and a predicted internal concentration was calculated for that ion.

Notice that, of all the ions shown in Table 6.1, only K^+ is at or near equilibrium. The anions NO_3^-, Cl^-, $H_2PO_4^-$, and SO_4^{2-} all have higher internal concentrations than predicted, indicating that their uptake is active. The cations Na^+, Mg^{2+}, and Ca^{2+} have lower internal concentrations than predicted; therefore, these ions enter the cell by diffusion down their electrochemical-potential gradients and then are actively exported.

The example shown in Table 6.1 is an oversimplification: Plant cells have several internal compartments, each of which can differ in its ionic composition from the others. The cytosol and the vacuole are the most important intracellular compartments that determine the ionic relations of plant cells. In mature plant cells, the central vacuole often occupies 90% or more of the cell's volume, and the cytosol is restricted to a thin layer around the periphery of the cell.

Because of its small volume, the cytosol of most angiosperm cells is difficult to assay chemically. For this reason, much of the early work on the ionic relations of

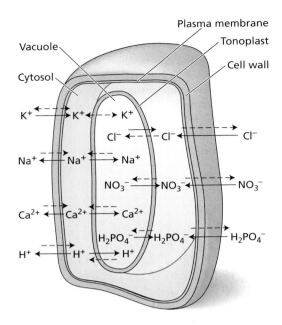

FIGURE 6.4 Ion concentrations in the cytosol and the vacuole are controlled by passive (dashed arrows) and active (solid arrows) transport processes. In most plant cells the vacuole occupies up to 90% of the cell's volume and contains the bulk of the cell solutes. Control of the ion concentrations in the cytosol is important for the regulation of metabolic enzymes. The cell wall surrounding the plasma membrane does not represent a permeability barrier and hence is not a factor in solute transport.

plants focused on certain green algae, such as *Chara* and *Nitella*, whose cells are several inches long and can contain an appreciable volume of cytosol. Figure 6.4 diagrams the conclusions from these studies and from related work with higher plants.

- Potassium is accumulated passively by both the cytosol and the vacuole. When extracellular K^+ concentrations are very low, K^+ may be taken up actively.
- Sodium is pumped actively out of the cytosol into the extracellular space and vacuole.
- Excess protons, generated by intermediary metabolism, are also actively extruded from the cytosol. This process helps maintain the cytosolic pH near neutrality, while the vacuole and the extracellular medium are generally more acidic by one or two pH units.
- Anions are taken up actively into the cytosol.
- Calcium is actively transported out of the cytosol at both the cell membrane and the vacuolar membrane, which is called the tonoplast (see Figure 6.4).

Many different ions permeate the membranes of living cells simultaneously, but K^+ and anions have the highest concentrations and largest permeabilities in plant cells. A modified version of the Nernst equation, the **Goldman equation**, includes all permeant ions (all ions for which mechanisms

TABLE 6.1
Comparison of observed and predicted ion concentrations in pea root tissue

Ion	Concentration in external medium (mmol L⁻¹)	Internal concentration (mmol L⁻¹)	
		Predicted	Observed
K^+	1	74	75
Na^+	1	74	8
Mg^{2+}	0.25	1340	3
Ca^{2+}	1	5360	2
NO_3^-	2	0.0272	28
Cl^-	1	0.0136	7
$H_2PO_4^-$	1	0.0136	21
SO_4^{2-}	0.25	0.00005	19

Source: Data from Higinbotham et al. 1967.

Note: The membrane potential was measured as –110 mV.

of transmembrane movement exist) and therefore gives a more accurate value for the diffusion potential in these cells. When permeabilities and ion gradients are known, it is possible to calculate a diffusion potential for the membrane from the Goldman equation. The diffusion potential calculated from the Goldman equation is termed the *Goldman diffusion potential* (for a detailed discussion of the Goldman equation, see **Web Topic 6.1**).

Proton transport is a major determinant of the membrane potential

In most eukaryotic cells, K^+ has both the greatest internal concentration and the highest membrane permeability, so the diffusion potential may approach E_K, the Nernst potential for K^+. In cells of some organisms, particularly in some mammalian cells such as nerve cells, the normal resting potential of the cell also may be close to E_K. This is not the case with plants and fungi, which often show experimentally measured membrane potentials (often −200 to −100 mV) that are much more negative than those calculated from the Goldman equation, which are usually only −80 to −50 mV. Thus, in addition to the diffusion potential, the membrane potential has a second component. The excess voltage is provided by the plasma membrane electrogenic H^+-ATPase.

Whenever an ion moves into or out of a cell without being balanced by countermovement of an ion of opposite charge, a voltage is created across the membrane. Any active transport mechanism that results in the movement of a net electric charge will tend to move the membrane potential away from the value predicted by the Goldman equation. Such a transport mechanism is called an *electrogenic pump* and is common in living cells.

The energy required for active transport is often provided by the hydrolysis of ATP. In plants we can study the dependence of the membrane potential on ATP by observing the effect of cyanide ($CN^−$) on the membrane potential (Figure 6.5). Cyanide rapidly poisons the mitochondria, and the cell's ATP consequently becomes depleted. As ATP synthesis is inhibited, the membrane potential falls to the level of the Goldman diffusion potential (see **Web Topic 6.1**).

Thus the membrane potentials of plant cells have two components: a diffusion potential and a component resulting from electrogenic ion transport (transport that results in the generation of a membrane potential) (Spanswick 1981). When cyanide inhibits electrogenic ion transport, the pH of the external medium increases while the cytosol becomes acidic because H^+ remains inside the cell. This is one piece of evidence that it is the active transport of H^+ out of the cell that is electrogenic.

As discussed earlier, a change in the membrane potential caused by an electrogenic pump will change the driving forces for diffusion of all ions that cross the membrane. For example, the outward transport of H^+ can create a driving force for the passive diffusion of K^+ into the cell. H^+ is transported electrogenically across the plasma membrane

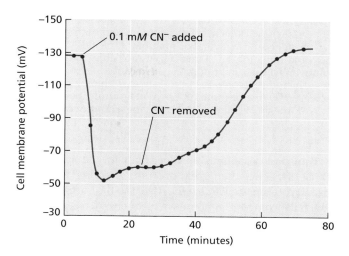

Figure 6.5 The membrane potential of a pea cell collapses when cyanide ($CN^−$) is added to the bathing solution. Cyanide blocks ATP production in the cells by poisoning the mitochondria. The collapse of the membrane potential upon addition of cyanide indicates that an ATP supply is necessary for maintenance of the potential. Washing the cyanide out of the tissue results in a slow recovery of ATP production and restoration of the membrane potential. (After Higinbotham et al. 1970.)

not only in plants but also in bacteria, algae, fungi, and some animal cells, such as those of the kidney epithelia.

ATP synthesis in mitochondria and chloroplasts also depends on a H^+-ATPase. In these organelles, this transport protein is sometimes called *ATP synthase* because it forms ATP rather than hydrolyzing it (see Chapter 11). The structure and function of membrane proteins involved in active and passive transport in plant cells will be discussed later.

Membrane Transport Processes

Artificial membranes made of pure phospholipids have been used extensively to study membrane permeability. When the permeability of artificial phospholipid bilayers for ions and molecules is compared with that of biological membranes, important similarities and differences become evident (Figure 6.6).

Both biological and artificial membranes have similar permeabilities for nonpolar molecules and many small polar molecules. On the other hand, as shown in Figure 6.6, biological membranes are much more permeable to ions, to some large polar molecules, such as sugars, and to water, than artificial bilayers are. The reason is that, unlike artificial bilayers, biological membranes contain **transport proteins** that facilitate the passage of selected ions and other polar molecules. The general term transport proteins encompasses three main categories of proteins, **channels**, **carriers**, and **pumps** (Figure 6.7), each of which will be described in more detail subsequently.

FIGURE 6.6 Typical values for the permeability, P, of a biological membrane to various substances, compared with those for an artificial phospholipid bilayer. For nonpolar molecules such as O_2 and CO_2, and for some small uncharged molecules such as glycerol, P values are similar in both systems. For ions and selected polar molecules, including water, the permeability of biological membranes is increased by one or more orders of magnitude, because of the presence of transport proteins. Note the logarithmic scale.

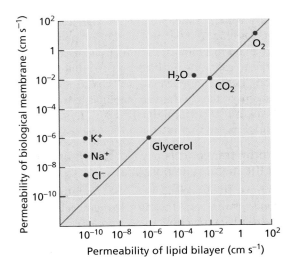

Transport proteins exhibit specificity for the solutes they transport, hence their great diversity in cells. The simple prokaryote *Haemophilus influenzae*, the first organism for which the complete genome was sequenced, has only 1743 genes, yet more than 200 of these genes (greater than 10% of the genome) encode various proteins involved in membrane transport. In Arabidopsis, out of a predicted 25,500 proteins, as many as 1800 (~7%) may execute transport functions (Schwacke et al. 2003).

Although a particular transport protein is usually highly specific for the kinds of substances it will transport, its specificity is often not absolute. For example, in plants a K^+ transporter on the plasma membrane may transport Rb^+ and Na^+ in addition to K^+, but K^+ is usually preferred. In contrast, most K^+ transporters are completely ineffective in transporting anions such as Cl^- or uncharged solutes such as sucrose. Similarly, a protein involved in the transport of neutral amino acids may move glycine, alanine, and valine with equal ease but not accept aspartic acid or lysine.

In the next several pages we will consider the structures, functions, and physiological roles of the various membrane transporters found in plant cells, especially on the plasma membrane and tonoplast. We begin with a discussion of the role of certain transporters (channels and carriers) in promoting the diffusion of solutes across membranes. We will then distinguish between primary and secondary active transport, and discuss the roles of the electrogenic H^+-ATPase and various symporters (proteins that transport two substances in the same direction simultaneously) in driving proton-coupled secondary active transport.

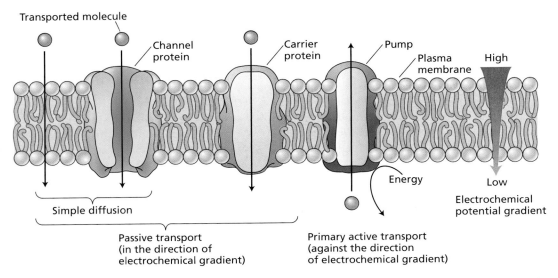

FIGURE 6.7 Three classes of membrane transport proteins: channels, carriers, and pumps. Channels and carriers can mediate the passive transport of solutes across membranes (by simple diffusion or facilitated diffusion), down the solute's gradient of electrochemical potential. Channel proteins act as membrane pores, and their specificity is determined primarily by the biophysical properties of the channel. Carrier proteins bind the transported molecule on one side of the membrane and release it on the other side. (The different types of carrier proteins are described in more detail in Figure 6.11). Primary active transport is carried out by pumps and uses energy directly, usually from ATP hydrolysis, to pump solutes against their gradient of electrochemical potential.

Channel transporters enhance diffusion across membranes

Channels are transmembrane proteins that function as selective pores, through which molecules or ions can diffuse across the membrane. The size of a pore and the density and nature of the surface charges on its interior lining determine its transport specificity. Transport through channels is always passive, and because the specificity of transport depends on pore size and electric charge more than on selective binding, channel transport is limited mainly to ions or water (Figure 6.8). As long as the channel pore is open, solutes that can penetrate the pore diffuse through it extremely rapidly: about 10^8 ions per second through each channel protein. Channels are not open all the time: Channel proteins have structures called **gates** that open and close the pore in response to external signals (see Figure 6.8B). Signals that can regulate channel activity include voltage changes, hormones, light, and posttranslational modifications such as phosphorylation. For example, voltage-gated channels open or close in response to changes in the membrane potential.

Individual ion channels can be studied in detail by the technique of patch clamp electrophysiology (see **Web Topic 6.2**), which can detect the electric current carried by ions diffusing through a single channel. Patch clamp studies reveal that for a given ion, such as K^+, a membrane has a variety of different channels. These channels may open in different voltage ranges, or in response to different signals, which may include K^+ or Ca^{2+} concentrations, pH, reactive oxygen species, and so on. This specificity enables the transport of each ion to be fine-tuned to the prevailing conditions. Thus the ion permeability of a membrane is a variable that depends on the mix of ion channels that are open at a particular time.

As we saw in the experiment of Table 6.1, the distribution of most ions is not close to equilibrium across the membrane. Anion channels always function to allow anions to diffuse only out of the cell, and other mechanisms are needed for anion uptake. Similarly, calcium channels can function only in the direction of calcium release into the cytosol, and calcium must be expelled by active transport. The exception is potassium, which can diffuse either inward or outward, depending on whether the membrane potential is more negative or more positive than E_K, the potassium equilibrium potential.

K^+ channels that open only at potentials more negative than the prevailing Nernst potential for K^+ are specialized for inward diffusion of K^+ and are known as **inwardly rectifying**, or simply inward, K^+ channels. Conversely, K^+ channels that open only at potentials positive of the Nernst potential for K^+ are **outwardly rectifying**, or outward, K^+ channels (see Figure 6.9 and **Web Essay 6.1**). Inward K^+ channels function in the accumulation of K^+ from the apoplast, as occurs, for example, during K^+ uptake by guard cells in the process of stomatal opening (see Figure 6.9). Various outward K^+ channels function in the closing of stomata, and in the release of K^+ into the xylem or apoplast.

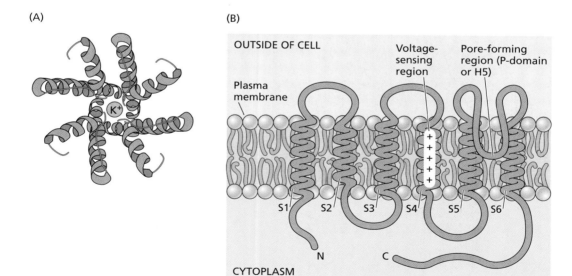

FIGURE 6.8 Models of K^+ channels in plants. (A) Top view of a channel, looking through the pore of the protein. Membrane-spanning helices of four subunits come together in an inverted teepee with the pore at the center. The pore-forming regions of the four subunits dip into the membrane, with a K^+ selectivity finger region formed at the outer (near) part of the pore (more details on the structure of this channel can be found in **Web Essay 6.1**). (B) Side view of the inwardly rectifying K^+ channel, showing a polypeptide chain of one subunit, with six membrane-spanning helices (S1–S6). The fourth helix contains positively charged amino acids and acts as a voltage sensor. The pore-forming region is a loop between helices 5 and 6. (A after Leng et al. 2002; B after Buchanan et al. 2000.)

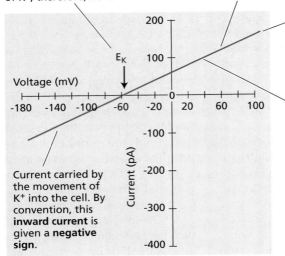

(A) Equilibrium or Nernst potential for K⁺: by definition, no net flux of K⁺, therefore, no current.

Current carried by the movement of K⁺ out of the cell. By convention, this **outward current** is given a **positive sign**.

The opening and closing or "**gating**" of these channels is not regulated by voltage. Therefore, current through the channel is a linear function of voltage.

The slope of the line ($\Delta I/DV$) gives the **conductance** of the channels mediating this K⁺ current.

Current carried by the movement of K⁺ into the cell. By convention, this **inward current** is given a **negative sign**.

$E_K = RT/ZF * \ln\{[K_{out}]/[K_{in}]\}$
$E_K = 0.025 * \ln\{10/100\}$
$E_K = -59$ mV

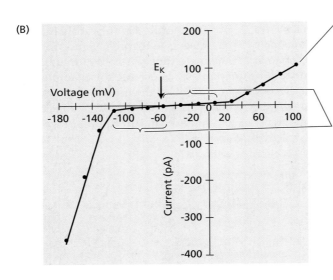

(B) This current/voltage relationship is produced by K⁺ movement through channels that are regulated ("gated") by voltage. Note that the I/V relationship is non-linear.

Little or no current over these voltage ranges because the channels are voltage-regulated and the effect of these voltages is to keep the channels in a closed state.

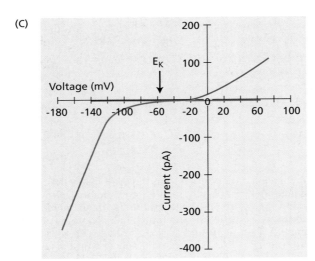

(C) Current response illustrated in (B) is shown here to arise from the activity of two molecularly distinct types of K⁺ channels. The outward K⁺ channels (red) are voltage-gated such that they open only at membrane potentials >E_K; thus these channels mediate K⁺ efflux from the cell. The inward K⁺ channels (blue) are voltage-gated such that they open only at membrane potentials <E_K; thus these channels mediate K⁺ uptake into the cell.

FIGURE 6.9 Current–voltage relationships. (A) A diagram showing the current that would result from K⁺ flux through a set of hypothetical plasma membrane K⁺ channels which were not voltage regulated, if the K⁺ concentration in the cytosol was 100 mM and if the extracellular K⁺ concentration was 10 mM. Note that the current would be linear, and that there would be zero current at the equilibrium (Nernst) potential for K⁺ (E_K). (B) Actual K⁺ current data from an Arabidopsis guard cell protoplast, with the same intracellular and extracellular K⁺ concentrations as in (A). These currents result from the activities of voltage-regulated K⁺ channels. Note that, again, there is zero net current at the equilibrium potential for K⁺. However, there is also zero net current over a broader voltage range: this is because the channels are closed over this voltage range in these conditions. When the channels are closed, no K⁺ can flow through them, hence zero current is observed over this voltage range. (C) The current–voltage relationship in (B) actually results from the activities of two sets of channels— the inwardly rectifying K⁺ channels and the outwardly rectifying K⁺ channels—which together produce the current–voltage relationship. (B after L. Perfus-Barbeoch and S. M. Assmann, unpublished data).

Carriers bind and transport specific substances

Unlike channels, **carrier** proteins do not have pores that extend completely across the membrane. In transport mediated by a carrier, the substance being transported is initially bound to a specific site on the carrier protein. This requirement for binding allows carriers to be highly selective for a particular substrate to be transported. Carriers therefore specialize in the transport of specific ions or organic metabolites. Binding causes a conformational change in the protein, which exposes the substance to the solution on the other side of the membrane. Transport is complete when the substance dissociates from the carrier's binding site.

Because a conformational change in the protein is required to transport individual molecules or ions, the rate of transport by a carrier is many orders of magnitude slower than that through a channel. Typically, carriers may transport 100 to 1000 ions or molecules per second, which is about 10^6 times slower than transport through a channel. The binding and release of a molecule at a specific site on a protein that occur in carrier-mediated transport are similar to the binding and release of molecules from an enzyme in an enzyme-catalyzed reaction. As will be discussed later in the chapter, enzyme kinetics have been used to characterize transport carrier proteins (for a detailed discussion on kinetics, see Chapter 2 on the web site).

Carrier-mediated transport (unlike transport through channels) can be either passive or active. Passive transport via a carrier is sometimes called **facilitated diffusion**, although it resembles diffusion only in that it transports substances down their gradient of electrochemical potential, without an additional input of energy. (The term "facilitated diffusion" might seem more appropriately applied to transport through channels, but historically it has not been used in this way.)

Primary active transport requires energy

To carry out active transport, a carrier must couple the energetically uphill transport of the solute with another, energy-releasing, event so that the overall free-energy change is negative. **Primary active transport** is coupled directly to a source of energy other than $\Delta \tilde{\mu}_j$, such as ATP hydrolysis, an oxidation–reduction reaction (the electron transport chain of mitochondria and chloroplasts), or the absorption of light by the carrier protein (bacteriorhodopsin in halobacteria).

The membrane proteins that carry out primary active transport are called **pumps** (see Figure 6.7). Most pumps transport ions, such as H⁺ or Ca^{2+}. However, as we will see later in this chapter, pumps belonging to the ATP-binding cassette (ABC) family of transporters can carry large organic molecules.

Ion pumps can be further characterized as either electrogenic or electroneutral. In general, **electrogenic transport** refers to ion transport involving the net movement of charge across the membrane. In contrast, **electroneutral transport**, as the name implies, involves no net movement of charge. For example, the Na⁺/K⁺-ATPase of animal cells pumps three Na⁺ ions out for every two K⁺ ions in, resulting in a net outward movement of one positive charge. The Na⁺/K⁺-ATPase is therefore an electrogenic ion pump. In contrast, the H⁺/K⁺-ATPase of the animal gastric mucosa pumps one H⁺ out of the cell for every one K⁺ in, so there is no net movement of charge across the membrane. Therefore, the H⁺/K⁺-ATPase is an electroneutral pump.

In the plasma membranes of plants, fungi, and bacteria, as well as in plant tonoplasts and other plant and animal endomembranes, H⁺ is the principal ion that is electrogenically pumped across the membrane. The **plasma membrane H⁺-ATPase** generates the gradient of electrochemical potential of H⁺ across the plasma membranes, while the **vacuolar H⁺-ATPase** and the **H⁺-pyrophosphatase** (H⁺-PPase) electrogenically pump protons into the lumen of the vacuole and the Golgi cisternae.

In plant plasma membranes, the most prominent pumps are those for H⁺ and Ca^{2+}, and the direction of pumping is outward. Therefore, another mechanism is needed to drive the active uptake of mineral nutrients such as NO_3^-, SO_4^{2-}, and PO_4^{2-}; the uptake of amino acids, peptides, and sucrose; and the export of Na⁺, which at high concentrations is toxic to plant cells. The other important way that solutes can be actively transported across a membrane against their gradient of electrochemical potential is by coupling the uphill transport of one solute to the downhill transport of another. This type of carrier-mediated cotransport is termed **secondary active transport** (Figure 6.10), and it is driven indirectly by pumps.

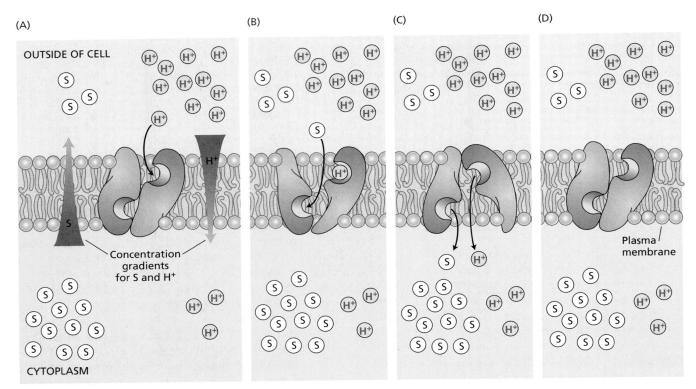

FIGURE 6.10 Hypothetical model for secondary active transport. In secondary active transport, the energetically uphill transport of one solute is driven by the energetically downhill transport of another solute. In the illustrated example, energy that was stored as proton motive force ($\Delta \tilde{\mu}_{H^+}$, symbolized by the red arrow on the right in A) is being used to take up a substrate (S) against its concentration gradient (left-hand red arrow). (A) In the initial conformation, the binding sites on the protein are exposed to the outside environment and can bind a proton. (B) This binding results in a conformational change that permits a molecule of S to be bound. (C) The binding of S causes another conformational change that exposes the binding sites and their substrates to the inside of the cell. (D) Release of a proton and a molecule of S to the cell's interior restores the original conformation of the carrier and allows a new pumping cycle to begin.

Secondary active transport uses stored energy

Protons are extruded from the cytosol by electrogenic H^+-ATPases operating in the plasma membrane and at the vacuole membrane. Consequently, a membrane potential and a pH gradient are created at the expense of ATP hydrolysis. This gradient of electrochemical potential for H^+, $\Delta \tilde{\mu}_{jH^+}$, or (when expressed in other units) the **proton motive force (PMF)**, or Δp, represents stored free energy in the form of the H^+ gradient (see **Web Topic 6.3**).

The proton motive force generated by electrogenic H^+ transport is used in secondary active transport to drive the transport of many other substances against their gradients of electrochemical potentials. Figure 6.10 shows how secondary transport may involve the binding of a substrate (S) and an ion (usually H^+) to a carrier protein, and a conformational change in that protein.

There are two types of secondary transport: symport and antiport. The example shown in Figure 6.10 is called **symport** (and the protein involved is called a *symporter*) because the two substances move in the same direction through the membrane (see also Figure 6.11A). **Antiport** (facilitated by a protein called an *antiporter*) refers to coupled transport in which the energetically downhill movement of protons drives the active (energetically uphill) transport of a solute in the opposite direction (see Figure 6.11B).

In both types of secondary transport, the ion or solute being transported simultaneously with the protons is moving against its gradient of electrochemical potential, so its transport is active. However, the energy driving this transport is provided by the proton motive force rather than directly by ATP hydrolysis.

Kinetic analyses can elucidate transport mechanisms

Thus far, we have described cellular transport in terms of its energetics. However, cellular transport can also be studied by use of enzyme kinetics, because transport involves the binding and dissociation of molecules at active sites on transport proteins (see Chapter 2 on the web site and **Web**

Figure 6.11 Two examples of secondary active transport coupled to a primary proton gradient. (A) In a symporter, the energy dissipated by a proton moving back into the cell is coupled to the uptake of one molecule of a substrate (e.g., a sugar) into the cell. (B) In an antiporter, the energy dissipated by a proton moving back into the cell is coupled to the active transport of a substrate (for example, a sodium ion) out of the cell. In both cases, the substrate under consideration is moving against its gradient of electrochemical potential. Both neutral and charged substrates can be transported by such secondary active transport processes.

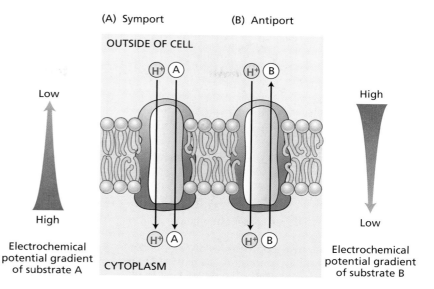

Topic 6.4). One advantage of the kinetic approach is that it gives new insights into the regulation of transport.

In kinetic experiments the effects of external ion (or other solute) concentrations on transport rates are measured. The kinetic characteristics of the transport rates can then be used to distinguish between different transporters. The maximum rate (V_{max}) of carrier-mediated transport, and often channel transport as well, cannot be exceeded, regardless of the concentration of substrate (Figure 6.12). V_{max} is approached when the substrate-binding site on the carrier is always occupied or when flux through the channel is maximal. The concentration of transporter, not the concentration of solute, becomes rate limiting. Thus V_{max} is an indicator of the number of molecules of the specific transport protein that are functioning in the membrane.

The constant K_m (which is numerically equal to the solute concentration that yields half the maximal rate of transport) tends to reflect the properties of the particular binding site (for a detailed discussion of K_m and V_{max}, see Chapter 2 on the web site). Low K_m values indicate high affinity of the transport site for the transported substance. Such values usually imply the operation of a carrier system. Higher values of K_m indicate a lower affinity of the transport site for the solute. The affinity is often so low that in practice V_{max} is never reached. In such cases, kinetics alone cannot distinguish between carriers and channels.

Often transport displays both high-affinity and low-affinity components when a plant cell or tissue is exposed to a wide range of solute concentrations. Figure 6.13 shows sucrose uptake by soybean cotyledon protoplasts as a function of the external sucrose concentration (Lin et al. 1984). Uptake increases sharply with concentration and begins to saturate at about 10 mM. At concentrations above 10 mM, uptake becomes linear and nonsaturable within the concentration range tested. Inhibition of ATP synthesis with metabolic poisons blocks the saturable component but not the linear one. The interpretation is that sucrose uptake at low concentrations is an active carrier-mediated process (sucrose–H$^+$ symport). At higher concentrations, sucrose enters the cells by diffusion down its concentration gradient and is therefore insensitive to metabolic poisons. However, additional research will be needed before we can conclude whether the nonsaturating component represents uptake by a carrier with very low affinity, or by a channel.

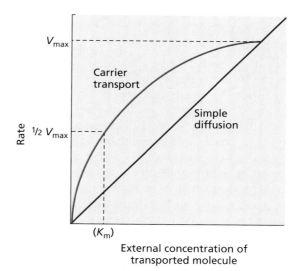

Figure 6.12 Carrier transport often shows saturation kinetics (V_{max}) (see Chapter 2 on the web site), because of saturation of a binding site. When channels are open, diffusion through them is ideally directly proportional to the concentration of the transported solute, or for an ion, to the difference in electrochemical potential across the membrane.

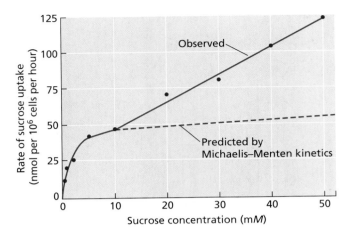

FIGURE 6.13 The transport properties of a solute can change at different solute concentrations. For example, at low concentrations (1 to 10 mM), the rate of uptake of sucrose by soybean cells shows saturation kinetics typical of carriers. A curve fitted to these data is predicted to approach a maximal rate (V_{max}) of 57 nmol per 10^6 cells per hour. Instead, at higher sucrose concentrations the uptake rate continues to increase linearly over a broad range of concentrations, suggesting the existence of other sucrose transporters, which might be carriers with very low affinity for the substrate. (After Lin et al. 1984.)

Membrane Transport Proteins

Numerous representative transport processes located on the plasma membrane and the tonoplast are illustrated in Figure 6.14. Typically, transport across a biological membrane is energized by one primary active transport system coupled to ATP hydrolysis. The transport of that ion—for example, H^+—generates an ion gradient and an electrochemical potential. Many other ions or organic substrates can then be transported by a variety of secondary active-transport proteins, which energize the transport of their respective substrates by simultaneously carrying one or two H^+ ions down their energy gradient. Thus H^+ ions circulate across the membrane, outward through the primary active transport proteins, and back into the cell through the secondary transport proteins. Most of the ionic gradients across membranes of higher plants are generated and maintained by electrochemical-potential gradients of H^+ (Tazawa et al. 1987). In turn, these H^+ gradients are generated by the electrogenic proton pumps. Evidence suggests that in plants, Na^+ is transported out of the cell by a Na^+–H^+ antiporter and that Cl^-, NO_3^-, $H_2PO_4^-$, sucrose, amino acids, and other substances enter the cell via specific proton symporters. In plants and fungi, sugars and amino acids are also taken up by symport with protons. What about K^+? At very low external concentrations, K^+ can be taken up by carriers, but at higher concentrations it can enter the cell by diffusion through specific K^+ channels.

However, even influx through channels is driven by the H^+-ATPase, in the sense that K^+ diffusion is driven by the membrane potential, which is maintained at a value more negative than the K^+ equilibrium potential by the action of the electrogenic H^+ pump. Conversely, K^+ efflux requires the membrane potential to be maintained at a value more positive than E_K, which can be achieved if efflux of Cl^- or other anions through anion channels is allowed.

We have seen in preceding sections that some transmembrane proteins operate as channels for the controlled diffusion of ions. Other membrane proteins act as carriers for other substances (mostly molecules and ions). Active transport utilizes carrier-type proteins that are energized directly by ATP hydrolysis or indirectly as symporters and antiporters. The latter systems use the energy of ion gradients (often a H^+ gradient) to drive the energetically uphill transport of another ion or molecule. In the pages that follow we will examine in more detail the molecular properties, cellular locations, and genetic manipulations of some of the transport proteins that mediate the movement of organic and inorganic nutrients, as well as water, across the plant cell membrane.

The genes for many transporters have been identified

Transporter gene identification has aided greatly in the elucidation of the molecular properties of transporter proteins. One way to identify transport genes is to screen plant complementary DNA (cDNA) libraries for genes that complement (i.e., compensate for) transport deficiencies in yeast. Many yeast transport mutants are known and have been used to identify corresponding plant genes by complementation. In the case of genes for ion channels, researchers have studied the behavior of the channel proteins by expressing the genes in oocytes of the toad *Xenopus*, which, because of their large size, are convenient for electrophysiological studies. Genes for both inwardly and outwardly rectifying K^+ channels have been cloned and studied in this way. As the number of sequenced genomes has increased, it has become increasingly common to identify putative transport genes by phylogenetic analysis, in which sequence comparison to genes encoding transporters of known function in another organism allows one to predict function in the organism of interest. Computer-aided prediction of molecular structures is also becoming a useful tool in assigning putative function to a gene product.

The emerging picture of plant transporter genes shows that a family of genes, rather than an individual gene, exists in the plant genome for most transport functions. Within a gene family, variations in transport characteristics such as K_m, in mode of regulation, and in differential tissue expression give plants a remarkable plasticity to acclimate to a broad range of environmental conditions.

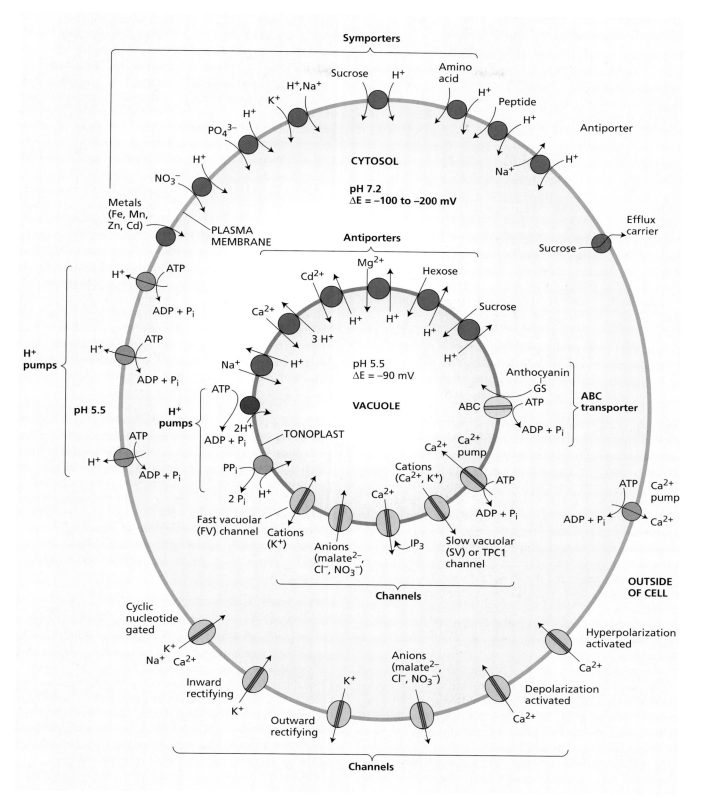

FIGURE 6.14 Overview of the various transport processes on the plasma membrane and tonoplast of plant cells.

Transporters exist for diverse nitrogen-containing compounds

Nitrogen is one of the three macronutrients, and nitrate (NO_3^-) uptake and its regulation is of great nutritional importance in plants. Transport of nitrate is an example that is also of interest because of its complexity. Kinetic analysis shows that nitrate transport, like the sucrose transport shown in Figure 6.13, has both high-affinity (low K_m) and low-affinity (high K_m) components. In contrast with sucrose, nitrate is negatively charged, and such an electric charge imposes an energy requirement for the transport of the nitrate ion at all concentrations. The energy is provided by symport with H^+. Nitrate transport is also strongly regulated according to nitrate availability: The enzymes required for nitrate transport, as well as for nitrate assimilation (see Chapter 12), are induced in the presence of nitrate in the environment, and uptake can also be repressed if nitrate accumulates in the cells.

Mutants in nitrate transport or nitrate reduction can be selected by growth in the presence of chlorate (ClO_3^-). Chlorate is a nitrate analog that is taken up and reduced in wild-type plants to the toxic product chlorite. If plants resistant to chlorate are selected, they are likely to show mutations that block nitrate transport or reduction. Several such mutations have been identified in Arabidopsis. The first transport gene identified in this way encodes an inducible nitrate–proton symporter that functions as a dual-affinity carrier that contributes to both high-affinity and low-affinity transport (Liu and Tsay 2003). As more genes for nitrate transport have been identified and characterized, it has become evident that high- and low-affinity transport each involve more than one gene product.

Peptide transporters provide another mechanism to move nitrogen across membranes. Peptide transporters are important for mobilizing nitrogen reserves during seed germination, and for distribution of nitrogenous compounds throughout the plant, via the vascular system. In the carnivorous pitcher plant (*Nepenthes alata*), high levels of expression of a peptide transporter were found in the pitcher, where the transporter presumably mediates uptake of peptides from digested insects into the internal tissue (Schulze et al. 1999).

Remarkably, the Arabidopsis genome encodes tenfold more peptide transporters than genomes of nonplant species, suggesting the importance of this group of transporters (Stacey et al. 2002). One family of transporters, which mediates di- and tripeptide uptake, also mediates the transport of nitrate and histidine and drives transport by coupling with the H^+ electrochemical gradient. A second family of peptide transporters, found only in plants and fungi, mediates uptake of tetra- and pentapeptides and is probably H^+-coupled as well. The third group of peptide transporters differs from the other two in that the transporter directly utilizes energy from ATP hydrolysis for transport; thus transport does not depend on a primary electrochemical gradient (see **Web Topic 6.5**). These transporters are members of the ATP-binding cassette (ABC) superfamily of proteins. ABC transporters are found on both the plasma membrane and in internal membranes.

The ABC superfamily is an extremely large protein family, and its members transport diverse substrates ranging from ions to macromolecules. For example, large metabolites such as flavonoids, anthocyanins, and secondary products of metabolism are sequestered in the vacuole via action of specific ABC transporters (Martinoia et al. 2002; Stacey et al. 2002).

Amino acids are another important category of nitrogen-containing compounds. Plasma membrane amino acid transporters of eukaryotes are divided into five superfamilies, three of which rely on the H^+ gradient for coupled amino acid uptake and are present in plants (Wipf et al. 2002). In general, amino acid transporters can provide high- or low-affinity transport and have overlapping substrate specificity. Many amino acid transporters show distinct tissue-specific expression patterns, suggesting specialized functions in different cell types. Amino acids comprise an important form in which nitrogen is distributed long distance in the plants, and thus it is not surprising that the expression patterns of many amino acid transporter genes include expression in vascular tissue.

Because plant hormones are frequently found conjugated to amino acids and peptides, transporters for these molecules may also be involved in the distribution of hormone conjugates throughout the plant body. The hormone auxin is derived from tryptophan, and the genes encoding auxin transporters are related to those of some other amino acid transporters. Proline is an amino acid that accumulates under salt stress. It lowers cellular water potential and allows water retention under stress conditions. Thus, amino acid and peptide transporters have important roles in addition to their function as distributors of nitrogen resources.

Cation transporters are diverse

CATION CHANNELS An estimated 56 genes in the Arabidopsis genome encode channels mediating cation uptake across the plasma membrane or intracellular membranes such as the tonoplast (Very and Sentenac 2002). Some of these channels are highly selective for specific ionic species, such as K^+, while others allow passage of a variety of cations, sometimes including the Na^+ ion, even though this ion is toxic when overaccumulated. As described in Figure 6.15, cation channels are categorized into six types based on their deduced structures and cation selectivity.

Of the six types of plant cation channels, the Shaker channels have been the most thoroughly characterized. These channels are named after a *Drosophila* K^+ channel whose mutation causes the flies to shake or tremble, hence the name. Plant Shaker channels are highly K^+ selective and can be either inwardly or outwardly rectifying, or

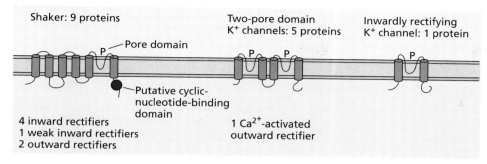

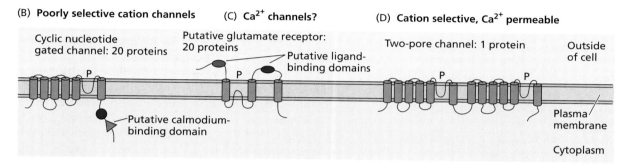

FIGURE 6.15 Six families of Arabidopsis cation channels. Some channels have been identified solely from sequence homology with channels of animals, while others have been experimentally verified. (A) K⁺ selective channels. (B) Poorly selective cation channels with activity regulated by binding of cyclic nucleotides. (C) Putative glutamate receptors. Based on measurements of cytosolic Ca^{2+} changes, these proteins are hypothesized to function as Ca^{2+}-permeable channels. (D) TPC1 is the sole two-pore channel of this type encoded in the Arabidopsis genome. TPC1 transports mono- and divalent cations, including Ca^{2+}. (After Very and Sentenac 2002.)

weakly rectifying (i.e., conducting much more current in one direction across the membrane than the other). Some members of the Shaker family mediate K⁺ uptake or efflux across the guard cell membrane, another has been identified as a major conduit for K⁺ uptake from the soil, another participates in K⁺ release to dead xylem vessels from the living stelar cells, and another has a role in K⁺ uptake in pollen, a process that promotes water influx and pollen tube elongation.

Some Shaker channels, such as those in roots, can function in high-affinity uptake, mediating K⁺ uptake at micromolar external K⁺ concentrations as long as the membrane potential is sufficiently hyperpolarized to drive this uptake (Hirsch et al. 1998). Thus, just as is the case for nitrate, it is not possible to assign high-affinity uptake roles to one set of cation transporters and low-affinity uptake roles to another, distinct set of cation transporters.

Unlike the strong voltage regulation exhibited by the majority of Shaker channels, the cyclic nucleotide–gated cation channels are only weakly voltage-dependent. Their activity is regulated primarily by the binding of cyclic nucleotides such as cGMP and probably also by the regulatory protein calmodulin. These channels can be permeable to K⁺, Ca^{2+}, or Na⁺. Their functions are largely unknown, but mutational analyses have implicated two of these cyclic nucleotide–gated cation channels in disease-resistance responses to bacterial pathogens (Clough et al. 2000; Balagué et al. 2003). Another interesting example is the glutamate receptor channels, which are homologous to a type of glutamate receptor in mammals that functions as glutamate-gated cation channels. Glutamate induces Ca^{2+} uptake into plant cells, suggesting that the glutamate receptor channels may be Ca^{2+}-permeable channels.

The two-pore domain channel (TPC) is another Ca^{2+}-permeable channel; it is also permeable to K⁺ and other cations, depending on cellular conditions. This channel is localized in the tonoplast membrane. TPC1 is expressed throughout the Arabidopsis plant, but it has been ascribed a special function in guard cells, where it is essential for stomatal closure in response to elevated external Ca^{2+} concentrations (Peiter et al. 2005). Calcium efflux from the vacuole into the cytosol is triggered in some cells, probably including guard cells, by inositol trisphosphate ($InsP_3$). $InsP_3$, which appears to act as a "second messenger" in certain signal transduction pathways, induces the opening of $InsP_3$-gated calcium channels on the tonoplast and endoplasmic reticulum. (For a more detailed description of these signal transduction pathways see Chapter 14 on the web site.)

CATION CARRIERS A variety of carriers also move cations into plant cells. There are two families of transporters that move K^+ across plant membranes: the KUP/HAK/KT family and the HKT family. The KUP family contains both high-affinity and low-affinity transporters, some of which also mediate Na^+ influx at high external Na^+ concentrations. The second family, the HKT transporters, can operate as K^+–Na^+ symporters, or as Na^+ channels under high external Na^+ concentrations. The functional roles of the HKT transporters remain mysterious, but they may play a role in redistribution of Na^+ within the plant under salt-stress conditions.

Irrigation increases soil salinity, and salinization of croplands is an increasing problem worldwide. While halophytic plants, such as those found in salt marshes, are adapted to a high Na^+ environment, such environments are deleterious to other, glycophytic, plant species, including the majority of crop species. Plants have evolved mechanisms to sequester salt in the vacuole, and to extrude it across the plasma membrane. Vacuolar Na^+ sequestration occurs by activity of Na^+–H^+ antiporters, which couple the energetically downhill movement of H^+ into the cytoplasm to drive Na^+ uptake into the vacuole. When the Arabidopsis *AtNHX1* Na^+–H^+ antiporter gene is overexpressed, it confers greatly increased salt tolerance in both Arabidopsis and crop species such as tomato (Apse et al. 1999; Zhang and Blumwald 2001).

The plasma membrane Na^+–H^+ antiporter was uncovered in a screen to identify Arabidopsis mutants that showed enhanced sensitivity to salt, hence this antiporter was named Salt Overly Sensitive or SOS1. SOS1 may function both to extrude Na^+ from the plant and to dilute root Na^+ concentrations via Na^+ extrusion to the transpiration stream, allowing eventual Na^+ sequestration in vacuoles of leaf mesophyll cells (Shi et al. 2002; Horie and Schroeder 2004).

Just as for Na^+, for Ca^{2+} there is a large free-energy gradient favoring its entry into the cytosol from both the apoplast and intracellular stores, and this entry is mediated by Ca^{2+}-permeable channels, which were described above. Calcium concentrations in the cell wall and the apoplast are usually in the millimolar range; in contrast, free cytosolic Ca^{2+} concentrations are kept in the hundreds of nanomolar (10^{-9} M) to one micromolar (10^{-6} M) range, against the large electrochemical-potential gradient for Ca^{2+} diffusion into the cell. Ca^{2+} efflux from the cytosol is achieved by Ca^{2+}-ATPases found at the plasma membrane (e.g., TPC [Sanders et al. 2002]) and in some endomembranes of plant cells such as the tonoplast and endoplasmic reticulum (see Figure 6.14) (see **Web Topic 8.10**).

Much of the Ca^{2+} inside the cell is stored in the central vacuole, where it is also sequestered via Ca^{2+}–H^+ antiporters (Hirschi et al. 1996), which use the electrochemical potential of the proton gradient to energize the accumulation of calcium into the vacuole. Ca^{2+} is not only a nutrient but also plays crucial signaling roles. Small fluctuations in cytosolic Ca^{2+} concentration drastically alter the activities of many enzymes, making calcium an important second messenger in signal transduction, and thus its cytosolic concentration is tightly regulated (Sanders et al. 2002).

Some anion transporters have been identified

Nitrate (NO_3^-), chloride (Cl^-), sulfate (SO_4^{2-}), and phosphate (PO_4^{2-}) are the major inorganic ions in plant cells, and malate^{2-} is the major organic anion (Barbier-Brygoo et al. 2000). The free energy gradient for all of these anions is in the direction of passive efflux. Several types of plant anion channels have been characterized by electrophysiological techniques, and most anion channels appear to be permeable to a variety of anions. In contrast to our detailed knowledge of K^+ transport mechanisms at the molecular level, it has proved to be more difficult to identify anion channel genes. Some research utilizing pharmacological inhibitors suggests that certain transporters related to the ABC family may be able to act as plasma membrane anion channels. In addition, one class of voltage-regulated chloride channels, first identified in the electric organ of the electric eel, has homologs encoded in the Arabidopsis genome (Barbier-Brygoo et al. 2000).

Compared to the relative lack of specificity of anion channels, anion carriers that mediate the energetically uphill transport of anions into plant cells exhibit preferences for particular anions. In addition to the transporters for nitrate uptake described above, plants have transporters for various organic anions, such as malate and citrate. As the third macronutrient, phosphate availability in the soil solution often limits plant growth. A family in Arabidopsis of about nine high-affinity plasma membrane phosphate transporters mediate phosphate uptake in symport with H^+. These transporters are expressed primarily in root tissues, and their expression is induced upon phosphate starvation (Rausch and Bucher 2002). Phosphate–H^+ symporters with lower affinity to phosphate have also been identified in plants, and have been localized to membranes of intracellular organelles such as plastids and mitochondria. Another group of phosphate transporters, the phosphate translocators, are localized in the inner plastid membrane, where they function to release phosphorylated carbon compounds derived from photosynthesis to the cytosol, in exchange for uptake of inorganic phosphate (Weber 2004) (see **Web Topic 8.10**).

Metals are transported by ZIP proteins

Several metals are essential nutrients in plants, although they are required only in trace amounts. One example is iron. Iron deficiency is the most common human nutritional disorder worldwide, so an increased understanding of how plants accumulate iron may also benefit efforts to improve the nutritional value of crops. Over 25 related

"ZIP" transporters mediate metal uptake in Arabidopsis and other species (Guerinot 2000). Metals are usually available at low concentration in the soil solution, and the ZIP transporters are high-affinity transporters. A given ZIP transporter typically can transport several different metals, such as iron, manganese, and zinc.

Several other families of high-affinity transporters mediate metal uptake into intracellular compartments; because overaccumulation of metals can be toxic to the plant, such sequestration mechanisms are important in maintaining metal concentrations at appropriate levels. Some of the metal transporters can also mediate the uptake of cadmium, which is undesirable in crop species, as cadmium is toxic to humans. However, this property may prove useful in phytoremediation, that is, the detoxifying of soils by uptake of contaminants into plants, which can then be removed and properly discarded (see Web Essay 26.2).

Aquaporins may have novel functions

Water channels, or "aquaporins," are a class of proteins that is relatively abundant in plant membranes (see Chapters 3 and 4). Many aquaporins reveal no ion currents when expressed in oocytes, indicating that they do not mediate ion transport, but when the osmolarity of the external medium is reduced, expression of these proteins results in swelling and bursting of the oocytes. The bursting results from rapid influx of water across the oocyte plasma membrane, which normally has very low water permeability. These results show that aquaporins form water channels in membranes (see Figures 3.13 and 4.5). Some aquaporin proteins also transport uncharged solutes, and it is currently debated whether aquaporins may act as conduits for carbon dioxide uptake into plant cells. It has also been hypothesized that aquaporins may function as sensors of gradients in osmotic potential and turgor pressure (Hill et al. 2004).

The existence of aquaporins was a surprise at first, because it was thought that the lipid bilayer is itself sufficiently permeable to water. Nevertheless, aquaporins are common in plant and animal membranes, and the Arabidopsis genome is predicted to encode approximately 35 aquaporins and related proteins (Luu and Maurel 2005). Aquaporin expression and activity appear to be regulated, likely by protein phosphorylation, in response to water availability (Tyerman et al. 2002). Aquaporins are also regulated by pH, calcium concentration, heteromerization, and reactive oxygen species (Luu and Maurel 2005).

Such regulation may account for the ability of plant cells to quickly alter their water permeability in response to circadian rhythm, and to stresses such as salt, chilling, drought, and flooding (anoxia). Regulation also occurs at the level of gene expression. Aquaporins are highly expressed in epidermal and endodermal cells, and in the xylem parenchyma, which may be critical points for control of water movement (Javot and Maurel 2002).

The plasma membrane H^+-ATPase has several functional domains

The outward, active transport of H^+ across the plasma membrane creates gradients of pH and electric potential that drive the transport of many other substances (ions and uncharged solutes) through the various secondary active-transport proteins. H^+-ATPase activity is also important for regulation of cytoplasmic pH and for the control of cell turgor, which drives organ movement, stomatal opening, and cell growth. Figure 6.16 illustrates how a membrane H^+-ATPase might work.

Plant and fungal plasma membrane H^+-ATPases and Ca^{2+}-ATPases are members of a class known as P-type ATPases, which are phosphorylated as part of the catalytic cycle that hydrolyzes ATP. Because of this phosphorylation step, the plasma membrane ATPases are strongly inhibited by orthovanadate (HVO_4^{2-}), a phosphate (HPO_4^{2-}) analog that competes with phosphate from ATP for the aspartic acid phosphorylation site on the enzyme.

Plasma membrane H^+-ATPases are encoded by a family of about ten genes (Sondergaard et al. 2004). The roles of each H^+-ATPase isoform are starting to be understood, based on information from gene expression patterns and functional analysis of Arabidopsis plants harboring null mutations in individual H^+-ATPase genes. Some H^+-ATPases exhibit cell-specific patterns of expression. For example, one H^+-ATPase gene is expressed specifically in the phloem, suggestive of a role in establishing the electrochemical gradient that drives sucrose uptake. When this H^+-ATPase gene is disrupted such that the corresponding H^+-ATPase protein is no longer produced, the plant exhibits stunted growth (Gottwald et al. 2000). Several H^+-ATPases are expressed in guard cells, where they energize the plasma membrane to drive solute uptake during stomatal opening (see Chapter 4).

In general, H^+-ATPase expression is high in cells with key functions in nutrient movement, including root endodermal cells and cells involved in nutrient uptake from the apoplast that surrounds the developing seed (Sondergaard et al. 2004). In cells in which multiple H^+-ATPases are coexpressed, they may be differentially regulated or may function redundantly, perhaps providing a "fail-safe" mechanism to this all-important transport function (Arango et al. 2003).

Figure 6.17 shows a model of the functional domains of a yeast plasma membrane H^+-ATPase, similar to those of plants. The protein has ten membrane-spanning domains that cause it to loop back and forth across the membrane. Some of the membrane-spanning domains make up the pathway through which protons are pumped. The catalytic domain, including the aspartic acid residue that becomes phosphorylated during the catalytic cycle, is on the cytosolic face of the membrane.

Like other enzymes, the plasma membrane H^+-ATPase is regulated by the concentration of substrate (ATP), pH,

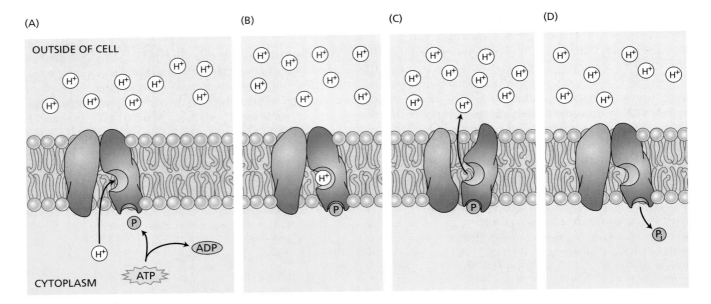

FIGURE 6.16 Hypothetical steps in the transport of a cation (such as H$^+$) against its chemical gradient by an electrogenic pump. The protein, embedded in the membrane, binds the cation on the inside of the cell (A) and is phosphorylated by ATP (B). This phosphorylation leads to a conformational change that exposes the cation to the outside of the cell and makes it possible for the cation to diffuse away (C). Release of the phosphate ion (P) from the protein into the cytosol (D) restores the initial configuration of the membrane protein and allows a new pumping cycle to begin.

temperature, and other factors. In addition, H$^+$-ATPase molecules can be reversibly activated or deactivated by specific signals, such as light, hormones, pathogen attack, and the like. This type of regulation is mediated by a specialized autoinhibitory domain at the C-terminal end of the polypeptide chain, which acts to regulate the activity of the proton pump (see Figure 6.17). If the autoinhibitory domain is removed by a protease, the enzyme becomes irreversibly activated (Palmgren 2001).

The autoinhibitory effect of the C-terminal domain can also be regulated by protein kinases and phosphatases that add or remove phosphate groups to serine or threonine residues on this domain. Phosphorylation recruits ubiquitous enzyme-modulating proteins called 14-3-3 proteins (Roberts 2003), which bind to the phosphorylated region and are thought to thus displace the autoinhibitory domain, leading to H$^+$-ATPase activation. The fungal toxin fusicoccin, which is a strong activator of the H$^+$-ATPase, activates this pump by increasing 14-3-3 binding affinity even in the absence of phosphorylation (Sondergaard et al. 2004). The activity of fusicoccin on the guard cell H$^+$-ATPases is so strong that it can lead to irreversible stomatal opening, wilting, and even plant death.

The tonoplast H$^+$-ATPase drives solute accumulation into vacuoles

Because plant cells increase their size primarily by taking up water into large, central vacuoles, the osmotic pressure of the vacuole must be maintained sufficiently high for water to enter from the cytoplasm. The tonoplast regulates the traffic of ions and metabolites between the cytosol and the vacuole, just as the plasma membrane regulates uptake into the cell. Tonoplast transport became a vigorous area of research following the development of methods for the isolation of intact vacuoles and tonoplast vesicles (see **Web Topic 6.6**). These studies led to the discovery of a new type of proton-pumping ATPase, which transports protons into the vacuole (see Figure 6.14).

The vacuolar H$^+$-ATPase (also called **V-ATPase**) differs both structurally and functionally from the plasma membrane H$^+$-ATPase (Kluge et al. 2003). The vacuolar ATPase is more closely related to the F-ATPases of mitochondria and chloroplasts (see Chapter 11). Because the hydrolysis of ATP by the vacuolar ATPase does not involve the formation of a phosphorylated intermediate, vacuolar ATPases are insensitive to vanadate, the inhibitor of plasma membrane ATPases discussed earlier. Vacuolar ATPases are specifically inhibited by the antibiotic bafilomycin, as well as by high concentrations of nitrate, neither of which inhibit plasma membrane ATPases. Use of these selective inhibitors makes it possible to identify different types of ATPases, and to assay their activity.

Vacuolar ATPases belong to a general class of ATPases that are present on the endomembrane systems of all eukaryotes. They are large enzyme complexes, about 750 kDa, composed of at least twelve different subunits (Kluge et al. 2003). These subunits are organized into a peripheral

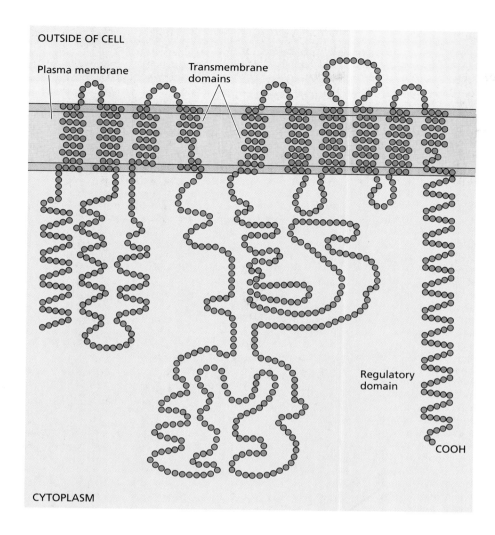

Figure 6.17 Two-dimensional representation of the plasma membrane H⁺-ATPase from yeast. The H⁺-ATPase has ten transmembrane segments. The regulatory domain is an autoinhibitory domain. Post-translational modifications that lead to displacement of the auto-inhibitory domain result in H⁺-ATPase activation. (After Palmgren 2001.)

catalytic complex, V_1, and an integral membrane channel complex, V_0 (Figure 6.18). Because of their similarities to F-ATPases, vacuolar ATPases are assumed to operate like tiny rotary motors (see Chapter 11).

Vacuolar ATPases are electrogenic proton pumps that transport protons from the cytoplasm to the vacuole and generate a proton motive force across the tonoplast. Electrogenic proton pumping accounts for the fact that the vacuole is typically 20 to 30 mV more positive than the cytoplasm, although it is still negative relative to the external medium. To maintain bulk electrical neutrality, anions such as Cl⁻ or malate²⁻ are transported from the cytoplasm into the vacuole through channels in the membrane (Barkla and Pantoja 1996). Without the simultaneous movement of anions along with the pumped protons, the charge buildup across the tonoplast would make the pumping of additional protons energetically impossible.

The conservation of bulk electrical neutrality by anion transport makes it possible for the vacuolar H⁺-ATPase to generate a large concentration (pH) gradient of protons across the tonoplast. This gradient accounts for the fact that the pH of the vacuolar sap is typically about 5.5, while the cytoplasmic pH is 7.0 to 7.5. Whereas the electrical component of the proton motive force drives the uptake of anions into the vacuole, the electrochemical-potential gradient for H⁺ ($\Delta\tilde{\mu}_{H^+}$) is harnessed to drive the uptake of cations and sugars into the vacuole via secondary transport (antiporter) systems (see Figure 6.14).

Although the pH of most plant vacuoles is mildly acidic (about 5.5), the pH of the vacuoles of some species is much lower—a phenomenon termed *hyperacidification*. Vacuolar hyperacidification is the cause of the sour taste of certain fruits (lemons) and vegetables (rhubarb). Some extreme examples are listed in Table 6.2. Biochemical studies with lemon fruits have suggested that the low pH of the lemon fruit vacuoles (specifically, those of the juice sac cells) is due to a combination of factors:

- The low permeability of the vacuolar membrane to protons permits a steeper pH gradient to build up.

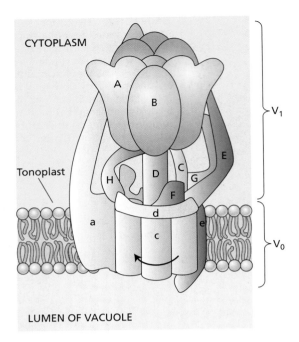

Figure 6.18 Model of the V-ATPase rotary motor. Many polypeptide subunits come together to make this complex enzyme. The V_1 catalytic complex is easily dissociated from the membrane, and contains the nucleotide-binding and catalytic sites. Components of V_1 are designated by uppercase letters. The intrinsic membrane complex mediating H^+ transport is designated V_0, and its subunits are given lowercase letters. It is proposed that ATPase reactions catalyzed by each of the A subunits, acting in sequence, drive the rotation of the shaft D and the six c subunits. The rotation of the c subunits relative to subunit a is thought to drive the transport of H^+ across the membrane. (After Kluge et al. 2003.)

TABLE 6.2
The vacuolar pH of some hyperacidifying plant species

Tissue	Species	pH[a]
Fruits		
	Lime (*Citrus aurantifolia*)	1.7
	Lemon (*Citrus limonia*)	2.5
	Cherry (*Prunus cerasus*)	2.5
	Grapefruit (*Citrus paradisi*)	3.0
Leaves		
	Rosette oxalis (*Oxalis deppei*)	1.3
	Wax begonia (*Begonia semperflorens*)	1.5
	Begonia 'Lucerna'	0.9 – 1.4
	Oxalis sp.	1.9 – 2.6
	Sorrel (*Rumex* sp.)	2.6
	Prickly pear (*Opuntia phaeacantha*)[b]	1.4 (6:45 A.M.) / 5.5 (4:00 P.M.)

Source: Data from Small 1946.

[a] The values represent the pH of the juice or expressed sap of each tissue, usually a good indicator of vacuolar pH.

[b] The vacuolar pH of the cactus *Opuntia phaeacantha* varies with the time of day. As will be discussed in Chapter 8, many desert succulents have a specialized type of photosynthesis, called crassulacean acid metabolism (CAM), that causes the pH of the vacuoles to decrease during the night.

- A specialized vacuolar ATPase is able to pump protons more efficiently (with less wasted energy) than normal vacuolar ATPases can (Müller et al. 1997).

- The accumulation of organic acids such as citric, malic, and oxalic acids helps maintain the low pH of the vacuole by acting as buffers.

H^+-pyrophosphatases also pump protons at the tonoplast

Another type of proton pump, an H^+-pyrophosphatase (H^+-PPase) (Rea et al. 1998), works in parallel with the vacuolar ATPase to create a proton gradient across the tonoplast (see Figure 6.14). This enzyme consists of a single polypeptide that has a molecular mass of 80 kDa. The H^+-PPase harnesses its energy from the hydrolysis of inorganic pyrophosphate (PP_i). Recent research suggests that this H^+-pyrophosphatase may also reside in other cellular membranes, and that it contributes to regulation of auxin transport (Li et al. 2005).

The free energy released by PP_i hydrolysis is less than that from ATP hydrolysis. However, the H^+-pyrophosphatase transports only one H^+ ion per PP_i molecule hydrolyzed, whereas the vacuolar ATPase appears to transport two H^+ ions per ATP hydrolyzed. Thus the energy available per H^+ ion transported appears to be the same, and the two enzymes seem to be able to generate comparable H^+ gradients. Interestingly, the plant H^+-PPase is not found in animals or yeast, although a similar enzyme is present in some bacteria and protists.

Ion Transport in Roots

Mineral nutrients absorbed by the root are carried to the shoot by the transpiration stream moving through the xylem (see Chapter 4). Both the initial uptake of nutrients and water and the subsequent movement of these substances from the root surface across the cortex and into the xylem are highly specific, well-regulated processes.

Ion transport across the root obeys the same biophysical laws that govern cellular transport. However, as we have seen in the case of water movement (see Chapter 4), the anatomy of roots imposes some special constraints on the pathway of ion movement. In this section we will discuss the pathways and mechanisms involved in the radial movement of ions from the root surface to the tracheary elements of the xylem.

Solutes move through both apoplast and symplast

Thus far, our discussion of cellular ion transport has not included the cell wall. In terms of the transport of small molecules, the cell wall is an open lattice of polysaccharides through which mineral nutrients diffuse readily. Because all plant cells are separated by cell walls, ions can diffuse across a tissue (or be carried passively by water flow) entirely through the cell wall space without ever entering a living cell. This continuum of cell walls is called the **extracellular space**, or **apoplast** (see Figure 4.4).

We can determine the apoplastic volume of a slice of plant tissue by comparing the uptake of ^{3}H-labeled water and ^{14}C-labeled mannitol. Mannitol is a nonpermeating sugar alcohol that diffuses within the extracellular space but cannot enter the cells. Water, on the other hand, penetrates both the cells and the cell walls. Measurements of this type usually show that 5 to 20% of the plant tissue volume is occupied by cell walls.

Just as the cell walls form a continuous phase, so do the cytoplasms of neighboring cells, collectively referred to as the **symplast**. Plant cells are interconnected by cytoplasmic bridges called plasmodesmata (see Chapter 1), cylindrical pores 20 to 60 nm in diameter (Figure 6.19). Each plasmodesma is lined with a plasma membrane and contains a narrow tubule, the desmotubule, that is a continuation of the endoplasmic reticulum.

In tissues where significant amounts of intercellular transport occur, neighboring cells contain large numbers of plasmodesmata, up to 15 per square micrometer of cell surface. Specialized secretory cells, such as floral nectaries and leaf salt glands, have high densities of plasmodesmata; so do the cells near root tips, where most nutrient absorption occurs.

By injecting dyes or by making electrical-resistance measurements on cells containing large numbers of plasmodesmata, investigators have shown that ions, water, and small solutes can move from cell to cell through these pores. Because each plasmodesma is partly occluded by the desmotubule and associated proteins (see Chapter 1), the movement of large molecules such as proteins through the plasmodesmata requires special mechanisms (Lucas and Lee 2004). Ions, on the other hand, appear to move symplastically through the entire plant by simple diffusion through plasmodesmata (see Chapter 4).

Ions cross both symplast and apoplast

Ion absorption by the roots (see Chapter 5) is more pronounced in the root hair zone than in the meristem and elongation zones. Cells in the root hair zone have completed their elongation but have not yet begun secondary growth. The root hairs are simply extensions of specific epidermal cells that greatly increase the surface area available for ion absorption.

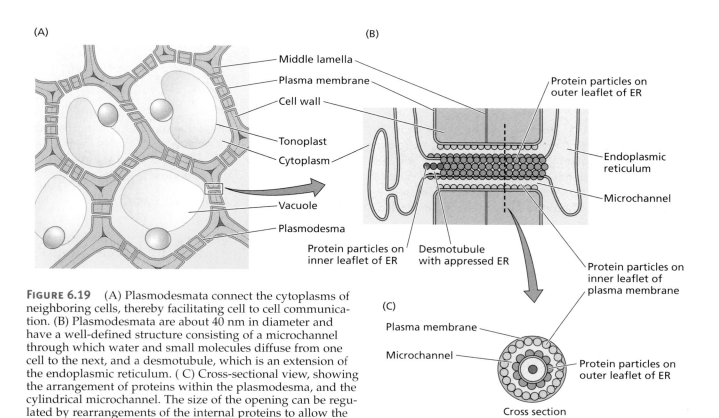

Figure 6.19 (A) Plasmodesmata connect the cytoplasms of neighboring cells, thereby facilitating cell to cell communication. (B) Plasmodesmata are about 40 nm in diameter and have a well-defined structure consisting of a microchannel through which water and small molecules diffuse from one cell to the next, and a desmotubule, which is an extension of the endoplasmic reticulum. (C) Cross-sectional view, showing the arrangement of proteins within the plasmodesma, and the cylindrical microchannel. The size of the opening can be regulated by rearrangements of the internal proteins to allow the passage of larger molecules. (C after Lucas and Lee 2004).

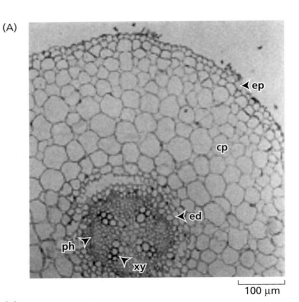
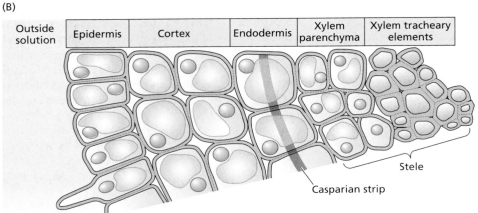

FIGURE 6.20 Tissue organization in roots. (A) Cross section through a soybean root showing the epidermis (ep); cortical parenchyma (cp); endodermis (ed); xylem (xy); and phloem (ph). (B) Schematic diagram of a root cross section, illustrating the cell layers through which solutes pass from the soil solution to the xylem tracheary elements. (A from Javot and Maurel 2002; B after Dunlop and Bowling 1971.)

An ion that enters a root may immediately enter the symplast by crossing the plasma membrane of an epidermal cell, or it may enter the apoplast and diffuse between the epidermal cells through the cell walls. From the apoplast of the cortex, an ion (or other solute) may either cross the plasma membrane of a cortical cell, thus entering the symplast, or diffuse radially all the way to the endodermis via the apoplast. The apoplast forms a continuous phase from the root surface through the cortex. However, in all cases, ions must enter the symplast before they can enter the stele, because of the presence of the Casparian strip. As discussed in Chapters 4 and 5, the Casparian strip is a suberized layer that forms rings around cells of the specialized endodermis (Figure 6.20) and effectively blocks the entry of water and mineral ions into the stele via the apoplast.

Once an ion has entered the stele through the symplastic connections across the endodermis, it continues to diffuse from cell to cell into the xylem. Finally, the ion reenters the apoplast as it diffuses into a xylem tracheid or vessel element. Again, the Casparian strip prevents the ion from diffusing back out of the root through the apoplast. The presence of the Casparian strip allows the plant to maintain a higher ionic concentration in the xylem than exists in the soil water surrounding the roots.

Xylem parenchyma cells participate in xylem loading

Once ions have been taken up into the symplast of the root at the epidermis or cortex, they must be loaded into the tracheids or vessel elements of the stele to be translocated to the shoot. The stele consists of dead tracheary elements and the living xylem parenchyma. Because the xylem tracheary elements are dead cells, they lack cytoplasmic continuity with surrounding xylem parenchyma. To enter the tracheary elements, the ions must exit the symplast by crossing a plasma membrane a second time.

The process whereby ions exit the symplast and enter the conducting cells of the xylem is called **xylem loading**. An early hypothesis, that ions leak out of the living cells of the stele in an unregulated fashion, has been largely discounted. Recent studies clearly indicate that xylem loading is a highly regulated process. Xylem parenchyma cells, like other living plant cells, maintain plasma membrane H^+-ATPase activity and a negative membrane potential. Transporters that specifically function in the unloading of solutes to the tracheary elements have been identified by electrophysiological and genetic approaches. The plasma membranes of xylem parenchyma cells contain proton pumps, aquaporins, and a variety of ion channels and carriers specialized for influx or efflux (Maathuis et al. 1997; De Boer and Volkov 2003).

In Arabidopsis xylem parenchyma, the stelar outwardly rectifying K^+ channel (SKOR) is expressed in cells of the pericycle and xylem parenchyma, where it functions as an efflux channel, transporting K^+ from the living cells out to the tracheary elements (Gaymard et al. 1998). In mutant Arabidopsis plants lacking the SKOR channel protein, or in plants in which SKOR has been pharmacologically inactivated, K^+ transport from the root to the shoot is severely reduced, confirming the function of this channel protein.

Several types of anion-selective channels have also been identified that participate in unloading of Cl^- and NO_3^- from the xylem parenchyma. Drought, abscisic acid (ABA) treatment, or elevation of cytosolic Ca^{2+} concentrations (which often occurs as a response to ABA), all reduce the activity of SKOR and anion channels, a response that could help maintain cellular hydration in the root under desiccating conditions.

Other, less-selective ion channels found in the plasma membrane of xylem parenchyma cells are permeable to K^+, Na^+, and anions. In addition, the SOS1 Na^+–H^+ antiporter appears to load Na^+ into the xylem, possibly serving to ameliorate Na^+ levels in the root symplasm. Other transport molecules have also been identified that mediate loading of boron, Mg^{2+}, and PO_4^{2-}. Thus, the flux of ions from the xylem parenchyma cells into the xylem tracheary elements is under tight metabolic control through regulation of plasma membrane H^+-ATPases, ion efflux channels, and carriers.

Summary

The movement of molecules and ions from one location to another is known as transport. Plants exchange solutes and water with their environment and among their tissues and organs. Both local and long-distance transport processes in plants are controlled largely by cellular membranes.

Forces that drive biological transport, which include concentration gradients, electric-potential gradients, and hydrostatic pressures, are integrated by an expression called the electrochemical potential. Transport of solutes down their free energy gradient (e.g., by diffusion) is known as passive transport. Movement of solutes against their free energy gradient is known as active transport and requires energy input.

The extent to which a membrane permits or restricts the movement of a substance is called membrane permeability. The permeability depends on the chemical properties of the particular solute and on the lipid composition of the membrane, as well as on the membrane proteins that facilitate the transport of specific substances.

When cations and anions move passively across a membrane at different rates, the electric potential that develops is called the diffusion potential. For each permeant ion, the distribution of that particular ionic species across a membrane that would occur at equilibrium is described by the Nernst equation. The Nernst equation shows that at equilibrium the difference in concentration of an ion between two compartments is balanced by the voltage difference between the compartments. The electrical effects of different ions diffusing simultaneously across a cell membrane are summed by the Goldman equation. Electrogenic pumps, which carry out active transport and carry a net charge, change the membrane potential from the value created by diffusion. A voltage difference, or membrane potential, is seen in all living cells because of the asymmetric ion distributions between the inside and outside of the cells. Membranes contain specialized proteins—channels, carriers, and pumps—that facilitate solute transport. Transport specificity is determined largely by the properties of channels and carriers. Channels are transport proteins that span the membrane, forming pores through which solutes diffuse down their gradient of electrochemical potentials. Carriers bind a solute on one side of the membrane and release it on the other side. A family of H^+-pumping ATPases provides the primary driving force for transport across the plasma membrane of plant cells. Two other kinds of electrogenic H^+ pumps serve this purpose at the tonoplast. Plant cells also have Ca^{2+}-pumping ATPases that participate in the regulation of intracellular calcium concentrations, as well as ATP-binding cassette transporters that use the energy of ATP to transport a great diversity of molecules. The gradient of electrochemical potential generated by H^+ pumping is used to drive the transport of other substances in a process called secondary transport.

Genome sequencing and genetic studies have revealed many genes, and their corresponding transport proteins, that account for the versatility of plant transport. Patch-clamp electrophysiology provides unique information on ion channels, and it enables measurement of the selectivity and gating of individual channel proteins.

Solutes move between cells either through the extracellular spaces (the apoplast) or from cytoplasm to cytoplasm (via the symplast). The cytoplasms in neighboring cells are connected by plasmodesmata, which facilitate symplastic transport. When an ion enters the root, it may be taken up into the cytoplasm of an epidermal cell, or it may diffuse through the apoplast into the root cortex and enter the symplast through a cortical or endodermal cell. From the symplast, the ion is loaded into the xylem and moves to the shoot in the transpiration stream.

Web Material

Web Topics

6.1 Relating the Membrane Potential to the Distribution of Several Ions across the Membrane: The Goldman Equation
The Goldman equation is used to calculate the membrane permeability of more than one ion.

6.2 Patch Clamp Studies in Plant Cells
Patch clamping is applied to plant cells for electrophysiological studies.

6.3 Chemiosmosis in Action
The chemiosmotic theory explains how electrical and concentration gradients are used to perform cellular work.

6.4 Kinetic Analysis of Multiple Transporter Systems
Application of principles of enzyme kinetics to transport systems provides an effective way to characterize different carriers.

6.5 ABC Transporters in Plants
ATP-binding cassette (ABC) transporters are a large family of active transport proteins energized directly by ATP.

6.6 Transport Studies with Isolated Vacuoles and Membrane Vesicles
Certain experimental techniques enable the isolation of tonoplasts and plasma membranes for study.

Web Essay

6.1 Potassium Channels
Several plant K$^+$ channels have been characterized.

Chapter References

Apse, M. P., Aharon, G. S., Snedden, W. A., and Blumwald, E. (1999) Salt tolerance conferred by overexpression of a vacuolar Na$^+$/H$^+$ antiport in Arabidopsis. *Science* 285: 1256–1258.

Arango, M., Gevaudant, F., Oufattole, M., and Boutry, M. (2003) The plasma membrane proton pump ATPase: The significance of gene subfamilies. *Planta* 216: 355–365.

Balagué, C., Lin, B., Alcon, C., Flottes, G., Malmstrom, S., Kohler, C., Neuhaus, G., Pelletier, G., Gaymard, F., and Roby, D. (2003) HLM1, an essential signaling component in the hypersensitive response, is a member of the cyclic nucleotide-gated channel ion channel family. *Plant Cell* 15: 365–379.

Barbier-Brygoo, H., Vinauger, M., Colcombet, J., Ephritikhine, G., Frachisse, J., and Maurel, C. (2000) Anion channels in higher plants: Functional characterization, molecular structure and physiological role. *Biochim. Biophys. Acta* 1465: 199–218.

Barkla, B. J., and Pantoja, O. (1996) Physiology of ion transport across the tonoplast of higher plants. *Annu. Rev. Plant Physiol. Plant Mol. Biol.* 47: 159–184.

Buchanan, B. B., Gruissem, W., and Jones, R. L., eds. (2000) *Biochemistry and Molecular Biology of Plants*. American Society of Plant Physiologists, Rockville, MD.

Clough, S. J., Fengler, K. A., Yu, I. C., Lippok, B., Smith, R. K., Jr., and Bent, A. F. (2000) The Arabidopsis *dnd1* "defense, no death" gene encodes a mutated cyclic nucleotide-gated ion channel. *Proc. Natl. Acad. Sci. USA* 97: 9323–9328.

De Boer, A. H., and Volkov, V. (2003) Logistics of water and salt transport through the plant: Structure and functioning of the xylem. *Plant Cell Environ.* 26: 87–101.

Dunlop, J., and Bowling, D. J. F. (1971) The movement of ions to the xylem exudate of maize roots. *J. Exp. Bot.* 22: 453–464.

Gaymard, F., Pilot, G., Lacombe, B., Bouchez, D., Bruneau, D., Boucherez, J., Michaux-Ferriere, N., Thibaud, J. B., and Sentenac, H. (1998) Identification and disruption of a plant shaker-like outward channel involved in K$^+$ release into the xylem sap. *Cell* 94: 647–655.

Gottwald, J. R., Krysan, P. J., Young, J. C., Evert, R. F., and Sussman, M. R. (2000) Genetic evidence for the in planta role of phloem-specific plasma membrane sucrose transporters. *Proc. Natl. Acad. Sci. USA* 97: 13979–13984.

Guerinot, M. L. (2000) The ZIP family of metal transporters. *Biochim. Biophys. Acta* 1465: 190–198.

Higinbotham, N., Etherton, B., and Foster, R. J. (1967) Mineral ion contents and cell transmembrane electropotentials of pea and oat seedling tissue. *Plant Physiol.* 42: 37–46.

Higinbotham, N., Graves, J. S., and Davis, R. F. (1970) Evidence for an electrogenic ion transport pump in cells of higher plants. *J. Membr. Biol.* 3: 210–222.

Hill, A. E., Shachar-Hill, B., and Shachar-Hill, Y. (2004) What are aquaporins for? *J. Membr. Biol.* 197: 1–32.

Hirsch, R. E., Lewis, B. D., Spalding, E. P., and Sussman, M. R. (1998) A role for the AKT1 potassium channel in plant nutrition. *Science* 280: 918–921.

Hirschi, K. D., Zhen, R.-G., Rea, P. A., and Fink, G. R. (1996) CAX1, an H$^+$/Ca^{2+} antiporter from *Arabidopsis*. *Proc. Natl. Acad. Sci. USA* 93: 8782–8786.

Horie, T., and Schroeder, J. I. (2004) Sodium transporters in plants. Diverse genes and physiological functions. *Plant Physiol.* 136: 2457–2462.

Javot, H., and Maurel, C. (2002) The role of aquaporins in root water uptake. *Ann. Bot. (Lond).* 90: 301–313.

Kluge, C., Lahr, J., Hanitzsch, M., Bolte, S., Golldack, D., and Dietz, K. J. (2003) New insight into the structure and regulation of the plant vacuolar H$^+$-ATPase. *J. Bioenerg. Biomembr.* 35: 377–388.

Leng, Q., Mercier, R. W., Hua, B. G., Fromm, H., and Berkowitz, G. A. (2002) Electrophysical analysis of cloned cyclic nucleotide-gated ion channels. *Plant Physiol.* 128: 400–410.

Li, J., Yang, H., Peer, W. A., Richter, G., Blakeslee, J., Bandyopadhyay, A., Titapiwantakun, B., Undurraga, S., Khodakovskaya, M., and Richards, E. L., et al. (2005) Arabidopsis H$^+$-PPase AVP1 regulates auxin-mediated organ development. *Science* 310: 121–125.

Lin, W., Schmitt, M. R., Hitz, W. D., and Giaquinta, R. T. (1984) Sugar transport into protoplasts isolated from developing soybean cotyledons. *Plant Physiol.* 75: 936–940.

Liu, K. H., and Tsay, Y. F. (2003) Switching between the two action modes of the dual-affinity nitrate transporter CHL1 by phosphorylation. *EMBO J.* 22: 1005–1013.

Lucas, W. J., and Lee, Y. J. (2004) Plasmodesmata as a supracellular control network in plants. *Nat. Rev. Mol. Cell Biol.* 5: 712–726.

Luu, D.-T., and Maurel, C. (2005) Aquaporins in a challenging environment: Molecular gears for adjusting plant water status. *Plant Cell Environ.* 28: 85–96.

Maathuis, F. J. M., Ichida, A. M., Sanders, D., and Schroeder, J. I. (1997) Roles of higher plant K$^+$ channels. *Plant Physiol.* 114: 1141–1149.

Martinoia, E., Klein, M., Geisler, M., Bovet, l., Forestier, C., Kolukisaoglu, U., Muller-Rober, B., and Schulz, B. (2002) Multifunctionality of plant ABC transporters—More than just detoxifiers. *Planta* 214: 345–355

Müller, M., Irkens-Kiesecker, U., Kramer, D., and Taiz, L. (1997) Purification and reconstitution of the vacuolar H$^+$-ATPases from lemon fruits and epicotyls. *J. Biol. Chem.* 272: 12762–12770.

Nobel, P. (1991) *Physicochemical and Environmental Plant Physiology*. Academic Press, San Diego, CA.

Palmgren, M. G. (2001) Plant plasma membrane H$^+$-ATPases: Powerhouses for nutrient uptake. *Annu. Rev. Plant Physiol. Plant Mol. Biol.* 52: 817–845.

Peiter, E., Maathuis, F. J., Mills, L. N., Knight, H., Pelloux, J., Hetherington, A. M., and Sanders, D. (2005) The vacuolar Ca^{2+}-activated channel TPC1 regulates germination and stomatal movement. *Nature* 434:404–408.

Rausch, C., and Bucher, M. (2002) Molecular mechanisms of phosphate transport in plants. *Planta.* 216:23-37.

Rea, P. A., Li, Z. S., Lu, Y. P., Drozdowicz, Y. M., and Martinoia, E. (1998) From vacuolar Gs-X pumps to multispecific ABC transporters. *Annu. Rev. Plant Physiol. Plant Mol. Biol.* 49: 727–760.

Roberts, M .R. (2003) 14-3-3 proteins find new partners in plant cell signalling. *Trends Plant Sci.* 8: 218–223.

Sanders, D., Pelloux, J., Brownlee, C., and Harper, J. F. (2002) Calcium at the crossroads of signaling. *Plant Cell.* 14 Suppl: S401–417.

Schulze, W., Frommer, W. B., and Ward, J. M. (1999) Transporters for ammonium, amino acids and peptides are expressed in pitchers of the carnivorous plant Nepenthes. *Plant J.* 17: 637–646.

Schwacke, R., Schneider, A., van der Graaff, E., Fischer, K., Catoni, E., Desimone, M., Frommer, W. B., Flugge, U. I., and Kunze R. (2003) ARAMEMNON, a novel database for Arabidopsis integral membrane proteins. *Plant Physiol.* 131: 16–26.

Shi, H., Quintero, F. J., Pardo, J. M., and Zhu, J. K. (2002) The putative plasma membrane Na$^+$/H$^+$ antiporter SOS1 controls long-distance Na$^+$ transport in plants. *Plant Cell.* 14: 465–477.

Small, J. (1946) *pH and Plants, an Introduction to Beginners*. D. Van Nostrand, New York.

Sondergaard, T. E., Schulz, A., and Palmgren, M. G. (2004) Energization of transport processes in plants: Roles of the plasma membrane H$^+$-ATPase. *Plant Physiol.* 136: 475–2482.

Spanswick, R. M. (1981) Electrogenic ion pumps. *Annu. Rev. Plant Physiol.* 32: 267–289.

Stacey, G., Koh, S., Granger, C., and Becker, J. M. (2002) Peptide transport in plants. *Trends Plant Sci.* 7: 257–263.

Tazawa, M., Shimmen, T., and Mimura, T. (1987) Membrane control in the *Characeae*. *Annu. Rev. Plant Physiol.* 38: 95–117.

Tyerman, S. D., Niemietz, C. M., and Bramley, H. (2002) Plant aquaporins: Multifunctional water and solute channels with expanding roles. *Plant Cell Envir.* 25: 173–194.

Very, A. A., and Sentenac, H. (2002) Cation channels in the Arabidopsis plasma membrane. *Trends Plant Sci.* 7: 168–175.

Weber, A. P., Schneidereit, J., and Voll, L. M. (2004) Using mutants to probe the in vivo function of plastid envelope membrane metabolite transporters. *J. Exp. Bot.* 55:1231-1244.

Wipf, D., Ludewig, U., Tegeder, M., Rentsch, D., Koch, W., and Frommer, W. B. (2002). Conservation of amino acid transporters in fungi, plants and animals. *Trends Biochem. Sci.* 27: 139–147.

Zhang, H. X., and Blumwald, E. (2001) Transgenic salt-tolerant tomato plants accumulate salt in foliage but not in fruit. *Nat. Biotechnol.* 19: 765–768.

Unit II

Biochemistry and Metabolism

Unit II BIOCHEMISTRY AND METABOLISM

7 Photosynthesis: The Light Reactions

8 Photosynthesis: Carbon Reactions

9 Photosynthesis: Physiological and Ecological Considerations

10 Translocation in the Phloem

11 Respiration and Lipid Metabolism

12 Assimilation of Mineral Nutrients

13 Secondary Metabolites and Plant Defense

Cross section through a cactus stem. Photograph by David McIntyre.

Chapter 7 | Photosynthesis: The Light Reactions

LIFE ON EARTH ULTIMATELY depends on energy derived from the sun. Photosynthesis is the only process of biological importance that can harvest this energy. In addition, a large fraction of the planet's energy resources results from photosynthetic activity in either recent or ancient times (fossil fuels). This chapter introduces the basic physical principles that underlie photosynthetic energy storage and the current understanding of the structure and function of the photosynthetic apparatus (Blankenship 2002).

The term *photosynthesis* means literally "synthesis using light." As we will see in this chapter, photosynthetic organisms use solar energy to synthesize carbon compounds that cannot be formed without the input of energy. More specifically, light energy drives the synthesis of carbohydrates and generation of oxygen from carbon dioxide and water:

$$6\ CO_2 + 6\ H_2O \rightarrow C_6H_{12}O_6 + 6\ O_2$$
Carbon dioxide — Water — Carbohydrate — Oxygen

Energy stored in these molecules can be used later to power cellular processes in the plant and can serve as the energy source for all forms of life.

This chapter deals with the role of light in photosynthesis, the structure of the photosynthetic apparatus, and the processes that begin with the excitation of chlorophyll by light and culminate in the synthesis of ATP and NADPH.

Photosynthesis in Higher Plants

The most active photosynthetic tissue in higher plants is the mesophyll of leaves. Mesophyll cells have many chloroplasts, which contain the specialized light-absorbing green pigments, the **chlorophylls**. In photosynthesis, the plant uses solar energy to oxidize water, thereby releasing oxygen, and to reduce carbon dioxide, thereby forming large carbon compounds, primarily sugars. The complex series of reactions that culminate in the reduction of CO_2 include the thylakoid reactions and the carbon fixation reactions.

The **thylakoid reactions** of photosynthesis take place in the specialized internal membranes of the chloroplast called thylakoids (see Chapter 1). The end products of these thylakoid reactions are the high-energy compounds ATP and NADPH, which are used for the synthesis of sugars in the **carbon fixation reactions**. These synthetic processes take place in the stroma of the chloroplasts, the aqueous region that surrounds the thylakoids. The thylakoid reactions, also often called the light reactions of photosynthesis, are the subject of this chapter; the carbon fixation reactions are discussed in Chapter 8.

In the chloroplast, light energy is converted into chemical energy by two different functional units called *photosystems*. The absorbed light energy is used to power the transfer of electrons through a series of compounds that act as electron donors and electron acceptors. The majority of electrons ultimately reduce $NADP^+$ to NADPH and oxidize H_2O to O_2. Light energy is also used to generate a proton motive force (see Chapter 6) across the thylakoid membrane, which is used to synthesize ATP.

General Concepts

In this section we will explore the essential concepts that provide a foundation for an understanding of photosynthesis. These concepts include the nature of light, the properties of pigments, and the various roles of pigments.

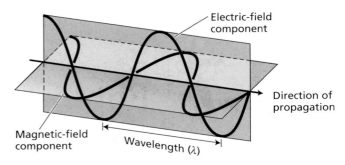

FIGURE 7.1 Light is a transverse electromagnetic wave, consisting of oscillating electric and magnetic fields that are perpendicular to each other and to the direction of propagation of the light. Light moves at a speed of 3×10^8 m s^{-1}. The wavelength (λ) is the distance between successive crests of the wave.

Light has characteristics of both a particle and a wave

A triumph of physics in the early twentieth century was the realization that light has properties of both particles and waves. A wave (Figure 7.1) is characterized by a **wavelength**, denoted by the Greek letter lambda (λ), which is the distance between successive wave crests. The **frequency**, represented by the Greek letter nu (ν), is the number of wave crests that pass an observer in a given time. A simple equation relates the wavelength, the frequency, and the speed of any wave:

$$c = \lambda \nu \qquad (7.1)$$

where c is the speed of the wave—in the present case, the speed of light (3.0×10^8 m s^{-1}). The light wave is a transverse (side-to-side) electromagnetic wave, in which both electric and magnetic fields oscillate perpendicularly to the direction of propagation of the wave and at 90° with respect to each other.

Light is also a particle, which we call a **photon**. Each photon contains an amount of energy that is called a **quan-**

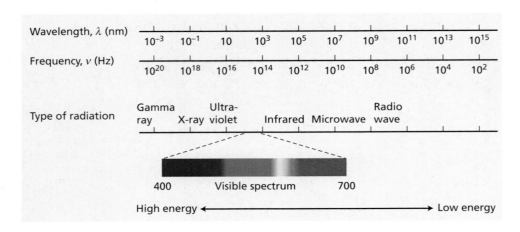

FIGURE 7.2 Electromagnetic spectrum. Wavelength (λ) and frequency (ν) are inversely related. Our eyes are sensitive to only a narrow range of wavelengths of radiation, the visible region, which extends from about 400 nm (violet) to about 700 nm (red). Short-wavelength (high-frequency) light has a high energy content; long-wavelength (low-frequency) light has a low energy content.

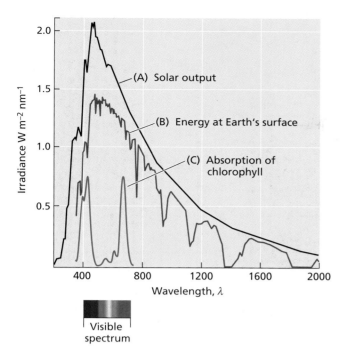

FIGURE 7.3 The solar spectrum and its relation to the absorption spectrum of chlorophyll. Curve A is the energy output of the sun as a function of wavelength. Curve B is the energy that strikes the surface of Earth. The sharp valleys in the infrared region beyond 700 nm represent the absorption of solar energy by molecules in the atmosphere, chiefly water vapor. Curve C is the absorption spectrum of chlorophyll, which absorbs strongly in the blue (about 430 nm) and the red (about 660 nm) portions of the spectrum. Because the green light in the middle of the visible region is not efficiently absorbed, most of it is reflected into our eyes and gives plants their characteristic green color.

wavelengths) is in the ultraviolet region of the spectrum, and light of slightly lower frequencies (or longer wavelengths) is in the infrared region. The output of the sun is shown in Figure 7.3, along with the energy density that strikes the surface of Earth. The absorption spectrum of chlorophyll *a* (green curve in Figure 7.3) indicates approximately the portion of the solar output that is utilized by plants.

An **absorption spectrum** (plural *spectra*) provides information about the amount of light energy taken up or absorbed by a molecule or substance as a function of the wavelength of the light. The absorption spectrum for a particular substance in a nonabsorbing solvent can be determined by a spectrophotometer as illustrated in Figure 7.4. Spectrophotometry, the technique used to measure the absorption of light by a sample, is more completely discussed in **Web Topic 7.1**.

When molecules absorb or emit light, they change their electronic state

Chlorophyll appears green to our eyes because it absorbs light mainly in the red and blue parts of the spectrum, so only some of the light enriched in green wavelengths (about 550 nm) is reflected into our eyes (see Figure 7.3).

The absorption of light is represented by Equation 7.3, in which chlorophyll (Chl) in its lowest-energy, or ground, state absorbs a photon (represented by $h\nu$) and makes a transition to a higher-energy, or excited, state (Chl*):

$$\text{Chl} + h\nu \rightarrow \text{Chl*} \quad (7.3)$$

The distribution of electrons in the excited molecule is somewhat different from the distribution in the ground-state molecule (Figure 7.5). Absorption of blue light excites the chlorophyll to a higher energy state than absorption of red light, because the energy of photons is higher when their wavelength is shorter. In the higher excited state, chlorophyll is extremely unstable, very rapidly gives up some of its energy to the surroundings as heat, and enters the lowest excited state, where it can be stable for a maximum of several nanoseconds (10^{-9} s). Because of this inherent instability of the excited state, any process that captures its energy must be extremely rapid.

In the lowest excited state, the excited chlorophyll has four alternative pathways for disposing of its available energy.

tum (plural *quanta*). The energy content of light is not continuous but rather is delivered in these discrete packets, the quanta. The energy (E) of a photon depends on the frequency of the light according to a relation known as Planck's law:

$$E = h\nu \quad (7.2)$$

where h is Planck's constant (6.626×10^{-34} J s).

Sunlight is like a rain of photons of different frequencies. Our eyes are sensitive to only a small range of frequencies—the visible-light region of the electromagnetic spectrum (Figure 7.2). Light of slightly higher frequencies (or shorter

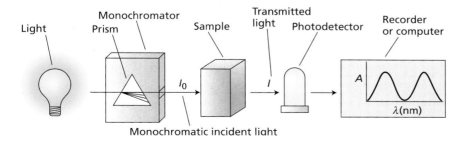

FIGURE 7.4 Schematic diagram of a spectrophotometer. The instrument consists of a light source, a monochromator that contains a wavelength selection device such as a prism, a sample holder, a photodetector, and a recorder or computer. The output wavelength of the monochromator can be changed by rotation of the prism; the graph of absorbance (A) versus wavelength (λ) is called a spectrum.

FIGURE 7.5 Light absorption and emission by chlorophyll. (A) Energy level diagram. Absorption or emission of light is indicated by vertical lines that connect the ground state with excited electron states. The blue and red absorption bands of chlorophyll (which absorb blue and red photons, respectively) correspond to the upward vertical arrows, signifying that energy absorbed from light causes the molecule to change from the ground state to an excited state. The downward-pointing arrow indicates fluorescence, in which the molecule goes from the lowest excited state to the ground state while re-emitting energy as a photon. (B) Spectra of absorption and fluorescence. The long-wavelength (red) absorption band of chlorophyll corresponds to light that has the energy required to cause the transition from the ground state to the first excited state. The short-wavelength (blue) absorption band corresponds to a transition to a higher excited state.

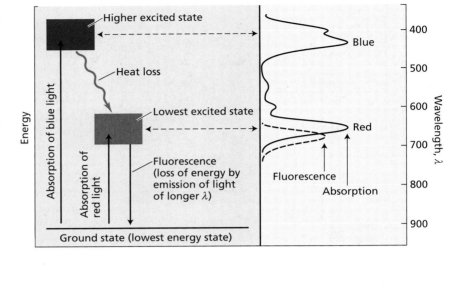

1. Excited chlorophyll can re-emit a photon and thereby return to its ground state—a process known as **fluorescence**. When it does so, the wavelength of fluorescence is slightly longer (and of lower energy) than the wavelength of absorption, because a portion of the excitation energy is converted into heat before the fluorescent photon is emitted. Chlorophylls fluoresce in the red region of the spectrum.

2. The excited chlorophyll can return to its ground state by directly converting its excitation energy into heat, with no emission of a photon.

3. Chlorophyll may participate in **energy transfer**, during which an excited chlorophyll transfers its energy to another molecule.

4. A fourth process is **photochemistry**, in which the energy of the excited state causes chemical reactions to occur. The photochemical reactions of photosynthesis are among the fastest known chemical reactions. This extreme speed is necessary for photochemistry to compete with the three other possible reactions of the excited state just described.

Photosynthetic pigments absorb the light that powers photosynthesis

The energy of sunlight is first absorbed by the pigments of the plant. All pigments active in photosynthesis are found in the chloroplast. Structures and absorption spectra of several photosynthetic pigments are shown in Figures 7.6 and 7.7, respectively. The chlorophylls and **bacteriochlorophylls** (pigments found in certain bacteria) are the typical pigments of photosynthetic organisms, but all organisms contain a mixture of more than one kind of pigment, each serving a specific function.

Chlorophylls a and b are abundant in green plants, and c and d are found in some protists and cyanobacteria. A number of different types of bacteriochlorophyll have been found; type a is the most widely distributed. **Web Topic 7.2** shows the distribution of pigments in different types of photosynthetic organisms.

All chlorophylls have a complex ring structure that is chemically related to the porphyrin-like groups found in hemoglobin and cytochromes (see Figure 7.6A). In addition, a long hydrocarbon tail is almost always attached to the ring structure. The tail anchors the chlorophyll to the hydrophobic portion of its environment. The ring structure contains some loosely bound electrons and is the part of the molecule involved in electron transitions and redox (reduction–oxidation) reactions.

The different types of **carotenoids** found in photosynthetic organisms are all linear molecules with multiple conjugated double bonds (see Figure 7.6B). Absorption bands in the 400 to 500 nm region give carotenoids their characteristic orange color. The color of carrots, for example, is due to the carotenoid β-carotene, whose structure and absorption spectrum are shown in Figures 7.6 and 7.7, respectively.

Carotenoids are found in all photosynthetic organisms, except for mutants incapable of living outside the laboratory. Carotenoids are integral constituents of the thylakoid membrane and are usually associated intimately with many of the proteins that make up the photosynthetic apparatus. The light absorbed by the carotenoids is transferred to chlorophyll for photosynthesis; because of this role they are called **accessory pigments**. Carotenoids also help protect the organism from damage caused by light.

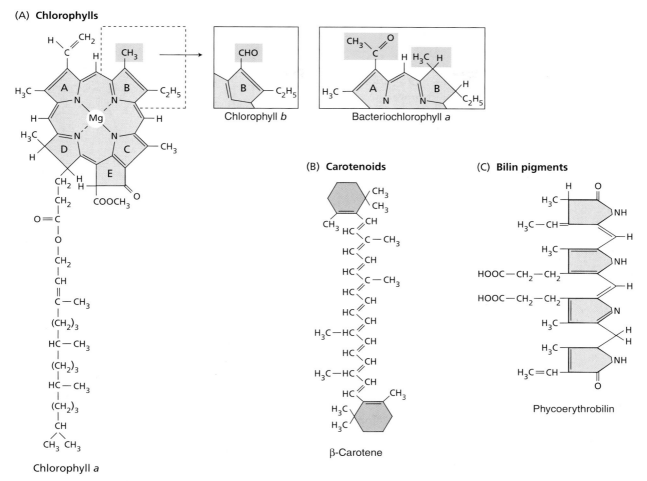

FIGURE 7.6 Molecular structure of some photosynthetic pigments. (A) The chlorophylls have a porphyrin-like ring structure with a magnesium atom (Mg) coordinated in the center and a long hydrophobic hydrocarbon tail that anchors them in the photosynthetic membrane. The porphyrin-like ring is the site of the electron rearrangements that occur when the chlorophyll is excited, and of the unpaired electrons when it is either oxidized or reduced. Various chlorophylls differ chiefly in the substituents around the rings and the pattern of double bonds. (B) Carotenoids are linear polyenes that serve as both antenna pigments and photoprotective agents. (C) Bilin pigments are open-chain tetrapyrroles found in antenna structures known as phycobilisomes that occur in cyanobacteria and red algae.

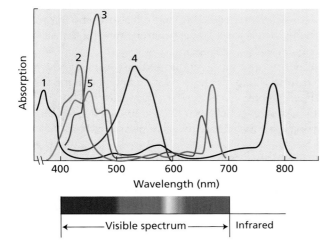

FIGURE 7.7 Absorption spectra of some photosynthetic pigments. Curve 1, bacteriochlorophyll a; curve 2, chlorophyll a; curve 3, chlorophyll b; curve 4, phycoerythrobilin; curve 5, β-carotene. The absorption spectra shown are for pure pigments dissolved in nonpolar solvents, except for curve 4, which represents an aqueous buffer of phycoerythrin, a protein from cyanobacteria that contains a phycoerythrobilin chromophore covalently attached to the peptide chain. In many cases the spectra of photosynthetic pigments in vivo are substantially affected by the environment of the pigments in the photosynthetic membrane. (After Avers 1985.)

Key Experiments in Understanding Photosynthesis

Establishing the overall chemical equation of photosynthesis required several hundred years and contributions by many scientists (literature references for historical developments can be found on the web site for this book). In 1771, Joseph Priestley observed that a sprig of mint growing in air in which a candle had burned out improved the air so that another candle could burn. He had discovered oxygen evolution by plants. A Dutchman, Jan Ingenhousz, documented the essential role of light in photosynthesis in 1779.

Other scientists established the roles of CO_2 and H_2O and showed that organic matter, specifically carbohydrate, is a product of photosynthesis along with oxygen. By the end of the nineteenth century, the balanced overall chemical reaction for photosynthesis could be written as follows:

$$6\,CO_2 + 6\,H_2O \xrightarrow{\text{Light, plant}} C_6H_{12}O_6 + 6\,O_2 \quad (7.4)$$

where $C_6H_{12}O_6$ represents a simple sugar such as glucose. As will be discussed in Chapter 8, glucose is not the actual product of the carbon fixation reactions. However, the energetics for the actual products is approximately the same, so the representation of glucose in Equation 7.4 should be regarded as a convenience but not taken literally.

The chemical reactions of photosynthesis are complex. In fact, at least 50 intermediate reaction steps have now been identified, and undoubtedly additional steps will be discovered. An early clue to the chemical nature of the essential chemical process of photosynthesis came in the 1920s from investigations of photosynthetic bacteria that did not produce oxygen as an end product. From his studies on these bacteria, C. B. van Niel concluded that photosynthesis is a redox process. This conclusion has served as a fundamental concept on which all subsequent research on photosynthesis has been based.

We now turn to the relationship between photosynthetic activity and the spectrum of absorbed light. We will discuss some of the critical experiments that have contributed to our present understanding of photosynthesis, and we will consider equations for essential chemical reactions of photosynthesis.

Action spectra relate light absorption to photosynthetic activity

The use of action spectra has been central to the development of our current understanding of photosynthesis. An **action spectrum** depicts the magnitude of a response of a biological system to light as a function of wavelength. For example, an action spectrum for photosynthesis can be constructed from measurements of oxygen evolution at different wavelengths (Figure 7.8). Often an action spectrum can identify the chromophore (pigment) responsible for a particular light-induced phenomenon.

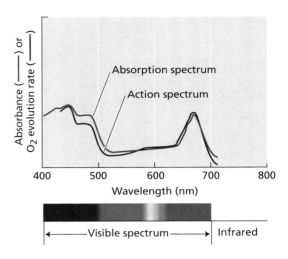

FIGURE 7.8 Action spectrum compared with an absorption spectrum. The absorption spectrum is measured as shown in Figure 7.4. An action spectrum is measured by plotting a response to light, such as oxygen evolution, as a function of wavelength. If the pigment used to obtain the absorption spectrum is the same as those that cause the response, the absorption and action spectra will match. In the example shown here, the action spectrum for oxygen evolution matches the absorption spectrum of intact chloroplasts quite well, indicating that light absorption by the chlorophylls mediates oxygen evolution. Discrepancies are found in the region of carotenoid absorption, from 450 to 550 nm, indicating that energy transfer from carotenoids to chlorophylls is not as effective as energy transfer between chlorophylls.

Some of the first action spectra were measured by T. W. Engelmann in the late 1800s (Figure 7.9). Engelmann used a prism to disperse sunlight into a rainbow that was allowed to fall on an aquatic algal filament. A population of O_2-seeking bacteria was introduced into the system. The bacteria congregated in the regions of the filaments that evolved the most O_2. These were the regions illuminated by blue light and red light, which are strongly absorbed by chlorophyll. Today, action spectra can be measured in room-sized spectrographs in which a huge monochromator bathes the experimental samples in monochromatic light. But the principle of the experiment is the same as that of Engelmann's experiments.

Action spectra were very important for the discovery of two distinct photosystems operating in O_2-evolving photosynthetic organisms. Before we introduce the two photosystems, however, we need to describe the light-gathering antennas and the energy needs of photosynthesis.

Photosynthesis takes place in complexes containing light-harvesting antennas and photochemical reaction centers

A portion of the light energy absorbed by chlorophylls and carotenoids is eventually stored as chemical energy via the formation of chemical bonds. This conversion of energy

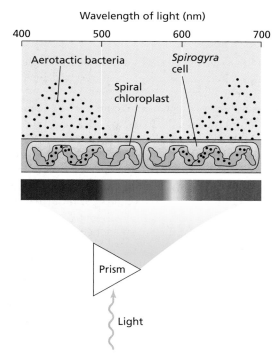

FIGURE 7.9 Schematic diagram of the action spectrum measurements by T. W. Engelmann. Engelmann projected a spectrum of light onto the spiral chloroplast of the filamentous green alga *Spirogyra* and observed that oxygen-seeking bacteria introduced into the system collected in the region of the spectrum where chlorophyll pigments absorb. This action spectrum gave the first indication of the effectiveness of light absorbed by accessory pigments in driving photosynthesis.

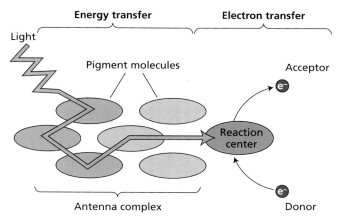

FIGURE 7.10 Basic concept of energy transfer during photosynthesis. Many pigments together serve as an antenna, collecting light and transferring its energy to the reaction center, where chemical reactions store some of the energy by transferring electrons from a chlorophyll pigment to an electron acceptor molecule. An electron donor then reduces the chlorophyll again. The transfer of energy in the antenna is a purely physical phenomenon and involves no chemical changes.

from one form to another is a complex process that depends on cooperation between many pigment molecules and a group of electron transfer proteins.

The majority of the pigments serve as an **antenna complex**, collecting light and transferring the energy to the **reaction center complex**, where the chemical oxidation and reduction reactions leading to long-term energy storage take place (Figure 7.10). Molecular structures of some of the antenna and reaction center complexes are discussed later in the chapter.

How does the plant benefit from this division of labor between antenna and reaction center pigments? Even in bright sunlight, a chlorophyll molecule absorbs only a few photons each second. If every chlorophyll had a complete reaction center associated with it, the enzymes that make up this system would be idle most of the time, only occasionally being activated by photon absorption. However, if many pigments can send energy into a common reaction center, the system is kept active a large fraction of the time.

In 1932, Robert Emerson and William Arnold performed a key experiment that provided the first evidence for the cooperation of many chlorophyll molecules in energy conversion during photosynthesis. They delivered very brief (10^{-5} s) flashes of light to a suspension of the green alga *Chlorella pyrenoidosa* and measured the amount of oxygen produced. The flashes were spaced about 0.1 s apart, a time that Emerson and Arnold had determined in earlier work was long enough for the enzymatic steps of the process to be completed before the arrival of the next flash. The investigators varied the energy of the flashes and found that at high energies the oxygen production did not increase when a more intense flash was given: The photosynthetic system was saturated with light (Figure 7.11).

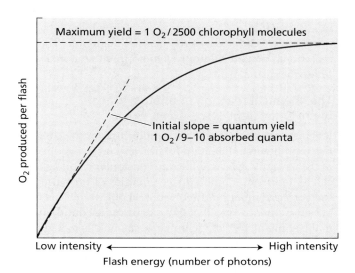

FIGURE 7.11 Relationship of oxygen production to flash energy, the first evidence for the interaction between the antenna pigments and the reaction center. At saturating energies, the maximum amount of O_2 produced is one molecule per 2500 chlorophyll molecules.

In their measurement of the relationship of oxygen production to flash energy, Emerson and Arnold were surprised to find that under saturating conditions, only one molecule of oxygen was produced for each 2500 chlorophyll molecules in the sample. We know now that several hundred pigments are associated with each reaction center and that each reaction center must operate four times to produce one molecule of oxygen—hence the value of 2500 chlorophylls per O_2.

The reaction centers and most of the antenna complexes are integral components of the photosynthetic membrane. In eukaryotic photosynthetic organisms, these membranes are found within the chloroplast; in photosynthetic prokaryotes, the site of photosynthesis is the plasma membrane or membranes derived from it.

The graph shown in Figure 7.11 permits us to calculate another important parameter of the light reactions of photosynthesis, the quantum yield. The **quantum yield** of photosynthesis (Φ) is defined as follows:

$$\Phi = \frac{\text{Number of photochemical products}}{\text{Total number of quanta absorbed}} \quad (7.5)$$

In the linear portion (low light intensity) of the curve, an increase in the number of photons stimulates a proportional increase in oxygen evolution. Thus the slope of the curve measures the quantum yield for oxygen production. The quantum yield for a particular process can range from 0 (if that process does not respond to light) to 1.0 (if every photon absorbed contributes to the process). A more detailed discussion of quantum yields can be found in **Web Topic 7.3**.

In functional chloroplasts kept in dim light, the quantum yield of photochemistry is approximately 0.95, the quantum yield of fluorescence is 0.05 or lower, and the quantum yields of other processes are negligible. The vast majority of excited chlorophyll molecules therefore lead to photochemistry.

The chemical reaction of photosynthesis is driven by light

It is important to realize that equilibrium for the chemical reaction shown in Equation 7.4 lies very far in the direction of the reactants. The equilibrium constant for Equation 7.4, calculated from tabulated free energies of formation for each of the compounds involved, is about 10^{-500}. This number is so close to zero that one can be quite confident that in the entire history of the universe no molecule of glucose has formed spontaneously from H_2O and CO_2 without external energy being provided. The energy needed to drive the photosynthetic reaction comes from light. Here's a simpler form of Equation 7.4:

$$CO_2 + H_2O \xrightarrow{\text{Light, plant}} (CH_2O) + O_2 \quad (7.6)$$

where (CH_2O) is one-sixth of a glucose molecule. About nine or ten photons of light are required to drive the reaction of Equation 7.6.

Although the photochemical quantum yield under optimum conditions is nearly 100%, the *efficiency* of the conversion of light into chemical energy is much less. If red light of wavelength 680 nm is absorbed, the total energy input (see Equation 7.2) is 1760 kJ per mole of oxygen formed. This amount of energy is more than enough to drive the reaction in Equation 7.6, which has a standard-state free-energy change of +467 kJ mol^{-1}. The efficiency of conversion of light energy at the optimal wavelength into chemical energy is therefore about 27%, which is remarkably high for an energy conversion system. Most of this stored energy is used for cellular maintenance processes; the amount diverted to the formation of biomass is much less (see Figure 9.2).

There is no conflict between the fact that the photochemical quantum efficiency (quantum yield) is nearly 1 (100%) and the energy conversion efficiency is only 27%. The *quantum efficiency* is a measure of the fraction of absorbed photons that engage in photochemistry; the *energy efficiency* is a measure of how much energy in the absorbed photons is stored as chemical products. The numbers indicate that almost all the absorbed photons engage in photochemistry, but only about a fourth of the energy in each photon is stored, the remainder being converted to heat.

Light drives the reduction of NADP and the formation of ATP

The overall process of photosynthesis is a redox chemical reaction, in which electrons are removed from one chemical species, thereby oxidizing it, and added to another species, thereby reducing it. In 1937, Robert Hill found that in the light, isolated chloroplast thylakoids reduce a variety of compounds, such as iron salts. These compounds serve as oxidants in place of CO_2, as the following equation shows:

$$4\ Fe^{3+} + 2\ H_2O \rightarrow 4\ Fe^{2+} + O_2 + 4\ H^+ \quad (7.7)$$

Many compounds have since been shown to act as artificial electron acceptors in what has come to be known as the Hill reaction. Their use has been invaluable in elucidating the reactions that precede carbon reduction. The demonstration of oxygen evolution linked to the reduction of artificial electron acceptors provided the first evidence that oxygen evolution could occur in the absence of carbon dioxide and led to the now accepted and proven idea that the oxygen in photosynthesis originates from water, not from carbon dioxide.

We now know that during the normal functioning of the photosynthetic system, light reduces nicotinamide adenine dinucleotide phosphate (NADP), which in turn serves as the reducing agent for carbon fixation in the Calvin cycle (see Chapter 8). ATP is also formed during the electron flow from water to NADP, and it, too, is used in carbon reduction.

The chemical reactions in which water is oxidized to oxygen, NADP is reduced, and ATP is formed are known as the *thylakoid reactions* because almost all the reactions up to NADP reduction take place within the thylakoids. The carbon fixation and reduction reactions are called the *stroma reactions* because the carbon reduction reactions take place in the aqueous region of the chloroplast, the stroma. Although this division is somewhat arbitrary, it is conceptually useful.

Oxygen-evolving organisms have two photosystems that operate in series

By the late 1950s, several experiments were puzzling the scientists who studied photosynthesis. One of these experiments, carried out by Emerson, measured the quantum yield of photosynthesis as a function of wavelength and revealed an effect known as the red drop (Figure 7.12). Note that while the quantum yield of photochemistry is nearly one, as discussed above, the actions of about ten photons are required to produce each molecule of O_2, so the overall maximum quantum yield of O_2 production is about 0.1.

If the quantum yield is measured for the wavelengths at which chlorophyll absorbs light, the values found throughout most of the range are fairly constant, indicating that any photon absorbed by chlorophyll or other pigments is as effective as any other photon in driving photosynthesis. However, the yield drops dramatically in the far-red region of chlorophyll absorption (greater than 680 nm).

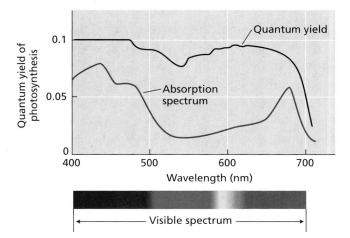

FIGURE 7.12 Red drop effect. The quantum yield of oxygen evolution (black curve) falls off drastically for far-red light of wavelengths greater than 680 nm, indicating that far-red light alone is inefficient in driving photosynthesis. The slight dip near 500 nm reflects the somewhat lower efficiency of photosynthesis using light absorbed by accessory pigments, carotenoids.

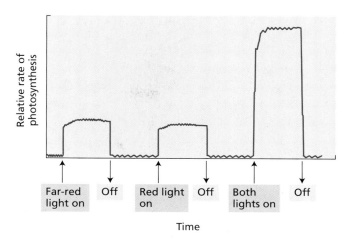

FIGURE 7.13 Enhancement effect. The rate of photosynthesis when red and far-red light are given together is greater than the sum of the rates when they are given apart. When it was demonstrated in the 1950s, the enhancement effect provided essential evidence in favor of the concept that photosynthesis is carried out by two photochemical systems working in tandem but with slightly different wavelength optima.

This drop cannot be caused by a decrease in chlorophyll absorption, because the quantum yield measures only light that has actually been absorbed. Thus, light with a wavelength greater than 680 nm is much less efficient than light of shorter wavelengths.

Another puzzling experimental result was the **enhancement effect**, also discovered by Emerson. He measured the rate of photosynthesis separately with light of two different wavelengths and then used the two beams simultaneously (Figure 7.13). When red and far-red light were given together, the rate of photosynthesis was greater than the sum of the individual rates, a startling and surprising observation. These and others observations were eventually explained by experiments performed in the 1960s (see **Web Topic 7.4**) that led to the discovery that two photochemical complexes, now known as **photosystems** I and II (**PSI** and **PSII**), operate in series to carry out the early energy storage reactions of photosynthesis.

Photosystem I preferentially absorbs far-red light of wavelengths greater than 680 nm; photosystem II preferentially absorbs red light of 680 nm and is driven very poorly by far-red light. This wavelength dependence explains the enhancement effect and the red drop effect. Another difference between the photosystems is that

- Photosystem I produces a strong reductant, capable of reducing $NADP^+$, and a weak oxidant.

- Photosystem II produces a very strong oxidant, capable of oxidizing water, and a weaker reductant than the one produced by photosystem I.

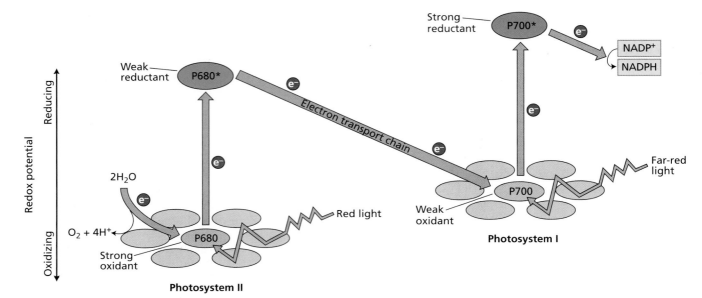

Figure 7.14 Z scheme of photosynthesis. Red light absorbed by photosystem II (PSII) produces a strong oxidant and a weak reductant. Far-red light absorbed by photosystem I (PSI) produces a weak oxidant and a strong reductant. The strong oxidant generated by PSII oxidizes water, while the strong reductant produced by PSI reduces $NADP^+$. This scheme is basic to an understanding of photosynthetic electron transport. P680 and P700 refer to the wavelengths of maximum absorption of the reaction center chlorophylls in PSII and PSI, respectively.

The reductant produced by photosystem II re-reduces the oxidant produced by photosystem I. These properties of the two photosystems are shown schematically in Figure 7.14.

The scheme of photosynthesis depicted in Figure 7.14, called the Z (for *zigzag*) *scheme*, has become the basis for understanding O_2-evolving (oxygenic) photosynthetic organisms. It accounts for the operation of two physically and chemically distinct photosystems (I and II), each with its own antenna pigments and photochemical reaction center. The two photosystems are linked by an electron transport chain.

Organization of the Photosynthetic Apparatus

The previous section explained some of the physical principles underlying photosynthesis, some aspects of the functional roles of various pigments, and some of the chemical reactions carried out by photosynthetic organisms. We now turn to the architecture of the photosynthetic apparatus and the structure of its components.

The chloroplast is the site of photosynthesis

In photosynthetic eukaryotes, photosynthesis takes place in the subcellular organelle known as the chloroplast. Figure 7.15 shows a transmission electron micrograph of a thin section from a pea chloroplast. The most striking aspect of the structure of the chloroplast is the extensive system of internal membranes known as **thylakoids**. All the chlorophyll is contained within this membrane system, which is the site of the light reactions of photosynthesis.

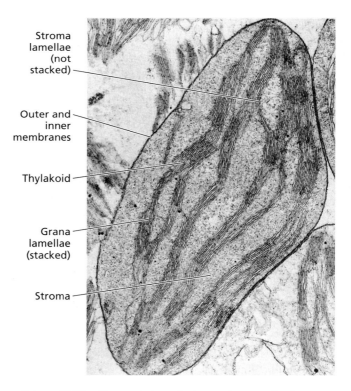

Figure 7.15 Transmission electron micrograph of a chloroplast from pea (*Pisum sativum*), fixed in glutaraldehyde and OsO_4, embedded in plastic resin, and thin-sectioned with an ultramicrotome. (14,500×) (Courtesy of J. Swafford.)

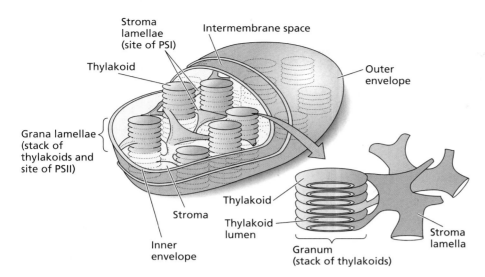

FIGURE 7.16 Schematic picture of the overall organization of the membranes in the chloroplast. The chloroplast of higher plants is surrounded by the inner and outer membranes (envelope). The region of the chloroplast that is inside the inner membrane and surrounds the thylakoid membranes is known as the stroma. It contains the enzymes that catalyze carbon fixation and other biosynthetic pathways. The thylakoid membranes are highly folded and appear in many pictures to be stacked like coins (the granum), although in reality they form one or a few large interconnected membrane systems, with a well-defined interior and exterior with respect to the stroma. The inner space within a thylakoid is known as the lumen. (After Becker 1986.)

The carbon reduction reactions, which are catalyzed by water-soluble enzymes, take place in the **stroma**, the region of the chloroplast outside the thylakoids. Most of the thylakoids appear to be very closely associated with each other. These stacked membranes are known as **grana lamellae** (singular *lamella*; each stack is called a *granum*), and the exposed membranes in which stacking is absent are known as **stroma lamellae**.

Two separate membranes, each composed of a lipid bilayer and together known as the **envelope**, surround most types of chloroplasts (Figure 7.16). This double-membrane system contains a variety of metabolite transport systems. The chloroplast also contains its own DNA, RNA, and ribosomes. Some of the chloroplast proteins are products of transcription and translation within the chloroplast itself, whereas most of the others are encoded by nuclear DNA, synthesized on cytoplasmic ribosomes, and then imported into the chloroplast. This remarkable division of labor, extending in many cases to different subunits of the same enzyme complex, will be discussed in more detail later in this chapter. For some dynamic structures of chloroplasts see **Web Essay 7.1**.

Thylakoids contain integral membrane proteins

A wide variety of proteins essential to photosynthesis are embedded in the thylakoid membranes. In many cases, portions of these proteins extend into the aqueous regions on both sides of the thylakoids. These **integral membrane proteins** contain a large proportion of hydrophobic amino acids and are therefore much more stable in a nonaqueous medium such as the hydrocarbon portion of the membrane (see Figure 1.5A).

The reaction centers, the antenna pigment–protein complexes, and most of the electron carrier proteins are all integral membrane proteins. In all known cases, integral membrane proteins of the chloroplast have a unique orientation within the membrane. Thylakoid membrane proteins have one region pointing toward the stromal side of the membrane and the other oriented toward the interior portion of the thylakoid, known as the *lumen* (see Figures 7.16 and 7.17).

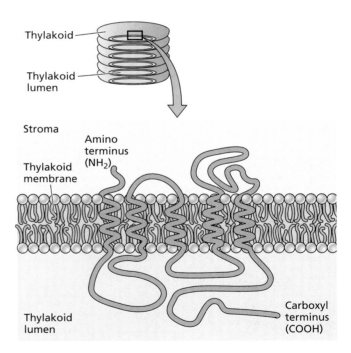

FIGURE 7.17 Predicted folding pattern of the D1 protein of the PSII reaction center. The hydrophobic portion of the membrane is traversed five times by the peptide chain rich in hydrophobic amino acid residues. The protein is asymmetrically arranged in the thylakoid membrane, with the amino (NH_2) terminus on the stromal side of the membrane and the carboxyl (COOH) terminus on the lumen side. (After Trebst 1986.)

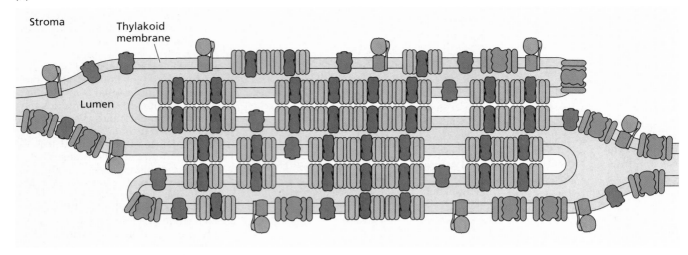

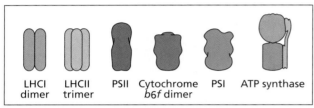

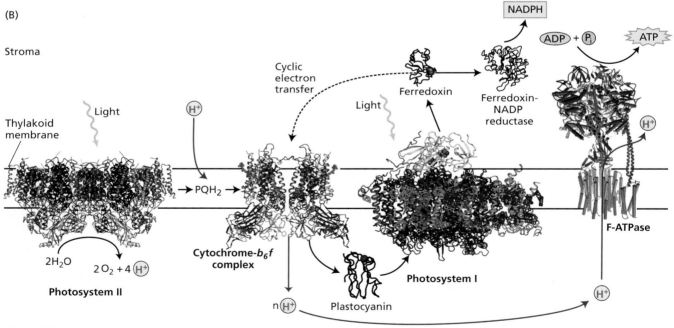

FIGURE 7.18 Organization and structure of the four major protein complexes of the thylakoid membrane. (A) Photosystem II is located predominantly in the stacked regions of the thylakoid membrane; photosystem I and ATP synthase are found in the unstacked regions protruding into the stroma. Cytochrome $b_6 f$ complexes are evenly distributed. This lateral separation of the two photosystems requires that electrons and protons produced by photosystem II be transported a considerable distance before they can be acted on by photosystem I and the ATP-coupling enzyme. (B) Structures of the four main protein complexes of the thylakoid membrane. Shown also are the two diffusible electron carriers—plastocyanin, which is located in the thylakoid lumen and plastohydroquinone (PQH_2) in the membrane. (A after Allen and Forsberg 2001; B after Nelson and Ben-Shem 2004).

The chlorophylls and accessory light-gathering pigments in the thylakoid membrane are always associated in a noncovalent but highly specific way with proteins, thereby forming pigment–protein complexes. Both antenna and reaction center chlorophylls are associated with proteins that are organized within the membrane so as to optimize energy transfer in antenna complexes and electron transfer in reaction centers, while at the same time minimizing wasteful processes.

Photosystems I and II are spatially separated in the thylakoid membrane

The PSII reaction center, along with its antenna chlorophylls and associated electron transport proteins, is located predominantly in the grana lamellae (Figure 7.18A) (Allen and Forsberg 2001). The PSI reaction center and its associated antenna pigments and electron transfer proteins, as well as the ATP synthase enzyme that catalyzes the formation of ATP, are found almost exclusively in the stroma lamellae and at the edges of the grana lamellae. The cytochrome b_6f complex of the electron transport chain that connects the two photosystems (see Figure 7.21) is evenly distributed between stroma and grana lamellae. The structures of all these complexes are shown in Figure 7.18B (Nelson and Ben-Shem 2004).

Thus the two photochemical events that take place in O_2-evolving photosynthesis are spatially separated. This separation implies that one or more of the electron carriers that function between the photosystems diffuses from the grana region of the membrane to the stroma region, where electrons are delivered to photosystem I. These diffusible carriers are plastocyanin (PC) and plastoquinone (PQ).

In PSII, the oxidation of two water molecules produces four electrons, four protons, and a single O_2 (see Equation 7.8, below). The protons produced by this oxidation of water must also be able to diffuse to the stroma region, where ATP is synthesized. The functional role of this large separation (many tens of nanometers) between photosystems I and II is not entirely clear, but is thought to improve the efficiency of energy distribution between the two photosystems (Allen and Forsberg 2001).

The spatial separation between photosystems I and II indicates that a strict one-to-one stoichiometry between the two photosystems is not required. Instead, PSII reaction centers feed reducing equivalents into a common intermediate pool of lipid-soluble electron carriers (plastoquinone), which will be described in detail later in the chapter. The PSI reaction centers remove the reducing equivalents from the common pool, rather than from any specific PSII reaction center complex.

Most measurements of the relative quantities of photosystems I and II have shown that there is an excess of photosystem II in chloroplasts. Most commonly, the ratio of PSII to PSI is about 1.5:1, but it can change when plants are grown in different light conditions. In contrast to the situation in chloroplasts of eukaryotic photosynthetic organisms, cyanobacteria usually have an excess of PSI over PSII.

Anoxygenic photosynthetic bacteria have a single reaction center

Non-O_2-evolving (anoxygenic) organisms contain only a single photosystem similar to either photosystem I or II. These simpler organisms have been very useful for detailed structural and functional studies that have contributed to a better understanding of oxygenic photosynthesis.

Reaction centers from purple photosynthetic bacteria were the first integral membrane proteins to have structures determined to high resolution (Deisenhofer and Michel, 1989) (see Figures 7.5.A and 7.5.B in **Web Topic 7.5**). Detailed analysis of these structures, along with the characterization of numerous mutants, has revealed many of the principles involved in the energy storage processes carried out by all reaction centers.

The structure of the purple bacterial reaction center is thought to be similar in many ways to that found in photosystem II from oxygen-evolving organisms, especially in the electron acceptor portion of the chain. The proteins that make up the core of the bacterial reaction center are relatively similar in sequence to their photosystem II counterparts, implying an evolutionary relatedness. A similar situation is found with respect to the reaction centers from the anoxygenic green sulfur bacteria and the heliobacteria, compared to photosystem I. The evolutionary implications of this pattern are discussed later in this chapter.

Organization of Light-Absorbing Antenna Systems

The antenna systems of different classes of photosynthetic organisms are remarkably varied, in contrast to the reaction centers, which appear to be similar in even distantly related organisms. The variety of antenna complexes reflects evolutionary adaptation to the diverse environments in which different organisms live, as well as the need in some organisms to balance energy input to the two photosystems (Grossman et al. 1995; Green and Durnford 1996; Green and Parson 2003).

Antenna systems function to deliver energy efficiently to the reaction centers with which they are associated (van Grondelle et al. 1994; Pullerits and Sundström 1996). The size of the antenna system varies considerably in different organisms, ranging from a low of 20 to 30 bacteriochlorophylls per reaction center in some photosynthetic bacteria, to generally 200 to 300 chlorophylls per reaction center in higher plants, to a few thousand pigments per reaction center in some types of algae and bacteria. The molecular structures of antenna pigments are also quite diverse, although all of them are associated in some way with the photosynthetic membrane.

The physical mechanism by which excitation energy is conveyed from the chlorophyll that absorbs the light to the

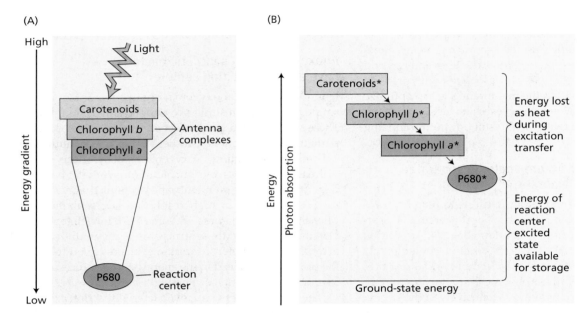

FIGURE 7.19 Funneling of excitation from the antenna system toward the reaction center. (A) The excited-state energy of pigments increases with distance from the reaction center; that is, pigments closer to the reaction center are lower in energy than those farther from the reaction center. This energy gradient ensures that excitation transfer toward the reaction center is energetically favorable and that excitation transfer back out to the peripheral portions of the antenna is energetically unfavorable. (B) Some energy is lost as heat to the environment by this process, but under optimal conditions almost all the excitations absorbed in the antenna complexes can be delivered to the reaction center. The asterisks denote an excited state.

reaction center is thought to be **fluorescence resonance energy transfer**, often abbreviated as FRET. By this mechanism the excitation energy is transferred from one molecule to another by a nonradiative process.

A useful analogy for resonance transfer is the transfer of energy between two tuning forks. If one tuning fork is struck and properly placed near another, the second tuning fork receives some energy from the first and begins to vibrate. As in resonance energy transfer in antenna complexes, the efficiency of energy transfer between the two tuning forks depends on their distance from each other and their relative orientation, as well as on their pitches or vibrational frequencies.

Energy transfer in antenna complexes is usually very efficient: Approximately 95 to 99% of the photons absorbed by the antenna pigments have their energy transferred to the reaction center, where it can be used for photochemistry. There is an important difference between energy transfer among pigments in the antenna and the electron transfer that occurs in the reaction center: Whereas energy transfer is a purely physical phenomenon, electron transfer involves chemical changes in molecules, that produces oxidized or reduced species.

The antenna funnels energy to the reaction center

The sequence of pigments within the antenna that funnel absorbed energy toward the reaction center has absorption maxima that are progressively shifted toward longer red wavelengths (Figure 7.19). This red shift in absorption maximum means that the energy of the excited state is somewhat lower nearer the reaction center than in the more peripheral portions of the antenna system.

As a result of this arrangement, when excitation is transferred, for example, from a chlorophyll b molecule absorbing maximally at 650 nm to a chlorophyll a molecule absorbing maximally at 670 nm, the difference in energy between these two excited chlorophylls is lost to the environment as heat.

For the excitation to be transferred back to the chlorophyll b, the energy lost as heat would have to be resupplied. The probability of reverse transfer is therefore smaller simply because thermal energy is not sufficient to make up the deficit between the lower-energy and higher-energy pigments. This effect gives the energy-trapping process a degree of directionality or irreversibility and makes the delivery of excitation to the reaction center more efficient. In essence, the system sacrifices some energy from each quantum so that nearly all of the quanta can be trapped by the reaction center.

Many antenna complexes have a common structural motif

In all eukaryotic photosynthetic organisms that contain both chlorophyll a and chlorophyll b, the most abundant

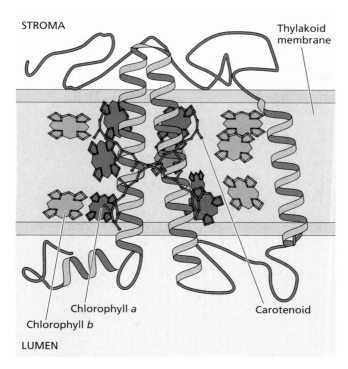

FIGURE 7.20 Two-dimensional view of the structure of the LHCII antenna complex from higher plants. The antenna complex is a transmembrane pigment protein, with three helical regions that cross the nonpolar part of the membrane. Approximately 14 chlorophyll a and b molecules are associated with the complex, as well as four carotenoids. Not all pigments are visible. The positions of several of the chlorophylls are shown, and two of the carotenoids form an X in the middle of the complex. In the membrane, the complex is trimeric and aggregates around the periphery of the PSII reaction center complex. (After Kühlbrandt et al. 1994; Liu et al. 2003.)

antenna proteins are members of a large family of structurally related proteins. Some of these proteins are associated primarily with photosystem II and are called **light-harvesting complex II (LHCII)** proteins; others are associated with photosystem I and are called *LHCI* proteins. These antenna complexes are also known as **chlorophyll *a*/*b* antenna proteins** (Paulsen 1995; Green and Durnford 1996; Green and Parson 2003).

The structure of one of the LHCII proteins has been determined (Figure 7.20) (Kühlbrandt et al. 1994; Liu et al. 2003). The protein contains three α-helical regions and binds 14 chlorophyll a and b molecules, as well as four carotenoids. The structure of the LHCI proteins is generally similar to that of the LHCII proteins (Ben-Shem et al. 2003). All of these proteins have significant sequence similarity and are almost certainly descendants of a common ancestral protein (Grossman et al. 1995; Green and Durnford 1996; Green and Parson 2003).

Light absorbed by carotenoids or chlorophyll b in the LHC proteins is rapidly transferred to chlorophyll a and then to other antenna pigments that are intimately associated with the reaction center. The LHCII complex is also involved in regulatory processes, which are discussed later in the chapter.

Mechanisms of Electron Transport

Some of the evidence that led to the idea of two photochemical reactions operating in series was discussed earlier in this chapter. Here we will consider in detail the chemical reactions involved in electron transfer during photosynthesis. We will discuss the excitation of chlorophyll by light and the reduction of the first electron acceptor, the flow of electrons through photosystems II and I, the oxidation of water as the primary source of electrons, and the reduction of the final electron acceptor (NADP$^+$). The chemiosmotic mechanism that mediates ATP synthesis will be discussed in detail later in the chapter (see "Proton Transport and ATP Synthesis in the Chloroplast").

Electrons from chlorophyll travel through the carriers organized in the "Z scheme"

Figure 7.21 shows a current version of the Z scheme, in which all the electron carriers known to function in electron flow from H$_2$O to NADP$^+$ are arranged vertically at their midpoint redox potentials (see **Web Topic 7.6** for further detail). Components known to react with each other are connected by arrows, so the Z scheme is really a synthesis of both kinetic and thermodynamic information. The large vertical arrows represent the input of light energy into the system.

Photons excite the specialized chlorophyll of the reaction centers (P680 for PSII, and P700 for PSI), and an electron is ejected. The electron then passes through a series of electron carriers and eventually reduces P700 (for electrons from PSII) or NADP$^+$ (for electrons from PSI). Much of the following discussion describes the journeys of these electrons and the nature of their carriers.

Almost all the chemical processes that make up the light reactions of photosynthesis are carried out by four major protein complexes: photosystem II, the cytochrome b_6f complex, photosystem I, and the ATP synthase. These four integral membrane complexes are vectorially oriented in the thylakoid membrane to function as follows (see Figures 7.18 and 7.22):

- Photosystem II oxidizes water to O$_2$ in the thylakoid lumen and in the process releases protons into the lumen.

- Cytochrome b_6f oxidizes plastohydroquinone (PQH$_2$) molecules that were reduced by PSII and delivers electrons to PSI. The oxidation of plastohydroquinone is coupled to proton transfer into the lumen from the stroma, generating a proton motive force.

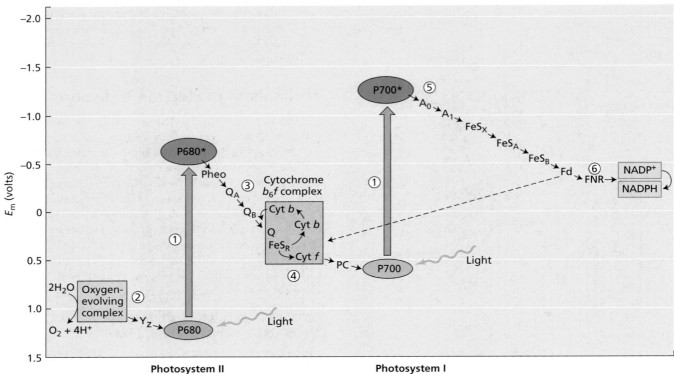

Figure 7.21 Detailed Z scheme for O_2-evolving photosynthetic organisms. The redox carriers are placed at their midpoint redox potentials (at pH 7). (1) The vertical arrows represent photon absorption by the reaction center chlorophylls: P680 for photosystem II (PSII) and P700 for photosystem I (PSI). The excited PSII reaction center chlorophyll, P680*, transfers an electron to pheophytin (Pheo). (2) On the oxidizing side of PSII (to the left of the arrow joining P680 with P680*), P680 oxidized by light is re-reduced by Y_z, that has received electrons from oxidation of water. (3) On the reducing side of PSII (to the right of the arrow joining P680 with P680*), pheophytin transfers electrons to the acceptors Q_A and Q_B, which are plastoquinones. (4) The cytochrome b_6f complex transfers electrons to plastocyanin (PC), a soluble protein, which in turn reduces P700$^+$ (oxidized P700). (5) The acceptor of electrons from P700* (A_0) is thought to be a chlorophyll, and the next acceptor (A_1) is a quinone. A series of membrane-bound iron–sulfur proteins (FeS_X, FeS_A, and FeS_B) transfers electrons to soluble ferredoxin (Fd). (6) The soluble flavoprotein ferredoxin–NADP reductase (FNR) reduces NADP$^+$ to NADPH, which is used in the Calvin cycle to reduce CO_2 (see Chapter 8). The dashed line indicates cyclic electron flow around PSI. (After Blankenship and Prince 1985.)

- Photosystem I reduces NADP$^+$ to NADPH in the stroma by the action of ferredoxin (Fd) and the flavoprotein ferredoxin–NADP reductase (FNR).
- ATP synthase produces ATP as protons diffuse back through it from the lumen into the stroma.

Energy is captured when an excited chlorophyll reduces an electron acceptor molecule

As discussed earlier, the function of light is to excite a specialized chlorophyll in the reaction center, either by direct absorption or, more frequently, via energy transfer from an antenna pigment. This excitation process can be envisioned as the promotion of an electron from the highest-energy filled orbital of the chlorophyll to the lowest-energy unfilled orbital (Figure 7.23). The electron in the upper orbital is only loosely bound to the chlorophyll and is easily lost if a molecule that can accept the electron is nearby.

The first reaction that converts electron energy into chemical energy—that is, the primary photochemical event—is the transfer of an electron from the excited state of a chlorophyll in the reaction center to an acceptor molecule. An equivalent way to view this process is that the absorbed photon causes an electron rearrangement in the reaction center chlorophyll, followed by an electron transfer process in which part of the energy in the photon is captured in the form of redox energy.

Immediately after the photochemical event, the reaction center chlorophyll is in an oxidized state (electron deficient, or positively charged), and the nearby electron acceptor molecule is reduced (electron rich, or negatively charged). The system is now at a critical juncture. The lower-energy orbital of the positively charged oxidized reaction center chlorophyll shown in Figure 7.23 has a vacancy and can accept an electron. If the acceptor molecule donates its electron back

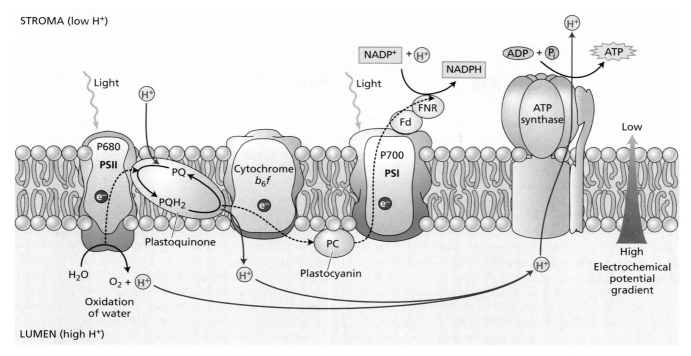

FIGURE 7.22 The transfer of electrons and protons in the thylakoid membrane is carried out vectorially by four protein complexes (see Figure 7.18B for structures). Water is oxidized and protons are released in the lumen by PSII. PSI reduces $NADP^+$ to NADPH in the stroma, via the action of ferredoxin (Fd) and the flavoprotein ferredoxin–NADP reductase (FNR). Protons are also transported into the lumen by the action of the cytochrome $b_6 f$ complex and contribute to the electrochemical proton gradient. These protons must then diffuse to the ATP synthase enzyme, where their diffusion down the electrochemical potential gradient is used to synthesize ATP in the stroma. Reduced plastoquinone (PQH_2) and plastocyanin transfer electrons to cytochrome $b_6 f$ and to PSI, respectively. The dashed lines represent electron transfer; solid lines represent proton movement.

to the reaction center chlorophyll, the system will be returned to the state that existed before the light excitation, and all the absorbed energy will be converted into heat.

This wasteful *recombination* process, however, does not appear to occur to any substantial degree in functioning reaction centers. Instead, the acceptor transfers its extra electron to a secondary acceptor and so on down the electron transport chain. The oxidized reaction center of the chlorophyll that had donated an electron is re-reduced by a secondary donor, which in turn is reduced by a tertiary donor. In plants, the ultimate electron donor is H_2O, and the ultimate electron acceptor is $NADP^+$ (see Figure 7.21).

The essence of photosynthetic energy storage is thus the initial transfer of an electron from an excited chlorophyll to an acceptor molecule, followed by a very rapid series of secondary chemical reactions that separate the positive and negative charges. These secondary reactions separate the

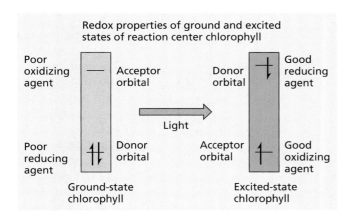

FIGURE 7.23 Orbital occupation diagram for the ground and excited states of reaction center chlorophyll. In the ground state the molecule is a poor reducing agent (loses electrons from a low-energy orbital) and a poor oxidizing agent (accepts electrons only into a high-energy orbital). In the excited state the situation is reversed, and an electron can be lost from the high-energy orbital, making the molecule an extremely powerful reducing agent. This is the reason for the extremely negative excited-state redox potential shown by P680* and P700* in Figure 7.21. The excited state can also act as a strong oxidant by accepting an electron into the lower-energy orbital, although this pathway is not significant in reaction centers. (After Blankenship and Prince 1985.)

charges to opposite sides of the thylakoid membrane in approximately 200 picoseconds (1 picosecond = 10^{-12} s).

With the charges thus separated, the reversal reaction is many orders of magnitude slower, and the energy has been captured. Each of the secondary electron transfers is accompanied by a loss of some energy, thus making the process effectively irreversible. The quantum yield for the production of stable products in purified reaction centers from photosynthetic bacteria has been measured as 1.0; that is, every photon produces stable products, and no reversal reactions occur.

Measured quantum requirements for O_2 production in higher plants under optimal conditions (low-intensity light) indicate that the values for the primary photochemical events are also very close to 1.0. The structure of the reaction center appears to be extremely fine-tuned for maximal rates of productive reactions and minimal rates of energy-wasting reactions.

The reaction center chlorophylls of the two photosystems absorb at different wavelengths

As discussed earlier in the chapter, PSI and PSII have distinct absorption characteristics. Precise measurements of absorption maxima were made possible by optical changes in the reaction center chlorophylls in the reduced and oxidized states. The reaction center chlorophyll is transiently in an oxidized state after losing an electron and before being re-reduced by its electron donor.

In the oxidized state, the strong light absorbance in the red region of the spectrum that is characteristic of chlorophylls is lost, or **bleached**. It is therefore possible to monitor the redox state of these chlorophylls by time-resolved optical absorbance measurements in which this bleaching is monitored directly (see **Web Topic 7.1**).

Using such techniques, it was found that the reaction center chlorophyll of photosystem I absorbs maximally at 700 nm in its reduced state. Accordingly, this chlorophyll is named **P700** (the P stands for *pigment*). The analogous optical transient of photosystem II is at 680 nm, so its reaction center chlorophyll is known as **P680**. Earlier, the reaction center bacteriochlorophyll from purple photosynthetic bacteria was identified as **P870**.

The X-ray structure of the bacterial reaction center (see Figures 7.5.A and 7.5.B in **Web Topic 7.5**) clearly indicates that P870 is a closely coupled pair or dimer of bacteriochlorophylls, rather than a single molecule. The primary donor of photosystem I, P700, is also a dimer of chlorophyll *a* molecules. Photosystem II also contains a dimer of chlorophylls, although the primary may not originate from these pigments. In the oxidized state, reaction center chlorophylls contain an unpaired electron. Molecules with unpaired electrons often can be detected by a magnetic-resonance technique known as **electron spin resonance (ESR)**. ESR studies, along with the spectroscopic measurements already described, have led to the discovery of many intermediate electron carriers in the photosynthetic electron transport system.

The photosystem II reaction center is a multisubunit pigment–protein complex

Photosystem II is contained in a multisubunit protein supercomplex (Figure 7.24) (Barber et al. 1999). In higher plants, the multisubunit protein supercomplex has two complete reaction centers and some antenna complexes. The core of the reaction center consists of two membrane proteins known as D1 and D2, as well as other proteins, as shown in Figure 7.25 and **Web Topic 7.7** (Zouni et al. 2001; Ferreira et al. 2004).

The primary donor chlorophyll, additional chlorophylls, carotenoids, pheophytins, and plastoquinones (two electron acceptors described in the following section) are bound to the membrane proteins D1 and D2. These proteins have some sequence similarity to the L and M peptides of purple bacteria. Other proteins serve as antenna complexes or are involved in oxygen evolution. Some, such as cytochrome b_{559}, have no known function but may be involved in a protective cycle around photosystem II.

Water is oxidized to oxygen by photosystem II

Water is oxidized according to the following chemical reaction (Hoganson and Babcock 1997):

$$2\,H_2O \rightarrow O_2 + 4\,H^+ + 4\,e^- \quad (7.8)$$

This equation indicates that four electrons are removed from two water molecules, generating an oxygen molecule and four hydrogen ions. (For more on oxidation–reduction reactions, see Chapter 2 on the web site and **Web Topic 7.6**.)

Water is a very stable molecule. Oxidation of water to form molecular oxygen is very difficult, and the photosynthetic oxygen-evolving complex is the only known biochemical system that carries out this reaction. Photosynthetic oxygen evolution is also the source of almost all the oxygen in Earth's atmosphere.

Many studies have provided a substantial amount of information about the process (see **Web Topic 7.7**). The protons produced by water oxidation are released into the lumen of the thylakoid, not directly into the stromal compartment (see Figure 7.22). They are released into the lumen because of the vectorial nature of the membrane and the fact that the oxygen-evolving complex is localized on the interior surface of the thylakoid. These protons are eventually transferred from the lumen to the stroma by translocation through ATP synthase. In this way, the protons released during water oxidation contribute to the electrochemical potential driving ATP formation.

It has been known for many years that manganese (Mn) is an essential cofactor in the water-oxidizing process (see Chapter 5), and a classic hypothesis in photosynthesis research postulates that Mn^{2+} ions undergo a series of oxidations—which are known as *S states*, and are labeled S_0,

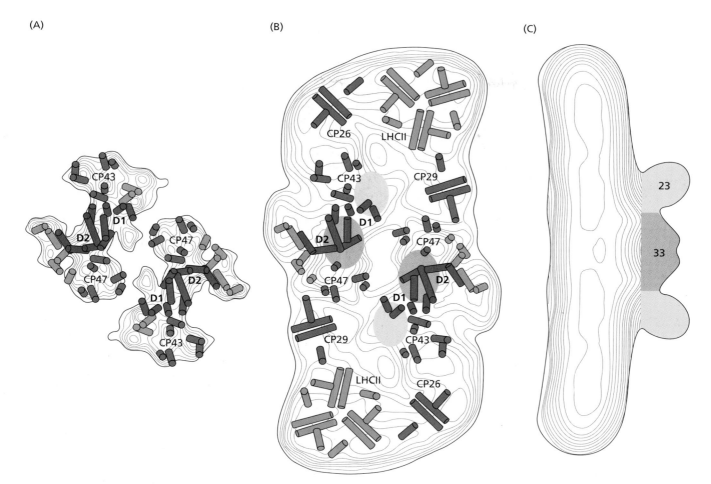

FIGURE 7.24 Structure of dimeric multisubunit protein supercomplex of photosystem II from higher plants, as determined by electron microscopy. The figure shows two complete reaction centers, each of which is a dimeric complex. (A) Helical arrangement of the D1 and D2 (red) and CP43 and CP47 (green) core subunits. (B) View from the lumenal side of the supercomplex, including additional antenna complexes, LHCII, CP26, and CP29, and an extrinsic oxygen-evolving complex, shown as orange and yellow circles. Unassigned helices are shown in gray. (C) Side view of the complex illustrating the arrangement of the extrinsic proteins of the oxygen-evolving complex. (After Barber et al. 1999.)

S_1, S_2, S_3, and S_4 (see **Web Topic 7.7**)—that are perhaps linked to H_2O oxidation and the generation of O_2. This hypothesis has received strong support from a variety of experiments, most notably X-ray absorption and ESR studies, both of which detect the manganese directly (Yachandra et al. 1996). Analytical experiments indicate that four Mn^{2+} ions are associated with each oxygen-evolving complex. Other experiments have shown that Cl^- and Ca^{2+} ions are essential for O_2 evolution (see **Web Topic 7.7**). The detailed chemical mechanism of the oxidation of water to O_2 is not yet well understood, but with structural information now available, rapid progress is being made in this area (McEvoy and Brudvig 2004).

One electron carrier, generally identified as Y_z, functions between the oxygen-evolving complex and P680 (see Figure 7.21). To function in this region, Y_z needs to have a very strong tendency to retain its electrons. This species has been identified as a radical formed from a tyrosine residue in the D1 protein of the PSII reaction center.

Pheophytin and two quinones accept electrons from photosystem II

Evidence from spectral and ESR studies indicates that pheophytin acts as an early acceptor in photosystem II, followed by a complex of two plastoquinones in close proximity to an iron atom. **Pheophytin** is a chlorophyll in which the central magnesium atom has been replaced by two hydrogen atoms. This chemical change gives pheophytin chemical and spectral properties that are slightly different from those of chlorophyll. The precise arrangement of the carriers in the electron acceptor complex is not known, but it is probably very similar to that of the reaction center of purple bacteria (for details, see Figure 7.5.B in **Web Topic 7.5**).

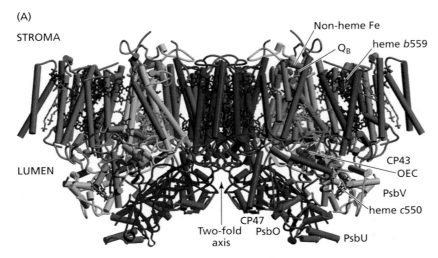
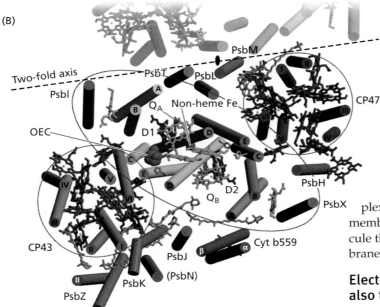
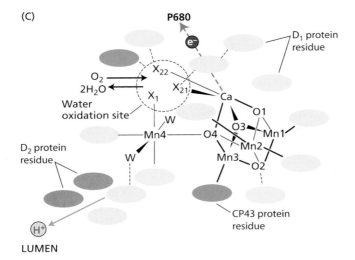

Figure 7.25 Structure of the photosystem II reaction center from the cyanobacterium *Thermosynechococcus elongatus*, resolved at 3.5 Å. The structure includes the D_1 (yellow) and D_2 (orange) core reaction center proteins, the CP43 (green) and CP47 (red) antenna proteins, cytochromes b_{559} and c_{550}, the extrinsic 33-kDa oxygen evolution protein PsbO (blue), and the pigments and other cofactors. (A) Side view parallel to the membrane plane. (B) View from the lumenal surface, perpendicular to the plane of the membrane. (C) Detail of the Mn-containing water-splitting complex. Proposed pathways for electron and proton transfer are indicated with green and blue arrows, respectively. W, water molecules; X_1, X_{21}, and X_{22} are water binding sites. (After Ferreira et al. 2004.)

Two plastoquinones (PQ_A and PQ_B) are bound to the reaction center and receive electrons from pheophytin in a sequential fashion (Okamura et al. 2000). Transfer of the two electrons to Q_B reduces it to PQ_B^{2-}, and the reduced PQ_B^{2-} takes two protons from the stroma side of the medium, yielding a fully reduced **plastohydroquinone** (PQH_2) (Figure 7.26). The plastohydroquinone then dissociates from the reaction center complex and enters the hydrocarbon portion of the membrane, where it in turn transfers its electrons to the cytochrome $b_6 f$ complex. Unlike the large protein complexes of the thylakoid membrane, plastohydroquinone is a small, nonpolar molecule that diffuses readily in the nonpolar core of the membrane bilayer.

Electron flow through the cytochrome $b_6 f$ complex also transports protons

The **cytochrome $b_6 f$ complex** is a large multisubunit protein with several prosthetic groups (Figure 7.27) (Cramer et al. 1996; Kurisu et al. 2003; Stroebel et al. 2003). It contains two *b*-type hemes and one *c*-type heme (**cytochrome *f***). In *c*-type cytochromes the heme is covalently attached to the peptide; in *b*-type cytochromes the chemically similar protoheme group is not covalently attached (see **Web Topic 7.8**). In addition, the complex contains a **Rieske iron–sulfur protein** (named for the scientist who discovered it), in which two iron atoms are bridged by two sulfur atoms. The functional roles of all these cofactors are reasonably well understood, as described below. However, the cytochrome $b_6 f$ complex also contains additional cofactors, including an additional heme, a chlorophyll, and a carotenoid, whose functions are yet to be resolved.

The structures of the cytochrome $b_6 f$ complex and the related cytochrome bc_1 suggest a mechanism for electron and proton flow. The precise way by which electrons and

(A)

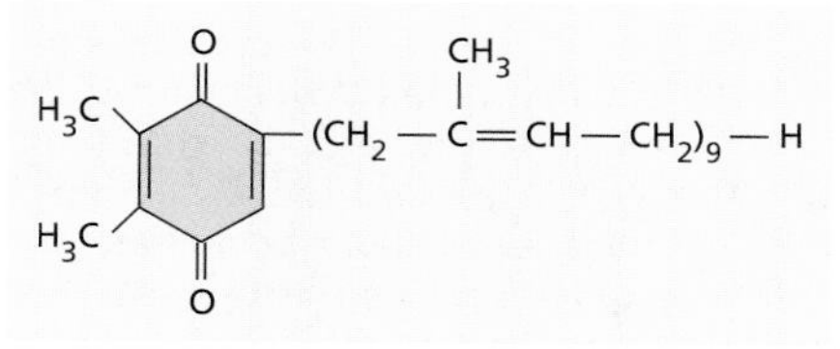

Plastoquinone

FIGURE 7.26 Structure and reactions of plastoquinones that operate in photosystem II. (A) The plastoquinone consists of a quinoid head and a long nonpolar tail that anchors it in the membrane. (B) Redox reactions of plastoquinone. The fully oxidized quinone (Q), anionic plastosemiquinone (Q•⁻), and reduced plastohydroquinone (PQH$_2$) forms are shown; R represents the side chain.

(B)

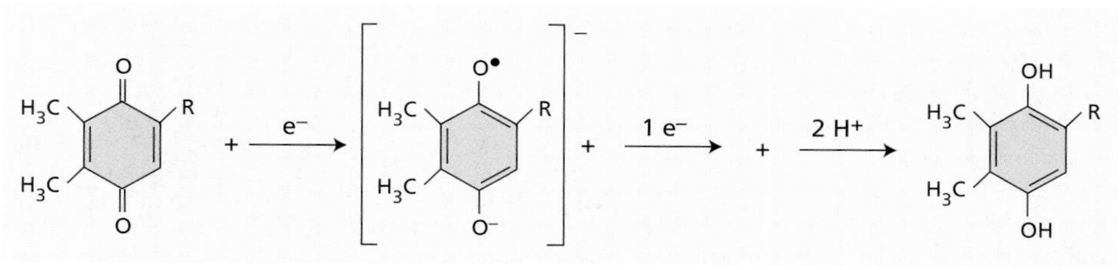

Quinone (Q) → Plastosemiquinone (Q•⁻) → Plastohydroquinone (QH$_2$)

protons flow through the cytochrome $b_6 f$ complex is not yet fully understood, but a mechanism known as the **Q cycle** accounts for most of the observations. In this mechanism, plastohydroquinone (PQH$_2$) is oxidized, and one of the two electrons is passed along a linear electron transport chain toward photosystem I, while the other electron goes through a cyclic process that increases the number of protons pumped across the membrane (Figure 7.28).

In the linear electron transport chain, the oxidized Rieske protein (**FeS$_R$**) accepts an electron from PQH$_2$ and transfers it to cytochrome f (see Figure 7.28A). Cytochrome f then transfers an electron to the blue-colored copper protein plastocyanin (PC), which in turn reduces oxidized P700 of PSI. In the cyclic part of the process (see Figure 7.28B), the plastosemiquinone (see Figure 7.26) transfers its other electron to one of the b-type hemes, releasing both of its protons to the lumenal side of the membrane.

The first b-type heme transfers its electron through the second b-type heme to an oxidized plastoquinone molecule, reducing it to the semiquinone form near the stromal surface of the complex. Another similar sequence of electron flow fully reduces the plastoquinone, which picks up

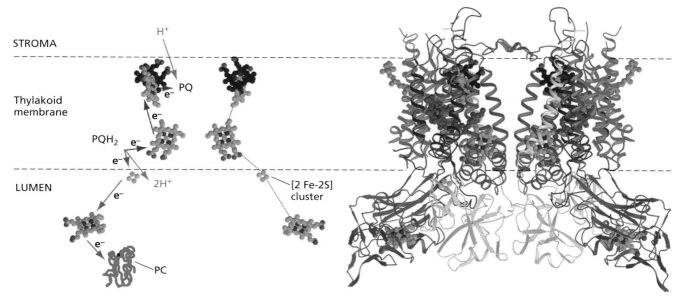

FIGURE 7.27 Structure of the cytochrome $b_6 f$ complex from cyanobacteria. Cofactor structure is on the left, and the protein structure is on the right. Cytochrome b_6 is shown in blue, cytochrome f in red. [2 Fe-2S] cluster, Rieske iron–sulfur protein; PC, plastocyanin; PQ, plastoquinone; PQH$_2$; plastohydroquinone. (After Kurisu et al. 2003.)

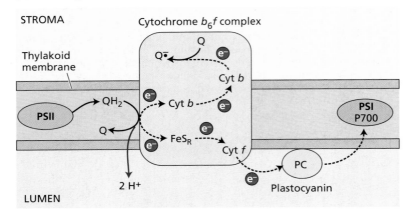

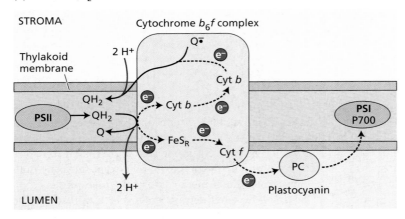

FIGURE 7.28 Mechanism of electron and proton transfer in the cytochrome b_6f complex. This complex contains two b-type cytochromes (Cyt b), a c-type cytochrome (Cyt c, historically called cytochrome f), a Rieske Fe–S protein (FeS_R), and two quinone oxidation–reduction sites. (A) The noncyclic or linear processes: A plastohydroquinone (PQH_2) molecule produced by the action of PSII (see Figure 7.26) is oxidized near the lumenal side of the complex, transferring its two electrons to the Rieske Fe–S protein and one of the b-type cytochromes and simultaneously expelling two protons to the lumen. The electron transferred to FeS_R is passed to cytochrome f (Cyt f) and then to plastocyanin (PC), which reduces P700 of PSI. The reduced b-type cytochrome transfers an electron to the other b-type cytochrome, which reduces a plastoquinone (PQ) to the plastosemiquinone (PQ•) state (see Figure 7.26). (B) The cyclic processes: A second PQH_2 is oxidized, with one electron going from FeS_R to PC and finally to P700. The second electron goes through the two b-type cytochromes and reduces the plastosemiquinone to the plastohydroquinone, at the same time picking up two protons from the stroma. Overall, four protons are transported across the membrane for every two electrons delivered to P700.

protons from the stromal side of the membrane and is released from the b_6f complex as plastohydroquinone.

The net result of two turnovers of the complex is that two electrons are transferred to P700, two plastohydroquinones are oxidized to the plastoquinone form, and one oxidized plastoquinone is reduced to the plastohydroquinone form. In the process of oxidizing the plastoquinones, four protons are transferred from the stromal to the lumenal side of the membrane.

By this mechanism, electron flow connecting the acceptor side of the PSII reaction center to the donor side of the PSI reaction center also gives rise to an electrochemical potential across the membrane, due in part to H⁺ concentration differences on the two sides of the membrane. This electrochemical potential is used to power the synthesis of ATP. The cyclic electron flow through the cytochrome b and plastoquinone increases the number of protons pumped per electron beyond what could be achieved in a strictly linear sequence.

Plastoquinone and plastocyanin carry electrons between photosystems II and I

The location of the two photosystems at different sites on the thylakoid membranes (see Figure 7.18) requires that at least one component be capable of moving along or within the membrane in order to deliver electrons produced by photosystem II to photosystem I. The cytochrome b_6f complex is distributed equally between the grana and the stroma regions of the membranes, but its large size makes it unlikely that it is the mobile carrier. Instead, plastoquinone or plastocyanin or possibly both are thought to serve as mobile carriers to connect the two photosystems.

Plastocyanin (PC) is a small (10.5 kDa), water-soluble, copper-containing protein that transfers electrons between the cytochrome b_6f complex and P700. This protein is found in the lumenal space (see Figure 7.28). In certain green algae and cyanobacteria, a c-type cytochrome is sometimes found instead of plastocyanin; which of these two proteins is synthesized depends on the amount of copper available to the organism.

The photosystem I reaction center reduces NADP⁺

The PSI reaction center complex is a large multisubunit complex (Figure 7.29) (Jordan et al. 2001; Ben-Shem et al. 2003). In contrast to PSII, a core antenna consisting of about 100 chlorophylls is an integral part of the PSI reaction center. The core antenna and P700 are bound to two proteins, PsaA and PsaB, with molecular masses in the range of 66 to 70 kDa (see **Web Topic 7.8**) (Brettel 1997; Chitnis 2001). The PSI reaction center complex from pea contains four LHCI complexes in addition to the core structure similar to that found in cyanobacteria (see Figure 7.29). The total number of chlorophyll molecules in this complex is nearly 200.

The core antenna pigments form a bowl surrounding the electron transfer cofactors, which are in the center of the complex. In their reduced form, the electron carriers that function in the acceptor region of photosystem I are all extremely strong reducing agents. These reduced species are very unstable and thus difficult to identify. Evidence indicates that one of these early acceptors is a chlorophyll molecule, and another is a quinone species, phylloquinone, also known as vitamin K_1.

Additional electron acceptors include a series of three membrane-associated iron–sulfur proteins, also known as **Fe–S centers**: **FeS$_X$**, **FeS$_A$**, and **FeS$_B$** (see Figure 7.29). Fe–S center X is part of the P700-binding protein; centers A and B reside on an 8-kDa protein that is part of the PSI reaction center complex. Electrons are transferred through centers A and B to **ferredoxin** (**Fd**), a small, water-soluble iron–sulfur protein (see Figures 7.21 and 7.29). The membrane-associated flavoprotein **ferredoxin–NADP reductase** (**FNR**) reduces $NADP^+$ to NADPH, thus completing the sequence of noncyclic electron transport that begins with the oxidation of water (Karplus et al. 1991).

In addition to the reduction of $NADP^+$, reduced ferredoxin produced by photosystem I has several other functions in the chloroplast, such as the supply of reductants to reduce nitrate and the regulation of some of the carbon-fixation enzymes (see Chapter 8).

Cyclic electron flow generates ATP but no NADPH

Some of the cytochrome $b_6 f$ complexes are found in the stroma region of the membrane, where photosystem I is located. Under certain conditions, **cyclic electron flow** is known to occur from the reducing side of photosystem I via plastohydroquinone and the $b_6 f$ complex and back to P700. This cyclic electron flow is coupled to proton pumping into the lumen, which can be utilized for ATP synthesis but does not oxidize water or reduce $NADP^+$. Cyclic electron flow is especially important as an ATP source in the bundle sheath chloroplasts of some plants that carry out C_4 carbon fixation (see Chapter 8).

Some herbicides block photosynthetic electron flow

The use of herbicides to kill unwanted plants is widespread in modern agriculture. Many different classes of herbicides have been developed, and they act by blocking amino acid, carotenoid, or lipid biosynthesis or by disrupting cell division. Other herbicides, such as dichlorophenyldimethylurea (DCMU) and paraquat, block photosynthetic electron flow (Figure 7.30). DCMU is also known as diuron. Paraquat has acquired public notoriety because of its use on marijuana crops.

Many herbicides, DCMU among them, act by blocking electron flow at the quinone acceptors of photosystem II,

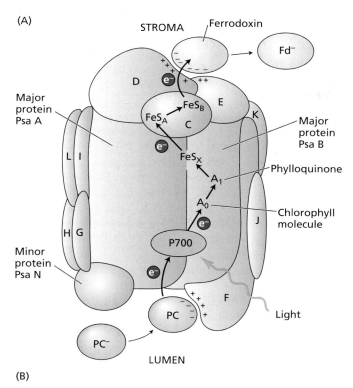

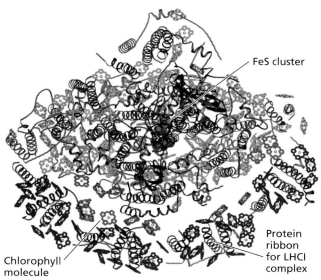

FIGURE 7.29 Structure of photosystem I. (A) Structural model of the PSI reaction center from higher plants. Components of the PSI reaction center are organized around two major core proteins, PsaA and PsaB. Minor proteins PsaC to PsaN are labeled C to N. Electrons are transferred from plastocyanin (PC) to P700 (see Figures 7.21 and 7.22) and then to a chlorophyll molecule (A_0), to phylloquinone (A_1), to the Fe–S centers FeS$_X$, FeS$_A$, and FeS$_B$, and finally to the soluble iron–sulfur protein ferredoxin (Fd). (B) Structure of the photosystem I reaction center complex from pea at 4.4 Å resolution, including the LHCI antenna complexes. This is viewed from the stromal side of the membrane. (A after Buchanan et al. 2000; B after Nelson and Ben-Shem 2004.)

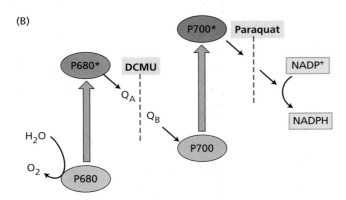

FIGURE 7.30 Chemical structure and mechanism of action of two important herbicides. (A) Chemical structure of dichlorophenyldimethylurea (DCMU) and methyl viologen (paraquat), two herbicides that block photosynthetic electron flow. DCMU is also known as diuron. (B) Sites of action of the two herbicides. DCMU blocks electron flow at the quinone acceptors of photosystem II, by competing for the binding site of plastoquinone. Paraquat acts by accepting electrons from the early acceptors of photosystem I.

by competing for the binding site of plastoquinone that is normally occupied by PQ_B. Other herbicides, such as paraquat, act by accepting electrons from the early acceptors of photosystem I and then reacting with oxygen to form superoxide, O_2^-, a species that is very damaging to chloroplast components, especially lipids.

Proton Transport and ATP Synthesis in the Chloroplast

In the preceding sections we learned how captured light energy is used to reduce $NADP^+$ to NADPH. Another fraction of the captured light energy is used for light-dependent ATP synthesis, which is known as **photophosphorylation**. This process was discovered by Daniel Arnon and his coworkers in the 1950s. In normal cellular conditions, photophosphorylation requires electron flow, although under some conditions electron flow and photophosphorylation can take place independently of each other. Electron flow without accompanying phosphorylation is said to be **uncoupled**.

It is now widely accepted that photophosphorylation works via the **chemiosmotic mechanism**. This mechanism was first proposed in the 1960s by Peter Mitchell. The same general mechanism drives phosphorylation during aerobic respiration in bacteria and mitochondria (see Chapter 11), as well as the transfer of many ions and metabolites across membranes (see Chapter 6). Chemiosmosis appears to be a unifying aspect of membrane processes in all forms of life.

In Chapter 6 we discussed the role of ATPases in chemiosmosis and ion transport at the cell's plasma membrane. The ATP used by the plasma membrane ATPase is synthesized by photophosphorylation in the chloroplast and oxidative phosphorylation in the mitochondrion. Here we are concerned with chemiosmosis and transmembrane proton concentration differences used to make ATP in the chloroplast.

The basic principle of chemiosmosis is that ion concentration differences and electric-potential differences across membranes are sources of free energy that can be utilized by the cell. As described by the second law of thermodynamics (see Chapter 2 on the web site for a detailed discussion), any nonuniform distribution of matter or energy represents a source of energy. Differences in **chemical potential** of any molecular species whose concentrations are not the same on opposite sides of a membrane provide such a source of energy.

The asymmetric nature of the photosynthetic membrane and the fact that proton flow from one side of the membrane to the other accompanies electron flow were discussed earlier. The direction of proton translocation is such that the stroma becomes more alkaline (fewer H^+ ions) and the lumen becomes more acidic (more H^+ ions) as a result of electron transport (see Figures 7.22 and 7.28).

Some of the early evidence supporting a chemiosmotic mechanism of photosynthetic ATP formation was provided by an elegant experiment carried out by André Jagendorf and coworkers (Figure 7.31). They suspended chloroplast thylakoids in a pH 4 buffer, and the buffer diffused across the membrane, causing the interior, as well as the exterior, of the thylakoid to equilibrate at this acidic pH. They then rapidly transferred the thylakoids to a pH 8 buffer, thereby creating a pH difference of 4 units across the thylakoid membrane, with the inside acidic relative to the outside.

They found that large amounts of ATP were formed from ADP and P_i by this process, with no light input or electron transport. This result supports the predictions of the chemiosmotic hypothesis, described in the paragraphs that follow.

Mitchell proposed that the total energy available for ATP synthesis, which he called the **proton motive force** (Δp), is the sum of a proton chemical potential and a transmembrane electric potential. These two components of the proton motive force from the outside of the membrane to the inside are given by the following equation:

$$\Delta p = \Delta E - 59(pH_i - pH_o) \qquad (7.9)$$

where ΔE is the transmembrane electric potential, and $pH_i - pH_o$ (or ΔpH) is the pH difference across the membrane. The constant of proportionality (at 25°C) is 59 mV per pH

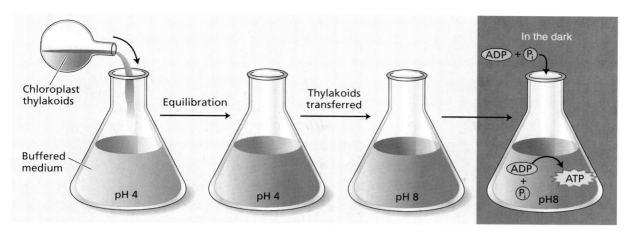

FIGURE 7.31 Summary of the experiment carried out by Jagendorf and coworkers. Isolated chloroplast thylakoids kept previously at pH 8 were equilibrated in an acid medium at pH 4. The thylakoids were then transferred to a buffer at pH 8 that contained ADP and P_i. The proton gradient generated by this manipulation provided a driving force for ATP synthesis in the absence of light. This experiment verified a prediction of the chemiosmotic theory stating that a chemical potential across a membrane can provide energy for ATP synthesis. (After Jagendorf 1967.)

unit, so a transmembrane pH difference of one pH unit is equivalent to a membrane potential of 59 mV.

In addition to the need for mobile electron carriers discussed earlier, the uneven distribution of photosystems II and I, and of ATP synthase at the thylakoid membrane (see Figure 7.18), poses some challenges for the formation of ATP. ATP synthase is found only in the stroma lamellae and at the edges of the grana stacks. Protons pumped across the membrane by the cytochrome b_6f complex or protons produced by water oxidation in the middle of the grana must move laterally up to several tens of nanometers to reach ATP synthase.

The ATP is synthesized by a large (400 kDa) enzyme complex known by several names: **ATP synthase**, **ATPase** (after the reverse reaction of ATP hydrolysis), and **CF$_o$–CF$_1$** (Boyer 1997). This enzyme consists of two parts: a hydrophobic membrane-bound portion called CF$_o$ and a portion that sticks out into the stroma called CF$_1$ (Figure 7.32). CF$_o$ appears to form a channel across the membrane through which protons can pass. CF$_1$ is made up of several peptides, including three copies of each of the α and β peptides arranged alternately much like the sections of an orange. Whereas the catalytic sites are located largely on the β polypeptide, many of the

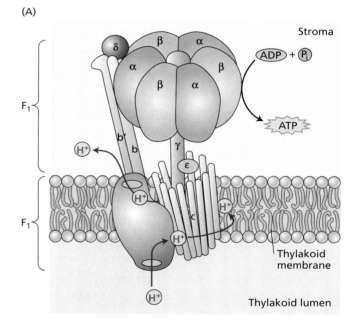

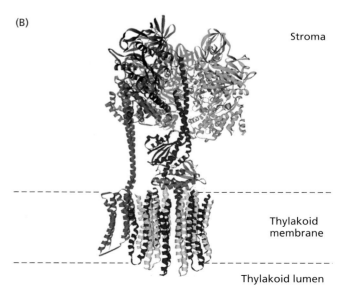

FIGURE 7.32 Subunit composition (A) and compiled crystal structure (B) of chloroplast F1F$_0$ ATP synthase. This enzyme consists of a large multisubunit complex, CF1, attached on the stromal side of the membrane to an integral membrane portion, known as CF$_0$. CF1 consists of five different polypeptides, with stoichiometry of α_3, β_3, γ, δ, ε. CF$_0$ contains probably four different polypeptides, with a stoichiometry of a, b, b', c_{14}. Protons from the lumen are transported by the rotating c polypeptide and ejected on the stroma side. (From W. Frasch unpublished results.)

(A) Purple bacteria

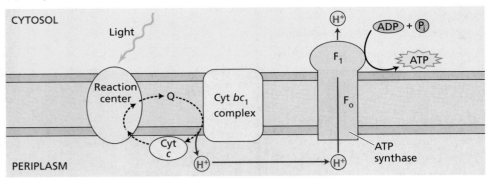

(B) Chloroplasts

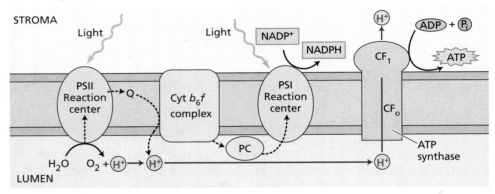

(C) Mitochondria

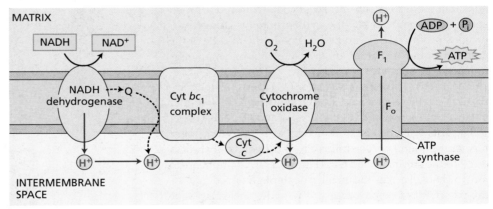

other peptides are thought to have primarily regulatory functions. CF_1 is the portion of the complex that synthesizes ATP.

The molecular structure of the mitochondrial ATP synthase has been determined by X-ray crystallography (Stock et al. 1999). Although there are significant differences between the chloroplast and mitochondrial enzymes, they have the same overall architecture and probably nearly identical catalytic sites. In fact, there are remarkable similarities in the way electron flow is coupled to proton translocation in chloroplasts, mitochondria, and purple bacteria (Figure 7.33). Another remarkable aspect of the mechanism of the ATP synthase is that the internal stalk and probably much of the CF_o portion of the enzyme rotate during catalysis (Yasuda et al. 2001). The enzyme is actually a tiny molecular motor (see **Web Topics 7.9 and 11.4**). Three molecules of ATP are synthesized for each rotation of the enzyme.

Direct microscopic imaging of the CF_o part of the chloroplast ATP synthase indicates that it contains 14 copies of the integral membrane subunit (Seelert et al. 2000). Each subunit can translocate one proton across the membrane

◀ **FIGURE 7.33** Similarities of photosynthetic and respiratory electron flow in bacteria, chloroplasts, and mitochondria. In all three, electron flow is coupled to proton translocation, creating a transmembrane proton motive force (Δp). The energy in the proton motive force is then used for the synthesis of ATP by ATP synthase. (A) A reaction center in purple photosynthetic bacteria carries out cyclic electron flow, generating a proton potential by the action of the cytochrome bc_1 complex. (B) Chloroplasts carry out noncyclic electron flow, oxidizing water and reducing $NADP^+$. Protons are produced by the oxidation of water and by the oxidation of PQH_2 (labeled "Q" in the illustration) by the cytochrome b_6f complex. (C) Mitochondria oxidize NADH to NAD^+ and reduce oxygen to water. Protons are pumped by the enzyme NADH dehydrogenase, the cytochrome bc_1 complex, and cytochrome oxidase. The ATP synthases in the three systems are very similar in structure.

each time the complex rotates. This suggests that the stoichiometry of protons translocated to ATP formed is 14/3, or 4.67. Measured values of this parameter are usually somewhat lower than this value and the reasons for this discrepancy are not yet understood.

Repair and Regulation of the Photosynthetic Machinery

Photosynthetic systems face a special challenge. They are designed to absorb large amounts of light energy and process it into chemical energy. At the molecular level, the energy in a photon can be damaging, particularly under unfavorable conditions. In excess, light energy can lead to the production of toxic species, such as superoxide, singlet oxygen, and peroxide, and damage can occur if the light energy is not dissipated safely (Horton et al. 1996; Asada 1999; Müller et al. 2001). Photosynthetic organisms therefore contain complex regulatory and repair mechanisms. Some of these mechanisms regulate energy flow in the antenna system, to avoid excess excitation of the reaction centers and ensure that the two photosystems are equally driven. Although very effective, these processes are not entirely fail-safe, and sometimes toxic compounds are produced. Additional mechanisms are needed to dissipate these compounds—in particular, toxic oxygen species. In this section we will examine how some of these processes work to protect the system against photodamage.

Despite these protective and scavenging mechanisms, damage can occur, and additional mechanisms are required to repair the system. Figure 7.34 provides an overview of the several levels of the regulation and repair systems.

Carotenoids serve as photoprotective agents

In addition to their role as accessory pigments, carotenoids play an essential role in **photoprotection**. The photosynthetic membrane can easily be damaged by the large amounts of energy absorbed by the pigments if this energy cannot be stored by photochemistry; this is why a pro-

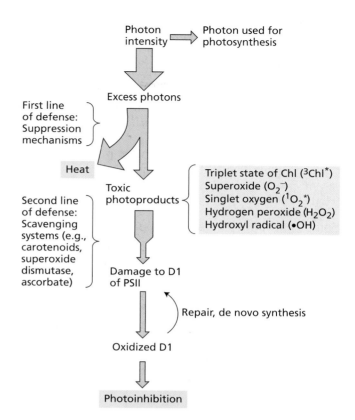

FIGURE 7.34 Overall picture of the regulation of photon capture and the protection and repair of photodamage. Protection against photodamage is a multilevel process. The first line of defense is suppression of damage by quenching of excess excitation as heat. If this defense is not sufficient and toxic photoproducts form, a variety of scavenging systems eliminate the reactive photoproducts. If this second line of defense also fails, the photoproducts can damage the D1 protein of photosystem II. This damage leads to photoinhibition. The D1 protein is then excised from the PSII reaction center and degraded. A newly synthesized D1 is reinserted into the PSII reaction center to form a functional unit. (After Asada 1999.)

tection mechanism is needed. The photoprotection mechanism can be thought of as a safety valve, venting excess energy before it can damage the organism. When the energy stored in chlorophylls in the excited state is rapidly dissipated by excitation transfer or photochemistry, the excited state is said to be **quenched**.

If the excited state of chlorophyll is not rapidly quenched by excitation transfer or photochemistry, it can react with molecular oxygen to form an excited state of oxygen known as **singlet oxygen** ($^1O_2^*$). The extremely reactive singlet oxygen goes on to react with and damage many cellular components, especially lipids. Carotenoids exert their photoprotective action by rapidly quenching the excited state of chlorophyll. The excited state of carotenoids does not have sufficient energy to form singlet oxygen, so it decays back to its ground state while losing its energy as heat.

Mutant organisms that lack carotenoids cannot live in the presence of both light and molecular oxygen—a rather difficult situation for an O_2-evolving photosynthetic organism. For non-O_2-evolving photosynthetic bacteria, mutants that lack carotenoids can be maintained under laboratory conditions if oxygen is excluded from the growth medium.

Recently carotenoids were found to play a role in nonphotochemical quenching, which is a second protective and regulatory mechanism (Holt et al. 2005).

Some xanthophylls also participate in energy dissipation

Nonphotochemical quenching, a major process regulating the delivery of excitation energy to the reaction center, can be thought of as a "volume knob" that adjusts the flow of excitations to the PSII reaction center to a manageable level, depending on the light intensity and other conditions. The process appears to be an essential part of the regulation of antenna systems in most algae and plants.

Nonphotochemical quenching is the quenching of chlorophyll fluorescence (see Figure 7.5) by processes other than photochemistry. As a result of nonphotochemical quenching, a large fraction of the excitations in the antenna system caused by intense illumination are quenched by conversion into heat (Krause and Weis 1991). Nonphotochemical quenching is thought to be involved in protecting the photosynthetic machinery against overexcitation and subsequent damage.

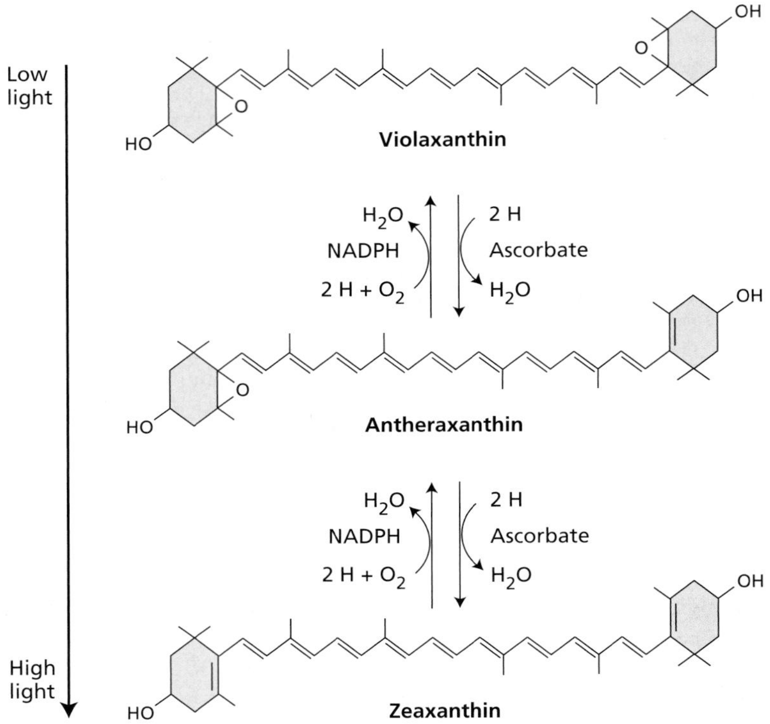

The molecular mechanism of nonphotochemical quenching is not well understood, and evidence suggests that there are several distinct quenching processes that may have different underlying mechanisms. It is clear that the pH of the thylakoid lumen and the state of aggregation of the antenna complexes are important factors. Three carotenoids, called **xanthophylls**, are involved in nonphotochemical quenching: violaxanthin, antheraxanthin, and zeaxanthin (Figure 7.35).

In high light, violaxanthin is converted into zeaxanthin, via the intermediate antheraxanthin, by the enzyme violaxanthin de-epoxidase. When light intensity decreases, the process is reversed. Binding of protons and zeaxanthin to light-harvesting antenna proteins is thought to cause conformational changes that lead to quenching and heat dissipation (Demmig-Adams and Adams 1992; Horton et al. 1996). Nonphotochemical quenching appears to be preferentially associated with a peripheral antenna complex of photosystem II, the PsbS protein (Li et al. 2000). Recent evidence suggests that a transient electron transfer process may be an important part of the molecular quenching mechanism (Holt et al. 2005).

FIGURE 7.35 Chemical structure of violaxanthin, antheraxanthin, and zeaxanthin. The highly quenched state of photosystem II is associated with zeaxanthin, the unquenched state with violaxanthin. Enzymes interconvert these two carotenoids, with antheraxanthin as the intermediate, in response to changing conditions, especially changes in light intensity. Zeaxanthin formation uses ascorbate as a cofactor, and violaxanthin formation requires NADPH. (After Pfündel and Bilger 1994.)

The photosystem II reaction center is easily damaged

Another effect that appears to be a major factor in the stability of the photosynthetic apparatus is photoinhibition, which occurs when excess excitation arriving at the PSII reaction center leads to its inactivation and damage (Long et al. 1994). **Photoinhibition** is a complex set of molecular processes, defined as the inhibition of photosynthesis by excess light.

As will be discussed in detail in Chapter 9, photoinhibition is reversible in early stages. Prolonged inhibition, however, results in damage to the system such that the PSII reaction center must be disassembled and repaired (Melis 1999). The main target of this damage is the D1 protein that makes up part of the PSII reaction center complex (see Figure 7.24). When D1 is damaged by excess light, it must be removed from the membrane and replaced with a newly synthesized molecule. The other components of the PSII reaction center are not damaged by excess excitation and are thought to be recycled, so the D1 protein is the only component that needs to be synthesized.

Photosystem I is protected from active oxygen species

Photosystem I is particularly vulnerable to damage from active oxygen species. The ferredoxin acceptor of PSI is a very strong reductant that can easily reduce molecular oxygen to form superoxide (O_2^-). This reduction competes with the normal channeling of electrons to the reduction of $NADP^+$ and other processes. Superoxide is one of a series of active oxygen species that can be very damaging to biological membranes, but when formed in this way it can be eliminated by the action of a series of enzymes, including superoxide dismutase and ascorbate peroxidase (Asada 1999).

Thylakoid stacking permits energy partitioning between the photosystems

The fact that photosynthesis in higher plants is driven by two photosystems with different light-absorbing properties poses a special problem. If the rate of delivery of energy to PSI and PSII is not precisely matched and conditions are such that the rate of photosynthesis is limited by the available light (low light intensity), the rate of electron flow will be limited by the photosystem that is receiving less energy. In the most efficient situation, the input of energy would be the same to both photosystems. However, no single arrangement of pigments would satisfy this requirement because at different times of day the light intensity and spectral distribution tend to favor one photosystem or the other (Allen and Forsberg 2001; Finazzi 2005).

This problem can be solved by a mechanism that shifts energy from one photosystem to the other in response to different conditions. Such a regulating mechanism has been shown to operate in different experimental conditions. The observation that the overall quantum yield of photosynthesis is nearly independent of wavelength (see Figure 7.12) strongly suggests that such a mechanism exists.

Thylakoid membranes contain a protein kinase that can phosphorylate a specific threonine residue on the surface of LHCII, one of the membrane-bound antenna pigment proteins described earlier in the chapter (see Figure 7.20). When LHCII is not phosphorylated, it delivers more energy to photosystem II, and when it is phosphorylated, it delivers more energy to photosystem I (Haldrup et al. 2001).

The kinase is activated when plastoquinone, one of the electron carriers between PSI and PSII, accumulates in the reduced state. Reduced plastoquinone accumulates when PSII is being activated more frequently than PSI. The phosphorylated LHCII then migrates out of the stacked regions of the membrane into the unstacked regions (see Figure 7.18), probably because of repulsive interactions with negative charges on adjacent membranes.

The lateral migration of LHCII shifts the energy balance toward photosystem I, which is located in the stroma lamellae, and away from photosystem II, which is located in the stacked membranes of the grana. This situation is called *state 2*. If plastoquinone becomes more oxidized because of excess excitation of photosystem I, the kinase is deactivated and the level of phosphorylation of LHCII is decreased by the action of a membrane-bound phosphatase. LHCII then moves back to the grana, and the system is in *state 1*. The net result is a very precise control of the energy distribution between the photosystems, allowing the most efficient use of the available energy.

Genetics, Assembly, and Evolution of Photosynthetic Systems

Chloroplasts have their own DNA, mRNA, and protein synthesis machinery, but many chloroplast proteins are encoded by nuclear genes and imported into the chloroplast. In this section we will consider the genetics, assembly, and evolution of the main chloroplast components.

Chloroplast, cyanobacterial, and nuclear genomes have been sequenced

The complete chloroplast genomes of many organisms have been sequenced. Chloroplast DNA is circular and ranges in size from 120 to 160 kilobases. In higher plants, the chloroplast genome contains coding sequences for approximately 120 proteins. Some of these DNA sequences code for proteins that are yet to be characterized.

The complete genome of several cyanobacteria, including *Synechocystis* (strain PCC 6803), and a number of algae and higher plants, including *Arabidopsis thaliana*, have been sequenced (Kotani and Tabata 1998; Arabidopsis Genome Initiative 2000). Genomes of important crop plants such as rice have also been completed (International Rice Genome Sequencing Project 2005). This genomic information is being complemented with data on gene expression using microarray techniques and high-resolution proteomics (Yamada et al. 2003; van Wijk 2004). Taken together, these powerful techniques promise to provide new insights into the mechanism of photosynthesis, as well as many other plant processes.

Chloroplast genes exhibit non-Mendelian patterns of inheritance

Chloroplasts and mitochondria reproduce by division rather than by **de novo synthesis**. This mode of reproduction is not surprising, since these organelles contain genetic information that is not present in the nucleus. During cell division, chloroplasts are divided between the two daughter cells. In most sexual plants, however, only the maternal plant contributes chloroplasts to the zygote. In these plants the normal Mendelian pattern of inheritance does not apply to chloroplast-encoded genes, because the offspring receive chloroplasts from only one parent. The result is **non-Mendelian**, or **maternal**, **inheritance**. Numerous traits are inherited in this way; one example is the herbicide-resistance trait discussed in Web Topic 7.10.

Many chloroplast proteins are imported from the cytoplasm

Chloroplast proteins can be encoded by either chloroplastic or nuclear DNA. The chloroplast-encoded proteins are synthesized on chloroplast ribosomes; the nucleus-encoded proteins are synthesized on cytoplasmic ribosomes and then transported into the chloroplast. Many nuclear genes contain introns—that is, base sequences that do not code for protein. The mRNA is processed to remove the introns, and the proteins are then synthesized in the cytoplasm.

The genes needed for chloroplast function are distributed in the nucleus and in the chloroplast genome with no evident pattern, but both sets are essential for the viability of the chloroplast. Some chloroplast genes are necessary for other cellular functions, such as heme and lipid synthesis. Control of the expression of the nuclear genes that code for chloroplast proteins is complex, involving light-dependent regulation mediated by both phytochrome (see Chapter 17) and blue light (see Chapter 18), as well as other factors (Bruick and Mayfield 1999; Wollman et al. 1999).

The transport of chloroplast proteins that are synthesized in the cytoplasm is a tightly regulated process (Chen and Schnell 1999). For example, the enzyme rubisco (see Chapter 8), which functions in carbon fixation, has two types of subunits, a chloroplast-encoded large subunit and a nucleus-encoded small subunit. Small subunits of rubisco are synthesized in the cytoplasm and transported into the chloroplast, where the enzyme is assembled.

In this and other known cases, the nucleus-encoded chloroplast proteins are synthesized as precursor proteins containing an N-terminal amino acid sequence known as a **transit peptide**. This terminal sequence directs the precursor protein to the chloroplast, facilitates its passage through both the outer and the inner envelope membranes, and is then clipped off. The electron carrier plastocyanin is a water-soluble protein that is encoded in the nucleus but functions in the lumen of the chloroplast. It therefore must cross three membranes to reach its destination in the lumen. The transit peptide of plastocyanin is very large and is processed in more than one step.

The biosynthesis and breakdown of chlorophyll are complex pathways

Chlorophylls are complex molecules exquisitely suited to the light absorption, energy transfer, and electron transfer functions that they carry out in photosynthesis (see Figure 7.6). Like all other biomolecules, chlorophylls are made by a biosynthetic pathway in which simple molecules are used as building blocks to assemble more complex molecules (Beale 1999; Eckhardt et al. 2004). Each step in the biosynthetic pathway is enzymatically catalyzed.

The chlorophyll biosynthetic pathway consists of more than a dozen steps (see Web Topic 7.11). The process can be divided into several phases (Figure 7.36), each of which can be considered separately, but in the cell are highly coordinated and regulated. This regulation is essential because free chlorophyll and many of the biosynthetic intermediates are damaging to cellular components. The damage results largely because chlorophylls absorb light efficiently, but in the absence of accompanying proteins, they lack a pathway for disposing of the energy, with the result that toxic singlet oxygen is formed.

The breakdown pathway of chlorophyll in senescent leaves is quite different from the biosynthetic pathway (Takamiya et al. 2000; Eckhardt et al. 2004). The first step is removal of the phytol tail by an enzyme known as chlorophyllase, followed by removal of the magnesium by magnesium de-chelatase. Next the porphyrin structure is opened by an oxygen-dependent oxygenase enzyme to form an open-chain tetrapyrrole.

The tetrapyrrole is further modified to form water-soluble, colorless products. These colorless metabolites are then exported from the senescent chloroplast and transported to the vacuole, where they are permanently stored. The chlorophyll metabolites are not further processed or recycled, although the proteins associated with them in the chloroplast are subsequently recycled into new proteins. The recycling of proteins is thought to be important for the nitrogen economy of the plant.

Complex photosynthetic organisms have evolved from simpler forms

The complicated photosynthetic apparatus found in plants and algae is the end product of a long evolutionary sequence. Much can be learned about this evolutionary process from analysis of simpler prokaryotic photosynthetic organisms, including the anoxygenic photosynthetic bacteria and the cyanobacteria.

The chloroplast is a semiautonomous cell organelle, with its own DNA and a complete protein synthesis apparatus. Many of the proteins that make up the photosynthetic apparatus, as well as all the chlorophylls and lipids, are synthesized in the chloroplast. Other proteins are imported from the cytoplasm and are encoded by nuclear genes. How did this curious division of labor come about? Most experts now agree that the chloroplast is the descendant of a symbiotic relationship between a cyanobacterium and a simple nonphotosynthetic eukaryotic cell. This type of relationship is called **endosymbiosis** (Cavalier-Smith 2000).

Originally the cyanobacterium was capable of independent life, but over time much of its genetic information needed for normal cellular functions was lost, and a substantial amount of information needed to synthesize the photosynthetic apparatus was transferred to the nucleus.

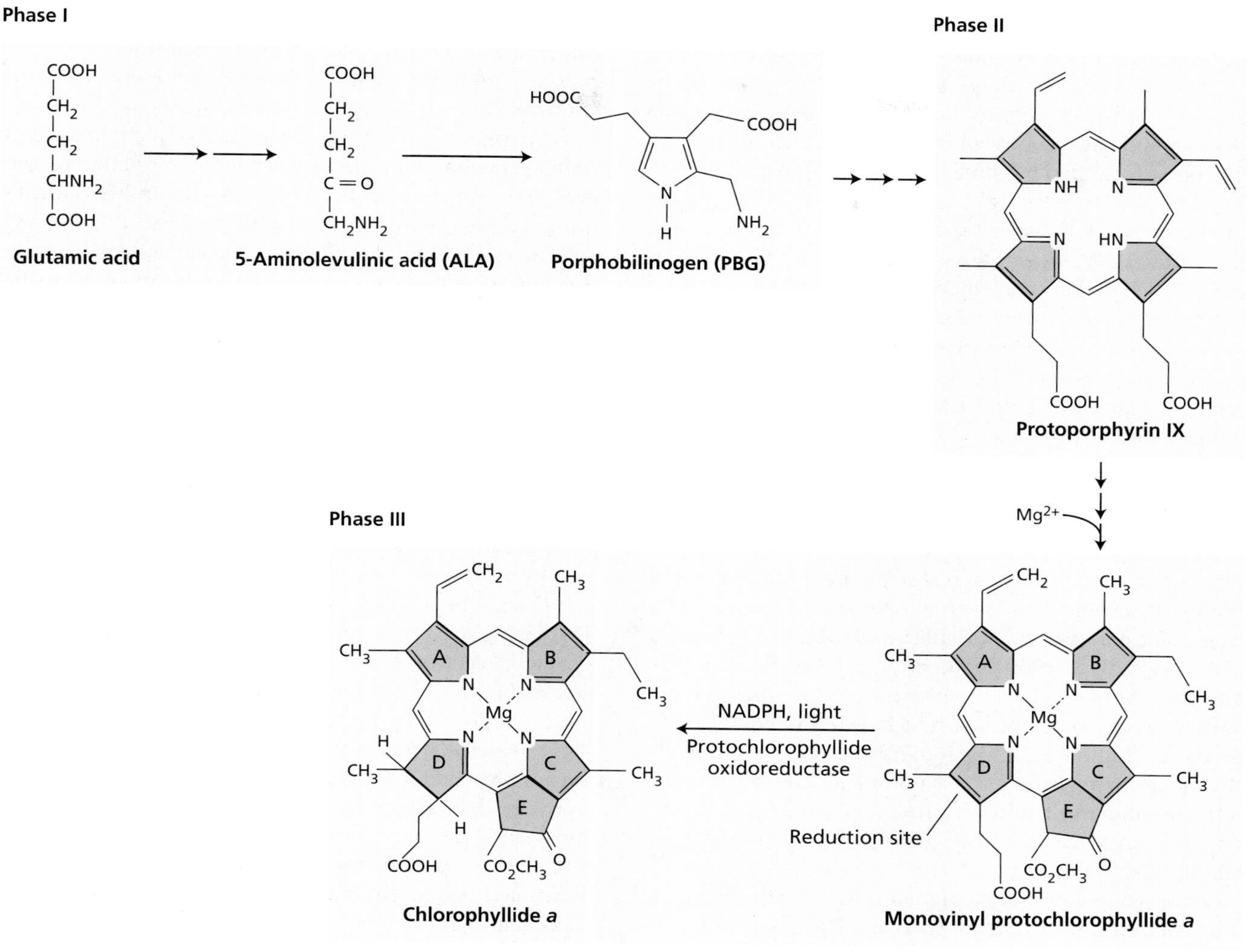

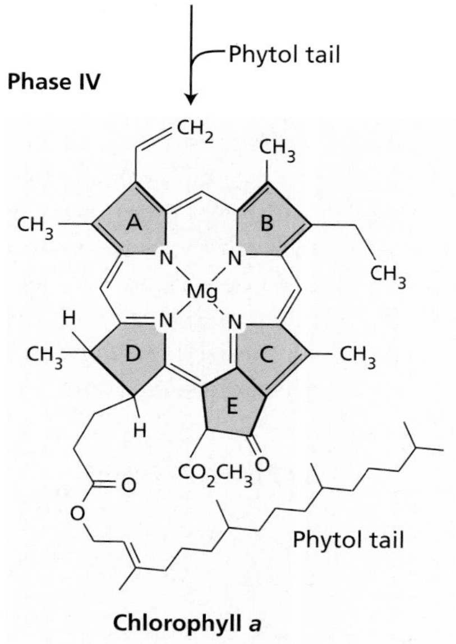

FIGURE 7.36 The biosynthetic pathway of chlorophyll. The pathway begins with glutamic acid, which is converted to 5-aminolevulinic acid (ALA). Two molecules of ALA are condensed to form porphobilinogen (PBG). Four PBG molecules are linked to form protoporphyrin IX. The magnesium (Mg) is then inserted, and the light-dependent cyclization of ring E, the reduction of ring D, and the attachment of the phytol tail complete the process. Many steps in the process are omitted in this figure.

So the chloroplast was no longer capable of life outside its host and eventually became an integral part of the cell.

In some types of algae, chloroplasts have arisen by endosymbiosis of eukaryotic photosynthetic organisms (Palmer and Delwiche 1996). In these organisms the chloroplast is surrounded by three and in some cases four membranes, which are thought to be remnants of the plasma membranes of the earlier organisms. Mitochondria are also thought to have originated by endosymbiosis in a separate event much earlier than chloroplast formation.

The answers to other questions related to the evolution of photosynthesis are less clear. These include the nature of the earliest photosynthetic systems, how the two photosystems became linked, and the evolutionary origin of the oxygen evolution complex (Blankenship and Hartman 1998; Xiong et al. 2000; Allen 2005).

Summary

Photosynthesis is the storage of solar energy carried out by plants, algae, and photosynthetic bacteria. Absorbed photons excite chlorophyll molecules, and these excited chlorophylls can dispose of this energy as heat, fluorescence, energy transfer, or photochemistry. Light is absorbed mainly in the antenna complexes, which comprise chlorophylls, accessory pigments, and proteins and are located at the thylakoid membranes of the chloroplast. The antenna system contains multisubunit protein complexes and hundreds or, in some organisms, thousands of chlorophylls and accessory pigments.

Photosynthetic antenna pigments transfer the energy to a specialized chlorophyll–protein complex known as a reaction center. The antenna complexes and the reaction centers are integral components of the thylakoid membrane. The reaction center initiates a complex series of chemical reactions that capture energy in the form of chemical bonds.

The relationship between the amount of absorbed quanta and the yield of a photochemical product made in a light-dependent reaction is given by the quantum yield. The quantum yield of the early steps of photosynthesis is approximately 0.95, indicating that nearly every photon that is absorbed yields a charge separation at the reaction center.

Plants, algae and cyanobacteria have two reaction centers, photosystem I and photosystem II, that function in series. The two photosystems are spatially separated: PSI is found exclusively in the nonstacked stroma membranes, PSII largely in the stacked grana membranes. The reaction center chlorophylls of PSI absorb maximally at 700 nm, those of PSII at 680 nm. Photosystems II and I carry out noncyclic electron transport, oxidize water to molecular oxygen, and reduce $NADP^+$ to NADPH. It is energetically very difficult to oxidize water to form molecular oxygen, and the photosynthetic oxygen-evolving system is the only known biochemical system that can oxidize water, thus providing the oxygen in Earth's atmosphere. The photooxidation of water is modeled by the five-step S state mechanism. Manganese is an essential cofactor in the water-oxidizing process, and the five S states appear to represent successive oxidized states of a manganese-containing enzyme.

A tyrosine residue of the D1 protein of the PSII reaction center functions as an electron carrier between the oxygen-evolving complex and P680. Pheophytin and two plastoquinones are electron carriers between P680 and the large cytochrome $b_6 f$ complex. Plastocyanin is the electron carrier between cytochrome $b_6 f$ and P700. The electron carriers that accept electrons from P700 are very strong reducing agents, and they include a quinone and three membrane-bound iron–sulfur proteins known as bound ferredoxins. The electron flow ends with the reduction of $NADP^+$ to NADPH by a membrane-bound, ferredoxin–NADP reductase.

A portion of the energy of photons is also initially stored as chemical potential energy, largely in the form of a pH difference across the thylakoid membrane. This energy is quickly converted into chemical energy during ATP formation by action of an enzyme complex known as the ATP synthase. The photophosphorylation of ADP by the ATP synthase is driven by a chemiosmotic mechanism. Photosynthetic electron flow is coupled to proton translocation across the thylakoid membrane, and the stroma becomes more alkaline and the lumen more acidic. This proton gradient drives ATP synthesis. NADPH and ATP formed by the light reactions provide the energy for carbon reduction.

Excess light energy can damage photosynthetic systems, and several mechanisms minimize such damage. Carotenoids work as photoprotective agents by rapidly quenching the excited state of chlorophyll. Changes in the phosphorylated state of antenna pigment proteins can change the energy distribution between photosystems I and II when there is an imbalance between the energy absorbed by each photosystem. The xanthophyll cycle also contributes to the dissipation of excess energy by nonphotochemical quenching.

Chloroplasts contain DNA and encode and synthesize some of the proteins that are essential for photosynthesis. Additional proteins are encoded by nuclear DNA, synthesized in the cytosol, and imported into the chloroplast. Chlorophylls are synthesized in a highly regulated biosynthetic pathway involving more than a dozen steps. Once synthesized, proteins and pigments are assembled into the thylakoid membrane.

Web Material

Web Topics

7.1 Principles of Spectrophotometry
Spectroscopy is a key technique to study light reactions.

7.2 The Distribution of Chlorophylls and Other Photosynthetic Pigments
The content of chlorophylls and other photosynthetic pigments varies among plant kingdoms.

7.3 Quantum Yield
Quantum yields measure how effectively light drives a photobiological process.

7.4 Antagonistic Effects of Light on Cytochrome Oxidation
Photosystems I and II were discovered in some ingenious experiments.

7.5 Structures of Two Bacterial Reaction Centers
X-ray diffraction studies resolved the atomic structure of the reaction center of photosystem II.

7.6 Midpoint Potentials and Redox Reactions
The measurement of midpoint potentials is useful for analyzing electron flow through photosystem II.

7.7 Oxygen Evolution
The S state mechanism is a valuable model for water splitting in PSII.

7.8 Photosystem I
The PSI reaction is a multiprotein complex.

7.9 ATP Synthase
The ATP synthase functions as a molecular motor.

7.10 Mode of Action of Some Herbicides
Some herbicides kill plants by blocking photosynthetic electron flow.

7.11 Chlorophyll Biosynthesis
Chlorophyll and heme share early steps of their biosynthetic pathways.

Web Essay

7.1 A Novel View of Chloroplast Structure
Stromules extend the reach of the chloroplasts.

Chapter References

Allen, J. F. (2005) A redox switch hypothesis for the origin of two light reactions in photosynthesis. *FEBS Lett.* 579: 963–968.

Allen, J. F., and Forsberg, J. (2001) Molecular recognition in thylakoid structure and function. *Trends Plant Sci.* 6: 317–326.

Arabidopsis Genome Initiative. (2000) Analysis of the genome sequence of the flowering plant *Arabidopsis thaliana*. *Nature* 408: 796–815.

Asada, K. (1999) The water–water cycle in chloroplasts: Scavenging of active oxygens and dissipation of excess photons. *Annu. Rev. Plant Physiol. Plant Mol. Biol.* 50: 601–639.

Avers, C. J. (1985) *Molecular Cell Biology*. Addison-Wesley, Reading, MA.

Barber, J., Nield, N., Morris, E. P., and Hankamer, B. (1999) Subunit positioning in photosystem II revisited. *Trends Biochem. Sci.* 24: 43–45.

Beale, S. I. (1999) Enzymes of chlorophyll biosynthesis. *Photosynth. Res.* 60: 43–73.

Becker, W. M. (1986) *The World of the Cell*. Benjamin/Cummings, Menlo Park, CA.

Ben-Shem, A., Frolow, F., and Nelson, N. Crystal structure of plant photosystem I. *Nature* 426: 630–635.

Blankenship, R. E. (2002) *Molecular Mechanisms of Photosynthesis*. Blackwell Science, Oxford.

Blankenship, R. E., and Hartman, H. (1998) The origin and evolution of oxygenic photosynthesis. *Trends Biochem. Sci.* 23: 94–97.

Blankenship, R. E., and Prince, R. C. (1985) Excited-state redox potentials and the Z scheme of photosynthesis. *Trends Biochem. Sci.* 10: 382–383.

Boyer, P. D. (1997) The ATP synthase: A splendid molecular machine. *Annu. Rev. Biochem.* 66: 717–749.

Brettel, K. (1997) Electron transfer and arrangement of the redox cofactors in photosystem I. *Biochim. Biophys. Acta* 1318: 322–373.

Bruick, R. K., and Mayfield, S. P. (1999) Light-activated translation of chloroplast mRNAs. *Trends Plant Sci.* 4: 190–195.

Buchanan, B. B., Gruissem, W., and Jones, R. L., eds. (2000) *Biochemistry and Molecular Biology of Plants*. American Society of Plant Physiologists, Rockville, MD.

Cavalier-Smith, T. (2000) Membrane heredity and early chloroplast evolution. *Trends Plant Sci.* 5: 174–182.

Chen, X., and Schnell, D. J. (1999) Protein import into chloroplasts. *Trends Cell Biol.* 9: 222–227.

Chitnis, P. R. (2001) Photosystem I: Function and physiology. *Annu. Rev. Plant Physiol. Plant Mol. Biol.* 52: 593–626.

Cramer, W. A., Soriano, G. M., Ponomarev, M., Huang, D., Zhang, H., Martinez, S. E., and Smith, J. L. (1996) Some new structural aspects and old controversies concerning the cytochrome b_6f complex of oxygenic photosynthesis. *Annu. Rev. Plant Physiol. Plant Mol. Biol.* 47: 477–508.

Deisenhofer, J., and Michel, H. (1989) The photosynthetic reaction center from the purple bacterium *Rhodopseudomonas viridis*. *Science* 245: 1463–1473.

Demmig-Adams, B., and Adams, W. W., III. (1992) Photoprotection and other responses of plants to high light stress. *Annu. Rev. Plant Physiol. Plant Mol. Biol.* 43: 599–626.

Eckhardt, U., Grimm, B. and Hortensteiner, S. (2004) Recent advances in chlorophyll biosynthesis and breakdown in higher plants. *Photosynth. Res.* 56: 1–14.

Ferreira, K. N., Iverson, T. M., Maghlaoui, K., Barber, J. and Iwata, S. (2004) Architecture of the photosynthetic oxygen-evolving center. *Science* 303: 1831–1838.

Finazzi, G. (2005) The central role of the green alga *Chlamydomonas reinhardtii* in revealing the mechanism of state transitions. *J. Exp. Bot.* 56: 383–388.

Green, B. R., and Durnford, D. G. (1996) The chlorophyll-carotenoid proteins of oxygenic photosynthesis. *Annu. Rev. Plant Physiol. Plant Mol. Biol.* 47: 685–714.

Green, B. R. and Parson, W. W., Eds. (2003) *Light-Harvesting Antennas in Photosynthesis*. Kluwer Academic Publishers, Dordrecht.

Grossman, A. R., Bhaya, D., Apt, K. E., and Kehoe, D. M. (1995) Light-harvesting complexes in oxygenic photosynthesis: Diversity, control, and evolution. *Annu. Rev. Genet.* 29: 231–288.

Haldrup, A., Jensen, P. E., Lunde, C., and Scheller, H. V. (2001) Balance of power: A view of the mechanism of photosynthetic state transitions. *Trends Plant Sci.* 6: 301–305.

Hoganson, C. W., and Babcock, G. T. (1997) A metalloradical mechanism for the generation of oxygen from water in photosynthesis. *Science* 277: 1953–1956.

Holt, N. E., Zigmantas, D. Valkunas, L., Li, X. P., Niyogi, K. K., and Fleming, G. R. (2005) Carotenoid cation formation and the regulation of photosynthetic light harvesting. *Science* 307: 433–436.

Horton, P., Ruban, A. V., and Walters, R. G. (1996) Regulation of light harvesting in green plants. *Annu. Rev. Plant Physiol. Plant Mol. Biol.* 47: 655–684.

International Rice Genome Sequencing Project. (2005) The map-based sequence of the rice genome. *Nature* 436: 793–800.

Jagendorf, A. T. (1967) Acid-based transitions and phosphorylation by chloroplasts. *Fed. Proc. Am. Soc. Exp. Biol.* 26: 1361–1369.

Jordan, P., Fromme, P., Witt, H. T., Klukas, O., Saenger, W., and Krauss, N. (2001) Three-dimensional structure of cyanobacterial photosystem I at 2.5 Å resolution. *Nature* 411: 909–917.

Karplus, P. A., Daniels, M. J., and Herriott, J. R. (1991) Atomic structure of ferredoxin-$NADP^+$ reductase: Prototype for a structurally novel flavoenzyme family. *Science* 251: 60–66.

Kotani, H., and Tabata, S. (1998) Lessons from sequencing of the genome of a unicellular cyanobacterium, *Synechocystis* sp. PCC6803. *Annu. Rev. Plant Physiol. Plant Mol. Biol.* 49: 151–171.

Krause, G. H., and Weis, E. (1991) Chlorophyll fluorescence and photosynthesis: The basics. *Annu. Rev. Plant Physiol. Plant Mol. Biol.* 42: 313–350.

Kühlbrandt, W., Wang, D. N., and Fujiyoshi, Y. (1994) Atomic model of plant light-harvesting complex by electron crystallography. *Nature* 367: 614–621.

Kurisu, G. Zhang, H. M., Smith, J. L., and Cramer, W. A. (2003) Structure of cytochrome b_6f complex of oxygenic photosynthesis: Tuning the cavity. *Science* 302: 1009–1014.

Li, X. P., Bjorkman, O., Shih, C., Grossman, A. R., Rosenquist, M., Jansson, S., and Niyogi, K. K. (2000) A pigment-binding protein essential for regulation of photosynthetic light harvesting. *Nature* 403: 391–395.

Liu, Z. F., Yan, H. C., Wang, K. B., Kuang, T. Y., Zhang, J. P., Gui, L. L., An, X. M., and Chang, W. R. (2004) Crystal structure of spinach major light harvesting complex at 2.72 A reslolution. *Nature* 428: 287–292.

Long, S. P., Humphries, S., and Falkowski, P. G. (1994) Photoinhibition of photosynthesis in nature. *Annu. Rev. Plant Physiol. Plant Mol. Biol.* 45: 633–662.

McEvoy, J. P., and Brudvig, G. W. (2004 Structure-based mechanism of photosynthetic water oxidation. *Physical Chem. Chemical Phys.* 6: 4754–4763.

Melis, A. (1999) Photosystem-II damage and repair cycle in chloroplasts: What modulates the rate of photodamage in vivo? *Trends Plant Sci.* 4: 130–135.

Müller, P., Li, X.-P., and Niyogi, K. K. (2001) Non-photochemical quenching: A response to excess light energy. *Plant Physiol.* 125: 1558–1566.

Nelson, N., and Ben-Shem, A. (2004) The complex architecture of oxygenic photosynthesis. *Nat. Rev. Mol. Cell Biol.* 5: 971–982.

Okamura, M. Y., Paddock, M. L., Graige, M. S., and Feher, G. (2000) Proton and electron transfer in bacterial reaction centers. *Biochim. Biophys. Acta* 1458: 148–163.

Palmer, J. D., and Delwiche, C. F. (1996) Second-hand chloroplasts and the case of the disappearing nucleus. *Proc. Natl. Acad. Sci. USA* 93: 7432–7435.

Paulsen, H. (1995) Chlorophyll *a/b*-binding proteins. *Photochem. Photobiol.* 62: 367–382.

Pullerits, T., and Sundström, V. (1996) Photosynthetic light-harvesting pigment-protein complexes: Toward understanding how and why. *Acc. Chem. Res.* 29: 381–389.

Seelert, H., Poetsch, A., Dencher, N. A., Engel, A., Stahlberg, H., and Muller, D. J. (2000) Structural biology. Proton-powered turbine of a plant motor. *Nature* 405: 418-419.

Stock, D., Leslie, A. G. W., and Walker, J. E. (1999) Molecular architecture of the rotary motor in ATP synthase. *Science* 286: 1700–1705.

Stroebel, D., Choquet, Y. Popot, J. L., and Picot, D. (2003) An atypical heme in the cytochrome b_6f complex. *Nature* 426: 413–418.

Takamiya, K.-I., Tsuchiya, T., and Ohta, H. (2000) Degradation pathway(s) of chlorophyll: What has gene cloning revealed? *Trends Plant Sci.* 5: 426–431.

Tommos, C., and Babcock, G. T. (1999) Oxygen production in nature: A light-driven metalloradical enzyme process. *Acc. Chem. Res.* 37: 18–25.

Trebst, A. (1986) The topology of the plastoquinone and herbicide binding peptides of photosystem II in the thylakoid membrane. *Z. Naturforsch. Teil C.* 240–245.

van Grondelle, R., Dekker, J. P., Gillbro, T., and Sundström, V. (1994) Energy transfer and trapping in photosynthesis. *Biochim. Biophys. Acta* 1187: 1–65.

van Wijk, K. J. (2004) Plastid proteomics. *Plant Physiol. Biochem.* 42: 963–977.

Wollman, F.-A., Minai, L., and Nechushtai, R. (1999) The biogenesis and assembly of photosynthetic proteins in thylakoid membranes. *Biochim. Biophys. Acta* 1411: 21–85.

Xiong, J., Fisher, W., Inoue, K., Nakahara, M., and Bauer, C. E. (2000) Molecular evidence for the early evolution of photosynthesis. *Science* 289: 1724–1730.

Yachandra, V. K., Sauer, K., and Klein, M. P. (1996) Manganese cluster in photosynthesis: Where plants oxidize water to dioxygen. *Chem. Rev.* 96: 2927–2950.

Yamada, K., Lim, J., Dale, J. M., Chen, H., Shinn, P., Palm, C. J., Southwick, A. M., Wu, H. C., Kim, C., Nguyen, M., et al. (2003) Empirical analysis of transcriptional activity in the *Arabidopsis* genome. *Science* 302: 842–846.

Yasuda, R., Noji, H., Yoshida, M., Kinosita, K., and Itoh, H. (2001) Resolution of distinct rotational substeps by submillisecond kinetic analysis of F 1-ATPase. *Nature* 410: 898–904.

Zouni, A., Witt, H.-T., Kern, J., Fromme, P., Krauss, N., Saenger, W., and Orth, P. (2001) Crystal structure of photosystem II from *Synechococcus elongatus* at 3.8 Å resolution. *Nature* 409: 739–743.

Chapter 8 | Photosynthesis: Carbon Reactions

IN CHAPTER 5 WE DISCUSSED the requirements of plants for essential elements such as minerals, carbon, and oxygen in order to grow and complete their life cycles. Because living organisms interact with one another and their environments, these essential elements cycle through the biosphere. Energy must be supplied to keep these cycles operational, because the amount of matter in the biosphere remains constant. Otherwise, increasing entropy dictates that the flow of matter would ultimately stop. Sunlight provides the energy that operates these cycles and, in particular, the energy that fuels the buildup by photosynthetic organisms of large reserves of organic carbon. A significant proportion of these reserves is subsequently used to drive the metabolism of other essential elements not only in photosynthetic organisms but also in other forms of life. Recent estimates indicate that about 200 billion tons of CO_2 are converted to biomass each year, of which about 40% originates from the activities of marine phytoplankton.

In Chapter 7 we saw how the energy associated with the photochemical oxidation of water to molecular oxygen drives the generation of adenosine triphosphate (ATP) and reduced pyridine nucleotide (NADPH) through reactions taking place in the chloroplast thylakoid membrane. Subsequently, the consumption of ATP and NADPH is coupled to reactions in which the reduction of CO_2 to carbohydrates is catalyzed by enzymes found in the stroma, the soluble phase of chloroplasts (Figure 8.1).

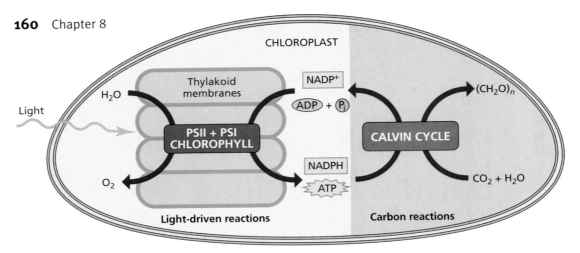

FIGURE 8.1 The light and carbon reactions of photosynthesis in chloroplasts of vascular plants. In thylakoid membranes, the excitation of chlorophyll in the photosynthetic electron transport system (PSII + PSI) by light elicits the formation of ATP and NADPH (see Chapter 7). In the stroma, both ATP and NADPH are consumed by the Calvin cycle in a series of enzyme-driven reactions that reduce CO_2 to carbohydrates (triose phosphates).

These stromal reactions were long thought to be independent of light and, as a consequence, were referred to as the *dark reactions*. However, because these stroma-localized reactions depend on products of the photochemical processes, and are also directly regulated by light, they are more properly referred to as the *carbon reactions of photosynthesis*.

In this chapter we will examine the cyclic reactions that accomplish the fixation and reduction of CO_2, then consider how the phenomenon of photorespiration catalyzed by the carboxylating enzyme alters the efficiency of photosynthesis. This chapter will also describe biochemical mechanisms that concentrate CO_2, allowing plants to mitigate the impact of photorespiration: CO_2 pumps, C_4 metabolism, and crassulacean acid metabolism (CAM). We will close the chapter with a consideration of the synthesis of starch and sucrose.

The Calvin Cycle

Autotrophic organisms have the ability to use energy from physical and chemical sources to transform the carbon of atmospheric CO_2 (carbon oxidation number: +4) into organic compounds that are compatible with the needs of the cell (e.g., —CHOH—, carbon oxidation number: 0).

The most important pathway of autotrophic CO_2 fixation is the Calvin cycle (also called the reductive pentose phosphate cycle), which is found in many prokaryotes and in all photosynthetic eukaryotes, from the most primitive alga to the most advanced angiosperms. This metabolic pathway reduces CO_2 to carbohydrates via the photosynthetic carbon reduction cycle originally described for C_3 species. Other metabolic pathways associated with the photosynthetic fixation of CO_2, such as the C_4 photosynthetic carbon assimilation cycle and the photorespiratory carbon oxidation cycle, are either auxiliary to or dependent on the basic Calvin cycle.

In this section we will examine how CO_2 is fixed via the Calvin cycle through the use of ATP and NADPH that is generated by the light reactions (see Figure 8.1), and how the Calvin cycle is regulated.

The Calvin cycle has three stages: carboxylation, reduction, and regeneration

The Calvin cycle was elucidated in a series of elegant experiments by M. Calvin, A. Benson, J. A. Bassham, and their colleagues in the 1950s. Calvin received a Nobel Prize for this work in 1961 (see **Web Topic 8.1**). In the Calvin cycle, CO_2 and water from the environment are enzymatically combined with a five-carbon acceptor molecule (ribulose-1,5-bisphosphate) to generate two molecules of a three-carbon intermediate. This intermediate (3-phosphoglycerate) is reduced to carbohydrate by enzymatic reactions driven by the ATP and NADPH generated photochemically. The cycle is completed by regeneration of the ribulose-1,5-bisphosphate.

The Calvin cycle proceeds in three stages (Figure 8.2):

1. *Carboxylation* of the CO_2 acceptor ribulose-1,5-bisphosphate, which forms two molecules of 3-phosphoglycerate, the first stable intermediate of the Calvin cycle

2. *Reduction* of 3-phosphoglycerate, which yields glyceraldehyde-3-phosphate, a carbohydrate

3. *Regeneration* of the CO_2 acceptor ribulose-1,5-bisphosphate from glyceraldehyde-3-phosphate

Carbon dioxide contains one of the most oxidized forms of carbon found in nature (+4). The carbon of the first stable intermediate, 3-phosphoglycerate, is more reduced (+3), and it is further reduced in the glyceraldehyde-3-phosphate product (+1). Overall, the reactions of the Calvin cycle reduce atmospheric carbon for its incorporation into organic compounds that can be used by the cell.

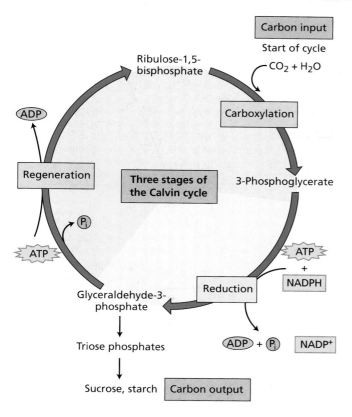

FIGURE 8.2 The Calvin cycle proceeds in three stages: carboxylation, in which CO_2 is covalently linked to a carbon skeleton; reduction, in which carbohydrate is formed at the expense of the photochemically derived ATP and reducing equivalents in the form of NADPH; and regeneration, which restores the CO_2 acceptor ribulose-1,5-bisphosphate. At steady state, the carbon entering the cycle as CO_2 equals the carbon leaving the cycle as triose phosphates.

The carboxylation of ribulose-1,5-bisphosphate is catalyzed by the enzyme rubisco

In the first step of the Calvin cycle, carboxylation, three molecules of CO_2 react with three molecules of ribulose-1,5-bisphosphate to yield six molecules of 3-phosphoglycerate (Figure 8.3 and Table 8.1, reaction 1), a reaction catalyzed by the chloroplast enzyme ribulose bisphosphate carboxylase/oxygenase, referred to as rubisco (see **Web Topic 8.2**). Specifically, for each CO_2 added to the carbon 2 of ribulose-1,5-bisphosphate, one unstable, enzyme-bound intermediate is produced that is then hydrolyzed to yield two molecules of the stable product 3-phosphoglycerate (Figure 8.4). Two features of this carboxylase reaction

TABLE 8.1
Reactions of the Calvin cycle

Enzyme	Reaction
1. Ribulose-1,5-bisphosphate carboxylase–oxygenase (rubisco)	3 Ribulose-1,5-bisphosphate + 3 CO_2 + 3 H_2O → 6 3-phosphoglycerate + 6 H^+
2. 3-Phosphoglycerate kinase	6 3-Phosphoglycerate + 6 ATP → 6 1,3-bisphosphoglycerate + 6 ADP
3. NADP–glyceraldehyde-3-phosphate dehydrogenase	6 1,3-Bisphosphoglycerate + 6 NADPH + 6 H^+ → 6 glyceraldehyde-3-phosphate + 6 $NADP^+$ + 6 P_i
4. Triose phosphate isomerase	2 Glyceraldehyde-3-phosphate ↔ 2 dihydroxyacetone-3-phosphate
5. Aldolase	Glyceraldehyde-3-phosphate + dihydroxyacetone-3-phosphate → fructose-1,6-bisphosphate
6. Fructose-1,6-bisphosphate phosphatase	Fructose-1,6-bisphosphate + H_2O → fructose-6-phosphate + P_i
7. Transketolase	Fructose-6-phosphate + glyceraldehyde-3-phosphate → erythrose-4-phosphate + xylulose-5-phosphate
8. Aldolase	Erythrose-4-phosphate + dihydroxyacetone-3-phosphate → sedoheptulose-1,7-bisphosphate
9. Sedoheptulose-1,7-bisphosphate phosphatase	Sedoheptulose-1,7-bisphosphate + H_2O → sedoheptulose-7-phosphate + P_i
10. Transketolase	Sedoheptulose-7-phosphate + glyceraldehyde-3-phosphate → ribose-5-phosphate + xylulose-5-phosphate
11a. Ribulose-5-phosphate epimerase	2 Xylulose-5-phosphate → 2 ribulose-5-phosphate
11b. Ribose-5-phosphate isomerase	Ribose-5-phosphate → ribulose-5-phosphate
12. Ribulose-5-phosphate kinase	3 Ribulose-5-phosphate + 3 ATP → 3 ribulose-1,5-bisphosphate + 3 ADP + 3 H^+
Net: 3 CO_2 + 5 H_2O + 6 NADPH + 9 ATP → glyceraldehyde-3-phosphate + 6 $NADP^+$ + 3 H^+ + 9 ADP + 8 P_i	

Note: P_i stands for inorganic phosphate.

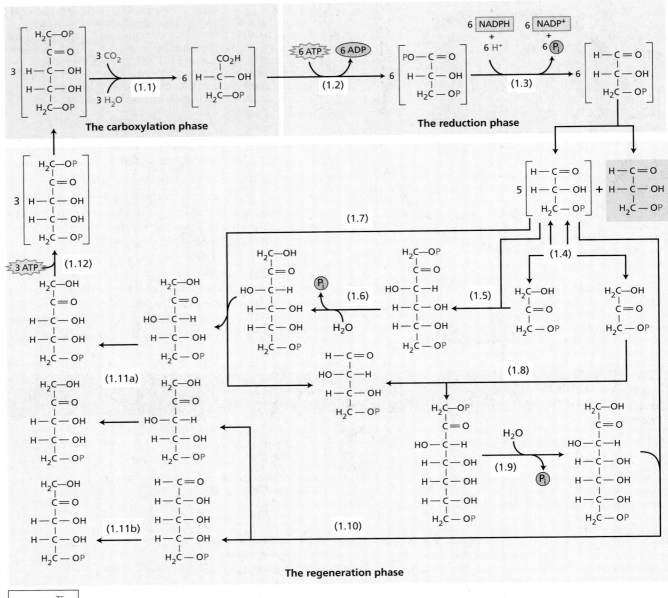

strongly favor the forward reaction: the large negative change in free energy associated with the carboxylation of ribulose-1,5-bisphosphate (see Chapter 2 on the web site for a discussion of free energy), and the affinity of rubisco for CO_2 that ensures the rapid carboxylation at the low concentrations of CO_2 found in photosynthetic cells.

Next in the Calvin cycle is the reduction phase (see Figure 8.3 and Table 8.1, reactions 2 and 3). Here, two reactions further reduce the carbon in the 3-phosphoglycerate formed in the carboxylation stage:

1. The ATP generated in the light reactions first phosphorylates 3-phosphoglycerate at the carboxylic group, yielding a mixed anhydride, 1,3-bisphosphoglycerate, in a reaction catalyzed by 3-phosphoglycerate kinase (see Table 8.1, reaction 2).

2. Then 1,3-bisphosphoglycerate is reduced to glyceraldehyde-3-phosphate with the energy from NADPH, also generated by the light reactions (see Table 8.1, reaction 3). The chloroplast enzyme NADP–glyceraldehyde-3-phosphate dehydrogenase catalyzes this step.

Note that this chloroplast enzyme is similar to that of glycolysis (which will be discussed in Chapter 11), except that NADP rather than NAD is the coenzyme. The NADP-linked form of the enzyme is synthesized during chloro-

◀ **FIGURE 8.3** The Calvin cycle. The carboxylation of three molecules of ribulose-1,5-bisphosphate yields six molecules of 3-phosphoglycerate (*carboxylation phase*). After the phosphorylation of the carboxylic group, 1,3-bisphosphoglycerate is reduced to six molecules of glyceraldehyde-3-phosphate with the concurrent release of six molecules of inorganic phosphate (*reduction phase*). From this total of six, one molecule of glyceraldehyde-3-phosphate accounts for the net assimilation of the three molecules of CO_2 (in yellow), while the other five undergo a series of reactions that finally regenerate the starting three molecules of ribulose-1,5-bisphosphate (*regeneration phase*). See Table 8.1 for a description of each numbered reaction.

regeneration of the CO_2 acceptor ribulose-1,5-bisphosphate. In this regeneration phase, three molecules of ribulose-1,5-bisphosphate (3 molecules × 5 carbons/molecule = 15 carbons total) are formed by reactions that reshuffle the carbons from five molecules of glyceraldehyde-3-phosphate (5 molecules × 3 carbons/molecule = 15 carbons) (see Figure 8.3). Concurrently, the sixth molecule of glyceraldehyde-3-phosphate (1 molecule × 3 carbons/molecule = 3 carbons total) represents the net assimilation of three molecules of CO_2 and becomes available for the carbon metabolism of the plant. This reshuffling consists of reactions 4 through 12 in Figure 8.3 and Table 8.1:

plast development (greening), and this form is preferentially used in biosynthetic reactions.

Operation of the Calvin cycle requires the regeneration of ribulose-1,5-bisphosphate

To prevent the depletion of Calvin cycle intermediates, the continued uptake of atmospheric CO_2 requires the constant

1. *Two molecules* of glyceraldehyde-3-phosphate are converted via triose phosphate isomerase to dihydroxyacetone-3-phosphate in an isomerization reaction (Table 8.1, reaction 4).

2. One molecule of dihydroxyacetone-3-phosphate then undergoes aldol condensation with a *third molecule* of glyceraldehyde-3-phosphate, a reaction catalyzed by

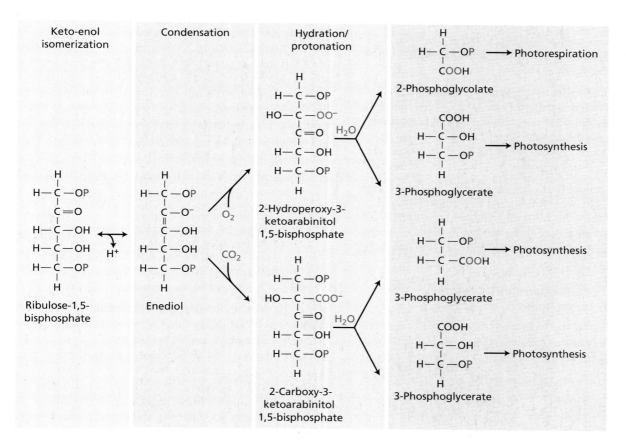

FIGURE 8.4 The carboxylation and the oxygenation of ribulose-1,5-bisphosphate by rubisco. The binding of ribulose-1,5-bisphosphate to rubisco facilitates the formation of an enzyme-bound enediol intermediate that can be attacked by CO_2 or O_2 at the carbon 2, yielding a six (2-carboxy-3-ketoarabinitol-1,5-bisphosphate) or a five (2-hydroperoxy-3-ketoarabinitol-1,5-bisphosphate) carbon reactive intermediate, respectively. The hydration of these intermediates at the carbon 3 triggers the cleavage of the carbon–carbon bond between the carbon 2 and the carbon 3, yielding two molecules of 3-phosphoglycerate (carboxylase activity) or one molecule of 2-phosphoglycolate and one molecule of 3-phosphoglyerate (oxygenase activity).

aldolase, to give fructose-1,6-bisphosphate (Table 8.1, reaction 5).

3. Fructose-1,6-bisphosphate occupies a key position in the cycle and is hydrolyzed to fructose-6-phosphate (Table 8.1, reaction 6).

4. A two-carbon unit (comprising the carbon 1 and carbon 2 of the fructose-6-phosphate) is transferred via the enzyme transketolase to a *fourth molecule* of glyceraldehyde-3-phosphate to give xylulose-5-phosphate. The remaining carbons (carbons 3–6 of the fructose-6-phosphate) form erythrose-4-phosphate (Table 8.1, reaction 7).

5. Erythrose-4-phosphate then combines, via aldolase, with the remaining molecule of dihydroxyacetone-3-phosphate to yield the seven-carbon sugar sedoheptulose-1,7-bisphosphate (Table 8.1, reaction 8).

6. Sedoheptulose-1,7-bisphosphate is then hydrolyzed by way of a specific phosphatase to give sedoheptulose-7-phosphate (Table 8.1, reaction 9).

7. Sedoheptulose-7-phosphate donates a two-carbon unit to the *fifth (and last) molecule* of glyceraldehyde-3-phosphate, via transketolase, producing xylulose-5-phosphate, and the remaining carbons (carbons 3–7) of sedoheptulose-7-phosphate become ribose-5-phosphate (Table 8.1, reaction 10).

8. The two molecules of xylulose-5-phosphate are converted to two molecules of ribulose-5-phosphate by a ribulose-5-phosphate epimerase (Table 8.1, reaction 11a), while a third molecule of ribulose-5-phosphate originates from ribose-5-phosphate by the action of ribose-5-phosphate isomerase (Table 8.1, reaction 11b).

9. Finally, ribulose-5-phosphate kinase (also called phosphoribulokinase) catalyzes the phosphorylation of three molecules of ribulose-5-phosphate with ATP, thus regenerating the three needed molecules of the initial CO_2 acceptor, ribulose-1,5-bisphosphate (Table 8.1, reaction 12).

The Calvin cycle regenerates its own biochemical components

The carboxylation and reduction phases of the Calvin cycle can be summarized as follows:

$$3\ CO_2 + 3\ \text{ribulose-1,5-bisphosphate} + 3\ H_2O +$$
$$6\ NADPH + 6\ H^+ + 6\ ATP$$
$$\downarrow$$
$$6\ \text{Triose phosphates} + 6\ NADP^+ + 6\ ADP + 6\ P_i$$

When leaves are kept for long periods in darkness (e.g., at night), the stromal concentration of most biochemical intermediates of the Calvin cycle is low. Therefore, when transferred to the light, leaves must regenerate these intermediates if the Calvin cycle is to operate. At this stage, stromal triose phosphates and most of the assimilated CO_2 are committed to the production of cycle intermediates, and as this production commences, assimilation of CO_2 by the Calvin cycle accelerates. The importance of this buildup of metabolites is shown by experiments in which previously darkened leaves or isolated chloroplasts are illuminated. In such experiments, CO_2 fixation starts only after a lag, called the *induction period*, and the rate of photosynthesis increases with time in the first few minutes after the onset of illumination. This increase in the rate of photosynthesis during the induction period is due both to an increase in the concentration of intermediates of the Calvin cycle and to the activation of enzymes by light (discussed later). Thus, in the induction period, the six triose phosphates are used for the regeneration of ribulose-1,5-bisphosphate.

In contrast, when photosynthesis reaches a *steady state*, the assimilation of atmospheric CO_2 increases the reserves of plant carbohydrates. At this stage, five-sixths of triose phosphates contribute to the regeneration of the CO_2 acceptor molecule ribulose-1,5-bisphosphate, while one-sixth are used for building up either starch in the chloroplast or the sucrose concentration in the cytosol (see Figure 8.3).

The Calvin cycle uses energy very efficiently

A constant input of energy, provided by ATP and NADPH, is required to keep the Calvin cycle functioning to fix CO_2. During photosynthetic CO_2 assimilation, every molecule of CO_2 fixed by the Calvin cycle into carbohydrates consumes two molecules of NADPH and three molecules of ATP (see Table 8.1). Thus, we can calculate the efficiency of the Calvin cycle by comparing the energy expended in the synthesis of carbohydrates from CO_2 to that derived from the oxidation of NADPH and the hydrolysis of ATP. To this end, the following points must be taken into account:

- The energy released by the total oxidation of one mole of a hexose to CO_2, 2804 kJ, is the minimum amount of energy needed for the synthesis of a hexose.

- The synthesis of one mole of the hexose fructose-6-phosphate from six moles of CO_2 consumes 3126 kJ in the oxidation of 12 moles of NADPH (12 moles × 217 kJ/mole) and the hydrolysis of 18 moles of ATP (18 moles × 29 kJ/mole) (see Table 8.1).

Hence, the thermodynamic efficiency of the Calvin cycle is close to 90% [(2804/3126) × 100]. A further inspection of these calculations reveals that the conversion of CO_2 to carbohydrates is mostly a reductive process, because the bulk of the required energy comes from NADPH. The oxidation of 12 moles of NADPH releases 2604 kJ, while the hydrolysis of 18 moles of ATP supplies 522 kJ; thus, 83% [(2604/3126) × 100] of the total energy used in the Calvin cycle comes from the reductant NADPH.

Alternatively, we can compute the maximum overall thermodynamic efficiency of photosynthesis if we know the energy content of the incident light and the minimum quantum requirement (moles of quanta absorbed per mole of CO_2 fixed; see Chapter 7). Red light at 680 nm contains 175 kJ per quantum mole of photons. The minimum quantum requirement is usually calculated to be eight photons per mole of CO_2 fixed, although numbers obtained experimentally are close to nine or ten (see Chapter 7). Therefore, the minimum light energy needed to reduce six moles of CO_2 to one mole of hexose is approximately $6 \times 8 \times 175$ kJ = 8400 kJ. Given that the synthesis of a mole of hexose requires at least 2804 kJ, the maximum overall thermodynamic efficiency of photosynthesis is about 33% [(2804/8400) × 100].

Thus, the theoretical efficiency of the Calvin cycle is 90% when thermodynamic calculations rely on the energy supplied by the substrates, but that of photosynthesis overall is only 33% when calculations rely on energy from incident light. This difference indicates that a large proportion of light energy is lost in the generation of ATP and NADPH by the light reactions (see Chapter 7) rather than during the operation of the Calvin cycle. The performance of plants under normal growing conditions in the field is actually far below these theoretical values.

In general, the developmental stage of plants and environmental factors (e.g., water, mineral nutrients, and ambient temperature) conspire to lower the yield of photosynthesis. The conversion efficiency for most crops (e.g., potatoes, soybeans, wheat, rice, and corn) typically ranges from 0.1 to 0.4%. For sugarcane in Texas and Hawaii, where nearly year-round active growth is possible under optimal conditions, the value approaches 2%.

The Calvin cycle does not occur in all autotrophic cells. In some anaerobic bacteria autotrophic growth proceeds via one of the following alternatives:

- The ferredoxin-mediated synthesis of organic acids from acetyl– and succinyl–CoA derivatives via the reversal of the citric acid cycle (the reductive carboxylic acid cycle of green sulfur bacteria)
- The glyoxylate-producing cycle (the hydroxypropionate pathway of green nonsulfur bacteria)
- The linear pathway characterized by the CO_2-fixing enzyme carbon monoxide dehydrogenase (the acetyl–CoA route of acetogenic and methanogenic bacteria)

Thus, although the Calvin cycle is quantitatively the most important pathway of autotrophic CO_2 fixation, autotrophy is not obligatorily linked to photosynthesis.

Regulation of the Calvin Cycle

The high energy efficiency of the Calvin cycle indicates the existence of careful regulatory mechanisms ensuring that all intermediates in the cycle are present at adequate concentrations and that the cycle is turned off to optimize energy use when not needed, in the dark. In general, the level and specific activity of enzymes control catalytic rates such that the concentration of cycle metabolites is adjusted in response to metabolic needs.

Changes in gene expression and protein biosynthesis determine the content of enzymes in cell compartments. In particular, the amount of each enzyme present in the chloroplast stroma is regulated by mechanisms that control the concerted expression of nuclear and chloroplast genomes (Maier et al. 1995; Purton 1995). Nucleus-encoded enzymes are translated as precursors on 80S ribosomes in the cytosol and are subsequently transported into the plastid, whereas plastid-encoded proteins are translated in situ on prokaryotic-like 70S ribosomes.

Most of the regulation between nucleus and plastid is anterograde (the signal proceeds forward from nucleus to plastid)—that is, nuclear gene products control the transcription and translation of plastid genes. For instance, rubisco consists of eight nucleus-encoded small subunits and eight plastid-encoded large subunits. The subunit stoichiometry of the stromal holoenzyme (S_8L_8) is partially controlled by the abundance of small subunits that start the translation of the large-subunit mRNA. Specific photoreceptors (e.g., phytochrome or blue-light receptor) are particularly relevant in this nucleus–plastid interaction because, after the perception of light signals, they trigger the expression of stromal enzymes encoded by the nuclear genome (Neff et al. 2000). However, in other cases (e.g., synthesis of chlorophyll proteins), regulation can be retrograde—i.e., the signal flows from the plastid to the nucleus.

In contrast to the slow changes in catalytic rates elicited by the synthesis of enzymes, posttranslational modifications effect changes in specific activity within minutes (Wolosiuk et al. 1993). Two general mechanisms modify the kinetic properties of enzymes:

1. Changes in covalent bonds that result in a chemically modified enzyme, such as reduction of disulfides or carbamylation of amino groups

2. Modifications of noncovalent interactions due, for example, to the binding of metabolites or changes in the ionic composition of the cellular milieu (e.g., pH)

Particularly relevant to the second mechanism is the binding of stromal enzymes to thylakoid membranes, a higher level of organization that enhances the efficiency of the Calvin cycle by channeling substrates and products inside the complex.

Light regulates the Calvin cycle

The light-dependent modulation of stromal enzymes, which is linked to photosynthetic electron transport, changes the activity of target enzymes within minutes of the onset of illumination. As a consequence, light–dark modulation provides an on–off switch for five key enzymes in the Calvin cycle:

1. Rubisco
2. Fructose-1,6-bisphosphate phosphatase
3. Sedoheptulose-1,7-bisphosphate phosphatase
4. Ribulose-5-phosphate kinase (phosphoribulokinase)
5. NADP–glyceraldehyde-3-phosphate dehydrogenase

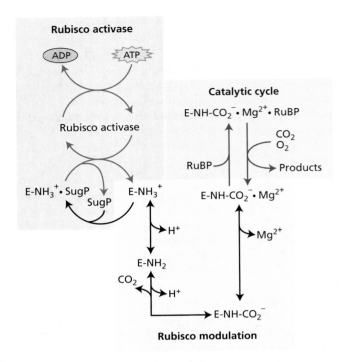

FIGURE 8.5 CO_2 can function as an activator (modulation) or a substrate (catalysis) for rubisco. Modulation: The activation of rubisco (E) involves the formation of a carbamate–Mg^{2+} complex (E–NH–CO_2^-·Mg^{2+}) at the enzyme's active site. In the stroma of illuminated chloroplasts, the increase of both the pH (lower H^+ concentration) and the concentration of Mg^{2+} facilitates the formation of the active form of rubisco. In the catalytic cycle, the active form of rubisco combines with ribulose-1,5-bisphosphate and subsequently with either the *substrate* CO_2, or O_2, starting the carboxylase or the oxygenase activity, respectively (see Figure 8.4). In the rubisco activase-mediated cycle, the tight binding of sugar phosphates such as ribulose-1,5-bisphosphate, either hinders carbamylation or halts the binding of substrates to the carbamylated enzyme. The hydrolysis of ATP by rubisco activase elicits a conformational changes in rubisco that reduces its binding affinity for sugar phophates. RuBP, ribulose-1,5-bisphosphate; SugP, sugar phosphate.

The activity of rubisco increases in the light

George Lorimer and colleagues found that rubisco is activated when an *activator* CO_2 (a different molecule from the *substrate* CO_2 that has become fixed in ribulose-1,5-bisphosphate) reacts slowly with the ε-NH_2 group of a specific lysine within the active site of the enzyme ("rubisco modulation" in Figure 8.5). The resulting carbamate derivative (a new anionic site) then rapidly binds Mg^{2+} to yield the activated enzyme. Two protons are released during the formation of the ternary rubisco–CO_2–Mg^{2+} complex; thus, activation is promoted by an increase in both pH and Mg^{2+} concentration. Accordingly, the light-triggered changes in stromal pH and Mg^{2+} concentration contribute to the activation of rubisco in illuminated chloroplasts (see below). In the active state, rubisco binds the *substrate* CO_2, which reacts with ribulose-1,5-bisphosphate, finally releasing two molecules of 3-phosphoglycerate ("catalytic cycle" in Figures 8.4 and 8.5) (Cleland et al. 1998).

The tight binding of sugar phosphate–like molecules, such as ribulose-1,5-bisphosphate, to rubisco prevents carbamylation. However, the interaction of rubisco with an associated protein, rubisco activase, in a reaction that requires ATP, brings about a structural change in rubisco that releases the sugar phosphate and prepares the enzyme for activation via carbamylation and metal binding ("rubisco activase" in Figure 8.5 and Web Topic 8.3) (Salvucci and Ogren 1996). Congruent with the view that rubisco activase removes sugar phosphate–like molecules from inactive forms of rubisco, Arabidopsis mutants lacking rubisco activase exhibit severely impaired photosynthesis at atmospheric levels of CO_2.

Rubisco activase is a member of a family of proteins that exhibit an ATPase activity associated with a variety of cellular responses that include chaperone-like functions. Many plant species contain two polypeptides of rubisco activase (about 42 and 47 kDa) originating from the alternative splicing of a unique pre-mRNA that produces two polypeptides that are identical except for the presence of extra amino acid residues at the C terminus of the longer form. The binding of ATP to rubisco activase elicits the oligomerization of these polypeptides (14–16 subunits), whose subsequent association with rubisco enhances the assimilation of CO_2. Interestingly, the longer isoform is subject to redox regulation through a thioredoxin-dependent reduction of a disulfide bond located at the C-terminal extension (Zhang and Portis 1999) (see below).

Rubisco is also regulated by the natural sugar phosphate 2-carboxyarabinitol-1-phosphate [H_2O_3P-O-CH_2-C(CO_2^-)OH-CHOH-CHOH-CH_2OH], which closely resembles the six-carbon transition intermediate of the carboxylation reaction (i.e., the 2-carboxy-3-ketoarabinitol-1,5-bisphosphate in Figure 8.4). Present at low concentrations in leaves of many species, this inhibitor is found at high concentrations in darkened leaves of legumes such as soybean and bean. The 2-carboxyarabinitol-1-phosphate binds to rubisco and accumulates at night, but it is removed in the morning through the action of rubisco activase. Disappearance of the inhibitor is enhanced by the action of a specific phosphatase that hydrolyzes its phosphate residue.

The ferredoxin–thioredoxin system regulates the Calvin cycle

In addition to rubisco, light controls the activity of four other enzymes of the Calvin cycle via the ferredoxin–thioredoxin system (ferredoxin, ferredoxin–thioredoxin reductase, thioredoxin). This oxidation–reduction mechanism, identified by Bob Buchanan and colleagues (Buchanan 1980; Wolosiuk et al. 1993; Buchanan and Balmer 2005; Mora-Garcia et al. 2006), uses the product of the photosynthetic electron transport system (reduced ferredoxin) for the modulation of enzyme activity (fructose-1,6-bisphosphate phosphatase,

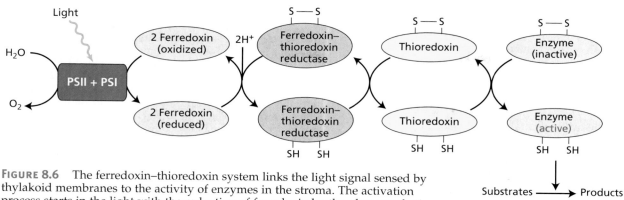

FIGURE 8.6 The ferredoxin–thioredoxin system links the light signal sensed by thylakoid membranes to the activity of enzymes in the stroma. The activation process starts in the light with the reduction of ferredoxin by the photosynthetic electron transport system (PSII + PSI) (see Chapter 7). The reduced ferredoxin, together with two protons, are used to reduce a catalytically active disulfide (—S—S—) group of the iron–sulfur enzyme ferredoxin–thioredoxin reductase, which, in turn, reduces the highly specific disulfide (—S—S—) bond of the small regulatory protein thioredoxin (see Web Topic 8.4 for details). The reduced form (—SH HS—) of thioredoxin then reduces the critical disulfide bond of a target enzyme, triggering its conversion to a catalytically active state. At this stage, the target enzyme catalyzes the transformation of substrates into products.

sedoheptulose-1,7-bisphosphate phosphatase, ribulose-5-phosphate kinase, and NADP–glyceraldehyde-3-phosphate dehydrogenase) (Figure 8.6). Thus, the ferredoxin–thioredoxin system links light absorbed by chlorophyll in the thylakoid membranes to metabolic activity in the chloroplast stroma.

The reducing power from reduced ferredoxin is transferred to thioredoxin by the iron–sulfur enzyme ferredoxin–thioredoxin reductase. The reduced thioredoxin, in turn, reduces a target enzyme, thereby altering its catalytic activity; biosynthetic enzymes are activated and degradative enzymes are deactivated. Initially associated with the Calvin cycle, there is recent evidence that the ferredoxin–thioredoxin system controls a spectrum of chloroplast processes (Buchanan and Balmer 2005; Mora-Garcia et al. 2006). New findings also suggest that thioredoxin may link the activity of a number of chloroplast enzymes to oxidants such as oxygen and reactive oxygen species.

The resolution of the crystal structures of components of the ferredoxin–thioredoxin system and target enzymes has provided valuable information about the mechanisms involved. Thioredoxin is the common reductant for chloroplast enzymes, but the disulfide bond of a target enzyme has structural and thermodynamic features that distinguish it from the others (Table 8.2) (see Web Topic 8.4). Moreover,

TABLE 8.2
Enzymes of the Calvin cycle modulated by the ferredoxin–thioredoxin system

Enzyme	$E_m{}^a$	Regulatory site[b]	Residues	Species	Source
Fructose-1,6-bisphosphate phosphatase	–305 mV	—ECX$_{19}$CIVNVCQ—	153 and 173	Pea	Chiadmi et al. 1999; Schurmann and Jacquot 2000
Sedoheptulose-1,7-bisphosphathate phosphatase	–300 mV	—SCGGTACV—	52 and 57	Wheat	Dunford et al. 1999
Ribulose-5-phosphate kinase	–295 mV	—GCX$_{38}$CL—	16 and 55	Spinach	Porter et al. 1988; Brandes et al. 1996; Hirasawa et al. 1999
Glyceraldehyde-3-phosphate dehydrogenase	ND	—FCX$_{10}$CK—(B subunit)	364 and 375	Pea	Brinkman et al. 1989

[a]Midpoint potential for the reaction $\varepsilon\text{-}(S)_2 + 2H^+ + 2e^- \leftrightarrow \varepsilon\text{-}(SH)_2$, where -(S)$_2$ represents the disulfide bond in the protein ε that links the cysteine residues depicted in bold in the column "Regulatory site"

[b]Amino acid sequence that holds the cysteine residues functional in redox processes. X_n represents the number amino acid residues (n) that separates redox active cysteines.

in some cases, chloroplast metabolites and ions act jointly with thioredoxin in the regulation of stromal enzymes.

Inactivation of target enzymes observed upon darkening appears to take place by a reversal of the reduction (activation) pathway. Oxygen transforms the reduced thioredoxin (—SH HS—) to the oxidized state (—S—S—), which in turn converts the reduced target enzyme to the oxidized state, leading to the original catalytic state (see Figure 8.6) (see Web Topic 8.4). Since the original work with chloroplasts, thioredoxin has been found to function throughout biology and is gaining increased importance in medical research; for example, it has featured in extensive studies on cancer.

Light-dependent ion movements regulate Calvin cycle enzymes

Concurrent with the posttranslational modification of chloroplast enzymes, light causes reversible changes in the ionic composition of the stroma, which, in turn, influences catalytic activities. Upon illumination, the flow of protons from the stroma into the lumen of the thylakoids is coupled to the release of Mg^{2+} from the intrathylakoid space to the stroma. These ion fluxes decrease the stromal concentration of H^+ (the pH increases from 7 to 8) and increase that of Mg^{2+}. Upon darkening, these changes in the ionic composition of the chloroplast stroma are reversed.

Several Calvin cycle enzymes (rubisco, fructose-1,6-bisphosphate phosphatase, sedoheptulose-1,7-bisphosphate phosphatase, and ribulose-5-phosphate kinase) are more active at pH 8 than at pH 7 and require Mg^{2+} as a cofactor for catalysis. Hence, these light-dependent ion fluxes enhance the activity of key enzymes of the Calvin cycle (Heldt 1979).

The C_2 Oxidative Photosynthetic Carbon Cycle

An important feature of rubisco is its capacity to catalyze both the carboxylation and oxygenation of ribulose-1,5-bisphosphate (see Figure 8.4) (Miziorko and Lorimer 1983). The latter capability initiates a series of physiological events in which light-dependent oxygen uptake is associated with CO_2 evolution in photosynthetically active leaves. This process, known as **photorespiration**, works in a diametrically opposite direction to photosynthesis, causing a loss of the CO_2 fixed by the Calvin cycle (Ogren 1984; Leegood et al. 1995).

In this section we will describe the C_2 oxidative photosynthetic cycle—the reactions that result in the partial recovery of carbon lost through photorespiration.

Photosynthetic CO_2 fixation and photorespiratory oxygenation are competing reactions

The incorporation of one molecule of O_2 into the 2,3-enediol isomer of ribulose-1,5-bisphosphate generates an unstable intermediate that rapidly splits into 2-phosphoglycolate and 3-phosphoglycerate (see Figures 8.4 and 8.7, and Table 8.3, reaction 1). The ability to catalyze the oxygenation of ribulose-1,5-bisphosphate is a property of all rubiscos, regardless of taxonomic origin. Even the rubisco from autotrophic anaerobic bacteria catalyzes the oxygenase reaction when exposed to oxygen in vitro. As alternative substrates for rubisco, CO_2 and O_2 compete for reaction with ribulose-1,5-bisphosphate, because carboxylation and oxygenation occur within the same active site of the enzyme.

The C_2 oxidative photosynthetic cycle acts as a scavenger operation to recover fixed carbon lost during photorespiration by the rubisco oxygenase reaction (Web Topic 8.5). The 2-phosphoglycolate formed in the chloroplast by oxygenation of ribulose-1,5-bisphosphate is rapidly hydrolyzed to glycolate by a specific chloroplast phosphatase (see Figure 8.7 and Table 8.3, reaction 2). The subsequent metabolism of glycolate involves the cooperation of two other organelles, peroxisomes and mitochondria (see Chapter 1) (Tolbert 1981).

Glycolate leaves the chloroplast via a specific transporter protein in the envelope inner membrane and diffuses to the peroxisome. There, a flavin mononucleotide–dependent oxidase, glycolate oxidase (see Figure 8.7 and Table 8.3, reaction 3), catalyzes the oxidation of glycolate by O_2 to produce H_2O_2 and glyoxylate. The former product is destroyed in the peroxisome by the action of catalase, releasing O_2 (see Table 8.3, reaction 4), while the latter

FIGURE 8.7 The main reactions of the photorespiratory cycle. Operation of the C_2 oxidative photosynthetic cycle involves the cooperative interaction among three organelles: chloroplasts, peroxisomes, and mitochondria. In the chloroplast, the oxygenase activity of rubisco yields two molecules of 2-phosphoglycolate, which, upon the action of phosphoglycolate phosphatase, releases two molecules of inorganic phosphate with the concurrent formation of two molecules of glycolate. The latter (four carbons) flow from the chloroplast to the peroxisome. At the same time, glutamate leaves the chloroplast for transamination in the peroxisome. Glycerate, 2-oxoglutarate, and NH_4^+ return to the chloroplast in a process that recovers part of the carbon and all of the nitrogen lost in photorespiration. First, glycerate is phosphorylated to 3-phosphoglycerate and incorporated back into the Calvin cycle. Second, the inorganic nitrogen (NH_4^+) and 2-oxoglutarate are used to recover the nitrogen lost in the exported glutamate. In the peroxisomal and mitochondrial reactions of the C_2 oxidative photosynthetic cycle, glycolate enters the peroxisome and it is successively transformed to glyoxylate and glycine. Two molecules of glycine are transported to the mitochondrion, where the successive action of the glycine decarboxylase complex and serine hydroxymethyl transferase yields a molecule of serine (three carbons) with the concurrent release of CO_2 (one carbon) and NH_4^+. Serine is transported to the peroxisome and transformed to glycerate. The latter flows into the chloroplast where it is phosphorylated to 3-phosphoglycerate and incorporated into the Calvin cycle. In addition to the flow of carbon and nitrogen, the uptake of oxygen in the peroxisome supports a short oxygen cycle coupled to oxidative reactions. The circuits of carbon, nitrogen, and oxygen are indicated in black, red, and blue, respectively. See Table 8.3 for a description of each numbered reaction.

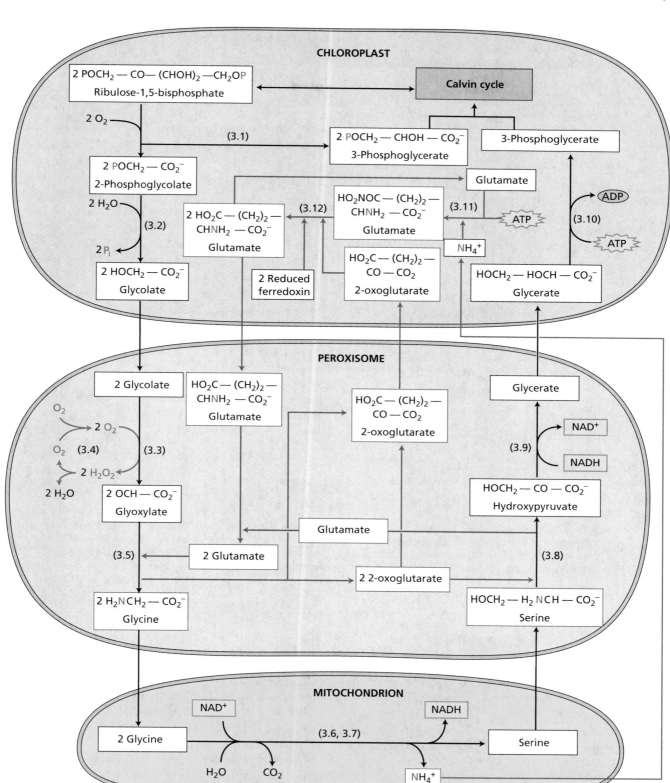

TABLE 8.3
Reactions of the C_2 oxidative photosynthetic cycle

Reaction	Enzyme	Location
1. 2 Ribulose-1,5-bisphosphate + 2 O_2 → 2 2-phosphoglycolate + 2 3-phosphoglycerate	Rubisco	Chloroplast
2. 2 2-Phosphoglycolate + 2 H_2O → 2 glycolate + 2 P_i	Phosphoglycolate phosphatase	Chloroplast
3. 2 Glycolate + 2 O_2 → 2 glyoxylate + 2 H_2O_2	Glycolate oxidase	Peroxisome
4. 2 H_2O_2 → 2 H_2O + O_2	Catalase	Peroxisome
5. 2 Glyoxylate + 2 glutamate → 2 glycine + 2 2-oxoglutarate	Glyoxylate–glutamate aminotransferase	Peroxisome
6. Glycine + NAD^+ + glycine decarboxylase complex (GDC) [GDC] → CO_2 + NH_4^+ + NADH + methylene-GDC	GDC	Mitochondrion
7. Methylene-GDC + glycine + H_2O → serine + GDC	Serine hydroxymethyl transferase	Mitochondrion
8. Serine + 2-oxoglutarate → hydroxypyruvate + glutamate	Serine aminotransferase	Peroxisome
9. Hydroxypyruvate + NADH + H^+ → glycerate + NAD^+	Hydroxypyruvate reductase	Peroxisome
10. Glycerate + ATP → 3-phosphoglycerate + ADP	Glycerate kinase	Chloroplast
11. Glutamate + NH_4^+ + ATP → glutamine + ADP + P_i	Glutamine synthetase	Chloroplast
12. 2-Oxoglutarate + glutamine + 2 Fd_{red} + 2 H^+ → 2 glutamate + 2 Fd_{oxid}	Ferredoxin-dependent glutamate synthase (GOGAT)	Chloroplast

Net reactions of the C_2 oxidative photosynthetic cycle

2 Ribulose-1,5-bisphosphate + 3 O_2 + H_2O + glutamate
↓
Glycerate + 2 3-phosphoglycerate + NH_4^+ + CO_2 + 2 P_i + 2-oxoglutarate

Reactions in the chloroplasts restore the molecule of glutamate:

2-Oxoglutarate + NH_4^+ + [(2 Fd_{red} + 2 H^+), ATP] → glutamate + H_2O + [(2 Fd_{oxid}), ADP + P_i]

and the molecule of 3-phosphoglycerate:

Glycerate + ATP → 3-phosphoglycerate + ADP

Hence, the consumption of three moles of atmospheric oxygen in the C_2 oxidative photosynthetic cycle (two in the oxygenase activity of rubisco and one in peroxisomal oxidations) elicits the release of one mole of CO_2 and the use of two moles of ATP and two moles of reducing equivalents (2 Fd_{red} + 2 H^+) for the purpose of incorporating two-carbon skeletons back into the Calvin cycle, restoring glutamate from NH_4^+ and 2-oxoglutarate.

---

undergoes transamination with glutamate, yielding the amino acid glycine (see Table 8.3, reaction 5).

Glycine leaves the peroxisome and enters the mitochondrion (see Figure 8.7) where two enzymes, the glycine decarboxylase complex (GDC) and serine hydroxymethyltransferase, sequentially catalyze the conversion of two molecules of glycine and one of NAD^+ to one molecule each of serine, NADH, NH_4^+, and CO_2. In the first reaction, a methyl group of glycine is transferred to the GDC complex, forming methylene GDC. The methyl group is transferred to the second glycine to form serine (see Table 8.3, reactions 6 and 7). The newly formed serine diffuses from the mitochondrion to the peroxisome, where it is converted by transamination to hydroxypyruvate (see Table 8.3, reaction 8) which, in turn, is transformed to glycerate (see Table 8.3, reaction 9) via an NADH-dependent reductase. A malate–oxaloacetate shuttle transfers NADH from the cytoplasm into the peroxisome, thus maintaining an adequate concentration of NADH for this reaction. Finally, glycerate reenters the chloroplast, where it is phosphorylated to yield 3-phosphoglycerate (see Figure 8.7 and Table 8.3, reaction 10).

In parallel with these reactions, the NH_4^+ released in the oxidation of glycine (see Table 8.3, reaction 6) diffuses rapidly from the matrix of the mitochondrion to chloroplasts, where glutamine synthetase drives the ATP-dependent incorporation to glutamate, yielding glutamine (see Figure 8.7 and Table 8.3, reaction 11). Subsequently, the ferredoxin-dependent glutamate synthase catalyzes the reaction of glutamine with 2-oxoglutarate, which leads to the production of two molecules of glutamate (see Figure 8.7 and Table 8.3, reaction 12).

The activity of rubisco and glycolate oxidase incorporate two molecules of O_2 each (see Figure 8.7 and Table 8.3, reactions 1 and 3), but catalase releases one molecule of O_2 from H_2O_2 (see Figure 8.7 and Table 8.3, reaction 4). In consequence, a total of three molecules of O_2 are reduced when two molecules of ribulose-1,5-bisphosphate enter the C_2 oxidative photosynthetic cycle.

In photorespiration, carbon, nitrogen, and oxygen atoms circulate simultaneously through the three cycles. In the first cycle, carbon exits the chloroplast as two molecules of glycolate and returns as one molecule of glycerate, leaving a molecule of CO_2 in the mitochondrion. In the second cycle, nitrogen exits the chloroplast as one molecule of glutamate and returns as one molecule of NH_4^+ ultimately associated with one molecule of 2-oxoglutarate (see Figure 8.7). In consequence, total nitrogen remains unchanged, because the formation of inorganic nitrogen (NH_4^+) in the mitochondrion is balanced by the synthesis of glutamine in the chloroplast. In the third cycle, the O_2-linked cycle, reactions catalyzed by rubisco in the chloroplast and by glycolate oxidase in the peroxisome incorporate O_2, giving an oxidative character to the whole process.

Overall, two molecules of phosphoglycolate (four carbons), lost from the Calvin cycle by the oxygenation of ribulose-1,5-bisphosphate, are converted into one molecule of 3-phosphoglycerate (three carbons) and one CO_2 (one carbon). In other words, 75% of the carbon lost by the oxygenation of ribulose-1,5-bisphosphate is recovered by the C_2 oxidative photosynthetic cycle and returned to the Calvin cycle (Lorimer 1981; Sharkey 1988). This decreased efficiency can be measured as an increase in the quantum requirement for CO_2 fixation under photorespiratory conditions (air with high O_2 and low CO_2) as opposed to nonphotorespiratory conditions (low O_2 and high CO_2).

In vivo, the balance between the Calvin cycle and the C_2 oxidative photosynthetic cycle is determined mainly by three factors, one inherent to the plant (the kinetic properties of rubisco) and two linked to the environment (temperature and the concentration of substrates, CO_2 and O_2). An increment of the external temperature:

- Modifies kinetic constants of rubisco, increasing the rate of oxygenation more than that of carboxylation (Ku and Edwards 1978)
- Lowers the concentration of CO_2 in a solution in equilibrium with air more than that of O_2 (see **Web Topic 8.6**)

Both features explain the increase of photorespiration (oxygenation) relative to photosynthesis (carboxylation) in response to higher temperatures. Overall, then, increasing the temperature progressively tilts the balance away from the Calvin cycle and toward the C_2 oxidative photosynthetic cycle (see Chapter 9).

Photorespiration depends on the photosynthetic electron transport system

The operation of the C_2 oxidative photosynthetic cycle is linked to the concurrent functioning of the photosynthetic electron transport system (Figure 8.8). For salvaging 2-phosphoglycolate by conversion to 3-phosphoglycerate, photophosphorylation provides the ATP necessary for the conversion of glycerate to 3-phosphoglycerate (see Table 8.3, reaction 10), while the consumption of NADH by hydroxypyruvate reductase (see Table 8.3, reaction 9) is counterbalanced by its production by glycine decarboxylase (see Table 8.3, reaction 6).

The photosynthetic electron transport system supplies the ATP and reduced ferredoxin needed for salvaging NH_4^+ through its uptake into glutamate via glutamine synthetase (see Table 8.3, reaction 11) and ferredoxin-dependent glutamine–2-oxoglutarate aminotransferase (glutamate synthase; GOGAT) (see Table 8.3, reaction 12), respectively.

In summary,

$$\begin{array}{c} 2 \text{ Ribulose-1,5-bisphosphate} + 3\, O_2 + H_2O + ATP + \\ [2\, Fd_{red} + 2\, H^+ + ATP] \\ \downarrow \\ 3 \text{ 3-Phosphoglycerate} + CO_2 + 2\, P_i + ADP + \\ [2\, Fd_{oxid} + ADP + P_i] \end{array}$$

Photosynthetic carbon metabolism in intact leaves reflects the integrated balance between two mutually opposing cycles, the Calvin cycle and the C_2 oxidative photosynthetic cycle, interlocked with the photosynthetic electron transport system, which provides a supply of ATP and reducing equivalents (reduced ferredoxin) (see Figure 8.8).

The biological function of photorespiration is under investigation

Although the C_2 oxidative photosynthetic cycle recovers as glycerate 75% of the carbon originally lost from the Calvin cycle as 2-phosphoglycolate, we are still left with the question of the adaptive value of photorespiration. One possible answer is that the formation of 2-phosphoglycolate is a consequence of the chemistry of the carboxylation reaction that generates an intermediate reactive species with both CO_2 and O_2. On this basis, while ratios of CO_2 to O_2 in air higher than today would have had little consequence in early evolutionary times, the low $CO_2:O_2$ ratios prevalent in modern times promote the recovery, through photorespiration, of some of the carbon lost as 2-phosphoglycolate.

An alternative view is that photorespiration constitutes a protective, supportive mechanism in chloroplasts that reduces the injury caused by reactive oxygen species generated especially under conditions of high light intensity and low intercellular CO_2 (e.g., when stomata are closed because of water stress). Evidence from work with transgenic plants is consistent with photorespiration protecting C_3 plants from photooxidation and photoinhibition (Kozaki and Takeba 1996). Free oxygen radicals react strongly with biological macromolecules and thereby alter their

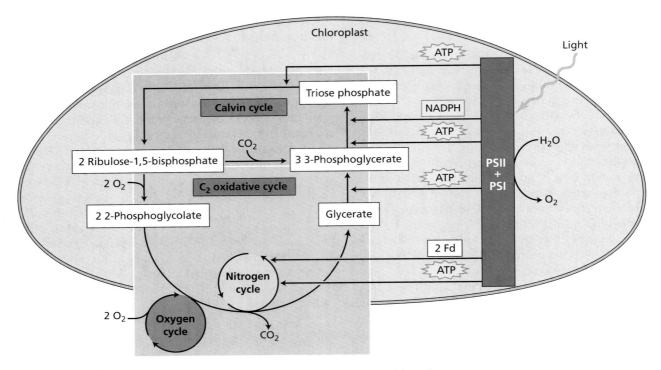

FIGURE 8.8 Dependence of the C_2 oxidative photosynthetic cycle on chloroplast metabolism. The functioning of the C_2 oxidative photosynthetic cycle requires the Calvin cycle for the regeneration of the starting ribulose-1,5-bisphosphate, and the photosynthetic electron transport system for (i) ATP needed to convert glycerate to 3-phosphoglycerate and (ii) ATP and reducing equivalents needed to restore glutamate from NH_4^+ and 2-oxoglutarate. The "nitrogen cycle," which also involves the peroxisome and the mitochondrion, is required for the removal of NH_4^+ and the supply of glutamate. The "oxygen cycle" takes place in the peroxisome and is required for the removal of H_2O_2.

function. Under high illumination and low atmospheric CO_2, the C_2 oxidative photosynthetic cycle would dissipate the excess of both ATP and reducing power from the light reactions, thus preventing damage to the photosynthetic apparatus. Arabidopsis mutants that are unable to photorespire grow normally under high (2%) CO_2, but they die rapidly when transferred to normal air.

It can also be argued that the recovery of carbon as 3-phosphoglycerate by the Calvin cycle is an essential feature of photorespiration; these photorespiration-deficient mutants may die from the accumulation of metabolites into dead-end pathways. In line with this alternative, Arabidopsis mutants lacking glycerate kinase accumulate glycerate and concurrently are not viable in normal air but viable in atmospheres with elevated proportions of CO_2 (Boldt et al. 2005).

Finally, a recent study has also linked photorespiration to nitrate assimilation via a mechanism still under investigation (Rachmilevitch et al. 2004). In short, while there is emerging evidence that photorespiration positively affects plants, a full understanding of the process is still not at hand.

CO_2-Concentrating Mechanisms

Many photosynthetic organisms do not photorespire at all or do so to only a limited extent. These organisms have normal rubiscos, and their lack of photorespiration is a consequence of mechanisms that concentrate CO_2 in the rubisco environment and thereby suppress the oxygenation reaction.

In the following sections we will discuss different mechanisms for concentrating CO_2 at the site of carboxylation:

- CO_2 pumps at the plasma membrane
- C_4 photosynthetic carbon fixation (C_4)
- Crassulacean acid metabolism (CAM)

The first of these CO_2-concentrating mechanisms is present in aquatic plants, but has been studied most extensively in prokaryotic cyanobacteria and eukaryotic algae. The last two mechanisms are found in some angiosperms and involve "add-ons" to the Calvin cycle. Plants with C_4 metabolism are often found in hot environments, and CAM plants are typical of desert environments.

I. CO_2 and HCO_3^- Pumps

In response to changing concentrations of HCO_3^- in aqueous environments, many aquatic organisms develop effective mechanisms for concentrating CO_2 and thus improving carboxylation by the relatively inefficient rubisco (Ogawa and Kaplan 1987; Giordano et al. 2005). When cyanobacteria and algae are grown in air enriched with 5% CO_2 and are subsequently transferred to a low-CO_2 medium, they exhibit symptoms typical of photorespiration (O_2 inhibition of photosynthesis at low concentrations of CO_2). But if the cells are grown in air containing 0.03% CO_2, they rapidly develop the ability to concentrate inorganic carbon (CO_2 plus HCO_3^-) internally.

Under these low-CO_2 conditions, the cells no longer photorespire, because they accumulate HCO_3^- in the cytosol using both HCO_3^- and CO_2 pumps associated with the operation of NAD(P)H–dehydrogenase complexes at the plasma membrane. The ionic character of HCO_3^- further enhances its active uptake, because lipid membranes are much less permeable to these species than to uncharged molecules. Recent evidence indicates that the CO_2 pump associated with green algae is a specific gas channel that, surprisingly, is highly homologous to the Rheus protein, known to be important in erythrocytes (Soupene et al. 2004). This work thus not only helps clarify an old question in plant physiology, but also defines the function of the Rheus protein, a long-standing mystery protein in medicine.

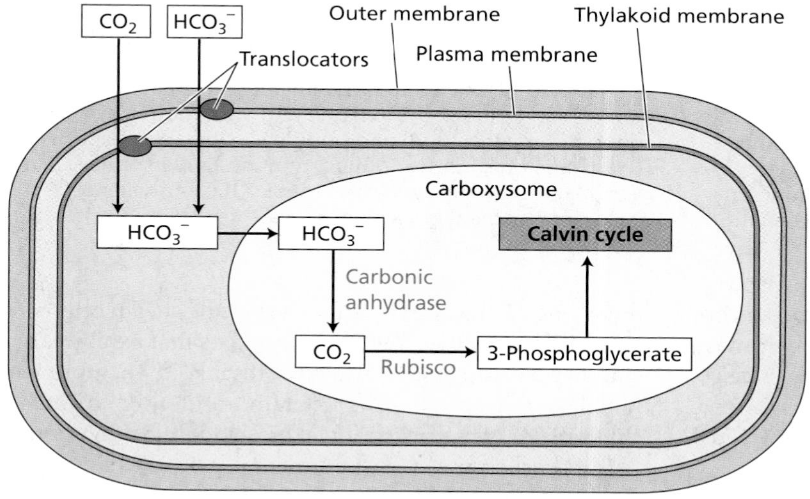

Central to this process is a specialized carbonic anhydrase that, by speeding up the conversion $H_2CO_3 \leftrightarrow CO_2 + H_2O$ elevates the concentration of CO_2 around rubisco localized inside specific compartments. In prokaryotic cyanobacteria, cytosolic HCO_3^- is transferred to a rubisco-containing microcompartment, the carboxysome, for the final fixation via the Calvin cycle (Figure 8.9) (Badger and Price 2003). This particular protein microbody consists of a protein coat that keeps most of the cellular rubisco sequestered. HCO_3^- diffuses into the carboxysome across the proteinaceous shell and, once inside the compartment, yields CO_2 via a specialized carbonic anhydrase. A poorly understood diffusion barrier, perhaps the protein shell, restricts the diffusion of CO_2 out of the carboxysome and, in so doing, increases the concentration of CO_2 that surrounds the active site of rubisco within the carboxysome.

The metabolic consequence of this mechanism for CO_2 enrichment around rubisco is the suppression of oxygenation and, as a consequence, of photorespiration. The energetic cost of this adaptation is the additional ATP needed, at the level of the NAD(P)H–dehydrogenase complexes, for pumping CO_2.

II. The C_4 Carbon Cycle

The oxygenase activity of rubisco limits significantly the efficiency of photosynthetic carbon assimilation in vascular plants, particularly under warmer temperatures or water stress. To minimize oxygenase activity and the concurrent loss of carbon through photorespiration, C_4 photosynthesis evolved as one of the major carbon-concentrating mechanisms used by vascular plants to compensate for limitations associated with low atmospheric CO_2.

Malate and aspartate are carboxylation products of the C_4 cycle

Early labeling of four-carbon acids was first observed in the late 1950s in $^{14}CO_2$ labeling studies with sugarcane and maize by H. P. Kortschack and Y. Karpilov, respectively. When leaves were exposed to $^{14}CO_2$ for a few seconds in the light, 70 to 80% of the label was found in the four-carbon acids malate and aspartate—a pattern very different from the one observed in leaves that photosynthesize solely via the Calvin cycle. In pursuing these initial observations, M. D. Hatch and C. R. Slack elucidated what is now known as the C_4 photosynthetic carbon cycle (C_4 cycle) (Figure 8.10). They established that malate and aspartate are the first stable, detectable intermediates of photosynthesis in leaves of sugarcane, and

FIGURE 8.9 The cyanobacterial CO_2-concentrating mechanism. Translocators located in the thylakoid membrane and the plasma membrane (green ovals) pump CO_2 and HCO_3^- into the cytosol and the thylakoid of a cyanobacterium. The diffusional resistance to efflux and the internal gradient of HCO_3^- drive the inorganic carbon to the carboxysome. The carboxysomal carbonic anhydrase catalyzes the interconversion between HCO_3^- and CO_2 and, in so doing, increases the concentration of CO_2 around rubisco, facilitating the carboxylation of ribulose-1,5-bisphosphate.

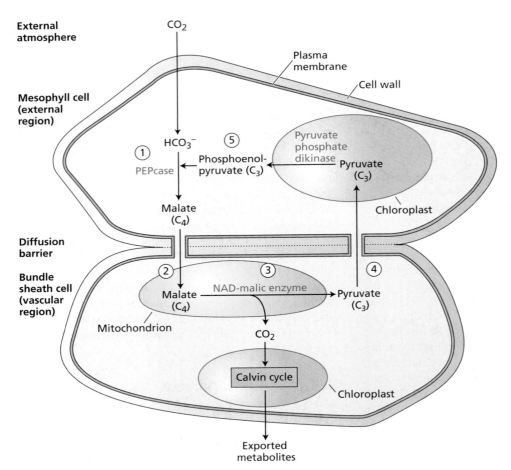

Figure 8.10 The C_4 photosynthetic carbon cycle involves four successive stages in two different compartments of leaf cells. (1) In the periphery of mesophyll cells ("external region"), close to the external environment, the HCO_3^- originating from atmospheric CO_2 reacts with a three-carbon compound, phosphoenolpyruvate, via the enzyme phosphoenolpyruvate carboxylase (PEPCase), yielding a four-carbon acid. (2) The four-carbon acid flows across a diffusion barrier to the bundle sheath cells (vascular region), close to vascular connections. (3) The decarboxylation of the four-carbon acid via a decarboxylating enzyme (e.g., NAD–malic enzyme) yields a high concentration of CO_2 around rubisco that facilitates its assimilation through the Calvin cycle. (4) The residual three-carbon acid (e.g., pyruvate) flows back to the region in contact with the external atmosphere. (5) The acceptor of the atmospheric CO_2, phosphoenolpyruvate is regenerated via the enzyme pyruvate–phosphate dikinase. In most land plants, the external and vascular regions are found in the mesophyll and bundle sheath cells, respectively (Kranz anatomy). A few others, typified by *Borszczowia aralocaspica* and *Bienertia cycloptera*, contain the equivalents of these compartments in a single cell.

that carbon 4 of malate subsequently becomes carbon 1 of 3-phosphoglycerate (Hatch and Slack 1966). The primary carboxylation in these leaves is catalyzed not by rubisco, but by phosphoenolpyruvate carboxylase (PEPCase) (Table 8.4, reaction 1) (see **Web Essay 8.1**) (Chollet et al. 1996; Vidal and Chollet 1997).

Two different types of cells participate in the C_4 cycle

Since the seminal studies of the 1950s and the 1960s, the C_4 pathway of photosynthesis has typically been found in leaves of higher plants with vascular tissues surrounded by two distinctive photosynthetic cell types: an internal ring of **bundle sheath cells**, which is wrapped with an outer ring of **mesophyll cells**. The bundle sheath cells contain centrifugally arranged chloroplasts that exhibit large starch granules and unstacked thylakoid membranes, whereas mesophyll cells contain randomly arranged chloroplasts with stacked thylakoids and little or no starch.

This particular dimorphic structure, named Kranz anatomy (*Kranz*, German for "wreath"), ensures the cell-specific compartmentalization of enzymes that is essential for providing CO_2 to the C_4 pathway (Figure 8.11). The manner in which carbon is transferred from carbon 4 of malate to carbon 1 of 3-phosphoglycerate became clear when the role of mesophyll and bundle sheath cells was elucidated. The participating enzymes occur in one of the two cell types. PEPCase and pyruvate–phosphate dikinase are restricted to

TABLE 8.4
Reactions of the C_4 photosynthetic carbon cycle

Enzyme	Reaction
1. PEPCase	Phosphoenolpyruvate + HCO_3^- → oxaloacetate + P_i
2. NADP–malate dehydrogenase	Oxaloacetate + NADPH + H^+ → malate + $NADP^+$
3. Aspartate aminotransferase	Oxaloacetate + glutamate → aspartate + 2-oxoglutarate
4. NAD(P)–malic enzyme	Malate + $NAD(P)^+$ → pyruvate + CO_2 + NAD(P)H + H^+
5. Phosphoenolpyruvate carboxykinase	Oxaloacetate + ATP → phosphoenolpyruvate + CO_2 + ADP
6. Alanine aminotransferase	Pyruvate + glutamate → alanine + 2-oxoglutarate
7. Pyruvate–phosphate dikinase	Pyruvate + P_i + ATP → phosphoenolpyruvate + AMP + PP_i
8. Adenylate kinase	AMP + ATP → 2 ADP
9. Pyrophosphatase	PP_i + H_2O → 2 P_i

Note: P_i and PP_i stand for inorganic phosphate and pyrophosphate, respectively.

(A) C_4 monocot

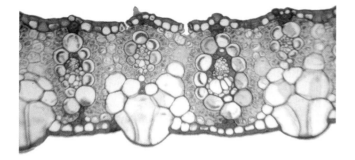

(B) C_3 monocot

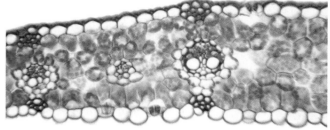

(C) C_4 dicot

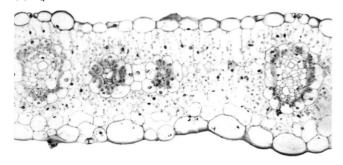

(D) C_4 leaf

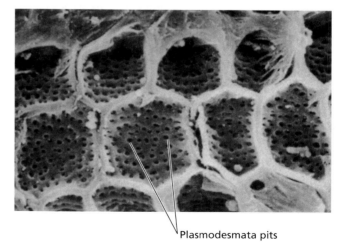

Plasmodesmata pits

FIGURE 8.11 Anatomic differences between C_3 and C_4 plants (Kranz anatomy). Cross-sections of leaves, showing the anatomic differences between C_3 and C_4 plants. (A) A C_4 monocot, sugarcane (*Saccharum officinarum*). (135×) (B) A C_3 monocot, a grass (*Poa* sp.). (240×). (C) A C_4 dicot, *Flaveria australasica* (family Asteraceae). (740×) The bundle sheath cells are large in C_4 leaves (A and C), and no mesophyll cell is more than two or three cells away from the nearest bundle sheath cell. These anatomic features are absent in the C_3 leaf (B). (D) Scanning electron micrograph of a C_4 leaf from *Triodia irritans*, showing the plasmodesmata pits in the bundle sheath cell walls through which metabolites of the C_4 carbon cycle are thought to be transported. (A and B © David Webb; C courtesy of Athena McKown; D from Craig and Goodchild 1977.)

mesophyll cells, whereas decarboxylases and the enzymes of the complete Calvin cycle are confined to bundle sheath cells. With this knowledge, Hatch and Slack described the basic model of the C_4 cycle in four stages (see Figure 8.10 and Table 8.4):

1. Fixation of CO_2 by the carboxylation of phosphoenolpyruvate in the mesophyll cells to form oxaloacetate (see Table 8.4, reaction 1). The oxaloacetate is subsequently reduced to malate by NADP–malate dehydrogenase (see Table 8.4, reaction 2) or converted to aspartate by transamination with glutamate (see Table 8.4, reaction 3).

2. Transport of the four-carbon acids (malate or aspartate) to bundle sheath cells that surround the vascular bundles.

3. Decarboxylation of the four-carbon acids and generation of CO_2, which is then reduced to carbohydrate via the Calvin cycle. In some C_4 plants, prior to this reaction an aspartate aminotransferase catalyzes the conversion of aspartate back to oxaloacetate (see Table 8.4, reaction 3). Different subtypes of C_4 plants have co-opted alternative decarboxylases for the release of CO_2 from organic acids whose action effectively reduces the oxygenase reaction of rubisco by increasing the proportion of CO_2 relative to O_2 (see Table 8.4, reactions 4 and 5) (see **Web Topic 8.7**).

4. Transport of the three-carbon backbone (pyruvate or alanine) formed by the decarboxylation step back to the mesophyll cells and regeneration of the CO_2 acceptor. When pyruvate is the available three-carbon acid, phosphoenolpyruvate is generated through the action of the pyruvate–phosphate dikinase (see Table 8.4, reaction 7). At this stage, an additional molecule of ATP is required for the transformation of AMP to ADP catalyzed by the adenylate kinase (see Table 8.4, reaction 8). When alanine is the three-carbon compound exported by the bundle sheath cells, the formation of pyruvate by the action of alanine aminotransferase precedes the phosphorylation by pyruvate–phosphate dikinase (see Table 8.4, reaction 6).

The recent finding of C_4 photosynthesis in vascular plants devoid of Kranz anatomy (Edwards et al. 2004); has shown that a much greater diversity in modes of C_4 carbon fixation exists than previously thought. *Borszczowia aralocaspica* (which grows in central Asia) and *Bienertia cycloptera* (which grows from east Anatolia eastward to Pakistan) perform C_4 photosynthesis within single chlorenchyma cells by intracellularly partitioning enzymes and chloroplasts into two compartments located in different parts of *the same cell*. Similarly, diatoms, important marine photoautotrophic protists, also perform C_4 photosynthesis in a single cell.

In summary, shuttling of metabolites between two compartments is driven by diffusion gradients not only *between* but also *within* the cells. The C_4 cycle thus effectively drives CO_2 from the external region close to the atmosphere into chloroplasts for CO_2 fixation via the rubisco located at the internal region (see Figure 8.10). Studies of a PEPCase-deficient mutant of *Amaranthus edulis* clearly showed that the lack of an effective mechanism for concentrating CO_2 in bundle sheath cells markedly enhances photorespiration in a C_4 plant (Dever et al. 1996).

The C_4 cycle concentrates CO_2 in the chloroplasts of bundle sheath cells

Originally described for tropical grasses, the C_4 cycle is now known to occur in 18 families of both monocotyledons and dicotyledons, and it is particularly prominent in Gramineae (Poaceae) (corn, millet, sorghum, sugarcane), Chenopodiaceae (*Atriplex*), and Cyperaceae (sedges) (Edwards and Walker 1983). There are differences in leaf anatomy between plants that have a four-carbon cycle (C_4 plants) and those that photosynthesize solely via the Calvin photosynthetic cycle (C_3 plants). A cross section of a typical C_3 leaf reveals one major cell type that has chloroplasts, the mesophyll. In contrast, a typical C_4 leaf has the two distinct chloroplast-containing cell types already discussed, mesophyll and bundle sheath cells (Figure 8.11).

In addition to anatomical differences, Kranz cells exhibit large differences in enzyme composition. PEPCase and rubisco are located in mesophyll and bundle sheath cells, respectively, while decarboxylases are found in different intracellular compartments of bundle sheath cells: NADP–malic enzyme in chloroplasts, NAD–malic enzyme in mitochondria, and phosphoenolpyruvate carboxykinase in the cytosol. Despite these biochemical differences, starch, the transitory storage product of photosynthesis (see below), accumulates predominantly in bundle sheath chloroplasts.

The chloroplasts in mesophyll and bundle sheath cells of C_4 plants also exhibit differences in ultrastructure. PSII activity and the accompanying linear electron flow to PSI, which yield both NADPH and ATP, are generally found in chloroplasts containing extensive grana, whereas the PSI-mediated cyclic electron flow, which produces mostly ATP, is present in chloroplasts low in grana. In general, this dimorphism of chloroplasts correlates with the energy requirements of the different biochemical systems of C_4 photosynthesis. Species using C_4 compounds of the NADP–malic enzyme type are predominantly malate formers that require more reducing power for the conversion of oxaloacetate to malate than do species that are aspartate formers (see Table 8.4, reaction 2). As a consequence, mesophyll chloroplasts have well-developed grana, while bundle sheath chloroplasts are deficient in grana. On the other hand, species of the NAD–malic enzyme type, in which aspartate is the primary product of CO_2 fixation, require mainly ATP, which is produced by PSI cyclic electron flow, to drive the conversion of pyruvate to phosphoenolpyruvate in mesophyll chloroplasts (see Table 8.4, reaction 7). Accordingly, the grana content of chloroplasts of these mesophyll cells is low.

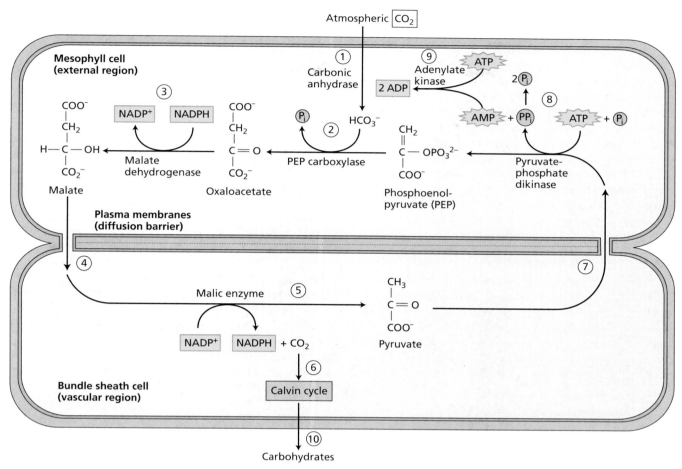

FIGURE 8.12 The C_4 photosynthetic pathway in leaves with Kranz anatomy. The uptake of CO_2 by mesophyll cells (1) provides the HCO_3^- for the primary carboxylation catalyzed by the enzyme phosphoenolpyruvate carboxylase (2). Upon the reduction of oxaloacetate to malate (3), the four-carbon acid diffuses to bundle sheath cells (4) for decarboxylation (5). The CO_2 is taken up by the Calvin cycle located at the chloroplasts of bundle sheath cells (6) while the three-carbon metabolite pyruvate (7) diffuses back to mesophyll cells. Finally, the phosphorylation of pyruvate restores phosphoenolpyruvate (8), the primary acceptor of CO_2. The consumption of two molecules of ATP per mole of fixed CO_2 (8 and 9) drives the cycle in the direction of the arrows, thus pumping CO_2 from the atmosphere to the Calvin cycle. The assimilated carbon leaves the organelle (10) and, after transformation to sucrose in the cytoplasm, enters the vascular tissues for sustaining growth or carbohydrate storage in other parts of the plant. For differences among plant species in reactions sustaining the C_4 photosynthetic pathways see **Web Topic 8.7**.

There is considerable anatomic variation in the arrangement of the bundle sheath cells with respect to mesophyll cells and vascular tissue. In all cases, the operation of the C_4 cycle requires the cooperative effort of both cell types. The transport process facilitated by the plasmodesmata connecting the two cell types generates a much higher concentration of CO_2 in bundle sheath cells (the vascular region) than in mesophyll cells (the "external" region) (Figure 8.12). This elevated concentration of CO_2 at the carboxylation site of rubisco results in the suppression of the ribulose-1,5-bisphosphate oxygenation and hence of photorespiration.

Our understanding of the regulation of C_4 photosynthesis advanced considerably with the application of recombinant DNA technology to plant metabolism. The use of transgenic plants with altered ratios of PEPCase to rubisco provided an opportunity to test whether the relative levels of these enzymes determine the efficiency with which C_4 photosynthesis operates. The antisense suppression of rubisco in *Flaveria* progressively reduced the content of the enzyme in leaves without a concurrent effect on the levels of PEPCase (Siebke et al. 1997; von Caemmerer et al. 1997). Although these transgenic plants exhibited higher concentrations of CO_2 in bundle sheath cells, the increased CO_2 leaked from the bundle sheath and photosynthetic efficiency declined. In contrast, the high leakage of HCO_3^- to mesophyll cells through plasmodesmata, together with the saturation of photosynthesis at low light observed in *Flaveria* expressing tobacco carbonic anhydrase in bundle sheath cells, indicated that a high resistance to CO_2 diffusion in the bundle sheath is essential for efficient C_4 photosynthesis (Ludwig et al. 1998).

The existence of C_4-like photosynthesis in cells surrounding the vascular bundles of C_3 plants suggests, however, that fewer modifications are needed to evolve the former pathway. Some C_3 plants (e.g., tobacco) exhibit features of C_4 photosynthesis in cells that surround the xylem and phloem, and use carbon supplied as malate from the vascular system (Hibberd and Quick 2002). Moreover, the presence of enzymes specific for the C_4 pathway [NAD(P)–malic enzyme and phosphoenolpyruvate carboxykinase] in these cells indicates that essential biochemical components required for C_4 photosynthesis are already present in C_3 plants. These studies are in line with the polyphyletic evolution of C_4 photosynthesis.

The C_4 cycle also concentrates CO_2 in single cells

The Asian plants *Borszczowia aralocaspica* and *Bienertia cycloptera* can perform complete C_4 photosynthesis within single chlorenchyma cells (Sage 2002; Edwards et al. 2004). As noted above, this property changed the long-standing view that plant C_4 photosynthesis relies exclusively on the concerted action of two types of cells. The intracellular partitioning of enzymes and organelles in two well-defined compartments of the cytosol can also contribute to increased concentrations of CO_2 around rubisco (Figure 8.13).

For example, confocal microscopy of chloroplast autofluorescence of *Borszczowia* leaves reveals a dense layer of chloroplasts close to vascular bundles and fewer chloroplasts near the leaf surface in contact with the atmosphere. The assimilation of atmospheric CO_2 takes place at the chlorenchyma cells, where cytosol PEPCase catalyzes the incorporation of HCO_3^- into phosphoenolpyruvate to form oxaloacetate, the precursor of malate and aspartate. The four-carbon acids diffuse across a cytoplasmic space devoid of organelles to the region of the cell proximal to vascular connections.

In mitochondria located at the vascular region, the NAD–malic enzyme decarboxylates four-carbon acids, releasing CO_2 and pyruvate. The former is captured, via rubisco, by the Calvin cycle localized in chloroplasts surrounding the mitochondria. The pyruvate diffuses back to the region in contact with the atmosphere where the pyruvate–phosphate dikinase, located in chloroplasts, converts pyruvate to phosphoenolpyruvate. In both *Bienertia* and *Borszczowia*, the chloroplasts containing rubisco are often positioned near mitochondria in which the NAD–malic enzyme generates CO_2 from malate. Hence, these parti-

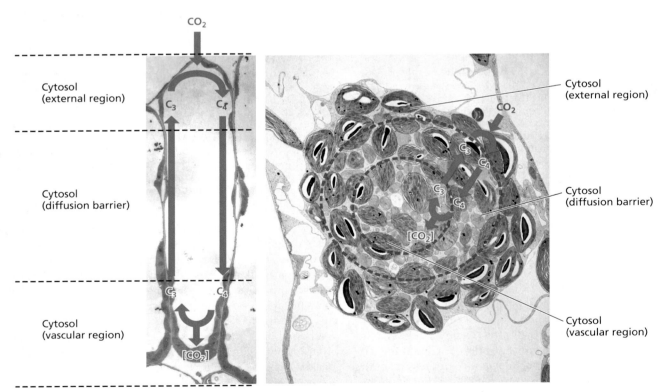

FIGURE 8.13 Single-cell C_4 photosynthesis. Diagrams of the C_4 cycle are superimposed on electron micrographs of *Borszczowia aralocaspica* (A) and *Bienertia cycloptera* (B). Studies on the localization of key photosynthetic enzymes also indicate two dimorphic chloroplasts located at different cytoplasmic compartments having photosynthetic functions analogous to mesophyll and bundle sheath cells in Kranz NAD–malic enzyme–type C_4 plants. (From Edwards et al. 2004.)

ular cells mimic intracellularly the proposed function of membranes and liquid diffusion paths of the Kranz anatomy, evolving the positioning of organelles as an adaptive response to ensure a high concentration of CO_2 for the function of rubisco.

DIATOMS Diatoms are important marine photoautotrophic protists that play a central role in food webs and carbon cycling. Short-term labeling experiments with the marine diatom *Thalassiosira weissflogii* revealed the initial incorporation of CO_2 into four-carbon acids and the subsequent transfer of carbon to 3-phosphoglycerate and sugars. A C_4 pathway that performs photosynthesis at low concentrations of CO_2 might seems redundant with the known CO_2 concentrating mechanism of marine diatoms. However, genes encoding PEPCase, pyruvate–phosphate dikinase, and phosphoenolpyruvate carboxykinase have been identified in the recently sequenced genome of *Thalassiosira pseudonana*. Moreover, a specific inhibitor of PEPCase (i.e., 3,3-dichloro-2-dihydroxyphosphinoylmethyl-2-propenoate) lowers by more than 90% whole-cell photosynthesis in *Thalassiosira weissflogii* cells acclimated to low CO_2 (10 µM). However, the subsequent increase of CO_2 to 150 µM or decrease of O_2 to 80 µM restores photosynthesis to the control levels (Reinfelder et al. 2004). Diatoms are a fine example of photosynthetic organisms that have the capacity to use different CO_2-concentrating mechanisms in response to environmental fluctuations.

The C_4 cycle has higher energy demand than the Calvin cycle

Thermodynamics tells us that work must be done to establish and maintain the CO_2 concentration gradients (for a detailed discussion of thermodynamics, see Chapter 2 on the web site). This principle also applies to the operation of the C_4 cycle. From a summation of the reactions involved (Table 8.5), we calculate that the regeneration of the primary acceptor, phosphoenolpyruvate, consumes two additional molecules of ATP to drive two endergonic reactions—catalyzed by pyruvate–phosphate dikinase (see Table 8.4, reaction 7) and adenylate kinase (see Table 8.4, reaction 8)—and is complemented with the exergonic hydrolysis of one molecule of pyrophosphate to two molecules of inorganic phosphate, catalyzed by pyrophosphatase (see Figure 8.12 and Table 8.4, reaction 9). Thus the total energy requirement for fixing CO_2 by the combined action of C_4 and Calvin cycles (calculated in Tables 8.5 and 8.1, respectively) is five ATP plus two NADPH per CO_2 fixed.

Because of this higher energy demand, C_4 plants photosynthesizing under nonphotorespiratory conditions (high CO_2 and low O_2) require more quanta of light per CO_2 than C_3 leaves do. In normal air, the quantum requirement of C_3 plants changes with factors that affect the balance between photosynthesis and photorespiration, such as temperature. By contrast, owing to the mechanisms built in to avoid photorespiration, the quantum requirement of C_4 plants remains relatively constant under different environmental conditions (see Figure 9.9)

Light regulates the activity of key C_4 enzymes

In addition to supplying ATP and NADPH, light is essential for the operation of the C_4 cycle for the regulation of several specific enzymes. Thus, the activities of PEPCase, NADP–malate dehydrogenase, and pyruvate–phosphate dikinase (see Table 8.4, reactions 1, 2, and 7, respectively) are regulated in response to variations in photon flux density by two different mechanisms: reduction–oxidation of thiol groups and phosphorylation–dephosphorylation.

NADP–malate dehydrogenase is regulated via the ferredoxin–thioredoxin system (see Figure 8.6). The enzyme is reduced (activated) upon illumination of leaves and is oxidized (deactivated) upon darkening. On the other hand, PEPCase is activated by a light-dependent phosphorylation–dephosphorylation mechanism yet to be characterized. The third regulatory member of the C_4 pathway, pyruvate–phosphate dikinase, is rapidly deactivated by an

TABLE 8.5
Energetics of the C_4 photosynthetic carbon cycle

Phosphoenolpyruvate + H_2O + NADPH + CO_2	→ malate + $NADP^+$ + P_i (mesophyll)
Malate + $NADP^+$	→ pyruvate + NADPH + CO_2 (bundle sheath)
Pyruvate + P_i + ATP	→ phosphoenolpyruvate + AMP + PP_i (mesophyll)
PP_i + H_2O	→ 2 P_i (mesophyll)
AMP + ATP	→ 2 ADP (mesophyll)
Net: CO_2 (mesophyll) + 2 ATP + 2 H_2O	→ CO_2 (bundle sheath) + 2 ADP + 2 P_i

Cost of concentrating CO_2 within the bundle sheath cell: 2 ATP per CO_2

Notes: As shown in reaction 1 of Table 8.4, the H_2O and CO_2 shown in the first line of this table stand for HCO_3^- + H^+. The cell type where the reaction takes place is indicated in parentheses.

P_i and PP_i stand for inorganic phosphate and pyrophosphate, respectively.

unusual ADP-dependent phosphorylation of the enzyme that occurs when the photon flux density drops (Burnell and Hatch 1985). Activation is accomplished by the phosphorolytic cleavage of this phosphate group. Both reactions, phosphorylation and dephosphorylation, appear to be catalyzed by a single regulatory protein.

In hot, dry climates, the C_4 cycle reduces photorespiration and water loss

Two features of the C_4 cycle overcome the deleterious effects of high temperature on photosynthesis. First, the affinity of PEPCase for its substrate, HCO_3^-, is sufficiently high that the enzyme is saturated by HCO_3^- in equilibrium with air levels of CO_2. Furthermore, oxygen is not a competitor in the reaction, because the substrate is HCO_3. This high activity of PEPCase enables C_4 plants to reduce the stomatal aperture at high temperatures and thereby conserve water while fixing CO_2 at rates equal to or greater than those of C_3 plants.

The second beneficial feature is a result of high concentration of CO_2 in bundle sheath cells, which minimizes the operation of the C_2 oxidative photosynthetic cycle (Maroco et al. 1998). These attributes enable C_4 plants to photosynthesize more efficiently at high temperatures than C_3 plants, likely accounting for the relative abundance of C_4 plants in drier, hotter climates. Expanding the alternatives to survive under changing environments, some plants exhibit properties intermediate between strictly C_3 and C_4 species.

III. Crassulacean Acid Metabolism (CAM)

Many plants that inhabit arid environments with seasonal water availability, including commercially important plants such as pineapple (*Ananas comosus*), agave (*Agave* spp.), cacti (Cactaceae), and orchids (Orchidaceae), exhibit a third mechanism for concentrating CO_2 at the site of rubisco. This important alternative of photosynthetic carbon fixation was named the crassulacean acid metabolism (CAM) to honor the initial observations that were made with *Bryophyllum calycinum*, a succulent member of the Crassulaceae (Cushman 2001). CAM is generally associated with anatomical features that minimize water loss such as thick cuticles, low surface-to-volume ratios, large vacuoles, and reduced size and frequency of stomatal opening. Typically, a CAM plant loses 50 to 100 g of water for every gram of CO_2 gained, compared with 250 to 300 g for C_4 plants and 400 to 500 g for C_3 plants (see Chapter 4). Thus, CAM plants have a competitive advantage in dry environments.

In C_4 plants, the formation of C_4 acids in one compartment (e.g., mesophyll cells) is spatially separated from their decarboxylation and refixation of the resulting CO_2 by the Calvin cycle in another compartment (e.g., bundle sheath cells). In CAM plants, the formation of the C_4 acids is both spatially and temporally separated (Figure 8.14). At night, the cytosol PEPCase captures CO_2 into oxaloacetate using the phosphoenolpyruvate formed via the glycolytic breakdown of stored carbohydrates (see Table 8.4, reaction 1). A cytosol NAD–malate dehydrogenase converts the oxaloacetate to malate, which in turn is stored in the vacuole (see Table 8.4, reaction 2).

During the day, the stored malate is transported to the chloroplast and decarboxylated by mechanisms similar to those in C_4 plants; that is, via a cytosol NADP–malic enzyme, a mitochondrial NAD–malic enzyme, or a mitochondrial phosphoenolpyruvate carboxykinase (see Table 8.4, reactions 4 and 5). The released CO_2 is refixed by the Calvin cycle, while the complementary three-carbon acids are thought to be converted first to triose phosphates, then to starch or sucrose via gluconeogenesis. Whether triose phosphates flow to chloroplasts to be stored as starch or to the vacuole to be accumulated as sucrose or hexose depends on the plant species. These processes ensure the formation of substrates for the next nocturnal carboxylation as well as for plant growth.

In summary, the temporal separation of carboxylation mediated by PEPCase and rubisco optimizes the carbon acquisition and the photosynthetic performance in limiting environments.

The stomata of CAM plants open at night and close during the day

CAM plants growing in deserts such as cacti achieve high water-use efficiency by opening their stomata during the cool nights and closing them during the hot, dry days. Closing the stomata during the day minimizes water loss but, because H_2O and CO_2 share the same diffusion pathway, CO_2 must then be taken up at night. The accumulation of substantial amounts of malic acid, equivalent to the amount of CO_2 fixed at night, has long been recognized as a nocturnal acidification of the leaf (Bonner and Bonner 1948). This four-carbon acid is stored in large vacuoles that are a typical, but not obligatory, anatomic feature of the leaf cells of CAM plants.

With the onset of day, the stomata close and thereby minimize the loss of water and further uptake of CO_2. The leaf cells deacidify as the reserves of vacuolar malic acid are decarboxylated by the action of NADP–malic enzyme (Drincovich et al. 2001). Because the stomata are closed, the released CO_2 cannot escape from the leaf and instead is fixed and converted to carbohydrate by the Calvin cycle. The elevated internal concentration of CO_2 not only suppresses the photorespiratory oxygenation of ribulose-1,5-bisphosphate that favors carboxylation, but also assists in the stomatal closure that helps conserve water (see **Web Topic 8.8**).

Some CAM plants change the pattern of CO_2 uptake in response to environmental conditions

In many species, transitional periods of net CO_2 uptake occur at the start and the end of the day, by PEPCase- and

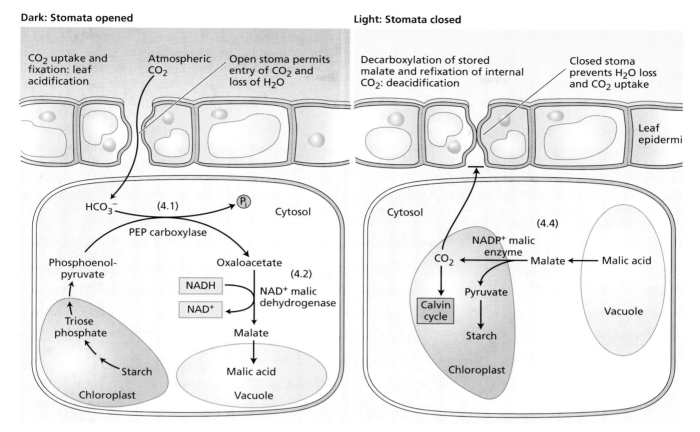

FIGURE 8.14 Crassulacean acid metabolism (CAM). Temporal separation of CO_2 uptake from photosynthetic reactions: CO_2 uptake and fixation take place at night, and decarboxylation and refixation of the internally released CO_2 occur during the day. The adaptive advantage of CAM is the reduction of water loss by transpiration, achieved by stomatal closure during the day. See Table 8.4 for a description of numbered sections.

rubisco-mediated carboxylation, interspersed between the nocturnal uptake of atmospheric CO_2 and the diurnal decarboxylation of organic acids. Genotypic and developmental attributes, and environmental factors such as light intensity and water availability, modulate the extent to which the biochemical and physiological capacity of CAM plants is expressed. The concerted actions of the following alter the proportion of CO_2 assimilated via PEPCase at night or by rubisco during the day (net CO_2 assimilation):

- Stomatal behavior
- Fluctuations in organic acid and storage carbohydrate accumulation
- Activity of primary (PEPCase) and secondary (rubisco) carboxylating enzymes
- Activity of decarboxylating enzymes and
- Synthesis and breakdown of C_3 carbon skeletons

Many CAM representatives show longer-term regulation and are able to adjust their pattern of CO_2 uptake to environmental conditions. Aizoaceae, Crassulaceae, Portulaceae, and Vitaceae are among the families that use CAM when water is scarce but undergo a gradual transition to C_3 when water is abundant. Other environmental conditions, such as salinity, temperature, and intensity or quality of light, also contribute to the extent of CAM induction in these species. This form of regulation requires the expression of numerous CAM genes in response to stress signals (Adams et al. 1998; Cushman 2001).

The advantage of high efficiency in the use of water likely accounts for the extensive diversification and speciation of CAM plants in water-limited environments. However, paradoxically CAM species are also found among aquatic vascular plants. Perhaps this mechanism enhances the acquisition of inorganic carbon from aquatic environments where low concentrations of CO_2 limit photosynthesis (Cushman 2001). Hence, the restriction in the availability of CO_2, caused either by the water-conserving closure of stomata in arid lands or to the high diffusional resistance of aquatic habitats, may both be responsible for the evolution of CAM.

Starch and Sucrose

The photosynthetic assimilation of atmospheric CO_2 by leaves yields sucrose and starch as end products of two gluconeogenic pathways that are physically separated: sucrose in the cytosol and starch in chloroplasts. Under illumination, the disaccharide sucrose is continuously exported from the leaf cytosol to nonphotosynthetic parts of the plant, while the polysaccharide starch concurrently accumulates as granules in chloroplasts (Figure 8.15). The onset of darkness not only stops the assimilation of carbon, but also starts the degradation of the chloroplast starch to maintain the exportation of sucrose. The large difference in the level of the stromal starch, elicited by the transition from light to dark, led to the usual name *transitory starch* for the polysaccharide stored in chloroplasts.

Sugars produced by photosynthesis move first from the site of synthesis (the mesophyll) to the vascular tissues (the phloem). Although sucrose is the main compound by which the photoassimilated carbon is continuously exported from leaves to the nonphotosynthetic parts of the plant (see Figure 8.15), the phloem of some species also transports sucrose–galactosyl oligosaccharides—raffinose, stachyose, verbascose (*Cucurbita* and Arabidopsis), sorbitol (apple), or mannitol (celery). In nutritionally dependent organs, exported sugars are used as source of energy for growth (stems, young leaves) and building blocks of storage polysaccharides (tubers, grains). Whereas starch is the principal reserve carbohydrate in most plants, other polysaccharides, mainly fructans, are found as reserves in vegetative tissues of both monocotyledons and dicotyledons (see **Web Topic 8.9**).

The regulation of growth by light and sugars ensures the optimal use of carbon and energy resources in carbohydrate-exporting and carbohydrate-importing tissues (see Figure 8.15). Moreover, this type of control drives the adap-

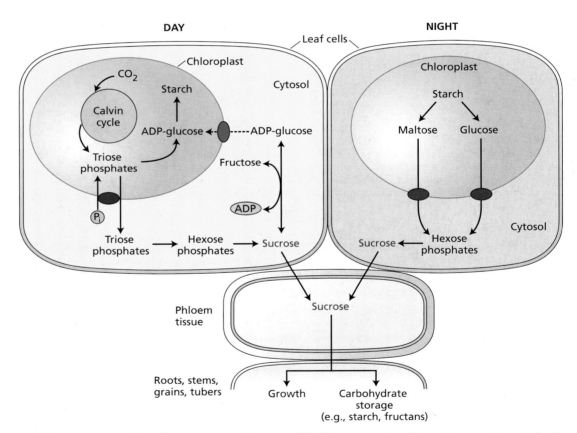

Figure 8.15 Carbon mobilization in vascular plants. During the day, the carbon assimilated photosynthetically is either incorporated into starch in the chloroplast or exported to the cytosol for the synthesis of sucrose. External and internal stimuli control the partitioning between starch and sucrose. ADP–glucose is the glucosyl donor for the elongation of the polysaccharide chain of starch. At night, the degradation of the starch granule supplies the carbon necessary for maintaining production of sucrose. The nocturnal cleavage of the glucosidic bond releases both maltose and glucose, which flow across the chloroplast envelope to feed the hexose phosphate pool and contribute to high rates of sucrose synthesis. Green, blue, and red ovals represent the P_i, ADPG, and maltose–glucose translocators, respectively. As a consequence of diurnal photosynthesis and nocturnal breakdown, the levels of chloroplast starch are maximal during the day and minimal at night. In all cases, sucrose links the assimilation of inorganic carbon in leaves to the utilization of organic carbon in stems, roots, seeds, and tubers.

tation of carbon metabolism to changing environmental conditions and to the availability of other nutrients. In general, low sugar status enhances photosynthesis, mobilization of reserves, and export, whereas abundant sugar promotes growth and carbohydrate storage.

Chloroplast starch is synthesized during the day and degraded at night

Starch is the main storage carbohydrate in plants; its abundance as a naturally occurring compound is surpassed only by cellulose. Starch is a complex polysaccharide composed of two entities: the essentially linear amylose and the highly branched amylopectin. Although α-D-1,4 glycosidic bonds link glucose moieties in both polymers, only amylopectin contains an appreciable proportion of α-D-1,6 glycosidic bonds. This relative distribution of amylose and amylopectin is important in food and in commercial (paper, glue) uses of starch. Branched starches—that is, those containing a large proportion of amylopectin (waxy starch)—have a firmer texture and greater strength than do linear starches. In chloroplasts and seed endosperm, amylose and amylopectin are organized in relatively dense granules (from 0.1 to over 50 μm in diameter) with an internal lamellar structure, resulting in alternating semicrystalline and amorphous layers.

Starch is synthesized in the chloroplast

Electron micrograph studies showing prominent starch deposits, together with enzyme localization studies, leave no doubt that the chloroplast is the site of starch synthesis in leaves. The biosynthesis of the α-D-1,4 glycosidic linkages of amylose proceeds through three successive steps: initiation, elongation, and termination of the polysaccharide chain. At specific stages in the elongation process, the linear amylose chain acquires novel α-D-1,6 glycosidic linkages leading to the formation of amylopectin. Numerous studies have improved the understanding of the main steps of polysaccharide elongation and branching, but knowledge of initiation and termination is limited.

Although it is a controversial issue (Baroja-Fernandez et al. 2005; Neuhaus et al. 2005), most of the ADP-glucose, the precursor that provides the glucosyl moiety in photosynthetically active leaves, is synthesized by the action of ADP-glucose pyrophosphorylase (Table 8.6, reaction 1). The elongation of starch proceeds via an enzyme, starch synthase, that catalyzes the transfer of the glucosyl moiety of ADP-glucose to the nonreducing end of a preexisting α-D-1,4-glucan primer, retaining the anomeric configuration of the glycosidic bond (i.e., the glucosyl moiety has an α configuration, as in the ADP-glucose donor) (see Table 8.6, reaction 3). Although multiple isoforms of starch synthases have been found in oxygenic eukaryotes, ranging from green algae to almost all plant tissues, plant starch synthase can be broadly divided into two groups. The first group, primarily associated with the granule matrix, is involved in amylose synthesis and is responsible for the extension of long glucans within the amylopectin fraction. The second group of starch synthases, whose distribution between the stroma and starch granules varies with species, tissues, and developmental stages, is principally confined to amylopectin biosynthesis.

The formation of amylopectin requires the participation of starch-branching enzymes that transfer a segment of an α-D-1,4-glucan to the carbon 6 of glucosyl moieties in the same glucan (see Table 8.6, reaction 4). Like starch synthase, starch-branching enzymes comprise various isoforms that differ not only in the length of the glucan chain transferred, but in their expression in different tissues and developmental stages. Although starch-branching isoforms are present both in the soluble stroma and the particulate starch granules of chloroplasts, the factors and mechanisms involved in their partitioning are not known.

The synthesis of starch depends not only on synthases and branching enzymes, but also on the participation of two groups of enzymes that hydrolyze α-D-1,6 linkages, the isoamylases and the limit-dextrinases (pullulanases). It remains unknown whether the debranching activity is responsible for the removal of inappropriately positioned branches (preamylopectin) generated at the surface of the growing starch granules or for the removal of soluble glucans from the stroma.

Starch degradation requires phosphorylation of amylopectin

In the last decade, creative molecular biological approaches in the construction of transgenic plants combined with biochemical analyses and genome sequence information have introduced a radically new picture of the pathway that results in nocturnal degradation of leaf starch (Figure 8.16) (Lloyd et al. 2005; Smith et al. 2005). Although most studies were performed with *Arabidopsis thaliana*, the activities of the main starch-cleaving enzymes have been identified in leaves of numerous species. The release of soluble glucans from the dynamic pool of transitory starch requires the prior incorporation of phosphoryl groups to the polysaccharide, mainly amylopectin. Relevant to this reaction is glucan–water dikinase, an enzyme that transfers the β-phosphate of ATP to positions 6 and 3 of glucosyl moieties of amylopectin.

$$\text{AM-P-P-P*} (\text{ATP}) + (\text{Glucan})\text{-OH} + H_2O \rightarrow$$
$$\text{AMP} + (\text{Glucan})\text{-O-P*} + P_i$$

Although the occurrence of phosphoryl groups in leaf starch is low (one phosphoryl group per 2000 glucosyl residues in *A. thaliana*), changes in the amount or the activity of glucan–water dikinase in transgenic plants dramatically lowers not only the incorporation of phosphoryl groups into amylopectin but also the rate of starch degradation. As a consequence, the content of starch in mature leaves of these transgenic lines—named *starch excess 1*

TABLE 8.6
Reactions of starch synthesis in chloroplasts

1. *ADP–glucose pyrophosphorylase*

   Glucose 1-phosphate + ATP → ADP–glucose + PP_i

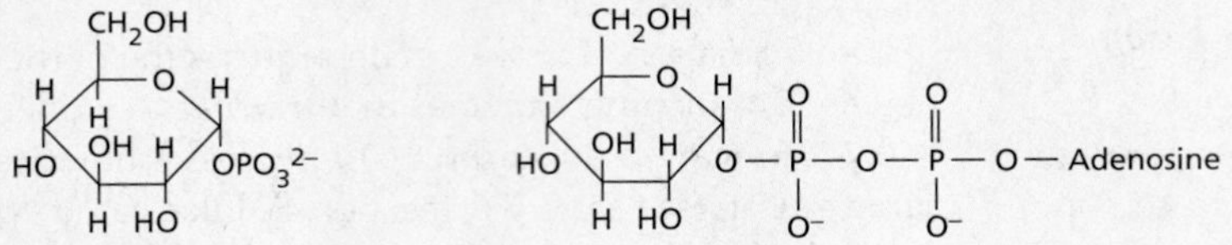

2. *Pyrophosphatase*

   $PP_i + H_2O \rightarrow 2\ P_i + 2\ H^+$

3. *Starch synthase*

   ADP-glucose + (α-D-1,4 glucosyl)$_n$ → ADP + (α-D-1,4 glucosyl)$_{n+1}$

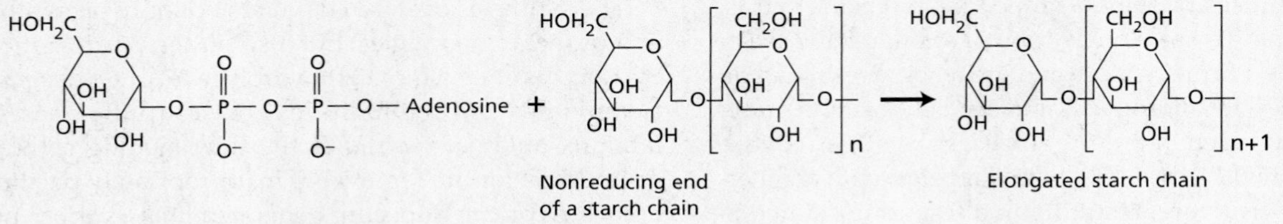

4. *Starch branching enzyme*

   (α-D-1,4 glucosyl)$_{(n+1)}$-(α-D-1,4 glucosyl)-(α-D-1,4 glucosyl)$_m$ → (α-D-1,4 glucosyl)$_n$-[(α-D-1,6 glucosyl)-(α-D-1,4 glucosyl)$_m$]

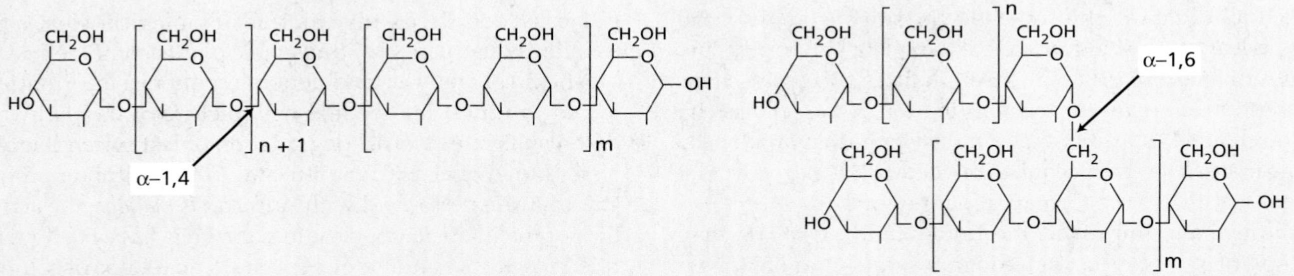

Note: P_i and PP_i stand for inorganic phosphate and pyrophosphate, respectively.

(*sex1*)—is up to seven times greater than in wild-type leaves (Yu et al. 2001). The activity of glucan–water dikinase was recently found to depend on reduction by thioredoxin in a reaction that also altered its starch-binding properties (Mikkelsen et al. 2005).

Vascular plants contain a second enzyme, phosphoglucan–water dikinase, that catalyzes a reaction similar to glucan–water dikinase but uses instead a phosphorylated glucan as substrate (see Figure 8.16).

$$AM\text{-}P\text{-}P\text{-}P^* \ (ATP) + (P\text{-glucan})\text{-}OH + H_2O \rightarrow$$
$$AMP + (P\text{-glucan})\text{-}O\text{-}P^* + P_i$$

In vitro studies revealed that the recombinant phosphoglucan–water dikinase adds additional phosphoryl groups to the phosphorylated, but not the unphosphorylated, form of amylopectin. But more important, transgenic lines lacking phosphoglucan–water dikinase also contain increased levels of starch but, at variance with *sex1* mutants, do not exhibit an altered content of phosphorylated amylopectin. Why the incorporation of phosphoryl groups in the leaf starch becomes essential for starch degradation is so far unknown.

The formation of linear glucans results from the action of amylases and debranching enzymes on the branched

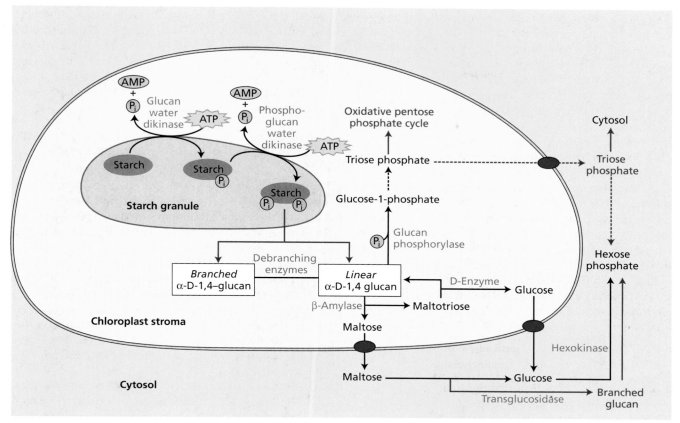

FIGURE 8.16 Nocturnal starch degradation in *A. thaliana* leaves. The release of soluble glucans from the starch granule at night requires the prior phosphorylation of the polysaccharide via a glucan–water dikinase and a phosphoglucan–water dikinase. At this stage, debranching enzymes transform the branched starch into linear glucans, which in turn can be converted into maltose via the β-amylolysis catalyzed by the chloroplastic β-amylase. Residual maltotriose is transformed to maltopentaose and glucose via the disproportionating enzyme (D-enzyme). Two pumps in the chloroplast envelope, one for maltose and another for glucose, facilitate the flow of these products of starch degradation to the cytosol. The utilization of maltose in the leaf cytosol proceeds via a transglucosidase that transfers one glucosyl moiety to a branched glucan, concurrently releasing a glucose. The cytosolic glucose can be phosphorylated via a hexokinase to glucose-6-phosphate for incorporation into the pool of hexose phosphates.

starch (see Figure 8.16). Initially, it was thought that α-amylases attack the surface of the insoluble starch granule and release soluble glucans that become ready for further cleavage. However, nocturnal starch degradation proceeds at a normal rate in leaves of transgenic *A. thaliana* that lack the three α-amylase isozymes, providing a good reason to suppose that this family of glucan hydrolases is ineffective.

On the other hand, debranching enzymes are essential for the complete breakdown of starch granules, because between 4 and 5% of amylopectin linkages are α-D-1,6 branch points. *A. thaliana* genes encode four proteins that may hydrolyze α-1,6 bonds in glucans. The absence of one of them, a limit-dextrinase (pullulanase), does not affect starch cleavage at night. The interpretation of data obtained with mutants lacking the three isoamylases, however, is complicated, because their reduction or elimination elicits abnormal starch synthesis.

Two mechanisms accomplish further cleavage of the α-D-1,4 glycosidic bond of linear glucans (see Figure 8.16).

1. Hydrolysis, catalyzed by β-amylases:

   Linear $[glucose]_n + H_2O \rightarrow$ linear $[glucose]_{n-2}$ + maltose

2. Phosphorolysis, catalyzed by glucan phosphohydrolases:

   Linear $[glucose]_n + P_i \rightarrow$ linear $[glucose]_{n-1}$ + glucose-1-phosphate

Although chloroplasts of vascular plants contain both enzymes, it is reasonable to assume that only β-amylases are necessary for sustaining normal starch degradation at night. Mutants of *A. thaliana* devoid of chloroplast β-amylase and glucan phosphohydrolase have reduced and normal rates of starch degradation, respectively (Zeeman et al.

2004). Under stress conditions, a minor portion of transitory starch may be degraded by the phosphorolytic pathway, yielding glucose-1-phosphate that can be subsequently converted into triose phosphates. However, in contrast to their photosynthetic use in daylight, triose phosphates are not used efficiently in the production of sucrose at night, because cytosol fructose-1,6-bisphosphatase is inactivated by high levels of fructose-2,6-bisphosphate (see below). Apparently, glucan phosphorylases supply glucose-1-phosphate for the nocturnal functioning of the oxidative pentose phosphate cycle and for photorespiration.

At the same time that α-amylase breaks down starch in chloroplasts, appreciable amounts of the disaccharide maltose are produced and exported from the chloroplasts. Further, the exhaustive action of β-amylases produce low amounts of maltotriose that cannot be further processed by this enzyme (see Figure 8.16). To avoid the accumulation of maltotriose in the nocturnal degradation of starch, a disproportionating enzyme (D-enzyme; an enzyme that alters oligosaccharide chain length) catalyzes the following transformation:

$$2\,[\text{Glucose}]_3 \rightarrow [\text{glucose}]_5 + \text{glucose}$$
$$(\text{Maltotriose}) \quad\quad (\text{maltopentaose})$$

The glucose released is exported through a glucose transporter of the inner membrane of the chloroplast envelope (Critchley et al. 2001). In support of this view, lower rates of starch degradation and a concurrent increase in the level of maltotriose are observed with transgenic plants deficient in the disproportionating enzyme.

Maltose is not metabolized in the stroma, because neither a maltose phosphorylase [maltose + P_i → glucose-1-phosphate + glucose] nor an α-glucosidase [maltose + H_2O → 2 glucose] appears to be functional in chloroplasts. But more important, the inner chloroplast membrane contains a protein that facilitates the transport of maltose across the envelope (Niittylä et al. 2004). Although α-D-1,4-linked oligosaccharides competitively inhibit the transport of maltose, they are not translocated. Transgenic plants in which the transporter locus *MEX1* is altered accumulate maltose at night to levels much higher than do wild-type leaves. Analysis of the *MEX1* gene reveals that homologs not only are present in angiosperms, gymnosperms, and mosses, but are expressed in both photosynthetic and nonphotosynthetic tissues, indicating that maltose export is a common feature of starch-storing plastids.

The utilization of maltose in the leaf cytosol follows a biochemical pathway unsuspected before the use of transgenic plants. Transgenic lines devoid of a putative transglucosidase degrade starch poorly and accumulate maltose to levels much higher than those of wild-type plants (Chia et al. 2004). The transglucosylation reaction catalyzed by this enzyme transfers one glucosyl moiety from maltose to branched glucans and releases a molecule of glucose. The phosphorylation of the latter via hexokinase adds the resulting sugar phosphate to the hexose phosphate pool for the subsequent conversion to sucrose. The molecular nature of the cytosol branched glucan, however, remains unknown. The processes, uncovered for transitory starch in leaves, are not immediately applicable to other storage organs, such as cereal endosperm, in which levels of starch do not change in a dynamic manner.

Triose phosphates synthesized in the chloroplast build up the pool of hexose phosphates in the cytosol

The major factors that modulate the partitioning of the assimilated carbon between the chloroplast and the cytosol are the relative concentrations of inorganic phosphate and triose phosphates. Two products of the Calvin cycle, dihydroxyacetone phosphate and glyceraldehyde-3-phosphate, are rapidly interconverted by triose phosphate isomerases in the plastid and cytosol (Table 8.7, reaction 1) (Lüttge and Higinbotham 1979; Flügge and Heldt 1991). To drive the exchange of chloroplast triose phosphates for cytosol phosphate, the plastid envelope harbors a protein complex located at the inner membrane—the triose phosphate translocator (see Table 8.7, reaction 2) (see **Web Topic 8.10**) (Weber et al. 2005).

During active photosynthesis on a sunny day, the triose phosphate translocator mediates the export of fixed carbon, in the form of triose phosphates, from the chloroplast to the cytosol. In turn, the phosphate released from biosynthetic processes in the cytosol is translocated back into the chloroplast, where it is used to replenish ATP and other phosphorylated metabolites, thereby sustaining photosynthetic electron transport and the Calvin cycle. This counterexchange balances the triose phosphate-to-phosphate ratio in both the stroma and the cytosol, reflecting the metabolic status of these two compartments. Transformed plants with impaired triose phosphate translocator activity fail to exhibit pronounced growth differences, but do show drastic alterations in carbon metabolism. These transgenic lines accelerate the turnover of starch and the export of neutral sugars from the chloroplast stroma, compensating for the lack of triose phosphates for sucrose biosynthesis in the cytosol.

The accumulation of triose phosphates in the cytosol increases the formation of fructose-1,6-bisphosphate from dihydroxyacetone phosphate and glyceraldehyde-3-phosphate catalyzed by the cytosol aldolase (Figure 8.17 and Table 8.7, reaction 3). The reaction is quite sensitive to the concentration of triose phosphates in the cytosol ($\Delta G^{0\prime} = -24$ kJ/mol).

Given the rapid interconversion catalyzed by triose phosphate isomerase, the K_{eq} = [dihydroxyacetone phosphate] × [glyceraldehyde-3-phosphate]/[fructose-1,6-bisphosphate] transforms into K_{eq} = [triose phosphate]^2/[fructose-1,6-bisphosphate], implying that the concentration of fructose-1,6-bisphosphate varies exponentially in response

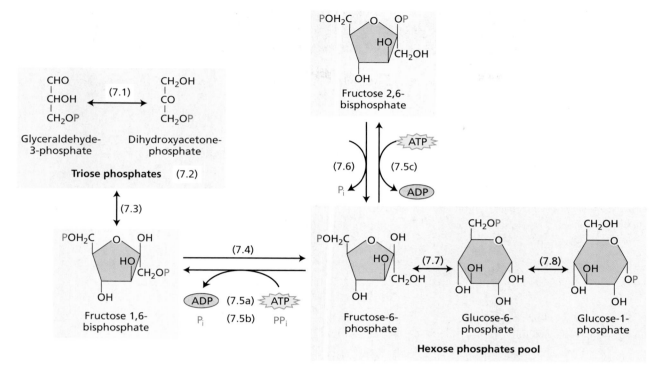

FIGURE 8.17 Interconversion of hexose phosphates. The fructose-1,6-bisphosphate, formed from triose phosphates by the action of aldolase, is cleaved at the carbon 1 position by cytosolic fructose-1,6-bisphosphate phosphatase, structurally and functionally different from the chloroplast counterpart. Fructose-6-phosphate constitutes a starting substrate for three transformations. First, vascular plants employ two alternative phosphorylation reactions of fructose-6-phosphate at the carbon 1 position of the furanose ring: (i) the classical ATP-dependent phosphofructokinase (compare with the glycolytic pathway) and (ii) a pyrophosphate-dependent phosphofructokinase that catalyzes the readily reversible phosphorylation of fructose-6-phosphate using pyrophosphate as substrate. Second, fructose-6-phosphate 2-kinase catalyzes the ATP-dependent phosphorylation of fructose-6-phosphate to fructose-2,6-bisphosphate and, in turn, fructose-2,6-bisphosphate phosphatase catalyzes the hydrolysis of fructose-2,6-bisphosphate, releasing the phosphoryl group and again yielding fructose-6-phosphate. Third, hexose phosphate isomerase favors the isomerization of fructose-6-phosphate to glucose-6-phosphate. See Table 8.7 for a description of numbered sections.

to changes in the concentration of triose phosphates. Hence, a constant input of triose phosphates from photosynthetically active chloroplasts biases the aldolase reaction in the cytosol of leaf cells toward the formation of fructose-1,6-bisphosphate. On the other hand, the aldol cleavage of fructose-1,6-bisphosphate to dihydroxyacetone phosphate and glyceraldehyde-3-phosphate takes place when the proportion of fructose-1,6-bisphosphate is high in relation to triose phosphates, for example in glycolysis.

Fructose-1,6-bisphosphate is subsequently hydrolyzed at the carbon 1 position by the action of cytosol fructose-1,6-bisphosphatase, yielding fructose 6-phosphate and phosphate ($\Delta G^{0\prime} = -16.7$ kJ/mol) (see Figure 8.17 and Table 8.7, reaction 4). The cytosol enzyme shows important differences from its chloroplast counterpart that participates in the Calvin cycle both in the primary structure and the regulatory properties. Like the enzyme from heterotrophic organisms, the cytosol fructose-1,6-bisphosphate phosphatase lacks the characteristic peptide insert that enables the chloroplast counterpart to be regulated through reduction by thioredoxin. Moreover, in addition to AMP, cytosol fructose-1,6-bisphosphate phosphatase is strongly inhibited by an important cytosol metabolite, fructose-2,6-bisphosphate. Fructose-6-phosphate can proceed to different destinations through the phosphorylation of (i) the carbon 1, which restores fructose-1,6-bisphosphate (see Table 8.7, reactions 5a and b), or (ii) the carbon 2, which yields fructose-2,6-bisphosphate (see Table 8.7, reaction 5c). However, the cytosolic concentration of fructose-6-phosphate is kept close to equilibrium with glucose-6-phosphate and glucose-1-phosphate by readily reversible reactions catalyzed by hexose phosphate isomerase ($\Delta G^{0\prime} = 8.7$ kJ/mol) and phosphoglucomutase ($\Delta G^{0\prime} = 7.3$ kJ/mol) (see Table 8.7, reactions 7 and 8). These three sugar monophosphates are collectively designated hexose phosphates (see Figure 8.17).

The impaired conversion of triose phosphates to hexose phosphates in illuminated transgenic plants lacking either the triose phosphate translocator or the cytosolic fructose-1,6-bisphosphate phosphatase drastically reduces the export of carbon from the chloroplast to the cytosol and

TABLE 8.7
Reactions of sucrose biosynthesis from triose phosphates formed photosynthetically

1. *Triose phosphate isomerase*
   Dihydroxyacetone-3-phosphate ↔ glyceraldehyde-3-phosphate

2. *Phosphate-triose phosphate translocator*
   Triose phosphate (chloroplast) + P_i (cytosol) ↔ triose phosphate (cytosol) + P_i (chloroplast)

3. *Fructose-1,6-bisphosphate aldolase*
   Dihydroxyacetone 3-phosphate + glyceraldehyde-3-phosphate → fructose-1,6-bisphosphate

4. *Fructose-1,6-bisphosphate phosphatatse*
   Fructose-1,6-bisphosphate + H_2O → fructose-6-phosphate + P_i

5a. *Fructose-6-phosphate 1-kinase (phosphofructokinase)*
   Fructose-6-phosphate + ATP → fructose-1,6-bisphosphate + ADP

5b. *PP_i-linked phosphofructokinase*
   Fructose-6-phosphate + PP_i → fructose-1,6-bisphosphate + P_i

5c. *Fructose-6-phosphate 2-kinase*
   Fructose-6-phosphate + ATP → fructose-2,6-bisphosphate + ADP

6. *Fructose-2,6-bisphosphate phosphatatse*
   Fructose-2,6-bisphosphate + H_2O → fructose-6-phosphate + P_i

TABLE 8.7 (continued)
Reactions of sucrose biosynthesis from triose phosphates formed photosynthetically

7. *Hexose phosphate isomerase*

   Fructose-6-phosphate → glucose-6-phosphate

8. *Phosphoglucomutase*

   Glucose-6-phosphate → glucose-1-phosphate

9. *UDP-glucose pyrophosphorylase*

   Glucose-1-phosphate + UTP → UDP-glucose + PP_i

10. *Sucrose-6^F-phosphate synthase*

    UDP-glucose + fructose-6-phosphate → UDP + sucrose-6^F-phosphate

11. *Sucrose-6^F-phosphate phosphatase*

    Sucrose-6^F-phosphate + H_2O → sucrose + P_i

Note: The P_i translocator (reaction 2) facilitates the exchange between triose phosphates and P_i across the chloroplast inner envelope membrane; all other enzymes catalyze cytosol reactions.

P_i and PP_i stand for inorganic phosphate and pyrophosphate, respectively.

promotes unusually high levels of plastid starch (Zrenner et al. 1996; Hausler et al. 1998).

Fructose-6-phosphate can be converted to fructose-1,6-bisphosphate by two different enzymes

In contrast to animals, land plants exhibit two routes for the conversion of fructose-6-phosphate to fructose-1,6-bisphosphate.

First, like most other eukaryotes, plants have a phosphofructokinase that uses ATP to irreversibly phosphorylate fructose-6-phosphate at the carbon 1 position (see Figure 8.17 and Table 8.7, reaction 5a). Two strong inhibitors, phosphate and phosphoenolpyruvate, together with the relative concentrations of ATP, ADP, and AMP, appear to regulate the activity of the plant phosphofructokinase. However, in contrast to animals, the plant enzyme is not appreciably modified by fructose-2,6-bisphosphate.

Second, the pyrophosphate-dependent phosphofructokinase that catalyzes the readily reversible phosphorylation of fructose-6-phosphate with pyrophosphate, yielding fructose-1,6-bisphosphate, constitutes an additional mechanism for the biosynthesis of fructose-1,6-bisphosphate in the cytosol of leaf cells (see Figure 8.17 and Table 8.7, reaction 5b). This enzyme, a heterotetramer composed of two α (regulatory) and two β (catalytic) subunits, possesses high affinity for the activator fructose-2,6-bisphosphate.

Once again, gene deletion experiments with transformed tobacco plants illustrate the remarkable flexibility of plants. Transformed plants can grow without a functional pyrophosphate-dependent fructose-6-phosphate kinase, indicating that, under certain conditions, phosphofructokinase alone accounts for the conversion of fructose-6-phosphate to fructose-1,6-bisphosphate (Paul et al. 1995).

Fructose-2,6-bisphosphate is an important regulatory compound

Plants phosphorylate fructose-6-phosphate in an additional way: ATP is used to phosphorylate the carbon 2 of the sugar moiety, yielding fructose-2,6-bisphosphate (see Figure 8.17 and Table 8.7, reaction 5c). The synthesis and cleavage of fructose-2,6-bisphosphate is catalyzed by a unique bifunctional enzyme confined to the cytosol, the fructose-6-phosphate 2-kinase–fructose-2,6-bisphosphate phosphatase (Table 8.7, reactions 5c and 6). Fructose-6-phosphate 2-kinase–fructose-2,6-bisphosphate phosphatase not only transfers the γ-phosphoryl group of ATP to the carbon 2 position of fructose-6-phosphate, but also can release the same group through hydrolysis. Although the quaternary structure of the plant enzyme (a homotetramer with a subunit of 83 kDa) differs from that of the mammalian counterpart (a homodimer with a subunit of 45–50 kDa), native and recombinant forms of the plant enzyme possess many regulatory properties found in the animal enzyme. The kinase activity is stimulated by phosphate, fructose-6-phosphate, and pyruvate and is inhibited by 3-phosphoglycerate, triose phosphates, phosphoenolpyruvate, and pyrophosphate. On the other hand, the phosphatase activity is inhibited by phosphate, fructose-6-phosphate, and 6-phosphogluconate. In consequence, phosphate and fructose-6-phosphate facilitate the synthesis of fructose-2,6-bisphosphate and concurrently halt the release of the phosphate at carbon 2.

Fructose-2,6-bisphosphate plays an important role in sucrose biosynthesis (Nielsen et al. 2004). A tightly controlled concentration of this important regulatory metabolite governs the functioning of the gluconeogenic pathway by inhibiting fructose-1,6-bisphosphate phosphatase and its activation of the pyrophosphate-dependent fructose-6-phosphate kinase.

Analysis of transgenic and mutant plants further supports the significant contribution of this metabolite to the regulation of carbon partitioning during photosynthesis. Leaves with decreased levels of fructose-2,6-bisphosphate partition more carbon into sucrose, and increased levels of fructose 2,6-bisphosphate impair the formation of sucrose. In contrast, sucrose synthesis is not affected by levels of fructose-2,6-bisphosphate at night, because maltose and hexoses originating from the nocturnal mobilization of chloroplast starch build up the pool of hexose phosphates in the cytosol. In the dark, the cytosol fructose-1,6-bisphosphate phosphatase does not contribute to the control of the hexose phosphate pool, minimizing in consequence the effect of fructose-2,6-bisphosphate.

HEXOSE PHOSPHATE INTERCONVERSION The cytosol hexose phosphate isomerase catalyzes the transformation of fructose-6-phosphate to glucose-6-phosphate, which in turn yields glucose-1-phosphate through the action of phosphoglucomutase (see Figure 8.17 and Table 8.7, reactions 7 and 8). Estimation of the in vivo concentration of these three sugar phosphates indicates that the interconversion is poised near equilibrium. Thus, the direction of carbon flow is dictated by the metabolic demands of the plant. Carbon enters the pool of hexose phosphates via either the gluconeogenic pathway (the photosynthetic assimilation of atmospheric CO_2) or the phosphorylation of free hexoses derived from the nocturnal breakdown of starch. Carbon leaves the pool of hexose phosphates via glycolysis, sucrose and polysaccharide synthesis, or the pentose phosphate pathway.

The hexose phosphate pool is regulated by fructose-2,6-bisphosphate

Under illumination, phosphate is taken up by the chloroplast in a reaction that is coupled to the transport of triose phosphates to the cytosol. Fructose-2,6-bisphosphate is a key control molecule in sensing the status of this transport pathway (Huber 1986; Stitt 1990). A high ratio of triose phosphates to phosphate in the cytosol, typical of photosynthetically active leaves, suppresses the formation of fructose-2,6-bisphosphate because triose phosphates strongly inhibit, and phosphate activates, fructose-6-phosphate 2-kinase activity. Concurrently, phosphate inhibits fructose-2,6-bisphosphate phosphatase activity. A low concentration of fructose-2,6-bisphosphate does not impair the hydrolysis of fructose-1,6-bisphosphate by fructose-1,6-bisphosphate phosphatase, thus contributing to enhance the rate of hexose phosphate formation.

The status of the hexose phosphate pool controls the intracellular content of fructose-2,6-bisphosphate, because fructose-6-phosphate inhibits fructose-2,6-bisphosphate phosphatase and activates fructose-6-phosphate 2-kinase. In turn, fructose 2,6-bisphosphate impairs the hydrolysis of fructose 1,6-bisphosphate by fructose-1,6-bisphosphate phosphatase. Thus, a high level of fructose-6-phosphate increases the concentration of fructose-2,6-bisphosphate which, in turn, limits the formation of fructose-6-phosphate by inhibiting fructose-1,6-bisphosphate phosphatase.

In summary, fructose-2,6-bisphosphate modulates the formation of the hexose phosphate pool in response not only to the photosynthate provided by the triose phosphate translocator, but also to the demand of the hexose phosphate pool itself.

Sucrose is continuously synthesized in the cytosol

Photosynthesis is most active in mature leaves. The photosynthate produced is transported, primarily as sucrose, to meristems and developing organs such as growing leaves, roots, flowers, fruits, and seeds (see Figure 8.15). The concentration of sucrose in the cytosol of leaves is largely dependent on the rates of:

- Photosynthesis, because triose phosphates are exported from the chloroplast to the leaf cytosol

- Export of carbon from leaves, because sucrose fulfills the energy demands of other tissues

Cell fractionation studies in which the organelles are physically separated from one another to analyze intrinsic enzyme activities have shown that sucrose is synthesized in the cytosol from triose phosphates by the pathways depicted in Figures 8.17 and 8.18, using the reactions described in Table 8.7.

The conversion of hexose phosphates to sugar nucleotides precedes the formation of sucrose and starch. Members of a family of enzymes, the nucleoside diphosphate–sugar pyrophosphorylases, catalyze the reaction between a nucleotide and a hexose-1-phosphate, and yield the respective sugar-nucleotide and pyrophosphate:

$$\text{NTP} + \text{hexose-1-phosphate} \leftrightarrow \text{NDP–sugar} + \text{PP}_i$$

In the cytosol, a specific UDP–glucose pyrophosphorylase produces UDP–glucose from UTP and glucose-1-phosphate (see Table 8.7, reaction 9). Given that this reaction is not far from equilibrium ($\Delta G^{0\prime} = -2.88$ kJ/mol), the in vivo

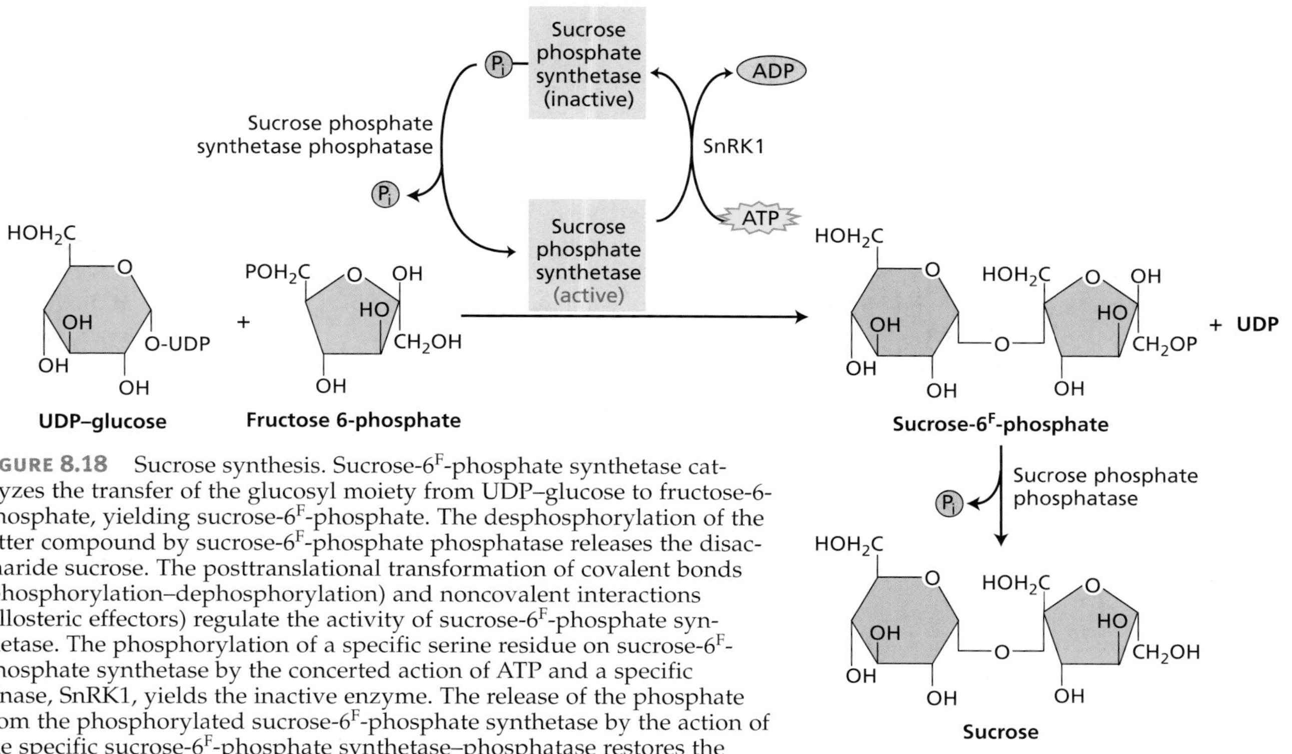

FIGURE 8.18 Sucrose synthesis. Sucrose-6^F-phosphate synthetase catalyzes the transfer of the glucosyl moiety from UDP–glucose to fructose-6-phosphate, yielding sucrose-6^F-phosphate. The desphosphorylation of the latter compound by sucrose-6^F-phosphate phosphatase releases the disaccharide sucrose. The posttranslational transformation of covalent bonds (phosphorylation–dephosphorylation) and noncovalent interactions (allosteric effectors) regulate the activity of sucrose-6^F-phosphate synthetase. The phosphorylation of a specific serine residue on sucrose-6^F-phosphate synthetase by the concerted action of ATP and a specific kinase, SnRK1, yields the inactive enzyme. The release of the phosphate from the phosphorylated sucrose-6^F-phosphate synthetase by the action of the specific sucrose-6^F-phosphate synthetase–phosphatase restores the basal activity. The 6^F notation in sucrose-6^F-phosphate indicates that sucrose is phosphorylated at the carbon 6 of the fructose moiety.

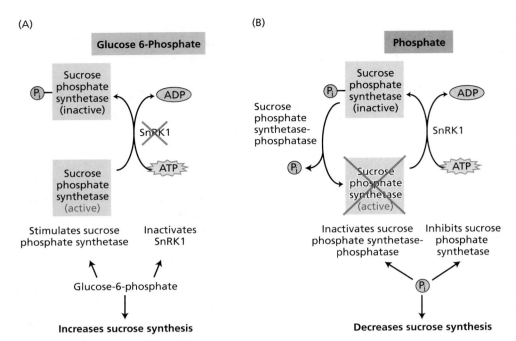

FIGURE 8.19 Regulation of sucrose synthesis. Effect of glucose-6-phosphate and phosphate. The transition of leaves from dark to light increases the concentration of glucose-6-phosphate and concurrently decreases the concentration of phosphate. Both events contribute to enhance the synthesis of sucrose. As indicated, (A) glucose-6-phosphate precludes the phosphorylation of sucrose-6^F-phosphate synthetase by inactivating the kinase SnRK1 and stimulating the active form of sucrose-6^F-phosphate synthetase, and (B) phosphate inactivates sucrose-6^F-phosphate synthetase–phosphatase and inhibits the activity of sucrose-6^F-phosphate synthetase.

equilibrium is largely dependent on processes that utilize pyrophosphate, thereby driving the reaction to the right. Because of the absence of an inorganic pyrophosphatase in the leaf cytosol, the pyrophosphate is used by other enzymes in transphosphorylation reactions (see the pyrophosphate-dependent phosphofructokinase in the previous section).

Two consecutive reactions complete the synthesis of sucrose from UDP–glucose. Sucrose-6^F-phosphate synthase—6^F indicates that sucrose is phosphorylated at the carbon 6 of the fructose moiety—first catalyzes the formation of sucrose-6^F-phosphate from fructose-6-phosphate and UDP–glucose (see Table 8.7, reaction 10). Subsequently, sucrose-6^F-phosphate phosphatase releases inorganic phosphate from sucrose-6^F-phosphate, yielding sucrose (see Table 8.7, reaction 11) (Huber and Huber 1996; Lund et al. 2000).

The close association of the reversible formation of sucrose-6^F-phosphate ($\Delta G^{0\prime}$ = –5.7 kJ/mol) with its irreversible hydrolysis ($\Delta G^{0\prime}$ = –16.5 kJ/mol), renders the pathway of sucrose synthesis essentially irreversible in vivo. Sucrose-6^F-phosphate synthetase and sucrose-6^F-phosphate phosphatase may exist as a supramolecular complex. The complex has an enzymatic activity that is higher than those of the isolated constituent enzymes due to substrate channeling, the process in which the intermediate produced by one enzyme is transferred to the next enzyme without complete mixing with the bulk phase (Salerno et al. 1996).

The sucrose-6^F-phosphate synthetase reaction has been identified as a key step in the control of sucrose synthesis (Huber and Huber 1996). The enhancement of sucrose-6^F-phosphate synthetase levels in transgenic plants has a dramatic impact on photosynthesis and leaf carbohydrate metabolism in several species. For example, the leaves of plants overexpressing sucrose-6^F-phosphate synthetase show increased leaf sucrose-to-starch ratios as well as higher maximal photosynthetic rates. Moreover, tomato plants expressing the gene encoding sucrose-6^F-phosphate synthetase have increased biomass compared to controls.

Sucrose-6^F-phosphate synthetase is subject to a complex system of regulation involving direct control via metabolite effectors and posttranslational modulation of activity via protein phosphorylation. On one hand, the active form of sucrose-6^F-phosphate synthetase is stimulated by glucose-6-phosphate and inhibited by phosphate (Figure 8.19). On the other hand, the active enzyme is inactivated in the dark by phosphorylation of a specific serine residue via a protein kinase, likely SnRK1, and activated in the light by dephosphorylation via a protein phosphatase. Cytosolic metabolites further potentiate the effect of phosphorylation, because glucose-6-phosphate inhibits the kinase, and phosphate inhibits the phosphatase. Thus, increased levels of hexose phosphates and decreased levels of phosphate in the cytosol, brought about by high rates of photosynthesis, increase the synthesis of sucrose. Conversely, the enzyme is inefficient when the level of hexose phosphates falls in response to lower rates of photosynthesis, a process that is accompanied by an increase in the level of cytosol phosphate.

Summary

The reduction of CO_2 to carbohydrate via the carbon-linked reactions of photosynthesis is coupled to the consumption of NADPH and ATP synthesized by the light reactions of thylakoid membranes. Photosynthetic eukaryotes reduce CO_2 via the Calvin cycle, which takes place in the stroma, or soluble phase, of chloroplasts. Here, CO_2 and water are combined with ribulose-1,5-bisphosphate to form two molecules of 3-phosphoglycerate, which are reduced and converted to carbohydrate. The continued operation of the cycle is ensured by the regeneration of ribulose-1,5-bisphosphate. The Calvin cycle consumes two molecules of NADPH and three molecules of ATP for every CO_2 fixed and, provided these substrates, has a thermodynamic efficiency close to 90%.

Several light-dependent systems act jointly to regulate the Calvin cycle: changes in ions (Mg^{2+} and H^+), effector metabolites (enzyme substrates), and protein-mediated systems (rubisco activase, ferredoxin–thioredoxin system).

The ferredoxin–thioredoxin control system plays a versatile role by linking light to the regulation of other chloroplast processes, such as carbohydrate breakdown, photophosphorylation, fatty acid biosynthesis, and mRNA translation. Control of these reactions by light separates opposing processes of biosynthesis from degradation, and thereby minimizes the waste of resources that would occur if the processes operated concurrently.

Rubisco, the enzyme that catalyzes the carboxylation of ribulose-1,5-bisphosphate, also acts as an oxygenase. In both cases the enzyme must be carbamylated to be fully active. The carboxylation and oxygenation reactions take place at the active site of rubisco. When reacting with oxygen, rubisco produces 2-phosphoglycolate and 3-phosphoglycerate from ribulose-1,5-bisphosphate rather than two 3-phosphoglycerates as occurs with CO_2, thereby decreasing the efficiency of photosynthesis.

The C_2 oxidative photosynthetic carbon cycle rescues the carbon lost as 2-phosphoglycolate by rubisco oxygenase activity. The dissipative effects of photorespiration are avoided in some plants by mechanisms that concentrate CO_2 at the carboxylation sites in the chloroplast. These mechanisms include a C_4 photosynthetic carbon cycle, CAM metabolism, and the "CO_2 pumps" of algae and cyanobacteria.

The carbohydrates synthesized by the Calvin cycle are converted into the storage forms of energy and carbon, sucrose and starch. Sucrose, the transportable form of carbon and energy in most plants, is synthesized in the cytosol, and its synthesis is regulated by phosphorylation of sucrose phosphate synthase. Starch is synthesized in the chloroplast. The balance between the biosynthetic pathways for sucrose and starch is determined by the relative concentrations of metabolite effectors (orthophosphate, fructose-6-phosphate, 3-phosphoglycerate, and dihydroxyacetone phosphate).

These metabolite effectors function in the cytosol by way of the enzymes synthesizing and degrading fructose-2,6-bisphosphate, the regulatory metabolite that plays a primary role in controlling the partitioning of photosynthetically fixed carbon between sucrose and starch. Two of these effectors, 3-phosphoglycerate and orthophosphate, also act on starch synthesis in the chloroplast by allosterically regulating the activity of ADP–glucose pyrophosphorylase. In this way the synthesis of starch from triose phosphates during the day can be separated from its breakdown, which is required to provide energy to the plant at night.

Web Material

Web Topics

8.1 How the Calvin Cycle Was Elucidated
Experiments carried out in the 1950s led to the discovery of the path of CO_2 fixation.

8.2 Rubisco: A Model Enzyme for Studying Structure and Function
As the most abundant enzyme on Earth, rubisco was obtained in quantities sufficient for elucidating its structure and catalytic properties.

8.3 Rubisco Activase
Rubisco is unique among Calvin cycle enzymes in its regulation by a specific protein, rubisco activase.

8.4 Thioredoxins
First found to regulate chloroplast enzymes, thioredoxins are now known to play a regulatory role in all types of cells.

8.5 Operation of the C_2 Oxidative Photosynthetic Carbon Cycle
The enzymes of the C_2 oxidative photosynthetic carbon cycle are localized in three different organelles.

8.6 Carbon Dioxide: Some Important Physicochemical Properties
Plants have adapted to the properties of CO_2 by altering the reactions catalyzing its fixation.

8.7 Three Variations of C_4 Metabolism
Certain reactions of the C_4 photosynthetic pathway differ among plant species.

8.8 Photorespiration in CAM Plants
During the day, stomatal closing and photosynthesis in CAM leaves lead to very high intracellular concentrations of both oxygen and carbon dioxide. These unusual conditions pose interesting adaptive challenges to CAM leaves.

8.9 Fructans
Fructans are fructose polymers often found as reserve carbohydrates in plants.

8.10 Chloroplast Phosphate Translocators
Chloroplast phosphate translocators are antiports that catalyze a strict 1:1 exchange of orthophosphate with other metabolites between the chloroplast and the cytosol.

Web Essay

8.1 Modulation of Phosphoenolpyruvate Carboxylase in C₄ and CAM Plants
The CO_2-fixing enzyme phosphoenolpyruvate carboxylase is regulated differently in C_4 and CAM species.

Chapter References

Adams, P., Nelson, D. E., Yamada, S., Chmara, W., Jensen, R. G., Bohnert, H. J., and Griffiths, H. (1998) Tansley Review No. 97; Growth and development of *Mesembryanthemum crystallinum*. *New Phytol.* 138: 171–190.

Badger, M. R., and Price, G. D. (2003) CO_2 concentrating mechanisms in cyanobacteria: Molecular components, their diversity and evolution. *J. Exp. Bot.* 54: 609–622.

Baroja-Fernandez, E., Muñoz, F. J., and Pozueta-Romero, J. (2005) Response to Neuhaus et al.: No need to shift the paradigm on the metabolic pathway to transitory starch in leaves. *Trends Plant Sci.* 10: 156–158.

Boldt, R., Edner, C., Kolukisaoglu, U., Hagemann, M., Weckwerth, W., Wienkoop, S., Morgenthal, K., and Bauwe, H. (2005) D-Gycerate 3-kinase, the last unknown enzyme in the photorespiratory cycle in Arabidopsis, belongs to a novel kinase family. *Plant Cell* 17: 2413–2420.

Bonner, W., and Bonner, J. (1948) The role of carbon dioxide in acid formation by succulent plants. *Am. J. Bot.* 35: 113–117.

Brandes, H. K., Hartman, F. C., Lu, T. Y., and Larimer, F. W. (1996) Efficient expression of the gene for spinach phosphoribulokinase in *Pichia pastoris* and utilization of the recombinant enzyme to explore the role of regulatory cysteinyl residues by site-directed mutagenesis. *J. Biol. Chem.* 271: 6490–6496.

Brinkmann, H., Cerff, R., Salomon, M., and Soll, J. (1989) Cloning and sequence analysis of cDNAs encoding the cytosolic precursors of subunits GapA and GapB of chloroplast glyceraldehyde-3-phosphate dehydrogenase from pea and spinach. *Plant Mol. Biol.* 13: 81–94.

Buchanan, B. B. (1980) Role of light in the regulation of chloroplast enzymes. *Annu. Rev. Plant Physiol.* 31: 341–374.

Buchanan, B. B., and Balmer, Y. (2005) Redox regulation: A broadening horizon. *Annu. Rev. Plant Biol.* 56:187–220.

Burnell, J. N., and Hatch, M. D. (1985) Light–dark modulation of leaf pyruvate, P_i dikinase. *Trends Biochem. Sci.* 10: 288–291.

Chia, T., Thorneycroft, D., Chapple, A., Messerli, G., Chen, J., Zeeman, S. C., Smith, S. M., and Smith, A. M. (2004) A cytosolic glucosyl-transferase is required for conversion of starch to sucrose in *Arabidopsis* leaves at night. *Plant J.* 37: 853–863.

Chiadmi M., Navaza A., Miginiac-Maslow M., Jacquot J. P., and Cherfils J. (1999) Redox signalling in the chloroplast: Structure of oxidized pea fructose-1,6-bisphosphate phosphatase. *Embo. J.* 18(23): 6809–6815.

Chollet, R., Vidal, J., and O'Leary, M. H. (1996) Phosphoenolpyruvate carboxylase: A ubiquitous, highly regulated enzyme in plants. *Annu. Rev. Plant Physiol. Plant Mol. Biol.* 47: 273–298.

Cleland, W. W., Andrews, T. J., Gutteridge, S., Hartman, F. C., and Lorimer, G. H. (1998) Mechanism of rubisco: The carbamate as general base. *Chem. Rev.* 98: 549–561.

Craig, S., and Goodchild, D. J. (1977) Leaf ultrastructure of *Triodia irritans*: A C_4 grass possessing an unusual arrangement of photosynthetic tissues. *Aust. J. Bot.* 25: 277–290.

Critchley, J. H., Zeeman, S. C., Takaha, T., Smith, A. M., and Smith, S. M. (2001) A critical role for disproportionating enzyme in starch breakdown is revealed by a knock-out mutant in *Arabidopsis*. *Plant J.* 26: 89–100.

Cushman, J. C. (2001) Crassulacean acid metabolism: A plastic photosynthetic adaptation to arid environments. *Plant Physiol.* 127: 1439–1448.

Dever, L. V., Bailey, K. J., Lacuesta, M., Leegood, R. C., and Lea P. J. (1996) The isolation and characterization of mutants of the C_4 plant *Amaranthus edulis*. *Comptes Rendus Acad. Sci. Ser. III. Life Sci.* 319: 919–959.

Drincovich, M. F., Casati, P., and Andreo, C. S. (2001) NADP-malic enzyme from plants: A ubiquitous enzyme involved in different metabolic pathways. *FEBS Lett.* 490: 1–6.

Dunford, R. P., Catley M. A., Raines, C. A., Lloyd, J. C., and Dyer, T. A. (1998) Purification of active chloroplast sedoheptulose-1,7-bisphosphatase expressed in *Escherichia coli*. *Protein Expr. Purif.* 14: 139–145.

Edwards, G. E., and Walker, D. (1983) *C3, C4: Mechanisms and Cellular and Environmental Regulation of Photosynthesis*. University of California Press, Berkeley.

Edwards, G. E., Franceschi, V. R., and Voznesenskaya, E. V. (2004) Single-cell C4 photosynthesis versus the dual-cell (Kranz) paradigm. *Annu. Rev. Plant Biol.* 55: 173–196.

Flügge, U. I., and Heldt, H. W. (1991) Metabolite translocators of the chloroplast envelope. *Annu. Rev. Plant Physiol. Plant Mol. Biol.* 42: 129–144.

Giordano, M., Beardall, J., and Raven, J. A (2005) CO_2 concentrating mechanisms in algae: Mechanisms, environmental modulation, and evolution. *Annu. Rev. Plant Biol.* 56: 99–131.

Hatch, M. D., and Slack, C. R. (1966) Photosynthesis by sugarcane leaves. A new carboxylation reaction and the pathway of sugar formation. *Biochem. J.* 101: 103–111.

Hausler, R. E., Schlieben, N. H., Schulz, B., and Flugge, U. I. (1998) Compensation of decreased triose phosphate/phosphate translocator activity by accelerated starch turnover and glucose transport in transgenic tobacco. *Planta* 204: 366–376.

Heldt, H. W. (1979) Light-dependent changes of stromal H⁺ and Mg^{2+} concentrations controlling CO_2 fixation. In *Photosynthesis II (Encyclopedia of Plant Physiology*, New Series, vol. 6) M. Gibbs and E. Latzko, eds. Springer, Berlin, pp. 202–207.

Hibberd, J. M., and Quick, P. (2002) Characteristics of C4 photosynthesis in stems and petioles of C3 flowering plants. *Nature* 415: 451–454.

Hirasawa M., Schurmann P., Jacquot J. P., Manieri W., Jacquot P., Keryer E., et al. (1999) Oxidation-reduction properties of chloroplast thioredoxins, ferredoxin:thioredoxin reductase, and thioredoxin f-regulated enzymes. *Biochemistry* 38: 5200–5205.

Huber, S. C. (1986) Fructose-2,6-bisphosphate as a regulatory metabolite in plants. *Annu. Rev. Plant Physiol.* 37: 233–246.

Huber, S. C., and Huber, J. L. (1996) Role and regulation of sucrose-phosphate synthase in higher plants. *Annu. Rev. Plant Physiol. Plant Mol. Biol.* 47: 431–444.

Kozaki, A., and Takeba, G. (1996) Photorespiration protects C_3 plants from photooxidation. *Nature* 384: 557–560.

Ku, S. B., and Edwards, G. E. (1978) Oxygen inhibition of photosynthesis. III. Temperature dependence of quantum yield and its relation to O_2/CO_2 solubility ratio. *Planta* 140: 1–6.

Leegood, R. C., Lea, P. J., Adcock, M. D., and Haeusler, R. D. (1995) The regulation and control of photorespiration. *J. Exp. Bot.* 46: 1397–1414.

Lloyd, J. R., Kossmann, J., and Ritte, G. (2005) Leaf starch degradation comes out of the shadows. *Trends Plant Sci.* 10: 130–137.

Lorimer, G. H. (1981) The carboxylation and oxygenation of ribulose 1,5-bisphosphate: The primary events in photosynthesis and photorespiration. *Annu. Rev. Plant Physiol.* 32 349–383.

Ludwig, M., von Caemmerer, S., Price, G. D., Badger, M. R., and Furbank, R. T. (1998) Expression of tobacco carbonic anhydrase in the C_4 dicot *Flaveria bidentis* leads to increased leakiness of the bundle sheath and a defective CO_2 concentrating mechanism. *Plant Physiol.* 117: 1071–1081.

Lund, J. E., Ashton, A. R., Hatch, M. D., and Heldt, H. W. (2000) Purification, molecular cloning, and sequence analysis of sucrose-6^F-phosphate phosphohydrolase from plants. *Proc. Natl. Acad. Sci. USA* 97: 12914–12919.

Lüttge, U., and Higinbotham, N. (1979) *Transport in Plants.* Springer-Verlag, New York.

Maier, R. M., Neckermann, K., Igloi, G. L., and Koessel, H. (1995) Complete sequence of the maize chloroplast genome: Gene content, hotspots of divergence and fine tuning of genetic information by transcript editing. *J. Mol. Biol.* 251: 614–628.

Maroco, J. P., Ku, M. S. B., Lea P. J., Dever, L. V., Leegood, R. C., Furbank, R. T., and Edwards, G. E. (1998) Oxygen requirement and inhibition of C_4 photosynthesis: An analysis of C_4 plants deficient in the C_3 and C_4 cycles. *Plant Physiol.* 116: 823–832.

Mikkelsen, R., Mutenda, K. E., Mant, A., Schurmann, P., and Blennow, A. (2005) Alpha-glucan, water dikinase (GWD): A plastidic enzyme with redox-regulated and coordinated catalytic activity and binding affinity. *Proc. Natl. Acad. Sci. USA* 102: 1785–1790.

Miziorko, H. M., and Lorimer G. H. (1983) Ribulose-1,5-bisphosphate oxygenase. *Annu. Rev. Biochem.* 52: 507–535.

Mora-Garcia, S., Stolowicz, F., and Wolosiuk, R. A. (2006) Redox signal transduction in plant metabolism. Control of primary metabolism in plants. *Annual Plant Reviews*, Vol. 22, Blackwell Publishing, Oxford, UK, pp. 150–186.

Neff, M. M., Fankhauser, C., and Chory, J. (2000). Light: An indicator of time and place. *Genes Dev.* 14, 257–271.

Neuhaus, H. E., Hausler, R. E., and Sonnewald, U. (2005) No need to shift the paradigm on the metabolic pathway to transitory starch in leaves. *Trends Plant Sci.* 10: 154–156.

Nielsen, T. H., Rung, J. H., and Villadsen, D. (2004) Fructose-2,6-bisphosphate: A traffic signal in plant metabolism. *Trends Plant Sci.* 9: 556–563.

Niittyla, T., Messerli, G., Trevisan, M., Chen, J., Smith, A. M., and Zeeman, S. C. (2004) A previously unknown maltose transporter essential for starch degradation in leaves. *Science* 303: 87–89.

Ogawa, T., and Kaplan, A. (1987) The stoichiometry between CO_2 and H^+ fluxes involved in the transport of inorganic carbon in cyanobacteria. *Plant Physiol.* 83: 888–891.

Ogren, W. L. (1984) Photorespiration: Pathways, regulation and modification. *Annu. Rev. Plant Physiol.* 35: 415–422.

Paul, M., Sonnewald, U., Hajirezaei, M., Dennis, D., and Stitt, M. (1995) Transgenic tobacco plants with strongly decreased expression of pyrophosphate: Fructose-6-phosphate 1-phosphotransferase do not differ significantly from wild type in photosynthate partitioning, plant growth or their ability to cope with limiting phosphate, limiting nitrogen and suboptimal temperatures. *Planta* 196: 277–283.

Porter M. A, Stringer, C. D., and Hartman, F. C. (1988) Characterization of the regulatory thioredoxin site of phosphoribulokinase. *J. Biol. Chem.* 263(1): 123–129.

Purton, S. (1995) The chloroplast genome of *Chlamydomonas. Sci. Prog.* 78: 205–216.

Rachmilevitch, S., Cousins, A. B., and Bloom, A. J. (2004) Nitrate assimilation in plant shoots depends on photorespiration. *Proc. Natl. Acad. Sci. USA* 101: 11506–11510.

Reinfelder, J. R., Milligan, A. J., and Morel, F. M. M. (2004) The role of the C_4 pathway in carbon accumulation and fixation in a marine diatom. *Plant Physiol.* 135: 2106–2111.

Sage, R. F. (2002) C_4 photosynthesis in terrestrial plants does not require Kranz anatomy. *Trends Biochem. Sci.* 7: 283–285.

Salerno, G. L., Echeverria, E., and Pontis, H. G. (1996) Activation of sucrose-phosphate synthase by a protein factor/sucrose-phosphate phosphatase. *Cell. Mol. Biol.* 42: 665–672.

Salvucci, M. E., and Ogren, W. L. (1996) The mechanism of rubisco activase: Insights from studies of the properties and structure of the enzyme. *Photosynth. Res.* 47: 1–11.

Sharkey, T.D. (1988) Estimating the rate of photorespiration in leaves. *Physiol. Plant.* 73: 147–152.

Siebke, K., von Caemmerer, S., Badger, M., and Furbank, R. T. (1997) Expressing an *rbcS* antisense gene in transgenic *Flaveria bidentis* leads to an increased quantum requirement for CO_2 fixed in photosystems I and II. *Plant Physiol.* 115: 1163–1174.

Smith, A. M., Zeeman, S. C., and Smith, S. M. (2005) Starch degradation. *Annu. Rev. Plant Biol.* 56: 73–97.

Soupene, E., Inwood, W., and Kustu, S. (2004) Lack of the Rhesus protein Rh1 impairs growth of the green alga *Chlamydomonas reinhardtii* at high CO_2. *Proc. Natl. Acad. Sci. USA* 101: 7787–7792.

Stitt, M. (1990) Fructose 2,6-bisphosphate as a regulatory molecule in plants. *Annu. Rev. Plant Physiol. Plant Mol. Biol.* 41: 153–185.

Tolbert, N. E. (1981) Metabolic pathways in peroxisomes and glyoxysomes. *Annu. Rev. Biochem.* 50: 133–157.

Vidal, J., and Chollet, R. (1997) Regulatory phosphorylation of C_4 PEP carboxylase. *Trends Plant Sci.* 2: 230–237.

von Caemmerer, S., Millgate, A., Farquhar, G. D., and Furbank, R. T. (1997) Reduction of ribulose-1,5-bisphosphate carboxylase/oxygenase by antisense RNA in the C_4 plant *Flaveria bidentis* leads to reduced assimilation rates and increased carbon isotope discrimination. *Plant Physiol.* 113: 469–477.

Weber, A. P. M., Schwacke, R., and Flugge U-I. (2005) Solute transporters of the plastid envelope membrane. *Annu. Rev. Plant Biol.* 56: 133–164.

Wolosiuk, R. A., Ballicora, M. A., and Hagelin, K. (1993) The reductive pentose phosphate cycle for photosynthetic carbon dioxide assimilation: Enzyme modulation. *FASEB J.* 7: 622–637.

Yu, T. S., Kofler H., Hausler, R. E., Hille, D., Flugge, U. I., Zeeman, S. C., Smith, A. M., Kossmann, J., Lloyd, J., Ritte, G., et al. (2001) The Arabidopsis sex1 mutant is defective in the R1 protein, a general regulator of starch degradation in plants, and not in the chloroplast hexose transporter. *Plant Cell* 13: 1907–1918.

Zeeman, S. C., Thorneycroft, D., Schupp, N., Chapple, A., Weck M., Dunstan, H., Haldimann, P., Bechtold, N., Smith, A. M., and Smith, S. M. (2004) Plastidial α-glucan phosphorylase is not required for starch degradation in *Arabidopsis* leaves but has a role in the tolerance of abiotic stress. *Plant Physiol.* 135: 849–858.

Zhang, N., and Portis, A. R. (1999) Mechanism of light regulation of rubisco: A specific role for the larger rubisco activase isoform involving reductive activation by thioredoxin-f. *Proc. Natl. Acad. Sci. USA* 96: 9438–9443.

Zrenner, R., Krause, K. P., Apel, P., and Sonnewald, U. (1996) Reduction of the cytosolic fructose–1,6-bisphosphatase in transgenic potato plants limits photosynthetic sucrose biosynthesis with no impact on plant growth and tuber yield. *Plant J.* 9: 671–681.

Chapter 9 | Photosynthesis: Physiological and Ecological Considerations

THE CONVERSION OF SOLAR ENERGY to the chemical energy of organic compounds is a complex process that includes electron transport and photosynthetic carbon metabolism (see Chapters 7 and 8). Earlier discussions of the photochemical and biochemical reactions of photosynthesis should not overshadow the fact that, under natural conditions, the photosynthetic process takes place in intact organisms that are continuously responding to internal and external changes. This chapter addresses some of the photosynthetic responses of the intact leaf to its environment. Additional photosynthetic responses to different types of stress are covered in Chapter 26.

The impact of the environment on photosynthesis is of interest to plant physiologists, ecologists, and agronomists. From a physiological standpoint, we wish to understand how photosynthetic rate responds directly to environmental factors such as light, ambient CO_2 concentrations, and temperature, or indirectly, through the effects of stomatal control, to environmental factors such as humidity and soil moisture. The dependence of photosynthetic processes on environment is also important to agronomists because plant productivity, and hence crop yield, depend strongly on prevailing photosynthetic rates in a dynamic environment. To the ecologist, the fact that photosynthetic rates and capacities show differences in different environments is of great interest in terms of adaptation.

In studying the environmental dependence of photosynthesis, a central question arises: How many environmental

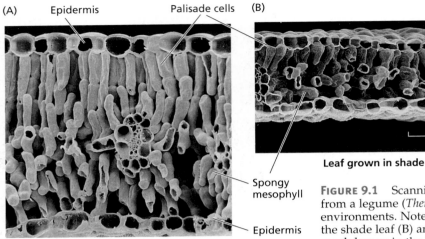

FIGURE 9.1 Scanning electron micrographs of the leaf anatomy from a legume (*Thermopsis montana*) grown in different light environments. Note that the sun leaf (A) is much thicker than the shade leaf (B) and that the palisade (columnlike) cells are much longer in the leaves grown in sunlight. Layers of spongy mesophyll cells can be seen below the palisade cells. (Courtesy of T. Vogelmann.)

factors can limit photosynthesis at one time? The British plant physiologist F. F. Blackman hypothesized in 1905 that, under any particular conditions, the rate of photosynthesis is limited by the slowest step, the so-called *limiting factor*.

The implication of this hypothesis is that at any given time, photosynthesis can be limited either by light or by CO_2 concentration, for instance, but not by both factors. This hypothesis has had a marked influence on the approach used by plant physiologists to study photosynthesis—that is, varying one factor and keeping all other environmental conditions constant. In the intact leaf, three major metabolic steps have been identified as important for optimal photosynthetic performance:

- Rubisco activity
- Regeneration of ribulose bisphosphate (RuBP)
- Metabolism of the triose phosphates

The first two steps are the most important under natural conditions.

Farquhar and Sharkey (1982) added a new dimension to our understanding of photosynthesis by pointing out that we should think of the controls over photosynthetic rate in leaves through "supply" and "demand" functions. The biochemical activities referred to above take place in the palisade cells and spongy mesophyll of the leaf (Figure 9.1); they describe the "demand" by photosynthetic metabolism in the cells for CO_2 as a substrate. However, the actual rate of CO_2 "supply" to these cells is controlled by stomatal guard cells located on the epidermal portions of the leaf. These supply and demand functions associated with photosynthesis take place in different cells. It is the coordinated actions of "demand" by photosynthetic cells and "supply" by guard cells that determine the leaf photosynthetic rate.

In the following sections, we will focus on how naturally occurring variations in light and temperature influence photosynthesis in leaves and how leaves in turn adjust or acclimate to variations in light and temperature. In addition, we will consider the impacts of atmospheric carbon dioxide, a major factor that influences photosynthesis and one that is rapidly increasing in concentration as humans continue to burn fossil fuels for energy uses.

Light, Leaves, and Photosynthesis

Scaling up from the chloroplast (the focus of Chapters 7 and 8) to the leaf adds new levels of complexity to photosynthesis. At the same time, the structural and functional properties of the leaf make possible other levels of regulation.

We will start by examining how leaf anatomy and leaf orientation control the absorption of light for photosynthesis. Then we will describe how chloroplasts and leaves acclimate to their light environment. We will see that the photosynthetic response of leaves grown under different light conditions also reflects an acclimation capacity to growth under a different light environment. We will also see that there are limits in the extent to which photosynthesis in a species can acclimate to very different light environments.

It will become clear that multiple environmental factors can influence photosynthesis. For example, consider that both the amount of light and the amount of CO_2 determine the photosynthetic response of leaves. In some situations involving these two factors, photosynthesis is limited by an inadequate supply of light or CO_2. In other situations, absorption of too much light can cause severe problems, and special mechanisms protect the photosynthetic system from excessive light. While plants have multiple levels of acclimation control over photosynthesis that allow them to grow successfully in constantly changing environments, there are ultimately limits to possible acclimations to sun versus shade, high temperature versus low temperature, and high water stress versus low water stress environments.

Units in the Measurement of Light

Think of the different ways in which leaves are exposed to different spectra and quantities of light that result in photosynthesis. Plants grown outdoors are exposed to sunlight, and the spectrum of that sunlight will depend on whether it is measured in full sunlight or under the shade of a canopy. Plants grown indoors may receive either incandescent or fluorescent lighting, each of which is different from sunlight. To account for these differences in spectral quality and quantity, we need uniformity in how we measure and express the light that impacts photosynthesis.

Three light parameters are especially important in the measurement of light: (1) spectral quality, (2) amount, and (3) direction. Spectral quality was discussed in Chapter 7 (see Figures 7.2 and 7.3, and Web Topic 7.1). A discussion of the amount and direction of light reaching the plant requires consideration of the geometry of the part of the plant that receives the light: Is the plant organ flat or cylindrical?

Flat, or planar, light sensors are best suited for flat leaves. The light reaching the plant can be measured as energy, and the amount of energy that falls on a flat sensor of known area per unit time is quantified as **irradiance** (see Table 9.1). Units can be expressed in terms of energy, such as watts per square meter (W m^{-2}). Time (seconds) is contained within the term watt: 1 W = 1 joule (J) s^{-1}.

Light can also be measured as the number of incident **quanta** (singular *quantum*). In this case, units can be expressed in moles per square meter per second (mol m^{-2} s^{-1}), where *moles* refers to the number of photons (1 mol of light = 6.02 × 10^{23} photons, Avogadro's number). This measure is called **photon irradiance**. Quanta and energy units can be interconverted relatively easily, provided that the wavelength of the light, λ is known. The energy of a photon is related to its wavelength as follows:

$$E = \frac{hc}{\lambda}$$

where c is the speed of light (3 × 10^8 m s^{-1}), h is Planck's constant (6.63 × 10^{-34} J s), and λ is the wavelength of light, usually expressed in nm (1 nm = 10^{-9} m). From this equation it can be shown that a photon at 400 nm has twice the energy of a photon at 800 nm (see Web Topic 9.1).

Now let's turn our attention to the direction of light. Light can strike a flat surface directly from above or obliquely. When light deviates from perpendicular, irradiance is proportional to the cosine of the angle at which the light rays hit the sensor (Figure 9.2).

There are many examples in nature in which the light-intercepting object is not flat (e.g., complex shoots, whole plants, chloroplasts). In addition, in some situations light

TABLE 9.1
Concepts and units for the quantification of light

	Energy measurements (W m^{-2})	Photon measurements (mol m^{-2}s^{-1})
Flat light sensor	Irradiance	Photon irradiance
	Photosynthetically active radiation (PAR, 400-700 nm, energy units)	PAR (quantum units)
	—	Photosynthetic photon flux density (PPFD)
Spherical light sensor	Fluence rate (energy units)	Fluence rate (quantum units)
	Scalar irradiance	Quantum scalar irradiance

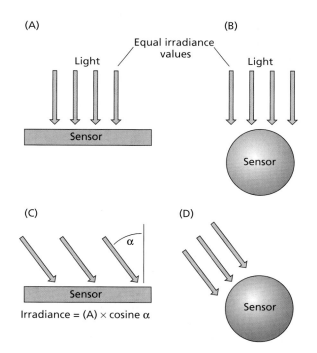

FIGURE 9.2 Flat and spherical light sensors. Equivalent amounts of collimated light strike a flat irradiance–type sensor (A) and a spherical sensor (B) that measure fluence rate. With collimated light, A and B will give the same light readings. When the light direction is changed 45°, the spherical sensor (D) will measure the same quantity as in B. In contrast, the flat irradiance sensor (C) will measure an amount equivalent to the irradiance in A multiplied by the cosine of the angle α in C. (After Björn and Vogelmann 1994.)

can come from many directions simultaneously (e.g., direct light from the sun plus the light that is reflected upward from sand, soil, or snow). In these situations it makes more sense to measure light with a spherical sensor that takes measurements omnidirectionally (from all directions).

The term for this omnidirectional measurement is **fluence rate** (see Table 9.1) (Rupert and Letarjet 1978), and this quantity can be expressed in watts per square meter (W m^{-2}) or moles per square meter per second (mol m^{-2} s^{-1}). The units clearly indicate whether light is being measured as energy (W) or as photons (mol).

In contrast to a flat sensor's sensitivity, the sensitivity to light of a spherical sensor is independent of direction (see Figure 9.2). Depending on whether the light is collimated (rays are parallel) or diffuse (rays travel in random directions), values for fluence rate versus irradiance measured with a flat or a spherical sensor can provide different values (for a detailed discussion, see Björn and Vogelmann [1994]).

Photosynthetically active radiation (**PAR**, 400–700 nm) may also be expressed in terms of energy (W m^{-2}) or quanta (mol m^{-2} s^{-1}) (McCree 1981). Note that PAR is an irradiance-type measurement. In research on photosynthesis, when PAR is expressed on a quantum basis, it is given the special term **photosynthetic photon flux density** (**PPFD**). However, it has been suggested that the term *density* be discontinued, because within the International System of Units (SI units, where *SI* stands for *Système International*), *density* can mean area or volume.

In summary, when choosing how to quantify light, it is important to match sensor geometry and spectral response with that of the plant. Flat, cosine-corrected sensors are ideally suited to measure the amount of light that strikes the surface of a leaf; spherical sensors are more appropriate in other situations, such as in studies of a chloroplast suspension or a branch from a tree (see Table 9.1).

How much light is there on a sunny day? What is the relationship between PAR irradiance and PAR fluence rate? Under direct sunlight, PAR irradiance and fluence rate are both about 2000 μmol m^{-2} s^{-1}, although higher values can be measured at high altitudes. The corresponding value in energy units is about 900 W m^{-2}.

Leaf anatomy maximizes light absorption

Roughly 1.3 kW m^{-2} of radiant energy from the sun reaches Earth, but only about 5% of this energy can be converted into carbohydrates by a photosynthesizing leaf (Figure 9.3). The reason this percentage is so low is that a major fraction of the incident light is of a wavelength either too short or too long to be absorbed by the photosynthetic pigments (see Figure 7.3). Of the absorbed light energy, a significant fraction is lost as heat, and a smaller amount is lost as fluorescence (see Chapter 7).

Recall from Chapter 7 that radiant energy from the sun consists of many different wavelengths of light. Only photons of wavelengths from 400 to 700 nm are utilized in photosynthesis, and about 85 to 90% of this PAR is absorbed by the leaf; the remainder is either reflected at the leaf surface or transmitted through the leaf (see Figure 9.4). Because chlorophyll absorbs very strongly in the blue and the red regions of the spectrum (see Figure 7.3), the transmitted and reflected light are vastly enriched in green—hence the green color of vegetation.

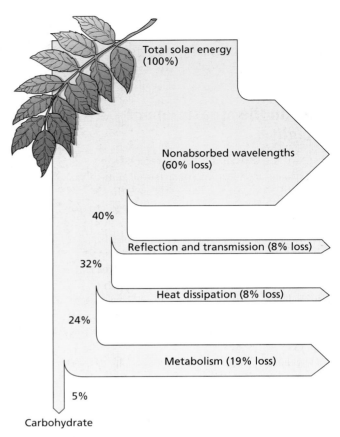

Figure 9.3 Conversion of solar energy into carbohydrates by a leaf. Of the total incident energy, only 5% is converted into carbohydrates.

The anatomy of the leaf is highly specialized for light absorption (Terashima and Hikosaka 1995). The outermost cell layer, the epidermis, is typically transparent to visible light, and the individual cells are often convex. Convex epidermal cells can act as lenses and focus light so that the amount reaching some of the chloroplasts can be many times greater than the amount of ambient light (Vogelmann et al. 1996). Epidermal focusing is common among herbaceous plants and is especially prominent among tropical plants that grow in the forest understory, where light levels are very low.

Below the epidermis, the top layers of photosynthetic cells are called **palisade cells**; they are shaped like pillars that stand in parallel columns one to three layers deep (see

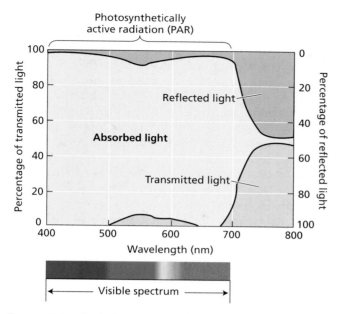

FIGURE 9.4 Optical properties of a bean leaf. Shown here are the percentages of light absorbed, reflected, and transmitted, as a function of wavelength. The transmitted and reflected green light in the wave band at 500 to 600 nm gives leaves their green color. Note that most of the light above 700 nm is not absorbed by the leaf. (After Smith 1986.)

reflect and refract the light, thereby randomizing its direction of travel. This phenomenon is called *light scattering*.

Light scattering is especially important in leaves because the multiple reflections between cell–air interfaces greatly increase the length of the path over which photons travel, thereby increasing the probability for absorption. In fact, photon path lengths within leaves are commonly four times or more longer than the thickness of the leaf (Richter and Fukshansky 1996). Thus the palisade cell properties that allow light to pass through, and the spongy mesophyll cell properties that are conducive to light scattering, result in more uniform light absorption throughout the leaf.

Some environments, such as deserts, have so much light that it is potentially harmful to leaves. In these environments leaves often have special anatomic features, such as hairs, salt glands, and epicuticular wax that increase the reflection of light from the leaf surface, thereby reducing light absorption (Ehleringer et al. 1976). Such adaptations can decrease light absorption by as much as 40%, minimizing heating and other problems associated with the absorption of too much light.

Plants compete for sunlight

Plants normally compete for sunlight. Held upright by stems and trunks, leaves configure a canopy that absorbs light and influences photosynthetic rates and growth beneath them.

As we will see, leaves that are shaded by other leaves experience lower light levels and have much lower photosynthetic rates. Some plants have very thick leaves that transmit little, if any, light. Other plants, such as those of the dandelion (*Taraxacum* sp.), have a rosette growth habit, in which leaves grow radially very close to each other on a very short stem, thus preventing the growth of any leaves below them.

Trees with their leaves high above the ground surface represent an outstanding adaptation for light interception. The elaborate branching structure of trees vastly increases the interception of sunlight. Very little PAR penetrates the canopy of many forests; almost all of it is absorbed by leaves (Figure 9.5).

Another feature of the shady habitat is **sunflecks**, patches of sunlight that pass through small gaps in the leaf canopy and move across shaded leaves as the sun moves. In a dense forest, sunflecks can change the photon flux impinging on a leaf in the forest floor more than tenfold within seconds. For some of these leaves, sunflecks contain nearly 50% of the total light energy available during the day, but this critical energy is available for only a few minutes now and then in a very high dose.

Sunflecks also play a role in the carbon metabolism of lower leaves in dense crops that are shaded by the upper leaves of the plant. Rapid responses of both the photosynthetic apparatus and the stomata to sunflecks have been of

Figure 9.1). Some leaves have several layers of columnar palisade cells, and we may wonder how efficient it is for a plant to invest energy in the development of multiple cell layers when the high chlorophyll content of the first layer would appear to allow little transmission of the incident light to the leaf interior. In fact, more light than might be expected penetrates the first layer of palisade cells because of the sieve effect and light channeling.

The **sieve effect** is due to the fact that chlorophyll is not uniformly distributed throughout cells but instead is confined to the chloroplasts. This packaging of chlorophyll results in shading between the chlorophyll molecules and creates gaps between the chloroplasts, where light is not absorbed—hence the reference to a sieve. Because of the sieve effect, the total absorption of light by a given amount of chlorophyll in a palisade cell is less than the light absorbed by the same amount of chlorophyll in a solution.

Light channeling occurs when some of the incident light is propagated through the central vacuole of the palisade cells and through the air spaces between the cells, an arrangement that facilitates the transmission of light into the leaf interior (Vogelmann 1993).

Below the palisade layers is the **spongy mesophyll**, where the cells are very irregular in shape and are surrounded by large air spaces (see Figure 9.1). The large air spaces generate many interfaces between air and water that

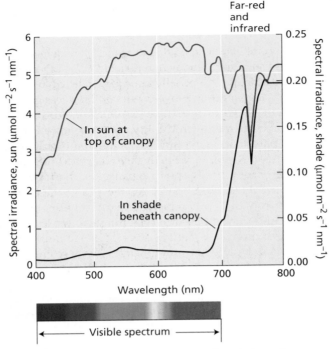

Figure 9.5 The spectral distribution of sunlight at the top of a canopy and under the canopy. For unfiltered sunlight, the total irradiance was 1900 μmol m^{-2} s^{-1}; for shade, 17.7 μmol m^{-2} s^{-1}. Most of the photosynthetically active radiation was absorbed by leaves in the canopy. (After Smith 1994.)

substantial interest to plant physiologists and ecologists (Pearcy et al. 1997), because they represent unique physiological responses specialized in the capture of short bursts of sunlight.

Leaf angle and leaf movement can control light absorption

How do leaves influence the light levels within a canopy? The angle of the leaf relative to the sun will determine the amount of sunlight incident upon it in a manner identical to that shown for the flat light sensor in Figure 9.2. If the sun is directly overhead, a horizontal leaf (such as the flat sensor in Figure 9.2A) will receive much more sunlight than a leaf at a steeper angle. Under natural conditions, leaves exposed to full sunlight at the top of the canopy tend to have steep leaf angles so that less than the maximum amount of sunlight is incident on the leaf blade; this allows more sunlight to penetrate into the canopy. It is common to see that the angle of different leaves within a canopy decrease (become more horizontal) with increasing depth into a canopy.

Leaves have the highest light absorption when the leaf blade, or lamina, is perpendicular to the incident light. Some plants control light absorption by **solar tracking** (Koller 2000); that is, their leaves continuously adjust the orientation of their laminae such that they remain perpendicular to the sun's rays (Figure 9.6). Many crop and wild species have leaves capable of solar tracking, including alfalfa, cotton, soybean, bean, and lupine.

Solar-tracking leaves keep a nearly vertical position at sunrise, facing the eastern horizon, where the sun will rise. The leaf blades then lock on to the rising sun and follow its movement across the sky with an accuracy of ±15° until sunset, when the laminae are nearly vertical, facing the west, where the sun will set. During the night the leaf takes a horizontal position and reorients just before dawn so that it faces the eastern horizon in anticipation of another sunrise. Leaves track the sun only on clear days, and they stop when a cloud obscures the sun. In the case of intermittent cloud cover, some leaves can reorient as rapidly as 90° per hour and thus can catch up to the new solar position when the sun emerges from behind a cloud (Koller 1990).

Solar tracking is a blue-light response (see Chapter 18), and the sensing of blue light in solar-tracking leaves occurs in specialized regions. In species of *Lavatera* (Malvaceae), the photosensitive region is located in or near the major leaf veins (Koller 1990). In lupines (*Lupinus*, Fabaceae), leaves consist of five or more leaflets, and the photosensi-

Figure 9.6 Leaf movement in sun-tracking plants. (A) Initial leaf orientation in the lupine *Lupinus succulentus*. (B) Leaf orientation 4 hours after exposure to oblique light. The direction of the light beam is indicated by the arrows. Movement is generated by asymmetric swelling of a pulvinus, found at the junction between the lamina and the petiole. In natural conditions, the leaves track the sun's trajectory in the sky. (From Vogelmann and Björn 1983, courtesy of T. Vogelmann.)

tive region is the pulvinus and is located in the basal part of each leaflet lamina (see Figure 9.6).

In many cases, leaf orientation is controlled by a specialized organ called the **pulvinus** (plural *pulvini*), found at the junction between the blade and petiole. The pulvinus contains motor cells that change their osmotic potential and generate mechanical forces that determine laminar orientation. In other plants, leaf orientation is controlled by small mechanical changes along the length of the petiole and by movements of the younger parts of the stem.

Some solar-tracking plants can also move their leaves such that they avoid full exposure to sunlight, thus minimizing heating and water loss. Building on the term **heliotropism** (bending toward the sun), which is often used to describe sun-induced leaf movements, these sun-avoiding leaves are called *paraheliotropic*, and leaves that maximize light interception by solar tracking are called *diaheliotropic*. Some plant species have leaves that can display diaheliotropic movements when they are well watered and paraheliotropic movements when they experience water stress.

Since full sunlight usually exceeds the amount of light that can be utilized for photosynthesis, what advantage is gained by solar tracking? By keeping leaves perpendicular to the sun, solar-tracking plants are able to maintain maximum photosynthetic rates throughout the day, including early morning and late afternoon. Moreover, air temperature is lower during the early morning and late afternoon, so that water stress is lower. Solar tracking therefore gives an advantage to plants with short growing periods, as in rain-fed crops such as pinto beans. Diaheliotropic solar tracking appears to be a feature common to wild plants that are short lived and must complete their life cycle before the onset of drought (Ehleringer and Forseth 1980). Similarly, paraheliotropic leaves are able to regulate the amount of sunlight incident on the leaf to a nearly constant value, although the amount of incident sunlight is often only one-half to two-thirds of full sunlight.

Plants acclimate and adapt to sun and shade

Shady habitats can receive less than 20% of the PAR available in an exposed habitat; deep shade habitats receive less than 1% of the PAR at the top of the canopy. Some plants have enough developmental plasticity to respond to a range of light regimes, growing as sun plants in sunny areas and as shade plants in shady habitats. We call this **acclimation**, a process whereby the newly produced leaf has a set of biochemical and morphological characteristics that are best suited for a particular environment.

In some plant species, individual leaves that develop under very sunny or very shady environments are often unable to persist when transferred in the other type of habitat (see Figure 9.5). In this case, the mature leaf will be abscised and a new leaf develop better suited for the new environment. For example, what happens when you take a plant that has developed indoors and transfer it outdoors? After some time, if it's the right type of plant, it develops a new set of leaves better suited to high sunlight. However, other species of plants are not able to acclimate when transferred from sunny to shade environments, but instead they are **adapted** to a sunny environment or to a shade environment. For example, when plants adapted to deep shade conditions are transferred into full sunlight, the leaves experience chronic photoinhibition and leaf bleaching, and they eventually die, as will be discussed later in this chapter.

Sun and shade leaves have some contrasting biochemical characteristics:

- Shade leaves have more total chlorophyll per reaction center, have a higher ratio of chlorophyll b to chlorophyll a, and are usually thinner than sun leaves.

- Sun leaves have more rubisco and a larger pool of xanthophyll cycle components than shade leaves (see Chapter 7).

Contrasting anatomic characteristics can also be found in leaves of the same plant that are exposed to different light regimes. Figure 9.1 shows some anatomic differences between a leaf grown in the sun and a leaf grown in the shade. Sun-grown leaves are thicker and have longer palisade cells than leaves grown in the shade. Even different parts of a single leaf show adaptations to their light microenvironment (Terashima 1992).

These morphological and biochemical modifications are associated with specific functions. Far-red light is absorbed primarily by PSI, and altering the ratio of PSI to PSII or changing the light-harvesting antennae associated with the photosystems makes it possible to maintain a better balance of energy flow through the two photosystems (Melis 1996). These adaptations are found in nature; some shade plants show a 3:1 ratio of photosystem II to photosystem I reaction centers, compared with the 2:1 ratio found in sun plants (Anderson 1986). Other shade plants, rather than altering the ratio of PSI to PSII, add more antenna chlorophyll to PSII. These adaptations appear to enhance light absorption and energy transfer in shady environments, where far-red light is more abundant.

Sun and shade plants also differ in their dark respiration rates, and these differences alter the relationship between respiration and photosynthesis, as we'll see a little later in this chapter.

Photosynthetic Responses to Light by the Intact Leaf

Light is a critical resource for plants that can limit growth and reproduction. The photosynthetic properties of the leaf provide valuable information about plant adaptations to their light environment. In this section we describe typical photosynthetic responses to light as measured in light-

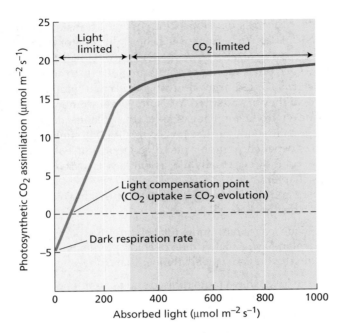

FIGURE 9.7 Response of photosynthesis to light in a C_3 plant. In darkness, respiration causes a net efflux of CO_2 from the plant. The light compensation point is reached when photosynthetic CO_2 assimilation equals the amount of CO_2 evolved by respiration. Increasing light above the light compensation point proportionally increases photosynthesis, indicating that photosynthesis is limited by the rate of electron transport, which in turn is limited by the amount of available light. This portion of the curve is referred to as light-limited. Further increases in photosynthesis are eventually limited by the carboxylation capacity of rubisco or the metabolism of triose phosphates. This part of the curve is referred to as CO_2-limited.

The photon flux at which different leaves reach the light compensation point varies with species and developmental conditions. One of the more interesting differences is found between plants grown in full sunlight and those grown in the shade (Figure 9.8). Light compensation points of sun plants range from 10 to 20 µmol m^{-2} s^{-1}; corresponding values for shade plants are 1 to 5 µmol m^{-2} s^{-1}.

The values for shade plants are lower because respiration rates in shade plants are very low, so little net photosynthesis suffices to bring the net rates of CO_2 exchange to zero. Low respiratory rates represent a basic response that allows shade plants to survive in light-limited environments through their ability to achieve positive CO_2 uptake rates at lower values of PAR than sun plants.

Increasing photon flux above the light compensation point results in a proportional increase in photosynthetic rate (see Figure 9.7), yielding a linear relationship between photon flux and photosynthetic rate. Such a linear relationship comes about because photosynthesis is light limited at those levels of incident light, so more light stimulates proportionately more photosynthesis.

In this linear portion of the curve, the slope of the line reveals the **maximum quantum yield** of photosynthesis for

response curves. We also consider how an important feature of light-response curves, the light compensation point, explains contrasting physiological properties of sun and shade plants. The section then continues with descriptions of how leaves respond to excess light.

Light-response curves reveal photosynthetic properties

Measuring net CO_2 fixation in intact leaves at increasing photon flux allows us to construct light-response curves (Figure 9.7) that provide useful information about the photosynthetic properties of leaves. In the dark there is no photosynthetic carbon assimilation; instead CO_2 is given off by the plant because of mitochondrial respiration (see Chapter 11). By convention, CO_2 assimilation is negative in this part of the light-response curve. As the photon flux increases, photosynthetic CO_2 assimilation increases linearly until it equals CO_2 release by mitochondrial respiration. The point at which photosynthetic CO_2 uptake exactly balances CO_2 release is called the **light compensation point**.

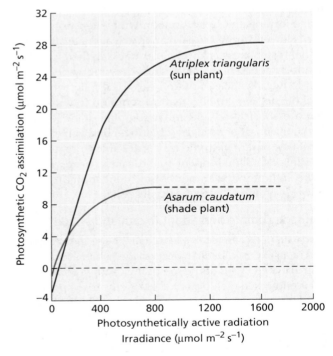

FIGURE 9.8 Light-response curves of photosynthetic carbon fixation in sun and shade plants. Triangle orache (*Atriplex triangularis*) is a sun plant, and a wild ginger (*Asarum caudatum*) is a shade plant. Typically, shade plants have a low light compensation point and have lower maximal photosynthetic rates than sun plants. The dashed line has been extrapolated from the measured part of the curve. (After Harvey 1979.)

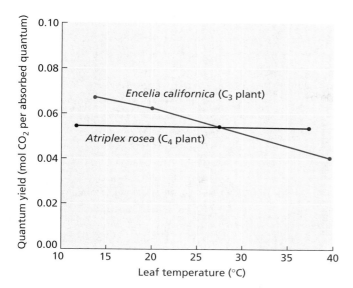

FIGURE 9.9 The quantum yield of photosynthetic carbon fixation in a C_3 plant and in a C_4 plant as a function of leaf temperature. In normal air, photorespiration increases with temperature in C_3 plants, and the energy cost of net CO_2 fixation increases accordingly. This higher energy cost is expressed in lower quantum yields at higher temperatures. Because of the CO_2-concentrating mechanisms of C_4 plants, photorespiration is low in these plants, and the quantum yield does not show a temperature dependence. Note that at lower temperatures the quantum yield of C_3 plants is higher than that of C_4 plants, indicating that photosynthesis in C_3 plants is more efficient at lower temperatures. (After Ehleringer and Björkman 1977.)

the leaf. Recall that quantum yield is the ratio of a given light-dependent product (in this case, CO_2 assimilation) to the number of absorbed photons (see Equation 7.5).

Photosynthetic quantum yield can be expressed on either a CO_2 or O_2 basis. Recall from Chapter 7 that the quantum yield of photochemistry is about 0.95. However, the photosynthetic quantum yield of an integrated process such as photosynthesis is lower whether it is measured in chloroplasts (organelles) or whole leaves. Based on the biochemistry discussed in Chapter 8, we expect the maximum quantum yield for photosynthesis to be 0.125 for C_3 plants (one CO_2 molecule fixed per eight photons absorbed).

Under today's atmospheric conditions (380 ppm CO_2, 21% O_2), the quantum yields for CO_2 of C_3 and C_4 leaves are similar and vary between 0.04 and 0.06 mole of CO_2 per mole of photons. In C_3 plants the reduction from the theoretical maximum is primarily caused by energy loss through photorespiration. In C_4 plants the reduction is caused by the energy requirements of the CO_2-concentrating mechanism. If C_3 leaves are exposed to low O_2 concentrations, photorespiration is inhibited and the quantum yield increases to about 0.09 mole of CO_2 per mole of photons. If C_4 leaves are exposed to low O_2 concentrations, the quantum yields for CO_2 fixation remain constant at about 0.05 mole of CO_2 per mole of photons, irrespective of whether leaves are exposed to a current (21%) or low O_2 concentration. This is because the carbon-concentrating mechanism in C_4 photosynthesis effectively eliminates CO_2 evolution via photorespiration.

Quantum yield varies with temperature and CO_2 concentration because of their effect on the ratio of the carboxylase and oxygenase reactions of rubisco (see Chapter 8). Below 30°C in today's environment, quantum yields of C_3 plants are generally higher than those of C_4 plants; above 30°C, the situation is usually reversed (see Figure 9.9). Despite their different growth habitats, sun and shade plants show very similar quantum yields, because the basic biochemical processes that determine quantum yield are the same for these two types of plants.

At higher photon fluxes, the photosynthetic response to light starts to level off (see Figure 9.10) and eventually reaches *saturation*. Once the saturation point is reached, further increases in photon flux no longer affect photosynthetic rates, indicating that factors other than incident light, such as electron transport rate, rubisco activity, or the metabolism of triose phosphates, have become limiting to photosynthesis.

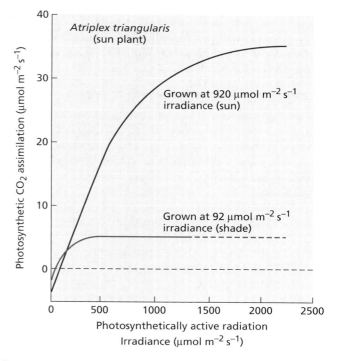

FIGURE 9.10 Light-response of photosynthesis of a sun plant grown under sun or shade conditions. The upper curve represents an *A. triangularis* leaf grown at an irradiance ten times higher than that of the lower curve. In the leaf grown at the lower light levels, photosynthesis saturates at a substantially lower irradiance, indicating that the photosynthetic properties of a leaf depend on its growing conditions. The dashed red line has been extrapolated from the measured part of the curve. (After Björkman 1981.)

After the saturation point, photosynthesis is commonly referred to as *CO₂ limited* (see Figure 9.8), reflecting the inability of the Calvin cycle enzymes to keep pace with the absorbed light energy that is producing ATP and NADPH. Light saturation levels for shade plants are substantially lower than those for sun plants. These levels usually reflect the maximum photon flux to which the leaf was exposed during growth.

The light-response curve of most leaves saturates between 500 and 1000 µmol m^{-2} s^{-1}—well below full sunlight (which is about 2000 µmol m^{-2} s^{-1}). Although individual leaves are rarely able to utilize full sunlight, whole plants usually consist of many leaves that shade each other, so only a small fraction of a tree's leaves are exposed to full sun at any given time of the day. The rest of the leaves receive subsaturating photon fluxes in the form of small patches of light that pass through gaps in the leaf canopy or in the form of light transmitted through other leaves. Because the photosynthetic response of the intact plant is the sum of the photosynthetic activity of all the leaves, only rarely is photosynthesis saturated with light at the level of the whole plant.

Light-response curves of individual trees and of the forest canopy show that photosynthetic rate increases with photon flux and photosynthesis usually does not saturate, even in full sunlight (Figure 9.11). Along these lines, crop productivity is related to the total amount of light received during the growing season, and given enough water and nutrients, the more light a crop receives, the higher the biomass (Ort and Baker 1988).

Leaves must dissipate excess light energy

When exposed to excess light, leaves must dissipate the surplus absorbed light energy so that it does not harm the photosynthetic apparatus (Figure 9.12). There are several routes for energy dissipation involving *nonphotochemical quenching* (see Chapter 7), which is the quenching of chlorophyll fluorescence by mechanisms other than photochemistry. The most important example involves the transfer of absorbed light energy away from electron transport toward heat production. Although the molecular mechanisms are not yet fully understood, the xanthophyll cycle appears to be an important avenue for dissipation of excess light energy (see **Web Essay 9.1**).

THE XANTHOPHYLL CYCLE Recall from Chapter 7 that the xanthophyll cycle, which comprises the three carotenoids violaxanthin, antheraxanthin, and zeaxanthin, is involved in the dissipation of excess light energy in the leaf (see Figure 7.36). Under high light, violaxanthin is converted to antheraxanthin and then to zeaxanthin. Note that the two aromatic rings of violaxanthin have a bound oxygen atom in them, antheraxanthin has one, and zeaxanthin has none (again, see Figure 7.36). Experiments have shown that zeaxanthin is the most effective of the three xanthophylls in heat dissipation, and antheraxanthin is only half as effective.

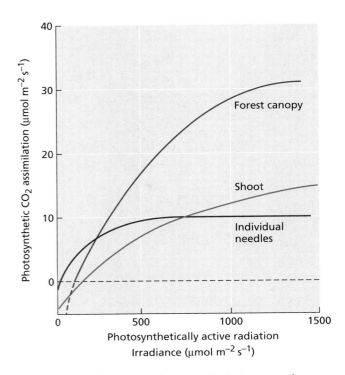

FIGURE 9.11 Changes in photosynthesis (expressed on a per-square-meter basis) in individual needles, a complex shoot, and a forest canopy of Sitka spruce (*Picea sitchensis*) as a function of irradiance. Complex shoots consist of groupings of needles that often shade each other, similar to the situation in a canopy where branches often shade other branches. As a result of shading, much higher irradiance levels are needed to saturate photosynthesis. The dashed portion of the forest canopy trace has been extrapolated from the measured part of the curve. (After Jarvis and Leverenz 1983.)

Whereas the levels of antheraxanthin remain relatively constant throughout the day, the zeaxanthin content increases at high irradiances and decreases at low irradiances.

In leaves growing under full sunlight, zeaxanthin and antheraxanthin can make up 60% of the total xanthophyll cycle pool at maximal irradiance levels attained at midday (Figure 9.13). In these conditions a substantial amount of excess light energy absorbed by the thylakoid membranes can be dissipated as heat, thus preventing damage to the photosynthetic machinery of the chloroplast (see Chapter 7). The fraction of light energy that is dissipated depends on irradiance, species, growth conditions, nutrient status, and ambient temperature (Demmig-Adams and Adams 1996).

THE XANTHOPHYLL CYCLE IN SUN AND SHADE Leaves that grow in full sunlight contain a substantially larger xanthophyll pool than do shade leaves, so they can dissipate higher amounts of excess light energy. Nevertheless, the xanthophyll cycle also operates in plants that grow in the low light of the forest understory, where they are only occasion-

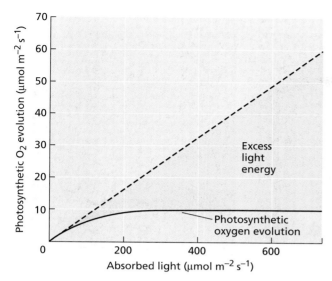

FIGURE 9.12 Excess light energy in relation to a light-response curve of photosynthetic evolution. The broken line shows theoretical oxygen evolution in the absence of any rate limitation to photosynthesis. At levels of photon flux up to 150 µmol m^{-2} s^{-1}, a shade plant is able to utilize the absorbed light. Above 150 µmol m^{-2} s^{-1}, however, photosynthesis saturates, and an increasingly larger amount of the absorbed light energy must be dissipated. At higher irradiances there is a large difference between the fraction of light used by photosynthesis versus that which must be dissipated (excess light energy). The differences are much greater in a shade plant than in a sun plant. (After Osmond 1994.)

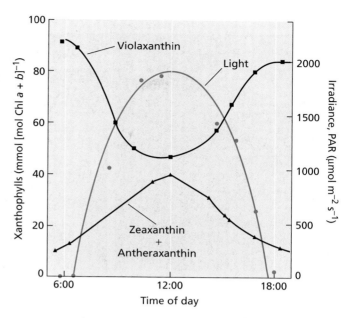

FIGURE 9.13 Diurnal changes in xanthophyll content as a function of irradiance in sunflower (*Helianthus annuus*). As the amount of light incident to a leaf increases, a greater proportion of violaxanthin is converted to antheraxanthin and zeaxanthin, thereby dissipating excess excitation energy and protecting the photosynthetic apparatus. (After Demmig-Adams and Adams 1996.)

ally exposed to high light when sunlight passes through gaps in the overlying leaf canopy, forming sunflecks (described earlier in the chapter). Exposure to one sunfleck results in the conversion of much of the violaxanthin in the leaf to zeaxanthin. In contrast to typical leaves, in which violaxanthin levels increase again when irradiances drop, the zeaxanthin formed in shade leaves of the forest understory is retained and protects the leaf against exposure to subsequent sunflecks.

The xanthophyll cycle is also found in species such as conifers, the leaves of which remain green during winter, when photosynthetic rates are very low yet light absorption remains high. Contrary to the diurnal cycling of the xanthophyll pool observed in the summer, zeaxanthin levels remain high all day during the winter. Presumably this mechanism maximizes dissipation of light energy, thereby protecting the leaves against photooxidation during winter (Adams et al. 2001).

In addition to protecting the photosynthetic system against high light, the xanthophyll cycle may help protect against high temperatures. Chloroplasts are more tolerant of heat when they accumulate zeaxanthin (Havaux et al. 1996). Thus, plants may employ more than one biochemical mechanism to guard against the deleterious effect of excess heat.

CHLOROPLAST MOVEMENTS An alternative means of reducing excess light energy is to move the chloroplasts so that they are no longer exposed to high light. Chloroplast movement is widespread among algae, mosses, and leaves of higher plants (Haupt and Scheuerlein 1990). If chloroplast orientation and location are controlled, leaves can regulate how much of the incident light is absorbed. Under low light (Figure 9.14B), chloroplasts gather at the cell surfaces parallel to the plane of the leaf so that they are aligned perpendicularly to the incident light—a position that maximizes absorption of light.

Under high light (Figure 9.14C), the chloroplasts move to the cell surfaces that are parallel to the incident light, thus avoiding excess absorption of light. Such chloroplast rearrangement can decrease the amount of light absorbed by the leaf by about 15% (Gorton et al. 1999). Chloroplast movement in leaves is a typical blue-light response (see Chapter 18). Blue light also controls chloroplast orientation in many of the lower plants, but in some algae, chloroplast movement is controlled by phytochrome (Haupt and Scheuerlein 1990). In leaves, chloroplasts move along actin microfilaments in the cytoplasm, and calcium regulates their movement (Tlalka and Fricker 1999).

LEAF MOVEMENTS Plants have evolved structural features that reduce the excess light load on leaves during high sunlight periods, especially when transpiration and its cooling

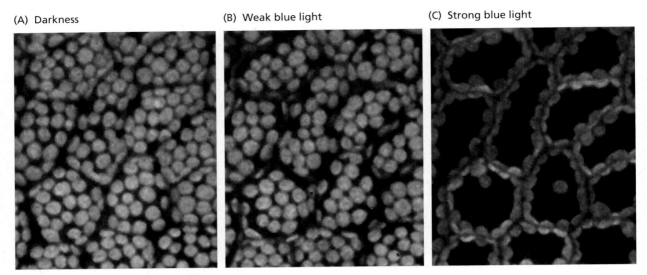

FIGURE 9.14 Chloroplast distribution in photosynthesizing cells of the duckweed *Lemna*. These surface views show the same cells under three conditions: (A) darkness, (B) weak blue light, and (C) strong blue light. In A and B, chloroplasts are positioned near the upper surface of the cells, where they can absorb maximum amounts of light. When the cells are irradiated with strong blue light (C), the chloroplasts move to the side walls, where they shade each other, thus minimizing the absorption of excess light. (Courtesy of M. Tlalka and M. D. Fricker.)

effects are reduced because of water stress. These features often involve changes in the leaf orientation relative to the incoming sunlight. For example, paraheliotropic leaves track the sun but at the same time can reduce incident light levels by folding leaflets together so that the leaf lamina become nearly parallel to the sun's rays. Another common feature is wilting, whereby a leaf droops to a vertical orientation, again effectively reducing the incident heat load and reducing transpiration and incident light levels.

Absorption of too much light can lead to photoinhibition

Recall from Chapter 7 that when leaves are exposed to more light than they can utilize (see Figure 9.12), the reaction center of PSII is inactivated and often damaged in a phenomenon called **photoinhibition**. The characteristics of photoinhibition in the intact leaf depend on the amount of light to which the plant is exposed (Figure 9.15). The two types of photoinhibition are dynamic photoinhibition and chronic photoinhibition (Osmond 1994).

Under moderate excess light, **dynamic photoinhibition** is observed. Quantum efficiency decreases (contrast the slopes of the curves in Figure 9.15), but the maximum photosynthetic rate remains unchanged. Dynamic photoinhibition is caused by the diversion of absorbed light energy toward heat dissipation—hence the decrease in quantum efficiency. This decrease is often temporary, and quantum efficiency can return to its initial higher value when photon flux decreases below saturation levels.

Chronic photoinhibition results from exposure to high levels of excess light that damage the photosynthetic system and decrease both quantum efficiency and maximum

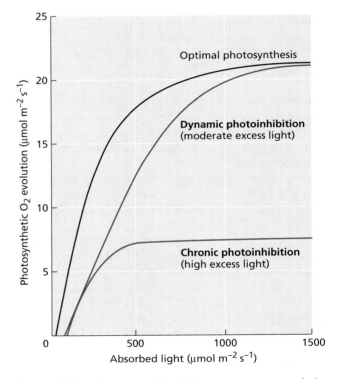

FIGURE 9.15 Changes in the light-response curves of photosynthesis caused by photoinhibition. Exposure to moderate levels of excess light can decrease quantum efficiency (reduced slope of curve) without reducing maximum photosynthetic rate, a condition called dynamic photoinhibition. Exposure to high levels of excess light leads to chronic photoinhibition, where damage to the chloroplast decreases both quantum efficiency and maximum photosynthetic rate. (After Osmond 1994.)

photosynthetic rate (see Figure 9.15). Chronic photoinhibition is associated with damage and replacement of the D1 protein from the reaction center of PSII (see Chapter 7). In contrast to dynamic photoinhibition, these effects are relatively long-lasting, persisting for weeks or months.

Early researchers of photoinhibition interpreted all decreases in quantum efficiency as damage to the photosynthetic apparatus. It is now recognized that short-term decreases in quantum efficiency seem to reflect protective mechanisms (see Chapter 7), whereas chronic photoinhibition represents actual damage to the chloroplast resulting from excess light or a failure of the protective mechanisms.

How significant is photoinhibition in nature? Dynamic photoinhibition appears to occur normally at midday, when leaves are exposed to maximum amounts of light and there is a corresponding reduction in carbon fixation. Photoinhibition is more pronounced at low temperatures, and it becomes chronic under more extreme climatic conditions.

Studies of natural willow populations, and of crops such as oilseed rape (*Brassica napus*) and maize (*Zea mays*), have shown that the cumulative effects of a daily depression in photosynthetic rates caused by photoinhibition decrease biomass by 10% at the end of the growing season (Long et al. 1994). This may not seem a particularly large effect at any one moment in time, but compounded daily or over an entire growing season, it could be significant in natural plant populations competing for limited resources—conditions under which any reduction in carbon allocated to reproduction can adversely affect survival and reproductive success.

Photosynthetic Responses to Temperature

Photosynthesis (CO_2 uptake) and transpiration (H_2O loss) share a common pathway. That is, CO_2 diffuses into the leaf, and H_2O diffuses out, through the stomatal opening regulated by the guard cells. While these are independent processes, vast quantities of water are lost during photosynthetic periods, with the molar ratio of H_2O loss to CO_2 uptake often reaching 250 to 500. This high water loss rate also removes heat from leaves and keeps them relatively cool under full sunlight conditions. Since photosynthesis is a temperature-dependent process, it is important to remember this linkage between two processes influenced by the degree of stomatal opening. As we will see, stomatal opening influences both leaf temperature and the extent of transpiration water loss.

Leaves must dissipate vast quantities of heat

The heat load on a leaf exposed to full sunlight is very high. In fact, a leaf with an effective thickness of water of 300 μm would warm up to a very high temperature if all available solar energy were absorbed and no heat were lost. However, this does not occur because leaves absorb only about 50% of the total solar energy (300–3,000 nm), with most of the absorption occurring in the visible portion of the spectrum (see Figure 9.4). Yet the amount of the sun's

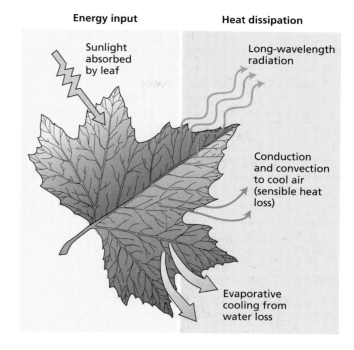

FIGURE 9.16 The absorption and dissipation of energy from sunlight by the leaf. The imposed heat load must be dissipated in order to avoid damage to the leaf. The heat load is dissipated by emission of long-wavelength radiation, by sensible heat loss to the air surrounding the leaf, and by the evaporative cooling caused by transpiration.

energy absorbed by leaves is still enormous, and this heat load is dissipated by the emission of long-wave radiation (at about 10,000 nm), by sensible (i.e., perceptible) heat loss, and by evaporative (or latent) heat loss (Figure 9.16):

- Radiative heat loss: All objects emit radiation in proportion to their temperature. However, the maximum wavelength is inversely proportional to its temperature, and leaf temperatures are low enough that the wavelengths emitted are not visible to the human eye.

- Sensible heat loss: Air circulation around the leaf removes heat from the leaf surfaces if the temperature of the leaf is higher than that of the air; the heat is convected from the leaf to the air.

- Latent heat loss: Evaporative heat loss occurs because the evaporation of water requires energy. Thus, as water evaporates from a leaf (transpiration), it withdraws large amounts of heat from the leaf and cools it. The human body is cooled by the same principle, through perspiration.

Sensible heat loss and evaporative heat loss are the most important processes in the regulation of leaf temperature, and the ratio of the two is called the **Bowen ratio** (Campbell 1977):

$$\text{Bowen ratio} = \frac{\text{Sensible heat loss}}{\text{Evaporative heat loss}}$$

In well-watered crops, transpiration (see Chapter 4), and hence water evaporation from the leaf, are high, so the Bowen ratio is low (see Web Topic 9.2). Conversely, when evaporative cooling is limited, the Bowen ratio is large. For example, in a water-stressed crop, partial stomatal closure reduces evaporative cooling and the Bowen ratio is increased. The amount of evaporative heat loss (and thus the Bowen ratio) is influenced by the degree to which stomata remain open.

Plants with very high Bowen ratios conserve water, but also endure very high leaf temperatures. However, the high temperature difference between the leaf and air does increase the amount of sensible heat loss. Reduced growth is usually correlated with these high Bowen ratios, because a high Bowen ratio is indicative of at least partial stomatal closure.

Photosynthesis is temperature sensitive

When photosynthetic rate is plotted as a function of temperature in a leaf with C_3 photosynthesis under ambient CO_2 concentrations, the curve has a characteristic bell shape (Figure 9.17A). The ascending arm of the curve represents a temperature-dependent stimulation of enzymatic activities; the flat top portion of the curve represents a temperature range over which temperature is optimum for photosynthesis; the descending arm is associated with temperature-sensitive deleterious effects, some of which are reversible while others are not.

Temperature affects all biochemical reactions of photosynthesis as well as membrane integrity in chloroplasts, so it is not surprising that the responses to temperature are complex. We can gain insight into the underlying mechanisms by comparing photosynthetic rates of C_3 leaves in air at normal and at high CO_2 concentrations. At high CO_2 (see Figure 9.17B), there is an ample supply of CO_2 at the carboxylation sites, and the rate of photosynthesis is limited primarily by biochemical reactions connected with electron transport (see Chapter 7). In these conditions, temperature changes have large effects on fixation rates.

At ambient CO_2 concentrations (see Figure 9.17A), photosynthesis is limited by the activity of rubisco, and the response reflects two conflicting processes: an increase in carboxylation rate as the temperature rises and a decrease in the affinity of rubisco for CO_2 (see Chapter 8). There is also evidence that rubisco activity decreases at high temperatures because of temperature effects on rubisco activase (see Chapter 8). These opposing effects dampen the temperature response of photosynthesis at ambient CO_2 concentrations.

By contrast, when photosynthetic rate is plotted as a function of temperature in a leaf with C_4 photosynthesis, the curves have a bell shape in both cases (see Figure 9.17), since photosynthesis is CO_2-saturated (as was discussed in Chapter 8). This is one of the reasons that leaves of C_4 plants tend to have a higher photosynthetic temperature optimum than do leaves of C_3 plants when grown under common conditions.

At low temperatures, photosynthesis can also be limited by other factors such as phosphate availability at the chloroplast (Sage and Sharkey 1987). When triose phosphates are exported from the chloroplast to the cytosol, an equimolar amount of inorganic phosphate is taken up via translocators in the chloroplast membrane.

If the rate of triose phosphate utilization in the cytosol decreases, phosphate uptake into the chloroplast is inhibited and photosynthesis becomes phosphate limited (Geiger and Servaites 1994). Starch synthesis and sucrose synthesis decrease rapidly with temperature, reducing the demand for triose phosphates and causing the phosphate limitation observed at low temperatures.

The highest photosynthetic rates seen in temperature responses represent the so-called *optimal temperature response*. When these temperatures are exceeded, photosynthetic rates decrease again. It has been argued that this optimal temperature is the point at which the capacities of the various steps of photosynthesis are optimally balanced, with some of the steps becoming limiting as the temperature decreases or increases. What factors are associated

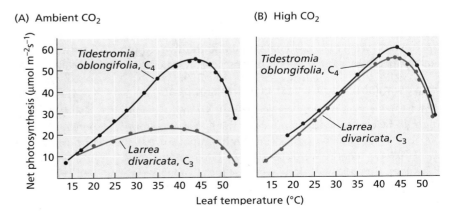

FIGURE 9.17 Changes in photosynthesis as a function of temperature at normal atmospheric CO_2 concentrations (A) and at high CO_2 concentrations, which saturate photosynthetic CO_2 assimilation (B). Photosynthesis depends strongly on temperature at saturating CO_2 concentrations. Note the significantly higher photosynthetic rates at saturating CO_2 concentrations. (After Berry and Björkman 1980).

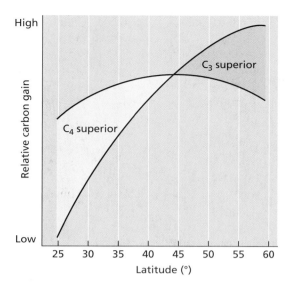

Figure 9.18 The relative rates of photosynthetic carbon gain predicted for identical C_3 and C_4 grass canopies as a function of latitude across the Great Plains of North America (After Ehleringer 1978).

with the decline in photosynthesis beyond the temperature optimum? Respiration rates increase as a function of temperature, but they are not the primary reason for the sharp decrease in net photosynthesis at high temperatures. Rather, membrane-bound electron transport processes become unstable at high temperatures, cutting off the supply of reducing power and leading to a sharp overall decrease in photosynthesis.

Optimal temperatures have strong genetic (adaptation) and environmental (acclimation) components. Plants of different species growing in habitats with different temperatures have different optimal temperatures for photosynthesis, and plants of the same species, grown at different temperatures and then tested for their photosynthetic responses, show temperature optima that correlate with the temperature at which they were grown. Plants growing at low temperatures maintain higher photosynthetic rates at low temperatures than plants grown at high temperatures.

These changes in photosynthetic properties in response to temperature play an important role in plant adaptations to different environments. Plants are remarkably plastic in their adaptations to temperature. In the lower temperature range, plants growing in alpine areas are capable of net CO_2 uptake at temperatures close to 0°C; at the other extreme, plants living in Death Valley, California, have optimal rates of photosynthesis at temperatures approaching 50°C.

Figure 9.9 shows changes in quantum yield as a function of temperature in a C_3 plant and in a C_4 plant. In the C_4 plant the quantum yield or light-use efficiency remains constant with temperature, reflecting typical low rates of photorespiration. In the C_3 plant the quantum yield decreases with temperature, reflecting a stimulation of photorespiration by temperature and an ensuing higher energy demand per net CO_2 fixed. While quantum yield effects are most expressed under light-limited conditions, a similar pattern is reflected in photorespiration rates under high light as a function of temperature. The combination of reduced quantum yield and increased photorespiration leads to expected differences in the photosynthetic capacities of C_3 and C_4 plants in habitats with different temperatures. The predicted relative rates of primary productivity of C_3 and C_4 grasses along a latitudinal transect in the Great Plains of North America from southern Texas in the USA to Manitoba in Canada (Ehleringer 1978) are shown in Figure 9.18. This decline in C_3 relative to C_4 productivity moving southward very closely parallels the actual abundances of plants with these pathways in the Great Plains: C_3 species are more common above 45°N, and C_4 species dominate below 40°N (Figure 9.18) (Web Topic 9.3).

Photosynthetic Responses to Carbon Dioxide

We have discussed how plant growth and leaf anatomy are influenced by light and temperature. Now we turn our attention to how CO_2 concentration affects photosynthesis. CO_2 diffuses from the atmosphere into leaves—first through stomata, then through the intercellular air spaces, and ultimately into cells and chloroplasts. In the presence of adequate amounts of light, higher CO_2 concentrations support higher photosynthetic rates. The reverse is also true: Low CO_2 concentrations can limit the amount of photosynthesis.

In this section we will discuss the concentration of atmospheric CO_2 in recent history, and its availability for carbon-fixing processes. Then we'll consider the limitations that CO_2 places on photosynthesis and the impact of the CO_2-concentrating mechanisms of C_4 plants.

Atmospheric CO_2 concentration keeps rising

Carbon dioxide is a trace gas in the atmosphere, presently accounting for about 0.038%, or 380 parts per million (ppm), of air. The partial pressure of ambient CO_2 (c_a) varies with atmospheric pressure and is approximately 38 pascals (Pa) at sea level (see Web Topic 9.4). Water vapor usually accounts for up to 2% of the atmosphere and O_2 for about 21%. The bulk of the atmosphere is nitrogen, at 77%.

The current atmospheric concentration of CO_2 is almost twice the concentration that has prevailed during most of the last 420,000 years, as measured from air bubbles trapped in glacial ice in Antarctica (Figure 9.19A). Today's atmospheric CO_2 is likely higher than Earth has experienced in the last 2 million years. Except for the last 200 years, atmospheric CO_2 concentrations during the recent geological past are thought to have been low, meaning that the plants in the world today evolved in a low-CO_2 world. The available evidence indicates that high CO_2 concentra-

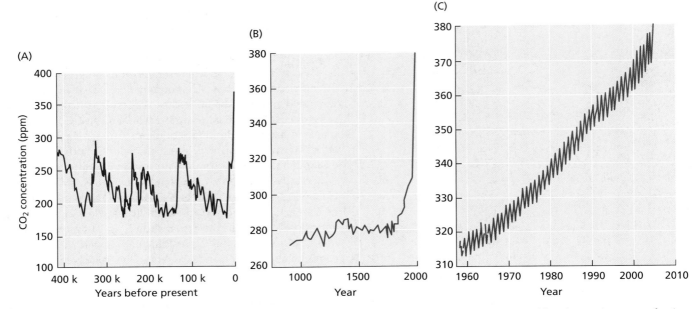

FIGURE 9.19 Concentration of atmospheric CO_2 from 420,000 years ago to the present. (A) Past atmospheric CO_2 concentrations, determined from bubbles trapped in glacial ice in Antarctica, were much lower than current levels. (B) In the last 1000 years, the rise in CO_2 concentration coincides with the Industrial Revolution and the increased burning of fossil fuels. (C) Current atmospheric concentrations of CO_2 measured at Mauna Loa, Hawaii, continue to rise. The wavy nature of the trace is caused by change in atmospheric CO_2 concentrations associated with seasonal changes in relative balance between photosynthesis and respiration rates. Each year the highest CO_2 concentration is observed in May, just before the Northern Hemisphere growing season, and the lowest concentration is observed in October. (After Barnola et al. 1994, Keeling and Whorf 1994, Neftel et al. 1994, and Keeling et al. 1995.)

tions (greater than 1,000 ppm) have not existed on Earth since the warm Cretaceous, over 70 million years ago. Thus, until the dawn of the Industrial Revolution, the geological trend over the past 50 to 70 million years had been one of decreasing atmospheric CO_2 concentrations (**Web Topic 9.5**).

The current CO_2 concentration of the atmosphere is increasing by about 1 to 3 ppm each year, primarily because of the burning of fossil fuels such as coal, oil, and natural gas (see Figure 9.19C). Since 1958, when C. David Keeling began systematic measurements of CO_2 in the clean air at Mauna Loa, Hawaii, atmospheric CO_2 concentrations have increased by more than 20% (Keeling et al. 1995). By 2100 the atmospheric CO_2 concentration could reach 600 to 750 ppm unless fossil fuel emissions are controlled (**Web Topic 9.6**).

THE GREENHOUSE EFFECT The consequences of this increase in atmospheric CO_2 are under intense scrutiny by scientists and government agencies, particularly because of predictions that the **greenhouse effect** is altering the world's climate. The term *greenhouse effect* refers to the resulting warming of Earth's climate, which is caused by the trapping of long-wavelength radiation by the atmosphere.

A greenhouse roof transmits visible light, which is absorbed by plants and other surfaces inside the greenhouse. The absorbed light energy is converted to heat, and part of it is re-emitted as long-wavelength radiation. Because glass transmits long-wavelength radiation very poorly, this radiation cannot leave the greenhouse through the glass roof, and the greenhouse heats up.

Certain gases in the atmosphere, particularly CO_2 and methane, play a role similar to that of the glass roof in a greenhouse. The increased CO_2 concentration and temperature associated with the greenhouse effect can influence photosynthesis. At current atmospheric CO_2 concentrations, photosynthesis in C_3 plants is CO_2 limited (as we will discuss later in the chapter), but this situation could change as atmospheric CO_2 concentrations continue to rise. Under laboratory conditions, most C_3 plants grow 30 to 60% faster when CO_2 concentration is doubled (to 600–750 ppm), and the growth rate changes depend on nutrient status (Bowes 1993).

CO_2 diffusion to the chloroplast is essential to photosynthesis

For photosynthesis to occur, carbon dioxide must diffuse from the atmosphere into the leaf and into the carboxylation site of rubisco. Because diffusion rates depend on concentration gradients in leaves (see Chapters 3 and 6), appropriate gradients are needed to ensure adequate diffusion of CO_2 from the leaf surface to the chloroplast.

The cuticle that covers the leaf is nearly impermeable to CO_2, so the main port of entry of CO_2 into the leaf is the stomatal pore. The same path is traveled in the reverse

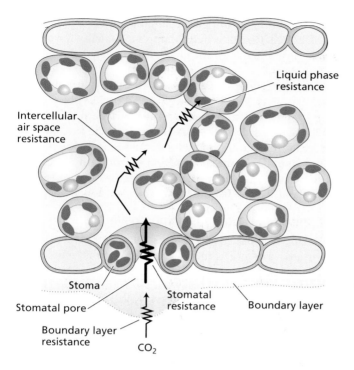

FIGURE 9.20 Points of resistance to the diffusion of CO_2 from outside the leaf to the chloroplasts. The stomatal pore is the major point of resistance to CO_2 diffusion.

The boundary layer consists of relatively unstirred air at the leaf surface, and its resistance to diffusion is called the **boundary layer resistance**. The magnitude of the boundary layer resistance decreases with leaf size and wind speed. The boundary layer resistance to water and CO_2 diffusion is physically related to the boundary layer resistance to sensible heat loss discussed earlier.

Smaller leaves have a lower boundary layer resistance to CO_2 and water diffusion, and to sensible heat loss. Leaves of desert plants are usually small, facilitating sensible heat loss. The large leaves often found in the shade of the humid Tropics can have large boundary layer resistances, but these leaves can dissipate the radiation heat load by evaporative cooling made possible by the abundant water supply in these habitats.

After diffusing through the boundary layer, CO_2 enters the leaf through the stomatal pores, which impose the next type of resistance in the diffusion pathway, the **stomatal resistance**. Under most conditions in nature, in which the air around a leaf is seldom completely still, the boundary layer resistance is much smaller than the stomatal resistance, and the main limitation to CO_2 diffusion is imposed by the stomatal resistance.

There is also a resistance to CO_2 diffusion in the air spaces that separate the substomatal cavity from the walls of the mesophyll cells, called the **intercellular air space resistance**. This resistance is also usually small—causing a drop of 0.5 Pa or less in partial pressure of CO_2, compared with the 38 Pa outside the leaf.

The resistance to CO_2 diffusion of the liquid phase in C_3 leaves—the **liquid phase resistance**, also called **mesophyll resistance**—encompasses diffusion from the intercellular leaf spaces to the carboxylation sites in the chloroplast. This point of resistance to CO_2 diffusion has been calculated as approximately one-tenth of the combined boundary layer resistance and stomatal resistance when the stomata are fully open. This low resistance value can be attributed to the localization of chloroplasts near the cell periphery, which minimizes the distance that CO_2 must diffuse through liquid to reach carboxylation sites within the chloroplast.

Because the stomatal pores usually impose the largest resistance to CO_2 uptake and water loss in the diffusion pathway, this single point of regulation provides the plant with an effective way to control gas exchange between the leaf and the atmosphere. In experimental measurements of gas exchange from leaves, the boundary layer resistance and the intercellular air space resistance are often ignored, and the stomatal resistance is used as the single parameter describing the gas phase resistance to CO_2 (see **Web Topic 9.4**).

direction by H_2O. CO_2 diffuses through the pore into the substomatal cavity and into the intercellular air spaces between the mesophyll cells. This portion of the diffusion path of CO_2 into the chloroplast is a gaseous phase. The remainder of the diffusion path to the chloroplast is a liquid phase, which begins at the water layer that wets the walls of the mesophyll cells and continues through the plasma membrane, the cytosol, and the chloroplast. (For the properties of CO_2 in solution, see **Web Topic 8.6**.)

Each portion of this diffusion pathway imposes a resistance to CO_2 diffusion, so the supply of CO_2 for photosynthesis meets a series of different points of resistance (Figure 9.20). An evaluation of the magnitude of each point of resistance is helpful for understanding CO_2 limitations to photosynthesis.

The sharing of the stomatal entry pathway by CO_2 and water presents the plant with a functional dilemma. In air of high relative humidity, the diffusion gradient that drives water loss is about 50 times larger than the gradient that drives CO_2 uptake. In drier air, this difference can be even larger. Therefore, a decrease in stomatal resistance through the opening of stomata facilitates higher CO_2 uptake but is unavoidably accompanied by substantial water loss.

Recall from Chapter 4 that the gas phase of CO_2 diffusion into the leaf can be divided into three components—the boundary layer, the stomata, and the intercellular spaces of the leaf—each of which imposes a resistance to CO_2 diffusion (see Figure 9.20).

Patterns of light absorption generate gradients of CO_2 fixation

We have discussed how leaf anatomy is specialized for capturing light and how it also facilitates the internal diffusion

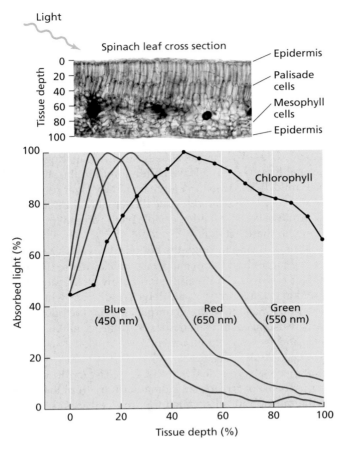

FIGURE 9.21 Distribution of absorbed light in spinach sun leaves. Irradiation with blue, green or red light results in different profiles of absorbed light in the leaf. The micrograph above the graph shows a cross section of a spinach leaf, with rows of palisade cells occupying nearly half of the leaf thickness. The shapes of the curves are in part a result of the unequal distribution of chlorophyll within the leaf tissues. (After Nishio et al. 1993 and Vogelmann and Han 2000; micrograph courtesy of T. Vogelmann.)

absorbs poorly in the green (again, see Figure 7.5), yet green light is very effective in supplying energy for photosynthesis in the tissues within the leaf depleted of blue and red photons.

The capacity of the leaf tissue for photosynthetic CO_2 assimilation depends to a large extent on its rubisco content. In spinach (*Spinacea oleracea*) and fava bean (*Vicia faba*), rubisco content starts out low at the top of the leaf, increases toward the middle, and then decreases again toward the bottom, similar to the distribution of chlorophyll in a leaf as shown in Figure 9.21. As a result, the distribution of photosynthetic carbon fixation within the leaf is bell-shaped.

CO_2 imposes limitations on photosynthesis

For many crops, such as tomatoes, lettuce, cucumbers, and roses growing in greenhouses under optimal water and nutrient nutrition, the carbon dioxide enrichment in the greenhouse environment above natural atmospheric levels results in increased productivity. Expressing photosynthetic rate as a function of the partial pressure of CO_2 in the intercellular air space (c_i) within the leaf (see **Web Topic 9.4**) makes it possible to evaluate limitations to photosynthesis imposed by CO_2 supply. At very low intercellular CO_2 concentrations, photosynthesis is strongly limited by the low CO_2.

Increasing intercellular CO_2 to the concentration at which these two processes balance each other defines the **CO_2 compensation point**, at which the net efflux of CO_2 from the leaf is zero (Figure 9.22). This concept is analogous to that of the light compensation point discussed earlier in

of CO_2, but where inside an individual leaf do maximum rates of photosynthesis occur? In most leaves, light is preferentially absorbed at the upper surface, whereas CO_2 enters through the lower surface. Given that light and CO_2 enter from opposing sides of the leaf, does photosynthesis occur uniformly within the leaf tissues, or is there a gradient in photosynthesis across the leaf?

For most leaves, once CO_2 has diffused through the stomata, internal CO_2 diffusion is rapid, so limitations on photosynthetic performance within the leaf are imposed by factors other than internal CO_2 supply. When white light enters the upper surface of a leaf, blue and red photons are preferentially absorbed by chloroplasts near the irradiated surface (Figure 9.21), owing to the strong absorption bands of chlorophyll in the blue and red regions of the spectrum (see Figure 7.5). Green light, on the other hand, penetrates deeper into the leaf. Compared to blue and red, chlorophyll

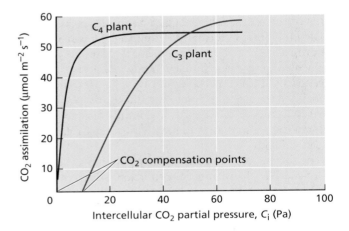

FIGURE 9.22 Changes in photosynthesis as a function of intercellular CO_2 concentrations in Arizona honeysweet (*Tidestromia oblongifolia*), a C_4 plant, and creosote bush (*Larrea divaricata*), a C_3 plant. Photosynthetic rate is plotted against calculated intercellular partial pressure of CO_2 inside the leaf (see Equation 5 in **Web Topic 9.4**). The partial pressure at which CO_2 assimilation is zero defines the CO_2 compensation point. (After Berry and Downton 1982.)

the chapter: *The CO_2 compensation point reflects the balance between photosynthesis and respiration as a function of CO_2 concentration, and the light compensation point reflects that balance as a function of photon flux under constant O_2 concentration.*

In C_3 plants, increasing atmospheric CO_2 above the compensation point stimulates photosynthesis over a wide concentration range (see Figure 9.22). At low to intermediate CO_2 concentrations, photosynthesis is limited by the carboxylation capacity of rubisco. At high CO_2 concentrations, photosynthesis becomes limited by the capacity of the Calvin cycle to regenerate the acceptor molecule ribulose-1,5-bisphosphate, which depends on electron transport rates. However, photosynthesis continues to increase with CO_2 because carboxylation replaces oxygenation on rubisco. By regulating stomatal conductance, most leaves appear to regulate their c_i (internal partial pressure for CO_2) such that it is at an intermediate concentration between limitations imposed by carboxylation capacity and limits in the capacity to regenerate ribulose-1,5-bisphosphate.

A plot of CO_2 assimilation as a function intercellular partial pressures of CO_2 tells us how photosynthesis is regulated by CO_2, independent of the functioning of stomata (see Figure 9.22). Inspection of such a plot for C_3 and C_4 plants reveals interesting differences between the two pathways of carbon metabolism:

- In C_4 plants, photosynthetic rates saturate at c_i values of about 15 Pa, reflecting the effective CO_2-concentrating mechanisms operating in these plants (see Chapter 8).
- In C_3 plants, increasing c_i levels continue to stimulate photosynthesis over a much broader CO_2 range.
- In C_4 plants, the CO_2 compensation is zero or nearly zero, reflecting their very low levels of photorespiration (see Chapter 8).
- In C_3 plants, the CO_2 compensation point is about 10 Pa, reflecting CO_2 production because of photorespiration (see Chapter 8).

These responses indicate that C_3 plants may benefit more from ongoing increases in today's atmospheric CO_2 concentrations (see Figure 9.19). In contrast, photosynthesis in C_4 plants is CO_2-saturated at low concentrations, and as a result C_4 plants do not benefit much from increases in atmospheric CO_2 concentrations. In fact, the ancestral photosynthetic pathway is C_3 photosynthesis, and C_4 photosynthesis is a derived pathway. During geologically historical time periods when atmospheric CO_2 concentrations were very much higher than they are today, CO_2 diffusion through stomata into C_3 leaves would have resulted in higher c_i values and therefore higher photosynthetic rates. While C_3 photosynthesis is typically CO_2-diffusion limited today, C_3 plants still account for nearly 70% of the world's primary productivity. The evolution of C_4 photosynthesis is one biochemical adaptation to overcome a CO_2-limited

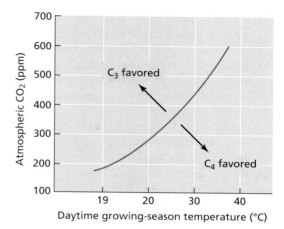

FIGURE 9.23 The combination of atmospheric carbon dioxide levels and daytime growing season temperatures that are predicted to favor C_3 versus C_4 grasses. At any point in time, the Earth is at a single atmospheric carbon dioxide concentration, resulting in the expectation that C_4 plants would be most common in the hottest growing–season habitats. (After Ehleringer et al. 1997.)

atmosphere. Our current understanding is that C_4 photosynthesis may have evolved recently, some 10 to 15 million years ago.

If the ancestral Earth more than 50 million years ago was one of elevated atmospheric CO_2 concentrations that were well above current atmospheric conditions, under what atmospheric conditions might we have expected that C_4 photosynthesis should become a major photosynthetic pathway found in the Earth's ecosystems? Ehleringer et al. (1997) suggest that C_4 photosynthesis first became a prominent component of terrestrial ecosystems in the warmest growing regions of the Earth when global CO_2 concentrations decreased below some critical and as-yet unknown threshold CO_2 concentration (Figure 9.23). That is, the negative impacts of high photorespiration and CO_2 limitation on C_3 photosynthesis would be greatest under warm to hot growing conditions, especially when atmospheric CO_2 is reduced. The C_4-favorable growing areas would have been located in those geographic regions with the warmest temperatures. C_4 plants would have been most favored during periods of the Earth's history when CO_2 levels were lowest. In today's world, these regions are the subtropical grasslands and savannas. There are now extensive data to indicate that C_4 photosynthesis was more prominent during the glacial periods when atmospheric CO_2 levels were below 200 ppm than it is today (see Figure 9.19). Other factors may have contributed to the expansion of C_4 plants, but certainly low atmospheric CO_2 was one important factor favoring their geographic expansion.

Because of the CO_2-concentrating mechanisms in C_4 plants, CO_2 concentration at the carboxylation sites within C_4 chloroplasts is often saturating for rubisco activity. As a result, plants with C_4 metabolism need less rubisco than C_3 plants to achieve a given rate of photosynthesis, and require less nitrogen to grow (von Caemmerer 2000).

In addition, the CO_2-concentrating mechanism allows the leaf to maintain high photosynthetic rates at lower c_i values, which require lower rates of stomatal conductance for a given rate of photosynthesis. Thus, C_4 plants can use water and nitrogen more efficiently than C_3 plants can. On the other hand, the additional energy cost of the concentrating mechanism (see Chapter 8) makes C_4 plants less efficient in their utilization of light. This is probably one of the reasons that most shade-adapted plants in temperate regions are C_3 plants.

Crassulacean Acid Metabolism

Many cacti, orchids, bromeliads, and other succulent plants with crassulacean acid metabolism (CAM) have stomatal activity patterns that contrast with those found in C_3 and C_4 plants. CAM plants open their stomata at night and close them during the day, exactly the opposite of the pattern observed in guard cells in leaves of C_3 and C_4 plants (Figure 9.24). At night, atmospheric CO_2 diffuses into CAM plants where it is combined with phosphoenolpyruvate and fixed into malate (see Chapter 8).

The ratio of water loss to CO_2 uptake is much lower in CAM plants than it is in either C_3 or C_4 plants. This is because stomata are open only at night when temperatures are lower and humidities higher than daytime conditions, both of which contribute to a lower transpiration rate.

The main photosynthetic constraint on CAM metabolism is that the capacity to store malic acid is limited, and this limitation restricts the total amount of CO_2 uptake. However, some CAM plants are able to enhance total photosynthesis during wet conditions by fixing CO_2 via the Calvin cycle at the end of the day, when temperature gradients are less extreme. In water-limited conditions, stomata then only open at night.

Cladodes (flattened stems) of cacti can survive after detachment from the plant for several months without water. Their stomata are closed all the time, and the CO_2 released by respiration is refixed into malate. This process, which has been called *CAM idling*, also allows the intact plant to survive for prolonged drought periods while losing remarkably little water.

Carbon isotope ratio variations reveal different photosynthetic pathways

We can learn more about the different photosynthetic pathways in plants by measuring their chemical composition. That is, the stable isotopes of carbon atoms in a leaf contain useful information about photosynthesis. The two stable isotopes of carbon are ^{12}C and ^{13}C, differing only in composition by the addition of an additional neutron in ^{13}C.

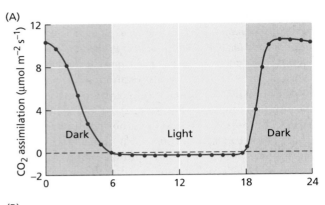

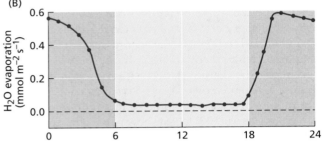

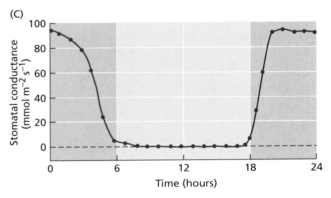

FIGURE 9.24 Photosynthetic carbon assimilation, evaporation, and stomatal conductance of a CAM plant, the cactus *Opuntia ficus-indica*, during a 24-hour period. The whole plant was kept in a gas exchange chamber in the laboratory. The dark period is indicated by shaded areas. Three parameters were measured over the study period: (A) photosynthetic rate, (B) water loss, and (C) stomatal conductance. In contrast to plants with C_3 or C_4 metabolism, CAM plants open their stomata and fix CO_2 at night. (After Gibson and Nobel 1986.)

How do we measure the carbon isotopes of plants?

Atmospheric CO_2 contains the naturally occurring stable carbon isotopes ^{12}C and ^{13}C in the proportions 98.9% and 1.1%, respectively. $^{14}CO_2$ is radioactive and is present in small quantities ($10^{-10}\%$). The chemical properties of $^{13}CO_2$ are identical to those of $^{12}CO_2$, but plants assimilate less $^{13}CO_2$ than $^{12}CO_2$. In other words, leaves discriminate against the heavier isotope of carbon during photosynthesis, and therefore they have smaller $^{13}C/^{12}C$ ratios than are found in atmospheric CO_2.

The $^{13}C/^{12}C$ isotope composition is measured by use of a mass spectrometer, which yields the following ratio:

$$R = \frac{^{13}CO_2}{^{12}CO_2} \qquad (9.1)$$

The **carbon isotope ratio** of plants, $\delta^{13}C$, is quantified on a per mil (‰) basis:

$$\delta^{13}C\,^0/_{00} = \left(\frac{R_{sample}}{R_{standard}} - 1\right) \times 1000 \qquad (9.2)$$

where the standard represents the carbon isotopes contained in a fossil belemnite from the Pee Dee limestone formation of South Carolina. The $\delta^{13}C$ of atmospheric CO_2 has a value of −8‰, meaning that there is less ^{13}C in atmospheric CO_2 than is found in the carbonate of the belemnite standard.

What are some typical values for carbon isotope ratios of plants? C_3 plants have a $\delta^{13}C$ value of about −28‰; C_4 plants have an average value of −14‰ (Farquhar et al. 1989). Both C_3 and C_4 plants have less ^{13}C than does CO_2 in the atmosphere, which means that they discriminate against ^{13}C during the photosynthetic process. Cerling et al. (1997) provided $\delta^{13}C$ data for a large number of C_3 and C_4 plants from around the world (Figure 9.25). What becomes clear from this figure is that there is a wide spread of $\delta^{13}C$ values in C_3 and C_4 plants values from averages of −28‰ and −14‰, respectively. These $\delta^{13}C$ variations actually reflect the consequences of small variations in physiology associated with changes in stomatal conductance in different environmental conditions. Thus, $\delta^{13}C$ values can be used both to distinguish between C_3 and C_4 photosynthesis and then to further reveal details about stomatal conditions for plants grown in different environmental conditions (such as the tropics versus deserts).

Differences in carbon isotope ratio are easily detectable with mass spectrometers that allow for very precise measurements of the abundance of ^{12}C and ^{13}C in either different molecules or different tissues. Many of our foods are products of C_3 plants, such as wheat (*Triticum aestivum*), rice (*Oryza sativa*), potatoes (*Solanum tuberosum*), and beans (*Phaseoulus* spp.). Yet many of our most productive crops are C_4 plants, such as corn (maize; *Zea mays*), sugarcane (*Saccharum officinarum*), and sorghum (*Sorghum bicolor*). Carbohydrates extracted from each of these foods may be chemically identical, but they are C_3–C_4 distinguishable on the basis of their $\delta^{13}C$ values. For example, measuring the $\delta^{13}C$ values of table sugar (sucrose) makes it possible to determine if the sucrose came from sugar beet (*Beta vulgaris*; a C_3 plant) or sugarcane (a C_4 plant) (see **Web Topic 9.7**).

Why are there carbon isotope ratio variations in plants?

What is the physiological basis for ^{13}C depletion in plants relative to CO_2 in the atmosphere? It turns out that both the diffusion of CO_2 into the leaf and the carboxylation selectivity for $^{12}CO_2$ play a role.

CO_2 diffuses from air outside of the leaf to the carboxylation sites within leaves in both C_3 and C_4 plants. Because $^{12}CO_2$ is lighter than $^{13}CO_2$, it diffuses slightly faster toward the carboxylation site, creating an effective diffusion fractionation factor of −4.4‰. Thus, we would expect leaves to have a more negative $\delta^{13}C$ value simply because of this diffusion effect. Yet this factor alone is not sufficient to explain the $\delta^{13}C$ values of C_3 plants as shown in Figure 9.25.

The initial carboxylation event is a determining factor in the carbon isotope ratio of plants. Rubisco represents the first carboxylation reaction in C_3 photosynthesis and has an intrinsic discrimination value against ^{13}C of −30‰. By contrast, PEP carboxylase, the primary CO_2 fixation enzyme of C_4 plants, has a much smaller isotope discrimination effect—about −2‰. Thus, the inherent difference between the two carboxylating enzymes contributes to the different isotope ratio differences observed in C_3 and C_4 plants (Farquhar et al. 1989).

Other physiological characteristics of plants affect its carbon isotope ratio. One primary factor is the partial pressure of CO_2 in the intercellular air spaces of leaves (c_i). In C_3 plants the potential isotope discrimination by rubisco of −30‰ is not fully expressed during photosynthesis because the availability of CO_2 at the carboxylation site becomes a limiting factor restricting the discrimination by rubisco. Greater discrimination against $^{13}CO_2$ occurs when c_i is high, as when stomata are open. Yet open stomata also

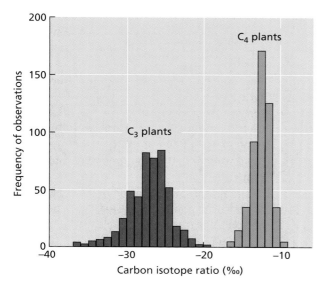

FIGURE 9.25 Frequency histograms for the observed carbon isotope ratios in C_3 and C_4 taxa from around the world. (After Cerling et al. 1997.)

facilitate water loss. Thus, lower ratios of photosynthesis to transpiration are correlated with greater discrimination against ^{13}C (Ehleringer et al. 1993). When leaves are exposed to water stress, stomata tend to close, reducing c_i values. As a consequence, C_3 plants grown under water stress conditions tend to have more positive carbon isotope ratios.

Measuring $\delta^{13}C$ in fossil, carbonate-containing soils and fossil teeth makes it possible to reconstruct the photosynthetic pathways present at times in the ancient past. These approaches have been used to determine that C_4 photosynthesis developed and became prevalent about 6 million years ago and to reconstruct the diets of ancient and modern animals (see **Web Topic 9.8**).

CAM plants can have $\delta^{13}C$ values that are very near those of C_4 plants. In CAM plants that fix CO_2 at night via PEP carboxylase, $\delta^{13}C$ is expected to be similar to that of C_4 plants. However, when some CAM plants are well watered, they can switch to C_3 mode by opening their stomata and fixing CO_2 during the day via rubisco. Under these conditions the isotope composition shifts more toward that of C_3 plants. Thus the $\delta^{13}C$ values of CAM plants reflect how much carbon is fixed via the C_3 pathway versus the C_4 pathway.

Summary

Photosynthetic activity in the intact leaf is an integral process that depends on many biochemical reactions. Different environmental factors can limit photosynthetic rates.

Leaf anatomy is highly specialized for light absorption, and the properties of palisade and mesophyll cells ensure light absorption throughout the leaf. In addition to the anatomic features of the leaf, chloroplast movements within cells and solar tracking by the leaf blade help maximize light absorption. Light transmitted through upper leaves is absorbed by leaves growing beneath them.

Many properties of the photosynthetic apparatus change as a function of the available light, including the light compensation point, which is higher in sun leaves than in shade leaves. The linear portion of the light-response curve for photosynthesis provides a measure of the quantum yield of photosynthesis in the intact leaf. In temperate areas, quantum yields of C_3 plants are generally higher than those of C_4 plants.

Sunlight imposes on the leaf a substantial heat load, which is dissipated back into the air by long-wavelength radiation, by sensible heat loss, or by evaporative heat loss. Increasing CO_2 concentrations in the atmosphere are increasing the heat load on the biosphere. This process could cause damaging changes in the world's climate, but it could also reduce the CO_2 limitations on photosynthesis. At high photon flux, photosynthesis in most plants is CO_2 limited, but the limitation is substantially lower in C_4 and CAM plants because of their CO_2-concentrating mechanisms.

Diffusion of CO_2 into the leaf is constrained by a series of different points of resistance. The largest resistance is usually that imposed by the stomata, so modulation of stomatal apertures provides the plant with an effective means of controlling water loss and CO_2 uptake. Both stomatal and non-stomatal factors affect CO_2 limitations on photosynthesis.

Temperature responses of photosynthesis reflect the temperature sensitivity of the biochemical reactions of photosynthesis and are most pronounced at high CO_2 concentrations. Because of the role of photorespiration, the quantum yield is strongly dependent on temperature in C_3 plants but is nearly independent of temperature in C_4 plants.

Leaves growing in cold climates can maintain higher photosynthetic rates at low temperatures than leaves growing in warmer climates. Leaves grown at high temperatures perform better at high temperatures than leaves grown at low temperatures do. Functional changes in the photosynthetic apparatus in response to prevailing temperatures in the environment have an important effect on the ability of plants to live in diverse habitats.

The carbon isotope ratios of leaves can be used to distinguish photosynthetic pathway differences among different plant species. Within each photosynthetic pathway, variations in the carbon isotope ratios reveal information about environmental factors such as water stress.

Web Material

Web Topics

9.1 Working with Light
Amount, direction, and spectral quality are important parameters for the measurement of light.

9.2 Heat Dissipation from Leaves: The Bowen Ratio
Sensible heat loss and evaporative heat loss are the most important processes in the regulation of leaf temperature.

9.3 The Geographic Distributions of C_3 and C_4 Plants
The geographic distribution of C_3 and C_4 plants corresponds closely with growing season temperature in today's world.

9.4 Calculating Important Parameters in Leaf Gas Exchange
Gas exchange methods allow us to measure photosynthesis and stomatal conductance in the intact leaf.

9.5 Prehistoric Changes in Atmospheric CO_2
Over the past 800,000 years, atmospheric CO_2 levels changed between 180 ppm (glacial periods) and 280 ppm (interglacial periods) as Earth moved between ice ages.

9.6 Projected Future Increases in Atmospheric CO_2
In 2005 atmospheric CO_2 reached 379 ppm and is expected to reach 400 ppm by 2015.

9.7 Using Carbon Isotopes to Detect Adulteration in Foods
Carbon isotopes are frequently used to detect the substitution of C_4 sugars into C_3 food products, such as the introduction of sugar cane into honey to increase yield.

9.8 Reconstruction of the Expansion of C_4 Taxa
The $\delta^{13}C$ of animal teeth faithfully record the carbon isotope ratios of food sources and can be used to reconstruct the abundances of C_3 and C_4 plants eaten by mammalian grazers.

Web Essay

9.1 The Xanthophyll Cycle
Molecular and biophysical studies are revealing the role of the xanthophyll cycle on the photoprotection of leaves.

Chapter References

Adams, W. W., Demmig-Adams, B., Rosenstiel, T. N., and Ebbert, V. (2001) Dependence of photosynthesis and energy dissipation activity upon growth form and light environment during the winter. *Photosynth. Res.* 67: 51–62.

Anderson, J. M. (1986) Photoregulation of the composition, function, and structure of thylakoid membranes. *Annu. Rev. Plant Physiol.* 37: 93–136.

Barnola, J. M., Raynaud, D., Lorius, C., and Korothevich, Y. S. (1994) Historical CO_2 record from the Vostok ice core. In *Trends '93: A Compendium of Data on Global Change* (ORNL/CDIAC-65), T. A. Boden, D. P. Kaiser, R. J. Sepanski, and F. W. Stoss, eds., Carbon Dioxide Information Center, Oak Ridge National Laboratory, Oak Ridge, TN, pp. 7–10.

Berry, J., and Björkman, O. (1980) Photosynthetic response and adaptation to temperature in higher plants. *Annu. Rev. Plant Physiol.* 31: 491–543.

Berry, J. A., and Downton, J. S. (1982) Environmental regulation of photosynthesis. In *Photosynthesis: Development, Carbon Metabolism and Plant Productivity*, Vol. II, Govindjee, ed., Academic Press, New York, pp. 263–343.

Björkman, O. (1981) Responses to different quantum flux densities. In *Encyclopedia of Plant Physiology*, New Series, Vol. 12A, O. L. Lange, P. S. Nobel, C. B. Osmond, and H. Zeigler, eds., Springer, Berlin, pp. 57–107.

Björn, L. O., and Vogelmann, T. C. (1994) Quantification of light. In *Photomorphogenesis in Plants*, 2nd ed., R. E. Kendrick and G. H. M. Kronenberg, eds., Kluwer, Dordrecht, Netherlands, pp. 17–25.

Bowes, G. (1993) Facing the inevitable: Plants and increasing atmospheric CO_2. *Annu. Rev. Plant Physiol. Plant Mol. Biol.* 44: 309–332.

Campbell, G. S. (1977) *An Introduction to Environmental Biophysics*. Springer-Verlag, New York.

Cerling, T. E., Harris, J. M., MacFadden, B. J., Leakey, M. G., Quade, J., Eisenmann, V., and Ehleringer, J. R. (1997) Global vegetation change through the Miocene–Pliocene boundary. *Nature* 389: 153–158.

Demmig-Adams, B., and Adams, W. (1996) The role of xanthophyll cycle carotenoids in the protection of photosynthesis. *Trends Plant Sci.* 1: 21–26.

Ehleringer, J. R. (1978) Implications of quantum yield differences on the distributions of C_3 and C_4 grasses. *Oecologia* 31: 255–267.

Ehleringer, J. R., and Björkman, O. (1977) Quantum yields for CO_2 uptake in C_3 and C_4 plants. *Plant Physiol.* 59: 86–90.

Ehleringer, J. R., and Forseth, I. (1980) Solar tracking by plants. *Science* 210: 1094–1098.

Ehleringer, J. R., Björkman, O., and Mooney, H. A. (1976) Leaf pubescence: Effects on absorptance and photosynthesis in a desert shrub. *Science* 192: 376–377.

Ehleringer, J. R., Sage, R. F., Flanagan, L. B., and Pearcy, R. W. (1991) Climate change and the evolution of C_4 photosynthesis. *Trends Ecol. Evol.* 6: 95–99.

Ehleringer, J. R., Hall, A. E., and Farquhar, G. D. (eds.) (1993) *Stable Isotopes and Plant Carbon-Water Relations*. Academic Press, San Diego, CA.

Ehleringer, J. R., Cerling, T. E., and Helliker, B. R. (1997) C_4 photosynthesis, atmospheric CO_2, and climate. *Oecologia* 112: 285–299.

Farquhar, G. D., and Sharkey, T. D. (1982) Stomatal conductance and photosynthesis. *Annu. Rev. Plant Physiol.* 33: 317–345.

Farquhar, G. D., Ehleringer, J. R., and Hubick, K. T. (1989) Carbon isotope discrimination and photosynthesis. *Annu. Rev. Plant Physiol. Plant Mol. Biol.* 40: 503–538.

Geiger, D. R., and Servaites, J. C. (1994) Diurnal regulation of photosynthetic carbon metabolism in C_3 plants. *Annu. Rev. Plant Physiol. Plant Mol. Biol.* 45: 235–256.

Gibson, A. C., and Nobel, P. S. (1986) *The Cactus Primer*. Harvard University Press, Cambridge, MA.

Gorton, H. L., Williams, W. E., and Vogelmann, T. C. (1999) Chloroplast movement in *Alocasia macrorrhiza*. *Physiol. Plant.* 106: 421–428.

Harvey, G. W. (1979) Photosynthetic performance of isolated leaf cells from sun and shade plants. *Carnegie Inst. Washington Yearbook* 79: 161–164.

Haupt, W., and Scheuerlein, R. (1990) Chloroplast movement. *Plant Cell Environ.* 13: 595–614.

Havaux, M., Tardy, F., Ravenel, J., Chanu, D., and Parot, P. (1996) Thylakoid membrane stability to heat stress studied by flash spectroscopic measurements of the electrochromic shift in intact potato leaves: Influence of the xanthophyll content. *Plant Cell Environ.* 19: 1359–1368.

Jarvis, P. G., and Leverenz, J. W. (1983) Productivity of temperate, deciduous and evergreen forests. In *Encyclopedia of Plant Physiology*, New Series, Vol. 12D, O. L. Lange, P. S. Nobel, C. B. Osmond, and H. Zeigler, eds., Springer, Berlin, pp. 233–280.

Keeling, C. D., and Whorf, T. P. (1994) Atmospheric CO_2 records from sites in the SIO air sampling network. In *Trends '93: A Compendium of Data on Global Change* (ORNL/CDIAC-65), T. A. Boden, D. P. Kaiser, R. J. Sepanski, and F. W. Stoss, eds., Carbon Dioxide Information Center, Oak Ridge National Laboratory, Oak Ridge, TN, pp. 16–26.

Keeling, C. D., Whorf, T. P., Wahlen, M., and Van der Plicht, J. (1995) Interannual extremes in the rate of rise of atmospheric carbon dioxide since 1980. *Nature* 375: 666–670.

Koller, D. (1990) Light-driven leaf movements. *Plant Cell Environ.* 13: 615–632.

Koller, D. (2000) Plants in search of sunlight. *Adv. Bot. Res.* 33: 35–131.

Long, S. P., Humphries, S., and Falkowski, P. G. (1994) Photoinhibition of photosynthesis in nature. *Annu. Rev. Plant Physiol. Plant Mol. Biol.* 45: 633–662.

McCree, K. J. (1981) Photosynthetically active radiation. In *Encyclopedia of Plant Physiology*, New Series, Vol. 12A, O. L. Lange, P. S. Nobel, C. B. Osmond, and H. Zeigler, eds., Springer, Berlin, pp. 41–55.

Melis, A. (1996) Excitation energy transfer: Functional and dynamic aspects of Lhc (cab) proteins. In *Oxygenic Photosynthesis: The Light Reactions*, D. R. Ort and C. F. Yocum, eds., Kluwer, Dordrecht, Netherlands, pp. 523–538.

Neftel, A., Friedle, H., Moor, E., Lötscher, H., Oeschger, H., Siegenthaler, U., and Stauffer, B. (1994) Historical CO_2 record from the Siple Station ice core. In *Trends '93: A Compendium of Data on Global Change* (ORNL/CDIAC-65), T. A. Boden, D. P. Kaiser, R. J. Sepanski, and F. W. Stoss, eds., Carbon Dioxide Information Center, Oak Ridge National Laboratory, Oak Ridge, TN, pp. 11–15.

Nishio, J. N., Sun, J., and Vogelmann, T. C. (1993) Carbon fixation gradients across spinach leaves do not follow internal light gradient. *Plant Cell* 5: 953–961.

Ort, D. R., and Baker, N. R. (1988) Consideration of photosynthetic efficiency at low light as a major determinant of crop photosynthetic performance. *Plant Physiol. Biochem.* 26: 555–565.

Osmond, C. B. (1994) What is photoinhibition? Some insights from comparisons of shade and sun plants. In *Photoinhibition of Photosynthesis: From Molecular Mechanisms to the Field*. N. R. Baker and J. R. Bowyer, eds., BIOS Scientific, Oxford, pp. 1–24.

Pearcy, R. W., Gross, L. J., and He, D. (1997) An improved dynamic model of photosynthesis for estimation of carbon gain in sunfleck light regimes. *Plant Cell Environ.* 20: 411–424.

Richter, T., and Fukshansky, L. (1996) Optics of a bifacial leaf: 2. Light regime as affected by leaf structure and the light source. *Photochem. Photobiol.* 63: 517–527.

Rupert, C. S., and Letarjet, R. (1978) Toward a nomenclature and dosimetric scheme applicable to all radiations. *Photochem. Photobiol.* 28: 3–5.

Sage, R. F., and Sharkey, T. D. (1987) The effect of temperature on the occurrence of O_2 and CO_2 insensitive photosynthesis in field grown plants. *Plant Physiol.* 84: 658–664.

Smith, H. (1986). The perception of light quality. In *Photomorphogenesis in Plants*, R. E. Kendrick and G. H. M. Kronenberg, eds., Nijhoff, Dordrecht, Netherlands, pp. 187–217.

Smith, H. (1994). Sensing the light environment: The functions of the phytochrome family. In *Photomorphogenesis in Plants*, 2nd ed., R. E. Kendrick and G. H. M. Kronenberg, eds., Nijhoff, Dordrecht, Netherlands, pp. 377–416.

Terashima, I. (1992) Anatomy of non-uniform leaf photosynthesis. *Photosynth. Res.* 31: 195–212.

Terashima, I., and Hikosaka, K. (1995) Comparative ecophysiology of leaf and canopy photosynthesis. *Plant Cell Environ.* 18: 1111–1128.

Tlalka, M., and Fricker, M. (1999) The role of calcium in blue-light-dependent chloroplast movement in *Lemna trisulca* L. *Plant J.* 20: 461–473.

Vogelmann, T. C. (1993) Plant tissue optics. *Annu. Rev. Plant Physiol. Plant Mol. Biol.* 44: 231–251.

Vogelmann, T. C., and Björn, L. O. (1983) Response to directional light by leaves of a sun-tracking lupine (*Lupinus succulentus*). *Physiol. Plant.* 59: 533–538.

Vogelmann, T. C., and Han, T. (2000) Measurement of gradients of absorbed light in spinach leaves from chlorophyll fluorescence profiles. *Plant Cell Environ.* 23: 1303–1311.

Vogelmann, T. C., Bornman, J. F., and Yates, D. J. (1996) Focusing of light by leaf epidermal cells. *Physiol. Plant.* 98: 43–56.

von Caemmerer, S. (2000) *Biochemical Models of Leaf Photosynthesis*. CSIRO, Melbourne, Australia.

Chapter 10 | Translocation in the Phloem

SURVIVAL ON LAND poses some serious challenges to terrestrial plants; foremost among these challenges is the need to acquire and retain water. In response to such environmental pressures, plants evolved roots and leaves. Roots anchor the plant and absorb water and nutrients; leaves absorb light and exchange gases. As plants increased in size, the roots and leaves became increasingly separated from each other in space. Thus, systems evolved for long-distance transport that allowed the shoot and the root to efficiently exchange products of absorption and assimilation.

You will recall from Chapters 4 and 6 that the xylem is the tissue that transports water and minerals from the root system to the aerial portions of the plant. The **phloem** is the tissue that translocates the products of photosynthesis from mature leaves to areas of growth and storage, including the roots. Sugars are the most significant transported products of photosynthesis.

The phloem also transmits signals between sources and sinks in the form of regulatory molecules, and redistributes water and various compounds throughout the plant body. All of these molecules appear to move with the transported sugars. The compounds to be redistributed, some of which initially arrive in the mature leaves via the xylem, can be either transferred out of the leaves without modification or metabolized before redistribution.

The discussion that follows emphasizes translocation in the phloem of angiosperms, because most of the research has been conducted on that group of plants. Gymnosperms will

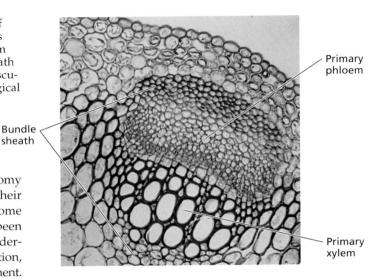

FIGURE 10.1 Transverse section of a vascular bundle of trefoil, a clover (*Trifolium*). (130×) The primary phloem is toward the outside of the stem. Both the primary phloem and the primary xylem are surrounded by a bundle sheath of thick-walled sclerenchyma cells, which isolate the vascular tissue from the ground tissue. (© J. N. A. Lott/Biological Photo Service.)

be compared briefly to angiosperms in terms of the anatomy of their conducting cells and possible differences in their mechanism of translocation. First we will examine some aspects of translocation in the phloem that have been researched extensively and are thought to be well understood, including the pathway and patterns of translocation, materials translocated in the phloem, and rates of movement.

In the second part of the chapter we will explore aspects of translocation in the phloem that need further investigation. These areas include phloem loading and unloading and the allocation and partitioning of photosynthetic products. Finally, we will explore an area of intensive research at present, the role of the phloem as a transport pathway for signaling molecules such as proteins and RNA.

Pathways of Translocation

The two long-distance transport pathways—the phloem and the xylem—extend throughout the plant body. The phloem is generally found on the outer side of both primary and secondary vascular tissues (Figures 10.1 and 10.2). In plants with secondary growth the phloem constitutes the inner bark.

The cells of the phloem that conduct sugars and other organic materials throughout the plant are called **sieve elements**. *Sieve element* is a comprehensive term that includes both the highly differentiated **sieve tube elements** typical of the angiosperms and the relatively unspecialized **sieve cells** of gymnosperms. In addition to sieve elements, the phloem tissue contains companion cells (discussed below) and parenchyma cells (which store and release food molecules). In some cases the phloem tissue also includes fibers and sclereids (for protection and strengthening of the tissue) and laticifers (latex-containing cells). However, only the sieve elements are directly involved in translocation.

The small veins of leaves and the primary vascular bundles of stems are often surrounded by a **bundle sheath** (see Figure 10.1), which consists of one or more layers of compactly arranged cells. (You will recall the bundle sheath cells involved in C_4 metabolism discussed in Chapter 8.) In the vascular tissue of leaves, the bundle sheath surrounds the small veins all the way to their ends, isolating the veins from the intercellular spaces of the leaf.

We will begin our discussion of translocation pathways with the experimental evidence demonstrating that the sieve elements are the conducting cells in the phloem. Then we will examine the structure and physiology of these unusual plant cells.

Sugar is translocated in phloem sieve elements

Early experiments on phloem transport date back to the nineteenth century, indicating the importance of long-dis-

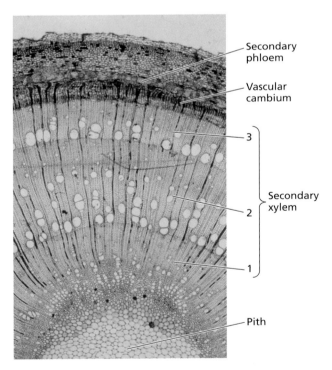

FIGURE 10.2 Transverse section of a 3-year-old stem of an ash (*Fraxinus excelsior*) tree. (27×) The numbers 1, 2, and 3 indicate growth rings in the secondary xylem. The old secondary phloem has been crushed by expansion of the xylem. Only the most recent (innermost) layer of secondary phloem is functional. (© P. Gates/Biological Photo Service.)

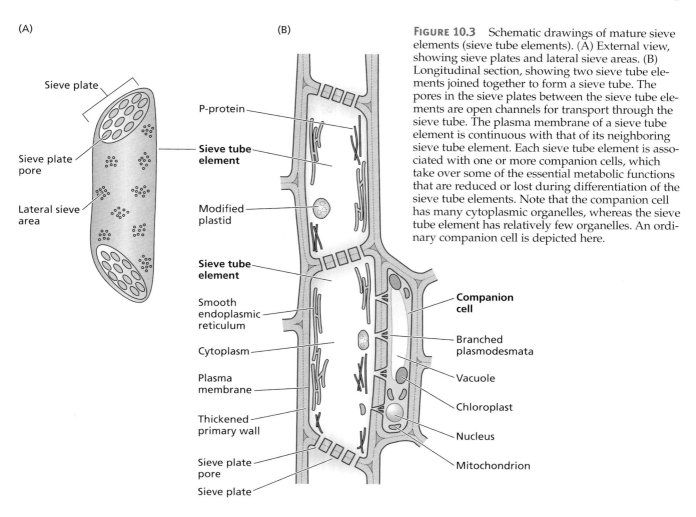

FIGURE 10.3 Schematic drawings of mature sieve elements (sieve tube elements). (A) External view, showing sieve plates and lateral sieve areas. (B) Longitudinal section, showing two sieve tube elements joined together to form a sieve tube. The pores in the sieve plates between the sieve tube elements are open channels for transport through the sieve tube. The plasma membrane of a sieve tube element is continuous with that of its neighboring sieve tube element. Each sieve tube element is associated with one or more companion cells, which take over some of the essential metabolic functions that are reduced or lost during differentiation of the sieve tube elements. Note that the companion cell has many cytoplasmic organelles, whereas the sieve tube element has relatively few organelles. An ordinary companion cell is depicted here.

tance transport in plants (see Web Topic 10.1). These classical experiments demonstrated that removal of a ring of bark around the trunk of a tree, which removes the phloem, effectively stops sugar transport from the leaves to the roots without altering water transport through the xylem. When radioactive compounds became available, $^{14}CO_2$ was used to show that sugars made in the photosynthetic process are translocated through the phloem sieve elements (see Web Topic 10.1).

Mature sieve elements are living cells specialized for translocation

Detailed knowledge of the ultrastructure of sieve elements is critical to any discussion of the mechanism of translocation in the phloem. Mature sieve elements are unique among living plant cells (Figures 10.3 and 10.4). They lack many structures normally found in living cells, even the

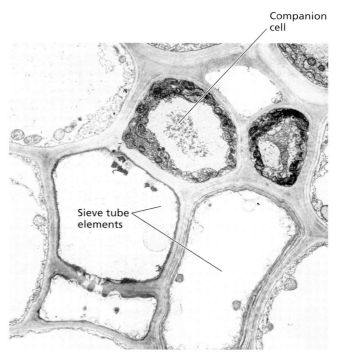

FIGURE 10.4 Electron micrograph of a transverse section of ordinary companion cells and mature sieve tube elements. (3600×) The cellular components are distributed along the walls of the sieve tube elements, where they offer less resistance to mass flow. (From Warmbrodt 1985.)

FIGURE 10.5 Sieve elements and open sieve plate pores. Open pores provide a low-resistance pathway for transport between sieve elements. (A) Electron micrograph of a longitudinal section of two mature sieve elements (sieve tube elements), showing the wall between the sieve elements (called a sieve plate) in the hypocotyl of winter squash (*Cucurbita maxima*). (3685×) (B) The inset shows sieve plate pores in face view. (4280×) In both images A and B, the sieve plate pores are open—that is, unobstructed by P-protein. (From Evert 1982.)

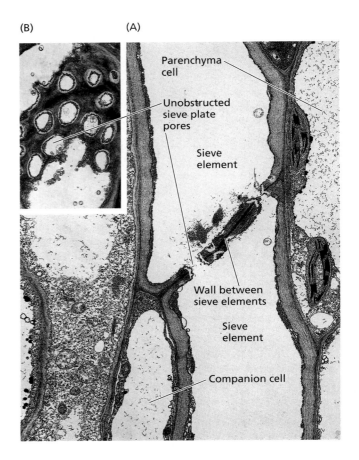

undifferentiated cells from which mature sieve elements are formed. For example, sieve elements lose their nuclei and tonoplasts (vacuolar membrane) during development. Microfilaments, microtubules, Golgi bodies, and ribosomes are also generally absent from the mature cells. In addition to the plasma membrane, organelles that are retained include somewhat modified mitochondria, plastids, and smooth endoplasmic reticulum. The walls are nonlignified, though they are secondarily thickened in some cases.

Thus the sieve elements have a cellular structure different from that of tracheary elements of the xylem, which are dead at maturity, lack a plasma membrane, and have lignified secondary walls. As we will see, living cells are critical to the mechanism of translocation in the phloem.

Large pores in cell walls are the prominent feature of sieve elements

Sieve elements (sieve cells and sieve tube elements) have characteristic sieve areas in their cell walls, where pores interconnect the conducting cells (see Figure 10.5). The sieve area pores range in diameter from less than 1 μm to approximately 15 μm. Unlike sieve areas of gymnosperms, the sieve areas of angiosperms can differentiate into **sieve plates** (see Figure 10.5 and Table 10.1).

Sieve plates have larger pores than the other sieve areas in the cell and are generally found on the end walls of sieve tube elements, where the individual cells are joined together to form a longitudinal series called a **sieve tube** (see Figure 10.3). Furthermore, the sieve plate pores of sieve tube elements are open channels that allow transport between cells (see Figure 10.5).

In contrast, all of the sieve areas are more or less the same in gymnosperms such as conifers. The pores of gymnosperm sieve areas meet in large median cavities in the middle of the wall. Smooth endoplasmic reticulum (SER) covers the sieve areas (Figure 10.6) and is continuous through the sieve pores and median cavity, as indicated by ER-specific staining. Observation of living material with confocal laser scanning microscopy confirms that the observed distribution of SER is not an artifact of fixation (Schulz 1992).

Damaged sieve elements are sealed off

Sieve-element sap is rich in sugars and other organic molecules. (*Sap* is a general term used to refer to the fluid con-

TABLE 10.1
Characteristics of the two types of sieve elements in seed plants

Sieve tube elements found in angiosperms

1. Some sieve areas are differentiated into sieve plates; individual sieve tube elements are joined together into a sieve tube.
2. Sieve plate pores are open channels.
3. P-protein is present in all dicots and many monocots.
4. Companion cells are sources of ATP and perhaps other compounds and, in some species, are transfer cells or intermediary cells.

Sieve cells found in gymnosperms

1. There are no sieve plates; all sieve areas are similar.
2. Pores in sieve areas appear blocked with membranes
3. There is no P-protein.
4. Albuminous cells sometimes function as companion cells.

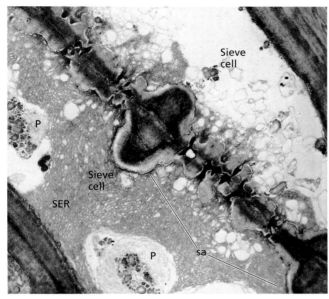

FIGURE 10.6 Electron micrograph showing a sieve area (sa) linking two sieve cells of a conifer (*Pinus resinosa*). Smooth endoplasmic reticulum (SER) covers the sieve area on both sides and is also found within the pores and the extended median cavity. Such obstructed pores would result in a high resistance to solution flow between sieve cells. P, plastid (From Schulz 1990.)

tents of plant cells.) These molecules represent an energy investment for the plant, and their loss must be prevented when sieve elements are damaged. Short-term sealing mechanisms involve sap proteins, while the principal long-term mechanism for preventing sap loss entails closing sieve-plate pores with a glucose polymer.

The main phloem proteins involved in sealing damaged sieve elements are structural proteins called **P-proteins** (see Figure 10.3B) (Clark et al. 1997). (In classical literature, P-protein was called *slime*.) The sieve tube elements of most angiosperms, including all dicots and many monocots, are rich in P-protein. However, P-protein is absent in gymnosperms. It occurs in several different forms (tubular, fibrillar, granular, and crystalline), depending on the species and maturity of the cell.

In immature cells, P-protein is most evident as discrete bodies in the cytosol known as **P-protein bodies**. P-protein bodies may be spheroidal, spindle-shaped, or twisted and coiled. They generally disperse into tubular or fibrillar forms during cell maturation.

P-proteins have been characterized at the molecular level (Dinant et al. 2003). For example, P-proteins from the genus *Cucurbita* consist of two major proteins: PP1, a phloem protein that forms filaments, and PP2, a phloem lectin associated with the filaments. Both PP1 and PP2 are synthesized in companion cells (discussed in the next section) and transported via the plasmodesmata to the sieve elements, where they associate to form P-protein filaments and P-protein bodies (Clark et al. 1997). PP2 may serve to anchor PP1 in the sieve element.

P-protein appears to function in sealing off damaged sieve elements by plugging up the sieve plate pores. Sieve tubes are under very high internal turgor pressure, and the sieve elements in a sieve tube are connected through open sieve plate pores. When a sieve tube is cut or punctured, the release of pressure causes the contents of the sieve elements to surge toward the cut end, from which the plant could lose much sugar-rich phloem sap if there were no sealing mechanism. When surging occurs, however, P-protein is trapped on the sieve plate pores, helping to seal the sieve element and to prevent further loss of sap. The sieve-element organelles and sometimes the ER, on the other hand, appear to be anchored to each other and to the sieve-element plasma membrane (Ehlers et al. 2000).

Another mechanism for blocking wounded sieve tubes occurs in plants in the legume family (Fabaceae). These plants contain large crystalloid protein bodies that do not disperse during development. However, following damage or osmotic shock, the protein bodies rapidly disperse and block the sieve tube. The process is reversible and controlled by calcium (Knoblauch et al. 2001). While other plant families have protein bodies that do not disperse during development, it is not known whether other plants have similar mechanisms for plugging wounded sieve tubes.

A longer-term solution to sieve tube damage is the production of **callose** in the sieve pores. Callose, a β-1,3-glucan, is synthesized by an enzyme in the plasma membrane and is deposited between the plasma membrane and the cell wall. Callose is synthesized in functioning sieve elements in response to damage and other stresses, such as mechanical stimulation and high temperatures, or in preparation for normal developmental events, such as dormancy. The deposition of **wound callose** in the sieve pores efficiently seals off damaged sieve elements from surrounding intact tissue. As the sieve elements recover from damage, the callose disappears from these pores.

Companion cells aid the highly specialized sieve elements

Each sieve tube element is usually associated with one or more **companion cells** (see Figures 10.3B, 10.4, and 10.5). The division of a single mother cell forms the sieve tube element and the companion cell. Numerous plasmodesmata (see Chapter 1) penetrate the walls between sieve tube elements and their companion cells; the plasmodesmata are often complex and branched on the companion cell side. The presence of abundant plasmodesmata suggests a close functional relationship between a sieve element and its companion cell, an association that is demonstrated by the rapid exchange of solutes, such as fluorescent dyes, between the two cells.

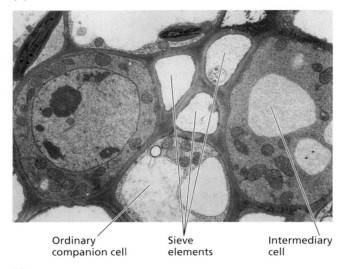

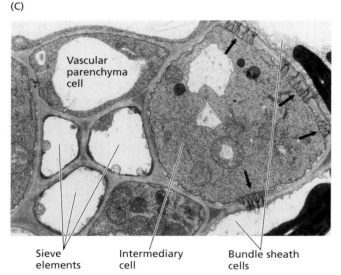

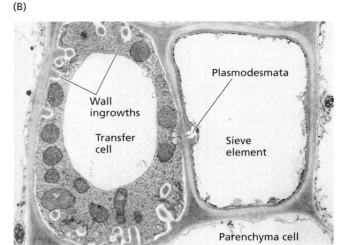

FIGURE 10.7 Electron micrographs of companion cells in minor veins of mature leaves. (A) Three sieve elements abut two intermediary cells and a more lightly stained ordinary companion cell in a minor vein from scarlet monkey flower (*Mimulus cardinalis*). (6585×) (B) A sieve element adjacent to a transfer cell with numerous wall ingrowths in pea (*Pisum sativum*). (8020×) Such ingrowths greatly increase the surface area of the transfer cell's plasma membrane, thus increasing the transfer of materials from the mesophyll to the sieve elements. (C) A typical intermediary cell with numerous fields of plasmodesmata (arrows) connecting it to neighboring bundle sheath cells. These plasmodesmata are branched on both sides, but the branches are longer and narrower on the intermediary cell side. Minor-vein phloem was taken from heartleaf maskflower (*Alonsoa warscewiczii*). (4700×) (A and C from Turgeon et al. 1993, courtesy of R. Turgeon; B from Brentwood and Cronshaw 1978.)

Companion cells play a role in the transport of photosynthetic products from producing cells in mature leaves to the sieve elements in the minor (small) veins of the leaf. They also take over some of the critical metabolic functions, such as protein synthesis, that are reduced or lost during differentiation of the sieve elements. In addition, the numerous mitochondria in companion cells may supply energy as ATP to the sieve elements.

There are at least three different types of companion cells in the minor veins of mature, exporting leaves: "ordinary" companion cells, transfer cells, and intermediary cells. All three cell types have dense cytoplasm and abundant mitochondria.

Ordinary companion cells (Figure 10.7A) have chloroplasts with well-developed thylakoids and a cell wall with a smooth inner surface. Relatively few plasmodesmata connect this type of companion cell to any of the surrounding cells except its own sieve element. As a result, the symplast of the sieve element and its companion cell is relatively, if not entirely, symplastically isolated from that of surrounding cells.

Transfer cells are similar to ordinary companion cells, except for the development of fingerlike wall ingrowths, particularly on the cell walls that face away from the sieve element (see Figure 10.7B). These wall ingrowths greatly increase the surface area of the plasma membrane, thus increasing the potential for solute transfer across the membrane.

Because of the scarcity of cytoplasmic connections to surrounding cells and the wall ingrowths in transfer cells, the ordinary companion cell and the transfer cell are thought to be specialized for taking up solutes from the apoplast or cell wall space. Xylem parenchyma cells can also be modified as transfer cells, probably serving to retrieve and reroute solutes moving in the xylem, which is also part of the apoplast.

Though ordinary companion cells and transfer cells are relatively isolated symplastically from surrounding cells, there are some plasmodesmata in the walls of these cells. The function of these plasmodesmata is not known. The fact that they are present indicates that they must have a function, and an important one, since the cost of having them is high: they are the avenues by which viruses become systemic in the plant. They are, however, difficult to study because they are so inaccessible.

Intermediary cells appear well suited for taking up solutes via cytoplasmic connections (see Figure 10.7C). Intermediary cells have numerous plasmodesmata connecting them to bundle sheath cells. Although the presence of many plasmodesmatal connections to surrounding cells is their most characteristic feature, intermediary cells are also distinctive in having numerous small vacuoles, as well as poorly developed thylakoids and a lack of starch grains in the chloroplasts.

In general, ordinary companion cells and transfer cells are found in plants where transport sugars enter the apoplast during the movement of sugars from mesophyll cells to sieve elements. In these plants ordinary companion cells and transfer cells transfer sugars from the apoplast to the symplast of the sieve elements and companion cells in the source. Intermediary cells, on the other hand, function in symplastic transport of sugars from mesophyll cells to sieve elements in plants where no apoplastic step appears to occur in the source leaf.

Patterns of Translocation: Source to Sink

Sap in the phloem is not translocated exclusively in either an upward or a downward direction, and translocation in the phloem is not defined with respect to gravity. Rather, sap is translocated from areas of supply, called *sources*, to areas of metabolism or storage, called *sinks*.

Sources include any exporting organs, typically mature leaves, that are capable of producing photosynthate in excess of their own needs. The term *photosynthate* refers to products of photosynthesis. Another type of source is a storage organ during the exporting phase of its development. For example, the storage root of the biennial wild beet (*Beta maritima*) is a sink during the growing season of the first year, when it accumulates sugars received from the source leaves. During the second growing season the same root becomes a source; the sugars are remobilized and utilized to produce a new shoot, which ultimately becomes reproductive.

Sinks include any nonphotosynthetic organs of the plant and organs that do not produce enough photosynthetic products to support their own growth or storage needs. Roots, tubers, developing fruits, and immature leaves, which must import carbohydrate for normal development, are all examples of sink tissues. Both girdling and labeling studies support the source-to-sink pattern of translocation in the phloem.

Source-to-sink pathways follow anatomic and developmental patterns

Although the overall pattern of transport in the phloem can be stated simply as source-to-sink movement, the specific pathways involved are often more complex. Not all sources supply all sinks on a plant; rather, certain sources preferentially supply specific sinks. In the case of herbaceous plants, such as sugar beet (*Beta vulgaris*) and soybean (*Glycine max*), the following generalizations can be made.

PROXIMITY The proximity of the source to the sink is a significant factor. The upper mature leaves on a plant usually provide photosynthates to the growing shoot tip and young, immature leaves; the lower leaves supply predominantly the root system. Intermediate leaves export in both directions, bypassing the intervening mature leaves.

DEVELOPMENT The importance of various sinks may shift during plant development. Whereas the shoot and root apices are usually the major sinks during vegetative growth (and in that order), seeds and fruits generally become the dominant sinks during reproductive development, particularly for adjacent and other nearby leaves.

VASCULAR CONNECTIONS Source leaves preferentially supply sinks with which they have direct vascular connections (Figure 10.8B). In the shoot system, for example, a given leaf is generally connected via the vascular system to other leaves directly above or below it on the stem. Such a vertical row of leaves is called an *orthostichy*. The number of internodes between leaves on the same orthostichy varies with the species. Figure 10.8A shows the three-dimensional structure of the phloem in an internode of dahlia (*Dahlia pinnata*).

MODIFICATION OF TRANSLOCATION PATHWAYS Interference with a translocation pathway by wounding or pruning can alter the patterns established by proximity and vascular connections that have been outlined here. In the absence of direct connections between source and sink, vascular interconnections, called **anastomoses** (singular *anastomosis*) (see Figure 10.8A), can provide an alternative pathway. In sugar beet, for example, removing source leaves from one side of the plant can bring about cross-transfer of photosynthates to young leaves (sink leaves) on the pruned side (see Figure 10.8C). Removal of the lower source leaves on a plant can force the upper source leaves to translocate materials to the roots, and removal of the upper source leaves can force lower source leaves to translocate materials to the upper parts of the plant.

The plasticity of the translocation pathway depends on the extent of the interconnections between vascular bun-

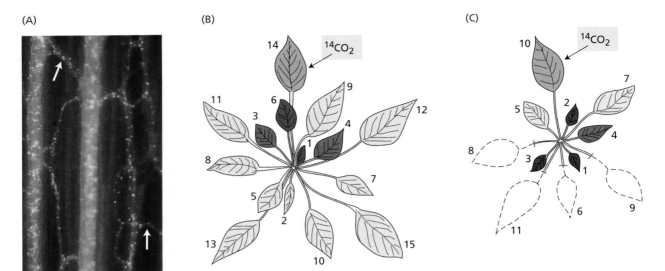

FIGURE 10.8 Source-to-sink patterns of phloem translocation. (A) Longitudinal view of a typical three-dimensional structure of the phloem in a thick section (from an internode of dahlia [*Dahlia pinnata*]), viewed here after clearing, staining with aniline blue, and observing under an epifluorescent microscope. The sieve plates are seen as numerous small dots because of the yellow staining of callose in the sieve areas. Two large longitudinal vascular bundles are prominent. This staining reveals the delicate sieve tubes forming the phloem network; two phloem anastomoses are marked by arrows. (B) Distribution of radioactivity from a single labeled source leaf in an intact plant. The distribution of radioactivity in leaves of a sugar beet plant (*Beta vulgaris*) was determined 1 week after $^{14}CO_2$ was supplied for 4 hours to a single source leaf (arrow). The degree of radioactive labeling is indicated by the intensity of shading of the leaves. Leaves are numbered according to their age; the youngest, newly emerged leaf is designated 1. The ^{14}C label was translocated mainly to the sink leaves directly above the source leaf (that is, sink leaves on the same orthostichy as the source; for example, leaves 1 and 6 are sink leaves directly above source leaf 14). (C) Same as B, except all source leaves on the side of the plant opposite the labeled leaf were removed 24 hours before labeling. Sink leaves on both sides of the plant now receive ^{14}C-labeled assimilates from the source. (A courtesy of R. Aloni; B and C based on data from Joy 1964.)

dles and thus on the species and organs studied. In some species, the leaves on a branch with no fruits cannot transport photosynthate to the fruits on an adjacent defoliated branch. But in other plants, such as soybean, photosynthate is transferred readily from a partly defruited side to a partly defoliated side.

Materials Translocated in the Phloem

Water is the most abundant substance in the phloem. Dissolved in the water are the translocated solutes, including carbohydrates, amino acids and proteins, hormones, and some inorganic ions. Carbohydrates are the most significant and concentrated solutes in phloem sap (Table 10.2), with sucrose being the sugar most commonly transported in sieve elements. There is always some sucrose in sieve element sap, and it can reach concentrations of 0.3 to 0.9 M.

Nitrogen is found in the phloem largely in amino acids and amides, especially glutamate and aspartate and their respective amides, glutamine and asparagine. Reported levels of amino acids and organic acids vary widely, even for the same species, but they are usually low compared with carbohydrates. See **Web Topic 10.2** for more information on nitrogen transport in the phloem.

Some inorganic solutes move in the phloem, including potassium, magnesium, phosphate, and chloride (see Table

TABLE 10.2
The composition of phloem sap from castor bean (*Ricinus communis*), collected as an exudate from cuts in the phloem

Component	Concentration (mg mL⁻¹)
Sugars	80.0–106.0
Amino acids	5.2
Organic acids	2.0–3.2
Protein	1.45–2.20
Potassium	2.3–4.4
Chloride	0.355–0.675
Phosphate	0.350–0.550
Magnesium	0.109–0.122

Source: Hall and Baker 1972.

TABLE 10.3
Functions of some soluble proteins in the sieve element–companion cell complexes

Function	Examples
Sugar metabolism	Sucrose synthase
Sugar transport across membranes	ATPases and sucrose carriers
Membrane water permeability	Aquaporins
Protein degradation	Ubiquitin
Protein disulfide reduction that regulates their functions	Thioredoxin h and glutaredoxin
Protein phosphorylation	Protein kinases
Protein folding	Chaperones
Antioxidant defense	A suite of enzymes and radical scavengers
Protection of phloem proteins from degradation plus defense against pathogens and phloem-feeding insects	Protease inhibitors

Source: Hayashi et al. 2000; Walz et al. 2004.

10.2). In contrast, nitrate, calcium, sulfur, and iron are relatively immobile in the phloem.

Almost all the endogenous plant hormones, including auxin, gibberellins, cytokinins, and abscisic acid (see Chapters 19, 20, 21, and 23), have been found in sieve elements. The long-distance transport of hormones is thought to occur at least partly in the sieve elements. Nucleotide phosphates have also been found in phloem sap.

A variety of proteins and RNAs occur in phloem sap, in relatively low concentrations. Proteins found in the phloem include structural P-proteins such as PP1 and PP2 (involved in the sealing of wounded sieve elements), as well as a number of water-soluble proteins. The functions of some of these proteins are listed in Table 10.3; many are related to stress and defense reactions (Walz et al. 2004). RNAs found in the phloem include mRNAs (Doering-Saad et al. 2002), pathogenic RNAs, and small regulatory RNA molecules.

We will continue the discussion of phloem content with a look at the methods used to identify materials translocated in the phloem and an examination of the translocated sugars. The possible roles of RNAs and proteins as signal molecules will be discussed at the end of the chapter.

Phloem sap can be collected and analyzed

The collection of phloem sap has been experimentally challenging (see **Web Topic 10.3**). A few species exude phloem sap from wounds that sever sieve elements, making it possible to collect small samples of the exuded sap. The initial samples may, however, be contaminated by the contents of surrounding damaged cells. A preferable approach is to use an aphid stylet as a "natural syringe." This method yields relatively pure sap from the sieve elements and companion cells (Doering-Saad et al. 2002).

Aphids are small insects that feed by inserting their mouthparts, consisting of four tubular stylets, into a single sieve element of a leaf or stem. Sap can be collected from aphid stylets cut from the body of the insect, usually with a laser, after the aphid has been anesthetized with CO_2. The high turgor pressure in the sieve element forces the cell contents through the stylet to the cut end, where they can be collected. Exudate from severed stylets provides a fairly accurate picture of the composition of phloem sap (see **Web Topic 10.3**). Exudation from severed stylets can continue for hours, suggesting that the aphid prevents the plant's normal sealing mechanisms from operating.

Sugars are translocated in nonreducing form

Results from analyses of collected sap indicate that the translocated carbohydrates are all nonreducing sugars. Reducing sugars, such as glucose and fructose, contain an exposed aldehyde or ketone group (Figure 10.9A). In a nonreducing sugar, such as sucrose, the ketone or aldehyde group is reduced to an alcohol or combined with a similar group on another sugar (see Figure 10.9B). Most researchers believe that the nonreducing sugars are the major compounds translocated in the phloem because they are less reactive than their reducing counterparts.

Sucrose is the most commonly translocated sugar; many of the other mobile carbohydrates contain sucrose bound to varying numbers of galactose molecules. Raffinose consists of sucrose and one galactose molecule, stachyose consists of sucrose and two galactose molecules, and verbascose consists of sucrose and three galactose molecules (see Figure 10.9B). Translocated sugar alcohols include mannitol and sorbitol.

FIGURE 10.9 Structures of compounds not normally translocated in the phloem (A) and of compounds commonly translocated in the phloem (B).

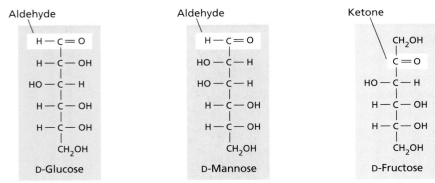

(A) **Reducing sugars, which are not generally translocated in the phloem**

The reducing groups are aldehyde (glucose and mannose) and ketone (fructose) groups.

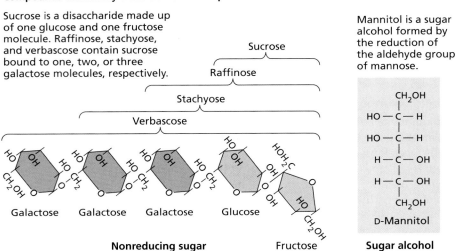

(B) **Compounds commonly translocated in the phloem**

Sucrose is a disaccharide made up of one glucose and one fructose molecule. Raffinose, stachyose, and verbascose contain sucrose bound to one, two, or three galactose molecules, respectively.

Mannitol is a sugar alcohol formed by the reduction of the aldehyde group of mannose.

Glutamic acid, an amino acid, and glutamine, its amide, are important nitrogenous compounds in the phloem, in addition to aspartate and asparagine.

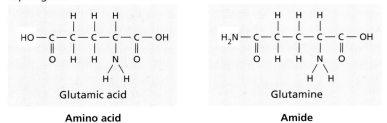

Species with nitrogen-fixing nodules also utilize ureides as transport forms of nitrogen.

Ureides

Rates of Movement

The rate of movement of materials in the sieve elements can be expressed in two ways: as **velocity**, the linear distance traveled per unit time, or as **mass transfer rate**, the quantity of material passing through a given cross section of phloem or sieve elements per unit time. Mass transfer rates based on the cross-sectional area of the sieve elements are preferred because the sieve elements are the conducting cells of the phloem. Values for mass transfer rate range from 1 to 15 g h^{-1} cm^{-2} of sieve elements (see Web Topic 10.4).

In early publications reporting on rates of transport in the phloem, the units of velocity were centimeters per hour (cm h^{-1}), and the units of mass transfer were grams per hour per square centimeter (g h^{-1} cm^{-2}) of phloem or sieve elements. However, the currently preferred units (SI units) are meters (m) or millimeters (mm) for length, seconds (s) for time, and kilograms (kg) for mass.

Both velocities and mass transfer rates can be measured with radioactive tracers. (Methods of measuring mass transfer rates are described in Web Topic 10.4.) In the simplest type of experiment for measuring velocity, ^{11}C- or ^{14}C-labeled CO_2 is applied for a brief period of time to a source leaf (pulse labeling), and the arrival of label at a sink tissue or at a particular point along the pathway is monitored with an appropriate detector.

In general, velocities measured by a variety of conventional techniques far exceed the rate of diffusion, averaging about 100 cm h^{-1} and ranging from 30 to 150 cm h^{-1}. A more recent measurement of velocity using NMR spectrometry yielded a value of 0.25 mm sec^{-1} (equivalent to 90 cm h^{-1}), which is remarkably close to the average obtained using older methods (Peuke et al. 2001). Transport velocities in the phloem are clearly quite high and well in excess of the rate of diffusion over long distances. Any proposed mechanism of phloem translocation must account for these high velocities.

The Pressure-Flow Model for Phloem Transport

The mechanism of phloem translocation in angiosperms is best explained by the pressure-flow model, which is consistent with most of the experimental and structural data currently available. The pressure-flow model explains phloem translocation as a flow of solution (mass flow or bulk flow) driven by an osmotically generated pressure gradient between source and sink. The pressure-flow model, predictions arising from mass flow, and supporting data will be described in this section. At the end of the section some remaining questions about the model will be briefly explored.

In early research on phloem translocation, both active and passive mechanisms were considered. All theories, both active and passive, assume an energy requirement in both sources and sinks. In sources, energy is necessary to move photosynthate from producing cells into the sieve elements. This movement of photosynthate is called *phloem loading*, and it is discussed in detail later in the chapter. In sinks, energy is essential for some aspects of movement from sieve elements to sink cells, which store or metabolize the sugar. This movement of photosynthate from sieve elements to sink cells is called *phloem unloading* and will also be discussed later.

The passive mechanisms of phloem transport further assume that energy is required in the sieve elements of the path between sources and sinks simply to maintain structures such as the cell plasma membrane and to recover sugars lost from the phloem by leakage. The pressure-flow model is an example of a passive mechanism. The active theories, on the other hand, postulate an additional expenditure of energy by path sieve elements in order to drive translocation itself (Zimmermann and Milburn 1975).

A pressure gradient drives translocation in the pressure-flow model

Diffusion is far too slow to account for the velocities of solute movement observed in the phloem. Translocation velocities average 1 m h^{-1}; the rate of diffusion is 1 m per 32 years! (See Chapter 3 for a discussion of diffusion velocities and the distances over which diffusion is an effective transport mechanism.)

The **pressure-flow model**, first proposed by Ernst Münch in 1930, states that a flow of solution in the sieve elements is driven by an osmotically generated *pressure gradient* between source and sink ($\Delta\Psi_p$). The pressure gradient is established as a consequence of phloem loading at the source and phloem unloading at the sink.

Recall from Chapter 3 (Equation 3.6) that $\Psi_w = \Psi_s + \Psi_p$; that is, $\Psi_p = \Psi_w - \Psi_s$. In source tissues, energy-driven phloem loading leads to an accumulation of sugars in the sieve elements, generating a low (negative) solute potential ($\Delta\Psi_s$) and causing a steep drop in the water potential ($\Delta\Psi_w$). In response to the water potential gradient, water enters the sieve elements and causes the turgor pressure (Ψ_p) to increase.

At the receiving end of the translocation pathway, phloem unloading leads to a lower sugar concentration in the sieve elements, generating a higher (more positive) solute potential in the sieve elements of sink tissues. As the water potential of the phloem rises above that of the xylem, water tends to leave the phloem in response to the water potential gradient, causing a decrease in turgor pressure in the sieve elements of the sink. Figure 10.10 illustrates the pressure-flow hypothesis.

If no cross-walls were present in the translocation pathway—that is, if the entire pathway were a single membrane-enclosed compartment—the different pressures at the source and sink would rapidly equilibrate. The presence of sieve plates greatly increases the resistance along the pathway and is thought to result in the generation and

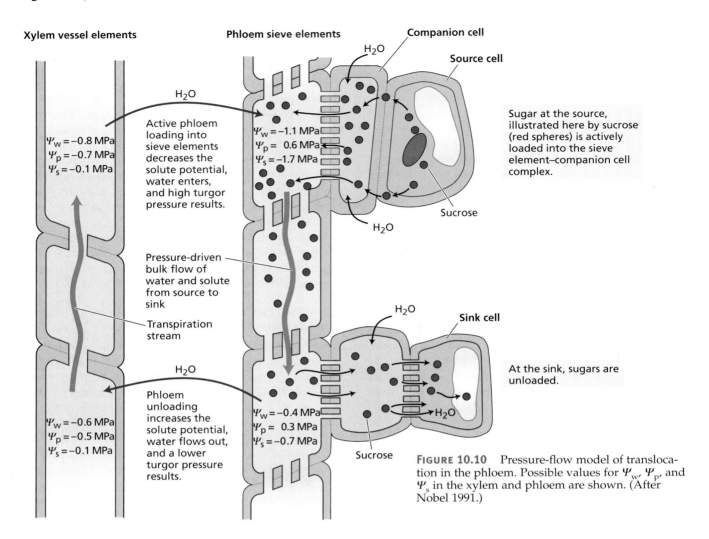

FIGURE 10.10 Pressure-flow model of translocation in the phloem. Possible values for Ψ_w, Ψ_p, and Ψ_s in the xylem and phloem are shown. (After Nobel 1991.)

maintenance of a considerable pressure gradient in the sieve elements between source and sink.

The phloem sap moves by mass flow rather than by osmosis. That is, no membranes are crossed during transport from one sieve tube to another, and solutes move at the same rate as the water molecules. Since this is the case, mass flow can occur from a source organ with a lower water potential to a sink organ with a higher water potential, or vice versa, depending on the identities of the source and sink organs. In fact, Figure 10.10 illustrates an example in which the flow is against the water potential gradient. Such water movement does not violate the laws of thermodynamics, because the movement is mass flow and not osmosis.

Under prevailing conditions in the sieve elements, the solute potential, Ψ_s, cannot contribute to the driving force for water movement, although it still influences the water potential. Water movement in the translocation pathway is driven by the pressure gradient according to the pressure-flow model, not by the water potential gradient. Of course, the passive, pressure-driven, long-distance translocation in the sieve tubes ultimately depends on the active, short-distance transport mechanisms involved in phloem loading and unloading. These active mechanisms are responsible for setting up the pressure gradient.

The predictions of mass flow have been confirmed

Some important predictions emerge from the model of phloem translocation as mass flow:

- The sieve plate pores must be unobstructed. If P-protein or other materials blocked the pores, the resistance to flow of the sieve element sap would be too great.

- True bidirectional transport (i.e., simultaneous transport in both directions) in a single sieve element cannot occur. A mass flow of solution precludes such bidirectional movement because a solution can flow in only one direction in a pipe at any one time. Solutes within the phloem can move bidirectionally, but in different sieve elements or at different times.

- Great expenditures of energy are not required in order to drive translocation in the tissues along the path. Therefore, treatments that restrict the supply of ATP in the path, such as low temperature, anoxia, and metabolic inhibitors, should not stop translocation. However, energy *is* required to maintain the structure of the sieve elements, to reload any sugars lost to the apoplast by leakage, and to regulate sieve element turgor.

- The pressure-flow hypothesis demands the presence of a positive pressure gradient. Turgor pressure must be higher in sieve elements of sources than in those of sinks, and the pressure difference must be large enough to overcome the resistance of the pathway and to maintain flow at the observed velocities.

The available evidence testing these predictions is consistent with mass flow and with the pressure-flow hypothesis.

Sieve plate pores are open channels

Ultrastructural studies of sieve elements are challenging because of the high internal pressure in these cells. When the phloem is excised or killed slowly with chemical fixatives, the turgor pressure in the sieve elements is released. The contents of the cell, particularly P-protein, surge toward the point of pressure release and, in the case of sieve tube elements, accumulate on the sieve plates. This accumulation is probably the reason that many earlier electron micrographs show sieve plates that are obstructed.

Newer, rapid freezing and fixation techniques provide reliable pictures of undisturbed sieve elements. Electron micrographs of sieve tube elements prepared by such techniques show that P-protein is usually found along the periphery of the sieve tube elements (see Figures 10.3, 10.4, and 10.5), or it is evenly distributed throughout the lumen of the cell. Furthermore, the pores contain P-protein in similar positions, lining the pore or in a loose network. The open condition of the pores, seen in many species, such as cucurbits, sugar beet, and bean (*Phaseolus vulgaris*) (e.g., see Figure 10.5), is consistent with mass flow.

In addition to obtaining the structural evidence provided by electron microscopy, it is important to determine whether the sieve plate pores are open in the intact tissue. The use of confocal laser scanning microscopy, which allows for the direct observation of translocation through living sieve elements, addresses this question (Knoblauch and van Bel 1998). Such experiments show that the sieve plate pores of living, translocating sieve elements are open (Figure 10.11).

There is no bidirectional transport in single sieve elements

Researchers have investigated bidirectional transport by applying two different radiotracers to two source leaves, one above the other. Each leaf receives one of the tracers, and a point between the two sources is monitored for the presence of both tracers.

Transport in two directions has often been detected in sieve elements of different vascular bundles in stems. Transport in two directions has also been seen in adjacent sieve elements of the same bundle in petioles. Bidirectional transport in adjacent sieve elements can occur in the petiole of a leaf that is undergoing the transition from sink to source and simultaneously importing and exporting photosynthates through its petiole. However, simultaneous bidirectional transport in a single sieve element has never been demonstrated.

The energy requirement for transport through the phloem pathway is small

In plants that can survive periods of low temperature, such as sugar beet, rapidly chilling a short segment of the petiole of a source leaf to approximately 1°C does not cause sustained inhibition of mass transport out of the leaf (Figure 10.12). Rather, there is a brief period of inhibition, after which transport slowly returns to the control rate. Chilling reduces respiration rate and both the synthesis and the consumption of ATP in the petiole by about 90%, at a time when translocation has recovered and is proceeding normally. These experiments show that the energy requirement for transport through the pathway of these herbaceous plants is small, consistent with mass flow.

Extreme treatments that inhibit all energy metabolism do inhibit translocation. For example, in bean, treating the petiole of a source leaf with a metabolic inhibitor (cyanide) inhibited translocation out of the leaf. However, examination of the treated tissue by electron microscopy revealed blockage of the sieve plate pores by cellular debris (Giaquinta and Geiger 1977). Clearly, these results do not bear on the question of whether energy is required for translocation along the pathway.

Pressure gradients are sufficient to drive a mass flow of phloem sap

Mass flow or bulk flow is the combined movement of all the molecules in a solution, driven by a pressure gradient. What are the pressure values in sieve elements, how can they be determined, and are they sufficient to drive mass flow through the phloem sieve elements?

Turgor pressure in sieve elements can be either calculated from the water potential and solute potential ($\Psi_p = \Psi_w - \Psi_s$) or measured directly. The most effective technique uses micromanometers or pressure transducers sealed over exuding aphid stylets (see Figure 10.2.A in **Web Topic 10.3**) (Wright and Fisher 1980). The data obtained are accurate because aphids pierce only a single sieve element, and the plasma membrane apparently seals well around the aphid stylet. When the turgor pressure of sieve elements is measured by this technique, the pressure at the source is higher than that at the sink.

In soybean, the observed pressure difference between source and sink has been shown to be sufficient to drive a

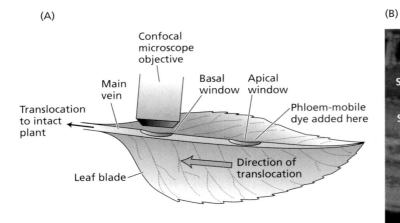
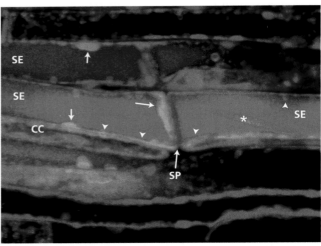
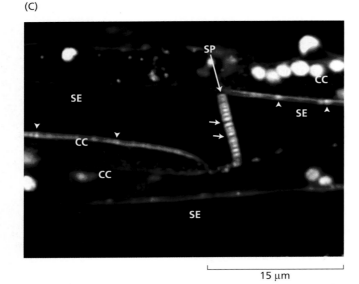

FIGURE 10.11 Translocation in living, functional sieve elements of a leaf attached to an intact broad bean (*Vicia faba*) plant. (A) Two windows were sliced parallel to the epidermis on the lower side of the main vein of a mature leaf, exposing the phloem tissue. The objective of the laser confocal microscope was positioned over the basal window. A phloem-mobile fluorescent dye was added at the apical window. If translocation occurred, the dye would become visible in the microscope at the basal window of the leaf. In this way it could be demonstrated that the sieve elements being observed were alive and functional. (B) Phloem tissue of bean doubly stained with a locally applied fluorescent dye (red) that primarily stains membranes, and a different fluorescent dye (green) that is translocated. The presence of the green dye at the basal, observation window indicates that translocation has occurred from the apical, application window. The sieve elements were alive and functional. Protein (arrows) deposited against the plasma membrane and the sieve plate does not impede translocation. A crystalline P-protein body (asterisk) is stained by the green dye. Plastids (arrowheads) are evenly distributed around the periphery of the sieve element. (C) Phloem tissue of bean stained only with the locally applied fluorescent dye that stains membranes. Arrows indicate sieve plate pores, which are not occluded. CC, companion cell; SP, sieve plate. (B and C from Knoblauch and van Bel 1998; courtesy of A. van Bel.)

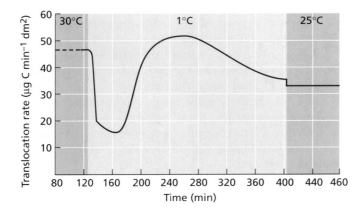

FIGURE 10.12 Loss of metabolic energy resulting from the chilling of the leaf petiole partially reduces the rate of translocation in sugar beet, although translocation rates recover with time. The fact that translocation recovers when ATP production and utilization are largely inhibited by chilling indicates that the energy requirement for translocation in the path is small. $^{14}CO_2$ was supplied to a source leaf, and a 2-cm portion of its petiole was chilled to 1°C. Translocation was monitored by the arrival of ^{14}C at a sink leaf. (1 dm [decimeter] = 0.1 meter) (After Geiger and Sovonick 1975.)

solution through the pathway by mass flow, taking into account the path resistance (caused mainly by the sieve plate pores), the path length, and the velocity of translocation (Fisher 1978). The actual pressure difference between source and sink was calculated from the water potential and solute potential to be 0.41 MPa, and the pressure difference required for translocation by pressure flow was calculated to be 0.12 to 0.46 MPa. Thus the observed pressure difference appears to be sufficient to drive mass flow through the phloem.

We can therefore conclude that all the experiments and data described here are consistent with the operation of mass flow in angiosperm phloem. The lack of an energy requirement in the pathway and the presence of open sieve plate pores provide definitive evidence for a mechanism in which the path phloem is relatively passive. The failure to detect bidirectional transport or motility proteins, as well as the positive data on pressure gradients, is in accord with current interpretations of the pressure-flow hypothesis.

Significant questions about the pressure-flow model still exist

Does translocation in gymnosperms involve a different mechanism? Although mass flow explains translocation in angiosperms, it may not be sufficient for gymnosperms. Very little physiological information on gymnosperm phloem is available, and speculation about translocation in these species is based almost entirely on interpretations of electron micrographs. As discussed previously, the sieve cells of gymnosperms are similar in many respects to sieve tube elements of angiosperms, but the sieve areas of sieve cells are relatively unspecialized and do not appear to consist of open pores (see Figure 10.6).

The pores in gymnosperms are filled with numerous membranes that are continuous with the smooth endoplasmic reticulum adjacent to the sieve areas. Such pores are clearly inconsistent with the requirements of mass flow. Although these electron micrographs might be artifactual and fail to show conditions in the intact tissue, translocation in gymnosperms might involve a different mechanism—a possibility that requires further investigation.

Is the pressure gradient large or small in angiosperm sieve elements? The size of the pressure gradient predicted by any mathematical model of phloem transport depends on the assumptions made while developing the model and on the data put into the model. Results from one recent model indicate that the phloem transport system works most efficiently with relatively small turgor differences between sources and sinks (Thompson 2006). The model does not contradict Münch's hypothesis but deemphasizes pressure differences as controlling flow. In this view, pressure differences in the sieve elements result from osmotically generated mass flow, rather than drive mass flow.

The mathematical expression of this model, called osmoregulatory flow, has several significant outcomes. One is that turgor can be globally regulated under conditions of a small gradient; that is, turgor in all sieve elements and companion cells in the path can be regulated simultaneously in the same way. Information in the form of pressure waves would be rapidly transmitted in this system, resulting in an efficient sieve tube that could rapidly transmit information on changes in the pressure or concentration of sap over long distances.

What kinds of measurements might enable us to distinguish between a mass flow with large pressure differences between sources and sinks and one with small gradients? One obvious answer is to measure pressure gradients between sources and sinks. Such studies have been relatively few in number and restricted to small herbaceous plants. More measurements on a wider variety of plants are needed, necessitating the development of techniques that can measure turgor differences along the same transport stream, ideally in large plants such as trees. Developing such techniques will be a huge technical challenge.

In addition, more data are required for use with mathematical models, for example, on the anatomy of sieve elements and sieve plates in different species, sieve tube length, and whole-plant phloem architecture. While the model of phloem transport with small turgor gradients is more difficult to imagine than the traditional picture, the questions it poses should lead to some exciting discussions in the future and to more studies on whole-plant phloem transport.

Phloem Loading

Several transport steps are involved in the movement of photosynthate from the mesophyll chloroplasts to the sieve elements of mature leaves:

1. Triose phosphate formed by photosynthesis during the day (see Chapter 8) is transported from the chloroplast to the cytosol, where it is converted to sucrose. During the night, carbon from stored starch exits the chloroplast primarily in the form of maltose and is converted to sucrose. (Other transport sugars are later synthesized from sucrose in some species.)

2. Sucrose moves from producing cells in the mesophyll to cells in the vicinity of the sieve elements in the smallest veins of the leaf (Figure 10.13). This **short-distance transport** pathway usually covers a distance of only a few cell diameters.

3. In the process called **phloem loading**, sugars are transported into the sieve elements and companion cells. In most of the species studied so far, sugars become more concentrated in the sieve elements and companion cells than in the mesophyll. Note that with respect to loading, the sieve elements and companion cells are often considered a functional unit, called the *sieve element–companion cell complex*. Once inside the

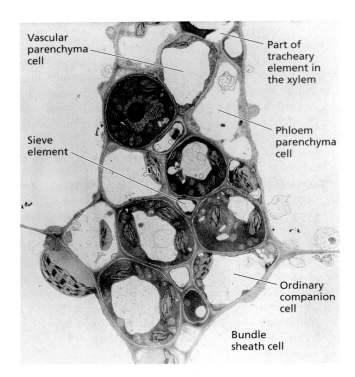

FIGURE 10.13 Electron micrograph showing the relationship between the various cell types of a small vein in a source leaf of sugar beet (5000×). Photosynthetic cells (mesophyll cells) surround the compactly arranged cells of the bundle sheath layer. Photosynthate from the mesophyll must move a distance equivalent to only several cell diameters before being loaded into the sieve elements. Movement from the mesophyll to the sieve elements is thus known as short-distance transport. (From Evert and Mierzwa 1985, courtesy of R. Evert.)

sieve elements, sucrose and other solutes are translocated away from the source, a process known as **export**. Translocation through the vascular system to the sink is referred to as **long-distance transport**.

As discussed earlier, the processes of loading at the source and unloading at the sink provide the driving force for long-distance transport and are thus of considerable basic, as well as agricultural, importance. A thorough understanding of these mechanisms should provide the basis of technology aimed at enhancing crop productivity by increasing the accumulation of photosynthate by edible sink tissues, such as cereal grains.

Phloem loading can occur via the apoplast or symplast

We have seen that solutes (mainly sugars) in source leaves must move from the photosynthesizing cells in the mesophyll to the sieve elements. The initial short-distance pathway is always symplastic (Figure 10.14). However, sugars might move entirely through the symplast (cytoplasm) to the sieve elements via the plasmodesmata (see Figure

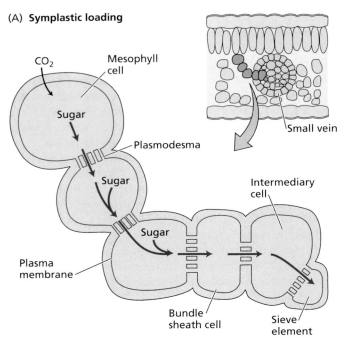

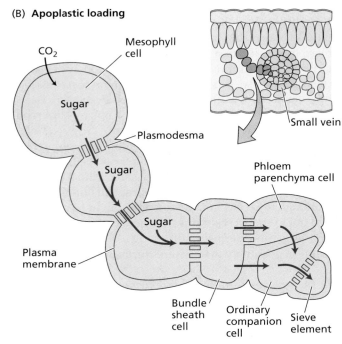

FIGURE 10.14 Schematic diagram of pathways of phloem loading in source leaves. (A) In the totally symplastic pathway, sugars move from one cell to another in the plasmodesmata, all the way from the mesophyll to the sieve elements. (B) In the partly apoplastic pathway, sugars initially move through the symplast but enter the apoplast just prior to loading into the companion cells and sieve elements. Sugars loaded into the companion cells are thought to move through plasmodesmata into the sieve elements.

10.14A), or they might enter the apoplast prior to phloem loading (see Figure 10.14B). (See Figure 4.4 for a general description of the symplast and apoplast.) In the case of apoplastic loading the sugars enter the apoplast quite near the sieve element–companion cell complex. Sugars are then actively transported from the apoplast into the sieve elements and companion cells by an energy-driven, selective transporter located in the plasma membranes of these cells. Efflux into the apoplast is highly localized, probably into the walls of phloem parenchyma cells. The apoplastic and symplastic routes are used in different species.

Early research on phloem loading focused on the apoplastic pathway. Apoplastic phloem loading leads to three basic predictions (Grusak et al. 1996):

1. Transported sugars should be found in the apoplast.

2. In experiments in which sugars are supplied to the apoplast, the exogenously supplied sugars should accumulate in sieve elements and companion cells.

3. Inhibition of sugar uptake from the apoplast should result in inhibition of export from the leaf.

Many studies devoted to testing these predictions have provided solid evidence for apoplastic loading in several species (see **Web Topic 10.5**).

Sucrose uptake in the apoplastic pathway requires metabolic energy

In source leaves, sugars become more concentrated in the sieve elements and companion cells than in the mesophyll. This difference in solute concentration, found in most of the species studied, can be demonstrated through measurement of the osmotic potential (Ψ_s) of the various cell types in the leaf (see Chapter 3).

In sugar beet, the osmotic potential of the mesophyll is approximately –1.3 MPa, and the osmotic potential of the sieve elements and companion cells is about –3.0 MPa (Geiger et al. 1973). Most of this difference in osmotic potential is thought to result from accumulated sugar, specifically sucrose because sucrose is the major transport sugar in this species. Experimental studies have also demonstrated that both externally supplied sucrose and sucrose made from photosynthetic products accumulate in the sieve elements and companion cells of the minor veins of sugar beet source leaves (Figure 10.15).

The fact that sucrose is at a higher concentration in the sieve element–companion cell complex than in surrounding cells indicates that sucrose is actively transported against its chemical-potential gradient. The dependence of sucrose accumulation on active transport is supported by the fact that treating source tissue with respiratory inhibitors both decreases ATP concentration and inhibits loading of exogenous sugar. In contrast, other metabolites, such as organic acids and hormones, may enter sieve elements passively (see **Web Topic 10.6**).

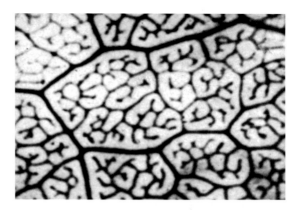

FIGURE 10.15 This autoradiograph shows that labeled sugar moves from the apoplast into sieve elements and companion cells against its concentration gradient. A solution of ^{14}C-labeled sucrose was applied for 30 minutes to the upper surface of a sugar beet leaf that had previously been kept in darkness for 3 hours. The leaf cuticle was removed to allow penetration of the solution to the interior of the leaf. Label accumulates in the small veins, sieve elements, and companion cells of the source leaf, indicating the ability of these cells to transport sucrose against its concentration gradient. (From Fondy 1975, courtesy of D. Geiger.)

Phloem loading in the apoplastic pathway involves a sucrose–H$^+$ symporter

A sucrose–H$^+$ symporter is thought to mediate the transport of sucrose from the apoplast into the sieve element–companion cell complex. Recall from Chapter 6 that symport is a secondary transport process that uses the energy generated by the proton pump (see Figure 6.11A). The energy dissipated by protons moving back into the cell is coupled to the uptake of a substrate, in this case sucrose (Figure 10.16).

Data from a number of studies support the operation of a sucrose–H$^+$ symporter in phloem loading:

- Proton-pumping ATPases, localized by immunological techniques, have been found in the plasma membranes of companion cells (*Arabidopsis thaliana*) and in transfer cells (broad bean, *Vicia faba*). In transfer cells, the H$^+$-ATPase molecules are most concentrated in the plasma membrane infoldings that face the bundle sheath and phloem parenchyma cells (for details, see **Web Topic 10.7**).

- The distribution of the H$^+$-ATPases in companion cells appears to be correlated with the distribution of a sucrose–H$^+$ symporter called *SUC2* (Arabidopsis, Table 10.4; broad-leaved plantain, *Plantago major*, see **Web Topic 10.7**). Thus, the H$^+$-ATPase that supplies the driving force for sucrose uptake and the sucrose transporter that utilizes it have been shown to be located in the same cells in some species.

- Reducing transporter activity inhibits transport from source leaves. Potato plants (*Solanum tuberosum*) trans-

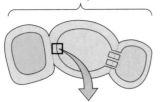

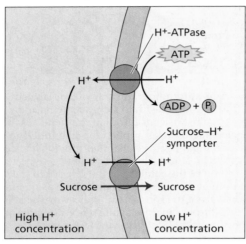

FIGURE 10.16 ATP-dependent sucrose transport in sieve element loading. In the cotransport model of sucrose loading into the symplast of the sieve element–companion cell complex, the plasma membrane ATPase pumps protons out of the cell into the apoplast, establishing a high proton concentration there. The energy in this proton gradient is then used to drive the transport of sucrose into the symplast of the sieve element–companion cell complex through a sucrose–H$^+$ symporter.

formed with antisense DNA to *SUT1*, another sucrose–H$^+$ symporter, showed reduced transporter activity, a reduction in root and tuber growth, and accumulation of starch and lipids in source leaves (Schulz et al. 1998).

Several sucrose–H$^+$ symporters have been cloned and localized in the phloem (see Table 10.4). The carriers are found in plasma membranes of either sieve elements or companion cells. (Carriers found in sieve element membranes include SUT1, SUT2, and SUT4, in potato and tomato [*Lycopersicum esculentum*]. SUC 2 has been found in the companion cell membranes of Arabidopsis and plantain.) Work with SUT1 has shown that the messenger RNAs for symporters found in the sieve element membrane are synthesized in the companion cells (Kuhn et al. 1997; Vaughn et al. 2002). This finding agrees with the fact that sieve elements lack nuclei. The symporter protein is probably also synthesized in the companion cells, since ribosomes do not appear to persist in mature sieve elements.

All sucrose transporters fall into one of three large subfamilies; the roles played by the many of the carriers are still being investigated (Kuhn 2003). SUT1 and SUC2 appear to be the major sucrose transporters in phloem loading, into either companion cells or sieve elements. The functions of other sucrose transporters, such as SUT2, remain to be discovered by future experimentation.

REGULATING SUCROSE LOADING The mechanisms that regulate the loading of sucrose from the apoplast to the sieve elements by the sucrose–H$^+$ symporter are not completely clear. Possible regulatory factors include the following:

- The solute potential or, more likely, the turgor pressure of the sieve elements. A decrease in sieve element turgor below a certain threshold would lead to a compensatory increase in loading. Although many publications suggest that turgor pressure regulates membrane transport, very few have actually demonstrated such control (Daie and Wyse 1985).

- Sucrose concentration in the apoplast. High sucrose concentrations in the apoplast would increase phloem loading. A number of investigators have reported a direct correlation between export and source-leaf sucrose concentration or photosynthetic rate, although few have measured the sucrose concentration in the apoplast at the site of loading.

- The available number of symporter protein molecules. Data suggest that sucrose levels in source leaves of sugar beet regulate the concentration of sucrose–H$^+$ symporter molecules (SUT1) in the leaf, which in turn could regulate loading. The transcription of *SUT1* declines in sugar beet leaves fed sucrose via the

TABLE 10.4
Selected sucrose transporters from dicotyledons and their functions

Transporter gene family	Plant examples	Name	Location/function
SUT1	Potato, tomato, tobacco	SUT1	SE/loading
	Arabidopsis, plantain	SUC2	CC/Loading
	Plantain	SUC1	SE of petioles/retrieval
SUT2	Potato, tomato, plantain	SUT2	SE/function unknown
	Arabidopsis		Bundle sheath in Arabidopsis/function unknown
SUT4	Arabidopsis	SUT4	Unknown
	Potato, tomato		SE

Source: Kuhn 2003.

Note: CC, companion cell; SE, sieve element

transpiration stream. Both SUT1 and its mRNA are degraded rapidly, so a decrease in transcription results in a decline in SUT1 protein. Uptake of sucrose into plasma membrane vesicles purified from the leaves declines at the same sucrose concentration (Vaughn et al. 2002), suggesting that the level of the symporter has declined. Protein phosphorylation has been demonstrated to play a role in these regulatory events (Ransom-Hodgkins et al. 2003). The levels of SUT1 transporter mRNA and protein are also regulated diurnally, being lower after 15 hours of darkness than after a light treatment.

Other studies have shown that sucrose efflux into the apoplast is enhanced by potassium availability in the apoplast, suggesting that a better nutrient supply increases translocation to sinks and enhances sink growth.

Phloem loading is symplastic in plants with intermediary cells

Many results point to apoplastic phloem loading in species that have ordinary companion cells or transfer cells in the minor veins, and that transport only sucrose. On the other hand, a symplastic pathway has become evident in species that transport raffinose and stachyose in the phloem, in addition to sucrose, and that have intermediary cells in the minor veins. Some examples of such species are common coleus (*Coleus blumei*), squash (*Cucurbita pepo*), and melon (*Cucumis melo*).

The operation of a symplastic pathway requires the presence of open plasmodesmata between the different cells in the pathway. Many species have numerous plasmodesmata at the interface between the sieve element–companion cell complex and the surrounding cells (see Figure 10.7C).

The polymer-trapping model explains symplastic loading

The composition of sieve element sap is generally different from the solute composition in tissues surrounding the phloem. This difference indicates that certain sugars are specifically selected for transport in the source leaf. The involvement of symporters in apoplastic phloem loading provides a clear mechanism for selectivity, because symporters are specific for certain sugar molecules. Symplastic loading, in contrast, depends on the diffusion of sugars from the mesophyll to the sieve elements via the plasmodesmata. It is more difficult to envision how diffusion through plasmodesmata during symplastic loading could be selective for certain sugars.

Furthermore, data from several species showing symplastic loading indicate that sieve elements and companion cells have a higher osmotic content than the mesophyll. How could diffusion-dependent symplastic loading account for the observed selectivity for transported molecules and the accumulation of sugars against a concentration gradient?

The **polymer-trapping model** (Figure 10.17) has been developed to address these questions (Turgeon and Gowan 1990). This model states that the sucrose synthesized in the mesophyll diffuses from the bundle sheath cells into the intermediary cells through the abundant plasmodesmata that connect the two cell types. In the intermediary cells, raffinose and stachyose (polymers made of three and four hexose sugars, respectively; see Figure 10.9B) are synthesized from the transported sucrose and from galactose. Because of the anatomy of the tissue and the relatively large size of raffinose and stachyose, the polymers cannot diffuse back into the bundle sheath cells, but they can diffuse into the sieve element. Sucrose can continue to diffuse into the intermediary cells, because its synthesis in the mesophyll and its utilization in the intermediary cells maintain the concentration gradient (see Figure 10.17).

The polymer-trapping model makes three predictions:

1. Sucrose should be more concentrated in the mesophyll than in the intermediary cells.
2. The enzymes for raffinose and stachyose synthesis should be preferentially located in the intermediary cells.
3. The plasmodesmata linking the bundle sheath cells and the intermediary cells should exclude molecules larger than sucrose. Plasmodesmata between the intermediary cells and sieve elements must be wider to allow passage of raffinose and stachyose.

Many studies support the polymer-trapping model. For instance, the enzymes required to synthesize stachyose from sucrose are localized in intermediary cells. In melon, raffinose and stachyose are present in high concentrations in intermediary cells, but not in mesophyll cells.

The type of phloem loading is correlated with plant family and with climate

As discussed earlier, the operation of apoplastic and symplastic phloem-loading pathways is correlated with the transport sugar, the type of companion cell in the minor veins, and the number of plasmodesmata connecting the sieve elements and companion cells to the surrounding photosynthetic cells (Table 10.5) (van Bel et al. 1992):

- Species showing apoplastic phloem loading translocate sucrose almost exclusively and have either ordinary companion cells or transfer cells in the minor veins. These species usually possess few connections between the sieve element–companion cell complex and the surrounding cells.

- Species having symplastic phloem loading translocate oligosaccharides such as raffinose in addition to sucrose, have intermediary-type companion cells in the minor veins, and possess abundant connections between the sieve element–companion cell complex and the surrounding cells.

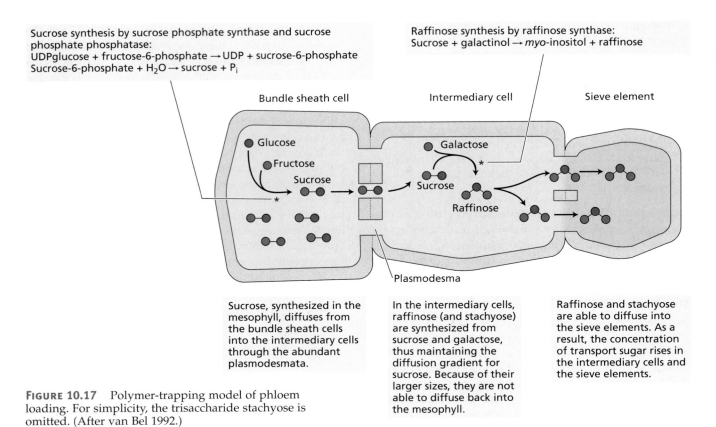

FIGURE 10.17 Polymer-trapping model of phloem loading. For simplicity, the trisaccharide stachyose is omitted. (After van Bel 1992.)

TABLE 10.5
Patterns in apoplastic and symplastic loading

	Apoplastic loading	Symplastic loading
Transport sugar	Sucrose	Oligosaccharides in addition to sucrose
Type of companion cell in the minor veins	Ordinary companion cells or transfer cells	Intermediary cells
Number of plasmodesmata connecting the sieve elements and companion cells to surrounding cells	Few	Abundant

Source: Drawings after van Bel et al. 1992.

Note: Some species may load both apoplastically and symplastically, since different types of companion cells can be found within the veins of a single species.

Which of these characteristics is the best predictor of symplastic loading? The number of plasmodesmata linking the sieve element–companion cell complex and surrounding cells has been considered to be diagnostic for loading type. However, some species that had been classified as symplastic loaders on the basis of plasmodesmatal frequencies have now been shown to load apoplastically. The type of transport sugar (sucrose alone or sucrose plus larger oligosaccharides) may be a more universal predictor of loading type than plasmodesmatal frequencies. Plants with abundant plasmodesmata in the minor vein phloem also tend to possess abundant plasmodesmata between mesophyll cells, so plasmodesmatal frequencies may be related to some other plant characteristic (Turgeon and Medville 2004).

A number of species have more than one type of companion cell in their minor veins. For example, coleus has both intermediary cells and ordinary companion cells. It has been suggested that the symplastic and apoplastic pathways may coexist in some species, simultaneously or at different times, in different sieve elements in the same vein or in sieve elements in veins of different sizes.

Future research will likely reveal new loading pathways or combinations of pathways (Turgeon 2006). Certainly, the evolution of different loading types and the environmental pressures related to their evolution will continue to be important research areas in the future as loading pathways are clarified in more species. At present, apoplastic loading is thought to be the ancestral condition, while symplastic loading is the derived condition. What adaptive advantage is conferred on the symplastic loaders? Only more research will provide the answers to these questions.

Phloem Unloading and Sink-to-Source Transition

Now that we have learned about the events leading up to the export of sugars from sources, let's take a look at **import** into sinks, such as developing roots, tubers, and reproductive structures. In many ways the events in sink tissues are simply the reverse of the events in sources. The following steps are involved in the import of sugars into sink cells.

1. *Phloem unloading.* This is the process by which imported sugars leave the sieve elements of sink tissues.
2. *Short-distance transport.* After unloading, the sugars are transported to cells in the sink by means of a short-distance transport pathway. This pathway has also been called *post–sieve element transport.*
3. *Storage and metabolism.* In the final step, sugars are stored or metabolized in sink cells.

In this section we will discuss the following questions: Are phloem unloading and short-distance transport symplastic or apoplastic? Is sucrose hydrolyzed during the process? Do phloem unloading and subsequent steps require energy? Finally, we will examine the transition process by which a young, importing leaf becomes an exporting source leaf.

Phloem unloading and short-distance transport can occur via symplastic or apoplastic pathways

In sink organs, sugars move from the sieve elements to the cells that store or metabolize them. Sinks vary widely from growing vegetative organs (root tips and young leaves) to storage tissues (roots and stems) to organs of reproduction and dispersal (fruits and seeds). Because sinks vary so greatly in structure and function, there is no single scheme of phloem unloading and short-distance transport. Differences in import pathways due to differences in sink types are emphasized in this section; however, the pathway often depends on the stage of sink development as well.

As in sources, the sugars may move entirely through the symplast via the plasmodesmata in sinks, or they may enter the apoplast at some point. Figure 10.18 diagrams the several possible pathways in sinks. Both unloading and the short-distance pathway appear to be completely symplastic in some young dicot leaves, such as sugar beet and tobacco (*Nicotiana tabacum*) (see Figure 10.18A). Evidence for the symplastic pathway includes insensitivity to *p*-chloromercuribenzenesulfonic acid (PCMBS), a reagent that inhibits the transport of sucrose across plasma membranes but does not permeate the symplastic pathway. Meristematic and elongating regions of primary root tips also appear to unload symplastically. Sufficient plasmodesmata exist in these pathways to support symplastic unloading.

While symplastic import predominates in most sink tissues, part of the short-distance pathway is apoplastic in some sink organs (see Figure 10.18B). The apoplastic step could be located at the site of unloading itself (type 1 in Figure 10.18B) (Zhang et al. 2004) or farther removed from the sieve elements (type 2). This arrangement (type 2), typical of developing seeds, appears to be the most common in apoplastic pathways.

An apoplastic step is required in developing seeds because there are no symplastic connections between the maternal tissues and the tissues of the embryo. Sugars exit the sieve elements (phloem unloading) via a symplastic pathway and are transferred from the symplast to the apoplast at some point removed from the sieve element–companion cell complex (type 2 in Figure 10.18B). The apoplastic step permits membrane control over the substances that enter the embryo, because two membranes must be crossed in the process.

When an apoplastic step occurs in the import pathway, the transport sugar can be partly metabolized in the apoplast, or it can cross the apoplast unchanged (see **Web Topic 10.8**). For example, sucrose can be hydrolyzed into

(A) **Symplastic phloem unloading and short-distance transport**

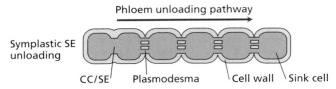

(B) **Apoplastic phloem unloading and short-distance transport**

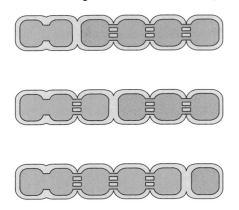

FIGURE 10.18 Pathways for phloem unloading and short-distance transport. The sieve element–companion cell complex (CC/SE) is considered a single functional unit. The presence of plasmodesmata is assumed to provide functional symplastic continuity. An absence of plasmodesmata between cells indicates an apoplastic transport step. (A) Symplastic phloem unloading and short-distance transport. All steps are symplastic. (B) Apoplastic phloem unloading and short-distance transport.

Type 1: This short-distance pathway is designated apoplastic because one step, phloem unloading from the sieve element–companion cell complex occurs in the apoplast. Once the sugars are taken back up into the symplast of adjoining cells, transport is symplastic.

Type 2: These pathways also have an apoplastic step. However, phloem unloading from the sieve element–companion cell complex is symplastic. The apoplastic step occurs later in the pathways. The upper figure (type 2A) shows an apoplastic step close to the sieve element–companion cell complex; the lower figure (type 2B), an apoplastic step that is further removed.

glucose and fructose in the apoplast by invertase, a sucrose-splitting enzyme, and glucose and/or fructose would then enter the sink cells. As we will discuss later, such sucrose-cleaving enzymes play a role in the control of phloem transport by sink tissues.

Transport into sink tissues requires metabolic energy

Inhibitor studies have shown that import into sink tissues is energy dependent. Growing leaves, roots, and storage sinks, in which carbon is stored in starch or protein, utilize symplastic phloem unloading and short-distance transport. Transport sugars are used as substrate for respiration and are metabolized into storage polymers and into compounds needed for growth. Sucrose metabolism results in a low sucrose concentration in the sink cells, thus maintaining a concentration gradient for sugar uptake. No membranes are crossed during sugar uptake into the sink cells, and transport through the plasmodesmata is passive because transport sugars move from a high concentration in the sieve elements to a low concentration in the sink cells. Metabolic energy is thus required in these sink organs for respiration and for biosynthesis reactions.

In apoplastic import, sugars must cross at least two membranes: the plasma membrane of the cell that is exporting the sugar, and the plasma membrane of the sink cell. When sugars are transported into the vacuole of the sink cell, they must also traverse the tonoplast. As discussed earlier, transport across membranes in an apoplastic pathway may be energy dependent. While some evidence indicates that both efflux and uptake of sucrose can be active (see **Web Topic** 10.9), the transporters have yet to be isolated and characterized. A sucrose–H$^+$ symporter important in phloem loading (SUT1) has been found in some sink tissues, for example, potato tubers. The symporter may function in sucrose retrieval from the apoplast, in import, or both.

The transition of a leaf from sink to source is gradual

Leaves of dicots such as tomato or bean begin their development as sink organs. A transition from sink to source status occurs later in development, when the leaf is approximately 25% expanded, and it is usually complete when the leaf is 40 to 50% expanded.

Export from the leaf begins at the tip or apex of the blade and progresses toward the base until the whole leaf becomes a sugar exporter. During the transition period, the tip exports sugar, while the base imports it from the other source leaves (Figure 10.19).

The maturation of leaves is accompanied by a large number of functional and anatomic changes, resulting in a reversal of transport direction from importing to exporting. In general, the cessation of import and the initiation of export are independent events (Turgeon 1984). In albino leaves of tobacco, which have no chlorophyll and therefore are incapable of photosynthesis, import stops at the same developmental stage as in green leaves, even though export is not possible. Therefore some change besides the initiation of export must occur in developing leaves of tobacco that causes them to cease importing sugars.

Sugars are unloaded and loaded almost entirely via different veins in tobacco (Figure 10.20) (Roberts et al. 1997),

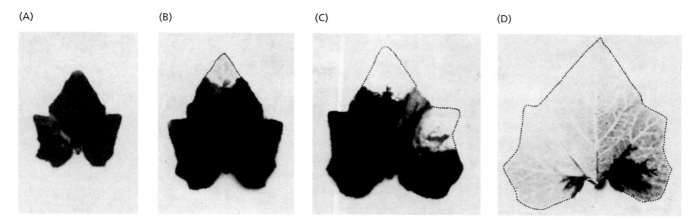

Figure 10.19 Autoradiographs of a leaf of summer squash (*Cucurbita pepo*), showing the transition of the leaf from sink to source status. In each case, the leaf imported ^{14}C from the source leaf on the plant for 2 hours. Label is visible as black accumulations. (A) The entire leaf is a sink, importing sugar from the source leaf. (B–D) The base is still a sink. As the tip of the leaf loses the ability to unload and stops importing sugar (as shown by the loss of black accumulations), it gains the ability to load and to export sugar. (From Turgeon and Webb 1973.)

contributing to the fact that import cessation and export initiation are two separate events. The minor veins that are eventually responsible for most of the loading in tobacco and other *Nicotiana* species do not mature until about the time import ceases and cannot play a role in unloading.

The change that stops import must thus involve blockage of unloading from the large veins at some point in the development of mature leaves. Factors that could account for the cessation of unloading include plasmodesmatal closure, a decrease in plasmodesmatal frequency, or another change in symplastic continuity. Experimental data have shown that both plasmodesmatal closure and elimination of plasmodesmata can occur.

Export of sugars begins when phloem loading has accumulated sufficient photosynthate in the sieve elements to drive translocation out of the leaf. Thus, export is initiated when:

- The unloading pathway is closed.
- Minor veins responsible for loading have matured.
- The leaf is synthesizing photosynthate in sufficient quantity that some is available for export.

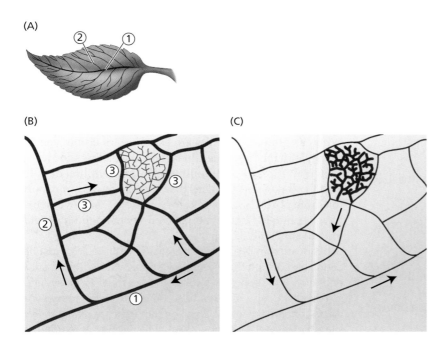

Figure 10.20 Division of labor in the veins of a tobacco leaf shown in (A). When the leaf is immature and still in its sink phase (B), photosynthate is imported from mature leaves and distributed (arrows) throughout the blade (or lamina) via the larger, major veins (thicker lines). The major veins are numbered, with the midrib being the first-order vein. The imported photosynthate unloads from the same major veins into the mesophyll. The smallest, minor veins are shown within the areas enclosed by the third-order veins. The minor veins do not function in import and unloading because they are immature. In a source leaf (C), import has ceased, and export has begun. Photosynthate loads into the minor veins (thicker lines), while the larger veins serve only in export (arrows); they can no longer unload. Although (B) is drawn to scale from an autoradiograph, (C) is not to scale or in correct proportions, since the lamina grows considerably as the leaf matures. (From Turgeon 2006.)

- The sucrose-synthesizing genes are being expressed.
- The sucrose–H$^+$ symporter is in place in the plasmalemma of the sieve element–companion cell complex.

In leaves of plants such as sugar beet and tobacco, the ability to accumulate exogenous [^{14}C]sucrose in the sieve element–companion cell complex is acquired as the leaves undergo the sink-to-source transition, suggesting that the symporter required for loading has become functional. In developing leaves of Arabidopsis, expression of the symporter that is thought to transport sugars during loading begins in the tip and proceeds to the base during a sink-to-source transition. The same basipetal pattern is seen in the development of export capacity.

Photosynthate Distribution: Allocation and Partitioning

The photosynthetic rate determines the total amount of fixed carbon available to the leaf. However, the amount of fixed carbon available for translocation depends on subsequent metabolic events. The regulation of the distribution of fixed carbon into various metabolic pathways is termed **allocation**.

The vascular bundles in a plant form a system of "pipes" that can direct the flow of photosynthates to various sinks: young leaves, stems, roots, fruits, or seeds. However, the vascular system is often highly interconnected, forming an open network that allows source leaves to communicate with multiple sinks. Under these conditions, what determines the volume of flow to any given sink? The differential distribution of photosynthates within the plant is termed **partitioning**.

After giving an overview of allocation and partitioning, we will examine the coordination of starch and sucrose synthesis. We will conclude by discussing how sinks compete, how sink demand might regulate photosynthetic rate in the source leaf, and how sources and sinks communicate with each other.

Allocation includes storage, utilization, and transport

The carbon fixed in a source cell can be used for storage, metabolism, and transport:

- Synthesis of storage compounds. Starch is synthesized and stored within chloroplasts and, in most species, is the primary storage form that is mobilized for translocation during the night. Plants that store carbon primarily as starch are called starch storers.
- Metabolic utilization. Fixed carbon can be utilized within various compartments of the photosynthesizing cell to meet the energy needs of the cell or to provide carbon skeletons for the synthesis of other compounds required by the cell.
- Synthesis of transport compounds. Fixed carbon can be incorporated into transport sugars for export to various sink tissues. A portion of the transport sugar can also be stored temporarily in the vacuole (see **Web Topic 10.8**).

Allocation is also a key process in sink tissues. Once the transport sugars have been unloaded and enter the sink cells, they can remain as such or can be transformed into various other compounds. In storage sinks, fixed carbon can be accumulated as sucrose or hexose in vacuoles or as starch in amyloplasts. In growing sinks, sugars can be utilized for respiration and for the synthesis of other molecules required for growth.

Various sinks partition transport sugars

Sinks compete for the photosynthate being exported by the sources. Such competition determines the distribution of transport sugars among the various sink tissues of the plant (partitioning), at least in the short term. The processes of allocation, including the ability of a sink to store or metabolize imported sugar, affects its ability to compete for available sugars. In this way, partitioning and allocation interact with each other.

Of course, events in sources and sinks must be synchronized. Partitioning determines the patterns of growth, and such growth must be balanced between shoot growth (photosynthetic productivity) and root growth (water and mineral uptake). So an additional level of control lies in the interaction between areas of supply and demand. Turgor pressure in the sieve elements could be an important means of communication between sources and sinks, acting to coordinate rates of loading and unloading. Chemical messengers are also important in communicating the status of one organ to the other organs in the plant. Such chemical messengers include plant hormones and nutrients, such as potassium and phosphate, and even the transport sugars themselves. Recent findings suggest that macromolecules (RNA and protein) may also play a role in photosynthate partitioning, perhaps by influencing transport through plasmodesmata.

Attainment of higher yields of crop plants is one goal of research on photosynthate allocation and partitioning. Whereas grains and fruits are examples of edible yields, total yield includes inedible portions of the shoot. An understanding of partitioning enables plant breeders to select and develop varieties that have improved transport to edible portions of the plant. Significant improvements have been made in the ratio of commercial or edible yield to total shoot yield.

Allocation and partitioning in the whole plant must be coordinated such that increased transport to edible tissues does not occur at the expense of other essential processes and structures. Crop yield will also be improved if photosynthates that are normally "lost" by the plant are retained. For example, losses due to nonessential respiration or exudation from roots could be reduced. In the latter case, care must be taken not to disrupt essential processes outside the plant, such as growth of beneficial microbial species in the vicinity of the root that obtain nutrients from the root exudate.

Source leaves regulate allocation

Increases in the rate of photosynthesis in a source leaf generally result in an increase in the rate of translocation from the source. Control points for the allocation of photosynthate (Figure 10.21) include the allocation of triose phosphates to the following processes:

- Regeneration of intermediates in the C_3 photosynthetic carbon reduction cycle (the Calvin cycle; see Chapter 8)
- Starch synthesis
- Sucrose synthesis, as well as distribution of sucrose between transport and temporary storage pools

Various enzymes operate in the pathways that process the photosynthate, and the control of these steps is complex (Geiger and Servaites 1994).

During the day the rate of starch synthesis in the chloroplast must be coordinated with sucrose synthesis in the cytosol. Triose phosphates (glyceraldehyde-3-phosphate and dihydroxyacetone phosphate) produced in the chloroplast by the Calvin cycle (see Chapter 8) can be used for either starch or sucrose synthesis. Sucrose synthesis in the cytoplasm diverts triose phosphate away from starch synthesis and storage. For example, it has been shown that when the demand for sucrose by other parts of a soybean plant is high, less carbon is stored as starch by the source leaves. The key enzymes involved in the regulation of sucrose synthesis in the cytoplasm and of starch synthesis in the chloroplast are sucrose phosphate synthase and fructose-1,6-bisphosphatase in the cytoplasm and ADP-glucose pyrophosphorylase in the chloroplast (see Figure 10.21) (see Chapter 8).

However, there is a limit to the amount of carbon that normally can be diverted from starch synthesis in species that store carbon primarily as starch. Studies of allocation between starch and sucrose under different conditions suggest that a fairly steady rate of translocation throughout the 24-hour period is a priority for most plants.

The use of mutants and transgenic plants enables us to ask a new set of questions about allocation. For example, what happens when one of the competing processes, sucrose or starch synthesis, is inhibited or enhanced? Starch-deficient tobacco mutants synthesize only trace amounts of starch but are able to compensate for a lack of stored carbon by doubling the rate of sucrose synthesis and export during the day and by switching most of their growth to the day (Geiger et al. 1995).

In contrast, potato plants accumulate four times more starch during the daytime when the phosphate translocator activity (see Figure 10.21) is decreased 30% due to anti-

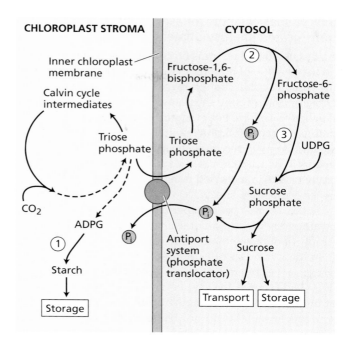

FIGURE 10.21 A simplified scheme for starch and sucrose synthesis during the day. Triose phosphate, formed in the Calvin cycle, can either be utilized in starch formation in the chloroplast or transported into the cytosol in exchange for inorganic phosphate (P_i) via the phosphate translocator in the inner chloroplast membrane. The outer chloroplast membrane is porous to small molecules and is omitted here for clarity. In the cytosol, triose phosphate can be converted to sucrose for either storage in the vacuole or transport. Key enzymes involved are starch synthetase (1), fructose-1,6-bisphosphatase (2), and sucrose phosphate synthase (3). The second and third enzymes, along with ADP-glucose pyrophosphorylase, which forms adenosine diphosphate glucose (ADPG), are regulated enzymes in sucrose and starch synthesis (see Chapter 8). UDPG, uridine diphosphate glucose. (After Preiss 1982.)

sense repression of its RNA. Under these conditions, fixed carbon cannot exit the chloroplast as well as in the control, and more starch is synthesized. These plants respond to increased starch accumulation by exporting more of their fixed carbon during the night, when a different transporter is operating between the chloroplast and the cytoplasm.

Like the potato plants just described, tobacco plants with an 80% reduction of phosphate-translocator activity also accumulate more starch in the chloroplasts and utilize the nighttime translocator to transport fixed carbon out of the chloroplast. However, they use the starch reserves to synthesize translocated sugar during the day, rather than during the night. Plants in which the phosphate translocator activity is inhibited generally show no obvious changes in growth. They simply bypass the phosphate translocator and utilize alternate means to move carbon out of the chloroplast. These and other results reveal the amazing flexibility of plants (Flugge et al. 2003).

Sink tissues compete for available translocated photosynthate

As discussed earlier, translocation to sink tissues depends on the position of the sink in relation to the source and on the vascular connections between source and sink. Another factor determining the pattern of transport is competition between sinks. For example, young leaves might compete with roots for photosynthates in the translocation stream. Competition has been shown by numerous experiments in which removal of a sink tissue from a plant generally results in increased translocation to alternative, and hence competing, sinks.

In the reverse type of experiment, the source supply can be altered while the sink tissues are left intact. When the supply of photosynthates from sources to competing sinks is suddenly and drastically reduced by shading of all the source leaves but one, the sink tissues become dependent on a single source. In sugar beet and bean plants, the rates of photosynthesis and export from the single remaining source leaf usually do not change over the short term (approximately 8 hours; Fondy and Geiger 1980). However, the roots receive less sugar from the single source, while the young leaves receive relatively more. Shading in general decreases partitioning to roots. Presumably, the young leaves can deplete the sugar content of the sieve elements more readily and thus increase the pressure gradient and the rate of translocation toward themselves.

Treatments such as making the sink water potential more negative increase the pressure gradient and enhance transport to the sink. Treatment of the root tips of pea (*Pisum sativum*) seedlings with 350 mM mannitol solutions increased the import of [^{14}C]sucrose over the short term by more than 300%, presumably because of a turgor decrease in the sink cells (Schulz 1994).

Sink strength depends on sink size and activity

The ability of a sink to mobilize photosynthate toward itself is often described as **sink strength**. Sink strength depends on two factors—sink size and sink activity—as follows:

$$\text{Sink strength} = \text{sink size} \times \text{sink activity}$$

Sink size is the total biomass of the sink tissue, and **sink activity** is the rate of uptake of photosynthates per unit biomass of sink tissue. Altering either the size or the activity of the sink results in changes in translocation patterns. For example, the ability of a pea pod to import carbon depends on the dry weight of that pod as a proportion of the total number of pods (Jeuffroy and Warembourg 1991).

Changes in sink activity can be complex, because various activities in sink tissues can potentially limit the rate of uptake by the sink. These activities include unloading from the sieve elements, metabolism in the cell wall, uptake from the apoplast, and metabolic processes that use the photosynthate in either growth or storage.

Conventional treatments to manipulate sink strength were often nonspecific. For example, cooling a sink tissue, which inhibits any activities that require metabolic energy, often results in a decrease in the speed of transport toward the sink. More recent experiments take advantage of our ability to specifically over- or underexpress enzymes related to sink activity, for example, those involved in sucrose metabolism in the sink. The two major enzymes that split sucrose are acid invertase and sucrose synthase, both of which can catalyze the first step in sucrose utilization. We will use invertase activity to illustrate the relationship between the activity of sucrose-splitting enzymes and sink demand.

Acid invertase, which cleaves sucrose into a fructose and a glucose molecule, can be found bound to the cell wall or accumulated in the vacuole. Carrot plants (*Daucus carota*) display altered sucrose partitioning when the cell-wall invertase activity in the roots is decreased due to antisense suppression. Partitioning to the roots is decreased, reducing taproot development. Since these plants also have more leaves than the controls, their dry weight leaf-to-root ratio is greatly increased. Both sucrose and starch levels are increased severalfold in the leaves of the transformed plants. Most of the plants show an opposite change in carbohydrate concentrations in the roots. Since unloading into carrot storage roots is most likely apoplastic, decreasing the formation of glucose and fructose in the cell wall would prevent sugar uptake into the sink storage tissues, causing these profound changes in the partitioning patterns in the plant (Tang et al. 1999).

Genes for invertase and sucrose synthase are often expressed at different times during sink development. In bean pods and corn (maize; *Zea mays*) kernels, changes in invertase activity are found to precede changes in photosynthate import. These results point to a key role of invertase and sucrose synthase in controlling import patterns, both during the genetic program of sink development and during responses to environmental stresses (see **Web Topic 10.8**).

The source adjusts over the long term to changes in the source-to-sink ratio

If all but one of the source leaves of a soybean plant are shaded for an extended period (e.g., 8 days), many changes occur in the single remaining source leaf. These changes include a decrease in starch concentration and increases in photosynthetic rate, rubisco activity, sucrose concentration, transport from the source, and orthophosphate concentration (Thorne and Koller 1974). These data indicate that, besides the observed short-term changes in the distribution of photosynthate among different sinks, the metabolism of the source adjusts to the altered conditions in longer-term experiments.

Photosynthetic rate (the net amount of carbon fixed per unit leaf area per unit time) often increases over several days when sink demand increases, and it decreases when sink demand decreases. An accumulation of photosynthate (sucrose or hexoses) in the source leaf can account for the linkage between sink demand and photosynthetic rate in starch-storing plants (see **Web Topic 10.10**). Sugars act as signaling molecules that regulate many metabolic and developmental processes in plants. In general, carbohydrate depletion enhances the expression of genes for photosynthesis, reserve mobilization, and export processes, while abundant carbon resources favor genes for storage and utilization (Koch 1996; Halford and Paul 2003).

Sucrose or hexoses that would accumulate as a result of decreased sink demand are well known to repress photosynthetic genes. Interestingly, the genes for invertase and sucrose synthase (discussed in the previous section) and genes for sucrose–H^+ symporters (discussed in the section above, *Regulating sucrose loading*) are also among those regulated by carbohydrate supply.

The Transport of Signaling Molecules

Besides its major function in the long-distance transport of photosynthate, the phloem is also a conduit for the transport of signaling molecules from one part of the organism to another. Such long-distance signals coordinate the activities of sources and sinks and regulate plant growth and development. As indicated earlier, the signals between sources and sinks might be physical or chemical. Physical signals such as turgor change could be transmitted rapidly via the interconnecting system of sieve elements. Molecules traditionally considered to be chemical signals, such as proteins and plant hormones, are found in the phloem sap, as are mRNAs and small RNAs, which have more recently been added to the list of signal molecules. The translocated carbohydrates themselves may also act as signals.

Turgor pressure and chemical signals coordinate source and sink activities

Turgor pressure may play a role in coordinating the activities of sources and sinks. For example, if phloem unloading were rapid under conditions of rapid sugar utilization at the sink tissue, turgor pressures in the sieve elements of sinks would be reduced, and this reduction would be transmitted to the sources. If loading were controlled in part by turgor in the sieve elements of the source, loading would increase in response to this signal from the sinks. The opposite response would be seen when unloading was slow in the sinks. Some data suggest that cell turgor can modify the activity of the proton-pumping ATPase at the plasma membrane and therefore alter transport rates.

Shoots produce growth regulators such as auxin (see Chapter 19), which can be rapidly transported to the roots via the phloem, and roots produce cytokinins (see Chapter 21), which move to the shoots through the xylem. Gibberellins (GA) and abscisic acid (ABA) (see Chapters 20 and 23) are also transported throughout the plant in the vascular system. Plant hormones play a role in regulating source–sink relationships. They affect photosynthate partitioning in part by controlling sink growth, leaf senescence, and other developmental processes.

Loading of sucrose has been shown to be stimulated by exogenous auxin but inhibited by ABA in some source tissues, while exogenous ABA enhances, and auxin inhibits, sucrose uptake by some sink tissues. Active transporters in plasma membranes are obvious targets for regulation of apoplastic loading and unloading by hormones, for example by effects on the levels of transporter proteins. Other potential sites of hormone regulation of unloading include tonoplast transporters, enzymes for metabolism of incoming sucrose, wall extensibility, and plasmodesmatal permeability in the case of symplastic unloading (see the next section).

As indicated earlier, carbohydrate levels can influence the expression of genes encoding photosynthesis component genes, as well as genes involved in sucrose hydrolysis. Many genes have been shown to be responsive to sugar depletion and abundance (Koch 1996). Thus, not only is sucrose transported in the phloem, but sucrose or its metabolites can also act as signals that modify the activities of sources and sinks. As discussed earlier, sucrose–H^+ symporter mRNA declines in sugar-beet source leaves fed exogenous sucrose through the xylem. The decline in symporter mRNA is accompanied by a loss of symporter activity in plasma membrane vesicles isolated from the leaves. A working model includes the following steps:

1. Decreased sink demand leads to high sucrose levels in the vascular tissue.
2. High sucrose levels lead to down-regulation of the symporter in the source.
3. Decreased loading results in increased sucrose concentrations in the source.

Increased sucrose concentrations in the source can result in a lower photosynthetic rate (see **Web Topic 10.10**). An increase of starch accumulation in source leaves of plants transformed with antisense DNA to the sucrose–H^+ symporter SUT1 also supports this model (Schulz et al. 1998).

Sugars and other metabolites have been shown to interact with hormonal signals to control and integrate many plant processes. Gene expression in some source–sink systems responds to both sugar and hormonal signals (Thomas and Rodriguez 1994).

Signal molecules in the phloem regulate growth and development

It has long been known that viruses can move in the phloem, traveling as complexes of proteins and nucleic acids or as intact virus particles. More recently, endogenous RNA molecules and proteins have been found in phloem sap, and at least some of these could be signal molecules or could generate phloem-mobile signals.

To be assigned a signaling role in plants, a macromolecule must meet a number of significant criteria (Oparka and Santa Cruz 2000):

- The macromolecule must move from source to sink in the phloem.

- The macromolecule must be able to leave the sieve element–companion cell complex in sink tissues. Alternatively, the macromolecule might trigger the formation of a second signal that transmits information to the sink tissues surrounding the phloem; that is, it might initiate a signal cascade.

- Perhaps most importantly, the macromolecule must be able to modify the functions of specific cells in the sink.

How well do various macromolecules in the phloem meet these criteria?

PROTEINS MAY SERVE AS SIGNAL MOLECULES Proteins synthesized in companion cells can clearly enter the sieve elements through the plasmodesmata that connect the two cell types. As noted earlier, the P-proteins in cucurbit sap, PP1 and PP2, appear to be synthesized in companion cells. The plasmodesmata connecting the companion cells and sieve elements must thus allow these macromolecules to move across them. Furthermore, subunits of PP1 and PP2 move with the translocation stream to sink tissues.

Subunits of P-proteins from cucumber (*Cucumis sativus*) can move across graft unions from a cucumber stock (basal graft partner) to a pumpkin (*Cucurbita maxima*) scion (upper graft partner). One experiment showed that the smaller PP2 protein is able to move from the sieve elements to companion cells of the scion stem; the larger PP1 was not detected in the companion cells. Neither protein was able to move beyond the sieve element–companion cell complex (Golecki et al. 1999). Furthermore, no definite signaling function has yet been established for either protein.

Other proteins move in the translocation stream from sources to sinks. For example, passive movement of proteins from companion cells to sieve elements has been demonstrated in Arabidopsis and tobacco plants. These plants were transformed with the gene for green fluorescent protein (GFP) from jellyfish, under control of the *SUC2* promoter from Arabidopsis. The SUC2 sucrose–H⁺ symporter is synthesized within the companion cells, so proteins expressed under the control of its promoter, including GFP, are also synthesized in the companion cells. GFP, which is localized by its fluorescence after excitation with blue light, moves through plasmodesmata from companion cells into sieve elements of source leaves (Figure 10.22A) and migrates within the phloem to sink tissues.

Finally, the jellyfish green fluorescent protein is unloaded symplastically through the plasmodesmata into sink tissues, such as seed coats, anthers, root tips, and mesophyll cells in

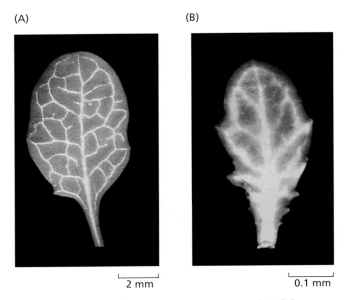

FIGURE 10.22 GFP fluorescence in source and sink leaves from transgenic Arabidopsis plants expressing GFP under control of the *SUC2* promoter. (A) GFP is synthesized in companion cells and moves into the sieve elements of the source, as indicated by the bright fluorescence in the veins. (B) Free GFP is imported into sink leaves and moves into the surrounding mesophyll. Because GFP has moved into surrounding tissues, the veins are no longer distinctly delineated, and GFP fluorescence is much more diffuse. Even though the source leaf in (A) appears to be the same size as the sink leaf in (B), the source leaf is actually much larger. The scales in (A) and (B) are different, as indicated by the bars on the photographs. (From Stadler et al. 2005.)

importing leaves (see Figure 10.22B) (Imlau et al. 1999, Stadler et al. 2005). Because jellyfish GFP is unlikely to possess specific sequences for interaction with plasmodesmatal structures, its movement into and out of sieve elements is likely to occur by passive diffusion. Some soluble fusion proteins between GFP and other proteins (ubiquitin, for example), expressed in Arabidopsis companion cells, show similar transport patterns. However, only free GFP is able to move symplastically into sink tissues of the root (Stadler et al. 2005).

Phloem transport of proteins that modify cellular functions has also been demonstrated. The PRms protein is a pathogenesis-related protein that is involved in the defense response of maize against fungal pathogens. This protein has been located in plasmodesmata between companion cells and sieve elements in transgenic tobacco plants. Localization was accomplished using an anti-PRms antibody. PRms protein moves into and through the phloem as well, as shown by its presence in tobacco phloem exudate and its movement across graft unions. PRms can move across graft unions from a transgenic stock to a nontransgenic control (Bortolotti et al. 2005). Unloading of the protein has not yet been demonstrated, however, nor is its exact mechanism of action known.

Clearly, proteins can be transported from the companion cells in the source through the intervening sieve elements to sink companion cells. However, little evidence exists for a similar movement of proteins synthesized outside the companion cells. Other signals from outside the sieve element–companion cell complex may give rise to the production of mobile proteins in the companion cells.

RNAS MAY SERVE AS SIGNAL MOLECULES RNAs transported in the phloem consist of endogenous mRNAs, pathogenic RNAs, and small RNAs associated with gene silencing (see Chapter 14 on the web site). At least some of these RNAs appear to travel in the phloem as complexes of RNA and protein (ribonucleoproteins [RNPs]) (Gomez et al. 2005). RNPs in the phloem resemble the complexes formed by viral RNAs with their movement proteins. Viral "movement proteins" interact directly with plasmodesmata to allow the passage of viral nucleic acids between cells.

Pathogenic RNAs—for example, viroids—are found in the phloem exudate of cucumber in a complex with cucumber PP2, a phloem structural protein well known for its ability to seal off damaged sieve elements. In addition, PP2 and the viroid RNA are simultaneously translocated from a cucumber stock to a pumpkin scion. Following transport, the viroid is capable of causing disease symptoms in the pumpkin scion (Gomez and Pallas 2004).

Posttranscriptional gene silencing (see Chapter 14 on the web site) is one system that clearly demonstrates the role of RNAs in long-distance signaling in the phloem (van Bel 2003). RNA silencing of gene expression causes degradation of target mRNAs, for example, excess or foreign mRNA transcribed from transgenes or viral nucleic acid. When the transgenes are homologous to host genes, the host genes are also silenced. RNA silencing is triggered by the presence of double-stranded RNAs, induced by the virus or transgene. The double-stranded RNAs are processed by proteins called Dicer-like proteins to form small RNAs called short interfering RNAs (siRNAs). These siRNAs are incorporated into complexes that carry out the degradation of the target mRNAs. In this way, the gene is "silenced."

In plants, RNA silencing plays a role not only in defense processes against foreign mRNAs but also in developmental processes. Small RNAs called microRNAs (miRNAs) are formed similarly to siRNAs and likewise are incorporated into complexes that mediate degradation of mRNA or inhibit translation of mRNA (Yu and Kumar 2003) (see Chapter 16).

Small RNAs appear to meet all the requirements for signaling molecules: entry into sieve elements and transport in the phloem to sink tissues, where functional changes are triggered. Small RNAs shown to be authentic regulatory RNAs (siRNA and miRNA) are found in phloem sap, and their levels respond to growth conditions and viral infection (Yoo et al. 2004). Furthermore, small RNAs can be transported through the phloem. An siRNA complementary to a viral coat-protein gene sequence was found in the sap of a wild-type cucumber scion that had been grafted onto a squash stock silenced for transgene coat-protein expression. Silencing occurred in the tissues of a scion from a nonsilenced line that had been grafted onto a stock from a silenced squash line; that is, the levels of coat protein mRNA were reduced in the nonsilenced line to the same low levels as found in the silenced line (Yoo et al. 2004).

Finally, the RNA transported in the phloem can cause visible changes in the sink. *FLOWERING LOCUS T (FT)* mRNA appears to be a significant component of the floral stimulus that moves from leaf to apex in response to conditions inductive for flowering (see Chapter 25). Changes in phenotype (appearance) due to the transport of RNA to a scion has been shown in a number of grafting experiments: the changes include chlorosis in tobacco due to posttranscriptional gene silencing and changes in leaf morphology in tomato due to transport of a mutant homeobox mRNA. In both cases, RNA movement occurred in an acropetal (toward the apex) direction, but movement in the phloem was not confirmed. However, mRNA for a regulator of gibberellic acid responses (called GAI) was localized to sieve elements and companion cells of pumpkin and was found in pumpkin phloem sap. Transgenic tomato plants expressing a mutant version of the regulator gene were dwarf and dark green. The mRNA for the mutant regulator was localized to sieve elements, was able to be transported across graft unions into wild type scions, and was unloaded into apical tissues. As a result, the mutant phenotype developed in new growth on the wild-type scion (Haywood et al. 2005).

PLASMODESMATA PLAY A ROLE IN SIGNALING Plasmodesmata have been implicated in nearly every aspect of phloem translocation, from loading to long-distance transport (pores in sieve areas and sieve plates are modified plasmodesmata) to allocation and partitioning. What role might plasmodesmata play in macromolecular signaling in the phloem?

The mechanism of plasmodesmatal transport (called trafficking) can be either passive (nontargeted) or selective and regulated (Cilia and Jackson 2004). When a molecule moves passively, its size must be smaller than the *size exclusion limit* (*SEL*) of the plasmodesmata. As indicated earlier, green fluorescent protein moves passively through plasmodesmata. In contrast, when a molecule moves in a selective fashion, it must possess a trafficking signal or be targeted in some other way to the plasmodesmata. The transport of viral movement proteins and developmental transcription factors probably occurs by means of a selective mechanism. Viral movement proteins have been shown in a number of cases to use endogenous cellular proteins to "target" the plasmodesmata as their destination. Once at the plasmodesmata, movement proteins and/or the cellular proteins act to increase the SEL of the plasmodesmata to allow the viral genome to move between cells.

The endogenous proteins commandeered by viruses to facilitate their specific movement to and through plasmodesmata are thought to carry out similar functions for endogenous macromolecules. Proteins have been identified from phloem exudate that bind RNA and increase the SEL of plasmodesmata between mesophyll cells (for example, PP2 and CmPP16); in addition, proteins have been identified in plasmodesmata that interact with the exudate proteins (for example, NtNCAP1). The control over plasmodesmatal transport exerted by NtNCAP1 appears to be specific and selective (Ding et al. 2003). It is believed that proteins such as CmPP16 are involved in macromolecular trafficking into sieve elements and long-distance transport in the phloem. Recently, CmPP16 from pumpkin was introduced into rice sieve elements through aphid stylets; the transport of CmPP16 to roots appeared to be selective (Aoki et al. 2005). Although it has not been shown to increase plasmodesmatal permeability, the maize pathogenesis-related protein (PRms) described earlier increases sucrose efflux from source leaves of transgenic tobacco plants. Allocation into sucrose increases, while allocation into starch decreases in such plants (Murillo et al. 2003).

It is fitting to end this chapter with research topics that will continue to engage plant physiologists of the future: regulation of growth and development via the transport of endogenous RNA and protein signals, the nature of the proteins that facilitate the transport of signals through plasmodesmata, and the possibility of targeting signals to specific sinks in contrast to mass flow. Many other potential areas of inquiry have been indicated in this chapter as well, such as the mechanism of phloem transport in gymnosperms, the possibility of additional modes of phloem loading, and the nature of phloem unloading. As always in science, an answer to one question generates more questions!

Summary

Translocation in the phloem is the movement of the products of photosynthesis from mature leaves to areas of growth and storage. The phloem also transmits chemical signals between sources and sinks and redistributes water and various compounds throughout the plant body.

Some aspects of phloem translocation have been well established by extensive research over many years. These include the following:

- *The pathway of translocation.* Sugars and other organic materials are conducted throughout the plant in the phloem, specifically in cells called sieve elements. Sieve elements display a variety of structural adaptations that make them well suited for transport.

- *Patterns of translocation.* Materials are translocated in the phloem from sources (areas of photosynthate supply) to sinks (areas of metabolism or storage of photosynthate). Sources are usually mature leaves. Sinks include organs such as roots and immature leaves and fruits.

- *Materials translocated in the phloem.* The translocated solutes are mainly carbohydrates, and sucrose is the most commonly translocated sugar. Phloem sap also contains other organic molecules, such as amino acids and proteins, RNAs, and plant hormones, as well as inorganic ions.

- *Rates of movement.* Rates of movement in the phloem are quite rapid, well in excess of rates of diffusion. Velocities average 1 m h^{-1}, and mass transfer rates range from 1 to $15 \text{ g h}^{-1}\text{cm}^{-2}$ of sieve elements.

Other aspects of phloem translocation require further investigation, and most of these are being studied intensively at the present time. These aspects include the following:

- *Phloem loading and unloading.* Transport of sugars into and out of the sieve elements is called phloem loading and unloading, respectively. In some species, sugars must enter the apoplast of the source leaf before loading. In these plants, loading into the sieve elements requires metabolic energy, provided in the form of a proton gradient. In other species, the whole pathway from the photosynthesizing cells to the sieve elements occurs in the symplast of the source leaf. In either case, phloem loading is specific for the transported sugar. Phloem unloading requires metabolic energy, but the transport pathway, the site of metabolism of transport sugars, and the site where energy is expended vary with the organ and species.

- *Photosynthate allocation and partitioning.* Allocation is the regulation of the quantities of fixed carbon that are channeled into various metabolic pathways. In sources, the regulatory mechanisms of allocation determine the quantities of fixed carbon that will be stored (usually as starch), metabolized within the cells of the source, or immediately transported to sink tissues. In sinks, transport sugars are allocated to growth processes or to storage. Partitioning is the differential distribution of photosynthates within the whole plant. Partitioning mechanisms determine the quantities of fixed carbon delivered to specific sink tissues. Phloem loading and unloading, and photosynthate allocation and partitioning, are of great research interest because of their roles in crop productivity.

- *The transport of signal molecules.* The phloem is also thought to transport signaling molecules from one part of the organism to another. Such long-distance signals coordinate the activities of sources and sinks and regulate plant growth and development. The signals between sources and sinks might be physical, such as changes in turgor pressure, or chemical, such as RNA and protein molecules.

Web Material

Web Topics

10.1 Classical Studies on Phloem Transport
Classical experiments illustrate some basic properties of phloem transport.

10.2 Nitrogen Transport in Soybean
Nitrogen compounds synthesized in the roots are transferred from the xylem to the phloem.

10.3 Sampling Phloem Sap
Aphid stylets are optimally suited to sample phloem sap.

10.4 Monitoring Traffic on the Sugar Freeway
Sugar transport rates in the phloem are measured with radioactive tracers.

10.5 Evidence for Apoplastic Loading of Sieve Elements
Transgenic plants have provided experimental support for apoplastic loading.

10.6 Some Substances Enter the Phloem by Diffusion
Substances such as plant hormones might enter the phloem by diffusion.

10.7 Localization of the Sucrose–H$^+$ Symporter in the Phloem of Apoplastic Loaders
The sucrose–H$^+$ symporter of companion cells has been localized using fluorescent dyes.

10.8 Sugars in the Phloem
The transport, allocation, and metabolism of phloem sugars are tightly regulated.

10.9 Energy Requirements for Unloading in Developing Seeds and Storage Organs
Unloading of seed storage sugars into the embryo is mediated by active transporters.

10.10 Possible Mechanisms Linking Sink Demand and Photosynthetic Rate in Starch Storers
Photosynthate accumulation decreases the photosynthetic rate.

Chapter References

Aoki, K., Suzui, N., Fujimaki, S., Dohmae, N., Yonekura-Sakakibara, K., Fujiwara, T., Hayashi, H. Yamaya, T., and Sakakibara, H. (2005) Destination-selective long-distance movement of phloem proteins. *Plant Cell* 17: 1801–1814.

Bortolotti, C., Murillo, I., Fontanet, P., Coca, M., and San Segundo, B. (2005) Long-distance transort of the maize pathogenesis-related PRms protein through the phloem in transgenic tobacco plants. *Plant Sci.* 168: 813–821.

Brentwood, B., and Cronshaw, J. (1978) Cytochemical localization of adenosine triphosphatase in the phloem of *Pisum sativum* and its relation to the function of transfer cells. *Planta* 140: 111–120.

Cilia, M. L. and Jackson, D. (2004) Plasmodesmata form and function. *Curr. Opin. Cell Biol.* 16: 500–506.

Clark, A. M., Jacobsen, K. R., Bostwick, D. E., Dannenhoffer, J. M., Skaggs, M. I., and Thompson, G. A. (1997) Molecular characterization of a phloem-specific gene encoding the filament protein, phloem protein 1 (PP1), from *Cucurbita maxima*. *Plant J.* 12: 49–61.

Daie, J., and Wyse, R. E. (1985) Evidence on the mechanism of enhanced sucrose uptake at low cell turgor in leaf discs of *Phaseolus coccineus* cultivar scarlet. *Physiol. Plant.* 64: 547–552.

Dinant, S., Clark, A. M., Zhu, Y., Vilaine, F., Palauqui, J.-C., Kusiak, C., and Thompson, G. A. (2003) Diversity of the superfamily of phloem lectins (phloem protein 2) in angiosperms. *Plant Physiol.* 131: 114–128.

Ding, B., Itaya, A., and Qi, Y. (2003) Symplasmic protein and RNA traffic: Regulatory points and regulatory factors. *Curr. Opin. Plant Biol.* 6: 596–602.

Doering-Saad, C., Newbury, H. J., Bale, J. S., and Pritchard, J. (2002) Use of aphid stylectomy and RT-PCR for the detection of transporter mRNAs in sieve elements. *J. Exp. Bot.* 53: 631–637.

Ehlers, K., Knoblauch, M. and van Bel, A. J. E. (2000) Ultrastructural features of well-preserved and injured sieve elements: Minute clamps keep the phloem transport conduits free for mass flow. *Protoplasma* 214: 80–92.

Evert, R. F. (1982) Sieve-tube structure in relation to function. *BioScience* 32: 789–795.

Evert, R. F., and Mierzwa, R. J. (1985) Pathway(s) of assimilate movement from mesophyll cells to sieve tubes in the *Beta vulgaris* leaf. In *Phloem Transport. Proceedings of an International Conference on Phloem Transport, Asilomar, CA*, J. Cronshaw, W. J. Lucas, and R. T. Giaquinta, eds. Liss, New York, pp. 419–432.

Fisher, D. B. (1978) An evaluation of the Münch hypothesis for phloem transport in soybean. *Planta* 139: 25–28.

Flugge, U.-I., Hausler, R. E., Ludewig, F., and Fischer, K. (2003) Functional genomics of phosphate antiport systems of plastids. *Physiol. Plant.* 118: 475–482.

Fondy, B. R. (1975) Sugar selectivity of phloem loading in *Beta vulgaris, vulgaris* L. and *Fraxinus americanus, americana* L. Thesis, University of Dayton, Dayton, OH.

Fondy, B. R., and Geiger, D. R. (1980) Effect of rapid changes in sink–source ratio on export and distribution of products of photosynthesis in leaves of *Beta vulgaris* L. and *Phaseolus vulgaris* L. *Plant Physiol.* 66: 945–949.

Geiger, D. R., and Servaites, J. C. (1994) Diurnal regulation of photosynthetic carbon metabolism in C_3 plants. *Annu. Rev. Plant Physiol. Plant Mol. Biol.* 45: 235–256.

Geiger, D. R., and Sovonick, S. A. (1975) Effects of temperature, anoxia and other metabolic inhibitors on translocation. In *Transport in Plants*, 1: *Phloem Transport* (Encyclopedia of Plant Physiology, New Series, Vol. 1), M. H. Zimmerman and J. A. Milburn, eds., Springer, New York, pp. 256–286.

Geiger, D. R., Giaquinta, R. T., Sovonick, S. A., and Fellows, R. J. (1973) Solute distribution in sugar beet leaves in relation to phloem loading and translocation. *Plant Physiol.* 52: 585–589.

Geiger, D. R., Shieh, W.-J., and Yu, X.-M. (1995) Photosynthetic carbon metabolism and translocation in wild-type and starch-deficient mutant *Nicotiana sylvestris* L. *Plant Physiol.* 107: 507–514.

Giaquinta, R. T., and Geiger, D. R. (1977) Mechanism of cyanide inhibition of phloem translocation. *Plant Physiol.* 59: 178–180.

Golecki, B., Schulz, A., and Thompson, G. A. (1999) Translocation of structural P proteins in the phloem. *Plant Cell* 11: 127–140.

Gomez, G., and Pallas, V. (2004) A long-distance translocatable phloem protein from cucumber forms a ribonucleoprotein complex in vivo with *Hop stunt viroid* RNA. *J. Virology* 78: 10104–10110.

Gomez, G., Torres, H., and Pallas, V. (2005) Identification of translocatable RNA-binding phloem proteins from melon, potential components of the long-distance RNA transport system. *Plant J.* 41: 107–116.

Grusak, M. A., Beebe, D. U., and Turgeon, R. (1996) Phloem loading. In *Photoassimilate Distribution in Plants and Crops:*

Source–Sink Relationships, E. Zamski and A. A. Schaffer, eds., Dekker, New York, pp. 209–227.

Halford, N. G. and Paul, M. J. (2003) Carbon metabolite sensing and signaling. *Plant Biotech. J.* 1: 381–398.

Hall, S. M., and Baker, D. A. (1972) The chemical composition of *Ricinus* phloem exudate. *Planta* 106: 131–140.

Hayashi, H., Fukuda, A., Suzui, N., and Fujimaki, S. (2000) Proteins in the sieve element–companion cell complexes: Their detection, localization and possible functions. *Aust. J. Plant Physiol.* 27: 489–496.

Haywood, V., Yu, T.-S., Huang, N.-C., and Lucas, W. J. (2005) Phloem long-distance trafficking of *GIBBERELLIC ACID-INSENSITIVE* RNA regulates leaf development. *Plant J.* 42: 49–68.

Imlau, A., Truernit, E., and Sauer, N. (1999) Cell-to-cell and long-distance trafficking of the green fluorescent protein in the phloem and symplastic unloading of the protein into sink tissues. *Plant Cell* 11: 309–322.

Jeuffory, M.-H., and Warembourg, F. R. (1991) Carbon transfer and partitioning between vegetative and reproductive organs in *Pisum sativum* L. *Plant Physiol.* 97: 440–448.

Joy, K. W. (1964) Translocation in sugar beet. I. Assimilation of $^{14}CO_2$ and distribution of materials from leaves. *J. Exp. Bot.* 15: 485–494.

Knoblauch, M., and van Bel, A. J. E. (1998) Sieve tubes in action. *Plant Cell* 10: 35–50.

Knoblauch, M., Peters, W. S., Ehlers, K., and van Bel, A. J. E. (2001) Reversible calcium-regulated stopcocks in legume sieve tubes. *Plant Cell* 13: 1221–1230.

Koch, K. E. (1996) Carbohydrate-modulated gene expression in plants. *Annu. Rev. Plant Physiol. Plant Mol. Biol.* 47: 509–540.

Kuhn, C. (2003) A comparison of the sucrose transporter systems of different plant species. *Plant Biol.* 5: 215–232.

Kuhn, C., Franceschi, V. R., Schulz, A., Lemoine, R., and Frommer, W. B. (1997) Macromolecular trafficking indicated by localization and turnover of sucrose transporters in enucleate sieve elements. *Science* 275: 1298–1300.

Münch, E. (1930) *Die Stoffbewegungen in der Pflanze*. Gustav Fischer, Jena, Germany.

Murillo, I., Roca, R., Bortolotti, C., and San Segundo, B. (2003) Engineering photoassimilate partitioning in tobacco plants improves growth and productivity and provides pathogen resistance. *Plant J.* 36: 330–341.

Nobel, P. S. (1991) *Physicochemical and Environmental Plant Physiology*. Academic Press, San Diego, CA.

Oparka, K. J., and Santa Cruz, S. (2000) The great escape: Phloem transport and unloading of macromolecules. *Annu. Rev. Plant Physiol. Plant Mol. Biol.* 51: 323–347.

Peuke, A. D., Rokitta, M., Zimmermann, U., Schreiber, L., and Haase, A. (2001) Simultaneous measurement of water flow velocity and solute transport in xylem and phloem of adult plants of *Ricinus communis* over a daily time course by nuclear magnetic resonance spectrometry. *Plant Cell Environ.* 24: 491–503.

Preiss, J. (1982) Regulation of the biosynthesis and degradation of starch. *Annu. Rev. Plant Physiol.* 33: 431–454.

Ransom-Hodgkins, W. D., Vaughn, M. W., and Bush, D. R. (2003) Protein phosphorylation plays a key role in sucrose-mediated transcriptional regulation of a phloem-specific proton-sucrose symporter. *Planta* 217: 483–489.

Roberts, A. G., Santa Cruz, S., Roberts, I. M., Prior, D. A. M., Turgeon, R., and Oparka, K. J. (1997) Phloem unloading in sink leaves of *Nicotiana benthamiana*: Comparison of a fluorescent solute with a fluorescent virus. *Plant Cell* 9: 1381–1396.

Schulz, A. (1990) Conifers. In *Sieve Elements: Comparative Structure, Induction and Development*. H.-D. Behnke and R. D. Sjolund, eds. Springer-Verlag, Berlin.

Schulz, A. (1992) Living sieve cells of conifers as visualized by confocal, laser-scanning fluorescence microscopy. *Protoplasma* 166: 153–164.

Schulz, A. (1994) Phloem transport and differential unloading in pea seedlings after source and sink manipulations. *Planta* 192: 239–248.

Schulz, A., Kuhn, C., Riesmeier, J. W., and Frommer, W. B. (1998) Ultrastructural effects in potato leaves due to antisense-inhibition of the sucrose transporter indicate an apoplasmic mode of phloem loading. *Planta* 206: 533–543.

Stadler, R., Wright, K. M., Lauterbach, C., Amon, G., Gahrtz, M., Feuerstein, A., Oparka, K. J., and Sauer, N. (2005) Expression of GFP-fusions in Arabidopsis companion cells reveals nonspecific protein trafficking into sieve elements and identifies a novel post-phloem domain in roots. *Plant J.* 41: 319–331.

Tang, G.-Q., Luscher, M., and Sturm, A. (1999) Antisense repression of vacuolar and cell wall invertase in transgenic carrot alters early plant development and sucrose partitioning. *Plant Cell* 11: 177–189.

Thomas, B. R., and Rodriguez, R. L. (1994) Metabolite signals regulate gene expression and source/sink relations in cereal seedlings. *Plant Physiol.* 106: 1235–1239.

Thompson, M. V. (2006) Phloem: The long and the short of it. *Trends Plant Sci.* 11: 26–32.

Thorne, J. H., and Koller, H. R. (1974) Influence of assimilate demand on photosynthesis, diffusive resistances, translocation, and carbohydrate levels of soybean leaves. *Plant Physiol.* 54: 201–207.

Turgeon, R. (1984) Termination of nutrient import and development of vein loading capacity in albino tobacco leaves. *Plant Physiol.* 76: 45–48.

Turgeon, R. (2006) Phloem loading: How leaves gain their independence. *BioScience* 56: 15–24.

Turgeon, R., and Gowan, E. (1990) Phloem loading in *Coleus blumei* in the absence of carrier-mediated uptake of export sugar from the apoplast. *Plant Physiol.* 94: 1244–1249.

Turgeon, R., and Medville, R. (2004) Phloem loading. A reevaluation of the relationship between plasmodesmatal frequencies and loading strategies. *Plant Physiol.* 136: 3795–3803.

Turgeon, R., and Webb, J. A. (1973) Leaf development and phloem transport in *Cucurbita pepo*: Transition from import to export. *Planta* 113: 179–191.

Turgeon, R., Beebe, D. U., and Gowan, E. (1993) The intermediary cell: Minor-vein anatomy and raffinose oligosaccharide synthesis in the Scrophulariaceae. *Planta* 191: 446–456.

van Bel, A. J. E. (1992) Different phloem-loading machineries correlated with the climate. *Acta Bot. Neerl.* 41: 121–141.

van Bel, A. J. E. (2003) The phloem, a miracle of ingenuity. *Plant, Cell Environ.* 26: 125–149.

van Bel, A. J. E., Gamalei, Y. V., Ammerlaan, A., and Bik, L. P. M. (1992) Dissimilar phloem loading in leaves with symplasmic or apoplasmic minor-vein configurations. *Planta* 186: 518–525.

Vaughn, M. W., Harrington, G. N., and Bush, D. R. (2002) Sucrose-mediated transcriptional regulation of sucrose symporter activity in the phloem. *Proc. Natl. Acad. Sci. USA* 99: 10876–10880.

Walz, C., Giavalisco, P., Schad, M., Juenger, M., Klose, J., and Kehr, J. (2004) Proteomics of cucurbit phloem exudate reveals a network of defence proteins. *Phytochemistry* 65: 1795–1804.

Warmbrodt, R. D. (1985) Studies on the root of *Hordeum vulgare* L.—Ultrastructure of the seminal root with special reference to the phloem. *Am. J. Bot.* 72: 414–432.

Wright, J. P., and Fisher, D. B. (1980) Direct measurement of sieve tube turgor pressure using severed aphid stylets. *Plant Physiol.* 65: 1133–1135.

Yoo, B.-C., Kragler, F., Varkonyl-Gasic, E., Haywood, V., Archer-Evans, S., Lee, Y. M., Lough, T. J., and Lucas, W. J. (2004) A systemic small RNA signaling system in plants. *Plant Cell* 16: 1979–2000.

Yu, H., and Kumar, P. P. (2003) Post-transcriptional gene silencing in plants by RNA. *Plant Cell Rep.* 22: 167–174.

Zhang, L.-Y., Peng, Y.-B., Pelleschi-Travier, S., Fan, Y., Lu, Y.-F., Lu, Y.-M., Gao, X.-P., Shen, Y.-Y., Delrot, S., and Zhang, D.-P. (2004) Evidence for apoplasmic phloem unloading in developing apple fruit. *Plant Physiol.* 135: 574–586.

Zimmermann, M. H., and Milburn, J. A., eds. (1975) *Transport in Plants, 1: Phloem Transport* (Encyclopedia of Plant Physiology, New Series, Vol. 1). Springer, New York.

Chapter 11 | Respiration and Lipid Metabolism

PHOTOSYNTHESIS PROVIDES the organic building blocks that plants (and nearly all other life) depend on. Respiration, with its associated carbon metabolism, releases the energy stored in carbon compounds in a controlled manner for cellular use. At the same time it generates many carbon precursors for biosynthesis. In the first part of this chapter we will review respiration in its metabolic context, emphasizing the interconnections and the special features that are peculiar to plants. We will also relate respiration to recent developments in our understanding of the biochemistry and molecular biology of plant mitochondria.

In the second part of the chapter we will describe the pathways of lipid biosynthesis that lead to the accumulation of fats and oils, which many plants use as energy and carbon storage. We will also examine lipid synthesis and the influence of lipids on membrane properties. Finally, we will discuss the catabolic pathways involved in the breakdown of lipids and the conversion of the degradation products to sugars that occurs during the germination of fat-storing seeds.

Overview of Plant Respiration

Aerobic (oxygen-requiring) respiration is common to nearly all eukaryotic organisms, and in its broad outlines, the respiratory process in plants is similar to that found in animals and lower eukaryotes. However, some specific aspects of plant respiration distinguish it from its animal counterpart. **Aerobic respiration** is the biological process by which reduced organic compounds are mobilized and subsequently

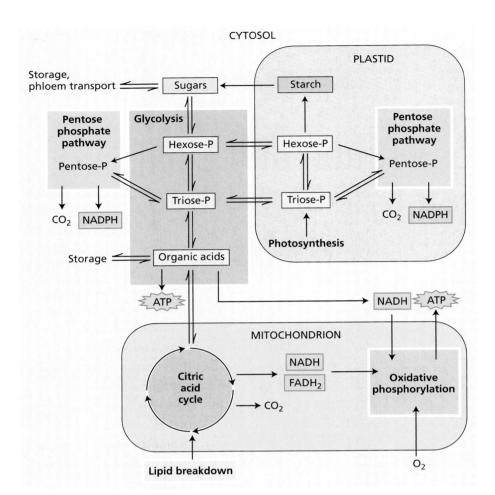

Figure 11.1 Overview of respiration. Substrates for respiration are generated by other cellular processes and enter the respiratory pathways. Glycolysis and the pentose phosphate pathways in the cytosol and plastid convert sugars to organic acids, via hexose phosphates and triose phosphates, generating NADH or NADPH and ATP. The organic acids are oxidized in the mitochondrial citric acid cycle, and the NADH and FADH$_2$ produced provide the energy for ATP synthesis by the electron transport chain and ATP synthase in oxidative phosphorylation. In gluconeogenesis, carbon from lipid breakdown is broken down in the glyoxysomes, metabolized in the citric acid cycle, and then used to synthesize sugars in the cytosol by reverse glycolysis.

oxidized in a controlled manner. During respiration, free energy is released and transiently stored in a compound, adenosine triphosphate (ATP), which can be readily utilized for the maintenance and development of the plant.

Glucose is most commonly cited as the substrate for respiration. However, in a functioning plant cell reduced carbon is mainly derived from sources such as the disaccharide sucrose, triose phosphates from photosynthesis, fructose-containing polymers (fructans), and other sugars, as well as lipids (primarily triacylglycerols), organic acids, and on occasion, proteins (Figure 11.1).

From a chemical standpoint, plant respiration can be expressed as the oxidation of the 12-carbon molecule sucrose and the reduction of 12 molecules of O$_2$:

$$C_{12}H_{22}O_{11} + 13\ H_2O \rightarrow 12\ CO_2 + 48\ H^+ + 48\ e^-$$

$$12\ O_2 + 48\ H^+ + 48\ e^- \rightarrow 24\ H_2O$$

giving the following net reaction:

$$C_{12}H_{22}O_{11} + 12\ O_2 \rightarrow 12\ CO_2 + 11\ H_2O$$

This reaction is the reversal of the photosynthetic process; it represents a coupled redox reaction in which sucrose is completely oxidized to CO$_2$ while oxygen serves as the ultimate electron acceptor and is reduced to water in the process. The standard free-energy decrease for the reaction as written is 5760 kJ (1380 kcal) per mole (342 g) of sucrose oxidized. The controlled release of this free energy, along with its coupling to the synthesis of ATP, is the primary, although by no means only, role of respiratory metabolism.

To prevent damage (incineration) of cellular structures, the cell mobilizes the large amount of free energy released in the oxidation of sucrose in a series of step-by-step reactions. These reactions can be grouped into four major processes: glycolysis, the citric acid cycle (also termed Krebs cycle), the reactions of the pentose phosphate pathway, and oxidative phosphorylation. The substrates of respiration enter the respiratory process at different points in the pathways, as summarized in Figure 11.1:

- **Glycolysis** involves a series of reactions carried out by soluble enzymes located in both the cytosol and the plastid. A sugar—for example, sucrose—is partly oxidized via six-carbon sugar phosphates (hexose phosphates) and three-carbon sugar phosphates (triose phosphates) to produce an organic acid—for example, pyruvate. The process yields a small amount of energy as ATP and reducing power in the form of a reduced pyridine nucleotide, NADH.

- In the **pentose phosphate pathway**, also located both in the cytosol and the plastid, the six-carbon glucose-6-phosphate is initially oxidized to the five-carbon ribulose-5-phosphate. The carbon is lost as CO$_2$, and reducing power is conserved in the form of two molecules of another reduced pyridine nucleotide,

NADPH. In following near-equilibrium reactions of the pentose phosphate pathway, ribulose-5-phosphate is converted into sugars containing three to seven carbon atoms.

- In the **citric acid cycle**, pyruvate is oxidized completely to CO_2. This generates the major amount of reducing power (16 NADH + 4 $FADH_2$ equivalents per sucrose) from the breakdown of sucrose. With one exception (succinate dehydrogenase), these reactions are carried out by enzymes located in the internal aqueous compartment, or matrix, of the mitochondrion. As we will discuss later, succinate dehydrogenase is localized in the inner of the two mitochondrial membranes.

- In **oxidative phosphorylation**, electrons are transferred along an electron transport chain consisting of a collection of electron transport proteins bound to the inner of the two mitochondrial membranes. This system transfers electrons from NADH (and related species)—produced during glycolysis, the pentose phosphate pathway, and the citric acid cycle—to oxygen. This electron transfer releases a large amount of free energy, much of which is conserved through the synthesis of ATP from ADP and P_i (inorganic phosphate) catalyzed by the enzyme ATP synthase. Collectively the redox reactions of the electron transport chain and the synthesis of ATP are called oxidative phosphorylation. This final stage completes the oxidation of sucrose.

Nicotinamide adenine dinucleotide (NAD^+/NADH) is an organic cofactor (coenzyme) associated with many enzymes that catalyze cellular redox reactions. NAD^+ is the oxidized form of the cofactor, and it undergoes a reversible two-electron reaction that yields NADH (Figure 11.2):

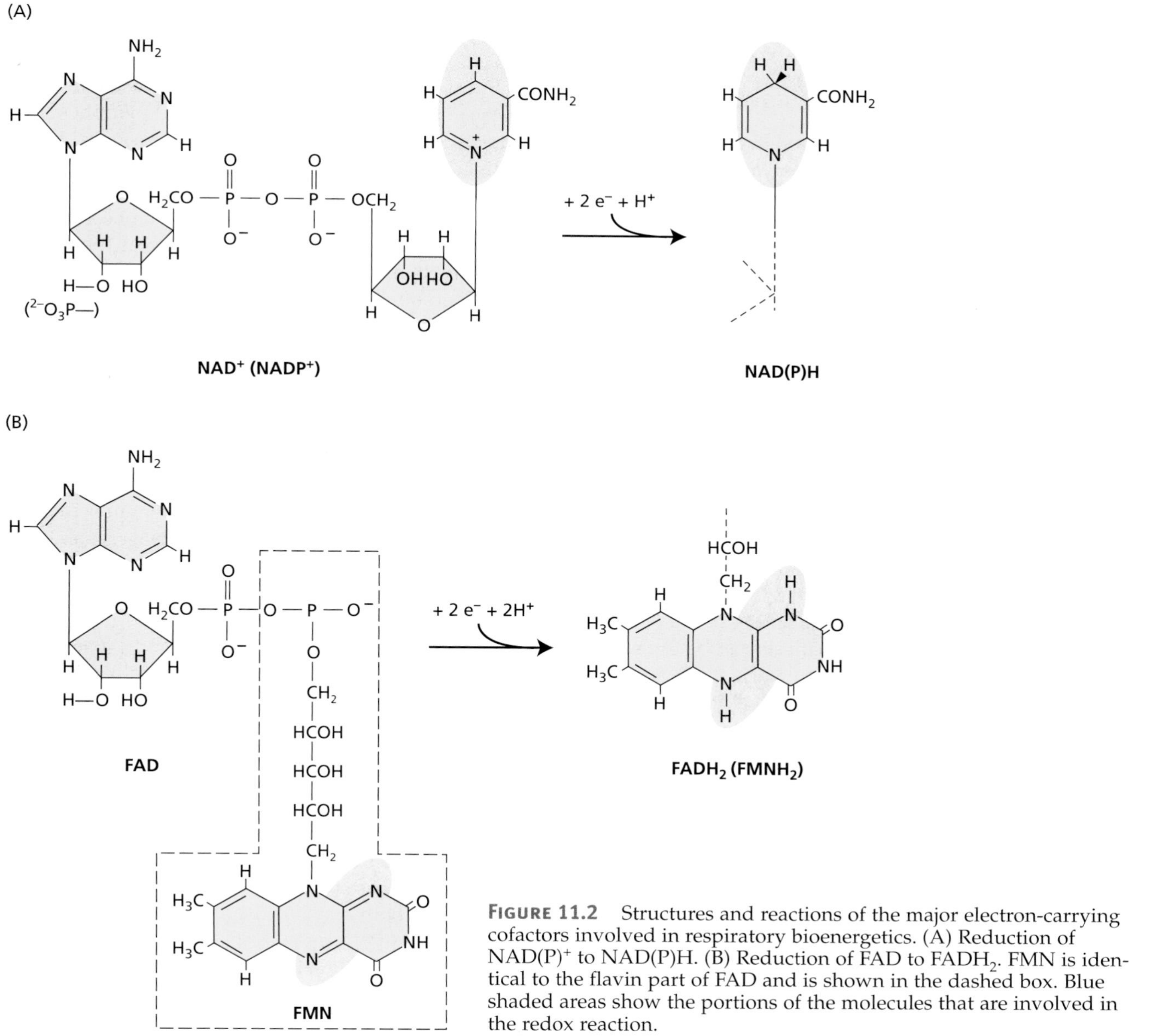

FIGURE 11.2 Structures and reactions of the major electron-carrying cofactors involved in respiratory bioenergetics. (A) Reduction of $NAD(P)^+$ to NAD(P)H. (B) Reduction of FAD to $FADH_2$. FMN is identical to the flavin part of FAD and is shown in the dashed box. Blue shaded areas show the portions of the molecules that are involved in the redox reaction.

$$NAD^+ + 2\,e^- + H^+ \rightarrow NADH$$

The standard reduction potential for this redox couple is about –320 mV, which makes it a relatively strong reductant (i.e., electron donor). NADH is thus a good molecule in which to conserve the free energy carried by electrons released during the stepwise oxidations of glycolysis and the citric acid cycle. A related compound, nicotinamide adenine dinucleotide phosphate ($NADP^+/NADPH$), functions in redox reactions of photosynthesis (see Chapter 8) and of the oxidative pentose phosphate pathway; it also takes part in mitochondrial metabolism (Møller and Rasmusson 1998). This will be discussed later in the chapter.

The oxidation of NADH by oxygen via the electron transport chain releases free energy (220 kJ mol^{-1}, or 52 kcal mol^{-1}) that drives the synthesis of ATP (the details of which we will see later). We can formulate a more complete picture of respiration as related to its role in cellular energy metabolism by coupling the following two reactions:

$$C_{12}H_{22}O_{11} + 12\,O_2 \rightarrow 12\,CO_2 + 11\,H_2O$$

$$60\,ADP + 60\,P_i \rightarrow 60\,ATP + 60\,H_2O$$

Keep in mind that not all the carbon that enters the respiratory pathway ends up as CO_2. Many respiratory intermediates are the starting points for pathways that assimilate nitrogen into organic form, pathways that synthesize nucleotides and lipids, and many others.

Glycolysis: A Cytosolic and Plastidic Process

In the early steps of glycolysis (from the Greek words *glykos*, "sugar," and *lysis*, "splitting"), carbohydrates are converted to hexose phosphates, which are then split into two triose phosphates. In a subsequent energy-conserving phase, the triose phosphates are oxidized and rearranged to yield two molecules of pyruvate, an organic acid. Besides preparing the substrate for oxidation in the citric acid cycle, glycolysis yields a small amount of chemical energy in the form of ATP and NADH.

When molecular oxygen is unavailable—for example, in plant roots in flooded soils—glycolysis can be the main source of energy for cells. For this to work, the **fermentation pathways**, which are carried out in the cytosol, reduce pyruvate to recycle the NADH produced by glycolysis. In this section we will describe the basic glycolytic and fermentative pathways, emphasizing features that are specific for plant cells. In the following section we will discuss the pentose phosphate pathway, another pathway for sugar oxidation in plants.

Glycolysis converts carbohydrates into pyruvate, producing NADH and ATP

Glycolysis occurs in all living organisms (prokaryotes and eukaryotes). The principal reactions associated with the classic glycolytic and fermentative pathways in plants are almost identical to those of animal cells (Figure 11.3). However, plant glycolysis has unique regulatory features, as well as a parallel partial glycolytic pathway in plastids and alternative enzymatic routes for several cytosolic steps. In animals the substrate of glycolysis is glucose and the end product pyruvate. Because sucrose is the major translocated sugar in most plants and is therefore the form of carbon that most nonphotosynthetic tissues import, sucrose (not glucose) can be argued to be the true sugar substrate for plant respiration. The end products of plant glycolysis include another organic acid, malate.

In the early steps of glycolysis, sucrose is split into its two monosaccharide units—glucose and fructose—which can readily enter the glycolytic pathway. Two pathways for the splitting of sucrose are known in plants, both of which also take part in the unloading of sucrose from the phloem (see Chapter 10).

In most plant tissues sucrose synthase, localized in the cytosol, degrades sucrose by combining it with UDP to produce fructose and UDP-glucose. UDP-glucose pyrophosphorylase then converts UDP-glucose and pyrophosphate (PP_i) into UTP and glucose-6-phosphate (see Figure 11.3). In some tissues, invertases present in the cell wall, vacuole, or cytosol hydrolyze sucrose to its two component hexoses (glucose and fructose). The hexoses are then phosphorylated by the hexokinase that uses ATP. Whereas the sucrose synthase reaction is close to equilibrium, the invertase reaction releases sufficient energy to be essentially irreversible.

In plastids, a partial glycolysis occurs that produces metabolites for biosynthetic reactions there, but can also supply substrates for glycolysis in the cytoplasm. Starch is both synthesized and catabolized only in plastids, and carbon obtained from starch degradation (for example, in a chloroplast at night) enters the glycolytic pathway in the cytosol primarily as glucose (see Chapter 8). In the light, photosynthetic products can also directly enter the glycolytic pathway as triose phosphate (Hoefnagel et al. 1998). So glycolysis resembles a funnel with an initial phase collecting carbon from different cellular sources, depending on physiological conditions.

In the initial phase of glycolysis, each hexose unit is phosphorylated twice and then split, eventually producing two molecules of triose phosphate. This series of reactions consumes two to four molecules of ATP per sucrose unit, depending on whether the sucrose is split by sucrose synthase or invertase. These reactions also include two of the three essentially irreversible reactions of the glycolytic pathway that are catalyzed by hexokinase and phosphofructokinase (see Figure 11.3). As we will see later, the phosphofructokinase reaction is one of the control points of glycolysis in both plants and animals.

THE ENERGY-CONSERVING PHASE OF GLYCOLYSIS The reactions discussed thus far transfer carbon from the various substrate pools to triose phosphates. Once glyceraldehyde-3-phosphate is formed, the glycolytic pathway can begin to extract usable energy in the energy-conserving phase. The enzyme glyceraldehyde-3-phosphate dehydrogenase catalyzes the oxidation of the aldehyde to a carboxylic acid, reducing NAD^+ to NADH. This reaction releases sufficient free energy to allow the phosphorylation (using inorganic phosphate) of glyceraldehyde-3-phosphate to produce 1,3-bisphosphoglycerate. The phosphorylated carboxylic acid on carbon 1 of 1,3-bisphosphoglycerate (see Figure 11.3) has a large standard free energy of hydrolysis (–49.3 kJ mol^{-1}, or –11.8 kcal mol^{-1}). Thus, 1,3-bisphosphoglycerate is a strong donor of phosphate groups.

In the next step of glycolysis, catalyzed by phosphoglycerate kinase, the phosphate on carbon 1 is transferred to a molecule of ADP, yielding ATP and 3-phosphoglycerate. For each sucrose entering the pathway, four ATPs are generated by this reaction—one for each molecule of 1,3-bisphosphoglycerate.

This type of ATP synthesis, traditionally referred to as **substrate-level phosphorylation**, involves the direct transfer of a phosphate group from a substrate molecule to ADP, to form ATP. As we will see, ATP synthesis by substrate-level phosphorylation is mechanistically distinct from ATP synthesis by ATP synthases involved in the oxidative phosphorylation in mitochondria (which will be described later in this chapter) or photophosphorylation in chloroplasts (see Chapter 7).

In the subsequent two reactions, the phosphate on 3-phosphoglycerate is transferred to carbon 2 and then a molecule of water is removed, yielding the compound phosphoenolpyruvate (PEP). The phosphate group on PEP has a high standard free energy of hydrolysis (–61.9 kJ mol^{-1}, or –14.8 kcal mol^{-1}), which makes PEP an extremely good phosphate donor for ATP formation. Using PEP as substrate, the enzyme pyruvate kinase catalyzes a second substrate-level phosphorylation to yield ATP and pyruvate. This final step, which is the third essentially irreversible step in glycolysis, yields four additional molecules of ATP for each sucrose molecule that enters the pathway.

Plants have alternative glycolytic reactions

The sequence of reactions leading to the formation of pyruvate from glucose occurs in all organisms that carry out glycolysis. In addition, organisms can operate this pathway in the opposite direction to synthesize sugar from organic acids. This process is known as **gluconeogenesis**.

Gluconeogenesis is not common in plants, but it does operate in the seeds of some plants, such as castor bean (*Ricinus communis*) and sunflower, that store a significant quantity of their carbon reserves in the form of oils (triacylglycerols). After the seed germinates, much of the oil is converted by gluconeogenesis to sucrose, which is then used to support the growing seedling. In the initial phase of glycolysis, gluconeogenesis overlaps with the pathway for synthesis of sucrose from photosynthetic triose phosphate described in Chapter 8, which is typical for plants.

Because the glycolytic reaction catalyzed by ATP-dependent phosphofructokinase is essentially irreversible (see Figure 11.3), an additional enzyme, fructose-1,6-bisphosphatase, converts fructose-1,6-bisphosphate to fructose-6-phosphate and P_i during gluconeogenesis. ATP-dependent phosphofructokinase and fructose-1,6-bisphosphatase represent a major control point of carbon flux through the glycolytic/gluconeogenic pathways of both plants and animals, as well as in sucrose synthesis in plants (see Chapter 8).

In plants, the interconversion of fructose-6-phosphate and fructose-1,6-bisphosphate is made more complex by the presence of an additional (cytosolic) enzyme, a PP_i-dependent phosphofructokinase (pyrophosphate:fructose-6-phosphate 1-phosphotransferase), which catalyzes the following reversible reaction (see Figure 11.3):

$$\text{Fructose-6-P} + PP_i \leftrightarrow \text{fructose-1,6-}P_2 + P_i$$

where P represents phosphate and P_2 bisphosphate. PP_i-dependent phosphofructokinase is found in the cytosol of most plant tissues at levels that are considerably higher than those of the ATP-dependent phosphofructokinase (Kruger 1997). Suppression of the PP_i-dependent phosphofructokinase in transgenic potato has indicated that it contributes to glycolytic flux, but that it is not essential for plant survival, indicating that other enzymes can take over its function. That different pathways serve a similar function and can therefore replace each other without a clear loss in function is called metabolic redundancy; this is a common feature in plant metabolism.

The reaction catalyzed by the PP_i-dependent phosphofructokinase is readily reversible, but it is unlikely to operate in sucrose synthesis (Dennis and Blakely 2000). Like ATP-dependent phosphofructokinase and fructose bisphosphatase, this enzyme appears to be regulated by fluctuations in cell metabolism (discussed later in the chapter), suggesting that under some circumstances operation of the glycolytic pathway in plants differs from that in many other organisms (see Web Essay 11.1).

At the end of the glycolytic sequence, plants have alternative pathways for metabolizing PEP. In one pathway PEP is carboxylated by the ubiquitous cytosolic enzyme PEP carboxylase to form the organic acid oxaloacetate (OAA). The OAA is then reduced to malate by the action of malate dehydrogenase, which uses NADH as the source of electrons, and this performs a role similar to that of the dehydrogenases during fermentative metabolism (see Figure 11.3). The resulting malate can be stored by export to the vacuole or transported to the mitochondrion, where it can enter the citric acid cycle. Thus the operation of pyruvate kinase and PEP carboxylase can produce alternative organic acids—pyruvate or malate—for mitochondrial respiration, although pyruvate dominates in most tissues.

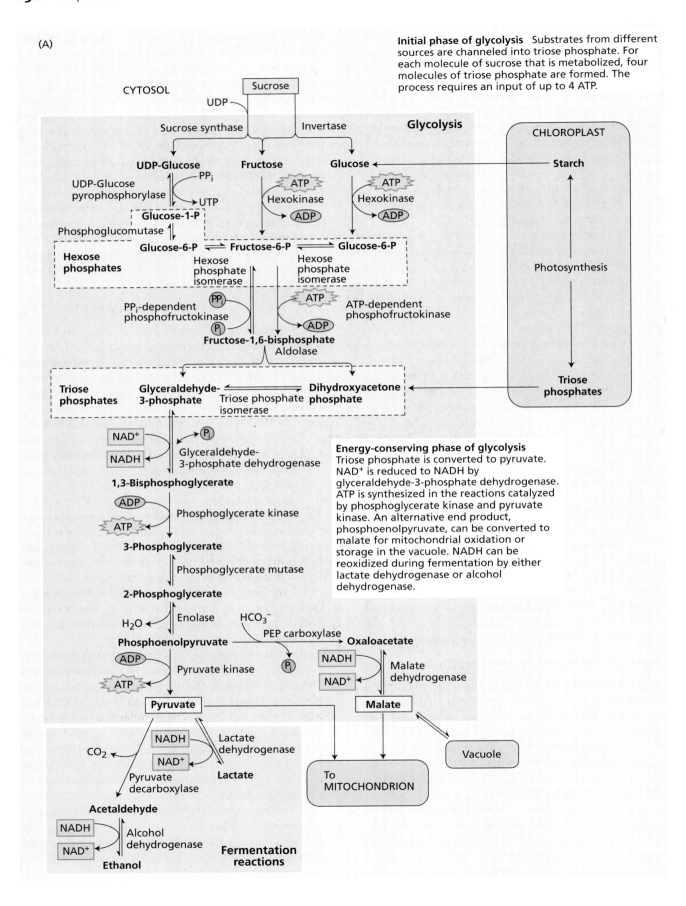

(A) **Initial phase of glycolysis** Substrates from different sources are channeled into triose phosphate. For each molecule of sucrose that is metabolized, four molecules of triose phosphate are formed. The process requires an input of up to 4 ATP.

Energy-conserving phase of glycolysis Triose phosphate is converted to pyruvate. NAD$^+$ is reduced to NADH by glyceraldehyde-3-phosphate dehydrogenase. ATP is synthesized in the reactions catalyzed by phosphoglycerate kinase and pyruvate kinase. An alternative end product, phosphoenolpyruvate, can be converted to malate for mitochondrial oxidation or storage in the vacuole. NADH can be reoxidized during fermentation by either lactate dehydrogenase or alcohol dehydrogenase.

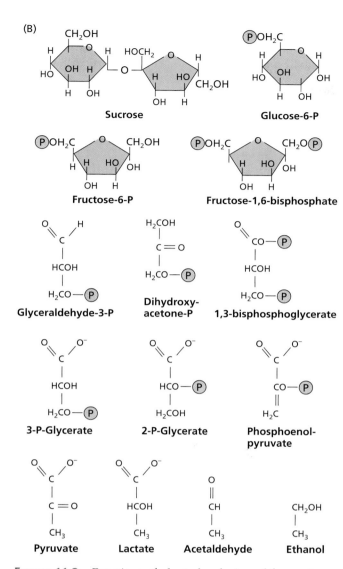

FIGURE 11.3 Reactions of plant glycolysis and fermentation. (A) In the main pathway, sucrose is oxidized via hexose phosphates and triose phosphates to the organic acid pyruvate, but plants also carry out alternative reactions. All the enzymes have been measured at levels sufficient to support the respiration rates in intact plant tissues, and flux through the pathway has been observed in vivo. The double arrows denote reversible reactions; the single arrows, essentially irreversible reactions. (B) The structures of the intermediates. P, phosphate.

In the absence of O_2, fermentation regenerates the NAD+ needed for glycolysis

In the absence of oxygen, the citric acid cycle and oxidative phosphorylation cannot function. Glycolysis thus cannot continue to operate because the cell's supply of NAD+ is limited, and once all the NAD+ becomes tied up in the reduced state (NADH), the glyceraldehyde-3-phosphate dehydrogenase comes to a halt. To overcome this problem, plants and other organisms can further metabolize pyruvate by carrying out one or more forms of **fermentative metabolism** (see Figure 11.3).

In alcoholic fermentation (common in plants, but more widely known in brewer's yeast), the two enzymes, pyruvate decarboxylase and alcohol dehydrogenase, act on pyruvate, ultimately producing ethanol and CO_2 and oxidizing NADH in the process. In lactic acid fermentation (common in mammalian muscle but also found in plants), the enzyme lactate dehydrogenase uses NADH to reduce pyruvate to lactate, thus regenerating NAD+.

Under some circumstances, plant tissues may be subjected to low (hypoxic) or zero (anoxic) concentrations of ambient oxygen, forcing them to carry out fermentative metabolism. The best-studied example involves flooded or waterlogged soils in which the diffusion of oxygen is sufficiently reduced to cause root tissues to become hypoxic.

In corn the initial response to low oxygen is lactic acid fermentation, but the subsequent response is alcoholic fermentation. Ethanol is thought to be a less toxic end product of fermentation because it can diffuse out of the cell, whereas lactate accumulates and promotes acidification of the cytosol. In numerous other cases plants function under near-anaerobic conditions by carrying out some form of fermentation.

Fermentation does not liberate all the energy available in each sugar molecule

Before we leave the topic of glycolysis, we need to consider the efficiency of fermentation. *Efficiency* is defined here as the energy conserved as ATP relative to the energy potentially available in a molecule of sucrose. The standard free-energy change ($\Delta G^{0\prime}$) for the complete oxidation of sucrose is –5760 kJ mol^{-1} (1380 kcal mol^{-1}). The value of $\Delta G^{0\prime}$ for the synthesis of ATP is 32 kJ mol^{-1} (7.7 kcal mol^{-1}). However, under the nonstandard conditions that normally exist in both mammalian and plant cells, the synthesis of ATP requires an input of free energy of approximately 50 kJ mol^{-1} (12 kcal mol^{-1}). (For a discussion of free energy, see Chapter 2 on the web site.)

Given the net synthesis of four molecules of ATP for each sucrose molecule that is converted to ethanol (or lactate), the efficiency of anaerobic fermentation is only about 4%. Most of the energy available in sucrose remains in the reduced by-product of fermentation: lactate or ethanol. During aerobic respiration, the pyruvate produced by glycolysis is transported into mitochondria, where it is further oxidized, resulting in a much more efficient utilization of the free energy originally available in the sucrose.

Because of the low efficiency of energy conservation under fermentation, an increased rate of glycolysis is needed to sustain the ATP production necessary for cell survival. This is called the *Pasteur effect* after the French microbiologist Louis Pasteur, who first noted it when yeast switched from aerobic respiration to anaerobic alcoholic fermentation. The higher rates of glycolysis result from changes in glycolytic metabolite levels, as well as from increased expression of genes encoding enzymes of glycolysis and fermentation (Sachs et al. 1996).

Plant glycolysis is controlled by its products

In vivo, glycolysis appears to be regulated at the level of fructose-6-phosphate phosphorylation and PEP turnover. In contrast to animals, AMP and ATP are not major effectors of plant phosphofructokinase and pyruvate kinase. The cytosolic concentration of PEP, which is a potent inhibitor of the plant ATP-dependent phosphofructokinase, is a more important regulator of plant glycolysis.

This inhibitory effect of PEP on phosphofructokinase is strongly decreased by inorganic phosphate, making the cytosolic ratio of PEP to P_i a critical factor in the control of plant glycolytic activity. Pyruvate kinase and PEP carboxylase, the enzymes that metabolize PEP in the last steps of glycolysis (see Figure 11.3), are in turn sensitive to feedback inhibition by citric acid cycle intermediates and their derivatives, including malate, citrate, 2-oxoglutarate, and glutamate.

In plants, therefore, the control of glycolysis comes from the "bottom up" (as discussed later in the chapter) with primary regulation at the level of PEP metabolism by pyruvate kinase and PEP carboxylase and secondary regulation exerted by PEP at the conversion of fructose-6-phosphate to fructose-1,6-bisphosphate (see Figure 11.3). In animals, the primary control operates at the phosphofructokinase, and secondary control at the pyruvate kinase.

One conceivable benefit of bottom-up control of glycolysis is that it permits plants to control net glycolytic flux to pyruvate independently of related metabolic processes such as the Calvin cycle and sucrose–triose phosphate–starch interconversion (Plaxton 1996). Another benefit of this control mechanism is that glycolysis may adjust to the demand for biosynthetic precursors.

The presence of two enzymes metabolizing PEP in plant cells—pyruvate kinase and PEP carboxylase—has consequences for the control of glycolysis that are not quite clear. Although the two enzymes are inhibited by similar metabolites, the PEP carboxylase can under some conditions perform a bypass reaction around the pyruvate kinase. The resulting malate can then enter the mitochondrial citric acid cycle.

Experimental support for multiple pathways of PEP metabolism comes from the study of transgenic tobacco plants with less than 5% of the normal level of cytosolic pyruvate kinase in their leaves (Plaxton 1996). In these plants, rates of both leaf respiration and photosynthesis were unaffected relative to controls having wild-type levels of pyruvate kinase. However, reduced root growth in the transgenic plants indicated that the pyruvate kinase reaction could not be circumvented without some detrimental effects.

The regulation of the conversion of fructose-6-phosphate to fructose-1,6-bisphosphate is also complex. Fructose-2,6-bisphosphate, another hexose bisphosphate, is present at varying levels in the cytosol (see Chapter 8). It markedly inhibits the activity of cytosolic fructose-1,6-bisphosphatase, but stimulates the activity of PP_i-dependent phosphofructokinase. These observations suggest that fructose-2,6-bisphosphate plays a central role in partitioning flux between ATP-dependent and PP_i-dependent pathways of fructose phosphate metabolism at the crossing point between sucrose synthesis and glycolysis.

Understanding the dynamics of the regulation of glycolysis requires the study of temporal changes in metabolite levels (Givan 1999). Methods of rapid extraction and simultaneous analyses of many metabolites are now available—for example, mass spectrometry—an approach called *metabolic profiling* (see **Web Essay 11.2**).

The pentose phosphate pathway produces NADPH and biosynthetic intermediates

The glycolytic pathway is not the only route available for the oxidation of sugars in plant cells. Sharing common metabolites, the **oxidative pentose phosphate pathway** (also known as the *hexose monophosphate shunt*) can also accomplish this task (Figure 11.4). The reactions are carried out by soluble enzymes present in the cytosol and in plastids. Generally, the pathway in plastids predominates over that in the cytosol (Dennis et al. 1997).

The first two reactions of this pathway involve the oxidative events that convert the six-carbon glucose-6-phosphate to a five-carbon sugar, ribulose-5-phosphate, with loss of a CO_2 molecule and generation of two molecules of NADPH (not NADH). The remaining reactions of the pathway convert ribulose-5-phosphate to the glycolytic intermediates glyceraldehyde-3-phosphate and fructose-6-phosphate. Because glucose-6-phosphate can be regenerated from glyceraldehyde-3-phosphate and fructose-6-phosphate by glycolytic enzymes, for six turns of the cycle we can write the reaction as follows:

$$6 \text{ glucose-6-P} + 12 \text{ NADP}^+ + 7 \text{ H}_2\text{O} \rightarrow$$
$$5 \text{ glucose-6-P} + 6 \text{ CO}_2 + P_i + 12 \text{ NADPH} + 12 \text{ H}^+$$

The net result is the complete oxidation of one glucose-6-phosphate molecule to CO_2 (five molecules are regenerated) with the concomitant synthesis of 12 NADPH molecules.

Studies of the release of $^{14}CO_2$ from isotopically labeled glucose indicate that glycolysis is the more dominant breakdown pathway, accounting for 80 to 95% of the total

FIGURE 11.4 Reactions of the oxidative pentose phosphate pathway in higher plants. The first two reactions—which are oxidizing reactions—are essentially irreversible and supply NADPH to the cytoplasm and to plastids in the absence of photosynthesis. The downstream part of the pathway is reversible (denoted by double-headed arrows), so it can supply five-carbon substrates for biosynthesis also when the oxidizing reactions are inhibited, for example in the chloroplast in the light. P, phosphate.

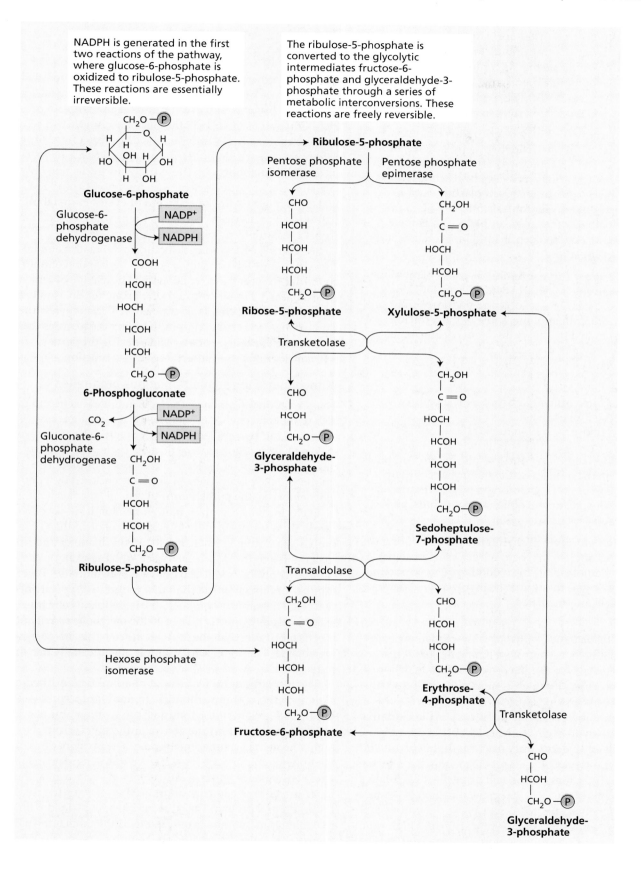

carbon flux in most plant tissues. However, the pentose phosphate pathway does contribute to the flux, and developmental studies indicate that its contribution increases as plant cells develop from a meristematic to a more differentiated state (Ap Rees 1980). The oxidative pentose phosphate pathway plays several roles in plant metabolism:

- **NADPH supply for biosynthetic redox reactions.** The product of the two oxidative steps is NADPH, and this NADPH is thought to drive reductive steps associated with various biosynthetic reactions that occur in the cytosol. In nongreen plastids, such as amyloplasts, and in chloroplasts functioning in the dark, the pathway may also supply NADPH for biosynthetic reactions such as lipid biosynthesis and nitrogen assimilation.

- **NADPH supply for respiration.** Because plant mitochondria are able to oxidize cytosolic NADPH via an NADPH dehydrogenase localized on the external surface of the inner membrane, some of the reducing power generated by this pathway may contribute to cellular energy metabolism; that is, electrons from NADPH may end up reducing O_2 and generating ATP.

- **Supply of biosynthesis substrates.** The pathway produces ribose-5-phosphate, a precursor of the ribose and deoxyribose needed in the synthesis of RNA and DNA, respectively. Another intermediate in this pathway, the four-carbon erythrose-4-phosphate, combines with PEP in the initial reaction that produces plant phenolic compounds, including the aromatic amino acids and the precursors of lignin, flavonoids, and phytoalexins (see Chapter 13).

- **Generation of Calvin cycle intermediates.** During the early stages of greening, before leaf tissues become fully photoautotrophic, the oxidative pentose phosphate pathway is thought to be involved in generating Calvin cycle intermediates.

CONTROL OF THE OXIDATIVE PATHWAY The oxidative pentose phosphate pathway is controlled by the initial reaction of the pathway catalyzed by glucose-6-phosphate dehydrogenase, the activity of which is markedly inhibited by a high ratio of NADPH to $NADP^+$.

In the light, however, little operation of the oxidative pathway is likely to occur in the chloroplast, because the end products of the pathway, fructose-6-phosphate and glyceraldehyde-3-phosphate, are being synthesized by the Calvin cycle. Thus, mass action will drive the nonoxidative interconversions of the pathway in the direction of pentose synthesis. Moreover, glucose-6-phosphate dehydrogenase will be inhibited during photosynthesis by the high ratio of NADPH to $NADP^+$ in the chloroplast, as well as by a reductive inactivation involving the ferredoxin–thioredoxin system (see Chapter 8).

The Citric Acid Cycle: A Mitochondrial Matrix Process

During the nineteenth century, biologists discovered that in the absence of air, cells produce ethanol or lactic acid, whereas in the presence of air, cells consume O_2 and produce CO_2 and H_2O. In 1937 the German-born British biochemist Hans A. Krebs reported the discovery of the **citric acid cycle**—also called the *tricarboxylic acid cycle* or *Krebs cycle*. The elucidation of the citric acid cycle not only explained how pyruvate is broken down to CO_2 and H_2O; it also highlighted the key concept of cycles in metabolic pathways. For his discovery, Hans Krebs was awarded the Nobel Prize in physiology or medicine in 1953.

Because the citric acid cycle occurs in the matrix of mitochondria, we will begin with a general description of mitochondrial structure and function, the knowledge of which was obtained mainly through experiments on isolated mitochondria (see **Web Topic 11.1**). We will then review the steps of the citric acid cycle, emphasizing the features that are specific to plants. For all plant-specific properties, we will consider how they affect respiratory function.

Mitochondria are semiautonomous organelles

The breakdown of sucrose to pyruvate releases less than 25% of the total energy in sucrose; the remaining energy is stored in the two molecules of pyruvate. The next two stages of respiration (the citric acid cycle and oxidative phosphorylation—i.e., electron transport coupled to ATP synthesis) take place within an organelle enclosed by a double membrane, the **mitochondrion** (plural *mitochondria*).

In electron micrographs, plant mitochondria—whether in situ or in vitro—usually look spherical or rodlike (Figure 11.5) and range from 0.5 to 1.0 µm in diameter and up to 3 µm in length (Douce 1985) (see **Web Essay 11.3**). With some exceptions, plant cells have a substantially lower number of mitochondria than that found in a typical animal cell. The number of mitochondria per plant cell varies, and it is usually directly related to the metabolic activity of the tissue, reflecting the mitochondrial role in energy metabolism. Guard cells, for example, are unusually rich in mitochondria.

The ultrastructural features of plant mitochondria are similar to those of mitochondria in nonplant tissues (see Figure 11.5). Plant mitochondria have two membranes: a smooth **outer membrane** that completely surrounds a highly invaginated **inner membrane**. The invaginations of the inner membrane are known as **cristae** (singular *crista*). As a consequence of the greatly enlarged surface area, the inner membrane can contain more than 50% of the total mitochondrial protein. The aqueous phase contained within the inner membrane is referred to as the mitochondrial **matrix** (plural *matrices*), and the region between the two mitochondrial membranes is known as the **intermembrane space**.

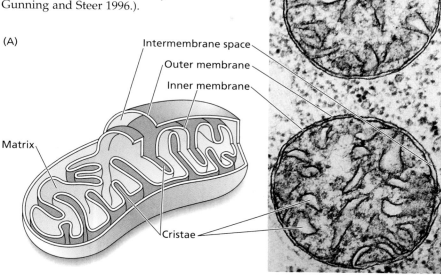

FIGURE 11.5 Structure of plant mitochondria. (A) Three-dimensional representation of a mitochondrion, showing the invaginations of the inner membrane, called cristae, as well as the location of the matrix and intermembrane spaces (see also Figure 11.10). (B) Electron micrograph of mitochondria in a mesophyll cell of broad bean (*Vicia faba*). Typically, individual mitochondria are 1 to 3 μm long in plant cells, which means that they are substantially smaller than nuclei and plastids. (B from Gunning and Steer 1996.).

Intact mitochondria are osmotically active; that is, they take up water and swell when placed in a hypo-osmotic medium. Most inorganic ions and charged organic molecules are not able to diffuse freely into the matrix space. The inner membrane is the osmotic barrier; the outer membrane is permeable to solutes that have a molecular mass of less than approximately 10,000 Daltons (Da)—that is, most cellular metabolites and ions, but not proteins. The lipid fraction of both membranes is primarily made up of phospholipids, 80% of which are either phosphatidylcholine or phosphatidylethanolamine.

Like chloroplasts, mitochondria are semiautonomous organelles, because they contain ribosomes, RNA, and DNA, which encodes a limited number of mitochondrial proteins. Plant mitochondria are thus able to carry out the various steps of protein synthesis and to transmit their genetic information. Mitochondria proliferate through division by fission of preexisting mitochondria and not through de novo biogenesis of the organelle.

Pyruvate enters the mitochondrion and is oxidized via the citric acid cycle

As already noted, the citric acid cycle is also known as the tricarboxylic acid cycle, because of the importance of the tricarboxylic acids citric acid (citrate) and isocitric acid (isocitrate) as early intermediates (Figure 11.6). This cycle constitutes the second stage in respiration and takes place in the mitochondrial matrix. Its operation requires that the pyruvate generated in the cytosol during glycolysis be transported through the impermeable inner mitochondrial membrane via a specific transport protein (as will be described shortly).

Once inside the mitochondrial matrix, pyruvate is decarboxylated in an oxidation reaction catalyzed by the enzyme pyruvate dehydrogenase. The products are NADH (from NAD^+), CO_2, and acetic acid in the form of acetyl-CoA, in which a thioester bond links the acetic acid to a sulfur-containing cofactor, coenzyme A (CoA) (see Figure 11.6). Pyruvate dehydrogenase exists as a large complex of several enzymes that catalyze the overall reaction in a three-step process: decarboxylation, oxidation, and conjugation to CoA.

In the next reaction the enzyme citrate synthase, formally the first enzyme in the citric acid cycle, combines the acetyl group of acetyl-CoA with a four-carbon dicarboxylic acid (oxaloacetate, OAA) to give a six-carbon tricarboxylic acid (citrate). Citrate is then isomerized to isocitrate by the enzyme aconitase.

The following two reactions are successive oxidative decarboxylations, each of which produces one NADH and releases one molecule of CO_2, yielding a four-carbon molecule, succinyl-CoA. At this point, three molecules of CO_2 have been produced for each pyruvate that entered the mitochondrion, or 12 CO_2 for each molecule of sucrose oxidized.

During the remainder of the citric acid cycle, succinyl-CoA is oxidized to OAA, allowing the continued operation of the cycle. Initially the large amount of free energy available in the thioester bond of succinyl-CoA is conserved through the synthesis of ATP from ADP and P_i via a substrate-level phospho-

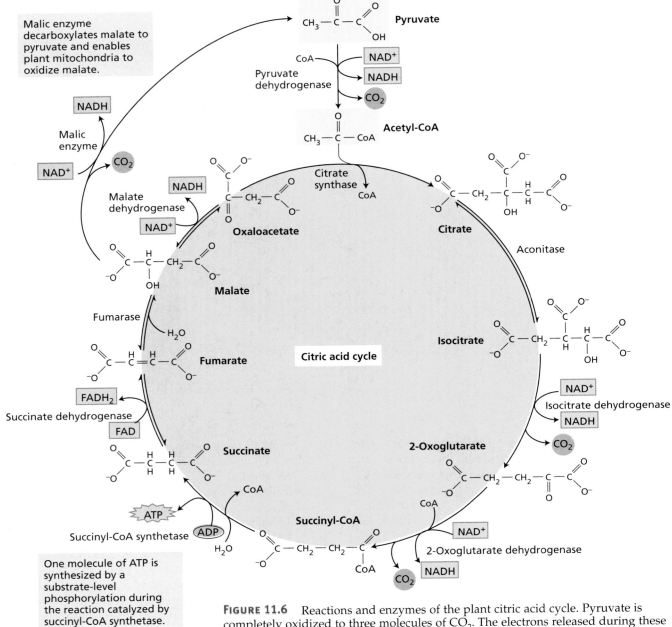

FIGURE 11.6 Reactions and enzymes of the plant citric acid cycle. Pyruvate is completely oxidized to three molecules of CO_2. The electrons released during these oxidations are used to reduce four molecules of NAD^+ to NADH and one molecule of FAD to $FADH_2$.

rylation catalyzed by succinyl-CoA synthetase. (Recall that the free energy available in the thioester bond of acetyl-CoA was used to form a carbon–carbon bond in the step catalyzed by citrate synthase.) The resulting succinate is oxidized to fumarate by succinate dehydrogenase, which is the only membrane-associated enzyme of the citric acid cycle and also part of the electron transport chain (which is the next major topic to be discussed in this chapter).

The electrons and protons removed from succinate end up not on NAD^+ but on another cofactor involved in redox reactions: flavin adenine dinucleotide (FAD). FAD is covalently bound to the active site of succinate dehydrogenase and undergoes a reversible two-electron reduction to produce $FADH_2$ (see Figure 11.2B).

In the final two reactions of the citric acid cycle, fumarate is hydrated to produce malate, which is subsequently oxidized by malate dehydrogenase to regenerate OAA and produce another molecule of NADH. The OAA produced is now able to react with another acetyl-CoA and continue the cycling.

The stepwise oxidation of one molecule of pyruvate in the mitochondrion gives rise to three molecules of CO_2, and much of the free energy released during these oxidations is conserved in the form of four NADH and one

FADH$_2$. In addition, one molecule of ATP is produced by a substrate-level phosphorylation during the citric acid cycle.

All the enzymes associated with the citric acid cycle are found in plant mitochondria. Some of them may be associated in multienzyme complexes, which would facilitate movement of metabolites between the enzymes.

The citric acid cycle of plants has unique features

The citric acid cycle reactions outlined in Figure 11.6 are not all identical to those carried out by animal mitochondria. For example, the step catalyzed by succinyl-CoA synthetase produces ATP in plants and GTP in animals.

A feature of the plant citric acid cycle that is absent in many other organisms is the significant activity of NAD$^+$ malic enzyme, which has been found in the matrix of all plant mitochondria analyzed to date. This enzyme catalyzes the oxidative decarboxylation of malate:

$$\text{Malate} + \text{NAD}^+ \rightarrow \text{pyruvate} + \text{CO}_2 + \text{NADH}$$

The presence of NAD$^+$ malic enzyme enables plant mitochondria to operate alternative pathways for the metabolism of PEP derived from glycolysis (see **Web Essay 11.1**). As already described, malate can be synthesized from PEP in the cytosol via the enzymes PEP carboxylase and malate dehydrogenase (see Figure 11.3). For degradation, malate is transported into the mitochondrial matrix, where NAD$^+$ malic enzyme can oxidize it to pyruvate. This reaction makes possible the complete net oxidation of citric acid cycle intermediates such as malate (Figure 11.7A) or citrate (see Figure 11.7B) (Oliver and McIntosh 1995). Many plant tissues, not only those that carry out crassulacean acid metabolism (see Chapter 8), store significant amounts of malate or other organic acids in their vacuoles. Degradation of malate via mitochondrial malic enzyme is important for regulating the level of organic acids in the cells—for example, during fruit ripening.

Instead of being degraded, the malate produced via the PEP carboxylase can replace citric acid cycle intermediates used in biosynthesis. Reactions that can replenish intermediates in a metabolic cycle are known as *anaplerotic*. For example, export of 2-oxoglutarate for nitrogen assimilation in the chloroplast will cause a shortage of malate needed in the citrate synthase reaction. This malate can be replaced through the PEP carboxylase pathway (see Figure 11.7C).

Mitochondrial Electron Transport and ATP Synthesis

ATP is the energy carrier used by cells to drive living processes, and chemical energy conserved during the citric acid cycle in the form of NADH and FADH$_2$ (redox equivalents with high-energy electrons) must be converted to ATP to perform useful work in the cell. This O$_2$-dependent process, called **oxidative phosphorylation**, occurs in the inner mitochondrial membrane.

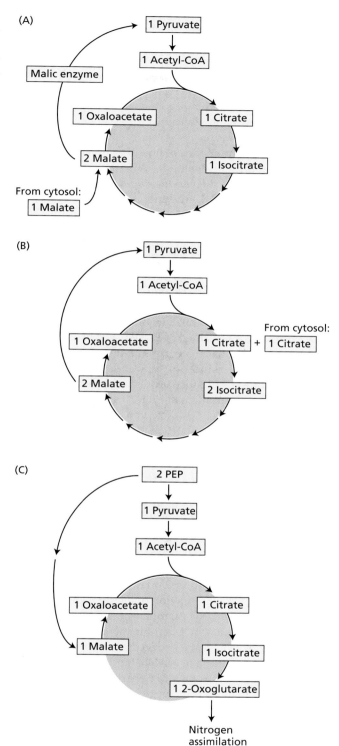

FIGURE 11.7 Malic enzyme and PEP carboxylase provide plants with metabolic flexibility for the metabolism of PEP and pyruvate. Malic enzyme converts malate to pyruvate and makes it possible for plant mitochondria to oxidize both malate (A) and citrate (B) to CO$_2$ without involving pyruvate delivered by glycolysis. The joint action of PEP carboxylase and pyruvate kinase can convert glycolytic PEP to 2-oxoglutarate, which is used for nitrogen assimilation (C).

In this section we will describe the process by which the energy level of the electrons is lowered in a stepwise fashion and conserved in the form of an electrochemical proton gradient across the inner mitochondrial membrane. Although fundamentally similar in all aerobic cells, the electron transport chain of plants (and fungi) contains multiple NAD(P)H dehydrogenases and an alternative oxidase not found in mammalian mitochondria.

We will also examine the enzyme that uses the energy of the proton gradient to synthesize ATP: the F_oF_1-ATP synthase. After examining the various stages in the production of ATP, we will summarize the energy conservation steps at each stage, as well as the regulatory mechanisms that coordinate the different pathways.

The electron transport chain catalyzes a flow of electrons from NADH to O_2

For each molecule of sucrose oxidized through glycolysis and the citric acid cycle pathways, four molecules of NADH are generated in the cytosol and 16 molecules of NADH plus four molecules of $FADH_2$ (associated with succinate dehydrogenase) are generated in the mitochondrial matrix. These reduced compounds must be reoxidized or the entire respiratory process will come to a halt.

The electron transport chain catalyzes an electron flow from NADH (and $FADH_2$) to oxygen, the final electron acceptor of the respiratory process. For the oxidation of NADH, the overall two-electron transfer can be written as follows:

$$NADH + H^+ + \tfrac{1}{2}O_2 \rightarrow NAD^+ + H_2O$$

From the reduction potentials for the NADH–NAD^+ pair (–320 mV) and the H_2O–$\tfrac{1}{2}O_2$ pair (+810 mV), it can be calculated that the standard free energy released during this overall reaction ($-nF\Delta E0'$) is about 220 kJ mol^{-1} (52 kcal mol^{-1}) per two electrons. (For a detailed discussion on standard free energy see Chapter 2 on the web site). Because the succinate–fumarate reduction potential is higher (+30 mV), only 152 kJ mol^{-1} (36 kcal mol^{-1}) of energy is released for each two electrons generated during the oxidation of succinate. The role of the electron transport chain is to bring about the oxidation of NADH (and $FADH_2$) and, in the process, utilize some of the free energy released to generate an electrochemical proton gradient, $\Delta \tilde{\mu}_{H^+}$, across the inner mitochondrial membrane.

The electron transport chain of plants contains the same set of electron carriers found in mitochondria from other organisms (Figure 11.8) (Siedow 1995; Siedow and Umbach 1995). The individual electron transport proteins are organized into four multiprotein complexes (identified by Roman numerals I through IV), all of which are localized in the inner mitochondrial membrane:

COMPLEX I (NADH DEHYDROGENASE) Electrons from NADH generated in the mitochondrial matrix during the citric acid cycle are oxidized by complex I (an NADH dehydrogenase). The electron carriers in complex I include a tightly bound cofactor (flavin mononucleotide [FMN], which is chemically similar to FAD; see Figure 11.2B) and several iron–sulfur centers. Complex I then transfers these electrons to ubiquinone. Four protons are pumped from the matrix to the intermembrane space for every electron pair passing through the complex.

Ubiquinone, a small lipid-soluble electron and proton carrier, is located within the inner membrane. It is not tightly associated with any protein, and it can diffuse within the hydrophobic core of the membrane bilayer.

COMPLEX II (SUCCINATE DEHYDROGENASE) Oxidation of succinate in the citric acid cycle is catalyzed by this complex, and the reducing equivalents are transferred via the $FADH_2$ and a group of iron–sulfur proteins into the ubiquinone pool. This complex does not pump protons.

COMPLEX III (CYTOCHROME BC_1 COMPLEX) This complex oxidizes reduced ubiquinone (ubiquinol) and transfers the electrons via an iron–sulfur center, two b-type cytochromes (b_{565} and b_{560}), and a membrane-bound cytochrome c_1 to cytochrome c. Four protons per electron pair are pumped out of the matrix by complex III using a mechanism called the Q-cycle (see **Web Topic 11.2**.)

Cytochrome c is a small protein loosely attached to the outer surface of the inner membrane and serves as a mobile carrier to transfer electrons between complexes III and IV.

COMPLEX IV (CYTOCHROME C OXIDASE) This complex contains two copper centers (Cu_A and Cu_B) and cytochromes a and a_3. Complex IV is the terminal oxidase and brings about the four-electron reduction of O_2 to two molecules of H_2O. Two protons are pumped out of the matrix per electron pair (see Figure 11.8).

Both structurally and functionally, ubiquinone and the cytochrome bc_1 complex are very similar to plastoquinone and the cytochrome b_6f complex, respectively, in the photosynthetic electron transport chain (see Chapter 7).

Reality may be more complex than the description above implies. Plant respiratory complexes contain a number of plant-specific subunits the function of which is still unknown. Several of the complexes contain subunits that participate in functions other than electron transport, such as protein import. Finally, several of the complexes appear to be present in supercomplexes, instead of freely mobile in the membrane, although the functional significance of these supercomplexes is not clear (Millar et al. 2005).

Some electron transport enzymes are unique to plant mitochondria

In addition to the set of electron carriers described in the previous section, plant mitochondria contain some components not found in mammalian mitochondria (see Figure

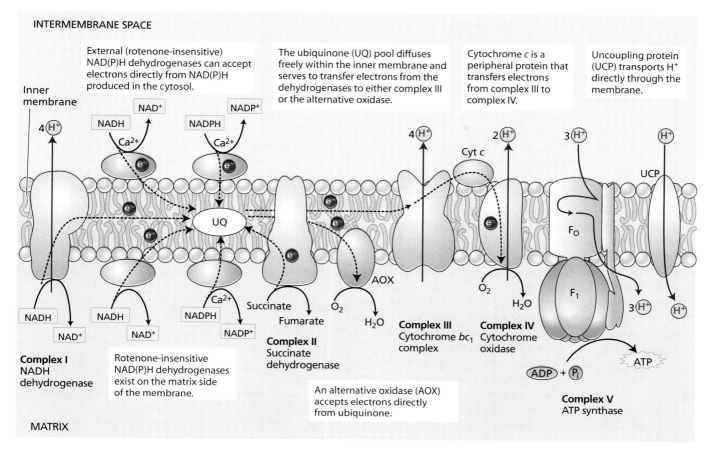

FIGURE 11.8 Organization of the electron transport chain and ATP synthesis in the inner membrane of plant mitochondria. In addition to the four standard protein complexes found in nearly all other mitochondria, the electron transport chain of plant mitochondria contains five additional enzymes (depicted in green). None of these additional enzymes pumps protons. Additionally, the uncoupling protein directly bypasses the ATP synthase by allowing passive proton influx. This multiplicity of bypasses in plants, where animals have only the uncoupling protein, gives a greater flexibility to plant energy coupling (see **Web Topic 11.3**). Specific inhibitors—rotenone for complex I, antimycin for complex III, cyanide for complex IV, and salicylhydroxamic acid (SHAM) for the alternative oxidase—are important tools used to investigate mitochondrial electron transport. Some also have commercial uses, such as rotenone used as an insecticide and to remove unwanted fish from lakes. Because plants have the alternative pathways, they can survive exposure to inhibitors of the respiratory complexes.

11.8). Note that none of these additional enzymes pump protons and that energy conservation is therefore lower whenever they are used:

- Two NAD(P)H dehydrogenases, both Ca^{2+}-dependent, attached to the outer surface of the inner membrane facing the intermembrane space can oxidize cytosolic NADH and NADPH. Electrons from these external NAD(P)H dehydrogenases—$ND_{ex}(NADH)$ and $ND_{ex}(NADPH)$—enter the main electron transport chain at the level of the ubiquinone pool (see **Web Topic 11.3**) (Møller 2001; Rasmusson et al. 2004).

- Plant mitochondria have two pathways for oxidizing matrix NADH. Electron flow through complex I, described in the previous section, is sensitive to inhibition by several compounds, including rotenone and piericidin. In addition, plant mitochondria have a rotenone-resistant dehydrogenase, $ND_{in}(NADH)$, for the oxidation of NADH derived from citric acid cycle substrates. The role of this pathway may well be as a bypass being engaged when complex I is overloaded (Møller and Rasmusson 1998; Møller 2001; Rasmusson et al. 2004), such as under photorespiratory conditions, as we will see shortly (see also **Web Topic 11.3**).

- An NADPH dehydrogenase, $ND_{in}(NADPH)$, is present on the matrix surface. Very little is known about this enzyme.

- Most, if not all, plants have an "alternative" respiratory pathway for the reduction of oxygen. This pathway involves the so-called alternative oxidase that, unlike cytochrome c oxidase, is insensitive to inhibition by cyanide, azide, or carbon monoxide (see **Web Topic 11.3**).

The nature and physiological significance of these plant-specific enzymes will be considered more fully later in the chapter.

ATP synthesis in the mitochondrion is coupled to electron transport

In oxidative phosphorylation, the transfer of electrons to oxygen via complexes I to IV is coupled to the synthesis of ATP from ADP and P_i via the ATP synthase (complex V). The number of ATPs synthesized depends on the nature of the electron donor.

In experiments conducted on isolated mitochondria, electrons derived from internal (matrix) NADH give ADP:O ratios (the number of ATPs synthesized per two electrons transferred to oxygen) of 2.4 to 2.7 (Table 11.1). Succinate and externally added NADH each give values in the range of 1.6 to 1.8, while ascorbate, which serves as an artificial electron donor to cytochrome *c*, gives values of 0.8 to 0.9. Results such as these (for both plant and animal mitochondria) have led to the general concept that there are three sites of energy conservation along the electron transport chain, at complexes I, III, and IV.

The experimental ADP:O ratios agree quite well with the values calculated on the basis of the number of H^+ pumped by complexes I, III, and IV and the cost of 4 H^+ for synthesizing one ATP (see next section and Table 11.1). For instance, electrons from external NADH pass only complexes III and IV, so a total of 6 H^+ are pumped, giving 1.5 ATP (when the alternative oxidase pathway is not used).

The mechanism of mitochondrial ATP synthesis is based on the chemiosmotic hypothesis, described in **Web Topic 6.3** and Chapter 7, which was first proposed in 1961 by Nobel laureate Peter Mitchell as a general mechanism of energy conservation across biological membranes (Nicholls and Ferguson 2002). According to the chemiosmotic theory, the orientation of electron carriers within the mitochondrial inner membrane allows for the transfer of protons across the inner membrane during electron flow. Numerous studies have confirmed that mitochondrial electron transport is associated with a net transfer of protons from the mitochondrial matrix to the intermembrane space (see Figure 11.8) (Whitehouse and Moore 1995).

Because the inner mitochondrial membrane is impermeable to protons, an electrochemical proton gradient can build up. As discussed in Chapters 6 and 7, the free energy associated with the formation of an electrochemical proton gradient ($\Delta \tilde{\mu}_{H^+}$, also referred to as a proton motive force, Δp, when expressed in units of volts) is made up of an electric transmembrane potential component (ΔE) and a chemical-potential component (ΔpH) according to the following equation:

$$\Delta p = \Delta E - 59 \Delta pH \text{ (at } 25°C\text{)}$$

where

$$\Delta E = E_{inside} - E_{outside}$$

and

$$\Delta pH = pH_{inside} - pH_{outside}$$

ΔE results from the asymmetric distribution of a charged species (H^+) across the membrane, and ΔpH is due to the proton concentration difference across the membrane. Because protons are translocated from the mitochondrial matrix to the intermembrane space, the resulting ΔE across the inner mitochondrial membrane is negative.

As this equation shows, both ΔE and ΔpH contribute to the proton motive force in plant mitochondria, although ΔE is consistently found to be of greater magnitude, probably because of the large buffering capacity of both cytosol and matrix, which prevents large pH changes. This situation contrasts that in the chloroplast, where almost all of the proton motive force across the thylakoid membrane is due to a proton gradient (see Chapter 7).

The free-energy input required to generate $\Delta \tilde{\mu}_{H^+}$ comes from the free energy released during electron transport. How electron transport is coupled to proton translocation is not well understood in all cases. Because of the low permeability (conductance) of the inner membrane to protons, the proton electrochemical gradient is reasonably stable once generated, and the free energy $\Delta \tilde{\mu}_{H^+}$ can be utilized to carry out chemical work (ATP synthesis). The $\Delta \tilde{\mu}_{H^+}$ is coupled to the synthesis of ATP by an additional protein complex associated with the inner membrane, the F_oF_1-ATP synthase.

The **F_oF_1-ATP synthase** (also called *complex V*) consists of two major components, F_1 and F_o (see Figure 11.8). **F_1** is a peripheral membrane protein complex that is composed of at least five different subunits and contains the catalytic site for converting ADP and P_i to ATP. This complex is attached to the matrix side of the inner membrane. **F_o** (sub-

TABLE 11.1
Theoretical and experimental ADP:O ratios in isolated plant mitochondria

Substrate	ADP:O ratio	
	Theoretical[a]	Experimental
Malate	2.5	2.4–2.7
Succinate	1.5	1.6–1.8
NADH (external)	1.5	1.6–1.8
Ascorbate	1.0[b]	0.8–0.9

[a]It is assumed that complexes I, III, and IV pump 4, 4, and 2 H^+ per 2 electrons, respectively; that the cost of synthesizing one ATP and exporting it to the cytosol is 4 H^+ (Brand 1994); and that the nonphosphorylating pathways are not active.

[b]Cytochrome *c* oxidase pumps only two protons when it is measured with ascorbate as electron donor. However, two electrons move from the outer surface of the inner membrane (where the electrons are donated) across the inner membrane to the inner, matrix side. As a result, 2 H^+ are consumed on the matrix side. This means that the net movement of H^+ and charges is equivalent to the movement of a total of 4 H^+, giving an ADP:O ratio of 1.0.

script "o" for oligomycin-sensitive) is an integral membrane protein complex that consists of at least three different polypeptides that form the channel through which protons cross the inner membrane.

The passage of protons through the channel is coupled to the catalytic cycle of the F_1 component of the ATP synthase, allowing the ongoing synthesis of ATP and the simultaneous utilization of the $\Delta\tilde{\mu}_{H^+}$. For each ATP synthesized, 3 H^+ pass through the F_o from the intermembrane space to the matrix down the electrochemical proton gradient.

A high-resolution X-ray structure of most of the F_1 complex of the mammalian mitochondrial ATP synthase supports a "rotational model" for the catalytic mechanism of ATP synthesis (Abrahams et al. 1994) (see **Web Topic 11.4**). The structure and function of the mitochondrial ATP synthase is similar to that of the CF_0–CF_1 ATP synthase in photophosphorylation (see Chapter 7).

The operation of a chemiosmotic mechanism of ATP synthesis has several implications. First, the true site of ATP formation on the mitochondrial inner membrane is the ATP synthase, not complex I, III, or IV. These complexes serve as sites of energy conservation whereby electron transport is coupled to the generation of a $\Delta\tilde{\mu}_{H^+}$.

Second, the chemiosmotic theory explains the action mechanism of uncouplers, a wide range of chemically unrelated compounds (including 2,4-dinitrophenol and p-trifluoromethoxycarbonylcyanide phenylhydrazone [FCCP]) that decrease mitochondrial ATP synthesis but stimulate the rate of electron transport (see **Web Topic 11.5**). All of these compounds make the inner membrane leaky to protons, which prevents the buildup of a sufficiently large $\Delta\tilde{\mu}_{H^+}$ to drive ATP synthesis.

In experiments on isolated mitochondria, higher rates of electron flow (measured as the rate of oxygen uptake in the presence of a substrate such as succinate) are observed upon addition of ADP (referred to as *state 3*) than in its absence (Figure 11.9). ADP provides a substrate that stimulates dissipation of the $\Delta\tilde{\mu}_{H^+}$ through the F_oF_1-ATP synthase during ATP synthesis. Once all the ADP has been converted to ATP, the $\Delta\tilde{\mu}_{H^+}$ builds up again and reduces the rate of electron flow (*state 4*). The ratio of the rates with and without ADP (state 3:state 4) is referred to as the *respiratory control ratio*.

Transporters exchange substrates and products

The electrochemical proton gradient also plays a role in the movement of the organic acids of the citric acid cycle and of substrates and products of ATP synthesis in and out of mitochondria. Although ATP is synthesized in the mitochondrial matrix, most of it is used outside the mitochondrion, so an efficient mechanism is needed for moving ADP in and ATP out of the organelle.

Adenylate transport involves another inner-membrane protein, the ADP/ATP (adenine nucleotide) transporter, which catalyzes an exchange of ADP and ATP across the

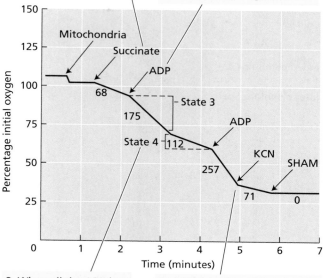

FIGURE 11.9 Regulation of respiratory rate by ADP during succinate oxidation in isolated mitochondria from mung bean (*Vigna radiata*). A limited amount of ADP is normally present in vivo, so the situation is intermediate between state 3 and state 4. The numbers below the traces are the rates of oxygen uptake expressed as O_2 consumed (nmol min^{-1} mg $protein^{-1}$). (Data courtesy of Steven J. Stegink.)

inner membrane (Figure 11.10). The movement of the more negatively charged ATP^{4-} out of the mitochondria in exchange for ADP^{3-}—that is, one net negative charge out—is driven by the electric-potential gradient (ΔE, positive outside) generated by proton pumping.

The uptake of inorganic phosphate (P_i) involves an active phosphate transporter protein that uses the proton gradient component (ΔpH) of the proton motive force to drive the electroneutral exchange of P_i^- (in) for OH^- (out). As long as a ΔpH is maintained across the inner membrane, the P_i content within the matrix will remain high. Similar reasoning applies to the uptake of pyruvate, which is driven by the electroneutral exchange of pyruvate for OH^-, leading to continued uptake of pyruvate from the cytosol (see Figure 11.10).

270 Chapter 11

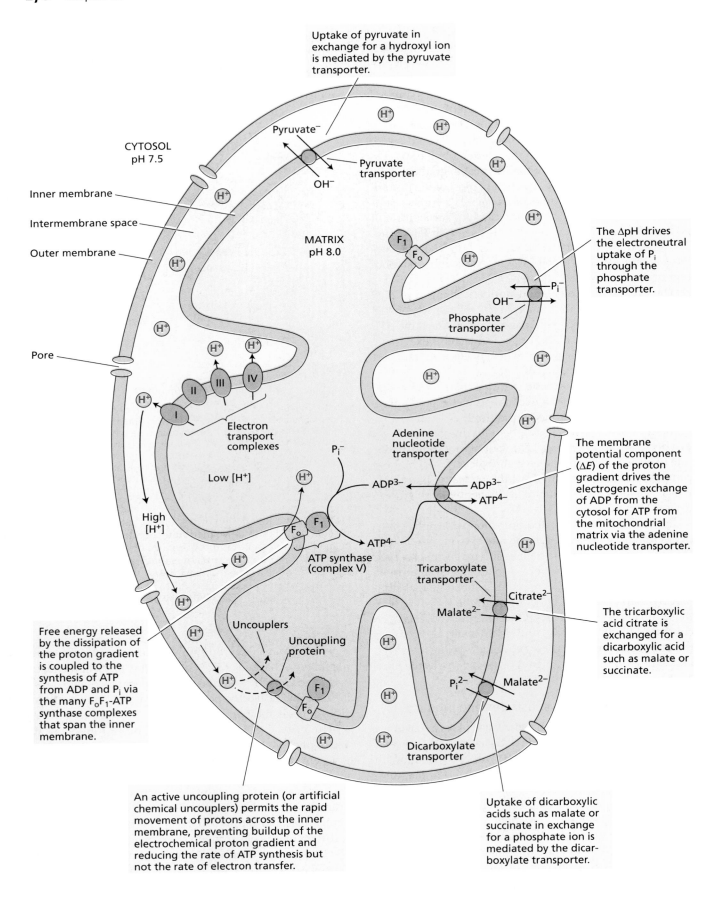

◀ **FIGURE 11.10** Transmembrane transport in plant mitochondria. An electrochemical proton gradient $\Delta\tilde{\mu}_{H^+}$, consisting of a membrane potential (ΔE, –200 mV, negative inside) and a ΔpH (alkaline inside), is established across the inner mitochondrial membrane during electron transport, as outlined in the text. Specific metabolites are moved across the inner membrane by specialized proteins called transporters or carriers. (After Douce 1985.)

The total energetic cost of taking up one phosphate and one ADP into the matrix and exporting one ATP is the movement of 1 H$^+$ from the intermembrane space into the matrix:

- One OH$^-$ out in exchange for the phosphate ion, which is the same as 1 H$^+$ in, so this electroneutral exchange will consume the proton gradient, but not the transmembrane potential
- One negative charge out (ADP^{3-} entering the matrix in exchange for ATP^{4-} leaving), which is the same as one positive charge in, so this will lower only the transmembrane potential

This proton should also be included in calculation of the cost of synthesizing one ATP. Thus the total cost is 3 H$^+$ used by the ATP synthase plus 1 H$^+$ for the exchange across the membrane, or a total of 4 H$^+$.

The inner membrane also contains transporters for dicarboxylic acids (malate or succinate) exchanged for P$_i^{2-}$ and for the tricarboxylic acid citrate exchanged for malate (see Figure 11.10 and **Web Topic 11.5**).

Aerobic respiration yields about 60 molecules of ATP per molecule of sucrose

The complete oxidation of a sucrose molecule leads to the net formation of:

- Eight molecules of ATP by substrate-level phosphorylation (four from glycolysis and four from the citric acid cycle)
- Four molecules of NADH in the cytosol
- 16 molecules of NADH plus four molecules of FADH$_2$ (via succinate dehydrogenase) in the mitochondrial matrix

On the basis of theoretical ADP:O values (see Table 11.1), a total of approximately 52 molecules of ATP will be generated per molecule of sucrose by oxidative phosphorylation. The complete aerobic oxidation of sucrose (including substrate-level phosphorylation) results in a total of about 60 ATPs synthesized per sucrose molecule (Table 11.2).

Using 50 kJ mol^{-1} (12 kcal mol^{-1}) as the actual free energy of formation of ATP in vivo, we find that about 3010 kJ mol^{-1} (720 kcal mol^{-1}) of free energy is conserved in the form of ATP per mole of sucrose oxidized during aerobic respiration. This amount represents about 52% of the standard free energy available from the complete oxidation of sucrose; the rest is lost as heat. This is a vast improvement over the conversion of only 4% of the energy available in sucrose to ATP that is associated with fermentative metabolism.

Several subunits of respiratory complexes are encoded by the mitochondrial genome

The genetic system of plant mitochondria differs not only from that in the nucleus and the chloroplast, but also from the ones found in the mitochondria of animals, protozoans, or even fungi. Most notably, processes involving RNA differ between plant mitochondria and mitochondria from most other organisms (see **Web Topic 11.6**). Major differences are found in:

- RNA splicing (for example, special introns are present)
- RNA editing (where the nucleotide sequence is changed)
- Signals regulating RNA stability
- Translation (plant mitochondria use the universal genetic code, whereas mitochondria in other eukaryotes have deviating codons)

The size of the plant mitochondrial genome varies substantially even between closely related plant species, but at 200 to 2400 kilobase pairs (kbp), it is always much larger than the compact and uniform 16 kbp genome found in mammalian mitochondria. The size differences are due mainly to the presence of much noncoding sequence, including numerous introns, in plant mitochondrial DNA (mtDNA). Mammalian mtDNA encodes only 13 proteins,

TABLE 11.2
The maximum yield of cytosolic ATP from the complete oxidation of sucrose to CO_2 via aerobic glycolysis and the citric acid cycle

Part reaction	ATP per sucrose[a]
Glycolysis	
4 substrate-level phosphorylations	4
4 NADH	4 × 1.5 6
Citric acid cycle	
4 substrate-level phosphorylations	4
4 FADH$_2$	4 × 1.5 6
16 NADH	16 × 2.5 40
Total	60

Source: Adapted from Brand 1994.

Note: Cytosolic NADH is assumed oxidized by the external NADH dehydrogenase. The nonphosphorylating pathways are assumed not to be engaged.

[a]Calculated using the theoretical values from Table 11.1

in contrast to the 35 known proteins encoded by Arabidopsis mtDNA (Marienfeld et al. 1999). Both plant and mammalian mitochondria encode rRNAs and tRNAs.

The genes of mtDNA can be divided into two main groups: those needed for expression of mitochondrial genes (tRNA, rRNA, and ribosome proteins) and those for oxidative phosphorylation complexes. Plant mtDNA encodes subunits for complexes I to IV, the ATP synthase, and proteins that take part in cytochrome biogenesis. The mitochondrially encoded subunits are essential for the activity of the respiratory complexes, a feature also evident in the sequence similarity to their bacterial homologs. Except for the proteins encoded by mtDNA, all mitochondrial proteins (possibly more than 2000) are encoded by nuclear DNA—for example, all proteins in the citric acid cycle (Millar et al. 2005). These nuclear-encoded mitochondrial proteins are synthesized by cytosolic ribosomes and imported via translocators in the outer and inner mitochondrial membrane. Therefore, oxidative phosphorylation is dependent on expression of genes located in two separate genomes. Any change in expression in response to a stimulus or for developmental reasons must be coordinated.

Whereas the expression of nuclear genes for mitochondrial proteins appears to be regulated as are other nuclear genes, much less is known about the expression of mitochondrial genes. The circular chromosome of plant mtDNA is normally split into several smaller subgenomic segments, and genes can be down-regulated by decreased copy number for a segment of the mtDNA (Leon et al. 1998). The gene promoters in mtDNA are of several kinds and show different transcriptional activity. However, a main control of mitochondrial gene expression appears to take place at the posttranslational level, by degradation of excess polypeptides (McCabe et al. 2000).

Plants have several mechanisms that lower the ATP yield

As we have seen, a complex machinery is required for high efficiency of energy conservation in oxidative phosphorylation. So it is perhaps surprising that plant mitochondria have several functional proteins that reduce this efficiency (see **Web Topic 11.3**). Probably plants are less limited by the energy supply (sunlight) than by other factors in the environment (e.g., access to nitrogen or phosphate). As a consequence, metabolic flexibility may be more important than energetic efficiency.

In the following subsections we will discuss the role of the nonphosphorylating mechanisms and their possible usefulness in the life of the plant.

THE ALTERNATIVE OXIDASE If cyanide (1 mM) is added to actively respiring animal tissues, cytochrome c oxidase is inhibited and the respiration rate quickly drops to less than 1% of its initial level. However, most plant tissues display a level of cyanide-resistant respiration that can represent 10 to 25%—and in some tissues up to 100%—of the uninhibited control rate. The enzyme responsible for this oxygen uptake has been identified as a cyanide-resistant oxidase component of the plant mitochondrial electron transport chain called the **alternative oxidase** (Vanlerberghe and McIntosh 1997) (see Figure 11.8 and **Web Topic 11.3**).

Electrons feed off the main electron transport chain into the alternative pathway at the level of the ubiquinone pool (see Figure 11.8). The alternative oxidase, the only component of the alternative pathway, catalyzes a four-electron reduction of oxygen to water and is specifically inhibited by several compounds, most notably salicylhydroxamic acid (SHAM).

When electrons pass to the alternative pathway from the ubiquinone pool, two sites of proton pumping (at complexes III and IV) are bypassed. Because there is no energy conservation site in the alternative pathway between ubiquinone and oxygen, the free energy that would normally be conserved as ATP is lost as heat when electrons are shunted through the alternative pathway.

How can a process as seemingly energetically wasteful as the alternative pathway contribute to plant metabolism? One example of the functional usefulness of the alternative oxidase is its activity during floral development in certain members of the Araceae (the arum family)—for example, the voodoo lily (*Sauromatum guttatum*). Just before pollination, tissues of the inflorescence exhibit a dramatic increase in the rate of respiration via the alternative pathway. As a result, the temperature of the upper appendix increases by as much as 25°C over the ambient temperature. During this extraordinary burst of heat production, certain amines, indoles, and terpenes are volatilized, and the plant therefore gives off a putrid odor that attracts insect pollinators. Salicylic acid, a phenolic compound related to aspirin (see Chapter 13), has been identified as the chemical signal responsible for initiating this thermogenic event in the voodoo lily (Raskin et al. 1989) (see **Web Essay 11.4**).

In most plants, however, both the respiratory rates and the rate of cyanide-resistant respiration are too low to generate sufficient heat to raise the temperature significantly, so what other role(s) does the alternative pathway play? To answer that question we need to consider the regulation of the alternative oxidase: Its transcription is often specifically induced, for example by various types of abiotic stress. The activity of the alternative oxidase, which functions as a dimer, is regulated by reversible oxidation/reduction of an intermolecular sulfhydryl bridge, by the reduction level of the ubiquinone pool, and by pyruvate. The first two ensure that the enzyme is most active under reducing conditions, while the latter ensures that the enzyme has high activity when there is plenty of substrate for the citric acid cycle.

If the respiration rate exceeds the cell's demand for ATP (i.e., if ADP levels are very low), the reduction level in the mitochondrion will be high and activate the alternative oxidase. Thus the alternative oxidase makes it possible for the mitochondrion to adjust the relative rates of ATP production and synthesis of carbon skeletons for use in biosynthetic reactions.

Another possible function of the alternative pathway is in the response of plants to a variety of stresses (phosphate deficiency, chilling, drought, osmotic stress, and so on), many of which can inhibit mitochondrial respiration (Wagner and Krab 1995) (see Chapter 25 and Web Essay 11.5). By draining off electrons from the electron transport chain, the alternative pathway prevents a potential overreduction of the ubiquinone pool (see Figure 11.8), which, if left unchecked, can lead to the generation of destructive reactive oxygen species such as superoxide anions and hydroxyl radicals. In this way, the alternative pathway may lessen the detrimental effects of stress on respiration (Wagner and Krab 1995; Møller 2001) (see Web Essay 11.5).

THE UNCOUPLING PROTEIN A protein found in the inner membrane of mammalian mitochondria, the **uncoupling protein**, can dramatically increase the proton permeability of the membrane and thus act as an uncoupler. As a result, less ATP and more heat is generated. Heat production appears to be one of the uncoupling protein's main functions in mammalian cells.

It has long been thought that the alternative oxidase in plants and the uncoupling protein in mammals were simply two different means of achieving the same end. It was therefore surprising when a protein similar to the uncoupling protein was discovered in plant mitochondria (Vercesi et al. 1995; Laloi et al. 1997). This protein is stress induced and, like the alternative oxidase, may function to prevent overreduction of the electron transport chain (see Web Topic 11.3 and Web Essay 11.5). It remains unclear, however, why plant mitochondria require both mechanisms.

THE INTERNAL, ROTENONE-INSENSITIVE NADH DEHYDROGENASE, ND$_{in}$(NADH) This is one of the multiple NAD(P)H dehydrogenases found in plant mitochondria (see Figure 11.8). It has been suggested to work as a non-proton-pumping bypass when complex I is overloaded. Complex I has a higher affinity for NADH (ten times lower K_m), than ND$_{in}$(NADH). At lower NADH levels in the matrix, typically when ADP is available (state 3), complex I will dominate, whereas when ADP is rate limiting (state 4), NADH levels will increase and ND$_{in}$(NADH) will be more active. The physiological importance of this enzyme is, however, still unclear.

Mitochondrial respiration is controlled by key metabolites

The substrates of ATP synthesis—ADP and P$_i$—appear to be key regulators of the rates of glycolysis in the cytosol, as well as of the citric acid cycle and oxidative phosphorylation in the mitochondria. Control points exist at all three stages of respiration; here we will give just a brief overview of some major features.

The best-characterized site of regulation of the citric acid cycle is at the pyruvate dehydrogenase complex, which is reversibly phosphorylated by a regulatory kinase and a phosphatase. Pyruvate dehydrogenase is inactive in the phosphorylated state, and the regulatory kinase is inhibited by pyruvate, allowing the enzyme to be active when substrate is available (Figure 11.11).

The citric acid cycle oxidations, and subsequently respiration, are dynamically controlled by the cellular level of adenine nucleotides. As the cell's demand for ATP in the cytosol decreases relative to the rate of synthesis of ATP in the mitochondria, less ADP is available, and the electron transport chain operates at a reduced rate (see Figure 11.10). This slowdown could be signaled to citric acid cycle enzymes through an increase in matrix NADH, which inhibits the activity of several citric acid cycle dehydrogenases (Oliver and McIntosh 1995).

The buildup of citric acid cycle intermediates and their derivates, such as citrate and glutamate, inhibits the action of cytosolic pyruvate kinase, increasing the cytosolic PEP concentration, which in turn reduces the rate of conversion

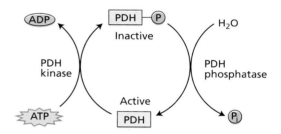

Pyruvate + CoA + NAD$^+$ ⟶ Acetyl-CoA + CO$_2$ + NADH

Effect on PDH activity	Mechanism
Activating	
Pyruvate	Inhibits kinase
ADP	Inhibits kinase
Mg^{2+} (or Mn^{2+})	Stimulates phosphatase
Inactivating	
NADH	Inhibits PDH Stimulates kinase
Acetyl CoA	Inhibits PDH Stimulates kinase
NH$_4^+$	Inhibits PDH Stimulates kinase

FIGURE 11.11 Metabolic regulation of pyruvate dehydrogenase (PDH) activity, directly and by reversible phosphorylation. Pyruvate dehydrogenase forms the entry point to the citric acid cycle, and this regulation adjusts the activity of the cycle to the cellular demand.

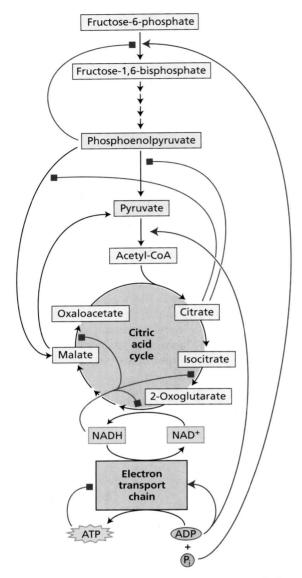

Figure 11.12 Concept of bottom-up regulation of plant respiration. Several substrates for respiration (e.g., ADP) stimulate enzymes in early steps of the pathways (green arrows). In contrast, accumulation of products (e.g., ATP) inhibits (red squares) earlier reactions in a stepwise fashion. For instance, ATP inhibits the electron transport chain leading to an accumulation of NADH. NADH inhibits citric acid cycle enzymes such as isocitrate dehydrogenase and 2-oxoglutarate dehydrogenase. Then, citric acid cycle intermediates such as citrate inhibit the PEP-metabolizing enzymes in the cytosol. Finally, PEP inhibits the conversion of fructose-6-phosphate to fructose-1,6-bisphosphate and restricts carbon feeding into glycolysis. In this way, respiration can be up- or down-regulated in response to changing demands for either of its products, ATP and organic acids.

of fructose-6-phosphate to fructose-1,6-bisphosphate, thus inhibiting glycolysis.

In summary, plant respiratory rates are controlled from the "bottom up" by the cellular level of ADP (Figure 11.12).

Figure 11.13 Glycolysis, the pentose phosphate pathway, and the citric acid cycle contribute precursors to many biosynthetic pathways in higher plants. The pathways shown illustrate the extent to which plant biosynthesis depends on the flux of carbon through these pathways and emphasize the fact that not all the carbon that enters the glycolytic pathway is oxidized to CO_2.

ADP initially regulates the rate of electron transfer and ATP synthesis, which in turn regulates citric acid cycle activity, which, finally, regulates the rate of the glycolytic reactions. The bottom-up model allows the respiratory carbon pathways to adjust to the demands for biosynthetic building blocks, thereby increasing respiratory flexibility.

Respiration is tightly coupled to other pathways

Glycolysis, the pentose phosphate pathway, and the citric acid cycle are linked to several other important metabolic pathways, some of which will be covered in greater detail in Chapter 13. The respiratory pathways are central to the production of a wide variety of plant metabolites, including amino acids, lipids and related compounds, isoprenoids, and porphyrins (Figure 11.13). Indeed, much of the reduced carbon that is metabolized by glycolysis and the citric acid cycle is diverted to biosynthetic purposes and not oxidized to CO_2. Mitochondria also carry out steps in the biosynthesis of coenzymes necessary for many metabolic enzymes in other cell compartments (see **Web Essay 11.6**).

Respiration in Intact Plants and Tissues

Many rewarding studies of plant respiration and its regulation have been carried out on isolated organelles and on cell-free extracts of plant tissues. But how does this knowledge relate to the function of the whole plant in a natural or agricultural setting?

In this section we will examine respiration and mitochondrial function in the context of the whole plant under a variety of conditions. First, we will explore what happens when green tissues are exposed to light: Respiration and photosynthesis operate simultaneously and interact in complex ways. Next we will discuss different rates of tissue respiration, which may be under developmental control, as well as the very interesting case of cytoplasmic male sterility. Finally, we will look at the influence of various environmental factors on respiration rates.

Plants respire roughly half of the daily photosynthetic yield

Many factors can affect the respiration rate of an intact plant or of its individual organs. Relevant factors include the species and growth habit of the plant, the type and age of the specific organ, and environmental variables such as the external oxygen concentration, temperature, and nutri-

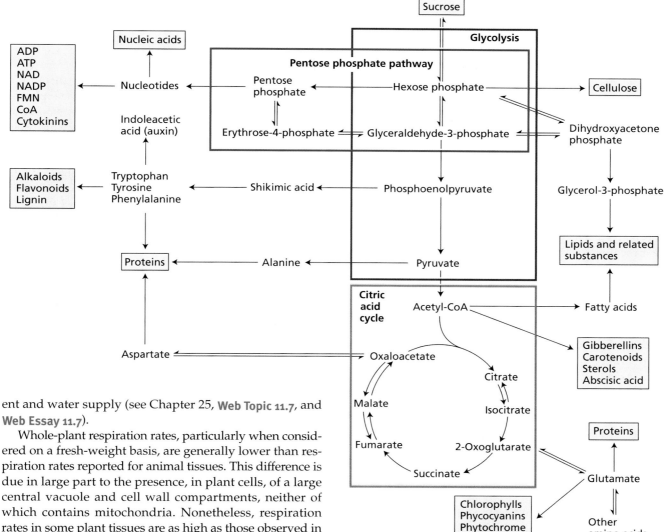

ent and water supply (see Chapter 25, **Web Topic 11.7**, and **Web Essay 11.7**).

Whole-plant respiration rates, particularly when considered on a fresh-weight basis, are generally lower than respiration rates reported for animal tissues. This difference is due in large part to the presence, in plant cells, of a large central vacuole and cell wall compartments, neither of which contains mitochondria. Nonetheless, respiration rates in some plant tissues are as high as those observed in actively respiring animal tissues, so the plant respiratory process is not inherently slower than in animals. In fact, isolated plant mitochondria respire as fast as or faster than mammalian mitochondria.

Although plants generally have low respiration rates, the contribution of respiration to the overall carbon economy of the plant can be substantial (see **Web Topic 11.7**). Whereas only green tissues photosynthesize, all tissues respire, and they do so 24 hours a day. Even in photosynthetically active tissues, respiration, if integrated over the entire day, can represent a substantial fraction of gross photosynthesis. A survey of several herbaceous species indicated that 30 to 60% of the daily gain in photosynthetic carbon was lost to respiration, although these values tended to decrease in older plants (Lambers 1985).

Young trees lose roughly a third of their daily photosynthate as respiration, and this loss can double in older trees as the ratio of photosynthetic to nonphotosynthetic tissue decreases. In tropical areas, 70 to 80% of the daily photosynthetic gain can be lost to respiration because of the high dark respiration rates associated with elevated night temperatures. See **Web Topic 11.7** for a discussion of how crop yield is affected by changes in respiration.

Respiration operates during photosynthesis

Mitochondria are involved in the metabolism of photosynthesizing leaves. The glycine generated by photorespiration is oxidized to serine in the mitochondrion, a reaction involving mitochondrial oxygen consumption (see Chapter 8). At the same time, mitochondria in photosynthesizing tissue also carry out normal mitochondrial respiration (i.e., not involving photorespiration). Relative to the maximum rate of photosynthesis, rates of mitochondrial respiration measured in green tissues in the light are far slower, generally by a factor of 6- to 20-fold. Given that rates of photorespiration can often reach 20 to 40% of the gross photosynthetic rate, the normal mitochondrial respiration also operates at rates well below the rate of photorespiration.

The activity of pyruvate dehydrogenase, one of the ports of entry into the citric acid cycle, decreases in the light to 25% of the activity in darkness (Budde and Randall 1990). Consistently, the overall rate of mitochondrial respiration decreases in the light, but the extent of the decrease remains uncertain at present. It is clear, however, that the mitochondrion is a major supplier of ATP to the cytosol (e.g., for driving biosynthetic pathways) even in illuminated leaves (Krömer 1995).

Another role of mitochondrial respiration during photosynthesis is to supply precursors for biosynthetic reactions—for example, by formation of 2-oxoglutarate needed for nitrogen assimilation (see Figures 11.7C and 11.12). This reaction also produces NADH in the matrix, linking the process to oxidative phosphorylation or to nonphosphorylating respiratory chain activities (Hoefnagel et al. 1998; Noctor and Foyer 1998).

Additional evidence for the involvement of mitochondrial respiration in photosynthesis has been obtained in studies with mitochondrial mutants defective in respiratory complexes. Compared to the wild type, these plants have slower leaf development and photosynthesis, because changes in levels of redox-active metabolites are communicated between mitochondria and chloroplasts, negatively affecting photosynthetic function (Noctor et al. 2004).

Different tissues and organs respire at different rates

A useful rule of thumb is that the greater the overall metabolic activity of a given tissue, the higher its respiration rate. Developing buds usually show very high rates of respiration (on a dry-weight basis), and respiration rates of vegetative tissues usually decrease from the point of growth (e.g., the leaf tip in dicotyledons and the leaf base in monocotyledons) to more differentiated regions. A well-studied example is the growing barley leaf (Thompson et al. 1998). In mature vegetative tissues, stems generally have the lowest respiration rates, and leaf and root respiration varies with the plant species and the conditions under which the plants are growing. Low availability of soil nutrients will, for example, increase the demand for respiratory ATP production in the root.

When a plant tissue has reached maturity, its respiration rate will either remain roughly constant or decrease slowly as the tissue ages and ultimately senesces. An exception to this pattern is the marked rise in respiration, known as the *climacteric*, that accompanies the onset of ripening in many fruits (e.g., avocado, apple, and banana) and senescence in detached leaves and flowers. Both ripening and the climacteric respiratory rise are triggered by the endogenous production of ethylene, or by an exogenous application of ethylene (see Chapter 22). In general, ethylene-induced respiration is associated with an active cyanide-resistant alternative pathway, but the role of this pathway in ripening is not clear (Tucker 1993).

Mitochondrial function is crucial during pollen development

A physiological feature directly linked to the plant mitochondrial genome is a phenomenon known as **cytoplasmic male sterility**, or **cms**. Plant lines that display cytoplasmic male sterility do not form viable pollen—hence the designation *male sterility*. The term *cytoplasmic* here refers to the fact that this trait is transmitted in a non-Mendelian fashion; the *cms* genotype is always maternally inherited with the mitochondrial genome. Because a stable male sterile line can facilitate the production of hybrid seed stock, *cms* is an important trait in plant breeding. For this use, *cms* traits that produce no major effects throughout the plant's life cycle, except for male sterility, have been found for many species.

All plants carrying the *cms* trait that have been characterized at the molecular level show the presence of distinct rearrangements in their mtDNA, relative to wild-type plants. These rearrangements create novel translated reading frames and have been strongly correlated with *cms* phenotypes in various systems. Nuclear restorer genes can overcome the effects of the mtDNA rearrangements and restore fertility to plants with the *cms* genotype. Such restorer genes are essential for the commercial utilization of *cms* if seeds are the harvested product.

An interesting consequence of the use of the *cms* gene occurred in the late 1960s, at which time 85% of the hybrid feed corn grown in the United States was derived from the use of a *cms* line of maize (*Zea mays*) called *cms*-T (Texas). In *cms*-T maize, the mtDNA rearrangements give rise to a unique 13-kDa protein, URF13 (Levings and Siedow 1992). How the URF13 protein acts to bring about male sterility is not known, but in the late 1960s a disease appeared, caused by a race of the fungus *Bipolaris maydis* (also called *Cochliobolus heterostrophus*). This specific race synthesizes a compound (HmT-toxin) that specifically interacts with the URF13 protein to produce pores in the inner mitochondrial membrane, with the result that selective permeability is lost.

The interaction between HmT-toxin and URF13 made *Bipolaris maydis* race T a particularly virulent pathogen on *cms*-T maize and led to an epidemic in the corn-growing regions of the United States that was known as southern corn leaf blight. As a result of this epidemic, the use of *cms*-T in the production of hybrid maize was discontinued. Alternative, toxin-resistant *cms* traits are nowadays in use, but the majority of hybrid corn production is done by mechanical detasseling, which prevents self-pollination.

As compared to other organs, the amount of mitochondria per cell and the expression of respiratory proteins are very high in developing anthers, where pollen development is an energy-demanding process (Huang et al. 1994). Male sterility is a common phenotype of mutations in mitochondrial genes for subunits of the complexes of oxidative phosphorylation (Vedel et al. 1999). Such mutants can be viable because of the existence of the alternative nonphosphorylating respiratory pathways.

Programmed cell death (PCD) is part of normal anther development. There are now indications that mitochondria are involved in plant PCD and that PCD is premature in anthers of *cms* sunflower (see **Web Essay 11.8**).

Environmental factors alter respiration rates

Many environmental factors can alter the operation of metabolic pathways and respiratory rates. Here we will examine the roles of environmental oxygen (O_2), temperature, and carbon dioxide (CO_2).

OXYGEN Oxygen can affect plant respiration because of its role as a substrate in the overall process. At 25°C, the equilibrium concentration of O_2 in an air-saturated (21% O_2), aqueous solution is about 250 μM. The K_m value for oxygen in the reaction catalyzed by cytochrome c oxidase is well below 1 μM, so there should be no apparent dependence of the respiration rate on external O_2 concentrations (see Chapter 2 on the web site for a discussion of K_m). However, respiration rates decrease if the atmospheric oxygen concentration is below 5% for whole tissues or below 2 to 3% for tissue slices. These findings show that oxygen diffusion through the aqueous phase in the tissue imposes a limitation on plant respiration.

The diffusion limitation imposed by an aqueous phase emphasizes the importance of the intercellular air spaces found in plant tissues for oxygen availability in the mitochondria. If there were no gaseous diffusion pathway throughout the plant, the cellular respiration rates of many plants would be limited by an insufficient oxygen supply (see **Web Essay 11.4**).

WATER SATURATION/LOW O_2 Diffusion limitation is even more significant when plant organs are growing in an aqueous medium. When plants are grown hydroponically, the solutions must be aerated vigorously to keep oxygen levels high in the vicinity of the roots. The problem of oxygen supply also arises with plants growing in very wet or flooded soils (see Chapter 25).

Some plants, particularly trees, have a restricted geographic distribution because of the need to maintain a supply of oxygen to their roots. For instance, dogwood (*Cornus florida*) and tulip tree poplar (*Liriodendron tulipifera*) can survive only in well-drained, aerated soils, because their roots cannot tolerate more than a limited exposure to a flooded condition. On the other hand, many plant species are adapted to grow in flooded soils. Herbaceous species such as rice and sunflower often rely on a network of intercellular air spaces (aerenchyma) running from the leaves to the roots to provide a continuous, gaseous pathway for the movement of oxygen to the flooded roots.

Limitation in oxygen supply can be more severe for trees having very deep roots that grow in wet soils. Such roots must survive on anaerobic (fermentative) metabolism or develop structures that facilitate the movement of oxygen to the roots. Examples of such structures are outgrowths of the roots, called *pneumatophores*, that protrude out of the water and provide a gaseous pathway for oxygen diffusion into the roots. Pneumatophores are found in *Avicennia* and *Rhizophora*, trees that grow in mangrove swamps under continuously flooded conditions.

TEMPERATURE Respiration typically increases substantially with temperatures between 0 and 30°C and reaches a plateau at 40 to 50°C. At higher temperatures it again decreases because of inactivation of the respiratory machinery. The increase in respiration rate for every 10°C increase in temperature is commonly called the temperature coefficient, Q_{10}. This coefficient describes how respiration responds to short-term temperature changes and varies with plant development and external factors. On a longer time scale, plants will become acclimated to low temperatures by increasing the respiratory capacity, so that ATP production can be continued (Atkin et al. 2003).

Low temperatures are utilized to retard postharvest respiration rates during the storage of fruits and vegetables. However, complications may arise from such storage. For instance, when potato tubers are stored at temperatures above 10°C, respiration and ancillary metabolic activities are sufficient to allow sprouting. Below 5°C, respiration rates and sprouting are reduced in most tissues, but the breakdown of stored starch and its conversion to sucrose impart an unwanted sweetness to the tubers. As a compromise, potatoes are stored at 7 to 9°C, which prevents the breakdown of starch while minimizing respiration and germination.

CO_2 CONCENTRATION It is common practice in the commercial storage of fruits to take advantage of the effects of atmospheric oxygen and temperature on respiration, and to store fruits at low temperatures under 2 to 3% oxygen and 3 to 5% CO_2. The reduced temperature lowers the respiration rate, as does the reduced oxygen. Low levels of oxygen are used instead of anoxic conditions to avoid lowering tissue oxygen tensions to the point where fermentative metabolism sets in. Carbon dioxide has a limited direct inhibitory effect on the respiration rate at the artificially high concentration of 3 to 5%.

The atmospheric CO_2 concentration is normally 360 ppm, but it is increasing as a result of human activities, and it is projected to double, to 700 ppm, before the end of the twenty-first century (see Chapter 9). The atmospheric exchange of CO_2 by plant photosynthesis and respiration is much larger than the flux caused by burning of fossil fuels. Therefore, effects of elevated CO_2 on plant respiration would severely affect the predictions for future global atmospheric changes. Recent studies have shown that 700 ppm CO_2 does not directly affect plant respiration, but measurements on whole ecosystems inside artificial biospheres indicate that respiration still may decrease per biomass unit with increased carbon dioxide. The mechanism behind this effect is not yet clarified, and it is at present not possible to fully predict the potential importance of plants as a sink for released CO_2 (Gonzales-Meler et al. 2004).

Lipid Metabolism

Whereas animals use fats for energy storage, plants use them mainly for carbon storage. Fats and oils are important storage forms of reduced carbon in many seeds, including those of agriculturally important species such as soybean, sunflower, peanut, and cotton. Oils often serve a major storage function in nondomesticated plants that produce small seeds. Some fruits, such as olives and avocados, also store fats and oils.

In this final part of the chapter we describe the biosynthesis of two types of glycerolipids: the *triacylglycerols* (the fats and oils stored in seeds) and the *polar glycerolipids* (which form the lipid bilayers of cellular membranes) (Figure 11.14). We will see that the biosynthesis of triacylglycerols and polar glycerolipids requires the cooperation of two organelles: the plastids and the endoplasmic reticulum. Plants can also use fats and oils for energy production. We will thus examine the complex process by which germinating seeds obtain metabolic energy from the oxidation of fats and oils.

Fats and oils store large amounts of energy

Fats and oils belong to the general class termed *lipids*, a structurally diverse group of hydrophobic compounds that are soluble in organic solvents and highly insoluble in water. Lipids represent a more reduced form of carbon than carbohydrates, so the complete oxidation of 1 g of fat or oil (which contains about 40 kJ, or 9.3 kcal, of energy) can produce considerably more ATP than the oxidation of 1 g of starch (about 15.9 kJ, or 3.8 kcal). Conversely, the biosynthesis of fats, oils, and related molecules, such as the phospholipids of membranes, requires a correspondingly large investment of metabolic energy.

Other lipids are important for plant structure and function but are not used for energy storage. These include waxes, which make up the protective cuticle that reduces water loss from exposed plant tissues, and terpenoids (also known as isoprenoids), which include carotenoids involved in photosynthesis and sterols present in many plant membranes (see Chapter 13).

Triacylglycerols are stored in oil bodies

Fats and oils exist mainly in the form of triacylglycerols (*acyl* refers to the fatty acid portion), in which fatty acid molecules are linked by ester bonds to the three hydroxyl groups of glycerol (see Figure 11.14).

The fatty acids in plants are usually straight-chain carboxylic acids having an even number of carbon atoms. The carbon chains can be as short as 12 units and as long as 20, but more commonly they are 16 or 18 carbons long. *Oils* are liquid at room temperature, primarily because of the presence of unsaturated bonds in their component fatty acids; *fats*, which have a higher proportion of saturated fatty acids, are solid at room temperature. The major fatty acids in plant lipids are shown in Table 11.3.

The composition of fatty acids in plant lipids varies with the species. For example, peanut oil is about 9% palmitic acid, 59% oleic acid, and 21% linoleic acid, and cottonseed oil is 25% palmitic acid, 15% oleic acid, and 55% linoleic acid. The biosynthesis of these fatty acids will be discussed shortly.

Triacylglycerols in most seeds are stored in the cytoplasm of either cotyledon or endosperm cells in organelles known as **oil bodies** (also called *spherosomes* or *oleosomes*) (see Chapter 1). Oil bodies have an unusual membrane barrier that separates the triacylglycerols from the aqueous cytoplasm. A single layer of phospholipids (i.e., a half-bilayer) surrounds the oil body with the hydrophilic ends of the phospholipids exposed to the cytosol and the hydrophobic acyl hydrocarbon chains facing the triacylglycerol interior (see Chapter 1). The oil body is stabilized by the presence of specific proteins, called *oleosins*, that coat the surface and prevent the phospholipids of adjacent oil bodies from coming in contact and fusing.

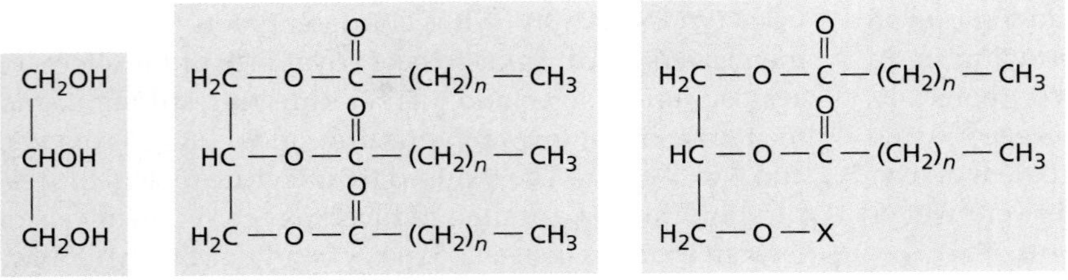

Glycerol — Triacylglycerol (the major stored lipid) — Glycerolipid

$X = H$	Diacylglycerol (DAG)
$X = HPO_3^{2-}$	Phosphatidic acid
$X = PO_3^{2-} — CH_2 — CH_2 — \overset{+}{N}(CH_3)_3$	Phosphatidylcholine
$X = PO_3^{2-} — CH_2 — CH_2 — NH_2$	Phosphatidylethanolamine
X = galactose	Galactolipids

FIGURE 11.14 Structural features of triacylglycerols and polar glycerolipids in higher plants. The carbon chain lengths of the fatty acids, which always have an even number of carbons, range from 12 to 20 but are typically 16 or 18. Thus, the value of n is usually 14 or 16.

TABLE 11.3
Common fatty acids in higher plant tissues

Name[a]	Structure
Saturated Fatty Acids	
Lauric acid (12:0)	$CH_3(CH_2)_{10}CO_2H$
Myristic acid (14:0)	$CH_3(CH_2)_{12}CO_2H$
Palmitic acid (16:0)	$CH_3(CH_2)_{14}CO_2H$
Stearic acid (18:0)	$CH_3(CH_2)_{16}CO_2H$
Unsaturated Fatty Acids	
Oleic acid (18:1)	$CH_3(CH_2)_7CH=CH(CH_2)_7CO_2H$
Linoleic acid (18:2)	$CH_3(CH_2)_4CH=CH-CH_2-CH=CH(CH_2)_7CO_2H$
Linolenic acid (18:3)	$CH_3CH_2CH=CH-CH_2-CH=CH-CH_2-CH=CH-(CH_2)_7CO_2H$

[a] Each fatty acid has a numerical abbreviation. The number before the colon represents the total number of carbons; the number after the colon is the number of double bonds.

This unique membrane structure for oil bodies results from the pattern of triacylglycerol biosynthesis. Triacylglycerol synthesis is completed by enzymes located in the membranes of the endoplasmic reticulum (ER), and the resulting fats accumulate between the two monolayers of the ER membrane bilayer. The bilayer swells apart as more fats are added to the growing structure, and ultimately a mature oil body buds off from the ER (Napier et al. 1996).

Polar glycerolipids are the main structural lipids in membranes

As outlined in Chapter 1, each membrane in the cell is a bilayer of *amphipathic* (i.e., having both hydrophilic and hydrophobic regions) lipid molecules in which a polar head group interacts with the aqueous phase while hydrophobic fatty acid chains form the center of the membrane. This hydrophobic core prevents random diffusion of solutes between cell compartments and thereby allows the biochemistry of the cell to be organized.

The main structural lipids in membranes are the polar glycerolipids (see Figure 11.14), in which the hydrophobic portion consists of two 16-carbon or 18-carbon fatty acid chains esterified to positions 1 and 2 of a glycerol backbone. The polar head group is attached to position 3 of the glycerol. There are two categories of polar glycerolipids:

1. **Glyceroglycolipids**, in which sugars form the head group (Figure 11.15A)
2. **Glycerophospholipids**, in which the head group contains phosphate (see Figure 11.15B)

Plant membranes have additional structural lipids, including sphingolipids and sterols (see Chapter 13), but these are minor components. Other lipids perform specific roles in photosynthesis and other processes. Included among these lipids are chlorophylls, plastoquinone, carotenoids, and tocopherols, which together account for about one-third of the lipids in plant leaves.

Figure 11.15 shows the nine major glycerolipid classes in plants, each of which can be associated with many different fatty acid combinations. The structures shown in Figure 11.15 illustrate some of the more common molecular species.

Chloroplast membranes, which account for 70% of the membrane lipids in photosynthetic tissues, are dominated by glyceroglycolipids; other membranes of the cell contain glycerophospholipids (Table 11.4). In nonphotosynthetic tissues, phospholipids are the major membrane glycerolipids.

Fatty acid biosynthesis consists of cycles of two-carbon addition

Fatty acid biosynthesis involves the cyclic condensation of two-carbon units in which acetyl-CoA is the precursor. In plants, fatty acids are synthesized exclusively in the plastids; in animals, fatty acids are synthesized primarily in the cytosol.

TABLE 11.4
Glycerolipid components of cellular membranes

	Lipid composition (percentage of total)		
	Chloroplast	Endoplasmic reticulum	Mitochondrion
Phosphatidylcholine	4	47	43
Phosphatidylethanolamine	—	34	35
Phosphatidylinositol	1	17	6
Phosphatidylglycerol	7	2	3
Diphosphatidylglycerol	—	—	13
Monogalactosyldiacylglycerol	55	—	—
Digalactosyldiacylglycerol	24	—	—
Sulfolipid	8	—	—

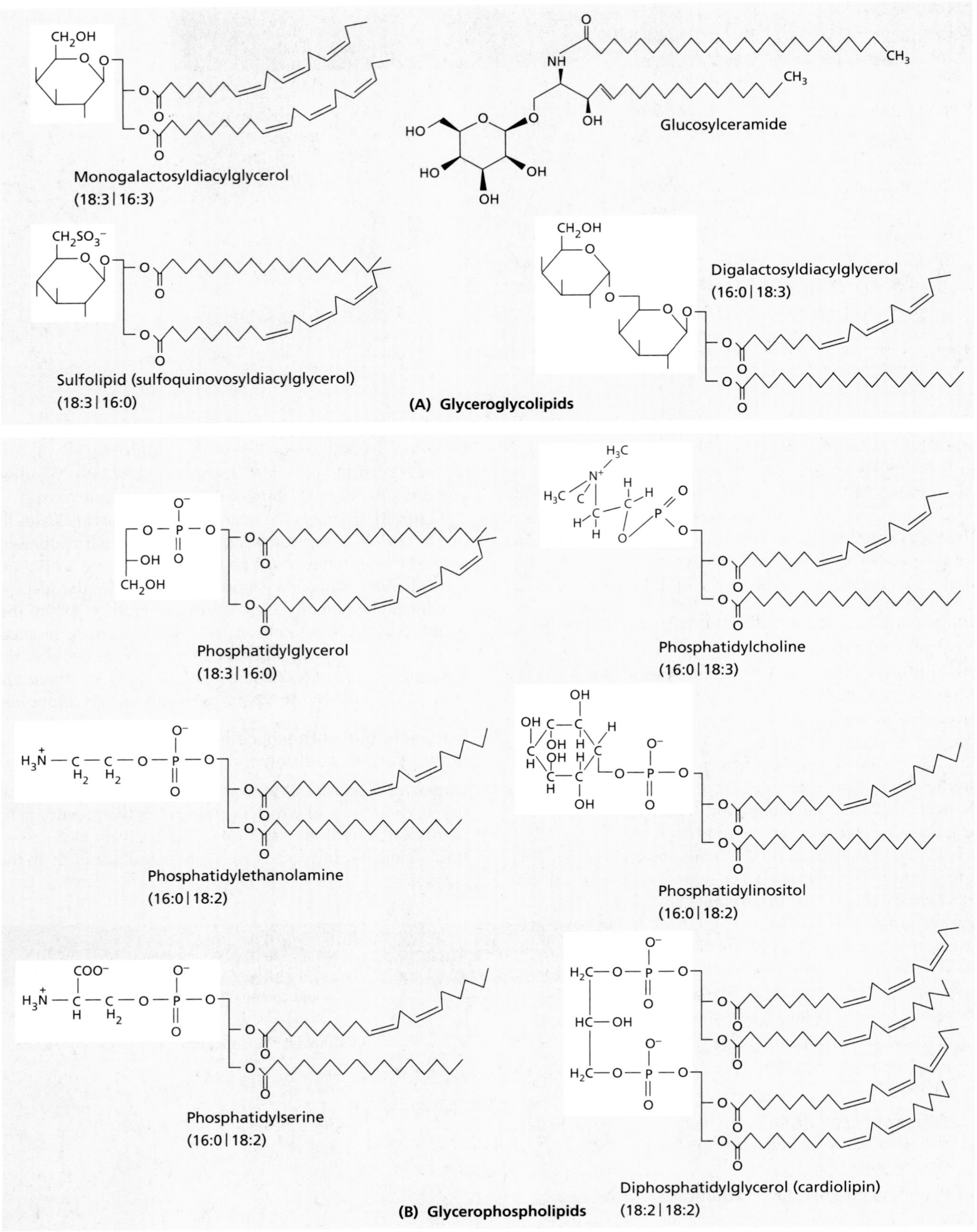

Monogalactosyldiacylglycerol
(18:3 | 16:3)

Glucosylceramide

Sulfolipid (sulfoquinovosyldiacylglycerol)
(18:3 | 16:0)

Digalactosyldiacylglycerol
(16:0 | 18:3)

(A) Glyceroglycolipids

Phosphatidylglycerol
(18:3 | 16:0)

Phosphatidylcholine
(16:0 | 18:3)

Phosphatidylethanolamine
(16:0 | 18:2)

Phosphatidylinositol
(16:0 | 18:2)

Phosphatidylserine
(16:0 | 18:2)

Diphosphatidylglycerol (cardiolipin)
(18:2 | 18:2)

(B) Glycerophospholipids

◀ **FIGURE 11.15** Major polar lipids of plant membranes: glyceroglycolipids and a sphingolipid (A) and glycerophospholipids (B). At least six different fatty acids may be attached to the glycerol backbone. One of the more common molecular species is shown for each lipid. The numbers given below each name refer to the number of carbons (number before the colon) and the number of double bonds (number after the colon).

The enzymes of the pathway are thought to be held together in a complex that is collectively referred to as *fatty acid synthase*. The complex probably allows the series of reactions to occur more efficiently than it would if the enzymes were physically separated from each other. In addition, the growing acyl chains are covalently bound to a low-molecular-weight, acidic protein called **acyl carrier protein** (**ACP**). When conjugated to the acyl carrier protein, the fatty acid chain is referred to as **acyl-ACP**.

The first committed step in the pathway (i.e., the first step unique to the synthesis of fatty acids) is the synthesis of malonyl-CoA from acetyl-CoA and CO_2 by the enzyme acetyl-CoA carboxylase (Figure 11.16) (Sasaki et al. 1995). The tight regulation of acetyl-CoA carboxylase appears to control the overall rate of fatty acid synthesis (Ohlrogge and Jaworski 1997). The malonyl-CoA then reacts with ACP to yield malonyl-ACP in the following four steps:

1. In the first cycle of fatty acid synthesis, the acetate group from acetyl-CoA is transferred to a specific cysteine of *condensing enzyme* (3-ketoacyl-ACP synthase) and then combined with malonyl-ACP to form acetoacetyl-ACP.

2. Next the keto group at carbon 3 is removed (reduced) by the action of three enzymes to form a new acyl chain (butyryl-ACP), which is now four carbons long (see Figure 11.16).

3. The four-carbon acid and another molecule of malonyl-ACP then become the new substrates for condensing enzyme, resulting in the addition of another

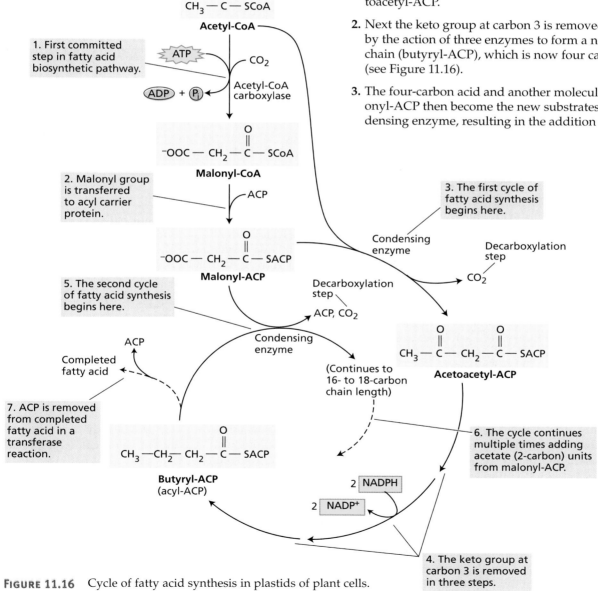

FIGURE 11.16 Cycle of fatty acid synthesis in plastids of plant cells.

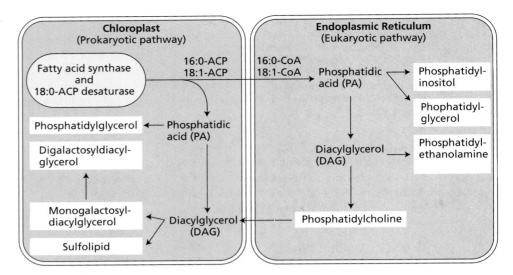

FIGURE 11.17 The two pathways for glycerolipid synthesis in the chloroplast and endoplasmic reticulum of Arabidopsis leaf cells. The major membrane components are shown in boxes. Glycerolipid desaturates in the chloroplast, and enzymes in the ER convert 16:0 and 18:1 fatty acids to the more highly unsaturated fatty acids shown in Figure 11.15.

two-carbon unit to the growing chain, and the cycle continues until 16 or 18 carbons have been added.

4. Some 16:0-ACP is released from the fatty acid synthase machinery, but most molecules that are elongated to 18:0-ACP are efficiently converted to 18:1-ACP by a desaturase enzyme. The repetition of this sequence of events makes 16:0-ACP and 18:1-ACP the major products of fatty acid synthesis in plastids (Figure 11.17).

Fatty acids may undergo further modification after they are linked with glycerol to form glycerolipids. Additional double bonds are placed in the 16:0 and 18:1 fatty acids by a series of desaturase isozymes. Desaturase isozymes are integral membrane proteins found in the chloroplast and the endoplasmic reticulum (ER). Each desaturase inserts a double bond at a specific position in the fatty acid chain, and the enzymes act sequentially to produce the final 18:3 and 16:3 products (Ohlrogge and Browse 1995).

Glycerolipids are synthesized in the plastids and the ER

The fatty acids synthesized in the plastid are next used to make the glycerolipids of membranes and oil bodies. The first steps of glycerolipid synthesis are two acylation reactions that transfer fatty acids from acyl-ACP or acyl-CoA to glycerol-3-phosphate to form **phosphatidic acid**.

The action of a specific phosphatase produces **diacylglycerol** (**DAG**) from phosphatidic acid. Phosphatidic acid can also be converted directly to phosphatidylinositol or phosphatidylglycerol; DAG can give rise to phosphatidylethanolamine or phosphatidylcholine (see Figure 11.17).

The localization of the enzymes of glycerolipid synthesis reveals a complex and highly regulated interaction between the chloroplast, where fatty acids are synthesized, and other membrane systems of the cell. In simple terms, the biochemistry involves two pathways referred to as the *prokaryotic* (or chloroplast) pathway and the *eukaryotic* (or ER) pathway:

1. In chloroplasts, the **prokaryotic pathway** utilizes the 16:0- and 18:1-ACP products of chloroplast fatty acid synthesis to synthesize phosphatidic acid and its derivatives. Alternatively, the fatty acids may be exported to the cytoplasm as CoA esters.

2. In the cytoplasm, the **eukaryotic pathway** uses a separate set of acyltransferases in the ER to incorporate the fatty acids into phosphatidic acid and its derivatives.

A simplified version of this two-pathway model is depicted in Figure 11.17.

In some higher plants, including Arabidopsis and spinach, the two pathways contribute almost equally to chloroplast lipid synthesis. In many other angiosperms, however, phosphatidylglycerol is the only product of the prokaryotic pathway, and the remaining chloroplast lipids are synthesized entirely by the eukaryotic pathway.

The biochemistry of triacylglycerol synthesis in oilseeds is generally the same as described for the glycerolipids. 16:0- and 18:1-ACP are synthesized in the plastids of the cell and exported as CoA thioesters for incorporation into DAG in the endoplasmic reticulum (see Figure 11.17).

The key enzymes in oilseed metabolism (not shown in Figure 11.17) are acyl-CoA:DAG acyltransferase and PC:DAG acyltransferase, which catalyze triacylglycerol synthesis (Dahlqvist et al. 2000). As noted earlier, triacylglycerol molecules accumulate in specialized subcellular structures—the oil bodies—from which they can be mobilized during germination and converted to sugar.

Lipid composition influences membrane function

A central question in membrane biology is the functional reason behind lipid diversity. Each membrane system of the cell has a characteristic and distinct complement of lipid types, and within a single membrane each class of lipids has a distinct fatty acid composition. Our understanding of a membrane is one in which lipids make up the fluid, semipermeable bilayer that is the matrix for the functional membrane proteins.

Since this bulk lipid role could be satisfied by a single unsaturated species of phosphatidylcholine, obviously such a simple model is unsatisfactory. Why is lipid diversity needed? One aspect of membrane biology that might offer answers to this central question is the relationship between lipid composition and the ability of organisms to adjust to temperature changes (Wolter et al. 1992). For example, chill-sensitive plants experience sharp reductions in growth rate and development at temperatures between 0 and 12°C (see Chapter 25). Many economically important crops, such as cotton, soybean, maize, rice, and many tropical and subtropical fruits, are classified as chill sensitive. In contrast, most plants that originate from temperate regions are able to grow and develop at chilling temperatures and are classified as chill-resistant plants.

It has been suggested that because of the decrease in lipid fluidity at lower temperatures, the primary event of chilling injury is a transition from a liquid-crystalline phase to a gel phase in the cellular membranes. According to this proposal, this transition would result in alterations in the metabolism of chilled cells and lead to injury and death of the chill-sensitive plants. The degree of unsaturation of the fatty acids would determine the temperature at which such damage occurred.

Recent research, however, suggests that the relationship between membrane unsaturation and plant responses to temperature is more subtle and complex (see **Web Topic 11.8**). The responses of Arabidopsis mutants with increased saturation of fatty acids to low temperature appear quite distinct from what is predicted by the chilling sensitivity hypothesis, suggesting that normal chilling injury may not be strictly related to the level of unsaturation of membrane lipids.

On the other hand, experiments with transgenic tobacco plants that are chill sensitive show opposite results. The transgenic expression of exogenous genes in tobacco has been used specifically to decrease the level of saturated phosphatidylglycerol or to bring about a general increase in membrane unsaturation. In each case, damage caused by chilling was alleviated to some extent.

These new findings make it clear that the extent of membrane unsaturation or the presence of particular lipids, such as disaturated phosphatidylglycerol, can affect the responses of plants to low temperature. As discussed in **Web Topic 11.8**, more work is required to fully understand the relationship between lipid composition and membrane function.

Membrane lipids are precursors of important signaling compounds

Plants, animals, and microbes all use membrane lipids as precursors for compounds that are used for intracellular or long-range signaling. For example, jasmonate derived from linolenic acid (18:3) activates plant defenses against insects and many fungal pathogens. In addition, jasmonate regulates other aspects of plant growth, including the development of anthers and pollen (Stintzi and Browse 2000). **Phosphatidylinositol-4,5-bisphosphate (PIP$_2$)** is the most important of several phosphorylated derivatives of phosphatidylinositol known as *phosphoinositides*. In animals, receptor-mediated activation of phospholipase C leads to the hydrolysis of PIP$_2$ to inositol trisphosphate (IP$_3$) and diacylglycerol, which both act as intracellular secondary messengers.

The action of IP$_3$ in releasing Ca^{2+} into the cytoplasm (through calcium-sensitive channels in the tonoplast and other membranes) and thereby regulating cellular processes has been demonstrated in several plant systems, including the stomatal guard cells (Schroeder et al. 2001). Information about other types of lipid signaling in plants is becoming available through biochemical and molecular genetic studies of phospholipases (Wang 2001) and other enzymes involved in the generation of these signals.

Storage lipids are converted into carbohydrates in germinating seeds

After germinating, oil-containing seeds metabolize stored triacylglycerols by converting lipids to sucrose. Plants are not able to transport fats from the endosperm to the root and shoot tissues of the germinating seedling, so they must convert stored lipids to a more mobile form of carbon, generally sucrose. This process involves several steps that are located in different cellular compartments: oil bodies, glyoxysomes, mitochondria, and cytosol.

OVERVIEW: LIPIDS TO SUCROSE The conversion of lipids to sucrose in oilseeds is triggered by germination, and it begins with the hydrolysis of triacylglycerols stored in the oil bodies to free fatty acids, followed by oxidation of the fatty acids to produce acetyl-CoA (Figure 11.18). The fatty acids are oxidized in a type of peroxisome called a **glyoxysome**, an organelle enclosed by a single bilayer membrane that is found in the oil-rich storage tissues of seeds. Acetyl-CoA is metabolized in the glyoxysome (see Figure 11.18A) to produce succinate, which is transported from the glyoxysome to the mitochondrion, where it is converted first to oxaloacetate and then to malate. The process ends in the cytosol with the conversion of malate to glucose via gluconeogenesis, and then to sucrose.

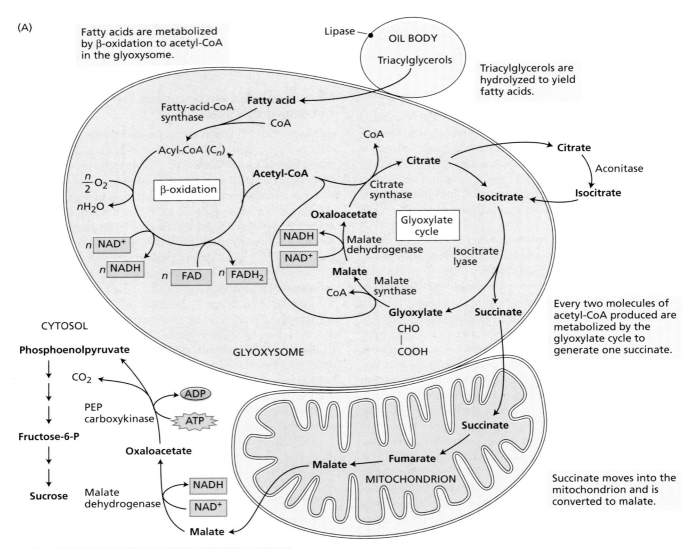

(A) Fatty acids are metabolized by β-oxidation to acetyl-CoA in the glyoxysome.

Triacylglycerols are hydrolyzed to yield fatty acids.

Every two molecules of acetyl-CoA produced are metabolized by the glyoxylate cycle to generate one succinate.

Succinate moves into the mitochondrion and is converted to malate.

Malate is transported into the cytosol and oxidized to oxaloacetate, which is converted to phosphoenolpyruvate by the enzyme PEP carboxykinase. The resulting PEP is then metabolized to produce sucrose via the gluconeogenic pathway.

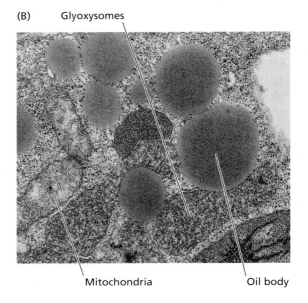

◀ **FIGURE 11.18** The conversion of fats to sugars during germination in oil-storing seeds. (A) Carbon flow during fatty acid breakdown and gluconeogenesis (refer to Figures 11.2, 11.3, and 11.6 for structures). (B) Electron micrograph of a cell from the oil-storing cotyledon of a cucumber seedling, showing glyoxysomes, mitochondria, and oleosomes. (B courtesy of R. N. Trelease.)

Although some of this fatty acid–derived carbon is diverted to other metabolic reactions in certain oilseeds, in castor bean the process is so efficient that each gram of lipid metabolized results in the formation of 1 g of carbohydrate, which is equivalent to a 40% recovery of free energy in the form of carbon bonds ($[15.9 \text{ kJ}/40 \text{ kJ}] \times 100 = 40\%$).

LIPASE-MEDIATED HYDROLYSIS The initial step in the conversion of lipids to carbohydrate is the breakdown of triacylglycerols stored in the oil bodies by the enzyme lipase, which, at least in castor bean endosperm, is located on the half-membrane that serves as the outer boundary of the oil body. The lipase hydrolyzes triacylglycerols to three molecules of fatty acid and glycerol. Corn and cotton also contain a lipase activity in the oil body, but peanut, soybean, and cucumber show lipase activity in the glyoxysome instead. During the breakdown of lipids, oil bodies and glyoxysomes are generally in close physical association (see Figure 11.18B).

β-OXIDATION OF FATTY ACIDS After hydrolysis of the triacylglycerols, the resulting fatty acids enter the glyoxysome, where they are activated by conversion to fatty-acyl-CoA by the enzyme fatty-acyl-CoA synthase. Fatty-acyl-CoA is the initial substrate for the β-oxidation series of reactions, in which C_n fatty acids (fatty acids composed of n number of carbons) are sequentially broken down to $n/2$ molecules of acetyl-CoA (see Figure 11.18A). This reaction sequence involves the reduction of $\frac{1}{2} O_2$ to H_2O and the formation of one NADH and one $FADH_2$ for each acetyl-CoA produced.

In mammalian tissues, the four enzymes associated with β-oxidation are present in the mitochondrion; in plant seed storage tissues, they are localized exclusively in the glyoxysome. Interestingly, in plant vegetative tissues (e.g., mung bean hypocotyl and potato tuber), the β-oxidation reactions are localized in a related organelle, the peroxisome (see Chapter 1).

THE GLYOXYLATE CYCLE The function of the glyoxylate cycle is to convert two molecules of acetyl-CoA to succinate. The acetyl-CoA produced by β-oxidation is further metabolized in the glyoxysome through a series of reactions that make up the glyoxylate cycle (see Figure 11.18A). Initially, the acetyl-CoA reacts with oxaloacetate to give citrate, which is then transferred to the cytoplasm for isomerization to isocitrate by aconitase. Isocitrate is reimported into the peroxisome and converted to malate by two reactions that are unique to the glyoxylate pathway:

1. First, isocitrate (C_6) is cleaved by the enzyme isocitrate lyase to give succinate (C_4) and glyoxylate (C_2). This succinate is exported to the mitochondria.

2. Next, malate synthase combines a second molecule of acetyl-CoA with glyoxylate to produce malate.

Malate is then oxidized by malate dehydrogenase to oxaloacetate, which can combine with another acetyl-CoA to continue the cycle (see Figure 11.18A). The glyoxylate produced keeps the cycle operating in the glyoxysome, but the succinate is exported to the mitochondria for further processing.

THE MITOCHONDRIAL ROLE Moving from the glyoxysomes to the mitochondria, the succinate is converted to malate by the normal citric acid cycle reactions. The resulting malate can be exported from the mitochondria in exchange for succinate via the dicarboxylate transporter located in the inner mitochondrial membrane. Malate is then oxidized to oxaloacetate by malate dehydrogenase in the cytosol, and the resulting oxaloacetate is converted to carbohydrate.

This conversion requires circumventing the irreversibility of the pyruvate kinase reaction (see Figure 11.3) and is facilitated by the enzyme PEP carboxykinase, which utilizes the phosphorylating ability of ATP to convert oxaloacetate to PEP and CO_2 (see Figure 11.18A). From PEP, gluconeogenesis can proceed to the production of glucose, as described earlier. Sucrose is the final product of this process, and the primary form of reduced carbon translocated from the cotyledons to the growing seedling tissues. Not all seeds quantitatively convert fat to sugar (see **Web Topic 11.9**).

Summary

In plant respiration, reduced cellular carbon generated during photosynthesis is oxidized to CO_2 and water, and this oxidation is coupled to the synthesis of ATP. Respiration takes place in three main stages: glycolysis, the citric acid (or Krebs) cycle, and oxidative phosphorylation. The latter comprises the electron transport chain and ATP synthesis.

In glycolysis, carbohydrate is converted to pyruvate in the cytosol, and a small amount of ATP is synthesized via substrate-level phosphorylation. Pyruvate is subsequently oxidized within the mitochondrial matrix through the citric acid cycle, generating a large number of reducing equivalents in the form of NADH and $FADH_2$.

In the third stage, oxidative phosphorylation, electrons from NADH and $FADH_2$ pass through the electron transport chain in the inner mitochondrial membrane to reduce oxygen. The chemical energy is conserved in the form of an electrochemical proton gradient, which is created by the coupling of electron flow to proton pumping from the matrix to the intermembrane space. This energy is then converted into chemical energy in the form of ATP by the F_oF_1-ATP synthase, also located in the inner membrane, which couples

ATP synthesis from ADP and P_i to the flow of protons back into the matrix down their electrochemical gradient.

Aerobic respiration in plants has several unique features, including the presence of alternative pathways, that are not coupled to ATP turnover, in glycolysis and the electron transport chain. Carbohydrates can also be oxidized via the oxidative pentose phosphate pathway, in which the reducing power is produced in the form of NADPH mainly for biosynthetic purposes. Numerous glycolytic and citric acid cycle intermediates also provide the starting material for a multitude of biosynthetic pathways. Consistent with the multiple functions, respiration is regulated mainly by cellular demand for ATP and biosynthesis intermediates via control points in the electron transport chain, the citric acid cycle, and glycolysis.

More than 50% of the daily photosynthetic yield can be respired by a plant, but many factors can affect the respiration rate observed at the whole-plant level. These factors include the nature and age of the plant tissue and environmental factors such as light, oxygen concentration, temperature, and CO_2 concentration.

Lipids play a major role in plants: Amphipathic lipids serve as the primary nonprotein components of plant membranes; fats and oils are an efficient storage form of reduced carbon, particularly in seeds. Glycerolipids play important roles as structural components of membranes. Fatty acids are synthesized in plastids using acetyl-CoA. Fatty acids from the plastid can be transported to the ER, where they are further modified.

Membrane function may be influenced by the lipid composition. The degree of unsaturation of the fatty acids influences the sensitivity of plants to cold but does not seem to be involved in normal chilling injury. On the other hand, certain membrane lipid breakdown products, such as jasmonic acid, can act as signaling agents in plant cells.

Triacylglycerol is synthesized in the ER and accumulates within the phospholipid bilayer, forming oil bodies. During germination in oil-storing seeds, the stored lipids are metabolized to carbohydrate in a series of reactions that involve a metabolic sequence known as the glyoxylate cycle. This cycle takes place in glyoxysomes, and subsequent steps occur in the mitochondria. The reduced carbon generated during lipid breakdown in the glyoxysomes is ultimately converted to carbohydrate in the cytosol by gluconeogenesis.

Web Material

Web Topics

11.1 Isolation of Mitochondria
Methods for the isolation of intact, functional mitochondria have been developed.

11.2 The Q-Cycle Explains How Protons Can Be Pumped across the Inner Mitochondrial Membrane
The best known example is the coupling of electron transport to proton pumping in complex III.

11.3 Multiple Energy-Conservation Bypasses in Oxidative Phosphorylation of Plant Mitochondria
With molecular characterization, physiological roles of the enigmatic "energy-wasting" pathways of respiration are being uncovered.

11.4 F_oF_1-ATP Synthases: The World's Smallest Rotary Motors
Rotation of the γ subunit brings about the conformational changes that couples proton flux to ATP synthesis.

11.5 Transport Into and Out of Plant Mitochondria
Plant mitochondria operate different transport mechanisms.

11.6 The Genetic System in Plant Mitochondria Has Many Special Features
The mitochondrial genome encodes about 40 mitochondrial proteins.

11.7 Does Respiration Reduce Crop Yields?
Crop yield is correlated to low respiration in a way that is not understood.

11.8 The Lipid Composition of Membranes Affects the Cell Biology and Physiology of Plants
Lipid mutants are expanding our understanding of the ability of organisms to adapt to temperature changes.

11.9 Utilization of Oil Reserves in Cotyledons
In some species, only part of the stored lipid in the cotyledons is exported as carbohydrate.

Web Essays

11.1 Metabolic Flexibility Helps Plants Survive Stress
The ability of plants to carry out a metabolic step in different ways increases plant survival under stress.

11.2 Metabolic Profiling of Plant Cells
Metabolic profiling complements genomics and proteomics.

11.3 Mitochondrial Dynamics: When Form Meets Function
New microscopy methods have shown that mitochondria dynamically change shape in vivo.

11.4 Temperature Regulation by Thermogenic Flowers
In thermogenic flowers, such as the *Arum* lilies, temperature can increase up to 20°C above their surroundings.

11.5 Reactive Oxygen Species (ROS) and Plant Mitochondria
The production of damaging reactive oxygen species is an unavoidable consequence of aerobic respiration.

11.6 Coenzyme Synthesis in Plant Mitochondria
Pathways for synthesis of coenzymes are often split between organelles.

11.7 The Role of Respiration in Desiccation Tolerance
Respiration has both positive and negative effects on the survival of plant cells under water stress.

11.8 Balancing Life and Death: The Role of the Mitochondrion in Programmed Cell Death
Programmed cell death is an integral part of the life cycle of plants, often directly involving mitochondria.

Chapter References

Abrahams, J. P., Leslie, A. G. W., Lutter, R., and Walker, J. E. (1994) Structure at 2.8 Å resolution of F_1-ATPase from bovine heart mitochondria. *Nature* 370: 621–628.

Ap Rees, T. (1980) Assessment of the contributions of metabolic pathways to plant respiration. In *The Biochemistry of Plants*, Vol. 2, D. D. Davies, ed., Academic Press, New York, pp. 1–29.

Atkin, O. K., and Tjoelker, M. G. (2003) Thermal acclimation and the dynamic response of plant respiration to temperature. *Trends Plant Sci.* 8: 343–351.

Brand, M. D. (1994) The stoichiometry of proton pumping and ATP synthesis in mitochondria. *Biochemist* 16 (4): 20–24.

Budde, R. J. A., and Randall, D. D. (1990) Pea leaf mitochondrial pyruvate dehydrogenase complex is inactivated in vivo in a light-dependent manner. *Proc. Natl. Acad. Sci. USA* 87: 673–676.

Dahlqvist, A., Stahl, U., Lenman, M., Banas, A., Lee, M., Sandager, L., Ronne, H., and Stymne, S. (2000) Phospholipid:diacylglycerol acyltransferase: An enzyme that catalyzes the acyl-CoA-independent formation of triacylglycerol in yeast and plants. *Proc. Natl. Acad. Sci. USA* 97: 6487–6492.

Dennis, D. T., and Blakely, S. D. (2000) Carbohydrate metabolism. In *Biochemistry & Molecular Biology of Plants*, B. Buchanan, W. Gruissem, and R. Jones, eds., American Society of Plant Physiologists, Rockville, MD, pp. 630–674.

Dennis, D. T., Huang, Y., and Negm, F. B. (1997) Glycolysis, the pentose phosphate pathway and anaerobic respiration. In *Plant Metabolism*, 2nd ed., D. T. Dennis, D. H. Turpin, D. D. Lefebvre, and D. B. Layzell, eds., Longman, Singapore, pp. 105–123.

Douce, R. (1985) *Mitochondria in Higher Plants: Structure, Function, and Biogenesis*. Academic Press, Orlando, FL.

Givan, C. V. (1999) Evolving concepts in plant glycolysis: Two centuries of progress. *Biol. Rev.* 74: 277–309.

Gonzalez-Meler, M. A., Taneva, L., and Trueman, R. J. (2004) Plant respiration and elevated atmospheric CO_2 concentration: Cellular responses and global significance. *Ann. Bot.* 94: 647–656.

Gunning, B. E. S., and Steer, M. W. (1996) *Plant Cell Biology: Structure and Function of Plant Cells*. Jones and Bartlett, Boston.

Hoefnagel, M. H. N., Atkin, O. K., and Wiskich, J. T. (1998) Interdependence between chloroplasts and mitochondria in the light and the dark. *Biochim. Biophys. Acta* 1366: 235–255.

Huang, J., Struck, F., Matzinger, D. F., and Levings, C. S. (1994) Flower-enhanced expression of a nuclear-encoded mitochondrial respiratory protein is associated with changes in mitochondrion number. *Plant Cell* 6: 439–448.

Krömer, S. (1995) Respiration during photosynthesis. *Annu. Rev. Plant Physiol. Plant Mol. Biol.* 46: 45–70.

Kruger, N. J. (1997) Carbohydrate synthesis and degradation. In *Plant Metabolism*, 2nd ed., D. T. Dennis, D. H. Turpin, D. D. Lefebvre, and D. B. Layzell, eds., Longman, Singapore, pp. 83–104.

Laloi, M., Klein, M., Riesmeier, J. W., Müller-Röber, B., Fleury, C., Bouillaud, F., and Ricquier, D. (1997) A plant cold-induced uncoupling protein. *Nature* 389: 135–136.

Lambers, H. (1985) Respiration in intact plants and tissues. Its regulation and dependence on environmental factors, metabolism and invaded organisms. In *Higher Plant Cell Respiration* (Encyclopedia of Plant Physiology, New Series, Vol. 18), R. Douce and D. A. Day, eds., Springer, Berlin, pp. 418–473.

Leon, P., Arroyo, A., and Mackenzie, S. (1998) Nuclear control of plastid and mitochondrial development in higher plants. *Annu. Rev. Plant Physiol. Plant Mol. Biol.* 49: 453–480.

Levings, C. S., III, and Siedow, J. N. (1992) Molecular basis of disease susceptibility in the Texas cytoplasm of maize. *Plant Mol. Biol.* 19: 135–147.

Marienfeld, J., Unseld, M., and Brennicke, A. (1999) The mitochondrial genome of Arabidopsis is composed of both native and immigrant information. *Trends Plant Sci.* 4: 495–502.

McCabe, T. C., Daley, D., and Whelan, J. (2000) Regulatory, developmental and tissue aspects of mitochondrial biogenesis in plants. *Plant Biol.* 2: 121–135.

Millar, A. H., Heazlewood, J. L., Kristensen, B. K., Braun, H.-P., and Møller, I. M. (2005) The plant mitochondrial proteome. *Trends Plant Sci.* 10: 36–43.

Møller, I. M. (2001) Plant mitochondria and oxidative stress. Electron transport, NADPH turnover and metabolism of reactive oxygen species. *Annu. Rev. Plant Physiol. Plant Mol. Biol.* 52: 561–591.

Møller, I. M., and Rasmusson, A. G. (1998) The role of NADP in the mitochondrial matrix. *Trends Plant Sci.* 3: 21–27.

Napier, J. A., Stobart, A. K., and Shewry, P. R. (1996) The structure and biogenesis of plant oil bodies: The role of the ER membrane and the oleosin class of proteins. *Plant Mol. Biol.* 31: 945–956.

Nicholls, D. G., and Ferguson, S. J. (2002) *Bioenergetics 3*, 3rd ed. Academic Press, San Diego, CA.

Noctor, G., and Foyer, C. H. (1998) A re-evaluation of the ATP:NADPH budget during C3 photosynthesis: A contribution from nitrate assimilation and its associated respiratory activity? *J. Exp. Bot.* 49: 1895–1908.

Noctor, G., Dutilleul, C., De Paepe, R., and Foyer, C. H. (2004) Use of mitochondrial electron transport mutants to evaluate the effects of redox state on photosynthesis, stress tolerance and the integration of carbon/nitrogen metabolism. *J. Exp. Bot.* 55: 49–57.

Ohlrogge, J. B., and Browse, J. A. (1995) Lipid biosynthesis. *Plant Cell* 7: 957–970.

Ohlrogge, J. B., and Jaworski, J. G. (1997) Regulation of fatty acid synthesis. *Annu. Rev. Plant Physiol. Plant Mol. Biol.* 48: 109–136.

Oliver, D. J., and McIntosh, C. A. (1995) The biochemistry of the mitochondrial matrix. In *The Molecular Biology of Plant Mitochondria*, C. S. Levings III and I. Vasil, eds., Kluwer, Dordrecht, Netherlands, pp. 237–280.

Plaxton, W. C. (1996) The organization and regulation of plant glycolysis. *Annu. Rev. Plant Physiol. Plant Mol. Biol.* 47: 185–214.

Raskin, I., Turner, I. M., and Melander, W. R. (1989) Regulation of heat production in the inflorescences of an *Arum* lily by endogenous salicylic acid. *Proc. Natl. Acad. Sci. USA* 86: 2214–2218.

Rasmusson, A. G., Soole, K. L., and Elthon, T. E. (2004) Alternative NAD(P)H dehydrogenases of plant mitochondria. *Annu. Rev. Plant Biol.* 55: 23–39.

Sachs, M. M., Subbaiah, C. C., and Saab, I. N. (1996) Anaerobic gene expression and flooding tolerance in maize. *J. Exp. Bot.* 47: 1–15.

Sasaki, Y., Konishi, T., and Nagano, Y. (1995) The compartmentation of acetyl-coenzyme A carboxylase in plants. *Plant Physiol.* 108: 445–449.

Schroeder, J. I., Allen, G. J., Hugouvieux, V., Kwak, J. M., and Waner, D. (2001) Guard cell signal transduction. *Annu. Rev. Plant Physiol. Plant Mol. Biol.* 52: 627–658.

Siedow, J. N. (1995) Bioenergetics: The plant mitochondrial electron transfer chain. In *The Molecular Biology of Plant Mitochondria*, C. S. Levings III and I. Vasil, eds., Kluwer, Dordrecht, Netherlands, pp. 281–312.

Siedow, J. N., and Umbach, A. L. (1995) Plant mitochondrial electron transfer and molecular biology. *Plant Cell* 7: 821–831.

Stintzi, A., and Browse, J. (2000) The Arabidopsis male-sterile mutant, *opr3*, lacks the 12-oxophytodienoic acid reductase required for jasmonate synthesis. *Proc. Natl. Acad. Sci. USA* 97: 10625–10630.

Thompson, P., Bowsher, C. G., and Tobin, A. K. (1998) Heterogeneity of mitochondrial protein biogenesis during primary leaf development in barley. *Plant Physiol.* 118: 1089–1099.

Tucker, G. A. (1993) Introduction. In *Biochemistry of Fruit Ripening*, G. Seymour, J. Taylor, and G. Tucker, eds., Chapman & Hall, London, pp. 1–51.

Vanlerberghe, G. C., and McIntosh, L. (1997) Alternative oxidase: From gene to function. *Annu. Rev. Plant Physiol. Plant Mol. Biol.* 48: 703–734.

Vedel, F., Lalanne, É., Sabar, M., Chétrit, P., and De Paepe, R. (1999) The mitochondrial respiratory chain and ATP synthase complexes: Composition, structure and mutational studies. *Plant Physiol. Biochem.* 37: 629–643.

Vercesi, A. E., Martins I. S., Silva, M. P., and Leite, H. M. F. (1995) PUMPing plants. *Nature* 375: 24.

Wagner, A. M., and Krab, K. (1995) The alternative respiration pathway in plants: Role and regulation. *Physiol. Plant.* 95: 318–325.

Wang, X. (2001) Plant phospholipases. *Annu. Rev. Plant Physiol. Plant Mol. Biol.* 52: 211–231.

Whitehouse, D. G., and Moore, A. L. (1995) Regulation of oxidative phosphorylation in plant mitochondria. In *The Molecular Biology of Plant Mitochondria*, C. S. Levings III and I. K. Vasil, eds., Kluwer, Dordrecht, Netherlands, pp. 313–344.

Wolter, F. P., Schmidt, R., and Heinz, E. (1992) Chilling sensitivity of *Arabidopsis thaliana* with genetically engineered membrane lipids. *EMBO J.* 11: 4685–4692.

Chapter 12 | Assimilation of Mineral Nutrients

HIGHER PLANTS ARE AUTOTROPHIC organisms that can synthesize their organic molecular components out of inorganic nutrients obtained from their surroundings. For many mineral nutrients, this process involves absorption from the soil by the roots (see Chapter 5) and incorporation into the organic compounds that are essential for growth and development. This incorporation of mineral nutrients into organic substances such as pigments, enzyme cofactors, lipids, nucleic acids, and amino acids is termed **nutrient assimilation**.

Assimilation of some nutrients—particularly nitrogen and sulfur—requires a complex series of biochemical reactions that are among the most energy-requiring reactions in living organisms.

- In nitrate (NO_3^-) assimilation, the nitrogen in NO_3^- is converted to a higher-energy form in nitrite (NO_2^-), then to a yet–higher-energy form in ammonium (NH_4^+), and finally into the amide nitrogen of glutamine. This process consumes the equivalent of 12 ATPs per nitrogen (Bloom et al. 1992).

- Plants such as legumes form symbiotic relationships with nitrogen-fixing bacteria to convert molecular nitrogen (N_2) into ammonia (NH_3). Ammonia (NH_3) is the first stable product of natural fixation; at physiological pH, however, ammonia is protonated to form the ammonium ion (NH_4^+). The process of biological nitrogen fixation, together with the subsequent assimilation of NH_3 into an amino acid, consumes about 16 ATPs per nitrogen (Pate and Layzell 1990; Vande Broek and Vanderleyden 1995).

- The assimilation of sulfate (SO_4^{2-}) into the amino acid cysteine via the two pathways found in plants consumes about 14 ATPs (Hell 1997).

For some perspective on the enormous energies involved, consider that if these reactions run rapidly in reverse—say, from NH_4NO_3 (ammonium nitrate) to N_2—they become explosive, liberating vast amounts of energy as motion, heat, and light. Nearly all explosives (e.g., nitroglycerin, TNT, and gunpowder) are based on the rapid oxidation of nitrogen or sulfur compounds.

Assimilation of other nutrients, especially the macronutrient and micronutrient cations (see Chapter 5), involves the formation of complexes with organic compounds. For example, Mg^{2+} associates with chlorophyll pigments, Ca^{2+} associates with pectates within the cell wall, and Mo^{6+} associates with enzymes such as nitrate reductase and nitrogenase. These complexes are highly stable, and removal of the nutrient from the complex may result in total loss of function.

This chapter outlines the primary reactions through which the major nutrients (nitrogen, sulfur, phosphate, cations such as Mg^{2+} and K^+, and oxygen) are assimilated and discusses the organic products of these reactions. We emphasize the physiological implications of the required energy expenditures and introduce the topic of symbiotic nitrogen fixation. Plants serve as the major conduit through which nutrients pass from inert geophysical domains into dynamic biological ones; this chapter thus highlights the vital role of plant nutrient assimilation in the human diet.

Nitrogen in the Environment

Many biochemical compounds present in plant cells contain nitrogen (see Chapter 5). For example, nitrogen is found in the nucleoside phosphates and amino acids that form the building blocks of nucleic acids and proteins, respectively. Only the elements oxygen, carbon, and hydrogen are more abundant in plants than nitrogen. Most natural and agricultural ecosystems show dramatic gains in productivity after fertilization with inorganic nitrogen, attesting to the importance of this element.

In this section we will discuss the biogeochemical cycle of nitrogen, the crucial role of nitrogen fixation in the conversion of molecular nitrogen into ammonium and nitrate, and the fate of nitrate and ammonium in plant tissues.

Nitrogen passes through several forms in a biogeochemical cycle

Nitrogen is present in many forms in the biosphere. The atmosphere contains vast quantities (about 78% by vol-

TABLE 12.1
The major processes of the biogeochemical nitrogen cycle

Process	Definition	Rate (10^{12} g yr^{-1})[a]
Industrial fixation	Industrial conversion of molecular nitrogen to ammonia	80
Atmospheric fixation	Lightning and photochemical conversion of molecular nitrogen to nitrate	19
Biological fixation	Prokaryotic conversion of molecular nitrogen to ammonia	170
Plant acquisition	Plant absorption and assimilation of ammonium or nitrate	1200
Immobilization	Microbial absorption and assimilation of ammonium or nitrate	N/C
Ammonification	Bacterial and fungal catabolism of soil organic matter to ammonium	N/C
Anammox	Anaerobic ammonium oxidation: bacterial conversion of ammonium and nitrate to molecular nitrogen	N/C
Nitrification	Bacterial (*Nitrosomonas* sp.) oxidation of ammonium to nitrite and subsequent bacterial (*Nitrobacter* sp.) oxidation of nitrite to nitrate	N/C
Mineralization	Bacterial and fungal catabolism of soil organic matter to mineral nitrogen through ammonification or nitrification	N/C
Volatilization	Physical loss of gaseous ammonia to the atmosphere	100
Ammonium fixation	Physical embedding of ammonium into soil particles	10
Denitrification	Bacterial conversion of nitrate to nitrous oxide and molecular nitrogen	210
Nitrate leaching	Physical flow of nitrate dissolved in groundwater out of the topsoil and eventually into the oceans	36

Note: Terrestrial organisms, the soil, and the oceans contain about 5.2×10^{15} g, 95×10^{15} g, and 6.5×10^{15} g, respectively, of organic nitrogen that is active in the cycle. Assuming that the amount of atmospheric N_2 remains constant (inputs = outputs), the *mean residence time* (the average time that a nitrogen molecule remains in organic forms) is about 370 years [(pool size)/(fixation input) = (5.2×10^{15} g + 95×10^{15} g)/(80×10^{12} g yr^{-1} + 19×10^{12} g yr^{-1} + 170×10^{12} g yr^{-1})] (Schlesinger 1997).

[a] N/C, not calculated.

ume) of molecular nitrogen (N$_2$) (see Chapter 9). For the most part, this large reservoir of nitrogen is not directly available to living organisms. Acquisition of nitrogen from the atmosphere requires the breaking of an exceptionally stable triple covalent bond between two nitrogen atoms (N≡N) to produce ammonia (NH$_3$) or nitrate (NO$_3^-$). These reactions, known as **nitrogen fixation**, can be accomplished by both industrial and natural processes.

Under elevated temperature (about 200°C) and high pressure (about 200 atmospheres) and in the presence of a metal catalyst (usually iron), N$_2$ combines with hydrogen to form ammonia. The extreme conditions are required to overcome the high activation energy of the reaction. This nitrogen fixation reaction, called the *Haber–Bosch process*, is a starting point for the manufacture of many industrial and agricultural products. Worldwide industrial production of nitrogen fertilizers amounts to more than 85×10^{12} g yr^{-1} (FAOSTAT 2005).

Natural processes, which fix about 190×10^{12} g yr^{-1} of nitrogen (Table 12.1), are the following (Schlesinger 1997):

- *Lightning.* Lightning is responsible for about 8% of the nitrogen fixed. Lightning converts water vapor and oxygen into highly reactive hydroxyl free radicals, free hydrogen atoms, and free oxygen atoms that attack molecular nitrogen (N$_2$) to form nitric acid (HNO$_3$). This nitric acid subsequently falls to Earth with rain.

- *Photochemical reactions.* Approximately 2% of the nitrogen fixed derives from photochemical reactions between gaseous nitric oxide (NO) and ozone (O$_3$) that produce nitric acid (HNO$_3$).

- *Biological nitrogen fixation.* The remaining 90% results from biological nitrogen fixation, in which bacteria or blue-green algae (cyanobacteria) fix N$_2$ into ammonium (NH$_4^+$).

From an agricultural standpoint, biological nitrogen fixation is critical, because industrial production of nitrogen fertilizers seldom meets agricultural demand (FAOSTAT 2005).

Once fixed in ammonium or nitrate, nitrogen enters a biogeochemical cycle and passes through several organic or inorganic forms before it eventually returns to molecular nitrogen (Figure 12.1; see also Table 12.1). The ammonium (NH$_4^+$) and nitrate (NO$_3^-$) ions that are generated through fixation or released through decomposition of soil organic matter become the object of intense competition among plants and microorganisms. To remain competitive, plants have developed mechanisms for scavenging these ions rapidly from the soil solution (see Chapter 5). Under the elevated soil concentrations that occur after fertilization, the absorption of ammonium and nitrate by the roots may exceed the capacity of a plant to assimilate these ions, leading to their accumulation within the plant's tissues.

Unassimilated ammonium or nitrate may be dangerous

Ammonium, if it accumulates to high levels in living tissues, is toxic to both plants and animals. Ammonium dissipates transmembrane proton gradients (Figure 12.2) that are required for both photosynthetic and respiratory electron transport (see Chapters 7 and 11) and for sequestering metabolites in the vacuole (see Chapter 6). Because high levels of ammonium are dangerous, animals have developed a

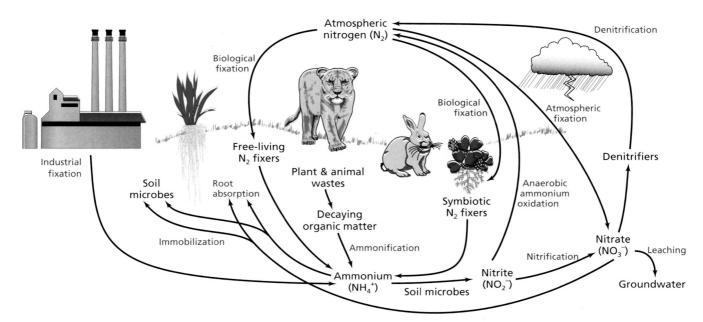

FIGURE 12.1 Nitrogen cycles through the atmosphere as it changes from a gaseous form to reduced ions before being incorporated into organic compounds in living organisms. Some of the steps involved in the nitrogen cycle are shown.

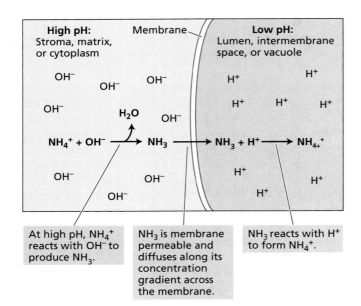

Figure 12.2 NH_4^+ toxicity can dissipate pH gradients. The left side represents the stroma, matrix, or cytoplasm, where the pH is high; the right side represents the lumen, intermembrane space, or vacuole, where the pH is low; and the membrane represents the thylakoid, inner mitochondrial, or tonoplast membrane for a chloroplast, mitochondrion, or root cell, respectively. The net result of the reaction shown is that both the OH^- concentration on the left side and the H^+ concentration on the right side have been diminished; that is, the pH gradient has been dissipated. (After Bloom 1997.)

strong aversion to its smell. The active ingredient in smelling salts, a medicinal vapor released under the nose of a person who has fainted to revive them, is ammonium carbonate. Plants assimilate ammonium near the site of absorption or generation and rapidly store any excess in their vacuoles, thus avoiding toxic effects on membranes and the cytosol.

In contrast to ammonium, plants can store high levels of nitrate, or they can translocate it from tissue to tissue without deleterious effect. Yet if livestock and humans consume plant material that is high in nitrate, they may suffer methemoglobinemia, a disease in which the liver reduces nitrate to nitrite, which combines with hemoglobin and renders the hemoglobin unable to bind oxygen. Humans and other animals may also convert nitrate into nitrosamines, which are potent carcinogens. Some countries limit the nitrate content in plant materials sold for human consumption.

In the next sections we will discuss the process by which plants assimilate nitrate into organic compounds via the enzymatic reduction of nitrate first into nitrite, next into ammonium, and then into amino acids.

Nitrate Assimilation

Plant roots actively absorb nitrate from the soil solution via several low- and high-affinity nitrate–proton cotransporters (Crawford and Forde 2002). Plants eventually assimilate most of this nitrate into organic nitrogen compounds. The first step of this process is the reduction of nitrate to nitrite in the cytosol (Oaks 1994). The enzyme **nitrate reductase** catalyzes this reaction:

$$NO_3^- + NAD(P)H + H^+ + 2\ e^- \rightarrow$$
$$NO_2^- + NAD(P)^+ + H_2O \qquad (12.1)$$

where NAD(P)H indicates NADH or NADPH. The most common form of nitrate reductase uses only NADH as an electron donor; another form of the enzyme that is found predominantly in nongreen tissues such as roots can use either NADH or NADPH (Warner and Kleinhofs 1992).

The nitrate reductases of higher plants are composed of two identical subunits, each containing three prosthetic groups: flavin adenine dinucleotide (FAD), heme, and a molybdenum complexed to an organic molecule called a *pterin* (Campbell 1999).

A pterin (fully oxidized)

Nitrate reductase is the main molybdenum-containing protein in vegetative tissues, and one symptom of molybdenum deficiency is the accumulation of nitrate that results from diminished nitrate reductase activity (Mendel 2005).

Comparison of the amino acid sequences for nitrate reductase from several species with those of other well-characterized proteins that bind FAD, heme, or molybdenum has led to a multiple-domain model for nitrate reductase; a simplified three-domain model is shown in Figure 12.3. The FAD-binding domain accepts two electrons from NADH or NADPH. The electrons then pass through the

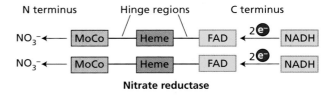

Nitrate reductase

Figure 12.3 A model of the nitrate reductase dimer, illustrating the three binding domains whose polypeptide sequences are similar in eukaryotes: molybdenum complex (MoCo), heme, and FAD. The NADH binds at the FAD-binding region of each subunit and initiates a two-electron transfer from the carboxyl (C) terminus, through each of the electron transfer components, to the amino (N) terminus. Nitrate is reduced at the molybdenum complex near the amino terminus. The polypeptide sequences of the hinge regions are highly variable among species.

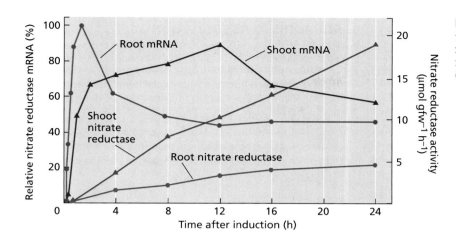

FIGURE 12.4 Stimulation of nitrate reductase activity follows the induction of nitrate reductase mRNA in shoots and roots of barley; gfw, grams fresh weight. (After Kleinhofs et al. 1989.)

heme domain to the molybdenum complex, where they are transferred to nitrate.

Many factors regulate nitrate reductase

Nitrate, light, and carbohydrates influence nitrate reductase at the transcription and translation levels (Sivasankar and Oaks 1996). In barley seedlings, nitrate reductase mRNA was detected approximately 40 minutes after addition of nitrate, and maximum levels were attained within 3 hours (Figure 12.4). In contrast to the rapid mRNA accumulation, there was a gradual linear increase in nitrate reductase activity, reflecting the slower synthesis of the protein.

In addition, the protein is subject to posttranslational modification (involving a reversible phosphorylation) that is analogous to the regulation of sucrose phosphate synthase (see Chapters 8 and 10). Light, carbohydrate levels, and other environmental factors stimulate a protein phosphatase that dephosphorylates a key serine residue in the hinge 1 region of nitrate reductase (between the molybdenum complex and heme-binding domains) and thereby activates the enzyme.

Operating in the reverse direction, darkness and Mg^{2+} stimulate a protein kinase that phosphorylates the same serine residues, which then interact with a 14-3-3 inhibitor protein, and thereby inactivate nitrate reductase (Kaiser et al. 1999). *Regulation of nitrate reductase activity through phosphorylation and dephosphorylation provides more rapid control than can be achieved through synthesis or degradation of the enzyme (minutes versus hours).*

Nitrite reductase converts nitrite to ammonium

Nitrite (NO_2^-) is a highly reactive, potentially toxic ion. Plant cells immediately transport the nitrite generated by nitrate reduction (see Equation 12.1) from the cytosol into chloroplasts in leaves and plastids in roots. In these organelles, the enzyme nitrite reductase reduces nitrite to ammonium according to the following overall reaction:

$$NO_2^- + 6\ Fd_{red} + 8\ H^+ + 6\ e^- \rightarrow \\ NH_4^+ + 6\ Fd_{ox} + 2\ H_2O \quad (12.2)$$

where Fd is ferredoxin, and the subscripts *red* and *ox* stand for *reduced* and *oxidized*, respectively. Reduced ferredoxin is derived from photosynthetic electron transport in the chloroplasts (see Chapter 7) and from NADPH generated by the oxidative pentose phosphate pathway in nongreen tissues (see Chapter 11).

Chloroplasts and root plastids contain different forms of the enzyme, but both forms consist of a single polypeptide containing two prosthetic groups: an iron–sulfur cluster (Fe_4S_4) and a specialized heme (Siegel and Wilkerson 1989). These groups act together to bind nitrite and reduce it to ammonium. Although no nitrogen compounds of intermediate redox states accumulate, a small percentage (0.02–0.2%) of the nitrite reduced is released as nitrous oxide (N_2O), a greenhouse gas (Smart and Bloom 2001). The electron flow through ferredoxin, Fe_4S_4, and heme can be represented as in Figure 12.5.

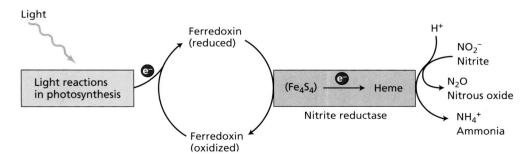

FIGURE 12.5 Model for coupling of photosynthetic electron flow, via ferredoxin, to the reduction of nitrite by nitrite reductase. The enzyme contains two prosthetic groups, Fe_4S_4 and heme, which participate in the reduction of nitrite to ammonium.

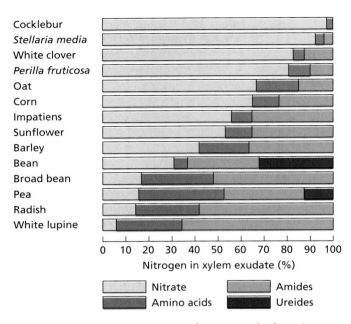

FIGURE 12.6 Relative amounts of nitrate and other nitrogen compounds in the xylem sap of various plant species. The plants were grown with their roots exposed to nitrate solutions, and xylem sap was collected by severing the stem. Note the presence of ureides in bean and pea; only legumes of tropical origin export nitrogen in such compounds. (After Pate 1983.)

Nitrite reductase is encoded in the nucleus and synthesized in the cytoplasm with an N-terminal transit peptide that targets it to the plastids (Wray 1993). Elevated concentrations of NO_3^- or exposure to light induce the transcription of nitrite reductase mRNA. Accumulation of the end products in the process—asparagine and glutamine—repress this induction.

Both roots and shoots assimilate nitrate

In many plants, when the roots receive small amounts of nitrate, nitrate is reduced primarily in the roots. As the supply of nitrate increases, a greater proportion of the absorbed nitrate is translocated to the shoot and assimilated there (Marschner 1995). Even under similar conditions of nitrate supply, the balance between root and shoot nitrate metabolism—as indicated by the proportion of nitrate reductase activity in each of the two organs or by the relative concentrations of nitrate and reduced nitrogen in the xylem sap—varies from species to species.

In plants such as the cocklebur (*Xanthium strumarium*), nitrate metabolism is restricted to the shoot; in other plants, such as white lupine (*Lupinus albus*), most nitrate is metabolized in the roots (Figure 12.6). Generally, species native to temperate regions rely more heavily on nitrate assimilation by the roots than do species of tropical or subtropical origins.

Ammonium Assimilation

Plant cells avoid ammonium toxicity by rapidly converting the ammonium generated from nitrate assimilation or photorespiration (see Chapter 8) into amino acids. The primary pathway for this conversion involves the sequential actions of glutamine synthetase and glutamate synthase (Lea et al. 1992). In this section we will discuss the enzymatic processes that mediate the assimilation of ammonium into essential amino acids, and the role of amides in the regulation of nitrogen and carbon metabolism.

Converting ammonium to amino acids requires two enzymes

Glutamine synthetase (GS) combines ammonium with glutamate to form glutamine (Figure 12.7A):

$$\text{Glutamate} + NH_4^+ + ATP \rightarrow \text{glutamine} + ADP + P_i \quad (12.3)$$

This reaction requires the hydrolysis of one ATP and involves a divalent cation such as Mg^{2+}, Mn^{2+}, or Co^{2+} as a cofactor. Plants contain two classes of GS, one in the cytosol and the other in root plastids or shoot chloroplasts. The cytosolic forms are expressed in germinating seeds or in the vascular bundles of roots and shoots and produce glutamine for intracellular nitrogen transport. The GS in root plastids generates amide nitrogen for local consumption; the GS in shoot chloroplasts reassimilates photorespiratory NH_4^+ (Lam et al. 1996). Light and carbohydrate levels alter the expression of the plastid forms of the enzyme, but they have little effect on the cytosolic forms.

Elevated plastid levels of glutamine stimulate the activity of **glutamate synthase** (also known as *glutamine:2-oxo-glutarate aminotransferase*, or **GOGAT**). This enzyme transfers the amide group of glutamine to 2-oxoglutarate, yielding two molecules of glutamate (see Figure 12.7A). Plants contain two types of GOGAT: One accepts electrons from NADH; the other accepts electrons from ferredoxin (Fd):

$$\text{Glutamine} + \text{2-oxoglutarate} + NADH + H^+ \rightarrow$$
$$\text{2 glutamate} + NAD^+ \quad (12.4)$$

$$\text{Glutamine} + \text{2-oxoglutarate} + Fd_{red} \rightarrow$$
$$\text{2 glutamate} + Fd_{ox} \quad (12.5)$$

The NADH type of the enzyme (NADH-GOGAT) is located in plastids of nonphotosynthetic tissues such as roots or vascular bundles of developing leaves. In roots, NADH-GOGAT is involved in the assimilation of NH_4^+ absorbed from the rhizosphere (the soil near the surface of the roots); in vascular bundles of developing leaves, NADH-GOGAT assimilates glutamine translocated from roots or senescing leaves.

The ferredoxin-dependent type of glutamate synthase (Fd-GOGAT) is found in chloroplasts and serves in photorespiratory nitrogen metabolism. Both the amount of protein and

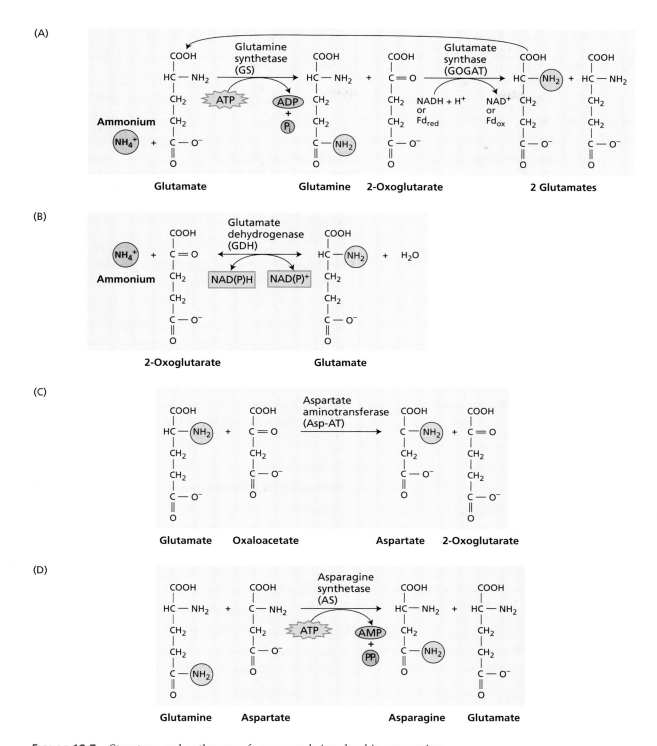

FIGURE 12.7 Structure and pathways of compounds involved in ammonium metabolism. Ammonium can be assimilated by one of several processes. (A) The GS-GOGAT pathway that forms glutamine and glutamate. A reduced cofactor is required for the reaction: ferredoxin in green leaves and NADH in nonphotosynthetic tissue. (B) The GDH pathway that forms glutamate using NADH or NADPH as a reductant. (C) Transfer of the amino group from glutamate to oxaloacetate to form aspartate (catalyzed by aspartate aminotransferase). (D) Synthesis of asparagine by transfer of an amino acid group from glutamine to aspartate (catalyzed by asparagine synthesis).

its activity increase with light levels. Roots, particularly those under nitrate nutrition, have Fd-GOGAT in plastids. Fd-GOGAT in the roots presumably functions to incorporate the glutamine generated during nitrate assimilation.

Ammonium can be assimilated via an alternative pathway

Glutamate dehydrogenase (GDH) catalyzes a reversible reaction that synthesizes or deaminates glutamate (Figure 12.7B):

$$\text{2-Oxoglutarate} + NH_4^+ + NAD(P)H \leftrightarrow \text{glutamate} + H_2O + NAD(P)^+ \quad (12.6)$$

An NADH-dependent form of GDH is found in mitochondria, and an NADPH-dependent form is localized in the chloroplasts of photosynthetic organs. Although both forms are relatively abundant, they cannot substitute for the GS–GOGAT pathway for assimilation of ammonium, and their primary function is in deaminating glutamate during the reallocation of nitrogen (see Figure 12.7B).

Transamination reactions transfer nitrogen

Once assimilated into glutamine and glutamate, nitrogen is incorporated into other amino acids via transamination reactions. The enzymes that catalyze these reactions are known as aminotransferases. An example is **aspartate aminotransferase (Asp-AT)**, which catalyzes the following reaction (Figure 12.7C):

$$\text{Glutamate} + \text{oxaloacetate} \rightarrow \text{aspartate} + \text{2-oxoglutarate} \quad (12.7)$$

in which the amino group of glutamate is transferred to the carboxyl group of aspartate. Aspartate is an amino acid that participates in the malate–aspartate shuttle to transfer reducing equivalents from the mitochondrion and chloroplast into the cytosol (see Chapter 11) and in the transport of carbon from the mesophyll to the bundle sheath for C_4 carbon fixation (see Chapter 8). All transamination reactions require pyridoxal phosphate (vitamin B_6) as a cofactor.

Aminotransferases are found in the cytoplasm, chloroplasts, mitochondria, glyoxysomes, and peroxisomes. The aminotransferases in chloroplasts may have a significant role in amino acid biosynthesis, because plant leaves or isolated chloroplasts exposed to radioactively labeled carbon dioxide rapidly incorporate the label into glutamate, aspartate, alanine, serine, and glycine.

Asparagine and glutamine link carbon and nitrogen metabolism

Asparagine, isolated from asparagus as early as 1806, was the first amide to be identified (Lam et al. 1996). It serves not only as a protein precursor, but as a key compound for nitrogen transport and storage because of its stability and high nitrogen-to-carbon ratio (2 N to 4 C for asparagine, versus 2 N to 5 C for glutamine or 1 N to 5 C for glutamate).

The major pathway for asparagine synthesis involves the transfer of the amide nitrogen from glutamine to asparagine (Figure 12.7D):

$$\text{Glutamine} + \text{aspartate} + ATP \rightarrow \text{asparagine} + \text{glutamate} + AMP + PP_i \quad (12.8)$$

Asparagine synthetase (AS), the enzyme that catalyzes this reaction, is found in the cytosol of leaves and roots and in nitrogen-fixing nodules (see the next section). In maize roots, particularly those under potentially toxic levels of ammonia, ammonium may replace glutamine as the source of the amide group (Sivasankar and Oaks 1996).

High levels of light and carbohydrate—conditions that stimulate plastid GS and Fd-GOGAT—inhibit the expression of genes coding for AS and the activity of the enzyme. The opposing regulation of these competing pathways helps balance the metabolism of carbon and nitrogen in plants (Lam et al. 1996). Conditions of ample energy (i.e., high levels of light and carbohydrates) stimulate GS (see Equation 12.3) and GOGAT (see Equations 12.4 and 12.5), inhibit AS, and thus favor nitrogen assimilation into glutamine and glutamate, compounds that are rich in carbon and participate in the synthesis of new plant materials.

In contrast, energy-limited conditions inhibit GS and GOGAT, stimulate AS, and thus favor nitrogen assimilation into asparagine, a compound that is rich in nitrogen and sufficiently stable for long-distance transport or long-term storage.

Amino Acid Biosynthesis

Humans and most animals cannot synthesize certain amino acids—histidine, isoleucine, leucine, lysine, methionine, phenylalanine, threonine, tryptophan, valine and arginine (the young only; adults can synthesize arginine)—and thus must obtain these so-called essential amino acids from their diet. In contrast, plants synthesize all the 20 or so amino acids found in proteins. The nitrogen-containing amino group, as discussed in the previous section, derives from transamination reactions with glutamine or glutamate. The carbon skeleton for amino acids derive from 3-phosphoglycerate, phosphoenolpyruvate, or pyruvate generated during glycolysis, or from α-ketoglutarate or oxaloacetate generated in the citric acid cycle (Figure 12.8). Parts of these pathways required for synthesis of the essential amino are appropriate targets for herbicides ("Roundup," Chapter 13), because they are missing from animals, so substances that block these pathways are lethal to plants but in low concentrations do not injure animals.

Biological Nitrogen Fixation

Biological nitrogen fixation accounts for most of the conversion of atmospheric N_2 into ammonium, and thus serves as the key entry point of molecular nitrogen into the

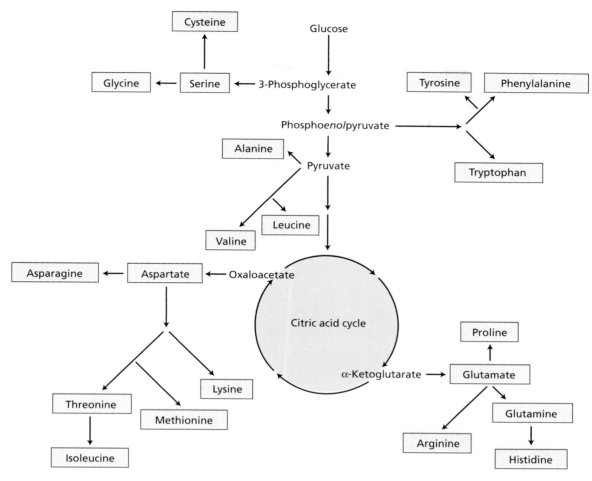

FIGURE 12.8 Biosynthetic pathways for the carbon skeletons of the 20 standard amino acids.

biogeochemical cycle of nitrogen (see Figure 12.1). In this section we will describe the symbiotic relations between nitrogen-fixing organisms and higher plants, the specialized structures that form in roots when infected by nitrogen-fixing bacteria, the genetic and signaling interactions that regulate nitrogen fixation by symbiotic prokaryotes and their hosts, and the properties of the nitrogenase enzymes that fix nitrogen.

Free-living and symbiotic bacteria fix nitrogen

Some bacteria, as stated earlier, can convert atmospheric nitrogen into ammonium (Table 12.2). Most of these nitrogen-fixing prokaryotes live in the soil, generally independent of other organisms. A few form symbiotic associations with higher plants in which the prokaryote directly provides the host plant with fixed nitrogen in exchange for other nutrients and carbohydrates (top portion of Table 12.2). Such symbioses occur in nodules that form on the roots of the plant and contain the nitrogen-fixing bacteria.

The most common type of symbiosis occurs between members of the plant family Leguminosae and soil bacteria of the genera *Azorhizobium*, *Bradyrhizobium*, *Photorhizobium*, *Rhizobium*, and *Sinorhizobium* (collectively called **rhizobia**; Table 12.3 and Figure 12.9). Another common type of symbiosis occurs between several woody plant species, such as alder trees, and soil bacteria of the genus *Frankia*; these plants are known as **actinorhizal** plants. Still other types of nitrogen-fixing symbioses involve the South American herb *Gunnera* and the tiny water fern *Azolla*, which form associations with the cyanobacteria *Nostoc* and *Anabaena*, respectively (see Table 12.2 and Figure 12.10).

Nitrogen fixation requires anaerobic conditions

Because nitrogen fixation involves the transfer of large amounts of energy, the **nitrogenase** enzymes that catalyze these reactions have sites that facilitate the high-energy exchange of electrons. Oxygen, being a strong electron acceptor, can damage these sites and irreversibly inactivate nitrogenase, so nitrogen must be fixed under anaerobic conditions. Each of the nitrogen-fixing organisms listed in Table 12.2 either functions under natural anaerobic condi-

TABLE 12.2
Examples of organisms that can carry out nitrogen fixation

Symbiotic nitrogen fixation	
Host plant	**N-fixing symbionts**
Leguminous: legumes, *Parasponia*	*Azorhizobium, Bradyrhizobium, Photorhizobium, Rhizobium, Sinorhizobium*
Actinorhizal: alder (tree), *Ceanothus* (shrub), *Casuarina* (tree), *Datisca* (shrub)	*Frankia*
Gunnera	*Nostoc*
Azolla (water fern)	*Anabaena*
Sugarcane	*Acetobacter*
Free-living nitrogen fixation	
Type	**N-fixing genera**
Cyanobacteria (blue-green algae)	*Anabaena, Calothrix, Nostoc*
Other bacteria	
Aerobic	*Azospirillum, Azotobacter, Beijerinckia, Derxia*
Facultative	*Bacillus, Klebsiella*
Anaerobic	
Nonphotosynthetic	*Clostridium, Methanococcus* (archaebacterium)
Photosynthetic	*Chromatium, Rhodospirillum*

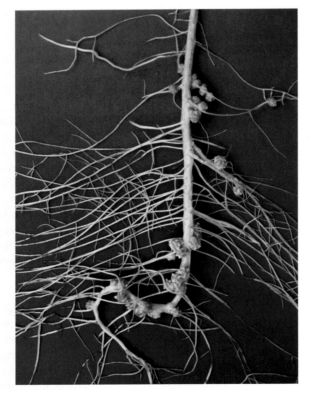

FIGURE 12.9 Root nodules on soybean. The nodules are a result of infection by *Rhizobium japonicum*. (© Courtesy of R. F. Denison.)

tions or can create an internal anaerobic environment in the presence of oxygen.

In cyanobacteria, anaerobic conditions are created in specialized cells called *heterocysts* (see Figure 12.10). Heterocysts are thick-walled cells that differentiate when filamentous cyanobacteria are deprived of NH_4^+. These cells lack photosystem II, the oxygen-producing photosystem of chloroplasts (see Chapter 7), so they do not generate oxygen (Burris 1976). Heterocysts appear to represent an adaptation for nitrogen fixation, in that they are widespread among aerobic cyanobacteria that fix nitrogen.

Cyanobacteria can fix nitrogen under anaerobic conditions such as those that occur in flooded fields. In Asian countries, nitrogen-fixing cyanobacteria of both the heterocyst and nonheterocyst types are a major means for maintaining an adequate nitrogen supply in the soil of rice fields. These microorganisms fix nitrogen when the fields are flooded and die as the fields dry, releasing the fixed nitrogen to the soil. Another important source of available nitrogen in flooded rice fields is the water fern *Azolla*, which associates with the cyanobacterium *Anabaena*. The *Azolla–Anabaena* association can fix as much as 0.5 kg of atmospheric nitrogen per hectare per day, a rate of fertilization that is sufficient to attain moderate rice yields.

Free-living bacteria that are capable of fixing nitrogen are aerobic, facultative, or anaerobic (see Table 12.2, bottom):

- *Aerobic* nitrogen-fixing bacteria such as *Azotobacter* are thought to maintain reduced oxygen conditions

TABLE 12.3
Associations between host plants and rhizobia

Plant host	Rhizobial symbiont
Parasponia (a nonlegume, formerly called *Trema*)	*Bradyrhizobium* spp.
Soybean (*Glycine max*)	*Bradyrhizobium japonicum* (slow-growing type); *Sinorhizobium fredii* (fast-growing type)
Alfalfa (*Medicago sativa*)	*Sinorhizobium meliloti*
Sesbania (aquatic)	*Azorhizobium* (forms both root and stem nodules; the stems have adventitious roots)
Bean (*Phaseolus*)	*Rhizobium leguminosarum* bv. *phaseoli*; *Rhizobium tropicii*; *Rhizobium etli*
Clover (*Trifolium*)	*Rhizobium leguminosarum* bv. *trifolii*
Pea (*Pisum sativum*)	*Rhizobium leguminosarum* bv. *viciae*
Aeschenomene (aquatic)	*Photorhizobium* (photosynthetically active rhizobia that form stem nodules, probably associated with adventitious roots)

(microaerobic conditions) through their high levels of respiration (Burris 1976). Others, such as *Gloeothece*, evolve O_2 photosynthetically during the day and fix nitrogen during the night.

- *Facultative* organisms, which are able to grow under both aerobic and anaerobic conditions, generally fix nitrogen only under anaerobic conditions.

- For *anaerobic* nitrogen-fixing bacteria, oxygen does not pose a problem, because it is absent in their habitat. These anaerobic organisms can be either photosynthetic (e.g., *Rhodospirillum*), or nonphotosynthetic (e.g., *Clostridium*).

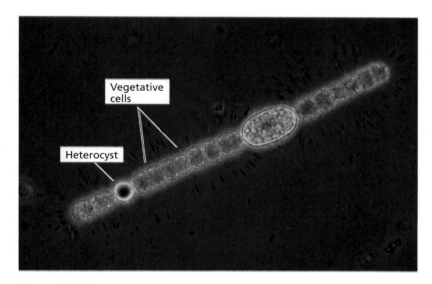

FIGURE 12.10 A heterocyst in a filament of the nitrogen-fixing cyanobacterium *Anabaena*. The thick-walled heterocysts, interspaced among vegetative cells, have an anaerobic inner environment that allows cyanobacteria to fix nitrogen in aerobic conditions. (© Paul W. Johnson/Biological Photo Service.)

Symbiotic nitrogen fixation occurs in specialized structures

Symbiotic nitrogen-fixing prokaryotes dwell within **nodules**, the special organs of the plant host that enclose the nitrogen-fixing bacteria (see Figure 12.9). In the case of *Gunnera*, these organs are existing stem glands that develop independently of the symbiont. In the case of legumes and actinorhizal plants, the nitrogen-fixing bacteria induce the plant to form root nodules.

Grasses can also develop symbiotic relationships with nitrogen-fixing organisms, but in these associations root nodules are not produced. Instead, the nitrogen-fixing bacteria seem to colonize plant tissues or anchor to the root surfaces, mainly around the elongation zone and the root hairs (Reis et al. 2000). For example, the nitrogen-fixing bacterium *Acetobacter diazotrophicus* lives in the apoplast of stem tissues in sugarcane and may provide its host with sufficient nitrogen to grant independence from nitrogen fertilization (Dong et al. 1994). The potential for applying *Azospirillum* to corn and other grains has been explored, but *Azospirillum* seems to fix little nitrogen when associated with plants (Vande Broek and Vanderleyden 1995).

Legumes and actinorhizal plants regulate gas permeability in their nodules, maintaining a level of oxygen within the nodule that can support respiration but is sufficiently low to avoid inactivation of the nitrogenase (Kuzma et al. 1993). Gas permeability increases in the light and decreases under drought or upon exposure to nitrate. The mechanism for regulating gas permeability is not yet known.

Nodules contain an oxygen-binding heme protein called **leghemoglobin**. Leghemoglobin is present in the cytoplasm of infected nodule cells at high concentrations (700 µM in soybean nodules) and gives the nodules a pink color. The host plant produces the globin portion of leghemoglobin in response to infection by the bacteria (Marschner 1995); the bacterial symbiont produces the heme portion. Leghemoglobin has a high affinity for oxygen (a K_m of about 0.01 µM), about ten times higher than the β chain of human hemoglobin.

Although leghemoglobin was once thought to provide a buffer for nodule oxygen, more recent studies indicate that it stores only enough oxygen to support nodule respiration for a few seconds (Denison and Harter 1995). Its function is to help transport oxygen to the respiring symbiotic bacterial cells in a manner analogous to hemoglobin transporting oxygen to respiring tissues in animals (Ludwig and de Vries 1986). To continue aerobic respiration under such conditions, the bacteroid uses a specialized electron transport chain (see Chapter 11) in which the terminal oxidase has an even higher affinity for oxygen, a K_m of about 0.007 µM (Preisig et al. 1996).

Establishing symbiosis requires an exchange of signals

The symbiosis between legumes and rhizobia is not obligatory. Legume seedlings germinate without any association with rhizobia, and they may remain unassociated throughout their life cycle. Rhizobia also occur as free-living organisms in the soil. Under nitrogen-limited conditions, however, the symbionts seek out one another through an elaborate exchange of signals. This signaling, the subsequent infection process, and the development of nitrogen-fixing nodules involve specific genes in both the host and the symbionts.

Plant genes specific to nodules are called *nodulin* (*Nod*) genes; rhizobial genes that participate in nodule formation are called *nodulation* (*nod*) genes (Heidstra and Bisseling 1996). The *nod* genes are classified as common *nod* genes or host-specific *nod* genes. The common *nod* genes—*nodA*, *nodB*, and *nodC*—are found in all rhizobial strains; the host-specific *nod* genes—such as *nodP*, *nodQ*, and *nodH*; or *nodF*, *nodE*, and *nodL*—differ among rhizobial species and determine the host range. Only one of the *nod* genes, the regulatory *nodD*, is constitutively expressed, and as we will explain in detail, its protein product (NodD) regulates the transcription of the other *nod* genes.

The first stage in the formation of the symbiotic relationship between the nitrogen-fixing bacteria and their host is migration of the bacteria toward the roots of the host plant. This migration is a chemotactic response mediated by chemical attractants, especially (iso)flavonoids and betaines, secreted by the roots. These attractants activate the rhizobial NodD protein, which then induces transcription of the other *nod* genes (Phillips and Kapulnik 1995). The promoter region of all *nod* operons, except that of *nodD*, contains a highly conserved sequence called the *nod* box. Binding of the activated NodD to the *nod* box induces transcription of the other *nod* genes.

Nod factors produced by bacteria act as signals for symbiosis

The *nod* genes, which NodD activates, code for nodulation proteins, most of which are involved in the biosynthesis of Nod factors. **Nod factors** are lipochitin oligosaccharide signal molecules, all of which have a chitin β-1→4-linked *N*-acetyl-D-glucosamine backbone (varying in length from three to six sugar units) and a fatty acid chain on the C-2 position of the nonreducing sugar (Figure 12.11).

Three of the *nod* genes (*nodA*, *nodB*, and *nodC*) encode enzymes (NodA, NodB, and NodC, respectively) that are required for synthesizing this basic structure (Stokkermans et al. 1995):

1. NodA is an *N*-acyltransferase that catalyzes the addition of a fatty acyl chain.
2. NodB is a chitin-oligosaccharide deacetylase that removes the acetyl group from the terminal nonreducing sugar.
3. NodC is a chitin-oligosaccharide synthase that links *N*-acetyl-D-glucosamine monomers.

Host-specific *nod* genes that vary among rhizobial species are involved in the modification of the fatty acyl chain or the addition of groups important in determining host specificity (Carlson et al. 1995):

- NodE and NodF determine the length and degree of saturation of the fatty acyl chain; those of *Rhizobium leguminosarum* bv. *viciae* and *R. meliloti* result in the synthesis of an 18:4 and a 16:2 fatty acyl group, respectively. (Recall from Chapter 11 that the number before the colon gives the total number of carbons in the fatty acyl chain, and the number after the colon gives the number of double bonds.)

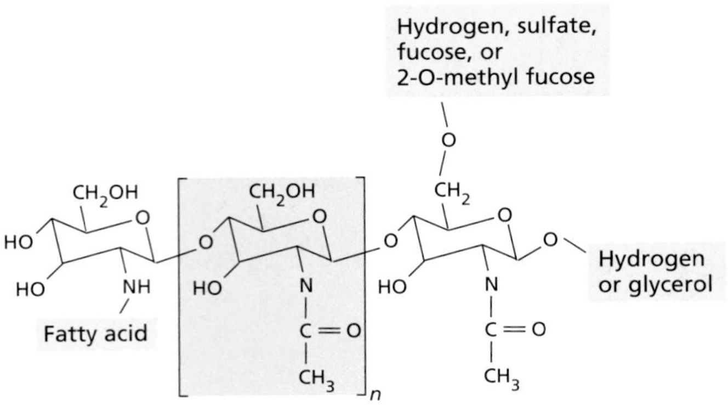

FIGURE 12.11 Nod factors are lipochitin oligosaccharides. The fatty acid chain typically has 16 to 18 carbons. The number of repeated middle sections (*n*) is usually 2 or 3. (After Stokkermans et al. 1995.)

- Other enzymes, such as NodL, influence the host specificity of Nod factors through the addition of specific substitutions at the reducing or nonreducing sugar moieties of the chitin backbone.

A particular legume host responds to a specific Nod factor. The legume receptors for Nod factors appear to involve special lectins (sugar-binding proteins) produced in the root hairs (van Rhijn et al. 1998; Etzler et al. 1999). Nod factors activate these lectins, increasing their hydrolysis of phosphoanhydride bonds of nucleoside di- and triphosphates. This lectin activation directs particular rhizobia to appropriate hosts and facilitates attachment of the rhizobia to the cell walls of a root hair.

Nodule formation involves phytohormones

Two processes—infection and nodule organogenesis—occur simultaneously during root nodule formation. During the infection process, rhizobia that are attached to the root hairs release Nod factors that induce a pronounced curling of the root hair cells (Figure 12.12A and B). The rhizobia become enclosed in the small compartment formed by the curling. The cell wall of the root hair degrades in these regions, also in response to Nod factors, allowing the bacterial cells direct access to the outer surface of the plant plasma membrane (Geurts and Bissling 2002).

The next step is formation of the **infection thread** (Figure 12.12C), an internal tubular extension of the plasma membrane that is produced by the fusion of Golgi-derived membrane vesicles at the site of infection. The thread grows at its tip by the fusion of secretory vesicles to the end of the tube. Deeper into the root cortex, near the xylem, cortical cells dedifferentiate and start dividing, forming a distinct area within the cortex, called a *nodule primordium*, from which the nodule will develop. The nodule primordia form opposite the protoxylem poles of the root vascular bundle (Timmers et al. 1999) (See **Web Topic 12.1**).

Different signaling compounds, acting either positively or negatively, control the position of nodule primordia. The nucleoside uridine diffuses from the stele into the cortex in the protoxylem zones of the root and stimulates cell division (Geurts and Bisseling 2002). Ethylene is synthesized in the region of the pericycle, diffuses into the cortex, and blocks cell division opposite the phloem poles of the root.

The infection thread filled with proliferating rhizobia elongates through the root hair and cortical cell layers, in the direction of the nodule primordium. When the infection thread reaches specialized cells within the nodule, its tip fuses with the plasma membrane of the host cell, releasing bacterial cells that are packaged in a membrane derived from the host cell plasma membrane (see Figure 12.12D). Branching of the infection thread inside the nodule enables the bacteria to infect many cells (see Figure 12.12E and F) (Mylona et al. 1995).

At first the bacteria continue to divide, and the surrounding membrane increases in surface area to accommodate this growth by fusing with smaller vesicles. Soon thereafter, upon an undetermined signal from the plant, the bacteria stop dividing and begin to enlarge and to differentiate into nitrogen-fixing endosymbiotic organelles called **bacteroids**. The membrane surrounding the bacteroids is called the *peribacteroid membrane*.

The nodule as a whole develops such features as a vascular system (which facilitates the exchange of fixed nitrogen produced by the bacteroids for nutrients contributed by the plant) and a layer of cells to exclude O_2 from the root nodule interior. In some temperate legumes (e.g., peas), the nodules are elongated and cylindrical because of the presence of a *nodule meristem*. The nodules of tropical legumes, such as soybeans and peanuts, lack a persistent meristem and are spherical (Rolfe and Gresshoff 1988).

The nitrogenase enzyme complex fixes N_2

Biological nitrogen fixation, like industrial nitrogen fixation, produces ammonia from molecular nitrogen. The overall reaction is

$$N_2 + 8\ e^- + 8\ H^+ + 16\ ATP \rightarrow$$
$$2\ NH_3 + H_2 + 16\ ADP + 16\ P_i \quad (12.9)$$

Note that the reduction of N_2 to $2\ NH_3$, a six-electron transfer, is coupled to the reduction of two protons to evolve H_2. The **nitrogenase enzyme complex** catalyzes this reaction (Dixon and Kahn 2004).

The nitrogenase enzyme complex can be separated into two components—the Fe protein and the MoFe protein—neither of which has catalytic activity by itself (Figure 12.13):

- The Fe protein is the smaller of the two components and has two identical subunits of 30 to 72 kDa each, depending on the organism. Each subunit contains an iron–sulfur cluster (4 Fe and 4 S^{2-}) that participates in the redox reactions involved in the conversion of N_2 to NH_3. The Fe protein is irreversibly inactivated by O_2 with typical half-decay times of 30 to 45 seconds (Dixon and Wheeler 1986).

- The MoFe protein has four subunits, with a total molecular mass of 180 to 235 kDa, depending on the species. Each subunit has two Mo–Fe–S clusters. The MoFe protein is also inactivated by oxygen, with a half-decay time in air of 10 minutes.

In the overall nitrogen reduction reaction (see Figure 12.13), ferredoxin serves as an electron donor to the Fe protein, which in turn hydrolyzes ATP and reduces the MoFe protein. The MoFe protein then can reduce numerous substrates (Table 12.4), although under natural conditions it reacts only with N_2 and H^+. One of the reactions catalyzed by nitrogenase, the reduction of acetylene to ethylene, is used in estimating nitrogenase activity (see **Web Topic 12.2**).

The energetics of nitrogen fixation is complex. The production of NH_3 from N_2 and H_2 is an exergonic reaction

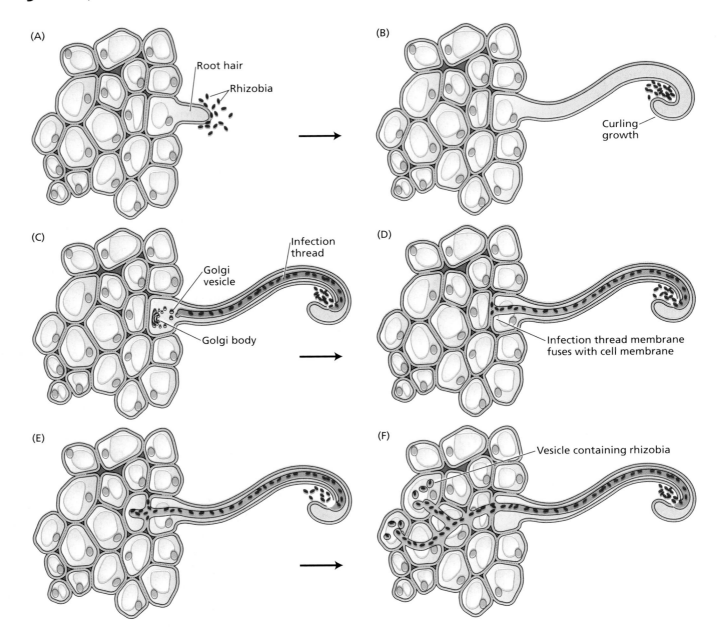

FIGURE 12.12 The infection process during nodule organogenesis. (A) Rhizobia bind to an emerging root hair in response to chemical attractants sent by the plant. (B) In response to factors produced by the bacteria, the root hair exhibits abnormal curling growth, and rhizobia cells proliferate within the coils. (C) Localized degradation of the root hair wall leads to infection and formation of the infection thread from Golgi secretory vesicles of root cells. (D) The infection thread reaches the end of the cell, and its membrane fuses with the plasma membrane of the root hair cell. (E) Rhizobia are released into the apoplast and penetrate the compound middle lamella to the subepidermal cell plasma membrane, leading to the initiation of a new infection thread, which forms an open channel with the first. (F) The infection thread extends and branches until it reaches target cells, where vesicles composed of plant membrane that enclose bacterial cells are released into the cytosol.

TABLE 12.4
Reactions catalyzed by nitrogenase

$N_2 \rightarrow NH_3$	Molecular nitrogen fixation
$N_2O \rightarrow N_2 + H_2O$	Nitrous oxide reduction
$N_3^- \rightarrow N_2 + NH_3$	Azide reduction
$C_2H_2 \rightarrow C_2H_4$	Acetylene reduction
$2\,H^+ \rightarrow H_2$	H_2 production
$ATP \rightarrow ADP + P_i$	ATP hydrolytic activity

Source: After Burris 1976.

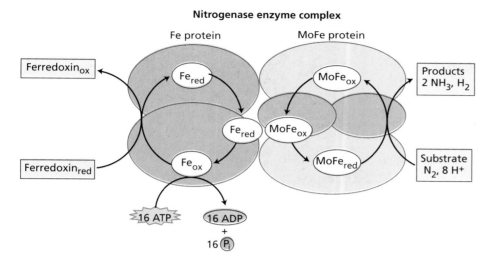

FIGURE 12.13 The reaction catalyzed by nitrogenase. Ferredoxin reduces the Fe protein. Binding and hydrolysis of ATP to the Fe protein is thought to cause a conformational change of the Fe protein that facilitates the redox reactions. The Fe protein reduces the MoFe protein, and the MoFe protein reduces the N_2. (After Dixon and Wheeler 1986; Buchanan et al. 2000.)

(see Chapter 2 on the web site for a discussion of exergonic reactions), with a $\Delta G^{0'}$ (change in free energy) of –27 kJ mol^{-1}. However, industrial production of NH_3 from N_2 and H_2 is *endergonic*, requiring a very large energy input because of the activation energy needed to break the triple bond in N_2. For the same reason, the enzymatic reduction of N_2 by nitrogenase also requires a large investment of energy (see Equation 12.9), although the exact changes in free energy are not yet known.

Calculations based on the carbohydrate metabolism of legumes show that a plant consumes 12 g of organic carbon per gram of N_2 fixed (Heytler et al. 1984). On the basis of Equation 12.9, the $\Delta G^{0'}$ for the overall reaction of biological nitrogen fixation is about –200 kJ mol^{-1}. Because the overall reaction is highly exergonic, ammonium production is limited by the slow operation (number of N_2 molecules reduced per unit time is about 5 s^{-1}) of the nitrogenase complex (Ludwig and de Vries 1986). To compensate for this slow turnover time, the bacteroid synthesizes large amounts of nitrogenase (up to 20% of the total protein in the cell).

Under natural conditions, substantial amounts of H$^+$ are reduced to H_2 gas, and this process can compete with N_2 reduction for electrons from nitrogenase. In rhizobia, 30 to 60% of the energy supplied to nitrogenase may be lost as H_2, diminishing the efficiency of nitrogen fixation. Some rhizobia, however, contain hydrogenase, an enzyme that can split the H_2 formed and generate electrons for N_2 reduction, thus improving the efficiency of nitrogen fixation (Marschner 1995).

Amides and ureides are the transported forms of nitrogen

The symbiotic nitrogen-fixing prokaryotes release ammonia that, to avoid toxicity, must be rapidly converted into organic forms in the root nodules before being transported to the shoot via the xylem. Nitrogen-fixing legumes can be divided into amide exporters or ureide exporters depending on the composition of the xylem sap. Amides (principally the amino acids asparagine or glutamine) are exported by temperate-region legumes, such as pea (*Pisum*), clover (*Trifolium*), broad bean (*Vicia*), and lentil (*Lens*).

Ureides are exported by legumes of tropical origin, such as soybean (*Glycine*), kidney bean (*Phaseolus*), peanut (*Arachis*), and southern pea (*Vigna*). The three major ureides are allantoin, allantoic acid, and citrulline (Figure 12.14). Allantoin is synthesized in peroxisomes from uric acid, and allantoic acid is synthesized from allantoin in the endoplasmic reticulum. The site of citrulline synthesis from the amino acid ornithine has not yet been determined. All

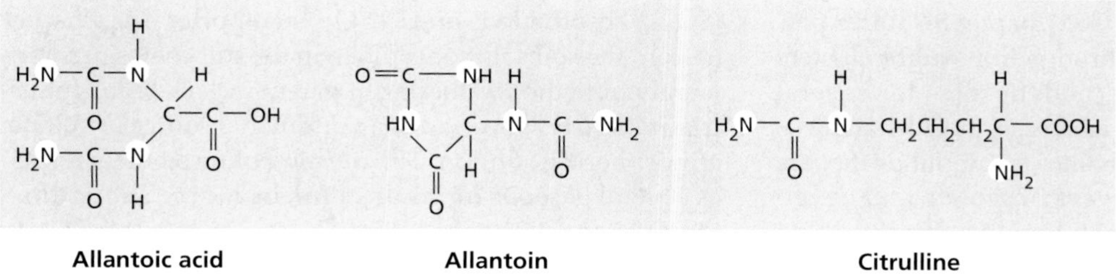

Allantoic acid   **Allantoin**   **Citrulline**

FIGURE 12.14 The major ureide compounds that are used to transport nitrogen from sites of fixation to sites where their deamination will provide nitrogen for amino acid and nucleoside synthesis.

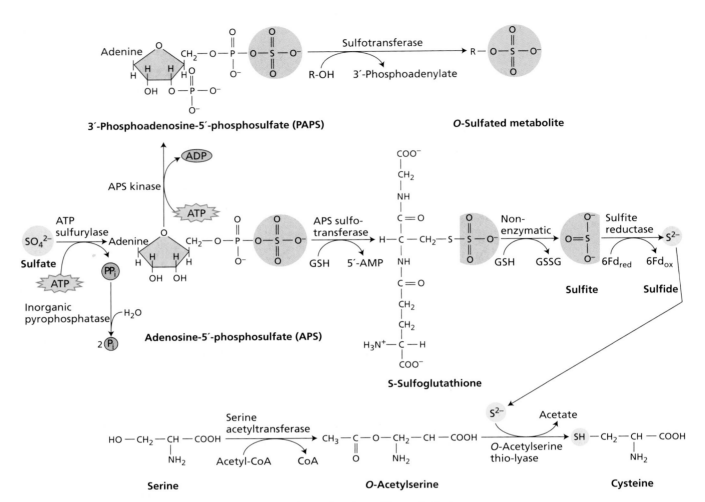

FIGURE 12.15 Structure and pathways of compounds involved in sulfur assimilation. The enzyme ATP sulfurylase cleaves pyrophosphate from ATP and replaces it with sulfate. Sulfide is produced from APS through reactions involving reduction by glutathione and ferredoxin. The sulfide or thiosulfide reacts with O-acetylserine to form cysteine. Fd, ferredoxin; GSH, glutathione, reduced; GSSG, glutathione, oxidized.

three compounds are ultimately released into the xylem and transported to the shoot, where they are rapidly catabolized to ammonium. This ammonium enters the assimilation pathway described earlier.

Sulfur Assimilation

Sulfur is among the most versatile elements in living organisms (Hell 1997). Disulfide bridges in proteins play structural and regulatory roles (see Chapter 8). Sulfur participates in electron transport through iron–sulfur clusters (see Chapters 7 and 11). The catalytic sites for several enzymes and coenzymes, such as urease and coenzyme A, contain sulfur. Secondary metabolites (compounds that are not involved in primary pathways of growth and development) that contain sulfur range from the rhizobial Nod factors discussed in the previous section to the antiseptic alliin in garlic and the anticarcinogen sulforaphane in broccoli.

The versatility of sulfur derives in part from the property that it shares with nitrogen: *multiple stable oxidation states*. In this section, we discuss the enzymatic steps that mediate sulfur assimilation, and the biochemical reactions that catalyze the reduction of sulfate into the two sulfur-containing amino acids, cysteine and methionine.

Sulfate is the absorbed form of sulfur in plants

Most of the sulfur in higher-plant cells derives from sulfate (SO_4^{2-}) absorbed via an H^+–SO_4^{2-} symporter (see Chapter 6) from the soil solution. Sulfate in the soil comes predominantly from the weathering of parent rock material. Industrialization, however, adds an additional source of sulfate: atmospheric pollution. The burning of fossil fuels releases several gaseous forms of sulfur, including sulfur dioxide (SO_2) and hydrogen sulfide (H_2S), which find their way to the soil in rain.

When dissolved in water, SO_2 is hydrolyzed to become sulfuric acid (H_2SO_4), a strong acid, which is the major source of acid rain. Plants can also metabolize sulfur dioxide taken up in the gaseous form through their stomata. Nonetheless, prolonged exposure (more than 8 hours) to high atmospheric concentrations (greater than 0.3 ppm) of SO_2 causes extensive tissue damage because of the formation of sulfuric acid.

Sulfate assimilation requires the reduction of sulfate to cysteine

The first step in the synthesis of sulfur-containing organic compounds is the reduction of sulfate to the amino acid cysteine (Figure 12.15). Sulfate is very stable and thus needs to be activated before any subsequent reactions may proceed. Activation begins with the reaction between sulfate and ATP to form 5′-adenylylsulfate (which is sometimes referred to as adenosine-5′-phosphosulfate and thus is abbreviated APS) and pyrophosphate (PP_i) (see Figure 12.15):

$$SO_4^{2-} + Mg\text{-}ATP \rightarrow APS + PP_i \quad (12.10)$$

The enzyme that catalyzes this reaction, ATP sulfurylase, has two forms: The major one is found in plastids, and a minor one is found in the cytoplasm (Leustek et al. 2000). The activation reaction is energetically unfavorable. To drive this reaction forward, the products APS and PP_i must be converted immediately to other compounds. PP_i is hydrolyzed to inorganic phosphate (P_i) by inorganic pyrophosphatase according to the following reaction:

$$PP_i + H_2O \rightarrow 2\, P_i \quad (12.11)$$

The other product, APS, is rapidly reduced or sulfated. Reduction is the dominant pathway (Leustek et al. 2000).

The reduction of APS is a multistep process that occurs exclusively in the plastids. First, APS reductase transfers two electrons, apparently from reduced glutathione (GSH), to produce sulfite (SO_3^{2-}):

$$APS + 2\, GSH \rightarrow SO_3^{2-} + 2\, H^+ + GSSG + AMP \quad (12.12)$$

where GSSG stands for oxidized glutathione. (The *SH* in GSH and the *SS* in GSSG stand for S—H and S—S bonds, respectively.)

Second, sulfite reductase transfers six electrons from ferredoxin (Fd_{red}) to produce sulfide (S^{2-}):

$$SO_3^{2-} + 6\, Fd_{red} \rightarrow S^{2-} + 6\, Fd_{ox} \quad (12.13)$$

The resultant sulfide then reacts with *O*-acetylserine (OAS) to form cysteine and acetate. The *O*-acetylserine that reacts with S^{2-} is formed in a reaction catalyzed by serine acetyltransferase:

$$\text{Serine} + \text{acetyl-CoA} \rightarrow \text{OAS} + \text{CoA} \quad (12.14)$$

The reaction that produces cysteine and acetate is catalyzed by OAS thiol-lyase:

$$\text{OAS} + S^{2-} \rightarrow \text{cysteine} + \text{acetate} \quad (12.15)$$

The sulfation of APS, localized in the cytosol, is the alternative pathway. First, APS kinase catalyzes a reaction of APS with ATP to form 3′-phosphoadenosine-5′-phosphosulfate (PAPS):

$$APS + ATP \rightarrow PAPS + ADP \quad (12.16)$$

Sulfotransferases then may transfer the sulfate group from PAPS to various compounds, including choline, brassinosteroids, flavonol, gallic acid glucoside, glucosinolates, peptides, and polysaccharides (Leustek and Saito 1999).

Sulfate assimilation occurs mostly in leaves

The reduction of sulfate to cysteine changes the oxidation number of sulfur from +6 to –4, thus entailing the transfer of 10 electrons. Glutathione, ferredoxin, NAD(P)H, or *O*-acetylserine may serve as electron donors at various steps of the pathway (see Figure 12.15).

Leaves are generally much more active than roots in sulfur assimilation, presumably because photosynthesis provides reduced ferredoxin, and photorespiration generates serine, that may stimulate the production of *O*-acetylserine (see Chapter 8). Sulfur assimilated in leaves is exported via the phloem to sites of protein synthesis (shoot and root apices, and fruits) mainly as glutathione (Bergmann and Rennenberg 1993):

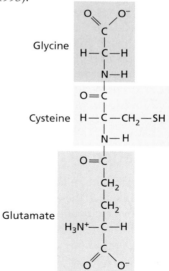

Reduced glutathione

Glutathione also acts as a signal that coordinates the absorption of sulfate by the roots and the assimilation of sulfate by the shoot.

Methionine is synthesized from cysteine

Methionine, the other sulfur-containing amino acid found in proteins, is synthesized in plastids from cysteine (see **Web Topic 12.3** for further detail). After cysteine and methionine are synthesized, sulfur can be incorporated into proteins and a variety of other compounds, such as acetyl-CoA and *S*-adenosylmethionine. The latter compound is important in the synthesis of ethylene (see Chapter 22) and in reactions involving the transfer of methyl groups, as in lignin synthesis (see Chapter 13).

Phosphate Assimilation

Phosphate (HPO_4^{2-}) in the soil solution is readily absorbed by plant roots via an H^+–HPO_4^{2-} symporter (see Chapter 6) and incorporated into a variety of organic compounds, including sugar phosphates, phospholipids, and nucleotides. The main entry point of phosphate into assimilatory pathways occurs during the formation of ATP, the energy "currency" of the cell. In the overall reaction for this process, inorganic phosphate is added to the second phosphate group in adenosine diphosphate to form a phosphate ester bond.

In mitochondria, the energy for ATP synthesis derives from the oxidation of NADH by oxidative phosphorylation (see Chapter 11). ATP synthesis is also driven by light-dependent photophosphorylation in the chloroplasts (see Chapter 7). In addition to these reactions in mitochondria and chloroplasts, reactions in the cytosol such as glycolysis also assimilate phosphate.

Glycolysis incorporates inorganic phosphate into 1,3-bisphosphoglyceric acid, forming a high-energy acyl phosphate group. This phosphate can be donated to ADP to form ATP in a substrate-level phosphorylation reaction (see Chapter 11). Once incorporated into ATP, the phosphate group may be transferred via many different reactions to form the various phosphorylated compounds found in higher-plant cells.

Cation Assimilation

Cations taken up by plant cells form complexes with organic compounds in which the cation becomes bound to the complex by noncovalent bonds (for a discussion of noncovalent bonds, see Chapter 2 on the web site). Plants assimilate macronutrient cations such as potassium, magnesium, and calcium, as well as micronutrient cations such as copper, iron, manganese, cobalt, sodium, and zinc, in this manner. In this section we will describe coordination bonds and electrostatic bonds, which mediate the assimilation of several cations that plants require as nutrients, and the special requirements for the absorption of iron by roots and subsequent assimilation of iron within plants.

Cations form noncovalent bonds with carbon compounds

The noncovalent bonds formed between cations and carbon compounds are of two types: coordination bonds and electrostatic bonds. In the formation of a coordination complex, several oxygen or nitrogen atoms of a carbon compound donate unshared electrons to form a bond with the cation nutrient. As a result, the positive charge on the cation is neutralized.

Coordination bonds typically form between polyvalent cations and carbon molecules—for example, complexes between copper and tartaric acid (Figure 12.16A) or magnesium and chlorophyll *a* (Figure 12.16B). The nutrients that are assimilated as coordination complexes include copper, zinc, iron, and magnesium. Calcium can also form coordination complexes with the polygalacturonic acid of cell walls (Figure 12.16C).

Electrostatic bonds form because of the attraction of a positively charged cation for a negatively charged group such as carboxylate (—COO^-) on a carbon compound. Unlike the situation in coordination bonds, the cation in an electrostatic bond retains its positive charge. Monovalent cations such as potassium (K^+) can form electrostatic bonds with the carboxylic groups of many organic acids (Figure 12.17A). Nonetheless, much of the potassium that is accumulated by plant cells and functions in osmotic regulation and enzyme activation remains in the cytosol and the vacuole as the free ion. Divalent ions such as calcium form electrostatic bonds with pectates (Figure 12.17B) and the carboxylic groups of polygalacturonic acid (see Chapter 15).

In general, cations such as magnesium (Mg^{2+}) and calcium (Ca^{2+}) are assimilated by the formation of both coordination complexes and electrostatic bonds with amino acids, phospholipids, and other negatively charged molecules.

Roots modify the rhizosphere to acquire iron

Iron is important in iron–sulfur proteins (see Chapter 7) and as a catalyst in enzyme-mediated redox reactions (see Chapter 5), such as those of nitrogen metabolism discussed earlier. Plants obtain iron from the soil, where it is present primarily as ferric iron (Fe^{3+}) in oxides such as $Fe(OH)^{2+}$, $Fe(OH)_3$, and $Fe(OH)_4^-$. At neutral pH, ferric iron is highly insoluble. To absorb sufficient amounts of iron from the soil solution, roots have developed several mechanisms that increase iron solubility and thus its availability (Figure 12.18). These mechanisms include:

- Soil acidification, which increases the solubility of ferric iron.

- Reduction of ferric iron to the more soluble ferrous form (Fe^{2+}).

- Release of compounds that form stable, soluble complexes with iron (Marschner 1995). Recall from Chapter 5 that such compounds are called iron chelators (see Figure 5.2).

Roots generally acidify the soil around them. They extrude protons during the absorption and assimilation of cations, particularly ammonium, and release organic acids, such as

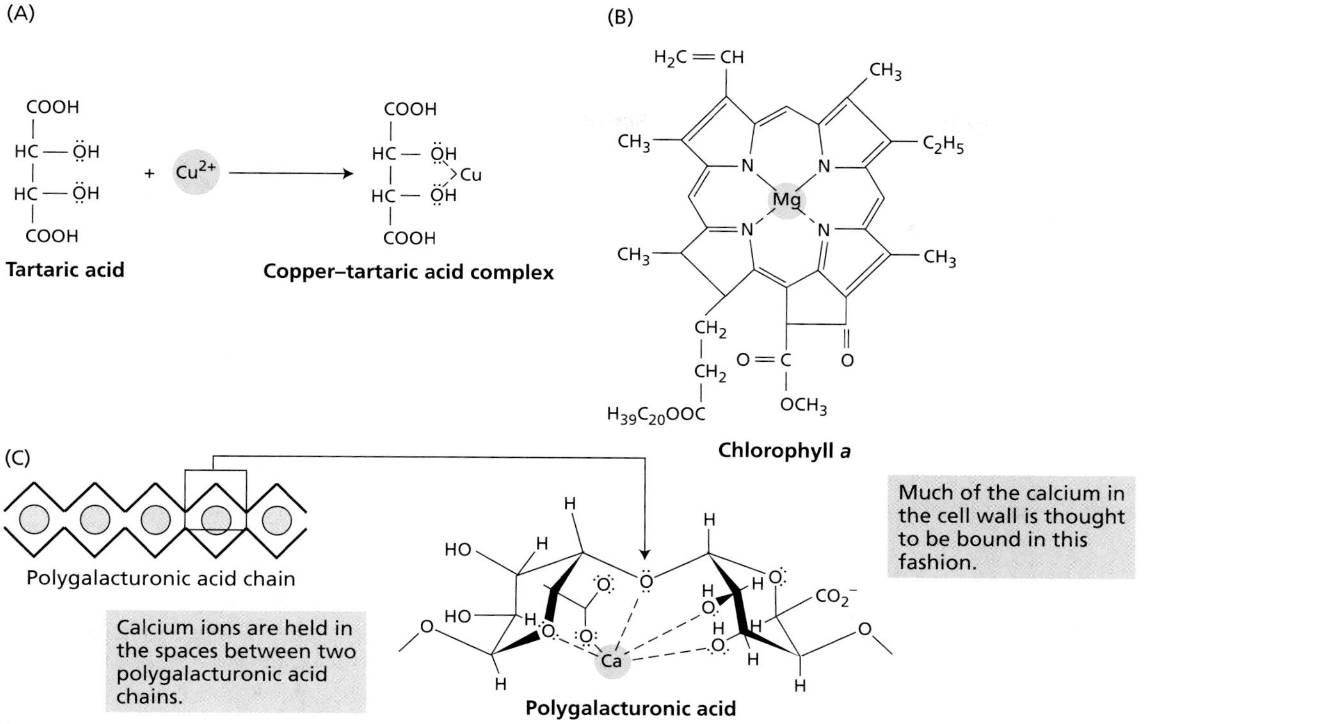

FIGURE 12.16 Examples of coordination complexes. Coordination complexes form when oxygen or nitrogen atoms of a carbon compound donate unshared electron pairs (represented by dots) to form a bond with a cation. (A) Copper ions share electrons with the hydroxyl oxygens of tartaric acid. (B) Magnesium ions share electrons with nitrogen atoms in chlorophyll *a*. Dashed lines represent a coordination bond between unshared electrons from the nitrogen atoms and the magnesium cation. (C) The "egg box" model of the interaction of polygalacturonic acid, a major constituent of pectins in cell walls, and calcium ions. At right is an enlargement of a single calcium ion forming a coordination complex with the hydroxyl oxygens of the galacturonic acid residues. (After Rees 1977.)

malic acid and citric acid, that enhance iron and phosphate availability (see Figure 5.4). Iron deficiencies stimulate the extrusion of protons by roots. In addition, plasma membranes in roots contain an enzyme, called *iron-chelating reductase*, that reduces ferric iron (Fe^{3+}) to the ferrous (Fe^{2+}) form, with NADH or NADPH serving as the electron donor (Figure 12.18A). The activity of this enzyme increases under iron deprivation.

Several compounds secreted by roots form stable chelates with iron. Examples include malic acid, citric acid, phenolics, and piscidic acid. Grasses produce a special class of iron chelators called *siderophores*. Siderophores are made of amino acids that are not found in proteins, such as mugineic acid, and form highly stable complexes with Fe^{3+}. Root cells of grasses have Fe^{3+}–siderophore transport systems in their plasma membrane that bring the chelate into

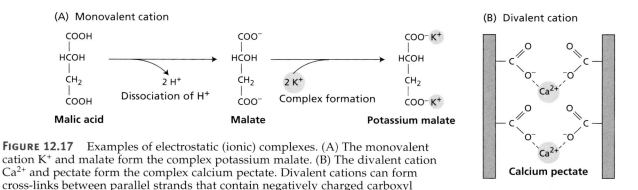

FIGURE 12.17 Examples of electrostatic (ionic) complexes. (A) The monovalent cation K^+ and malate form the complex potassium malate. (B) The divalent cation Ca^{2+} and pectate form the complex calcium pectate. Divalent cations can form cross-links between parallel strands that contain negatively charged carboxyl groups. Calcium cross-links play a structural role in the cell walls.

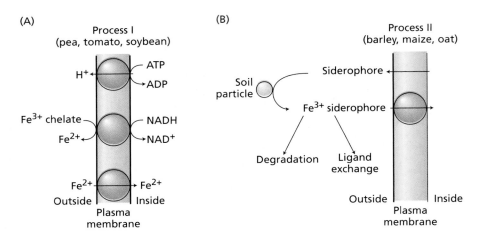

FIGURE 12.18 Two processes through which plants roots absorb iron. (A) A process common to dicots such as pea, tomato, and soybean. The chelates include organic compounds such as malic acid, citric acid, phenolics, and piscidic acid. (B) A process common to grasses such as barley, maize, and oat. After the grass excretes the siderophore and it removes iron soil particles, the complex may degrade and release the iron to the soil, exchange iron for another ligand, or be transported into the root. (After Guerinot and Yi 1994.)

the cytoplasm. Under iron deficiency, grass roots release more siderophores into the soil and increase the capacity of their Fe^{3+}–siderophore transport system (Figure 12.18B).

Iron forms complexes with carbon and phosphate

After the roots absorb iron or an iron chelate, they oxidize it to a ferric form and translocate much of it to the leaves as an electrostatic complex with citrate.

Once in the leaves, iron undergoes an important assimilatory reaction through which it is inserted into the porphyrin precursor of heme groups found in the cytochromes located in chloroplasts and mitochondria (see Chapter 7). This reaction is catalyzed by the enzyme ferrochelatase (Figure 12.19) (Jones 1983). Most of the iron in the plant is found in heme groups. In addition, iron–sulfur proteins of the electron transport chain (see Chapter 7) contain non-heme iron covalently bound to the sulfur atoms of cysteine residues in the apoprotein. Iron is also found in Fe_2S_2 centers, which contain two irons (each complexed with the sulfur atoms of cysteine residues) and two inorganic sulfides.

Free iron (iron that is not complexed with carbon compounds) may interact with oxygen to form superoxide anions (O_2^-), which can damage membranes by degrading unsaturated lipid components. Plant cells may limit such damage by storing surplus iron in an iron–protein complex called **phytoferritin** (Bienfait and Van der Mark 1983). Phytoferritin consists of a protein shell with 24 identical subunits forming a hollow sphere that has a molecular mass of about 480 kDa. Within this sphere is a core of 5400 to 6200 iron atoms present as a ferric oxide–phosphate complex.

How iron is released from phytoferritin is uncertain, but breakdown of the protein shell appears to be involved. The level of free iron in plant cells regulates the de novo biosynthesis of phytoferritin (Lobreaux et al. 1992). Interest in phytoferritin is high because iron in this protein-bound form may be highly available to humans, and foods rich in phytoferritin such as soybean may address dietary anemia problems (Welch and Graham 2004).

Oxygen Assimilation

Respiration accounts for the bulk (about 90%) of the oxygen (O_2) assimilated by plant cells (see Chapter 11). Another major pathway for the assimilation of O_2 into organic compounds involves the incorporation of O_2 from water (see reaction 1 in Table 8.1). A small proportion of oxygen can be directly assimilated into organic compounds in the process of *oxygen fixation*.

In oxygen fixation, molecular oxygen is added directly to an organic compound in reactions carried out by enzymes known as *oxygenases*. Recall from Chapter 8 that oxygen is directly incorporated into an organic compound during photorespiration in a reaction that involves the oxygenase activity of ribulose-1,5-bisphosphate carboxylase/oxygenase (rubisco), the enzyme of CO_2 fixation (Ogren 1984). The first stable product that contains oxygen originating from molecular oxygen is 2-phosphoglycolate.

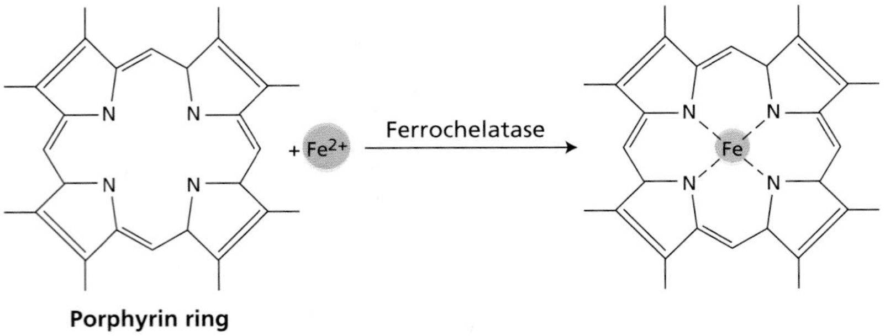

FIGURE 12.19 The ferrochelatase reaction. The enzyme ferrochelatase catalyzes the insertion of iron into the porphyrin ring to form a coordination complex. See Figure 7.36 for illustration of the biosynthesis of the porphyrin ring.

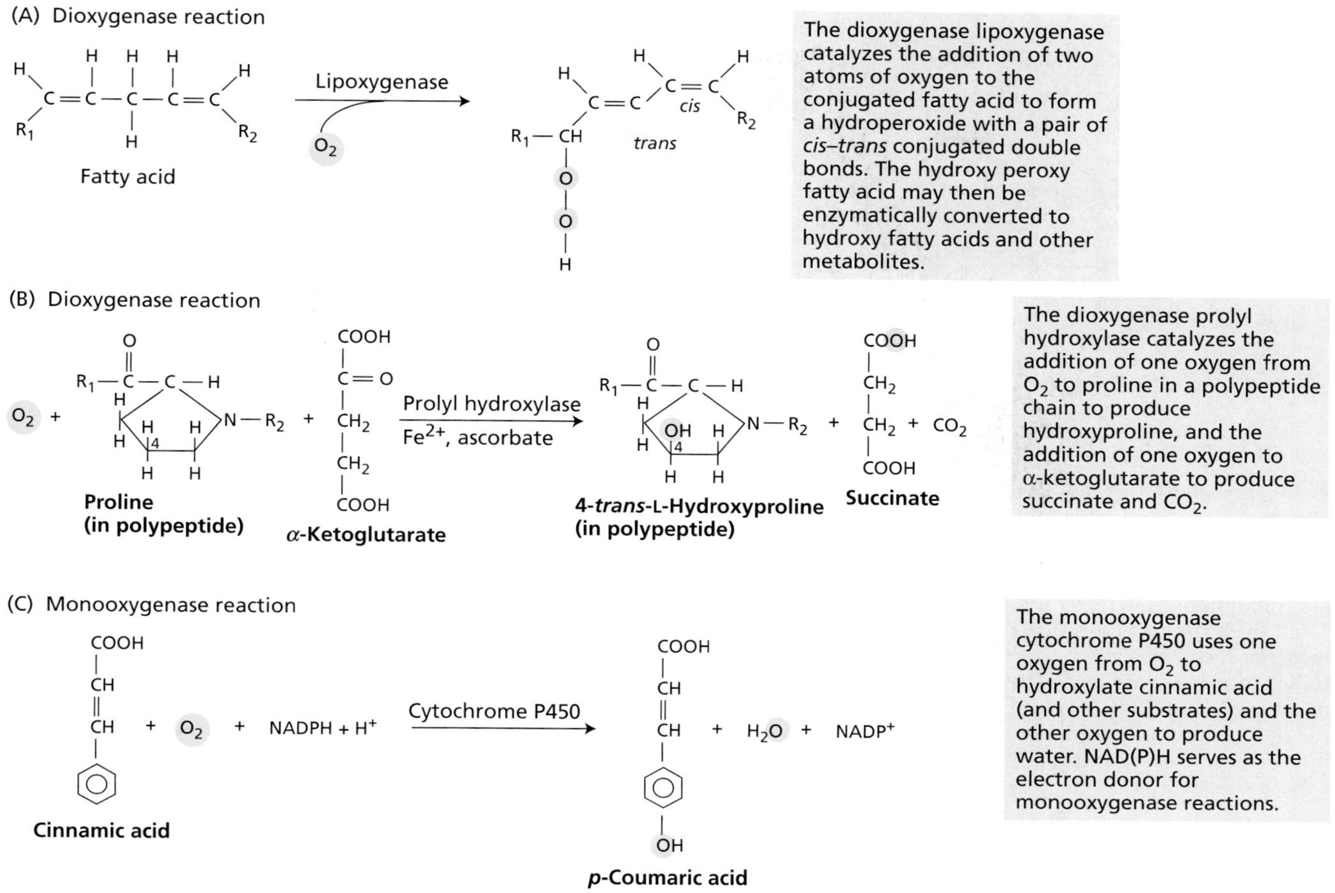

FIGURE 12.20 Examples of the two types of oxygenase reactions in cells of higher plants, dioxygenase (A and B) and monooxygenase (C).

In general, oxygenases are classified as dioxygenases or monooxygenases, according to the number of atoms of oxygen that are transferred to a carbon compound in the catalyzed reaction. In **dioxygenase** reactions, both oxygen atoms are incorporated into one or two carbon compounds (Figure 12.20A and B). Examples of dioxygenases in plant cells are lipoxygenase, which catalyzes the addition of two atoms of oxygen to unsaturated fatty acids (see Figure 12.20A), and prolyl hydroxylase, the enzyme that converts proline to the less common amino acid hydroxyproline (see Figure 12.20B).

Hydroxyproline is an important component of the cell wall protein extensin (see Chapter 15). The synthesis of hydroxyproline from proline differs from the synthesis of all other amino acids in that the reaction occurs after the proline has been incorporated into protein, and it is therefore a posttranslational modification reaction. Prolyl hydroxylase is localized in the endoplasmic reticulum, suggesting that most proteins containing hydroxyproline are found in the secretory pathway.

Monooxygenases add one of the atoms in molecular oxygen to a carbon compound; the other oxygen atom is converted into water. Monooxygenases are sometimes referred to as *mixed-function oxidases* because of their ability to catalyze simultaneously both the oxygenation reaction and the oxidase reaction (reduction of oxygen to water). The monooxygenase reaction also requires a reduced substrate (NADH or NADPH) as an electron donor, according to the following equation:

$$A + O_2 + BH_2 \rightarrow AO + H_2O + B$$

where A represents an organic compound and B represents the electron donor.

An important monooxygenase in plants is the family of heme proteins collectively called cytochrome P450, which catalyzes the hydroxylation of cinnamic acid to *p*-coumaric acid (Figure 12.20C). In monooxygenases, the oxygen is first activated by being combined with the iron atom of the heme group; NADPH serves as the electron donor. The mixed-function oxidase system is localized on the endoplasmic reticulum and is capable of oxidizing a variety of substrates, including mono- and diterpenes and fatty acids.

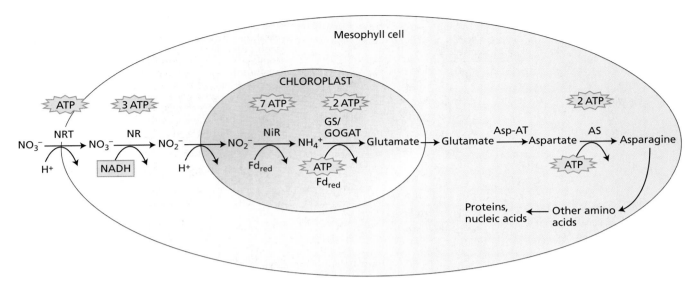

FIGURE 12.21 Summary of the processes involved in the assimilation of mineral nitrogen in the leaf. Nitrate translocated from the roots through the xylem is absorbed by a mesophyll cell via one of the nitrate–proton symporters (NRT) into the cytoplasm. There it is reduced to nitrite via nitrate reductase (NR). Nitrite is translocated into the stroma of the chloroplast along with a proton. In the stroma, nitrite is reduced to ammonium via nitrite reductase (NiR) and this ammonium is converted into glutamate via the sequential action of glutamine synthetase (GS) and glutamate synthase (GOGAT). Once again in the cytoplasm, the glutamate is transaminated to aspartate via aspartate aminotransferase (Asp-AT). Finally, asparagine synthetase (AS) converts aspartate into asparagine. The approximate amounts of ATP equivalents are given above each reaction.

The Energetics of Nutrient Assimilation

Nutrient assimilation generally requires large amounts of energy to convert stable, low-energy inorganic compounds into high-energy organic compounds. For example, the reduction of nitrate to nitrite and then to ammonium requires the transfer of about ten electrons and accounts for about 25% of the total energy expenditures in both roots and shoots (Bloom 1997). Consequently, a plant may use one-fourth of its energy to assimilate nitrogen, a constituent that accounts for less than 2% of the total dry weight of the plant.

Many of these assimilatory reactions occur in the stroma of the chloroplast, where they have ready access to powerful reducing agents such as NADPH, thioredoxin, and ferredoxin generated during photosynthetic electron transport. This process—coupling nutrient assimilation to photosynthetic electron transport—is called **photoassimilation** (Figure 12.21).

Photoassimilation and the Calvin cycle occur in the same compartment, but only when photosynthetic electron transport generates reductant in excess of the needs of the Calvin cycle—for example, under conditions of high light and low CO_2—does photoassimilation proceed (Robinson 1988). High levels of CO_2 inhibit nitrate assimilation in the

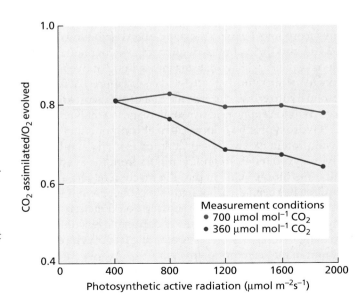

FIGURE 12.22 The assimilatory quotient (AQ = CO_2 assimilated/O_2 evolved) of wheat seedlings as a function of light level (photosynthetic active radiation). Nitrate photoassimilation is directly related to assimilatory quotient, because transfer of electrons to nitrate and nitrite during photoassimilation increases O_2 evolution from the light-dependent reactions of photosynthesis, while CO_2 assimilation by the light-independent reactions continues at similar rates. Therefore, plants that are photoassimilating nitrate exhibit a lower AQ. In measurements carried out at ambient CO_2 concentrations (360 µmol mol^{-1}, red trace), the AQ decreases as a function of incident radiation, indicating that photoassimilation rates increased. At elevated CO_2 concentrations (700 µmol mol^{-1}, blue trace) the AQ remains constant at all light levels used, indicating that the CO_2-fixing reactions are competing for reductant and inhibiting photoassimilation. (After Bloom et al. 2002.)

shoots of C_3 plants (Figure 12.22) (see **Web Essay 12.1**) and in the bundle sheath cells (see Chapter 8) of C_4 plants (Becker et al. 1993).

The mechanisms that regulate the partitioning of reductant between the Calvin cycle and photoassimilation warrant investigation, because atmospheric levels of CO_2 are expected to double during this century (see Chapter 9), so this phenomenon will affect plant–nutrient relations.

Summary

Nutrient assimilation is the process by which nutrients acquired by plants are incorporated into the carbon constituents necessary for growth and development. These processes often involve chemical reactions that are highly energy intensive and thus may depend directly on reductant generated through photosynthesis.

For nitrogen, assimilation is but one in a series of steps that constitute the nitrogen cycle. The nitrogen cycle encompasses the various states of nitrogen in the biosphere and their interconversions. The principal sources of nitrogen available to plants are nitrate (NO_3^-) and ammonium (NH_4^+).

The nitrate absorbed by roots is assimilated in either roots or shoots, depending on nitrate availability and plant species. In nitrate assimilation, nitrate is reduced to nitrite (NO_2^-) in the cytosol via the enzyme nitrate reductase; then nitrite is reduced to ammonium in root plastids or chloroplasts via the enzyme nitrite reductase.

Ammonium, derived either from root absorption or generated through nitrate assimilation or photorespiration, is converted to glutamine and glutamate through the sequential actions of glutamine synthetase and glutamate synthase, which are located in the cytosol and root plastids or chloroplasts.

Once assimilated into glutamine or glutamate, nitrogen may be transferred to many other organic compounds through various reactions, including the transamination reactions. Interconversion between glutamine and asparagine by asparagine synthetase balances carbon metabolism and nitrogen metabolism within a plant.

Many plants form a symbiotic relationship with nitrogen-fixing bacteria that contain an enzyme complex, nitrogenase, that can reduce atmospheric nitrogen to ammonia. Legumes and actinorhizal plants form associations with rhizobia and *Frankia*, respectively. These associations result from a finely tuned interaction between the symbiont and the host plant that involves the recognition of specific signals, the induction of a specialized developmental program within the plant, the uptake of the bacteria by the plant, and the development of nodules, unique organs that house the bacteria within plant cells. Some nitrogen-fixing prokaryotic microorganisms do not form symbiotic relationships with higher plants but benefit plants by enriching the nitrogen content of the soil.

Like nitrate, sulfate (SO_4^{2-}) must be reduced by assimilation. In sulfate reduction, an activated form of sulfate called 5′-adenylylsulfate (APS) forms. Sulfide (S^{2-}), the end product of sulfate reduction, does not accumulate in plant cells, but is instead rapidly incorporated into the amino acids cysteine and methionine.

Phosphate (HPO_4^{2-}) is present in a variety of compounds found in plant cells, including sugar phosphates, lipids, nucleic acids, and free nucleotides. The initial product of its assimilation is ATP, which is produced by substrate-level phosphorylations in the cytosol, oxidative phosphorylation in the mitochondria, and photophosphorylation in the chloroplasts.

Whereas the assimilation of nitrogen, sulfur, and phosphorus requires the formation of covalent bonds with carbon compounds, many macro- and micronutrient cations (e.g., K^+, Mg^{2+}, Ca^{2+}, Cu^{2+}, Fe^{3+}, Mn^{2+}, Co^{2+}, Na^+, Zn^{2+}) simply form complexes. These complexes may be held together by electrostatic bonds or coordination bonds.

Iron assimilation may involve chelation, oxidation–reduction reactions, and the formation of complexes. In order to store large amounts of iron, plant cells synthesize phytoferritin, an iron storage protein. An important function of iron in plant cells is to act as a redox component in the active site of enzymes, often as an iron–porphyrin complex. Iron is inserted into a porphyrin group in the ferrochelatase reaction.

In addition to being utilized in respiration, molecular oxygen can be assimilated in the process of oxygen fixation, the direct addition of oxygen to organic compounds. This process is catalyzed by enzymes known as oxygenases, which are classified as monooxygenases or dioxygenases.

Nutrient assimilation requires large amounts of energy to convert stable, low-energy, inorganic compounds into high-energy organic compounds. A plant may use one-fourth of its energy to assimilate nitrogen. Plants use energy from photosynthesis to assimilate inorganic compounds in a process called photoassimilation.

Web Material

Web Topics

12.1 Development of a Root Nodule
Nodule primordia form opposite to the protoxylem poles of the root vascular bundles.

12.2 Measurement of Nitrogen Fixation
Acetylene reduction is used as an indirect measurement of nitrogen reduction.

12.3 The Synthesis of Methionine
Methionine is synthesized in plastids from cysteine.

Web Essay

12.1 Elevated CO₂ and Nitrogen Photoassimilation In leaves grown under high CO_2 concentrations, CO_2 inhibits nitrogen photoassimilation because it competes for reductant.

Chapter References

Becker, T. W., Perrot-Rechenmann, C., Suzuki, A., and Hirel, B. (1993) Subcellular and immunocytochemical localization of the enzymes involved in ammonia assimilation in mesophyll and bundle-sheath cells of maize leaves. *Planta* 191: 129–136.

Bergmann, L., and Rennenberg, H. (1993) Glutathione metabolism in plants. In *Sulfur Nutrition and Assimilation in Higher Plants. Regulatory, Agricultural and Environmental Aspects*, L. J. De Kok, I. Stulen, H. Rennenberg, C. Brunold, and W. E. Rauser, eds., SPB Academic Publishing, The Hague, Netherlands, pp. 109–123.

Bienfait, H. F., and Van der Mark, F. (1983) Phytoferritin and its role in iron metabolism. In *Metals and Micronutrients: Uptake and Utilization by Plants*, D. A. Robb and W. S. Pierpoint, eds., Academic Press, New York, pp. 111–123.

Bloom, A. J. (1997) Nitrogen as a limiting factor: Crop acquisition of ammonium and nitrate. In *Ecology in Agriculture*, L. E. Jackson, ed., Academic Press, San Diego, CA, pp. 145–172.

Bloom, A. J., Smart, D. R., Nguyen, D. T., and Searles, P. S. (2002) Nitrogen assimilation and growth of wheat under elevated carbon dioxide. *Proc. Natl. Acad. Sci. USA* 99: 1730–1735.

Bloom, A. J., Sukrapanna, S. S., and Warner, R. L. (1992) Root respiration associated with ammonium and nitrate absorption and assimilation by barley. *Plant Physiol.* 99: 1294–1301.

Buchanan, B., Gruissem, W., and Jones, R., eds. (2000) *Biochemistry and Molecular Biology of Plants*. American Society of Plant Physiologists. Rockville, MD.

Burris, R. H. (1976) Nitrogen fixation. In *Plant Biochemistry*, 3rd ed., J. Bonner and J. Varner, eds., Academic Press, New York, pp. 887–908.

Campbell, W. H. (1999) Nitrate reductase structure, function and regulation: Bridging the gap between biochemistry and physiology. *Annu. Rev. Plant Physiol. Plant Mol. Biol.* 50: 277–303.

Carlson, R. W., Forsberg, L. S., Price, N. P. J., Bhat, U. R., Kelly, T. M., and Raetz, C. R. H. (1995) The structure and biosynthesis of *Rhizobium leguminosarum* lipid A. In *Progress in Clinical and Biological Research*, Vol. 392: *Bacterial Endotoxins: Lipopolysaccharides from Genes to Therapy: Proceedings of the Third Conference of the International Endotoxin Society, held in Helsinki, Finland, on August 15-18, 1994*, J. Levin et al., eds., John Wiley and Sons, New York, pp. 25–31.

Crawford, N. M., and Forde, B. J. 2002. Molecular and developmental biology of inorganic nitrogen nutrition. In: *The Arabidopsis Book*, Somerville, C. and Meyerowitz, E., eds. American Society of Plant Physiologists, Rockville, MD, pp. doi/10.1199/tab.0011, http://www.aspb.org/publications/arabidopsis/.

Denison, R. F., and Harter, B. L. (1995) Nitrate effects on nodule oxygen permeability and leghemoglobin. *Plant Physiol.* 107: 1355–1364.

Dixon, R. O. D., and Wheeler, C. T. (1986) *Nitrogen Fixation in Plants*. Chapman and Hall, New York.

Dixon, R., and Kahn, D. (2004) Genetic regulation of biological nitrogen fixation. *Nat. Rev. Microbiol.* 2: 621–631.

Dong, Z., Canny, M. J., McCully, M. E., Roboredo, M. R., Cabadilla, C. F., Ortega, E., and Rodes, R. (1994) A nitrogen-fixing endophyte of sugarcane stems: A new role for the apoplast. *Plant Physiol.* 105: 1139–1147.

Etzler, M. E., Kalsi, G., Ewing, N. N., Roberts, N. J., Day, R. B., and Murphy, J. B. (1999) A nod factor binding lectin with apyrase activity from legume roots. *Proc. Natl. Acad. Sci. USA* 96: 5856–5861.

FAOSTAT. (2001) *Agricultural Data*. Food and Agricultural Organization of the United Nations, Rome.

Geurts, R., and Bisseling, T. (2002) Rhizobium nod factor perception and signalling. *Plant Cell* 14: S239–S249.

Guerinot, M. L., and Yi, Y. (1994) Iron: Nutritious, noxious, and not readily available. *Plant Physiol.* 104: 815–820.

Heidstra, R., and Bisseling, T. (1996) Nod factor-induced host responses and mechanisms of Nod factor perception. *New Phytol.* 133: 25–43.

Hell, R. (1997) Molecular physiology of plant sulfur metabolism. *Planta* 202: 138–148.

Heytler, P. G., Reddy, G. S., and Hardy, R. W. F. (1984) In vivo energetics of symbiotic nitrogen fixation in soybeans. In *Nitrogen Fixation and CO₂ Metabolism*, P. W. Ludden and I. E. Burris, eds., Elsevier, New York, pp. 283–292.

Jones, O. T. G. (1983) Ferrochelatase. In *Metals and Micronutrients: Uptake and Utilization by Plants*, D. A. Robb and W. S. Pierpoint, eds., Academic Press, New York, pp. 125–144.

Kaiser, W. M., Weiner, H., and Huber, S. C. (1999) Nitrate reductase in higher plants: A case study for transduction of environmental stimuli into control of catalytic activity. *Physiol. Plant.* 105: 385–390.

Kleinhofs, A., Warner, R. L., Lawrence, J. M., Melzer, J. M., Jeter, J. M., and Kudrna, D. A. (1989) Molecular genetics of nitrate reductase in barley. In *Molecular and Genetic Aspects of Nitrate Assimilation*, J. L. Wray and J. R. Kinghorn, eds., Oxford Science, New York, pp. 197–211.

Kuzma, M. M., Hunt, S., and Layzell, D. B. (1993) Role of oxygen in the limitation and inhibition of nitrogenase activity and respiration rate in individual soybean nodules. *Plant Physiol.* 101: 161–169.

Lam, H.-M., Coschigano, K. T., Oliveira, I. C., Melo-Oliveira, R., and Coruzzi, G. M. (1996) The molecular-genetics of nitrogen assimilation into amino acids in higher plants. *Annu. Rev. Plant Physiology Plant Mol. Biol.* 47: 569–593.

Lea, P. J., Blackwell, R. D., and Joy, K. W. (1992) Ammonia assimilation in higher plants. In *Nitrogen Metabolism of Plants* (Proceedings of the Phytochemical Society of Europe 33), K. Mengel and D. J. Pilbeam, eds., Clarendon, Oxford, pp. 153–186.

Leustek, T., and Saito, K. (1999) Sulfate transport and assimilation in plants. *Plant Physiol.* 120: 637–643.

Leustek, T., Martin, M. N., Bick, J.-A., and Davies, J. P. (2000) Pathways and regulation of sulfur metabolism revealed through molecular and genetic studies. *Annu. Rev. Plant Physiol. Plant Mol. Biol.* 51: 141–165.

Lobreaux, S., Massenet, O., and Briat, J.-F. (1992) Iron induces ferritin synthesis in maize plantlets. *Plant Mol. Biol.* 19: 563–575.

Ludwig, R. A., and de Vries, G. E. (1986) Biochemical physiology of *Rhizobium* dinitrogen fixation. In *Nitrogen Fixation*, Vol. 4: *Molecular Biology*, W. I. Broughton and S. Puhler, eds., Clarendon, Oxford, pp. 50–69.

Marschner, H. (1995) *Mineral Nutrition of Higher Plants*, 2nd ed. Academic Press, London.

Mendel, R. R. (2005) Molybdenum: Biological activity and metabolism. *Dalton Transactions* 2005: 3404–3409.

Mylona, P., Pawlowski, K., and Bisseling, T. (1995) Symbiotic nitrogen fixation. *Plant Cell* 7: 869–885.

Oaks, A. (1994) Primary nitrogen assimilation in higher plants and its regulation. *Can. J. Bot.* 72: 739–750.

Ogren, W. L. (1984) Photorespiration: Pathways, regulation, and modification. *Annu. Rev. Plant Physiol.* 35: 415–442.

Pate, J. S. (1983) Patterns of nitrogen metabolism in higher plants and their ecological significance. In *Nitrogen as an Ecological Factor: The 22nd Symposium of the British Ecological Society,*

Oxford 1981, J. A. Lee, S. McNeill, and I. H. Rorison, eds., Blackwell, Boston, pp. 225–255.

Pate, J. S., and Layzell, D. B. (1990) Energetics and biological costs of nitrogen assimilation. In *The Biochemistry of Plants*, Vol. 16: *Intermediary Nitrogen Metabolism*, B. J. Miflin and P. J. Lea, eds., Academic Press, San Diego, CA, pp. 1–42.

Phillips, D. A., and Kapulnik, Y. (1995) Plant isoflavonoids, pathogens and symbionts. *Trends Microbiol.* 3: 58–64.

Preisig, O., Zufferey, R., Thony-Meyer, L., Appleby, C. A., and Hennecke, H. (1996) A high-affinity cbb3-type cytochrome oxidase terminates the symbiosis-specific respiratory chain of *Bradyrhizobium japonicum*. *J. Bacteriol.* 178: 1532–1538.

Rees, D. A. (1977) *Polysaccharide Shapes*. Chapman and Hall, London.

Reis, V. M., Baldani, J. I., Baldani, V. L. D., and Dobereiner, J. (2000) Biological dinitrogen fixation in Gramineae and palm trees. *Crit. Rev. Plant Sci.* 19: 227–247.

Robinson, J. M. (1988) Spinach leaf chloroplast carbon dioxide and nitrite photoassimilations do not compete for photogenerated reductant: Manipulation of reductant levels by quantum flux density titrations. *Plant Physiol.* 88: 1373–1380.

Rolfe, B. G., and Gresshoff, P. M. (1988) Genetic analysis of legume nodule initiation. *Annu. Rev. Plant Physiol. Plant Mol. Biol.* 39: 297–320.

Schlesinger, W. H. (1997) *Biogeochemistry: An Analysis of Global Change*, 2nd ed. Academic Press, San Diego, CA.

Siegel, L. M., and Wilkerson, J. Q. (1989) Structure and function of spinach ferredoxin-nitrite reductase. In *Molecular and Genetic Aspects of Nitrate Assimilation*, J. L. Wray and J. R. Kinghorn, eds., Oxford Science, Oxford, pp. 263–283.

Sivasankar, S., and Oaks, A. (1996) Nitrate assimilation in higher plants—The effect of metabolites and light. *Plant Physiol. Biochem.* 34: 609–620.

Smart, D. R., and Bloom, A. J. (2001) Wheat leaves emit nitrous oxide during nitrate assimilation. *PNAS* 98: 7875–7878.

Stokkermans, T. J. W., Ikeshita, S., Cohn, J., Carlson, R. W., Stacey, G., Ogawa, T., and Peters, N. K. (1995) Structural requirements of synthetic and natural product lipo-chitin oligosaccharides for induction of nodule primordia on *Glycine soja*. *Plant Physiol.* 108: 1587–1595.

Timmers, A. C. J., Auriac, M.-C., and Truchet, G. (1999) Refined analysis of early symbiotic steps of the Rhizobium-Medicago: Interaction in relation with microtubular cytoskeleton rearrangements. *Development* 126: 3617–3628.

Vande Broek, A., and Vanderleyden, J. (1995) Review: Genetics of the *Azospirillum*-plant root association. *Crit. Rev. Plant Sci.* 14: 445–466.

van Rhijn, P., Goldberg, R. B., and Hirsch, A. M. (1998) *Lotus corniculatus* nodulation specificity is changed by the presence of a soybean lectin gene. *Plant Cell* 10: 1233–1249.

Warner, R. L., and Kleinhofs, A. (1992) Genetics and molecular biology of nitrate metabolism in higher plants. *Physiol. Plant.* 85: 245–252.

Welch, R. M., and Graham, R. D. (2004) Breeding for micronutrients in staple food crops from a human nutrition perspective. *J. Exp. Bot.* 55: 353–364.

Wray, J. L. (1993) Molecular biology, genetics and regulation of nitrite reduction in higher plants. *Physiol. Plant.* 89: 607–612.

Chapter 13 | Secondary Metabolites and Plant Defense

IN NATURAL HABITATS, plants are surrounded by an enormous number of potential enemies. Nearly all ecosystems contain a wide variety of bacteria, viruses, fungi, nematodes, mites, insects, mammals, and other herbivorous animals. By their nature, plants cannot avoid these herbivores and pathogens simply by moving away; they must protect themselves in other ways.

The cuticle (a waxy outer layer) and the periderm (secondary protective tissue), besides retarding water loss, provide barriers to bacterial and fungal entry. In addition, a group of plant compounds known as secondary metabolites defend plants against a variety of herbivores and pathogenic microbes. Secondary compounds may serve other important functions as well, such as structural support, as in the case of lignin, or pigments, as in the case of the anthocyanins.

In this chapter we will discuss some of the mechanisms by which plants protect themselves against both herbivory and pathogenic organisms. We will begin with a discussion of the three classes of compounds that provide surface protection to the plant: cutin, suberin, and waxes. Next we will describe the structures and biosynthetic pathways for the three major classes of secondary metabolites: terpenes, phenolics, and nitrogen-containing compounds. Induced plant defenses against insect herbivore damage will be discussed as an example of the important ecological functions of secondary metabolites. Finally, we will examine specific plant responses to pathogen attack, the genetic control of host–pathogen interactions, and cell signaling processes associated with infection.

Cutin, Waxes, and Suberin

All plant parts exposed to the atmosphere are coated with layers of lipid material that reduce water loss and help block the entry of pathogenic fungi and bacteria. The principal types of coatings are cutin, suberin, and waxes (Figure 13.1). Cutin is found on most aboveground parts; suberin is present on underground parts, woody stems, and healed wounds. Waxes are associated with both cutin and suberin.

Cutin, waxes, and suberin are made up of hydrophobic compounds

Cutin is a macromolecule, a polymer consisting of many long-chain fatty acids that are attached to each other by ester linkages, creating a rigid three-dimensional network. Cutin is formed from 16:0 and 18:1 fatty acids* with hydroxyl or epoxide groups situated either in the middle of the chain or at the end opposite the carboxylic acid function (Figure 13.1A).

Cutin is a principal constituent of the **cuticle**, a multilayered secreted structure that coats the outer cell walls of the epidermis on the aerial parts of all herbaceous plants (Figure 13.2). The cuticle is composed of a top coating of wax, a thick middle layer containing cutin embedded in wax (the cuticle proper), and a lower layer formed of cutin and wax blended with the cell wall substances pectin, cellulose,

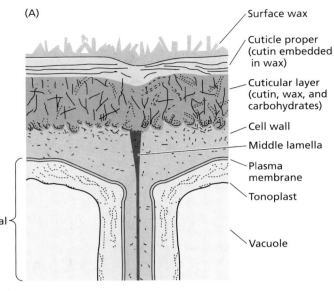

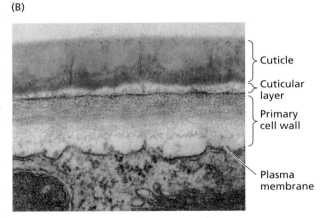

FIGURE 13.2 (A) Schematic drawing of the structure of the plant cuticle, the protective covering on the epidermis of leaves and young stems at the stage of full leaf expansion. (B) Electron micrograph of the cuticle of a glandular cell from a young leaf (*Lamium* sp.), showing the presence of the cuticle layers indicated in A, except for surface waxes, which are not visible. (51,000×) (A, after Jeffree 1996; B, from Gunning and Steer 1996.)

(A) Hydroxy fatty acids that polymerize to make **cutin**:

$$HOCH_2(CH_2)_{14}COOH$$

$$CH_3(CH_2)_8\underset{\underset{OH}{|}}{CH}(CH_2)_5COOH$$

(B) Common **wax** components:

Straight-chain alkanes $CH_3(CH_2)_{27}CH_3$

$CH_3(CH_2)_{29}CH_3$

Fatty acid ester $CH_3(CH_2)_{22}\overset{\overset{O}{\|}}{C}—O(CH_2)_{25}CH_3$

Long-chain fatty acid $CH_3(CH_2)_{22}COOH$

Long-chain alcohol $CH_3(CH_2)_{24}CH_2OH$

(C) Hydroxy fatty acids that polymerize along with other constituents to make **suberin**:

$$HOCH_2(CH_2)_{14}COOH$$

$$HOOC(CH_2)_{14}COOH \text{ (a dicarboxylic acid)}$$

FIGURE 13.1 Constituents of (A) cutin, (B) waxes, and (C) suberin.

and other carbohydrates (the cuticular layer). Recent research suggests that, in addition to cutin, the cuticle may contain a second lipid polymer, made up of long-chain hydrocarbons, that has been named *cutan* (Jeffree 1996).

Waxes are not macromolecules, but complex mixtures of long-chain acyl lipids that are extremely hydrophobic. The most common components of wax are straight-chain alkanes and alcohols of 25 to 35 carbon atoms (see Figure 13.1B). Long-chain aldehydes, ketones, esters, and free fatty

---

*Recall from Chapter 11 that the nomenclature for fatty acids is X:Y, where X is the number of carbon atoms and Y is the number of *cis* double bonds.

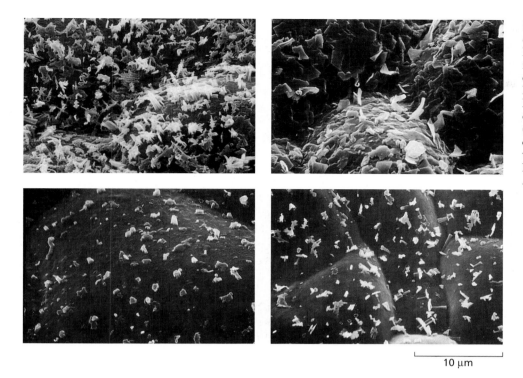

FIGURE 13.3 Surface wax deposits, which form the top layer of the cuticle, adopt different forms. These scanning electron micrographs show the leaf surfaces of two different lines of *Brassica oleracea*, which differ in wax crystal structure. (From Eigenbrode et al. 1991, courtesy of S. D. Eigenbrode, with permission from the Entomological Society of America.)

acids are also found. The waxes of the cuticle are synthesized by epidermal cells. They leave the epidermal cells as droplets that pass through pores in the cell wall by an unknown mechanism. The top coating of cuticle wax often crystallizes in an intricate pattern of rods, tubes, or plates (Figure 13.3). Certain patterns of these microstructures enhance water repellency by increasing the roughness of the wax surface. In this way, water droplets form instantly on contact and carry away contaminating particles, cleansing the plant's surface (Neinhuis et al. 1992). This phenomenon was first described for the leaves of the leguminous lotus *Lotus japonicus**, and for this reason it is sometimes referred to as the "Lotus effect."

Suberin is a polymer whose structure is poorly understood. Like cutin, suberin is formed from hydroxy or epoxy fatty acids joined by ester linkages. However, suberin differs from cutin in that it has dicarboxylic acids (see Figure 13.1C), more long-chain components, and a significant proportion of phenolic compounds as part of its structure.

Suberin is a cell wall constituent found in many locations throughout the plant. We have already noted its presence in the Casparian strip of the root endodermis, which forms a barrier between the apoplast of the cortex and the stele (see Chapter 4). Suberin is a principal component of the outer cell walls of all underground organs and is associated with the cork cells of the **periderm**, the tissue that forms the outer bark of stems and roots during secondary growth of woody plants. Suberin also forms at sites of leaf abscission and in areas damaged by disease or wounding.

Cutin, waxes, and suberin help reduce transpiration and pathogen invasion

Cutin, suberin, and their associated waxes form barriers between the plant and its environment that function to keep water in and pathogens out. The cuticle is very effective at limiting water loss from aerial parts of the plant, but it does not block transpiration completely, because even with the stomata closed, some water is lost. The thickness of the cuticle varies with environmental conditions. Plant species native to arid areas typically have thicker cuticles than do plants from moist habitats, but plants from moist habitats often develop thick cuticles when grown under dry conditions.

The cuticle and suberized tissue are both important in excluding fungi and bacteria, although they do not appear to be as important in pathogen resistance as some of the other defenses we will discuss in this chapter. Many fungi penetrate directly through the plant surface by mechanical means. Others produce *cutinase*, an enzyme that hydrolyzes cutin and thus facilitates entry into the plant.

Secondary Metabolites

Plants produce a large, diverse array of organic compounds that appear to have no direct function in growth and development. These substances are known as **secondary metabolites**, *secondary products*, or *natural products*. Secondary metabolites have no generally recognized, direct roles in the

---

*Not to be confused with the aquatic lotuses of the Nymphaeaceae family.

processes of photosynthesis, respiration, solute transport, translocation, protein synthesis, nutrient assimilation, or differentiation, or the formation of carbohydrates, proteins, and lipids discussed elsewhere in this book.

Secondary metabolites also differ from primary metabolites (amino acids, nucleotides, sugars, acyl lipids) in having a restricted distribution in the plant kingdom. That is, particular secondary metabolites are often found in only one plant species or related group of species, whereas primary metabolites are found throughout the plant kingdom.

Secondary metabolites defend plants against herbivores and pathogens

For many years the adaptive significance of most plant secondary metabolites was unknown. These compounds were thought to be simply functionless end products of metabolism, or metabolic wastes. Study of these substances was pioneered by organic chemists of the nineteenth and early twentieth centuries who were interested in them because of their importance as medicinal drugs, poisons, flavors, and industrial materials.

More recently, many secondary metabolites have been suggested to have important ecological functions in plants:

- They protect plants against being eaten by herbivores (herbivory) and against being infected by microbial pathogens.
- They serve as attractants (smell, color, taste) for pollinators and seed-dispersing animals.
- They function as agents of plant–plant competition and plant–microbe symbioses.

The ability of plants to compete and survive is therefore profoundly affected by the ecological functions of their secondary metabolites.

Secondary metabolism is also relevant to agriculture. The very defense compounds that increase the reproductive fitness of plants by warding off fungi, bacteria, and herbivores may also make them undesirable as food for humans. Many important crop plants have been artificially selected to produce relatively low levels of these compounds (which, of course, can make them more susceptible to insects and disease).

In the remainder of this chapter we will discuss the major types of plant secondary metabolites, their biosynthesis, and what is known about their functions in the plant, particularly their roles in defense.

Secondary metabolites are divided into three major groups

Plant secondary metabolites can be divided into three chemically distinct groups: terpenes, phenolics, and nitrogen-containing compounds. Figure 13.4 shows in simplified form the pathways involved in the biosynthesis of secondary metabolites and their interconnections with primary metabolism.

Terpenes

The **terpenes**, or *terpenoids*, constitute the largest class of secondary products. The diverse substances of this class are generally insoluble in water. They are biosynthesized from acetyl-CoA or glycolytic intermediates. After discussing the biosynthesis of terpenes, we'll examine how they act to repel herbivores and how some herbivores circumvent the toxic effects of terpenes.

Terpenes are formed by the fusion of five-carbon isoprene units

All terpenes are derived from the union of five-carbon elements (also called C_5 *units*) that have the branched carbon skeleton of isopentane:

$$\begin{array}{c} H_3C \\ \diagdown \\ CH - CH_2 - CH_3 \\ \diagup \\ H_3C \end{array}$$

The basic structural elements of terpenes are sometimes called **isoprene units**, because terpenes can decompose at high temperatures to give isoprene:

$$\begin{array}{c} H_3C \\ \diagdown \\ CH - CH = CH_2 \\ \diagup \\ H_2C \end{array}$$

Thus all terpenes are occasionally referred to as *isoprenoids*.

Terpenes are classified by the number of C_5 units they contain, although extensive metabolic modifications can sometimes make it difficult to pick out the original five-carbon residues. Ten-carbon terpenes, which contain two C_5 units, are called *monoterpenes*; 15-carbon terpenes (three C_5 units) are *sesquiterpenes*; and 20-carbon terpenes (four C_5 units) are *diterpenes*. Larger terpenes include *triterpenes* (30 carbons), *tetraterpenes* (40 carbons), and *polyterpenoids* ($[C_5]_n$ carbons, where $n > 8$).

There are two pathways for terpene biosynthesis

Terpenes are biosynthesized from primary metabolites in at least two different ways. In the well-studied **mevalonic acid pathway**, three molecules of acetyl-CoA are joined together stepwise to form mevalonic acid (Figure 13.5). This key six-carbon intermediate is then pyrophosphorylated, decarboxylated, and dehydrated to yield **isopentenyl diphosphate (IPP)***. IPP is the activated five-carbon building block of terpenes.

IPP also can be formed from intermediates of glycolysis or the photosynthetic carbon reduction cycle via a separate set of reactions called the **methylerythritol phosphate (MEP) pathway** that operates in chloroplasts and other plastids (Lichtenthaler 1999). Glyceraldehyde-3-phosphate and two carbon atoms derived from pyruvate condense to

---

*IPP is the abbreviation for isopentenyl *pyro*phosphate, an earlier name for this compound. The other pyrophosphorylated intermediates in this pathway are also now referred to as *di*phosphates.

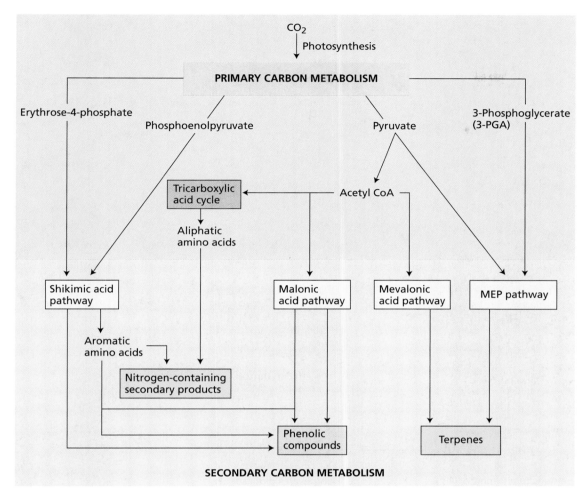

FIGURE 13.4 A simplified view of the major pathways of secondary-metabolite biosynthesis and their interrelationships with primary metabolism.

form the five-carbon intermediate 1-deoxy-D-xylulose-5-phosphate. After rearrangement and reduction of this intermediate to 2-C-methyl-D-erythritol 4-phosphate (MEP), it is eventually converted to IPP (Figure 13.5).

Isopentenyl diphosphate and its isomer combine to form larger terpenes

Isopentenyl diphosphate and its isomer, dimethylallyl diphosphate (DPP), are the activated five-carbon building blocks of terpene biosynthesis that join together to form larger molecules. First IPP and DPP react to give geranyl diphosphate (GPP), the 10-carbon precursor of nearly all the monoterpenes (see Figure 13.5). GPP can then link to another molecule of IPP to give the 15-carbon compound farnesyl diphosphate (FPP), the precursor of nearly all the sesquiterpenes. Addition of yet another molecule of IPP gives the 20-carbon compound geranylgeranyl diphosphate (GGPP), the precursor of the diterpenes. Finally, FPP and GGPP can dimerize to give the triterpenes (C_{30}) and the tetraterpenes (C_{40}), respectively.

Some terpenes have roles in growth and development

Certain terpenes have a well-characterized function in plant growth or development and so can be considered primary rather than secondary metabolites. For example, the *gibberellins*, an important group of plant hormones, are diterpenes. *Brassinosteroids*, another class of plant hormones with growth-regulating functions, originate from triterpenes.

Sterols are triterpene derivatives that are essential components of cell membranes, which they stabilize by interacting with phospholipids (see Chapter 11). The red, orange, and yellow carotenoids are tetraterpenes that function as accessory pigments in photosynthesis and protect photosynthetic tissues from photooxidation (see Chapter 7). The hormone abscisic acid (see Chapter 23) is a C_{15} terpene produced by degradation of a carotenoid precursor.

Long-chain polyterpene alcohols known as *dolichols* function as carriers of sugars in cell wall and glycoprotein synthesis (see Chapter 15). Terpene-derived side chains, such as the phytol side chain of chlorophyll (see Chapter 7),

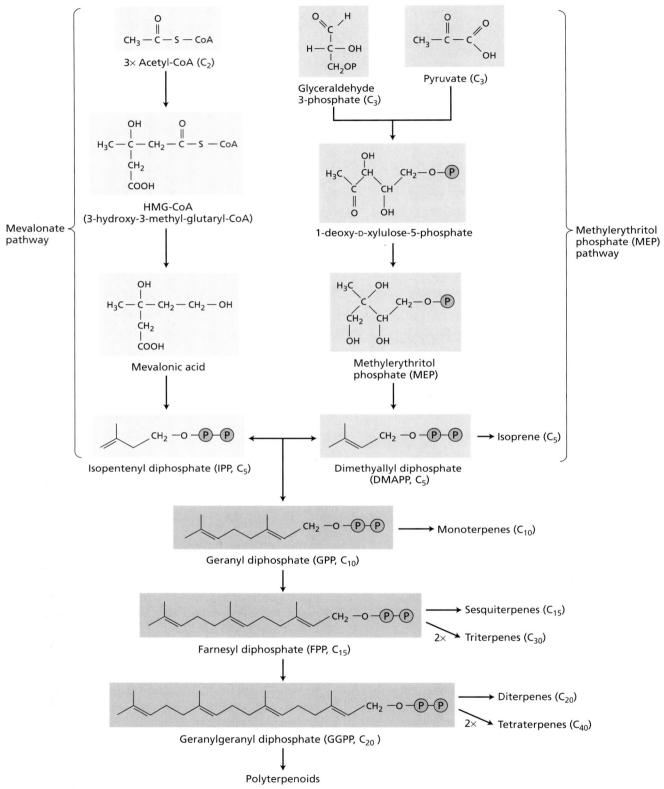

FIGURE 13.5 Outline of terpene biosynthesis. The basic 5-carbon units of terpenes are synthesized by two different pathways. The phosphorylated intermediates, IPP and DMAPP, are combined to make 10-carbon, 15-carbon, and larger terpenes.

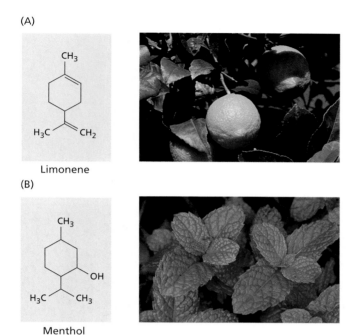

FIGURE 13.6 Structures of limonene (A) and menthol (B). These two well-known monoterpenes serve as defenses against insects and other organisms that feed on these plants. (A, photo © Calvin Larsen/Photo Researchers, Inc.; B, photo © David Sieren/Visuals Unlimited.)

help anchor certain molecules in membranes. Thus various terpenes have important primary roles in plants, the vast majority are secondary metabolites presumed to be involved in defense.

Terpenes defend against herbivores in many plants

Terpenes are toxins and feeding deterrents to many plant-feeding insects and mammals; thus they appear to play important defensive roles in the plant kingdom (Gershenzon and Croteau 1992). For example, the monoterpene esters called **pyrethroids** that occur in the leaves and flowers of *Chrysanthemum* species show striking insecticidal activity. Both natural and synthetic pyrethroids are popular ingredients in commercial insecticides because of their low persistence in the environment and their negligible toxicity to mammals.

In conifers such as pine and fir, monoterpenes accumulate in resin ducts found in the needles, twigs, and trunk. These compounds are toxic to numerous insects, including bark beetles, which are serious pests of conifer species throughout the world. Many conifers respond to bark beetle infestation by producing additional quantities of monoterpenes (Trapp and Croteau 2001).

Many plants contain mixtures of volatile monoterpenes and sesquiterpenes, called **essential oils**, that lend a characteristic odor to their foliage. Peppermint, lemon, basil,

and sage are examples of plants that contain essential oils. The chief monoterpene constituent of peppermint oil is menthol; that of lemon oil is limonene (Figure 13.6).

Essential oils have well-known insect repellent properties. They are frequently found in glandular hairs that project outward from the epidermis and serve to "advertise" the toxicity of the plant, repelling potential herbivores even before they take a trial bite. In the glandular hairs, the terpenes are stored in a modified extracellular space in the cell wall (Figure 13.7). Essential oils can be extracted from plants by steam distillation and are important commercially in flavoring foods and making perfumes (see **Web Essay 13.1**).

Among the nonvolatile terpene antiherbivore compounds are the **limonoids**, a group of triterpenes (C_{30}) well known as bitter substances in citrus fruit. Perhaps the most powerful deterrent to insect feeding known is *azadirachtin* (Figure 13.8A), a complex limonoid from the neem tree (*Azadirachta indica*) of Africa and Asia. Azadirachtin is a feeding deterrent to some insects at doses as low as 50 parts per billion, and it exerts a variety of toxic effects (Aerts and Mordue 1997). It has considerable potential as a commercial insect control agent because of its low toxicity to mammals, and several preparations containing azadirachtin are now being marketed in North America and India.

The **phytoecdysones**, first isolated from the common fern, *Polypodium vulgare*, are a group of plant steroids that have the same basic structure as insect molting hormones (Figure 13.8B). Ingestion of phytoecdysones by insects dis-

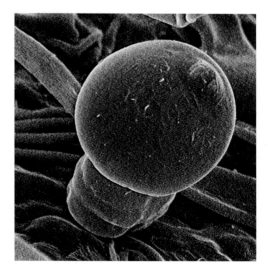

FIGURE 13.7 Monoterpenes and sesquiterpenes are commonly found in glandular hairs on the plant surface. This scanning electron micrograph shows a glandular hair on a young leaf of spring sunflower (*Balsamorhiza sagittata*). Terpenes are thought to be synthesized in the cells of the hair and are stored in the rounded cap at the top. This "cap" is an extracellular space that forms when the cuticle and a portion of the cell wall pull away from the remainder of the cell. (1105×) (© J. N. A. Lott/Biological Photo Service.)

FIGURE 13.8 Structure of two triterpenes, azadirachtin (A), and α-ecdysone (B), which serve as powerful feeding deterrents to insects. (A, photo © Inga Spence/Visuals Unlimited; B, photo © Wally Eberhart/Visuals Unlimited.)

(A) Azadirachtin, a limonoid

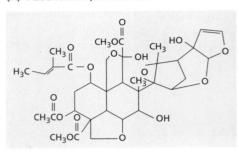

(B) α-Ecdysone, an insect molting hormone

rupts molting and other developmental processes, often with lethal consequences. In addition, phytoecdysones were recently found to have a defensive function against plant-parasitic nematodes (Soriano et al. 2004).

Triterpenes that are active against vertebrate herbivores include cardenolides and saponins. **Cardenolides** are glycosides (compounds containing an attached sugar or sugars) that taste bitter and are extremely toxic to higher animals. In humans, they have dramatic effects on the heart muscle through their influence on Na^+/K^+-activated ATPases. In carefully regulated doses, they slow and strengthen the heartbeat. Cardenolides extracted from foxglove (*Digitalis*) are prescribed to millions of patients for the treatment of some types of heart disease.

Saponins are steroid and triterpene glycosides, so named because of their soaplike properties. The presence of both lipid-soluble (the steroid or triterpene) and water-soluble (the sugar) elements in one molecule gives saponins detergent properties, and they form a soapy lather when shaken with water. The toxicity of saponins is thought to be a result of their ability to form complexes with sterols. Saponins may interfere with sterol uptake from the digestive system or disrupt cell membranes after being absorbed into the bloodstream. (See **Web Topic 13.1** for more information about structures of triterpenes.)

Phenolic Compounds

Plants produce a large variety of secondary products that contain a phenol group—a hydroxyl functional group on an aromatic ring:

These substances are classified as phenolic compounds. Plant **phenolics** are a chemically heterogeneous group of nearly 10,000 individual compounds: Some are soluble only in organic solvents, some are water-soluble carboxylic acids and glycosides, and others are large, insoluble polymers.

In keeping with their chemical diversity, phenolics play a variety of roles in the plant. After giving a brief account of phenolic biosynthesis, we will discuss several principal groups of phenolic compounds and what is known about their roles in the plant. Many serve as defense compounds against herbivores and pathogens. Others function in mechanical support, in attracting pollinators and fruit dispersers, in absorbing harmful ultraviolet radiation, or in reducing the growth of nearby competing plants.

Phenylalanine is an intermediate in the biosynthesis of most plant phenolics

Plant phenolics are biosynthesized by several different routes and thus constitute a heterogeneous group from a metabolic point of view. Two basic pathways are involved: the shikimic acid pathway and the malonic acid pathway (Figure 13.9). The shikimic acid pathway participates in the biosynthesis of most plant phenolics. The malonic acid pathway, although an important source of phenolic secondary products in fungi and bacteria, is of less significance in higher plants.

The **shikimic acid pathway** converts simple carbohydrate precursors derived from glycolysis and the pentose

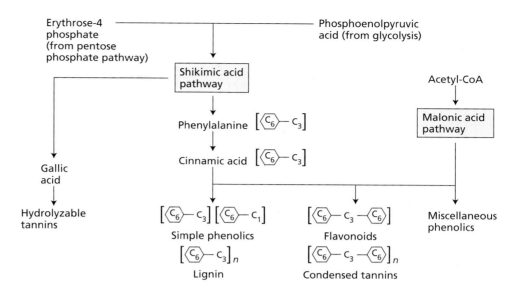

FIGURE 13.9 Plant phenolics are biosynthesized in several different ways. In higher plants, most phenolics are derived at least in part from phenylalanine, a product of the shikimic acid pathway. Formulas in brackets indicate the basic arrangement of carbon skeletons:

indicates a benzene ring, and C_3 is a three-carbon chain. More detail on the pathway from phenylalanine onward is given in Figure 13.10.

phosphate pathway to the aromatic amino acids (see **Web Topic 13.2**) (Herrmann and Weaver 1999). One of the pathway intermediates is shikimic acid, which has given its name to this whole sequence of reactions. The well-known, broad-spectrum herbicide glyphosate (available commercially as Roundup) kills plants by blocking a step in this pathway (see Chapter 2 on the web site). The shikimic acid pathway is present in plants, fungi, and bacteria but is not found in animals. Animals have no way to synthesize the three aromatic amino acids—phenylalanine, tyrosine, and tryptophan—which are therefore essential nutrients in animal diets.

The most abundant classes of secondary phenolic compounds in plants are derived from phenylalanine via the elimination of an ammonia molecule to form cinnamic acid (Figure 13.10). This reaction is catalyzed by **phenylalanine ammonia lyase** (**PAL**), perhaps the most studied enzyme in plant secondary metabolism. PAL is situated at a branch point between primary and secondary metabolism, so the reaction that it catalyzes is an important regulatory step in the formation of many phenolic compounds.

The activity of PAL is increased by environmental factors, such as low nutrient levels, light (through its effect on phytochrome), and fungal infection. The point of control appears to be the initiation of transcription. Fungal invasion, for example, triggers the transcription of messenger RNA that codes for PAL, thus increasing the amount of PAL in the plant, which then stimulates the synthesis of phenolic compounds.

The regulation of PAL activity in many plant species is made more complex by the existence of multiple PAL-encoding genes, some of which are expressed only in specific tissues or only under certain environmental conditions (Logemann et al. 1995).

Reactions subsequent to that catalyzed by PAL lead to the addition of more hydroxyl groups and other substituents. *Trans*-cinnamic acid, *p*-coumaric acid, and their derivatives are simple phenolic compounds called **phenylpropanoids** because they contain a benzene ring:

and a three-carbon side chain. Phenylpropanoids are important building blocks of the more complex phenolic compounds discussed later in this chapter.

Now that the biosynthetic pathways leading to most widespread phenolic compounds have been determined, researchers have turned their attention to studying how these pathways are regulated. In some cases, specific enzymes, such as PAL, are important in controlling flux through the pathway. Several transcription factors have been shown to regulate phenolic metabolism by binding to the promoter regions of certain biosynthetic genes and activating transcription. Some of these factors activate the transcription of large groups of genes (Jin and Martin 1999).

Some simple phenolics are activated by ultraviolet light

Simple phenolic compounds are widespread in vascular plants and appear to function in different capacities. Their structures include the following:

- Simple phenylpropanoids, such as *trans*-cinnamic acid, *p*-coumaric acid, and their derivatives, such as caffeic acid, which have a basic phenylpropanoid carbon skeleton (Figure 13.11A):

- Phenylpropanoid lactones (cyclic esters) called *coumarins*, also with a phenylpropanoid skeleton (see Figure 13.11B)

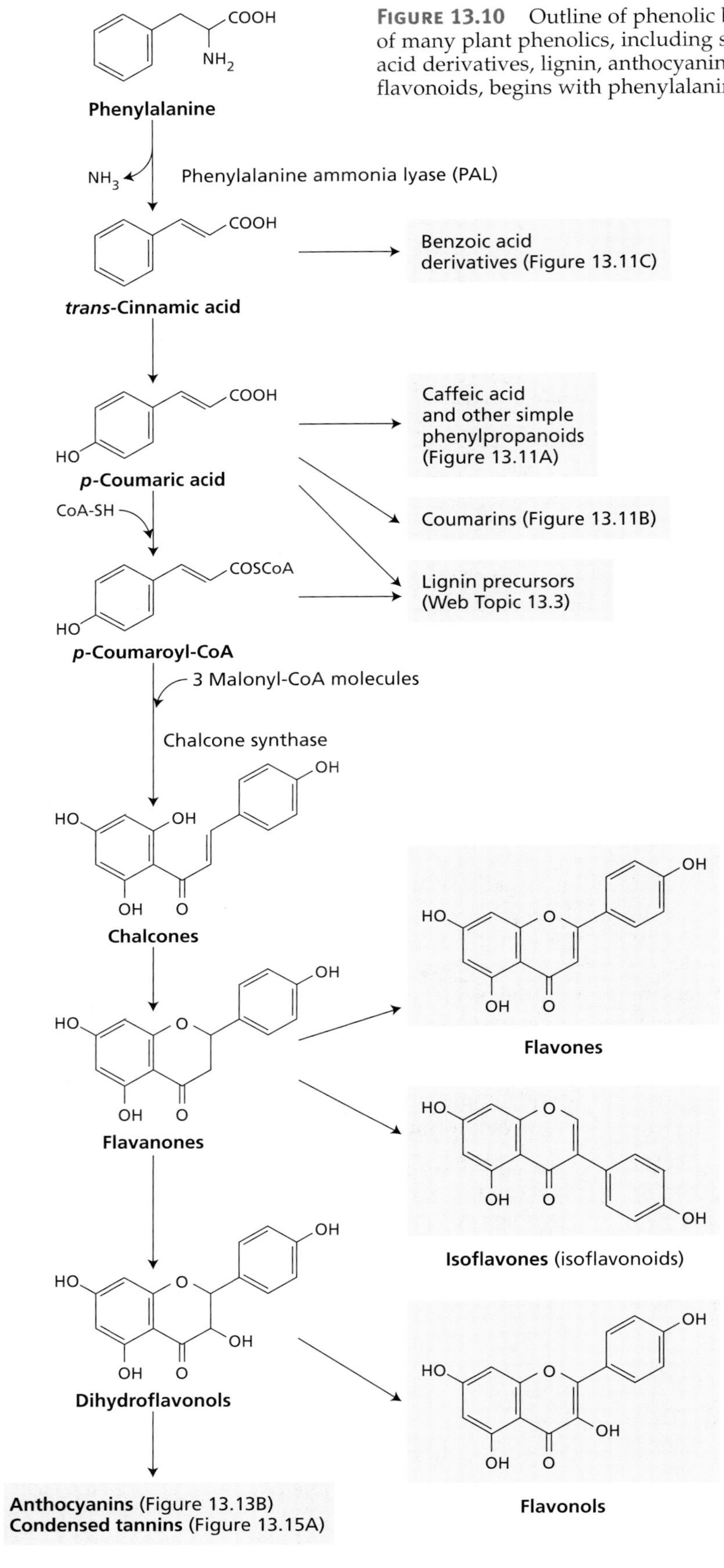

FIGURE 13.10 Outline of phenolic biosynthesis from phenylalanine. The formation of many plant phenolics, including simple phenylpropanoids, coumarins, benzoic acid derivatives, lignin, anthocyanins, isoflavones, condensed tannins, and other flavonoids, begins with phenylalanine.

- Benzoic acid derivatives, which have a carbon skeleton formed from phenylpropanoids by the cleavage of a two-carbon fragment from the side chain (see Figure 13.11C) (see also Figure 13.10):

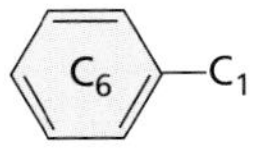

C_6—C_1

As with many other secondary products, plants can elaborate on the basic carbon skeleton of simple phenolic compounds to make more complex products.

Many simple phenolic compounds have important roles in plants as defenses against insect herbivores and fungi. Of special interest is the phototoxicity of certain coumarins called **furanocoumarins**, which have an attached furan ring (see Figure 13.11B).

These compounds are not toxic until they are activated by light. Sunlight in the ultraviolet A (UV-A) region (320–400 nm) causes some furanocoumarins to become activated to a high-energy electron state. Activated furanocoumarins can insert themselves into the double helix of DNA and bind to the pyrimidine bases cytosine and thymine, thus blocking transcription and repair and leading eventually to cell death.

Phototoxic furanocoumarins are especially abundant in members of the Umbelliferae family, including celery, parsnip, and parsley. In celery, the level of these compounds can increase about 100-fold if the plant is stressed or diseased. Celery pickers, and even some grocery shoppers, have been known to develop skin rashes from handling stressed or diseased celery. Some insects have adapted to survive on plants that contain furanocoumarins and other phototoxic compounds by living in silken webs or rolled-up leaves, which screen out the activating wavelengths (Sandberg and Berenbaum 1989).

The release of phenolics into the soil may limit the growth of other plants

From leaves, roots, and decaying litter, plants release a variety of primary and secondary metabolites into the environment. Investigation of the effects of these compounds on neighboring plants is the study of **allelopathy**.

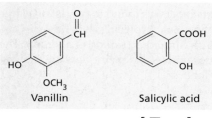

FIGURE 13.11 Simple phenolic compounds play a great diversity of roles in plants. (A) Caffeic acid and ferulic acid may be released into the soil and inhibit the growth of neighboring plants. (B) Psoralen is a furanocoumarin that exhibits phototoxicity to insect herbivores. (C) Salicylic acid is a plant growth regulator that is involved in systemic resistance to plant pathogens.

If a plant can reduce the growth of nearby plants by releasing chemicals into the soil, it may increase its access to light, water, and nutrients and thus its evolutionary fitness. Generally speaking, the term *allelopathy* has come to be applied to the harmful effects of plants on their neighbors, although a precise definition also includes beneficial effects.

Simple phenylpropanoids and benzoic acid derivatives are frequently cited as having allelopathic activity. Compounds such as caffeic acid and ferulic acid (see Figure 13.11A) occur in soil in appreciable amounts and have been shown in laboratory experiments to inhibit the germination and growth of many plants (Inderjit et al. 1995).

Allelopathy is currently of great interest because of its potential agricultural applications (Kruse et al. 2000). Reductions in crop yields caused by weeds or residues from the previous crop may in some cases be a result of allelopathy. An exciting future prospect is the development of crop plants genetically engineered to be allelopathic to weeds. (See **Web Essay 13.2**.)

Lignin is a highly complex phenolic macromolecule

After cellulose, the most abundant organic substance in plants is **lignin**, a highly branched polymer of phenylpropanoid groups:

that plays both primary and secondary roles. The precise structure of lignin is not known, because it is difficult to extract lignin from plants, where it is covalently bound to cellulose and other polysaccharides of the cell wall.

Lignin is generally formed from three different phenylpropanoid alcohols: coniferyl, coumaryl, and sinapyl, alcohols that are synthesized from phenylalanine via various cinnamic acid derivatives. The phenylpropanoid alcohols are joined into a polymer through the action of enzymes that generate free-radical intermediates. The proportions of the three monomeric units in lignin vary among species, plant organs, and even layers of a single cell wall. In the polymer, there are often multiple C—C and C—O—C bonds in each phenylpropanoid alcohol unit, resulting in a complex structure that branches in three dimensions. Unlike polymers such as starch, rubber, or cellulose, the units of lignin do not appear to be linked in a simple, repeating way. However, recent research suggests that a guiding protein may bind the individual phenylpropanoid units during lignin biosynthesis, giving rise to a scaffold that then directs the formation of a large, repeating unit (Davin and Lewis 2000; Hatfield and Vermerris 2001). (See **Web Topic 13.3** for the partial structure of a hypothetical lignin molecule.)

Lignin is found in the cell walls of various types of supporting and conducting tissue, notably the tracheids and vessel elements of the xylem. It is deposited chiefly in the thickened secondary wall but can also occur in the primary wall and middle lamella in close contact with the celluloses and hemicelluloses already present. The mechanical rigidity of lignin strengthens stems and vascular tissue, allowing upward growth and permitting water and minerals to be conducted through the xylem under negative pressure without collapse of the tissue. Because lignin is such a key component of water transport tissue, the ability to make lignin must have been one of the most important adaptations permitting primitive plants to colonize dry land.

Besides providing mechanical support, lignin has significant protective functions in plants. Its physical toughness deters feeding by animals, and its chemical durability makes it relatively indigestible to herbivores. By bonding to cellulose and protein, lignin also reduces the digestibility of these substances. Lignification blocks the growth of pathogens and is a common response to infection or wounding.

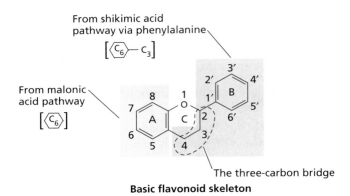

FIGURE 13.12 Basic flavonoid carbon skeleton. Flavonoids are biosynthesized from products of the shikimic acid and malonic acid pathways. Positions on the flavonoid ring system are numbered as shown.

There are four major groups of flavonoids

The **flavonoids** are one of the largest classes of plant phenolics. The basic carbon skeleton of a flavonoid contains 15 carbons arranged in two aromatic rings connected by a three-carbon bridge:

$$C_6 - C_3 - C_6$$

This structure results from two separate biosynthetic pathways: the shikimic acid pathway and the malonic acid pathway (Figure 13.12).

Flavonoids are classified into different groups, primarily on the basis of the degree of oxidation of the three-carbon bridge. We will discuss four of the groups shown in Figure 13.10: the anthocyanins, the flavones, the flavonols, and the isoflavones.

The basic flavonoid carbon skeleton may have numerous substituents. Hydroxyl groups are usually present at positions 4, 5, and 7, but they may also be found at other positions. Sugars are very common as well; in fact, the majority of flavonoids exist naturally as glycosides.

Whereas both hydroxyl groups and sugars increase the water solubility of flavonoids, other substituents, such as methyl ethers or modified isopentyl units, make flavonoids lipophilic (hydrophobic). Different types of flavonoids perform very different functions in the plant, including pigmentation and defense.

Anthocyanins are colored flavonoids that attract animals

In addition to predator–prey interactions, there are mutualistic associations among plants and animals. In return for the reward of ingesting nectar or fruit pulp, animals perform extremely important services for plants as carriers of pollen and seeds. Secondary metabolites are involved in these plant–animal interactions, helping to attract animals to flowers and fruit by providing visual and olfactory signals.

The colored pigments of plants are of two principal types: carotenoids and flavonoids. *Carotenoids*, as we have already seen, are yellow, orange, and red terpenoid compounds that also serve as accessory pigments in photosynthesis (see Chapter 7). *Flavonoids* are phenolic compounds that include a wide range of colored substances.

The most widespread group of pigmented flavonoids is the **anthocyanins**, which are responsible for most of the red, pink, purple, and blue colors observed in plant parts. By coloring flowers and fruits, the anthocyanins are vitally important in attracting animals for pollination and seed dispersal.

Anthocyanins are glycosides that have sugars at position 3 (Figure 13.13A) and sometimes elsewhere. Without their sugars, anthocyanins are known as **anthocyanidins** (Figure 13.13B). Anthocyanin color is influenced by many factors, including the number of hydroxyl and methoxyl groups in ring B of the anthocyanidin (see Figure 13.13A), the presence of aromatic acids esterified to the main skeleton, and the pH of the cell vacuole in which these compounds are stored. Anthocyanins may also exist in supramolecular complexes along with chelated metal ions and flavone copigments. The blue pigment of dayflower (*Commelina communis*) was found to consist of a large complex of six anthocyanin molecules, six flavones, and two associated magnesium ions (Kondo et al. 1992). The most common anthocyanidins and their colors are shown in Figure 13.13 and Table 13.1.

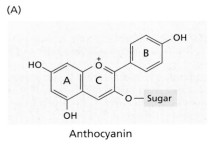

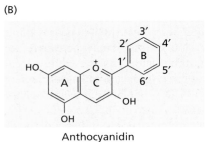

FIGURE 13.13 The structures of anthocyanins (A) and anthocyanidin (B). The colors of anthocyanidins depend in part on the substituents attached to ring B (see Table 13.1). An increase in the number of hydroxyl groups shifts absorption to a longer wavelength and gives a bluer color. Replacement of a hydroxyl group with a methoxyl group (—OCH$_3$) shifts absorption to a slightly shorter wavelength, resulting in a redder color.

TABLE 13.1
Effects of ring substituents on anthocyanidin color

Anthocyanidin	Substituents	Color
Pelargonidin	4'— OH	Orange red
Cyanidin	3'— OH, 4'— OH	Purplish red
Delphinidin	3'— OH, 4'— OH, 5'— OH	Bluish purple
Peonidin	3'— OCH$_3$, 4'— OH	Rosy red
Petunidin	3'— OCH$_3$, 4'— OH, 5'— OCH$_3$	Purple

Considering the variety of factors affecting anthocyanin coloration and the possible presence of carotenoids as well, it is not surprising that so many different shades of flower and fruit color are found in nature. The evolution of flower color may have been governed by selection pressures for different sorts of pollinators, which often have different color preferences.

Color, of course, is just one type of signal used to attract pollinators to flowers. Volatile chemicals, particularly monoterpenes, frequently provide attractive scents.

Flavonoids may protect against damage by ultraviolet light

Two other major groups of flavonoids found in flowers are **flavones** and **flavonols** (see Figure 13.10). These flavonoids generally absorb light at shorter wavelengths than do anthocyanins, so they are not visible to the human eye. However, insects such as bees, which see farther into the ultraviolet range of the spectrum than humans do, may respond to flavones and flavonols as attractant cues (Figure 13.14). Flavonols in a flower often form symmetric patterns of stripes, spots, or concentric circles called *nectar guides* (Lunau 1992). These patterns may be conspicuous to insects and are thought to help indicate the location of pollen and nectar.

Flavones and flavonols are not restricted to flowers; they are also present in the leaves of all green plants. These two classes of flavonoids function to protect cells from excessive UV-B radiation (280–320 nm), because they accumulate in the epidermal layers of leaves and stems and absorb light strongly in the UV-B region while allowing the visible (photosynthetically active) wavelengths to pass through uninterrupted. In addition, exposure of plants to increased UV-B light has been demonstrated to increase the synthesis of flavones and flavonols.

Arabidopsis thaliana mutants that lack the enzyme chalcone synthase produce no flavonoids. Lacking flavonoids, these plants are much more sensitive to UV-B radiation than wild-type individuals are, and they grow very poorly under normal conditions. When shielded from UV light, however, they grow normally (Li et al. 1993). A group of simple phenylpropanoid esters are also important in UV protection in Arabidopsis.

Other functions of flavonoids have recently been discovered. For example, flavones and flavonols secreted into the

(A)

(B)

FIGURE 13.14 Black-eyed Susan (*Rudbeckia* sp.) as seen by humans (A) and as it might appear to honeybees (B). (A) To humans, the flowers have yellow rays and a brown central disc. (B) To bees, the tips of the rays appear "light yellow," the inner portion of the rays "dark yellow," and the central disc "black." Ultraviolet-absorbing flavonols are found in the inner parts of the rays but not in the tips. The distribution of flavonols in the rays and the sensitivity of insects to part of the UV spectrum contribute to the "bull's-eye" pattern seen by honeybees, which presumably helps them locate pollen and nectar. Special lighting was used to simulate the spectral sensitivity of the honeybee visual system. (Courtesy of Thomas Eisner.)

soil by legume roots mediate the interaction of legumes and nitrogen-fixing symbionts, a phenomenon described in Chapter 12. As will be discussed in Chapter 19, recent work suggests that flavonoids also play a regulatory role in plant development as modulators of polar auxin transport.

Isoflavonoids have antimicrobial activity

The **isoflavonoids** (isoflavones) are a group of flavonoids in which the position of one aromatic ring (ring B) is shifted (see Figure 13.10). Isoflavonoids are found mostly in legumes and have several different biological activities. Some, such as the rotenoids, have strong insecticidal actions; others have anti-estrogenic effects. For example, sheep grazing on clover rich in isoflavonoids often suffer from infertility. The isoflavonoid ring system has a three-dimensional structure similar to that of steroids (see Figure 13.8B), allowing these substances to bind to estrogen receptors. Isoflavonoids may also be responsible for the anticancer benefits of food prepared from soybeans.

In the past few years, isoflavonoids have become best known for their role as *phytoalexins*, antimicrobial compounds synthesized in response to bacterial or fungal infection that help limit the spread of the invading pathogen. Phytoalexins are discussed in more detail later in this chapter (see **Web Essay 13.3**).

Tannins deter feeding by herbivores

A second category of plant phenolic polymers with defensive properties, besides lignins, is the **tannins**. The term *tannin* was first used to describe compounds that could convert raw animal hides into leather in the process known as tanning. Tannins bind the collagen proteins of animal hides, increasing their resistance to heat, water, and microbes.

There are two categories of tannins: condensed and hydrolyzable. **Condensed tannins** are compounds formed by the polymerization of flavonoid units (Figure 13.15A). They are common constituents of woody plants. Because condensed tannins can often be hydrolyzed to anthocyanidins by treatment with strong acids, they are sometimes called *pro-anthocyanidins*.

Hydrolyzable tannins are heterogeneous polymers containing phenolic acids, especially gallic acid, and simple sugars (Figure 13.15B). They are smaller than condensed tannins and may be hydrolyzed more easily; only dilute acid is needed. Most tannins have molecular masses between 600 and 3000 Da.

Tannins are general toxins that significantly reduce the growth and survivorship of many herbivores when added to their diets. In addition, tannins act as feeding repellents to a great diversity of animals. Mammals such as cattle, deer, and apes characteristically avoid plants or parts of plants with high tannin contents. Unripe fruits, for instance, frequently have very high tannin levels, which deter feeding on the fruits until they are mature enough for dispersal.

Although crop plants generally produce fewer secondary metabolites, there are exceptions. Humans often prefer a certain level of astringency in tannin-containing foods, such as apples, blackberries, tea, and grapes. Recently, polyphenols (tannins) in red wine were shown to block the formation of endothelin-1, a signaling molecule that makes blood vessels constrict (Corder et al. 2001). This effect of wine tannins may account for the often-touted health benefits of red wine, especially the reduction in the risk of heart disease associated with moderate red wine consumption.

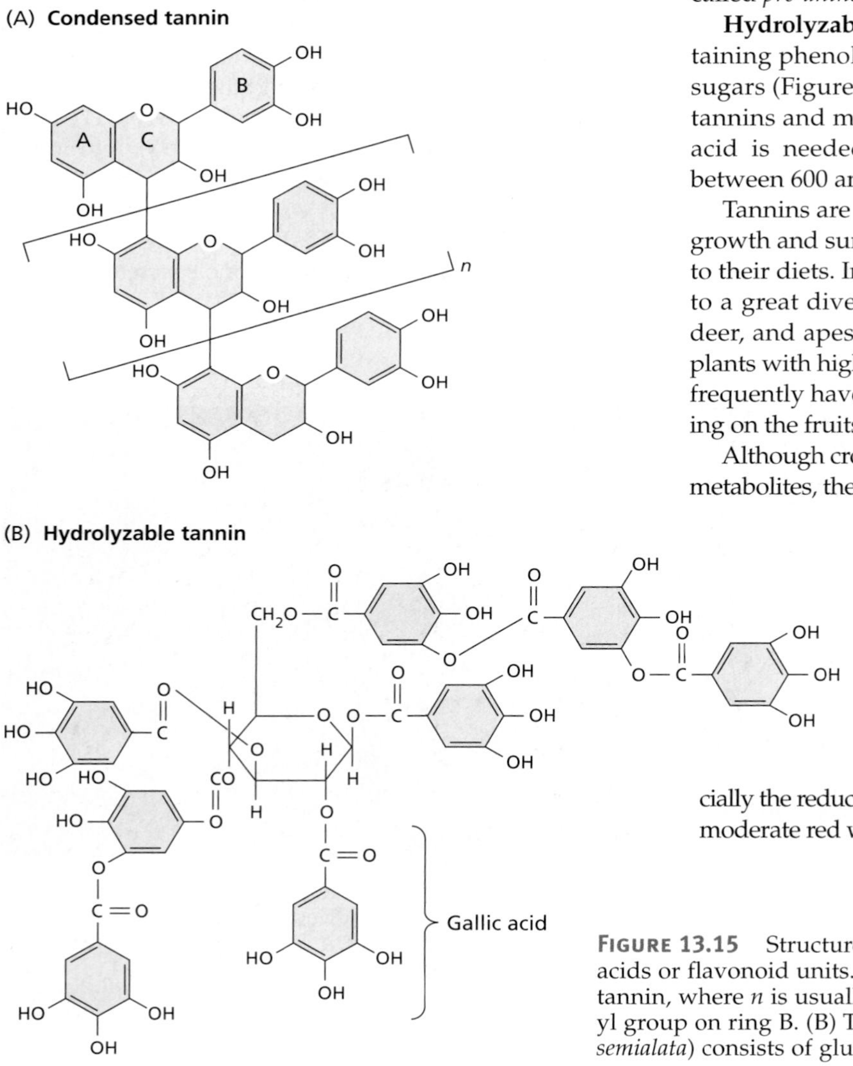

FIGURE 13.15 Structure of some tannins formed from phenolic acids or flavonoid units. (A) The general structure of a condensed tannin, where n is usually 1 to 10. There may also be a third hydroxyl group on ring B. (B) The hydrolyzable tannin from sumac (*Rhus semialata*) consists of glucose and eight molecules of gallic acid.

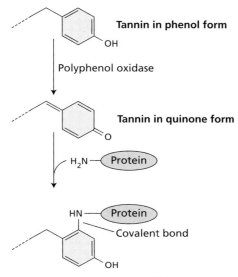

FIGURE 13.16 Proposed mechanisms for the interaction of tannins with proteins. (A) Hydrogen bonds may form between the phenolic hydroxyl groups of tannins and electronegative sites on the protein. (B) Phenolic hydroxyl groups may bind covalently to proteins following activation by oxidative enzymes, such as polyphenol oxidase.

Moderate amounts of specific polyphenolics may have health benefits for humans, but the defensive properties of most tannins are due to their toxicity, which is generally attributed to their ability to bind proteins nonspecifically. It has long been thought that plant tannins complex proteins in the guts of herbivores by forming hydrogen bonds between their hydroxyl groups and electronegative sites on the protein (Figure 13.16A).

More recent evidence indicates that tannins and other phenolics can also bind to dietary protein in a covalent fashion (Figure 13.16B). The foliage of many plants contains enzymes that oxidize phenolics to their corresponding quinone forms in the guts of herbivores (Felton et al. 1989). Quinones are highly reactive electrophilic molecules that readily react with the nucleophilic —NH_2 and —SH groups of proteins (see Figure 13.16B) (Appel 1993). By whatever mechanism protein–tannin binding occurs, this process has a negative impact on herbivore nutrition. Tannins can inactivate herbivore digestive enzymes and create complex aggregates of tannins and plant proteins that are difficult to digest.

Herbivores that habitually feed on tannin-rich plant material appear to possess some interesting adaptations to remove tannins from their digestive systems. For example, some mammals, such as rodents and rabbits, produce salivary proteins with a very high proline content (25–45%) that have a high affinity for tannins. Secretion of these proteins is induced by ingestion of food with a high tannin content and greatly diminishes the toxic effects of tannins (Butler 1989). The large number of proline residues gives these proteins a very flexible, open conformation and a high degree of hydrophobicity, which facilitate binding to tannins.

Plant tannins also serve as defenses against microorganisms. For example, the nonliving heartwood of many trees contains high concentrations of tannins that help prevent fungal and bacterial decay.

Nitrogen-Containing Compounds

A large variety of plant secondary metabolites have nitrogen in their structure. Included in this category are such well-known antiherbivore defenses as alkaloids and cyanogenic glycosides, which are of considerable interest because of their toxicity to humans and their medicinal properties. Most nitrogenous secondary metabolites are biosynthesized from common amino acids.

In this section we will examine the structure and biological properties of various nitrogen-containing secondary metabolites, including alkaloids, cyanogenic glycosides, glucosinolates, and nonprotein amino acids.

Alkaloids have dramatic physiological effects on animals

The **alkaloids** are a large family of more than 15,000 nitrogen-containing secondary metabolites found in approximately 20% of the species of vascular plants. The nitrogen atom in these substances is usually part of a **heterocyclic ring**, a ring that contains both nitrogen and carbon atoms. As a group, alkaloids are best known for their striking pharmacological effects on vertebrate animals.

As their name would suggest, most alkaloids are alkaline. At pH values commonly found in the cytosol (pH 7.2) or the vacuole (pH 5–6), the nitrogen atom is protonated; hence, alkaloids are positively charged and are generally water soluble.

Alkaloids are usually synthesized from one of a few common amino acids—in particular, lysine, tyrosine, and tryptophan. However, the carbon skeleton of some alkaloids contains a component derived from the terpene pathway. Table 13.2 lists the major alkaloid types and their amino acid precursors. Several different types, including nicotine and its relatives (Figure 13.17), are derived from ornithine, an intermediate in arginine biosynthesis. The B vitamin nicotinic acid (niacin) is a precursor of the pyridine (six-membered) ring of this alkaloid; the pyrrolidine (five-membered) ring of nicotine arises from ornithine (Figure

TABLE 13.2
Major types of alkaloids, their amino acid precursors, and well-known examples of each type

Alkaloid class	Structure	Biosynthetic precursor	Examples	Human uses
Pyrrolidine	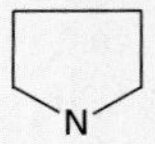 (pyrrolidine ring)	Ornithine (aspartate)	Nicotine	Stimulant, depressant, tranquilizer
Tropane	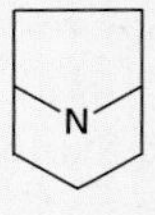 (tropane ring)	Ornithine	Atropine	Prevention of intestinal spasms, antidote to other poisons, dilation of pupils for examination
			Cocaine	Stimulant of the central nervous system, local anesthetic
Piperidine	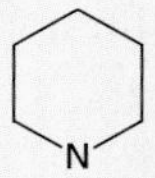 (piperidine ring)	Lysine (or acetate)	Coniine	Poison (paralyzes motor neurons)
Pyrrolizidine	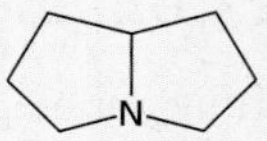 (pyrrolizidine ring)	Ornithine	Retrorsine	None
Quinolizidine	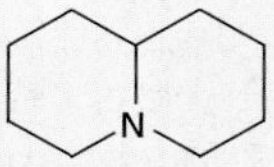 (quinolizidine ring)	Lysine	Lupinine	Restoration of heart rhythm
Isoquinoline	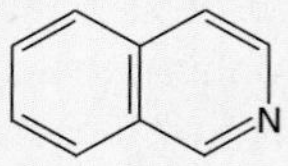 (isoquinoline ring)	Tyrosine	Codeine	Analgesic (pain relief), treatment of coughs
			Morphine	Analgesic
Indole	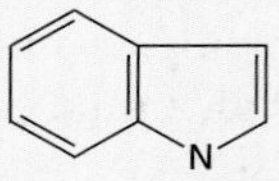 (indole ring)	Tryptophan	Psilocybin	Hallucinogen
			Reserpine	Treatment of hypertension, treatment of psychoses
			Strychnine	Rat poison, treatment of eye disorders

13.18). Nicotinic acid is also a constituent of NAD^+ and $NADP^+$, which serve as electron carriers in metabolism.

The role of alkaloids in plants has been a subject of speculation for at least 100 years. Alkaloids were once thought to be nitrogenous wastes (analogous to urea and uric acid in animals), nitrogen storage compounds, or growth regulators, but there is little evidence to support any of these functions. Most alkaloids are now believed to function as defenses against predators, especially mammals, because of their general toxicity and deterrence capability (Hartmann 1992).

Large numbers of livestock deaths are caused by the ingestion of alkaloid-containing plants. In the United States, a significant percentage of all grazing livestock animals are poisoned each year by consumption of large quantities of alkaloid-containing plants such as lupines (*Lupinus*), larkspur (*Delphinium*), and groundsel (*Senecio*). This phenomenon may be due to the fact that domestic animals, unlike wild animals, have not been subjected to natural selection for the avoidance of toxic plants. Indeed,

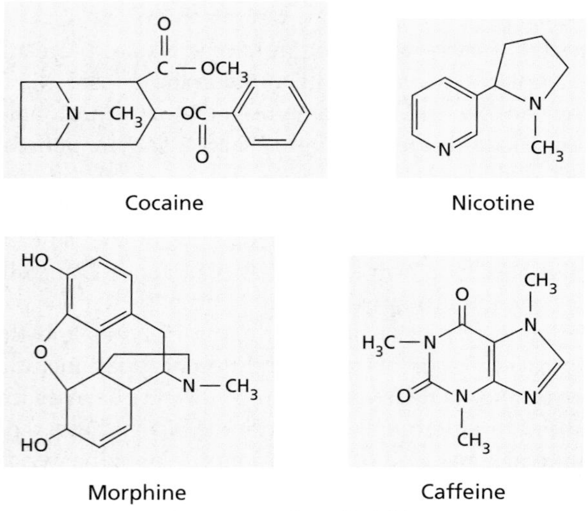

Representative alkaloids
Cocaine
Nicotine
Morphine
Caffeine

FIGURE 13.17 Examples of alkaloids, a diverse group of secondary metabolites that contain nitrogen, usually as part of a heterocyclic ring. Caffeine is a purine-type alkaloid similar to the nucleic acid bases adenine and guanine. The pyrrolidine (five-membered) ring of nicotine arises from ornithine; the pyridine (six-membered) ring is derived from nicotinic acid.

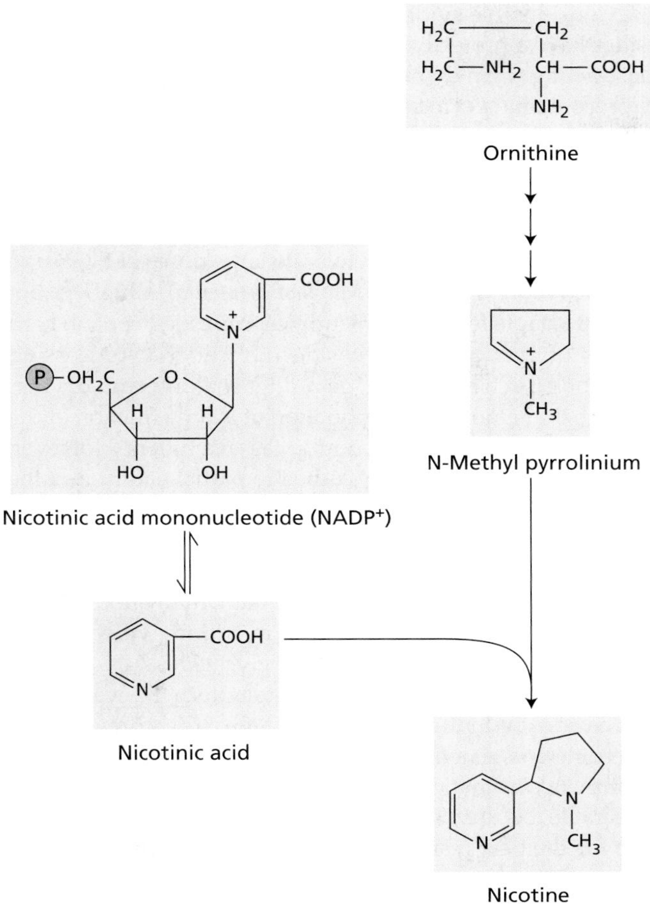

some livestock actually seem to prefer alkaloid-containing plants to less-harmful forage.

Nearly all alkaloids are also toxic to humans when taken in sufficient quantity. For example, strychnine, atropine, and coniine (from poison hemlock, *Conium maculatum*) are classic alkaloid poisoning agents. At lower doses, however, many are useful pharmacologically. Morphine, codeine, and scopolamine are just a few of the plant alkaloids currently used in medicine. Other alkaloids, including cocaine, nicotine, and caffeine (see Figure 13.17), have widespread nonmedical uses as stimulants or sedatives.

On a cellular level, the mode of action of alkaloids in animals is quite variable. Many alkaloids interfere with components of the nervous system, especially the chemical transmitters; others affect membrane transport, protein synthesis, or miscellaneous enzyme activities.

One group of alkaloids, the pyrrolizidine alkaloids, illustrates how herbivores can become adapted to tolerate plant defensive substances and even use them in their own defense (Hartmann 1999). Within plants, pyrrolizidine alkaloids occur naturally as nontoxic N-oxides. In the alkaline digestive tracts of some insect herbivores, however, they are quickly reduced to uncharged, hydrophobic tertiary alkaloids (Figure 13.19), which easily pass through membranes and are toxic. Nevertheless, some herbivores, such as cinnabar moth (*Tyria jacobeae*), have developed the ability to reconvert tertiary pyrrolizidine alkaloids to the nontoxic N-oxide form immediately after its absorption from the digestive tract. These herbivores may then store the N-oxides in their bodies as defenses against their own predators.

Not all of the alkaloids that appear in plants are produced by the plant itself. Many grasses harbor endogenous fungal symbionts that grow in the apoplast and synthesize a variety of different types of alkaloids. Grasses with fungal symbionts often grow faster and are better defended against insect and mammalian herbivores than those with-

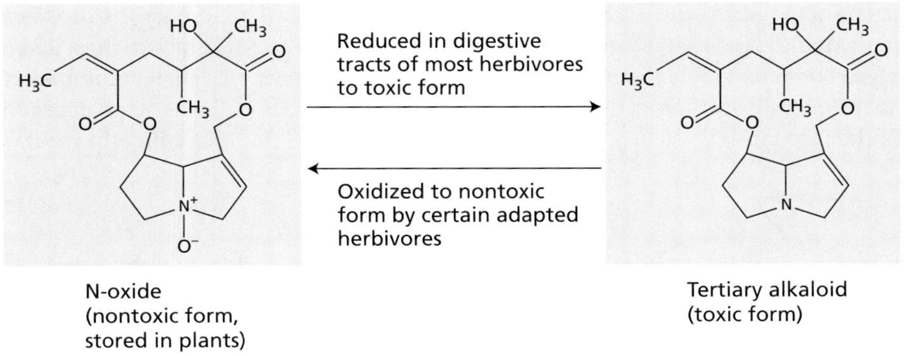

FIGURE 13.18 Nicotine biosynthesis begins with the biosynthesis of the nicotinic acid (niacin) from aspartate and glyceraldehyde-3-phosphate. Nicotinic acid is also a component of NAD^+ and $NADP^+$, important participants in biological oxidation–reduction reactions. The five-membered ring of nicotine is derived from ornithine, an intermediate in arginine biosynthesis.

FIGURE 13.19 Two forms of pyrrolizidine alkaloids occur in nature: the N-oxide form and the tertiary alkaloid. The nontoxic N-oxide found in plants is reduced to the toxic tertiary form in the digestive tracts of most herbivores. However, some adapted herbivores can convert the toxic tertiary alkaloid back to the nontoxic N-oxide. These forms are illustrated here for the alkaloid senecionine, found in species of ragwort (*Senecio*).

out symbionts. Unfortunately, certain grasses with symbionts, such as tall fescue, are important pasture grasses that may become toxic to livestock when their alkaloid content is too high. Efforts are under way to breed tall fescue with alkaloid levels that are not poisonous to livestock but still provide protection against insects (see **Web Essay 13.4**).

Cyanogenic glycosides release the poison hydrogen cyanide

Various nitrogenous protective compounds other than alkaloids are found in plants. Two groups of these substances—cyanogenic glycosides and glucosinolates—are not in themselves toxic but are readily broken down to give off poisons, some of which are volatile, when the plant is crushed. Cyanogenic glycosides release the well-known poisonous gas hydrogen cyanide (HCN).

The breakdown of cyanogenic glycosides in plants is a two-step enzymatic process. Species that make cyanogenic glycosides also make the enzymes necessary to hydrolyze the sugar and liberate HCN:

1. In the first step the sugar is cleaved by a glycosidase, an enzyme that separates sugars from other molecules to which they are linked (Figure 13.20).

2. In the second step the resulting hydrolysis product, called an α-hydroxynitrile or cyanohydrin, can decompose spontaneously at a low rate to liberate HCN. This second step can be accelerated by the enzyme hydroxynitrile lyase.

Cyanogenic glycosides are not normally broken down in the intact plant, because the glycoside and the degradative enzymes are spatially separated in different cellular compartments or in different tissues. In sorghum, for example, the cyanogenic glycoside dhurrin is present in the vacuoles of epidermal cells, while the hydrolytic and lytic enzymes are found in the mesophyll (Poulton 1990).

Under ordinary conditions this compartmentation prevents decomposition of the glycoside. When the leaf is damaged, however, as during herbivore feeding, the cell contents of different tissues mix and HCN forms. Cyanogenic glycosides are widely distributed in the plant kingdom and are frequently encountered in legumes, grasses, and species of the rose family.

Considerable evidence indicates that cyanogenic glycosides have a protective function in certain plants. HCN is a fast-acting toxin that inhibits metalloproteins, such as the iron-containing cytochrome oxidase, a key enzyme of mitochondrial respiration. The presence of cyanogenic glycosides deters feeding by insects and other herbivores, such as snails and slugs. As with other classes of secondary metabolites, however, some herbivores have adapted to feed on cyanogenic plants and can tolerate large doses of HCN.

The tubers of cassava (*Manihot esculenta*), a high-carbohydrate staple food in many tropical countries, contain high levels of cyanogenic glycosides. Traditional processing methods, such as grating, grinding, soaking, and drying, lead to the removal or degradation of a large fraction of the cyanogenic glycosides present in cassava tubers. However, chronic cyanide poisoning leading to partial paralysis of the limbs is still widespread in regions where cassava is a major food source, because the traditional detoxification methods employed to remove cyanogenic glycosides from cassava are not completely effective. In addition, many populations that consume cassava have poor nutrition, which aggravates the effects of the cyanogenic glycosides.

Efforts are currently under way to reduce the cyanogenic glycoside content of cassava through both conventional breeding and genetic engineering approaches. However, the complete elimination of cyanogenic glycosides may not be desirable, because these substances are probably responsible for the pest resistance of cassava stored for very long periods of time.

Glucosinolates release volatile toxins

A second class of plant glycosides, called the **glucosinolates**, or mustard oil glycosides, break down to release defensive substances. Found principally in the Brassicaceae and related plant families, glucosinolates break down to produce the compounds responsible for the smell and taste of vegetables such as cabbage, broccoli, and radishes.

Glucosinolate breakdown is catalyzed by a hydrolytic enzyme, called a *thioglucosidase* or *myrosinase*, that cleaves glucose from its bond with the sulfur atom (Figure 13.21). The resulting aglycone, the nonsugar portion of the molecule, rearranges and loses the sulfate to give pungent and chemically reactive products, including isothiocyanates

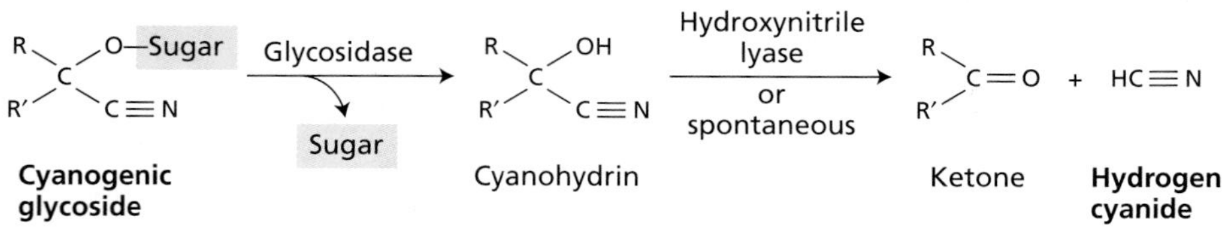

FIGURE 13.20 Enzyme-catalyzed hydrolysis of cyanogenic glycosides to release hydrogen cyanide. R and R′ represent various alkyl or aryl substituents. For example, if R is phenyl, R′ is hydrogen, and the sugar is the disaccharide β-gentiobiose, the compound is amygdalin (the common cyanogenic glycoside found in the seeds of almonds, apricots, cherries, and peaches).

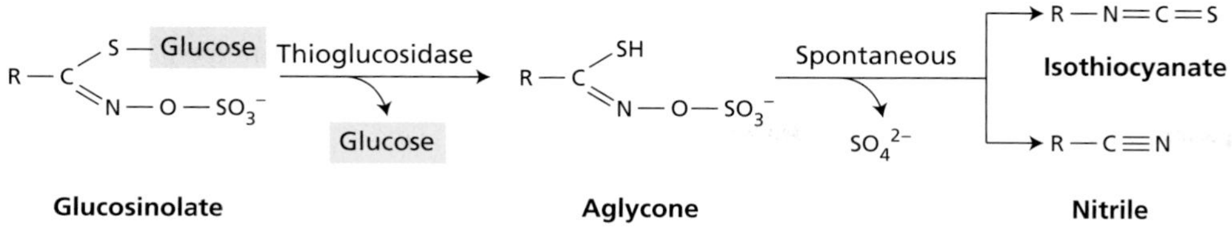

FIGURE 13.21 Hydrolysis of glucosinolates to mustard-smelling volatiles. R represents various alkyl or aryl substituents. For example, if R is $CH_2 = CH—CH_2^-$, the compound is sinigrin, a major glucosinolate of black mustard seeds and horseradish roots.

and nitriles, depending on the conditions of hydrolysis. These products function in defense as herbivore toxins and feeding repellents. Like cyanogenic glycosides, glucosinolates are stored in the intact plant separately from the enzymes that hydrolyze them, and they are brought into contact with these enzymes only when the plant is crushed.

As with other secondary metabolites, certain animals are adapted to feed on glucosinolate-containing plants without ill effects. For adapted herbivores, such as the cabbage butterfly, glucosinolates serve as adult stimulants for feeding and egg laying, and the isothiocyanates produced after glucosinolate hydrolysis act as volatile attractants (Renwick et. al. 1992). In addition, their caterpillars can redirect the glucosinolate hydrolysis reaction to the production of the less toxic nitriles (Wittstock et al. 2004) (see Web Essay 13.5).

Most of the recent research on glucosinolates in plant defense has concentrated on rape, or canola (*Brassica napus*), a major oil crop in both North America and Europe. Plant breeders have tried to lower the glucosinolate levels of rapeseed so that the high-protein seed meal remaining after oil extraction can be used as animal food. The first low-glucosinolate varieties tested in the field were unable to survive because of severe pest problems. However, more recently developed varieties with low glucosinolate levels in seeds but high glucosinolate levels in leaves are more resistant to pests and still provide a protein-rich seed residue for animal feeding.

Nonprotein amino acids defend against herbivores

Plants and animals incorporate the same 20 amino acids into their proteins. However, many plants also contain unusual amino acids, called **nonprotein amino acids**, that are not incorporated into proteins but are present instead in the free form and act as protective substances. Nonprotein amino acids are often very similar to common protein amino acids. Canavanine, for example, is a close analog of arginine, and azetidine-2-carboxylic acid has a structure very much like that of proline (Figure 13.22).

Nonprotein amino acids exert their toxicity in various ways. Some block the synthesis or uptake of protein amino acids; others, such as canavanine, can be mistakenly incorporated into proteins. After ingestion, canavanine is recognized by the herbivore enzyme that normally binds arginine to the arginine transfer RNA molecule, so it becomes incorporated into proteins in place of arginine. The usual result is a nonfunctional protein, because either its tertiary structure or its catalytic site is disrupted. Canavanine is less basic than arginine and may alter the ability of an enzyme to bind substrates or catalyze chemical reactions (Rosenthal 1991).

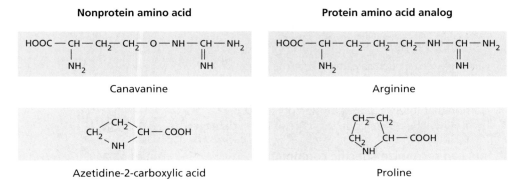

FIGURE 13.22 Nonprotein amino acids and their protein amino acid analogs. The nonprotein amino acids are not incorporated into proteins but are defensive compounds found in free form in plant cells.

Plants that synthesize nonprotein amino acids are not susceptible to the toxicity of these compounds. The jack bean (*Canavalia ensiformis*), which synthesizes large amounts of canavanine in its seeds, has protein-synthesizing machinery that can discriminate between canavanine and arginine, and it does not incorporate canavanine into its own proteins. Some insects that specialize on plants containing nonprotein amino acids have similar biochemical adaptations.

Induced Plant Defenses against Insect Herbivores

Plants have developed a wide variety of defense strategies against insect herbivory. These strategies can be divided into two categories, *constitutive defense responses* and *induced defense responses*. **Constitutive defense responses** are defense mechanisms that are always present. They are often species-specific and may exist as stored compounds, conjugated compounds (to reduce toxicity), or precursors of active compounds that can easily be activated upon damage. Most of the secondary metabolites serving as defense compounds that have been described so far in this chapter are constitutive defense responses. In some cases, however, the same insecticidal defense compounds are involved in both constitutive and induced responses. **Induced defense responses** are initiated only after actual damage occurs. In principle, induced defense responses require a smaller investment of plant resources then constitutive mechanisms, but they must be activated quickly to be effective.

Three categories of insect herbivores can cause varying degrees of damage to the plant:

1. *Phloem feeders*, such as aphids and whiteflies, cause little damage to the epidermis and mesophyll cells. The plant defense response to phloem feeders more closely resembles the response to *pathogens* rather than to herbivores. (Although the amount of direct injury to the plant is low, when these insects serve as vectors for plant viruses they can cause much greater damage.)

2. *Cell content feeders*, such as mites and thrips, are piercing/sucking insects that cause an intermediate amount of physical damage to plant cells.

3. *Chewing insects*, such as caterpillars (the larvae of moths and butterflies), grasshoppers, and beetles, cause the most significant damage to plants. In the discussion that follows, our definition of "insect herbivory" will be restricted to this type of insect damage.

Plants can recognize specific components of insect saliva

Plant responses to damage by insect herbivores involves both a wound response and recognition of certain insect-derived compounds called **elicitors**. Although repeated mechanical wounding can induce responses similar to those caused by insect herbivory in some plants, some molecules in insect saliva can serve as enhancers of this stimulus. In addition, such insect-derived elicitors can trigger signaling pathways systemically, thereby initiating defense responses in distal regions of the plant in anticipation of further damage.

The elicitors present in insect saliva and involved in insect herbivory have been identified as *fatty acid-amino acid conjugates* (or *fatty acid amides*) (Alborn et al. 1997). These compounds have been shown to elicit a response closely resembling the response to chewing insects, as opposed to wounding alone. The biosynthesis of these conjugates strongly depends on the plant as the source of the fatty acids linolenic acid (18:3) and linoleic acid (18:2). After the insect ingests plant tissue containing these fatty acids, an enzyme in its gut conjugates the plant-derived fatty acid to an insect-derived amino acid, typically glutamine. In some caterpillars the resulting conjugate of linolenic acid and glutamine is further processed by the introduction of a hydroxyl group at position 17 of linolenic acid (Figure 13.23). This

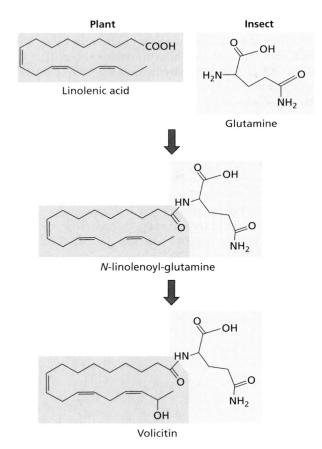

Figure 13.23 Concerted biosynthesis of elicitors from plant and insect precursors. Plant-derived fatty acids, in this case linolenic acid, are conjugated to an insect-derived amino acid (glutamine) to form a fatty acid–amino acid conjugate with elicitor activity. Hydroxylation at C-17 leads to the production of volicitin [*N*-(17-hydroxylinolenoyl)-L-glutamine].

compound, N-(17-hydroxylinolenoyl)-L-glutamine, was named **volicitin** for its potential to induce volatile secondary metabolites in corn plants (*Zea mays*) (Pare et al. 1998).

When plants recognize elicitors from the insect saliva, a complex signal transduction network is activated. A major signaling pathway involved in most plant defenses against insect herbivores is the *octadecanoid pathway*, which leads to the production of jasmonic acid.

Jasmonic acid is a plant hormone that activates many defense responses

Jasmonic acid (JA) levels rise steeply in response to insect herbivore damage and trigger the production of many proteins involved in plant defenses. The structure and biosynthesis of jasmonic acid have intrigued plant biologists because of the parallels to some *eicosanoids* that are central to inflammatory responses and other physiological processes in mammals (see Chapter 14 on the web site). In plants, jasmonic acid is synthesized from linolenic acid (18:3), which is released from membrane lipids and then converted to JA through a pathway also referred to as the *octadecanoid signaling pathway* as outlined in Figure 13.24. Two organelles participate in jasmonate biosynthesis, the chloroplast and peroxisome. In the chloroplast an intermediate derived from linolenic acid is cyclized and then transported to the peroxisome, where enzymes of the β-oxidation pathway (see Chapter 11) complete the conversion to jasmonic acid (see Figure 13.24).

Jasmonic acid is known to induce the transcription of a host of genes involved in plant defense metabolism. Among the genes induced are those that encode important key enzymes in all major pathways for secondary metabolites. The mechanisms for this gene activation are slowly becoming clear. For example, recent research on the Madagascar periwinkle (*Catharanthus roseus*), which makes some valuable anticancer alkaloids, identified a transcription factor that responds to jasmonic acid by activating the expression of several genes encoding alkaloid biosynthetic genes (van der Fits and Memelink 2000). This transcription factor also activates the genes of certain primary metabolic pathways that provide precursors for alkaloid formation, so it appears to be a master regulator of metabolism in Madagascar periwinkle.

Direct demonstration of the role of jasmonic acid in insect resistance has come from research with mutant lines of Arabidopsis that produce only low levels of JA (McConn et al. 1997). Such mutants are easily killed by insect pests, such as fungus gnats, that normally do not damage Arabidopsis. Application of exogenous jasmonic acid restores resistance nearly to the levels of the wild-type plant.

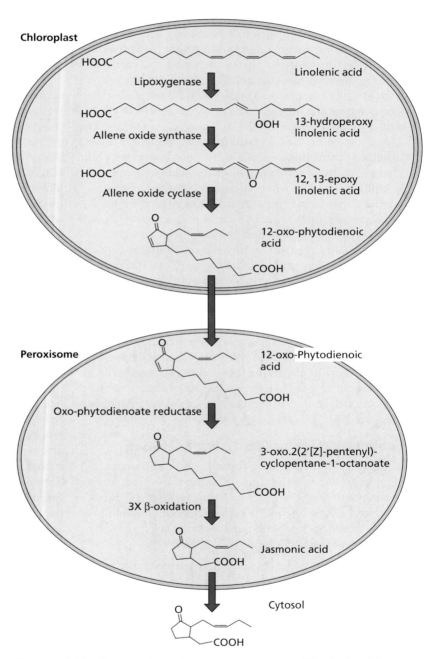

FIGURE 13.24 Steps in the pathway for conversion of linolenic acid (18:3) to jasmonic acid. The first three enzymatic steps occur in the chloroplast, resulting in the cyclized product 12-oxo-phytodienoic acid. This intermediate is transported to the peroxisome, where it is first reduced and then converted to jasmonic acid by β-oxidation.

Several other signaling compounds—including ethylene, salicylic acid, and methyl salicylate—are also induced by insect herbivory. Often, a concerted action of these signaling compounds is necessary for the full activation of induced defense responses.

Some plant proteins inhibit herbivore digestion

Among the diverse components of plant defense arsenals induced by jasmonic acid are proteins that interfere with herbivore digestion. For example, some legumes synthesize **α-amylase inhibitors** that block the action of the starch-digesting enzyme α-amylase. Other plant species produce **lectins**, defensive proteins that bind to carbohydrates or carbohydrate-containing proteins. After ingestion by an herbivore, lectins bind to the epithelial cells lining the digestive tract and interfere with nutrient absorption (Peumans and Van Damme 1995).

The best-known antidigestive proteins in plants are the **proteinase inhibitors**. Found in legumes, tomatoes, and other plants, these substances block the action of herbivore proteolytic enzymes. After entering the herbivore's digestive tract, they hinder protein digestion by binding tightly and specifically to the active site of protein-hydrolyzing enzymes such as trypsin and chymotrypsin. Insects that feed on plants containing proteinase inhibitors suffer reduced rates of growth and development that can be offset by supplemental amino acids in their diet.

The defensive role of proteinase inhibitors has been confirmed by experiments with transgenic tobacco. Plants that had been transformed to accumulate increased levels of proteinase inhibitors suffered less damage from insect herbivores than did untransformed control plants (Johnson et al. 1989). As with glucosinolates, some insect herbivores have become adapted to plant proteinase inhibitors by production of digestive proteinases resistant to inhibition (Jongsma et al. 1995, Opperta et al. 2005).

Herbivore damage induces systemic defenses

In tomatoes, insect feeding leads to the rapid accumulation of proteinase inhibitors throughout the plant, even in undamaged areas far from the initial feeding site (Schilmiller and Howe 2005). The systemic production of proteinase inhibitors in young tomato plants is triggered by a complex sequence of events:

1. Wounded tomato leaves synthesize **prosystemin**, a large (200 amino acid) precursor protein.

2. Prosystemin is proteolytically processed to produce the short (18 amino acid) polypeptide called **systemin**.

3. Systemin is released from damaged cells into the apoplast.

4. In adjacent intact tissue (phloem parenchyma) systemin binds to its receptor, a leucine-rich repeat (LRR) protein with protein kinase activity, on the plasma membrane. (See Chapter 14 on the web site and **Web Essay 13.6**.)

5. The activated systemin receptor becomes phosphorylated and activates a phospholipase A_2 (PLA_2).

6. The activated PLA_2 generates the signal that initiats jasmonic acid (JA) biosynthesis.

7. JA is then transported through the phloem to systemic parts of the plant by an unknown mechanism.

8. In target tissues JA is released and eventually activates the expression of genes that encode proteinase inhibitors (Figure 13.25).

Recently the systemin receptor of tomato has been shown to be identical to the receptor for *brassinolide*, a steroidal plant hormone (see Chapter 24). The systemin receptor thus appears to be a dual-function receptor. Since the *systemin signaling pathway* occurs exclusively in solanaceous plants, it has been proposed that early in the evolution of the Solanaceae, the polypeptide that would eventually become systemin evolved in ways that allowed binding to the ubiquitous brassinolide receptor, thereby activating a potential defense signaling pathway (Scheer and Ryan 2002). Systemic signaling responses have also been detected in many other plant species, including cotton, corn, and lima bean, but the mechanisms are still unclear.

Herbivore-induced volatiles have complex ecological functions

The induction and release of volatile organic compounds (hereafter referred to as *volatiles*) in response to insect herbivore damage provides an excellent example of the complex ecological functions of secondary metabolites in nature. The emitted combination of molecules is often specific for each insect herbivore species and typically includes representatives from the three major pathways of secondary metabolism: the terpenes, alkaloids, and phenolics discussed above. In addition, in response to mechanical damage all plants also emit lipid-derived products such as **green-leaf volatiles**, a mixture of 6-carbon aldehydes, alcohols, and esters. The ecological functions of these volatiles are manifold. Often, they attract natural enemies of the attacking insect herbivore—predators or parasites—that utilize the volatile cues to find their prey or host for their offspring. Volatiles released by the leaf during moth oviposition (egg-laying) can act as repellents to other female moths, thereby preventing further egg deposition and herbivory. In addition, many of these compounds, although volatile, remain attached to the surface of the leaf and serve as feeding deterrents because of their taste.

Plants have the ability to distinguish among various insect herbivore species and to respond differentially. For example, *Nicotiana attenuata*, a wild tobacco that grows in the deserts of the Great Basin in the western United States,

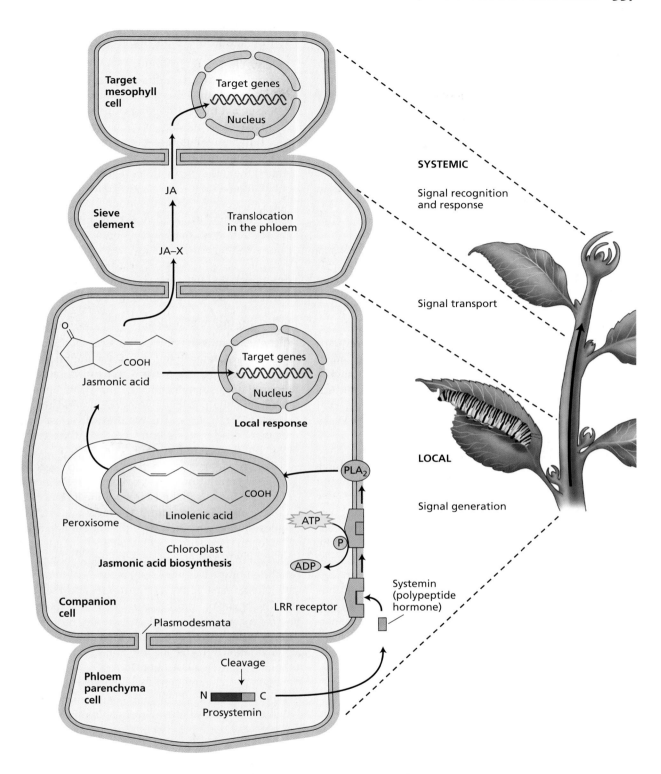

FIGURE 13.25 Proposed systemin signaling pathway for the rapid induction of proteinase inhibitor biosynthesis in wounded tomato plants. Wounded tomato leaves (bottom of figure) synthesize prosystemin in phloem parenchyma cells, and the prosystemin is proteolytically processed to systemin. Systemin is released from phloem parenchyma cells and binds to receptors on the plasma membrane of adjacent companion cells, activating a signaling cascade which results in the biosynthesis of jasmonic acid (JA). JA is then transported via sieve elements, possibly in a conjugated form (JA-X), to unwounded leaves. There, JA initiates a signaling pathway in target mesophyll cells, resulting in the activation of proteinase inhibitor genes. Plasmodesmata facilitate the spread of the signal at various steps in the pathway.

typically produces higher levels of nicotine following herbivory. However, when it is attacked by nicotine-tolerant caterpillars, there is no increase in nicotine. Instead, volatile terpenes are released that attract insect predators of the caterpillars (Karban and Baldwin 1997) (see **Web Essay 13.7**). Clearly, wild tobacco and other plants must have ways of determining what type of insect herbivore is damaging their foliage. Herbivores might signal their presence by the type of damage they inflict or the distinctive chemical compounds they release in their oral secretions.

There is an interesting twist on the role of volatiles in plant protection. Certain volatiles emitted by infested plants can serve as signals for neighboring plants to initiate expression of defense-related genes. In addition to several terpenoids, green-leaf volatiles act as potent signals in this process (Arimura et al. 2000). For example, when corn plants (*Z. mays*) were exposed to green-leaf volatiles, jasmonic acid and JA-related gene expression were rapidly induced. More important, however, was the finding that exposure to green-leaf volatiles primed corn plant defenses to respond more strongly to subsequent attacks by insect herbivores (Engelberth et al. 2004). Green-leaf volatiles have been shown to prime or sensitize the defensive mechanisms of a variety of plant species, including the induction of phytoalexins and other antimicrobial compounds (discussed in the next section) (see **Web Essay 13.8**).

In spite of all the chemical mechanisms plants have evolved to protect themselves, herbivorous insects have acquired mechanisms for circumventing or overcoming these plant defenses by the process of *reciprocal evolutionary change between plant and insect*, a type of co-evolution. These adaptations, like plant defense responses, can be either **constitutive** (always active) or **induced** (activated by the plant). Constitutive adaptations are more widely distributed among specialist herbivorous insects, which can feed only on a few plant species, whereas induced adaptations are more likely to be found among insects that are dietary generalists. Although it is not always obvious, in most natural environments plant-insect interactions have led to a standoff in which each can develop and survive under suboptimal conditions.

Plant Defense against Pathogens

Even though they lack an immune system as complex as that in animals, plants are surprisingly resistant to diseases caused by the fungi, bacteria, viruses, and nematodes that are ever present in the environment. In this section we will examine the diverse array of mechanisms that plants have evolved to resist infection, including the production of antimicrobial agents and a type of programmed cell death (see Chapter 16) called the *hypersensitive response*. Finally, we will discuss a special type of plant immunity called *systemic acquired resistance*.

Some antimicrobial compounds are synthesized before pathogen attack

Several classes of secondary metabolites that we have already discussed have strong antimicrobial activity when tested in vitro; thus they have been proposed to function as defenses against pathogens in the intact plant. Among these are saponins, a group of triterpenes thought to disrupt fungal membranes by binding to sterols.

Experiments utilizing genetic approaches have demonstrated the role of saponins in defense against pathogens of oat (Papadopoulou et al. 1999). Mutant oat lines with reduced saponin levels had much less resistance to fungal pathogens than did wild-type oats. Interestingly, one fungal strain that normally grows on oats was able to detoxify one of the principal saponins in the plant. However, mutants of this strain that could no longer detoxify saponins failed to infect oats, but could grow successfully on wheat that did not contain any saponins.

Infection induces additional antipathogen defenses

After being infected by a pathogen, plants deploy a broad spectrum of defenses against invading microbes. A common defense is the **hypersensitive response**, in which cells immediately surrounding the infection site die rapidly, depriving the pathogen of nutrients and preventing its spread. After a successful hypersensitive response, a small region of dead tissue is left at the site of the attempted invasion, but the rest of the plant is unaffected.

The hypersensitive response is often preceded by the rapid accumulation of reactive oxygen species and nitric oxide (NO). Cells in the vicinity of the infection synthesize a burst of toxic compounds formed by the reduction of molecular oxygen, including the superoxide anion ($O_2^{\bullet-}$), hydrogen peroxide (H_2O_2), and the hydroxyl radical ($\bullet OH$). An NADPH-dependent oxidase located at the plasma membrane (Figure 13.26) is thought to produce $O_2^{\bullet-}$, which in turn is converted into $\bullet OH$ and H_2O_2.

The hydroxyl radical is the strongest oxidant of these active oxygen species and can initiate radical chain reactions with a range of organic molecules, leading to lipid peroxidation, enzyme inactivation, and nucleic acid degradation (Lamb and Dixon 1997). Active oxygen species may contribute to cell death as part of the hypersensitive response or act to kill the pathogen directly.

A rapid spike of **nitric oxide** (**NO**) production accompanies the oxidative burst in infected leaves (Delledonne et al. 1998). NO, which acts as a second messenger in many signaling pathways in animals and plants (see Chapter 14 on the web site) is synthesized from the amino acid arginine by the enzyme *NO synthase*. An increase in the cytosolic calcium concentration appears to be required for the activation of NO synthase during the response. An increase in *both* NO and reactive oxygen species is required for the activation of the hypersensitive response: Increasing only one of these signals has little effect on the induction of cell death.

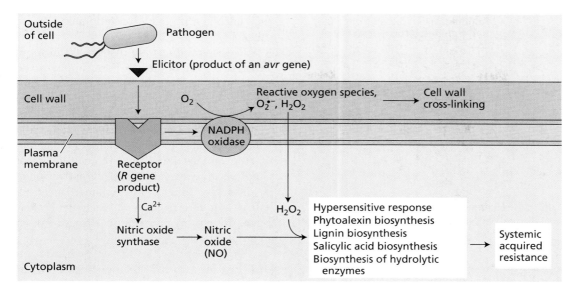

FIGURE 13.26 Many types of antipathogen defense are induced by infection. Fragments of pathogen molecules called elicitors initiate a complex signaling pathway leading to the activation of defensive responses. Some bacterial protein elicitors are injected directly into the cell, where they interact with R gene products. A burst in oxidation activity and nitric oxide production stimulates the hypersensitive response and other defense mechanisms.

Many species react to fungal or bacterial invasion by synthesizing lignin or callose (see Chapter 10). These polymers are thought to serve as barriers, walling off the pathogen from the rest of the plant and physically blocking its spread. A related response is the modification of cell wall proteins. Certain proline-rich proteins of the wall become oxidatively cross-linked after pathogen attack in an H_2O_2-mediated reaction (see Figure 13.26) (Bradley et al. 1992). This process strengthens the walls of the cells in the vicinity of the infection site, increasing their resistance to microbial digestion.

Another defensive response to infection is the formation of hydrolytic enzymes that attack the cell wall of the pathogen. An assortment of glucanases, chitinases, and other hydrolases are induced by fungal invasion. Chitin, a polymer of *N*-acetylglucosamine residues, is a principal component of fungal cell walls. These hydrolytic enzymes belong to a group of proteins that are closely associated with pathogen infection and so are known as *pathogenesis-related (PR) proteins*.

PHYTOALEXINS Perhaps the best-studied response of plants to bacterial or fungal invasion is the synthesis of **phytoalexins**. Phytoalexins are a chemically diverse group of secondary metabolites with strong antimicrobial activity that accumulate around the site of infection.

Phytoalexin production appears to be a common mechanism of resistance to pathogenic microbes in a wide range of plants. However, different plant families employ different types of secondary products as phytoalexins. For example, in leguminous plants, such as alfalfa and soybeans, isoflavonoids are common phytoalexins, whereas in solanaceous plants, such as potato, tobacco, and tomato, various sesquiterpenes are produced as phytoalexins (Figure 13.27).

Phytoalexins are generally undetectable in the plant before infection, but they are synthesized very rapidly after microbial attack because of the activation of new biosynthetic pathways. The point of control is usually the initiation of gene transcription. Thus, plants do not appear to store any of the enzymatic machinery required for phytoalexin synthesis. Instead, soon after microbial invasion they begin transcribing and translating the appropriate mRNAs and synthesizing the enzymes de novo (see **Web Essay 13.3**).

Although phytoalexins accumulate in concentrations that have been shown to be toxic to pathogens in bioassays, the defensive significance of these compounds in the intact plant is not fully known. Recent experiments on genetically modified plants and pathogens have provided the first direct proof of phytoalexin function in vivo. For example, tobacco transformed with a gene catalyzing the biosynthesis of the phenylpropanoid phytoalexin resveratrol become much more resistant to a fungal pathogen than nontransformed control plants (Hain et al. 1993). Similarly, resistance of Arabidopsis to a fungal pathogen depends on the tryptophan-derived phytoalexin camalexin, because mutants deficient in camalexin production were more susceptible than the wild-type. In other experiments, pathogens transformed with genes encoding phytoalexin-degrading enzymes were able to infect plants normally resistant to them (Kombrink and Somssich 1995).

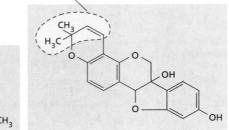

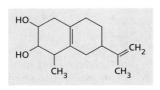

FIGURE 13.27 Structure of some phytoalexins—secondary metabolites with antimicrobial properties that are rapidly synthesized after microbial infection.

Some plants recognize specific substances released from pathogens

Within a species, individual plants often differ greatly in their resistance to microbial pathogens. These differences often lie in the speed and intensity of a plant's reactions. Resistant plants respond more rapidly and more vigorously to pathogens than do susceptible plants. Hence it is important to learn how plants sense the presence of pathogens and initiate defense.

Researchers have isolated more than 20 plant resistance genes, known as **R genes**, that function in defense against fungi, bacteria, and nematodes. Most of the R genes are thought to encode protein receptors that recognize and bind specific molecules originating from pathogens. This binding alerts the plant to the pathogen's presence (see Figure 13.26). These specific pathogen-derived elicitors include proteins, peptides, sterols, and polysaccharide fragments arising from the pathogen cell wall or outer membrane, or a secretion process (Boller 1995).

The R gene products themselves are nearly all proteins with a leucine-rich domain that is repeated inexactly several times in the amino acid sequence (see Chapter 14 on the web site). Such domains may be involved in elicitor binding and pathogen recognition. In addition, the R gene product is equipped to initiate signaling pathways that activate the various modes of antipathogen defense. Some R genes encode a nucleotide-binding site that binds ATP or GTP; others encode a protein kinase domain (Young 2000).

R gene products are distributed in more than one place in the cell. Some appear to be attached to the outside of the plasma membrane, where they could rapidly detect elicitors; others are in the cytoplasm, where they detect either pathogen molecules that are injected into the cell or other metabolic changes indicating pathogen infection. R genes constitute one of the largest gene families in plants and are often clustered together in the genome. The structures of R gene clusters may help generate R gene diversity by promoting exchange between chromosomes.

Studies of plant disease have revealed complex patterns of host relationships between plants and pathogen strains. Plant species are generally susceptible to the attack of certain pathogen strains, but resistant to others. This specificity is thought to be determined by interaction between the products of host R genes and pathogen *avr* (*avirulence*) **genes** believed to encode specific elicitors. According to current thinking, successful resistance requires the elicitor, a product of the pathogen *avr* gene, to be rapidly recognized by a host plant receptor, the product of an R gene. Despite their name, *avr* genes appear to encode factors that promote infection.

Exposure to elicitors induces a signal transduction cascade

Within a few minutes after pathogen elicitors have been recognized by an R gene, complex signaling pathways are set in motion that lead eventually to defense responses (see Figure 13.26). A common early element of these cascades is a transient change in the ion permeability of the plasma membrane. R gene activation stimulates an influx of Ca^{2+} and H^+ ions into the cell and an efflux of K^+ and Cl^- ions (Nürnberger and Scheel 2001). The influx of Ca^{2+} activates the oxidative burst that may act directly in defense (as already described), as well as signaling other defense reactions. Other components of pathogen-stimulated signal transduction pathways include nitric oxide, mitogen-activated protein (MAP) kinases, calcium-dependent protein kinases, jasmonic acid, and salicylic acid (discussed in the next section).

A single encounter with a pathogen may increase resistance to future attacks

When a plant survives the infection of a pathogen at one site, it often develops increased resistance to subsequent attacks at sites throughout the plant and enjoys protection against a wide range of pathogen species. This phenome-

Unit III
Growth and Development

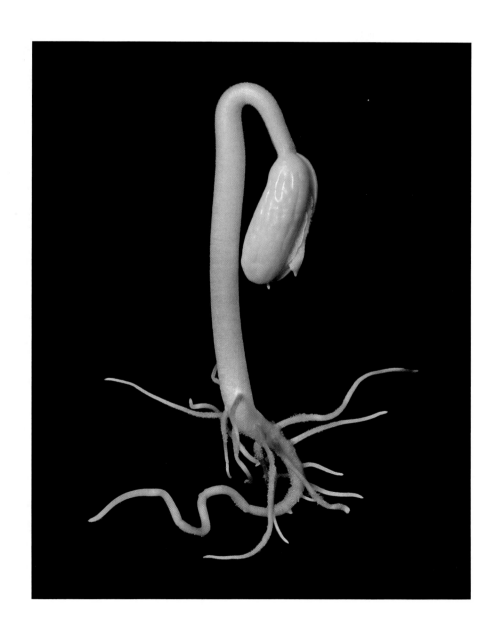

Unit III GROWTH AND DEVELOPMENT

14 *Available at www.plantphys.net*
   Gene Expression and Signal Transduction
15 Cell Walls: Structure, Biogenesis, and Expansion
16 Growth and Development
17 Phytochrome and Light Control of Plant Development
18 Blue-Light Responses: Stomatal Movements and Morphogenesis
19 Auxin: The Growth Hormone
20 Gibberellins: Regulators of Plant Height
21 Cytokinins: Regulators of Cell Division
22 Ethylene: The Gaseous Hormone
23 Abscisic Acid: A Seed Maturation and Antistress Signal
24 Brassinosteriods
25 The Control of Flowering
26 Stress Physiology

Etiolated garden bean seedling. Photograph by David McIntyre.

Web Chapter 14 | Gene Expression and Signal Transduction

Content available at www.plantphys.net

Genome Size, Organization, and Complexity
Most plant haploid genomes contain 20,000 to 30,000 genes

Prokaryotic Gene Expression
DNA-binding proteins regulate transcription in prokaryotes

Eukaryotic Gene Expression
Eukaryotic nuclear transcripts require extensive processing
Various posttranscriptional regulatory mechanisms have been identified
Transcription in eukaryotes is controlled by *cis*-acting regulatory sequences
Transcription factors contain specific structural motifs
Homeodomain proteins are a special class of helix-turn-helix proteins
Eukaryotic genes can be coordinately regulated
Small RNAs are posttranscriptional repressors of gene expression
The ubiquitin pathway regulates protein turnover

Signal Transduction in Prokaryotes
Bacteria employ two-component regulatory systems to sense extracellular signals
Osmolarity is detected by a two-component system
Related two-component systems have been identified in eukaryotes

Signal Transduction in Eukaryotes
Two classes of hormonal signals define two classes of receptors
Most steroid receptors act as transcription factors
Cell surface receptors can interact with G proteins
Heterotrimeric G proteins cycle between active and inactive forms
Activation of adenylyl cyclase increases the level of cyclic AMP
Activation of phospholipase C initiates the IP_3 pathway

Each living cell contains a set of instructions for building the entire organism, consisting of genes linearly arranged in the form of chromosomes. This fundamental concept in biology began with Mendel's genetic studies with garden peas in 1865, and culminated with Watson and Crick's discovery of the structure of DNA in 1953. But the story did not end there. A new field of molecular biology arose focused on the structure, replication, and expression of genes. Genes encode proteins, and elucidation of the elaborate machinery involved in transcription and translation was one of the early triumphs of the new field of molecular biology. More recently, molecular biologists have sought to understand how gene expression is regulated, for it turns out that the genetic "instructions" found on the chromosomes are incomplete without a full complement of regulatory proteins from the cytoplasm to direct their activity. In this chapter we will review basic concepts in gene expression in prokaryotes and eukaryotes.

While molecular biologists were studying cell function from the gene outward, developmental biologists were tracking the signals that regulate development, both external and internal, from the "skin" inward. They discovered that developmental signals, such as light or hormones, involve specific receptors and typically require amplification in the form of "second messengers." Ultimately these second messengers regulate the activities of crucial processes, such as membrane transport or gene expression, which bring about the physiological or developmental response. Thus developmental and molecular biologists approach the same problem from opposite directions. The second part of this chapter provides an overview of various signaling mechanisms found

IP$_3$ opens calcium channels on the ER and on the tonoplast
Cyclic ADP-ribose mediates intracellular Ca^{2+} release independently of IP$_3$ signaling
Some protein kinases are activated by calcium–calmodulin complexes
Plants contain calcium-dependent protein kinases
Diacylglycerol activates protein kinase C
Phospholipase A$_2$ and phospholipase D generate other membrane-derived signaling agents
In vertebrate vision, a heterotrimeric G protein activates cyclic GMP phosphodiesterase
Nitric oxide gas stimulates the synthesis of cGMP
Cell surface receptors may have catalytic activity
Ligand binding to receptor tyrosine kinases induces auto-phosphorylation
Intracellular signaling proteins that bind to RTKs are activated by phosphorylation
Ras recruits Raf to the plasma membrane
The activated MAP kinase enters the nucleus
Plant receptor kinases are structurally similar to animal receptor tyrosine kinases

Summary

in living cells. The models presented are derived mainly from animal and microbial systems, in which they were first discovered. Related mechanisms in plants will be discussed in the various chapters of the text devoted to development, light, and hormones.

Chapter 15 | Cell Walls: Structure, Biogenesis, and Expansion

PLANT CELLS, UNLIKE ANIMAL CELLS, are surrounded by a thin but mechanically strong cell wall. This wall consists of a complex mixture of polysaccharides and other polymers that are secreted by the cell and are assembled into an organized network linked together by both covalent and noncovalent bonds. Plant cell walls also contain structural proteins, enzymes, phenolic polymers, and other materials that modify the wall's physical and chemical characteristics.

The cell walls of prokaryotes, fungi, algae, and plants are distinct from each other in chemical composition and microscopic structure, yet they share three common functions: regulating cell volume, determining cell shape, and protecting the protoplast. As we will see, however, plant cell walls have acquired additional functions that are not apparent in the walls of other organisms. Consistent with these diverse functions, the structure and composition of plant cell walls are complex and variable.

In addition to these biological functions, the plant cell wall is important in human economics. As a natural product, the plant cell wall is used commercially in the form of paper, textiles, fibers (cotton, flax, hemp, and others), charcoal, lumber, and other wood products. Another major use of plant cell walls is in the form of structurally modified polysaccharides used to make synthetic fibers (such as rayon), plastics, films, coatings, adhesives, gels, and thickeners in a huge variety of products.

As the largest reservoir of organic carbon in nature, the plant cell wall also takes part in the processes of carbon flow through ecosystems. Ruminants and other foraging animals are able to digest cellulose with the aid of gut microbes that make the enzymes capable of digesting the cell wall. The organic substances that make up humus in the soil and that enhance soil structure and fertility are derived from cell walls. Finally, as an important source of roughage in our diet, the plant cell wall is a significant factor in human health and nutrition.

We begin this chapter with a description of the general structure and composition of plant cell walls and the mechanisms of the biosynthesis and secretion of cell wall materials. We then turn to the role of the primary cell wall in cell expansion. The mechanisms of tip growth will be contrasted with those of diffuse growth, particularly with respect to the establishment of cell polarity and the control of the rate of cell expansion. Finally, we will describe the dynamic changes in the cell wall that often accompany cell differentiation, along with the role of cell wall fragments as signaling molecules.

The Structure and Synthesis of Plant Cell Walls

Without a cell wall, plants would be very different organisms from what we know. Indeed, the plant cell wall is essential for many processes in plant growth, development, maintenance, and reproduction:

- Cell walls determine the mechanical strength of plant structures, allowing many plants to grow to great heights.

- Cell walls "glue" cells together, preventing them from sliding past one another. This constraint on cellular movement contrasts markedly with the situation in animal cells, and it dictates the way in which plants develop (see Chapter 16).

- A tough outer coating enclosing the cell, the cell wall acts as a cellular "exoskeleton" that controls cell shape and allows high turgor pressures to develop.

- Plant morphogenesis depends largely on the control of cell wall properties, because the expansive growth of plant cells is limited principally by the ability of the cell wall to expand.

- The cell wall is required for normal water balance in plants, because the wall determines the relationship between the cell turgor pressure and cell volume (see Chapter 3).

- The bulk flow of water in the xylem requires a mechanically tough wall that resists collapse by the negative pressure in the xylem.

- The wall acts as a diffusion barrier that limits the size of macromolecules that can reach the plasma membrane from outside, and it is a major structural barrier to pathogen invasion.

Much of the carbon assimilated in photosynthesis is channeled into polysaccharides that make up the wall. During specific phases of development, these polymers may be hydrolyzed into their constituent sugars to be scavenged by the cell and used to make new polymers. This phenomenon is most notable in many seeds, in which wall polysaccharides of the endosperm or cotyledons function primarily as food reserves. Furthermore, oligosaccharide components of the cell wall may act as important signaling molecules during cell differentiation and during recognition of pathogens and symbionts.

The diversity of functions of the plant cell wall requires a diverse and complex plant cell wall structure. In this section we will begin with a brief description of the morphology and basic architecture of plant cell walls. Then we will discuss the organization, composition, and synthesis of primary and secondary cell walls.

Plant cell walls have varied architecture

Stained sections of plant tissues reveal that the cell wall is not uniform, but varies greatly in appearance and composition in different cell types (Figure 15.1). Cell walls of the cortical parenchyma are generally thin (~100 nm) and have few distinguishing features. In contrast, the walls of some

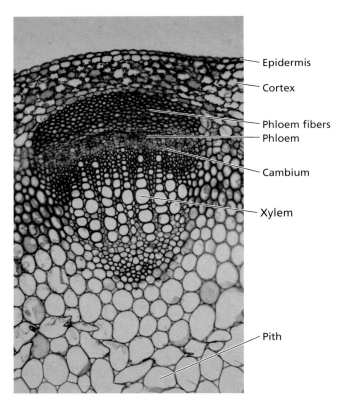

Figure 15.1 Cross-section of a stem of clover (*Trifolium*), showing cells with varying wall morphology. Note the highly thickened walls of the phloem fibers. (Photo © James Solliday/Biological Photo Service.)

specialized cells, such as epidermal cells, collenchyma, phloem fibers, xylem tracheary elements, and other forms of sclerenchyma have thicker, multilayered walls. Often these walls are intricately sculpted and are impregnated with other substances, such as lignin, cutin, suberin, waxes, silica, or structural proteins, that alter the walls' chemical and physical properties.

The individual sides of a wall surrounding a cell may also vary in thickness, impregnated substances, sculpting, and frequency of pitting and plasmodesmata. For example, the outer wall of the epidermis lacks plasmodesmata, is impregnated with cutin and waxes, and is much thicker than the other walls of the cell. In guard cells, the wall adjacent to the stomatal pore is much thicker than the walls on the other sides of the cell. Such variations in wall architecture for a single cell reflect the cell's polarity and differentiated functions and arise from targeted secretion of wall components to the cell surface.

Despite this diversity in cell wall morphology, cell walls commonly are classified into two major types: primary walls and secondary walls. **Primary walls** are extensible walls formed by growing cells and are usually considered to be relatively unspecialized and similar in molecular architecture in all cell types. Nevertheless, the appearance of primary walls can show wide variation. Some primary walls, such as those of the onion bulb parenchyma, are very thin and architecturally simple (Figure 15.2). Other primary walls, such as those found in collenchyma or in the epidermis (Figure 15.3), may be much thicker and consist of multiple layers.

Secondary walls are the cell walls that form after cell growth (enlargement) has ceased. Secondary walls may become highly specialized in structure and composition, reflecting the differentiated state of the cell. Xylem cells, such as those found in wood, are notable for possessing highly thickened secondary walls that are strengthened and waterproofed by **lignin** (see Chapter 13).

A thin layer of material, the **middle lamella** (plural *lamellae*), can usually be seen at the junction where the walls of neighboring cells come into contact. The composition of the middle lamella differs from the rest of the wall in that it is high in pectin and contains proteins different from those in the bulk of the wall. Its origin can be traced to the cell plate that formed during cell division.

As we saw in Chapter 1, the cell wall is usually penetrated by tiny membrane-lined channels, called **plasmodesmata** (singular *plasmodesma*), which connect neighboring cells. Plasmodesmata function in communication between cells by allowing passive transport of small molecules and active transport of proteins and nucleic acids between the cytoplasm of adjacent cells.

The primary cell wall is composed of cellulose microfibrils embedded in a polysaccharide matrix

In primary cell walls, cellulose microfibrils are embedded in a highly hydrated matrix (Figure 15.4). This structure

(A)
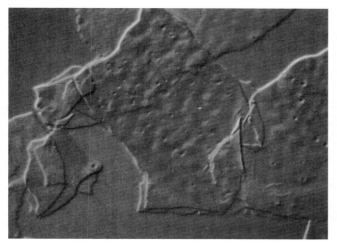
20 µm

(B)
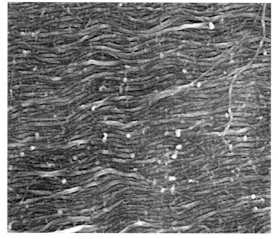
200 nm

FIGURE 15.2 Two views of primary cell walls. (A) This surface view of cell wall fragments from onion parenchyma was taken using Nomarski optics. Note that the wall looks like a very thin sheet with small surface depressions; these depressions may be pit fields, places where plasmodesmatal connections between cells are concentrated. (B) This surface view of a cell wall from a growing cucumber hypocotyl was visualized by scanning electron microscopy. Note the fibrous texture of the wall and the more or less parallel orientation of the fibrils, which are oriented transverse to the long axis of the cell. The fibrils are cellulose microfibrils coated with matrix polymers. (A from McCann et al. 1990; B from Marga et al. 2005.)

FIGURE 15.3 Electron micrograph of the outer epidermal cell wall from the growing region of a bean hypocotyl. Multiple layers are visible within the wall. The inner layers are thicker and more defined than the outer layers, because the outer layers are the older regions of the wall and have been stretched and thinned by cell expansion. (From Roland et al. 1982.)

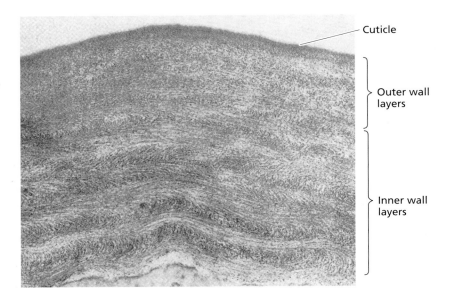

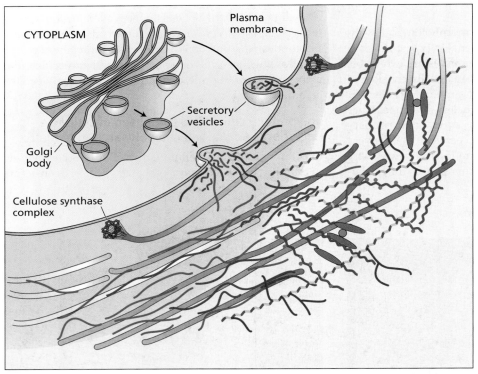

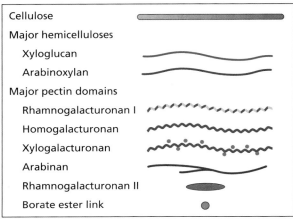

FIGURE 15.4 Schematic diagram of the major structural components of the primary cell wall and their likely arrangement. Cellulose microfibrils (grey rods) are synthesized at the cell surface and are coated with hemicelluloses (blue and purple strands) which bind microfibrils to one another. Pectins (red, yellow, and green strands) form an interlocking matrix that controls microfibril spacing and wall porosity. Pectins and hemicelluloses are synthesized in the Golgi apparatus and delivered to the wall via vesicles that fuse with the plasma membrane and thus deposit these polymers to the cell surface. For clarity, the hemicellulose–cellulose network is emphasized on the left, and the pectin network is emphasized on the right. (After Cosgrove 2005.)

TABLE 15.1
Structural components of plant cell walls

Class	Examples
Cellulose	Microfibrils of (1→4)β-D-glucan
Matrix polysaccharides	
Pectins	Homogalacturonan
	Rhamnogalacturonan
	Arabinan
	Galactan
Hemicelluloses	Xyloglucan
	Xylan
	Glucomannan
	Arabinoxylan
	Callose (1→3)β-D-glucan
	(1→3,1→4)β-D-glucan [grasses only]
Lignin	(see Chapter 13)
Structural proteins	(see Table 15.2)

provides both strength and flexibility. In cell walls, the **matrix** (plural *matrices*) consists of two major groups of polysaccharides, usually called hemicelluloses and pectins, plus a small amount of structural protein. The matrix polysaccharides consist of a variety of polymers that may vary according to cell type and plant species (Table 15.1).

These polysaccharides are named after the principal sugars they contain. For example, a *glucan* is a polymer made up of glucose, a *xylan* is a polymer made up of xylose, a *galactan* is made from galactose, and so on. *Glycan* is the general term for a polymer made up of sugars.

For branched polysaccharides, the backbone of the polysaccharide is usually indicated by the last part of the name. For example, *xyloglucan* has a glucan backbone (a linear chain of glucose residues) with xylose sugars attached to it in the side chains; *glucuronoarabinoxylan* has a xylan backbone (made up of xylose subunits) with glucuronic acid and arabinose side chains. However, a compound name does not necessarily imply a branched structure. For example, *glucomannan* is the name given to a polymer containing both glucose and mannose in its backbone.

Cellulose microfibrils are relatively stiff structures that contribute to the strength and structural bias of the cell wall. The individual glucans that make up the microfibril are closely aligned and bonded to each other to make a highly ordered (**crystalline**) ribbon that excludes water and is relatively inaccessible to enzymatic attack. As a result, cellulose is very strong and very stable and resists degradation.

Hemicelluloses are flexible polysaccharides that characteristically bind to the surface of cellulose. They may form tethers that bind cellulose microfibrils together into a cohesive network (see Figure 15.4), or they may act as a slippery coating to prevent direct microfibril–microfibril contact. Another term for these molecules is *cross-linking glycans*, but in this chapter we'll use the more traditional term *hemicelluloses*. As described later, the term *hemicellulose* includes several different kinds of polysaccharides.

Pectins form a hydrated gel phase in which the cellulose–hemicellulose network is embedded. They act as hydrophilic filler to prevent aggregation and collapse of the cellulose network. They also determine the porosity of the cell wall to macromolecules. Like hemicelluloses, pectins include several different kinds of polysaccharides.

The precise role of wall **structural proteins** is uncertain, but they may add mechanical strength to the wall and assist in the proper assembly of other wall components.

The primary wall is composed of approximately 25% cellulose, 25% hemicelluloses, and 35% pectins, with perhaps 2 to 5% structural protein, on a dry-weight basis. However, large deviations from these values may be found among species. For example, the walls of grass coleoptiles consist of 60 to 70% hemicelluloses, 20 to 25% cellulose, and only about 10% pectins. Cereal endosperm walls are mostly (about 85%) hemicelluloses. Xylem secondary walls typically contain much higher cellulose contents. The composition of the cell wall can also change substantially during and after cell elongation (Gibeaut et al. 2005).

In this chapter we will present a basic model of the primary wall, but be aware that plant cell walls are more diverse than this model suggests. The composition of matrix polysaccharides and structural proteins in walls varies significantly among different species and cell types (O'Neill and York 2003). Most notably, in grasses and related species the makeup of the major matrix polysaccharides differs from that of other land plants (Carpita 1996). Pectins are reduced in quantity in grass walls and replaced by hemicelluloses (specifically, mixed-link glucans and branched xylans known as glucuronoarabinoxylans—see the section on hemicelluloses later in this chapter).

The primary wall also contains water, located mostly in the matrix, which is about 75 to 80% water. The hydration state of the matrix is an important determinant of the physical properties of the wall; for example, removal of water makes the wall stiffer and less extensible. This stiffening effect of dehydration may play a role in growth inhibition by water deficits. We will examine the structure of each of the major polymers of the cell wall in more detail in the sections that follow.

Cellulose microfibrils are synthesized at the plasma membrane

Cellulose is a tightly packed microfibril composed of linear chains of (1→4)-linked β-D-glucose (for sugar structures, see Figure 15.5 and **Web Topic 15.1**). Because of the alternating spatial configuration of the glucosidic bond linking adjacent glucose residues, the repeating unit in cellulose is considered to be cellobiose, a (1→4)-linked β-D-glucose disaccharide.

Cellulose microfibrils are of indeterminate length and vary considerably in width and degree of order, depending

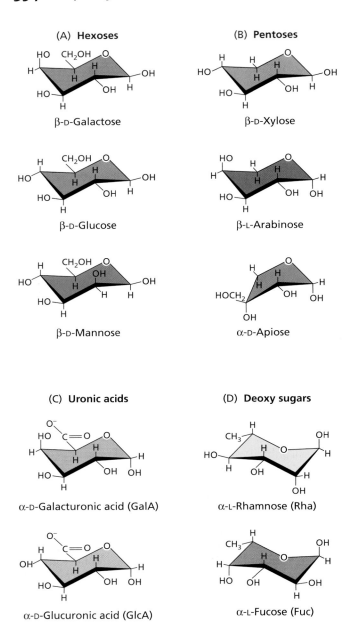

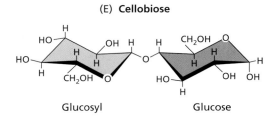

FIGURE 15.5 Conformational structures of sugars commonly found in plant cell walls. (A) Hexoses (six-carbon sugars). (B) Pentoses (five-carbon sugars). (C) Uronic acids (acidic sugars). (D) Deoxy sugars. (E) Cellobiose, showing the (1→4)β-D-linkage between two glucose residues in inverted orientation.

on the source. For instance, cellulose microfibrils in land plants are 2 to 5 nm wide, whereas those formed by algae may be up to 20 nm wide and may be more highly ordered (more crystalline) than those found in land plants (Sturcova et al. 2004). This variation in width corresponds to a variation in the number of parallel chains that make up the cross section of a microfibril—estimated to consist of as few as six individual chains in the crystalline core of the thinnest microfibrils to as many as 30 to 50 chains in larger ones.

The precise molecular structure of the cellulose microfibril is still a matter of debate. Some models of microfibril organization suggest that it has a substructure consisting of highly crystalline domains linked together by less organized, "amorphous" regions, while other models conceive of a solid crystalline core surrounded by a less organized layer (Figure 15.6) (Vietor et al. 2002). Within crystalline domains, adjacent glucans are highly ordered and firmly attached to each other by noncovalent bonding, such as hydrogen bonds and hydrophobic interactions. Native cellulose in plants is found in two variant crystalline forms, called allomorphs Iα and Iβ, which differ slightly in the way the parallel glucan chains are packed. These forms may be interconverted by chemical and physical treatments. The significance of these two crystalline forms is unclear at present.

The individual glucan chains in cellulose microfibrils are composed of 2000 to more than 25,000 glucose residues (Brown et al. 1996). These chains are long enough (about 1–5 μm) to extend through multiple crystalline and amorphous regions within a microfibril.

The extensive noncovalent bonding between adjacent glucans within a cellulose microfibril gives this structure remarkable properties. Cellulose has a high tensile strength, equivalent to that of steel. Cellulose is also insoluble, chemically stable, and relatively immune to chemical and enzymatic attack. These properties make cellulose an excellent structural material for building a strong cell wall.

Evidence from electron microscopy indicates that cellulose microfibrils are synthesized by large, ordered protein complexes, called *particle rosettes* or *terminal complexes*, that are embedded in the plasma membrane (Figure 15.7) (Kimura et al. 1999). These rosettes are made up of six subunits, each of which is believed to contain six units of **cellulose synthase**, the enzyme that synthesizes the individual (1→4)β-D-glucans that make up the microfibril (see **Web Topic 15.2**).

Cellulose synthases in higher plants are encoded by a gene family named ***CesA*** (*Ce*llulose *s*ynthase *A*) (Pear et al. 1996; Arioli et al. 1998), which has ten members in Arabidopsis (Doblin et al. 2002). The *CesA* family is part of a larger ***Csl*** (*C*ellulose *s*ynthase-*l*ike) superfamily, whose other families are named *CslA* through *H*. ***CslA*** genes have been found to encode synthases for a hemicellulose, specifically (1→4)β-D-mannan (Dhugga et al. 2004; Liepman et al. 2005), and it is hypothesized that the other *Csl* families

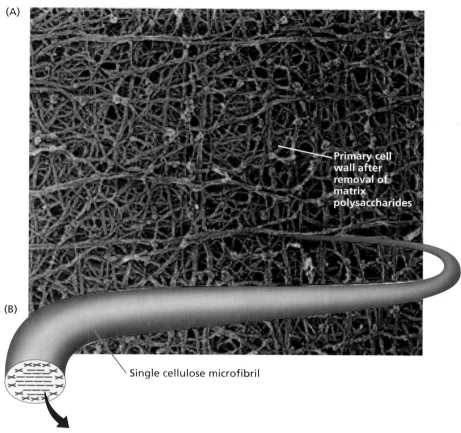
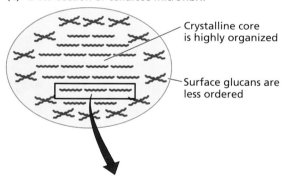
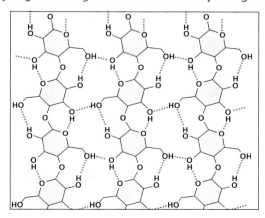

Figure 15.6 A structural model of a cellulose microfibril. The microfibril has regions of high crystallinity intermixed with less ordered regions. Some hemicelluloses may also be trapped within the microfibril and bound to the surface. (A) Scanning EM of the primary cell wall from onion parenchyma, after matrix polysaccharides have been extracted. Note its fibrillar texture, which arises from layers of cellulose microfibrils. (B) A single cellulose microfibril composed of two to four dozen (1→4)β-D-glucan chains tightly bonded to each other to form a crystalline ribbon. (C) Cross-section of a cellulose microfibril, illustrating one model of cellulose structure, with a crystalline core of highly ordered (1→4)β-D-glucans surrounded by a less organized layer. (D) The crystalline regions of cellulose have precise alignment of glucans, with hydrogen bonding within, but not between, layers of (1→4)β-D-glucans. (After McCann et al. 2001 and Matthews et al. 2006.)

encode enzymes that synthesize the backbones of other hemicelluloses. These synthases are **sugar-nucleotide polysaccharide glycosyltransferases**, which transfer monosaccharides from sugar nucleotides to the growing end of the polysaccharide chain.

Cellulose synthase, which is located on the cytoplasmic side of the plasma membrane, transfers a glucose residue from a sugar nucleotide donor to the growing glucan chain. The sugar donor is uridine diphosphate D-glucose (UDP-glucose). There is evidence that the glucose in the UDP-glucose used for cellulose synthesis is obtained from sucrose, a disaccharide composed of fructose and glucose (Amor et al. 1995; Salnikov et al. 2001). According to this idea, the enzyme **sucrose synthase** acts as a metabolic channel to transfer glucose taken from sucrose, via UDP-glucose, to the growing cellulose chain (Figure 15.8).

Sterol-glucosides (sterols linked to a chain of one or more glucose residues) (Figure 15.9) are hypothesized to serve as the primers, or initial acceptors, that start the elongation of the glucan chain (Peng et al. 2002). After the chain has elongated sufficiently, the sterol may be clipped from the glucan by a membrane-bound endoglucanase, allowing the growing glucan chain to be extruded through the membrane to the exterior of the cell, where, together with other glucan chains, it forms a crystalline ribbon and binds hemicelluloses to form a

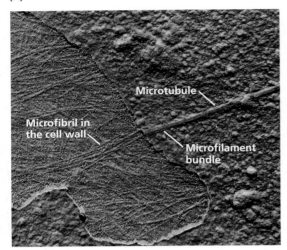

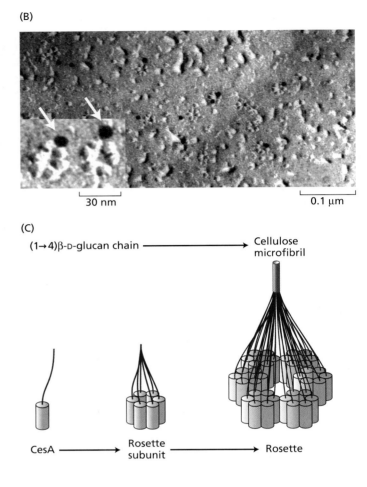

FIGURE 15.7 Cellulose synthesis by the cell. (A) Electron micrograph showing newly synthesized cellulose microfibrils immediately exterior to the plasma membrane. (B) Freeze-fracture labeled replicas showing reactions with antibodies against cellulose synthase. A field of labeled rosettes (arrows) with seven clearly labeled rosettes and one unlabeled rosette. The inset shows an enlarged view of two selected particle rosettes (terminal complexes) with immunogold labeling of CesA. The gold nanoparticles are the dark circles indicated with arrows. (C) Schematic model of the relationships between the particle rosettes (far right) and the single CesA proteins (far left). Six CesA proteins (encoded by three different genes) are thought to make one particle, which assembles to form a hexamer that is microscopically visible and that produces 36 glucan strands that form the ordered microfibril. (A from Gunning and Steer 1996; B from Kimura et al. 1999; C after Doblin et al. 2002.)

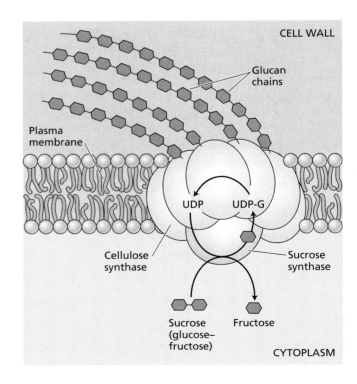

FIGURE 15.8 Model of cellulose synthesis by a multisubunit complex containing cellulose synthase. Glucose residues are donated to the growing glucan chains by UDP-glucose (UDP-G). Sucrose synthase may act as a metabolic channel to transfer glucose taken from sucrose to UDP-glucose, or UDP-glucose may be obtained directly from the cytoplasm. (After Amor et al. 1995.)

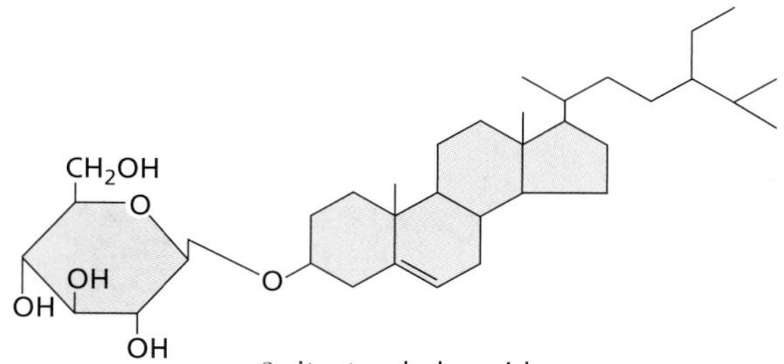

Cell Walls: Structure, Biogenesis, and Expansion **357**

FIGURE 15.9 Structure of a sterol glucoside that may act as an initial primer (i.e., acceptor for glucan chain elongation) for cellulose synthesis. This β-sitosterol glucoside consists of a sterol (right) linked to a glucose residue (left). Additional glucose residues are added to the one shown here, forming a glucan used in the formation of a cellulose microfibril.

strong and resilient network (see Figure 15.4). It is also possible that some hemicellulose gets entrapped in the microfibril as it forms (Hayashi 1989); this may have the effect of creating disorder in the crystalline microfibril and also anchoring the microfibril to the matrix.

Matrix polymers are synthesized in the Golgi and secreted via vesicles

The matrix is a highly hydrated phase in which the cellulose microfibrils are embedded. The matrix polysaccharides are synthesized by membrane-bound glycosyltransferases in the Golgi apparatus and are delivered to the cell wall via exocytosis of tiny vesicles (see Figure 15.4) (**Web Topic 15.3**). As described above, genes in the *Csl* superfamily encode glycosyltransferases for synthesis of the backbone of some of the wall polysaccharides. Additional sugar residues may be added as branches to the polysaccharide backbone by other sets of glycosyltransferases (Scheible and Pauly 2004).

Unlike cellulose, which forms a crystalline microfibril, the matrix polysaccharides are much less ordered and are often described as amorphous. This noncrystalline character is a consequence of the structure of these polysaccharides—their branching and their nonlinear conformation. Nevertheless, studies using infrared spectroscopy and nuclear magnetic resonance (NMR) indicate partial order in the orientation of hemicelluloses and pectins in the cell wall, probably as a result of a physical tendency for these polymers to become aligned along the long axis of cellulose (Wilson et al. 2000).

Hemicelluloses are matrix polysaccharides that bind to cellulose

Hemicelluloses are a heterogeneous group of polysaccharides (Figure 15.10) that are bound tightly to the wall. Typically they are solubilized from depectinated walls by the use of a strong alkali (2–4 M NaOH). Several kinds of hemicelluloses are found in plant cell walls, and walls from different tissues and different species vary in their hemicellulose composition.

In the primary wall of dicotyledons, the most abundant hemicellulose is **xyloglucan** (see Figure 15.10A). Like cellulose, this polysaccharide has a backbone of (1→4)-linked β-D-glucose residues. Unlike cellulose, however, xyloglucan has short side chains that contain xylose, galactose, and often, though not always, a terminal fucose. Primary wall xyloglucans are typically fucosylated, but those of storage cell walls are not. There is taxonomic variation as well; for example, solanaceous and mint species have arabinose side groups, and no galactose or fucose.

By interfering with the linear alignment of the glucan backbones with one another, these side chains prevent the assembly of xyloglucan into a crystalline microfibril. Because xyloglucans are longer (about 50–500 nm) than the spacing between cellulose microfibrils (20–40 nm), they have the potential to link several microfibrils together.

Varying with the developmental state and plant species, the hemicellulose fraction of the wall may also contain large amounts of other physiologically important polysaccharides—for example, **glucuronoarabinoxylans** (see Figure 15.10B) and **glucomannans**. Secondary walls typically contain little xyloglucan and more xylans and glucomannans, which also bind tightly to cellulose. The cell walls of grasses contain relatively small amounts of xyloglucan and pectin, which are replaced by glucuronoarabinoxylan and (1→3,1→4)β-D-glucan (also called "mixed-link glucan").

Pectins are gel-forming components of the matrix

Like the hemicelluloses, pectins constitute a heterogeneous group of polysaccharides (Figure 15.11), characteristically containing acidic sugars such as galacturonic acid and neutral sugars such as rhamnose, galactose, and arabinose. Pectins are the most soluble of the wall polysaccharides; they can be extracted with hot water or with calcium chelators. In the wall, pectins are very large and complex molecules composed of different pectic polysaccharide domains believed to be linked together by covalent and noncovalent bonds.

Some pectic polysaccharide domains, such as *homogalacturonan*, have a relatively simple primary structure (see Figure 15.11A). This polysaccharide, also called *polygalacturonic acid*, is a (1→4)-linked polymer of α-D-glucuronic acid residues. Figure 15.12 shows a triple-fluorescence-labeled section of tobacco stem parenchyma cells showing the distribution of cellulose and pectic homogalacturonan, which are concentrated in different parts of the cell wall.

Another abundant pectin polysaccharide is **rhamnogalacturonan I (RG I)**, which has a long backbone of alternating rhamnose and galacturonic acid residues (see Figure 15.11B). This molecule is very large and is believed to carry long side-chains of arabinans and galactans.

Much further up on the scale of molecular complexity is a highly branched pectic polysaccharide called **rhamnogalacturonan II (RG II)** (see Figure 15.11E), which contains a homogalacturonan backbone decorated with side chains

FIGURE 15.10 Partial structures of common hemicelluloses. (For details on carbohydrate nomenclature, see **Web Topic 15.1**.) (A) Xyloglucan has a backbone of (1→4)-linked β-D-glucose (Glc), with (1→6)-linked branches containing α-D-xylose (Xyl). In some cases galactose (Gal) and fucose (Fuc) are added to the xylose side chains. (B) Glucuronoarabinoxylans have a (1→4)-linked backbone of β-D-xylose (Xyl). They may also have side chains containing arabinose (Ara), 4-O-methylglucuronic acid (4-O-Me-α-D-GlcA), or other sugars. (From Carpita and McCann 2000.)

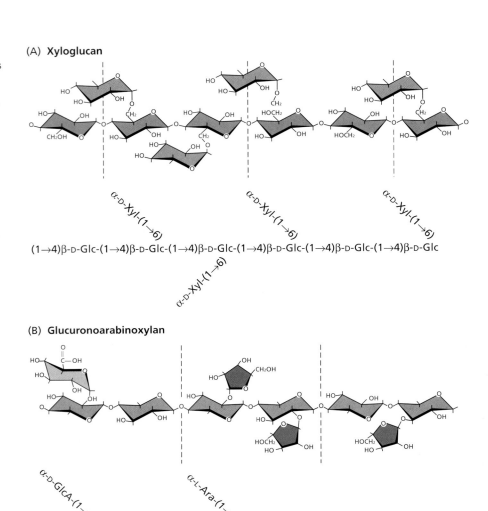

comprising at least ten different sugars in a complicated pattern of linkages. Although RG I and RG II have similar names, *they have very different structures*. RG II units are cross-linked in the wall by borate diesters (Ishii et al. 1999), and this cross-linking is important for the structure and the mechanical strength of cell walls. For example, Arabidopsis mutants that synthesize an altered RG II show substantial growth abnormalities, apparently resulting from an unstable borate cross-link (O'Neill et al. 2001; Ryden et al. 2003). Lack of RG II cross-linking by borate leads to radical swelling of the cell wall, increases in wall porosity, and mechanical weakening of the wall (Ryden et al. 2003; O'Neill et al. 2004).

It is believed that in the wall these pectic polysaccharides are covalently linked to one another, but the exact linkage has not been established. Figures 15.13A and B illustrate two models for how the various pectin domains may be linked to one another, but neither model is firmly established.

FIGURE 15.11 Partial structures of the most common pectins. (A) Homogalacturonan, also known as polygalacturonic acid or pectic acid, is made up of (1→4)-linked α-D-galacturonic acid (GalA) with occasional rhamnosyl residues that put a kink in the chain. The carboxyl residues are often methyl esterified. (B) Rhamnogalacturonan I (RG I) is a very large pectin, with a backbone of alternating (1→4)α-D-galacturonic acid (GalA) and (1→2)α-D-rhamnose (Rha). Side chains are attached to rhamnose and are composed principally of arabinans (C), galactans, and arabinogalactans (D). These side chains may be short or quite long. The galacturonic acid residues are often methyl esterified. (E) Rhamnogalacturonan II (RG II) is a very complicated, but compact, polysaccharide made up of 11 different sugar residues. It is dimerized via a borate ester. (From Carpita and McCann 2000.)

Cell Walls: Structure, Biogenesis, and Expansion **359**

(A) Homogalacturonan (HGA)

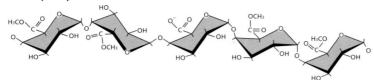

(B) Rhamnogalacturonan I (RG I)

(C) 5-Arabinan

(D) Type I arabinogalactan

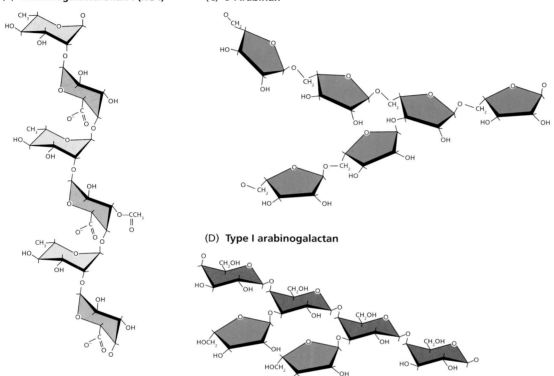

(E) Rhamnogalacturonan II (RG II) dimer cross-linked by borate diester bonds

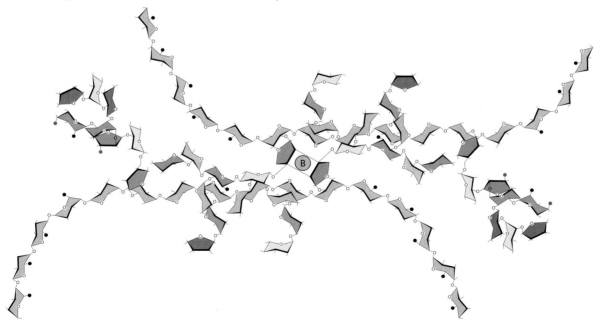

FIGURE 15.12 Triple-fluorescence-labeled section of tobacco stem showing the primary cell walls of three adjacent parenchyma cells bordering an intercellular space. The blue color is calcofluor (staining of cellulose), and the red and green colors indicate the binding of two monoclonal antibodies to different epitopes (immunologically distinct regions) of pectic homogalacturonan. (Courtesy of W. Willats.)

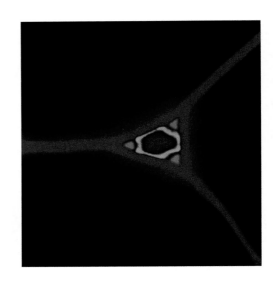

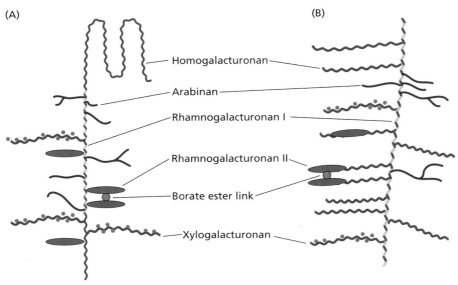

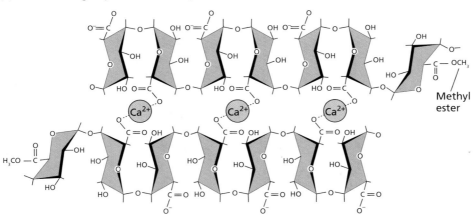

FIGURE 15.13 Pectin structure. (A and B) Two proposed models of how the various pectic polysaccharides may be linked together to form a huge macromolecule in the cell wall. In the traditional model (A), the backbone is made up of alternating homogalacturonan regions and rhamnogalacturonan I (RG I) regions. RG I can be highly branched with the other pectin polysaccharides. In an alternative model (B), RG I forms the backbone and all the other pectins, including homogalacturonan, are side branches. (C) Formation of a pectin network involves ionic bridging of the nonesterified carboxyl groups (COO^-) by calcium ions. When blocked by methyl-esterified groups, the carboxyl groups cannot participate in this type of interchain network formation. Likewise, the presence of side chains on the backbone interferes with network formation. (A, B after Vincken et al. 2003; C after Carpita and McCann 2000.)

TABLE 15.2 Structural proteins of the cell wall

Class of cell wall proteins	Percentage carbohydrate	Localization typically in:
HRGP (hydroxyproline-rich glycoprotein)	~55	Phloem, cambium, sclereids
PRP (proline-rich protein)	~0–20	Xylem, fibers, cortex
GRP (glycine-rich protein)	0	Xylem

Pectins typically form gels—loose networks formed by highly hydrated polymers. Pectins are what make fruit jams and jellies "gel," or solidify. In pectic gels, the charged carboxyl (COO^-) groups of neighboring pectin chains are linked together via Ca^{2+}, which forms a tight complex with pectin. A large calcium-bridged network may thus form, as illustrated in Figure 15.13C.

Pectins are subject to modifications that may alter their conformation and linkage in the wall. Many of the acidic residues are esterified with methyl, acetyl, and other unidentified groups during biosynthesis in the Golgi apparatus. Such esterification masks the charges of carboxyl groups and prevents calcium bridging between pectins, thereby reducing the gel-forming character of the pectin.

Once the pectin has been secreted into the wall, the ester groups may be removed by pectin esterases found in the wall, thus unmasking the charges of the carboxyl groups and increasing the ability of the pectin to form a rigid gel. By creating free carboxyl groups, de-esterification also increases the electric-charge density in the wall, which in turn may influence the concentration of ions in the wall and the activities of wall enzymes. In addition to being connected by calcium bridging, pectins may be linked to each other by various covalent bonds, including ester linkages between phenolic residues such as ferulic acid (see Chapter 13).

Structural proteins become cross-linked in the wall

In addition to the major polysaccharides described in the previous section, the cell wall contains several classes of structural proteins. These proteins usually are classified according to their predominant amino acid composition—for example, hydroxyproline-rich glycoprotein (HRGP), glycine-rich protein (GRP), proline-rich protein (PRP), and so on (Table 15.2). Some wall proteins have sequences that are characteristic of more than one class. Many structural proteins of the wall have highly repetitive primary structures that form simple helical rods, and some are highly glycosylated (Figure 15.14).

In vitro extraction studies have shown that newly secreted wall structural proteins are relatively soluble, but they become more and more insoluble during cell maturation or in response to wounding. The biochemical nature of the insolubilization process is uncertain, however.

Wall structural proteins vary greatly in their abundance, depending on cell type, maturation, and previous stimulation. Wounding, pathogen attack, and treatment with elicitors (molecules that activate plant defense responses; see Chapter 13) increase expression of the genes that code for many of these proteins. In histological studies, wall structural proteins are often localized to specific cell and tissue types. For example, HRGPs are associated mostly with cambium, phloem parenchyma, and various types of sclerenchyma. GRPs and PRPs are most often localized to xylem vessels and fibers and thus are more characteristic of a differentiated cell wall.

In addition to the structural proteins already listed, cell walls contain **arabinogalactan proteins** (**AGPs**), which usually amount to less than 1% of the dry mass of the wall (Schultz et al. 2000). These water-soluble proteins are very heavily glycosylated. More than 90% of the mass of AGPs may be sugar residues—primarily galactose and arabinose (Figure 15.15) (Gaspar et al. 2001). Multiple AGP forms are found in plant tissues, either in the wall or associated with

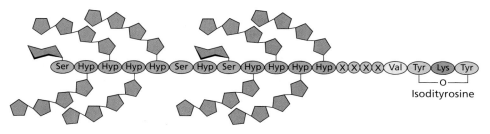

Tomato HRGP (extensive glycosylation)

FIGURE 15.14 A repeated hydroxyproline-rich motif from a molecule of HRGP from tomato, showing extensive glycosylation and the formation of intramolecular isodityrosine bonds. (After Carpita and McCann 2000.)

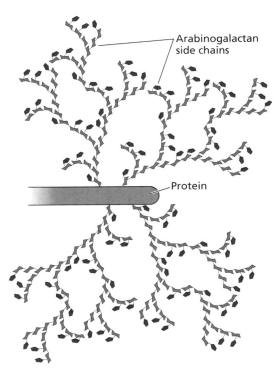

Figure 15.15 A highly branched arabinogalactan molecule. (After Carpita and McCann 2000.)

the plasma membrane (via a GPI anchor), and they display tissue- and cell-specific expression patterns.

AGPs may function in cell adhesion and in cell signaling during cell differentiation. As evidence for the latter idea, treatment of suspension cultures with exogenous AGPs or with agents that specifically bind AGPs is reported to influence cell proliferation and embryogenesis. AGPs are also implicated in the growth, nutrition, and guidance of pollen tubes through stylar tissues, as well as in other developmental processes (Gaspar et al. 2001). Finally, AGPs may also function as a kind of polysaccharide chaperone within secretory vesicles to reduce spontaneous association of newly synthesized polysaccharides until they are secreted to the cell wall.

New primary walls are assembled during cytokinesis

Primary walls originate de novo during the final stages of cell division, when the newly formed **cell plate** separates the two daughter cells and solidifies into a stable wall that is capable of bearing a physical load from turgor pressure.

The cell plate forms when Golgi vesicles and ER cisternae aggregate in the spindle midzone area of a dividing cell. This aggregation is organized by the **phragmoplast**, a complex assembly of microtubules, membranes, and vesicles that forms during late anaphase or early telophase (see Chapter 1). The membranes of the vesicles fuse with each other and with the lateral plasma membrane to become the new plasma membrane separating the daughter cells. The contents of the vesicles are the precursors from which the new middle lamella and the primary wall are assembled.

After a wall forms, it can grow and mature through a process that may be outlined as follows:

Synthesis → secretion → assembly →
expansion (in growing cells) →
cross-linking and secondary wall formation

The synthesis and secretion of the major wall polymers were described earlier. Here we will consider the assembly and expansion of the wall.

After their secretion into the extracellular space, the wall polymers must be assembled into a cohesive structure; that is, the individual polymers must attain the physical arrangement and bonding relationships that are characteristic of the wall. Although the details of wall assembly are not understood, the prime candidates for this process are self-assembly and enzyme-mediated assembly.

SELF-ASSEMBLY Self-assembly is an attractive model because it is mechanistically simple. Wall polysaccharides possess a marked tendency to aggregate spontaneously into organized structures. For example, isolated cellulose can be dissolved in strong solvents and then extruded to form stable fibers, called rayon.

Similarly, hemicelluloses may be dissolved in strong alkali; when the alkali is removed, these polysaccharides aggregate into concentric, ordered networks that resemble the native wall at the ultrastructural level. This tendency to aggregate can make the separation of hemicellulose into its component polymers technically difficult. In contrast, pectins are more soluble and tend to form dispersed, isotropic (randomly arranged) networks (gels). These observations indicate that the wall polymers have an inherent ability to aggregate into partly ordered structures.

ENZYME-MEDIATED ASSEMBLY In addition to self-assembly, wall enzymes may take part in putting the wall together. A prime candidate for enzyme-mediated wall assembly is **xyloglucan endotransglucosylase** (**XET**). This enzyme, which belongs to a large family of enzymes named **xyloglucan endotransglucosylase/hydrolases** (**XHTs**), has the ability to cut the backbone of a xyloglucan and to join one end of the cut xyloglucan with the free end of an acceptor xyloglucan (Figure 15.16). Such a transfer reaction integrates newly synthesized xyloglucans into the wall (Thompson and Fry 2001; Rose et al. 2002).

Other wall enzymes that might aid in assembly of the wall include glycosidases, pectin methyl esterases, and various oxidases. Some glycosidases remove the side chains of hemicelluloses. This "debranching" activity increases the tendency of hemicelluloses to adhere to the surface of cellulose microfibrils. Pectin methyl esterases hydrolyze the methyl esters that block the carboxyl groups of pectins. By unblocking the carboxyl groups, these enzymes increase the

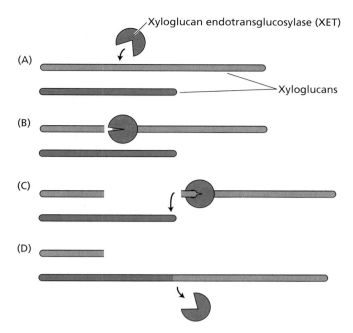

FIGURE 15.16 Action of xyloglucan endotransglucosylase (XET) to cut and stitch xyloglucan polymers into new configurations. Two xyloglucan chains are shown in (A) with two distinct patterns to emphasize their rearrangement. XET binds to the middle of one xyloglucan, cuts it (B), and transfers one end to the end of a second xyloglucan (C), resulting in one shorter and one longer xyloglucan (D). (After Smith and Fry 1991.)

concentration of acidic groups on the pectins and enhance the ability of pectins to form a Ca^{2+}-bridged gel network.

Oxidases such as peroxidase may catalyze cross-linking between phenolic groups (tyrosine, phenylalanine, ferulic acid) in wall proteins, pectins, and other wall polymers. Such oxidative cross-links tie lignin subunits together in complex ways (Figure 15.17), and they may likewise link other wall components together.

Secondary walls form in some cells after expansion ceases

After wall expansion ceases, cells sometimes continue to synthesize a **secondary wall**. Secondary walls are often quite thick, as in tracheids, fibers, and other cells that provide mechanical support to the plant (Figure 15.18). The rigid secondary wall in xylem is also important for preventing collapse of the water-conducting cells during periods of high water tension due to high transpiration.

Often such secondary walls are multilayered and differ in structure and composition from the primary wall. For example, the secondary walls in wood contain xylans rather than xyloglucans, as well as a higher proportion of cellulose. The orientation of the cellulose microfibrils may be more neatly aligned parallel to each other in secondary walls than in primary walls. Secondary walls are often (but not always) impregnated with lignin.

Lignin is a phenolic polymer with a complex, irregular pattern of linkages that link the aromatic alcohol subunits together (see Chapter 13). These subunits are synthesized from phenylalanine and are secreted to the wall, where they are oxidized in place by the enzymes peroxidase and laccase. As lignin forms in the wall, it displaces water from the matrix and forms a hydrophobic network that bonds tightly to cellulose and prevents wall enlargement (see Figure 15.17).

Lignin adds significant mechanical strength to cell walls and makes the walls hydrophobic, reducing the susceptibility of walls to attack by hydrolytic enzymes from pathogens. Lignin also reduces the digestibility of plant material by animals and interferes with the pulping process (conversion of wood into free fibers) for paper making. Current efforts at genetic engineering of lignin content and structure may improve the digestibility and nutritional content of plants used as animal fodder, as well as increase the value of cell walls for the production of paper and biofuel (ethanol used in autos).

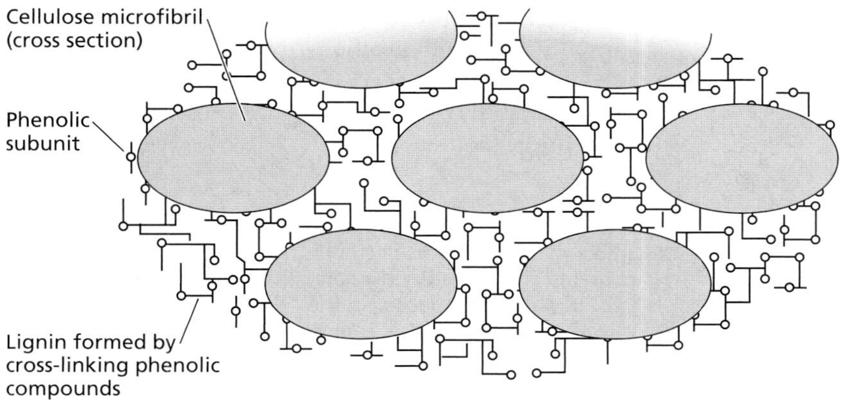

FIGURE 15.17 Diagram illustrating how the phenolic subunits of lignin infiltrate the space between cellulose microfibrils, where they become cross-linked. (Other components of the matrix are omitted from this diagram.)

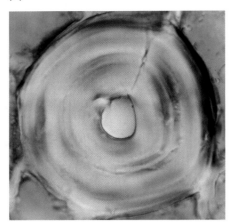

FIGURE 15.18 (A) Cross section of a *Podocarpus* sclereid, in which multiple layers in the secondary wall are visible. (B) Diagram of the cell wall organization often found in tracheids and other cells with thick secondary walls. Three distinct layers (S_1, S_2, and S_3) are formed interior to the primary wall. (Photo © David Webb.)

Patterns of Cell Expansion

During plant cell enlargement, new wall polymers are continuously synthesized and secreted at the same time that the preexisting wall is expanding. Wall expansion may be highly localized (as in the case of **tip growth**) or evenly distributed over the wall surface (**diffuse growth**) (Figure 15.19). Tip growth is characteristic of root hairs and pollen tubes; it is strongly regulated by components of the cytoskeleton, especially actin microfilaments (see **Web Essay 15.1**). Most of the other cells in the plant body exhibit diffuse growth. Cells such as fibers, some sclereids, and trichomes grow in a pattern that is intermediate between diffuse growth and tip growth.

Even in cells with diffuse growth, however, different parts of the wall may enlarge at different rates or in different directions. For example, in cortical cells of the stem, the end walls grow much less than side walls. This difference may be due to structural or enzymatic variations in specific walls or to variations in the stresses borne by different walls. As a consequence of this uneven pattern of wall expansion, plant cells may assume irregular forms.

Microfibril orientation influences growth directionality of cells with diffuse growth

During growth, the loosened cell wall is extended by physical forces generated from cell turgor pressure. Turgor pressure creates an outward-directed force, equal in all directions. The directionality of growth is determined largely by the structure of the cell wall—in particular, the orientation of cellulose microfibrils.

When cells first form in the meristem, they are isodiametric; that is, they have equal diameters in all directions. If the orientation of cellulose microfibrils in the primary cell wall were **isotropic** (randomly arranged), the cell would grow equally in all directions, expanding radially to generate a sphere (Figure 15.20A). In most plant cell walls, however, the arrangement of cellulose microfibrils is **anisotropic** (nonrandom).

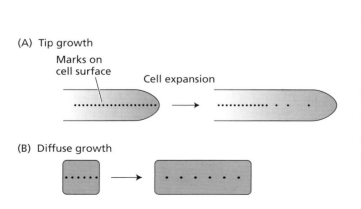

FIGURE 15.19 The cell surface expands differently during tip growth and diffuse growth. (A) Expansion of a tip-growing cell is confined to an apical dome at one end of the cell. If marks are placed on the cell surface and the cell is allowed to continue to grow, only the marks that were initially within the apical dome grow farther apart. Root hairs and pollen tubes are examples of plant cells that exhibit tip growth. (B) If marks are placed on the surface of a diffuse-growing cell, the distance between all the marks increases as the cell grows. Most cells in multicellular plants grow by diffuse growth.

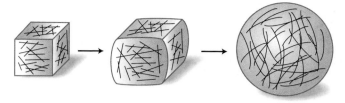

(A) Randomly oriented cellulose microfibrils

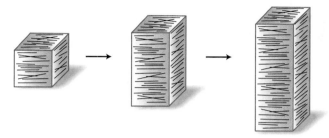

(B) Transverse cellulose microfibrils

FIGURE 15.20 The orientation of newly deposited cellulose microfibrils determines the direction of cell expansion. (A) If the cell wall is reinforced by randomly oriented cellulose microfibrils, the cell will expand equally in all directions, forming a sphere. (B) When most of the reinforcing cellulose microfibrils have the same orientation, the cell expands at right angles to the microfibril orientation and is constrained in the direction of the reinforcement. Here the microfibril orientation is transverse, so cell expansion is longitudinal.

Cellulose microfibrils are synthesized mainly in the lateral walls of cylindrical, enlarging cells such as cortical and vascular cells of stems and roots, or the giant internode cells of the filamentous green alga *Nitella*. Moreover, the cellulose microfibrils are deposited circumferentially (transversely) in these lateral walls, at right angles to the long axis of the cell. The circumferentially arranged cellulose microfibrils have been likened to hoops in a barrel, restricting growth in girth and promoting growth in length (see Figure 15.20B). However, because individual cellulose microfibrils do not actually form closed hoops around the cell, a more accurate analogy would be the glass fibers in fiberglass.

Fiberglass is a *complex composite material*, composed of an amorphous resin matrix reinforced by discontinuous strengthening elements, in this case glass fibers. In complex composites, rod-shaped crystalline elements exert their maximum reinforcement of the matrix in the direction parallel to their orientation, and their minimum reinforcement perpendicular to their orientation. The reinforcement of the wall is greater in the parallel direction because the matrix must physically scrape along the entire length of the fibers for lateral displacement to occur.

In contrast, when the material is stretched in the perpendicular direction, the matrix polymers need only slip over the diameters of the fibrous elements, resulting in little or no strengthening of the matrix. Because the glass fibers in fiberglass are randomly arranged, fiberglass is equally strong in all directions; that is, it is mechanically isotropic.

Plant cell walls, like fiberglass, are complex composite materials, composed of an amorphous phase and crystalline elements. Unlike fiberglass, however, the microfibril strengthening elements of a typical primary cell wall are transversely oriented, rendering the wall structurally and mechanically *anisotropic*. For this reason, growing plant cells tend to elongate, and they increase only minimally in girth.

Cell wall deposition continues as cells enlarge. According to the **multinet growth hypothesis**, each successive wall layer is stretched and thinned during cell expansion, so the microfibrils would be expected to be passively reoriented in the longitudinal direction—that is, in the direction of growth. Consistent with the multinet growth hypothesis, electron micrographs of the cell wall indicate that the most recently deposited, inner wall layers have transversely oriented cellulose microfibrils, while the microfibrils of the older, outer wall layers have a more disordered arrangement.

In a study to test the ability of cell wall microfibrils to passively reorient in response to wall tension, frozen-thawed cucumber hypocotyl segments were mechanically stretched under conditions that mimicked normal growth, and the effect of this extension on the orientation of the cellulose microfibrils of the wall was examined. Surprisingly, stretching the wall longitudinally by 20 to 30% failed to alter the transverse angle of the microfibrils of the inner wall surface, suggesting that the microfibrils separated from each other in a coordinate fashion, unlike extension of fiberglass-like material (Marga et al. 2005). These results implicate a selective loosening of the tethers that hold microfibrils together, rather than a generalized loosening of the matrix.

In contrast to the microfibrils of the inner wall layer, it is clear from electron micrographs that those of the outer wall layers do reorient longitudinally, presumably because of nonspecific entanglements in the matrix. Thus, the multinet model holds for the outer wall layers. However, because of thinning and fragmentation of the older wall layers as the wall extends, these outer layers have much less influence on the direction of cell expansion than do the newly deposited inner layers. Accordingly, the inner one-fourth of the wall bears nearly all the stress due to turgor pressure and determines the directionality of cell expansion (see **Web Topic 15.4**).

So far we have considered only a simple pattern of diffuse growth. So-called "pavement cells" in the epidermis of many dicot leaves, however, present a more complicated situation. These cells are highly lobed, creating an interlocking pattern resembling a jigsaw puzzle (Figure 15.21A, B). This pattern of interdigitating cell wall expansion combines aspects of diffuse growth and tip growth and requires the action of GTP-binding proteins called ROP (*Rho*-related GTPase from *p*lants) GTPases, and their activating proteins called RICs (*ROP*-*i*nteracting *C*RIB motif-containing

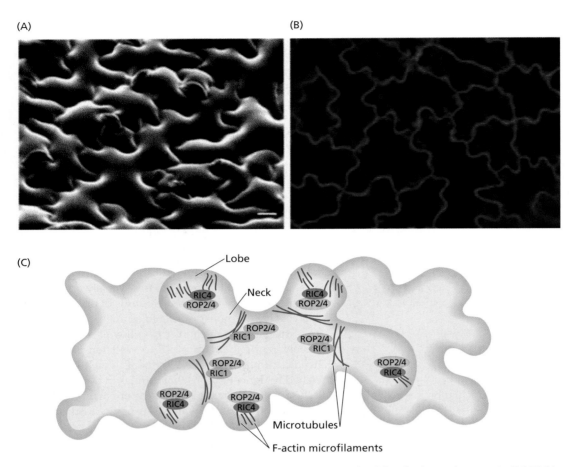

FIGURE 15.21 Interdigitating cell growth of leaf pavement cells and its regulation by ROP GTPases. (A) Scanning electron micrograph of pavement cells from an Arabidopsis leaf. Note the jigsaw puzzle-like appearance. (B) Immunofluorescent image of pavement cells shows more clearly the lobes and indentations formed by interdigitated cells. (C) A model to explain the role of ROP GTPases and their effectors (RICs) in leaf morphogenesis. ROP2/4 GTPases, when activated by RIC4, promote actin microfilament formation in regions of growing lobes, whereas when activated by RIC1 they promote microtubule bundling at neck regions. These cytoskeletal changes somehow act as signals to direct the direction of wall growth. (A courtesy of Dan Szymanski; B from Settleman 2005, courtesy of J. Settleman; C after Fu et al. 2005.)

proteins) (see Figure 15.21C). These proteins organize the cytoskeleton (actin microfilaments and tubulin microtubules) which then modify wall growth (Fu et al. 2005). How does the cytoskeleton influence wall growth? This topic is covered in the next section.

Cortical microtubules influence the orientation of newly deposited microfibrils

Newly deposited cellulose microfibrils usually are coaligned with microtubule arrays in the cytoplasm, close to the plasma membrane (Figure 15.22) (Baskin 2001; Baskin et al. 2004). A striking example occurs in xylem vessel elements, where bands of cortical microtubules mark the sites of secondary wall thickenings and also the sites of cellulose synthase A (CesA) localization (Gardiner et al. 2003). Moreover, experimental disruptions of microtubule organization with drugs or by genetic defects often leads to disorganized wall structure and disorganized growth. For example, several drugs bind to tubulin, the subunit protein of microtubules, causing them to depolymerize. When growing roots are treated with a microtubule-depolymerizing drug, such as oryzalin, the region of elongation expands laterally, becoming bulbous and tumorlike (Figure 15.23). This disrupted growth is due to the isotropic expansion of the cells; that is, they enlarge like a sphere instead of elongating. The drug-induced destruction of microtubules in the growing cells interferes with the transverse deposition of cellulose. Cellulose microfibrils continue to be synthesized in the absence of microtubules, but they are deposited randomly and consequently the cells expand equally in all directions. These observations have led to the suggestion that microtubules serve as tracks that guide or direct the movement of CesA complexes as they synthesize microfibrils. However, the relationship between the overall microtubule and microfibril ori-

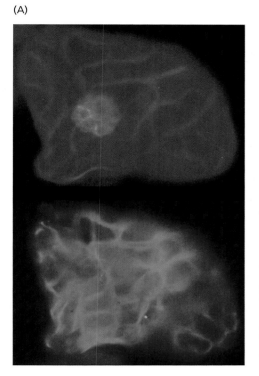
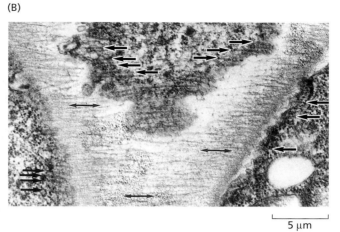

FIGURE 15.22 The orientation of microtubules in the cortical cytoplasm mirrors the orientation of newly deposited cellulose microfibrils in the walls of cells that are elongating. (A) The arrangement of microtubules can be revealed with fluorescently labeled antibodies to the microtubule protein tubulin. In this differentiating tracheary element from a *Zinnia* cell suspension culture, the pattern of microtubules (green) mirrors the orientation of the cellulose microfibrils in the wall, as shown by calcofluor staining (blue). (B) The alignment of cellulose microfibrils in the cell wall can sometimes be seen in grazing sections prepared for electron microscopy, as in this micrograph of a developing sieve tube element in a root of *Azolla* (a water fern). The longitudinal axis of the root and the sieve tube element runs vertically. Both the wall microfibrils (double-headed arrows) and the cortical microtubules (single-headed arrows) are aligned transversely (see also Figure 15.7). (A courtesy of Robert W. Seagull; B courtesy of A. Hardham.)

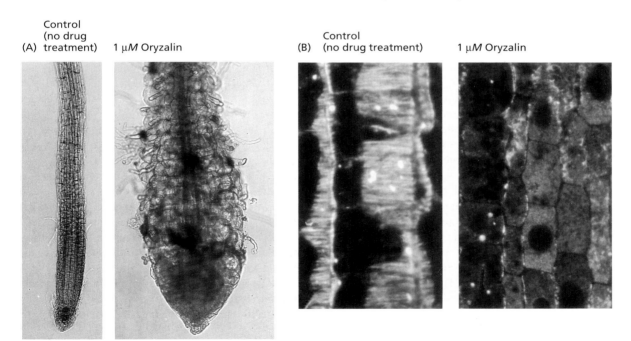

FIGURE 15.23 The disruption of cortical microtubules results in a dramatic increase in radial cell expansion and a concomitant decrease in elongation. (A) Root of Arabidopsis seedling treated with the microtubule-depolymerizing drug oryzalin (1 μM) for 2 days before this photomicrograph was taken. The drug has altered the polarity of growth. (B) Microtubules were visualized by means of an indirect immunofluorescence technique and an antitubulin antibody. Whereas cortical microtubules in the control are oriented at right angles to the direction of cell elongation, very few microtubules remain in roots treated with 1 μM oryzalin. (From Baskin et al. 1994, courtesy of T. Baskin.)

entations is not always so clear, and thus other mechanisms of controlling growth anisotropy are still being sought (Sugimoto et al. 2003; Baskin et al. 2004) (see **Web Essay 15.2**).

The Rate of Cell Elongation

Plant cells typically expand 10- to 100-fold in volume before reaching maturity. In extreme cases, cells may enlarge more than 10,000-fold in volume (e.g., xylem vessel elements) compared with their meristematic initials. The cell wall undergoes this profound expansion without losing its mechanical integrity and without becoming thinner. Thus, newly synthesized polymers are integrated into the wall without destabilizing it. Exactly how this integration is accomplished is uncertain, although self-assembly and xyloglucan endotransglucosylase (XET) play important roles, as already described.

This integrating process may be particularly critical for rapidly growing root hairs, pollen tubes, and other specialized cells that exhibit tip growth, in which the region of wall deposition and surface expansion is localized to the hemispherical dome at the apex of the tube-like cell, and cell expansion and wall deposition must be closely coordinated.

In rapidly growing cells with tip growth, the wall doubles its surface area and is displaced to the nonexpanding part of the cell within minutes. This is a much greater rate of wall expansion than is typically found in cells with diffuse growth, and tip-growing cells are therefore susceptible to wall thinning and bursting. Although diffuse growth and tip growth appear to be different growth patterns, both types of wall expansion must have analogous, if not identical, processes of polymer integration, stress relaxation, and wall polymer creep.

Many factors influence the rate of cell wall expansion. Cell type and age are important developmental factors. So, too, are hormones such as auxin and gibberellin. Environmental conditions such as light and water availability may likewise modulate cell expansion. These internal and external factors most likely modify cell expansion by altering the way in which the cell wall is loosened, so that it yields (stretches irreversibly) differently. In this context we speak of the *yielding properties* of the cell wall.

In this section we will first examine the biomechanical and biophysical parameters that characterize the yielding properties of the wall. For cells to expand at all, the rigid cell wall must be loosened in some way. The type of wall loosening involved in plant cell expansion is termed *stress relaxation*.

According to the acid growth hypothesis for auxin action (see Chapter 19), one mechanism that causes wall stress relaxation and wall yielding is cell wall acidification, resulting from proton extrusion across the plasma membrane. Cell wall loosening is enhanced at acidic pH. A little later we will explore the biochemical basis for acid-induced wall loosening and stress relaxation, including the role of a special class of wall-loosening proteins called *expansins*.

As the cell approaches its maximum size, its growth rate diminishes and finally ceases altogether. At the end of this section we will consider the process of cell wall rigidification that leads to the cessation of growth.

Stress relaxation of the cell wall drives water uptake and cell elongation

Because the cell wall is the major mechanical restraint that limits cell expansion, much attention has been given to its physical properties. As a hydrated polymeric material, the plant cell wall has physical properties that are intermediate between those of a solid and those of a liquid. We call these **viscoelastic**, or **rheological** (flow), **properties**. Walls of cells that are growing are generally less rigid than those of nongrowing cells, and under appropriate conditions they exhibit a long-term irreversible stretching, or **yielding**, that is lacking or nearly lacking in nongrowing walls.

Stress relaxation is a crucial concept for understanding how cell walls enlarge (Cosgrove 1997). The term *stress* is used here in the mechanical sense, as force per unit area. Wall stresses arise as an inevitable consequence of cell turgor. The turgor pressure in growing plant cells is typically between 0.3 and 1.0 megapascals (MPa). Turgor pressure stretches the cell wall and generates a counterbalancing physical stress or tension in the wall. Because of cell geometry (a large pressurized volume contained by a thin wall), this wall tension is equivalent to 10 to 100 MPa of tensile stress—a very large stress indeed.

This simple fact has important consequences for the mechanics of cell enlargement. Whereas animal cells can change shape in response to cytoskeleton-generated forces, such forces are negligible compared with the turgor-generated forces that are resisted by the plant cell wall. To change shape, plant cells must thus control the direction and rate of wall expansion, which they do by depositing cellulose in a biased orientation (which determines the directionality of cell wall expansion) and by selectively loosening the bonding between cell wall polymers. This biochemical loosening enables cellulose microfibrils and their associated matrix polysaccharides to slip by each other, thereby increasing the wall surface area. At the same time, such loosening reduces the physical stress in the wall.

Wall stress relaxation is crucial because it allows growing plant cells to reduce their turgor and water potentials, which enables them to absorb water and to expand. Without stress relaxation, wall synthesis would only thicken the wall, not expand it. During secondary-wall deposition in nongrowing cells, for example, stress relaxation does not occur.

The rate of cell expansion is governed by two growth equations

When plant cells enlarge before maturation, the increase in volume is generated mostly by water uptake. This water ends up mainly in the vacuole, which takes up an ever larger proportion of the cell volume as the cell grows. Here we will describe how growing cells regulate their water uptake and how this uptake is coordinated with wall yielding.

Water uptake by growing cells is a passive process. There are no active water pumps; instead the growing cell is able to lower the water potential inside the cell so that water is taken up spontaneously in response to a water potential difference, without direct energy expenditure.

We define the water potential difference, $\Delta\Psi_w$ (expressed in megapascals, MPa), as the water potential outside the cell minus the water potential inside (see Chapters 3 and 4). The rate of uptake also depends on the surface area of the cell (A, in square meters) and the permeability of the plasma membrane to water (Lp, in meters per second per MPa).

Membrane Lp is a measure of how readily water crosses the membrane, and it is a function of the physical structure of the membrane and the activity of aquaporins (see Chapter 3). Thus we have the rate of water uptake in volume units, $\Delta V/\Delta t$, expressed in cubic meters per second. Assuming that a growing cell is in contact with pure water (with zero water potential), then

$$\text{Rate of water uptake} = A \times Lp\,(\Delta\Psi_w) \\ = A \times Lp\,(\Psi_o - \Psi_i) \qquad (15.1)$$

This equation states that the rate of water uptake depends only on the cell area, membrane permeability to water, cell turgor, and osmotic potential.

Equation 15.1 is valid for both growing and nongrowing cells in pure water. But how can we account for the fact that growing cells can continue to take up water for a long time, whereas nongrowing cells soon cease water uptake?

In a nongrowing cell, water absorption would increase cell volume, causing the protoplast to push harder against the cell wall, thereby increasing cell turgor pressure, Ψ_p. This increase in Ψ_p would increase cell water potential Ψ_w, quickly bringing $\Delta\Psi_w$ to zero. Water uptake would then cease.

In a growing cell, $\Delta\Psi_w$ is prevented from reaching zero because the cell wall is "loosened": it yields irreversibly to the forces generated by turgor and thereby reduces simultaneously the wall stress and the cell turgor. This process is called **stress relaxation**, and it is the crucial physical difference between growing and nongrowing cells.

Stress relaxation can be understood as follows. In a turgid cell, the cell contents push against the wall, causing the wall to stretch elastically (i.e., reversibly) and giving rise to a counterforce, wall stress. In a growing cell, biochemical loosening enables the wall to yield inelastically (irreversibly) to the wall stress. Because water is nearly incompressible, only an infinitesimal expansion of the wall is needed to reduce cell turgor pressure and, simultaneously, wall stress. Thus, *stress relaxation is a decrease in wall stress with nearly no change in wall dimensions.*

As a consequence of wall stress relaxation, the cell water potential is reduced and water flows into the cell, causing a measurable extension of the cell wall and increasing cell surface area and volume. Sustained growth of plant cells entails simultaneous stress relaxation of the wall (which tends to reduce turgor pressure) and water absorption (which tends to increase turgor pressure).

Empirical evidence has shown that wall relaxation and expansion depend on turgor pressure. As turgor is reduced, wall relaxation and growth slow down. Growth usually ceases before turgor reaches zero. The turgor value at which growth ceases is called the **yield threshold** (usually represented by the symbol Y). This dependence of cell wall expansion on turgor pressure is expressed in the following equation:

$$GR = m(\Psi_p - Y) \qquad (15.2)$$

where GR is the cell growth rate, and m is the coefficient that relates growth rate to the turgor in excess of the yield threshold. The coefficient m is usually called **wall extensibility** and is *the slope of the line relating growth rate to turgor pressure.*

Under conditions of steady-state growth, GR in Equation 15.2 is the same as the rate of water uptake in Equation 15.1. That is, the increase in the volume of the cell equals the volume of water taken up. The two equations are plotted in Figure 15.24. Note that the two processes of wall expansion

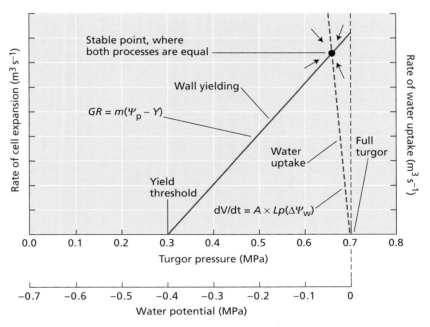

FIGURE 15.24 Graphic representation of the two equations that relate water uptake and cell expansion to cell turgor pressure and cell water potential. The values for the rates of cell expansion and water uptake are arbitrary. Steady-state growth is attained only at the point where the two equations intersect. Any imbalance between water uptake and wall expansion will result in changes in cell turgor and bring the cell back to this stable point of intersection between the two processes.

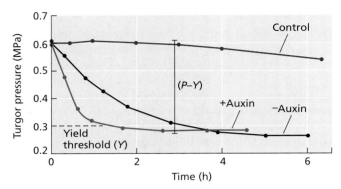

Figure 15.25 Reduction of cell turgor pressure (water potential) by stress relaxation. In this experiment, the excised stem segments from growing pea seedlings were incubated in solution with or without auxin, then blotted dry and sealed in a humid chamber. Cell turgor pressure (P) was measured at various time points. The segments treated with auxin rapidly reduced their turgor to the yield threshold (Y), as a result of rapid wall relaxation. The segments without auxin showed a slower rate of relaxation. The control segments were treated the same as the group treated with auxin, except that they remained in contact with a drop of water, which prevented wall relaxation. (After Cosgrove 1985.)

and water uptake show opposing reactions to a change in turgor. For example, an increase in turgor increases wall extension but reduces water uptake. Under normal conditions, turgor is dynamically balanced in a growing cell exactly at the point where the two lines intersect. At this point both equations are satisfied, and water uptake is exactly matched by enlargement of the wall chamber.

This intersection point in Figure 15.24 is the steady-state condition, and any deviations from this point will cause transient imbalances between the processes of water uptake and wall expansion. The result of these imbalances is that turgor will return to the point of intersection, the point of dynamic steady state for the growing cell.

The regulation of cell growth—for example, by hormones or by light—typically is accomplished by regulation of the biochemical processes that regulate wall loosening and stress relaxation. Such changes can be measured as a change in m or in Y.

The water uptake that is induced by wall stress relaxation enlarges the cell and tends to restore wall stress and turgor pressure to their equilibrium values, as we have shown. However, if growing cells are physically prevented from taking up water, wall relaxation progressively reduces cell turgor. This situation may be detected, for example, by turgor measurements with a *pressure probe* or by water potential measurements with a *psychrometer* or a *pressure chamber* (see **Web Topic 3.6**). Figure 15.25 shows the results of such an experiment.

Acid-induced growth is mediated by expansins

An important characteristic of growing cell walls is that they extend much faster at acidic pH than at neutral pH (Rayle and Cleland 1992). This phenomenon is called **acid growth**. In living cells, acid growth is evident when growing cells are treated with acid buffers or with the drug fusicoccin, which induces acidification of the cell wall solution by activating an H^+-ATPase in the plasma membrane.

An example of acid-induced growth can be found in the initiation of the root hair, where the local wall pH drops to a value of 4.5 at the time when the epidermal cell begins to bulge outward (Bibikova et al. 1998). Auxin-induced growth is also associated with wall acidification, but it is probably not sufficient to account for the entire growth induction by this hormone (see Chapter 19), and other wall-loosening processes may be involved. Some work, for example, implicates the production of hydroxyl radicals in wall loosening during auxin-induced growth (Schopfer 2001). Nevertheless, this pH-dependent mechanism of wall extension appears to be an evolutionarily conserved process common to all land plants (Cosgrove 2000) and involved in a variety of growth processes.

Acid growth may also be observed in isolated cell walls which lack normal cellular, metabolic, and synthetic processes. Such an observation requires the use of an extensometer to put the walls under tension and to measure the pH-dependent wall creep (Figure 15.26).

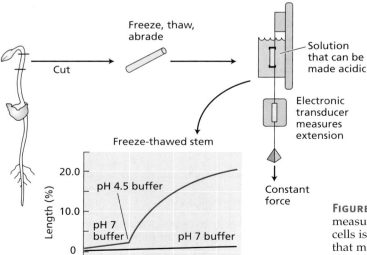

Figure 15.26 Acid-induced extension of isolated cell walls, measured in an extensometer. The wall sample from killed cells is clamped and put under tension in an extensometer that measures the length with an electronic transducer attached to a clamp. When the solution surrounding the wall is replaced with an acidic buffer (e.g., pH 4.5), the wall extends irreversibly in a time-dependent fashion (it creeps).

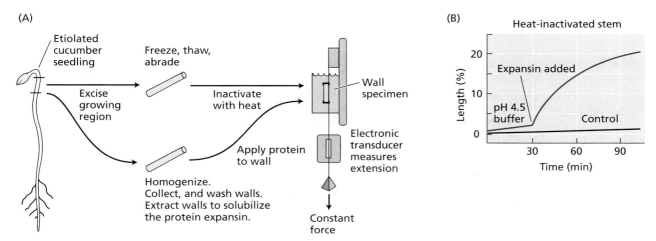

FIGURE 15.27 Scheme for the reconstitution of extensibility of isolated cell walls. (A) Cell walls are prepared as in Figure 15.26 and briefly heated to inactivate the endogenous acid extension response. To restore this response, proteins are extracted from growing walls and added to the solution surrounding the wall. (B) Addition of proteins containing expansins restores the acid extension properties of the wall. (After Cosgrove 1997.)

The term **creep** refers to a time-dependent irreversible extension, typically the result of slippage of wall polymers relative to one another. When growing walls are incubated in neutral buffer (pH 7) and clamped in an extensometer, the walls extend briefly when tension is applied, but extension soon ceases. When transferred to an acidic buffer (pH 5 or less), the wall begins to extend rapidly, in some instances continuing for many hours.

This acid-induced creep is characteristic of walls from growing cells, but it is not observed in mature (nongrowing) walls. When walls are pretreated with heat, proteases, or other agents that denature proteins, they lose their acid growth ability. Such results indicate that acid growth is not due simply to the physical chemistry of the wall (e.g., a weakening of the pectin gel), but is catalyzed by one or more wall proteins.

The idea that proteins are required for acid growth was confirmed in reconstitution experiments in which heat-inactivated walls were restored to nearly full acid growth responsiveness by addition of proteins extracted from growing walls (Figure 15.27). The active components proved to be a group of proteins that were named **expansins** (Cosgrove 2000). These proteins catalyze the pH-dependent extension and stress relaxation of cell walls. They are effective in catalytic amounts (about 1 part protein per 5000 parts wall, by dry weight).

The molecular basis for expansin action on wall rheology is still uncertain, but most evidence indicates that expansins cause wall creep by loosening noncovalent adhesion between wall polysaccharides. Binding studies suggest that expansins may act at the interface between cellulose and one or more hemicelluloses.

With the completion of the Arabidopsis genome, we now know that expansins belong to a large superfamily of genes, divided into two major expansin families, *α*-**expansins** (**EXPA**) and *β*-**expansins** (**EXPB**), plus two smaller families of unknown function (Sampedro and Cosgrove 2005). The two kinds of expansins act on different polymers of the cell wall and work coordinately during cell growth, fruit ripening, and other situations where wall loosening occurs (Cosgrove 2000).

Glucanases and other hydrolytic enzymes may modify the matrix

Several types of experiments implicate endo-(1→4)β-D-glucanases (**EGases**) in cell wall loosening, especially during auxin-induced cell elongation (see Chapter 19). These enzymes cut (1→4)β-D-glucans randomly along their backbone and use water to complete hydrolysis. Matrix glucans such as xyloglucan show enhanced hydrolysis and turnover in excised segments when growth is stimulated by auxin. Interference with this hydrolytic activity by antibodies or lectins reduces growth in excised segments (Hoson 1993).

EGase activity is associated with growing tissues, and glucanase treatments may stimulate cell growth. Such results support the idea that glucanases promote wall stress relaxation and expansion. However, plant EGases have not been found to cause extension of isolated cell walls, but may instead act indirectly by weakening the cell wall in a way that promotes expansin-mediated polymer creep.

Plants contain two gene families that encode enzymes with EGase activity. The first family consists of the large **XTH family** discussed earlier, some members of which have xyloglucan endotransglucosylase activity (see Figure 15.16), while others have hydrolase activity. XTH enzymes may affect wall mechanics via their action on xyloglucan (Takeda et al. 2002). It has also been suggested that some enzymes in this group may act on other substrates (e.g. xylans), but this has not been experimentally confirmed.

The second family of plant EGase is also a large family (Libertini et al. 2004) with diverse presumptive functions, including wall softening during fruit ripening and abscission, cellulose formation (i.e., trimming nascent glucans for microfibril formation), and growth. These enzymes may act on xyloglucan or on the noncrystalline regions of cellulose microfibrils (Park et al. 2003).

Structural changes accompany the cessation of wall expansion

The growth cessation that occurs during cell maturation is generally irreversible and is typically accompanied by a reduction in wall extensibility, as measured by various biophysical methods. These physical changes in the wall might come about by (1) a reduction in wall-loosening processes, (2) an increase in wall cross-linking, or (3) an alteration in the composition of the wall, making for a more rigid structure or one less susceptible to wall loosening. There is some evidence for each of these ideas (Cosgrove 1997).

Several modifications of the maturing wall may contribute to wall rigidification:

- Newly secreted matrix polysaccharides may be altered in structure so as to form tighter complexes with cellulose or other wall polymers, or they may be resistant to wall-loosening activities.
- Removal of $(1\rightarrow3,1\rightarrow4)\beta$-D-glucan in grass cell walls is coincident with growth cessation in these walls and may cause wall rigidification.
- De-esterification of pectins, leading to more-rigid pectin gels, is similarly associated with growth cessation in both grasses and dicotyledons.
- Cross-linking of phenolic groups in the wall (such as tyrosine residues in HRGPs, ferulic acid residues attached to pectins, and lignin) generally coincides with wall maturation and is believed to be mediated by peroxidase, a putative wall rigidification enzyme.

Thus, many structural changes in the wall occur during and after cessation of growth, and it has not yet been possible to identify the significance of individual processes for cessation of wall expansion.

Wall Degradation and Plant Defense

The plant cell wall is not simply an inert, static exoskeleton. In addition to acting as a mechanical restraint, the wall serves as an extracellular matrix that interacts with cell surface proteins, providing positional and developmental information. It contains numerous enzymes and smaller molecules that are biologically active and that can modify the physical properties of the wall, sometimes within seconds. In some cases, wall-derived molecules can also act as signals to inform the cell of environmental conditions, such as the presence of pathogens. This is an important aspect of the defense response of plants (see Chapter 13).

Walls may also be substantially modified long after growth has ceased. For instance, the cell wall may be massively degraded, such as occurs in ripening fruit or in the endosperm of germinating seeds. In cells that make up the abscission zones of leaves and fruits (see Chapter 22), the middle lamella may be selectively degraded, with the result that the cells become unglued and separate. Cells may also separate selectively during the formation of intercellular air spaces, during the emergence of the root from germinating seeds, and during other developmental processes. Plant cells may also modify their walls during pathogen attack as a form of defense.

In the sections that follow we will consider two types of dynamic changes that can occur in mature cell walls: hydrolysis and oxidative cross-linking. We will also discuss how fragments of the cell wall released during pathogen attack, or even during normal cell wall turnover, may act as cellular signals that influence metabolism and development.

Enzymes mediate wall hydrolysis and degradation

Hemicelluloses and pectins may be modified and broken down by a variety of enzymes that are found naturally in the cell wall. This process has been studied in greatest detail in ripening fruit, in which softening is thought to be the result of disassembly of the wall (Rose and Bennett 1999). Glucanases and related enzymes may hydrolyze the backbone of hemicelluloses. Xylosidases and related enzymes may remove the side branches from xyloglucan (particularly xyloglucan fragments, or oligomers). Transglycosylases may cut and join hemicelluloses together. Such enzymatic changes may alter the physical properties of the wall, for example, by changing the viscosity of the matrix or by altering the tendency of the hemicelluloses to stick to cellulose.

Messenger RNAs for expansin are expressed in ripening tomato fruit, suggesting that they play a role in wall disassembly (Rose et al. 1997). Similarly, softening fruits express high levels of pectin methyl esterase, which hydrolyzes the methyl esters from pectins. This hydrolysis makes the pectin more susceptible to subsequent hydrolysis by pectinases and related enzymes. The presence of these and related enzymes in the cell wall indicates that walls are capable of significant modification during development.

Oxidative bursts accompany pathogen attack

When plant cells are wounded or treated with certain low molecular weight elicitors (see Chapter 13), they activate a defense response that results in the production of high concentrations of hydrogen peroxide, superoxide radicals, and other active oxygen species in the cell wall. This "oxidative burst" appears to be part of a defense response against invading pathogens (Brisson et al. 1994; Otte et al. 2001) (see Chapter 13).

Active oxygen species may directly attack the pathogenic organisms, and they may indirectly deter subsequent invasion by the pathogenic organisms by causing a rapid cross-linking of phenolic components of the cell wall. In tobacco stems, for example, proline-rich structural proteins of the wall become rapidly insolubilized upon wounding or elicitor treatment, and this cross-linking is associated with an oxidative burst and with a mechanical stiffening of the cell walls.

Wall fragments can act as signaling molecules

Degradation of cell walls can result in the production of biologically active fragments 10 to 15 residues long, called **oligosaccharins**, that may be involved in natural developmental responses and in defense responses (see **Web Topic 15.5**). Some of the reported physiological and developmental effects of oligosaccharins include stimulation of phytoalexin synthesis, oxidative bursts, ethylene synthesis, membrane depolarization, changes in cytoplasmic calcium, induced synthesis of pathogen-related proteins such as chitinase and glucanase, other systemic and local "wound" signals, and alterations in the growth and morphogenesis of isolated tissue samples (John et al. 1997).

The best-studied examples are oligosaccharide elicitors produced during pathogen invasion (see Chapter 13). For example, the fungus *Phytophthora* secretes an endopolygalacturonase (a type of pectinase) during its attack on plant tissues. As this enzyme degrades the pectin component of the plant cell wall, it produces pectin fragments—**oligogalacturonans**—that elicit multiple defense responses by the plant cell (Figure 15.28). The oligogalacturonans that are 10 to 13 residues long are most active in these responses.

Plant cell walls also contain a β-D-glucanase that attacks the β-D-glucan that is specific to the fungal cell wall. When this enzyme attacks the fungal wall, it releases glucan oligomers with potent elicitor activity. The wall components serve in this case as part of a sensitive system for the detection of pathogen invasion.

Plants and microbes also possess inhibitory proteins that block the activity of the each other's degradative enzymes (York et al. 2004). For example, plants have inhibitory proteins that inhibit or otherwise modify the activity of microbial (but not plant) polygalacturonases, glucanases, and xylanases, presumably to thwart microbial attacks. A similar trick is played out by some plant pathogens, which secrete proteins that inhibit plant defense enzymes.

Summary

The architecture, mechanics, and function of plants depend crucially on the structure of the cell wall. The wall is secreted and assembled as a complex structure that varies in form and composition as the cell differentiates. Primary cell walls are synthesized in actively growing cells, and secondary cell walls are deposited in certain cells, such as xylem vessel elements and sclerenchyma, after cell expansion ceases.

The basic model of the primary wall is of a network of cellulose microfibrils embedded in a matrix of hemicelluloses, pectins, and structural proteins. Cellulose microfibrils are highly ordered arrays of glucan chains synthesized at the surface of the cell by protein complexes called particle rosettes. These complexes contain three different cellulose synthase (CesA) isoforms that associate with each other to form a hexameric subunit. Matrix polysaccharides are synthesized in the Golgi apparatus and secreted via vesicles. Hemicelluloses bind microfibrils together, and pectins form hydrophilic gels that can become cross-linked by calcium ions. Wall assembly is partly spontaneous, but also may be mediated by enzymes. For example, xyloglucan endotransglucosylase (XET) has the ability to carry out transglycosylation reactions that integrate newly synthesized xyloglucans into the wall.

Secondary walls differ from primary walls in that they contain a higher percentage of cellulose, they have different hemicelluloses, and lignin replaces pectins in the matrix. Secondary walls can also become highly thickened, sculpted, and embedded with specialized structural proteins.

In diffuse-growing cells, growth directionality is determined by wall structure, in particular the orientation of the cellulose microfibrils, which in turn is determined by the orientation of microtubules in the cytoplasm. Upon leaving the meristem, plant cells typically elongate greatly. Cell enlargement is limited by the ability of the cell wall to undergo polymer creep, which in turn is controlled in a complex way by the adhesion of wall polymers to one another and by the influence of pH on wall-loosening proteins such as expansins, glucanases, and other enzymes.

According to the acid growth hypothesis, proton extrusion by the plasma membrane H^+-ATPase acidifies the wall, activating the protein

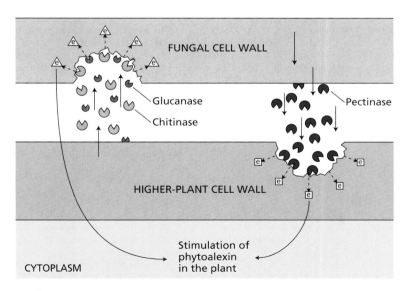

FIGURE 15.28 Scheme for the production of oligosaccharins during fungal invasion of plant cells. Enzymes secreted by the plant, such as chitinase and glucanase, attack the fungal wall, releasing oligosaccharins that elicit the production of defense compounds (phytoalexins) in the plant. Similarly, fungal pectinase releases biologically active oligosaccharins from the plant cell wall. (After Brett and Waldron 1996.)

expansin. Expansins induce stress relaxation of the wall by loosening the bonds holding microfibrils together. The cessation of cell elongation appears to be due to cell wall rigidification caused by an increase in the number of cross-links.

Hydrolytic enzymes may degrade mature cell walls completely or selectively during fruit ripening, seed germination, and the formation of abscission layers. Cell walls can also undergo oxidative cross-linking in response to pathogen attack. In addition, pathogen attack may release cell wall fragments, and certain wall fragments have been shown to be capable of acting as cell signaling agents.

Web Material

Web Topics

15.1 Terminology for Polysaccharide Chemistry
A brief review of terms used to describe the structures, bonds, and polymers in polysaccharide chemistry is provided.

15.2 Molecular Model for the Synthesis of Cellulose and Other Wall Polysaccharides That Consist of a Disaccharide Repeat
A model is presented for the polymerization of cellobiose units into glucan chains by the enzyme cellulose synthase.

15.3 Matrix Components of the Cell Wall
The secretion of xyloglucan and glycosylated proteins by the Golgi can be demonstrated at the ultrastructural level.

15.4 The Mechanical Properties of Cell Walls: Studies with *Nitella*
Experiments have demonstrated that the inner 25% of the cell wall determines the directionality of cell expansion.

15.5 Structure of Biologically Active Oligosaccharins
Some cell wall fragments have been demonstrated to have biological activity.

Web Essays

15.1 Calcium Gradients and Oscillations in Growing Pollen Tube
Calcium plays a role in regulating pollen tube tip growth.

15.2 Microtubules, Microfibrils, and Growth Anisotropy
The orientations of microtubules and/or microfibrils are not always correlated with the directionality of growth.

Chapter References

Amor, Y., Haigler, C. H., Johnson, S., Wainscott, M., and Delmer, D. P. (1995) A membrane-associated form of sucrose synthase and its potential role in synthesis of cellulose and callose in plants. *Proc. Natl. Acad. Sci. USA* 92: 9353–9357.

Arioli, T., Peng, L., Betzner, A. S., Burn, J., Wittke, W., Herth, W., Camilleri, C., Hofte, H., Plazinski, J., Birch, R., et al. (1998) Molecular analysis of cellulose biosynthesis in Arabidopsis. *Science* 279: 717–720.

Baskin, T. I. (2001) On the alignment of cellulose microfibrils by cortical microtubules: A review and a model. *Protoplasma* 215: 150–171.

Baskin, T. I., Beemster, G. T., Judy-March, J. E., and Marga, F. (2004) Disorganization of cortical microtubules stimulates tangential expansion and reduces the uniformity of cellulose microfibril alignment among cells in the root of Arabidopsis. *Plant Physiol.* 135: 2279–2290.

Baskin, T. I., Wilson, J. E., Cork, A., and Williamson, R. E. (1994) Morphology and microtubule organization in Arabidopsis roots exposed to oryzalin or taxol. *Plant Cell Physiol.* 35: 935–942.

Bibikova, T. N., Jacob, T., Dahse, I., and Gilroy, S. (1998) Localized changes in apoplastic and cytoplasmic pH are associated with root hair development in *Arabidopsis thaliana*. *Development* 125: 2925–2934.

Brett, C. T., and Waldron, K. (1996) *Physiology and Biochemistry of Plant Cell Walls*. Chapman and Hall, London.

Brisson, L. F., Tenhaken, R., and Lamb, C. (1994) Function of oxidative cross-linking of cell wall structural proteins in plant disease resistance. *Plant Cell* 6: 1703–1712.

Brown, R. M., Jr., Saxena, I. M., and Kudlicka, K. (1996) Cellulose biosynthesis in higher plants. *Trends Plant Sci.* 1: 149–155.

Carpita, N. C. (1996) Structure and biogenesis of the cell walls of grasses. *Annu. Rev. Plant Physiol. Plant Mol. Biol.* 47: 445–476.

Carpita, N. C., and McCann, M. (2000) The cell wall. In *Biochemistry and Molecular Biology of Plants*, B. B. Buchanan, W. Gruissem, and R. L. Jones, eds., American Society of Plant Biologists, Rockville, MD, pp. 52–108.

Cosgrove, D. J. (1985) Cell wall yield properties of growing tissues. Evaluation by in vivo stress relaxation. *Plant Physiol.* 78: 347–356.

Cosgrove, D. J. (1997) Relaxation in a high-stress environment: The molecular bases of extensible cell walls and cell enlargement. *Plant Cell* 9: 1031–1041.

Cosgrove, D. J. (2000) Loosening of plant cell walls by expansins. *Nature* 407: 321–326.

Cosgrove, D. J. (2005) Growth of the plant cell wall. *Nat. Rev. Mol. Cell Biol.* 6: 850–861.

Dhugga, K. S., Barreiro, R., Whitten, B., Stecca, K., Hazebroek, J., Randhawa, G. S., Dolan, M., Kinney, A. J., Tomes, D., Nichols, S., et al. (2004) Guar seed beta-mannan synthase is a member of the cellulose synthase super gene family. *Science* 303: 363–366.

Doblin, M. S., Kurek, I., Jacob-Wilk, D., and Delmer, D. P. (2002) Cellulose biosynthesis in plants: From genes to rosettes. *Plant Cell Physiol.* 43: 1407–1420.

Fu, Y., Gu, Y., Zheng, Z., Wasteneys, G., and Yang, Z. (2005) Arabidopsis interdigitating cell growth requires two antagonistic pathways with opposing action on cell morphogenesis. *Cell* 120: 687–700.

Gardiner, J. C., Taylor, N. G., and Turner, S. R. (2003) Control of cellulose synthase complex localization in developing xylem. *Plant Cell* 15: 1740–1748.

Gaspar, Y., Johnson, K. L., McKenna, J. A., Bacic, A., and Schultz, C. J. (2001) The complex structures of arabinogalactan-proteins and the journey towards understanding function. *Plant Mol. Biol.* 47: 161–176.

Gibeaut, D. M., Pauly, M., Bacic, A., and Fincher, G. B. (2005) Changes in cell wall polysaccharides in developing barley (*Hordeum vulgare*) coleoptiles. *Planta* 221: 729–738.

Gunning, B. E. S., and Steer, M. (1996) *Plant Cell Biology: Structure and Function*. Bartlet and Jones Publishers, Boston.

Hayashi, T. (1989) Xyloglucans in the primary cell wall. *Annu. Rev. Plant Physiol. Plant Mol. Biol.* 40: 139–168.

Hoson, T. (1993) Regulation of polysaccharide breakdown during auxin-induced cell wall loosening. *J. Plant Res.* 103: 369–381.

Ishii, T., Matsunaga, T., Pellerin, P., O'Neill, M. A., Darvill, A., and Albersheim, P. (1999) The plant cell wall polysaccharide rhamnogalacturonan II self-assembles into a covalently cross-linked dimer. *J. Biol. Chem.* 274: 13098–13104.

John, M., Röhrig, H., Schmidt, J., Walden, R., and Schell, J. (1997) Cell signalling by oligosaccharides. *Trends Plant Sci.* 2: 111–115.

Kimura, S., Laosinchai, W., Itoh, T., Cui, X. J., Linder, C. R., and Brown, R. M., Jr. (1999) Immunogold labeling of rosette terminal cellulose-synthesizing complexes in the vascular plant *Vigna angularis*. *Plant Cell* 11: 2075–2085.

Libertini, E., Li, Y., and McQueen-Mason, S. J. (2004) Phylogenetic analysis of the plant endo-beta-1,4-glucanase gene family. *J. Mol. Evol.* 58: 506–515.

Liepman, A. H., Wilkerson, C. G., and Keegstra, K. (2005) Expression of cellulose synthase-like (Csl) genes in insect cells reveals that CslA family members encode mannan synthases. *Proc. Natl. Acad. Sci. USA* 102: 2221–2226.

Marga, F., Grandbois, M., Cosgrove, D. J., and Baskin, T. I. (2005) Cell wall extension results in the coordinate separation of parallel microfibrils: Evidence from scanning electron microscopy and atomic force microscopy. *Plant J.* 43: 181–190.

Matthews, J. F., Skopec, C. E., Mason, P. E., Zuccato, P., Torget, R. W., Sugiyama, J., Himmel, M. E., and Brady, J. W. (2006) Computer simulation studies of microcrystalline cellulose B. *Carbohydrate Res.* 341:138–152.

McCann, M. C., Roberts, K., and Carpita, N. C. (2001) Plant cell growth and elongation. In *Encyclopedia of Life Sciences*. John Wiley & Sons, Ltd. Chichester. http://www.els.net/ [doi:10.1038/npr.els.0001688]

McCann, M. C., Wells, B., and Roberts, K. (1990) Direct visualization of cross-links in the primary plant cell wall. *J Cell Sci.* 96: 323–334.

O'Neill, M. A., and York, W. S. (2003) The composition and structure of plant primary cell walls. In *The Plant Cell Wall*, J. K. C. Rose, ed, Blackwell, Oxford, pp. 1–54.

O'Neill, M. A., Eberhard, S., Albersheim, P., and Darvill, A. G. (2001) Requirement of Borate Cross-Linking of Cell Wall Rhamnogalacturonan II for Arabidopsis Growth. *Science* 294: 846–849.

O'Neill, M. A., Ishii, T., Albersheim, P., and Darvill, A. G. (2004) Rhamnogalacturonan II: Structure and function of a borate cross-linked cell wall pectic polysaccharide. *Annu. Rev. Plant Biol.* 55: 109–139.

Otte, O., Pachten, A., Hein, F., and Barz, W. (2001) Early elicitor-induced events in chickpea cells: Functional links between oxidative burst, sequential occurrence of extracellular alkalinisation and acidification, K^+/H^+ exchange and defence-related gene activation. *Z. Naturforsch.* 56: 65–76.

Park, Y. W., Tominaga, R., Sugiyama, J., Furuta, Y., Tanimoto, E., Samejima, M., Sakai, F., and Hayashi, T. (2003) Enhancement of growth by expression of poplar cellulase in *Arabidopsis thaliana*. *Plant J.* 33: 1099–1106.

Pear, J. R., Kawagoe, Y., Schreckengost, W. E., Delmer, D. P., and Stalker, D. M. (1996) Higher plants contain homologs of the bacterial celA genes encoding the catalytic subunit of cellulose synthase. *Proc. Natl. Acad. Sci. USA* 93: 12637–12642.

Peng, L., Kawagoe, Y., Hogan, P., and Delmer, D. (2002) Sitosterol-beta-glucoside as primer for cellulose synthesis in plants. *Science* 295: 147–150.

Rayle, D. L., and Cleland, R. E. (1992) The acid growth theory of auxin-induced cell elongation is alive and well. *Plant Physiol.* 99: 1271–1274.

Roland, J. C., Reis, D., Mosiniak, M., and Vian, B. (1982) Cell wall texture along the growth gradient of the mung bean hypocotyl: Ordered assembly and dissipative processes. *J. Cell Sci.* 56: 303–318.

Rose, J. K. C., and Bennett, A. B. (1999) Cooperative disassembly of the cellulose-xyloglucan network of plant cell walls: Parallels between cell expansion and fruit ripening. *Trends Plant Sci.* 4: 176–183.

Rose, J. K., Braam, J., Fry, S. C., and Nishitani, K. (2002) The XTH family of enzymes involved in xyloglucan endotransglucosylation and endohydrolysis: Current perspectives and a new unifying nomenclature. *Plant Cell Physiol.* 43: 1421–1435.

Rose, J. K., Lee, H. H., and Bennett, A. B. (1997) Expression of a divergent expansin gene is fruit-specific and ripening- regulated. *Proc. Natl. Acad. Sci. USA* 94: 5955–5960.

Ryden, P., Sugimoto-Shirasu, K., Smith A. C., Findlay, K., Reiter, W. D., and McCann, M. C. (2003) Tensile properties of Arabidopsis cell walls depend on both a xyloglucan cross-linked microfibrillar network and rhamnogalacturonan II-borate complexes. *Plant Physiol.* 132: 1033–1040.

Salnikov, V. V., Grimson, M. J., Delmer, D. P., and Haigler, C. H. (2001) Sucrose synthase localizes to cellulose synthesis sites in tracheary elements. *Phytochem.* 57: 823–833.

Sampedro, J., and Cosgrove, D. J. (2005) The expansin superfamily. *Genome Biol.* 6: 242.

Scheible, W. R., and Pauly, M. (2004) Glycosyltransferases and cell wall biosynthesis: Novel players and insights. *Curr. Opin. Plant Biol.* 7: 285–295.

Schopfer, P. (2001) Hydroxyl radical-induced cell-wall loosening in vitro and in vivo: Implications for the control of elongation growth. *Plant J.* 28: 679–688.

Schultz, C. J., Johnson, K. L., Currie, G., and Bacic, A. (2000) The classical arabinogalactan protein gene family of Arabidopsis. *Plant Cell* 12: 1751–1768.

Settleman, J. (2005) Intercalating Arabidopsis leaf cells: A jigsaw puzzle of lobes, necks, ROPs, and RICs. *Cell* 120: 570–572.

Smith, R. C., and Fry, S. C. (1991) Endotransglycosylation of xyloglucans in plant cell suspension cultures. *Biochem. J.* 279: 529–535.

Sturcova, A., His, I., Apperley, D. C., Sugiyama, J., and Jarvis, M. C. (2004) Structural details of crystalline cellulose from higher plants. *Biomacromolecules* 5: 1333–1339.

Sugimoto, K., Himmelspach, R., Williamson, R. E., and Wasteneys, G. O. (2003) Mutation or drug-dependent microtubule disruption causes radial swelling without altering parallel cellulose microfibril deposition in Arabidopsis root cells. *Plant Cell* 15: 1414–1429.

Takeda, T., Furuta, Y., Awano, T., Mizuno, K., Mitsuishi, Y., and Hayashi, T. (2002) Suppression and acceleration of cell elongation by integration of xyloglucans in pea stem segments. *Proc. Natl. Acad. Sci. USA* 99: 9055–9060.

Thompson, J. E., and Fry, S. C. (2001) Restructuring of wall-bound xyloglucan by transglycosylation in living plant cells. *Plant J.* 26: 23–34.

Vietor, R. J., Newman, R. H., Ha, M. A., Apperley, D. C., and Jarvis, M. C. (2002) Conformational features of crystal-surface cellulose from higher plants. *Plant J.* 30: 721–731.

Vincken, J. P., Schols, H. A., Oomen, R. J., McCann, M. C., Ulvskov, P., Voragen, A. G., and Visser, R. G. (2003) If homogalacturonan were a side chain of rhamnogalacturonan I. Implications for cell wall architecture. *Plant Physiol.* 132: 1781–1789.

Wilson, R. H., Smith, A. C., Kacurakova, M., Saunders, P. K., Wellner, N., and Waldron, K. W. (2000) The mechanical properties and molecular dynamics of plant cell wall polysaccharides studied by Fourier-transform infrared spectroscopy. *Plant Physiol.* 124: 397–405.

York, W. S., Qin, Q., and Rose, J. K. (2004) Proteinaceous inhibitors of endo-beta-glucanases. *Biochim. Biophys. Acta* 1696: 223–233.

Chapter 16 | Growth and Development

HIKERS WHO VISIT THE HIGH, arid White Mountain Range just north of Death Valley in east central California have the opportunity to view an unusual grove of gnarled and weather-beaten old bristlecone pine trees. The most ancient of these provides a remarkable chronicle that began long ago, before the pyramids were completed. The character of over 4000 annual growth rings (viewed with the benefit of a nondestructive coring device) offers a glimpse of savage storms, freezing winters, and droughts endured, leaving one to ponder how a single cell contained within the nascent seed might have been guided to divide, grow, and develop into such an ancient organism.

Biologists who seek to understand such questions are generally concerned with two broad issues: the morphological and anatomical changes that take place during development and the molecular and biochemical mechanisms underlying these changes. The first issue involves a detailed description of development, the complex process by which the size, composition, and organization of an organism change during its life history. The vegetative phase of development begins with embryogenesis, which initiates plant development. Embryogenesis establishes the basic plant body plan and forms the meristems that generate additional organs in the adult. Unlike animal development, plant development is an ongoing process. Vegetative meristems are highly repetitive—they produce the same or similar structures over and over again—and their activity can continue indefinitely, a phenomenon known as *indeterminate growth*. Some long-lived trees, such as bristlecone pines and the California redwoods, continue to grow for thousands of years. Others, particularly annual plants, may cease vegetative development with the initiation of flowering after only a few weeks or months of growth.

When the adult plant undergoes a transition from vegetative to reproductive development, culminating in the production of a zygote, the process begins again. At some point in the life cycle the adult plant undergoes senescence and dies, and this, too, is a developmentally controlled, genetically determined process. Even bristlecone pine trees eventually pass from the scene.

How does the organism change over time? As an organism grows, are there corresponding increases its complexity, and if so, how can this organizational complexity be described? Is growth accompanied by cell division and differentiation? How is this cellular behavior related to the formation of specific tissues? How do the types of cells and their relationships to others change over time?

With a detailed account of development formulated from answers to these questions, the biologist is in a position to address the second issue, the nature of the underlying mechanisms. The general nature and sequence of developmental changes are to a large extent predictable, leading one to conclude that these changes must reflect the presence of underlying genetic programs. What is the nature of these programs, and what features allow these programs to respond to environmental cues? How are complex patterns of cellular differentiation regulated in space and time to yield a predictable organization consisting of a diverse array of tissues and cell types?

This chapter explores these questions, both descriptive and mechanistic, drawing examples from both the classic and the contemporary literature. The chapter first considers basic anatomical features of plant growth, and how these may have evolved in response to an autotrophic, photosynthetic lifestyle. With this background, an overview of key stages in the development of the sporophyte is provided, beginning with the zygote and its transformation into an embryo, before discussing programs that enable indeterminate patterns of vegetative growth. A more detailed treatment of each stage is then provided, focusing on key developmental processes and the nature of the underlying molecular mechanisms.

Complex patterns of the determinate type of growth that characterizes reproductive development are discussed in Chapter 25. Quantitative aspects of plant growth analysis are presented in **Web Topic 16.1**. Two classic works, *Patterns of Plant Development* (Steeves and Sussex 1989) and *Plant Anatomy* (Esau 1965), provide rich sources of relevant background information on the basic form and development of plants. Valuable perspectives can also be gained from the study of plant morphology (Gifford and Foster 1987). Here, systematic comparisons across different taxonomic groups point toward the nature of basic programs of development that have been adapted to produce a variety of basic plant forms.

Overview of Plant Growth and Development

An essential aspect of almost all land plants is their sedentary lifestyle. By virtue of their ability to photosynthesize, favorably positioned plants can readily obtain both the energy and the nutrients they need to grow and survive. Relieved of the need to move, plants have never evolved the sort of anatomical complexity that enables mobility in their animal counterparts. In its place, one finds a relatively rigid anatomy adapted to the capture of light energy and nutrients. As a consequence, plant cells, unlike animal cells, are firmly attached to their neighbors in a relatively inflexible, often woody, matrix. The rigid nature of this anatomy is also reflected by the manner in which plants grow. Cells are added progressively through the activity of structures called meristems. By contrast, the early development of animals is characterized by cell migration to new positions to form specific tissues.

While the sedentary habit of plants allows a relatively simple organization (Figure 16.1), this lack of mobility presents a significant challenge. Because plants are unable to relocate to optimal habitats, they must instead adapt to their local environments. While this adaptation can occur on a physiological level, it may also be achieved through a flexible pattern of development. Unlike animals, whose basic adult body plan is established during embryogenesis, plants elaborate their forms throughout their lives through programs of vegetative development (Steeves and Sussex 1989). Here again, meristems provide a means of adapting to changing conditions by maintaining reservoirs of undetermined cells. Through the regulated proliferation of these cells and their recruitment into tissues and organs, plants are able to produce complex, but often variable forms that are best adapted to their local environment.

Sporophytic development can be divided into three major stages

All seed plants pass through three stages during the development of the sporophyte: embryogenesis, vegetative development, and reproductive development.

EMBRYOGENESIS The term embryogenesis describes the process by which a single cell is transformed into a multicellular entity having a characteristic but typically rudimentary organization. In most seed plants, embryogenesis takes place within the confines of the ovule, a specialized structure borne within the carpel of the flower. The ovule serves several related functions, beginning with providing an environment in which the female gametophyte can develop. In flowering plants, the development of this gametophyte, also known as the embryo sac, is limited to the production of a small number of cells, including the egg cell and two polar nuclei. Fertilization (syngamy) occurs when the pollen sperm cell fuses with the egg cell to create

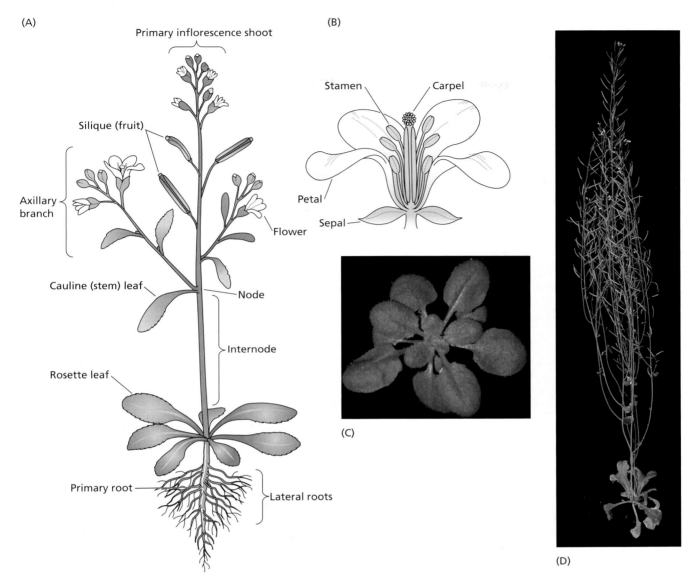

Figure 16.1 *Arabidopsis thaliana*. (A) Drawing of a mature Arabidopsis plant showing the various organs. (B) Drawing of a flower showing the floral organs. (C) An immature vegetative plant consisting of basal rosette leaves (root system not shown). (D) A mature plant after most of the flowers have matured and the siliques have developed. (C and D courtesy of Caren Chang.)

the zygote (Figure 16.2). As such, the zygote has twice the chromosome number, or ploidy, compared to either of the gametophyte cells from which it derives. At the same time, a second sperm cell combines with the two polar nuclei of the embryo sac to initiate the formation of the triploid endosperm that will ultimately support the growth of the embryo. Together, these two coordinated syngamy events are referred to as double fertilization.

This overall sequence of embryonic development is highly predictable, perhaps reflecting the need for the embryo to be effectively packaged within the maternally derived integuments to form the seed. Given the progressive nature of growth and development, careful examination of how the mature embryo develops from a single cell provides one of the clearest examples of basic patterning processes in plants (Figure 16.3). Among these processes are those responsible for establishing polarity, by which the activities of cells differentiate according to their positions in the embryo. Within this spatial framework, groups of cells become functionally specialized to form epidermal, cortical, and vascular tissues. Certain groups of cells, known as apical meristems, are established at the growing points of the shoot and root and thus enable the elaboration of additional tissues and organs during subsequent

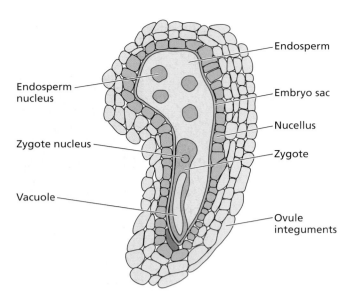

Figure 16.2 Arabidopsis ovule containing the embryo sac at about 4 hours after double fertilization. The embryo sac is bounded by the maternally derived nucellus. The zygote exhibits a marked polarization. The terminal half of the zygote has dense cytoplasm and a single large nucleus, while a large central vacuole occupies the basal half of the cell. At this stage, the embryo sac surrounding the zygote also contains four endosperm nuclei.

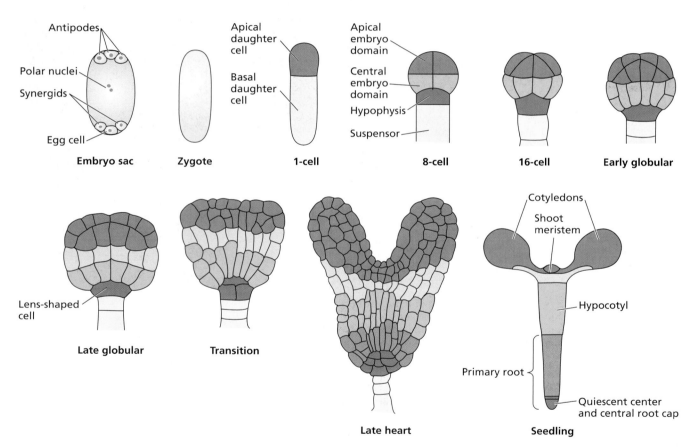

Figure 16.3 Cellular characteristics of Arabidopsis embryogenesis. A series of successive stages are shown to illustrate how specific cells in the young embryo contribute to specific anatomically defined features of the seedling. Clonally related groups of cells (cells that can be traced back to a common progenitor) are indicated by distinct colors. Following the asymmetric division of the zygote, the smaller apical daughter cell divides to form an eight-celled proembryo consisting of two tiers of four cells each. The upper tier gives rise to the shoot apical meristem and most of the flanking cotyledon primordia. The lower tier produces the hypocotyl and some of the cotyledon, the embryonic root, and the upper cells of the root apical meristem. The basal daughter cell of the zygote produces a file of nonembryonic cells that make up the suspensor, which attaches the embryo to the embryo sac. The uppermost cell of the suspensor adjacent to the proembryo, the hypophysis (blue), becomes part of the embryo. The hypophysis divides to form the quiescent center (discussed later in the chapter) and the stem cells that form the root cap.

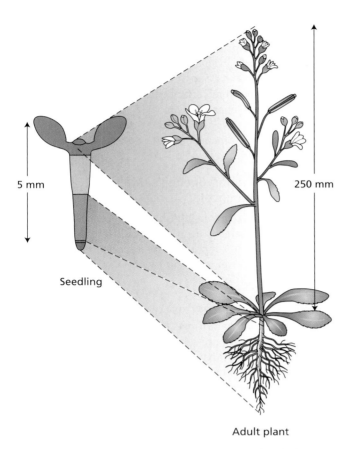

FIGURE 16.4 Postembryonic development of *Arabidopsis thaliana*. As illustrated, the majority of tissues formed during postembryonic growth derive from small groups of cells contained within the primary root and shoot apical meristems.

vegetative growth. At the conclusion of embryogenesis, a number of physiological changes occur to enable the embryo to withstand long periods of dormancy and harsh environmental conditions (see **Web Topic 16.2**).

VEGETATIVE DEVELOPMENT With germination, the embryo breaks its dormant state and, by mobilizing stored reserves, the seedling (Figure 16.4) commences a period of vegetative growth. Depending on the species, germination can depend on a variety of factors, including moisture levels, extended cold, heat, and light. Drawing initially on reserves stored in its cotyledons (e.g., beans) or in endosperm (e.g., grasses), the plant builds on its rudimentary form through the activity of the root and shoot apical meristems. Through photomorphogenesis (see Chapter 17) and further development of the shoot, the seedling becomes photosynthetically competent, thus enabling further vegetative growth. Unlike animals, this growth is often indeterminate—not predetermined, but subject to variability with no definite end point. Indeterminate growth is characterized by reiterated programs of lateral organ development that allow the plant to elaborate an architecture best suited to the local environment.

REPRODUCTIVE DEVELOPMENT After a period of vegetative growth, plants respond to a combination of intrinsic and extrinsic cues, including size, temperature, and photoperiod, to undergo the transition to reproductive development. In flowering plants, this transition involves the formation of specialized floral meristems that give rise to flowers. The processes by which floral meristems first are specified and then develop to produce a stereotyped sequence of organ formation have provided some of the best-studied examples of plant development and are described in detail in Chapter 25.

Development can be analyzed at the molecular level

While anatomically focused analyses of plants can provide a detailed account of the sequences of changes that occur during the life of an organism, including cellular behaviors, they do not reveal a complete picture of underlying mechanisms. In recent years, methods by which specific classes of molecules can be localized and quantified have offered valuable insights into the possible nature of these mechanisms. The far-reaching insights offered by such methods are not unexpected, since characteristic patterns of development reflect genetic programs whose expression is ultimately mediated by the production and interactions of such molecules. Such detailed descriptions of how molecular interactions correlate with specific developmental events can often suggest models for underlying mechanisms.

Although molecular models of development can be further refined and to some degree tested by increasingly detailed descriptive analyses, the most critical tests are typically those that examine causal or interdependent relationships between various elements of the mechanism. For example, based on the observation that shoots are a major source of auxin, one might suppose that this hormone could potentially influence the position at which leaves emerge on the shoot apex.

As will be discussed later in this chapter, this type of hypothetical dependency relationship can be tested in a number of ways. For example, one could artificially perturb the levels of the growth hormone auxin (see Chapter 19), and assess whether this alters the position of leaf initiation. However, even if auxin were found to have this activity in this artificial situation, one must then ask whether it would have a similar function in a normal plant.

Similar functional analyses can be applied to other developmental processes. In each case, the consequences of perturbing or altering the system are assessed and used to infer causal relationships between components of the system. A range of methods can be used to disrupt the developmental process, including chemical and environmental treatments. Classic studies have utilized surgical approaches to physically disrupt normal signaling pathways. More recent genetic approaches have been based on comparisons of individuals in which the activity, or expression, of one or more specific genes differ.

In the following sections, we apply different analytical perspectives to consider critical processes that guide the development of the plant, first discussing embryonic stages, then those relating to vegetative growth. As will be seen, each approach has its strengths and weaknesses. Together, these examples illustrate how detailed models for developmental mechanisms can be formulated on the basis of descriptive data, then tested and refined using a range of genetic and experimental approaches.

Embryogenesis: The Origins of Polarity

In seed plants, embryogenesis transforms a single-celled zygote into the considerably more complex individual contained in the mature seed. As such, embryogenesis is likely to encompass a range of developmental processes by which the basic plant architecture is established, including the elaboration of basic forms (**morphogenesis**), the associated formation of functionally organized structures (**organogenesis**), and the **differentiation** of cells contained within various tissues (**histogenesis**). An essential feature of this basic plant architecture is the presence of apical meristems at the tips of the shoot and root axes (see Figure 16.3). In addition, there must be additional mechanisms that mediate the complex physiological responses of the embryo, including its ability to withstand prolonged periods of inactivity (**dormancy**) and to recognize and interpret environmental cues that signal the plant to resume growth (**germination**).

The progressive nature of embryo growth is clearly illustrated by comparing the size and complexity of the embryo over time. Its basic form, and how that form changes, can be described by a variety of anatomical and cytological approaches. Comparisons involving cellular detail, which reflect division behavior or functional specialization, typically involve microscopic examinations of whole or sectioned specimens. Subcellular characteristics, cytoplasmic density, organelle size and distribution, and physiological traits can provide valuable clues to regionalized activities. This approach can be extended further by immunological and biochemical methods that allow a more detailed understanding of the molecular architecture of cells and their physiological status. Finally, exquisitely sensitive methods have been developed by which these characteristics can be related to the expression of specific genes at the RNA or protein level.

In the sections that follow we will examine the phenomenon of patterning during embryogenesis at several levels. We begin with an anatomical comparison of embryogenesis in dicots and monocots. Next we address the problem of how polarity is established in the embryo, and the importance of position-dependent information for cell differentiation. The growth hormone auxin (see Chapter 19) plays a crucial role in shaping the embryo and in maintaining the apical–basal polarity of the plant. As we will see, many of the genes that control embryogenesis are involved in auxin action. Genes have also been identified that regulate radial patterning during embryogenesis. Finally, the important role of plasmodesmata during embryogenesis will be discussed.

The pattern of embryogenesis differs in dicots and monocots

Anatomical comparisons highlight differences in the pattern of embryogenesis seen among different seed plant groups, such as those seen between monocots and dicots. Arabidopsis and rice provide two contrasting examples from each group.

ARABIDOPSIS EMBRYOGENESIS Perhaps by virtue of the small size of the Arabidopsis embryo, the patterns of cell division in this dicot are relatively simple and predictable (Mansfield and Briarty 1991). Five widely recognized stages are described in the following list, in the sequence in which they usually occur during embryogenesis:

1. *Zygotic* stage. A single-celled stage that follows fusion of the haploid egg and sperm and concludes with an asymmetric transverse division (Figure 16.5A).

2. *Globular* stage. After the first asymmetric zygotic division, the apical cell undergoes a series of divisions (see Figure 16.5B–D) to generate a spherical, eight-cell (**octant**) globular embryo by 30 hours after syngamy (see Figure 16.5C). Additional cell divisions increase the number of cells in the sphere (see Figure 16.5D) and create a discrete superficial layer of protoderm.

3. *Heart* stage. This stage forms by rapid cell division in two regions on either side of the future shoot apex. These two regions produce outgrowths that later become the cotyledons and give the embryo bilateral symmetry (see Figure 16.5E and F).

4. *Torpedo* stage. This stage forms as a result of cell elongation throughout the embryo axis and further development of the cotyledons (see Figure 16.5G).

5. *Mature* stage. Toward the end of embryogenesis, the embryo and seed lose water and become metabolically inactive as they enter dormancy (discussed in Chapter 23). Storage compounds accumulate in the cells at the mature stage (see Figure 16.5H).

RICE EMBRYOGENESIS Rice illustrates a distinct pattern of embryogenesis that is typical of many monocotyledonous plants (Figure 16.6). Like many plants, the patterns of division associated with embryogenesis are more variable than that seen in Arabidopsis. Nevertheless, it is possible to

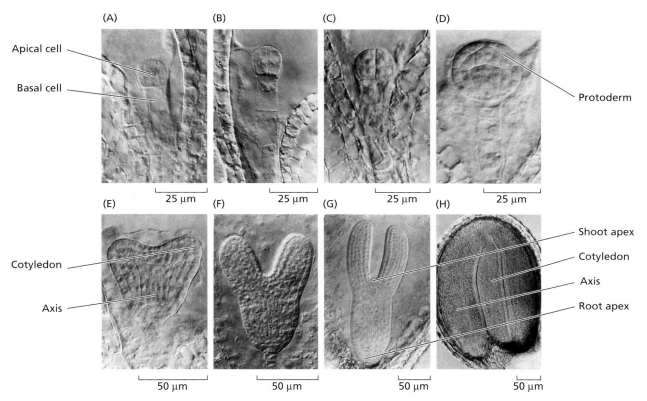

Figure 16.5 Arabidopsis embryogenesis is characterized by a precise pattern of cell division. Successive stages of embryogenesis are depicted here. (A) One-cell embryo after the first division of the zygote, which forms the apical and basal cells; (B) two-cell embryo; (C) eight-cell embryo; (D) early globular stage, which has developed a distinct protoderm (surface layer); (E) early heart stage; (F) late heart stage; (G) torpedo stage; and (H) mature embryo. (From West and Harada 1993; photographs taken by K. Matsudaira Yee; courtesy of John Harada, © American Society of Plant Biologists, reprinted with permission.)

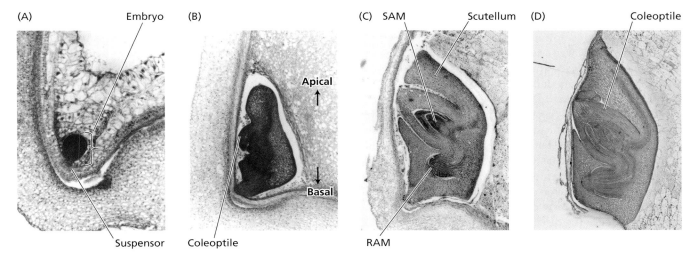

Figure 16.6 Rice (a monocot in the grass [Poaceae] family) embryonic development illustrating the globular stage (A), coleoptile stage (B), juvenile embryo stage (C), and mature embryo stage (D). The scutellum is a modified cotyledon specialized to absorb sugars released from the endosperm during germination. (Courtesy of Y. Nagato.)

describe embryogenesis in rice in terms of morphologically defined developmental stages as follows (Itoh et al. 2005):

1. *Zygotic* stage. This is the single-celled stage that follows fusion of the haploid egg and sperm.

2. *Globular* stage. This stage occurs 2–4 days after pollination (DAP). Following an initial horizontal division to create the apical and basal cells, a series of variable cell divisions create a multilayered, globular embryo (Figure 16.6A).

3. *Coleoptile* stage. At 5 DAP, the coleoptile (specialized tubular first leaf), shoot apical meristem (SAM), root apical meristem (RAM), and radicle (embryonic root) form (see Figure 16.6B).

4. *Juvenile vegetative* stage. At 6–10 DAP, the SAM initiates several vegetative leaves (see Figure 16.6C).

5. *Maturation* stage. During 11–20 DAP, maturation-related genes are expressed, which precedes onset of dormancy (see Figure 16.6D).

Even a superficial comparison of embryogenesis in Arabidopsis and rice illustrates certain obvious differences in size, shape, cell number, and division patterns. Despite these differences, several common themes emerge that can be generalized to all seed plants. Perhaps the most fundamental of these relate to **polarity**.

Beginning with the single-celled zygote, embryos become progressively more polarized throughout their development along two axes: an **apical–basal axis**, which runs between the tips of the embryonic shoot and root, and a **radial axis**, which extends from the center of the plant outwards. (For a discussion of the establishment of polarity in a simpler [algal] zygote, see **Web Topic 16.3** and **Web Essay 16.1**.)

In the following section, we consider how these axes are established and discuss how specific molecular processes may guide this development. The discussion focuses on Arabidopsis. The large variety of analyses that have been applied to this species attests to features that make it an especially attractive model, including a relatively small and completely sequenced diploid genome and a rapid pattern of growth and development that lends itself to genetic analysis. Another advantage relates to the simple and highly stereotyped cell divisions associated with early stages of embryo development. On the basis of changes to this simple pattern, factors that may influence embryo development can be more easily recognized. A graphic depiction of these cell divisions, provided in Figure 16.3, offers a convenient guide for the discussion that follows.

The axial polarity of the plant is established by the embryo

Seed plants exhibit an axial polarity in which the tissues and organs are arrayed in a precise order along a linear, or polarized, axis that extends from the shoot apical meristem (SAM) at one end to the root apical meristem (RAM) at the other. This asymmetry can be traced back to the earliest stage of somatic development, in which the zygote itself becomes polarized and elongates approximately threefold before its first division. The apical end of the zygote is densely cytoplasmic, in contrast to the basal half, which contains a large central vacuole. This first division of the zygote is also asymmetric and occurs at right angles to its long axis to create two cells—the apical and basal cells—each of which has a very distinct fate (see Figures 16.3, 16.5B and 16.7). The smaller, apical daughter cell is densely cytoplasmic, while the larger, basal cell inherits the large zygotic vacuole.

Nearly the entire embryo, and ultimately the mature plant, are derived from the smaller apical cell, which gives rise to the *proembryo*. The term **proembryo** refers to the earliest stages of plant embryo development before the protoderm and suspensor (see below) have fully developed. Two vertical divisions and one horizontal division of the apical cell generate the eight-celled (octant) globular embryo (see Figures 16.3 and 16.5B).

The basal cell also divides, but all of its divisions are horizontal, at right angles to the long axis. The result is a filament of six to nine cells known as the **suspensor**, which attaches the embryo to the vascular system of the plant. Only one of the basal cell derivatives contributes to the embryo. The basal cell derivative nearest the proembryo is known as the **hypophysis** (plural *hypophyses*), and in subsequent stages it will divide further to form the **columella**, or central part of the root cap, and an essential part of the root apical meristem known as the **quiescent center**, which will be discussed later in the chapter (see Figure 16.3).

Even though the embryo is roughly spherical throughout the globular stage of embryogenesis (see Figure 16.3), the cells derived from the apical and basal halves of the sphere will eventually have different identities and functions. As the embryo continues to grow and reaches the heart stage, its axial polarity becomes more distinct (see Figure 16.3), and three axial regions can readily be recognized:

1. The *apical region*, derived from the apical quartet of cells, gives rise to the cotyledons and shoot apical meristem.

2. The *middle region*, derived from the basal quartet of cells, gives rise to the hypocotyl, the root, and most of the root meristem.

3. The *hypophysis*, derived from the uppermost cell of the suspensor, gives rise to the rest of the root meristem, including the quiescent center and columella.

Position-dependent signaling guides embryogenesis

The reproducible patterns of cell division during early embryogenesis in Arabidopsis might suggest that a fixed sequence of division is an essential aspect of this phase of development. This consistency could be explained if the fate of individual cells within the embryo became fixed, or **determined**, early; once this fate was established, these cells would be committed to fixed programs of cell division, which would in turn produce a defined form.

This model for development could be likened to assembling a structure from a standard set of parts that are created by programmed patterns of cell division. Such **lineage-dependent** mechanisms have been well described in animal systems. However, this type of explanation seems difficult to reconcile with more variable patterns of cell division seen in other plants, including species closely related to Arabidopsis. It is possible that the relatively predictable pattern of cell division seen in Arabidopsis simply reflects the small size of its embryo, which places physical limits on how division can occur. Moreover, even for Arabidopsis, some limited variability in cell division behavior can be seen by following the fate of individual cells with fate mapping techniques (see Figure 16.7).

Given the variability of cell division observed during embryogenesis, it is likely that **position-dependent** signaling mechanisms play significant roles. Such mechanisms operate by modulating the behavior of cells in a manner that depends on the position of these cells within the developing embryo. This type of mechanism would explain how equivalent forms can arise through different patterns of cell division.

By one line of reasoning, such mechanisms could be expected to contain at least three functional elements:

- There must be a coordinate system by which unique positions within the developing structure can be specified.

- Individual cells must have the means to assess their position within this coordinate system.

- Cells must somehow respond to this information in an appropriate way, which may range from further division to terminal differentiation.

If such a mechanism was responsible for the formation of the embryo, we might expect several classes of embryo pattern mutants, including those that perturb the coordinate system; others that alter the ability of cells to assess their location within this system; and a third class that fails to differentiate accordingly (Willemsen and Scheres 2004).

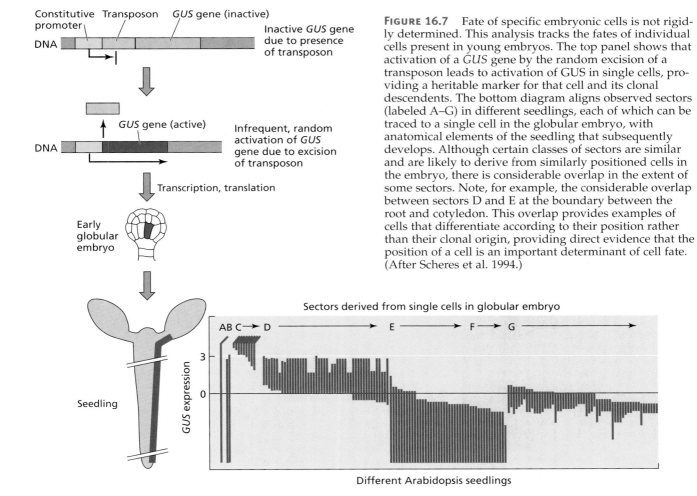

FIGURE 16.7 Fate of specific embryonic cells is not rigidly determined. This analysis tracks the fates of individual cells present in young embryos. The top panel shows that activation of a *GUS* gene by the random excision of a transposon leads to activation of GUS in single cells, providing a heritable marker for that cell and its clonal descendents. The bottom diagram aligns observed sectors (labeled A–G) in different seedlings, each of which can be traced to a single cell in the globular embryo, with anatomical elements of the seedling that subsequently develops. Although certain classes of sectors are similar and are likely to derive from similarly positioned cells in the embryo, there is considerable overlap in the extent of some sectors. Note, for example, the considerable overlap between sectors D and E at the boundary between the root and cotyledon. This overlap provides examples of cells that differentiate according to their position rather than their clonal origin, providing direct evidence that the position of a cell is an important determinant of cell fate. (After Scheres et al. 1994.)

Certain classes of embryo mutants in the fruit fly, *Drosophila*, are most easily explained by such a scheme, including those causing homeotic-like transformations, in which pattern elements are transformed into forms generally seen at other positions. Molecular analyses have strengthened such models by showing that the gene products affected by these mutants function in the predicted manner.

An alternative model for patterning supposes that there is no separate coordinate system per se. In its place, transmissible factors somehow achieve a localized distribution and thereby evoke patterned cellular responses directly without reference to more global system of positional information. Once again, *Drosophila* provides examples of behavior that is well explained by such schemes; for example, the patterning of elements that make up the *Drosophila* compound eye has been described in terms of complex, short-range signaling processes.

In the following discussion we consider the nature of mechanisms that guide axial patterning of the plant embryo—first by identifying critical elements of the process, and second by considering the evidence for how they may interact. In animals, chemical signals called *morphogens* play a key role in embryogenesis. **Morphogens** provide spatial information via a concentration gradient during embryonic development. As discussed in the following sections, the hormone auxin appears to act as a morphogen during plant embryogenesis.

Auxin may function as a morphogen during embryogenesis

Given the position-dependent nature of patterning processes during embryogenesis, it is reasonable to consider the possible involvement of plant hormones. The mobility of these small molecules and the physiological responses they evoke highlight their potential in this regard.

For many plant species, it is possible to induce the formation of embryos from somatic cells with synthetic growth media containing the plant growth hormone auxin: indole-3-acetic acid or its analogs (see Chapter 19). This regeneration behavior not only highlights the potential involvement of auxin, but also indicates the intrinsic character of embryogenic programs that can proceed in the absence of information supplied by the maternal plant. Later stages of embryogenesis appear to also involve the activity of auxin, as demonstrated by the effects of auxin or its inhibitors on immature embryos propagated in vitro (Figure 16.8) (Liu et al. 1993).

For auxin to function as a morphogen, it must not only have the ability to elicit specific responses in target tissues, but must also somehow achieve a graded distribution throughout the developing embryo. As described in more detail in Chapter 19, the spatial distribution of auxin reflects the complex interplay among factors acting act several levels, including those affecting auxin synthesis, transport, conjugation, and catabolism.

(A) Wild type plus *trans*-cinnamic acid

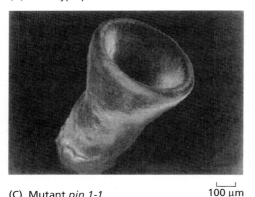

(B) Wild-type control

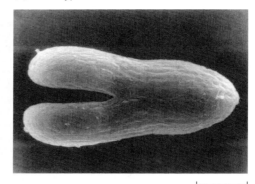

(C) Mutant *pin 1-1*

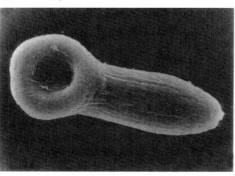

FIGURE 16.8 Evidence for a role for auxin in early embryonic development. The figure compares the altered morphology of a *Brassica juncea* embryo, obtained by culturing in vitro for 10 days in the presence of (A) *trans*-cinnamic acid (a flavonoid that has been shown to reduce auxin levels and inhibit its transport), (B) with a wild-type embryo, and (C) a mutant, *pin1-1* embryo. Note the similar failure in cotyledon separation caused by either the chemical inhibition of auxin in vitro or the disruption of auxin transport by mutations to the *PIN* gene. (From Liu et al. 1993.)

In general, sites where synthesis occurs are termed **sources**, and regions where accumulation, breakdown, or conversion to inactive forms occurs are termed **sinks**. Historically, source–sink relationships for auxin, and its associated movement, have been inferred by a variety of methods, including direct measurement of auxin levels in different tissues, immunodetection, and the use of radiolabeled molecules to track auxin synthesis, movement, and turnover.

Two additional methods have been developed that provide a highly detailed view of the distribution and movement of auxin in the plant. In the first method, control elements that are normally found upstream of auxin-inducible genes (such as the one represented in the synthetic promoter *DR5*) are fused to a reporter gene, such as that encoding β-glucuronidase (*GUS*) or green fluorescent protein (*GFP*). By assessing the activity of these reporters, it is possible to infer the relative concentration of auxin throughout the plant (see Chapter 19).

It should be noted, however, that expression of such auxin–reporter fusions appears to depend on the activity of another class of hormones, the brassinosteroids (see Chapter 24), suggesting that lack of expression does not in itself demonstrate low auxin levels. Nevertheless, the use of such auxin-dependent reporters has provided key insights into the dynamic behavior of auxin in the plant.

A second novel approach by which the directional movement of auxin in the developing plant can be inferred involves monitoring the distribution of proteins responsible for the intercellular movement of auxin. Molecular and biochemical analyses have established a close correlation between the asymmetric cellular localization of PINs, a family of auxin efflux proteins that govern the directional movement of auxin (see Chapter 19). By using immunolocalization methods, the asymmetric cellular distribution of these efflux proteins can be described, thus allowing the direction of auxin movement to be inferred.

Application of these methods has made it clear that polar auxin transport occurs during embryogenesis, and that this results in a patterned distribution of auxin across the developing embryo. Based on data derived from various analytical approaches, provisional maps can be developed that illustrate the both the directional movement of auxin and the relative auxin levels (Figure 16.9). With this representation, we can consider whether processes associated with the polar patterns of development can be correlated with a similarly patterned distribution of auxin.

Genes control apical–basal patterning

During the late 1980s and early 1990s, systematic genetic approaches that had been successfully applied to the development of animals such as *Drosophila* were applied for the first time to plants. In these so-called *forward genetic screens*, mutant individuals are first identified in which changes in the process of interest can be traced to alterations in specif-

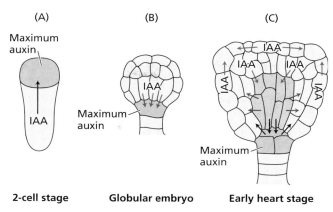

FIGURE 16.9 PIN-dependent auxin movement during early stages of embryogenesis. Auxin movement, as inferred from the asymmetric distribution of PIN proteins and the activity of *DR5* reporters, is depicted by arrows. Blue areas denote cells with maximum auxin concentrations.

ic genes. By analyzing the pattern with which the mutant trait is transmitted to its progeny with respect to other defined chromosomal genes, the chromosomal location, or map position, of the corresponding gene can be determined.

For plants such as Arabidopsis, this knowledge can ultimately be used to physically clone the gene through map-based cloning techniques. Together with knowledge of how these genes influence development, this information has been used to develop detailed models for processes that mediate embryogenesis (Laux et al. 2004).

A number of these mutant screens have been performed to gain insight into how the basic polarity of the embryo is established and how subsequent patterning occurs. It is unclear how the initial asymmetry between the apical and basal axis arises. Certain embryo mutants exhibit reversals in the normal polarity, pointing toward specific genes that are responsible for this early asymmetry. Although the attachment of the suspensor to the maternal plant might provide an obvious reference point, relatively normal embryos can be cultured from single cells in suspension culture as long as the appropriate balance of hormones, including auxin, is provided. Therefore, maternal cues may play a relatively insignificant role once an initial asymmetry is established.

Mutations that block very early stages of embryogenesis are termed *embryo lethal*. Given that these mutants are generally recessive, they can be detected by the presence of incompletely filled ovules on the parent plant. While the arrested development of these mutants is consistent with a role for the corresponding genes during embryogenesis, it can be difficult to distinguish whether this reflects a function specifically related to embryogenesis or a more general metabolic activity.

In an attempt to isolate mutants that specifically affect embryonic patterning processes, one group screened for *seedling defective* mutants capable of developing into mature seeds, but that displayed an abnormal organization when germinated and examined as seedlings (Mayer et al. 1991). Among these mutants were those in which the normal apical–basal morphology was disrupted, so that either the shoot apical meristem, the root apical meristem, or both, were missing. The nature of defects seen in these mutants suggests that the corresponding genes are required for establishing the normal apical–basal pattern (Figure 16.10).

Embryogenesis genes have diverse biochemical functions

After their initial phenotypic characterization, the cloning of these genes by map-based techniques has offered some insight into their molecular functions.

- *GURKE*, named for the cucumber-like shape of the mutant in which the cotyledons and shoot apical meristem have been deleted, encodes an acetyl-CoA carboxylase.
- *FACKEL*, in which the mutant lacks a hypocotyl, encodes a sterol C-14 reductase.
- *MONOPTEROS* (*MP*), necessary for the normal formation of basal elements such as the root and hypocotyl, encodes an auxin response factor (ARF) transcription factor (discussed below).
- *GNOM* (*GN*), required for the establishment of both the apical and basal terminal elements of the embryo, encodes a guanine nucleotide exchange factor (GEF).

Although cloning of these genes revealed the likely biochemical function of the proteins they each encode, how these activities help establish the normal apical–basal pattern of the embryo is not immediately obvious. Further molecular and genetic analyses have suggested that the effects of *MP* and *GN* can be understood in terms of their role in mediating the response to auxin.

Comparisons of the amino acid sequences of MP show it belongs to a subfamily of 22 genes that encode **auxin response factors**, which are thought to be responsible for the transcriptional activation of genes involved in auxin responses (see Chapter 19). In the absence of auxin, the activity of the ARF proteins is inhibited through their association with specific repressor proteins. Auxin responses are triggered when the targeted degradation of these repressors occurs and ARFs can activate their target genes (see Chapter 19).

MONOPTEROS activity is inhibited by a repressor protein

Several lines of evidence support the view that *MP* mediates auxin responses. In addition to

(A) Wild type vs. *gnom* mutant

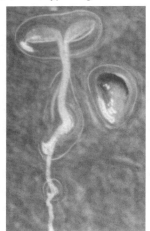

GNOM genes control apical-basal polarity

(B) Wild type vs. *monopteros* mutant

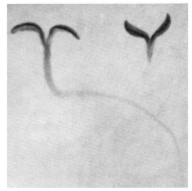

MONOPTEROS genes control formation of the primary root

(C) Schematic of mutant types

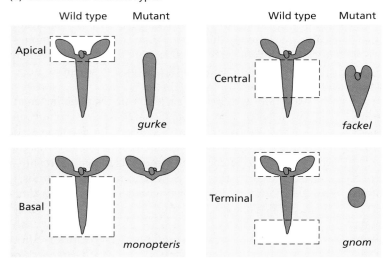

FIGURE 16.10 Genes essential for Arabidopsis embryogenesis have been identified from their mutant phenotypes. The development of mutant seedlings is contrasted here with that of the wild type at the same stage of development. (A) The *GNOM* gene helps establish apical–basal polarity. A plant homozygous for *gnom* is shown on the right. (B) The *MONOPTEROS* gene is necessary for basal patterning and formation of the primary root. Plants homozygous for the *monopteros* mutation have a hypocotyl, a normal shoot apical meristem, and cotyledons, but they lack the primary root. (C) Schematic of four deletion mutant types. The hatched regions of the wild type on the left are missing from each of the mutants on the right. (A from Willemsen et al. 1998; B from Berleth and Jürgens 1993; C from Mayer et al. 1991.)

causing the deletion of basal domains of the embryo (Figure 10B and C), *mp* mutants have defects in vascular patterning similar to those observed when auxin levels or movements are artificially blocked.

Further evidence has been provided by the characterization of *BODENLOS* (*BDL*), a gene whose mutant phenotype resembles that of *mp*, in which basal domains of the embryo are deleted (Hamann et al. 2002). Molecular cloning of the *BDL* gene shows that it encodes one of the previously discussed transcriptional repressor proteins. The mutant form of the gene appears to produce a protein resistant to auxin-induced degradation, and as such, likely blocks some forms of downstream auxin-induced gene expression.

Given the similarity between *mp* and *bd* phenotypes, one simple model is that the BODENLOS protein normally binds to the MP protein, repressing its function as a transcriptional activator. However, in tissues containing auxin, BODENLOS is degraded, thus allowing MP to bind promoter sequences to activate the transcription of specific auxin-regulated genes. (For more detail, see Figure 19.41.)

Gene expression patterns correlate with auxin

Given the defects in basal regions of the embryo seen in *mp* mutants, the foregoing model predicts that the higher auxin levels there normally activate *MP*-dependent expression of genes that direct basal programs of development. In support of this model are *DR5:GFP* expression data that suggest a high level of auxin in this region (Figure 16.11). Linked changes in the expression patterns of the *PIN* genes are also observed that likely account for the observed distribution of auxin (Friml et al. 2003; Blilou et al. 2005). At the two-cell stage, transcripts of *PIN7* accumulate in the basal cell and are asymmetrically distributed in the wall facing the apical cell (see Figure 16.11).

A likely outcome of the biased distribution of PIN7 protein is the accumulation of auxin in the apical cell. Later, during the development of the proembryo, the distribution of PIN proteins is reversed, with higher levels along the basal faces of cells. The outcome is a reversal in the directions of auxin movement. The spatial patterns of other genes involved in embryogenesis are also shown in Figure 16.11.

GNOM gene determines the distribution of efflux proteins

Recent studies extend this auxin-linked theme by suggesting that the *GN* gene product plays a critical role during embryogenesis by regulating the cellular distribution of PIN auxin efflux proteins (Geldner et al. 2003). It had been noted that chemical interference with auxin transport mimics many aspects of the *gn* mutant phenotype (see Figure 16.8), suggesting that the protein may normally be required for maintaining normal auxin fluxes.

Through its guanine nucleotide exchange factor (GEF) activity, GN is thought to facilitate the targeting of vesicles responsible for the polar deposition of PIN class proteins. The asymmetric cellular distribution of these PIN proteins, which facilitate auxin efflux, is thought to determine the directional movement of auxin in the developing embryo. In both *gn* mutants, and in wild-type plants treated with brefeldin A (BFA), an inhibitor of vesicle trafficking, PIN proteins no longer show polar localization (see Chapter 19). These effects of BFA were shown to result from the specific inhibition of GN (Geldner et al. 2003). The involvement of PIN proteins in apical–basal patterning is further reinforced by the early cell division defects that result from disrupting genes that encode PIN proteins themselves.

WOX genes represent a third set of players in the apical–basal patterning process (Haecker et al. 2004). Named for the homeobox DNA-binding sequence motif they share and their similarity to *WUSCHEL* (discussed later in relation to SAM function), transcripts from *WOX2* and *WOX9* accumulate in complementary positions in the apical and basal cells respectively. Evidence for their role in early patterning is provided by *wox2* mutants, which have abnormal cell divisions in the proembryo. It remains to be seen exactly what the relationship between *WOX* expression and other early patterning events might be.

A detailed picture of where and when in the embryo specific genes are expressed has been provided by in situ localization studies (Weijers and Jürgens 2005) (see Figure 16.11). However, many questions remain. Do auxin gradients enable the establishment of complementary *WOX* expression domains? Alternatively, perhaps the patterned expression of different *WOX* genes leads to the establishment of auxin gradients by affecting the expression of genes that encode PIN proteins. The former model is supported by the observation that mutations in *BDL* or *MP* lead to changes in the pattern of *WOX9* expression, suggesting that the expression of *WOX* genes is a downstream effect of regulated auxin levels.

From these few examples, it is apparent that a clear understanding of the functional relationships between different embryonic genes will ultimately require a more detailed accounting of genetic and physiological interactions.

Radial patterning establishes fundamental tissue layers

In addition to the distinctions among cells and tissues positioned along the apical–basal axis, differences can also be seen along a radial axis of the developing embryo. Initially, this radial axis is most clearly defined in geometric terms, extending perpendicularly (with respect to the longitudinal axis of the embryo) from the interior to the surface. In Arabidopsis, the radial pattern is established through predictable patterns of cell divisions in the globular embryo.

Differentiation of tissues along the radial axis is first observed in the globular embryo (see Figures 16.3 and 16.5) where **periclinal** divisions (cell divisions in which the new

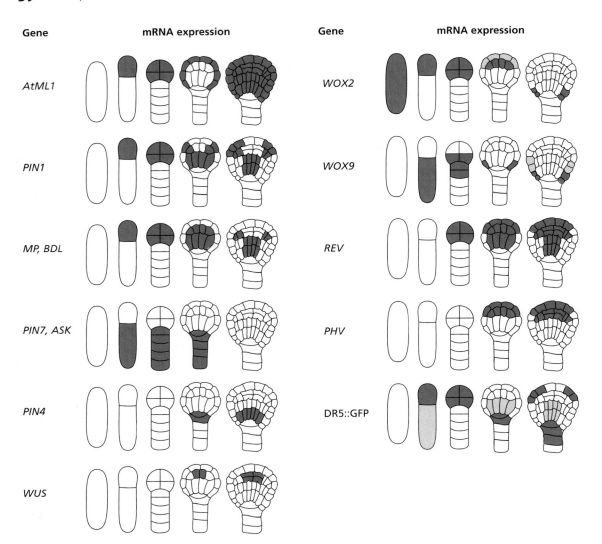

FIGURE 16.11 Early patterning of the embryo at the transcriptional level. The diagram provides a schematic presentation to illustrate how early steps in embryogenesis are accompanied by changes in spatially regulated patterns of transcription. For most genes, early expression has not been studied in detail. Uncharacterized stages are left blank. Different shades of blue represent different transcription levels, from high (dark blue) to low (white). Abbreviations: *ASK*, *Arabidopsis SHAGGY-related protein kinase*; *ATML1*, *Arabidopsis thaliana MERISTEM LAYER 1*; *BDL*, *BODENLOS*; *PHV*, *PHAVOLUTA*; *REV*, *REVOLUTA*; *WOX*, *WUSCHEL related homeobox*; *WUS*, *WUSCHEL*.

cross-walls form in a plane *parallel* to the surface of the organ) divide the embryo into three radially defined regions. The outermost cells form a one-cell-thick surface layer known as the **protoderm**. The protoderm covers both halves of the embryo and eventually differentiates into the epidermis. Cells that will later become the ground meristem underlie the protoderm. The ground meristem gives rise to the **cortex** (the ground tissue between the vascular system and the epidermis) and, in the root and hypocotyl, it also produces the **endodermis**, the layer of suberized cells that restricts water and ion movements into and out of the stele via the apoplast (see Chapter 4). The procambium is the inner core of elongated cells that generates the **vascular tissues** and, in the root, the **pericycle** (Figure 16.12).

As was seen for axial patterning, a precisely defined sequence of cell division does not appear essential for the establishment of basic radial pattern elements of the embryo. Comparison of early embryogenesis in different species reveals considerable variation in cell division patterns. In Arabidopsis, this flexibility is illustrated by mutations in the *FASS* gene (also known as *TONNEAU 2* [*TON2*]) (Torres-Ruiz and Jürgens 1994). Disruption of this gene, whose protein phosphatase activity normally regulates features of the cytoskeleton, clearly alters patterns of cell division, leading to extra cell division, but it does not block the establishment of basic radial pattern elements (Figure 16.13; see also **Web Essay 16.2** and **Web Topic 16.4** for discussion of how cell division planes are established).

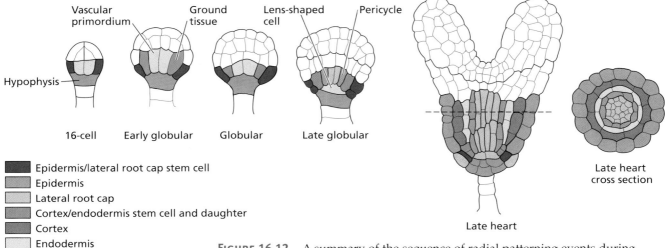

FIGURE 16.12 A summary of the sequence of radial patterning events during Arabidopsis embryogenesis. A series of five successive embryonic stages are shown in longitudinal cross section illustrate the origin of distinct tissues, beginning with the delineation of the protoderm (left) and ending with the formation of the vascular tissues (right). Note how the number of tissues increases through the actions of stem cells. A cross-sectional view of the basal portion of the late heart stage embryo is shown on the far right, and the level of the cross-section is shown by the line in the longitudinal section to the left.

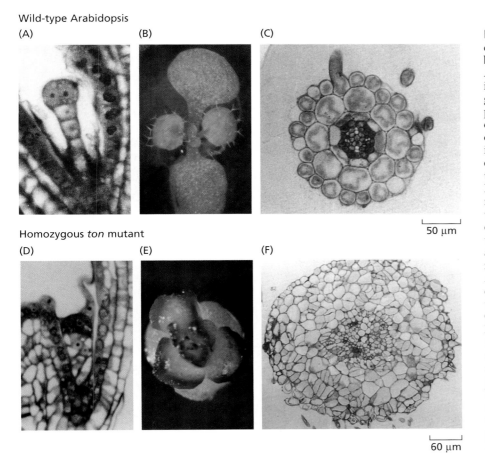

FIGURE 16.13 Extra cell divisions do not block the establishment of basic radial pattern elements. Arabidopsis plants with mutations in the *FASS* (alternatively, *TON2*) gene are unable to form a preprophase band of microtubules in cells at any stage of division. Plants carrying this mutation are highly irregular in their cell division and expansion planes, and as a result they are severely deformed. However, they continue to produce recognizable tissues and organs in their correct positions. Although the organs and tissues produced by these mutant plants are highly abnormal, the radial tissue pattern is not disturbed. Wild-type Arabidopsis: (A) early globular stage embryo; (B) seedling seen from the top; (C) cross-section of a root. Comparable stages of Arabidopsis homozygous for the *fass* mutation: (D) early embryogenesis; (E) mutant seedling seen from the top; (F) cross-section of the mutant root showing the random orientation of the cells, but a near wild-type tissue order; an outer epidermal layer covers a multicellular cortex, which in turn surrounds the vascular cylinder. (From Traas et al. 1995.)

Two genes regulate protoderm differentiation

Unlike axial patterning of the embryo, where auxin plays a key role, the molecular basis for radial patterning of the embryo is less clear. Although position-dependent mechanisms are almost certainly involved, their nature is obscure. Some clues can be drawn from detailed anatomical and molecular analyses of early embryogenesis that point to the possible nature and behavior of positional cues. One obvious and singular aspect of the radial axis is provided by the surface layer of the embryo, the protoderm.

Cells that make up the protoderm can be uniquely defined by their superficial position. Instead of being completely surrounded by other cells, protodermal cells have exposed walls that could, in theory, facilitate communication with the external environment. Alternatively, these superficial cells might act as boundary that restricts the movement of signals transmitted from cell to cell. In either case, the protoderm exhibits unique properties that differentiate it from the internal cell layers. For example, studies in *Citrus* have shown the presence of a cuticle layer on the surface of the embryo from the earliest zygotic stages through to maturity (Laux et al. 2004). This observation suggests that surface-associated factors might provide important cues for radial patterning.

Genetic studies have pointed toward two genes, *ARABIDOPSIS THALIANA MERISTEM LAYER 1* (*ATML1*) (Lu et al. 1996) and *PROTODERMAL FACTOR 2* (*PDF2*) (Abe et al. 2003), as having essential roles in establishing the epidermal identity of superficially positioned cells. Both genes encode homeodomain transcription factors and are expressed from early stages of embryogenesis in the outer cells of the embryo proper. This expression appears necessary for the establishment of normal epidermal identity, since the mutant plants have an abnormal epidermis in which cells display characteristics normally associated with mesophyll cells (Figure 16.14).

The regulatory activities of the protein products of both genes appears to be linked to their recognition of a specific, 8–base pair recognition sequence shared by the promoters of epidermis-specific genes. The binding of *ATML1* and *PDF2* to these promoter sequences promotes the transcription of the genes, whose active products lead to the differentiation of the epidermis. The *ATML1* and *PDF2* genes, themselves, contain this same recognition sequence, suggesting that their expression is maintained by a positive auto-feedback loop.

Cytokinin stimulates cell divisions for vascular elements

Genetic analysis has also revealed the identity of genes involved in the establishment of more internal tissues, including the vasculature and cortex. Arabidopsis mutants in which the *WOODEN LEG* (*WOL*) gene is disrupted fail to undergo a critical round of cell division that normally produces precursors for xylem and phloem (Figure 16.15). This defect leads to the development of a vascular system that contains xylem, but not phloem elements. *WOL* (also known as *CYTOKININ RECEPTOR 1*) encodes one of several receptors for cytokinin, implicating this hormone in the establishment of radial pattern elements (Mahonen et al. 2000; Inoue et al. 2001) (see Chapter 21). However, because these defects can be rescued by the *fass* mutation (that is, by making a *wol/fass* double mutant), which causes extra rounds of cell

(A) Wild type

(B) *atml1/pdf2* mutant

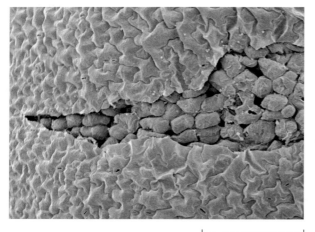
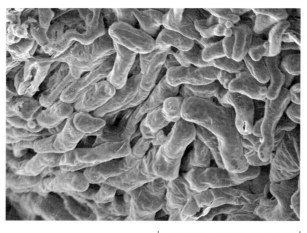

10 μm

10 μm

Figure 16.14 *ATML1* (*Arabidopsis thaliana MERISTEM LAYER 1*) and *PDF2* (*PROTODERMAL FACTOR 2*) are genes required for the establishment of a normal epidermis. (A) Wild type, partially peeled to reveal the underlying mesophyll. (B) Surface of the double mutant, *atml1/pdf2*. The epidermis of the mutant resembles the mesophyll of the wild type. (From Abe et al. 2003.)

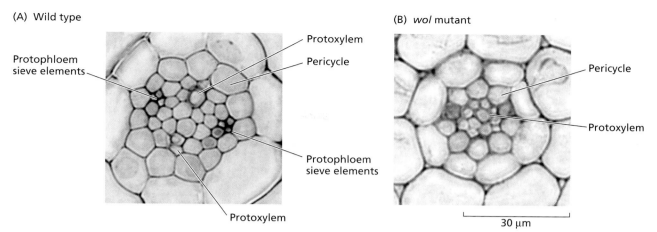

FIGURE 16.15 The cytokinin receptor encoded by the Arabidopsis *WOL* gene is required for normal phloem development. Comparison of wild type (left) and *wol* mutant (right) roots showing the absence of phloem elements in *wol* that is accompanied by an apparent decrease in the number of cell layers. (From Mähönen et al. 2000.)

division, it appears that the absence of phloem in *WOL* reflects the absence of an appropriately positioned precursor cell layer rather than the inability to specify phloem cell identity.

Two genes control the differentiation of cortical and endodermal tissues through intercellular communication

The interaction between genes involved in the differentiation of endodermal and cortical tissues illustrates the role that communication between adjacent cell layers plays in radial patterning processes. Two Arabidopsis genes, *SCARECROW* (*SCR*) (Di Laurenzio et al. 1996) and *SHORT-ROOT* (*SHR*) (Helariutta et al. 2000), are essential for the normal formation of cortical and endodermal cell layers, with mutations in either gene blocking the round of cell division that creates these layers (Figure 16.16). Both genes encode members of the structurally related family of GRAS transcription factors. The name GRAS is drawn from the first chracterized members of the family, including *GIBBERELLIN-INSENSITIVE* (*GAI*) and *REPRESSOR OF GA1–3* (*RGA*) and *SCR*. Given the type of protein they encode, both SCR and SHR could be expected to control the transcription of other "downstream" genes necessary for the formation of cortical and endodermal tissues.

In *scr* mutants, the single layer that remains exhibits characteristics of both the endodermis and cortex, suggesting that the mutant is still able to express these characters, but is unable to separate them into discrete layers. This interpretation is supported by the ability of *fass* to restore more normal growth patterns. Much like its ability to rescue *wol*, *fass* appears to compensate for the division defect of *scr*, and thus provides separate layers in which distinct endodermal and cortical traits can be expressed.

The mutant *shr*, by contrast, not only exhibits a cell division defect similar to *scr*, but also is defective in the specification of endodermal characteristics. In its absence, the single undivided layer takes on characteristics of the cortex, with no clear endodermal characteristics. This appar-

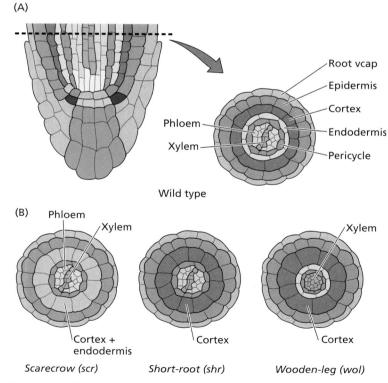

FIGURE 16.16 A comparison of normal and mutant radial root patterns shows the spatially defined functions of specific genes. (A) Wild-type root. Cell types are given in the key at bottom. (B) Defective root radial pattern of three Arabidopsis mutants, *scarecrow* (*scr*), *shortroot* (*shr*), and *wooden leg* (*wol*). (After Nakajima and Benfey 2002.)

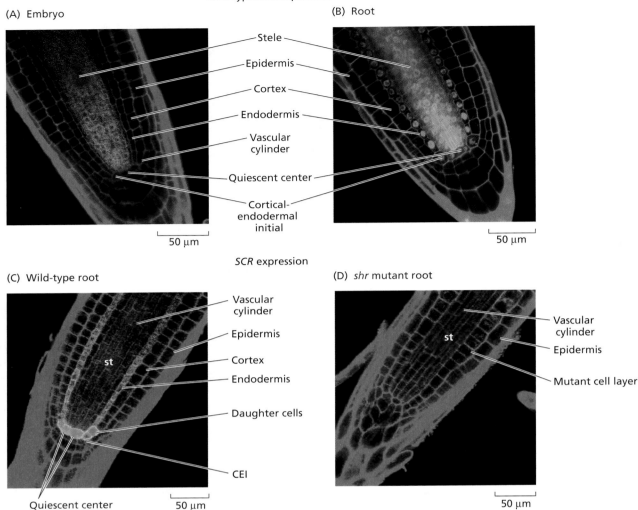

FIGURE 16.18 The *SHORTROOT* (*SHR*) and *SCARECROW* (*SCR*) genes in Arabidopsis control tissue patterning during root development. Localized patterns of gene expression are shown by confocal laser scanning microscopy visualization of green fluorescent protein (GFP). When the coding region of GFP is fused to gene promoters, the location of its fluorescence (greenish-yellow color) indicates tissues in which the promoter is most active, and where transcripts for the gene are likely to accumulate (A,C, and D). When GFP is fused in frame to the coding region of a gene (B), the fluorescence indicates where the protein gene product is likely to accumulate. A comparison of (A) and (B) indicates that while there is considerable overlap in signal from the promoter driven fusion versus the fusion to SHR protein, only the protein fusion can be detected in the endodermis suggesting the SHR protein is translated in the stele, then transported intercellularly to the endodermal layer. In panels (C) and (D), the fusion of the *SCR* promoter to GFP is used to show how *SCR* transcription appears to depend on *SHR* gene activity. (C) In wild-type roots, the *SCR* promoter is active in the quiescent center, endodermis, and cortical–endodermal stem cell (CEI), but is not active in the cortex, vascular cylinder, or epidermis. By contrast, the activity of *SCR* is markedly reduced in the *shr* mutant root (D), and now appears only in the mutant cell layer that has characteristics of both endodermis and cortex. (From Helariutta et al. 2000 and Nakajima et al. 2001).

ent requirement for *SHR* gene activity to specify endodermal traits is puzzling, since the expression of its mRNA is normally restricted to more internal provascular tissues.

An explanation of why *SHR* activity is required for normal specification of the endodermal layer was provided by a detailed study that described the distribution of the SHR protein (Nakajima et al. 2001). This study revealed that, although the mRNA for *SHR* is confined to layers of the vascular cylinder, the translated protein is transported into the adjacent external layer, where it somehow induces endodermal traits (Figure 16.17). As endodermal traits are not normally produced in the vascular layers, the SHR protein must interact with additional factors that are present in the layer that will develop as the endodermis, but which are absent from the more internal provascular tissues.

Intercellular communication is central to plant development

Although the foregoing discussion provides several examples of differentiation along the radial axis, the nature of the positional information that guides these processes and the rules that govern its transmission remain unclear. In some cases, as exemplified by SHR, **inductive effects**, in which the differentiation of one layer is *induced* by factors from adjacent cells, are evident. The movement of the SHR protein is mediated by *plasmodesmata*, membrane-lined tubes that penetrate the cell wall and connect the cytoplasms of adjacent cells (see Chapter 1). Plasmodesmata play an important role in the intercellular trafficking of macromolecules, such as proteins, whose expression can influence the determination of cell fate (Haywood et al. 2002).

The rules that govern the intercellular movement of macromolecules are complex, but studies of SHR protein fused to green fluorescent protein to facilitate tracking suggest that the expression pattern of the *SCR* gene has the potential to limit the movement of the *SHR* protein. For example, the SHR protein moves more readily between cell layers in a mutant *scr* background than it does in a wild-type background, although the mechanism by which SHR movement is regulated is unknown.

Furthermore, the rules governing macromolecular movements appear to change as the embryo develops. Studies monitoring the movement of soluble versus endoplasmic reticulum (ER)–localized GFP have shown that large, soluble macromolecules can move relatively freely within multicellular domains in a number of different types of developing tissues (Figure 16.18) (Kim et al. 2005). The existence of such domains highlights their potential for enabling regionalized patterns of cell fate determination, but the mechanisms for establishing such patterns remain undefined. This regionalized communication might enable patterning processes that operate on a supracellular level. Such communication could explain many characteristic patterns of plant development in which individual cells appear to differentiate according to their position rather than being guided by intrinsic factors that might reflect the cell's ancestry.

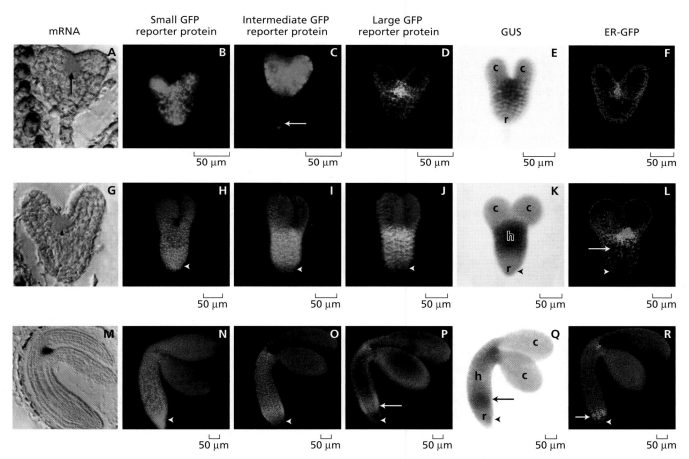

FIGURE 16.18 Potential for intercellular protein movement changes during development. Figures show the distribution of small, intermediate, and large GFP reporter proteins in embryos of different ages. All constructs are transcribed using a *STM* promoter which produces transcripts in a relatively small regions of the embryos as shown by in situ hybridization (A, G, and M) and fusions to the non-diffusible GUS (E, K, and Q) and ER-GFP reporters (F, L, and R). Small proteins appear to move readily in all stages while the mobility of larger protein is less and becomes more restricted in older embryos. Early heart (A–F), late heart (G–L), and midtorpedo (M–R) embryos for small GFP construct (B, H, and N), intermediate GFP contstruct (C, I, and O), and large GFP construct (D, J, and P) movement. Arrows indicate nucleus in suspensor cells (C), and ectopic expression of *STM* promoter in hypocotyls (L and P–R). Arrowheads indicate root. Abbreviations: c, cotyledons; h, hypocotyl; r, root. (From Kim et al. 2005.)

Shoot Apical Meristem

Meristem tissues play a crucial role throughout vegetative development by providing a source of undifferentiated cells that can be recruited to form various tissues and organs (see **Web Essay 16.2** for a historical overview of plant meristems).

In a broad sense, meristems can be defined as groups of cells that retain the capacity to proliferate and whose ultimate fate remains undetermined. A variety of meristem types, defined by their position in the plant, contribute to the development of the plant. The root and shoot **apical meristems** (RAM and SAM, respectively), enable the root and shoot systems to grow in an indeterminate manner. *Intercalary meristems*, as their name suggests, represent regions embedded in developing tissues that enable further localized growth, such as that associated with the base of elongating grass leaves. Small, superficial clusters of cells, known as *meristemoids*, give rise to structures such as stomata. The nature of mechanisms by which these types of meristems are established and maintained poses challenging questions that promise important insights into the control of plant development.

The shoot apical meristem provides one of the best examples of how a meristem is established and maintained, and its maintenance and complex behaviors pose intriguing questions for the developmental biologist. As we have discussed, the formation of the SAM appears to require a complex sequence of patterned gene expression. Beyond this, several questions can be asked. How does the SAM maintain its characteristic size and shape? How does the SAM balance the need to supply new cells for organogenesis with the need to perpetuate itself? How does the SAM recognize and adjust to diverse growth conditions?

In the following section we consider these questions from different perspectives, starting with a physical overview of the SAM and a description of its characteristic activities, followed by a consideration of the genetic programs that regulate these activities.

The shoot apical meristem forms at a position where auxin is low

The role that auxin plays in the initial establishment of the apical–basal axis extends to the formation of terminal domains of the embryo. These terminal domains ultimately function as the apical meristems of the shoot and root. Key to this activity is a means by which specific groups of cells are set apart from their neighbors and somehow maintained in a **pluripotent state**—that is, capable of differentiating into many cell types. As will be seen, molecular studies suggest that auxin does not specify this potential directly, but instead acts through intermediate programs of gene expression that are spatially limited and progressively define smaller domains that can function as **stem cells**. In this context, *stem cells* are undifferentiated cells that can both perpetuate themselves and give rise to daughter cells capable of differentiating into specialized cells.

The establishment of the shoot apical meristem in the embryo can be linked to auxin related gene activities. An initial surge of auxin into the apical region of the early embryo appears to induce the patterned activity of genes such as *MP*, which, as discussed previously, encodes an auxin response factor whose activity as a transcriptional activator facilitates further auxin flow by promoting vascular development (Jenik and Barton 2005). Indeed, the roles of the ARF transcription factors MP and the related NONPHOTOTROPIC HYPOCOTYL 4 (NPH4) are essential to this auxin-based patterning process: Embryos that are doubly mutant for these genes fail to produce cotyledons, and in this respect resemble mutants with defects in PIN-mediated auxin transport.

As we saw earlier in Figure 16.9, PIN-mediated auxin flow directs auxin to regions that flank the shoot apical meristem, while the *central zone* becomes relatively auxin-deficient (see Figure 16.19A). As a result of this patterning process, *MP* expression is relatively weak in the auxin deficient central region of the apex but stronger in flanking regions that will later develop into cotyledons.

In addition to their influence on vascular development, *MP*-like genes also appear instrumental in establishing the patterned expression of three related transcription factor genes, *CUP-SHAPED COTYLEDON (CUC) 1, 2,* and *3* (Jenik and Barton 2005). The activation of these three genes appears to constitute an early step toward defining the centrally positioned domain of cells that will later develop into the SAM (see Figure 16.19B and C). Embryos that are mutant for *MP* and *NPH4* show abnormal accumulations of *CUC* transcripts in cotyledonary regions, indicating that *MP* and *NPH4* expression normally helps limit the expression of *CUC* to the central region between the cotyledons. The role of auxin in the CUC-mediated process is supported by the observation that normal embryos treated with auxin transport inhibitors (see Figure 16.8) exhibit defects (cup-shaped cotyledons) similar to those observed among *cuc* mutants.

The expression of *CUC*-like genes in the central apical domain of the embryo provides an environment for further patterning processes, including the localized expression of the *SHOOT MERISTEMLESS (STM)* gene, which initially coincides with the stripe-like CUC expression domain, but later becomes focused in a central circular domain. This expression depends on *CUC* gene activities, as indicated by analysis of *cuc* mutant embryos in which no *STM* expression can be detected.

The expression of *STM* appears essential for establishment and maintenance of the shoot apical meristem, as demonstrated by the phenotypes of *STM* loss-of-function mutants, in which the SAM fails to form. Subsequent changes in the pattern of *CUC* gene expression, in turn, appear dependent on *STM* activity (Figure 16.20). The manner in which these genes achieve complementary expression patterns is unclear,

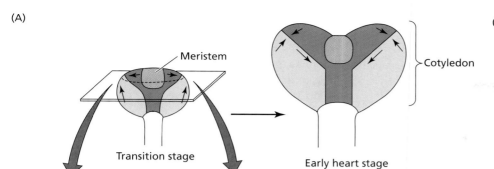

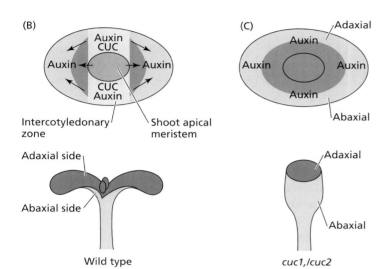

FIGURE 16.19 A model for auxin-dependent patterning of the shoot apical domain. (A) The direction of auxin transport (arrows) during the transition-stage and early heart-stage Arabidopsis embryos. (B and C) Cross-sections (as shown in A) through the apical domain of a wild-type or *CUP-SHAPED COTYLEDON* gene double mutant (*cuc1/cuc2*), showing the region in the embryo that will develop into the shoot apical meristem, the intercotyledonary zones, and the adaxial and abaxial domains of the cotyledon. In the wild type (B), the meristem and intercotyledonary zones have low auxin levels and, consequently, high CUC levels, whereas the opposite pattern is seen in the flanking cotyledon primordia. (C) In a *cuc* mutant, no separation occurs between the cotyledons, thus preventing the formation of a shoot apical meristem. (After Jenik and Barton 2005.)

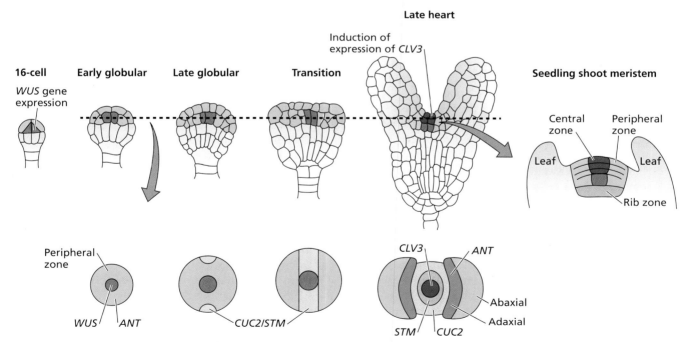

FIGURE 16.20 The formation of the apical domain involves a defined sequence of gene expression. The top row illustrates the early onset of internal *WUS* expression, which induces the expression of *CLV3* in adjacent external cell layers. The bottom row shows cross-sections at the level indicated by the dashed line at left, and emphasizes the gene expression patterns that demarcate the emerging cotyledonary and shoot apical domains. (After Laux et al. 2004.)

but presumably involves the transcriptional regulatory activity associated with their gene products.

Forming an embryonic SAM requires many genes

A host of other gene activities are required for the normal establishment of the embryonic shoot apical meristem. For example, mutants of the gene *WUSCHEL* (*WUS*), which encodes a homeodomain transcription factor, fail to produce a functional SAM (Mayer et al. 1998). *WUS* transcripts are expressed early in embryogenesis, from the 16-cell stage onward. In later stages, *WUS* expression become restricted to a small sub-apical domain situated directly beneath initial cells in the outer three cell layers, where it appears to play an essential role in maintaining their stem cell–like behavior as initial cells. Functional analyses of *WUS* and *STM*, discussed more fully in the next section, suggest that both genes play essential roles in maintaining the SAM during postembryonic development.

Another class of transcription factors, encoded by Class III homeodomain–leucine zipper (HD-Zip) gene family members, has recently been shown to play essential roles in establishing the SAM. In addition, Class III HD-Zip genes promote the differentiation of the upper epidermis and palisade parenchyma, referred to collectively as the **adaxial** cell layers of the leaf. We will return to the role of Class III HD-Zip genes in establishing the planar asymmetry of the leaf later in the chapter, in the section titled Vegetative Organogenesis (see also **Web Topic 16.5**).

Shoot apical meristems vary in size and shape

Located at the extreme tip of the shoot, the vegetative shoot apical meristem generates precursors that will develop into the stem and lateral organs such as leaves and branches. The SAM is normally concealed by **leaf primordia** and **immature leaves** that form at its base and envelop the meristem as they enlarge. Because organogenesis occurs in close proximity to the initial cells of the SAM, specific terminology has proven useful to distinguish these activities. In this context, the term **shoot apical meristem** refers specifically to the initial cells and their undifferentiated derivatives, but excludes adjacent regions of the apex that contain cells committed to particular developmental fates. The more inclusive term **shoot apex** (plural *apices*), refers to the apical meristem plus the most recently formed leaf primordia.

The size and shape of the SAM varies according to a number of parameters, including the species, developmental stage, and growth conditions (Steeves and Sussex 1989). Cycads have the largest SAM among vascular plants, measuring over 3 mm in diameter; at the other extreme, the SAM of Arabidopsis is less than 50 μm in diameter and contains only a few dozen cells.

Within a given species, significant variations in shape and size can also occur, ranging from flat to mounded structures. Such periodic changes in the size and shape of the shoot apical meristem are associated with successive rounds of leaf initiation. Further anatomical differences can occur in connection with changes in the rates of leaf initiation and overall growth associated with seasonal changes and the onset of dormancy.

The shoot apical meristem contains distinct zones and layers

The cellular organization of the shoot apical meristem provides a useful framework for a more detailed description of its growth and development and is best appreciated by microscopic examination of shoot apices. Longitudinal sections of shoot apices reveal an organizational pattern called **zonation**, which refers to a combination of cytohistological characteristics (Figure 16.21). The center of an active SAM contains a cluster of relatively large, highly vacuolate cells called the **central zone**. The central zone is somewhat comparable to the quiescent center of root meristems, which will be discussed later in the chapter. A flanking region (doughnut-shaped in cross-section) consisting of smaller, cytoplasmically dense cells is called the **peripheral zone**. A **rib zone** that lies beneath the central cell zone gives rise to the internal tissues of the stem.

These cytologically defined zones most likely reflect regionalized metabolic activity differences. The dense cytoplasm of the peripheral zone is probably linked to the increased growth and cell division observed in this region. The centrally located rib zone exhibits similar behavior and contributes cells that become the pith of the stem. By contrast, the central zone contains cells having low metabolic activity that act as apical initials, dividing slowly to replenish cells in the rib and peripheral zone populations (Bowman and Eshed 2000).

Anatomical studies also provide clues to how various tissues of the shoot might be derived from the **initials**, or stem cells, of the shoot apical meristem. In addition to revealing cytological zonation in the SAM, longitudinal sections illustrate its layered organization. This layered pattern is most apparent for the protoderm. These epidermal cells, and in many cases one or more layers of cells beneath them, owe their layered organization to highly consistent **anticlinal** cell divisions, in which new cell walls form in a plane that is *perpendicular* to the surface. As long as divisions are limited to the anticlinal plane, the integrity of the layer is maintained, with increases in cell number confined to that layer. In contrast, no layered organization is apparent for the more internally located cells, suggesting that divisions occur in a variety of planes, including **periclinal** divisions, in which new cross-walls are formed *parallel* to the surface. This variable division orientation contributes to increases in the volume of the internal tissues.

Groups of relatively stable initial cells have been identified

Our understanding of the relationships between initial cells in the apical meristem and their derivative tissues in the

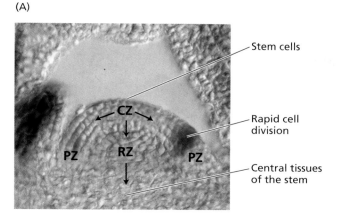

FIGURE 16.21 The Arabidopsis shoot apical meristem can be analyzed in terms of cytohistological zones or cell layers. (A) The shoot apical meristem has cytohistological zones that represent regions with different identities and functions. The central zone (CZ) contains the stem cells, which divide slowly but are the ultimate source of the tissues that make up the plant body. The peripheral zone (PZ), in which cells divide rapidly, surrounds the central zone and produces the leaf primordia. A rib zone (RZ) lies below the central zone and generates the central tissues of the stem. (B) The shoot apical meristem also has cell layers that contribute to specific tissues of the shoot. Most cell divisions are anticlinal in the outer, L1 and L2 layers, while the planes of cell divisions are more randomly oriented in the L3 layer. The outermost (L1) layer generates the shoot epidermis; the L2 and L3 layers generate internal tissues. (From Bowman and Eshed 2000.)

shoot has been facilitated by the use of **clonal analysis** or **fate mapping**, whereby the fates of individual cells can be traced with genetically determined markers (Poethig 1987).

In earlier studies, researchers applied the chemical colchicine to shoot apices to induce the occasional formation of polyploid cells that grow relatively normally, but can be easily recognized by their increased nuclear volume and cell size. Examination of sectioned shoot apices of such a plant, which had been treated and allowed to grow for a time, revealed sectors of polyploid cells that occurred in layers. Those sectors that appeared to include cells at the apex of the shoot were stably maintained, suggesting that a small number of permanent initials are positioned at the apex (Figure 16.22).

In most angiosperm species, more than one set of initials appears to be maintained. One set, positioned in the surface layer at the very tip of the shoot apex, divides anticlinally (perpendicular to the surface) to produce progenitors of the epidermis. A second set of adjacent, but more internal initials give rise to a subepidermal layer, while a third even more internal set gives rise to cortical tissues. The derivatives of these discrete sets of initials represent separate cell lineages, and as such, are designated **L1**, **L2**, and **L3**, respectively (see Figure 16.21). Some species, such as grasses, possess only two layers of apical initials and, therefore, have only L1 and L2, while other species may have more than three layers.

Broadly defined, the undetermined initials contained within the SAM are stem cells, sharing many similarities with their animal counterparts, including a relatively undifferentiated character and an indefinite capacity for division. Like their animal counterparts, considerable interest has focused on how these cells are maintained. Some important clues can be seen in the behavior of periclinal chimeras. **Periclinal chimeras** are plants which contain one or more genetically distinct layers that run parallel to the tissue surface. Such plants arise when a set of

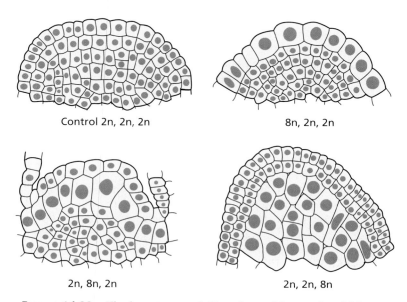

FIGURE 16.22 The long-term stability of cytochimeras in which one of the cell layers contains enlarged, polyploid nuclei demonstrates the presence of clonally distinct layers in the shoot apical meristem. (After Steeves and Sussex 1989.)

FIGURE 16.23 Periclinal chimeras demonstrate that the mesophyll tissue has more than a single clonal origin in English ivy (*Hedera helix*). These variegated leaves provide clues on the clonal origins of different tissues. A mutation in a gene essential for chloroplast development occurred in some of the initial cells of the meristem, and cells derived from these mutated stem cells lack chloroplasts and are white, while cells derived from other stem cells have normal chloroplasts and appear green. (Courtesy of S. Poethig.)

apical initials giving rise to a particular cell layer is genetically marked to give the layer a visually distinctive appearance (Figures 16.22 and 16.23). When layers are marked with a trait such as albinism, the resulting variegated foliage seen on such plants reflects the characteristic, but somewhat variable, contribution of specific cell lineages to different elements of the shoots.

In some cases, variegation is limited to one side or sector of the shoot, implying the existence of more than one initial for each layer, each contributing cells to a particular side. In other cases, the extent of variegation changes abruptly, suggesting that cells functioning as initials occasionally are displaced by neighboring cells. Once displaced from their unique apical position, these cells cease to function as initials, but instead differentiate because of their position in organogenic zones.

SAM function may require intercellular protein movement

Although the localized transcription of genes has offered important clues to the regulation of many developmental processes, there are an increasing number of examples where knowledge of the gene expression at the protein level has provided further crucial insights. For example, although the localization of transcripts of *PIN* genes offers some insight into large scale patterning of the embryo, it was only through protein localization studies that the asymmetric cellular distribution of PIN proteins could be determined, thus providing crucial insights into the basis of polar auxin transport. These analyses of proteins frequently make use of antibodies against the protein of interest, which can then be visualized by specific staining protocols. Alternatively, for those plants that can be transformed using the Ti plasmid of Agrobacterium, "tagged" forms of the protein can be introduced into the plant, which can be readily visualized by fluorescence microscopy (e.g. GFP tags), or by immunological techniques (epitope tags).

As noted earlier in the chapter, through the use of GFP-tagged proteins characteristic patterns of intercellular protein movement within the shoot apex have been established. By these studies, it appears that many transcription factor proteins have the ability to move *between* the cell layers of the meristem. By contrast, their ability to move *laterally* within an individual cell layer is more restricted. The preferential movement of proteins between cell layers probably reflects the greater number of active plasmodesmatal connections between cells of adjacent layers versus those that connect cells within the same layer.

Protein turnover may spatially restrict gene activity

In addition to the auxin-induced turnover of the repressor proteins that block auxin-induced gene expression, analogous targeted destruction of other proteins has been proposed to play an important role in restricting the expression of proteins that promote stem-cell like activity. One example is the Arabidopsis *UNUSUAL FLORAL ORGANS* (*UFO*) gene, or its *Antirrhinum* equivalent, *FIMBRIATA*, both of which are expressed in the peripheral regions that surround the central zone (Ingram et al. 1997). Both genes encode previously discussed F-box type proteins (components of ubiquitin ligase), which typically target specific proteins for destruction (see Chapter 14 on the web site). Mutants for these genes show various organ defects, suggesting that the genes may normally ensure that specific proteins that function in the central zone to confer meristematic properties are not expressed in surrounding organogenic regions.

Stem cell population is maintained by a transcriptional feedback loop

Given the ongoing diversion of stem cell derivatives into various tissues and organs of the shoot, we would expect that some mechanism monitors this process and compensates for any depletion of the stem cell population with an increased rate of cell division. The analysis of a group of functionally related genes, *CLAVATA* (*CLV*) *1*, *2*, and *3* reveals at least part of such a mechanism, and suggests involvement of a combination of both positive and negative feedback regulation.

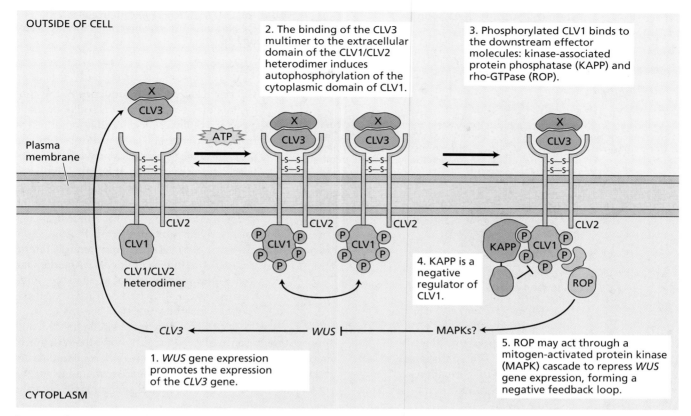

FIGURE 16.24 Model of the CLAVATA1/CLAVATA2 (CLV1/CLV2) receptor kinase signaling cascade forming a negative feedback loop with the *WUS* gene. (See Chapter 14 on the web site for further information about receptor kinase signaling pathways.) (After Clark 2001.)

All three *CLV* genes were first described in Arabidopsis in terms of their loss-of-function *clv* mutant phenotypes, in which the SAM was grossly enlarged, leading to an easily recognizable increase in the number of lateral organs produced, especially floral organs (Sharma et al. 2003). The similarity of the loss-of-function phenotypes for each of the three genes suggested that their gene products function interdependently, a prediction that has been supported by further analyses. Molecular cloning revealed that all *CLV* genes encode components of a membrane-spanning receptor kinase signaling complex (Figure 16.24). Plant genes are known to encode hundreds of such proteins, which enable cellular responses to extracellular signals.

Binding of a specific ligand (signaling molecule) to the exposed extracellular domain is thought to promote a kinase activity on the intracellular domain, and thus initiates an intracellular signaling cascade. *CLV1* encodes a leucine-rich repeat kinase (LRR kinase), which contains an extracellular ligand-binding domain, a membrane-spanning domain, and an intracellular kinase domain. *CLV2* encodes a closely related protein that lacks the kinase domain. The leucine-rich domains of both proteins are thought to enable the formation of a heterodimer complex between *CLV1* and *CLV2*, which appears essential for the kinase activity of *CLV1* (see Figure 16.24).

For kinase activity, the CLV1 and CLV2 complex requires a small secreted protein encoded by the *CLV3* gene that probably acts as a ligand (Rojo et al. 2002). Biochemical analyses have shown that, once synthesized, the water-soluble CLV3 protein is secreted into the apoplast, where it can diffuse in a concentration-dependent manner.

The **apoplast** consists mostly of the space occupied by the cell walls and intercellular spaces (see Chapter 4). Cell wall macromolecules are largely hydrophilic, and the wall contains passages between the macromolecules with an apparent pore size of 3.5 to 5.0 nm. Given this pore size, molecules with a mass of less than approximately 15 kDa can be expected to diffuse freely through the apoplast. With its molecular weight of approximately 11 kDa, the CLV3 protein could diffuse easily through the apoplast. Since the loss of gene activity in mutants is associated with an increase in meristem size, we can conclude that activation of *CLV* signaling by CLV3 normally limits the size of the SAM.

How does CLV kinase activity limit meristem growth? By one well-supported model, CLV kinase activity initiates a signaling cascade that ultimately leads to the repression

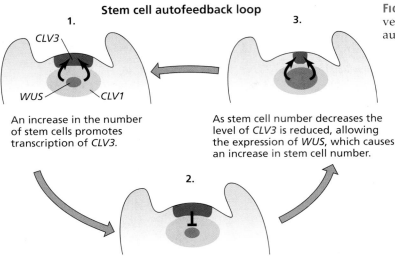

Figure 16.25 Maintenance of stem cells in the vegetative and inflorescence meristems involves an autofeedback loop.

of *WUSCHEL* (*WUS*) gene expression (Figure 16.25) (Schoof et al. 2000). (Recall that WUS, a homeodomain transcription factor, is required for the formation and maintenance of the shoot apical meristem, and mutants lacking WUS fail to produce a functional SAM.) Consistent with the CLV kinase repression model, WUS activity is increased in *clv* mutants, which leads to the increase in the size of the SAM. Thus, although the elements of this cascade are not completely defined, it is evident that the transcription of WUS is repressed by CLV kinase activity. Conversely, WUS transcription can be completely suppressed by overexpression of CLV3, leading to a phenocopy of the *wus* mutant phenotype in which the meristem is nonfunctional.

The foregoing observations suggest that the amount of CLV3 in the cells of the SAM may provide the SAM with a means of a measuring its own size, with high levels feeding back to limit WUS expression, this limiting meristem growth. This hypothesis raises the question of how CLV3 levels are regulated. Molecular analyses have determined that the transcription of the *CLV3* gene is promoted by WUS, and that meristem growth is controlled by a self-limiting positive/negative feedback loop. WUS expression stimulates the activity of the apical initials and promotes meristem growth, but at the same time induces an increase in CLV activity. The rise in CLV activity inhibits WUS expression, leading to an inhibition of meristem growth. By maintaining a dynamic equilibrium between active and inactive states, the SAM is able to maintain its population of stem cells at the genetically determined size.

Root Apical Meristem

Roots are adapted for growing through soil and absorbing water and mineral nutrients in the capillary spaces between soil particles. As a consequence of the constraints they encounter in the soil, roots have a pattern of lateral organ initiation different from that of shoots. Cells produced by the root apical meristem (RAM) divide, differentiate, and elongate as they are displaced away from the tip, much like their counterparts in the shoot. However, in contrast to the shoot, the region of the root where branches form is well separated from the tip. Only after cell elongation is complete do lateral organs form, thus avoiding shearing damage that would result if lateral appendages were pushed through a solid soil matrix by the root to which they were attached.

A further difference between the root and shoot is the presence of a root cap covering the root apical meristem. Unlike the shoot, in which epidermal initials are located on the surface layer of the meristem, initials for the root epidermis are located beneath several layers of cells that make up the root cap.

The growing RAM is faced with many of the same issues as the SAM is. Both must balance the production of undifferentiated cells with the recruitment of these cells into various tissues. However, given the distinct patterns of organogenesis associated with each, and the very distinct physical environments in which they grow, we would expect to find some differences in the behavior of their apical meristems. In the sections that follow we will highlight unique features of root development, beginning with the role of auxin in the establishment of the root apical meristem. After a review of the four developmental zones of the root tip we will discuss the different groups of initial cells found in the RAM that produce the different tissue types of the radial pattern of the root.

High auxin levels stimulate the formation of the root apical meristem

As we saw earlier, the central zone of the shoot apical meristem, which contains the initial cells, forms in the presence of low auxin concentrations as the result of PIN-mediated auxin flow toward the peripheral region, where leaf primordia develop. In contrast, the normal specification of the root apical meristem begins with PIN-mediated polar transport that is likely to cause *high* levels of auxin in the basal regions of the proembryo (Blilou et al. 2005). This high level of auxin in the cells leads to the derepression of two genes encoding ARF transcription factors, *MP* and its close relative *NPH4*, which activate additional gene expression. Among the genes activated are two, *PLETHORA*

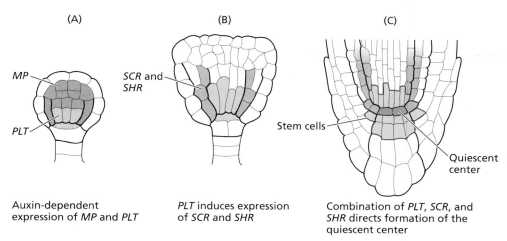

FIGURE 16.26 Model for the specification of stem cell identity in the root. (A) Early auxin-dependent expression of the *MONOPTEROS* (*MP*) and *NONPHO-TOTROPIC HYPOCOTYL 4* (*NPH4*) genes. *MP* and *NPH4* expression promote *PLETHORA* (*PLT*) expression in a basal domain. (B) *PLT* expression promotes the expression of *SCARECROW* (*SCR*) and *SHORTROOT* (*SHR*). (C) The combination of *PLT*, *SCR*, and *SHR* gene expression directs centrally positioned cells to become the quiescent center, which signals surrounding cells to maintain their stem cell activity. (After Aida et al. 2004.)

(*PLT*) *1* and *2*, whose products are transcription factors (Aida et al. 2004) (Figure 16.26). The two PLT transcription factors are then responsible for the activation of yet another set of genes that, together with genes such as *SCR* (Sabatini et al., 2003), specify the identity and function of the *quiescent center* of the root apical meristem. As discussed later in the chapter, the **quiescent center** (**QC**) is a central region of the RAM characterized by low mitotic activity.

Despite its low rate of cell division, the QC functions as the ultimate source of all the cells of the root. Thus the formation of the QC as a consequence of PLT action plays a critical role in the formation of the RAM. Studies in which the artificial activation of *PLT1* and *PLT2* alone induces the formation of the root apical meristem reinforce models in which auxin's effects are mediated through intermediate levels of transcriptional regulation (Leyser 2005). These steps are shown diagrammatically in Figure 16.27.

The root tip has four developmental zones

The basic features of root development can be analyzed by dividing the root into separate zones that display characteristic cellular behaviors. Although the boundaries are not sharp, four developmental zones can be distinguished in a root tip: the root cap, the meristematic zone, the elongation zone, and the maturation zone (Figure 16.28). In Arabidopsis, these four developmental zones occupy little more than the first millimeter of the root tip. The developing region is larger in other species, but growth is still confined to the tip. With the exception of the root cap, the boundaries of these zones overlap considerably. The zones can be described as follows:

- The **root cap** protects the apical meristem from mechanical injury as the root pushes its way through the soil. Root cap cells form by specialized root cap stem cells. As the root cap stem cells produce new cells, older cells are progressively displaced toward the tip, where they are eventually sloughed off. As root cap cells differentiate, they acquire the ability to perceive gravitational stimuli and secrete mucopolysaccharides (slime) that help the root penetrate the soil.

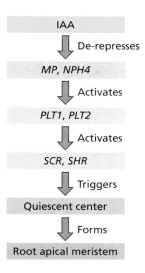

FIGURE 16.27 During embryogenesis, auxin stimulates root formation via several intermediate steps. The diagram illustrates the hierarchy of genes involved in organogenesis.

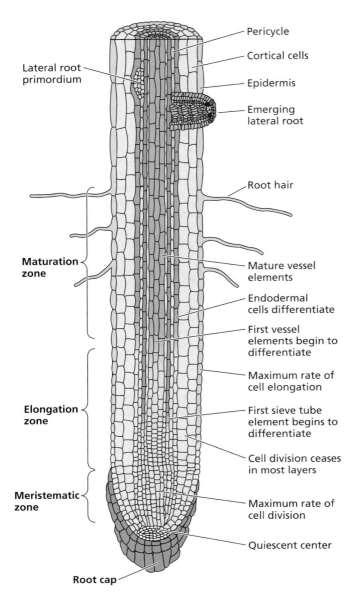

FIGURE 16.28 Simplified diagram of a primary root showing the root cap, the meristematic zone, the elongation zone, and the maturation zone. Cells in the meristematic zone have small vacuoles and expand and divide rapidly, generating many files of cells.

- The **meristematic zone** lies just under the root cap, and in Arabidopsis it is about a quarter of a millimeter long. The root meristem generates only one organ, the primary root. It produces no lateral appendages.

- The **elongation zone**, as its name implies, is the site of rapid and extensive cell elongation. Although some cells may continue to divide while they elongate within this zone, the rate of division decreases progressively to zero with increasing distance from the meristem.

- The **maturation zone** is the region in which cells acquire their differentiated characteristics. Cells enter the maturation zone after division and elongation have ceased. Differentiation may begin much earlier, but cells do not achieve the mature state until they reach this zone. The radial pattern of differentiated tissues becomes obvious in the maturation zone. Later in the chapter we will examine the differentiation and maturation of one of these cell types, the tracheary element.

Specific root initials produce different root tissues

Like the shoot, various tissues of the root in higher plants can be traced to apical initials. In species for which the root is relatively small, such as Arabidopsis, lineage relationships between cells that make up the root can inferred by direct observation (Dolan et al. 1993). Unlike shoots, where lateral organs conceal the apical meristem, the tips of growing roots are exposed and relatively transparent, thus facilitating studies that explore how cellular processes such as the orientation and frequency of cell division influence development. The most striking structural feature of the root tips of many species, when viewed in longitudinal section, is the presence of long files cells, which are generated by anticlinal divisions (that is, the walls that separate newly formed cells are oriented at right angles to the long axis of the root). Very close to the tip, however, periclinal divisions (new walls are parallel to the root surface) may occur, thus establishing new files of cells.

In the roots of most higher plants, cell files appear to converge on a central group of subapically positioned cells, the quiescent center. The QC of Arabidopsis consists of four to seven cells. Cells of the QC are distinguished not only by their unique position, but also by the low frequency of cell division (Figure 16.29), a condition not unlike the cells that make up the central zone of the SAM. Adjacent to the quiescent center are initial cells for various tissues. While the actual number of initials and their relationship to specific tissues vary among species, they share many properties with the initials of the SAM, including their relatively undifferentiated character and slow rates of cell division.

Some have argued that the distinction between the quiescent center and adjacent stem cells is somewhat artificial, given that cells of the QC may occasionally divide to replace adjacent initials. As such, the QC could be viewed as a group of pluripotent stem cells, whose derivatives give rise to all the tissues of the root. Lower vascular plants, such as the water fern *Azolla*, provide examples in which a single, centrally positioned *apical cell* functions in this manner, by retaining mitotic activity throughout vegetative development. (For a discussion of this work, see **Web Topic 16.6**.)

Root apical meristems contain several types of initials

The quiescent center is the ultimate source of all the tissues of the root. However, in species such as Arabidopsis the QC exhibits little mitotic activity during postembryonic

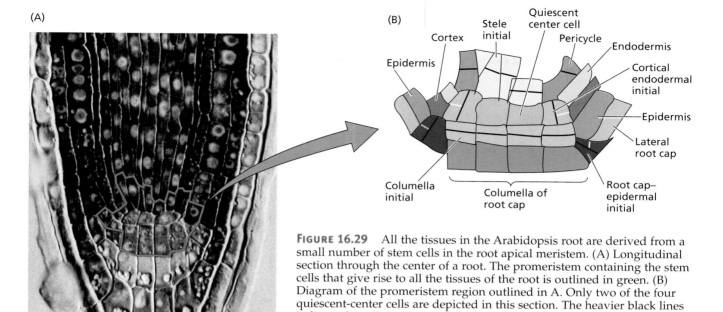

Figure 16.29 All the tissues in the Arabidopsis root are derived from a small number of stem cells in the root apical meristem. (A) Longitudinal section through the center of a root. The promeristem containing the stem cells that give rise to all the tissues of the root is outlined in green. (B) Diagram of the promeristem region outlined in A. Only two of the four quiescent-center cells are depicted in this section. The heavier black lines indicate the cell division planes that occur in the stem cells. White lines indicate the secondary cell divisions that occur in the cortical–endodermal and lateral root cap–epidermal stem cells. (A from Schiefelbein et al. 1997, courtesy of J. Schiefelbein, © the American Society of Plant Biologists, reprinted with permission.)

development. Therefore it is more useful to treat the cells *adjacent* to the QC as initials, since their slow but predictable patterns of division can be directly tied to the origin of specific root tissues. Four distinct sets of initials, all of which are adjacent to the QC, can be identified in terms of their position and the tissues they give rise to (see Figure 16.29B):

1. **Columella initials**. Located directly below (distal to) the QC, these initials give rise to the central portion of the root cap.

2. **Epidermal–lateral root cap initials**. Located to the side of the QC, these initials first divide anticlinally to set off daughter cells, which then divide periclinally to form two files of cells that will mature into the lateral root cap and epidermis.

3. **Cortical–endodermal initials**. Located interior and adjacent to the epidermal–lateral root initial, the cortical–endodermal initials divide anticlinally to set off daughter cells, which then divide periclinally to form the cortical and endodermal cell layers.

4. **Vascular initials**. Located directly above (proximal to) the QC, these initials give rise to the vascular system.

As in the shoot, the maintenance of the identity of the initials in the root appears to depend on interactions with adjacent cells. In a series of classic experiments, the role of cell–cell interactions on stem cell maintenance in the root was demonstrated by using focused laser beams to ablate (kill) single cells in the relatively transparent Arabidopsis root. Using this approach, it was shown that the cells of the QC help maintain adjacent initials in an indeterminate state. When the QC was removed, adjacent initial cells underwent abnormal division and precocious differentiation (see Figure 16.29B).

A positive/negative feedback loop involving the *WUS* gene and a gene related to *CLV3*, similar to that which occurs in the shoot apical meristem, may also be involved in maintaining the initials of the root apical meristem (Fiers et al. 2005). The specific identity of apical initials also appears to depend on intercellular interactions. In this case, information from more mature differentiated cells may provides cues that determine the characteristic behaviors of initials. This acropetal (toward the tip) flow of information was demonstrated by alterations in the normal division patterns of cortical–endodermal initials that resulted from ablating more mature derivative cells (Van den Berg et al. 1995).

Vegetative Organogenesis

The three-dimensional architecture of the shoot and root systems of most plants is elaborated gradually throughout the life of the individual. In addition to vertical growth by the axes (stem and root), lateral organs are produced. Shoots differ from roots in giving rise to leaves, but both roots and shoots exhibit branching patterns. In this section we will describe regulatory mechanisms involved in the emergence and patterning of leaf primordia on the shoot apical meristem. We will also discuss the contrasting modes of branch formation in shoots and roots.

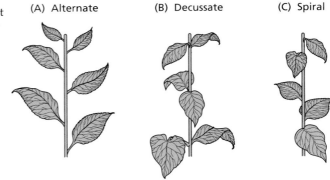

Figure 16.30 Three types of leaf arrangements (phyllotactic patterns) along the shoot axis. The same terms also are used for inflorescences and flowers.

(A) Alternate (B) Decussate (C) Spiral

Periclinal cell divisions initiate leaf primordia

The earliest overt sign of leaf development is the appearance of a bulge on the flank of the shoot apex. Periclinal cell divisions in subepidermal layers are closely associated with this event and signify the establishment of a new growth axis. The complex makeup of the leaf is most easily appreciated by observation of plants in which one or more apical initials are genetically marked to produce an easily recognized visible trait. For example, in a plant in which the initials for the L2 layer are marked with mutations that result in an albino phenotype, leaves exhibit **variegation** (Tilney-Basset 1986). The characteristic appearance of these variegated leaves—the irregular borders of green and white tissue—reflect the variable contribution of specific initials and their derivatives to the developing leaf (see Figure 16.23).

A longstanding question in plant biology is how the characteristic arrangement of leaves on the shoot, or **phyllotaxy**, is achieved. Three basic phyllotactic patterns, termed alternate, decussate, and spiral can be directly linked to the pattern in which leaf primordia are initiated on the shoot apical meristem (Figure 16.30). This pattern depends on a number of factors, including intrinsic factors that determine characteristic phyllotaxies for individual species. Mutations or environmental factors that lead to changes in meristem size or shape can also affect phyllotaxy, suggesting that diffusible factors may play an important role. In the following section the role of auxin in this process is discussed.

Local auxin concentrations in the SAM control leaf initiation

Studies involving the experimental manipulation of the shoot apex in Arabidopsis and tomato have shown that auxin can influence the position of leaf and flower initiation. For example, leaves can be induced to form at abnormal positions on the shoot apex by applying small quantities of auxin directly to the shoot apical meristem, suggesting that auxin normally acts as a key factor in promoting the initiation of leaves (see Figure 19.36). Related studies, in which the position of leaves can be altered by application of auxin-transport inhibitors, add further support to the idea that auxin levels influence leaf initiation.

Immunolocalization studies of PIN-type auxin efflux transporters have now provided compelling evidence that auxin movement is directly related to leaf initiation events (Reinhardt et al. 2003). Based on the pattern of PIN gene expression and the asymmetric distribution of the protein in the cell, it has been possible to infer the likely direction of auxin movements in the shoot apex. Detailed analyses of this sort indicate that leaf primordia are likely to form where auxin movements converge to create local concentration maxima (Figure 16.31).

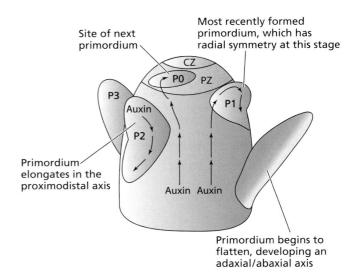

Figure 16.31 Sites of leaf formation are related to patterns of polar auxin transport. The figure depicts patterns of auxin movement (black arrows) inferred by asymmetric localization of PIN proteins. P0, P1, and P2 refer to the age of the leaf primordia, with P0 corresponding to the stage at which the leaf begins its overt development, while P1 and P2 represent increasingly older leaves. Leaf primordia are initiated where auxin accumulates. Acropetal (toward tip) movement of auxin is blocked at the boundary separating the central and peripheral zones (CZ and PZ, respectively), leading to increased auxin levels at this position and the initiation of a leaf (P0). The newly formed primordium (P1) acts as an auxin sink, thus preventing initiation of new leaves directly above. The displacement of a more mature leaf away from the PZ allows acropetal auxin movements to become reestablished, thus enabling the initiation of the P0 leaf. (After Friml et al. 2003.)

Three developmental axes describe the leaf's planar form

In many environments, the characteristic planar form of the leaf represents the most efficient means to intercept light, to promote gas exchange, and to enable highly regulated patterns of transpiration. How this planar architecture is established presents intriguing questions whose answers have begun to emerge through a variety of experimental approaches.

The most obvious question relates to how the planar geometry and tissue organization of the leaf arises. Much like the shoot, the leaf can be described by a proximal–distal axis—that is, one running from the base to the tip. However, unlike the shoot, where cells at the apex remain uncommitted, the cells in the leaf divide in more complex patterns, and all of the cells eventually differentiate to form a mature, determinate structure that ceases to grow.

Furthermore, in place of the single radial axis that runs perpendicular to the long axis of the shoot, from its center to the surface, the leaf has two axes: the first, termed the lateral axis, runs from edge to edge across the breadth of the leaf; the second axis, termed the adaxial–abaxial axis (or dorsiventral axis, the term adaxial referring to the upper surface of the leaf and abaxial to the lower), runs from the upper to the lower surface of the leaf (Figure 16.32).

Spatially regulated gene expression controls leaf pattern

Developmental analyses emphasize the complex patterns of cell division that accompany leaf development. The first observable signs of leaf initiation are periclinal divisions in subepidermal layers that define the new proximal–distal axis of the future leaf. Subsequently, divisions are observed throughout the primordium, but later are confined to more proximal regions as the leaf matures. Although the general distribution of cell division during leaf development is predictable, considerable variation is seen in the contribution that individual cells in the primordia make to the final leaf. The clearest examples of this are seen in variable size of sectors induced in fate mapping studies, or in the variable patterns of pigment variegation seen in previously discussed periclinal chimeras. Given the predictable architecture of the mature leaf, it is difficult to reconcile these variable patterns of cell division with hypothetical control mechanisms involving fixed programs of cell division. Instead, growth must be controlled by mechanisms that respond to positional cues that allow fine-tuning of cell division and expansion to achieve a predictable form.

Although the molecular nature of positional cues that guide leaf development remain elusive, several classes of genes that affect this process have been identified through genetic analyses. An early example is the *Antirrhinum PHANTASTICA* (*PHAN*) gene (equivalent to the Arabidopsis *ASYMMETRIC LEAVES 1* [*AS1*]), which is required for the establishment of the adaxial–abaxial axis. Disruption of the gene leads to the formation of filamentous leaves. The *PHAN* gene encodes a MYB family transcription factor (see Chapter 14 on the web site) that is expressed throughout the developing leaf primordia (Waites et al. 1998).

One model for the function of *AS1* (or *PHAN*) is that it somehow enables the specification of an adaxial identity on the upper surface of the leaf. Given that one activity of PHAN is to repress the expression of *KNOX* genes in developing leaves, such as *KNAT2* in Arabidopsis, the absence of *KNAT2* gene activity may be a key requirement for the establishment of adaxial (upper leaf surface) characters. Without *AS1*, a default abaxial (lower leaf surface) identity is expressed. The filamentous leaves that result are interpreted as resulting from the absence of a boundary in which abaxial and adaxial cell types are juxtaposed (Figure 16.33). By analogy with mechanisms that control wing growth in *Drosophila*, it has been proposed that this juxtaposition of distinct tissue types induces lateral growth to create the lamina.

In contrast to *PHAN/AS1*, which enables specification of adaxial cell types, a distinct class of genes, known as the YABBY* gene family (Bowman and Eshed, 2000), is essential for specification of abaxial identity. YABBY gene family members encode zinc finger proteins that likely function as transcription factors. The first member of this gene family identified, *CRABS CLAW* (*CRC*), was defined by the phenotype of its Arabidopsis loss-of-function mutant, in which the organization of the carpels is disturbed (Figure 16.34). The more general activity of this class of genes is

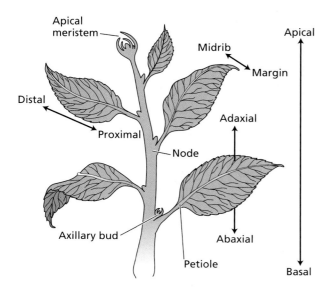

Figure 16.32 Diagram of a shoot showing the various axes along which development occurs.

---

*The YABBY gene family was named for the *crabs claw* mutant of Arabidopsis. "Yabby" is Australian slang for crayfish, a small crustacean.

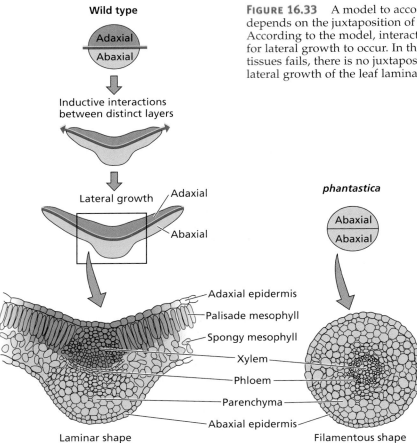

FIGURE 16.33 A model to account for the lateral growth of the leaf lamina that depends on the juxtaposition of adaxial (upper) and abaxial (lower) tissues. According to the model, interactions between these two tissue layers are required for lateral growth to occur. In the *phan* mutant, in which the specification of adaxial tissues fails, there is no juxtaposition of distinct tissue types, leading to a failure of lateral growth of the leaf lamina and the production of the mutant phenotype.

observed when mutations affecting several members of the Arabidopsis YABBY family are combined.

These multiple mutants show defects in both floral and vegetative leaf-like organs in which abaxial characters have been replaced by adaxial characters, suggesting that there is an overlap in the functions of members of the YABBY gene family. The abaxial-promoting activity of *YABBY* genes is further supported by the phenotypes of plants in which *YABBY* genes are overexpressed, causing the ectopic (in the wrong location) formation of abaxial tissues (see Figure 16.34).

A second class of genes required for specification of abaxial cell identity is represented by *KANADI* genes (Eshed et al. 2004). Like the YABBYs, the *KANADI* genes appear to have overlapping functions, with the most obvious loss of abaxial identity observed when loss-of-function mutations of these genes are combined

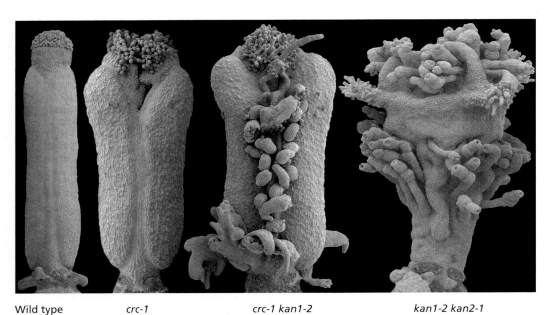

FIGURE 16.34 The normal establishment of adaxial–abaxial polarity requires *YABBY* and *KANADI* gene activities. A comparison of the wild type with single and double mutants of members of these two gene families (*crabs claw* [*crc*], is a mutant form of a *YABBY* gene) illustrates how normal organ polarity, such as that shown in these seed pods, depends on these gene activities. In mutants, one sees various degrees of abaxial tissues (the outer surface of the seed pod, or silique) being replaced by placental tissues, including ovules, that are normally found on the inner face of the carpels that form the silique. (From Bowman et al. 2002.)

(see Figure 16.34). Also like *YABBY* genes, the opposite effect (the abnormal formation of abaxial tissues) is observed when *KANADI*s are overexpressed.

MicroRNAs regulate the sidedness of the leaf

A third class of gene more directly involved in specifying adaxial identity encodes Class III HD-Zip proteins, named, as noted previously, for the presence of both a DNA-binding homeodomain and a leucine zipper protein–protein interaction domain (Emery et al. 2003). Analyses of proteins in this family indicate that their expression is closely coupled with the specification of adaxial identity—that is, the identities of the upper epidermal and palisade parenchyma tissues. Combinations of loss-of-function mutants for members of this gene family cause the tissues of the adaxial surfaces of leaves to differentiate into abaxial tissues—spongy mesophyll and lower epidermis—suggesting that members of the Class III HD-Zip family act redundantly to specify adaxial identity. This interpretation is also supported by gain-of-function mutant phenotypes, where increasing Class III HD-Zip gene products leads to the abnormal formation of adaxial tissues on the lower leaf surface (see Figure 16.34).

A spatially regulated pattern of gene expression is required for Class III HD-Zips to properly pattern organ formation and this requires the activity of microRNAs. **MicroRNAs** are short, 21- to 24-base RNA molecules that can regulate the activity of genes with complementary base sequences (Kidner and Martienssen, 2005; see also Chapter 14 on the web site). They can function by inhibiting gene transcription, triggering transcript degradation, or inhibiting translation. *PHABULOSA* (*PHB*) and *PHAVOLUTA* (*PHV*) are two Class III HD-Zip genes required for adaxial cell fate, and transcripts are normally localized to the shoot apical meristem and the adaxial side of leaf primordia. Mutations in *PHB* or *PHV* that alter the region of sequence complimentary to the microRNA, or mutations that disrupt the microRNA machinery, cause *PHB* and *PHV* transcripts to accumulate throughout the leaf primordium, resulting in radially symmetrical organs with adaxial identity on all sides.

While the foregoing discussion has highlighted a prominent role for different classes of transcription factors and described how their localized expression determines tissue identity, it is less clear how this localized expression is established. In some cases, the expression of one class of transcription factor appears activated by the expression of another. Other transcription factors appear to repress the expression of others. As these examples all involve regulation at the transcriptional level, it is possible that activation or repression could be mediated directly by the transcription factors themselves. Alternatively, regulation may require intermediate regulatory factors that remain to be identified in future analyses. A model for the interaction of transcription factor genes during leaf primordium formation is shown in Figure 16.35.

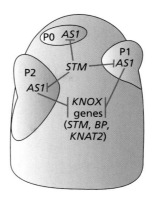

Figure 16.35 Lateral organ formation involves complex patterns of transcription factor interactions. The figure depicts the domains in which genes encoding specific transcription factors are transcribed. Inhibitory interactions between certain genes, inferred by phenotypic analysis of mutants, are also shown. The initial expression of *ASYMMETRIC LEAVES 1* (*AS1*) in P0 represses transcription of *SHOOT MERISTEMLESS* (*STM*) and other *KNOX* genes in leaf founder cells, thus permitting more determinate (restricted) programs associated with leaf development to proceed. Simultaneously, *STM* expression in other regions of the shoot apical meristem represses *AS1* activity, thus maintaining cells in a proliferative state.

Branch roots and shoots have different origins

Lateral roots begin their development some distance back from the root apical meristem (Malamy and Benfey 1997). Only after the derivatives of the RAM have ceased elongation do lateral root initials first appear. The first histological signs of root initiation are periclinal divisions among a small number of cells in the pericycle (see Chapter 1), leading to further divisions and growth in a plane perpendicular the parent root axis, with the new root tip rupturing through the endodermal and cortical layers of the parent root. Simultaneously, a new RAM becomes organized at the tip of the lateral root, enabling it to grow in an indeterminate manner (Figure 16.36). The pattern of lateral root formation and growth can be influenced by intrinsic factors such as auxin concentration, and by environmental factors such as soil nutrient availability.

Lateral organs of the shoot are initiated in a very different manner. Rather than drawing on cells from a single internal layer, lateral organs of the shoot incorporate cells from several clonally distinct layers. In most species, the predominant means of shoot branching occurs through the formation of axillary meristems. These are meristems that develop in the axils of leaf primordia and as such, the pattern of branch formation is directly related to the SAM activities that regulate phyllotaxy. Similar to roots, the growth of lateral branches can be influenced by intrinsic factors, such as hormones, and environmental factors such as light. The phenomenon of **apical dominance**, in which the terminal bud suppresses the growth of the axillary buds, is primarily regulated by auxin from the terminal

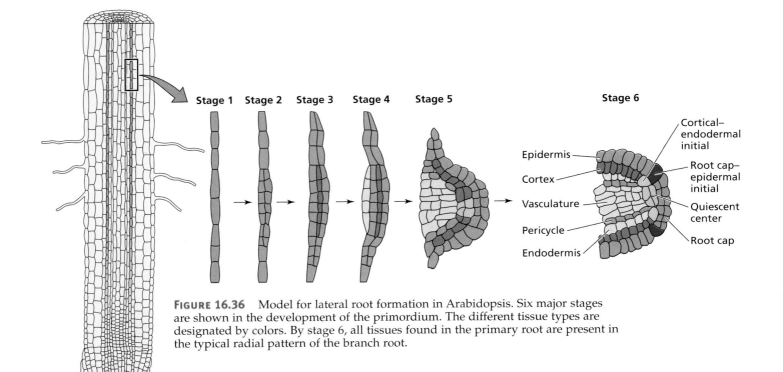

FIGURE 16.36 Model for lateral root formation in Arabidopsis. Six major stages are shown in the development of the primordium. The different tissue types are designated by colors. By stage 6, all tissues found in the primary root are present in the typical radial pattern of the branch root.

bud and an unidentified root signal, thought to be a carotenoid cleavage product. However, we will defer an in-depth discussion of the hormonal control of branching until Chapter 19.

Senescence and Programmed Cell Death

Every autumn, people who live in temperate regions can enjoy the beautiful color changes that precede the loss of leaves from deciduous trees. The leaves change color because changing day length and cooling temperatures trigger developmental processes that lead to leaf senescence and death. Senescence is distinct from necrosis, although both senescence and necrosis lead to death. **Necrosis** is death brought about by physical damage, poisons, or other external injury. In contrast, **senescence** is a normal, energy-dependent developmental process that is controlled by the plant's own genetic program. Leaves are genetically programmed to die, and their senescence can be initiated by environmental cues.

As new leaves are initiated from the shoot apical meristem, older leaves often are shaded and lose the ability to function efficiently in photosynthesis. Senescence recovers a portion of the valuable resources that the plant invested in leaf formation. During senescence, hydrolytic enzymes break down many cellular proteins, carbohydrates, and nucleic acids. The component sugars, nucleosides, and amino acids are then transported back into the plant via the phloem, where they will be reused for synthetic processes. Many minerals also are transported out of senescing organs back into the main body of the plant.

Senescence of plant organs is frequently associated with **abscission**, a process whereby specific cells in the petiole differentiate to form an abscission layer, allowing the senescent organ to separate from the plant. In Chapter 22 we will have more to say about the control of abscission by ethylene.

In this section we will examine the roles that senescence and programmed cell death play in plant development. We will see that there are many types of senescence, each with its own genetic program. Then, in Chapters 21 and 22, we will describe how cytokinins and ethylene can act as signaling agents that regulate plant senescence.

Plants exhibit various types of senescence

Senescence occurs in a variety of organs and in response to many different cues. Many annual plants, including major crop plants such as wheat, maize (corn; *Zea mays*), and soybeans, abruptly yellow and die following fruit production, even under optimal growing conditions. Senescence of the entire plant after a single reproductive cycle is called **monocarpic senescence** (Figure 16.37).

Other types of senescence include the following:

- Senescence of aerial shoots in herbaceous perennials
- Seasonal leaf senescence (as in deciduous trees)
- Sequential leaf senescence (in which the leaves die when they reach a certain age)
- Senescence (ripening) of fleshy fruits; senescence of dry fruits
- Senescence of storage cotyledons and floral organs (Figure 16.38)
- Senescence of specialized cell types (e.g., trichomes, tracheids, and vessel elements)

FIGURE 16.37 Monocarpic senescence in soybeans (*Glycine max*). The entire plant on the left underwent senescence after flowering and producing fruit (pods). The plant on the right remained green and vegetative because its flowers were continually removed. (Courtesy of L. Noodén.)

The triggers for the various types of senescence are different and can be internal, as in reproductive processes in monocarpic senescence, or external, such as day length and temperature in the autumnal leaf senescence of deciduous trees. Regardless of the initial stimulus, the different senescence patterns may share common internal programs in which a regulatory senescence gene initiates a cascade of secondary gene expression that eventually brings about senescence and death.

Senescence involves ordered cellular and biochemical changes

Because it is genetically encoded, senescence follows a predictable course of cellular events. On the cytological level, some organelles are destroyed while others remain active. The chloroplast is the first organelle to deteriorate during the onset of leaf senescence, with the destruction of thylakoid protein components and stromal enzymes.

In contrast to the rapid deterioration of chloroplasts, nuclei remain structurally and functionally intact until the late stages of senescence. Senescing tissues carry out catabolic processes that require the de novo synthesis of various hydrolytic enzymes, such as proteases, nucleases, lipases, and chlorophyll-degrading enzymes. The synthesis of these senescence-specific enzymes involves the activation of specific genes.

Not surprisingly, the levels of most leaf mRNAs decline significantly during the senescence phase, but the abundance of certain specific mRNA transcripts increases. Genes whose expression decreases during senescence are called **senescence down-regulated genes** (**SDGs**). SDGs include genes that encode proteins involved in photosynthesis. However, senescence involves much more than the simple switching off of photosynthesis genes.

Genes whose expression is induced during senescence are called **senescence-associated genes** (**SAGs**). SAGs include genes that encode hydrolytic enzymes, such as pro-

FIGURE 16.38 Stages of flower senescence in morning glory (*Ipomoea acuminata*). (Courtesy of Lee Taiz.)

teases, ribonucleases, and lipases, as well as enzymes involved in the biosynthesis of ethylene, such as l-aminocyclopropane-l-carboxylic acid (ACC) synthase and ACC oxidase (see Chapter 22). Another class of SAGs have secondary functions in senescence. These genes encode enzymes involved in the conversion or remobilization of breakdown products, such as glutamine synthetase, which catalyzes the conversion of ammonium to glutamine (see Chapter 12) and is responsible for nitrogen recycling from senescing tissues. Pectinases play important roles in cell wall breakdown during leaf abscission and fruit ripening.

Programmed cell death is a specialized type of senescence

Senescence can occur at the level of the whole plant, as in monocarpic senescence; at the organ level, as in leaf senescence; and at the cellular level, as in tracheary element differentiation. The process whereby individual cells activate an intrinsic senescence program is called **programmed cell death** (**PCD**). PCD plays an important part in animal development, in which the molecular mechanism has been studied extensively. PCD can be initiated by specific developmental signals, or by potentially lethal events, such as pathogen attack or errors in DNA replication during cell division. It involves the expression of a characteristic set of genes that orchestrate the dismantling of cellular components, ultimately resulting in cell death.

Much less is known about PCD in plants than in animals (Pennell and Lamb 1997). PCD in animals is usually accompanied by a distinct set of morphological and biochemical changes called **apoptosis** (plural *apoptoses*) (from a Greek word meaning "falling off," as in autumn leaves). During apoptosis, the cell nucleus condenses and the nuclear DNA fragments in a specific pattern caused by degradation of the DNA between nucleosomes (see Chapter 1).

Some plant cells, particularly in senescing tissues, exhibit similar cytological changes. PCD also appears to occur during the differentiation of xylem tracheary elements, during which the nuclei and chromatin degrade and the cytoplasm disappears. These changes result from the activation of genes that encode nucleases and proteases.

One of the important functions of PCD in plants is protection against pathogenic organisms. When a pathogenic organism infects a plant, signals from the pathogen cause the plant cells at the site of the infection to quickly accumulate high concentrations of toxic phenolic compounds and die. The dead cells form a small circular island of cell death called a **necrotic lesion**. The necrotic lesion isolates and prevents the infection from spreading to surrounding healthy tissues by surrounding the pathogen with a toxic and nutritionally depleted environment. This rapid, localized cell death due to pathogen attack is called the *hypersensitive response* (see Chapter 13).

The existence of Arabidopsis mutants that can mimic the effect of infection and trigger the entire cascade of events leading to the formation of necrotic lesions, even in the absence of the pathogen, has demonstrated that the hypersensitive response is a genetically programmed process rather than simple necrosis.

Summary

An increasingly detailed understanding of plant growth and development continues to emerge through a combination of experimental and analytical approaches. The ability to describe developmental changes in molecular terms and to correlate these changes with the activity of specific genes has provided the means to develop detailed and testable models for underlying mechanisms.

Although biological development is by nature an ongoing and complex process, it is possible to discern relatively discrete developmental programs that operate in a semiautonomous manner. For example, certain aspects of development, such as embryogenesis, can occur relatively normally in individuals that, by mutation, are unable to form normal vegetative leaves. Similarly, mutations in genes that alter floral development may have no effect on vegetative processes. By systematically cataloging the phenotypes of developmental mutants it is possible to define groups of functionally related genes.

Through a combination of genetic and molecular techniques, the relationships between functional components of a process can be revealed. We find, for example, that the levels and distribution of auxin affect a variety of processes, including the establishment of apical–basal polarity and the process of organogenesis. The asymmetric cellular distribution of PIN auxin efflux proteins appears to play a critical role in establishing this patterned distribution of auxin, both in the developing embryo and in subsequent vegetative development.

Two classes of proteins play antagonistic roles in mediating auxin responses: auxin response factors (ARFs), which act as transcriptional activators of auxin-induced genes, and repressor proteins that inhibit the activities of the ARF proteins. Finally, there are the auxin-responsive genes themselves, which presumably mediate specific auxin-directed growth responses.

In radial patterning, the roles of several classes of genes have been defined in terms of their localized, cell layer–specific expression, as well as their function as revealed by genetic analyses. Epidermis-specific markers can be detected at early stages of embryogenesis. These include the cuticle and transcription factors, such as *ARABIDOPSIS THALIANA MERISTEM LAYER 1* (*ATML1*) and *PROTODERMAL FACTOR 2* (*PDF2*), whose activities are required for normal epidermal development.

Another pair of related transcription factors are encoded by *SCARECROW* (*SCR*) and *SHORTROOT* (*SHR*), required for the normal development of the cortical and endodermal cell layers. The interaction between these two proteins provides an exceptionally clear example of inductive interactions between cells. In this case, the movement of SHR protein from vascular layers into adjacent outer layers that specifically express SCR appears to enable a partitioning of endodermal and cortical activities into separate layers.

The normal development of phloem and xylem is also linked to specific, genetically defined activities. Auxin-related gene activities, such as the two *ARF* genes, *MONOPTEROS* (*MP*) and the related *NONPHOTOTROPIC HYPOCOTYL 4* (*NPH4*), are also required for normal vascular development, adding to a large body of evidence that supports a role for auxin in vascular development. Finally, phloem defects in *wooden leg* (*wol*) mutants suggest that cytokinin also plays a role in normal vascular development.

For the indeterminate growth that distinguishes plants from animals, auxin once again presents itself as a key player. In the root, high concentrations of auxin appear to trigger the expression of key transcription factors, such as *PLETHORA* (*PLT*) 1 and 2, which in turn are responsible for establishing the quiescent center and associated apical initials. In the shoot, a relatively low concentration of auxin in the intercotyledonary regions of the embryo permits the expression of *CUP-SHAPED COTYLEDON* (*CUC*) genes—transcription factors responsible for the establishment of cell division-promoting *KNOX* gene activities such as *STM*.

Maintenance of apical meristem activities presents another well-studied example of genetic interactions, especially in the case of the shoot apical meristem, where *WUSCHEL* (*WUS*) and *CLAVATA* (*CLV*) genes appear to act in an autoregulatory feedback loop. Expression of the *WUS* gene, which encodes a homeodomain transcription factor, is thought to somehow promote stem cell activity, while at the same time promoting the transcription of *CLV3*. The CLV3 protein, in turn, activates the *CLV1* and *CLV2* leucine-rich repeat kinase complex, whose activity ultimately leads to the repression of *WUS* transcription. A similar feedback loop may operate to maintain the root apical meristem.

For the vast majority of developmental processes that have been modeled in detail, transcription factors play the most prominent roles. However, it is almost certain that, at some downstream level, other types of gene products, such as those that mediate specific patterns of cellular growth and differentiation, also play essential roles.

Senescence and programmed cell death are additional key aspects of plant development. Plants exhibit a variety of senescence phenomena. Leaves are genetically programmed to senesce and die. Senescence is an active developmental process that is controlled by the plant's genetic program and initiated by specific environmental or developmental cues.

Senescence is an ordered series of cytological and biochemical events. The expression of most genes is reduced during senescence, but the expression of some genes (senescence-associated genes, or SAGs) is initiated. The newly active genes encode various hydrolytic enzymes, such as proteases, ribonucleases, lipases, and enzymes involved in the biosynthesis of ethylene, which carry out the degradative processes as the tissues die, and help to recapture the dying tissues' nutrients for reuse by the living plant.

Programmed cell death (PCD) is a specialized type of senescence. One important function of PCD in plants is protection against pathogenic organisms in what is called the hypersensitive response, which has been demonstrated to be a genetically programmed process.

Web Material

Web Topics

16.1 The Analysis of Plant Growth
Basic methods to describe in detail the growth of plants over time provide an essential starting point for understanding underlying developmental programs.

16.2 Embryonic Dormancy
The ability of seeds to lie dormant for long periods and then germinate under favorable conditions reflects the activity of complex physiological programs.

16.3 Polarity of *Fucus* Zygotes
A wide variety of external gradients can polarize growth of cells that are initially apolar.

16.4 The Preprophase Band of Microtubules
Ultrastructural studies have elucidated the structure of the preprophase band of microtubules and its role in orienting the plane of cell division.

16.5 Genes Required for Shoot Apical Meristem Formation: Class III HD-Zip Genes
Molecular genetic analyses of a versatile family of transcription factors highlight diverse roles in vascular development, meristem activity and polarity determination.

16.6 *Azolla* Root Development
Anatomical studies of the root of the aquatic fern, Azolla, have provided insights into cell fate during root development.

Web Essays

16.1 Division Plane Determination in Plant Cells
Plant cells appear to utilize mechanisms different from those used by other eukaryotes to control their division planes.

16.2 Plant Meristems: An Historical Overview
Scientists have used many approaches to unraveling the secrets of plant meristems.

Chapter References

Abe, M., Katsumata, H., Komeda, Y., and Takahashi, T. (2003) Regulation of shoot epidermal cell differentiation by a pair of homeodomain proteins in Arabidopsis. *Development* 130: 635–643.

Aida, M., Beis, D., Heidstra, R., Willemsen, V., Blilou, I., Galinha, C., Nussaume, L., Noh, Y.-S., Amasino, R., and Scheres, B. (2004) The *PLETHORA* genes mediate patterning of the Arabidopsis root stem cell niche. *Cell* 119: 109–120.

Berleth, T., and Jürgens, G. (1993) The role of the *MONOPTEROS* gene in organising the basal body region of the Arabidopsis embryo. *Development* 118: 575–587.

Blilou, I., Xu, J., Wildwater, M., Willemsen, V., Paponov, I., Friml, J., Heidstra, R., Aida, M., Palme, K., and Scheres, B. (2005) The PIN auxin efflux facilitator network controls growth and patterning in Arabidopsis roots. *Nature* 433: 39–44.

Bowman, J. L., and Eshed, Y. (2000) Formation and maintenance of the shoot apical meristem. *Trends Plant Sci.* 5: 110–115.

Bowman, J. L., Eshed, Y., and Baum, S. F. (2002) Establishment of polarity in angiosperm lateral organs. *Trends Genet.* 18: 134–141.

Clark, S. E. (2001) Cell signaling at the shoot meristem. *Nature Rev. Mol. Cell. Biol.* 2: 276–284.

Di Laurenzio, L., Wysocka-Diller, J., Malamy, J. E., Pysh, L., Helariutta, Y., Freshour, G., Hahn, M. G., Feldmann, K. A., and Benfey, P. N. (1996) The *SCARECROW* gene regulates an asymmetric cell division that is essential for generating the radial organization of the Arabidopsis root. *Cell* 86: 423–433.

Dolan, L., Janmaat, K., Willemsen, V., Linstead, P., Poethig, S., Roberts, K., and Scheres, B. (1993) Cellular organisation of the *Arabidopsis thaliana* root. *Development* 119: 71–84.

Esau, K. (1965) *Plant Anatomy*. John Wiley and Sons, New York.

Emery, J. F., Floyd, S. K., Alvarez, J., Eshed, Y., Hawker, N. P., Izhaki, A., Baum, S. F., and Bowman, J. L. (2003) Radial patterning of Arabidopsis shoots by class III HD-ZIP and KANADI genes. *Curr. Biol.* 13: 1768–1774.

Eshed, Y., Izhaki, A., Baum, S. F., Floyd, S. K., and Bowman, J. L. (2004) Asymmetric leaf development and blade expansion in Arabidopsis are mediated by KANADI and YABBY activities. *Development* 131: 2997–3006.

Fiers, M., Golemiec, E., Xu, J., van der Geest, L., Heidstra, R., Stiekema, W., and Liu, C.-M. (2005) The 14-amino acid CLV3, CLE19, and CLE40 peptides trigger consumption of the root meristem in Arabidopsis through a CLAVATA2-dependent pathway. *Plant Cell* 17: 2542–2553.

Friml, J., Vieten, A., Sauer, M., Weijers, D., Schwarz, H., Hamann, T., Offringa, R., and Jürgens, G. (2003) Efflux-dependent auxin gradients establish the apical-basal axis of Arabidopsis. *Nature* 426: 147–153.

Geldner, N., Anders, N., Wolters, H., Keicher, J., Kornberger, W., Muller, P., Delbarre, A., Ueda, T., Nakano, A., and Jürgens, G. (2003) The Arabidopsis GNOM ARF-GEF mediates endosomal recycling, auxin transport, and auxin-dependent plant growth. *Cell* 112: 219–230.

Gifford, E. M., and Foster, C. E. (1987) *Morphology and Evolution of Vascular Plants*. W.H. Freeman and Company, New York.

Haecker, A., Gross-Hardt, R., Geiges, B., Sarkar, A., Breuninger, H., Herrmann, M., and Laux, T. (2004) Expression dynamics of *WOX* genes mark cell fate decisions during early embryonic patterning in *Arabidopsis thaliana*. *Development* 131: 657–668.

Hamann, T., Benkova, E., Baurle, I., Kientz, M., and Jürgens, G. (2002) The Arabidopsis *BODENLOS* gene encodes an auxin response protein inhibiting MONOPTEROS-mediated embryo patterning. *Genes Dev.* 16: 1610–1615.

Haywood, V., Kragler, F., and Lucas, W. J. (2002) Plasmodesmata: Pathways for protein and ribonucleoprotein signaling. *Plant Cell* 14: S303–325.

Helariutta, Y., Fukaki, H., Wysocka-Diller, J., Nakajima, K., Jung, J., Sena, G., Hauser, M. T., and Benfey, P. N. (2000) The *SHORT-ROOT* gene controls radial patterning of the Arabidopsis root through radial signaling. *Cell* 101: 555–567.

Ingram, G. C., Doyle, S., Carpenter, R., Schultz, E. A., Simon, R., and Coen, E. S. (1997) Dual role for fimbriata in regulating floral homeotic genes and cell division in *Antirrhinum*. *EMBO J.* 16: 6521–6534.

Inoue, T., Higuchi, M., Hashimoto, Y., Seki, M., Kobayashi, M., Kato, T., Tabata, S., Shinozaki, K., and Kakimoto, T. (2001) Identification of CRE1 as a cytokinin receptor from Arabidopsis. *Nature* 409: 1060–1063.

Itoh, J., Nonomura, K., Ikeda, K., Yamaki, S., Inukai, Y., Yamagishi, H., Kitano, H., and Nagato, Y. (2005) Rice plant development: From zygote to spikelet. *Plant Cell Physiol.* 46: 23–47.

Jenik, P. D., and Barton, M. K. (2005) Surge and destroy: The role of auxin in plant embryogenesis. *Development* 132: 3577–3585.

Kidner, C. A., and Martienssen, R. A. (2005) The developmental role of microRNA in plants. *Curr. Opin. Plant Biol.* 8: 38–44.

Kim, I., Kobayashi, K., Cho, E., and Zambryski, P. C. (2005) Subdomains for transport via plasmodesmata corresponding to the apical-basal axis are established during Arabidopsis embryogenesis. *Proc. Natl. Acad. Sci. USA* 102: 11945–11950.

Laux, T., Wurschum, T., and Breuninger, H. (2004) Genetic regulation of embryonic pattern formation. *Plant Cell* 16: S190–202.

Liu, C., Xu, Z., and Chua, N. H. (1993) Auxin polar transport is essential for the establishment of bilateral symmetry during early plant embryogenesis. *Plant Cell* 5: 621–630.

Lu, P., Porat, R., Nadeau, J. A., and O'Neill, D. (1996) Identification of a meristem L1 layer-specific gene in Arabidopsis that is expressed during embryonic pattern formation and defines a new class of homeobox genes. *Plant Cell* 8: 2155–2168.

Mahonen, A. P., Bonke, M., Kauppinen, L., Riikonen, M., Benfey, P. N., and Helariutta, Y. (2000) A novel two-component hybrid molecule regulates vascular morphogenesis of the Arabidopsis root. *Genes Dev.* 14: 2938–2943.

Malamy, J. E., and Benfey, P. N. (1997) Organization and cell differentiation in lateral roots of *Arabidopsis thaliana*. *Development* 124: 33–44.

Mansfield, S. G., and Briarty, L. G. (1991) Early embryogenesis in *Arabidopsis thaliana*: The developing embryo. *Can. J. Bot.* 69: 461–476.

Mayer, K. F., Schoof, H., Haecker, A., Lenhard, M., Jürgens, G., and Laux, T. (1998) Role of *WUSCHEL* in regulating stem cell fate in the Arabidopsis shoot meristem. *Cell* 95: 805–815.

Mayer, U., Torres Ruiz, R. A., Berleth, T., Misera, S., and Jürgens, G. (1991) Mutations affecting body organisation in the Arabidopsis embryo. *Nature* 353: 402–407.

Nakajima, K., and Benfey, P. N. (2002) Signaling in and out: control of cell division and differentiation in the shoot and root. *Plant Cell* 14 (Suppl): S265–276.

Nakajima, K., Sena, G., Nawy, T., and Benfey, P. N. (2001) Intercellular movement of the putative transcription factor SHR in root patterning. *Nature* 413: 307–311.

Pennell, R. I., and Lamb, C. (1997) Programmed cell death in plants. *Plant Cell* 9: 1157–1168.

Poethig, R. S. (1987) Clonal analysis of cell lineage patterns in plant development. *Am. J. Bot.* 74: 581–594.

Reinhardt, D., Pesce, E. R., Stieger, P., Mandel, T., Baltensperger, K., Bennett, M., Traas, J., Friml, J., and Kuhlemeier, C. (2003) Regulation of phyllotaxis by polar auxin transport. *Nature* 426: 255–260.

Rojo, E., Sharma, V. K., Kovaleva, V., Raikhel, N. V., and Fletcher, J. C. (2002) CLV3 is localized to the extracellular space, where it activates the Arabidopsis CLAVATA stem cell signaling pathway. *Plant Cell* 14: 969–977.

Sabatini, S., Heidstra, R., Wildwater, M., and Scheres, B. (2003) SCARECROW is involved in positioning the stem cell niche in the Arabidopsis root meristem. *Genes Dev.* 17: 354–358.

Schiefelbein, J. W., Masucci, J. D., and Wang, H. (1997) Building a root: The control of patterning and morphogenesis during root development. *Plant Cell* 9: 1089–1098.

Scheres, B., Wolkenfelt, H., Willemsen, V., Terlouw, M., Lawson, E., Dean, C., and Weisbeek, P. (1994) Embryonic origin of the Arabidopsis primary root and root meristem initials. *Development* 120: 2475–2487.

Schmitz, G., and Theres, K. (2005) Shoot and inflorescence branching. *Curr. Opin. Plant Biol.* 8: 506–511.

Schoof, H., Lenhard, M., Haecker, A., Mayer, K. F., Jürgens, G., and Laux, T. (2000) The stem cell population of Arabidopsis shoot meristems in maintained by a regulatory loop between the *CLAVATA* and *WUSCHEL* genes. *Cell* 100: 635–644.

Sharma, V. K., Carles, C., and Fletcher, J. C. (2003) Maintenance of stem cell populations in plants. *Proc. Natl. Acad. Sci. USA* 100 (Suppl 1): 11823–11829.

Steeves, T. A., and Sussex, I. M. (1989) *Patterns in Plant Development*. Cambridge University Press, Cambridge.

Tilney-Basset, R. A. E. (1986) *Plant Chimeras*. London: Edward Arnold.

Torres-Ruiz, R. A., and Jürgens, G. (1994) Mutations in the *FASS* gene uncouple pattern formation and morphogenesis in Arabidopsis development. *Development* 120: 2967–2978.

Traas, J., Bellini, C., Nacry, P., Kronenberger, J. Bouchez, D., and Caboche, M. (1995) Normal differentiation patterns in plants lacking microtubular preprophase bands. *Nature* 375: 676–677.

Van den Berg, C., Willemsen, V., Hage, W., Weisbeek, P., and Scheres, B. (1995) Cell fate in the Arabidopsis root meristem determined by directional signalling. *Nature* 378: 62–65.

Waites, R., Selvadurai, H. R., Oliver, I. R., and Hudson, A. (1998) The *PHANTASTICA* gene encodes a MYB transcription factor involved in growth and dorsoventrality of lateral organs in *Antirrhinum. Cell* 93: 779–789.

Weigel, D., and Jürgens, G. (2002) Stem cells that make stems. *Nature* 415: 751–754.

Weijers, D., and Jürgens, G. (2005) Auxin and embryo axis formation: The ends in sight. *Curr. Opin. Plant Biol.* 8: 1–6.

Willemsen, V., Wolkenfelt, H., de Vrieze, G., Weisbeek, P., and Scheres, B. (1998) The *HOBBIT* gene is required for formation of the root meristem in the Arabidopsis embryo. *Development* 125: 521–531.

Willemsen, V., and Scheres, B. (2004) Mechanisms of pattern formation in plant embryogenesis. *Annu. Rev. Genet.* 38: 587–614.

Chapter 17

Phytochrome and Light Control of Plant Development

HAVE YOU EVER DISSECTED your sandwich and wondered why the bean sprouts looked the way they do? Most edible sprouts (e.g., alfalfa and mung bean) are germinated and grown in the dark, where they undergo a special kind of development called **skotomorphogenesis** (from *skotos*, the Greek word for darkness). Such **etiolated seedlings** have elongated stems, folded cotyledons, and they fail to accumulate chlorophyll (Figure 17.1). Now imagine these seedlings growing in the soil rather than decorating your sandwich. Envision the elongating shoot pushing the delicate first leaves up through the soil using the apical hook to clear a path (see Figure 17.1D). When the seedling emerges from the soil and the limited energy reserves from the cotyledons (dicots) or seeds (monocots) are depleted, the seedling must begin making food for itself.

This transition from skotomorphogenesis to **photomorphogenesis** is an extremely rapid yet complex process. Within hours of applying a single flash of relatively dim light to a dark-grown bean seedling, several developmental changes occur: a decrease in the rate of stem elongation, the beginning of apical-hook straightening, and the initiation of the synthesis of pigments that are characteristic of green plants. Thus, light acts as a signal to induce a change in the form of the seedling, from one that facilitates growth beneath the soil to one that will enable the plant to efficiently harvest light energy and transform it into the essential sugars, proteins, and lipids necessary for growth of the plant and for the people and animals that eat plants.

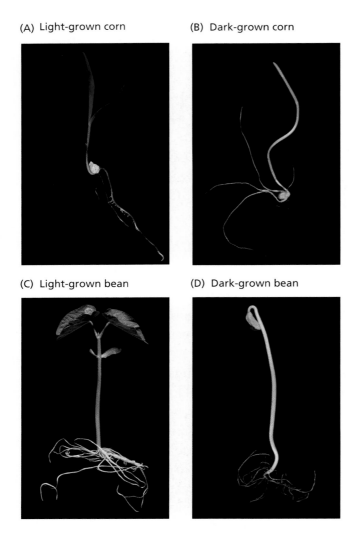

FIGURE 17.1 Corn (maize; *Zea mays*) (A and B) and kidney bean (*Phaseolus vulgaris*) (C and D) seedlings grown either in the light (A and C) or the dark (B and D). Symptoms of etiolation in corn, a monocot, include the absence of greening, reduction in leaf width, failure of leaves to unroll, and elongation of the coleoptile and mesocotyl. In bean, a dicot, etiolation symptoms include absence of greening, reduced leaf size, hypocotyl elongation, and maintenance of the apical hook. (Photos © M. B. Wilkins.)

cuss the biochemical and photochemical properties of phytochrome and the conformational changes induced by light. Different types of phytochromes are encoded by different members of a multigene family, and different phytochromes regulate distinct processes in the plant. These different phytochrome responses can be classified according to the amount and quality of light required to produce the effect. Over the past few years, tremendous progress has been made in defining the molecular events that underlie phytochrome responses, and these will be discussed in detail. Finally, we will examine ecological functions of phytochrome that enable plants to adapt to ever-changing environments.

The Photochemical and Biochemical Properties of Phytochrome

Phytochrome, a blue protein pigment with a molecular mass of about 125 kilodaltons (kDa), was not identified as a unique chemical species until 1959, mainly because of technical difficulties in isolating and purifying the protein. However, many of the biological properties of phytochrome had been established earlier in studies of whole plants.

The first clues regarding the role of phytochrome in plant development came from studies that began in the 1930s on red light–induced morphogenic responses, especially seed germination. The list of such responses is now enormous and includes one or more responses at almost every stage in the life history of a wide range of different green plants (Table 17.1).

A key breakthrough in the history of phytochrome was the discovery that the effects of *red light* (650–680 nm) on morphogenesis could be reversed by a subsequent irradiation with light of longer wavelengths (710–740 nm), called *far-red light*. This phenomenon was first demonstrated in germinating seeds, but was also observed in relation to stem and leaf growth, as well as floral induction (see Chapter 25). The initial observation, made in 1935, was that the germination of lettuce seeds is stimulated by red light and inhibited by far-red light (Flint 1936). But the real breakthrough came many years later, in 1952, when lettuce seeds were exposed to alternating treatments of red and far-red light. Nearly 100% of the seeds that received red light as the final treatment germinated; in seeds that received far-red light as the final treatment, however, germination was strongly inhibited (Figure 17.2) (Borthwick et al. 1952). This pivotal experiment demonstrated that the responses to red and far-red light were not merely opposite, they were also antagonistic.

Two interpretations of these results were possible. One is that there are two pigments, a red light–absorbing pigment and a far-red light–absorbing pigment, and the two pigments act antagonistically in the regulation of seed germination. Alternatively, there might be a single pigment that can exist in two interconvertible forms: a red light–absorbing form and a far-red light–absorbing form (Borthwick et al. 1952). The model chosen—the one-pig-

Among the different pigments that can promote photomorphogenic responses in plants, the most important are those that absorb red and blue light. The blue-light photoreceptors will be discussed in relation to guard cells and phototropism in Chapter 18. The focus of this chapter is **phytochrome**, a protein pigment that absorbs red and far-red light most strongly, but that also absorbs blue light. As we will see in this chapter and in Chapter 25, phytochrome plays a key role in light-regulated vegetative and reproductive development.

We begin with the discovery of phytochrome and the phenomenon of red/far-red photoreversibility. Next we will dis-

TABLE 17.1
Typical photoreversible responses induced by phytochrome in a variety of higher and lower plants

Group	Genus	Stage of development	Effect of red light
Angiosperms	*Lactuca* (lettuce)	Seed	Promotes germination
	Avena (oat)	Seedling (etiolated)	Promotes de-etiolation (e.g., leaf unrolling)
	Sinapis (mustard)	Seedling	Promotes formation of leaf primordia, development of primary leaves, and production of anthocyanin
	Pisum (pea)	Adult	Inhibits internode elongation
	Xanthium (cocklebur)	Adult	Inhibits flowering (photoperiodic response)
Gymnosperms	*Pinus* (pine)	Seedling	Enhances rate of chlorophyll accumulation
Pteridophytes	*Onoclea* (sensitive fern)	Young gametophyte	Promotes growth
Bryophytes	*Polytrichum* (moss)	Germling	Promotes replication of plastids
Chlorophytes	*Mougeotia* (alga)	Mature gametophyte	Promotes orientation of chloroplasts to directional dim light

ment model—was the more radical of the two because there was no precedent for such a photoreversible pigment. Several years later phytochrome was demonstrated in plant extracts for the first time, and its unique photoreversible properties were exhibited in vitro, confirming the prediction (Butler et al. 1959) (see Web Topic 17.1).

Phytochrome can interconvert between Pr and Pfr forms

In dark-grown or etiolated plants, phytochrome is present in a red light–absorbing form, referred to as **Pr**. This blue-colored inactive form is converted by red light to a far-red light–absorbing form called **Pfr**, which is blue-green. Pfr, in turn, can be converted back to Pr by far-red light.

Known as **photoreversibility**, this conversion/reconversion property is the most distinctive property of phytochrome, and it may be expressed in abbreviated form as follows:

$$\text{Pr} \underset{\text{Far-red light}}{\overset{\text{Red light}}{\rightleftarrows}} \text{Pfr}$$

The interconversion of the Pr and Pfr forms can be measured in vivo or in vitro. In fact, most of the spectral properties, such as absorption spectrum and photoreversibility,

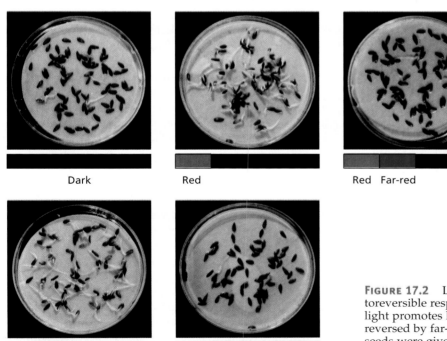

FIGURE 17.2 Lettuce seed germination is a typical photoreversible response controlled by phytochrome. Red light promotes lettuce seed germination, but this effect is reversed by far-red light. Imbibed (water-moistened) seeds were given alternating treatments of red followed by far-red light. The effect of the light treatment depended on the last treatment given. (Photos © M. B. Wilkins.)

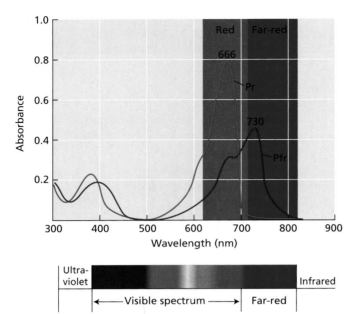

FIGURE 17.3 Absorption spectra of purified oat phytochrome in the Pr (green line) and Pfr (blue line) forms overlap. (After Vierstra and Quail 1983.)

of carefully purified phytochrome measured in vitro are the same as those observed in vivo.

It is important to note that the phytochrome pool is never fully converted to the Pfr or Pr forms following red or far-red irradiation, because the absorption spectra of the Pfr and Pr forms overlap. Thus, when Pr molecules are exposed to red light, most of them absorb the photons and are converted to Pfr, but some of the Pfr made also absorbs the red light and is converted back to Pr (Figure 17.3). The proportion of phytochrome in the Pfr form after saturating irradiation by red light is only about 85%. Similarly, the very small amount of far-red light absorbed by Pr makes it impossible to convert Pfr entirely to Pr by broad-spectrum far-red light. Instead, an equilibrium of 97% Pr and 3% Pfr is achieved. This equilibrium is termed the **photostationary state.**

In addition to absorbing red light, both forms of phytochrome absorb light in the blue region of the spectrum (see Figure 17.3). Therefore, phytochrome effects can be elicited also by blue light, which can convert Pr to Pfr and vice versa. Blue-light responses can also result from the action of one or more specific blue-light photoreceptors (see Chapter 18). Whether phytochrome is involved in a response to blue light is often determined by a test of the ability of far-red light to reverse the response, since only phytochrome-induced responses are reversed by far-red light.

Pfr is the physiologically active form of phytochrome

Because phytochrome responses are induced by red light, they could in theory result either from the appearance of Pfr or from the disappearance of Pr. In most cases studied, a quantitative relationship holds between the magnitude of the physiological response and the amount of Pfr generated by light, but no such relationship holds between the physiological response and the loss of Pr. Evidence such as this has led to the conclusion that Pfr is the physiologically active form of phytochrome.

The use of narrow waveband red and far-red light was central to the discovery and eventual isolation of phytochrome. However, a plant growing outdoors never is exposed to strictly "red" or "far-red" light, as is commonly used in laboratory-based photobiological experiments. In natural settings plants are exposed to a much broader spectrum of light, and it is under these conditions that phytochrome must work to regulate developmental responses to changes in the light environment.

Characteristics of Phytochrome-Induced Responses

The variety of different phytochrome responses in intact plants is extensive, in terms of both the kinds of responses (see Table 17.1) and the quantity of light needed to induce the responses. A survey of this variety will show how diversely the effects of a single photoevent—the absorption of light by Pr—are manifested throughout the plant. For ease of discussion, phytochrome-induced responses may be logically grouped into two types:

1. Rapid biochemical events
2. Slower morphological changes, including movements and growth

Some of the early biochemical reactions affect later developmental responses. The nature of these early biochemical events, which comprise signal transduction pathways, will be treated in detail later in the chapter. Here we will focus on the effects of phytochrome on whole-plant responses. As we will see, such responses can be classified into various types depending on the amount and duration of light required and on their action spectra.

Phytochrome responses vary in lag time and escape time

Morphological responses to the photoactivation of phytochrome may be observed visually after a *lag time*—the time between stimulation and observed response. The lag time may be as brief as a few minutes or as long as several weeks. The more rapid of these responses are usually reversible movements of organelles (see **Web Topic 17.2**) or reversible volume changes (swelling, shrinking) in cells, but even some growth responses are remarkably fast.

Red-light inhibition of the stem elongation rate of light-grown pigweed (*Chenopodium album*) is observed within 8 minutes after its relative level of Pfr is increased. Kinetic studies using Arabidopsis have confirmed this observation and further shown that phytochrome acts within minutes after exposure to red light (Parks and Spalding 1999). Longer lag times of several weeks are observed for the induction of flowering (see Chapter 25).

Information about the lag time for a phytochrome response helps researchers evaluate the kinds of biochemical events that could precede and cause the induction of that response. The shorter the lag time, the more limited the range of biochemical events that could have been involved.

Variety in phytochrome responses can also be seen in the phenomenon called **escape from photoreversibility**. Red light–induced events are reversible by far-red light for only a limited period of time, after which the response is said to have "escaped" from reversal control by light.

This phenomenon can be explained by a model based on the assumption that phytochrome-controlled morphological responses are the end result of a multi-step sequence of linked biochemical reactions in the responding cells. Early stages in the sequence may be fully reversible by removing Pfr, but at some point in the sequence a point of no return is reached beyond which the reactions proceed irreversibly toward the response. The escape time therefore represents the amount of time it takes before the overall sequence of reactions becomes irreversible; essentially, the time it takes for Pfr to complete its primary action. The escape time for different responses ranges from less than a minute to, remarkably, hours.

Phytochrome responses can be distinguished by the amount of light required

Phytochrome responses can be distinguished by the amount of light required to induce them. The amount of light is referred to as the **fluence**, which is defined as the number of photons impinging on a unit surface area. The most commonly used units for fluence are micromoles of quanta per square meter ($\mu mol\ m^{-2}$). In addition to the fluence, some phytochrome responses are sensitive to the **irradiance***, or *fluence rate*, of light. The units of irradiance in terms of photons are micromoles of quanta per square meter per second ($\mu mol\ m^{-2}\ s^{-1}$). (For definitions of these and other terms used in light measurement see Chapter 9 and **Web Topic 9.1**.)

Each phytochrome response has a characteristic range of light fluences over which the magnitude of the response is proportional to the fluence. As Figure 17.4 shows, these responses fall into three major categories based on the amount of light required: very low–fluence responses (VLFRs), low-fluence responses (LFRs), and high-irradiance responses (HIRs).

---

*Irradiance is sometimes loosely equated with light intensity. The term *intensity*, however, refers to light emitted by the source, whereas *irradiance* refers to light that is incident on the object.

Very low–fluence responses are nonphotoreversible

Some phytochrome responses can be initiated by fluences as low as 0.0001 $\mu mol\ m^{-2}$ (one-tenth of the amount of light emitted from a firefly in a single flash), and they saturate (i.e., reach a maximum) at about 0.05 $\mu mol\ m^{-2}$. For example, in dark-grown oat seedlings, red light can stimulate the growth of the coleoptile and inhibit the growth of the mesocotyl (the elongated axis between the coleoptile and the root) at such low fluences. Arabidopsis seeds can be induced to germinate with red light in the range of 0.001 to 0.1 $\mu mol\ m^{-2}$. These remarkable effects of vanishingly low levels of illumination are called **very low–fluence responses** (**VLFRs**).

The minute amount of light needed to induce VLFRs converts less than 0.02% of the total phytochrome to Pfr. Because the far-red light that would normally reverse a red-light effect converts only 97% of the Pfr to Pr (as discussed earlier), about 3% of the phytochrome remains as Pfr—significantly more than the 0.02% needed to induce VLFRs (Mandoli and Briggs 1984). In other words, far-red light cannot lower the Pfr concentration below 0.02%, so it is unable to inhibit VLFRs. The VLFR action spectrum matches the absorption spectrum of Pr, supporting the view that Pfr is the active form for these responses (Shinomura et al. 1996).

Ecological implications of the VLFR in seed germination are discussed in **Web Essay 17.1**.

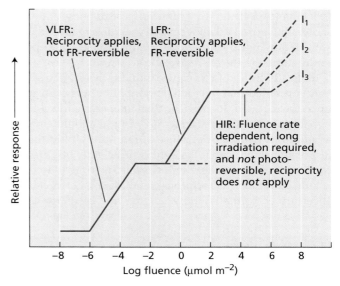

FIGURE 17.4 Three types of phytochrome responses, based on their sensitivities to fluence. The relative magnitudes of representative responses are plotted against increasing fluences of red light. Short light pulses activate VLFRs and LFRs. Because HIRs are also proportional to the irradiance, the effects of three different irradiances given continuously are illustrated ($I_1 > I_2 > I_3$). (After Briggs et al. 1984.)

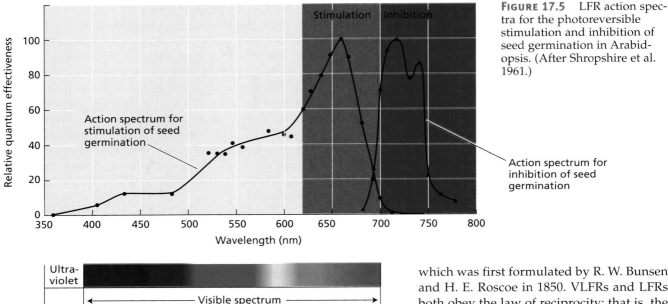

FIGURE 17.5 LFR action spectra for the photoreversible stimulation and inhibition of seed germination in Arabidopsis. (After Shropshire et al. 1961.)

Low-fluence responses are photoreversible

Another set of phytochrome responses cannot be initiated until the fluence reaches 1.0 µmol m^{-2}, and they are saturated at about 1000 µmol m^{-2}. These responses are referred to as **low-fluence responses** (**LFRs**), and they include most of the red/far-red photoreversible responses, such as the promotion of lettuce seed germination and the regulation of leaf movements, that are mentioned in Table 17.1. The LFR action spectrum for Arabidopsis seed germination is shown in Figure 17.5. LFR spectra include a main peak for stimulation in the red region (660 nm), and a major peak for inhibition in the far-red region (720 nm).

Both VLFRs and LFRs can be induced by brief pulses of light, provided that the total amount of light energy adds up to the required fluence. The total fluence is a function of two factors: the fluence rate (µmol m^{-2} s^{-1}) and the time of irradiation. Thus, a brief pulse of red light will induce a response, provided that the light is sufficiently bright, and conversely, very dim light will work if the irradiation time is long enough. This reciprocal relationship between fluence rate and time is known as the **law of reciprocity**, which was first formulated by R. W. Bunsen and H. E. Roscoe in 1850. VLFRs and LFRs both obey the law of reciprocity; that is, the magnitude of the response (whether, for instance, it be percent germination or degree of inhibition of hypocotyl elongation) is dependent on the product of the fluence rate and the time of irradiation.

High-irradiance responses are proportional to the irradiance and the duration

Phytochrome responses of the third type are termed **high-irradiance responses** (**HIRs**), several of which are listed in Table 17.2. HIRs require prolonged or continuous exposure to light of relatively high irradiance, and the response is proportional to the irradiance until the response saturates and additional light has no further effect (see **Web Topic 17.3**).

The reason that these responses are called high-irradiance responses rather than high-fluence responses is that they are proportional to irradiance (loosely speaking, the brightness of the light) rather than to fluence. HIRs saturate at much higher fluences than LFRs—at least 100 times higher—and are not photoreversible. Because neither continuous exposure to dim light nor transient exposure to bright light can induce HIRs, HIRs do not obey the law of reciprocity.

Many of the photoreversible LFRs listed in Table 17.1, particularly those involved in de-etiolation, also qualify as HIRs. For example, at low fluences the action spectrum for anthocyanin production in seedlings of white mustard (*Sinapis alba*) shows a single peak in the red region of the spectrum, the effect is reversible with far-red light, and the response obeys the law of reciprocity. However, if the dark-grown seedlings are instead exposed to high-irradiance light for several hours, the action spectrum now includes peaks in the far-red and blue regions, the effect is no longer photoreversible, and the response becomes proportional to the

TABLE 17.2
Some plant photomorphogenic responses induced by high irradiances

Synthesis of anthocyanin in various dicot seedlings and in apple skin segments
Inhibition of hypocotyl elongation in mustard, lettuce, and petunia seedlings
Induction of flowering in henbane (*Hyoscyamus*)
Plumular hook opening in lettuce
Enlargement of cotyledons in mustard
Production of ethylene in sorghum

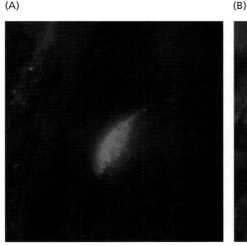
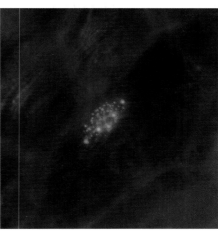

FIGURE 17.11 Nuclear localization of phy–GFP fusion proteins in epidermal cells of Arabidopsis hypocotyls. Transgenic Arabidopsis expressing phyA–GFP (A) or phyB–GFP (right) was observed under a fluorescence microscope. Only nuclei are visible. The plants were placed either under continuous far-red light (A) or white light (B) to induce the nuclear accumulation. The smaller bright green dots inside the nucleus in B are called "speckles." The significance of speckles is unknown. (From Yamaguchi et al. 1999, courtesy of A. Nagatani.)

Genetic Analysis of Phytochrome Function

Genetic screens have proven invaluable in defining light–signal transduction pathways in Arabidopsis. In a now classic study, Marteen Koornneef was one of the first to use genetics to identify components of light signaling pathways (Koornneef et al. 1980). To generate mutations, he soaked Arabidopsis seeds in a solution of ethane methyl sulfonate (EMS), which creates point mutations in DNA. The plants derived from the mutagenized seed were allowed to self-pollinate, and large pools of mutagenized seed were generated. The "families" were then grown under white light, and mutants with defects in light perception or response were identified as having a long hypocotyl (*hy*), similar to what would be observed if plants were grown in the dark. All of the mutations identified segregated as single recessive alleles.

Crosses made among the mutant plants defined five loci or complementation groups (*HY1–HY5*) that have served as a foundation for studies into light signaling in Arabidopsis. The cloning and sequence analysis of all five *HY* genes has led to the identification of several components of light-signal transduction, including genes necessary for phytochrome chromophore biosynthesis (*HY1* and *HY2*), the photoreceptor *PHYB* (*HY3*), the blue-light photoreceptor cryptochrome (*HY4*), and a light-induced transcription factor (*HY5*).

By combining genetic analysis, molecular biology, and detailed physiological studies, tremendous insights have been made into the mechanisms of light response over the past decade. Perhaps not surprisingly, these studies have revealed a complex interplay of light with plant signal transduction components that mediate multiple developmental decisions of the plant throughout its life cycle from seed germination to flowering.

In this section we will examine the diverse physiological functions and complex interactions of the five phytochrome genes as revealed by genetic analyses. In addition to their contrasting roles in VLFR and LFR responses, two types of high-irradiance responses have been identified, each dependent on a different phytochrome: the classical far-red light HIR of etiolated seedlings, which is mediated by phyA; and the red light (or white light) HIR of light-grown plants, which is mediated by phyB.

Phytochrome A mediates responses to continuous far-red light

No phytochrome gene mutations other than for phyB were found in the original *hy* collection, so the identification of phyA mutants required the development of more ingenious screens. As discussed previously, because the far-red HIRs were known to require light-labile (Type I) phytochrome, it was suspected that phyA must be the photoreceptor involved in the perception of continuous far-red light. If this is true, then the phyA mutants should fail to respond to continuous far-red light and grow tall and spindly under these light conditions. However, mutants lacking chromophore would also look like this because phyA can detect far-red light only when assembled with the chromophore into holophytochrome.

To select for just the phyA mutants, the seedlings that grew tall in continuous far-red light were then grown under continuous red light. The phyA-deficient mutants can grow normally under this regimen, but a chromophore-deficient mutant, which also lacks functional phyB, does not respond. The *phyA* mutant seedlings selected in this screen had no obvious phenotype when grown in normal white light, suggesting that phyA has no discernible role in sensing white light. This also explains why *phyA* mutants were not detected in the original long-hypocotyl screen. Thus, phyA appears to function primarily during de-etiolation in response to far-red light. It is also clear from the characterization of this mutant that none of the other phytochromes is sufficient for the perception of constant far-red light.

Phytochrome A also appears to be involved in the germination VLFR of Arabidopsis seeds in response to broad spectrum light. Thus, mutants lacking phyA cannot germinate in response to millisecond pulses of light, but they show a normal response to red light in the low-fluence range (Shinomura et al. 1996). This result demonstrates that phyA functions as the primary photoreceptor for this VLFR.

Phytochrome B mediates responses to continuous red or white light

The characterization of the *hy3* mutant revealed an important role for phyB in de-etiolation, since mutant seedlings grown in continuous white light had long hypocotyls. The *phyB* mutant is deficient in chlorophyll and in some mRNAs that encode chloroplast proteins, and it is impaired in its ability to respond to plant hormones.

In addition to white and red light-mediated HIR responses, phytochrome B also appears to regulate LFRs, such as photoreversible seed germination, the phenomenon that originally led to the discovery of phytochrome. Wild-type Arabidopsis seeds require light for germination, and the response shows red/far-red reversibility in the low-fluence range. Mutants that lack phyA respond normally to red light; mutants deficient in phyB are unable to respond to low-fluence red light (Shinomura et al. 1996). This experimental evidence strongly suggests that phyB mediates photoreversible seed germination.

Roles for phytochromes C, D, and E are emerging

Although phyA and phyB are the predominant forms of phytochrome in Arabidopsis, phyC, phyD, and phyE each play unique roles in regulating response to red and far-red light. The creation of double and triple mutants has made it possible to assess the relative role of each phytochrome in a given response. PhyD and phyE are structurally similar to phyB, but are not functionally redundant. Responses mediated by phyD and phyE include petiole and internode elongation and the control of flowering time (see Chapter 25). The characterization of *phyC* mutants in Arabidopsis has suggested a complex interplay between phyC, phyA, and phyB response pathways (Franklin et al. 2003; Monte et al. 2003). Loss-of-function mutants have revealed a minor role for phyC in seedling de-etiolation, cotyledon expansion and repression of flowering under nonpermissive (short day) conditions. This specialization in *PHY* gene function is likely to be important in fine-tuning phytochrome responses to daily and seasonal changes in light regimes. As we will discuss later in the chapter, phyB, phyD, and phyE also play an important ecological role in mediating shade avoidance responses.

Phy gene family interactions are complex

Clearly, genetic analysis has been a powerful tool in dissecting *PHY* gene function in Arabidopsis. There are, however, several limitations to this approach. For one, *phy* mutant phenotypes are often interpreted under the assumption that the activity of the other family members is unchanged. However, detailed molecular studies have shown that phyC and to some extent phyD accumulation are dependent on active phyB, as are *PHYA* transcript levels. Furthermore, transgenic Arabidopsis plants overexpressing the gene encoding PHYB have increased levels of phyA, phyC, and phyD in Arabidopsis (Hirschfeld et al. 1998). The interpretation of the genetic data is further complicated by the recent finding that the light-stable phy proteins can form both homo- and heterodimers (e.g., phyB-phyD) (Sharrock and Clack 2004). Thus, loss-of-function mutations in *PHYB* or plants overexpressing the PHYB apoprotein are likely to display phenotypes that are the consequence of an altered dynamic of many phytochrome interactions. Unraveling the complexities of phytochrome gene interactions thus represents a challenging problem for the future.

PHY gene functions have diversified during evolution

Although much of our understanding of the molecular events underlying phytochrome responses has been revealed through studies of Arabidopsis, the *PHY* gene family is rapidly evolving among the angiosperms (Mathews and Sharrock 1997). While most dicots have four subfamilies of phytochrome genes (*PHYA*, *PHYB/D*, *PHYC/F*, and *PHYE*), the monocots have only three (*PHYA*, *PHYB*, and *PHYC*). Through gene duplication/loss, genetic drift, and rapid diversification of *PHY* gene function, phytochrome signal transduction networks can be refashioned to suit the needs of plants in different habitats and under different selection pressures. Only when a detailed genetic analysis is conducted in a given species can we begin to identify the similarities and differences in phytochrome-regulated pathways.

For example, a phyA deficiency has little phenotypic effect on white light–grown Arabidopsis and rice, but loss of phyA function in pea results in a highly pleiotropic phenotype, including shortened internodes, delayed senescence, and increased yield under long days. In tomato, a loss of phyA and both copies of the duplicated phyB genes prevents chlorophyll accumulation in the fruits and greatly extends the length of the fruit-bearing cluster or "truss" (Figure 17.12). However, chlorophyll accumulates in the leaves of these plants, indicating that the role of phyA and phyB in chlorophyll accumulation is fruit-specific in tomato. The emerging picture from such cross-species comparisons indicates that, while the mode of action of phy family members may be highly conserved (e.g., phyA mediates VLFR and far-red HIR), the downstream effectors and, ultimately, the physiological responses mediated by these photoreceptors, may be quite different across taxa (Figure 17.13).

FIGURE 17.12 Phytochrome deficiencies alter growth and development in pea and tomato. (A) Pea plants with lesions in *phyA* exhibit delayed flowering and shortened internodes. (B) In tomato, mutations in *phyA* and both copies of *phyB* prevent chlorophyll accumulation in the fruits and greatly extend the length of the fruit-bearing cluster. (From Weller et al. 1997; Weller et al. 2000.)

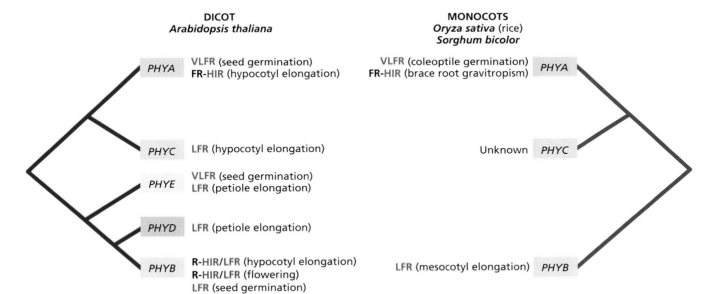

FIGURE 17.13 Differences in phytochrome gene family structure and function in the dicot *Arabidopsis thaliana* and the monocots *Oryza sativa* and *Sorghum bicolor*. PHY genes in monocots and dicots appear to utilize the same gene family members for VLFRs, LFRs, and HIRs (e.g., phyA is used for FR-HIR). However, the developmental phenomenon being regulated may be quite different (e.g., seed germination vs. coleoptile elongation for the phyA-mediated VLFR). Also note that monocots do not contain *PHYD* or *PHYE*.

Phytochrome Signaling Pathways

All phytochrome-regulated changes in plants begin with absorption of light by the pigment. After light absorption, the molecular properties of phytochrome are altered, probably affecting the interaction of the phytochrome protein with other cellular components that ultimately bring about changes in the growth, development, or position of an organ (see Table 17.1).

Molecular and biochemical techniques are helping to unravel the early steps in phytochrome action and the signal transduction pathways that lead to physiological or developmental responses. These responses fall into two general categories:

1. Ion fluxes, which cause relatively rapid turgor responses
2. Altered gene expression, which result in slower, long-term processes

In this section we will examine the effects of phytochrome on both membrane permeability and gene expression, as well as the possible chain of events constituting the signal transduction pathways that bring about these effects.

Phytochrome regulates membrane potentials and ion fluxes

Phytochrome can rapidly alter the properties of membranes, within seconds of a light pulse. Such rapid modulation has been measured in individual cells and has been inferred from the effects of red and far-red light on the surface potential of roots and oat (*Avena*) coleoptiles, where the lag between the production of Pfr and the onset of measurable potential changes is 4.5 s for hyperpolarization.

Changes in the bioelectric potential of cells imply changes in the flux of ions across the plasma membrane. Membrane isolation studies provide evidence that a small portion of the total phytochrome is tightly bound to various organellar membranes.

These findings led some workers to suggest that membrane-bound phytochrome represents the physiologically active fraction, and that all the effects of phytochrome on gene expression are initiated by changes in membrane permeability. On the basis of sequence analysis, however, it is now clear that phytochrome is a hydrophilic protein without membrane-spanning domains. The current view is that it may be associated with microtubules located directly beneath the plasma membrane, at least in the case of the alga *Mougeotia*, as described in **Web Topic 17.2**.

If phytochrome exerts its effects on membranes from some distance, no matter how small, involvement of a *second messenger* is implied, and calcium is a good candidate. Rapid changes in cytosolic free calcium have been implicated as second messengers in several signal transduction pathways, and there is evidence that calcium plays a role in chloroplast movement in *Mougeotia*.

Phytochrome regulates gene expression

As the term *photomorphogenesis* implies, plant development is profoundly influenced by light. Etiolation symptoms include spindly stems, small leaves (in dicots), and the absence of chlorophyll. Complete reversal of these symptoms by light involves major long-term alterations in metabolism that can be brought about only by changes in gene expression. Overall, the picture emerging for light-regulated plant promoters is similar to that for other eukaryotic genes: a collection of modular elements, the number, position, flanking sequences, and binding activities of which can lead to a wide range of transcriptional patterns. No single DNA sequence or binding protein is common to all phytochrome-regulated genes.

At first it may appear paradoxical that light-regulated genes have such a range of elements, any combination of which can confer light-regulated expression. However, this array of sequences allows for the differential light- and tissue-specific regulation of many genes through the action of multiple photoreceptors. (For a detailed discussion of promoter elements see **Web Topic 17.5**.)

The stimulation and repression of transcription by light can be very rapid, with lag times as short as 5 minutes. Using **DNA microarray analysis,** global patterns of gene expression in response to changes in light can be monitored. (For a discussion of methods for transcriptional analysis see **Web Topic 17.6**.) These studies have indicated that nuclear import triggers a transcriptional cascade involving thousands of genes that initiate photomorphogenic development. By monitoring gene expression profiles over time following a shift of plants from darkness to light, both early and late targets of *PHY* gene action were identified (Tepperman et al. 2001; Tepperman et al. 2004).

Some of the early gene products that are rapidly up-regulated following a shift from darkness to light are themselves transcription factors that activate the expression of other genes. The genes encoding these rapidly up-regulated proteins are called **primary response genes**. Expression of the primary response genes depends on *signal transduction pathways* (discussed next) and is independent of protein synthesis. In contrast, the expression of the late genes, or **secondary response genes**, requires the synthesis of new proteins. DNA microarray analyses have thus revealed the global reprogramming of plant gene expression that accompanies the transition from *skotomorphogenic* to *photomorphogenic* development.

Phytochrome interacting factors (PIFs) act early in phy signaling

Two techniques have been used extensively in recent years to identify interacting partners of plant proteins—two-hybrid library screens and co-immunoprecipitation (see **Web Topic 17.7** for descriptions). Using these two methods, several **phytochrome-interacting factors** (PIFs) have been identified in Arabidopsis (Wang and Deng 2004). Proteins that interact with *either* phyA or phyB define branch points in the phy signaling networks, whereas proteins that interact with *both* phyA and phyB are likely to represent points of convergence.

One of the most extensively characterized of these factors is **PIF3**, a basic helix-loop-helix (bHLH) transcription factor that interacts with both phyA and phyB (Ni et al. 1998) (see Chapter 14 on the web site). PIF3 and several related PIF or **PIF-like proteins** (**PILs**) are particularly notable because at least five members of this gene family selectively interact with phytochromes in their active Pfr conformations. The facts that these proteins are localized to the nucleus and can bind to DNA suggest an intimate association between phytochrome and gene transcription.

Detailed studies of PIF3 have revealed that it can act both as a positive and negative regulator of phytochrome signaling (Monte et al. 2004; Park et al. 2004). These apparently contradictory roles may be attributable to the fact that PIF3 can act as either an activator or a repressor of gene transcription depending on interactions with other transcriptional regulators. Furthermore, the PIF3 protein itself appears to be rapidly degraded in the light through a phytochrome-mediated pathway (Bauer et al. 2004). The phytochrome-induced rapid degradation of PIF3 and other proteins that act as negative regulators of phy responses may provide a mechanism for modulating light responses that is tightly coupled to the activities of phy proteins.

Phytochrome associates with protein kinases and phosphatases

In addition to nucleus-localized transcription factors, two-hybrid screens also identified cytosolic protein kinases as potential partners for phy proteins. **Phytochrome kinase substrate 1** (**PKS1**) is capable of interacting with phyA and phyB in both the active Pfr and inactive Pr form. This protein can accept a phosphate from phyA, suggesting that phosphorylation is an important part of the signal transduction cascade. The PKS1 phosphorylation is regulated by phytochrome both in the test tube and in the plant, with Pfr having a twofold higher level of activity than Pr. Overexpression studies and loss-of-function mutants of PKS1 and the closely related PKS2 suggest that these two molecules maintain balanced levels through a negative feedback loop. Molecular and genetic analyses suggest that these proteins act selectively to promote phyA-mediated VLFR (Lariguet et al. 2003).

Another protein kinase associated with phytochrome is **nucleoside diphosphate kinase 2** (**NDPK2**). Phytochrome A has been found to interact with this protein, and its kinase activity is increased about twofold when phyA is bound to it in the Pfr form. Association of NDPK2 with PfrA increases the ability of NDPK2 to convert GDP to GTP (Shen et al. 2005). This GTP could be used by *heterotrimeric G proteins* or one of many *monomeric GTP-binding proteins* in plants, including Rop, Arf, and Rab family members that are regulators of vesicle trafficking (reviewed in Chapter 14 on the web site). Defining the downstream targets of NDPK2 should help to illuminate the role of G proteins in phytochrome-mediated signal transduction (see **Web Topic 17.8**).

Recently, the phosphatase **phytochrome-associated protein phosphatase 5** (**PAPP5**) was identified as a factor that accentuates phy interactions with NDPK2 (Ryu et al. 2005). A possible model for the regulation of phy activity by phosphorylation is shown in Figure 17.14. In the dark, phytochrome in the Pr form is inactive and likely to be phosphorylated at serine residues in the N-terminal region. Absorption of red light induces a conformational change in phy that stimulates autophosphorylation of a serine residue in the hinge region and subsequent nuclear import;

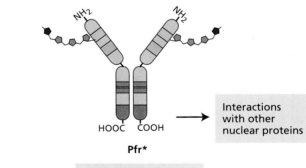

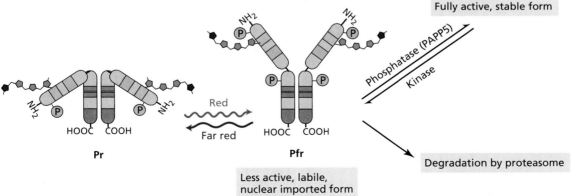

FIGURE 17.14 Phytochrome activity is modulated by phosphorylation status. Following activation by red light, the phy-associated phosphatase PAPP5 and as-yet unidentified kinases modulate phy activity in response to the intensity or quality of light. (After Ryu et al. 2005.)

it may also target phy for degradation (discussed below). Dephosphorylation of the serine residue in the hinge region enhances interaction of phy with downstream effectors such as NDPK2 and increases the stability of the protein in the light. The action of an unknown kinase and perhaps autophosphorylation serve to drive phy toward the less active, phosphorylated version that no longer interacts efficiently with its effector proteins.

Phytochrome-induced gene expression involves protein degradation

We often equate the initiation of a signal transduction cascade to flipping a switch that activates a process, like a phone ringing when an incoming call arrives. However, what would happen if the phone never stopped ringing? It would lose its utility as a signaling mechanism after the first call! In much the same way, termination or resetting of a pathway is just as important as the initiation of the event. As suggested above, protein degradation is emerging as a ubiquitous mechanism regulating many cellular processes, including light and hormone signaling, circadian rhythms, and flowering time (for example, see Chapters 19 and 25).

Genetic screens conducted independently by several groups identified mutants that exhibited light-grown phenotypes when grown in the dark, such as opened cotyledons, expanded leaves, and shortened hypocotyls. The genes identified in these screens were called *CONSTITUTIVE PHOTOMORPHOGENESIS* (*COP*), *DE-ETIOLATED* (*DET*), and *FUSCA* (for the red color of the anthocyanins that accumulated in light-grown seedlings). Cloning and genetic complementation revealed that many of these genes were allelic or part of the same complex, and they are collectively known as *COP/DET/FUS*.

The cloning of several *COP/DET/FUS* genes has revealed an essential role for protein degradation in the regulation of the light response. *COP1* encodes an E3 ubiquitin ligase that is essential for placing a small peptide tag known as *ubiquitin* onto proteins (see Chapter 1 and Chapter 14 on the web site). Once tagged by ubiquitin, the proteins are transported to the proteasome, a cellular garbage disposal that chews up proteins into their constituent amino acids. COP9 and several other COP proteins compose the **COP9 signalosome (CSN)**, which forms the lid of this garbage disposal, helping to determine which proteins enter the complex. As shown in Figure 17.15,

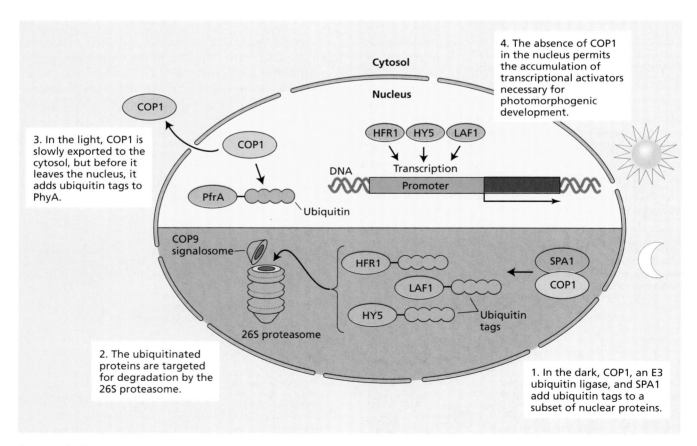

FIGURE 17.15 COP proteins regulate the turnover of proteins required for photomorphogenic development. During the night, COP1 enters the nucleus, and the COP1/SPA1 complex adds ubiquitin to a subset of transcriptional activators. The transcription factors are then degraded by the COP9 signalosome-proteasome complex. During the day, COP1 exits the nucleus, allowing the transcriptional activators to accumulate. Blue tails represent ubiquitin tags on proteins destined for the COP9 signalosome complex (CSN) that serves as the gatekeeper of the 26S proteasome.

COP1 has been shown to interact with several proteins involved in the light response, including the transcription factors HFR1, HY5, and LAF1, targeting them for degradation in the dark. It is likely that COP1 functions with the SUPPRESSOR OF phyA-1 (SPA1) protein to mediate the ubiquitination of these transcription factors.

In the light, COP1 is exported from the nucleus to the cytosol (see Figure 17.15), excluding it from interaction with many of the nucleus-localized transcription factors. These transcription factors can then bind to promoter elements in genes that mediate photomorphogenic development.

As mentioned above, phyA protein is highly unstable in light, and its degradation has long been linked to the sequential attachment of multiple copies of the small protein ubiquitin to a specific site on the target protein—a process called **ubiquitination** (Vierstra 1994). The discovery that COP1 can ubiquitinate phyA, thus targeting the protein for destruction, provides a satisfying link between COP1 function and the attenuation of phyA signaling in the light (Seo et al. 2004).

Furthermore, the relatively slow kinetics of nuclear–cytosolic partitioning of COP1 compared to phytochrome ensures that any phyA molecules that enter the nucleus in the light while COP1 is leaving will also be degraded. However, COP1 does not appear to mediate the destruction of PIF3 in the light, suggesting that other proteins are involved in the degradation of light-signaling components.

It is also unclear if the cytosolic pool of PfrA is ubiquitinated through an interaction with COP1 or with some other ubiquitin E3 ligase. There is increasing evidence for a high degree of specialization among ubiquitin E3 ligase family members, so it is possible that another E3 ligase participates in the degradation of cytosolic PfrA.

Circadian Rhythms

Various metabolic processes in plants, such as oxygen evolution and respiration, cycle alternately through high-activity and low-activity phases with a regular periodicity of about 24 hours. These rhythmic changes are referred to as **circadian rhythms** (from the Latin *circa diem*, meaning "approximately a day"). The **period** of a rhythm is the time that elapses between successive peaks or troughs in the cycle, and because the rhythm persists in the absence of external controlling factors, it is considered to be **endogenous**. (For a more detailed description of circadian rhythms, see Chapter 25.)

The endogenous nature of circadian rhythms suggests that they are governed by an internal pacemaker; this mechanism is called an **oscillator**. The endogenous oscillator is coupled to a variety of physiological processes. An important feature of the oscillator is that it is unaffected by temperature, which enables the clock to function normally under a wide variety of seasonal and climatic conditions. The clock is said to exhibit **temperature compensation**.

Light is a strong modulator of rhythms in both plants and animals. Although circadian rhythms that persist under controlled laboratory conditions usually have periods one or more hours longer or shorter than 24 hours, in nature their periods tend to be uniformly closer to 24 hours because of the synchronizing effects of light at daybreak, referred to as **entrainment**.

The molecular basis of the circadian rhythms has long fascinated both plant and animal biologists, and the isolation of clock mutants has been an important tool for the identification of clock genes in other organisms. Isolating clock mutants in plants requires a convenient assay that allows monitoring of the circadian rhythms of many thousands of individual plants to detect the rare abnormal phenotype.

To allow screening for clock mutants in Arabidopsis, the promoter region of the *LHCB* (also called *CHLOROPHYLL a/b BINDING* [*CAB*]) gene was fused to the gene that encodes luciferase, an enzyme that emits light in the presence of its substrate, luciferin. This reporter gene construct was then used to transform Arabidopsis with the Ti plasmid of *Agrobacterium* as a vector. Investigators were then able to monitor the temporal and spatial regulation of bioluminescence in individual seedlings in real time using a video camera (Millar et al. 1995). A total of 21 independent *toc* (*timing of CAB* [*LHCB*] *expression*) mutants were isolated, including both short-period and long-period lines. The *toc1* mutant in particular has been implicated in the core oscillator mechanism (Strayer et al. 2001).

Another important discovery came with the isolation and characterization of two MYB-related transcription factors, CIRCADIAN CLOCK–ASSOCIATED 1 (CCA1) and LATE ELONGATED HYPOCOTYLS (LHY). (For information on MYB, see Chapter 14 on the web site.) Loss-of-function mutants or constitutive overexpression of *CCA1* or *LHY* abolishes the circadian and phytochrome regulation of several genes, and physiological responses such as leaf movements become arrhythmic. These observations strongly suggest that *CCA1* and *LHY* are components of the circadian clock.

The circadian oscillator involves a transcriptional negative feedback loop

The circadian oscillators of cyanobacteria (*Synechococcus*), fungi (*Neurospora crassa*), insects (*Drosophila melanogaster*), and mouse (*Mus musculus*) have now been elucidated. In these four organisms, the oscillator is composed of several "clock genes" involved in a transcriptional–translational negative feedback loop.

So far, three major clock genes have been identified in Arabidopsis: *TOC1*, *LHY*, and *CCA1*. The protein products of these genes are all regulatory proteins. *TOC1* is not related to the clock genes of other organisms, suggesting that the plant oscillator is unique.

Models have been proposed that incorporate the findings of several genetic and molecular studies of circadian rhythms in Arabidopsis (Alabadi et al. 2001; Salome and McClung 2004). According to these models, light and the

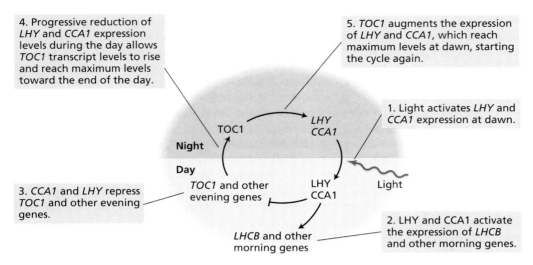

FIGURE 17.16 Circadian oscillator model showing the hypothetical interactions between the *TOC1* and *MYB* genes *LHY* and *CCA1*. Light at dawn increases *LHY* and *CCA1* expression. *LHY* and *CCA1* regulate other daytime and evening genes.

TOC1 regulatory protein activate *LHY* and *CCA1* expression at dawn (Figure 17.16). The increase in LHY and CCA1 represses the expression of the *TOC1* gene. Because TOC1 is a positive regulator of the *LHY* and *CCA1* genes, the repression of *TOC1* expression causes a progressive reduction in the levels of LHY and CCA1, which reach their minimum levels at the end of the day. As LHY and CCA1 levels decline, *TOC1* gene expression is released from inhibition. TOC1 reaches its maximum at the end of the day, when LHY and CCA1 are at their minimum. TOC1 then either directly or indirectly stimulates the expression of *LHY* and *CCA1*, and the cycle begins again.

A number of additional proteins help regulate this central oscillator. A protein kinase, CK2, can interact with and phosphorylate CCA1. The CK2 kinase is a multisubunit protein with serine/threonine kinase activity that, when mutated, changes the period of rhythmic expression of CCA1. The nuclear protein GIGANTEA (GI) is also required to maintain high levels of expression of LHY and CCA, although its mechanism of action is unknown. Degradation of TOC1 is mediated by the F-box protein ZTL. **F-boxes** are protein motifs that promote protein–protein interactions. F-box proteins were discovered as components of ubiquitin E3 ligase complexes, which target proteins for degradation by the 26S proteasome. ZTL protein levels peak at dusk and are lowest at dawn. Interestingly, ZTL is also structurally similar to the blue-light photoreceptor phototropin, suggesting that blue light at dawn may negatively regulate its activity.

The two MYB regulator proteins—LHY and CCA1—have dual functions. In addition to serving as components of the oscillator, they regulate the expression of other genes, such as *LHCB* and other "morning genes," and they repress genes expressed at night. Light reinforces the effect of the *TOC1* gene in promoting *LHY* and *CCA1* expression. This reinforcement represents the underlying mechanism of *entrainment*.

LIGHT REGULATION OF THE CIRCADIAN CLOCK To function properly, the oscillator must be entrained to the daily light/dark cycles of the external environment. In experiments designed to characterize the role of photoreceptors in this process, phytochrome-deficient mutants were crossed with lines carrying the luciferase reporter gene discussed above (Somers et al. 1998). The pace of the oscillator was slowed (i.e., period length increased) when *phyA* mutant plants were grown under dim red light, but not high-fluence red light. However, *phyB* mutants showed timing defects only under high-fluence red light. The blue-light photoreceptors CRY1 and CRY2 were required for blue light–mediated entrainment of the circadian clock.

These studies indicate that both phytochromes and cryptochromes entrain the circadian clock in Arabidopsis. This light input appears to be modulated by the genes *EARLY FLOWERING 3* (*ELF3*) and *TIME FOR COFFEE* (*TIC*). Mutations in *ELF3* stop the oscillations of the clock at dusk, whereas mutations in *TIC* stop the clock at dawn. The *elf3/tic* double mutant is completely arrhythmic, suggesting that *TIC* and *ELF* interact with different components of the clock at different phases in the rhythm (Hall et al. 2003).

GENE EXPRESSION AND CIRCADIAN RHYTHMS Phytochrome can also interact with circadian rhythms at the level of gene expression. The expression of genes in the *LHCB* family, encoding the light-harvesting chlorophyll *a/b*–binding proteins of photosystem II, is regulated at the transcriptional level by both circadian rhythms and phytochrome.

In leaves of pea and wheat, the level of *LHCB* mRNA has been found to oscillate during daily light–dark cycles, rising in the morning and falling in the evening. Since the rhythm persists even in continuous darkness, it appears to be a circadian rhythm. But phytochrome can perturb this cyclical pattern of expression.

When wheat plants are transferred from a cycle of 12 hours light and 12 hours dark to continuous darkness, the rhythm persists for a while, but it slowly *damps out* (i.e., reduces in amplitude until no peaks or troughs are discernible). If, however, the plants are given a pulse of red light before they are transferred to continuous darkness, no damping occurs (i.e., the levels of *LHCB* mRNA continue to oscillate as they do during the light–dark cycles).

In contrast, a far-red flash at the end of the day prevents the expression of *LHCB* in continuous darkness, and the effect of far-red is reversed by red light. Note that it is not the oscillator that damps out under constant conditions, but the coupling of the oscillator to the physiological event being monitored. Red light restores the coupling between the oscillator and the physiological process.

Microarray analyses have indicated that up to 6% of expressed genes in Arabidopsis are under the control of the circadian clock (Harmer et al. 2000). Interestingly, many genes that participate in similar cellular activities display rhythms with a similar phase. For instance, transcripts of many genes necessary for photosynthesis peak near the middle of the subjective day*, whereas transcripts required for cell wall biosynthesis peak near the middle of the subjective night. By carefully examining the sequence of promoter regions of these genes, a nine-nucleotide motif termed the **evening element** (AAAATATCT) was identified that appears to mediate expression of many genes whose expression peaks at the end of the subjective day.

CIRCADIAN RHYTHMS AND FITNESS Although circadian rhythms have long been known to play an essential role in photoperiodism during flowering (see Chapter 25), only recently has their importance in optimizing vegetative growth been tested experimentally (Dodd et al. 2005). Arabidopsis clock mutants with abnormally long or short periods were grown under artificial day–night cycles that either matched or were out of phase with oscillator periods. Plants whose circadian rhythms matched the light–dark cycle of the environment (**circadian resonance**) contained more chlorophyll and had greater biomass than plants whose clocks were out of phase with the environment. Moreover, when grown together in competition experiments, plants with correctly matched circadian clocks out-competed plants with mismatched clocks. Thus, circadian resonance enhances evolutionary fitness by promoting vegetative growth (photosynthesis and biomass) and reproductive development at optimal times.

Ecological Functions

Thus far we have discussed phytochrome-regulated responses as studied in the laboratory. However, phytochrome plays important ecological roles for plants growing in the environment. In the discussion that follows we will learn how phytochrome is involved in regulating various daily rhythms and how plants sense and respond to shading by other plants. We will also examine the specialized functions of the different phytochrome gene family members in these processes.

Phytochrome regulates the sleep movements of leaves

The sleep movements of leaves, referred to as **nyctinasty**, are a well-described example of a plant circadian rhythm that is regulated by light. In nyctinasty, leaves and/or leaflets extend horizontally (open) to face the light during the day and fold together vertically (close) at night (Figure 17.17). Nyctinastic leaf movements are exhibited by many legumes,

(A)

(B)

FIGURE 17.17 Thigmonastic (touch-sensitive) leaf movements of *Mimosa pudica*. Open leaflets (A) close in response to touch (B). Similar leaflet movements occur in a day–night cycle in nyctinastic species. (Courtesy of David McIntyre.)

---

*According to standardized circadian time (CT), the *subjective day* begins at CT = 0 hour of the 24-hour day–night cycle, while the *subjective night* starts at CT = 12 hours. The middle of the subjective day would therefore correspond to CT = 6 hours, and the middle of the subjective night would occur at CT = 18 hours.

FIGURE 17.18 Circadian rhythm in the diurnal movements of *Albizia* (a nyctinastic species) leaves. The leaves are elevated in the morning and lowered in the evening. In parallel with the raising and lowering of the leaves, the leaflets open and close. The rhythm persists at a lower amplitude for a limited time in total darkness.

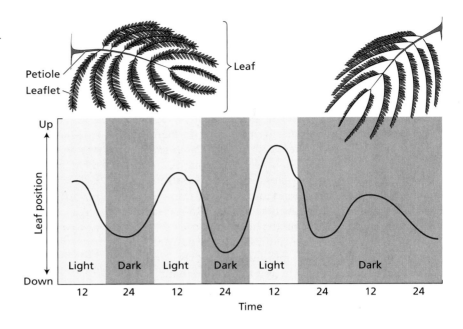

such as *Mimosa*, *Albizia*, and *Samanea*, as well as members of the Oxalis family. The change in leaf or leaflet angle is caused by rhythmic turgor changes in the cells of the **pulvinus** (plural *pulvini*), a specialized structure at the base of the petiole (see **Web Topic 17.9**).

Once initiated, the rhythm of opening and closing persists even in constant darkness, both in whole plants and in isolated leaflets (Figure 17.18). The phase of the rhythm (see Chapter 25), however, can be shifted by various exogenous signals, including red or blue light.

Light also directly affects movement: Blue light stimulates closed leaflets to open, and red light followed by darkness causes open leaflets to close. The leaflets begin to close within 5 minutes after being transferred to darkness, and closure is complete in 30 minutes. Because the effect of red light can be canceled by far-red light, phytochrome regulates leaflet closure.

The physiological mechanism of leaf movement is well understood. It results from turgor changes in cells located on opposite sides of the pulvinus, called **ventral motor cells** and **dorsal motor cells** (Figure 17.19). These changes in turgor pressure depend on K^+ and Cl^- fluxes across the plasma membranes of the dorsal and ventral motor cells. Leaflets close when the adaxial, or dorsal, motor cells accumulate K^+ and Cl^-, causing them to swell, while the abaxial, or ventral, motor cells release K^+ and Cl^-, causing them to shrink. Reversal of this process results in leaflet opening. Leaflet movement is therefore an example of a rapid response to phytochrome involving ion fluxes across membranes.

During phytochrome-mediated leaflet closure, the apoplastic pH of the dorsal motor cells (the cells that swell during leaflet closure) decreases, while the apoplastic pH of the ventral motor cells (the cells that shrink during leaflet closure) increases. Thus the plasma membrane H^+ pump of

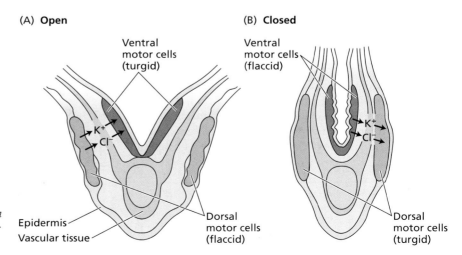

FIGURE 17.19 Ion fluxes between the flexor and extensor motor cells of *Albizia* pulvini regulate leaflet opening and closing. (After Galston 1994.)

the dorsal cells appears to be activated by darkness (provided that phytochrome is in the Pfr form), and the H⁺ pump of the ventral cells appears to be deactivated under the same conditions (see Figure 17.19). The reverse pattern of apoplastic pH change is observed during leaflet opening.

Studies have also been carried out on phytochrome regulation of K⁺ channels in isolated protoplasts (cells without their cell walls) of both dorsal and ventral motor cells from *Samanea* leaves (Kim et al. 1993). When the extracellular K⁺ concentration was raised, K⁺ entered the protoplasts and depolarized the membrane potential only if the K⁺ channels were open. When the dorsal and ventral motor cell protoplasts were transferred to constant darkness, the state of the K⁺ channels exhibited a circadian rhythmicity during a 21-hour incubation period, and the two cell types varied reciprocally, just as they do in vivo. That is, when the dorsal cell K⁺ channels were open, the ventral cell K⁺ channels were closed, and vice versa. Thus the circadian rhythm of leaf movements has its origins in the circadian rhythm of K⁺ channel opening.

On the basis of the evidence thus far, we can conclude that phytochrome brings about leaflet closure by regulating the activities of the primary proton pumps and the K⁺ channels of the dorsal and ventral motor cells. Although the effect is rapid, it is not instantaneous, and it is therefore unlikely to be due to a direct effect of phytochrome on the membrane. Instead, phytochrome acts indirectly via one or more signal transduction pathways.

Phytochrome enables plant adaptation to light quality changes

The presence of a red/far-red reversible pigment in all green plants, from algae to dicots, suggests that these wavelengths of light provide information that helps plants adjust to their environment. What environmental conditions change the relative levels of these two wavelengths of light in natural radiation?

The ratio of red light (R) to far-red light (FR) varies remarkably in different environments. This ratio can be defined as follows:

$$R = \frac{\text{Photon fluence rate in 10 nm band centered on 660 nm}}{\text{Photon fluence rate in 10 nm band centered on 730 nm}}$$

Table 17.3 compares both the total light intensity in photons (400–800 nm) and the R:FR values in eight natural environments. Both parameters vary greatly in different environments.

Compared with direct daylight, there is relatively more far-red light during sunset, under 5 mm of soil, or under the canopy of other plants (as on the floor of a forest). The canopy phenomenon results from the fact that green leaves absorb red light because of their high chlorophyll content, but are relatively transparent to far-red light.

Decreasing the R:FR ratio causes elongation in sun plants

An important function of phytochrome is that it enables plants to sense shading by other plants. Plants that increase stem extension in response to shading are said to exhibit a **shade avoidance response**. As shading increases, the R:FR ratio decreases. The greater proportion of far-red light converts more Pfr to Pr, and the ratio of Pfr to total phytochrome (Pfr/P_{total}) decreases. When simulated natural radiation was used to vary the far-red content, it was found that for so-called sun plants (plants that normally grow in an open-field habitat), stem extension rates were increased when they experienced higher far-red content (i.e., a lower Pfr:P_{total} ratio) (Figure 17.20).

In other words, simulated canopy shading (high levels of far-red light; low Pfr:P_{total} ratio) induced these plants to allocate more of their resources to growing taller. This correlation was not as strong for "shade plants," which normally grow in a shaded environment. Shade plants showed less reduction in their stem extension rate than did sun plants as they were exposed to higher R:FR values (see Figure 17.20). Thus there appears to be a systematic relationship between phytochrome-controlled growth and species habitat. Such results are taken as an indication of the involvement of phytochrome in shade perception.

For a "sun plant" or "shade-avoiding plant" there is a clear adaptive value in allocating its resources toward more rapid extension growth when it is shaded by another plant. In this way it can enhance its chances of growing above the

TABLE 17.3
Ecologically important light parameters

	Photon flux density (µmol m⁻² s⁻¹)	R:FR[a]
Daylight	1900	1.19
Sunset	26.5	0.96
Moonlight	0.005	0.94
Ivy canopy	17.7	0.13
Lakes, at a depth of 1 m		
Black Loch	680	17.2
Loch Leven	300	3.1
Loch Borralie	1200	1.2
Soil, at a depth of 5 mm	8.6	0.88

Source: Smith 1982, p. 493.

Note: The light intensity factor (400–800 nm) is given as the photon flux density, and phytochrome-active light is given as the R:FR ratio.

[a]Absolute values taken from spectroradiometer scans; the values should be taken to indicate the relationships between the various natural conditions and not as actual environmental means.

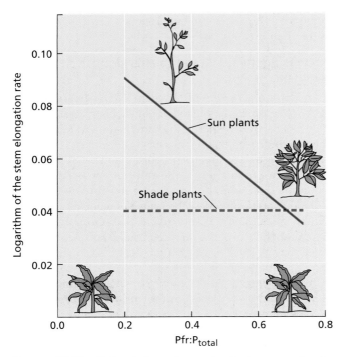

FIGURE 17.20 Role of phytochrome in shade perception in sun plants (solid line) versus shade plants (dashed line). (After Morgan and Smith 1979.)

avoidance responses, but phyD and phyE also contribute, particularly to petiole elongation. PhyA also plays a role by antagonizing the responses mediated by phyB, D, and E (see Figure 17.21) (discussed later in the chapter). Evidence is also emerging for the integration of a number of hormonal pathways in the control of this aspect of developmental plasticity, including auxin, gibberellins, and ethylene. Microarray analyses of shaded vs. unshaded plants have confirmed and extended the genetic studies (see **Web Topic 17.10**). For a discussion of how plants sense their neighbors using reflected light, see **Web Essay 17.3**.

Small seeds typically require a high R:FR ratio for germination

Light quality also plays a role in regulating the germination of some seeds. As discussed earlier, phytochrome was discovered in studies of light-dependent lettuce seed germination.

In general, large seeds, with their ample food reserves to sustain prolonged seedling growth in darkness (e.g., underground), do not require light for germination. However, light is required by the small seeds of many herbaceous and grassland species, many of which remain dormant, even while hydrated, if they are buried below the depth to which light penetrates. Even when such seeds are on or near the soil surface, their level of shading by the vegetation canopy (i.e., the R:FR ratio they receive) is likely to affect their germination. For example, it is well documented that far-red enrichment imparted by a leaf canopy inhibits germination in a range of small-seeded species.

For the small seeds of the tropical species trumpet tree (*Cecropia obtusifolia*) and Veracruz pepper (*Piper auritum*) planted on the floor of a deeply shaded forest, this inhibition can be reversed if a light filter is placed immediately

canopy and acquiring a greater share of unfiltered, photosynthetically active light. The price for favoring internode elongation is usually reduced leaf area and reduced branching, but at least in the short run this adaptation to canopy shade seems to work.

Genetic analyses of Arabidopsis have indicated that phyB plays the predominant role in mediating many shade

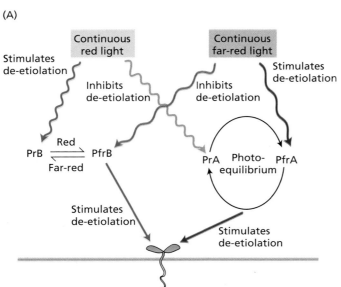

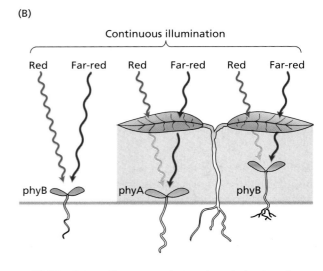

FIGURE 17.21 Mutually antagonistic roles of phyA and phyB. (After Quail et al. 1995.)

above the seeds that permits the red component of the canopy-shaded light to pass through while blocking the far-red component. Although the canopy transmits very little red light, the level is enough to stimulate the seeds to germinate, probably because most of the inhibitory far-red light is excluded by the filter and the R:FR ratio is very high. These seeds would also be more likely to germinate in spaces receiving sunlight through gaps in the canopy than in densely shaded spaces. The sunlight would help ensure that the seedlings became photosynthetically self-sustaining before their seed food reserves were exhausted.

Phytochrome interactions are important early in germination

As emphasized throughout this chapter, interactions among phytochrome proteins are complex. The actions of phyB, D, and E are often synergistic, whereas the action of phyA on phyB may be either synergistic or antagonistic, depending on the type of response and the tissue. These complex interactions help the plant to carefully monitor and respond to ever-changing light environments.

Figure 17.21 summarizes some of these interactions for the de-etiolation and shade avoidance responses. Continuous red light absorbed by phyB stimulates de-etiolation by maintaining high levels of PfrB. Continuous far-red light absorbed by PfrB prevents this stimulation by reducing the amount of PfrB. The stimulation of de-etiolation by phyA depends on the photostationary state of phytochrome (indicated in Figure 17.21A by the circular arrows). Continuous far-red light stimulates de-etiolation when absorbed by the phyA system; continuous red light inhibits the response.

The effects of phyA and phyB on seedling development in sunlight versus canopy shade (enriched in far-red light) are shown in Figure 17.21B. In open sunlight, which is enriched in red light compared with canopy shade, de-etiolation is mediated primarily by the phyB system (on the left in the figure). A seedling emerging under canopy shade, enriched in far-red light, initiates de-etiolation primarily through the phyA system (center). Because phyA is labile, however, the response is taken over by phyB (right). In switching over to phyB, the stem is released from growth inhibition (see Figure 17.21B), allowing for the accelerated rate of stem elongation that is part of the shade avoidance response.

Reducing shade avoidance responses can improve crop yields

Shade avoidance responses may be highly adaptive in a natural setting to help plants out-compete neighboring vegetation, but for many crop species a reallocation of resources from reproductive to vegetative growth can reduce crop yield. In recent years, yield gains in crops such as maize have largely come through the breeding of new maize varieties with a higher tolerance to crowding (which induces shade avoidance responses) than through increases in basic yield per plant. As a consequence, today's maize crops can be grown at higher densities than older varieties without suffering decreases in plant yield (Figure 17.22).

As the mechanism of shade avoidance has become better understood, the prospect of engineering crop plants with decreased shade avoidance has been opened up. For example, Robson and colleagues created transgenic tobacco plants that overexpressed phyA (Robson et al. 1996). When these plants were grown at high density, they failed to display a typical shade avoidance response and were actually shorter at high density than low. This result is predicted by the model shown in Figure 17.21B. Overexpression of phyA would lead to the persistence of phyA under far-red–rich shade beyond the early seedling stage, when phyB normally takes over. The persistence of phyA would cause an increase in the FR-HIR response (discussed earlier), which would counteract the phyB-mediated shade avoidance response.

Although much remains to be learned, such studies have shown that the manipulation of phytochromes or their downstream targets is a promising new approach to improving crop yield (Sawers et al. 2005).

FIGURE 17.22 Modern maize varieties are planted at high density. Traditionally, maize was grown by Native Americans in small hills or mounds, and each mound was separated by several feet. The plants were short and often produced multiple small ears. In contrast, modern hybrids are machine-planted in rows with little space between plants (typically 30,000 to 38,000 plants/acre). Although yield per plant has not increased for many years in commercial hybrids, overall yield increases have continued, largely through the better performance of plants at high planting density. That is, more plants can be grown in a given area. As shown in this image of a typical upstate New York cornfield, modern varieties have upright leaves that help the plants to capture sunlight energy under crowded conditions. (Courtesy of T. Brutnell.)

Phytochrome responses show ecotypic variation

To date, most of our understanding of light responses in any given model plant species has been derived from experiments performed on a limited number of varieties or accessions. For example, the complete genome of Arabidopsis has been sequenced only in the Columbia ecotype, and the genome of maize is being sequenced in a single inbred line, B73. As a result, plant research programs throughout the world have tended to focus on these varieties.

When considering the role of phytochromes in an ecological context, however, it is essential to examine a much broader germplasm collection. Surveys of the light responses in Arabidopsis and maize have revealed tremendous ecotypic variation, both in the physiology of their light responses and in their phytochrome gene families. For instance, the Wassilewskija (Ws) accession of Arabidopsis contains a naturally occurring deletion of the phyD gene, whereas an accession from Le Mans, France (Lm-2) carries a light-stable form of phyA that fails to mediate responses to continuous far-red light (Aukerman et al. 1997; Maloof et al. 2001). These studies indicate that variations in phytochrome responses may have some adaptive value. Determining how such variations contribute to fitness in diverse habitats is a challenge for the future.

Phytochrome action can be modulated

Can the actions of other photoreceptors modify the action of phytochrome? The isolation of the genes encoding the cryptochrome and phototropin photoreceptors (see Chapter 18) mediating blue light–regulated responses has made it possible to analyze whether these photoreceptors functionally overlap with phytochrome (Chory and Wu 2001). This possibility was suspected because mutations in the cryptochrome *CRY2* gene led to delayed flowering under continuous white light, and flowering time was also known to be under phytochrome control.

In Arabidopsis, continuous blue or far-red light treatment leads to promotion of flowering, and red light inhibits flowering. Far-red light acts through phyA, and the antagonistic effect of red light is produced by of phyB. One might expect the *cry2* mutant to be delayed in flowering, since blue light promotes flowering. However, *cry2* mutants flower at the same time as the wild type under either continuous blue or continuous red light. Delay is observed only if both blue and red light are given together. Therefore, *cry2* probably promotes flowering in blue light by repressing phyB function.

Additional experiments have confirmed that the other cryptochrome, CRY1, also interacts with phytochromes. Both CRY1 and CRY2 interact with phyA in vitro and can be phosphorylated in a phyA-dependent manner. Phosphorylation of CRY1 has also been demonstrated to occur in vivo in a red light–dependent manner. Indeed, the importance of cryptochromes as developmental regulators has been underscored by their subsequent discovery in animal systems, such as mouse and human.

Summary

The term *photomorphogenesis* refers to the dramatic effects of light on plant development and cellular metabolism. Red light exerts the strongest influence, and the effects of red light are often reversible by far-red light.

Phytochrome is the pigment involved in most photomorphogenic phenomena. Phytochrome exists in two forms: a red light–absorbing form (Pr) and a far-red light–absorbing form (Pfr). Phytochrome is synthesized in the dark in the Pr form. Absorption of red light by Pr converts it to Pfr, and absorption of far-red light by Pfr converts it to Pr. However, the absorption spectra of the two forms overlap in the red region of the spectrum, leading to an equilibrium between the two forms called a photostationary state.

Pfr is considered the active form that gives rise to the physiological response. Other factors in addition to light regulate the steady-state level of Pfr, including the expression level of the protein and its stability in the Pfr form.

Both rapid biochemical events, such as changes in membrane potentials, and slower physiological changes, such as germination, growth, and reproductive development, are mediated by phytochrome. These different processes are regulated by different cellular signaling intermediates and can be classified on the basis of lag time and escape from photoreversibility.

Phytochrome responses have been classified into very low–fluence, low-fluence, and high-irradiance responses (VLFRs, LFRs, and HIRs). These three types of responses differ not only in their fluence requirements but in other parameters, such as their escape times, action spectra, and photoreversibility.

Phytochrome is a large dimeric protein made up of two subunits. The monomer has a molecular mass of about 125 kDa and is covalently bound to a chromophore molecule, an open-chain tetrapyrrole called phytochromobilin. Phytochrome is a light-regulated serine/threonine kinase, and the C-terminal domain encodes a region that shows homology with histidine kinases from bacteria. Absorption of red or far-red light results in isomerization of the pyrrole ring of phytochromobilin, which mediates a conformational change and autophosphorylation of the *PHY* apoprotein. Several domains have been defined for phytochrome, including the bilin lyase domain that that is necessary for chromophore attachment and a PHY domain necessary for stabilizing the active form. Phosphorylation and dephosphorylation of the phy protein helps regulate its activity.

When activated by red light, phytochrome is translocated from the cytosol to the nucleus, where it interacts with transcriptional regulators. However, a small pool of phytochrome remains in the cytosol to mediate very rapid physiological changes.

Phytochrome is encoded by a family of divergent genes that encode two types of proteins: Type I and Type II. Type I, which is encoded by the *PHYA* gene, is abundant in etiolated tissue and mediates the VLFR and FR-HIRs. Type II phytochrome (encoded by the *PHYB*, *PHYC*, *PHYD*, and *PHYE* genes in Arabidopsis) is represented by a divergent gene family among angiosperms, suggesting a rapid evolutionary diversification in function.

Phytochrome B plays an important role in the detection of shade in plants adapted to high levels of sunlight; phytochrome A has a more limited role, mediating the far-red HIR during early greening. Phytochromes C, D, and E also have specific roles during limited phases of development, and these roles are partially redundant with those of phyA and phyB.

Phytochrome induces a variety of rapid responses, including chloroplast rotation in the alga *Mougeotia*, leaf closure during nyctinasty, and alterations in membrane potential. These responses involve rapid changes in membrane properties. The current view is that even these rapid effects of phytochrome involve cytosolic signal transduction pathways.

Phytochrome is also known to regulate the transcription of numerous genes in the nucleus. Many of the genes involved in greening, such as the nucleus-encoded gene for chlorophyll *a/b*–binding protein of the light-harvesting complex, are transcriptionally regulated by phytochrome (both phyA and phyB).

Phytochrome also represses the transcription of numerous genes. Microarray analysis has indicated that the expression of thousands of genes change in response to red or far-red light, amounting to about 10% of the total genome of Arabidopsis. Activation or repression of these genes is thought to be mediated by general transcription factors. In some cases, phytochrome in the Pfr form interacts directly with these factors. In addition, kinases play an important role in regulating phytochrome responses.

Protein degradation plays an important role in regulating phytochrome response. The *COP/DET/FUS* genes encode many proteins required for protein turnover via the 26S proteasome and interact with phytochrome and immediate targets of phytochrome to control light response.

Phytochromes also play an important role in entraining the circadian clock. The central oscillator requires the activity of a transcriptional negative feedback loop. Several other proteins ensure that the clock maintains rhythmic periodicity. Light input to the clock is also regulated by specific proteins. Phytochrome can also mediate changes in gene expression that are regulated by the circadian clock.

The functions of phytochromes on seed germination, leaf movements, and responses to vegetative shade have ecological effects. These responses are mediated through a complex and sometimes antagonistic action of phytochromes. Recent comparisons of light responses within and between species have helped explain the diversification of the phytochrome response, and suggest that plasticity in response to light has adaptive value.

Web Material

Web Topics

17.1 The Structure of Phytochromes
The purification and characterization of phytochrome as a homodimer are described.

17.2 Phytochrome and High-Irradiance Responses
Dual-wavelength experiments helped demonstrate the role of phytochrome in HIRs.

17.3 The Origins of Phytochrome as a Bacterial Two-Component Receptor
The discovery of bacterial phytochrome led to the identification of phytochrome as a protein kinase.

17.4 *Mougeotia*: A Chloroplast with a Twist
Microbeam irradiation experiments have been used to localize phytochrome in this filamentous green alga.

17.5 Regulation of Transcription by *cis*-Acting Sequences
Phytochrome response elements are described briefly.

17.6 Phytochrome Regulation of Gene Expression
Evidence shows that phytochrome regulates gene expression at the level of transcription.

17.7 Two-Hybrid Screens and Co-immunoprecipitation
Protein–protein interactions can be studied using both molecular-genetic and immunological techniques.

17.8 The Roles of G Proteins and Calcium in Phytochrome Responses
Evidence suggests that G proteins and calcium participate in phytochrome action.

17.9 Phytochrome Effects on Ion Fluxes
Phytochrome regulates ion fluxes across membranes by altering the activities of ion channels and the plasma membrane proton pump.

17.10 Microarray Studies on Shade Avoidance
DNA microarray analyses have helped to characterize both global and specific effects of variations in the R:FR ratio on gene expression.

Web Essays

17.1 Awakened By a Flash of Sunlight
When placed in the proper soil environment, seeds acquire extraordinary sensitivity to light, such that germination can be stimulated by less than 1 second of exposure to sunlight during soil cultivation.

17.2 Diversity of Phytochrome Chromophores
Bacterial and higher plant chromophores vary in their structure, attachment chemistries, and spectral properties. By replacing plant chromophores with bacterial chromophores, plants can be engineered to "see" different wavelengths of light.

17.3 Know Thy Neighbor Through Phytochrome
Plants can detect the proximity of neighbors through phytochrome perception of the R:FR of reflected light and produce adaptive morphological changes before being shaded by potential competitors.

Chapter References

Alabadi, D., Oyama, T., Yanovsky, M. J., Harmon, F. G., Mas, P., and Kay, S. A. (2001) Reciprocal regulation between *TOC1* and *LHY/CCA1* within the *Arabidopsis* circadian clock. *Science* 293: 880–883.

Andel, F., Hasson, K. C., Gai, F., Anfinrud, P. A., and Mathies, R. A. (1997) Femtosecond time-resolved spectroscopy of the primary photochemistry of phytochrome. *Biospectroscopy* 3: 421–433.

Aukerman, M. J., Hirschfeld, M., Wester, L., Weaver, M., Clack, T., Amasino, R. M., and Sharrock, R. A. (1997) A deletion in the PHYD gene of the Arabidopsis Wassilewskija ecotype defines a role for phytochrome D in red/far-red light sensing. *Plant Cell* 9: 1317–1326.

Bauer, D., Viczian, A., Kircher, S., Nobis, T., Nitschke, R., Kunkel, T., Panigrahi, K. C., Adam, E., Fejes, E., and Schafer, E., et al. (2004). Constitutive photomorphogenesis 1 and multiple photoreceptors control degradation of phytochrome interacting factor 3, a transcription factor required for light signaling in *Arabidopsis*. *Plant Cell* 16: 1433–1445.

Borthwick, H. A., Hendricks, S. B., Parker, M. W., Toole, E. H., and Toole, V. K. (1952) A reversible photoreaction controlling seed germination. *Proc. Natl. Acad. Sci. USA* 38: 662–666.

Briggs, W. R., Mandoli, D. F., Shinkle, J. R., Kaufman, L. S., Watson, J. C., and Thompson, W. F. (1984) Phytochrome regulation of plant development at the whole plant, physiological, and molecular levels. In *Sensory Perception and Transduction in Aneural Organisms*, G. Colombetti, F. Lenci, and P.-S. Song, eds., Plenum, New York, pp. 265–280.

Butler, W. L., Norris, K. H., Siegelman, H. W., and Hendricks, S. B. (1959) Detection, assay, and preliminary purification of the pigment controlling photosensitive development of plants. *Proc. Natl. Acad. Sci. USA* 45: 1703–1708.

Chen, M., Tao, Y., Lim, J., Shaw, A., and Chory, J. (2005) Regulation of phytochrome B nuclear localization through light-dependent unmasking of nuclear-localization signals. *Curr. Biol.* 15: 637–642.

Chory, J., and Wu, D. (2001) Weaving the complex web of signal transduction. *Plant Physiol.* 125: 77–80.

Dodd, A. N., Salathia, N., Hall, A., Kevei, E., Toth, R., Nagy, F., Hibberd, J. M., Millar, A. J., and Webb, A. A. (2005) Plant circadian clocks increase photosynthesis, growth, survival, and competitive advantage. *Science* 309: 630–633.

Flint, L. H. (1936) The action of radiation of specific wave-lengths in relation to the germination of light-sensitive lettuce seed. *Proc. Int. Seed Test. Assoc.* 8: 1–4.

Franklin, K. A., Davis, S. J., Stoddart, W. M., Vierstra, R. D., and Whitelam, G. C. (2003) Mutant analyses define multiple roles for phytochrome C in *Arabidopsis* photomorphogenesis. *Plant Cell* 15: 1981–1989.

Galston, A. (1994) *Life Processes of Plants*. Scientific American Library, New York.

Hall, A., Bastow, R. M., Davis, S. J., Hanano, S., McWatters, H. G., Hibberd, V., Doyle, M. R., Sung, S., Halliday, K. J., and Amasino, R. M., et al. (2003). The *TIME FOR COFFEE* gene maintains the amplitude and timing of *Arabidopsis* circadian clocks. *Plant Cell* 15: 2719–2729.

Harmer, S. L., Hogenesch, J. B., Straume, M., Chang, H. S., Han, B., Zhu, T., Wang, X., Kreps, J. A., and Kay, S. A. (2000) Orchestrated transcription of key pathways in *Arabidopsis* by the circadian clock. *Science* 290: 2110–2113.

Hartmann, K. M. (1967) Ein Wirkungsspecktrum der Photomorphogenese unter Hochenergiebedingungen und seine Interpretation auf der Basis des Phytochroms (Hypokotylwachstumshem-mung bei *Lactuca sativa* L.). *Z. Naturforsch.* 22b: 1172–1175.

Hirschfeld, M., Tepperman, J. M., Clack, T., Quail, P. H., and Sharrock, R. A. (1998). Coordination of phytochrome levels in phyB mutants of *Arabidopsis* as revealed by apoprotein-specific monoclonal antibodies. *Genetics* 149: 523–535.

Kim, H. Y., Cote, G. G., and Crain, R. C. (1993) Potassium channels in *Samanea*-Saman protoplasts controlled by phytochrome and the biological clock. *Science* 260: 960–962.

Koornneef, M., Rolff, E., and Spruitt, C. J. P. (1980) Genetic control of light-induced hypocotyl elongation in *Arabidopsis thaliana* L. *Z. Pflanzenphysiol.* 100: 147–160.

Lariguet, P., Boccalandro, H. E., Alonso, J. M., Ecker, J. R., Chory, J., Casal, J. J., and Fankhauser, C. (2003) A growth regulatory loop that provides homeostasis to phytochrome a signaling. *Plant Cell* 15: 2966–2978.

Li, L., and Lagarias, J. C. (1992) Phytochrome assembly—Defining chromophore structural requirements for covalent attachment and photoreversibility. *J. Biol. Chem.* 267: 19204–19210.

Maloof, J. N., Borevitz, J. O., Dabi, T., Lutes, J., Nehring, R. B., Redfern, J. L., Trainer, G. T., Wilson, J. M., Asami, T., and Berry, C. C., et al. (2001). Natural variation in light sensitivity of *Arabidopsis*. *Nat. Genet.* 29: 441–446.

Mandoli, D. F., and Briggs, W. R. (1984) Fiber optics in plants. *Sci. Am.* 251: 90–98.

Mathews, S., and Sharrock, R. A. (1997) Phytochrome gene diversity. *Plant Cell Environ.* 20: 666–671.

Matsushita, T., Mochizuki, N., and Nagatani, A. (2003) Dimers of the N-terminal domain of phytochrome B are functional in the nucleus. *Nature* 424: 571–574.

Millar, A. J., Carre, I. A., Strayer, C. A., Chua, N.-H., and Kay, S. A. (1995) Circadian clock mutants in *Arabidopsis* identified by luciferase imaging. *Science* 267: 1161–1163.

Monte, E., Alonso, J. M., Ecker, J. R., Zhang, Y., Li, X., Young, J., Austin-Phillips, S., and Quail, P. H. (2003) Isolation and characterization of phyC mutants in *Arabidopsis* reveals complex crosstalk between phytochrome signaling pathways. *Plant Cell* 15, 1962–1980.

Monte, E., Tepperman, J. M., Al-Sady, B., Kaczorowski, K. A., Alonso, J. M., Ecker, J. R., Li, X., Zhang, Y., and Quail, P. H. (2004) The phytochrome-interacting transcription factor, PIF3, acts early, selectively, and positively in light-induced chloroplast development. *Proc. Natl. Acad. Sci. USA* 101: 16091–16098.

Morgan, D. C., and Smith, H. (1979) A systematic relationship between phytochrome-controlled development and species habitat, for plants grown in simulated natural irradiation. *Planta* 145: 253–258.

Nakasako, M., Wada, M., Tokutomi, S., Yamamoto, K. T., Sakai, J., Kataoka, M., Tokunaga, F., and Furuya, M. (1990) Quaternary structure of pea phytochrome I dimer studied with small angle X-ray scattering and rotary-shadowing electron microscopy. *Photochem. Photobiol.* 52: 3–12.

Ni, M., Tepperman, J. M., and Quail, P. H. (1998) PIF3, a phytochrome-interacting factor necessary for normal photoinduced signal transduction, is a novel basic helix-loop-helix protein. *Cell* 95: 657–667.

Park, E., Kim, J., Lee, Y., Shin, J., Oh, E., Chung, W. I., Liu, J. R., and Choi, G. (2004) Degradation of phytochrome interacting factor 3 in phytochrome-mediated light signaling. *Plant Cell Physiol.* 45: 968–975.

Parks, B. M., and Spalding, E. P. (1999) Sequential and coordinated action of phytochromes A and B during *Arabidopsis* stem growth revealed by kinetic analysis. *Proc. Natl. Acad. Sci. USA* 96: 14142–14146.

Quail, P. H., Boylan, M. T., Parks, B. M., Short, T. W., Xu, Y., and Wagner, D. (1995) Phytochrome: Photosensory perception and signal transduction. *Science* 268: 675–680.

Robson, P. R., McCormac, A. C., Irvine, A. S., and Smith, H. (1996) Genetic engineering of harvest index in tobacco through overexpression of a phytochrome gene. *Nat. Biotechnol.* 14: 995–998.

Ryu, J. S., Kim, J. I., Kunkel, T., Kim, B. C., Cho, D. S., Hong, S. H., Kim, S. H., Fernandez, A. P., Kim, Y., and Alonso, J. M., et al. (2005) Phytochrome-specific type 5 phosphatase controls light signal flux by enhancing phytochrome stability and affinity for a signal transducer. *Cell* 120: 395–406.

Salome, P. A., and McClung, C. R. (2004) The *Arabidopsis thaliana* clock. *J. Biol. Rhythms* 19: 425–435.

Sawers, R. J., Sheehan, M. J., and Brutnell, T. P. (2005) Cereal phytochromes: targets of selection, targets for manipulation? *Trends Plant Sci.* 10, 138–143.

Seo, H. S., Watanabe, E., Tokutomi, S., Nagatani, A., and Chua, N. H. (2004) Photoreceptor ubiquitination by COP1 E3 ligase desensitizes phytochrome A signaling. *Genes Dev.* 18: 617–622.

Sharma, R. (2001) Phytochrome: A serine kinase illuminates the nucleus! *Curr. Sci.* 80: 178–188.

Sharrock, R. A., and Clack, T. (2004) Heterodimerization of type II phytochromes in *Arabidopsis. Proc. Natl. Acad. Sci. USA* 101: 11500–11505.

Sharrock, R. A., and Quail, P. H. (1989) Novel phytochrome sequences in *Arabidopsis thaliana*: Structure, evolution, and differential expression of a plant regulatory photoreceptor family. *Genes Dev.* 3: 1745–1757.

Shen, Y., Kim, J. I., and Song, P. S. (2005). NDPK2 as a signal transducer in the phytochrome-mediated light signaling. *J. Biol. Chem.* 280, 5740–5749.

Shinomura, T., Nagatani, A., Hanzawa, H., Kubota, M., Watanabe, M., and Furuya, M. (1996) Action spectra for phytochrome A- and B-specific photoinduction of seed germination in *Arabidopsis thaliana. Proc. Natl. Acad. Sci. USA* 93: 8129–8133.

Shropshire, W., Jr., Klein, W. H., and Elstad, V. B. (1961) Action spectra of photomorphogenic induction and photoinactivation of germination in *Arabidopsis thaliana. Plant Cell Physiol.* 2: 63–69.

Smith, H. (1974) *Phytochrome and Photomorphogenesis: An Introduction to the Photocontrol of Plant Development*. McGraw-Hill, London.

Smith, H. (1982) Light quality photoperception and plant strategy. *Annu. Rev. Plant Physiol.* 33: 481–518.

Somers D. E., Devlin, P. F., and Kay, S. A. (1998) Phytochromes and cryptochromes in the entrainment of the *Arabidopsis* circadian clock. *Science* 282: 1488–1494.

Strayer, C., Oyama, T., Schultz, T. F., Raman, R., Somer, D. E., Mas, P., Panda, S., Kreps, J. A., and Kay, S. A. (2001) Cloning of the *Arabidopsis* clock gene *TOC1*, an autoregulatory response regulator homolog. *Science* 289: 768–771.

Tepperman, J. M., Hudson, M. E., Khanna, R., Zhu, T., Chang, S. H., Wang, X., and Quail, P. H. (2004). Expression profiling of phyB mutant demonstrates substantial contribution of other phytochromes to red-light-regulated gene expression during seedling de-etiolation. *Plant J.* 38: 725–739.

Tepperman, J. M., Zhu, T., Chang, H. S., Wang, X., and Quail, P. H. (2001) Multiple transcription factor genes are early targets of phytochrome A signaling. *Proc. Natl. Acad. Sci. USA* 98: 9437–9442.

Vierstra, R. D. (1994) Phytochrome degradation. In *Photomorphogenesis in Plants*, 2nd ed., R. E. Kendrick and G. H. M. Kronenberg, eds., Martinus Nijhoff, Dordrecht, Netherlands, pp. 141–162.

Vierstra, R. D., and Quail, P. H. (1983) Purification and initial characterization of 124-kilodalton phytochrome from *Avena. Biochemistry* 22: 2498–2505.

Wagner, J. R., Brunzelle, J. S., Forest, K. T., and Vierstra, R. D. (2005) A light-sensing knot revealed by the structure of the chromophore binding domain on phytochrome. *Nature* 17: 325–321.

Wang, H., and Deng, X. W. (2004) Phytochrome signaling mechanism. In *The Arabidopsis Book*, C. R. Somerville, and E. M. Meyerowitz, eds. American Society of Plant Biologists, Rockville, MD.

Weller, J. L., Murfet, I. C., and Reid, J. C. (1997) Pea mutants with reduced sensitivity to far-red light define an important role for phytochrome A in day-length detection. *Plant Physiol.* 114: 1225–1236.

Weller, J. L., Schreuder, M. E. L., Smith, H., Koornneef, M., Kendrick, R. E. (2000) Physiological interactions of phytochromes A, B1 and B2 in the control of development in tomato. *Plant J.* 24: 345–356.

Yamaguchi, R., Nakamura, M., Mochizuki, N., Kay, S. A., and Nagatani, A. (1999) Light-dependent translocation of a phytochrome B-GFP fusion protein to the nucleus in transgenic *Arabidopsis. J. Cell Biol.* 145: 437–445.

Chapter 18 | Blue-Light Responses: Stomatal Movements and Morphogenesis

MOST OF US ARE FAMILIAR with the observation that the branches of house plants placed near a window grow toward the incoming light. This phenomenon, called *phototropism*, is an example of how plants alter their growth patterns in response to the direction of incident radiation. This response to light is intrinsically different from light trapping by photosynthesis. In photosynthesis, plants harness light and convert it into chemical energy (see Chapters 7 and 8). In contrast, phototropism is an example of the use of light as an *environmental signal*. There are two major families of plant responses to light signals: the phytochrome responses, which were covered in Chapter 17, and the **blue-light responses**.

Some blue-light responses were introduced in Chapter 9—for example, chloroplast movement within cells in response to incident photon fluxes, and sun tracking by leaves. As with the family of the phytochrome responses, there are numerous plant responses to blue light. Besides phototropism, they include inhibition of hypocotyl elongation, stimulation of chlorophyll and carotenoid synthesis, activation of gene expression, stomatal movements, phototaxis (the movement of motile unicellular organisms such as algae and bacteria toward or away from light), enhancement of respiration, and anion uptake in algae (Senger 1984). Blue-light responses have been reported in higher plants, algae, ferns, fungi, and prokaryotes.

Some responses, such as electrical events at the plasma membrane, can be detected within seconds of irradiation by blue light. More complex metabolic or morphogenetic responses, such as blue light–stimulated pigment biosynthesis in the fungus *Neurospora* or branching in the alga *Vaucheria*, might require minutes, hours, or even days (Horwitz 1994).

Both chlorophylls and phytochrome absorb blue light (400–500 nm) from the visible spectrum, and other chromophores and some amino acids, such as tryptophan, absorb light in the ultraviolet (250–400 nm) region. How then can we functionally distinguish specific responses to blue light? Specific blue-light responses can be distinguished from photosynthetic responses by using red light, which stimulates photosynthetic responses and not blue-light responses. Phytochrome responses can be distinguished from blue light responses by testing red/far-red reversibility, which affects phytochrome responses and not blue-light responses.

Another key distinction is that *many blue-light responses of higher plants share a characteristic action spectrum*. You will recall from Chapter 7 that an action spectrum is a graph of the magnitude of the observed light response as a function of wavelength (see **Web Topic 7.1** for a detailed discussion of spectroscopy and action spectra). The action spectrum of the response can be compared with the *absorption spectra* of candidate photoreceptors. A close correspondence between action and absorption spectra provides a strong indication that the pigment under consideration is the photoreceptor mediating the light response under study (see Figure 7.8).

Action spectra for blue light–stimulated phototropism, stomatal movements, inhibition of hypocotyl elongation, and other key blue-light responses share a characteristic "three-finger" fine structure in the 400 to 500 nm region (Figure 18.1) that is not observed in spectra for responses to light that are mediated by photosynthesis, phytochrome, or other photoreceptors (Cosgrove 1994).

In this chapter we will describe representative blue-light responses in plants: phototropism, inhibition of stem elongation, and stomatal movements. The stomatal responses to blue light are discussed in detail because of the importance of stomata in leaf gas exchange (see Chapter 9) and in plant acclimations and adaptations to their environment. We will also discuss blue-light photoreceptors and the signal transduction cascade that links light perception with the final expression of blue-light sensing in the organism.

The Photophysiology of Blue-Light Responses

Blue-light signals are utilized by the plant in many responses, allowing the plant to sense the presence of light and its direction. This section describes the major morphological, physiological, and biochemical changes associated with typical blue-light responses.

Blue light stimulates asymmetric growth and bending

Directional growth toward (or in special circumstances away from) the light, is called **phototropism**. It can be observed in fungi, ferns, and higher plants. Phototropism is a **photomorphogenetic** response that is particularly dramatic in dark-grown seedlings of both monocots and dicots. Unilateral light is commonly used in experimental studies, but phototropism can also be observed when a seedling is exposed to two unequally bright light sources from different directions (Figure 18.2), a condition that can occur in nature.

As it grows through the soil, the shoot of a grass is protected by a modified leaf that covers it, called a **coleoptile** (Figure 18.3; see also Figure 19.1). As discussed in detail in Chapter 19, unequal light perception in the coleoptile results in unequal concentrations of auxin in the lighted and shaded sides of the coleoptile, in turn causing unequal growth and bending.

Keep in mind that phototropic bending occurs only in *growing* organs, and that coleoptiles and shoots that have stopped elongating will not bend when exposed to unilateral light. In grass seedlings growing in soil under sunlight, coleoptiles stop growing as soon as the shoot has emerged from the soil and the first true leaf has pierced the tip of the coleoptile.

On the other hand, dark-grown, *etiolated* coleoptiles continue to elongate at high rates for several days and, depending on the species, can attain several centimeters in length. The dramatic phototropic response of these etiolated coleoptiles (see Figure 18.3) has made them a classic model for studies of phototropism (Firn 1994).

The action spectrum shown in Figure 18.1 was obtained through measurement of the angles of curvature from oat

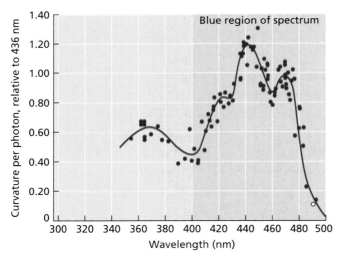

Figure 18.1 Action spectrum for blue light–stimulated phototropism in oat coleoptiles. An action spectrum shows the relationship between a biological response and the wavelengths of light absorbed. The "three-finger" pattern in the 400 to 500 nm region is characteristic of specific blue-light responses. (After Thimann and Curry 1960.)

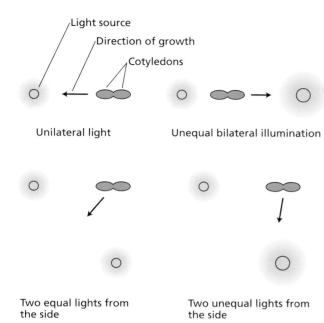

FIGURE 18.2 Relationship between direction of growth and unequal incident light. Cotyledons from a young seedling are shown as viewed from the top. The arrows indicate the direction of phototropic curvature. The diagrams illustrate how the direction of growth varies with the location and the intensity of the light source, but growth is always toward light. (After Firn 1994.)

coleoptiles that were irradiated with light of different wavelengths. The spectrum shows a peak at about 370 nm and the "three-finger" pattern in the 400 to 500 nm region discussed earlier. An action spectrum for phototropism in the dicot alfalfa (*Medicago sativa*) was found to be very similar to that of oat coleoptiles, suggesting that a common photoreceptor mediates phototropism in the two species.

Phototropism in sporangiophores of the mold *Phycomyces* has been studied to identify genes involved in phototropic responses. The sporangiophore consists of a sporangium (spore-bearing spherical structure) that develops on a stalk consisting of a single long cell. Growth in the sporangiophore is restricted to a growing zone just below the sporangium.

When irradiated with unilateral blue light, the sporangiophore bends toward the light, with an action spectrum similar to that of coleoptile phototropism (Cerda-Olmedo and Lipson 1987). These studies of *Phycomyces* have led to the isolation of many mutants with altered phototropic responses and the identification of several genes that are required for normal phototropism.

In recent years, phototropism of the stem of the small dicot Arabidopsis (Figure 18.4) has attracted much attention because of the ease with which advanced molecular techniques can be applied to Arabidopsis mutants. The genetics and the molecular biology of phototropism in Arabidopsis are discussed later in this chapter.

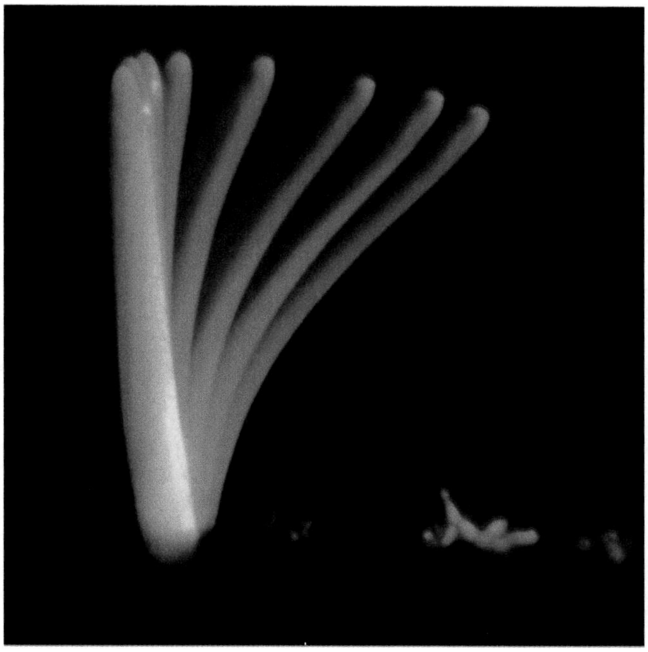

FIGURE 18.3 Time-lapse photograph of a corn coleoptile growing toward unilateral blue light given from the right. In the first image on the left, the coleoptile is about 3 cm. long. The consecutive exposures were made 30 minutes apart. Note the increasing angle of curvature as the coleoptile bends. (Courtesy of M. A. Quiñones.)

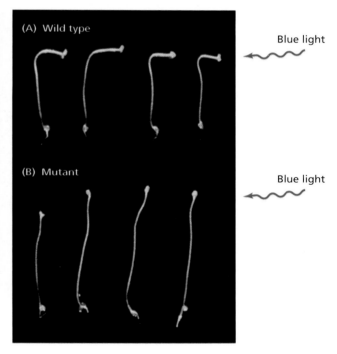

FIGURE 18.4 Phototropism in wild type (A) and mutant (B) Arabidopsis seedlings. Unilateral light was applied from the right. (Courtesy of Dr. Eva Huala.)

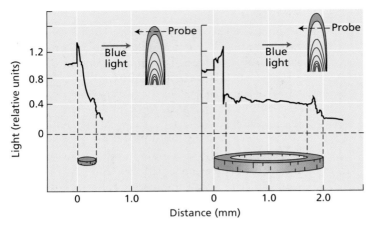

FIGURE 18.5 Distribution of transmitted, 450 nm blue light in an etiolated corn coleoptile. The diagram in the upper right of each frame shows the area of the coleoptile being measured by a fiber-optic probe. A cross section of the tissue appears at the bottom of each frame. The trace above it shows the amount of light sensed by the probe at each point. A sensing mechanism that depended on light gradients would sense the difference in the amount of light between the lighted and shaded sides of the coleoptile, and this information would be transduced into an unequal auxin concentration and bending. (After Vogelmann and Haupt 1985.)

How do plants sense the direction of the light signal?

Light gradients between lighted and shaded sides have been measured in coleoptiles of monocots and in hypocotyls of dicot seedlings irradiated with unilateral blue light. When a coleoptile is illuminated with 450-nm blue light, the ratio between the light that is incident to the surface of the illuminated side and the light that reaches the shaded side is 4:1 at the tip and the midregion of the coleoptile, and 8:1 at the base (Figure 18.5).

On the other hand, there is a *lens effect* in the sporangiophore of the mold *Phycomyces* irradiated with unilateral blue light, and as a result, the light measured at the distal cell surface of the sporangiophore is about twice the amount of light that is incident at the surface of the illuminated side. Light gradients and lens effects could play a role in how the bending organ senses the direction of the unilateral light (Vogelmann 1994).

Blue light rapidly inhibits stem elongation

The stems of seedlings growing in the dark elongate very rapidly, and the inhibition of stem elongation by light is a key morphogenetic response of the seedling emerging from the soil surface (see Chapter 17). The conversion of Pr to Pfr (the red- and far red–absorbing forms of phytochrome, respectively) in etiolated seedlings causes a phytochrome-dependent, sharp decrease in elongation rates (see Figure 17.1).

Note, however, that action spectra for the decrease in elongation rate show strong activity in the blue region, which cannot be explained by the absorption properties of phytochrome (see Figure 17.6). In fact, the 400- to 500-nm blue region of the action spectrum for the inhibition of stem elongation closely resembles that of phototropism (compare the action spectra in Figures 17.6 and 18.1).

It is possible to experimentally separate a reduction in elongation rates mediated by phytochrome from a reduction mediated by a specific blue-light response. If lettuce seedlings are given low fluence rates of blue light under a strong background of yellow light, their hypocotyl elongation rate is reduced by more than 50%. The background yellow light establishes a well-defined Pr:Pfr ratio (see Chapter 17). In such conditions, the low fluence rates of blue light added are too small to significantly change this ratio, ruling out a phytochrome effect on the reduction in elongation rate observed upon the addition of blue light. These results indicate that besides phytochrome, the elongation rate of the hypocotyl is also controlled by a specific blue light response.

A specific blue light–mediated hypocotyl response can also be distinguished from one mediated by phytochrome by their contrasting time courses. Whereas phytochrome-mediated changes in elongation rates can be detected within 8 to 90 minutes, depending on the species, blue-light responses are rapid, and can be measured within 15 to 30 s (Figure 18.6). Interactions between phytochrome and the blue light–dependent sensory transduction cascade in the regulation of elongation rates will be described later in the chapter.

Another fast response elicited by blue light is a depolarization of the membrane of hypocotyl cells that precedes the inhibition of growth rate (see Figure 18.6). The membrane depolarization is caused by the activation of anion channels (see Chapter 6), which facilitates the efflux of anions such as chloride. Application of an anion channel blocker prevents the blue light–dependent membrane depolarization and decreases the inhibitory effect of blue light on hypocotyl elongation (Spalding 2000).

Blue light regulates gene expression

Blue light also regulates the expression of genes involved in several important morphogenetic processes. Some of these light-activated genes have been studied in detail—for example, the genes that encode the enzyme chalcone synthase (which catalyzes the first committed step in flavonoid biosynthesis [see Chapter 13]), the small subunit of rubisco (see Chapter 8), and the proteins that bind chlorophylls *a* and *b* (see Chapter 7). Most of the studies on light-activated genes show sensitivity to both blue and red light, as well as red/far-red reversibility, implicating both phytochrome and specific blue-light responses.

The nuclear gene *SIG5*, one of six *SIG* nuclear genes in Arabidopsis that play a regulatory role in the transcription of the chloroplast gene *psbD*, which encodes the D2 subunit of the PSII reaction center (see Chapter 7), is specifically activated by blue light (Tsunoyama et al. 2004). In contrast,

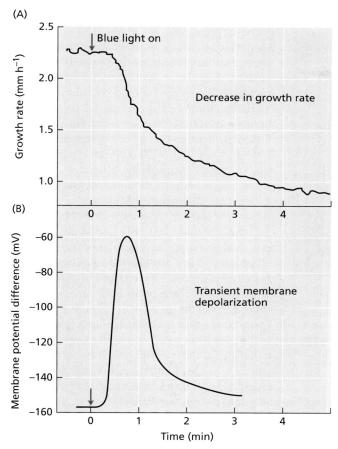

FIGURE 18.6 Blue light–induced (A) changes in elongation rates of etiolated cucumber seedlings and (B) transient membrane depolarization of hypocotyl cells. As the membrane depolarization (measured with intracellular electrodes) reaches its maximum, growth rate (measured with position transducers) declines sharply. Comparison of the two curves shows that the membrane starts to depolarize before the growth rate begins to decline, suggesting a cause–effect relation between the two phenomena. (After Spalding and Cosgrove 1989.)

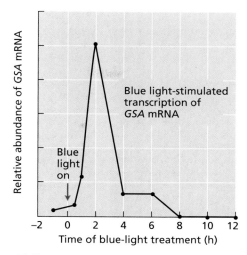

FIGURE 18.7 Time course of blue light–dependent gene expression in *Chlamydomonas reinhardtii*. The *GSA* gene encodes the enzyme glutamate-1-semialdehyde aminotransferase, which regulates an early step in chlorophyll biosynthesis. (After Matters and Beale 1995.)

the other five *SIG* genes are activated by both blue and red light.

Another well-documented instance of gene expression that is mediated solely by a blue light–sensing system involves the *GSA* gene in the photosynthetic unicellular alga *Chlamydomonas reinhardtii* (Im and Beale 2000) This gene encodes the enzyme glutamate-1-semialdehyde aminotransferase (GSA), a key enzyme in the chlorophyll biosynthesis pathway (see Chapter 7). The absence of phytochrome in *C. reinhardtii* simplifies the analysis of blue-light responses in this experimental system.

In synchronized cultures of *C. reinhardtii*, levels of *GSA* mRNA are strictly regulated by blue light, and 2 hours after the onset of illumination, *GSA* mRNA levels are 26-fold higher than they are in the dark (Figure 18.7). These blue light–mediated mRNA increases precede increases in chlorophyll content, indicating that chlorophyll biosynthesis is regulated by activation of the *GSA* gene.

Blue light stimulates stomatal opening

We now turn to the stomatal response to blue light. Stomata have a major regulatory role in gas exchange in leaves (see Chapter 9), and they can often affect yields of agricultural crops (see Chapter 26). Several characteristics of blue light–dependent stomatal movements make guard cells a valuable experimental system for the study of blue-light responses:

- The stomatal response to blue light is rapid and reversible, and it is localized in a single cell type, the guard cell.

- The stomatal response to blue light regulates stomatal movements throughout the life of the plant. This is unlike phototropism or hypocotyl elongation, which are functionally important only at early stages of development.

- The signal transduction cascade that links the perception of blue light with the opening of stomata is understood in considerable detail.

In the following sections we will discuss two central aspects of the stomatal response to light: the osmoregulatory mechanisms that drive stomatal movements, and the role of a blue light–activated H^+-ATPase in ion uptake by guard cells.

Light is the dominant environmental signal controlling stomatal movements in leaves of well-watered plants growing in a natural environment. Stomata open as light levels reaching the leaf surface increase, and close as they

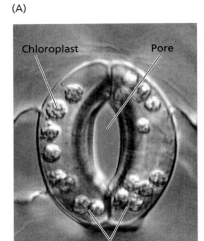
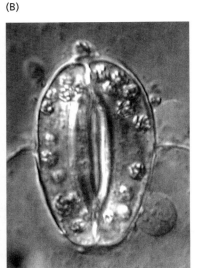

FIGURE 18.8 Light-stimulated stomatal opening in detached epidermis of *Vicia faba*. Open, light-treated stoma (A), is shown in the dark-treated, closed state in (B). Stomatal opening is quantified by microscopic measurement of the width of the stomatal pore. (Courtesy of E. Raveh.)

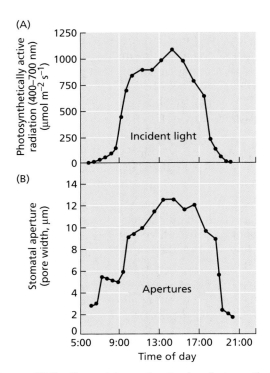

FIGURE 18.9 Stomatal opening tracks photosynthetic active radiation at the leaf surface. Stomatal opening in the lower surface of leaves of *V. faba* grown in a greenhouse, measured as the width of the stomatal pore (A), closely follows the levels of photosynthetically active radiation (400–700 nm) incident to the leaf (B), indicating that the response to light was the dominant response regulating stomatal opening. (After Srivastava and Zeiger 1995a.)

decrease (Figure 18.8). In greenhouse-grown leaves of broad bean (*Vicia faba*), stomatal movements closely track incident solar radiation at the leaf surface (Figure 18.9). This light dependence of stomatal movements has been documented in many species and conditions.

Early studies of the stomatal response to light showed that dichlorophenyldimethylurea (DCMU), an inhibitor of photosynthetic electron transport (see Figure 7.30), causes a partial inhibition of light-stimulated stomatal opening. These results indicate that photosynthesis in the guard cell chloroplast plays a role in light-dependent stomatal opening, but the observation that the inhibition was only partial pointed to a nonphotosynthetic component of the stomatal response to light. Detailed studies of the light response of stomata have shown that light activates two distinct responses of guard cells: photosynthesis in the guard cell chloroplast (see **Web Essay 18.1**), and a specific blue-light response.

Illumination with blue light cannot be used to properly characterize the specific stomatal response to blue light, because blue light simultaneously stimulates both the specific blue-light response and guard cell photosynthesis (for the photosynthetic response to blue light, see the action spectrum for photosynthesis in Figure 7.8). A clear-cut separation of the two light responses can be obtained in dual-beam experiments. High fluence rates of red light are used to *saturate* the photosynthetic response, and low photon fluxes of blue light are added after the response to the saturating red light has been completed (Figure 18.10). The addition of blue light causes substantial further stomatal opening that cannot be explained as a further stimulation of guard cell photosynthesis, because photosynthesis is saturated by the background red light.

An action spectrum for the stomatal response to blue light under background red illumination shows the three-finger pattern discussed earlier (Figure 18.11). This action spectrum, typical of blue-light responses and distinctly different from the action spectrum for photosynthesis, further indicates that, in addition to photosynthesis, guard cells respond specifically to blue light.

When guard cells are treated with cellulolytic enzymes that digest the cell walls, **guard cell protoplasts** are released. Guard cell protoplasts swell when illuminated with blue light (Figure 18.12), indicating that blue light is sensed within the guard cells proper. The swelling of guard cell protoplasts also illustrates how intact guard cells function. The light-stimulated uptake of ions and the accumulation of organic solutes decrease the cell's osmotic potential (increase the osmotic pressure). Water flows in as a result, leading to an increase in turgor that in guard cells with intact walls is mechanically transduced into an

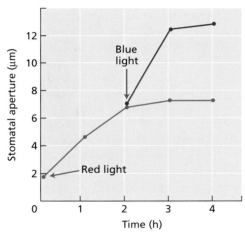

FIGURE 18.10 The response of stomata to blue light under a red-light background. Stomata from detached epidermis of common dayflower (*Commelina communis*) were treated with saturating photon fluxes of red light (red trace). In a parallel treatment, stomata illuminated with red light were also illuminated with blue light, as indicated by the arrow (blue trace). The increase in stomatal opening above the level reached in the presence of saturating red light indicates that a different photoreceptor system, stimulated by blue light, is mediating the additional increases in opening. (From Schwartz and Zeiger 1984.)

increase in stomatal apertures (see Chapter 4). In the absence of a cell wall, the blue light–mediated increase in osmotic pressure causes the guard cell protoplast to swell.

Blue light activates a proton pump at the guard cell plasma membrane

When guard cell protoplasts from broad bean (*V. faba*) are irradiated with blue light under background red-light illumination, the pH of the suspension medium becomes more acidic (Figure 18.13). This blue light–induced acidification is blocked by inhibitors that dissipate pH gradients, such as CCCP (discussed shortly), and by inhibitors of the pro-

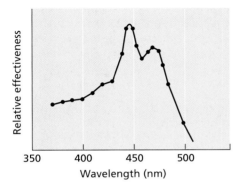

FIGURE 18.11 The action spectrum for blue light–stimulated stomatal opening (under a red-light background). (After Karlsson 1986.)

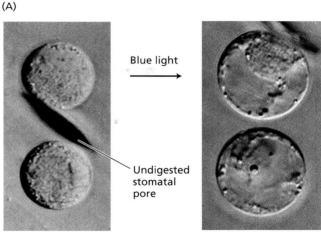

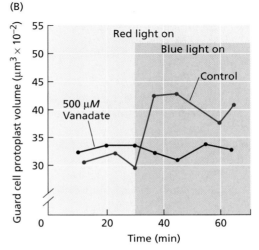

FIGURE 18.12 Blue light–stimulated swelling of guard cell protoplasts. (A) In the absence of a rigid cell wall, guard cell protoplasts of onion (*Allium cepa*) swell. (B) Blue light stimulates the swelling of guard cell protoplasts of broad bean (*V. faba*), and vanadate, an inhibitor of the H^+-ATPase, inhibits this swelling. Blue light stimulates ion and water uptake in the guard cell protoplasts, which in the intact guard cells provides a mechanical force that drives increases in stomatal apertures. (A from Zeiger and Hepler 1977; B after Amodeo et al. 1992.)

ton-pumping H^+-ATPase, such as vanadate (see Figure 18.12B) (see Chapter 6).

This indicates that *the acidification results from the activation by blue light of a proton-pumping ATPase in the guard cell plasma membrane* that extrudes protons into the protoplast suspension medium and lowers its pH. In the intact leaf, this blue-light stimulation of proton pumping lowers the pH of the apoplastic space surrounding the guard cells. The plasma membrane ATPase from guard cells has been isolated and extensively characterized (Kinoshita and Shimazaki 2001).

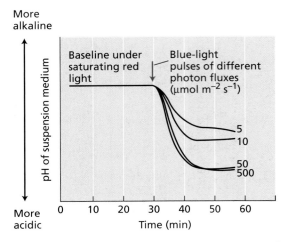

FIGURE 18.13 Acidification of a suspension medium of guard cell protoplasts of *V. faba* stimulated by a 30 s pulse of blue light. The acidification results from the stimulation of an H⁺-ATPase at the plasma membrane by blue light, and it is associated with protoplast swelling (see Figure 18.12). (After Shimazaki et al. 1986.)

ward electric current, which is abolished by the proton ionophore carbonyl cyanide *m*-chlorophenylhydrazone (CCCP). This proton ionophore makes the plasma membrane highly permeable to protons, thus precluding the formation of a proton gradient across the membrane and abolishing net proton efflux.

The relationship between proton pumping at the guard cell plasma membrane and stomatal opening is evident from the observations that fusicoccin stimulates both proton extrusion from guard cell protoplasts and stomatal opening, and that CCCP inhibits the fusicoccin-stimulated opening. The increase in proton-pumping rates as a function of fluence rates of blue light (see Figure 18.13) indicates that the increasing rates of blue photons in the solar radiation reaching the leaf cause a larger stomatal opening.

The close relationship among the number of incident blue-light photons, proton pumping at the guard cell plasma membrane, and stomatal opening further suggests that the blue-light response of stomata might function as a sensor of photon fluxes reaching the guard cell.

Pulses of blue light given under a saturating red-light background also stimulate an outward electric current from guard cell protoplasts (see Figure 18.14B). The acidification measurements shown in Figure 18.13 indicate that the outward electric current measured in patch clamp experiments is carried by protons.

Blue-light responses have characteristic kinetics and lag times

Some of the characteristics of the responses to blue-light pulses underscore important properties of blue-light responses: the persistence of the response after the light signal has been switched off, and a significant lag time separating the onset of the light signal and the beginning of the response.

In contrast to typical photosynthetic responses, which are activated very quickly after a "light on" signal, and cease when the light goes off (see, for instance, Figure 7.13), blue-light responses proceed at maximal rates for several minutes after application of the pulse (see Figure 18.14B). This property can be explained by a physiologically inactive form of the blue-light photoreceptor that is converted to an active form by blue light, with the active form reverting slowly to the physiologically inactive form in the absence of blue light

The activation of electrogenic pumps such as the proton-pumping ATPase can be measured in patch-clamping experiments as an outward electric current at the plasma membrane (see **Web Topic 6.2** for a description of patch clamping). A patch clamp recording of a guard cell protoplast treated with the fungal toxin fusicoccin, a well-characterized activator of plasma membrane ATPases, is shown in Figure 18.14A. Exposure to fusicoccin stimulates an out-

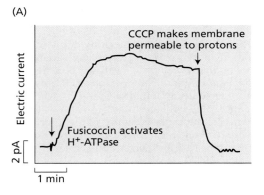

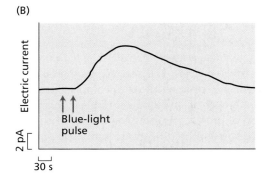

FIGURE 18.14 Activation of the H⁺-ATPase at the plasma membrane of guard cell protoplasts by fusicoccin and blue light can be measured as electric current in patch clamp experiments. (A) Outward electric current (measured in picoamps, pA) at the plasma membrane of a guard cell protoplast stimulated by the fungal toxin fusicoccin, an activator of the H⁺-ATPase. The current is abolished by the proton ionophore carbonyl cyanide *m*-chlorophenylhydrazone (CCCP). (B) Outward electric current at the plasma membrane of a guard cell protoplast stimulated by a blue-light pulse. These results indicate that blue light stimulates the H⁺-ATPase. (A after Serrano et al. 1988; B after Assmann et al. 1985.)

(Iino et al. 1985). The rate of the response to a blue-light pulse would thus depend on the time course of the reversion of the active form to the inactive one.

Another property of the response to blue-light pulses is a lag time, which lasts about 25 s in both the acidification response and the outward electric currents stimulated by blue light (see Figures 18.13 and 18.14). This amount of time is probably required for the signal transduction cascade to proceed from the photoreceptor site to the proton-pumping ATPase and for the proton gradient to form. Similar lag times have been measured for blue light–dependent inhibition of hypocotyl elongation, which was discussed earlier.

Blue light regulates osmotic relations of guard cells

Blue light modulates guard cell osmoregulation not only via its activation of proton pumping, but also via the stimulation of the synthesis of organic solutes. Before discussing these blue-light responses, let us briefly describe the major osmotically active solutes in guard cells.

The botanist Hugo von Mohl proposed in 1856 that turgor changes in guard cells provide the mechanical force for changes in stomatal apertures. The plant physiologist F. E. Lloyd hypothesized in 1908 that guard cell turgor is regulated by osmotic changes resulting from starch–sugar interconversions, a concept that led to a starch–sugar hypothesis of stomatal movements. The discovery of the changes in potassium concentrations in guard cells led to the modern theory of guard cell osmoregulation by potassium and its counterions.

Potassium concentration in guard cells increases severalfold when stomata open, from 100 mM in the closed state to 400 to 800 mM in the open state, depending on the plant species and the experimental conditions. In most species, these large concentration changes in K$^+$ are electrically balanced by varying amounts of the anions Cl$^-$ and malate^{2-} (Figure 18.15A) (Talbott et al. 1996). In species of the genus *Allium*, such as onion (*A. cepa*), K$^+$ ions are balanced solely by Cl$^-$.

Chloride ions are taken up into the guard cells during stomatal opening and extruded during stomatal closing. Malate, on the other hand, is synthesized in the guard cell cytosol, in a metabolic pathway that uses carbon skeletons generated by starch hydrolysis (see Figure 18.15B). The malate content of guard cells decreases during stomatal closing, but it remains to be established whether malate is catabolized in mitochondrial respiration or is extruded into the apoplast.

Potassium and chloride are taken up into guard cells via secondary transport mechanisms driven by the gradient of electrochemical potential for H$^+$, $\Delta\mu_{H^+}$, generated by the proton pump (see Chapter 6) discussed earlier in the chapter. Proton extrusion makes the electric-potential difference across the guard cell plasma membrane more negative; light-dependent hyperpolarizations as high as 64 mV have been measured (Roelfsema et al. 2001). In addition, proton pumping generates a pH gradient of about 0.5 to 1 pH unit.

The electrical component of the proton gradient provides a driving force for the passive uptake of potassium ions via voltage-regulated potassium channels (see Chapter 6) (Schroeder et al. 2001). Chloride is thought to be taken up through a proton–chloride symporter. Thus, blue light–dependent stimulation of proton pumping plays a key role in guard cell osmoregulation during light-dependent stomatal movements.

Guard cell chloroplasts (see Figure 18.8) contain large starch grains, and their starch content decreases during stomatal opening and increases during closing. Starch, an insoluble, high-molecular-weight polymer of glucose, does not contribute to the cell's osmotic potential, but the hydrolysis of starch into soluble sugars causes a decrease in the osmotic potential (or increase in osmotic pressure) of guard cells. In the reverse process, starch synthesis decreases the sugar concentration, resulting in an increase of the cell's osmotic potential, which the starch–sugar hypothesis predicted to be associated with stomatal closing.

With the discovery of the major role of potassium and its counterions in guard cell osmoregulation, the sugar–starch hypothesis was no longer considered important (Outlaw 1983). However, studies described in the next section have characterized a major osmoregulatory phase of guard cells in which sucrose is the dominant solute regulating guard cell osmoregulation.

Sucrose is an osmotically active solute in guard cells

Studies of daily courses of stomatal movements in intact leaves have shown that the potassium content in guard cells increases in parallel with early-morning opening, but it decreases in the early afternoon under conditions in which apertures continue to increase. In contrast, sucrose content increases slowly in the morning, and upon potassium efflux, sucrose becomes the dominant osmotically active solute in guard cells. Stomatal closing at the end of the day parallels a decrease in the sucrose content of guard cells (Figure 18.16) (Talbott and Zeiger 1998).

These osmoregulatory features indicate that stomatal opening is associated primarily with K$^+$ uptake, and closing is associated with a decrease in sucrose content (see Figure 18.16). The need for distinct potassium- and sucrose-dominated osmoregulatory phases is unclear, but it might underlie regulatory aspects of stomatal function. Potassium might be the solute of choice for the consistent daily opening that occurs at sunrise. The sucrose phase might be associated with the coordination of stomatal movements in the epidermis with rates of photosynthesis in the mesophyll.

Where do osmotically active solutes originate? Four distinct metabolic pathways that can supply osmotically active solutes to guard cells have been characterized (see Figure 18.15):

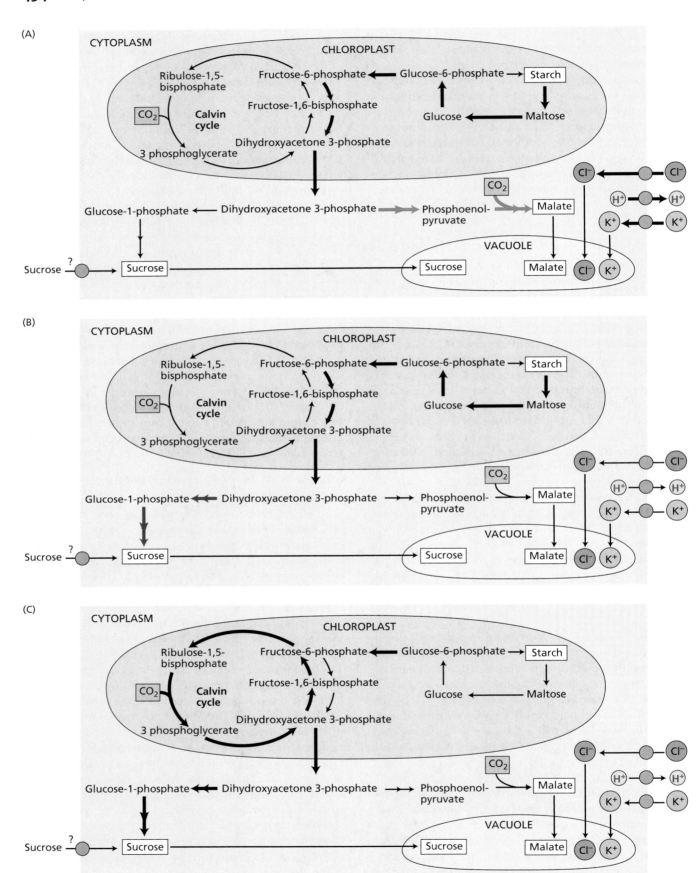

◀ **FIGURE 18.15** Three distinct osmoregulatory pathways in guard cells. The dark arrows identify the major metabolic steps of each pathway that lead to the accumulation of osmotically active solutes in the guard cells. (A) Potassium and its counterions. Potassium and chloride are taken up in secondary transport processes driven by a proton gradient; malate is formed from the hydrolysis of starch. (B) Accumulation of sucrose from starch hydrolysis. (C) Accumulation of sucrose from photosynthetic carbon fixation. The possible uptake of apoplastic sucrose is also indicated. (From Talbott and Zeiger 1998.)

1. The uptake of K^+ and Cl^- coupled to the biosynthesis of malate^{2-}
2. The production of sucrose from starch hydrolysis
3. The production of sucrose by photosynthetic carbon fixation in the guard cell chloroplast
4. The uptake of sucrose generated by mesophyll photosynthesis

Depending on environmental conditions, one or several pathways may be activated. For instance, red light–stimulated stomatal opening in detached epidermis kept under ambient CO_2 concentrations depends solely on sucrose generated by photosynthetic carbon fixation in the guard cell chloroplast, with no detectable K^+ uptake. The other osmoregulatory pathways can be selectively activated under different experimental conditions (see **Web Topic 18.1**). Current studies are unraveling the intricacy of guard cell osmoregulation in the intact leaf (Roelfsema et al. 2002; Outlaw 2003; Fan et al. 2004; Tallman 2004).

A remarkable feature emerging from these studies is the unusual plasticity of the guard cells. These plastic features include acclimations of the responses to blue light and CO_2, and daily changes in guard cell photosynthetic rates (Zeiger et al. 2002).

Blue-Light Photoreceptors

Experiments carried out by Charles Darwin and his son Francis in the nineteenth century determined that during phototropism, blue light is perceived in the coleoptile tip. Early hypotheses about blue-light photoreceptors focused on carotenoids and flavins. Despite active research efforts, no significant advances toward the identification of blue-light photoreceptors were made until the early 1990s. In the case of phototropism and the inhibition of stem elongation, progress resulted from the identification of mutants for key blue-light responses, and the subsequent isolation of the relevant gene.

In the following section we will describe three photoreceptors associated with blue-light responses: cryptochromes, phototropins, and zeaxanthin.

Cryptochromes are involved in the inhibition of stem elongation

The *hy4* mutant of Arabidopsis lacks the blue light–stimulated inhibition of hypocotyl elongation described earlier in the chapter. As a result of this genetic defect, *hy4* plants show an elongated hypocotyl when irradiated with blue light. Isolation of the *HY4* gene showed that it encodes a 75-kDa protein with significant sequence homology to microbial DNA **photolyase**, a blue light–activated enzyme that repairs pyrimidine dimers in DNA formed as a result of exposure to ultraviolet radiation (Ahmad and Cashmore 1993). In view of this sequence similarity, the HY4 protein, later renamed **cryptochrome 1 (CRY1)**, was proposed to be a blue-light photoreceptor mediating the inhibition of stem elongation.

Photolyases are pigment proteins that contain a flavin adenine dinucleotide (FAD; see Figure 11.2B) and a pterin. **Pterins** are light-absorbing, pteridine derivatives often found in pigmented cells of insects, fishes, and birds (see Chapter 12 for pterin structure). When expressed in the bacterium *Escherichia coli*, the CRY1 protein binds FAD and a pterin, but it lacks detectable photolyase activity. Despite intensive investigation, it has not been possible to obtain direct evidence for a role of flavins or pterins in cryptochrome functioning of intact cells (Ahmad et al. 2002)

Important evidence for a role of CRY1 in blue light–mediated inhibition of stem elongation comes from overexpression studies. Over-

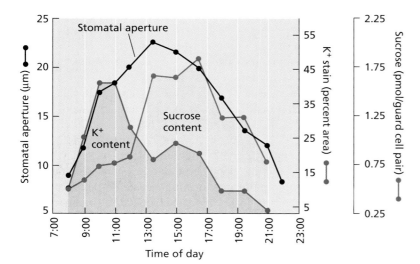

FIGURE 18.16 Daily course of changes in stomatal aperture, and in potassium and sucrose content, of guard cells from intact leaves of broad bean (*V. faba*). These results indicate that the changes in osmotic potential required for stomatal opening in the morning are mediated by potassium and its counterions, whereas the afternoon changes are mediated by sucrose. (After Talbott and Zeiger 1998.)

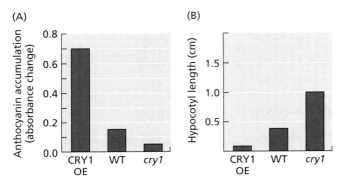

FIGURE 18.17 Blue light stimulates the accumulation of anthocyanin (A) and the inhibition of stem elongation (B) in transgenic and mutant seedlings of Arabidopsis. These bar graphs show a transgenic phenotype overexpressing the gene that encodes CRY1 (CRY1 OE), the wild type (WT), and *cry1* mutants. The enhanced blue-light response of the transgenic plant overexpressing the gene that encodes CRY1 demonstrates the important role of this gene product in stimulating anthocyanin biosynthesis and inhibiting stem elongation. (After Ahmad et al. 1998.)

expression of the CRY1 protein in transgenic tobacco or Arabidopsis plants results in a stronger blue light–stimulated inhibition of hypocotyl elongation than in the wild type, as well as increased production of anthocyanin, another well-known blue-light response (Figure 18.17). Thus, overexpression of CRY1 caused an enhanced sensitivity to blue light in transgenic plants.

A second gene product homologous to CRY1, named CRY2, has been isolated from Arabidopsis (Lin 2000). Both CRY1 and CRY2 appear to be ubiquitous throughout the plant kingdom. A major difference between them is that CRY2 is rapidly degraded in the light, whereas CRY1 is stable in light-grown seedlings.

Transgenic plants overexpressing the *cry2* gene show a small enhancement of the inhibition of hypocotyl elongation, indicating that unlike CRY1, CRY2 does not play a primary role in inhibiting stem elongation. On the other hand, transgenic plants overexpressing the gene that encodes CRY2 show a large increase in blue light–stimulated cotyledon expansion, yet another blue-light response. In addition, CRY1 is involved in the setting of the circadian clock in Arabidopsis (see Chapter 17), and both CRY1 and CRY2 play a role in the induction of flowering (see Chapter 25). Cryptochrome homologs regulate the circadian clock in *Drosophila*, mouse, and humans.

Both CRY1 and CRY2 are detectable in the nucleus upon illumination, and they interact with COP1, the ubiquitin ligase discussed in Chapter 17, both in vivo and in vitro. In addition, the activity of cryptochrome is influenced by its phosphorylation state. CRY1 and CRY2 have been shown to interact with phytochrome A in vivo, and to be phosphorylated by phytochrome A in vitro (see Chapter 17 and **Web Essay 18.2**). These studies indicate that both phosphory-

lation and protein degradation are important steps in the sensory transduction of cryptochrome mediated blue-light signals (Spalding and Folta 2005).

Some of the blue light–mediated cryptochrome functioning seems to occur in the cytoplasm, because one of the earliest detected defects in *cry1* mutant seedlings is impaired activation of anion channels at the plasma membrane (Spalding 2000). Activation of anion channels is an early step in the sensory transduction cascade mediating hypocotyls elongation. The mechanism involved in this blue light-dependent anion-channel activation is not yet known.

Phototropins mediate blue light–dependent phototropism and chloroplast movements

Arabidopsis mutants impaired in blue light–dependent phototropism of the hypocotyl have provided valuable information about cellular events preceding bending. One of these mutants, the *nph1* (*n*on*p*hototropic *h*ypocotyl) mutant, has been found to be genetically independent of the *hy4* (*cry1*) mutant discussed earlier: The *nph1* mutant lacks a phototropic response in the hypocotyl, but has normal blue light–stimulated inhibition of hypocotyl elongation, while *hy4* has the converse phenotype. The *nph1* gene was renamed *phot1*, and the protein it encodes was named **phototropin** (Briggs and Christie 2002).

The C-terminal half of phototropin is a serine/threonine kinase. The N-terminal half contains two similar domains, called LOV domains, of about 100 amino acids each. The LOV domains bind flavins and have sequence similarity to proteins involved in signaling in bacteria and mammals. These proteins are oxygen sensors in *E. coli* and *Azotobacter* and voltage sensors in potassium channels of *Drosophila* and vertebrates.

The N-terminal half of phototropin binds flavin mononucleotide (FMN) (see Figure 11.2B) and undergoes a blue light–dependent autophosphorylation reaction. This reaction resembles the blue light–dependent phosphorylation of a 120-kDa membrane protein found in growing regions of etiolated seedlings. The cellular events that follow the autophosphorylation of phototropin remain unknown.

Recent spectroscopic studies have shown that, in the dark, a FMN molecule is noncovalently bound to each LOV domain. Upon blue light illumination, the FMN molecule becomes covalently bound to a cysteine residue in the phototropin molecule, forming a cysteine-flavin covalent adduct (Figure 18.18) (Swartz et al. 2002). This reaction is reversed by a dark treatment.

The Arabidopsis genome contains a second gene, *phot2*, which is related to *phot1*. The *phot1* mutant lacks hypocotyl phototropism in response to low-intensity blue light (0.01–1 $\mu mol\ mol^{-2}\ s^{-1}$) but retains a phototropic response at higher intensities (1–10 $\mu mol\ m^{-2}\ s^{-1}$). The *phot2* mutant has a normal phototropic response, but the *phot1*/*phot2* double mutant is severely impaired at both low and high intensities. These data indicate that both *phot1* and *phot2* are

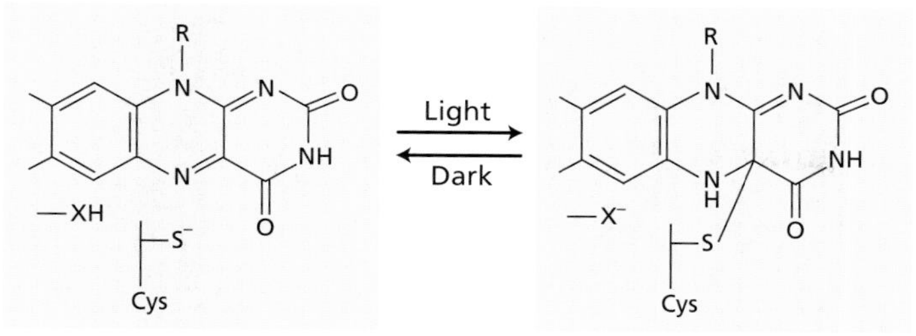

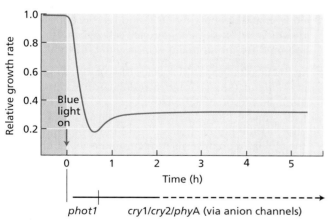

FIGURE 18.18 Adduct formation of FMN and a cysteine residue of phototropin protein upon blue-light irradiation. XH and X⁻ represent an unidentified proton donor and acceptor, respectively. (After Briggs and Christie 2002.)

involved in the phototropic response, with *phot2* functioning at high light fluence rates.

PHOTORECEPTOR INTERACTION IN THE INHIBITION OF STEM ELONGANTION High-resolution analysis of the changes in growth rate mediating the inhibition of hypocotyl elongation by blue light has provided valuable information about the interactions among phototropin, CRY1, CRY2, and the phytochrome PHYA (Parks et al. 2001). After a lag of 30 s, blue light–treated, wild-type Arabidopsis seedlings show a rapid decrease in elongation rates during the first 30 minutes, and then they grow very slowly for several days (Figure 18.19).

FIGURE 18.19 Sensory transduction cascade of blue light–stimulated inhibition of stem elongation in Arabidopsis. Elongation rates in the dark (0.25 mm h⁻¹) were normalized to 1. Within 30 s of the onset of blue-light irradiation, growth rates decreased and approached zero within 30 minutes, then continued at very reduced rates for several days. If blue light is applied to a *phot1* mutant, dark-growth rates remain unchanged for the first 30 minutes, indicating that the inhibition of elongation in the first 30 minutes is under phototropin control. Similar experiments with *cry1*, *cry2*, and *phyA* mutants indicate that the respective gene products control elongation rates at later times. (After Parks et al. 2001.)

Analysis of the same response in *phot1*, *cry1*, *cry2*, and *phyA* mutants has shown that suppression of stem elongation by blue light during seedling de-etiolation is initiated by *phot1*, with *cry1*, and to a limited extent *cry2*, modulating the response after 30 minutes. The slow growth rate of stems in blue light–treated seedlings is primarily a result of the persistent action of CRY1, and this is the reason that *cry1* mutants of Arabidopsis show a long hypocotyl, compared to the short hypocotyl of the wild type. There is also a role for phytochrome A in at least the early stages of blue light–regulated growth, because growth inhibition does not progress normally in *phyA* mutants.

BLUE LIGHT–ACTIVATED CHLOROPLAST MOVEMENT Leaves show an adaptive feature that can alter the intracellular distribution of their chloroplasts to control light absorption and prevent photodamage (see Figure 9.14). Chloroplast movement is triggered by a light signal, and the action spectrum for the response shows the "three finger" fine structure typical of specific blue-light responses. When the amount of light reaching the leaf is weak, chloroplasts gather at the upper and lower surfaces of the mesophyll cells (the "accumulation" response; see Figure 9.14B), thus maximizing light absorption.

Under strong light, the chloroplasts move to the cell surfaces that are parallel to the incident light (the "avoidance" response; see Figure 9.14C), thus minimizing light absorption. Recent studies have shown that mesophyll cells of the *phot1* mutant have a normal avoidance response and a rudimentary accumulation response. Cells from the *phot2* mutant show a normal accumulation response but lack the avoidance response. Cells from the *phot1/phot2* double mutant lack both the avoidance and accumulation responses (Wada et al. 2003). These results indicate that *phot2* plays a key role in the avoidance response, and that both *phot1* and *phot2* contribute to the accumulation response.

The carotenoid zeaxanthin mediates blue-light photoreception in guard cells

Recall from Chapters 7 and 9 that zeaxanthin is one of the three components of the xanthophyll cycle of chloroplasts, which protects photosynthetic pigments from excess excitation energy. In guard cells, however, the changes in zeaxanthin content as a function of incident radiation are distinctly different from the changes in mesophyll cells (Figure 18.20). For instance, in sun plants such as *V. faba*, zeaxanthin accumulation in the mesophyll begins at about 200 μmol m⁻² s⁻¹, and there is no detectable zeaxanthin in the early morning or late afternoon. In contrast, the zeaxanthin content in guard cells closely follows incident solar radiation at the leaf surface throughout the day, and it is nearly linearly proportional to incident photon fluxes in the early morning and late afternoon.

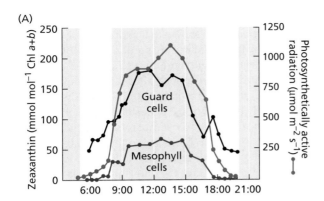

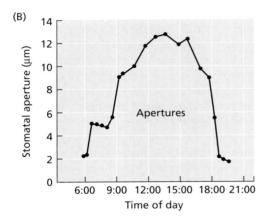

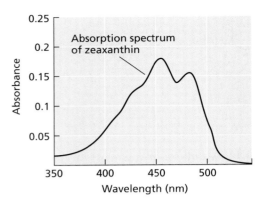

FIGURE 18.20 The zeaxanthin content of guard cells closely tracks photosynthetically active radiation and stomatal apertures. (A) Daily course of photosynthetic active radiation reaching the leaf surface, and of zeaxanthin content of guard cells (black tracing) and mesophyll cells (green tracing) of *V. faba* leaves grown in a greenhouse. The white areas within the graph highlight the contrasting sensitivity of the xanthophyll cycle in mesophyll and guard cell chloroplasts under the low irradiances prevailing early and late in the day. (B) Stomatal apertures in the same leaves used to measure guard cell zeaxanthin content. (After Srivastava and Zeiger 1995a.)

FIGURE 18.21 The absorption spectrum of zeaxanthin in ethanol.

Several studies indicate that zeaxanthin plays a central role in blue light-stimulated stomatal opening:

- The absorption spectrum of zeaxanthin (Figure 18.21) closely matches the action spectrum for blue light–stimulated stomatal opening (see Figure 18.11).

- In daily courses of stomatal opening in intact leaves grown in a greenhouse, incident radiation, zeaxanthin content of guard cells, and stomatal apertures are closely related (see Figure 18.20).

- The blue-light sensitivity of guard cells increases as a function of their zeaxanthin concentration. Zeaxanthin concentration in guard cells can be increased by a red-light treatment. When guard cells from epidermal peels illuminated with increasing fluence rates of red light are exposed to blue light, the resulting blue light–stimulated stomatal opening is linearly related to the fluence rate of background red-light irradiation and to zeaxanthin content (Srivastava and Zeiger 1995b). The same relationship among background red light, zeaxanthin content, and blue-light sensitivity has been found in blue light–stimulated phototropism of corn coleoptiles (see **Web Topic 18.2**).

- Blue light–stimulated stomatal opening is completely inhibited by 3 m*M* dithiothreitol (DTT), and the inhibition is concentration dependent. Zeaxanthin formation is blocked by DTT, a reducing agent that reduces S—S bonds to —SH groups and effectively inhibits the enzyme that converts violaxanthin into zeaxanthin. The specificity of the inhibition of blue light–stimulated stomatal opening by DTT, and its concentration dependence, indicate that guard cell zeaxanthin is required for the stomatal response to blue light.

- In the facultative CAM species *Mesembryanthemum crystallinum* (see Chapters 8 and 26), salt accumulation shifts its carbon metabolism from C_3 to CAM mode. In the C_3 mode, stomata accumulate zeaxanthin and show a blue-light response. CAM induction inhibits the ability of guard cells to accumulate zeaxanthin and to respond to blue light (Tallman et al. 1997).

STOMATA FROM *npq1* MUTANTS LACK A SPECIFIC BLUE LIGHT RESPONSE The Arabidopsis mutant *npq1* (*n*on*p*hotochemical *q*uenching) has a genetic lesion in the enzyme that converts violaxanthin to zeaxanthin (Figure 18.22) (Niyogi et al. 1998). Because of this mutation, neither mesophyll nor guard cell chloroplasts of *npq1* mutants accumulate zeaxanthin (Frechilla et al. 1999). Availability of this mutant made it possible to test the role of zeaxanthin in the stomatal response to blue light in guard cells in which zeaxanthin accumulation is genetically blocked.

As discussed earlier in the chapter, the blue-light sensitivity of guard cell photosynthesis (see Figure 18.10) requires that the specific stomatal response to blue light be tested

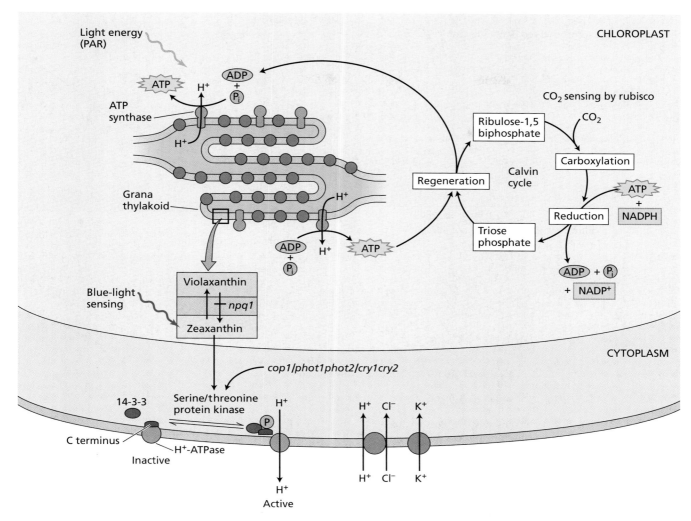

FIGURE 18.22 A sensory transduction cascade of blue light–stimulated stomatal opening.

under conditions in which guard cell photosynthesis cannot contribute to the opening response. Typically this is achieved by saturating the photosynthetic response with red light. Because saturating conditions vary with species and growth conditions, dose–response curves need to be obtained for each set of experimental conditions. In a recent study with *npq1* stomata irradiated with 100 µmol m^{-2} s^{-1} red light, shown to be saturating, a specific blue light response was clearly absent (Figure 18.23) (Talbott et al. 2003).

STOMATA FROM *phot1*/*phot2* HAVE A SMALL BUT CLEAR BLUE LIGHT RESPONSE Stomata from *phot1* and *phot2* mutants show suboptimal blue-light responses, and the response is markedly impaired in the *phot1*/*phot2* double mutant. Stomata from the double mutant fail to respond to blue light at fluence rates below 10 µmol m^{-2} s^{-1} (Kinoshita et al. 2001), and respond poorly at higher fluence rates (see Figure 18.23) (Talbott et al. 2003; Mao et al. 2005). Phototropin is widely believed to be a blue-light photoreceptor in guard cells; however, the response of the double mutant indicates that it retains a specific blue-light response despite the lack of phototropin.

BLUE LIGHT STIMULATES PHYTOCHROME-MEDIATED RESPONSES IN STOMATA Many studies seeking to define a role of phytochrome in stomatal movements have yielded mostly negative results. Recent findings, however, have shown that phytochrome modulates stomatal movements in the wild-type orchid *Paphiopedilum* and in *npq1*, the zeaxanthinless mutant of Arabidopsis.

Guard cells from *Paphiopedilum* lack chlorophyll. Because of that deficiency, *Paphiopedilum* stomata show blue light–specific opening, but lack an opening response driven by guard cell photosynthesis. Surprisingly, *Paphiopedilum* stomata open in response to low fluence rates of red and green light, and that opening can be reversed by far-red light, implicating phytochrome as the involved photoreceptor (Talbott et al. 2002).

Far-red reversible stomatal opening, stimulated by both blue and red light, can also be shown in *npq1*. Earlier stud-

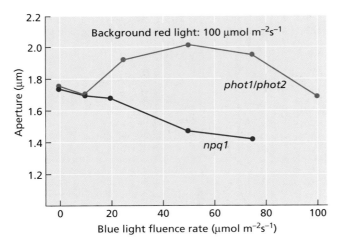

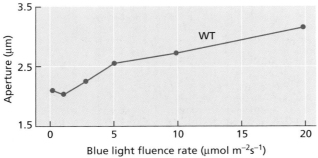

FIGURE 18.23 The blue light sensitivity of the zeaxanthin-less mutant *npq1*, and of the phototropinless double mutant *phot1/phot2*. Darkness is shown as zero fluence rate. Neither mutant shows opening when illuminated with 10 μmol m^{-2} s^{-1} blue light. The *phot1/phot2* mutant opens at higher fluence rates of blue light, whereas the *npq1* mutant closes, most likely because of the photoinhibition of the photosynthesis-driven opening caused by the higher fluence rates. Note the more reduced opening of *phot1/phot2* stomata, as compared with the wild-type (WT) response. (After Talbott et al. 2002.)

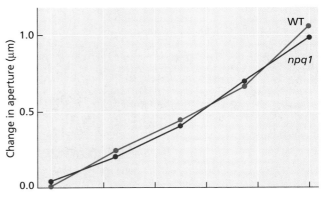

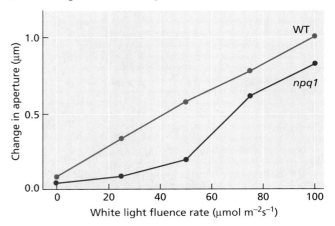

FIGURE 18.24 Stomata from the zeaxanthin-less mutant of Arabidopsis *npq1* lack a blue light response but show a phytochrome response. (A) Despite the lack of blue light response in *npq1*, wild-type and *npq1* stomata show indistinguishable dose–response curves to white light. (B) In the presence of 100 μmol m^{-2} s^{-1} of far-red light, the dose–response curve of wild-type stomata to white light remains the same but the response of *npq1* stomata is substantially reduced. These results indicate that phytochrome-dependent opening in *npq1* compensates for the absence of a blue-light response. (After Talbott et al. 2003.)

ies with *npq1* stomata have shown that, despite their lack of a specific blue-light response, they open normally in response to white light. However, if the experiments with white light are carried out in the presence of far-red light, the opening of wild-type stomata remains unaffected. In contrast, the opening of *npq1* is substantially reduced (Figure 18.24) (Talbott et al. 2003). These findings suggest that in both *Paphiopedilum* and *npq1*, phytochrome-dependent opening compensates for the absence of a major component of the stomatal response to light (guard cell photosynthesis in *Paphiopedilum* and a specific blue-light response in *npq1*).

Recent work with stomata from *cop1* mutants of Arabidopsis showed that they remain widely open in the dark (Mao et al. 2005). In addition, this study shows that CRY1, CRY2, PHOT1, and PHOT2 are positive modulators of the zeaxanthin-mediated stomatal response to blue light, whereas COP1 represses that modulation. Recall from Chapter 17 that COP1 and cryptochrome interact with phytochrome, both in vivo and in vitro. Taken together, these results point to a novel control of stomatal movement involving phytochrome, COP1, CRY1, CRY2, PHOT1, and PHOT2 (see Figure 18.22).

REGULATION OF BLUE LIGHT-STIMULATED STOMATAL OPENING
Several key steps in the sensory transduction cascade for blue light-stimulated stomatal opening have been characterized (see Figure 18.22). The C terminus of the H$^+$-ATPase (see Figure 6.17) has an autoinhibitory domain that regulates the activity of the enzyme. If this autoinhibitory domain is experimentally removed by a protease, the H$^+$-ATPase becomes irreversibly activated. The autoinhibitory domain of the C terminus is thought to lower the activity

of the enzyme by blocking its catalytic site. Conversely, fusicoccin appears to activate the enzyme by displacing the autoinhibitory domain away from the catalytic site (Kinoshita and Shimazaki 2001).

Upon blue-light irradiation, the H⁺-ATPase shows a lower K_m for ATP and a higher V_{max} (see Chapter 6), indicating that blue light activates the H⁺-ATPase. Activation of the enzyme involves the phosphorylation of serine and threonine residues of the C-terminal domain of the H⁺-ATPase. Blue light–stimulated proton pumping and stomatal opening are prevented by inhibitors of protein kinases, which might block phosphorylation of the H⁺-ATPase. As with fusicoccin, phosphorylation of the C-terminal domain appears also to displace the autoinhibitory domain of the C terminus from the catalytic site of the enzyme.

A **14-3-3 protein** has been found to bind to the phosphorylated C terminus of the guard cell H⁺-ATPase, but not the nonphosphorylated one. The 14-3-3 proteins are ubiquitous regulatory proteins in eukaryotic organisms. In plants, 14-3-3 proteins regulate transcription by binding to activators in the nucleus, and they regulate metabolic enzymes such as nitrate reductase.

Only one of the four 14-3-3 isoforms found in guard cells binds to the H⁺-ATPase, so the binding appears to be specific. The same 14-3-3 isoform binds to the guard cell H⁺-ATPase in response to both fusicoccin and blue-light treatments. The 14-3-3 protein seems to dissociate from the H⁺-ATPase upon dephosphorylation of the C-terminal domain. Recent studies have shown that guard cell phototropins are phosphorylated by blue light and that they bind the 14-3-3 protein upon phosphorylation (Kinoshita et al. 2003).

Proton-pumping rates of guard cells increase with fluence rates of blue light (see Figure 18.13), and the electrochemical gradient generated by the proton pump drives ion uptake into the guard cells, increasing turgor and turgor-mediated stomatal apertures. These processes define the major steps linking the activation of a serine/threonine protein kinase by blue light and blue light–stimulated stomatal opening (see Figure 18.22).

Green light reverses blue light-stimulated opening

A reversal of blue light-stimulated stomatal opening by green light has been recently discovered. This novel photobiological property of guard cells can be observed when green light is applied together with blue light in continuous light treatments, or after blue light in pulse treatments. Stomata in epidermal strips open in response to a 30-s blue-light pulse (Figure 18.25), but the opening is not observed if the blue-light pulse is followed by a green-light pulse. The opening is restored if the green pulse is followed by a second blue-light pulse, in a response analogous to the red/far-red reversibility of phytochrome responses (Frechilla et al. 2000).

The blue/green reversibility response has been reported in stomata of several species (see **Web Essay 18.3**). Stom-

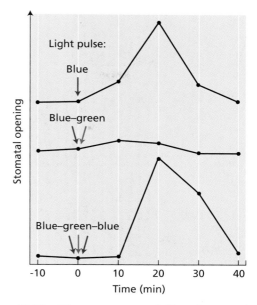

FIGURE 18.25 Blue–green reversibility of stomatal movements. Stomata open when given a 30 s blue-light pulse (1800 μmol m⁻² s⁻¹) under a background of continuous red light (120 μmol m⁻² s⁻¹). A green-light pulse (3600 μmol m⁻² s⁻¹) applied after the blue-light pulse blocks the blue-light response, and the opening is restored upon application of a second blue-light pulse given after the green-light pulse. (After Frechilla et al. 2000.)

ata from intact, attached leaves of Arabidopsis, grown in a growth chamber under blue, red, and green light, open when green light is turned off and close when the green light is turned on again (Figure 18.26) (Talbott et al. 2006). This stomatal sensitivity to green light is not observed if the leaves are illuminated with red and green light alone. This indicates that the response to green light ensues from an interaction between blue and green light, as observed in previous studies with isolated stomata in epidermal peels. The stomatal responses to green light in the intact leaf suggest that the green photons from solar radiation down-regulate the stomatal response to blue light under natural conditions.

The far red–reversible, blue light–stimulated stomatal opening observed with *npq1* stomata (see Figure 18.23) cannot be reversed by green light. In contrast, the blue light-stimulated opening observed with *phot1/phot2* stomata is green light-reversible (Talbott et al. 2003). In addition, the response to green light observed with intact leaves can be found in the *phot1/phot2* double mutant, but not in the zeaxanthinless *npq1* mutant (see Figure 18.26). These results indicate that the green reversal of the blue light response requires zeaxanthin but not phototropin.

The action spectrum for the green reversal of blue light–stimulated opening shows a maximum at 540 nm and two minor peaks at 490 and 580 nm. Such an action spectrum rules out the involvement of phytochrome or chlorophylls in the response. Rather, the action spectrum is

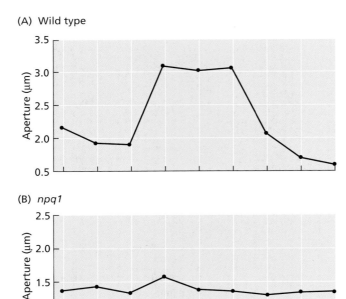

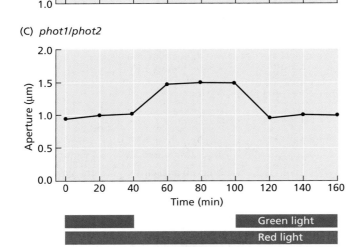

Figure 18.26 Stomata from intact, attached leaves of Arabidopsis grown in a growth chamber under blue, red, and green light open when green light is removed and close when green light is restored. Blue light is required for the expression of this stomatal sensitivity to green light. Stomata from the zeaxanthinless *npq1* mutant fail to respond to green light, whereas stomata from the *phot1/phot2* double mutant have a response similar to that of wild type. (After Talbott et al. 2006.)

remarkably similar to the action spectrum for blue light–stimulated stomatal opening (see Figure 18.11), but red-shifted (displaced toward the longer, red wave-band of the spectrum) by about 90 nm. Such spectral red shifts have been observed upon the isomerization of carotenoids in a protein environment. Spectroscopic studies have recently shown that green light is very effective in the isomerization of zeaxanthin (Milanowska and Gruszecki 2005). Isomerization of zeaxanthin changes the orientation of the molecule within the membrane, a transition that would be very effective as a transduction signal.

The blue/green reversal of stomatal movements, and the absorption spectrum changes elicited by blue and green light suggest that zeaxanthin is involved in a photocycle in which a physiologically inactive isomer is converted to an active isomer by blue light. The isomerization would start a sensory transduction cascade at the guard cell chloroplast. Green light would convert the active isomer into the inactive one, and thereby reverse the blue light–stimulated opening signal. Results from a study with guard cells treated with blue-light pulses further indicate that after a blue pulse, the active form slowly reverts to the inactive form in the dark (Iino et al. 1985).

The xanthophyll cycle confers plasticity to the stomatal responses to light

Zeaxanthin concentration in guard cells varies with the activity of the xanthophyll cycle. The enzyme that converts violaxanthin to zeaxanthin is an integral thylakoid protein showing a pH optimum of 5.2 (Yamamoto 1979). Acidification of lumen pH stimulates zeaxanthin formation, and lumen alkalinization favors violaxanthin formation.

Lumen pH depends on levels of incident photosynthetic active radiation (most effective at blue and red wavelengths; see Chapter 7), and on the rate of ATP synthesis, that dissipate the pH gradient across the thylakoid. Thus, photosynthetic activity in the guard cell chloroplast, lumen pH, zeaxanthin content, and blue-light sensitivity play an interactive role in the regulation of stomatal apertures.

Some unique properties of the guard cell chloroplast appear optimally geared for its sensory transducing function. Compared with their mesophyll counterparts, guard cell chloroplasts are enriched in photosystem II, and they have unusually high rates of photosynthetic electron transport and low rates of photosynthetic carbon fixation (Zeiger et al. 2002). These properties favor lumen acidification at low photon fluxes, and they explain zeaxanthin formation in the guard cell chloroplast early in the day (see Figure 18.20).

The regulation of zeaxanthin content by lumen pH, and the tight coupling between lumen pH and Calvin cycle activity in the guard cell chloroplast (see Figure 18.22), further suggest that sensing of carbon dioxide by the guard cell chloroplast is very likely to involve these processes (Zeiger et al. 2002).

Summary

Plants utilize light as a source of energy and as a signal that provides information about their environment. A large family of blue-light responses is used to sense light quantity and direction. These blue-light signals are transduced into electrical, metabolic, and genetic processes that allow plants to alter growth, development, and function in order to acclimate to changing environmental conditions. Blue-light responses include phototropism, stomatal movements, inhibition of stem elongation, gene activation, pigment biosynthesis, tracking of the sun by leaves, and chloroplast movements within cells.

Specific blue-light responses have a characteristic "three-finger" action spectrum in the 400 to 500 nm region.

The physiology of blue-light responses varies broadly. In phototropism, stems grow toward unilateral light sources by asymmetric growth on their shaded side. In the inhibition of stem elongation, perception of blue light depolarizes the membrane potential of elongating cells, and the rate of elongation rapidly decreases. In gene activation, blue light stimulates transcription and translation, leading to the accumulation of gene products that are required for the morphogenetic response to light.

Blue light–stimulated stomatal movements are driven by blue light–dependent changes in the osmoregulation of guard cells. Blue light stimulates an H^+-ATPase at the guard cell plasma membrane, and the resulting pumping of protons across the membrane generates an electrochemical-potential gradient that provides a driving force for ion uptake. Blue light also stimulates starch degradation and malate biosynthesis. Solute accumulation within guard cells leads to stomatal opening. Guard cells also utilize sucrose as a major osmotically active solute, and light quality can change the activity of different osmoregulatory pathways that modulate stomatal movements.

The two Arabidopsis genes *cry1* and *cry2* are involved in blue light–dependent inhibition of stem elongation, cotyledon expansion, anthocyanin synthesis, the control of flowering, and the setting of circadian rhythms. The gene products of *cry1* and *cry2* have sequence similarity to photolyase but no photolyase activity. The CRY1 protein, and to a lesser extent CRY2, accumulate in the nucleus and interact with the ubiquitin ligase COP1, both in vivo and in vitro. In addition, the activity of cryptochrome is influenced by its phosphorylation state. The CRY1 protein also regulates anion channel activity at the plasma membrane.

The protein phototropin has a major role in the regulation of phototropism. The C-terminal half of phototropin is a serine/threonine kinase, and the N-terminal half has two flavin-binding domains. Phototropin binds the flavin FMN and autophosphorylates in response to blue light. Phototropinless mutants are defective in phototropism and in chloroplast movements. The *phot1/phot2* double mutant, widely believed to play a role in the blue light photoreception of guard cells, has a well-defined response to blue light that is reversed by green light. These features are inconsistent with a role of phototropin in the blue light response of stomata.

The chloroplast carotenoid zeaxanthin has been implicated in blue-light photoreception in guard cells. Blue light–stimulated stomatal opening is blocked if zeaxanthin accumulation in guard cells is prevented by genetic or biochemical means. Manipulation of zeaxanthin content in guard cells makes it possible to regulate their response to blue light. The signal transduction cascade for the blue-light response of guard cells comprises blue-light perception in the guard cell chloroplast, transduction of the blue-light signal across the chloroplast envelope, activation of the H^+-ATPase, turgor buildup, and stomatal opening.

Green light can reverse the stomatal opening stimulated by blue light. The green light reversal is observed if green light is applied together with blue light in experiments under continuous illumination, or if a pulse of green light is applied after a pulse of blue light. The zeaxanthin-less mutant *npq1* does not show the green reversal, indicating that zeaxanthin is required for the response. Green light is very effective in the isomerization of zeaxanthin, suggesting that a physiologically active isomer of zeaxanthin absorbs blue light and mediates the blue light response, whereas green light reverses the isomerization reaction and regenerates the inactive zeaxanthin isomer.

Web Material

Web Topics

18.1 Guard Cell Osmoregulation and a Blue Light-Activated Metabolic Switch
Blue light controls major osmoregulatory pathways in guard cells and unicellular algae.

18.2 The Coleoptile Chloroplast
Both the coleoptile and the guard cell chloroplast specialize in sensory transduction.

Web Essays

18.1 Guard Cell Photosynthesis
Photosynthesis in the guard cell chloroplast shows unique regulatory features.

18.2 The Sensory Transduction of the Inhibition of Stem Elongation by Blue Light
The regulation of stem elongation rates by blue light has critical importance for plant development.

18.3 The Blue–Green Reversibility of the Blue-Light Response of Stomata
The blue–green reversal of stomatal movements is a remarkable photobiological response.

Chapter References

Ahmad, M., and Cashmore, A. R. (1993) HY4 gene of *A. thaliana* encodes a protein with characteristics of a blue light photoreceptor. *Nature* 366: 162–166.

Ahmad, M., Jarillo, J. A., Smirnova, O., and Cashmore, A. R. (1998) Cryptochrome blue light photoreceptors of *Arabidopsis* implicated in phototropism. *Nature* 392: 720–723.

Ahmad, M., Grancher, N., Heil, M., Black, R. C., Giovani, B., Galland, P., and Lardemer, D. (2002) Action spectrum for cryptochrome-dependent hypocotyl growth inhibition in Arabidopsis. *Plant Physiol.* 129: 774–785.

Amodeo, G., Srivastava, A., and Zeiger, E. (1992) Vanadate inhibits blue light–stimulated swelling of *Vicia* guard cell protoplasts. *Plant Physiol.* 100: 1567–1570.

Assmann, S. M., Simoncini, L., and Schroeder, J. I. (1985) Blue light activates electrogenic ion pumping in guard cell protoplasts of *Vicia faba*. *Nature* 318: 285–287.

Briggs, W. R., and Christie, J. M. (2002) Phototropins 1 and 2: Versatile plant blue-light receptors. *Trends Plant Sci.* 7: 204–210.

Cerda-Olmedo, E., and Lipson, E. D. (1987) *Phycomyces*. Cold Spring Harbor Laboratory, Cold Spring Harbor, NY.

Cosgrove, D. J. (1994) Photomodulation of growth. In *Photomorphogenesis in Plants*, 2nd ed., R. E. Kendrick and G. H. M. Kronenberg, eds., Kluwer, Dordrecht, Netherlands, pp. 631–658.

Fan, L. M., Zhao, Z., and Assmann, S. M. (2004) Guard cells: A dynamic signaling model. *Curr. Opin. Plant Biol.* 7: 537–546.

Firn, R. D. (1994) Phototropism. In *Photomorphogenesis in Plants*, 2nd ed., R. E. Kendrick and G. H. M. Kronenberg, eds., Kluwer, Dordrecht, Netherlands, pp. 659–681.

Frechilla, S., Zhu, J., Talbott, L. D., and Zeiger, E. (1999) Stomata from *npq1*, a zeaxanthin-less *Arabidopsis* mutant, lack a specific response to blue light. *Plant Cell Physiol.* 40: 949–954.

Frechilla, S., Talbott, L. D., Bogomolni, R. A., and Zeiger, E. (2000) Reversal of blue light-stimulated stomatal opening by green light. *Plant Cell Physiol.* 41: 171–176.

Horwitz, B. A. (1994) Properties and transduction chains of the UV and blue light photoreceptors. In *Photomorphogenesis in Plants*, 2nd ed., R. E. Kendrick and G. H. M. Kronenberg, eds., Kluwer, Dordrecht, Netherlands, pp. 327–350.

Iino, M., Ogawa, T., and Zeiger, E. (1985) Kinetic properties of the blue light response of stomata. *Proc. Natl. Acad. Sci. USA* 82: 8019–8023.

Im, C., and Beale, S. I. (2000) Identification of possible signal transduction components mediating light induction of the gsa gene for an early chlorophyll biosynthetic step in *Chlamydomonas reinhardtii*. *Planta* 210: 999–2005.

Karlsson, P. E. (1986) Blue light regulation of stomata in wheat seedlings. II. Action spectrum and search for action dichroism. *Physiol. Plant.* 66: 207–210.

Kinoshita, T., and Shimazaki, K. (2001) Analysis of the phosphorylation level in guard-cell plasma membrane H^+-ATPase in response to fusicoccin. *Plant Cell Physiol.* 42: 424–432.

Kinoshita, T., Doi, M., Suetsugu, N., Kagawa, T., Wada, M., and Shimazaki, K. (2001) *phot1* and *phot2* mediate blue light regulation of stomatal opening. *Nature* 414: 656–660.

Kinoshita, T., Emi, T., Tominaga, M., Sakamoto, K., Shigenaga, A., Doi, M., and Shimazaki, K. (2003) Blue-light- and phosphorylation-dependent binding of a 14-3-3 protein to phototropins in stomatal guard cells of broad bean. *Plant Physiol.* 133: 1453–1463.

Lin, C. (2000) Plant blue-light receptors. *Trends Plant Sci.* 5: 337–342.

Mao J., Zhang, Y. C., Sang, Y., Li, Q. H., and Yang, H. Q. (2005) A role for Arabidopsis cryptochromes and COP1 in the regulation of stomatal opening. *Proc. Natl. Acad. Sci. USA* 102: 12270–12275.

Matters, G. L., and Beale, S. I. (1995) Blue-light-regulated expression of genes for two early steps of chlorophyll biosynthesis in *Chlamydomonas reinhardtii*. *Plant Physiol.* 109: 471–479.

Milanowska, J. and Gruszecki, W. I. (2005) Heat-induced and light-induced isomerization of the xanthophyll pigment zeaxanthin. *J. Photochem. Photobiol. B.* 80: 178–186.

Niyogi, K. K., Grossman, A. R., and Björkman, O. (1998) *Arabidopsis* mutants define a central role for the xanthophyll cycle in the regulation of photosynthetic energy conversion. *Plant Cell* 10: 1121–1134.

Outlaw, W. H., Jr. (1983) Current concepts on the role of potassium in stomatal movements. *Physiol. Plant.* 59: 302–311.

Outlaw, W. H., Jr. (2003) Integration of cellular and physiological functions of guard cells. *Crit. Rev. Plant Sci.* 22: 503–529.

Parks, B. M., Folta, K. M., and Spalding, E. P. (2001) Photocontrol of stem growth. *Curr. Opin. Plant Biol.* 4: 436–440.

Roelfsema, M. R. G., Steinmeyer, R., Staal, M., and Hedrich, R. (2001) Single guard cell recordings in intact plants: Light-induced hyperpolarization of the plasma membrane. *Plant J.* 26: 1–13.

Roelfsema, R. G., Hanstein, S., Felle, H. H., and Hedrich, R. (2002) CO_2 provides an intermediate link in the red light response of guard cells. *Plant J.* 32: 65–75.

Schroeder, J. I., Allen, G. J., Hugouvieux, V., Kwak, J. M., and Waner, D. (2001) Guard cell signal transduction. *Annu. Rev. Plant Physiol. Plant Mol. Biol.* 52: 627–658.

Schwartz, A., and Zeiger, E. (1984) Metabolic energy for stomatal opening. Roles of photophosphorylation and oxidative phosphorylation. *Planta* 161: 129–136.

Senger, H. (1984) *Blue Light Effects in Biological Systems*. Springer, Berlin.

Serrano, E. E., Zeiger, E., and Hagiwara, S. (1988) Red light stimulates an electrogenic proton pump in *Vicia* guard cell protoplasts. *Proc. Natl. Acad. Sci. USA* 85: 436–440.

Shimazaki, K., Iino, M., and Zeiger, E. (1986) Blue light–dependent proton extrusion by guard cell protoplasts of *Vicia faba*. *Nature* 319: 324–326.

Spalding, E. P. (2000) Ion channels and the transduction of light signals. *Plant Cell Environ.* 23: 665–674.

Spalding, E. P., and Cosgrove, D. J. (1989) Large membrane depolarization precedes rapid blue-light induced growth inhibition in cucumber. *Planta* 178: 407–410.

Spalding, E. P. and Folta, K. M. (2005) Illuminating topics in plant photobiology. *Plant Cell Environ.* 28: 39–53.

Srivastava, A., and Zeiger, E. (1995a) Guard cell zeaxanthin tracks photosynthetic active radiation and stomatal apertures in *Vicia faba* leaves. *Plant Cell Environ.* 18: 813–817.

Srivastava, A., and Zeiger, E. (1995b) The inhibitor of zeaxanthin formation, dithiothreitol, inhibits blue-light-stimulated stomatal opening in *Vicia faba*. *Planta* 196: 445–449.

Swartz, T. E., Wenzel, P. J., Corchnoy, S. B., Briggs, W. R., and Bogomolni, R. A. (2002) Vibration spectroscopy reveals light-induced chromophore and protein structural changes in the LOV2 domains of the plant blue-light receptor phototropin 1. *Biochemistry* 41: 7183–7189.

Talbott, L. D., and Zeiger, E. (1998) The role of sucrose in guard cell osmoregulation. *J. Exp. Bot.* 49: 329–337.

Talbott, L. D., Srivastava, A., and Zeiger, E. (1996) Stomata from growth-chamber-grown *Vicia faba* have an enhanced sensitivity to CO_2. *Plant Cell Environ.* 19: 1188–1194.

Talbott, L. D., Shmayevich, I. J., Chung, Y., Hammad, J. W., and Zeiger, E. (2003) Blue light and phytochrome-mediated stomatal opening in the *npq1* and *phot1 phot2* mutants of Arabidopsis. *Plant Physiol.* 133: 1522–1529.

Talbott, L. D., Hammad, J. W., Harn, L. C, Nguyen, V., Patel, J., and Zeiger, E. (2006) Reversal by green light of blue light-stimulated stomatal opening in intact, attached leaves of Arabidopsis operates only in the potassium dependent, morning phase of movement. *Plant Cell Physiol.* 47: 249–257.

Tallman, G. (2004) Are diurnal patterns of stomatal movement the result of alternating metabolism of endogenous guard cell ABA and accumulation of ABA delivered to the apoplast around guard cells by transpiration? *J. Exp. Bot.* 55: 1963–1976.

Tallman, G., Zhu, J., Mawson, B. T., Amodeo, G., Nouhi, Z., Levy, K., and Zeiger, E. (1997) Induction of CAM in *Mesembryanthemum crystallinum* abolishes the stomatal response to blue light and light-dependent zeaxanthin formation in guard cell chloroplasts. *Plant Cell Physiol.* 38: 236–242.

Thimann, K. V., and Curry, G. M. (1960) Phototropism and phototaxis. In *Comparative Biochemistry*, Vol. 1, M. Florkin and H. S. Mason, eds., Academic Press, New York, pp. 243–306.

Tsunoyama, Y., Ishizaki, Y., Morikawa, K., Kobori, M., Nakahira, Y., Takeba, G., Toyoshima, Y., and Shiina, T. (2004) Blue light-induced transcription of plastid-encoded *psbD* gene is mediated by a nuclear-encoded transcription initiation factor, AtSig5. *Proc. Natl. Acad. Sci USA* 101: 3304–3309.

Vogelmann, T. C. (1994) Light within the plant. In *Photomorphogenesis in Plants*, 2nd ed., R. E. Kendrick and G. H. M. Kronenberg, eds., Kluwer, Dordrecht, Netherlands, pp. 491–533.

Vogelmann, T. C., and Haupt, W. (1985) The blue light gradient in unilaterally irradiated maize coleoptiles: Measurements with a fiber optic probe. *Photochem. Photobiol.* 41: 569–576.

Wada M., Kagawa T., and Sato, Y. (2003). Chloroplast movement. *Annu. Rev. Plant Biol.* 54: 455–468.

Yamamoto, H. Y. (1979) Biochemistry of the violaxanthin cycle in higher plants. *Pure Appl. Chem.* 51: 639–648.

Zeiger, E., and Hepler, P. K. (1977) Light and stomatal function: Blue light stimulates swelling of guard cell protoplasts. *Science* 196: 887–889.

Zeiger, E., Talbott, L. D., Frechilla, S., Srivastava, A., and Zhu, J. X. (2002) The guard cell chloroplast: A perspective for the twenty-first century. *New Phytol.* 153: 415–424.

Chapter 19
Auxin: The Growth Hormone

THE FORM AND FUNCTION of multicellular organisms would not be possible without efficient communication among cells, tissues, and organs. In higher plants, regulation and coordination of metabolism, growth, and morphogenesis often depend on chemical signals from one part of the plant to another. This idea originated in the nineteenth century with the German botanist Julius von Sachs (1832–1897).

Sachs proposed that chemical messengers are responsible for the formation and growth of different plant organs. He also suggested that external factors such as gravity could affect the distribution of these substances within a plant. Although Sachs did not know the identity of these chemical messengers, his ideas led to their eventual discovery.

Hormones are chemical messengers that are produced in one cell or tissue and modulate cellular processes in another cell by interacting with specific protein *receptors.* As is the case with animals, most plant hormones are synthesized in one tissue and act on specific target sites in another tissue at vanishingly low concentrations. Hormones that are transported to sites of action in tissues distant from their site of synthesis are referred to as *endocrine* hormones. Those that act on cells adjacent to the source of synthesis are referred to as *paracrine* hormones. Plant development is regulated by six major types of hormones: auxins, gibberellins, cytokinins, ethylene, abscisic acid, and brassinosteroids.

A variety of other signaling molecules that play roles in resistance to pathogens and defense against herbivores have also been identified in plants, including jasmonic acid, salicylic acid, and the polypeptide systemin (see Chapter 13). Other classes of molecules, such as flavonoids, function as both intracellular and localized extracellular modulators of signal transduction pathways, and a carotenoid cleavage product are throught to function in long distance signaling. Indeed, the list of plant hormones and hormone-like signaling agents keeps expanding.

The first signaling agent we will consider is the hormone auxin. Auxin was the first growth hormone to be studied in plants, and much of the early physiological work on the mechanism of plant cell expansion was carried out in relation to auxin action. Moreover, both auxin and cytokinin differ from the other plant hormones and signaling agents in one important respect: they are required for viability. Thus far, no mutants lacking either auxin or cytokinin have been found, suggesting that mutations that eliminate them are lethal. Whereas other plant hormones seem to act as on/off switches that regulate specific developmental processes, auxin and cytokinin appear to be required at some level more or less continuously.

We begin our discussion of auxins with a brief history of their discovery, followed by a description of their chemical structures and the methods used to detect auxins in plant tissues. A look at the pathways of auxin biosynthesis and the polar nature of auxin transport follows. We will then review the various developmental processes controlled by auxin, such as stem elongation, apical dominance, root initiation, fruit development, meristem development, and oriented, or *tropic*, growth. Finally, we will describe our current understanding of auxin signal transduction pathways, from receptor binding to gene expression.

The Emergence of the Auxin Concept

During the latter part of the nineteenth century, Charles Darwin and his son Francis studied plant growth phenomena involving tropisms. One of their interests was the bending of plants toward light. This phenomenon, which is caused by differential growth, is called **phototropism**. In some experiments the Darwins used seedlings of canary grass (*Phalaris canariensis*), in which, as in many other grasses, the youngest leaves are sheathed in a protective organ called the **coleoptile** (Figure 19.1).

Coleoptiles are very sensitive to light, especially to blue light (see Chapter 18). If illuminated on one side with a short pulse of dim blue light, they will bend (grow) toward the source of the light pulse within an hour. The Darwins found that the tip of the coleoptile perceived the light, for if they covered the tip with foil, the coleoptile would not bend. But the region of the coleoptile that is responsible for the bending toward the light, called the **growth zone**, is several millimeters below the tip.

Thus they concluded that some sort of signal is produced in the tip, travels to the growth zone, and causes the shaded side to grow faster than the illuminated side. The results of their experiments were published in 1881 in a volume entitled *The Power of Movement in Plants*.

There followed a long period of experimentation by many investigators on the nature of the growth stimulus in coleoptiles. This research culminated in the demonstration in 1926 by Frits Went of the presence of a growth-promoting chemical in the tip of oat (*Avena sativa*) coleoptiles. It was known that if the tip of a coleoptile was removed, coleoptile growth ceased. Previous workers had attempted to isolate and identify the growth-promoting chemical by grinding up coleoptile tips and testing the activity of the extracts. This approach failed because grinding up the tissue released into the extract inhibitory substances that normally were compartmentalized in the cell.

Went's major breakthrough was to avoid grinding by allowing the material to diffuse out of excised coleoptile tips directly into gelatin blocks. If placed asymmetrically on top of a decapitated coleoptile, these blocks could be tested for their ability to cause bending in the absence of a unilateral light source (see Figure 19.1). Because the substance promoted the elongation of the coleoptile sections (Figure 19.2), it was eventually named **auxin** from the Greek *auxein*, meaning "to increase" or "to grow." The next step was to chemically identify this substance and to understand its production, destruction, and action.

Identification, Biosynthesis, and Metabolism of Auxin

Went's studies with agar blocks demonstrated unequivocally that the growth-promoting "influence" diffusing from the coleoptile tip was a chemical substance. The fact that it was produced at one location and transported in minute amounts to its site of action qualified it as an authentic plant hormone.

In the years that followed, the chemical identity of the "growth substance" was determined, and because of its potential agricultural uses, many related chemical analogs were tested. This testing led to generalizations about the chemical requirements for auxin activity. In parallel with these studies, the agar block diffusion technique was used to study how auxin is transported in plant tissues. Technological advances, especially the use of radioactive isotopes as tracers, enabled plant biochemists to unravel the pathways of auxin biosynthesis and breakdown.

Our discussion begins with the chemical nature of auxin and continues with a description of its biosynthesis, transport, and metabolism. Increasingly powerful analytical methods and the application of molecular biological approaches have recently allowed scientists to identify

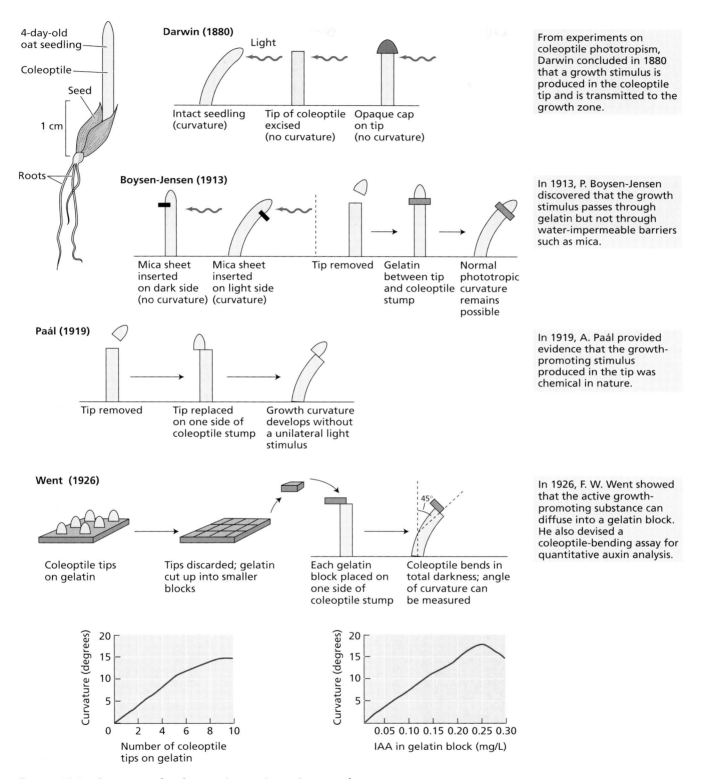

FIGURE 19.1 Summary of early experiments in auxin research.

(A)
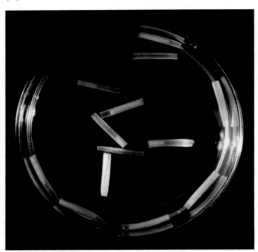

(B)
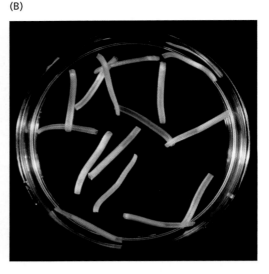

FIGURE 19.2 Auxin stimulates the elongation of oat coleoptile sections. These coleoptile sections were incubated for 18 hours in either water (A) or auxin (B). The yellow tissue inside the translucent coleoptile is the primary leaves. (Photos © M. B. Wilkins.)

auxin precursors and to study auxin turnover and distribution within the plant.

The principal auxin in higher plants is indole-3-acetic acid

In the mid-1930s it was determined that auxin is **indole-3-acetic acid** (**IAA**). Several other auxins in higher plants were discovered later (Figure 19.3), but IAA is by far the most abundant and physiologically important. Because the structure of IAA is relatively simple, academic and industrial laboratories were quickly able to synthesize a wide array of molecules with auxin activity. Some of these are used as herbicides in horticulture and agriculture (Figure 19.4) (synthetic auxins are discussed in **Web Topic 19.1**).

An early definition of auxins included all natural and synthetic chemical substances that stimulate elongation in coleoptiles and stem sections. However, auxins affect many developmental processes besides cell elongation. Thus auxins can be defined as compounds with biological activities similar to those of IAA, including the ability to promote cell elongation in coleoptile and stem sections, cell division in callus cultures in the presence of cytokinins, formation of adventitious roots on detached leaves and stems, and other developmental phenomena associated with IAA action.

Although they are chemically diverse, a common feature of all active auxins is a molecular distance of about 0.5 nm between a fractional positive charge on the aromatic ring and a negatively charged carboxyl group (see **Web Topic 19.2**).

Depending on the information that a researcher needs, the amounts and/or identity of auxins in biological samples can be determined by bioassay, mass spectrometry, or enzyme-linked immunosorbent assay, which is abbreviated as ELISA (see **Web Topic 19.3**).

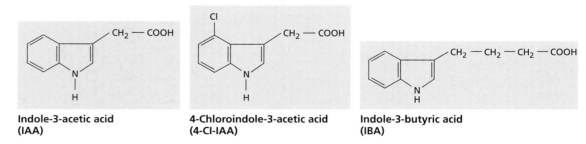

FIGURE 19.3 Structure of three natural auxins. Indole-3-acetic acid (IAA) occurs in all plants, but other related compounds in plants have auxin activity. Peas, for example, contain 4-chloroindole-3-acetic acid. Mustards and corn contain indole-3-butyric acid (IBA).

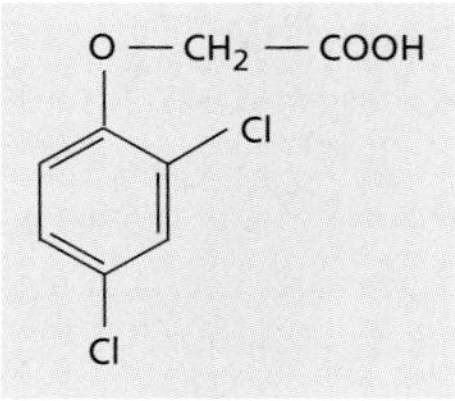

2,4-Dichlorophenoxyacetic acid (2,4-D)

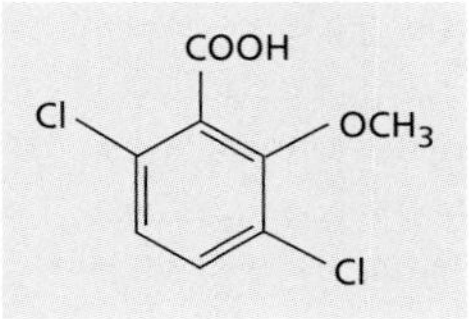

2-Methoxy-3, 6-dichlorobenzoic acid (dicamba)

FIGURE 19.4 Structures of two synthetic auxins. Most synthetic auxins are used as herbicides in horticulture and agriculture.

IAA is synthesized in meristems and young dividing tissues

IAA biosynthesis is associated with rapidly dividing and rapidly growing tissues, especially in shoots. Although virtually all plant tissues appear to be capable of producing low levels of IAA, shoot apical meristems and young leaves are the primary sites of auxin synthesis (Ljung et al. 2001). Root apical meristems are also important sites of auxin synthesis, especially as the roots elongate and mature, although the root is still dependent on the shoot for much of its auxin (Ljung et al. 2005). Young fruits and seeds contain high levels of auxin, but it is unclear whether this auxin is newly synthesized or transported from maternal tissues during development.

In very young leaf primordia of Arabidopsis, auxin accumulates at the tip. During leaf development, auxin accumulations are seen at the leaf margins, but gradually shift toward the base of the leaf and, later, to the central region of the lamina. The basipetal shift in auxin production correlates closely with, and is probably causally related to, the basipetal maturation sequence of leaf development and vascular differentiation (Aloni 2001).

The *GUS* (β-glucuronidase) reporter gene is a useful analytical tool because its activity and location in the tissue can be visualized by treating the tissue with a substrate that produces a blue color when hydrolyzed by the GUS enzyme (Ulmasov et al. 1997). By fusing the *GUS* reporter gene to a DNA promoter sequence that responds to auxin, and transforming Arabidopsis leaves with this construct in a Ti plasmid using *Agrobacterium*, it is possible to visualize the distribution of free auxin in young developing leaves. Wherever free auxin reaches a minimum threshold level, *GUS* expression occurs and can be detected as a blue color.

As shown in Figure 19.5, auxin accumulates at specific sites at the margins of young leaves. These sites represent future **hydathodes**—glandlike modifications of the ground and vascular tissues, typically at the leaf margins, that allow the release of liquid water (guttation fluid) through pores in the epidermis in the presence of root pressure (see

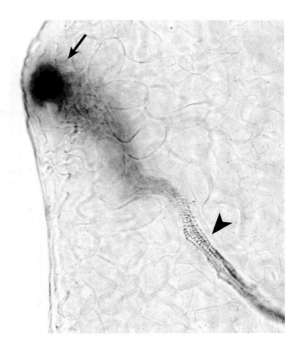

FIGURE 19.5 Detection of sites of auxin accumulation in a young leaf primordium of Arabidopsis using a synthetic auxin-sensitive *DR5* promoter driving a *GUS* reporter gene. During the early stages of hydathode differentiation, a center of auxin synthesis is evident as a concentrated dark blue *GUS* stain (arrow) in the lobes of the serrated leaf margin. A gradient of diluted GUS activity extends from the margin toward a differentiating vascular strand (arrowhead), which appears to function as a sink for the auxin flow originating in the lobe. (Courtesy of R. Aloni and C. I. Ullrich.)

Chapter 4). During early stages of hydathode differentiation a center of high auxin accumulation is evident as a concentrated dark blue GUS stain (arrow) in the lobes of young, serrated leaves of Arabidopsis, and these seem to be sites of auxin synthesis as well (Aloni et al. 2003). A trail of GUS activity, indicative of auxin diffusion, can be seen extending down to differentiating vessel elements in a developing vascular strand. We will return to the topic of auxin regulation of vascular differentiation later in the chapter.

Multiple pathways exist for the biosynthesis of IAA

IAA is structurally related to the amino acid *tryptophan*, and to the tryptophan precursor *indole-3-glycerol phosphate*, both of which can serve as precursors for IAA biosynthesis (Figure 19.6). Molecular genetic and radioisotope labeling studies have been used to identify the enzymes and intermediate molecules involved in tryptophan-dependent IAA biosynthesis, and the order in which they function. The three principal plant pathways and the bacterial pathway of tryptophan-dependent IAA biosynthesis are shown in Figure 19.7. Of these, the TAM and IPA pathways are probably the most common pathways in plants.

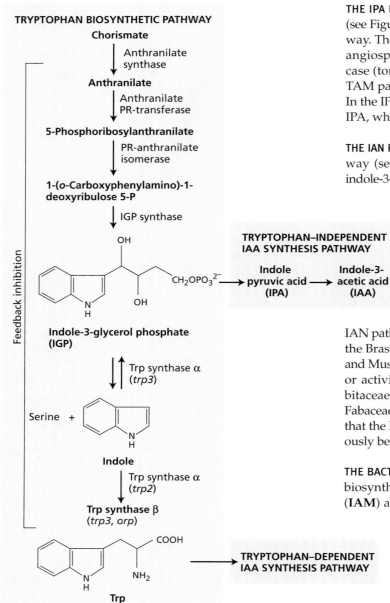

FIGURE 19.6 Tryptophan biosynthetic pathway provides precursors for IAA biosynthesis. In plants, tryptophan synthesis takes place in the chloroplast. Mutants discussed in **Web Topic 19.4** are indicated in parentheses. The branch-point precursor for tryptophan-independent auxin biosynthesis is indole-3-glycerol phosphate, and indole-3-pyruvic acid (IPA) serves as an intermediate.

THE TAM PATHWAY The **tryptamine (TAM) pathway** (see Figure 19.7B) functions in numerous plant species, including Arabidopsis. It begins with the decarboxylation of tryptophan to form TAM. A series of enzymatic reactions then converts tryptamine to indole-3-acetaldehyde (IAAld), which is then oxidized by a specific dehydrogenase to IAA.

THE IPA PATHWAY The **indole-3-pyruvate (IPA)** pathway (see Figure 19.7C), is found in plants lacking the TAM pathway. The precise distribution of the two pathways among angiosperms is not yet known. However, in at least one case (tomato [*Lycopersicon esculentum*]), both the IPA and TAM pathways appear to function (Nonhebel et al. 1993). In the IPA pathway, tryptophan is first deaminated to form IPA, which is then decarboxylated to form IAAld.

THE IAN PATHWAY In the **indole-3-acetonitrile (IAN)** pathway (see Figure 19.7A), tryptophan is first converted to indole-3-acetaldoxime (IAOx) and then to IAN, either directly or by multiple steps via indole-3-acetaldoxime N-oxide (IAOx N-oxide). Nitrilase enzymes then convert IAN to IAA. Four genes (*NIT1* through *NIT4*) that encode nitrilase enzymes have now been cloned from Arabidopsis. When *NIT2* was expressed in transgenic tobacco, the resultant plants acquired the ability to respond to IAN as an auxin by hydrolyzing it to IAA (Schmidt et al. 1996). The IAN pathway may be important in only three plant families: the Brassicaceae (mustard family), Poaceae (grass family), and Musaceae (banana family). However, nitrilase-like genes or activities have recently been identified in the Cucurbitaceae (squash family), Solanaceae (tobacco family), Fabaceae (legumes), and Rosaceae (rose family), suggesting that the IAN pathway may be more widespread that previously believed.

THE BACTERIAL PATHWAY Another tryptophan-dependent biosynthetic pathway—one that uses **indole-3-acetamide (IAM)** as an intermediate (see Figure 19.7D)—is used by various pathogenic bacteria, such as *Pseudomonas savastanoi* and *Agrobacterium tumefaciens*. This pathway involves the two enzymes tryptophan monooxygenase and IAM hydrolase. The auxins produced by these bacteria often elicit morphological changes in their plant hosts.

IAA can also be synthesized from indole-3-glycerol phosphate

In addition to the tryptophan-dependent pathways, genetic studies indicate that plants can synthesize IAA via one or more tryptophan-independent pathways (see Figure 19.6). A tryptophan-independent pathway of IAA biosynthesis had long been suspected because, in isotopically-labeled precursor feeding experiments, the low levels of conversion of labeled tryptophan to IAA in some plant tissues seemed inadequate to account for the amount of auxin being produced. However, demonstration of a tryptophan-independent pathway for auxin biosynthesis was only possible through the use of molecular genetic techniques. Perhaps the most striking of these studies involves the *orange pericarp (orp)* maize mutant (Figure 19.8), which harbors

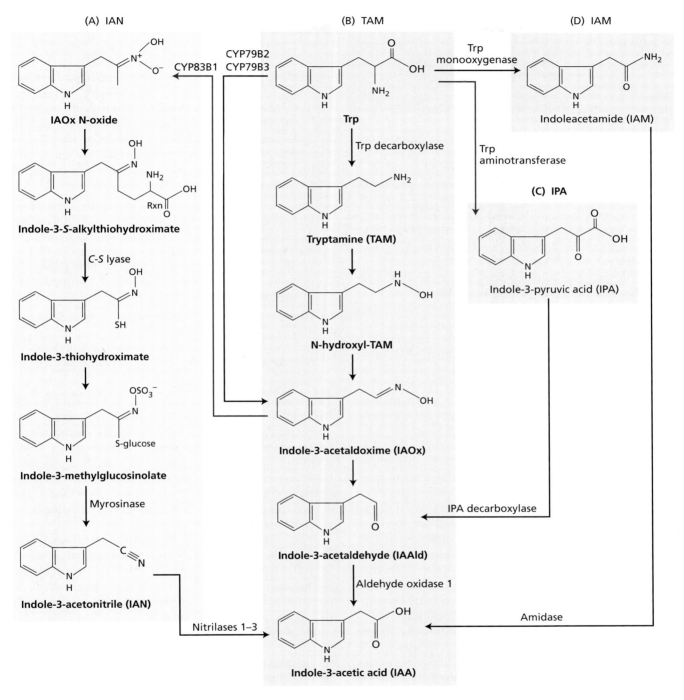

FIGURE 19.7 IAA biosynthetic pathways in plants and bacteria. The TAM and IPA pathways are the primary routes of IAA biosynthesis in plants. The IAN pathway may be limited to three families (Brassicaceae, Poaceae, and Musaceae). The IAM pathway functions in bacteria. The enzymes that catalyze specific reactions are indicated where known. The numbered CYPs are specific cytochrome p450 enzymes that catalyze steps in the IAN pathway, from tryptophan (Trp) to indole-3-acetaldoxime (IAOx), and from IAOx to IAOx N-oxide. Conversion of IAOx N-oxide to indole-3-thiohydroximate involves conjugation of a cysteine molecule followed by cleavage of a carbon sulfure (C-S) linkage C-S lyase. The enzyme myrosinase cleaves the cyanogen moiety from an indole-glucosinolate compound to form IAN (see Chapter 13). (After Woodward and Bartel 2005.)

FIGURE 19.8 The orange pericarp (*orp*) mutant of maize has mutations in both loci encoding tryptophan synthase β. As a result, the pericarps surrounding each mutant kernel (colored orange) accumulate glycosides of anthranilic acid and indole. The orange color is due to excess indole. (Courtesy of Jerry D. Cohen.)

mutations in both genes encoding the enzyme tryptophan synthase β. The *orp* mutant is a true tryptophan auxotroph, requiring exogenous tryptophan to survive. However, neither the *orp* seedlings nor the wild-type seedlings can convert tryptophan to IAA, even when the mutant seedlings are given enough tryptophan to reverse the lethal effects of the mutation.

Despite the block in tryptophan biosynthesis, the *orp* mutant contains amounts of IAA 50-fold higher than those of a wild-type plant (Wright et al. 1992). Significantly, when *orp* seedlings were fed [^{15}N]anthranilate (see Figure 19.6), the label subsequently appeared in IAA, but not in tryptophan. These results are the best experimental evidence for a tryptophan-independent pathway of IAA biosynthesis. Further studies established that the branch point for IAA biosynthesis is either indole-3-glycerol phosphate or its metabolite, indole. IAN and IPA are possible intermediates, but the immediate precursor of IAA in the tryptophan-independent pathway has not yet been identified.

The relative importance of the two IAA biosynthetic pathways (tryptophan-dependent versus tryptophan-independent) is poorly understood. In several plants it has been found that different IAA biosynthesis pathways function in different tissues and at different times in development. For example, during embryogenesis in carrot (*Daucus carota*), the tryptophan-dependent pathway is important very early in development, whereas the tryptophan-independent pathway appears to take over soon after the root–shoot axis is established. Some evidence suggests that tryptophan-dependent IAA biosynthesis is activated primarily in response to wounding, and that tryptophan-independent pathways may predominate under other conditions. The existence of multiple pathways for IAA biosynthesis makes it nearly impossible for plants to run out of auxin and is probably a reflection of the essential role of this hormone in plant development. (For more discussion of tryptophan-independent IAA biosynthesis, see **Web Topic 19.4**.)

Seeds and storage organs contain large amounts of covalently bound auxin

Apart from sites of synthesis, a significant proportion of auxin in tissues is covalently bound to both high and low molecular weight compounds. This is particularly the case in seeds and storage organs such as cotyledons. These conjugated, or "bound," auxins have been identified in all higher plants and are considered hormonally inactive. IAA is rapidly released from many, but not all, conjugates by enzymatic processes. Those conjugates that can release free auxin serve as reversible storage forms of the hormone. IAA conjugates are found in two categories:

- Low molecular weight compound–conjugated auxins include esters of IAA with a methyl group, glucose, or *myo*-inositol, and amide conjugates such as IAA-*N*-aspartate (Figure 19.9) Some of these low molecular weight conjugates function as hydrolyzable storage pools for IAA, while others function as intermediates in degradative processes.

- High molecular weight IAA conjugates include IAA-glucan (7–50 glucose units per IAA), IAA-peptides, and IAA-glycoproteins found in seeds. All of these conjugates appear to function in reversible storage of IAA.

The extent and nature of IAA conjugation depends on the tissue involved and the specific conjugating enzymes that are present. The best-studied reaction is the conjugation of IAA to glucose in the endosperm of maize (corn; *Zea mays*), but recent molecular genetic studies in Arabidopsis have enhanced our understanding of the regulation of IAA conjugation/deconjugation processes in vegetative tissues. The known IAA conjugates and the enzymes and intermediates involved are shown in Figure 19.9.

Indole-3-butyric acid (IBA) is another natural auxin. IBA and IBA conjugates also contribute to the pool of stored IAA. IBA-aminoacyl or glucosyl ester conjugates can be hydrolyzed to free IBA, which, in turn, can be enzymatically converted into IAA via β-oxidation in the peroxisome (see Figure 19.9A) (see Chapter 11).

Metabolism of conjugated auxin is a major factor in the regulation of the levels of free auxin. For example, during the germination of *Z. mays* seeds, IAA-*myo*-inositol is translocated from the endosperm to the coleoptile via the phloem. Most of the free IAA produced in coleoptile tips of *Z. mays* is believed to be derived from the hydrolysis of IAA-*myo*-inositol from the seed.

In addition, environmental stimuli such as light and gravity have been shown to influence both the rate of auxin conjugation (removal of free auxin) and the rate of release of free auxin (hydrolysis of conjugated auxin). The formation of conjugated auxins may serve other functions as

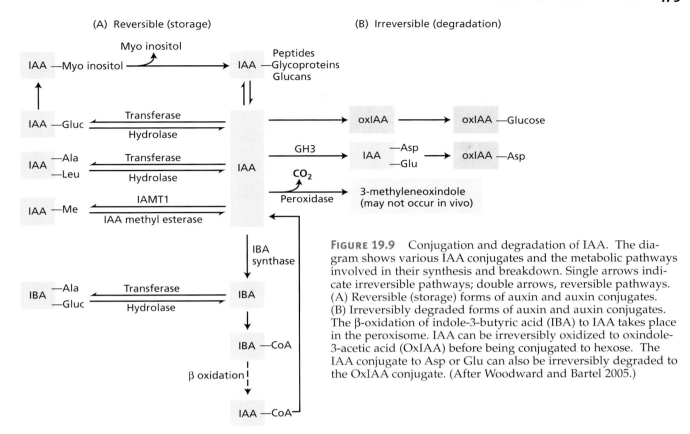

FIGURE 19.9 Conjugation and degradation of IAA. The diagram shows various IAA conjugates and the metabolic pathways involved in their synthesis and breakdown. Single arrows indicate irreversible pathways; double arrows, reversible pathways. (A) Reversible (storage) forms of auxin and auxin conjugates. (B) Irreversibly degraded forms of auxin and auxin conjugates. The β-oxidation of indole-3-butyric acid (IBA) to IAA takes place in the peroxisome. IAA can be irreversibly oxidized to oxindole-3-acetic acid (OxIAA) before being conjugated to hexose. The IAA conjugate to Asp or Glu can also be irreversibly degraded to the OxIAA conjugate. (After Woodward and Bartel 2005.)

well, including storage and protection against oxidative degradation.

IAA is degraded by multiple pathways

Hormones would be useless as developmental signals if they accumulated over time. Auxin catabolism ensures the degradation of active hormone when the concentration exceeds the optimal level or when the response to the hormone is complete. Like IAA biosynthesis, the enzymatic breakdown (oxidation) of IAA involves more than one pathway (see Figure 19.9B). Enzymatic decarboxylation of IAA to 3-methyleneoxindole was once thought to be the major degradative pathway, as peroxidases are ubiquitous in higher plants, and their ability to degrade IAA can be demonstrated in vitro. However, the physiological significance of the peroxidase pathway has been questioned. For example, no change in the IAA levels of transgenic plants was observed with either a tenfold increase in peroxidase expression or a tenfold repression of peroxidase activity (Normanly et al. 1995).

On the basis of isotopic labeling and metabolite identification, two other oxidative pathways are more likely to be involved in the controlled degradation of IAA (see Figure 19.9B). The end products of this pathway are oxindole-3-acetic acid (OxIAA) and, subsequently, OxIAA-glucose. The indole moiety of IAA is oxidized to form OxIAA. In another pathway, the IAA-aspartate conjugate is oxidized to OxIAA.

In vitro (especially in aqueous solution), IAA can be oxidized nonenzymatically when exposed to light, and its photodestruction in vitro can be promoted by plant pigments such as riboflavin. The products of auxin photooxidation have been isolated from plants, suggesting that photooxidation occurs in vivo as well as in vitro. However, the physiological role of auxin photooxidation is still unresolved (Geisler et al. 2005).

IAA partitions between the cytosol and the chloroplasts

The distribution of IAA and its metabolites has been studied in tobacco (*Nicotiana tabacum*) cells. About one-third of the IAA is found in the chloroplast, and the remainder is located in the cytosol. Tryptophan is synthesized in the chloroplast, and a number of IAA biosynthetic enzymes contain chloroplast-localizing sequences. As such, some IAA synthesis may take place in the chloroplast. However, if the conjugation of some of the IAA in the cytosol is taken into account, the pools of chloroplast and cytosolic free IAA appear to be at equilibrium, suggesting that chloroplast IAA pools are at least partially derived from cytosolic IAA pools (Sitbon et al. 1993). Alkaline trapping of anionic IAA may contribute to chloroplast accumulation (see the discussion of chemiosmotic auxin movement in the next section). The factors that regulate the steady-state concentration of free auxin in plant cells are diagrammatically summarized in **Web Topic 19.5**.

Auxin Transport

The main axes of shoots and roots, along with their branches, exhibit apex–base structural polarity, and this structural polarity is dependent on the polarity of auxin transport. Soon after Went developed the coleoptile curvature test for auxin, it was discovered that IAA moves mainly from the apical to the basal end (*basipetally*) in excised oat coleoptile sections. This type of unidirectional transport is termed **polar transport**. Auxin is the only plant growth hormone that has been clearly shown to be transported polarly, and transport of this hormone is found in almost all plants, including bryophytes and ferns. Recently, anatomical evidence of polar auxin flow was reported in 375 million year old fossil wood. In living woody plants, auxin "whirlpools" arise wherever polar transport is disrupted by the presence of obstacles such as buds and branches. As a result, the tracheary elements that differentiate in these regions form circular patterns. Identical circular patterns at the same positions in the wood (indicative of polar auxin transport) have now been detected in the fossil wood of a primitive gymnosperm dating to the Upper Devonian period (Rothwell and Lev-Yadun 2005).

Because the shoot apex serves as the primary source of auxin in the plant, polar transport has long been believed to be the principal cause of an auxin gradient extending from the shoot tip to the root tip. The longitudinal gradient of auxin from the shoot to the root affects various developmental processes, including embryonic development, stem elongation, apical dominance, wound healing, and leaf senescence. Some auxin transport also occurs in the phloem, and phloem-based movement, driven by "source–sink" translocation of sugars (see Chapter 10), contributes to the auxin transported *acropetally* (i.e., toward the tip) in the root. Thus, more than one mechanism is responsible for the distribution of auxin in the plant.

In the sections that follow we will discuss the cellular mechanisms underlying polar transport. We will also examine the cellular mechanisms that regulate polar transport and, later in the chapter, we will see how such regulatory mechanisms enable the plant to adapt to various environmental signals.

Polar transport requires energy and is gravity independent

Early studies of polar transport were carried out using the *donor–receiver agar block method* (Figure 19.10): An agar block containing radioisotope-labeled auxin (donor block) is placed on one end of a tissue segment, and a receiver block is placed on the other end. The movement of auxin through the tissue into the receiver block can be determined over time by measurement of the radioactivity in the receiver block. This method has been refined to allow for the deposition of much smaller droplets of radiolabeled auxin onto discrete surfaces of plants, improving the accuracy of transport studies over short distances (Murphy et al. 2000; Geisler et al. 2003).

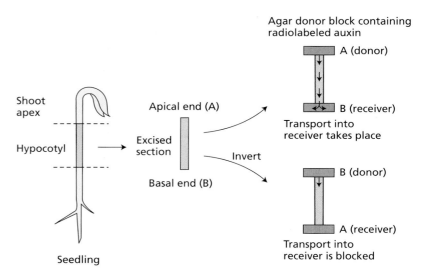

FIGURE 19.10 Donor–receiver block method for measuring polar auxin transport. The polarity of transport is independent of orientation with respect to gravity.

From a multitude of such studies, the general properties of polar IAA transport have emerged. Tissues differ in the degree of polarity of IAA transport. In coleoptiles, vegetative stems, and leaf petioles, basipetal transport predominates, whereas in the stelar tissues of the root auxin is transported acropetally. Polar transport is not affected by the orientation of the tissue (at least over short periods of time), so it is independent of gravity.

A demonstration of the lack of gravity effects on basipetal auxin transport is shown in Figure 19.11. When stem cuttings, in this case grape hardwood, are placed in a moist chamber, adventitious roots form at the basal ends of the cuttings, and shoots form at the apical ends, even when the cuttings are inverted. Roots form at the base because root differentiation is stimulated by auxin accumulation due to basipetal transport. Shoots tend to form at the apical end where the auxin concentration is lowest.

Polar transport proceeds in a cell-to-cell fashion, rather than via the symplast. That is, auxin exits the cell through the plasma membrane, diffuses across the compound middle lamella, and enters the next cell through its plasma membrane. The loss of auxin from cells is termed *auxin efflux*; the entry of auxin into cells is called *auxin uptake* or *influx*. The overall process requires metabolic energy, as evidenced by the sensitivity of polar transport to O_2 deprivation, sucrose depletion, and metabolic inhibitors.

FIGURE 19.11 Adventitious roots grow from the basal ends, and shoots grow from the apical ends, of grape hardwood cuttings, whether they are maintained in the inverted (the two cuttings on the left) or upright orientation (the cuttings on the right). The roots always form at the basal ends, because polar auxin transport is independent of gravity. (From Hartmann and Kester 1983.)

The velocity of polar auxin transport ranges from 2 to 20 cm h^{-1}, faster than the rate of diffusion (see **Web Topic 3.2**) but slower than phloem translocation rates (see Chapter 10). Higher rates of polar transport are observed in tissues immediately adjacent to the shoot and root apical meristems. Polar transport is also specific for active auxins, both natural and synthetic; other weak organic acids, inactive auxin analogs, and IAA conjugates are poorly transported. The specificity of polar transport indicates that auxin is recognized by protein carriers on the plasma membrane.

The major site of polar auxin transport in stems, leaves, and roots is the *vascular* parenchyma tissues, most likely the xylem. In vascular parenchyma, the overall direction of auxin transport in the plant is downward, toward the root tip. In other words, auxin is transported basipetally in the shoot and acropetally in the root. An exception is found in the coleoptiles of grasses, where basipetal polar transport occurs mainly in the *nonvascular* parenchyma tissues. In addition, auxin translocated in the phloem sieve tubes contributes to transport from shoot tissues to the growing root tip.

Basipetal auxin transport from the apex occurs in roots as well. In maize and Arabidopsis roots, for example, radiolabeled IAA applied to the root tip is transported basipetally for a distance of 2 to 8 mm. Basipetal auxin transport in the root occurs in the *epidermal* and *cortical tissues*, and as we shall see, it plays a central role in gravitropism.

A chemiosmotic model has been proposed to explain polar transport

The discovery of the chemiosmotic mechanism of solute transport in the late 1960s (see Chapter 6) led to the application of this model to polar auxin transport. According to the now generally accepted **chemiosmotic model** for polar auxin transport, auxin uptake is driven by the proton motive force ($\Delta E + \Delta pH$) across the plasma membrane, while auxin efflux is driven by the membrane potential, ΔE. (Proton motive force is described in more detail in Chapter 7.)

A crucial feature of the polar transport model is that the auxin efflux carriers are concentrated at the ends of the conducting cells (Figure 19.12). The evidence for each step in this model is considered separately in the discussion that follows.

AUXIN INFLUX The first step in polar transport is auxin influx. According to the model, auxin can enter plant cells by either of two mechanisms:

1. Passive diffusion of the protonated form (IAAH) from any direction across the phospholipid bilayer
2. Secondary active transport of the dissociated form (IAA$^-$) via a 2H$^+$–IAA$^-$ symporter

The dual pathway of auxin uptake arises because the passive permeability of the membrane to auxin depends strongly on the apoplastic pH.

The undissociated form of indole-3-acetic acid, in which the carboxyl group is protonated, is lipophilic and readily diffuses across lipid bilayer membranes. In contrast, the dissociated form of auxin is negatively charged and therefore does not cross membranes unaided. Because the plasma membrane H$^+$-ATPase normally maintains the cell wall solution at pH 5 to 5.5, about 25% of the auxin (pK$_a$ = 4.75) in the apoplast will be in the undissociated form and will diffuse passively across the plasma membrane down a concentration gradient. Experimental support for pH-dependent, passive auxin uptake was first provided by the demonstration that IAA uptake by plant cells increases as the extracellular pH is lowered from a neutral to a more acidic value.

A carrier-mediated, secondary active uptake mechanism has been shown to be saturable and specific for active auxins. In experiments in which the ΔpH and ΔE values of isolated membrane vesicles from zucchini (*Cucurbita pepo*) hypocotyls were manipulated artificially, the uptake of radiolabeled auxin was shown to be stimulated in the presence of a pH gradient, as in passive uptake, but also when the inside of the vesicle was negatively charged relative to the outside.

These and other experiments suggest that, at least in some tissues, an H$^+$–IAA$^-$ symporter cotransports two protons along with the auxin anion. This secondary active transport of auxin allows for greater auxin accumulation than simple diffusion does because it is driven across the membrane by the proton motive force.

A permease-type auxin uptake carrier, **AUX1**, related to bacterial amino acid carriers, functions in the leaf vascular tissue and root apices of Arabidopsis and other plants. In young leaves, AUX1 functions in the import of auxin into

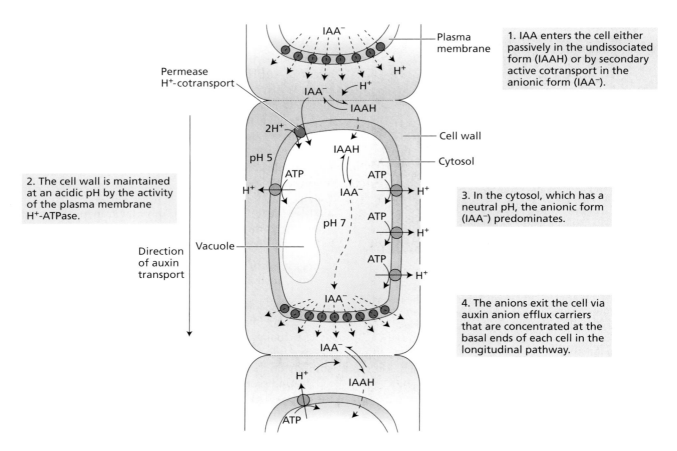

FIGURE 19.12 A simplified chemiosmotic model of polar auxin transport. Shown here is one elongated cell in a column of auxin-transporting cells. In smaller cells near the meristems, because of their high surface to volume ratio, back diffusion of IAA into cells is thought to require an additional energy-dependent efflux mechanism.

the vascular parenchyma polar transport stream. The roots of Arabidopsis *aux1* mutants exhibit agravitropic growth. This phenotype can be corrected by treatment with the synthetic auxin 1-naphthylene acetic acid (1-NAA), which readily crosses the lipid bilayer even in the absence of a carrier protein. This suggests that AUX1 and related proteins function as auxin influx carriers and that auxin influx is a limiting factor for root gravitropism.

AUX1 exhibits a polar localization at the ends of protophloem cells in the root apex (Figure 19.13A and C), where it is thought to function in the unloading of phloem-derived auxin moving acropetally toward the tip; but it also exhibits an apolar localization in 19.13B and root cap cells, where it appears to function in basipetal auxin transport. The significance of this basipetal redirection at the root tip will become apparent when we examine the mechanism of root gravitropism.

In addition to AUX1, an ATP-dependent mechanism may also contribute to auxin import in Arabidopsis root epidermal cells adjoining the apex. Active auxin transport mechanisms will be described following the discussion of auxin efflux according to the chemiosmotic model.

AUXIN EFFLUX Once IAA enters the cytosol, which has a pH of approximately 7.2, nearly all of it will dissociate to the anionic form. Because the membrane is less permeable to IAA$^-$ than to IAAH, IAA$^-$ will tend to accumulate in the cytosol. However, much of the auxin that enters the cell escapes via *auxin anion efflux carriers*. According to the chemiosmotic model, transport of IAA$^-$ out of the cell is driven by the inside negative membrane potential.

As noted earlier, the central feature of the chemiosmotic model for polar transport is that IAA$^-$ efflux is directed by polar localization of efflux carriers. The repetition of auxin uptake and preferential release from opposite poles of each cell in the pathway gives rise to the total polar transport effect. A family of proteins known as **PIN proteins** (named after the pin-shaped inflorescences formed by the *pin1* mutant of Arabidopsis; Figure 19.14A) are important components of auxin efflux carrier complexes. These proteins are localized precisely as the model would predict—aligned with the direction of auxin transport (see Figure 19.14B).

Mutational analysis and immunolocalization studies show that PIN proteins are required for auxin efflux and that different PIN family members mediate auxin efflux in

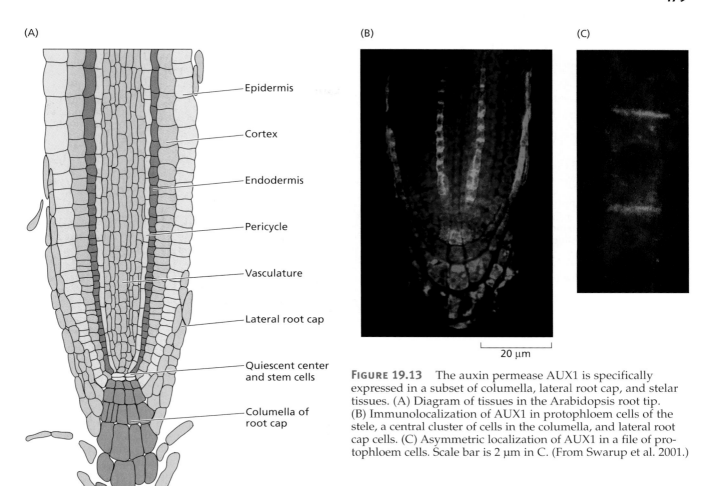

FIGURE 19.13 The auxin permease AUX1 is specifically expressed in a subset of columella, lateral root cap, and stelar tissues. (A) Diagram of tissues in the Arabidopsis root tip. (B) Immunolocalization of AUX1 in protophloem cells of the stele, a central cluster of cells in the columella, and lateral root cap cells. (C) Asymmetric localization of AUX1 in a file of protophloem cells. Scale bar is 2 μm in C. (From Swarup et al. 2001.)

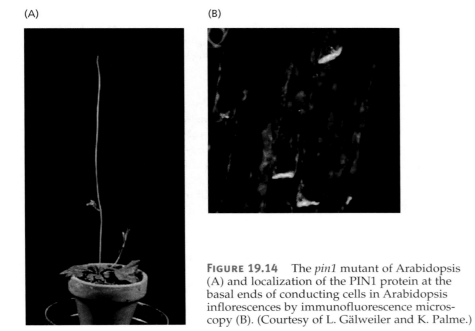

FIGURE 19.14 The *pin1* mutant of Arabidopsis (A) and localization of the PIN1 protein at the basal ends of conducting cells in Arabidopsis inflorescences by immunofluorescence microscopy (B). (Courtesy of L. Gälweiler and K. Palme.)

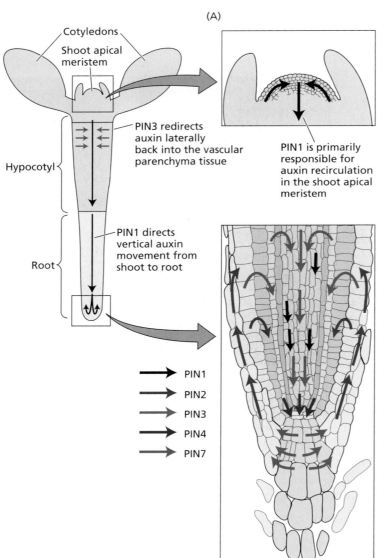
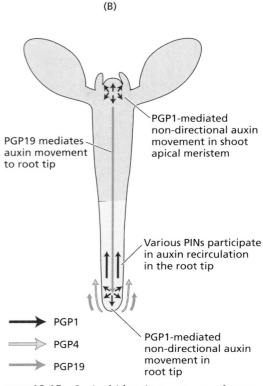

Figure 19.15 In Arabidopsis, two types of transport proteins direct the flow of auxin throughout the plant, PIN and PGP. (A) Directional auxin movement associated with the tissue-specific distribution of PIN efflux facilitator proteins. PIN1 mediates vertical transport of IAA from the shoot to the root. Since some lateral diffusion of auxin may occur, PIN3 is thought to redirect auxin back into vascular parenchymal tissue where polar transport takes place. The two insets show PIN1-mediated auxin movement in the shoot apical meristem (upper), and PIN-regulated auxin circulation in the root tip (lower). (B) Auxin flow associated with ATP-dependent PGP transport proteins. The multidirectional arrows at the shoot and root apices indicate non-directional auxin transport. However, when combined with polarly localized PIN proteins, directional transport occurs. (A, root model after Blilou et al. 2005.)

each tissue (Figure 19.15). PINs are integral membrane proteins that are similar to proteins of the major facilitator superfamily, which function in sugar transport and bacterial drug resistance. They are generally proteins with weak carrier transport activity that can facilitate transport by interacting with other proteins.

P-glycoproteins are also auxin transport proteins

Modeling of auxin transport in intact tissues suggests that polar auxin transport also requires an energy-dependent mechanism, especially in small cells of the apical meristems, where back-diffusion of auxin may occur. Moreover, molecular studies indicate that PIN-mediated auxin efflux involves interactions with other plasma membrane proteins, including both regulatory and transport proteins.

Plant cells contain ATP-dependent transporters belonging to the multidrug resistance/P-glycoprotein (MDR/PGP) family. **MDR/PGP proteins (PGPs)** are integral membrane proteins that have been shown to function as ATP-dependent hydrophobic anion carriers in cellular auxin efflux (Figure 19.16). Defective *PGP* genes in Arabidopsis, maize, and sorghum result in dwarf mutations of varying severity and in altered gravitropism and reduced auxin efflux (Figure 19.17).

There are 21 members of the PGP family in Arabidopsis and 17 in rice. Three members of the PGP family have been extensively characterized as participants in tissue-specific auxin transport (see Figure 19.15B). It is still unknown how many PGP family members play a role in auxin efflux; at least one family member has been shown to function in

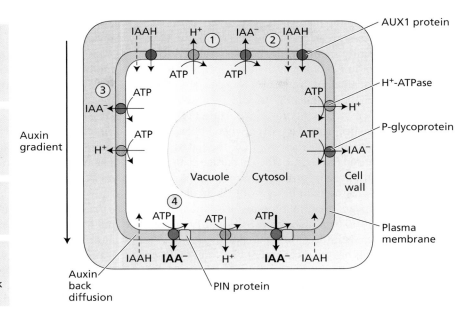

1. The plasma membrane H+-ATPase (purple) pumps protons into the apoplast. The acidity of apoplast affects the rate of auxin transport by altering the ratio of IAAH and IAA− present in the apoplast.

2. IAAH can enter the cell via proton symporters such as AUX1 (blue) or diffusion (dashed arrows). Once inside the cytosol, IAA is an anion, and may only exit the cell via active transport.

3. P-glycoproteins are localized nonpolarly on the plasma membrane and can drive active (ATP-dependent) auxin efflux.

4. Synergistically enhanced active polar transport occurs when polarly localized PIN proteins (yellow) associate with PGP proteins, overcoming the effects of back diffusion.

FIGURE 19.16 Model for polar auxin transport in small meristematic cells with significant back-diffusion of auxin due to a high surface-to-volume ratio.

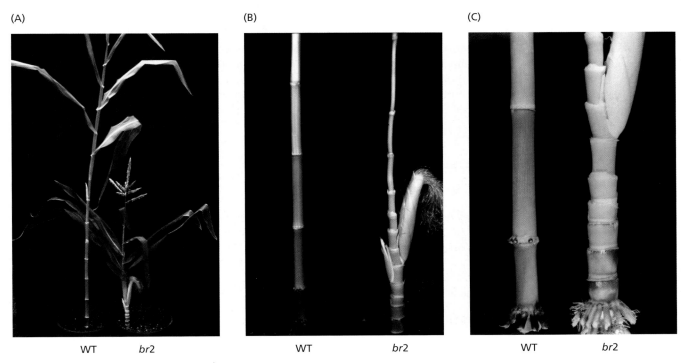

FIGURE 19.17 The *BR2* (*Brachytic 2*) gene encodes a P-glycoprotein required for normal auxin transport in corn, and *br2* mutants have short internodes. The mutant was created by insertional mutagenesis with the *Mutator* transposon. Unknown to the investigators, the *Mu8* transposon contained a fragment of the *BR2* gene. Expression of the *BR2* gene fragment was thought to produce interfering RNA (RNAi), which silenced BR2 expression (see Chapter 14 on the web site and Chapter 16). The *br2* mutants have normal tassels and ear (A and B), but compact lower stalks (B and C). (From Multani et al. 2003.)

alkaloid transport in the medicinal plant *Coptis japonica*, an Asian perennial evergreen (Ranunculaceae family). However, the PGP family member most closely related to the *Coptis* transporter in Arabidopsis, PGP4, appears to participate in auxin, not alkaloid, transport (Terasaka et al. 2005).

Unlike mammalian MDR/PGPs, which exhibit broad substrate specificity, plant auxin–transporting PGPs have relatively narrow specificity. Moreover, when PGPs interact with PIN proteins, the specificity of PGP-mediated auxin transport is enhanced (Noh et al. 2003, Blakeslee et al. 2005).

Although PGPs are uniformly rather than polarly distributed on the plasma membranes of cells in shoot and root apices, it appears that PIN proteins function synergistically with PGPs to establish the directionality of auxin transport. Studies with these proteins coexpressed in yeast and mamalian cells have indicated that PINs and PGPs can function both independently and synergistically to catalyze auxin transport (see **Web Topic 19.6**). However, in apical tissues it is the PIN proteins that confer directionality to PGP-mediated auxin efflux.

Inhibitors of auxin transport block auxin influx and efflux

Several compounds have been synthesized that can act as auxin transport inhibitors, including NPA (1-*N*-naphthylphthalamic acid), CPD (2-carboxyphenyl-3-phenylpropane-1,3-dione or "cyclopropyl propane dione"), TIBA (2,3,5-triiodobenzoic acid), and NOA (1-napthoxyacetic acid) (Figure 19.18). NPA, CPD, and TIBA are **auxin efflux inhibitors** (**AEIs**); they block polar transport by preventing auxin efflux. We can demonstrate this phenomenon by incorporating AEIs into either the donor or the receiver block in an auxin transport experiment. All three compounds inhibit auxin efflux into the receiver block, but they do not affect auxin uptake from the donor block. The reverse is seen with the influx inhibitor NOA.

Some AEIs, such as TIBA, have weak auxin activity and inhibit polar transport in part by competing with auxin at the efflux carrier site. Other AEIs, such as CPD and NPA, interfere with auxin transport by binding to a regulatory site. In Arabidopsis, PGPs have been shown to directly bind NPA (Noh et al. 2001; Murphy et al. 2002). In some tissues, NPA-binding proteins exhibit polar localization consistent with the localization of auxin efflux proteins (Jacobs and Gilbert 1983; Geisler et al. 2005).

In addition to binding to regulatory sites of auxin efflux carriers, AEIs can interfere with the trafficking of the plasma membrane proteins to which they bind, apparently by altering protein–protein interactions. The role of protein trafficking in regulating polar transport is discussed later in this section.

Auxin is also transported nonpolarly in the phloem

IAA originating in leaves can be transported to the rest of the plant nonpolarly via the phloem. Auxin, along with other components of phloem sap, can move from these leaves up or down the plant at velocities much higher than those of polar transport (see Chapter 10). Although loading and unloading of auxin into the phloem appears to be carrier-mediated, translocation in the phloem is largely passive and driven by source-sink forces. Long-distance auxin transport in the phloem appears to be important for controlling cambial cell divisions, callose accumulation or removal from sieve tube elements, and branch root formation. The phloem may make a major contribution to long-distance auxin translocation to the root.

Polar transport and phloem transport are not independent of each other. Studies with radiolabeled IAA suggest that

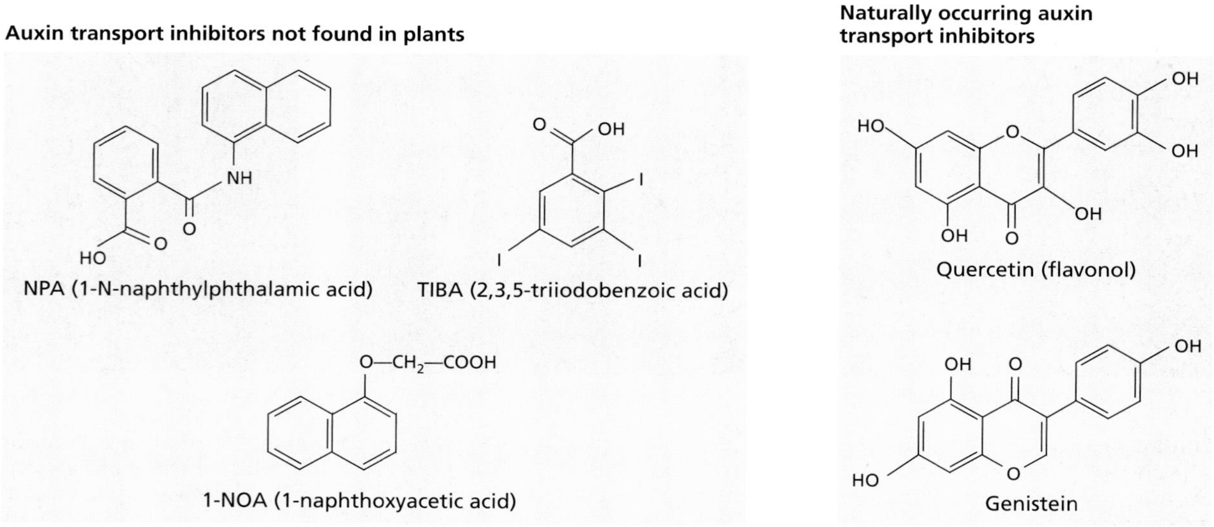

FIGURE 19.18 Structures of auxin transport inhibitors.

in pea (*Pisum sativum*), auxin can be transferred from the nonpolar phloem pathway to the polar transport pathway. This transfer takes place mainly in the immature tissues of the shoot apex. Similar results in Arabidopsis suggest that transfer of auxin from the phloem to polar transport streams are mediated by AUX1 and related proteins.

Auxin transport is regulated by multiple mechanisms

Expression of genes encoding auxin transport proteins is regulated in a tissue-specific fashion in response to developmental and environmental cues. Indeed, the expression of some auxin transport genes is regulated by auxin itself (Blakeslee et al. 2004; Peer et al. 2004). However, both efflux and influx appear to be regulated by posttranscriptional processes as well. Both protein phosphorylation and protein trafficking have been shown to play important roles in regulating polar transport.

PROTEIN PHOSPHORYLATION The phenotypes of auxin transport mutants can be mimicked by treating wild-type plants with inhibitors of protein kinases or phosphatases and by mutations in genes encoding protein kinases or phosphatases. The *pinoid* mutant in Arabidopsis has a phenotype similar to that of *pin1*. *PINOID* encodes a putative protein kinase, and alterations in PINOID expression result in altered auxin transport and subcellular localization of PIN proteins. The *rcn1* (*r*oot *c*urling in *N*PA *1*) mutant, which is insensitive to some effects of AEIs, harbors a mutation in a gene encoding a protein phosphatase subunit.

Flavonols such as quercetin and kaempferol, and isoflavones such as genestein, are naturally occurring plant compounds that inhibit the activity of certain kinases and phosphatases (see Chapter 13). Flavonols have been shown to displace AEIs from membrane binding sites, and flavonol-deficient mutants in Arabidopsis exhibit increased auxin transport and auxin-dependent growth. Arabidopsis mutants that overproduce flavonols exhibit decreased auxin transport (Murphy et al. 2000; Brown et al. 2001; Peer et al. 2004).

Flavonols are also inhibitors of auxin-transporting PGPs, as well as the associated NPA-binding membrane aminopeptidase APM1, which appears to function in the plasma membrane trafficking of PGP1, PGP4, and PGP19 (Murphy et al. 2002; Geisler et al. 2005). The mechanism of inhibition is suggested by the effects of flavonols on mammalian PGPs: both quercitin and kaempferol inhibit the hydrolysis of ATP to ADP and inorganic phosphate required for PGP transport proteins to undergo the conformational changes required for efflux activity; quercetin also inhibits phosphorylation of a key regulatory domain.

PROTEIN TRAFFICKING The polar localization of auxin transport proteins involves the secretion of vesicles to specific sites on the plasma membranes of auxin-conducting cells. Experiments with AEIs and cellular trafficking inhibitors have shown that plasma membrane localization of both PIN efflux facilitator proteins and AUX1 influx carriers is regulated by trafficking mechanisms that involve endocytotic cycling between the plasma membrane and endosomal compartments (Geldner et al. 2001, 2003; Swarup et al. 2001; Swarup and Bennet, 2003).

Endocytotic cycling was first demonstrated with PIN1 by Geldner et al. (2001). Before treatment, the PIN1 protein is localized at the lower ends of root vascular cells (Figure 19.19A). Treatment of Arabidopsis seedlings with brefeldin A (BFA), which causes Golgi vesicles and other endosomal compartments to aggregate near the nucleus, causes PIN1 to accumulate in abnormal intracellular compartments (see Figure 19.19B). When the BFA is washed out with buffer, the normal localization on the plasma membrane at the base of the cell is restored. But when cytochalasin D, an inhibitor of actin polymerization, is included in the buffer washout solution, normal relocalization of PIN1 to the plasma membrane is prevented.

(A) (B)

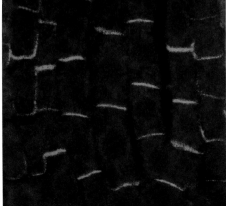
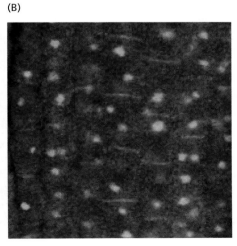

FIGURE 19.19 Auxin transport inhibitors block secretion of the auxin efflux carrier PIN1 to the plasma membrane. (A) Control, showing asymmetric localization of PIN1. (B) After treatment with brefeldin A (BFA), following a washout with buffer, PIN1 localization is restored as in A. When BFA is washed out with the auxin transport inhibitor TIBA, PIN1 is retained in endomembrane compartments as in B. (Photos courtesy of Klaus Palme 1999.)

These results indicate that PIN is rapidly cycled between the plasma membrane at the base of the cell and an unidentified endosomal compartment by an actin-dependent mechanism. Subsequently, GNOM, a protein involved in the regulation of vesicle trafficking, was shown to be the specific target of BFA (see Chapter 16). Several proteins, including a plasma membrane H$^+$-ATPase, are targeted to the membrane by the same BFA-sensitive mechanism that regulates PIN cycling.

When high concentrations of the auxin efflux inhibitors TIBA and NPA were added to the washout buffer in the experiments above, they prevented the normal relocalization of PIN on the plasma membrane. Replacement of synthetic AEIs in washout buffers with natural flavonol inhibitors of auxin transport also prevented plasma membrane relocalization of PIN1, although only in flavonoid-deficient mutants (Peer et al. 2004). TIBA and NPA also altered the endocytotic cycling of the plasma membrane H$^+$-ATPase and other proteins, suggesting either that some AEIs are general inhibitors of membrane cycling or that components of protein complexes containing PINs regulate specific cellular polar trafficking mechanisms. However, neither TIBA nor NPA alone causes PIN delocalization, even though they block auxin efflux. Therefore, NPA and TIBA must also directly inhibit the transport activity of efflux carrier complexes on the plasma membrane.

A simplified model of the effects of TIBA and NPA on PIN1 cycling and auxin efflux is shown in Figure 19.20. A more complete model, which incorporates additional regulatory steps, is presented in **Web Essay 19.1**.

Polar auxin transport is required for development

Polarized auxin movement is essential to the development of the basic shoot–root polarity of the plant. The presence alone of auxin is not sufficient; a polarized flow of auxin through the developing tissues is also necessary. Treatment of plant embryos with auxin transport inhibitors results in severe developmental abnormalities and loss of polar growth at both shoot and root apices. Studies of Arabidopsis mutants have demonstrated just how essential polar auxin transport is to early development: Severe *gnom* mutant alleles, which lack a key enzyme required for the normal cellular distribution of PIN proteins, exhibit a complete loss of apical-basal morphology (see Chapter 16). Double and triple *pin* mutants exhibit a similar loss of polar development. For a discussion of the developmental basis of plant polarity, see **Web Essay 19.2**.

To a large extent, the morphology of a plant depends on the directed movement of auxin via the polar transport system. Once auxin arrives at its destination, however, it initiates physiological responses that are determined by the target tissues. In the following section we will discuss the various physiological responses to auxin. At the end of the chapter we will examine the signaling pathways that bring about these responses.

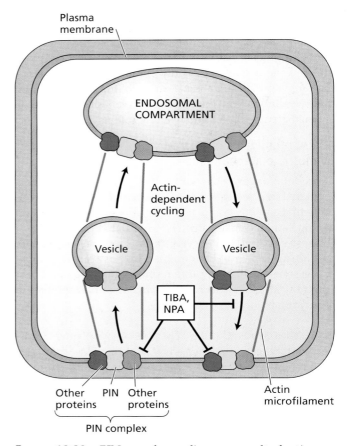

FIGURE 19.20 PIN complex cycling as a result of actin-dependent vesicle trafficking between the plasma membrane and an endosomal compartment. Auxin transport inhibitors TIBA and NPA interfere with the relocalization of PIN1 proteins to basal plasma membranes after a washout with brefeldin A (see Figure 19.19). This suggests that auxin transport inhibitors may act by interfering with PIN1 complex cycling.

Actions of Auxin: Cell Elongation

Auxin was discovered as the hormone involved in the bending of coleoptiles toward light. The coleoptile bends because of the unequal rates of cell elongation on its shaded versus its illuminated side (see Figure 19.1). The ability of auxin to regulate the rate of cell elongation has long fascinated plant scientists. In this section we will review the physiology of auxin-induced cell elongation, some aspects of which were discussed in Chapter 15.

Auxins promote growth in stems and coleoptiles, while inhibiting growth in roots

As we have seen, auxin synthesized in the shoot apex is transported basipetally to the tissues below. The steady supply of auxin arriving at the subapical region of the stem or coleoptile is required for the continued elongation of these cells. Because the level of endogenous auxin in the elongation region of a normal healthy plant is nearly optimal for growth, spraying the plant with exogenous auxin causes only a modest and short-lived stimulation in growth. Such spraying may even be inhibitory in the case of dark-grown seedlings, which are more sensitive to supraoptimal auxin concentrations than light-grown plants are.

However, when the endogenous source of auxin is removed by excision of sections containing the elongation zones, the growth rate rapidly decreases to a low basal rate. Such excised sections will often respond dramatically to exogenous auxin by rapidly increasing their growth rate back to the level in the intact plant.

In long-term experiments, treatment of excised sections of coleoptiles (see Figure 19.2) or dicot stems with auxin stimulates the rate of elongation of the section for up to 20 hours (Figure 19.21). The optimal auxin concentration for elongation growth in pea stems and oat coleoptiles is typically 10^{-6} to 10^{-5} M (see Figure 19.20). In other species, such as Arabidopsis, the optimum concentration is slightly lower. The inhibition beyond the optimal concentration is attributed mainly to auxin-induced ethylene biosynthesis. As we will see in Chapter 22, the gaseous hormone ethylene inhibits stem elongation in many species.

Auxin control of root elongation growth has been more difficult to demonstrate, perhaps because auxin induces the production of ethylene, which inhibits root growth as well. However, even if ethylene biosynthesis is specifically blocked, low concentrations (10^{-10} to 10^{-9} M) of auxin promote the growth of intact roots, whereas higher concentrations (10^{-6} M) inhibit growth. Thus, roots may require a minimum concentration of auxin to grow, but root growth is strongly inhibited by auxin concentrations that promote elongation in stems and coleoptiles.

The outer tissues of dicot stems are the targets of auxin action

Dicot stems are composed of many types of tissues and cells, only some of which may limit the growth rate. This point is illustrated by a simple experiment. When stem sections from growing regions of an etiolated dicot stem, such as pea, are split lengthwise and incubated in buffer, the two halves bend outward. This result indicates that in the absence of auxin the central tissues—including the pith, vascular tissues, and inner cortex—elongate at a faster rate than the outer tissues, consisting of the outer cortex and epidermis. Thus the outer tissues must be limiting the extension rate of the stem in the absence of auxin.

However, when the split sections are incubated in buffer plus auxin, the two halves now curve inward, demonstrating that the outer tissues of dicot stems are the primary targets of auxin action during cell elongation. Studies of auxin movement and auxin-responsive growth in Arabidopsis roots suggest that epidermal cells are the principal targets of auxin action in the root elongation zone as well.

To reach its targets in the elongating regions of dicot shoots, auxin derived from the shoot apex must be diverted laterally from the polar transport stream in vascular parenchyma cells to the outer shoot tissues. In contrast, all of the nonvascular tissues (epidermis plus mesophyll) are capable of transporting auxin as well as responding to it in coleoptiles.

This process of lateral auxin transport in dicots appears to be mediated by laterally oriented auxin efflux complexes in vascular parenchyma cells (Friml et al. 2002). The mechanism of lateral transport will be discussed later in the chapter in relation to plant bending responses.

The minimum lag time for auxin-induced growth is ten minutes

When a stem or coleoptile section is excised and inserted into a sensitive growth-measuring device, the growth response to auxin can be monitored at very high resolution. Without auxin in the medium, the growth rate declines rapidly. Addition of auxin markedly stimulates the growth rate after a lag period of only 10 to 12 minutes (see the inset in Figure 19.21). During this period biochemical, molecular, and cellular signaling events are occurring, as will be discussed later in the chapter. As is shown in Figure 19.22,

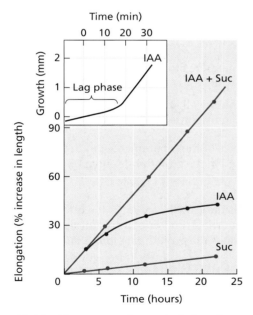

FIGURE 19.21 Time course for auxin-induced growth of oat *Avena* coleoptile sections. Growth is plotted as the percent increase in length. Auxin was added at time zero. When sucrose (Suc) is included in the medium, the response can continue for as long as 20 hours. Sucrose prolongs the growth response to auxin mainly by providing an osmotically active solute that can be taken up for the maintenance of turgor pressure during cell elongation. KCl can substitute for sucrose. The inset shows a short-term time course plotted with an electronic position-sensing transducer. In this graph, growth is plotted as the absolute length in millimeters versus time. The curve shows a lag time of about 15 minutes for auxin-stimulated growth to begin. (After Cleland 1995.)

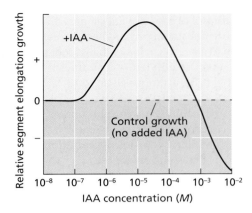

Figure 19.22 Typical dose–response curve for IAA-induced growth in pea stem or oat coleoptile sections. Elongation growth of excised sections of coleoptiles or young stems is plotted versus increasing concentrations of exogenous IAA. At higher concentrations (above 10^{-5} M), IAA becomes less and less effective; above about 10^{-4} M it becomes inhibitory, as shown by the fact that the curve falls below the dashed line, which represents growth in the absence of added IAA.

a threshold concentration of auxin must be reached to initiate this response. Beyond the optimum concentration, auxin becomes inhibitory.

Both oat (*Avena sativa*) coleoptiles and soybean (*Glycine max*) hypocotyls reach a maximum growth rate after 30 to 60 minutes of auxin treatment (Figure 19.23). This maximum represents a five- to tenfold increase over the basal rate. Oat coleoptile sections can maintain this maximum rate for up to 18 hours in the presence of osmotically active solutes such as sucrose or KCl.

The stimulation of growth by auxin requires energy, and metabolic inhibitors inhibit the response within minutes.

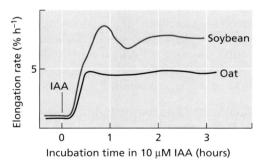

Figure 19.23 Comparison of the growth kinetics of oat coleoptile and soybean hypocotyl sections, incubated with 10 μM IAA and 2% sucrose. Growth is plotted as the rate at each time point, rather than the rate of the absolute length. The growth rate of the soybean hypocotyl oscillates after 1 hour, whereas that of the oat coleoptile is constant. (After Cleland 1995.)

Auxin-induced growth is also sensitive to inhibitors of protein synthesis such as cycloheximide, suggesting that protein synthesis is required for the response. Inhibitors of RNA synthesis also inhibit auxin-induced growth after a slightly longer delay.

Although the lag time for auxin-stimulated growth can be *lengthened* by lowering the temperature or by using low auxin concentrations so that it takes longer for auxin to diffuse into the tissue, the lag time cannot be *shortened* by raising the temperature, by using high auxin concentrations, or by abrading the waxy cuticle on the surface of the stem or coleoptile section to allow auxin to penetrate the tissue more rapidly. Thus the minimum lag time of 10 minutes is not determined by the time required for auxin to reach its site of action. Rather, the lag time reflects the time needed for the biochemical machinery of the cell to bring about the increase in the growth rate.

Auxin rapidly increases the extensibility of the cell wall

How does auxin cause a five- to tenfold increase in the growth rate in only 10 minutes? To understand the mechanism, we must first review the process of cell enlargement in plants (see Chapter 15). Plant cells expand in three steps:

1. Osmotic uptake of water across the plasma membrane is driven by the gradient in water potential ($\Delta \Psi_w$).

2. Turgor pressure builds up because of the rigidity of the cell wall.

3. Biochemical wall loosening occurs, allowing the cell to expand in response to turgor pressure.

The effects of these parameters on the growth rate are encapsulated in the growth rate equation:

$$GR = m\,(\Psi_p - Y)$$

where GR is the growth rate, Ψ_p is the turgor pressure, Y is the yield threshold, and m is the coefficient (*wall extensibility*) that relates the growth rate to the difference between Ψ_p and Y.

In principle, auxin could increase the growth rate by increasing m, increasing Ψ_p, or decreasing Y. Although extensive experiments have shown that auxin does not increase turgor pressure when it stimulates growth, conflicting results have been obtained regarding auxin-induced decreases in the yield threshold, Y. However, there is general agreement that auxin causes an increase in the wall extensibility parameter, m. This increase in m is mediated by hydrogen ions.

Auxin-induced proton extrusion increases cell extension

According to the widely accepted **acid growth hypothesis**, hydrogen ions act as the intermediate between auxin and cell wall loosening. The source of the hydrogen ions is the plasma membrane H^+-ATPase, whose activity is thought to increase in response to auxin. The acid growth hypothesis allows five main predictions:

1. Acid buffers alone should promote short-term growth, provided the cuticle has been abraded to allow the protons access to the cell wall.
2. Auxin should increase the rate of proton extrusion (wall acidification), and the kinetics of proton extrusion should closely match those of auxin-induced growth.
3. Neutral buffers should inhibit auxin-induced growth.
4. Compounds (other than auxin) that promote proton extrusion should stimulate growth.
5. Cell walls should contain a "wall-loosening factor" with an acidic pH optimum.

All five of these predictions have been confirmed. Acidic buffers cause a rapid and immediate increase in the growth rate, provided the cuticle has been abraded. Auxin stimulates proton extrusion into the cell wall after 10 to 15 minutes of lag time, consistent with the growth kinetics (Figure 19.24).

Auxin-induced growth has also been shown to be inhibited by neutral buffers, as long as the cuticle has been abraded. *Fusicoccin*, a fungal phytotoxin, stimulates both rapid proton extrusion and transient growth in stem and coleoptile sections (see Web Topic 19.7). And finally, wall-loosening proteins called *expansins* have been identified in the cell walls of a wide range of plant species (see Chapter 15). At acidic pH values, expansins loosen cell walls by weakening the hydrogen bonds between the polysaccharide components of the wall.

Auxin-induced proton extrusion may involve both activation and synthesis

In theory, auxin could increase the rate of proton extrusion by three possible mechanisms:

1. Activation of preexisting plasma membrane H^+-ATPases
2. Synthesis of new plasma membrane H^+-ATPases
3. Increased H^+-ATPase abundance on the plasma membrane

H^+-ATPASE ACTIVATION Adding auxin to isolated plasma membrane vesicles from tobacco cells in vitro causes a small stimulation (about 20%) of the ATP-driven proton-pumping activity, raising the possibility that auxin can directly activate the H^+-ATPase. A greater stimulation (about 40%) is observed if the living cells are treated with IAA prior to isolating the membranes, suggesting that a cellular factor may also be required. An auxin-binding protein (ABP1) has been linked to the direct activation of the plasma membrane H^+-ATPase in the presence of auxin, but, as discussed later in the chapter, how ABP1 mediates this response is unknown.

H^+-ATPASE ACTIVATION SYNTHESIS AND SECRETION The ability of protein synthesis inhibitors, such as cycloheximide, to rapidly inhibit auxin-induced proton extrusion and growth suggests that auxin might also stimulate proton pumping by increasing the synthesis of the H^+-ATPase. An increase in the amount of plasma membrane ATPase in *Z. mays* coleoptiles was detected immunologically after only 5 minutes of auxin treatment, and a doubling of the H^+-ATPase was observed after 40 minutes of treatment. A threefold stimulation by auxin of an mRNA for the plasma membrane H^+-ATPase was observed in the nonvascular tissues of the coleoptiles.

Plasma membrane H^+-ATPases appear to be dynamically cycled to and from the plasma membrane by the same auxin-sensitive mechanism that regulates the cellular trafficking of PIN1. Thus auxin may also up-regulate plasma membrane H^+-ATPases by increasing their residence time on the membrane.

In summary, the question of activation versus synthesis and secretion is still unresolved, and it is possible that auxin stimulates proton extrusion by all three mechanisms. Figure 19.25 summarizes some of the processes involved in auxin-induced cell wall loosening via proton extrusion. Auxin receptors shown to be involved in gene regulation, called TIR1/AFBs, will be discussed later in the chapter.

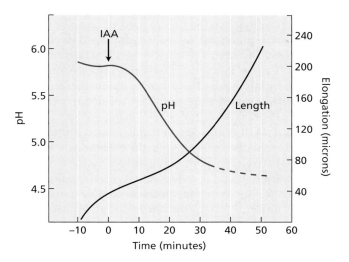

FIGURE 19.24 Kinetics of auxin-induced elongation and cell wall acidification in maize coleoptiles. The pH of the cell wall was measured with a pH microelectrode. Note the similar lag times (10 to 15 minutes) for both cell wall acidification and the increase in the rate of elongation. (After Jacobs and Ray 1976.)

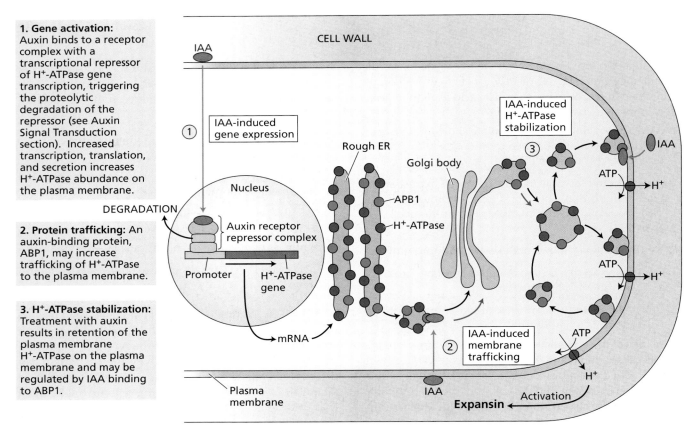

1. Gene activation: Auxin binds to a receptor complex with a transcriptional repressor of H^+-ATPase gene transcription, triggering the proteolytic degradation of the repressor (see Auxin Signal Transduction section). Increased transcription, translation, and secretion increases H^+-ATPase abundance on the plasma membrane.

2. Protein trafficking: An auxin-binding protein, ABP1, may increase trafficking of H^+-ATPase to the plasma membrane.

3. H^+-ATPase stabilization: Treatment with auxin results in retention of the plasma membrane H^+-ATPase on the plasma membrane and may be regulated by IAA binding to ABP1.

FIGURE 19.25 Current models for IAA-induced H^+ extrusion. In many plants, both of these mechanisms may operate. Regardless of how H^+ pumping is increased, acid-induced wall loosening is thought to be mediated by expansins.

Actions of Auxin: Phototropism and Gravitropism

Although plant growth can be influenced by many environmental factors, three main guidance systems control the orientation of the plant axis:

1. **Phototropism**, or growth with respect to light, is expressed in all shoots and some roots; it ensures that leaves will receive optimal sunlight for photosynthesis.

2. **Gravitropism**, growth in response to gravity, enables roots to grow downward into the soil and shoots to grow upward away from the soil, which is especially critical during the early stages of germination.

3. **Thigmotropism**, or growth with respect to touch, enables roots to grow around obstacles and is responsible for the ability of the shoots of climbing plants to wrap around other structures for support.

In this section we will examine the evidence that bending in response to light or gravity results from the lateral redistribution of auxin. We will also consider the cellular mechanisms involved in generating lateral auxin gradients during bending growth. Less is known about the mechanism of thigmotropism, although it, too, probably involves auxin gradients.

Phototropism is mediated by the lateral redistribution of auxin

As we saw earlier, Charles and Francis Darwin provided the first clue concerning the mechanism of phototropism by demonstrating that the sites of perception and differential growth (bending) are separate: Light is perceived at the tip, but bending occurs below the tip. The Darwins proposed that some "influence" that was transported from the tip to the growing region brought about the observed asymmetric growth response. This influence was later shown to be indole-3-acetic acid—auxin.

When a shoot is growing vertically, auxin is transported polarly from the growing tip to the elongation zone. The polarity of auxin transport from tip to base is developmentally determined and is independent of orientation with respect to gravity. However, auxin can also be transported laterally, and this lateral movement of auxin lies at the heart of a model for tropisms originally proposed separately by Nicolai Cholodny and Frits Went in the 1920s.

According to the Cholodny–Went model of phototropism, the tips of grass coleoptiles are sites of high auxin concentration and have two other specialized functions:

1. The perception of a unilateral light stimulus

2. The lateral transport of IAA in response to the phototropic stimulus

Thus, in response to a directional light stimulus, the auxin produced at the tip, instead of being transported basipetally, is transported laterally toward the shaded side.

Although phototropic mechanisms appear to be highly conserved across plant species, the precise sites of auxin production, light perception, and lateral transport have been difficult to define. In maize coleoptiles, auxin accumulates in the upper 1 to 2 mm of the tip. The zones of photosensing and lateral transport extend farther, within the upper 5 mm of the tip. The response is also strongly dependent on the light *fluence* (see **Web Topic 19.8**). Although similar zones of auxin synthesis/accumulation, light perception, and lateral transport are seen in the true shoots of all monocots and dicots examined to date, the localization of the primary polar transport mechanism in the vascular parenchyma tissues of dicots results in a lateral transport mechanism that is somewhat different from that seen in coleoptiles of monocots.

Two flavoproteins, *phototropins 1* and *2*, are the photoreceptors for the blue-light signaling pathway (see **Web Essay 19.3**) that induces phototropic bending in Arabidopsis hypocotyls and oat coleoptiles under both high- and low-fluence conditions (Briggs et al. 2001).

Phototropins are autophosphorylating protein kinases whose activity is stimulated by blue light. The action spectrum for blue-light activation of the kinase activity closely matches the action spectrum for phototropism, including the multiple peaks in the blue region. Phototropin 1 displays a lateral gradient in phosphorylation during exposure to low-fluence unilateral blue light.

In coleoptiles, the gradient in phototropin phosphorylation appears to induce the lateral movement of auxin to the shaded side of the coleoptile (see **Web Topic 19.8**). Once the auxin reaches the shaded side of the tip, it is transported

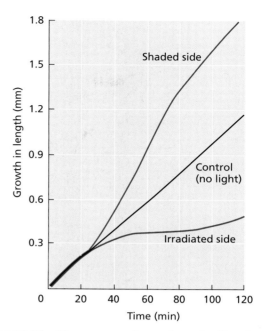

FIGURE 19.26 Time course of growth on the illuminated and shaded sides of a coleoptile responding to a 30-second pulse of unidirectional blue light. Control coleoptiles were not given a light treatment. (After Iino and Briggs 1984.)

basipetally to the elongation zone, where it stimulates cell elongation. The acceleration of growth on the shaded side and the slowing of growth on the illuminated side (called *differential growth*) result in curvature toward light (Figure 19.26).

Direct tests of the Cholodny–Went model using the agar block/coleoptile curvature bioassay have supported the model's prediction that auxin in coleoptile tips is transported laterally in response to unilateral light (Figure 19.27).

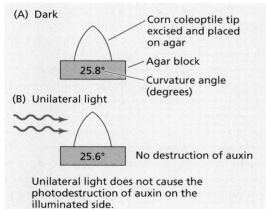
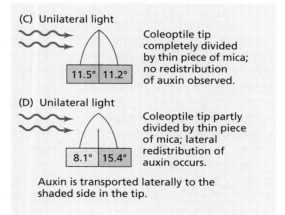

FIGURE 19.27 Evidence that the lateral redistribution of auxin is stimulated by unidirectional light in corn coleoptiles. The amount of auxin in the agar block is expressed as the angle of curvature the agar block induces when assayed by the coleoptile curvature bioassay (see Figure 19.1). (A) The amount of IAA diffusing into the block in darkness. (B) The amount of IAA diffusing into the block in unilateral light. (C) Lack of light-induced IAA migration to the shaded side in a coleoptile tip divided by mica barrier. (D) Light-induced migration of IAA to the shaded side in an intact tip.

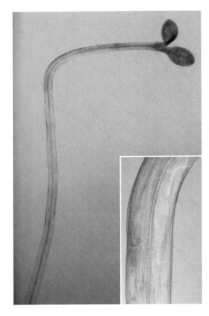
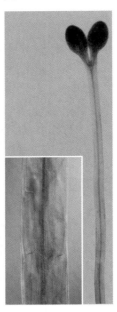

FIGURE 19.28 Lateral redistribution of auxin during phototropism can be visualized by transforming plants with the *DR5::GUS* reporter gene construct. (A) Lateral auxin gradients are formed in Arabidopsis hypocotyls during the bending response to unidirectional light. Auxin accumulation on the shaded side of the hypocotyls is indicated by the blue staining (inset). (B) Treatment with the auxin efflux inhibitor NPA blocks both phototropic bending and auxin redistribution. A similar redistribution of auxin occurs during gravitropism. (Photos courtesy of Klaus Palme.)

The total amount of auxin diffusing out of the tip (here expressed as the angle of curvature) is the same in the presence of unilateral light as in darkness (compare Figure 19.27A and B). This result indicates that, in coleoptiles, light does not cause the photodestruction of auxin on the illuminated side, as had been proposed by some investigators.

Acidification of the apoplast appears to play a role in phototropic growth: The apoplastic pH on the shaded side of phototropically bending stems or coleoptiles is more acidic than on the side facing the light. Decreased pH increases auxin transport by increasing both the rate of IAA entry into the cell and the chemiosmotic proton potential driven efflux mechanisms. Consistent with the acid growth hypothesis, this acidification would also be expected to enhance cellular elongation. Both processes—enhanced auxin uptake and increased cell elongation on the shaded side—would be expected to contribute to bending toward light.

Experiments in Arabidopsis suggest that phototropic auxin movement involves laterally-localized efflux complexes characterized by the PIN3 protein (Friml et al, 2003) and destabilization of basally-localized efflux complexes characterized by PIN1 (Blakeslee et al. 2004). The resulting loss of downward auxin movement is thought to increase the pool of auxin available for lateral redirection (Noh et al. 2003). These changes in auxin distribution can be visualized using the auxin-responsive reporter gene construct *DR5::GUS*, which utilizes the auxin-sensitive promoter, *DR5*, fused to the *GUS* reporter gene (Figure 19.28).

Gravitropism involves lateral redistribution of auxin

When dark-grown *Avena* seedlings are oriented horizontally, the coleoptiles bend upward in response to gravity. According to the Cholodny–Went model, auxin in a horizontally oriented coleoptile tip is transported laterally to the lower side, causing the lower side of the coleoptile to grow faster than the upper side. Early experimental evidence indicated that the tip of the coleoptile can perceive gravity and redistribute auxin to the lower side. For example, if coleoptile tips are oriented horizontally, a greater amount of auxin diffuses from the lower half than the upper half (Figure 19.29).

Tissues below the tip are able to respond to gravity as well. For example, when vertically oriented maize coleoptiles are decapitated by removal of the upper 2 mm of the tip and oriented horizontally, gravitropic bending occurs at a slow rate for several hours even without the tip. Application of IAA to the cut surface restores the rate of bending to normal levels. This finding indicates that both the perception of the gravitational stimulus and the lateral redistribution of auxin can occur in the tissues below the tip, although the tip is still required for auxin production.

Lateral redistribution of auxin is more difficult to demonstrate in shoot apical meristems than in coleoptiles because of the presence of auxin recirculation ("fountain effect"), similar to that in root tips, in developing leaf and apical shoot primordia (Benkova et al. 2003). However, mechanisms similar to those seen in phototropic bending are also involved in shoot gravitropic bending.

In soybean hypocotyls, gravitropism leads to a rapid asymmetry in the accumulation of a group of auxin-stimulated mRNAs called **SAURs** (*s*mall *a*uxin *u*p-regulated *R*NAs) (McClure and Guilfoyle 1989). In vertical seedlings, SAUR gene expression is symmetrically distributed. Within 20 minutes after the seedling is oriented horizontally, SAURs begin to accumulate on the lower half of the hypocotyl. Under these conditions, gravitropic bending first becomes evident after 45 minutes, well after the induction of the SAURs (see **Web Topic 19.9**). The existence of a lateral gradient in SAUR gene expression is indirect evidence for the existence of a lateral gradient in auxin detectable within 20 minutes of the gravitropic stimulus.

Dense plastids serve as gravity sensors

Unlike unidirectional light, gravity does not form a gradient between the upper and lower sides of an organ. All parts of the plant experience the gravitational stimulus equally. How do plant cells detect gravity? The only way

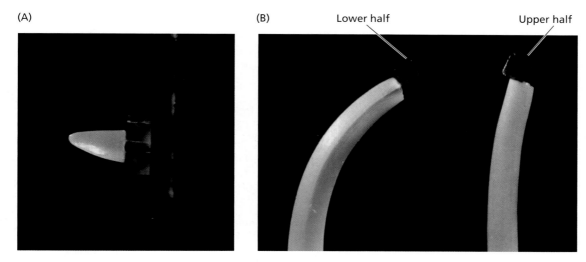

FIGURE 19.29 Auxin is transported to the lower side of a horizontally oriented oat coleoptile tip. (A) Auxin from the upper and lower halves of a horizontal tip is allowed to diffuse into two agar blocks. (B) The agar block from the lower half (left) induces greater curvature in a decapitated coleoptile than the agar block from the upper half (right). (Photo © M. B. Wilkins.)

that gravity can be sensed is through the motion of a falling or sedimenting body.

Obvious candidates for intracellular gravity sensors in plants are the large, dense amyloplasts that are present in gravity-sensing cells. These specialized amyloplasts (starch-containing plastids) are of sufficiently high density relative to the cytosol that they readily sediment to the bottom of the cell (Figure 19.30). Such amyloplasts that function as gravity sensors are called **statoliths**, and the specialized gravity-sensing cells in which they occur are called **statocytes**. Whether the statocyte is able to detect the downward motion of the statolith as it passes through the cytoskeleton or whether the stimulus is perceived only when the statolith comes to rest at the bottom of the cell has not yet been resolved.

SHOOTS AND COLEOPTILES In shoots and coleoptiles, gravity is perceived in the **starch sheath**, a layer of cells that surrounds the vascular tissues of the shoot. The starch sheath is continuous with the endodermis of the root, but unlike the endodermis it contains amyloplasts. Arabidopsis mutants lacking amyloplasts in the starch sheath display agravitropic shoot growth but normal gravitropic root growth.

As noted in Chapter 16, in the *scarecrow* (*scr*) mutant of Arabidopsis the cell layer from which the endodermis and the starch sheath are derived remains undifferentiated. As a result, the hypocotyl and inflorescence of the *scr* mutant are agravitropic, although the root exhibits a normal gravitropic response. On the basis of the phenotypes of these two mutants, we can conclude the following:

- The starch sheath is required for gravitropism in shoots.

- The root endodermis, which does not contain statoliths, is not required for gravitropism in roots.

ROOTS The site of gravity perception in primary roots is the root cap. Large, graviresponsive amyloplasts are located in the statocytes (see Figure 19.30A and B) in the central cylinder, or **columella**, of the root cap. Removal of the root cap from otherwise intact roots abolishes root gravitropism without inhibiting growth.

Precisely how the statocytes sense their falling statoliths is still poorly understood. According to one hypothesis, contact or pressure resulting from the amyloplast resting on the endoplasmic reticulum on the lower side of the cell triggers the response (see Figure 19.30C). The predominant form of endoplasmic reticulum in columella cells is of the tubular type, but an unusual form of ER, called "nodal ER," is also present. Nodal ER consists of five to seven rough ER sheets attached to a central nodal rod in a whorl, like petals on a flower. Nodal ER differs from the more typical tubular cortical ER cisternae and may play a role in the gravity response.

This **starch–statolith hypothesis** of gravity perception in roots is supported by several lines of evidence. Amyloplasts are the only organelles that consistently sediment in the columella cells of different plant species, and the rate of sedimentation correlates closely with the time required to perceive the gravitational stimulus. The gravitropic responses of starch-deficient mutants are generally much slower than those of wild-type plants. Nevertheless, starchless mutants exhibit some residual gravitropism, suggesting that although starch is required for a normal gravitropic response, starch-independent gravity perception mechanisms may also exist.

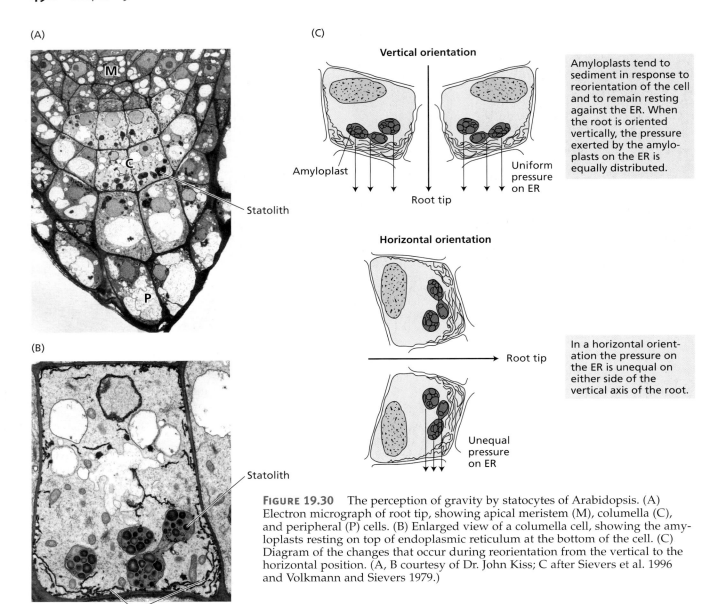

FIGURE 19.30 The perception of gravity by statocytes of Arabidopsis. (A) Electron micrograph of root tip, showing apical meristem (M), columella (C), and peripheral (P) cells. (B) Enlarged view of a columella cell, showing the amyloplasts resting on top of endoplasmic reticulum at the bottom of the cell. (C) Diagram of the changes that occur during reorientation from the vertical to the horizontal position. (A, B courtesy of Dr. John Kiss; C after Sievers et al. 1996 and Volkmann and Sievers 1979.)

Other organelles, such as nuclei, may be dense enough to act as statoliths. It may not even be necessary for a statolith to come to rest at the bottom of the cell. The cytoskeletal network may be able to detect a partial vertical displacement of an organelle. The structural components involved might consist of the meshwork of actin microfilaments that form part of the cytoskeleton of the central columella cells of the root cap. The actin network is assumed to be anchored to stretch-activated receptors on the plasma membrane. Stretch receptors in animal cells are typically mechanosensitive ion channels, and stretch-activated calcium channels have been demonstrated in plants. Pressure placed on elements of the actin meshwork by a sedimenting organelle could in this way alter ion channel activities. However, treatments with drugs that destabilize the microfilaments in statocytes appear to *promote* gravitropism rather than inhibit it. Thus the role of the actin microfilament network in gravity sensing is still unresolved.

GRAVITY PERCEPTION WITHOUT STATOLITHS? An alternative mechanism of gravity perception that does not involve statoliths of any kind has been proposed for the giant-celled freshwater alga *Chara*. See **Web Topic 19.10** for details.

Gravity sensing may involve pH and calcium as second messengers

When gravity-sensing mechanisms detect that the root or shoot axis is out of alignment with the gravity vector, signal transduction mechanisms involving second messengers transfer this information to initiate corrective differential

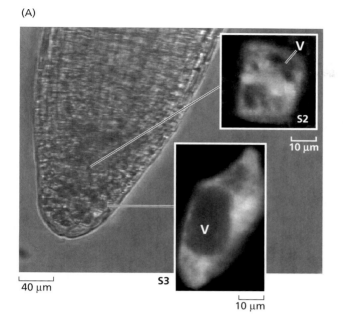
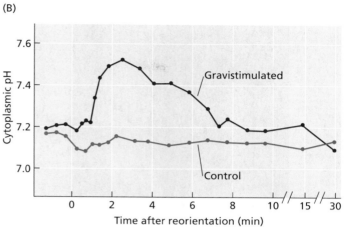
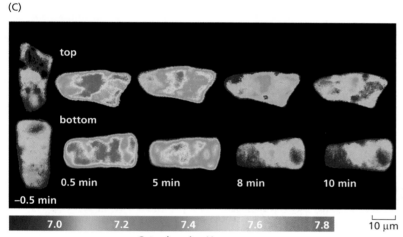

FIGURE 19.31 Experiments with a pH sensitive dye suggest that pH changes columella cells of the root cap are involved in gravitropic signal transduction. (A) Micrograph showing a magnification of the root tip and two columella cells at different levels (stories) of the root cap, labeled S2 (story 2) and S3 (story 3) (insets). The cytosols of the two columella cells are fluorescing because the cells have been microinjected with a pH-sensitive fluorescent dye. The vacuoles (V) contain no dye and therefore appear dark. (B) Cytoplasmic pH increases in less than 1 minute after gravistimulation. (C) Imaging of pH sensitive dyes in the response of the two columella cells to gravitropic stimulus. The color scale below was used to generate the data in (B). (From Fasano et al. 2001.)

growth. This process is called **gravistimulation**. A variety of experiments suggest that localized changes in pH and calcium gradients are part of that signaling.

Changes in intracellular pH can be detected early in root columella cells responding to gravity. When pH-sensitive dyes were used to monitor both intracellular and extracellular pH in Arabidopsis roots, rapid changes were observed after roots were rotated to a horizontal position (Fasano et al. 2001). Within 2 minutes of gravistimulation, the cytoplasmic pH of the columella cells of the root cap increased from 7.2 to 7.6, and the apoplastic pH declined from 5.5 to 4.5 (Figure 19.31). These changes preceded any detectable tropic curvature by about 10 minutes.

The alkalinization of the cytosol combined with the acidification of the apoplast suggests that activation of the plasma membrane H^+-ATPase is one of the initial events that mediate root gravity perception or signal transduction. The chemiosmotic model of auxin transport predicts that differential acidification of the apoplast and alkalinization of the cytosol would result in increased directional uptake and efflux of IAA from the affected cells.

Early physiological studies suggested that calcium release from storage pools may be involved in root gravitropic signal transduction. For example, treatment of maize roots with EGTA [ethylene glycol-bis(β-aminoethyl ether)-N,N,N′,N′-tetraacetic acid], a compound that can chelate (form a complex with) calcium ions, prevents calcium uptake by cells, and inhibits root gravitropism. Placing a block of agar that contains calcium ions on the side of the cap of a vertically oriented maize root induces the root to grow toward the side with the agar block (Figure 19.32). As in the case of [^{3}H]IAA, $^{45}Ca^{2+}$ is polarly transported to the

FIGURE 19.32 A maize root bending toward a calcium-containing agar block placed on one side of the cap. The result shows that an artificially imposed calcium gradient can override the normal gravitropic response. Although calcium does not appear to play a role in gravitropism itself, it may be part of a separate signaling pathway related to thigmotropism. (Courtesy of Michael L. Evans.)

lower half of the cap of a root stimulated by gravity. However, when sensitive calcium-responsive dyes are used to monitor cellular calcium levels, no changes in localized calcium distribution are observed after gravistimulation.

During thigmotropism in roots, a bending response that temporarily overrides gravitropism is initiated when the growing tip makes contact with a solid surface. In contrast to gravitropic bending, localized changes in internal calcium pools have been observed in root thigmotropic responses. However, because little is known about the role of auxin in thigmotropism, a relationship between calcium signaling and auxin-dependent thigmotropic bending has not been established. It is therefore possible that the bending response to calcium shown in Figure 19.32 has more to do with thigmotropism than with gravitropism.

Auxin is redistributed laterally in the root cap

In addition to protecting the sensitive cells of the apical meristem as the tip penetrates the soil, the root cap is the site of gravity perception. Because the cap is some distance away from the elongation zone where bending occurs, graviresponsive signaling events initiated in the root cap must induce production of a chemical messenger that modulates growth in the elongation zone. Microsurgery experiments in which half of the cap was removed showed that the cap supplies a root growth inhibitor to the lower side of the root during gravitropic bending (Figure 19.33).

Although root caps contain small amounts of IAA and abscisic acid (ABA) (see Chapter 23), IAA is more inhibitory to root growth than ABA when applied directly to the elongation zone, suggesting that IAA is the root cap inhibitor. Consistent with this conclusion, ABA-deficient Arabidopsis mutants have normal root gravitropism, whereas the roots of mutants defective in auxin transport, such as *aux1* and *agr1*, have impaired gravitropic responses. The *agr* (agravitropic) mutant lacks the PIN2 protein, which is localized at the basal (distal) end of epidermal and outer cortical cells that conduct auxin away from the root tip in Arabidopsis (see Figure 19.15).

The root cap and adjacent cells play an important role in the basipetal redirection of shoot-derived auxin and, apparently, the smaller amount of auxin synthesized in the root tip (Ljung et al. 2005). Auxin export from stelar tissues is thought to be mediated by PIN/PGP-mediated efflux and

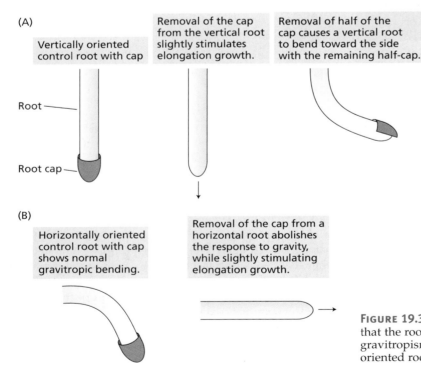

FIGURE 19.33 Microsurgery experiments demonstrating that the root cap produces an inhibitor that regulates root gravitropism. (A) Vertically oriented roots. (B) Horizontally oriented roots. (After Shaw and Wilkins 1973.)

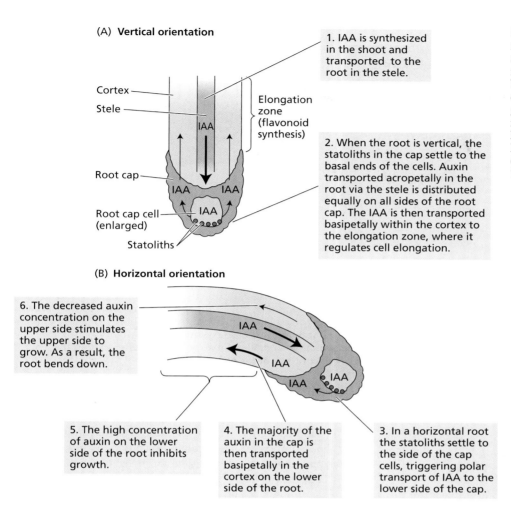

FIGURE 19.34 Proposed model for the redistribution of auxin during gravitropism in maize roots. (A) Equal distribution of auxin in a vertically oriented root results in straight growth downward. (B) Asymmetric auxin distribution in a horizontally oriented root leads to downward curvature. (After Hasenstein and Evans 1988.)

AUX1-mediated uptake. PIN and PGP proteins also mediate radial export for lateral redistribution.

The lateral root cap cells overlay the distal elongation zone (DEZ) of the root—the first region that responds to gravity. Rapid uptake of auxin into epidermal cells at the DEZ is mediated by AUX1 and PGP4. The importance of this uptake is demonstrated by experiments in which expression of AUX1 under the control of a lateral root cap/epidermal-specific promoter restored wild-type gravitropic bending to the *aux1* mutant, whereas expression of AUX1 in the columella alone did not. The auxin from the cap is taken up by the cortical and epidermal cells of the DEZ and transported basipetally through the elongation zone of the root. This basipetal transport, which is limited to the elongation zone, is facilitated by basally localized PIN proteins as well as by PGP proteins.

The PGP proteins are uniformly distributed in cells near the root apex and polarly localized in cells above the distal elongation zone (Geisler et al. 2005). An *auxin reflux loop* mediated by specific PIN proteins is thought to redirect auxin back into the acropetal stelar transport stream at the boundary of the elongation zone (see Figure 19.15). Auxin circulation at the growing tip may allow root growth to continue for a time independent of auxin from the shoot (Blilou et al. 2005).

According to the model for gravitropism, basipetal auxin transport in a vertically oriented root is equal on all sides (Figure 19.34A). When the root is oriented horizontally, however, the cap redirects most of the auxin to the lower side, thus inhibiting the growth of that lower side (see Figure 19.34B). Consistent with this model, the transport of [^{3}H]IAA across a horizontally oriented root cap is polar, with a preferential downward movement. The downward movement of auxin across a horizontal root cap has been confirmed using a reporter gene construct, *DR5::GFP*, consisting of green fluorescent protein (GFP) expressed under the control of the auxin-sensitive *DR5* promoter (Ottenschläger et al. 2003) (Figure 19.35).

One of the members of the PIN protein family, PIN3, is thought to participate in the redirection of auxin in roots that are displaced from the vertical orientation. In a vertically oriented root PIN3 is uniformly distributed around the columella cell, but when the root is placed on its side PIN3 is preferentially targeted to the lower side of the cell.

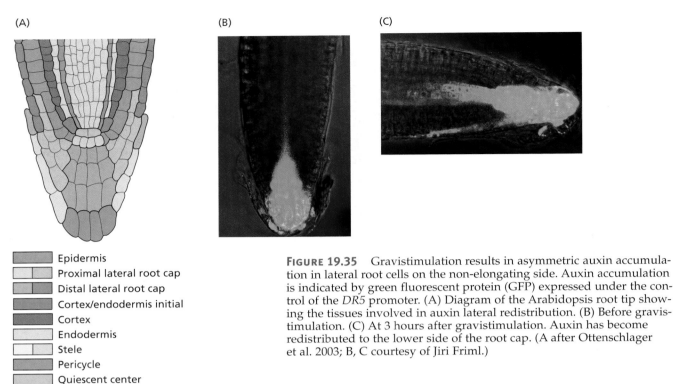

FIGURE 19.35 Gravistimulation results in asymmetric auxin accumulation in lateral root cells on the non-elongating side. Auxin accumulation is indicated by green fluorescent protein (GFP) expressed under the control of the *DR5* promoter. (A) Diagram of the Arabidopsis root tip showing the tissues involved in auxin lateral redistribution. (B) Before gravistimulation. (C) At 3 hours after gravistimulation. Auxin has become redistributed to the lower side of the root cap. (A after Ottenschlager et al. 2003; B, C courtesy of Jiri Friml.)

This reorientation of PIN3 is thought to accelerate auxin transport to the lower side of the cap. As noted earlier, some PIN proteins are rapidly cycled between the plasma membrane and intracellular secretory compartments. PIN3 may be localized by such a mechanism.

Developmental Effects of Auxin

Although originally discovered in relation to growth, auxin influences nearly every stage of a plant's life cycle from germination to senescence. As we have already seen, polar transport of auxin is required for the establishment of overall plant polarity during embryogenesis and is required for the maintenance of polarity in both normal and tropic growth. The polarity of auxin transport is established in the early stages of embryogenesis, and the cellular mechanisms that mediate polar auxin transport from the embryonic apical pole to the suspensor (point of maternal attachment) are maintained in mature tissues. For further discussion of developmental polarity, see Chapter 16 and **Web Essay 19.2**.

Because the effect that auxin produces depends on the identity of the target tissue, the response of a tissue to auxin is governed by its developmentally determined genetic program and is further influenced by the presence or absence of other signaling molecules. As we will see in this and subsequent chapters, interaction between two or more hormones is a recurring theme in plant development.

In this section we will examine some additional developmental processes regulated by auxin, including apical dominance, floral bud development, leaf arrangement (phyllotaxy), lateral root formation, vascular development, leaf abscission, and fruit formation. Throughout this discussion we assume that the mechanisms of auxin action are comparable in all cases, involving similar receptors and signal transduction pathways. The current state of our knowledge of auxin signaling pathways will be considered at the end of the chapter.

Auxin regulates apical dominance

In most higher plants, the growing apical bud inhibits the growth of lateral (axillary) buds—a phenomenon called **apical dominance**. Removal of the shoot apex (decapitation) results in the outgrowth of one or more of the lateral buds. Not long after the discovery of auxin, it was found that IAA could substitute for the apical bud in maintaining the inhibition of lateral buds of kidney bean (*Phaseolus vulgaris*) plants. This classic experiment of Kenneth V. Thimann and Folke Skoog in the 1920s is illustrated in Figure 19.36.

This result was soon confirmed for numerous other plant species, leading to the hypothesis that the outgrowth of the axillary bud is inhibited by auxin that is transported basipetally from the terminal bud. In support of this idea, a ring of the auxin transport inhibitor TIBA in lanolin paste (as a carrier) placed below the shoot apex released the axillary buds from inhibition.

(A) Terminal bud intact

(B) Terminal bud removed

(C) Auxin added to decapitated stem

FIGURE 19.36 Auxin suppresses the growth of axillary buds in bean (*Phaseolus vulgaris*) plants. (A) The axillary buds are suppressed in the intact plant because of apical dominance. (B) Removal of the terminal bud releases the axillary buds from apical dominance (arrows). (C) Applying IAA in lanolin paste (contained in the gelatin capsule) to the cut surface prevents the outgrowth of the axillary buds. (Photos © M. B. Wilkins.)

How does auxin from the shoot apex inhibit the outgrowth of lateral buds? Thimann and Skoog originally proposed that auxin synthesized in the shoot apex is transported basipetally to the axillary bud, which it enters to cause inhibition. According to the *direct-inhibition model*, axillary buds are more sensitive to auxin than other shoot tissues; hence, normal endogenous levels of auxin that promote shoot growth inhibit the outgrowth of axillary buds. To be consistent with the branching patterns seen in the shoots of numerous species, apical dominance should be weakest near the base of the shoot, where the auxin concentration is lowest.

A prediction of the direct-inhibition model is that the concentration of auxin in the axillary buds should decrease following decapitation of the shoot apex. However, the reverse appears to be the case. Measurements of auxin levels in axillary buds have shown that following decapitation the auxin content of the buds actually *increases*. In addition, application of auxin directly to the terminal bud, which raises the auxin concentration in the shoot, fails to inhibit normal axillary bud outgrowth. Furthermore, although basipetal auxin transport is required for apical dominance in many species, experiments with radiolabeled auxin have shown that the auxin does not enter the bud.

If auxin does not enter the bud, it must act remotely to suppress bud outgrowth, but where does auxin act? The answer to this question came from molecular studies using the *axr1* mutant of Arabidopsis (Booker et al. 2003). The AXR1 protein is related to the ubiquitin-activating enzyme (E1), which is required for auxin signaling (discussed at the end of the chapter). *axr1* mutants are unable to respond to auxin and, as a result, the phenotype of the mutant includes increased branching. Booker et al. (2003) found that normal apical dominance could be completely restored in the mutant by expressing the wild-type AXR1 protein exclusively in the xylem and the sclerenchyma cells between the vascular bundles (*interfascicular* sclerenchyma) of the stem. Furthermore, grafting studies between *axr1* and wild type demonstrated that auxin controls axillary bud growth by acting in the xylem and interfascicular sclerenchyma of the shoot.

OTHER BRANCHING SIGNALS Investigators have searched for other signaling agents that could interact with auxin to control axillary bud growth. Direct application of cytokinins to axillary buds stimulates bud growth in many species, overriding the inhibitory effect of the shoot apex. In kidney bean plants, decapitation results in an increase in cytokinin export to the shoot from the root via the xylem, and the effects of decapitation on cytokinin export can be reversed by auxin application to the cut surface. However, bud outgrowth following decapitation occurs prior to the increase in cytokinin export from the root, so cytokinins cannot be the signaling agent involved. As will be discussed in Chapter 21, the axillary bud appears to synthesize its own cytokinin when branching occurs.

Abscisic acid (see Chapter 23), which often serves as an inhibitor of plant growth, is present in axillary buds. In some species, when the shoot apex is removed, the ABA levels in the lateral buds decreases. However, the effect is not observed in all species, suggesting that although ABA may play a role in apical dominance in some species, it is not a universal mechanism for regulating apical dominance.

Analyses of a set of pea and Arabidopsis mutants with increased branching patterns have led to the identification of a carotenoid cleavage product that serves as a signaling agent in apical dominance. Grafting studies between mutant and wild-type plants have shown that this signal is produced by the roots and transported to the shoot, presumably via the xylem. Since mutants blocked in the synthesis of this compound exhibit increased branching, the carotenoid derivative, which has yet to be identified, somehow inhibits branching (Booker et al. 2004). How this root-derived signal interacts with auxin in the xylem and interfascicular sclerenchyma during apical dominance is the subject of ongoing research (see **Web Essay 19.4**).

Auxin transport regulates floral bud development and phyllotaxy

Treating Arabidopsis plants with the auxin transport inhibitor NPA causes abnormal leaf and flower development, suggesting that polar auxin transport in the inflorescence meristem is required for normal floral development. In Arabidopsis, the "pin-formed" mutant *pin1*, which lacks an auxin efflux carrier in shoot tissues, has abnormal flowers similar to those of NPA-treated plants (see Figure 19.14A).

The developing floral meristem depends on auxin being transported to it from subapical tissues. Auxin transport from these tissues also regulates leaf initiation and **phyllotaxy**, the pattern of leaf emergence from the shoot apex. In the absence of PIN1 and associated proteins, auxin movement to the meristem is impaired and the initiation of leaf and floral organ primordia is disrupted. Not surprisingly, therefore, leaf primordia can be induced to form on the meristem of a *pin1* mutant by applying a tiny spot of auxin in lanolin paste on the flank of the apical meristem (Figure 19.37)

Auxin promotes the formation of lateral and adventitious roots

Although elongation of the primary root is inhibited by auxin concentrations greater than $10^{-8}\,M$, initiation of lateral (branch) roots and adventitious roots is stimulated by high auxin levels. Lateral roots are commonly found above the elongation and root hair zone and originate from small groups of cells in the pericycle (see Chapter 16). Analyses of Arabidopsis mutants with aberrant patterns of root

(A)
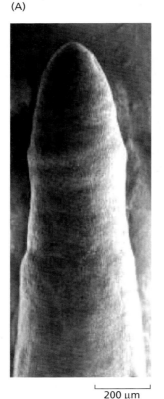

(B)

FIGURE 19.37 Scanning electron micrographs of leaf primordium induced by IAA on a vegetative *pin1* shoot apical meristem. (A) Untreated *pin1* inflorescence meristem fails to produce leaf primordia. (B) Leaf primordium induced on the inflorescence meristem of *pin1* mutant by placing a microdrop of IAA in lanoline paste on the side of the meristem. (A from Vernoux et al. 2000, B from Reinhardt et al. 2003).

branching, and a variety of physiological studies, have led to the following conclusions:

1. IAA transported acropetally (toward the tip) in the vascular parenchyma of the root is required to initiate cell division in the pericycle.

2. IAA is also required to promote cell division and maintain cell viability in the developing lateral root.

Adventitious roots (roots originating from nonroot tissue) can arise in a variety of tissue locations from clusters of mature cells that renew their cell division activity. These dividing cells develop into a root apical meristem in a manner somewhat analogous to the formation of a lateral root primordium. In horticulture, the stimulatory effect of auxin on the formation of adventitious roots has been very useful for the vegetative propagation of plants by cuttings.

Auxin induces vascular differentiation

New vascular tissues differentiate directly below developing buds and young growing leaves (see Figure 19.5), and removal of the young leaves prevents vascular differentiation. The ability of an apical bud to stimulate vascular differentiation can be demonstrated in tissue culture. When the apical bud is grafted onto a clump of undifferentiated cells, or *callus*, xylem and phloem differentiate beneath the graft.

Further evidence for the role of auxin in vascular differentiation comes from studies in which the auxin concentration is manipulated by localized transgenic expression of genes encoding auxin biosynthesis or degradation enzymes. Tissues with increased auxin levels develop more vascular tissue, while those that are depleted of auxin form new vascular tissue much more slowly after wounding. Vascular differentiation in some tissues appears to involve interaction with other hormones.

The relative amounts of xylem and phloem formed are regulated by the auxin concentration: High auxin concentrations induce the differentiation of both xylem and phloem, but only phloem differentiates at low auxin concentrations. Similarly, experiments on stem tissues have shown that low auxin concentrations induce phloem differentiation, whereas higher IAA levels induce xylem.

The regeneration of vascular tissue following wounding is also controlled by auxin produced by the young leaf directly above the wound site (Figure 19.38). Removal of the leaf prevents the regeneration of vascular tissue, and applied auxin can substitute for the leaf in stimulating regeneration.

Vascular differentiation is polar and occurs from leaves to roots. Mutant analysis in Arabidopsis indicates that normal auxin transport is required for vascular differentiation in leaves. In woody perennials, auxin produced by grow-

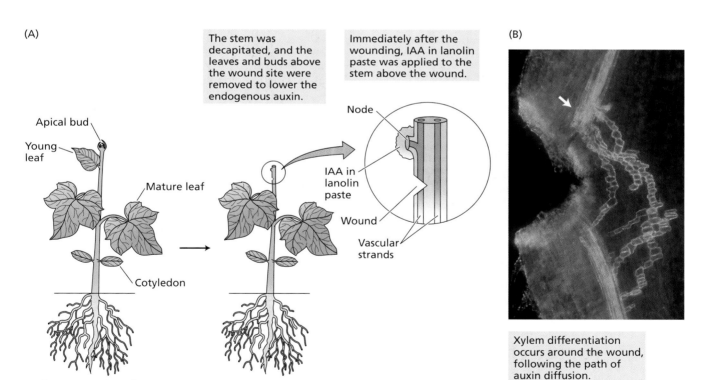

FIGURE 19.38 IAA-induced xylem regeneration around the wound in cucumber (*Cucumis sativus*) stem tissue. (A) Method for carrying out the wound regeneration experiment. (B) Fluorescence micrograph showing regenerating vascular tissue around the wound. The arrow indicates the wound site where auxin accumulates and xylem differentiation begins (B courtesy of R. Aloni.)

ing buds in the spring stimulates activation of the cambium in a basipetal direction. The new round of secondary growth begins at the smallest twigs and progresses downward toward the root tip.

Auxin delays the onset of leaf abscission

The shedding of leaves, flowers, and fruits from the living plant is known as **abscission**. These parts abscise in a region called the **abscission zone**, which is located near the base of the petiole of leaves. In most plants, leaf abscission is preceded by the differentiation of a distinct layer of cells, the **abscission layer**, within the abscission zone. During leaf senescence, the walls of the cells in the abscission layer are digested, which causes them to become soft and weak. The leaf eventually breaks off at the abscission layer as a result of stress on the weakened cell walls.

Auxin levels are high in young leaves, progressively decrease in maturing leaves, and are relatively low in senescing leaves when the abscission process begins. The role of auxin in leaf abscission can be readily demonstrated by excision of the blade from a mature leaf, leaving the petiole intact on the stem. Whereas removal of the leaf blade accelerates the formation of the abscission layer in the petiole, application of IAA in lanolin paste to the cut surface of the petiole prevents the formation of the abscission layer. (Lanolin paste alone does not prevent abscission.)

These results suggest the following:

- Auxin transported from the blade normally prevents abscission.
- Abscission is triggered during leaf senescence, when auxin is no longer being produced.

However, as will be discussed in Chapter 22, ethylene also plays a crucial role as a positive regulator of abscission.

Auxin promotes fruit development

Much evidence suggests that auxin is involved in the regulation of fruit development. Auxin is produced or mobilized from storage in pollen and in the endosperm and the embryo of developing seeds, and the initial stimulus for fruit growth may result from pollination.

Successful pollination initiates ovule growth, which is known as **fruit set**. After fertilization, fruit growth may depend on auxin from developing seeds. The endosperm may contribute auxin during the initial stage of fruit growth, and the developing embryo may take over as the main auxin source during the later stages.

Figure 19.39 shows the influence of auxin produced by the achenes of strawberry on the growth of the receptacle of strawberry.

Synthetic auxins have a variety of commercial uses

Auxins have been used commercially in agriculture and horticulture for more than 50 years. The early commercial uses included prevention of fruit and leaf drop, promotion of flowering in pineapple, induction of parthenocarpic fruit, thinning of fruit, and rooting of cuttings for plant propagation. Rooting is enhanced if the excised leaf or stem cutting is dipped in an auxin solution, which increases the initiation of adventitious roots at the cut end. This is the basis of commercial rooting compounds, which consist mainly of a synthetic auxin mixed with talcum powder.

In some plant species, seedless fruits may be produced naturally, or they may be induced by treatment of the unpollinated flowers with auxin. The production of such seedless fruits is called **parthenocarpy**. In stimulating the formation of parthenocarpic fruits, auxin may act primarily to induce fruit set, which in turn may trigger the endogenous production of auxin by certain fruit tissues to complete the developmental process.

Ethylene is also involved in fruit development, and some of the effects of auxin on fruiting may result from the promotion of ethylene synthesis. The control of ethylene in the commercial handling of fruit is discussed in Chapter 22.

In addition to these applications, synthetic auxins, such as 2,4-D and dicamba (see Figure 19.4 and **Web Topic 19.1**),

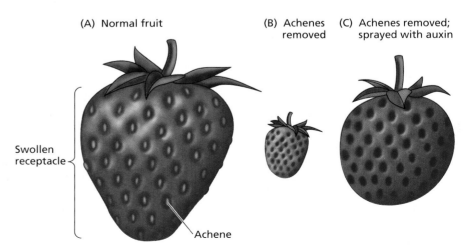

Figure 19.39 The strawberry "fruit" is actually a swollen receptacle whose growth is regulated by auxin produced by the "seeds," which are actually achenes, the true fruits. (A) When the achenes are present the receptacle enlarges and develops its characteristic flavor, sweetness, and red color. (B) When the achenes are removed, the receptacle fails to develop normally. (C) Spraying the receptacle minus its achenes with IAA restores normal growth and development. (After Galston 1994.)

are widely used as herbicides that induce excessive cell expansion and subsequent plant death. Synthetic auxins are so effective because they are not metabolized by the plant as quickly as is IAA. Some, like 2,4-D, are also transported more slowly than IAA, thus enhancing retention in shoots. Synthetic auxins are used by farmers for the control of dicot weeds, also called *broad-leaved weeds*, in commercial cereal fields, and by home gardeners for the control of weeds such as dandelions and daisies in lawns.

Monocots such as maize and grasses are less sensitive to artificial auxins, because monocots are able to inactivate synthetic auxins rapidly by conjugation, and they possess auxin binding and receptor proteins that bind synthetic auxins with lower affinity.

Auxin Signal Transduction Pathways

The ultimate goal of research on the molecular mechanism of hormone action is to reconstruct each step in the signal transduction pathway, from receptor binding to the physiological response. In the case of auxin, this would seem to be a particularly daunting task because auxin affects so many physiological and developmental processes. However, the initial steps in auxin signaling are surprisingly simple, and involve binding to a small group of receptors that regulate protein degradation via the ubiquitin-proteasome pathway (see Chapter 14 on the web site). Upon activation by auxin, the receptor-enzyme complex targets specific transcriptional repressors for hydrolysis, resulting in the activation (derepression) of auxin-responsive genes. While this mechanism appears to account for most auxin responses, a different type of auxin receptor protein may function in nontranscriptional activation and mobilization of plasma membrane H^+-ATPases to cause rapid cell wall acidification and cell elongation (see Figure 19.25). In this last section of the chapter, we will examine the signaling pathways involved in both types of auxin responses.

A ubiquitin E3 ligase subunit is an auxin receptor

The principal auxin receptors have been identified as "F-box" soluble proteins belonging to the **TIR1/AFB** family of proteins (Figure 19.40) (Dharmasiri et al. 2005a, 2005b; Kepinski et al. 2005). F-boxes are protein motifs about 50 amino acids long that function as sites of protein-protein interaction. **F-box proteins** were first discovered as subunits of a specific type of **ubiquitin E3 ligase complex**, called **SCF**, named after its three main subunits: *S*KP1, *c*ullin, and *F*-box protein. Like other ubiquitin ligases, SCF complexes catalyze the ATP-dependent, covalent addition of ubiquitin molecules to proteins targeted for proteolytic degradation (see Chapter 14 on the web site).

SCF complexes are found in all eukaryotes and play an important role in the regulation of protein abundance. The primary function of the F-box protein in the SCF complex is to bind the substrates for ubiquitin-mediated proteolysis. To carry out this function, F-box proteins typically have additional motifs besides the F-box that promote protein-protein interaction, such as leucine-rich repeat sequences (see Figure 19.40). The F-box motif itself serves to link the F-box protein to the other subunits of the SCF complex.

Mutant analyses in Arabidopsis led to the identification of an F-box protein, **TIR1** (*t*ransport *i*nhibitor *r*esponse 1), that appeared to be essential for auxin-dependent hypocotyl elongation and lateral root formation. TIR1 is a component of a specific E3 ubiquitin ligase complex, designated SCF^{TIR1}, that is required for auxin signaling in cells. The Arabidopsis homolog of SKP1 in the SCF^{TIR1} complex (see above) is called **ASK1**. The function of SCF^{TIR1} complexes is to tag constitutively expressed transcriptional regulators of auxin-induced genes with chains of ubiquitin proteins, thus targeting them for degradation by the proteasome. These regulators are called AUX/IAA proteins.

(A) The auxin TIR1 receptor protein

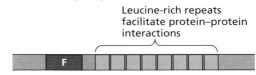

(B) The SCF^{TIR1} auxin receptor complex

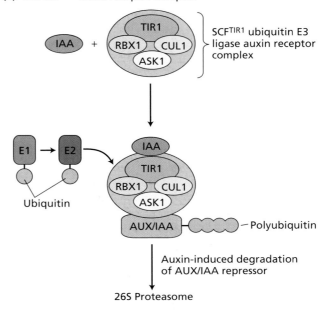

FIGURE 19.40 The auxin receptor. (A) The auxin receptor TIR1 has an F-box motif and leucine-rich repeat sequences that facilitate protein-protein interactions. (B) TIR1 functions as a subunit of the specific ubiquitin E3 ligase complex SCF^{TIR1}. When auxin binds to SCF^{TIR1} it activates it, resulting in the ubiquitination and proteolytic degradation of AUX/IAA repressor proteins.

The TIR1 protein binds auxin directly, and auxin binding to TIR1 promotes the association of the SCF^{TIR1} complex with the AUX/IAA protein. Thus far, three other F-box proteins related to TIR1 have also been shown to have auxin receptor activity. These proteins are called **auxin signaling F-box proteins** (**AFBs**), and like TIR1, they function in ubiquitin E3 ligase complexes.

Auxin-induced genes are negatively regulated by AUX/IAA proteins

Two families of transcriptional regulators participate in the TIR1 auxin signaling pathway: auxin response factors and AUX/IAA proteins. **Auxin response factors** (**ARFs**) are short-lived nuclear proteins that bind with specificity to TGTCTC **auxin response elements** (**AuxREs**) in the promoters of primary, or early, auxin-response genes. There are 23 different ARF proteins in Arabidopsis. The binding of ARFs to AuxREs results in the activation or repression of gene transcription, depending on the particular ARF involved. ARFs appear to be present on the promoters of early auxin genes regardless of the auxin status of the tissue.

AUX/IAA proteins are important secondary regulators of auxin-induced gene expression. There are 29 members of this family of short-lived small nuclear proteins in Arabidopsis. AUX/IAA proteins regulate gene transcription indirectly by binding to ARF protein bound to DNA. If the ARF in question is a transcriptional activator, the effect of the AUX/IAA protein is to repress transcription. Conversely, if the ARF bound to the AuxRE is a transcriptional repressor, the AUX/IAA protein functions as a transcriptional activator.

Auxin binding to SCF^{TIR1} stimulates AUX/IAA destruction

The short signaling pathway responsible for auxin-induced gene expression begins with the binding of auxin to the TIR1 subunit of the SCF^{TIR1} ubiquitin ligase complex. The binding of auxin to TIR1 induces a conformational change in SCF^{TIR1} that both enhances dimerization with AUX/IAA and activates E3 ligase activity (Figure 19.41A). Unlike other SCF complexes, no covalent modification is involved in SCF^{TIR1} activation by auxin. As a consequence, the AUX/IAA proteins are rapidly ubiquitinated and then hydrolyzed via the proteasome (see Figure 19.41B). In the absence of their negative regulators, ARF proteins either stimulate or repress gene expression, depending on the ARF.

It is not yet clear whether auxin binds its receptor, SCF^{TIR1}, in the cytosol or in the nucleus. One possibility is that inactive SCF^{TIR1} associates with AUX/IAA in the nucleus even in the absence of auxin. Consistent with this idea, auxin binds to SCF^{TIR1} with greater affinity in the presence of AUX/IAA than in its absence. Alternatively, auxin may bind SCF^{TIR1} in the cytosol and then migrate to the nucleus before associating with AUX/IAA. Research on these questions is currently ongoing.

Auxin-induced genes fall into two classes: early and late

Auxin-responsive genes whose expression is insensitive to protein synthesis inhibitors and involves the activation of preexisting transcription factors are called **primary response genes** or **early genes**. The time required for the expression of the early genes can be quite short, ranging from a few minutes to several hours. All of the primary response genes in auxin responses are induced via the SCF^{TIR1} signaling pathway.

In general, primary response genes have three main functions:

1. *Transcription factors.* Some of the early genes encode proteins that regulate the transcription of **secondary response genes**, or **late genes**, that are required for the long-term responses to the hormone. Because late genes require de novo protein synthesis, their expression can be blocked by protein synthesis inhibitors.

2. *Signaling.* Other early genes are involved in intercellular communication, or cell-to-cell signaling.

3. *Stress proteins.* Another group of early genes encode proteins involved in adaptations to stress.

Early response genes include the *AUX/IAA* genes, *SAUR* genes, and *GH3* genes. Transcription of another group of genes, such as those for glutathione S-transferases and genes involved in ethylene biosynthesis, is induced by higher levels of auxin and appears to be more associated with stress and wounding responses.

EARLY GENES FOR GROWTH AND DEVELOPMENT The expression of most *AUX/IAA* genes is stimulated by auxin within 5 to 60 minutes of hormone addition. AUX/IAA proteins have short half-lives (about 7 minutes), indicating that they are turning over rapidly.

Auxin stimulates the expression of *SAUR* genes within 2 to 5 minutes of treatment, and the response is insensitive to cycloheximide. The five *SAUR* genes of soybean are clustered together, contain no introns, and encode highly similar polypeptides of unknown function. Because of the rapidity of the response, expression of *SAUR* genes has proven to be a convenient probe for the lateral transport of auxin during photo- and gravitropism.

GH3 early-gene family members, identified in both soybean and Arabidopsis, are stimulated by auxin within 5 minutes. The GH3 protein functions in IAA conjugation (Staswick et al. 2005) (see Figure 19.9). Mutations in Arabidopsis *GH3*-like genes result in dwarfism (Nakazawa et al. 2001) and appear to function in light-regulated auxin responses (Hsieh et al. 2000). Because *GH3* expression is a good reflection of the presence of endogenous auxin, a syn-

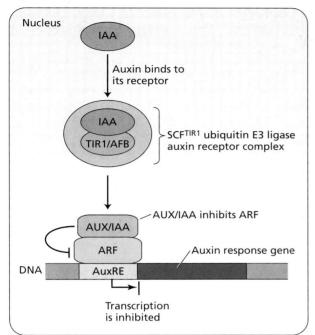

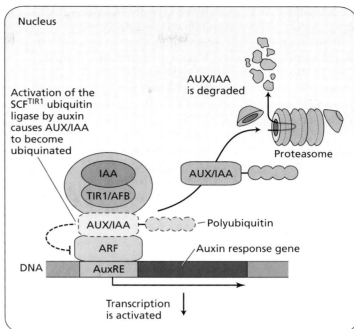

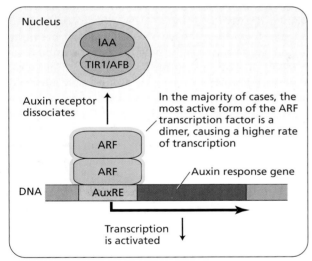

Figure 19.41 A model for auxin binding to TIR1/ABF auxin receptors and subsequent transcriptional activation of auxin response genes. (A) In the absence of auxin, AUX/IAA repressors inhibit the transcription of auxin-induced genes by binding to ARF transcriptional activators, locking them into an inactive state. Auxin binding to SCF$^{TIR/ABF}$ complexes promotes their association with AUX/IAA proteins. (B) Auxin-activated SCF$^{TIR/ABF}$ complexes attach ubiquitin molecules to AUX/IAA proteins, which promotes their destruction by the 26S proteasome. The removal and degradation of AUX/IAA proteins "unlocks" the ARF transcriptional activators. The ARF transcriptional activators bound to auxin response elements (AuxRE) stimulate the transcription of auxin-induced genes. (C) In most auxin-induced genes, two ARF proteins dimerize on the AuxRE, causing a further stimulation ("potentiation") of gene transcription.

thetic *GH3*-based reporter gene known as *DR5* is widely used in auxin bioassays (see Figure 19.5).

LATE GENES FOR STRESS ADAPTATIONS As mentioned earlier in the chapter, auxin is involved in stress responses, such as wounding. Several genes encoding glutathione-S-transferases (GSTs), a class of proteins stimulated by various stress conditions, are induced by elevated auxin concentrations. Likewise, ACC synthase, which is also induced by stress and is the rate-limiting step in ethylene biosynthesis (see Chapter 22), is induced by high levels of auxin.

Rapid auxin responses may involve a different receptor protein

As discussed earlier in the chapter, auxin induces cell elongation with a lag time of about 15 minutes, and auxin-induced increases in plasma membrane H$^+$-ATPase activity begin even sooner. Because these responses are probably too rapid to be mediated by transcriptional changes, they may not be regulated by the SCF^{TIR1} pathway.

ABP1 (*a*uxin-*b*inding *p*rotein *1*) has been identified as an auxin-binding protein required for auxin-dependent cell elongation, and it appears to function in some aspect of H$^+$-

ATPase activation. When added to tobacco protoplasts in vitro, purified ABP1 rapidly induces protoplast swelling and increased plasma membrane H^+-ATPase activity.

ABP1 homologs have been identified in a variety of monocot and dicot species. Knockout mutants of the *ABP1* gene in Arabidopsis exhibit embryo lethality, and less severe mutations result in altered development (Chen et al. 2001). ABP1 is localized primarily on the endoplasmic reticulum and other endomembrane compartments. It has been proposed that ABP1 may stimulate proton pumping by increasing the retention of H^+-ATPases and auxin efflux proteins on the plasma membrane.

However, ABP1 does not appear to be involved in other auxin-response pathways because expression of most auxin-responsive genes is not affected when protoplasts are incubated with anti-ABP1 antibodies, which block ABP1 activity. Thus ABP1-auxin interactions operate independently of the signaling pathway involving AUX/IAA degradation.

Summary

Auxin was the first hormone to be discovered in plants and is one of an expanding list of chemical signaling agents that regulate plant development. The most common naturally occurring form of auxin is indole-3-acetic acid (IAA). One of the most important roles of auxin in higher plants is the regulation of elongation growth in young stems and coleoptiles. Low levels of auxin are also required for root elongation, although at higher concentrations auxin acts as a root growth inhibitor.

Accurate measurement of the amount of auxin in plant tissues is critical for understanding the role of this hormone in plant physiology. Early coleoptile-based bioassays have been replaced by more accurate techniques, including physicochemical methods and immunoassay.

Regulation of growth in plants may depend in part on the amount of free auxin present in plant cells, tissues, and organs. There are two main pools of auxin in the cell: the cytosol and the chloroplasts. Levels of free auxin can be modulated by several factors, including the synthesis and breakdown of conjugated IAA, IAA metabolism, compartmentation, and polar auxin transport. Several pathways have been implicated in IAA biosynthesis, including tryptophan-dependent and tryptophan-independent pathways. Several degradative pathways for IAA have also been identified.

IAA is synthesized primarily in shoot and root apices and is transported polarly away from sites of synthesis. Polar transport is thought to occur mainly in the parenchyma cells associated with the vascular tissue in shoots and roots, but an additional, basipetal transport stream is found in root epidermal and cortical tissues. Polar auxin transport can be divided into two main processes: IAA influx/uptake and IAA efflux. In accord with the chemiosmotic model for polar transport, there are two modes of IAA influx: by a pH-dependent passive transport of the undissociated form, or by an active H^+ cotransport mechanism driven by the plasma membrane H^+-ATPase.

Auxin efflux is thought to occur preferentially at the basal ends of the transporting cells via anion efflux carrier complexes and to be driven by the membrane potential generated by the plasma membrane H^+-ATPase. Auxin efflux inhibitors can interrupt auxin transport directly by competing with auxin for the efflux channel pore or by binding to regulatory or structural proteins associated with the efflux channel. Auxin can be also transported via source–sink relationships in the phloem.

Auxin-induced cell elongation begins after a lag time of about 10 minutes. Auxin promotes elongation growth primarily by increasing cell wall extensibility. Auxin-induced wall loosening requires continuous metabolic input and can be mimicked in part by treatment with acidic buffers.

According to the acid growth hypothesis, one of the important actions of auxin is to induce cells to transport protons into the cell wall by stimulating the plasma membrane H^+-ATPase. Two mechanisms have been proposed for auxin-induced proton extrusion: direct activation of the proton pump and enhanced synthesis of the plasma membrane H^+-ATPase. The ability of protons to cause cell wall loosening is mediated by a class of proteins called expansins. Expansins loosen the cell wall by breaking hydrogen bonds between the polysaccharide components of the wall. In addition to proton extrusion, long-term auxin-induced growth involves the uptake of solutes and the synthesis and deposition of polysaccharides and proteins needed to maintain the acid-induced wall-loosening capacity.

Promotion of growth in stems and coleoptiles and inhibition of growth in roots are the best-studied physiological effects of auxins. Auxin-promoted differential growth in these organs is responsible for the responses to directional stimuli (i.e., light, gravity, and touch) called tropisms. According to the Cholodny–Went model, auxin is transported laterally to the shaded side during phototropism and to the lower side during gravitropism. Statoliths (starch-filled amyloplasts) in the statocytes are involved in the normal perception of gravity, but they are not absolutely required.

In addition to its roles in growth and tropisms, auxin plays central regulatory roles in apical dominance, lateral-root initiation, leaf abscission, vascular differentiation, floral bud formation, phyllotaxy, and fruit development. Commercial applications of auxins include rooting compounds and herbicides.

TIR1/AFB proteins function as primary auxin receptors and mediate auxin-dependent gene expression. AUX/IAA proteins, which function as transcriptional repressors, are the immediate targets of auxin-bound TIR1/AFB proteins, which accelerate AUX/IAA proteolytic degradation via a ubiquitin activation pathway.

Together with ARF proteins, AUX/IAA proteins mediate auxin-responsive gene expression. A second protein, ABP1, is located in primarily in the endoplasmic reticulum lumen, and appears to function in nontranscriptional activation of proton extrusion. Studies of the signal transduction pathways involved in auxin action have implicated other signaling intermediates, such as Ca^{2+}, intracellular pH, and kinases in auxin-mediated responses to environmental stimuli, such as light, touch, and gravity.

Auxin-induced genes fall into two categories: early and late. Induction of early genes by auxin does not require protein synthesis and is insensitive to protein synthesis inhibitors. The early genes fall into three functional classes: expression of the late genes (secondary response genes), stress adaptation, and intercellular signaling. The auxin response domains of the promoters of the auxin early genes have a composite structure in which an auxin-inducible response element is combined with a constitutive response element.

Web Material

Web Topics

19.1 Additional Synthetic Auxins
Biologically active synthetic auxins have surprisingly diverse structures.

19.2 The Structural Requirements for Auxin Activity
Comparisons of a wide variety of compounds that possess auxin activity have revealed common features at the molecular level that are essential for biological activity.

19.3 Auxin Measurement by Radioimmunoassay
Radioimmunoassay (RIA) allows the measurement of physiological levels (10^{-9} g = 1 ng) of IAA in plant tissues.

19.4 Evidence for the Tryptophan-Independent Biosynthesis of IAA
Additional experimental evidence for the tryptophan-independent biosynthesis of IAA is provided.

19.5 The Multiple Factors That Regulate Steady-State IAA Levels
The steady-state level of free IAA in the cytosol is determined by several interconnected processes, including synthesis, degradation, conjugation, compartmentation and transport.

19.6 Auxin Transport in Heterologous Systems
The activity of auxin transport proteins can be characterized when plant proteins are expressed in heterologous systems

19.7 The Mechanism of Fusicoccin Activation of the Plasma Membrane H^+-ATPase
Fusicoccin, a phytotoxin produced by the fungus *Fusicoccum amygdale*, causes membrane hyperpolarization and proton extrusion in nearly all plant tissues, and acts as a "super-auxin" in elongation assays.

19.8 The Fluence Response of Phototropism
The effect of light dose on phototropism is described, and a model explaining the phenomenon is presented.

19.9 Differential *SAUR* Gene Expression during Gravitropism
SAUR gene expression is used to detect the lateral auxin gradient during gravitropism.

19.10 Gravity Perception without Statoliths in *Chara*
The giant-celled freshwater alga *Chara* bends in response to gravity without any apparent statoliths.

Web Essays

19.1 Exploring the Cellular Basis of Polar Auxin Transport
Experimental evidence indicates that the polar transport of the plant hormone auxin is regulated at the cellular level. This implies that proteins involved in auxin transport must be asymmetrically distributed on the plasma membrane. How those transport proteins get to their destination is the focus of ongoing research.

19.2 Apical Basal Polarity Is Maintained in Mature Plants
Developmental biologist increasingly choose to apply the terminology applied to embryonic development to mature plants. The result can be a confusing use of terminology.

19.3 Phototropism: From Photoperception to Auxin-Dependent Changes in Gene Expression
How photoperception by phototropins is coupled to auxin signaling is the subject of this essay.

19.4 Multiple Signals Control Lateral Branching
Experimental evidence for multiple signals regulating lateral branding in pea and Arabidopsis.

Chapter References

Aloni, R. (2001) Foliar and axial aspects of vascular differentiation: Hypotheses and evidence. *J. Plant Growth Regul.* 20: 22–34.

Aloni, R., Schwalm, K., Langhans, M., and Ullrich, C. I. (2003) Gradual shifts in sites and levels of free-auxin production during leaf-primordium development and their role in vascular differentiation and leaf morphogenesis in *Arabidopsis*. *Planta* 216: 841–853.

Benkova, E., Michniewicz, M., Sauer, M., Teichmann, T., Seifertova, D., Jurgens, G., and Friml, J. (2003) Local, efflux-dependent auxin gradients as a common module for plant organ formation. *Cell* 114: 591–602.

Blakeslee, J. J., Peer, W. A., and Murphy, A. S. (2005) Auxin transport. *Curr. Opin. Plant Biol.* 8: 121–125.

Blilou, I., Xu, J., Wildwater, M., Willemsen, V., Paponov, I., Friml, J., Heidstra, R., Aida, M., Palme, K., and Scheres, B. (2005) The PIN auxin efflux facilitator network controls growth and patterning in *Arabidopsis* roots. *Nature* 433: 39–44.

Booker, J., Chatfield, S., and Leyser, O. (2003) Auxin acts in xylem-associated or medullary cells to mediate apical dominance. *Plant Cell* 15: 495–507.

Booker, J., Auldridge, M., Wills, S., McCarty, D., Klee, H., and Leyser, O. (2004) MAX3/CCD7 is a carotenoid cleavage dioxygenase required for the synthesis of a novel plant signaling molecule. *Curr. Biol.* 14: 1232–1238.

Briggs, W. R., Beck, C. F., Cashmore, A. R., Christie, J. M., Hughes, J., Jarillo, J. A., Kagawa, T., Kanegae, H., Liscum, E., Nagatani, A., et al. (2001) The phototropin family of photoreceptors. *Plant Cell* 13: 993–997.

Brown, D. E., Rashotte, A. M, Murphy, A. S., Normanly, J., Tague, B.W., Peer W. A., Taiz, L., and Muday, G. K. (2001) Flavonoids act as negative regulators of auxin transport in vivo in *Arabidopsis*. *Plant Physiol.* 126: 524–535.

Chen, J.-G., Ullah, H., Young, J. C., Sussman, M. R., and Jones, A. M. (2001) ABP1 is required for organized cell elongation and division in *Arabidopsis* embryogenesis. *Genes Dev.* 15: 902–911.

Cleland, R. E. (1995) Auxin and cell elongation. In *Plant Hormones and Their Role in Plant Growth and Development*, 2nd ed., P. J. Davies, ed., Kluwer, Dordrecht, Netherlands, pp. 214–227.

Dharmasiri, N., Dharmasiri, S., and Estelle, M. (2005a) The F-box protein TIR1 is an auxin receptor. *Nature* 436: 441–451.

Dharmasiri, N., Dharmasiri, S., Weijers, D., Lechner, E., Yamada, M., Hobbie, L., Ehrismann, J. S., Jurgens, G., and Estelle, M. (2005b) Plant development is regulated by a family of auxin receptor F box proteins. *Dev. Cell* 9: 109–119.

Fasano, J. M., Swanson, S. J., Blancaflor, E. B., Dowd, P. E., Kao, T. H., and Gilroy, S. (2001) Changes in root cap pH are required for the gravity response of the *Arabidopsis* root. *Plant Cell* 13: 907–921.

Friml, J., Wlśniewska, J., Benková, E., Mendgen, K., and Palme, K. (2002) Lateral relocation of auxin efflux regulator PIN3 mediates tropism in *Arabidopsis*. *Nature* 415: 806–809.

Galston, A. (1994) *Life Processes of Plants*. Scientific American Library, New York.

Geisler, M., Blakeslee, J. J., Bouchard, R., Lee, O. R., Vincenzetti, V., Bandyopadhyay, A., Peer, W. A., Bailly, A., Richards, E. L., Ejendal, K. F. K., et al. (2005) Cellular efflux of auxin mediated by the *Arabidopsis* MDR/PGP transporter AtPGP1. *Plant J.* 44: 179–194.

Geisler, M. Kolukisaoglu, H. U., Billion, K., Berger, J., Saal, B., Bouchard, R., Frangne, N., Koncz-Kalman, Z, Koncz, C., Dudler, R., et al. (2003) TWISTED DWARF1, a unique plasma membrane-anchored immunophilin-like protein, interacts with *Arabidopsis* multi-drug resistance-like transporters AtPGP1 and AtPGP19. *Mol Biol. Cell* 10: 4238–4249.

Geldner, N., Anders, N., Wolters, H., Keicher, J., Kornberger, W., Muller, P., Delbarre, A., Ueda, T., Nakano, A., and Jurgens, G. (2003) The *Arabidopsis* GNOM ARF-GEF mediates endosomal recycling, auxin transport, and auxin-dependent plant growth. *Cell* 112: 219–230.

Geldner, N., Friml, J., Stierhof, Y. D., Jurgens, G., and Palme, K. (2001) Auxin transport inhibitors block PIN1 cycling and vesicle trafficking. *Nature* 413: 425–428.

Hartmann, H. T., and Kester, D. E. (1983) *Plant Propagation: Principles and Practices*, 4th edition. Prentice-Hall, Inc., N.J.

Hasenstein, K. H., and Evans, M. L. (1988) Effects of cations on hormone transport in primary roots of *Zea mays*. *Plant Physiol.* 86: 890–894.

Hsieh, H. L., Okamoto, H, Wang, M. L., Ang, L. H., Matsui, M., Goodman, H., and Deng, X. W. (2000) *FIN219*, an Auxin-regulated gene, defines a link between phytochrome A and the downstream regulator COP1 in light control of *Arabidopsis* development. *Genes Dev.* 14: 1958–1970.

Iino, M., and Briggs, W. R. (1984) Growth distribution during first positive phototropic curvature of maize coleoptiles. *Plant Cell Environ.* 7: 97–104.

Imhoff, V., Muller, P., Guern, J., and Delbarre, A. (2000) Inhibitors of the carrier-mediated influx of auxin in suspension-cultured tobacco cells. *Planta* 210: 580–588.

Jacobs, M., and Gilbert, S. F. (1983) Basal localization of the presumptive auxin carrier in pea stem cells. *Science* 220: 1297–1300.

Jacobs M, and Ray, P. M. (1976) Rapid auxin-induced decrease in free space pH and its relationship to auxin-induced growth in maize and pea. *Plant Physiol.* 58: 203–209.

Kepinski S., and Leyser, O. (2005) The Arabidopsis F-box protein TIR1 is an auxin receptor. *Nature* 435: 446–451.

Ljung, K., Bhalerao, R. P., and Sandberg, G. (2001) Sites and homeostatic control of auxin biosynthesis in *Arabidopsis* during vegetative growth. *Plant J.* 29: 465–474.

Ljung, K., Hull, A. K., Celenza, J., Yamada, M., Estelle, M., Normanly, J., and Sandberg, G. (2005) Sites and regulation of auxin biosynthesis in *Arabidopsis* roots. *Plant Cell* 17: 1090–1104.

McClure, B. A., and Guilfoyle, T. (1989) Rapid redistribution of auxin-regulated RNAs during gravitropism. *Science* 243: 91–93.

Multani, D. S., Briggs, S., Chamberlin, M. A., Blakeslee, J. J., Murphy, A. S., and Johal, G. (2003) Control of plant height in maize by an ABC transporter of the multi-drug resistance class. *Science* 302: 81–84.

Murphy, A. S., Hoogner, K., Peer, W. A., and Taiz, L. (2002) Identification, purification, and molecular cloning of N-1 napthylphthalmic acid-binding plasma membrane-associated amino peptidases from *Arabidopsis*. *Plant Physiol.* 128: 935–950.

Murphy, A. S., Peer W. A., and Taiz, L. (2000) Regulation of auxin transport by aminopeptidases and endogenous flavonoids. *Planta* 211: 315–324.

Nakazawa, M., Yabe, N., Ishikawa, T., Yamamoto, Y. Y. Yoshizumi, T., Hasunuma, K., and Matsui, M. (2001) *DFL1*, an auxin-responsive *GH3* gene homologue, negatively regulates shoot cell elongation and lateral root formation, and positively regulates the light response of hypocotyls length. *Plant J.* 25: 213–221.

Noh, B., Bandyopadhyay, A., Peer, W. A., Spalding, E. P., and Murphy, A. S. (2003) Enhanced gravi- and phototropism in plant mdr mutants mislocalizing the auxin efflux protein PIN1. *Nature* 423: 999–1002.

Noh, B., Murphy, A. S., and Spalding, E. (2001) Multi-drug resistance-like genes of *Arabidopsis* required for auxin transport and auxin-mediated development. *Plant Cell* 13: 2551–2454.

Nonhebel, H. M., Cooney, T. P., and Simpson, R. (1993) The route, control and compartmentation of auxin synthesis. *Aust J. Plant Physiol.* 20: 527–539.

Normanly, J. P., Slovin, J., and Cohen, J. (1995) Rethinking auxin biosynthesis and metabolism. *Plant Physiol.* 107: 323–329.

Ottenschlager, I., Wolff, P., Wolverton, C., Bhalerao, R. P., Sandberg, G., Ishikawa, H., Evans, M., and Palme, K. (2003) Gravity-regulated differential auxin transport from columella to lateral root cap cells. *Proc. Natl. Acad. Sci. USA* 100: 2987–2991.

Palme, K., and Gälweiler, L. (1999) PIN-pointing the molecular basis of auxin transport. *Curr. Opin. Plant Biol.* 2: 375–381.

Peer, W. A., Bandyopadhyay, A., Blakeslee, J. J., Srinivas, M. N., Chen, R., Masson, P., and Murphy, A. S., (2004) Variation in PIN gene expression and protein localization in flavonoid mutants with altered auxin transport. *Plant Cell* 16: 1898–1911.

Reinhardt D., Pesce E.R., Stieger P., Mandel T., Baltensperger K., Bennett M., Traas J., Friml J., and Kuhlemeier, C. (2003) Regulation of phyllotaxis by polar auxin transport. *Nature* 426: 255–260.

Rothwell, G. W., and Lev-Yadun, S. (2005) Evidence of polar auxin flow in 375 million-year-old fossil wood. *Amer. J. Bot.* 92: 903–906.

Schmidt, R. C., Müller, A., Hain, R., Bartling, D., and Weiler, E. W. (1996) Transgenic tobacco plants expressing the *Arabidopsis thaliana* nitrilase II enzyme. *Plant J.* 9: 683–691.

Shaw, S., and Wilkins, M. B. (1973) The source and lateral transport of growth inhibitors in geotropically stimulated roots of *Zea mays* and *Pisum sativum*. *Planta* 109: 11–26.

Sievers, A., Buchen, B., and Hodick, D. (1996) Gravity sensing in tip-growing cells. *Trends Plant Sci.* 1: 273–279.

Staswick, P. E., Serban, B., Rowe, M., Tiryaki, I., Maldonado, M. T., Maldonado, M. C., and Suza, W. (2005) Characterization of an *Arabidopsis* enzyme family that conjugates amino acids to indole-3-acetic acid. *Plant Cell* 17: 616–627.

Swarup, R., and Bennett, M. (2003) Auxin transport: The fountain of life in plants? *Dev. Cell* 5: 824–826.

Swarup, R., Friml, J., Marchant, A., Ljung, K., Sandberg, G., Palme, K., and Bennett, M. (2001) Localization of the auxin permease AUX1 suggests two functionally distinct hormone transport pathways operate in the *Arabidopsis* root apex. *Genes Dev.* 15: 2648–2653.

Terasaka K., Blakeslee, J. J., Titapiwatanakun, B., Peer, W. A., Bandyopadhyaya, A., Makam, S. N., Lee, O. R., Richards, E. L., Murphy, A. S., Sato, F., et al. (2005) PGP4, an ATP binding cassette P-glycoprotein, catalyzes auxin transport in *Arabidopsis thaliana* roots. *Plant Cell* 17: 2922–2939.

Ulmasov, T., Murfett, J., Hagen, G., and Guilfoyle, T. J. (1997) Aux/IAA proteins repress expression of reporter genes containing natural and highly active synthetic auxin response elements. *Plant Cell.* 9: 1963–1971.

Vernoux, T., Kronenberger, J., Grandjean, O., Laufs, P., and Traas, J. (2000) *PIN-FORMED1* regulates cell fate at the periphery of the shoot apical meristem. *Development* 127: 5157–5165.

Volkmann, D., and Sievers, A. (1979) Graviperception in multicellular organs. In *Encyclopedia of Plant Physiology*, New Series, Vol. 7, W. Haupt and M. E. Feinleib, eds., Springer, Berlin, pp. 573–600.

Woodward, W., and Bartel, B. (2005) Auxin: Regulation, action, and interaction. *Annals Botany* 95: 707–735.

Wright, A. D., Moehlenkamp, C. A., Perrot, G. H., Neuffer, M. G., and Cone, K. C. (1992) The maize auxotrophic mutant *orange pericarp* is defective in duplicate genes for tryptophan synthase beta. *Plant Cell* 4: 711–719.

Chapter 20

Gibberellins: Regulators of Plant Height and Seed Germination

FOR NEARLY 30 YEARS after the discovery of auxin in 1927, and more than 20 years after its structural elucidation as indole-3-acetic acid, Western plant scientists tried to ascribe the regulation of all developmental phenomena in plants to the hormone auxin. However, as we will see in this and subsequent chapters, plant growth and development are regulated by several hormones acting both individually and in concert.

The second group of plant hormones to be characterized was the gibberellins (GAs). At least 136 naturally occurring GAs have been identified (MacMillan 2002), and their structures can be viewed at http://www.plant-hormones.info/gibberellin_nomenclature.htm. This website is frequently updated as naturally occurring GAs are newly characterized and named. Unlike the auxins, a diverse group of chemicals with similar biological properties, the GAs all share a similar chemical structure but relatively few of them have intrinsic biological activity. Many of the others are either metabolic precursors of the bioactive GAs or their deactivation products. There are often only a few bioactive GAs in any given plant, and their levels are generally correlated with stem length. Gibberellins play important roles in a variety of other physiological phenomena as well.

The biosynthesis of GAs is under strict genetic, developmental, and environmental control. Gibberellins are best known for their promotion of stem elongation, and GA-deficient mutants that have dwarf phenotypes have been isolated. Mendel's tall/dwarf alleles in peas are a famous example of a single gene locus that can control the level of bioactive GA and hence stem length. Such mutants have been useful in elucidating the complex pathways of GA biosynthesis, and in determining which of the GAs in a plant has intrinsic biological activity.

We begin this chapter by describing the discovery of GAs in a fungal pathogen of rice plants (*Oryza sativa*) and discussing their chemical structure. We then provide an overview of the physiological processes that are regulated by GAs. Underscoring their importance throughout a plant's life cycle, we see that GAs are involved in seed germination, shoot growth, transition to flowering, anther development, pollen tube growth, floral development, fruit set and subsequent growth, and seed development. We then examine biosynthesis of GAs and the role of factors that regulate the levels of bioactive GA in tissues or organs at specific developmental stages. In recent years, molecular genetic approaches have led to considerable progress in our understanding of the mechanisms of GA action. Among the important advances is the identification of a putative GA receptor from rice plants and of several intermediates in the GA signal transduction pathway.

Gibberellins: Their Discovery and Chemical Structure

A brief description of the discovery of gibberellins serves to underscore this groundbreaking work and helps to provide an explanation for the unusual terminology applied to GAs. After a consideration of their discovery, we will describe their chemical structures and numbering system. Finally, we'll explore the relation of specific chemical structures to biological action.

Gibberellins were discovered by studying a disease of rice

Although GAs first came to the attention of western scientists in the 1950s, they had been discovered much earlier in Japan. Rice farmers had long known of a fungal disease (termed "foolish seedling" or *bakanae* disease in Japanese) that caused rice plants to grow too tall and eliminated seed production. Plant pathologists found that these symptoms in rice were induced by a chemical or chemicals secreted by a pathogenic fungus, ***Gibberella fujikuroi***, that had infected the plants. Culturing this fungus in the laboratory and analyzing the culture filtrate enabled Japanese scientists in the 1930s to obtain impure crystals of two fungal "compounds" (they were later shown to be mixtures) with plant growth–promoting activity. One of these, because it was isolated from the fungus *Gibberella*, was named gibberellin A.

Gibberellic acid was first purified from *Gibberella* culture filtrates

In the 1950s two research groups—one at Imperial Chemical Industries in Britain, the other at the U.S. Department of Agriculture (USDA) in Illinois—elucidated the chemical structure of a compound that they had both purified from *Gibberella* culture filtrates and which they named *gibberellic acid*. At about the same time, scientists at Tokyo University separated and characterized three different gibberellins from the original gibberellin A sample, and named them gibberellin A_1, gibberellin A_2, and gibberellin A_3. The numbering system for gibberellins used for the past 50 years builds on this initial nomenclature of gibberellin A_1 (abbreviated GA_1), GA_2, and GA_3 as explained below. The Japanese scientists' GA_3 was later shown to be identical to the gibberellic acid isolated by the U.S. and British scientists. For this reason, GA_3 is also referred to as gibberellic acid.

It soon became evident that many different GAs were present in *Gibberella* cultures, although GA_3 was usually the principal component. (Thus GA_3 is the most frequently produced GA in commercial industrial-scale fermentations of *Gibberella*, for agronomic, horticultural and other scientific use). As GA_3 became available for scientific use, physiologists began testing it on a wide variety of plants. Spectacular responses were obtained in the elongation growth of dwarf and rosette plants, particularly in genetically dwarf peas (*Pisum sativum*), dwarf maize (corn; *Zea mays*) (Figure 20.1), and many rosette plants (Figure 20.2).

Figure 20.1 The effect of exogenous GA_1 on wild-type and dwarf (*d1*) maize. Gibberellin stimulates dramatic stem elongation in the dwarf mutant, but has little or no effect on the tall wild-type plant. (Courtesy of B. Phinney.)

FIGURE 20.2 Cabbage, a long-day plant, remains as a rosette in short days, but it can be induced to bolt and flower by applications of GA_3. In the case illustrated, giant flowering stalks were produced. (© Sylvan Wittwer/Visuals Unlimited.)

Because applications of fungus-derived GA_3 could increase the height of dwarf mutants, it was natural to ask whether wild-type plants contain their own endogenous GA. Bioassay of the extracts of a variety of plant species showed that GA-like substances* were indeed present. Higher concentrations were found in immature seeds (now known to be approximately 1 part per million) than in vegetative tissue (1–10 parts per billion). This made immature seeds the plant material of choice for GA extraction, but chemists still had to use tens of kilograms of seeds to obtain enough GA for chemical characterization. The first identification of a GA from a plant extract was the 1958 discovery of GA_1 from immature seeds of runner bean (*Phaseolus coccineus*). Nowadays, the availability of more-sensitive spectroscopic methods precludes the need to use such vast amounts of plant material to isolate and characterize new GAs.

As more and more GAs from *Gibberella* and different plant sources were characterized, a scheme was adopted in 1968 to number them (GA_1- GA_n) in chronological order of their discovery. Two organic chemists, Drs. Nobutaka Takahashi from Japan and Jake MacMillan from Britain, oversaw the naming and numbering so that only GAs that were naturally occurring and whose chemical structure had been conclusively determined could be assigned **A numbers**.* This naming system has now been taken over by Drs. Yuji Kamiya and Peter Hedden, preserving the historic link of Japanese and British scientists in the naming of these hormones.

All gibberellins are based on an *ent*-gibberellane skeleton

By definition, all GAs possess a tetracyclic (four-ringed) *ent*-gibberellane skeleton (containing 20 carbon atoms), or a 20-nor-*ent*-gibberellane skeleton (containing only 19 carbon atoms because carbon 20 is missing)†. Gibberellins that have the full diterpenoid complement of 20 carbon atoms are referred

ent-gibberellane

to as **C_{20}-GAs** (e.g., GA_{12}). The others that have only 19 carbons because they have lost carbon-20 (blue on structure) by metabolism are referred to as **C_{19}-GAs** (e.g., GA_9). In nearly all C_{19}-GAs, the carboxyl at C-4 forms a lactone at C-10 (red on structure). Other structural modifications

GA_{12} (C_{20}-GA)

GA_9 (C_{19}-GA)

---

*The term GA-like substance refers to a compound that shows activity in a GA bioassay, but has not been chemically characterized.

*Note that the "number" of a GA is simply a cataloging convenience, and there is no implied metabolic relationship between GAs with adjacent numbers.

†The prefix *ent* refers to the fact that the skeleton is derived from *ent*-kaurene, a tetracyclic hydrocarbon that is enantiomeric to the naturally occurring compound, kaurene.

include the insertion of additional functional groups, the position and stereochemistry of which can have profound effects on biological activity. Interestingly, and not surprisingly, the most biologically active GAs were among the first to be discovered, namely GA_1, GA_3, GA_4, and GA_7. These GAs, all of which have intrinsic or inherent stem growth-promoting activity, are C_{19}-GAs. They all possess a 4,10-lactone, a carboxylic acid (—COOH) at C-6, and a hydroxyl group (—OH) at C-3 in β-orientation. Although bioactive GAs

GA_4 R = H
GA_1 R = OH

GA_7 R = H
GA_3 R = OH

may possess other functionalities at various positions in the molecule, such as a double bond between C-1 and C-2 (as in GA_7 and GA_3) and/or an —OH group at C-13 (as in GA_1 and GA_3), it is a requirement that they *not* have a 2β-OH group that renders a GA inactive. (A further discussion of GA structure can be found in **Web Topic 20.1**.)

Although the molecular structure of a GA receptor has not yet been fully characterized, it is assumed that the strict structural requirements for bioactivity listed above reflect the necessary size and shape needed for binding to the receptor. GAs that do not fulfill these structural requirements, and thus are not *inherently* bioactive, may still have biological activity if they can be metabolized within the plant to a GA that does possess the required structural features.

Effects of Gibberellins on Growth and Development

Though they were originally discovered as the cause of disease symptoms of rice that resulted in internode elongation, endogenous GAs can influence a large number of developmental processes in addition to stem elongation. Many of these properties of GAs have been exploited in agriculture for decades, and GAs (or inhibitors of GA biosynthesis) have several important commercial uses.

Gibberellins can stimulate stem growth

Although applied GAs do not have dramatic effects on stem elongation of plants that are already "tall," they can promote internode elongation in genetically dwarf mutants, in "rosette" species, and in various members of the Poaceae (grass) family. Often GA_3 is used because it is most readily available, but the other bioactive GAs, such as GA_1, GA_4, and GA_7, are generally effective for internode elongation, too. Exogenous GA causes such extreme stem elongation in dwarf maize plants that they resemble the tallest varieties of the same species (see Figure 20.1).

Rosette species are plants in which the first-formed internodes do not elongate under certain growing conditions. This results in a compact cluster or rosette of leaves, as seen in members of the Brassicaceae (cabbage) family. Rosette formation is frequently observed in long-day plants grown in short-day conditions. Bolting (stem growth) and flowering will result if plants are treated with a bioactive GA, or are transferred to long days (see Figure 20.2). As discussed later, transfer of short-day grown spinach (*Spinacea oleracea*) plants to long-day conditions results in an increase in endogenous GA biosynthesis that is assumed to be important for both stem elongation and flowering. The application of GA_5 (see structure on page 533) can induce flowering but not stem growth in ryegrass (*Lolium temulentum*), allowing the two processes to be examined independently.

When GA_1 or GA_3 promotes internode elongation in members of the grass family, for example rice and wheat (*Triticum aestivum*), the target of GA action is the intercalary meristem—a meristem near the base of an internode that produces derivatives (new cells) above the meristematic cells. Deepwater rice, which is discussed later, is a particularly striking example; under field conditions, growth rates of up to 25 cm per day have been measured.

Gibberellins are also important for root growth. Extreme dwarf mutants of pea and Arabidopsis, in which GA biosynthesis is blocked, have shorter roots than wild-type plants, and GA application to the shoot enhances both shoot *and* root elongation (Yaxley et al. 2001; Fu and Harberd 2003).

Gibberellins regulate the transition from juvenile to adult phases

Many woody perennials do not flower or produce cones until they reach a certain stage of maturity; up to that stage they are said to be juvenile (see Chapter 25). The juvenile and mature stages may have different growth habits and leaf forms, as in English ivy (*Hedera helix*), in which the juvenile phase is a vine with lobed leaves, whereas the mature form is an erect shrub with a simple leaf form (refer to Figure 25.9). Applied GAs can regulate the juvenile to adult transition, although the nature of the effect depends on the species.

In many conifers, the juvenile phase, which can last up to 20 years, can be shortened by treatment with GA_3 or with a mixture of GA_4 and GA_7, and 2- or 3-year-old plants can be induced to enter the reproductive, cone-producing

(A) White spruce

(B) White spruce

(C) Giant sequoia seedling

Figure 20.3 Gibberellins induce conebud formation in juvenile conifers. (A) and (B) Female cones (already pollinated) developing on sapling-size, grafted propagules of white spruce (*Picea glauca*), stem-injected the previous summer with a mixture of GA_4/GA_7 in aqueous ethanol. (C) A 14-week-old seedling of giant sequoia (*Sequoiadendron giganteum*) that had been sprayed with an aqueous solution of GA_3 some 8 weeks earlier, showing the development of a female conebud. (Courtesy of S. D. Ross and R. P. Pharis.)

phase precociously (Figure 20.3). In contrast, GA_3 applied to English ivy can stop the mature phase from flowering and cause new growth at stem apices to revert to the juvenile form, with its characteristic lobed leaves and vine-like growth habit.

Gibberellins influence floral initiation and sex determination

As already noted, GAs can substitute for the long-day requirement for flowering in many plants, especially rosette species. The interaction of photoperiod and GAs in flowering is complex, and this subject is discussed later in this chapter and in Chapter 25.

In plants with unisexual rather than hermaphroditic flowers, sex determination is genetically regulated. However, it is also influenced by environmental factors such as photoperiod and nutritional status, and these environmental effects may be mediated by GAs. Just as in the case of the juvenile-to-adult transition, the nature of the effect of GA on sex determination can vary with species. In maize, for example, GAs suppress stamen development, resulting in female (pistillate) flowers. We will return to the stamen-suppressing effect of GA in maize later in the chapter when we discuss dwarfism. In dicots such as cucumber (*Cucumis sativus*), hemp (*Cannabis sativa*), and spinach, the opposite is observed: GAs promote the formation of staminate (male) flowers, and inhibitors of GA biosynthesis promote the formation of pistillate flowers.

Gibberellins promote pollen development and tube growth

Gibberellin-deficient dwarf mutants are known to have impaired anther development and pollen formation that can be rescued by treatment with bioactive GA. In mutants

FIGURE 20.4 Gibberellin induces growth in Thompson's seedless grapes. The bunch on the left is an untreated control. The bunch on the right was sprayed with GA$_3$ during fruit development. (© Sylvan Wittwer/Visuals Unlimited.)

in which GA response is blocked, defects in anther and pollen development cannot be reversed by GA treatment, so these mutants may be male-sterile. In addition, reducing the level of bioactive GA in Arabidopsis by overexpressing a GA deactivating enzyme severely inhibits pollen tube growth (Swain and Singh 2005). Thus GAs seem to be required for both the development of the pollen grain and the formation of the pollen tube.

Gibberellins promote fruit set and parthenocarpy

Gibberellin application can cause **fruit set** (the initiation of fruit growth following pollination) and growth of some fruits, sometimes even when auxin has no effect. For example, stimulation of fruit set by GA has been observed in pear (*Pyrus communis*). GA-induced fruit set may occur in the absence of pollination, resulting in parthenocarpic fruits (fruits without seeds). In grape (*Vitis vinifera*), the Thompson's seedless variety produces small seedless fruits that can be stimulated to enlarge by treatment with GA$_3$ (Figure 20.4). This treatment also reduces fungal infections that can be problematic in the compact clusters of untreated grapes. Both these effects of GAs on seedless grapes are exploited commercially.

Gibberellins promote seed development and germination

Some GA-deficient mutants, or transgenic plants with enhanced GA inactivation, have increased seed abortion. The failure of seeds to develop normally can be attributed to reduced levels of bioactive GAs in very young seeds. Treatment with GA will not restore normal seed development, because exogenous GA cannot enter the new seeds. However, the effect of GA deficiency on seed abortion can be negated by simultaneous expression of mutations that give a constitutive GA response (Swain and Singh 2005). Taken together, these results provide evidence for a role for GA in the early stages of seed development.

Some seeds, particularly those of wild plants, require light or cold treatment to induce germination. In such seeds this requirement (see Chapter 23) can often be overcome by application of bioactive GA. Since changes in GA levels are often, but not always, seen when seeds are chilled, GAs may represent a natural regulator of one or more of the processes involved in germination.

Gibberellin application also stimulates the production of numerous hydrolytic enzymes, notably α-amylase, by the aleurone layers of germinating cereal grains. Because germinating cereal grains are the principal system in which GA signal transduction pathways have been analyzed, this aspect will be treated in detail later in the chapter.

Commercial uses of gibberellins and GA biosynthesis inhibitors

The major commercial uses of GAs (typically GA$_3$) are to promote the growth of fruit crops, to stimulate the barley malting process in the beer-brewing industry, and to increase sugar yield in sugarcane. A description of the malting process can be found in **Web Topic 20.2**.

Inhibitors of GA biosynthesis have been useful for crops in which a *reduction* in plant height is desirable. For example, tallness is a disadvantage for cereal crops grown in cool, damp climates, as occur in Europe, where lodging can be a problem. (*Lodging*—the bending of stems to the ground caused by the weight of water collecting on the ripened heads—makes it difficult to harvest the grain with a combine harvester.) Shorter internodes reduce the tendency of the plants to lodge, increasing the yield of the crop. Even genetically dwarf wheat cultivars grown in Europe are sprayed with inhibitors of GA biosynthesis such as Cycocel to further reduce stem length and lodging.

In the field or greenhouse, tall plants are often difficult to manage. In floral crops, short, sturdy plants such as lilies, chrysanthemums, and poinsettias are desirable, and restrictions on elongation growth can be achieved by applications of inhibitors of GA biosynthesis. Such compounds are often used to control the size of container-grown ornamental plants in nurseries, greenhouses, and shade houses.

Further discussion of these topics can be found in **Web Topic 20.2**, together with the chemical structures of several commercially available inhibitors of GA biosynthesis in **Web Topic 20.1**.

Biosynthesis and Catabolism of Gibberellins

Gibberellins constitute a large family of tetracyclic diterpene acids synthesized via a terpenoid pathway. Early stages of that pathway are described in Chapter 13. In this section, covered in detail in **Web Topic 20.3**, we describe the later stages in the GA pathway. We also discuss the regulatory enzymes in the pathway, as well as the genes that encode them. A knowledge of GA biosynthesis and deactivation is important, as it contributes to our understanding of GA homeostasis.

Historically, much pioneering work on GA biosynthesis in plants was made using cell-free extracts from the liquid endosperm of pumpkin (*Cucurbita maxima*) seeds. These in vitro systems from pumpkin endosperm are relatively easy to prepare and have a high biosynthetic capacity. Many in vivo studies, utilizing intact pea seeds and pea and maize shoots as model systems, also helped to establish the major GA biosynthetic pathways in legumes and cereals.

The availability of pea and maize mutants with altered stem length was instrumental in determining the importance of GA_1 as the bioactive GA for stem growth in these species, as discussed later in this chapter. The isolation of stem length mutants of Arabidopsis was also important for cloning genes encoding enzymes in the GA biosynthetic pathway. Sequencing the Arabidopsis and rice genomes has led to the development of comprehensive databases that facilitate the rapid identification of new genes and proteins.

Fundamental to our understanding of how GAs control growth and development is the ability to identify the GAs present in our experimental plants. Systems of measurement using a biological response, called bioassays, were originally important for detecting GA-like activity in partly purified extracts and for assessing the biological activity of known GAs (Figure 20.5). However, bioassay of a plant extract does not give conclusive identification of the GAs within the extract, just an indication that it contains one or more GA-like substances. The use of bioassays for impure plant extracts has declined with the development of highly sensitive physical techniques such as mass spectrometry that allow precise identification and quantification of specific GAs from small amounts of tissue. This topic is discussed in **Web Topic 20.4**.

Gibberellins are synthesized via the terpenoid pathway

Terpenoids are compounds made up of five-carbon **isoprenoid** building blocks, joined head to tail. The GAs are diterpenoids that are formed from *four* such isoprenoid units. The GA biosynthetic pathway can be divided into three stages, each residing in a different cellular compartment (Figure 20.6) (Hedden and Phillips 2000). In *stage 1*, which occurs in plastids, four isoprenoid units are assembled to give a 20-carbon linear molecule, geranylgeranyl diphosphate (GGPP), which is then converted into a tetracyclic compound, *ent*-kaurene. In *stage 2*, which occurs on the plastid envelope and in the endoplasmic reticulum, *ent*-kaurene is converted, in a stepwise manner, to the first-formed GA, GA_{12}. The pathway to GA_{12} is essentially the same in all plant species studied so far. In *stage 3*, which occurs in the cytosol, GA_{12} is converted first into other C_{20}-GAs, and then into C_{19}-GAs, including the bioactive GA(s). Stage 3 differs not only from species to species, but even in different organs within one species, or in response to different environmental conditions. The pathway up to the bioactive GA (which, as discussed later, is GA_1 in cereals and legumes and GA_4 in Arabidopsis and cucurbits) is referred to as "biosynthesis," and later metabolic steps that involve deactivation of the bioactive GA are referred to as "catabolism." Gibberellin pathways (starting from GGPP, the first diterpenoid component of the pathway) are shown in Figure 20.6, and the entire pathway is described in **Web Topic 20.3**.

Some enzymes in the GA pathway are highly regulated

Work with biosynthetic mutants of Arabidopsis, pea, and maize (Figures 20.7 and 20.8) facilitated the cloning of

FIGURE 20.5 Gibberellin causes elongation of the leaf sheath of rice seedlings, and this response is used in the dwarf rice leaf sheath bioassay. Here, 4-day-old seedlings were treated with increasing amounts of GA_3 and allowed to grow for another 5 days. (Courtesy of P. Davies.)

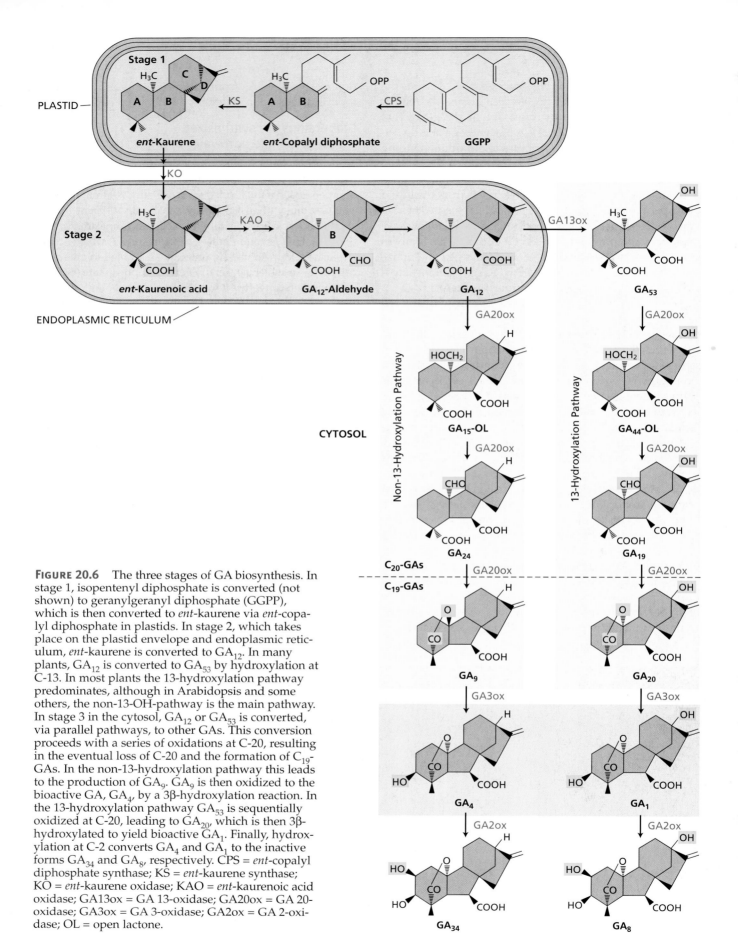

FIGURE 20.6 The three stages of GA biosynthesis. In stage 1, isopentenyl diphosphate is converted (not shown) to geranylgeranyl diphosphate (GGPP), which is then converted to *ent*-kaurene via *ent*-copalyl diphosphate in plastids. In stage 2, which takes place on the plastid envelope and endoplasmic reticulum, *ent*-kaurene is converted to GA_{12}. In many plants, GA_{12} is converted to GA_{53} by hydroxylation at C-13. In most plants the 13-hydroxylation pathway predominates, although in Arabidopsis and some others, the non-13-OH-pathway is the main pathway. In stage 3 in the cytosol, GA_{12} or GA_{53} is converted, via parallel pathways, to other GAs. This conversion proceeds with a series of oxidations at C-20, resulting in the eventual loss of C-20 and the formation of C_{19}-GAs. In the non-13-hydroxylation pathway this leads to the production of GA_9. GA_9 is then oxidized to the bioactive GA, GA_4, by a 3β-hydroxylation reaction. In the 13-hydroxylation pathway GA_{53} is sequentially oxidized at C-20, leading to GA_{20}, which is then 3β-hydroxylated to yield bioactive GA_1. Finally, hydroxylation at C-2 converts GA_4 and GA_1 to the inactive forms GA_{34} and GA_8, respectively. CPS = *ent*-copalyl diphosphate synthase; KS = *ent*-kaurene synthase; KO = *ent*-kaurene oxidase; KAO = *ent*-kaurenoic acid oxidase; GA13ox = GA 13-oxidase; GA20ox = GA 20-oxidase; GA3ox = GA 3-oxidase; GA2ox = GA 2-oxidase; OL = open lactone.

FIGURE 20.7 Phenotypes of wild-type and GA-deficient mutants of Arabidopsis, showing the position in the GA biosynthetic pathway that is blocked in each mutant. All mutant alleles (denoted by lower case notation of the wild-type alleles) are homozygous. Plants were grown in continuous light and are 7 weeks old. Note that the *ga1*, *ga2*, and *ga3* seedlings are sterile and have not produced siliques (seed pods). For abbreviations see Figure 20.6. (Courtesy of V. Sponsel.)

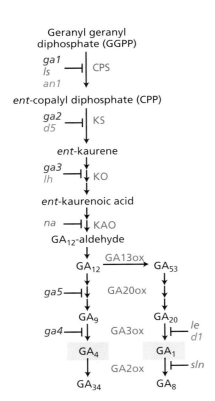

genes for many of the enzymes in GA biosynthesis and deactivation and enhanced our understanding of the pathway. Most notable from a regulatory standpoint are three enzymes in *stage 3* of the pathway. These are the GA 20-oxidase (GA20ox) and GA 3-oxidase (GA3ox) enzymes, which catalyze the steps prior to bioactive GA, and the GA 2-oxidase (GA2ox), which is involved in GA deactivation. All three of these enzymes are **dioxygenases**, and they have sequences in common with each other and with other enzymes utilizing 2-oxoglutarate and Fe^{2+} as cofactors. The common sequences represent the binding sites on the enzymes for 2-oxoglutarate and Fe^{2+}. We shall discuss each of these enzymes, before describing some of what is known of their regulation.

- **GA 20-oxidase** In Arabidopsis there is a small family of GA 20-oxidases, named AtGA20ox1 through

FIGURE 20.8 A portion of the GA biosynthetic pathway showing the metabolic steps that are blocked by known mutations (denoted by lower case notations of wild-type alleles) of Arabidopsis (green), pea (blue), and maize (orange). For abbreviations see Figure 20.6. As a historical note, the names *GA1–GA5* for five nonallelic loci in Arabidopsis that encode enzymes in the GA biosynthetic pathway were assigned long before the nature of the enzymes or their genes were known (Koornneef and van der Veen 1980). After genetic tests they were placed in sequence based on the anticipated order of action in the pathway.

AtGA20ox5. The main stem-expressed GA 20-oxidase (AtGA20ox1) is encoded by *GA5* (Phillips et al. 1995; Xu et al. 1995). Mutation in the *GA5* gene (*ga5*) results in a semidwarf, rather than extreme dwarf, phenotype (see Figure 20.7), presumably because of the overlapping functions of the other GA 20-oxidases in this plant. In contrast, *ga1*, *ga2*, and *ga3* are extreme dwarfs, since enzymes earlier in the pathway (*ent*-copalyl diphosphate synthase-CPS, *ent*-kaurene synthase-KS, and *ent*-kaurene oxidase-KO respectively) are encoded by single-copy genes (see Figure 20.7).

- **GA 3-oxidase** The main stem-expressed GA 3-oxidase in Arabidopsis (AtGA3ox1) is encoded by *GA4* (Chiang et al. 1995). Again, there is a small gene family, and the *ga4* mutant, like *ga5*, is a semidwarf rather than an extreme dwarf (see Figure 20.7). In pea, this enzyme is encoded by the *LE* gene (discussed later). In maize this enzyme is encoded by *D1*, and a recessive mutation in this gene gives semidwarf plants (see Figure 20.1).

- **GA 2-oxidase** There is no mutant phenotype known for the GA 2-oxidase in Arabidopsis. In pea this enzyme is encoded by *SLENDER* (*SLN*), so named because mutations in this gene cause young seedlings to be taller than wild-type ones. This is because GAs synthesized during seed maturation are not deactivated to the same extent as in wild-type seeds, and some potentially bioactive GAs remain in the mature seed. During germination they can be converted into bioactive GA_1, which can enhance internode elongation in the young seedling (Reid et al. 1992).

Gibberellin regulates its own metabolism

As explained in Chapter 19 for auxin, many factors are important for maintaining hormone homeostasis, including the relative balance between synthesis and deactivation. Part of a plant's response to bioactive GA is to depress GA biosynthesis and stimulate catabolism (deactivation), to prevent excessive stem elongation. For example, depression of biosynthesis is achieved through down-regulation (inhibition of expression) of the *GA20ox* and *GA3ox* genes encoding the last two enzymes in the formation of bioactive GA. This effect of GA on its own biosynthesis is termed **negative feedback regulation**. Enhanced GA deactivation is also important for maintaining GA homeostasis. It is achieved by up-regulating (stimulating) the expression of the *GA2ox* gene encoding an enzyme that deactivates GA. The ability of GA to promote the expression of genes involved in its own deactivation is termed **positive feed-forward regulation**.

The relative importance of each gene to this process varies with species and tissue (Hedden and Phillips 2000). This homeostatic mechanism allows changes in GA content as a result of developmental or environmental signals, but probably provides a mechanism to reestablish GA concentrations to a normal level afterwards.

Mutations that either severely reduce the levels of endogenous GAs, as in the GA-deficient dwarfs, or that prevent the GA response from taking place (considered later in the chapter), both alter this feedback and feed-forward regulation. This is because GA signal transduction does not occur in either type of dwarf mutant. Therefore, in the absence of a GA response, expression of both GA 20-oxidase and GA 3-oxidase is stimulated (up-regulated), while the expression of GA 2-oxidase is repressed (down-regulated). Thus dwarf mutants accumulate much higher levels of transcript for *GA20ox* and *GA3ox* genes, but have lower levels of *GA2ox* transcript than tall plants.

GA biosynthesis occurs at multiple cellular sites

All reactions in GA biosynthesis leading to a C_{19}-GA can be demonstrated in cell-free systems from seed or seed-parts, providing definitive evidence that developing seeds are sites of GA biosynthesis. In pea seeds, a surge in GA biosynthesis occurs soon after fertilization and is necessary for early seed and fruit growth. In later stages of seed maturation in many species, the level of GA accumulation is quite high.

In reporter gene studies, the promoter of a gene encoding, for example, a GA biosynthetic enzyme, is fused to a gene encoding β-glucuronidase (GUS) or green fluorescent protein (GFP). The GUS or GFP activity can be monitored by viewing the distribution or intensity of colored or fluorescent product, thereby giving a means to examine the activity of the GA biosynthesis gene to which the reporter is fused. Using these techniques, *GA1*, which encodes CPS (the first committed enzyme in GA biosynthesis), has been observed in immature seeds, shoot apices, root tips, and anthers of wild-type Arabidopsis plants (Figure 20.9) (Silverstone et al. 1997a). Not unexpectedly, these organs are sites known to be affected in *ga1* mutants, which exhibit seed dormancy, an extreme-dwarf growth habit, and male sterility. This study therefore indicates that the early stages of GA biosynthesis and GA action can occur in the same organs.

More-detailed studies of expression patterns in cells and tissues of germinating embryos of Arabidopsis have shown that *GA4* (encoding the GA 3-oxidase) is expressed in the cortex and endodermis (Yamaguchi et al. 2001). Since the GA 3-oxidase catalyses the final step in the synthesis of bioactive GA_4 (see Figure 20.6), we can therefore assume that GA_4 is produced in the cortex and endodermis. Using in situ hybridization to locate the products of genes known to be up-regulated by GA_4, Ogawa et al. (2003) showed that XTH5, an enzyme likely to be involved in wall loosening that is a prerequisite for cell expansion, is present in the cortex, endodermis, and radicle, whereas CP1, a cysteine proteinase, is present in the provasculature and the aleurone layer.

Thus, within germinating Arabidopsis embryos we see GA_4 *action* both in the same cells in which it is synthesized

(A) 5-day-old-seedling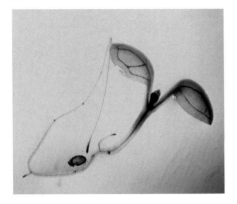

(B) 3-week-old seedling

(C) Open flower

(D) Mature silique with seeds

(E) Developing embryos

FIGURE 20.9 Histochemical analysis of Arabidopsis plants containing the *GA1* promoter-*GUS* gene fusion. *GA1* encodes CPS, the first committed enzyme in GA biosynthesis. *GA1* promoter activity (blue staining), indicating probable sites of GA biosynthesis, is visible throughout the life cycle. (From Silverstone et al. 1997a, courtesy of T-p. Sun.)

(cortex and endodermis) and in different cells (radicle, provasculature, and aleurone), implying that GA_4 or a component downstream of GA_4 in the response pathway must move from cell to cell. Similar extensive studies conducted by Kaneko et al. (2003) with germinating rice grains are discussed later in the chapter.

Environmental conditions can influence GA biosynthesis

Gibberellins play an important role in mediating the effects of environmental stimuli on plant development. For example, photoperiod and temperature can alter the levels of active GAs by affecting transcription of genes encoding specific enzymes in the GA pathway (Yamaguchi and Kamiya 2000). As mentioned earlier, several of the genes encoding enzymes in the pathway are members of small gene families, and it is apparent that individual members of a family can be regulated differently by environmental factors. Recently it has been determined that a rapid effect of far-red illumination on petiole elongation in Arabidopsis is a consequence of a 40-fold increase in transcription of *AtGA20ox2*, with only a minor effect on *AtGA20ox1* (Hisamatsu et al. 2005). The regulation of GA biosynthesis by environmental factors is covered in more detail in **Web Topic 20.5** and in the later section on flowering.

GA_1 and GA_4 have intrinsic bioactivity for stem growth

Seminal studies with GA biosynthetic mutants (also referred to as GA-deficient mutants) conducted in the 1980s achieved two important goals. Not only did they provide a way for the pathways of GA metabolism to be definitively established, but these studies also determined that GA_1 is the predominant bioactive GA for stem growth in pea and maize, and that its precursors have no intrinsic biological activity.

LE and *le* are two alleles of a gene that regulates tallness in peas, the genetic trait first investigated by Gregor Mendel in his pioneering study in 1866. If GA_{20} is applied to the *le* mutant of pea it is not bioactive, whereas GA_1 has

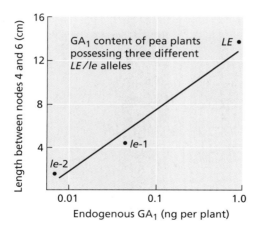

FIGURE 20.10 Stem elongation corresponds closely to the level of GA_1. Here the GA_1 content in peas with three different alleles at the *LE* locus is plotted against the internode elongation in plants with those alleles. The allele *le-2* is a more intense dwarfing allele of *LE* than is the original *le-1* allele. There is a close correlation between the GA_1 level and internode elongation. (After Ross et al. 1989.)

high biological activity and rescues the mutant phenotype (the plants grow tall). Gibberellin A_8 is also inactive. We can infer from this information, and from a knowledge of the GA metabolic pathway in pea ($GA_{20} \rightarrow GA_1 \rightarrow GA_8$), that GA_{20} is inactive unless it can be converted to GA_1 within the plant, and that *GA_1 has intrinsic bioactivity* because its metabolite, GA_8, is inactive (Ingram et al. 1984).

Metabolic studies using both stable and radioactive isotope–labeled GAs have demonstrated that the *LE* gene encodes an enzyme that 3β-hydroxylates GA_{20} to produce GA_1. At the same time, it was confirmed that tall stems contain more GA_1 than dwarf stems (Reid and Howell 1995). Mendel's *LE* gene was eventually cloned, and the recessive *le* allele (now known as *le-1*) was shown to have a single base change leading to a defective enzyme (Lester et al. 1997; Martin et al. 1997). Less GA_1 is produced in plants homozygous for the recessive allele than in wild-type plants. Since this is not a null mutation, and a partially active protein is produced, enough GA_1 is present for *le-1* seedlings to attain approximately 30% of the height of wild-type plants (Figure 20.10). Another *le* allele, *le-2*, (different mutation within the same locus) is more severe. It reduces the level of endogenous GA_1 even further, giving plants that are even shorter than those containing the original *le* allele (see Figure 20.10).

A study of other pea mutants has confirmed that the height of pea plants is directly correlated with the amount of endogenous GA_1. For example, *nana* mutants, which are almost completely devoid of GA_1 because the GA pathway is blocked at KAO (see Figure 20.8), achieve a stature of only about 1 cm at maturity (Figure 20.11). Furthermore, the *sln* mutant of pea is taller than wild-type plants because of impaired GA deactivation, which leads to young seedlings

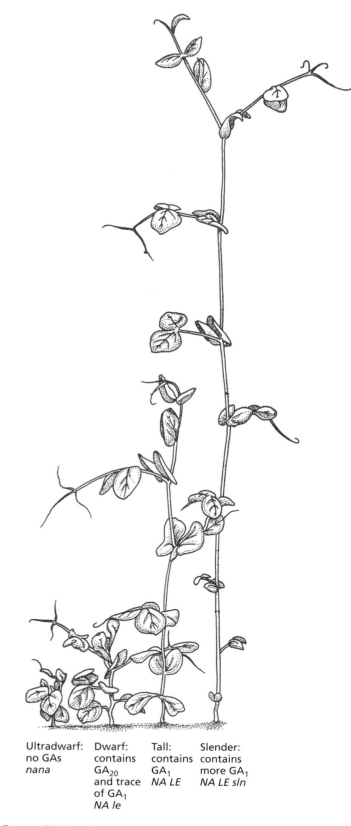

Ultradwarf: no GAs
nana

Dwarf: contains GA_{20} and trace of GA_1
NA le

Tall: contains GA_1
NA LE

Slender: contains more GA_1
NA LE sln

FIGURE 20.11 Phenotypes and genotypes of peas that differ in the GA content of their vegetative tissue. (All alleles are homozygous.) (After Davies 1995.)

having higher levels of endogenous GA_1 than wild-type plants (see Figure 20.11). A personal account of this research on pea stem growth can be found in **Web Essay 20.1**.

In maize, similar studies using GA-deficient dwarf mutants have determined that GA_1 is the bioactive GA regulating stem growth in this cereal monocot (Phinney 1984). Thus, the control of stem elongation by GA_1 occurs in both legumes and cereals and, in fact, has been shown to be quite widespread in other types of plants too.

In contrast, in a few species, including Arabidopsis and several members of the Cucurbitaceae (e.g., pumpkin and cucumber), GA_1 has less biological activity than its non-13-hydroxylated counterpart, GA_4. In these species the presence of a 13-OH group may inhibit the binding of GA to the GA receptor. In Arabidopsis GA_9 has low activity in the *ga4* mutant, and GA_{34} is inactive in wild-type plants. Therefore, in Arabidopsis, GA_4 is assumed to be the main biologically active GA.

Plant height can be genetically engineered

The identity of the bioactive GAs in crop plants, together with the characterization of key enzymes in GA biosynthesis and catabolism, has enabled genetic engineers to alter the levels of bioactive GA in crop plants, and thus affect plant height (Hedden and Phillips 2000). The conserved function of the GA20ox, GA3ox, and GA2ox enzymes between species extends the range of crop plants that can be manipulated. Several examples are given by Phillips (2004). For instance, the growth rate of quaking aspen (*Populus tremuloides*) seedlings can be enhanced by overexpression of a gene encoding an Arabidopsis GA 20-oxidase, with resulting increases in xylem fiber quantity and length that are desirable for paper manufacture (Eriksson et al. 2000).

As explained earlier, the desired effect is often to *decrease* growth. Reductions in GA_1 levels have been achieved in crops either by the transformation of plants with antisense constructs of the *GA20ox* or *GA3ox* genes, which encode the enzymes leading to the synthesis of GA_1, or by overexpressing the *GA2ox* gene, which is responsible for GA_1 deactivation. These approaches have been used to introduce more extreme dwarfing into wheat (Figure 20.12) and rice (Sakamoto et al. 2003).

Dwarf mutants often have other defects in addition to dwarfism

We have learned that bioactive GAs control many aspects of plant growth and development in addition to stem length. What happens in dwarf plants in which GA biosynthesis in stems is blocked by mutation? Clearly, many of these mutations give pleiotropic phenotypes. The severely dwarfed *ga1*, *ga2*, and *ga3* mutants of Arabidopsis have dormant seeds that cannot germinate unless treated with GA. Moreover, the plants are male-sterile unless treated with GA, because of a requirement for bioactive GA in anther and pollen development (see Figure 20.7).

FIGURE 20.12 Genetically engineered dwarf wheat plants. The untransformed wheat is shown on the extreme left. The three plants on the right were transformed with a *GA 2-oxidase* cDNA from bean under the control of a constitutive promoter, so that the endogenous active GA_1 was deactivated. The varying degrees of dwarfing reflect varying degrees of overexpression of the foreign gene. (From Hedden and Phillips 2000, courtesy of A. Phillips.)

In pea, mutations in genes encoding four enzymes in the GA pathway all give dwarf plants, but only one mutation (*lh-2*) leads to impaired seed development The *lh-2* mutation in the *LH* gene, which encodes KO (see Figure 20.8), reduces GA_1 levels in both stems *and* seeds (Davidson and Reid 2004). Thus this *lh-2* mutant, as well as being dwarf, also has significant seed abortion (Figure 20.13). Interesting-

FIGURE 20.13 Pods of wild-type (left) and the *lh-2* (right) mutant of pea, showing seed abortion in this GA-deficient mutant. (Courtesy of J. B. Reid.)

GA-deficient mutant plus GA

GA-deficient mutant minus GA

FIGURE 20.14 In corn, GAs normally suppress anther development, so that anthers develop in the ears of GA-deficient dwarf mutants. (Bottom) Unfertilized ear of the dwarf, GA-deficient, *an1* mutant showing conspicuous anthers. (Top) Anthers are suppressed in an ear from an *an1* plant that has been treated with GA_3. (Courtesy of M. G. Neuffer.)

ly, the *nana* mutant of pea, despite being blocked at KAO (see Figure 20.8) can still produce viable seeds with a normal complement of GAs even though it is a very extreme dwarf. This is because a second *NA* gene is expressed in pea seeds, but not in stems or roots (Davidson et al. 2003). The *nana* mutant therefore has impaired shoot and root growth but normal seed development.

In maize, GA-deficient dwarf mutants, e.g., *anther ear1* (*an1*) in which CPS is blocked (see Figure 20.8), are unusual in having anthers in the female flowers of the ears (Figure 20.14). The presence of anthers is a consequence of GA depletion since, as we saw earlier, GAs normally suppress stamen development in the ears. Thus, although dwarf mutants of our model species were initially selected because of impaired stem elongation, a number of them have pleiotropic phenotypes, underscoring the importance of GAs in many other aspects of growth and development.

Gibberellin Signaling: Significance of Response Mutants

Single-gene mutants impaired in their response to GA are valuable tools for identifying genes that encode possible GA receptors or components of signal transduction pathways. In general terms, factors that affect signal transduction can either be **positive** or **negative regulators**. Three main classes of mutants can be discerned.

1. A mutation that renders a positive regulator of GA signaling nonfunctional would give a dwarf phenotype. These loss-of-function mutations are recessive, and the mutants would not respond to applied GA because of the deficiency in an essential component of the GA signal transduction pathway.

2. A mutation that renders a negative regulator of GA action nonfunctional would have a tall phenotype. Again, these loss-of-function mutations are recessive.

3. A third class of mutants, in which a negative regulator is made constitutively active, also gives GA-nonresponsive dwarf plants, but in these cases the mutations are gain-of-function, and thus semidominant.

What sets these mutants apart from those with mutations that block enzymes in the GA biosynthetic pathway is that the GA-insensitive dwarf plants will not grow tall when treated with bioactive GA, and the constitutive extra-tall mutants (the phenotype is referred to as "slender") are still slender even if grown in the presence of inhibitors of GA biosynthesis. So, unlike the situation for GA biosynthetic mutants, the height of these GA response mutants is not proportional to the amount of endogenous bioactive GA. In fact, many of the GA-insensitive dwarf mutants contain abnormally high levels of GA_1, indicating that the normal feedback regulation of GA biosynthesis has been perturbed.

The response mutants have been extensively studied in Arabidopsis, starting with the *GA insensitive* (*gai-1*) mutant that was isolated in the mid-1980s (Koornneef et al. 1985). Sequencing some of the positive and negative regulators in Arabidopsis paved the way for their characterization in a number of important crop plants, including rice, barley (*Hordeum vulgare*), wheat, and maize. The discussion of GA signaling in the following sections is not treated historically, but the citations indicate the order in which the work was conducted.

Mutations of negative regulators of GA may produce slender or dwarf phenotypes

"Slender" mutants of rice (*slender rice 1* [*slr1*]) and barley (*slender 1* [*sln1*]) are perhaps the easiest of these response mutants to describe (Ikeda et al. 2001; Chandler et al. 2002). These plants, when homozygous for the recessive mutations, are excessively tall or slender, resembling plants that have been treated with a high dose of bioactive GA. Even when grown in the presence of inhibitors of GA biosynthesis these plants are slender. Thus the tallness is not due to the mutants containing very high levels of bioactive GA, but to the GA response being constitutively expressed.*

In these cereal slender mutants we can hypothesize that the signal transduction pathway is constantly "on" even in the absence of signal (GA). From these genetic results it is inferred that these loss-of-function mutants are missing a functional negative regulator, which, when present and functional, would repress the GA signal transduction pathway. Cloning the cereal *slender* genes revealed that they were orthologs[†] of a class of transcriptional regulators already known in Arabidopsis (Peng et al. 1997; Silverstone et al. 1997b). They belong to a subclass of the GRAS family of

---

*The *slender* mutants of rice and barley, which show a constitutive GA response, should not be confused with the slender mutant of pea, discussed earlier in this chapter, which has a defective GA 2-oxidase preventing the deactivation of bioactive GA_1.

[†]Genes in different species that evolved from a common ancestral gene and are therefore evolutionarily related.

transcriptional regulators, named after the first three members of the group: GIBBERELLIN-INSENSITIVE (GAI), REPRESSOR OF ga1-3 (RGA) and SCARECROW (SCR). These particular GRAS proteins have a domain at the N-terminal end in which the first five amino acids are aspartic acid (D), glutamic acid (E), leucine (L), leucine (L), and alanine (A), hence the notation "DELLA domain." However, unlike rice and barley, which each possess only one DELLA-domain protein, Arabidopsis has five. Arabidopsis plants with only one or two defective DELLA-domain proteins do not have a pronounced phenotype. Multiple defective DELLA-domain proteins are required to produce an excessively tall ("slender") phenotype in Arabidopsis.

In Arabidopsis, RGA was isolated as a suppressor of ga1 (Silverstone et al. 1997b, 1998). By mutagenizing a population of GA-deficient ga1 seeds in which GA biosynthesis is blocked at the CPS stage, a mutant (rga) was isolated in which the extreme dwarfism of ga1 was suppressed. These rga/ga1 plants are substantially taller than ga1 plants (Figure 20.15), even though they are GA-deficient. Thus, the rga mutation, which is recessive, is removing a negative regulator of GA signaling. Therefore, as in the sln1 and slr1 mutants, the GA response in rga/ga1 plants is "on."

RGA was shown to be 82% identical in amino acid sequence to GAI* (Peng et al. 1997; Silverstone et al. 1998). However, at that time only one mutation in gai was known (named gai-1 because it was the first allele of gai to be described), and this mutation gives a semidominant, GA-insensitive dwarf phenotype (Koornneef et al. 1985). This meant that the two mutations, gai-1 and rga, in two almost identical proteins, GAI and RGA, were giving quite different phenotypes! These mutants are depicted in Figure 20.15.

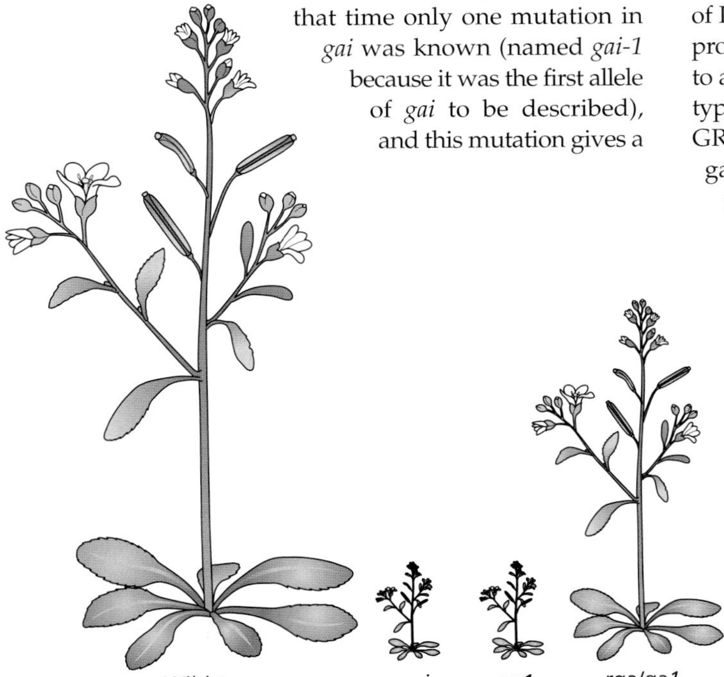

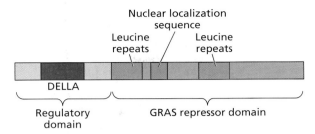

FIGURE 20.16 Domain structure of the RGA or GAI repressor protein showing the regulatory DELLA-domain and the repressor GRAS domain. Leucine repeat sequences and the nuclear localization signal are also shown. The gai-1 and rga-Δ17 mutations are in the DELLA domain, and the rga and gai-t6 mutations are in the GRAS domain.

When the sites of mutations were compared in the rga and gai-1 mutants they were shown to be in different parts of the two proteins—the gai-1 mutation results in a 17–amino acid deletion in the DELLA domain of GAI, whereas the rga mutation is in the repressor (GRAS) domain of RGA (Figure 20.16). When a 17–amino acid deletion was produced in the DELLA domain of wild-type RGA (referred to as rga-Δ17), it also produced an insensitive dwarf phenotype (Dill et al. 2001). Furthermore, a mutation (gai-t6) in the GRAS domain of GAI suppressed the effect on stem elongation of inhibitors of GA biosynthesis in a similar way to rga suppressing the effect of ga1 (Peng et al. 1997).

The alternative phenotypes are also seen in the slender mutations in cereals. For example, wild-type rice transformed with SLR1, which encodes a protein containing a 17–amino acid deletion in the DELLA domain, gives a GA-insensitive dwarf phenotype instead of the slender phenotype seen when mutations are in the GRAS domain (Ikeda et al. 2001). Similarly, the sln1d mutation in barley is a dominant, insensitive dwarf, in contrast to the original slender sln1c mutant with a constitutive GA response (slender) phenotype (Figure 20.17) (Chandler et al. 2002).

Negative regulators with DELLA domains have agricultural importance

The experiments described above provide a considerable body of evidence that the DELLA-domain proteins are important negative regulators of GA response in both dicots and monocots (see Olszewski et al. 2002). Additionally, in wheat the

FIGURE 20.15 Phenotypes of wild-type Arabidopsis and gai (GA-insensitive), ga1 (GA-deficient), and the rga (repressor of ga1) mutants. The latter is in a ga1 background, hence it is designated rga/ga1. This rga/ga1 double mutant is taller than the ga1 mutant, because the rga mutation negates the effect of RGA which is a negative regulator of the GA response. Two "negatives" make a "positive" so the GA response is "on" even in the absence of GA.

---

*Don't be confused by the similar-looking abbreviations: GAI/gai stands for GA-insensitive while GA1/ga1 are alleles of the gene encoding CPS.

Gibberellins signal the degradation of transcriptional repressors

To demonstrate the nuclear localization and repressor nature of RGA, an RGA-GFP fusion protein was expressed under the control of the RGA promoter (Silverstone et al. 2001). The effect of bioactive GA on levels and localization of the fusion protein was determined by immunoblotting and fluorescence microscopy. Roots were used because the autofluorescence of chlorophyll makes GFP analysis in shoots more difficult.

RGA-GFP was localized in nuclei of Arabidopsis root tips, and when the plants were treated with bioactive GA the GFP fluorescence disappeared. Moreover, if GA content was depleted by treatment with the GA biosynthesis inhibitor paclobutrazol, the nuclei retained intense green fluorescence, demonstrating persistence of RGA in the absence of GA (Figure 20.18). Confirming their earlier hypothesis on the vital importance of the DELLA domain, Silverstone et al. (2001) found that GFP-rga-Δ17 (the protein missing the crucial 17 amino acids in the DELLA domain) could not be degraded, and persisted in nuclei even after GA treatment.

In rice, GFP-labeled SLR1 is targeted to the nucleus and is also degraded in the presence of GA (Itoh et al. 2002). Thus, GA promotes the turnover of transcriptional repressors containing DELLA domains in both dicots and monocots.

F-box proteins target DELLA domain proteins for degradation

Dwarf mutants of rice and Arabidopsis have been identified that are defective in their abilities to degrade wild-type DELLA-domain proteins in the presence of bioactive GA. The *GA-insensitive dwarf 2* (*gid2*) and *sleepy 1* (*sly1*) recessive mutations are in orthologous proteins, and give rise to GA-insensitive dwarf phenotypes in rice (Sasaki et al. 2003) and Arabidopsis (McGinnis et al. 2003), respectively. These phenotypes suggest that the *GID2* and *SLY1* genes encode *positive* regulators of GA signaling. In fact, both genes encode proteins with conserved F-box domains that are components of E3-ubiquitin ligase complexes, also known as SCF complexes (Itoh et al. 2003) (see also Chapter 19). F-box components selectively recruit proteins into the SCF complex, leading to their modification by polyubiquitination and subsequent degradation by the 26S proteasome. (Further information on this topic can be found in Chapter 14 on the web site.)

These F-box proteins, GID2 and SLY1, target the DELLA-domain repressor proteins: SLR1 in rice and the RGA-type proteins in Arabidopsis. Thus we see more SLR1 accumulate in nuclei of *gid2* plants, and more RGA and GAI in *sly1* plants, than in their respective wild-type lines (Figure 20.19) (Sasaki et al. 2003; Dill et al. 2004). The characterization of GID2 and SLY1 has thus identified important upstream components of the DELLA-domain proteins in both monocot and dicot species. Moreover, we can now add GA signaling to the growing list of signaling pathways in plants

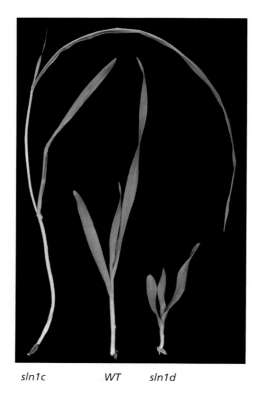

FIGURE 20.17 Demonstration of the opposite effects of two different mutations in the same *SLN1* repressor gene. Three shoots of 2-week-old barley seedlings are shown. The wild-type seedling is in the center. (Left) The *slnc* mutant has a GA-constitutive, slender phenotype associated with loss-of-function of the repressor protein due to a mutation in the GRAS repressor domain. (Right) The *sln1d* mutant is a gain-of-function dominant dwarf, because a mutation in the DELLA domain prevents the repressor protein from being degraded. (From Chandler et al. 2002, courtesy of P. M. Chandler.)

semidominant *Reduced height* mutations (*Rht-B1a* and *Rht-D1a*), and in maize the *d8* mutation, are both in the DELLA domains of the respective *GAI* orthologs, and both mutations give GA-insensitive dwarf plants (Peng et al. 1999).

Classical breeding of wheat and rice in the early and middle of the twentieth century paved the way for the "Green Revolution" of the 1960s—the introduction of high-yielding dwarf varieties of wheat and rice into Latin America and Southeast Asia to keep abreast of human population growth (Hedden 2003). Normal cereals grow too tall when close together in a field, especially with high levels of fertilizer. The result is unacceptably large losses in yield. The dissemination of wheat cultivars expressing *Rht* mutations, reducing GA response and giving shortened stems, was pivotal to the success of the Green Revolution. (More details of Dr. Norman Borlaug's Nobel Prize–winning work, and the elucidation of the molecular mechanisms that control GA-regulated cereal growth, can be found in **Web Essay 20.2**.)

FIGURE 20.18 The RGA protein is found in the cell nucleus, consistent with its identity as a transcription regulator, and its level is affected by the level of GA. (A) Plant cells were transformed with the gene for RGA fused to the gene for green fluorescent protein (GFP), allowing detection of RGA in the nucleus by fluorescence microscopy. (B) Effect of GA on RGA. A 2-hour pretreatment with GA causes the loss of RGA from the cell (top). When GA biosynthesis is inhibited by paclobutrazol, the RGA content in the nucleus increases (bottom). (From Silverstone et al. 2001.)

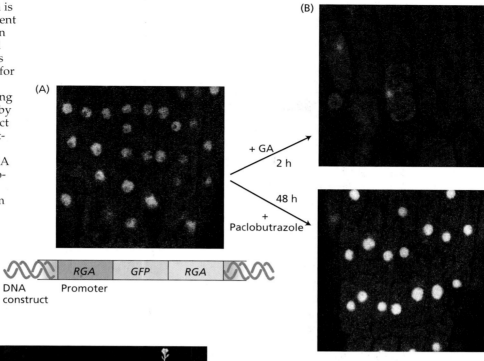

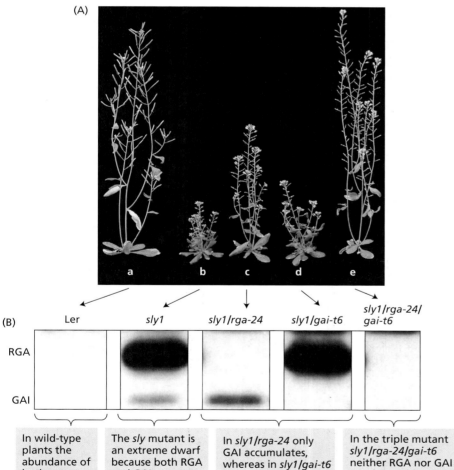

FIGURE 20.19 The SLY protein can degrade the RGA and GAI repressor proteins. (A) 42-day-old homozygous plants showing shoot phenotypes. (a) Wild-type (Ler), (b) *sly1*, (c) *sly1/rga-24* double mutant, (d) *sly1/gai-t6* double mutant, and (e) *sly1/rga-24/gai-t6* triple mutant. The *sly/rga* and *sly/gai* double mutants are taller than the *sly* single mutant, whereas the *sly/rga/gai* mutant shows complete suppression of the *sly* dwarf phenotype. (The *rga-24* and *gai-t6* single mutants are not shown, as their phenotypes are similar to wild-type.) (B) Blots showing the abundance of RGA and GAI protein in each mutant. The blots contain 20 μg of total proteins from rosette leaves of 24-day-old wild-type and mutant plants. The blots were probed with anti-RGA antibodies that also react with GAI, but with lower affinity. The results show a direct correlation between the accumulation of the RGA and GAI repressor proteins and the suppression of growth. (After Dill et al. 2004, and reproduced courtesy of T-p Sun.)

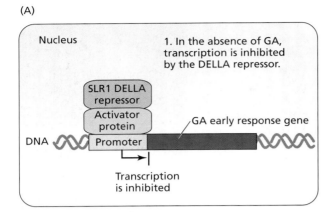
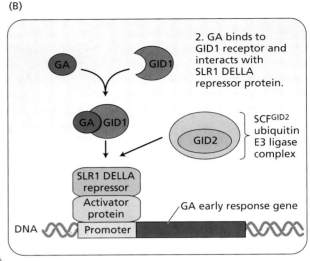
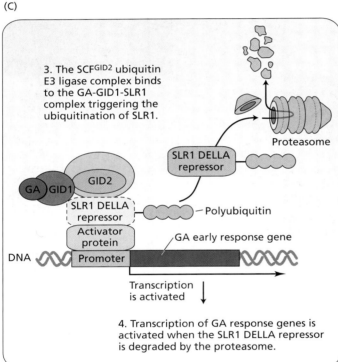

FIGURE 20.20 A model for GA binding to its receptor and the subsequent activation of gene expression leading to growth in rice. (A) In the absence of GA, the SLR1 DELLA-domain repressor blocks the transcription of GA-inducible genes, perhaps by binding and blocking the activity of a transcriptional activator (at present hypothetical). (B) Bioactive GA binds to a soluble receptor (GID1) in the nucleus and the GA-GID1 complex, then binds to the SLR1 repressor. (C) The GA-GID1-SLR1 complex associates with the GID2 F-box protein of the SCF^{GID2} ubiquitin ligase, activating it. SCF^{GID2} attaches ubiquitin molecules to SLR1, targeting it for degradation by the proteasome. The degradation of the DELLA-domain repressor protein "unlocks" the transcriptional activator, allowing transcription to proceed. Growth occurs as a result of GA-induced gene expression.

that are known to include proteasome-mediated degradation of targeted repressor proteins.

A possible GA receptor has been identified in rice

The newly characterized *gid1* mutation* in rice is a loss-of-function mutation producing GA-insensitive dwarf plants that are sterile. *GID1* has been cloned and is related to genes encoding *hormone-sensitive lipases* (Ueguchi-Tanaka et al. 2005). However, GID1 probably does not function as a lipase in rice.

*Although both *gid1* and *gid2* came out of a screen for GA-insensitive dwarfs, or GIDs, *GID1* and *GID2* are different loci and code for different proteins.

GID1 rapidly and reversibly binds biologically active GAs. Moreover, wild-type GID1 interacts with SLR1, the DELLA-domain protein repressor in rice, but only in the presence of bioactive GA (Figure 20.20A and B). This suggests that GA-binding to GID1 may be required to allow GID1 to bind SLR1.

Following its formation, the GA-GID1-SLR1 complex in rice is believed to interact with GID2, the F-box protein component of an SCF^{GID2} ubiquitin ligase complex, prior to the proteasome-mediated degradation of the SLR1 (see Figure 20.20C). SLR1 and other DELLA-domain repressor proteins do not have a known DNA-binding domain, so exactly how they repress GA response is not known. In one model, illustrated in Figure 20.20C, an intermediary pro-

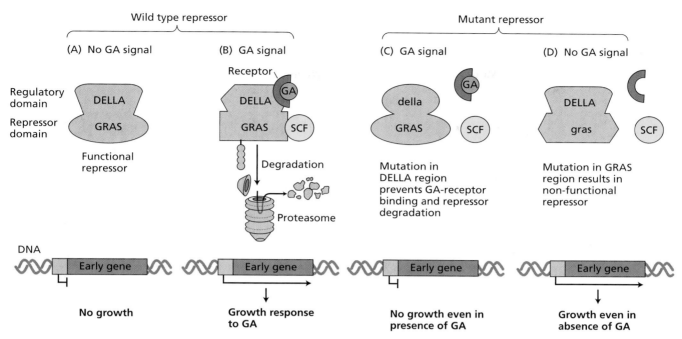

FIGURE 20.21 The current model of GA signaling during growth involves interactions between the GA receptor, a ubiquitin ligase SCF complex, and a DELLA-domain repressor protein. The GA repressor (GAI/RGA/SLR1/SLN1) proteins contain two domains: the regulatory domain (DELLA) and the repressor domain (GRAS). (A) The GRAS domain of the repressor protein is active in the absence of GA, blocking early GA-induced gene expression and giving a dwarf seedling. (B) GA bound to its receptor binds to the repressor protein and facilitates its association with the SCF ubiquitin ligase complex. The repressor protein is thus targeted for ubiquitination and degradation by the 26S proteasome. The destruction of the repressor proteins permits early GA-induced gene expression and seedling growth. (C) A mutation in the DELLA regulatory domain prevents it from binding the GA-receptor complex. Consequently the repressor cannot be degraded, and the mutant is a GA-insensitive dwarf. (D) Mutation in the GRAS domain gives a nonfunctional repressor, so the seedling can grow very tall (slender) even in the absence of GA.

tein is a hypothetical activator, which, when released from the restraint of SLR1, activates the transcription of a GA early response gene.

A summary of the status of the DELLA-domain repressor proteins relative to growth under various conditions is depicted in Figure 20.21. Several additional factors in GA signaling have not been covered in our discussion. For example SPINDLY (SPY), another negative regulator of GA action, is described briefly in **Web Topic 20.6**. (This and other components of GA signaling have been reviewed by Olszewski et al. 2002 and Sun and Gubler 2004.) **Web Essay 20.3** also describes the experimental approach to studying GA signaling in Arabidopsis.

Gibberellin Responses: The Cereal Aleurone Layer

We now turn our attention to the action of GA in cereal grains. In this section we will describe how studies in germinating grain of cereals have shed light on the site of GA biosynthesis, the location of putative receptor(s), and the signal transduction pathway that leads eventually to the transcriptional up-regulation of genes for α-amylase (Jacobson et al. 1995).

Cereal grains can be divided into three parts: the embryo, the endosperm, and the fused testa–pericarp (seed coat–fruit wall) (Figure 20.22A). The embryo, which will grow into the new seedling, has a specialized absorptive organ, the **scutellum**. The endosperm is composed of two tissues: the centrally located starchy endosperm and the **aleurone layer** (see Figure 20.22A).

The starchy endosperm, which is typically nonliving at maturity, consists of thin-walled cells filled with starch grains. Cells of the aleurone layer, which surrounds the starchy endosperm, have thick primary cell walls and contain large numbers of protein-storing vacuoles called protein bodies (see Figure 20.22B–D) (Bethke et al. 1997). *The sole function of the aleurone layer of the cereal grains appears to be the synthesis and release of hydrolytic enzymes into the starchy endosperm during germination.* As a consequence, the stored food reserves of the starchy endosperm are broken down, and the solubilized sugars, amino acids, and other products are transported to the growing embryo.

Two enzymes responsible for starch degradation are α- and β-amylase. α-Amylase (of which there are several isoforms) hydrolyzes starch chains internally to produce oligosaccharides consisting of α-1,4-linked glucose residues. β-Amylase degrades these oligosaccharides from the ends

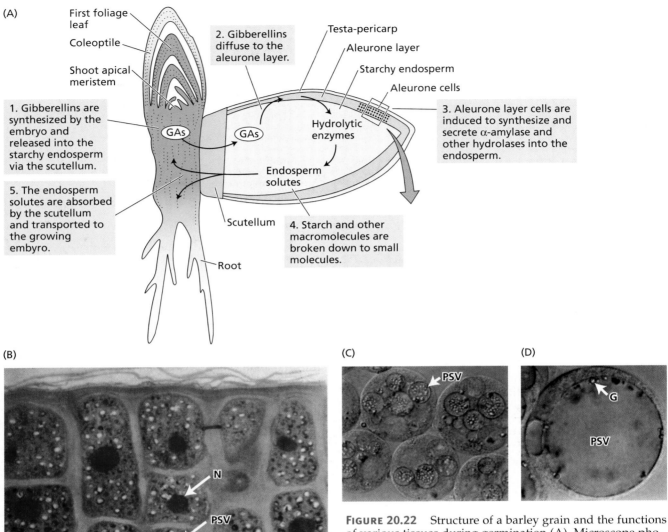

FIGURE 20.22 Structure of a barley grain and the functions of various tissues during germination (A). Microscope photos of the barley aleurone layer (B) and barley aleurone protoplasts at an early (C) and late (D) stage of amylase production. Protein storage vesicles (PSV) can be seen in each cell. G, phytin globoid; N, nucleus. (B–D from Bethke et al. 1997, courtesy of P. Bethke.)

to produce maltose, a disaccharide. Maltase then converts maltose to glucose. After completing the synthesis and secretion of hydrolytic enzymes, the aleurone cells undergo programmed cell death.

Since the 1960s, investigators have utilized isolated aleurone layers, or even aleurone cell protoplasts (see Figure 20.22C, D). The isolated aleurone layer, consisting of a homogeneous population of target cells, provides a unique opportunity to study the molecular aspects of GA action in the absence of nonresponding cell types.

GA is synthesized in the embryo

Experiments carried out in the 1960s confirmed Gottlieb Haberlandt's original 1890 observation that the secretion of starch-degrading enzymes by barley aleurone layers depends on the presence of the embryo. It was soon discovered that GA_3 could substitute for the embryo in stimulating starch degradation. When embryoless half-seeds were incubated in buffered solutions containing GA_3, secretion of α-amylase into the medium was greatly stimulated after an 8-hour lag period (relative to the control half-seeds incubated in the absence of GA_3).

The significance of the GA effect became clear when it was shown that the embryo synthesizes and releases GAs into the endosperm during germination. In rice, the precise location of GA biosynthesis within the germinating grain has been defined using *GUS* reporter genes for late-stage GA biosynthetic enzymes (Figure 20.23). *GA 20-oxidase* and

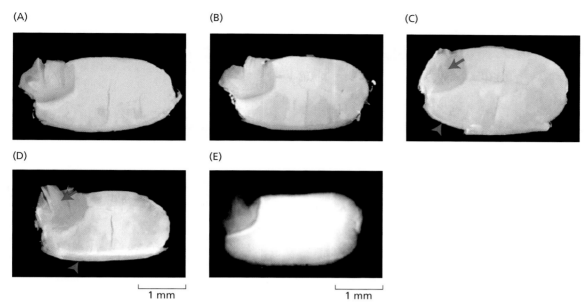

Figure 20.23 In germinating rice seeds, GA biosynthesis and signaling occur in the embryo, while only GA signaling occurs in the aleurone layer. Longitudinal sections of germinating rice grains, 24 hours from the start of imbibition. The sections were stained for GUS activity under the control of promoters for the rice (*Oryza sativa, Os*) genes encoding either GA biosynthetic enzymes (A and B) or GA signaling molecules (C–E). Arrows and arrowheads indicate GUS activity in embryos and aleurone layers, respectively. (A) *OsGA3ox1:GUS*, GA 3-oxidase (embryo only); (B) *OsGA20ox1:GUS*, GA 20-oxidase (embryo only); (C) *Gα:GUS*, α subunit of the heterotrimeric G protein (both embryo and aleurone layer); (D) *SLR1:GUS*, SLR1 DELLA repressor (both embryo and aleurone layer); (E) *OsGAMYB:GUS*, GAMYB transcription factor (both embryo and aleurone layer). (From Kaneko et al. 2003, 2004, courtesy of M. Matsuoka.)

GA 3-oxidase genes show tissue- and cell-specific patterns of expression (Kaneko et al. 2003), although this expression is confined to the epithelium and developing shoot tissues of the germinating embryo.

In contrast, the response genes, *Gα* and *SLR*, which we will be discussing in the next section, are not only expressed in the embryo but also in the aleurone layer (see Figure 20.23). Therefore, the embryo seems to be a site of GA biosynthesis and response, whereas the aleurone layer is a site of response only (see Figure 20.23). The "response" is not the same in both locations. In the aleurone it is the synthesis of α-amylase, whereas in the developing shoot it is cell division/elongation.

Aleurone cells may have two types of GA receptors

Several lines of evidence suggest that the site of GA perception in aleurone cells may be on the plasma membrane:

- GA_4 that has been covalently linked to agarose beads to prevent its passage through the plasma membrane can still induce α-amylase production in wild oat protoplasts.
- GA_3 injected directly into barley aleurone protoplasts appears to have no bioactivity.
- Evidence for the participation of heterotrimeric G proteins in GA signal transduction in wild oat and rice aleurone cells implicates a GA receptor on the plasma membrane.

However, as noted earlier, the GA receptor that regulates cell elongation in rice seedlings has been identified as the soluble protein, GID1. Could GID1 also function as a GA receptor in aleurone cells? If so, it would call into question the current model that the receptor is located on the plasma membrane.

How can the two sets of data be reconciled? It is possible that aleurone cells utilize both soluble and plasma membrane–bound GA receptors. Evidence for two types of hormone receptors has also been obtained for auxin (see Chapter 19) and abscisic acid (see Chapter 23). Given the great diversity of GA responses, the existence of multiple receptors may not be too surprising. Clearly, further research is needed to clarify the number and identity of GA receptors.

GA signaling requires several second messengers

Components of the GA signal transduction chain in aleurone cells have been identified in the plasma membrane, cytosol, and nucleus (Figure 20.24). Several lines of evidence indicate that biologically active GA, perceived at the plasma membrane and bound to a hypothetical receptor protein, interacts with a membrane-localized heterotrimeric G protein (Jones et al. 1998). (This evidence is discussed in **Web Topic 20.7**.) It is also known that GA signaling involves

530 Chapter 20

1. GA_1 from the embryo first binds to a hypothetical membrane receptor on the surface of an aleurone cell.

2. The cell-surface GA-receptor complex interacts with a heterotrimeric G protein, initiating two separate signal transduction chains.

3. A calcium-independent pathway involving cGMP and several other components results in the activation of an F-box protein, part of an SCF-ubiquitin ligase complex.

4. GA_1 may also enter the cell directly and bind to an alternate receptor protein, which is located mainly in the nucleus.

5. The activated F-box protein binds to a DELLA-domain repressor protein that is blocking the transcription of a *GAMYB* gene. The DELLA-domain repressor may be blocking the activity of a transcriptional activator.

6. The repressor is degraded via the SCF-ubiquitin ligase complex.

7. The degradation of the repressor allows the expression of *GAMYB* and other early response genes.

8. The newly synthesized GAMYB protein enters the nucleus and binds the promoters of α-amylase and genes encoding other hydrolytic enzymes.

9. Transcription of these genes is activated.

10. α-Amylase and other hydrolases are synthesized on the rough ER.

11. Proteins are secreted by the Golgi.

12. The secretory pathway requires GA-stimulation of a calcium-calmodulin-dependent pathway.

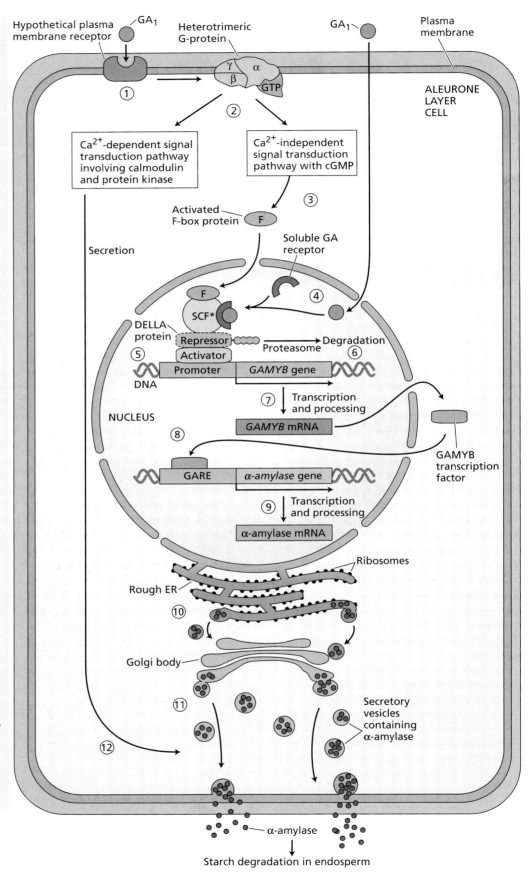

◀ **FIGURE 20.24** Composite model for the induction of α-amylase synthesis in barley aleurone layers by GA. A calcium-independent pathway induces *α-amylase* gene transcription; a calcium-dependent pathway is involved in α-amylase secretion.

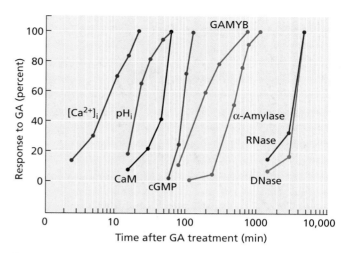

FIGURE 20.25 Following the addition of GA to barley aleurone protoplasts, a multiple-component signaling pathway is initiated. The timing of some of these events is shown. CaM = calmodulin. (After Bethke et al. 1997.)

both Ca^{2+}-dependent and Ca^{2+}-independent events (Bethke et al. 1997). One branch of the signaling pathway leads to rapid increases in intracellular calcium concentration (Figure 20.25), particularly in the peripheral cytoplasm (Figure 20.26), while the other branch leads to the induction of *α-amylase* gene expression (see Figure 20.24). Sustained lowering of Ca^{2+} levels for several hours does *not* affect α-amylase gene expression, but does inhibit α-amylase secretion. These results show that effects of GA on *α-amylase* gene expression are independent of Ca^{2+}, although intracellular Ca^{2+} concentration is important in α-amylase secretion. Ca^{2+} is probably also important in α-amylase formation, as it is a calcium-containing metalloenzyme. The mechanism for elevating intracellular Ca^{2+} concentration is presently unknown, although its source could be both extracellular (uptake from the apoplast) or intracellular (release from organelles).

Another potential second messenger in the GA signal transduction pathway in aleurone cells is cyclic GMP (cGMP), which increases in barley aleurone layers 2 hours after GA treatment (see Figure 20.25). Specific inhibitors that prevent the transient increase in cGMP also reduce the accumulation of *α-amylase* mRNA and that of the transcription factor GAMYB (discussed below). For this and other reasons, it has been suggested that cGMP may be a component of the signal transduction chain, provisionally placed downstream of the G proteins and upstream of MYB transcription factors (see Chapter 14 on the web site).

More recently, molecular studies have demonstrated that GA acts primarily by inducing the expression of the genes for α-amylase. It has been shown that GA_3 enhances the level of translatable mRNA for α-amylase in aleurone layers (Figure 20.27). Using isolated nuclei, investigators have also demonstrated that the increase in *α-amylase* gene expression is due to a stimulation of mRNA synthesis rather than to a decrease in mRNA turnover.

The purification of *α-amylase* mRNA, which is produced in relatively large amounts in aleurone cells, enabled the isolation of genomic clones containing both the structural gene for α-amylase and its upstream promoter sequences.

Gibberellins enhance the transcription of *α-amylase* mRNA

Even before molecular biological approaches were developed, there was already physiological and biochemical evidence that GA enhanced α-amylase production at the level of gene transcription (Jacobsen et al. 1995). The two main lines of evidence were:

1. GA_3-stimulated α-amylase production was shown to be blocked by inhibitors of transcription and translation.

2. Both stable- and radioactive isotope-labeling studies demonstrated that the stimulation of α-amylase activity by bioactive GA involved de novo synthesis of the enzyme from amino acids, rather than activation of preexisting enzyme.

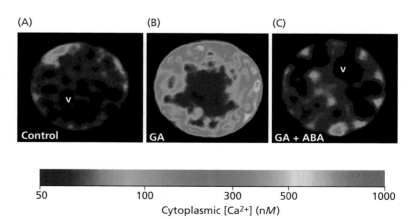

FIGURE 20.26 Increase in calcium in barley aleurone protoplasts following GA addition can be seen from this false-color image; vacuoles are indicated (v). The level of calcium corresponding to the colors is in the lower scale. (A) Untreated protoplast. (B) GA-treated protoplast. (C) Protoplast treated with both abscisic acid (ABA) and GA. Abscisic acid opposes the effects of GA in aleurone cells. (From Ritchie and Gilroy 1998.)

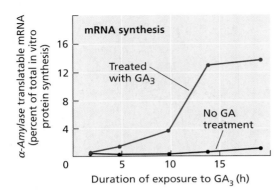

FIGURE 20.27 Gibberellin effects on mRNA synthesis. A gibberellin-induced increase in translatable α-*amylase* mRNA precedes the release of the α-amylase from the aleurone cells by several hours. (After Higgins et al. 1976.)

The partial deletion of known sequences of bases from α-amylase promoters of several cereals indicates that the sequences conferring GA responsiveness, termed **GA response elements** (**GARE**s), are located 200 to 300 base pairs upstream of the transcription start site. Identical GAREs were found in all cereal α-*amylase* promoters so far examined, and their presence was shown to be essential for the induction of α-*amylase* gene transcription by GA.

GAMYB is a positive regulator of α-*amylase* transcription

The sequence of the GARE in the α-*amylase* gene promoter (TAACAAA) is similar to that of the binding sites for MYB proteins, which are transcription factors in eukaryotes that share conserved DNA-binding domains.

In plants there is a large group of MYB proteins, referred to as R2R3-MYB factors. They are important in regulating a variety of primary and secondary metabolic pathways, including tryptophan and phenylpropanoid metabolism (see Chapter 13) and regulation of the circadian clock (see review by Stracke et al. 2001) (see Chapter 17). There are more than 120 R2R3-MYBs in Arabidopsis that can be divided into subclasses based on structural features. The single GAMYB in cereals is most closely related to subgroup 18 in Arabidopsis, of which three (AtMYB33, AtMYB65, and AtMYB101) can compensate for the loss of GAMYB in barley by transcriptionally activating α-*amylase* gene expression in aleurone (Gocal et al. 2001).

Testing the hypothesis that a MYB transcription factor turns on α-*amylase* gene expression, Gubler et al. (1995) conducted gel blot analysis of barley aleurone layers. They showed that the synthesis of *GAMYB* mRNA increases as early as 1 hour after GA treatment, preceding the increase in α-*amylase* mRNA by several hours. In fact, the peak accumulation of *GAMYB* RNA at 6 to 12 hours after GA application coincides with the maximal rate of α-*amylase* mRNA accumulation, consistent with GAMYB regulating α-*amylase* gene expression (Figure 20.28) (Gubler et al. 1995).

Cycloheximide, an inhibitor of translation, has no effect on the production of *GAMYB* mRNA, indicating that protein synthesis is not required for *GAMYB* expression. *GAMYB* can therefore be defined as a primary or early response gene. In contrast, the α-*amylase* gene is a secondary or late response gene, as indicated by the fact that its transcription is blocked by cycloheximide. Mutation in the GARE prevented GAMYB binding and α-*amylase* expression. Moreover, in the absence of GA, GAMYB can induce all the same responses as GA in aleurone cells, showing that GAMYB is necessary and sufficient for the enhancement of α-*amylase* expression.

DELLA domain proteins are rapidly degraded

How does GA application lead to the transcriptional activation of α-*amylase* by the GAMYB transcription factor? It is known that within 5 minutes of GA application to barley aleurone cells there is an effect on the negative regulator SLN1, a DELLA-domain repressor protein with homology to SLR1 in rice. Levels of a GFP-SLN1 fusion protein in nuclei decline, and indeed the protein is completely gone within 10 minutes of GA treatment (Gubler et al. 2002). However, GAMYB levels do not increase for 2 hours, suggesting that there are as-yet unidentified steps between SLN1 degradation and *GAMYB* transcription.

Drawing together our information for the cereal aleurone system, we can hypothesize the presence of GA receptor(s), one of which may be in the cell membrane, and the possible early involvement of a heterotrimeric G protein prior to signal transduction diverging into two branches (see Figure

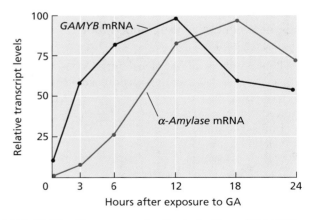

FIGURE 20.28 Time course for the induction of *GAMYB* and α-*amylase* mRNA by GA₃. The production of *GAMYB* mRNA precedes that of α-*amylase* mRNA by about 5 hours. This result is consistent with the role of *GAMYB* as an early GA response gene that regulates the transcription of the gene for α-amylase. In the absence of GA, the levels of both *GAMYB* and α-*amylase* mRNAs are negligible. (After Gubler et al. 1995.)

20.24). One branch leads via typical second messengers to the F-box protein-mediated (GID2-mediated in rice) degradation of DELLA-domain negative regulators (SLR1 in rice, SLN1 in barley) of the GA response. It is also possible that GID1, the soluble GA receptor discovered in rice, also functions as a GA receptor in aleurone cells. Given the importance of starch hydrolysis for germination in cereals, it would not be surprising if some functional redundancy were built into the system. If so, GA-binding to GID1 might also serve to activate F-box-mediated degradation of the SLR1 repressor.

As a consequence of SLR1 degradation, the expression of *GAMYB*, which encodes the sole MYB transcription factor in its class in cereals, is up-regulated. Finally, the GAMYB protein binds to a highly conserved GARE in the promoter of the gene for α-amylase, activating its transcription.

The other branch of the pathway, which requires Ca^{2+} accumulation, is responsible for the secretion of α-amylase from the aleurone cells into the starchy endosperm. Here, starch breakdown occurs by the action of α-amylase and other hydrolases, and sugars are transported to the growing embryo.

The involvement of DELLA-domain proteins in the regulation of the α-amylase response ties the known negative regulators of stem growth to those of the aleurone layer. As we shall see in subsequent sections of this chapter, after the first recognition of GAMYB as a *trans*-activator of a GA-inducible gene in aleurone cells, MYB transcription factors are now known to be positive regulators of GA responses in a variety of other tissues.

Gibberellin Responses: Flowering in Long-Day Plants

Gibberellins play a variety of roles in reproductive development. In the previous section we discussed processes involved in seed germination. Gibberellins also help to regulate the earliest stage of reproductive development: the transition from vegetative growth to flowering, especially the flowering of so-called long-day plants.

As we shall see in Chapter 25, many plants have evolved to flower in response to specific day lengths, a phenomenon known as **photoperiodism**. Most plants fall into three major categories: **short-day plants** (**SDPs**), **long-day plants** (**LDPs**), and **day-neutral plants** (**DNPs**). The perception of day length occurs in the leaves, not in the apical meristem. Thus a signal (termed "florigen") must be transmitted from the leaf to the apex for flowering to occur.

In this section we will begin with a brief overview of the different signaling pathways that can lead to flowering, including the GA pathway. We will then focus on the interaction between the GA pathway and photoperiodic pathway in LDPs. Next we will consider the role of GAMYB transcription factors in both floral initiation and floral development. Finally, we will briefly describe new findings that implicate microRNAs in the regulation of GAMYBs.

There are multiple independent pathways to flowering

The interaction of photoperiod and GAs in reproductive development is complex. Multiple independent pathways to flowering exist. In Arabidopsis, a facultative LDP, there is a facultative long-day (LD) pathway to flower initiation that is distinct from a GA-dependent pathway that promotes flowering under short-day (SD) conditions. Thus the *ga1* (GA-deficient) mutant will not flower in SD conditions unless treated with GA, whereas the same mutant will flower in LD conditions without GA treatment, although flowering is slightly delayed compared to wild-type plants.

The long day and gibberellin pathways interact

Despite the formal separation of the LD and GA pathways to flowering, there is an interaction between the two, because various steps in GA metabolism are altered by day length. In Arabidopsis, LDs enhance the expression of the GA biosynthesis gene, *GA20ox1*, leading to rapid stem extension (**bolting**) and accelerated flowering. In spinach, too, bolting requires the presence of bioactive GA(s), the synthesis of which is induced when plants are transferred from SDs to LDs (Lee and Zeevaart 2002).

In ryegrass (*Lolium temulentum*), which requires only one long photoperiod for induction, flowering occurs without much stem elongation, allowing the efficacy of GAs for both processes to be measured separately. GA_5 and GA_6 are potent inducers of flowering in *Lolium*.

Gibberellin A_5 levels have been shown to increase fivefold in illuminated leaves at the end of a single 16-hour day, and to double in the shoot apex, along with GA_6, after two long days (King et al. 2001). Increased GA_5 and GA_6 in the shoot apical meristem reflect their transport from the leaf to the apex via the phloem, and the rate of transport is consistent with the proposed movement of the flowering stimulus, florigen. Although recent genetic and molecular studies have identified specific mRNA molecules as the photoperiodic floral stimuli in both SDPs and LDPs (see Chapter 25), it is possible that GA can also serve as a floral stimulus in *Lolium*. GAs have been shown to regulate some of the genes controlling floral development in both *Lolium* and Arabidopsis (see **Web Topic 20.8**).

GAMYB regulates flowering and male fertility

We have seen that the GAMYB transcription factor activates α-*amylase* gene expression during the GA response of barley aleurone layers. We have also seen that the single GAMYB of cereals is closely related to three members of subgroup 18 of the R2R3-MYBs in Arabidopsis. To obtain some clues regarding the normal role(s) of these transcription factors, the expression of *MYB*-like genes was examined in Arabidopsis plants, at different stages in the life cycle and under different environmental conditions, by in situ hybridization. The highest expression of *AtMYB33* and *AtMYB65* (which have redundant functions) was observed in inflorescence apices 6 days after transfer of plants from SDs to LDs. Expression was also evident in some floral organs, particularly the locules of immature anthers.

As noted earlier, physiological studies have established the following features of GA-stimulated flowering in LDPs:

- Long days can induce stem and petiole elongation and eventual flowering of rosette species.
- Endogenous GAs increase following the transfer of plants from SDs to LDs.
- GA treatment of plants maintained in SDs can induce flowering.

When the expression of *AtMYB33* was examined in Arabidopsis, the increase observed in LDs could be mimicked by the application of GA_4 in SDs (Gocal et al. 2001). In both situations, enhanced *AtMYB33* expression accompanied floral inflorescence development. The meristem identity gene *LFY* is a downstream target of AtMYB33 (see Chapter 25). A DNA sequence element in the *LFY* promoter that is necessary for GA responsiveness contains a conserved MYB-binding sequence. Mutating this sequence abolishes AtMYB33 binding to, and the GA-responsiveness of, the *LFY* promoter. This work implicates a GAMYB–like transcription factor in another classic developmental response to GAs, that of flowering in Arabidopsis.

Loss-of-function mutants of the *GAMYB* in the SDP rice have also demonstrated the role of this transcription factor as a positive regulator of floral organ development (Kaneko et al. 2004). Rice plants lacking a functional *GAMYB* gene developed flowers with shrunken, white (instead of yellow) anthers that contained no pollen. In other flowers, even on the same plant, phenotypes were more pronounced, including grossly malformed pistils in the most severe phenotypes (Figure 20.29). In the LDP Arabidopsis, both *AtMYB33* and *AtMYB65* must be knocked out for male sterility to occur,

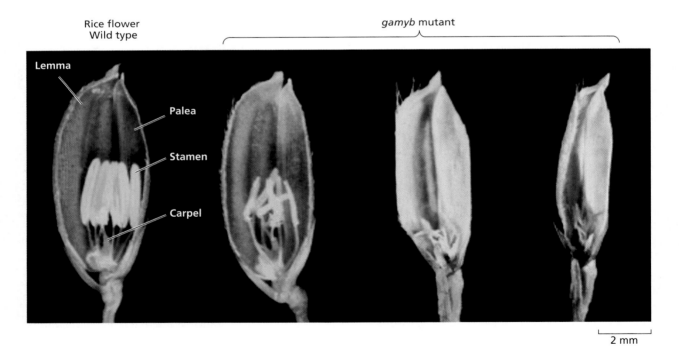

Figure 20.29 Phenotypes of floral organs of the *gamyb* mutants of rice. A wild-type flower is shown on the far left, and progressing toward the right are shown increasing severities of the mutant phenotype. In the absence of a functional *GAMYB* gene, flowers have shrunken, white (instead of yellow) anthers that lack pollen. More extreme phenotypes include white, shriveled bracts (lemma and palea) surrounding the flower and malformed carpels. (From Kaneko et al. 2004, courtesy of M. Matsuoka.)

because of functional redundancy between these two genes (Millar and Gubler 2005).

MicroRNAs regulate MYBs after transcription

MicroRNAs (miRNAs) are small gene-encoded RNAs containing about 20 to 25 nucleotides that regulate the expression of other genes post-transcriptionally. MiRNAs bind to ribonuclease complexes and serve as guides to target the ribonuclease to specific mRNA molecules. The mRNAs that hybridize to these miRNA guides are quickly digested by the ribonuclease complex, thus effectively silencing the expression of their target genes (for a more detailed discussion of miRNAs, see Chapter 14 on the web site).

Three miRNAs in Arabidopsis have been shown to be complementary to a conserved motif in the coding regions of *GAMYB–like* genes (Park et al. 2002; Rhoades et al. 2002). One of them, *miR159a*, can silence *AtMYB33*, and overexpressing or mutating *miR159a* can cause striking defects in plants. As previously discussed, *AtMYB33* is involved in floral organ development and flowering in Arabidopsis. Thus *miR159a* overexpressers, in which *AtMYB33* is silenced, are male-sterile and have delayed flowering (Achard et al. 2004).

Taken together, the work on both GA-induced α-amylase production and GA-induced flowering and floral development all support a role for MYB transcription factors as important intermediates of GA signaling.

Gibberellin Responses: Stem Growth

The effects of GAs on stem growth can be so dramatic that one might imagine it should be simple to determine how they act (see Figures 20.1 and 20.2). Unfortunately, this is not the case, because—as we have seen with auxin—much about plant cell growth is poorly understood. However, we do know some basic characteristics of GA-induced stem elongation. In the final section of the chapter we examine the distribution of GA within the shoot apical meristem (SAM) and the genetic interactions that maintain the identity of the SAM and allow the differentiation and growth of leaf primordia. Next we discuss studies aimed at elucidating the physiological and biochemical mechanisms for GA-stimulated cell elongation and cell division. Finally, we address briefly the interactions between auxin and GA in promoting cell elongation.

The shoot apical meristem interior lacks bioactive GA

For meristematic activity to be maintained in the SAM, GAs must be *excluded* from the inner cells. These interior cells must expand and divide in all planes to maintain the three-dimensional shape of the meristem. Bioactive GAs disrupt this process by promoting the deposition of transverse microfibrils in cell walls, causing elongation along a single (vertical) axis. When this occurs, the population of stem cells cannot be maintained, and the SAM disappears. The maintenance of the SAM thus depends on the exclusion of GA from the interior cells (Sakamoto et al. 2001). This exclusion is achieved through the suppression of GA biosynthesis in the inner cells by KNOX homeodomain transcription factors, encoded by *KNOTTED 1-LIKE HOMEOBOX (KNOX)* genes (see Chapter 16). KNOX proteins are required in inner cells of the SAM for the maintenance of meristematic activity, but they must be absent from derivative calls for the initiation of leaf primordia. Bioactive GAs must be excluded from the inner cells, but are required for leaf primordia initiation. This model for the mutually exclusive action of KNOX proteins and GAs was developed initially by studying the over-expression of KNOX genes.

In tobacco (*Nicotiana tabacum*), overexpression of the KNOX gene *NICOTIANA TABACUM HOMEOBOX 15 (NTH15)* interferes with the GA 20-oxidation step in GA biosynthesis, resulting in severely dwarfed shoots. The dwarfism can be relieved by application of GA_{20} or GA_1, but not by intermediates earlier in the pathway (Kusuba et al. 1998). NTH15 was shown to rapidly suppress the expression of the gene encoding GA 20-oxidase in tobacco, *Ntc12* (Sakamoto et al. 2001).

In situ hybridization of wild-type NTH15 and GA 20-oxidase transcripts showed their expression patterns to be mutually exclusive. *NTH15*, encoding the KNOX protein, is expressed in interior cells of the SAM, whereas GA 20-oxidase is expressed in the rib meristem and leaf primordia but *not* in the meristem interior (Figure 20.30). Mutating the NTH15-binding domain of the gene for GA 20-oxidase allowed the ectopic expression (expression in the wrong location) of the GA 20-oxidase in the interior cells of the SAM, confirming that in wild-type plants its absence from the interior cells is due to the KNOX protein suppressing *Ntc12* transcription.

GA-enhanced longitudinal cell expansion is actually required for the initiation of leaf primordia. For genes for GA 20-oxidase to be expressed in leaf primordia, *KNOX* gene expression must be suppressed in these organs. This is achieved in some plants by MYB transcription factors, which negatively regulate *KNOX* gene expression in the developing leaf primordia. This ensures a supply of bioactive GA to stimulate longitudinal cell expansion.

Gibberellins stimulate cell elongation and cell division

Gibberellins stimulate both cell elongation and cell division, as evidenced by increases in cell length and cell number in response to applications of bioactive GAs. Internodes of tall peas have more cells than those of dwarf peas, and the cells are longer. Mitosis increases markedly in the rib or ground meristem of rosette LDPs after treatment with bioactive GA. The dramatic stimulation of internode elongation in deepwater rice when submerged, or when treat-

FIGURE 20.30 Shoot apices of tobacco showing in situ hybridization of the KNOX protein NTH15 (A) and the GA 20-oxidase Ntc12 (B). Mutually exclusive expression is observed. NTH15 is expressed in the interior cells of the meristem but not in the flanks, while Ntc12 is expressed in the flanks of the meristem but not in the interior. (From Sakamoto et al. 2001.)

(A) Localization of *KNOX* gene expression

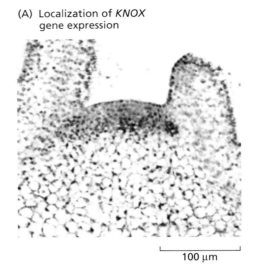

(B) Localization of GA *20-oxidase* gene expression

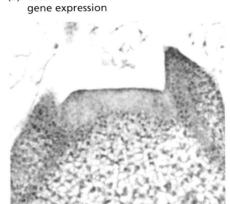

ed with GA, is due in part to increased cell division activity in the intercalary meristem (Figure 20.31). Because GA-induced cell elongation appears to precede GA-induced cell division, we begin our discussion with the role of GA in regulating cell elongation.

As discussed in Chapter 15, the elongation rate can be influenced by both cell wall extensibility and the osmotically driven rate of water uptake. Gibberellin has no effect on the osmotic parameters but has consistently been observed to cause an increase in both the mechanical extensibility of cell walls and the stress relaxation of the walls of living cells. An analysis of pea genotypes differing in GA_1 content or sensitivity showed that GA decreases the minimum force that will cause wall extension (the wall yield threshold). Thus, both GA and auxin seem to exert their effects by modifying cell wall properties.

In the case of auxin, cell wall loosening appears to be mediated in part by cell wall acidification (see Chapter 19). However, this does not appear to be the mechanism of GA action, since GA-stimulated increase in proton extrusion has not been demonstrated. On the other hand, GA is never present in tissues in the complete absence of auxin, and the effects of GA on growth may depend on auxin-induced wall acidification.

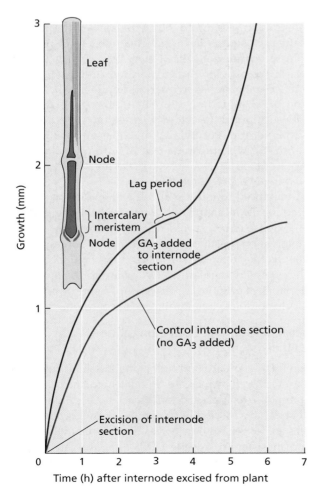

FIGURE 20.31 Continuous recording of the growth of the upper internode of deepwater rice in the presence or absence of exogenous GA_3. The control internode elongates at a constant rate after an initial growth burst during the first 2 hours after excision of the section. Addition of GA after 3 hours induces a sharp increase in the growth rate after a 40-minute lag period (upper curve). The difference in the initial growth rates of the two treatments is not statistically significant here, but reflects slight variation in experimental materials. The inset shows the internode section of the rice stem used in the experiment. The intercalary meristem just above the node responds to GA. (After Sauter and Kende 1992.)

do not grow in submerged conditions; overexpression of *OsEXP4* leads to taller rice plants (Figure 20.32) (Choi et al. 2003). Taken together, these results indicate that GA-induced cell elongation is at least in part mediated by the expansins.

GAs regulate the transcription of cell cycle kinases

The dramatic increase in growth rate of deepwater rice internodes with submergence is due partly to increased cell divisions in the intercalary meristem. To study the effect of GA on the cell cycle, researchers isolated nuclei from the intercalary meristem of deepwater rice and quantified the amount of DNA per nucleus (Sauter and Kende 1992). In submergence-induced plants, GA activates the transition from G_1 to S phase, leading to an increase in mitotic activity. The stimulation of cell division results from a GA-induced expression of the genes for several **cyclin-dependent protein kinases** (**CDKs**) (see Chapter 1). The transcription of these genes—first, those regulating the transition from G_1 to S phase, then those regulating the transition from G_2 to M phase—is induced in the intercalary meristem by GA (Fabian et al. 2000).

Auxin promotes GA biosynthesis and signaling

Although we often discuss the action of hormones as if they act singly, the net growth and development of the plant are the results of many combined signals. In addition, hormones can influence each other's biosynthesis and/or response so that the effects produced by one hormone may in fact be mediated by others.

For example, it has long been known that exogenous auxin induces ethylene biosynthesis. It is now evident that exogenous auxin can enhance GA biosynthesis in pea plants in which the site of auxin biosynthesis has been excised, or in which its transport is impeded by treatment with auxin transport inhibitors. If pea plants are decapitated, not only is the level of auxin lowered because its source has been removed, but the level of GA_1 in the upper stem drops sharply. This change can be shown to be an auxin effect, because replacing the bud with a supply of auxin restores the GA_1 level (Figure 20.33).

The presence of auxin has been shown to promote the transcription of *GA3ox* and to repress the transcription of *GA2ox*. Thus the apical bud promotes growth not only through the direct biosynthesis of auxin, but also through the auxin-induced biosynthesis of GA_1 (Figure 20.34) (Ross et al. 2000; Ross and O'Neill 2001). (This work is discussed in **Web Essay 20.3**.) Similarly, in young pea pods, excision of developing seeds retards pod elongation. Application of 4-chloro-IAA, which is an endogenous auxin in pea fruits, not only enhances pod growth but increases the transcription of both *GA20ox* and *GA3ox* genes (Ozga et al. 2003).

Other indications that auxin may be required for GA signaling came from studies of *ga1* (GA-deficient) Arabidopsis seedlings that had been either decapitated or treat-

FIGURE 20.32 Trangenic rice plants harboring sense and antisense constructs of *OsEXP4*. Shown are antisense (A), control (i.e., not transgenic) (B), and sense (C). (From Choi et al. 2003, courtesy of H. Kende.)

The typical lag time before GA-stimulated growth begins is longer for GA than it is for auxin; in deepwater rice it is about 40 minutes (see Figure 20.31), and in peas it is 2 to 3 hours (Yang et al. 1996). These longer lag times point to a growth-promoting mechanism distinct from that of auxin. Consistent with the existence of a separate GA-specific wall-loosening mechanism, the growth responses to applied GA and auxin are additive.

Various hypotheses have been explored regarding the mechanism of GA-stimulated stem elongation, and all have some experimental support, but as yet none provide a clear-cut answer. For example, there is evidence that the enzyme xyloglucan endotransglucosylase/hydrolase (XTH) is involved in GA-promoted wall extension (Xu et al. 1996). The function of XTH may be to facilitate the penetration of expansins into the cell wall. Expansins are cell wall proteins that cause wall loosening in acidic conditions by weakening hydrogen bonds between wall polysaccharides (see Chapter 15). Transcript levels of one particular expansin in rice, that encoded by *OsEXP4*, increases in deepwater rice within 30 minutes of GA treatment, or in response to submergence, both of which induce growth. Moreover, plants expressing an antisense version of *OsEXP4* are shorter and

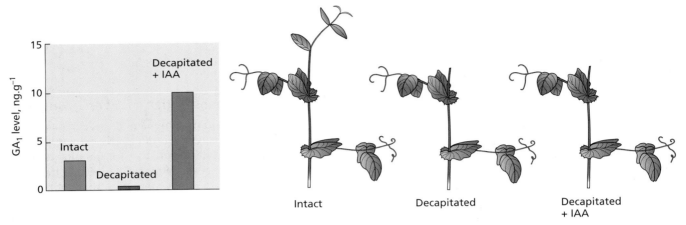

FIGURE 20.33 Decapitation reduces, and IAA (auxin) restores, endogenous GA$_1$ content in pea plants. (After Ross et al. 2000.)

ed with an auxin transport inhibitor (Fu and Harberd 2003). (This work is discussed in **Web Topic 20.9**.)

Our information on the effects of auxin on GA biosynthesis comes mainly from studies with pea, while our understanding of the effects of auxin on GA signaling is derived mainly from studies with Arabidopsis. Whether these two effects of auxin are widespread, and whether they complement one another in the same species is not yet known. Clearly, these are topics for ongoing study that will help to clarify the complex cross talk that appears to exist between these two growth-promoting hormones.

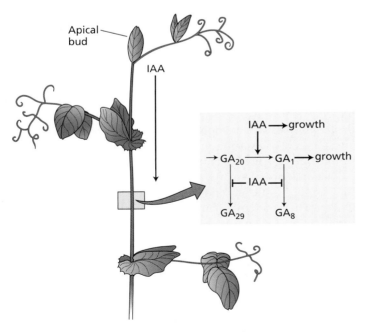

FIGURE 20.34 IAA (from the apical bud) promotes and is required for GA$_1$ biosynthesis in subtending internodes of pea. IAA also inhibits GA$_1$ breakdown. (After Ross and O'Neill 2001.)

Summary

Gibberellins are a family of compounds defined by their structure. Over 136 have been structurally identified so far, although only a few of them have intrinsic biological activity. Gibberellins affect many aspects of plant growth and development, although they are best known for their dramatic effects on internode elongation in grasses and dwarf and rosette species. Other physiological effects of GAs include roles in determination of flower sex, promotion of fruit growth, and germination of seed. Gibberellins have several commercial applications, mainly in enhancing the size of seedless grapes and in the malting of barley in the brewing industry. Inhibitors of GA biosynthesis are used as dwarfing agents.

Gibberellins are tetracyclic diterpenoid compounds made up of four isoprenoid units. The pathway to the first tetracyclic intermediate, *ent*-kaurene, occurs in plastids. Reactions converting *ent*-kaurene to GA$_{12}$—the precursor of all the other GAs— occur first on the plastid envelope and then on the endoplasmic reticulum, and are catalyzed by cytochrome P450 monooxygenases. In many species hydroxylation of GA$_{12}$ at C-13 takes place, giving GA$_{53}$.

Gibberellin A$_{12}$ and GA$_{53}$, both of which have 20 carbon atoms, are converted to other GAs by sequential oxidation of C-20, followed by the loss of this carbon atom. The products of these parallel pathways are GAs that contain only 19 carbon atoms, and so are termed C$_{19}$-GAs. This process is followed by hydroxylation at C-3 to give the growth-active GA$_4$ or GA$_1$. A subsequent hydroxylation at C-2 eliminates biological activity.

The steps after GA$_{12}$ or GA$_{53}$ occur in the cytoplasm. The genes that encode the three important dioxygenase enzymes in these parallel pathways have been cloned in many species. A GA 20-oxidase (GA20ox) catalyzes the steps between GA$_{12}$ and GA$_9$ and between GA$_{53}$ and GA$_{20}$, GA 3-oxidase (GA3ox) converts GA$_9$ into bioactive GA$_4$

and GA_{20} into bioactive GA_1, and GA 2-oxidase (GA2ox) deactivates the bioactive GAs. Dwarf plants can be produced by transformation of wild-type plants with antisense *GA20ox* or *GA3ox*, or by overexpressing *GA2ox*, since these strategies reduce the level of bioactive GA.

Gibberellin biosynthesis takes place in many parts of the plant, including germinating embryos, young seedlings, shoot apices, and developing seeds. Gibberellin action occurs in cells that may or may not have the ability to synthesize GAs—in the latter case, GAs must move from their sites of synthesis to their sites of action. Endogenous bioactive GA regulates its own synthesis by enhancing or inhibiting the transcription of the genes for the enzymes of GA biosynthesis and deactivation. Environmental factors such as photoperiod and temperature can modulate GA biosynthesis through effects on the transcription of the genes for the GA biosynthetic enzymes.

The most pronounced effect of applied GAs is stem elongation in dwarf and rosette plants. Normally, stem growth is directly correlated with the level of bioactive GA. However, in plants in which GA response is perturbed, the direct correlation between bioactive GA levels and growth is lost. Three types of GA response mutants have been useful in the identification of genes involved in the GA signaling pathway involved in stem growth: (1) recessive mutations resulting in the loss of a positive regulator (e.g., SLY in Arabidopsis and GID 2 in rice) leads to a GA nonresponsive dwarf phenotype; (2) recessive mutations leading to the loss of a negative regulator (e.g., SLR1 in rice and SLN1 in barley) leading to a tall or slender phenotype; and (3) semidominant gain-of-function mutations in which a negative regulator (e.g., GAI) is constitutively expressed, leading to a nonresponsive dwarf phenotype. These mutations have been instrumental in identifying components of the GA response pathway in both monocots and dicots, particularly a class of transcriptional regulators (GRAS proteins with regulatory DELLA domains) that must be degraded by GA-induced proteolysis for shoot growth to occur. The recent identification of a putative soluble GA receptor in rice seedlings (GID1) will advance our understanding of the chain of events between the perception of the GA signal and the manifestation of growth.

Bioactive GA, synthesized in embryos during imbibition, induces transcription of the genes encoding α-amylase in aleurone cells of cereal grain. This enzyme degrades starch stored within the endosperm, providing nourishment for the young seedling. There is some evidence, as yet not definitive, that a GA receptor is located on the surface of aleurone cells. Recent evidence also implicates the soluble GA receptor in α-amylase synthesis. GAMYB is a transcription factor that binds to a GA response element (GARE) in the upstream region of the *α-amylase* gene and enhances its transcription. AtMYB factors from Arabidopsis can rescue GAMYB-deficient cereal mutants. These AtMYBs, in Arabidopsis, seem to be important during the reproductive stage of the life cycle. The levels of these MYB transcription factors are regulated by microRNAs.

GAs must be excluded from the interior of shoot apical meristems for those cells to remain meristematic. However, GAs are necessary for the development of leaf primordia and for normal leaf expansion. In rib meristems proximal to the shoot apical meristem, and in intercalary meristems of grasses and cereals, GAs are important for cell expansion and cell division.

Gibberellins do not act alone; other hormones interact with them in a variety of different ways. For instance, auxins have been shown to regulate GA metabolism in some species, notably pea, and to be required for the GA-induced degradation of repressor proteins in Arabidopsis roots.

Web Material

Web Topics

20.1 Structures of Some Important Gibberellins and Their Precursors, Derivatives, and Biosynthetic Inhibitors
The chemical structures of various gibberellins and the inhibitors of their biosynthesis are presented.

20.2 Commercial Uses of Gibberellins
Gibberellins have roles in agronomy, horticulture, and the brewing industry.

20.3 Gibberellin Biosynthesis
The GA biosynthetic pathways and GA conjugates in plants are described.

20.4 Gas Chromatography–Mass Spectrometry of Gibberellins
Identification and quantitation of individual GAs are accomplished by gas chromatography–mass spectrometry.

20.5 Environmental Control of Gibberellin Biosynthesis
Light quality and duration are important in regulating GA metabolism.

20.6 Negative Regulators of GA Response
GA response factors such as SPINDLY are negative regulators of the GA response.

20.7 Second Messengers in GA Signaling in Cereal Aleurone Layers
GA signaling includes a role for certain second messengers.

20.8 Effects of GA on Flowering Genes
GA interacts with the floral meristem identity gene *Leafy*.

20.9 Auxin–GA Interaction
Auxin plays a role in GA signaling in Arabidopsis roots.

Web Essays

20.1 Gibberellins in Pea: From Mendel to Molecular Physiology
A personal description of GA research using pea, that most recently focuses on the interaction of environmental factors and other hormones on the regulation of GA biosynthesis in this plant.

20.2 Green Revolution Genes
High yielding dwarf varieties of wheat and rice introduced in the mid-1960s have altered GA response.

20.3 Ubiquitin becomes Ubiquitous in GA Signaling
Homologues of SLY1/GID2 and DELLAs have been found in just about every plant species that has an EST sequence database.

Chapter References

Achard, P., Herr, A., Baulcombe, D. C., and Harberd, N. P. (2004) Modulation of floral development by a gibberellin-regulated microRNA. *Development* 131: 3357–3365.

Bethke, P. C., Schuurink, R., and Jones, R. L. (1997) Hormonal signalling in cereal aleurone. *J. Exp. Bot.* 48: 1337–1356.

Chandler, P. M., Marion-Poll, A., Ellis, M., and Gubler, G. (2002) Mutants at the *Slender1* locus of barley cv Himalaya. Molecular and physiological characterization. *Plant Physiol.* 129: 181–190.

Chiang, H. H., Hwang, I., and Goodman, H. M. (1995) Isolation of Arabidopsis *GA4* locus. *Plant Cell* 7:195–201.

Choi, D., Lee, Y., Cho, H. T., and Kende, H. (2003) Regulation of expansin gene expression affects growth and development in transgenic rice plants. *Plant Cell* 15: 1386–1398.

Davidson, S. E., and Reid, J. B. (2004) The pea gene *LH* encodes *ent*-kaurene oxidase. *Plant Physiol.* 134: 1123–1134.

Davidson, S. E., Elliott, R. C., Helliwell, C. A., Poole, A. T., and Reid, J. B. (2003) The pea gene *NA* encodes *ent*-kaurenoic acid oxidase. *Plant Physiol.* 131: 335–344.

Davies, P. J. (1995) The plant hormones: Their nature, occurrence, and functions. In *Plant Hormones: Physiology, Biochemistry and Molecular Biology*, P. J. Davies, ed., Kluwer, Dordrecht, Netherlands, pp. 1–12.

Dill, A., Jung, H. S., and Sun, T-p. (2001) The DELLA motif is essential for gibberellin-induced degradation of RGA. *Proc. Natl. Acad. Sci. USA* 98: 14162–14167.

Dill, A., Thomas, S. G., Hu, J., Steber, C.M., and Sun, T-p. (2004) The Arabidopsis F-box protein SLEEPY1 targets gibberellin repressors for gibberellin-induced degradation. *Plant Cell* 16: 1392–1405.

Ericksson, M. E., Israelsson, M., Olsson, O., and Moritz, T. (2000) Increased gibberellin biosynthesis in transgenic trees promotes growth, biomass production and xylem fiber length. *Nature Biotechnol.* 18: 784–788.

Fabian, T., Lorbiecke, R., Umeda, M., and Sauter, M. (2000) The cell cycle genes *cycA1;1* and *cdc2Os-3* are coordinately regulated by gibberellin in plants. *Planta* 211: 376–383.

Fu, X., and Harberd, N. P. (2003) Auxin promotes Arabidopsis root growth by modulating gibberellin response. *Nature* 421: 740–743.

Gocal, G. F. W., Sheldon, C. C., Gubler, F., Moritz, T., Bagnall, D. J., MacMillan, C. P., Li, S. F., Parish, R. W., Dennis, E. S., Weigel, D., and King, R. W. (2001) *GAMYB*-like genes, flowering, and gibberellin signaling in Arabidopsis. *Plant Physiol.* 127: 1682–1693.

Gubler, F., Kalla, R., Roberts, J. K., and Jacobsen, J. V. (1995) Gibberellin-regulated expression of a *myb* gene in barley aleurone cells: Evidence of myb transactivation of a high-pI alpha-amylase gene promoter. *Plant Cell* 7: 1879–1891.

Gubler, F., Chandler, P. M., White, R. G., Llewellyn, D. J., and Jacobsen, J. V. (2002) Gibberellin signaling in barley aleurone cells. Control of SLN1 and GAMYB expression. *Plant Physiol.* 129: 191–200.

Hedden, P. (2003) The genes of the Green Revolution. *Trends Genetics* 19: 5–9.

Hedden, P., and Phillips, A. L. (2000) Gibberellin metabolism: New insights revealed by the genes. *Trends Plant Sci.* 5: 523–530.

Higgins, T. J. V., Zwar, J. A., and Jacobsen, J. V. (1976) Gibberellic acid enhances the level of translatable mRNA for a-amylase in barley aleurone layers. *Nature* 260: 166–169.

Hisamatsu, T., King, R. W., Helliwell, C. A., and Koshioka, M. (2005) The involvement of gibberellin 20-oxidase genes in phytochrome-regulated petiole elongation of Arabidopsis. *Plant Physiol.* 138: 1106–1116.

Ikeda, A., Ueguchi-Tanaka, M., Sonoda, Y., Kitano, H., Koshioka, M., Futsuhara, Y., Matsuoka, M., and Yamaguchi, J. (2001) *slender Rice*, a constitutive gibberellin response mutant, is caused by a null mutation of the *SLR1* gene, an ortholog of the height –regulating *GAI/RGA/RHT/D8*. *Plant Cell* 13: 999–1010.

Ingram, T. J., Reid, J. B., Murfet, I. C., Gaskin, P., Willis, C. L., and MacMillan, J. (1984) Internode length in *Pisum*: the *Le* gene controls the 3β–hydroxylation of gibberellin A_{20} to gibberellin A_1. *Planta* 160: 455–463.

Itoh, H., Ueguchi-Tanaka, M., Sato, Y., Ashikari, M., and Matsuoka, M. (2002) The gibberellin signaling pathway is regulated by the appearance and disappearance of SLENDER RICE1 in nuclei. *Plant Cell* 14: 57–70.

Itoh, H., Matsuoka, M., and Steber, C. (2003). A role for the ubiquitin-26S-proteasome pathway in gibberellin signaling. *Trends Plant Sci.* 8: 492–497.

Jacobsen, J. V., Gubler, F., and Chandler, P. M. (1995) Gibberellin action in germinated cereal grains. In *Plant Hormones: Physiology, Biochemistry and Molecular Biology*, P. J. Davies, ed., Kluwer, Dordrecht, Netherlands, pp. 246–271.

Jones, H. D., Smith, S. J., Desikan, R., Plakidou, D. S., Lovegrove, A., and Hooley, R. (1998) Heterotrimeric G proteins are implicated in gibberellin induction of a-amylase gene expression in wild oat aleurone. *Plant Cell* 10: 245–253.

Kaneko, M., Itoh, H., Inukai, Y., Sakamoto, T., Ueguchi-Tanaka, M., Ashikari, M., and Matsuoka, M. (2003) Where do gibberellin biosynthesis and gibberellin signaling occur in rice plants? *Plant J.* 35: 104–115.

Kaneko, M., Inukai, Y., Ueguchi-Tanaka, M., Itoh, H., Izawa, T., Kobayashi, Y., Hattori, T., Miyao, A., Hirochika, H., Ashikari, M., and Matsuoka, M. (2004) Loss-of-function mutations of the rice GAMYB gene impair alpha-amylase expression in aleurone and flower development. *Plant Cell* 16: 33–44.

King R. W., Moritz, T., Juntilla, O., and Evans, L. T. (2001) Long-day induction of flowering in *Lolium temulentum* L. involves sequential increases in specific gibberellins at the shoot apex. *Plant Physiol.* 127: 624–632.

Koornneef, M., and van der Veen, J. H. (1980) Induction and analysis of gibberellin sensitive mutants in *Arabidopsis thaliana* (L) Heynh. *Theor. Appl. Genet.* 58: 257–263.

Koornneef, M., Elgersma, A., Hanhart, C. J., van Loenen-Martinet, E. P., van Rijn, and Zeevaart, J.A. D. (1985) A gibberellin insensitive mutant of *Arabidopsis thaliana*. *Physiol. Plant.* 65: 33–39.

Kusuba, S., Kano-Murakami, Y., Matsuoka, M., Tamaoki, M., Sakamoto, T., Yamauchi, I., and Fukumoto, M. (1998) Alteration of hormone levels in transgenic tobacco plants overexpressing the rice homeobox gene *OSH1*. *Plant Physiol.* 116: 116–122.

Lee, D. J., and Zeevaart, J. A. D. (2002) Differential regulation of RNA levels of gibberellin dioxygenases by photoperiod in spinach. *Plant Physiol.* 130: 2085–2094.

Lester, D. R., Ross, J. J., Davies, P. J., and Reid, J. B. (1997) Mendel's stem length gene (*Le*) encodes a gibberellin 3b-hydroxylase. *Plant Cell* 9: 1435–1443.

MacMillan, J. (2002) Occurrence of gibberellins in vascular plants, fungi and bacteria. *J. Plant Growth Regul.* 20: 387–442.

Martin, D. N., Proebsting, W. M., and Hedden, P. (1997). Mendel's dwarfing gene: cDNAs from the *Le* alleles and function of the expressed proteins. *Proc. Natl. Acad. Sci. USA* 94: 8907–8911.

McGinnis, K. M., Thomas, S. G., Soule, J. D., Strader, L. C., Zale, J. M., Sun, T-p., and Steber, C. M. (2003) The Arabidopsis *SLEEPY1* gene encodes a putative F-box subunit of an SCF E3 ubiquitin ligase. *Plant Cell* 15: 1120–1130.

Millar, A. A., and Gubler, F. (2005) The Arabidopsis *GAMYB-like* genes, *MYB33* and *MYB 65*, are microRN-regulated genes that redundantly facilitate anther development. *Plant Cell* 17: 705–721.

Ogawa, M., Hanada, A., Yamauchi, Y., Kuwahara, A., Kamiya, Y., and Yamaguchi, S. (2003) Gibberellin biosynthesis and response during Arabidopsis seed germination. *Plant Cell* 15: 1591–1604.

Olszewski N., Sun, T-p., and Gubler, F. (2002) Gibberellin signaling: biosynthesis, catabolism, and response pathways. *Plant Cell* 14: S61–88.

Ozga, J. A., Yu, J., and Reinecke, D. M. (2003) Pollination-, development-, and auxin-specific regulation of gibberellin 3ß-hydroxylase gene expression in pea fruit and seeds. *Plant Physiol.* 131: 1137–1146.

Park, W. W., Li, J., Song, R., Messing, J., and Chen, X. (2002) CARPEL FACTORY, a Dicer homolog, and HEN1, a novel protein, act in microRNA metabolism in *Arabidopsis thaliana*. *Curr. Biol.* 2: 1484–1495.

Peng, J. R., Carol, P., Richards, D. E., King, K. E., Cowling, R. J., Murphy, G. P., and Harberd, N. P. (1997) The Arabidopsis *GAI* gene defines a signaling pathway that negatively regulates gibberellin responses. *Genes Devel.* 11: 3194–3205.

Peng, J. R., Richards, D. E., Hartley, N. M., Murphy, G. P., Flintham, J. E., Beales, J., Fish, L. J., Pelica, F., Sudhakar, D., Christou, P., Snape, J. W., Gale, M. D., and Harberd, N. P. (1999) 'Green revolution' genes encode mutant gibberellin response modulators. *Nature* 400: 256–261.

Phillips, A. L. (2004) Genetic and transgenic approaches to improving crop performance. In *Plant Hormones: Biosynthesis, signal transduction, action!* P. J. Davies, ed., Kluwer, Dordrecht, Netherlands, pp. 582–609.

Phillips, A. L., Ward, D. A., Uknes, S., Appleford, N. E. J., Lange, T., Huttley, A. K., Gaskin, P., Graebe, J. E., and Hedden, P. (1995) Isolation and expression of three gibberellin 20-oxidase cDNA clones from Arabidopsis. *Plant Physiol.* 108: 1049–1057.

Phinney, B. O. (1984) Gibberellin A_1, dwarfism and the control of shoot elongation in higher plants. In: *The Biosynthesis and Metabolism of Plant Hormones*. Crozier, A., and Hillman, J. R. (eds) vol 23, 17–41. Cambridge U.P., London

Reid, J. B., and Howell, S. H. (1995) Hormone mutants and plant development. In *Plant Hormones: Physiology, Biochemistry and Molecular Biology*, P. J. Davies, ed., Kluwer, Dordrecht, Netherlands, pp. 448–485.

Reid, J. B., Ross, J. J., and Swain, S. W. (1992) Internode length in *Pisum*. A new *slender* mutant with elevated levels of C_{19}-gibberellins. *Planta* 188: 462–467.

Rhoades, M. W., Reinhart, B. J., Lim, L. P., Burges, C. B., Bartel, B., and Bartel, D. P. (2002). Prediction of plant microRNA targets. *Cell* 110: 513–520.

Ritchie, S., and Gilroy, S. (1998) Tansley Review No. 100: Gibberellins: Regulating genes and germination. *New Phytol.* 140: 363–383.

Ross, J., and O'Neill, D. (2001) New interactions between classical plant hormones. *Trends Plant Sci.* 6: 2–4.

Ross, J. J., O'Neill, D. P., Smith, J. J., Kerckhoffs, L. H. J., and Elliott, R. C. (2000) Evidence that auxin promotes gibberellin A_1 biosynthesis in pea. *Plant J.* 21: 547–552.

Ross, J. J., Reid, J. B., Gaskin, P. and Macmillan, J. (1989) Internode length in *Pisum*. Estimation of GA_1 levels in genotypes *Le, le* and *le^d*. *Physiol. Plant.* 76: 173–176.

Sakamoto, T., Kamiya, N., Ueguchi-Tanaka, M., Iwahori, S., and Matsuoka, M. (2001) KNOX homeodomain protein directly suppresses the expression of a gibberellin biosynthetic gene in the tobacco shoot apical meristem. *Genes Dev.* 15: 581–590.

Sakamoto, T., Morinaka, Y., Ishiyama, K., Kobayashi, M., Itoh., H., Kayano, T., Iwahori, S., Matsuoaka, M., and Tanaka, H. (2003) Genetic manipulation of gibberellin metabolism in transgenic rice. *Nature* 21: 909–913.

Sasaki, A., Itoh, H., Gomi, K., Ueguchi-Tanaka, M., Ishayama, K., Kobayashi, M., Jeong, D.-H., An., G., Kitano, H., Ashikari, M., and Matsuoka, M. (2003) Accumulation of phosphorylated repressor for gibberellin signaling in an F box mutant. *Science* 299: 1896–1898.

Sauter, M., and Kende, H. (1992) Gibberellin-induced growth and regulation of the cell division cycle in deepwater rice. *Planta* 188: 362–368.

Silverstone, A. L., Chang, C.-W., Krol, E., and Sun, T-p. (1997a) Developmental regulation of the gibberellin biosynthetic gene *GA1* in *Arabidopsis thaliana*. *Plant J.* 12: 9–19.

Silverstone, A. L., Mak, P. Y. A., Casamitjana Martinez, E., and Sun, T-p. (1997b) The new *RGA* locus encodes a negative regulator of gibberellin responses in *Arabidopsis thaliana*. *Genetics* 146: 1087–1099.

Silverstone, A. L., Chang, C-w., Krol, E., and Sun, T-p. (1998) The Arabidopsis *RGA* encodes a transcriptional regulator repressing the gibberellin signal transduction pathway. *Plant Cell* 10, 155–169.

Silverstone, A. L., Jung, H. S., Dill, A., Kawaide, H., Kamiya, Y., and Sun, T-p. (2001) Repressing a repressor: Gibberellin-induced rapid reduction of the RGA protein in Arabidopsis. *Plant Cell* 13: 1555–1565.

Stracke, R., Werber, M., and Weisshaar, B. (2001) The *R2R3-MYB* gene family in *Arabidopsis thaliana*. *Curr. Opin. Plant Biol.* 4: 447–456

Sun, T-p., and Gubler, F. (2004) Molecular mechanism of gibberellin signaling in plants. *Annu. Rev. Plant Biol.* 55: 197–223.

Swain S. M., and Singh D. P. (2005) Tall tales from sly dwarves: novel functions of gibberellins in plant development. *Trends Plant Sci.* 10: 123–129.

Ueguchi-Tanaka, M., Ashikari, M., Nakajima, M., Itoh, H., Katoh, E., Kobayashi, M., Chow, T., Hsing, Y. C., Kitano, H., Yamaguchi, I., and Matsuoka, M. (2005) *GIBBERELLIN INSENSITIVE DWARF1* encodes a soluble receptor for gibberellin. *Nature* 437: 693–698.

Xu, W., Campbell, P., Vargheese, A. V., and Braam, J. (1996) The Arabidopsis XET-related gene family: Environmental and hormonal regulation of expression. *Plant J.* 9: 879–889.

Xu Y-L., Li, L., Wu, K., Peeters, A.. J. M., Gage, D. A., and Zeevaart, J. A. D. (1995) The *GA5* locus of *Arabidopsis thaliana* encodes a multi-functional gibberellin 20-oxidase: Molecular cloning and functional expression. *Proc. Natl. Acad. Sci. USA* 92: 6640–6644.

Yamaguchi, S., and Kamiya, Y. (2000) Gibberellin biosynthesis: Its regulation by endogenous and environmental signals. *Plant Cell Physiol.* 41: 251–257.

Yamaguchi, S., Kamiya, Y., and Sun, T-p. (2001) Distinct cell-specific expression patterns of early and late gibberellin biosynthetic genes during Arabidopsis seed germination. *Plant J.* 28: 443–453.

Yang, T., Davies, P. J., and Reid, J. B. (1996) Genetic dissection of the relative roles of auxin and gibberellin in the regulation of stem elongation in intact light-grown peas. *Plant Physiol.* 110: 1029–1034.

Yaxley, J. R., Ross, J. J., Sherriff, L. J., and Reid, J. B. (2001) Gibberellin biosynthesis mutations and root development in pea. *Plant Physiol.* 125: 627–633.

Chapter 21
Cytokinins: Regulators of Cell Division

THE CYTOKININS WERE DISCOVERED in the search for factors that stimulate plant cells to divide (i.e., undergo cytokinesis). Since their discovery, cytokinins have been shown to have effects on many other physiological and developmental processes, including leaf senescence, nutrient mobilization, apical dominance, the formation and activity of shoot apical meristems, floral development, the breaking of bud dormancy, and seed germination. Cytokinins also appear to mediate many aspects of light-regulated development, including chloroplast differentiation, the development of autotrophic metabolism, and leaf and cotyledon expansion.

Although cytokinins regulate many cellular processes, the control of cell division is central in plant growth and development and is considered diagnostic for this class of plant growth regulators. For these reasons we will preface our discussion of cytokinin function with a brief consideration of the roles of cell division in normal development, wounding, gall formation, and tissue culture.

Later in the chapter we will examine the regulation of plant cell proliferation by cytokinins. Then we will turn to cytokinin functions not directly related to cell division: chloroplast differentiation, the prevention of leaf senescence, and nutrient mobilization. Finally, we will consider the molecular mechanisms underlying cytokinin perception and signaling.

Cell Division and Plant Development

Plant cells form as the result of cell divisions in a primary or secondary meristem. Newly formed plant cells typically enlarge and differentiate, but once they have assumed their function—whether transport, photosynthesis, support, storage, or protection—usually they do not divide again during the life of the plant. In this respect they appear to be similar to animal cells, which are considered to be terminally differentiated.

However, this similarity to the behavior of animal cells is only superficial. Almost every type of plant cell that retains its nucleus at maturity has been shown to be capable of dividing. This property comes into play during such processes as wound healing and leaf abscission.

Differentiated plant cells can resume division

Under some circumstances, mature, differentiated plant cells may resume cell division in the intact plant. In many species, mature cells of the cortex and/or phloem resume division to form secondary meristems, such as the vascular cambium or the cork cambium. The abscission zone at the base of a leaf petiole is a region where mature parenchyma cells begin to divide again after a period of mitotic inactivity, forming a layer of cells with relatively weak cell walls where abscission can occur (see Chapter 22).

Wounding of plant tissues induces cell divisions at the wound site. Even highly specialized cells, such as phloem fibers and guard cells, may be stimulated by wounding to divide at least once. Wound-induced mitotic activity typically is self-limiting; after a few divisions the derivative cells stop dividing and redifferentiate. However, when the soil-dwelling bacterium *Agrobacterium tumefaciens* invades a wound, it can cause the neoplastic (tumor-forming) disease known as **crown gall**. This phenomenon is dramatic natural evidence of the mitotic potential of mature plant cells.

Without *Agrobacterium* infection, the wound-induced cell division would subside after a few days and some of the new cells would differentiate as a protective layer of cork cells or vascular tissue. However, *Agrobacterium* changes the character of the cells that divide in response to the wound, making them tumorlike. They do not stop dividing; rather they continue to divide throughout the life of the plant to produce an unorganized mass of tumor-like tissue called a **gall** (Figure 21.1). We will have more to say about this important disease later in this chapter.

Diffusible factors may control cell division

The considerations addressed in the previous section suggest that mature plant cells stop dividing because they no longer receive a particular signal, possibly a hormone, that is necessary for the initiation of cell division. The idea that cell division may be initiated by a diffusible factor originated with the Austrian plant physiologist G. Haberlandt,

FIGURE 21.1 Tumor that formed on a tomato stem infected with the crown gall bacterium, *Agrobacterium tumefaciens*. Two months before this photo was taken the stem was wounded and inoculated with a virulent strain of the crown gall bacterium. (From Aloni et al. 1998, courtesy of R. Aloni.)

who, in about 1913, demonstrated that vascular tissue contains a water-soluble substance or substances that will stimulate the division of wounded potato tuber tissue. The effort to determine the nature of this factor (or factors) led to the discovery of the cytokinins in the 1950s.

Plant tissues and organs can be cultured

Biologists have long been intrigued by the possibility of growing organs, tissues, and cells in culture on a simple nutrient medium, in the same way that microorganisms can be cultured in test tubes or on petri dishes. In the 1930s, Philip White demonstrated that tomato roots can be grown indefinitely in a simple nutrient medium containing only sucrose, mineral salts, and a few vitamins, with no added hormones (White 1934).

In contrast to roots, isolated stem tissues exhibit very little growth in culture without added hormones in the medium. Even if auxin is added, only limited growth may occur, and usually this growth is not sustained. Frequently this auxin-induced growth is due to cell enlargement only. The shoots of most plants cannot grow on a simple medium lacking hormones, even if the cultured stem tissue contains apical or lateral meristems, until adventitious roots form. Once the stem tissue has rooted, shoot growth resumes, but now as an integrated, whole plant.

These observations indicate that there is a difference in the regulation of cell division in root and shoot meristems. They also suggest that some root-derived factor(s) may regulate growth in the shoot.

Crown gall stem tissue is an exception to these generalizations. After a gall has formed on a plant, heating the plant to 42°C will kill the bacterium that induced gall formation. The plant will survive the heat treatment, and its gall tissue will continue to grow as a bacteria-free tumor (Braun 1958).

Tissues removed from these bacteria-free tumors grow on simple, chemically defined culture media that would not support the proliferation of normal stem tissue of the same species. However, these stem-derived tissues are not organized. Instead they grow as a mass of disorganized, relatively undifferentiated cells called **callus tissue**.

Callus tissue sometimes forms naturally in response to wounding, or in graft unions where stems of two different plants are joined. Crown gall tumors are a specific type of callus, whether they are growing attached to the plant or in culture. The finding that crown gall callus tissue can be cultured demonstrated that cells derived from stem tissues are capable of proliferating in culture and that contact with the bacteria may cause the stem cells to produce cell division–stimulating factors.

The Discovery, Identification, and Properties of Cytokinins

A great many substances were tested in an effort to initiate and sustain the proliferation of normal stem tissues in culture. Materials ranging from yeast extract to tomato juice were found to have a positive effect, at least with some tissues. However, culture growth was stimulated most dramatically when the liquid endosperm of coconut, also known as coconut milk, was added to the culture medium.

Philip White's nutrient medium, supplemented with an auxin and 10 to 20% coconut milk, will support the continued cell division of mature, differentiated cells from a wide variety of tissues and species, leading to the formation of callus tissue (Caplin and Steward 1948). This finding indicated that coconut milk contains a substance or substances that stimulate mature cells to enter and remain in the cell division cycle.

Eventually coconut milk was shown to contain the cytokinin *zeatin*, but this finding was not obtained until several years after the discovery of the cytokinins (Letham 1974). The first cytokinin to be discovered was the synthetic analog kinetin.

Kinetin was discovered as a breakdown product of DNA

In the 1940s and 1950s, Folke Skoog and coworkers at the University of Wisconsin tested many substances for their ability to initiate and sustain the proliferation of cultured tobacco pith tissue. They had observed that the nucleic acid base adenine had a slight promotive effect, so they tested the possibility that nucleic acids would stimulate division in this tissue. Surprisingly, aged or autoclaved herring sperm DNA had a powerful cell division–promoting effect.

After much work, a small molecule was identified from the autoclaved DNA and named **kinetin**. It was shown to be an adenine (or amino purine) derivative, 6-furfurylaminopurine (Miller et al. 1955):

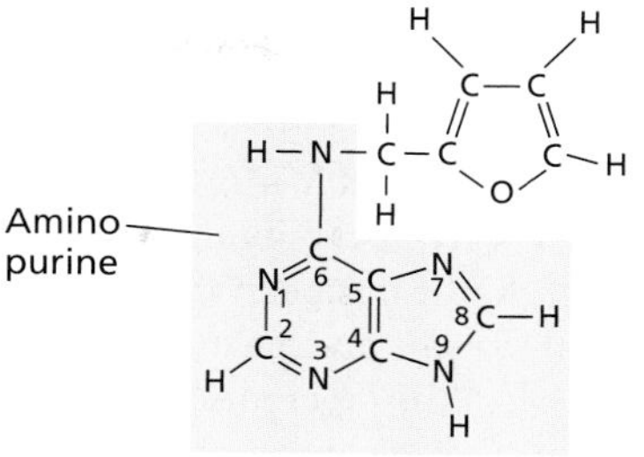

Kinetin

In the presence of an auxin, kinetin would stimulate tobacco pith parenchyma tissue to proliferate in culture. No kinetin-induced cell division occurs without auxin in the culture medium. (For more details, see **Web Topic 21.1**.)

Kinetin is not a naturally occurring plant growth regulator, and it does not occur as a base in the DNA of any species. It is a by-product of the heat-induced degradation of DNA, in which the deoxyribose sugar of adenosine is converted to a furfuryl ring and shifted from the 9 position to the 6 position on the adenine ring.

The discovery of kinetin was important because it demonstrated that cell division could be induced by a simple chemical substance. Of greater importance, the discovery of kinetin suggested that naturally occurring molecules with structures similar to that of kinetin regulate cell division activity within the plant. This hypothesis proved to be correct.

Zeatin was the first natural cytokinin discovered

Several years after the discovery of kinetin, extracts of the immature endosperm of maize (corn; *Zea mays*) were found to contain a substance that has the same biological effect as kinetin. This substance stimulates mature plant cells to divide when added to a culture medium along with an auxin. Letham (1973) isolated the molecule responsible for this activity and identified it as *trans*-6-(4-hydroxy-3-methylbut-2-enylamino)purine, which he called **zeatin**:

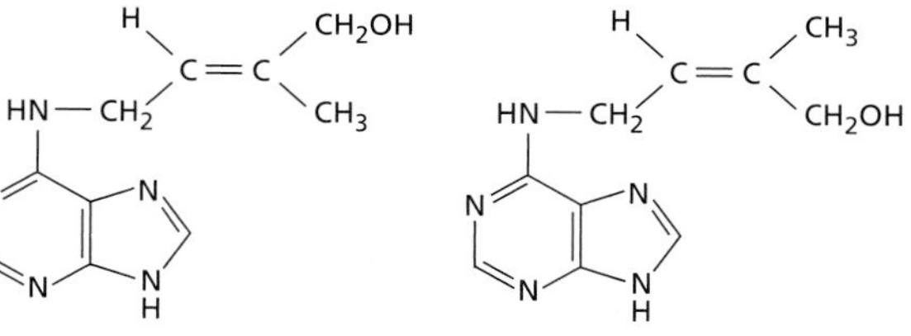

trans-Zeatin   *cis*-Zeatin
6-(4-Hydroxy-3-methylbut-2-enylamino)purine

The molecular structure of zeatin is similar to that of kinetin. Both molecules are adenine or aminopurine derivatives. Although they have different side chains, in both cases the side chain is attached to the 6 nitrogen of the aminopurine. Because the side chain of zeatin has a double bond, it can exist in either the *cis* or the *trans* configuration.

In higher plants, zeatin occurs in both the *cis* and the *trans* configurations, and these forms can be interconverted by an enzyme known as *zeatin isomerase*. Although the *trans* form of zeatin is much more active in biological assays, the *cis* form may also play important roles, as suggested by the fact that it has been found in high levels in a number of plant species and particular tissues. Furthermore, the maize cytokinin receptor binds to both *cis* and *trans* forms of zeatin with comparable affinity, and a gene encoding a glucosyl transferase enzyme specific to *cis*-zeatin has recently been cloned. Both of these observations provide further support that the *cis* isoform of zeatin may also play a biological role.

Since its discovery in immature maize endosperm, zeatin has been found in many plants and in some bacteria. It is generally the most prevalent active cytokinin in higher plants, but other substituted aminopurines that are active as cytokinins have been isolated from many plant and bacterial species. These aminopurines differ from zeatin in the nature of the side chain attached to the 6 nitrogen or in the attachment of a side chain to carbon 2:

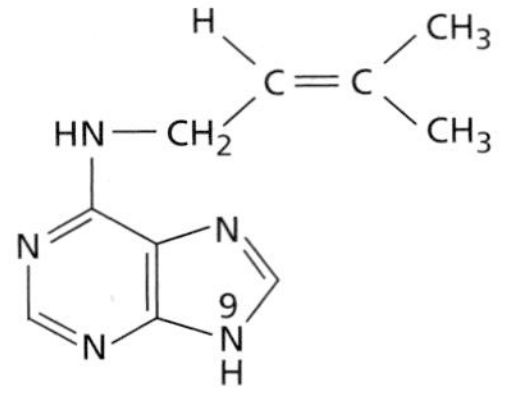

N^6-(Δ^2-Isopentenyl)-adenine (iP)

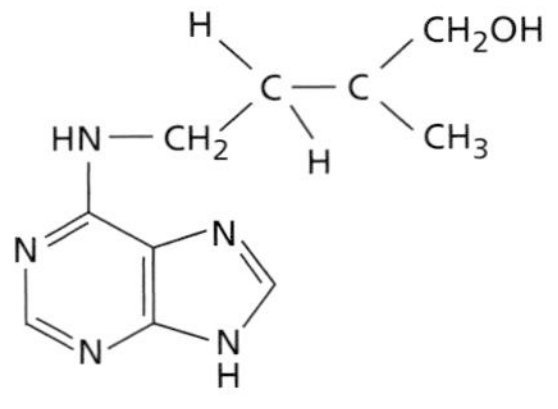

Dihydrozeatin (DZ)

In addition, these cytokinins can be present in the plant as a **riboside** (in which a ribose sugar is attached to the 9 nitrogen of the purine ring), a **ribotide** (in which the ribose sugar moiety contains a phosphate group), or a **glycoside** (in which a sugar molecule is attached to the 3, 7, or 9 nitrogen of the purine ring, or to the oxygen of the zeatin or dihydrozeatin side chain) (see **Web Topic 21.2**).

Some synthetic compounds can mimic or antagonize cytokinin action

Cytokinins are defined as compounds that have biological activities similar to those of *trans*-zeatin. These activities include the ability to do the following:

- Induce cell division in callus cells in the presence of an auxin
- Promote bud or root formation from callus cultures when in the appropriate molar ratios to auxin
- Delay senescence of leaves
- Promote expansion of dicot cotyledons

Many chemical compounds have been synthesized and tested for cytokinin activity. Analysis of these compounds provides insight into the structural requirements for activity. Nearly all compounds active as cytokinins are N^6-substituted aminopurines, such as benzyladenine (BA):

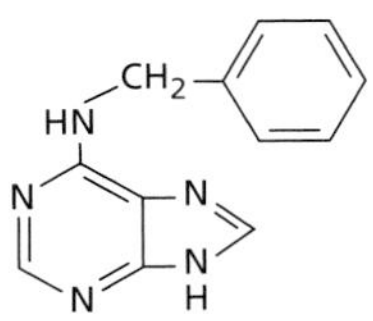

**Benzyladenine
(benzylaminopurine)
(BA)**

and all the naturally occurring cytokinins are aminopurine derivatives. There are also synthetic cytokinin compounds that have not been identified in plants, most notable of which are the diphenylurea-type cytokinins, such as thidiazuron, which is used commercially as a defoliant and an herbicide:

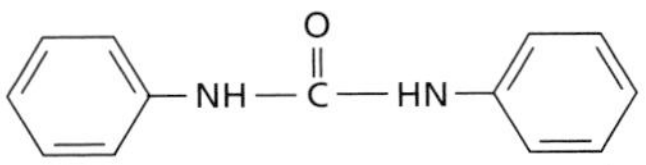

N,N'-Diphenylurea (nonamino purine with weak activity)

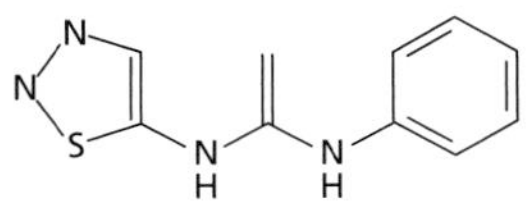

Thidiazuron

In the course of determining the structural requirements for cytokinin activity, investigators found that some molecules act as *cytokinin antagonists*:

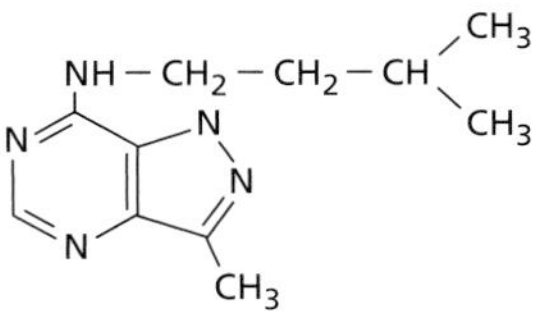

3-Methyl-7-(3-methylbutylamino)pyrazolo[4,3-D]pyrimidine

These molecules are able to block the action of cytokinins, and their effects may be overcome by the addition of more cytokinin. Naturally occurring molecules with cytokinin activity may be detected and identified primarily by physical methods (see **Web Topic 21.3**).

Cytokinins occur in both free and bound forms

Hormonal cytokinins are present as free molecules (not covalently attached to any macromolecule) in plants and certain bacteria. Free cytokinins have been found in a wide spectrum of angiosperms and probably are universal in this group of plants. They have also been found in algae, diatoms, mosses, ferns, and conifers.

The regulatory role of cytokinins has been demonstrated only in angiosperms, conifers, and mosses, but they may function to regulate the growth, development, and metabolism of all plants. Usually zeatin is the most abundant naturally occurring free cytokinin, but *dihydrozeatin* (*DZ*) and *isopentenyl adenine* (*iP*) also are commonly found in higher plants and bacteria. Numerous derivatives of these three cytokinins have been identified in plant extracts (see the structures illustrated in Figure 21.6).

Transfer RNA (tRNA) contains not only the four nucleotides used to construct all other forms of RNA, but also some unusual nucleotides in which the base has been modified. Some of these "hypermodified" bases act as cytokinins when the tRNA is hydrolyzed and tested in one of the cytokinin bioassays. Some plant tRNAs contain *cis*-zeatin as a hypermodified base. However, cytokinins are not confined to plant tRNAs. They are part of certain tRNAs from all organisms, from bacteria to humans. (For details, see **Web Topic 21.4**.)

The hormonally active cytokinin is the free base

It has been difficult to determine which species of cytokinin represents the active form of the hormone due to the interconversion among the various cytokinin species within the plant, but the recent identification of the cytokinin receptor CRE1 has allowed this question to be addressed directly. The relevant experiments have shown that the free-base form of *trans*-zeatin, but not its riboside or ribotide derivatives, binds directly to CRE1, indicating that the free base is the active form (Yamada et al. 2001). This is also likely true for dihydrozeatin and isopentenyl adenine.

Although the free-base forms are thought to be the hormonally active cytokinin, some other compounds have cytokinin activity, either because they are readily converted to zeatin, dihydrozeatin, or isopentenyl adenine, or because they release these compounds from other molecules, such as cytokinin glucosides. For example, tobacco cells in culture do not grow unless cytokinin ribosides supplied in the culture medium are converted to the free base.

In another example, excised radish cotyledons grow when they are cultured in a solution containing the cytokinin base benzyladenine (the N^6-substituted aminopurine cytokinin mentioned above). The cultured cotyledons readily take up the hormone and convert it to various BA glucosides, BA ribonucleoside, and BA ribonucleotide. When the cotyledons are transferred back to a medium lacking a cytokinin, their growth rate declines, as do the concentrations of BA, BA ribonucleoside, and BA ribonucleotide in the tissues. However, the level of the BA glucosides remains constant. This finding suggests that the glucosides cannot be the active form of the hormone.

Some plant pathogenic bacteria, fungi, insects, and nematodes secrete free cytokinins

Some bacteria and fungi are intimately associated with higher plants. Many of these microorganisms produce and secrete substantial amounts of cytokinins and/or cause the plant cells to synthesize plant hormones, including cytokinins (Akiyoshi et al. 1987). The cytokinins produced by microorganisms include *trans*-zeatin, [9R]iP, *cis*-zeatin, and their ribosides (Figure 21.2). Infection of plant tissues with these microorganisms can induce the tissues to divide and, in some cases, to form special structures, such as mycorrhizae, in which the microorganism can reside in a mutualistic relationship with the plant.

In addition to the crown gall bacterium, *A. tumefaciens*, other pathogenic bacteria may stimulate plant cells to divide. For example, *Corynebacterium fascians* is a major cause of the growth abnormality known as fasciation, and also as **witches' broom** (Figure 21.3). The shoots of plants infected by *C. fascians* resemble an old-fashioned straw

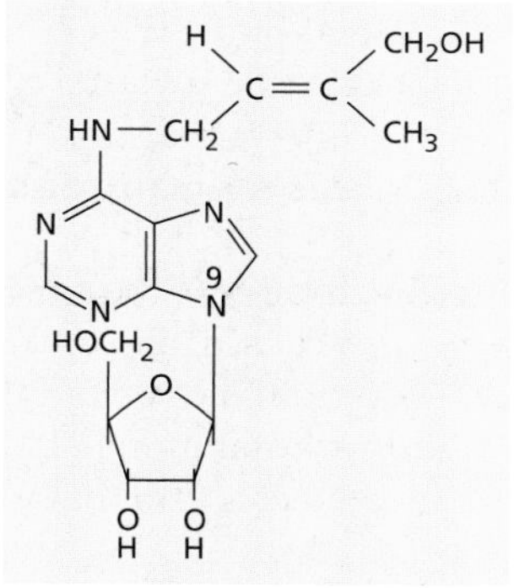

Ribosylzeatin (zeatin riboside)

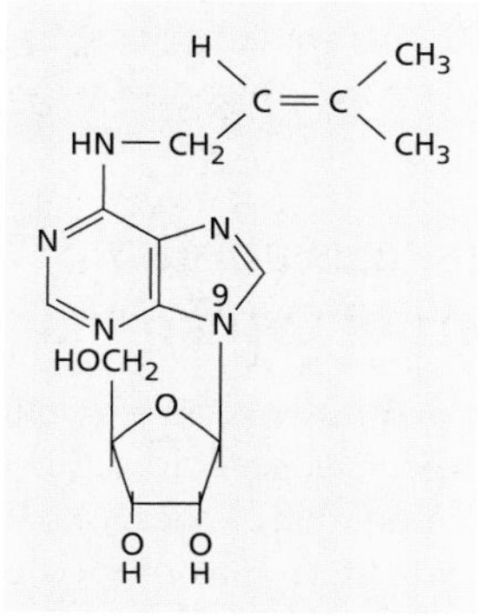

N^6-(Δ^2-Isopentenyl)adenosine ([9R]iP)

FIGURE 21.2 Structures of ribosylzeatin and N^6-(Δ^2-isopentenyl)adenosine ([9R]iP).

Figure 21.3 Witches' broom on balsam fir (*Abies balsamea*). (© Gregory K. Scott/Photo Researchers, Inc.)

broom because the lateral buds, which normally remain dormant, are stimulated by the bacterial cytokinin to grow (Hamilton and Lowe 1972). Fasciation can also occur spontaneously, and is the basis for many of the horticultural dwarf conifers.

Infection with a close relative of the crown gall organism, *Agrobacterium rhizogenes*, causes masses of roots instead of callus tissue to develop from the site of infection. *A. rhizogenes* is able to modify cytokinin metabolism in infected plant tissues through a mechanism that will be described later in this chapter.

Certain insects secrete cytokinins, which may play a role in the formation of galls utilized by these insects as feeding sites. Root-knot nematodes also produce cytokinins, which may be involved in manipulating host development to produce the giant cells from which the nematode feeds (Elzen 1983).

Biosynthesis, Metabolism, and Transport of Cytokinins

The side chains of naturally occurring cytokinins are chemically related to rubber, carotenoid pigments, the plant hormones gibberellin and abscisic acid, and some of the plant defense compounds known as phytoalexins. All of these compounds are constructed, at least in part, from isoprene units (see Chapter 13).

Isoprene is similar in structure to the side chains of zeatin and iP (see the structures illustrated in Figure 21.6). These cytokinin side chains are synthesized from an isoprene derivative. Large molecules of rubber and the carotenoids are constructed by the polymerization of many isoprene units; cytokinins contain just one of these units. The precursor(s) for the formation of these isoprene structures are either mevalonic acid or pyruvate plus 3-phosphoglycerate, depending on which pathway is involved (see Chapter 13). These precursors are converted to the biological isoprene unit dimethylallyl diphosphate (DMAPP).

Crown gall cells have acquired a gene for cytokinin synthesis

Bacteria-free tissues from crown gall tumors proliferate in culture without the addition of any hormones to the culture medium. Crown gall tissues contain substantial amounts of both auxin and free cytokinins. Furthermore, when radioactively labeled adenine is fed to periwinkle (*Vinca rosea*) crown gall tissues, it is incorporated into both zeatin and zeatin riboside, demonstrating that gall tissues contain the cytokinin biosynthetic pathway. Control stem tissue, which has not been transformed by *Agrobacterium*, does not incorporate labeled adenine into cytokinins.

During infection by *A. tumefaciens*, plant cells incorporate bacterial DNA into their chromosomes. The virulent strains of *Agrobacterium* contain a large plasmid known as the **Ti plasmid**. Plasmids are circular pieces of extrachromosomal DNA that are not essential for the life of the bacterium. However, plasmids frequently contain genes that enhance the ability of the bacterium to survive in special environments.

A small portion of the Ti plasmid, known as the **T-DNA**, is incorporated into the nuclear DNA of the host plant cell (Figure 21.4) (Chilton et al. 1977). T-DNA carries genes necessary for the biosynthesis of *trans*-zeatin and auxin, as well as a member of a class of unusual carbon- and nitrogen-containing amino acid derivatives called *opines* (Figure 21.5). Opines are not synthesized by plants except after crown gall transformation.

The T-DNA gene involved in cytokinin biosynthesis—known as the *ipt** gene—encodes an **isopentenyl transferase (IPT)** enzyme that transfers the isopentenyl group from 1-hydroxy-2-methyl-2-(E)-butenyl 4-diphosphate (HMBDP) to adenosine monophosphate (AMP) to form tZRMP (*trans*-zeatin riboside 5'-monophosphate) (Figure 21.6) (Akiyoshi et al. 1984; Barry et al. 1984; Sakakibara et al. 2005). The *ipt* gene has been called the *tmr* locus because, when *inactivated* by mutation, it results in "rooty" *tumors*. This conversion route is somewhat similar to the pathway for cytokinin synthesis that has been postulated for normal tissue (see Figure 21.6). However, one important difference is that the side chain used by the bacterial enzyme is distinct (HMBDP for the bacterial vs. DMAPP for the plant enzyme), which results in the bacterial enzyme directly

---

*Bacterial genes, unlike plant genes, are written in lowercase italics.

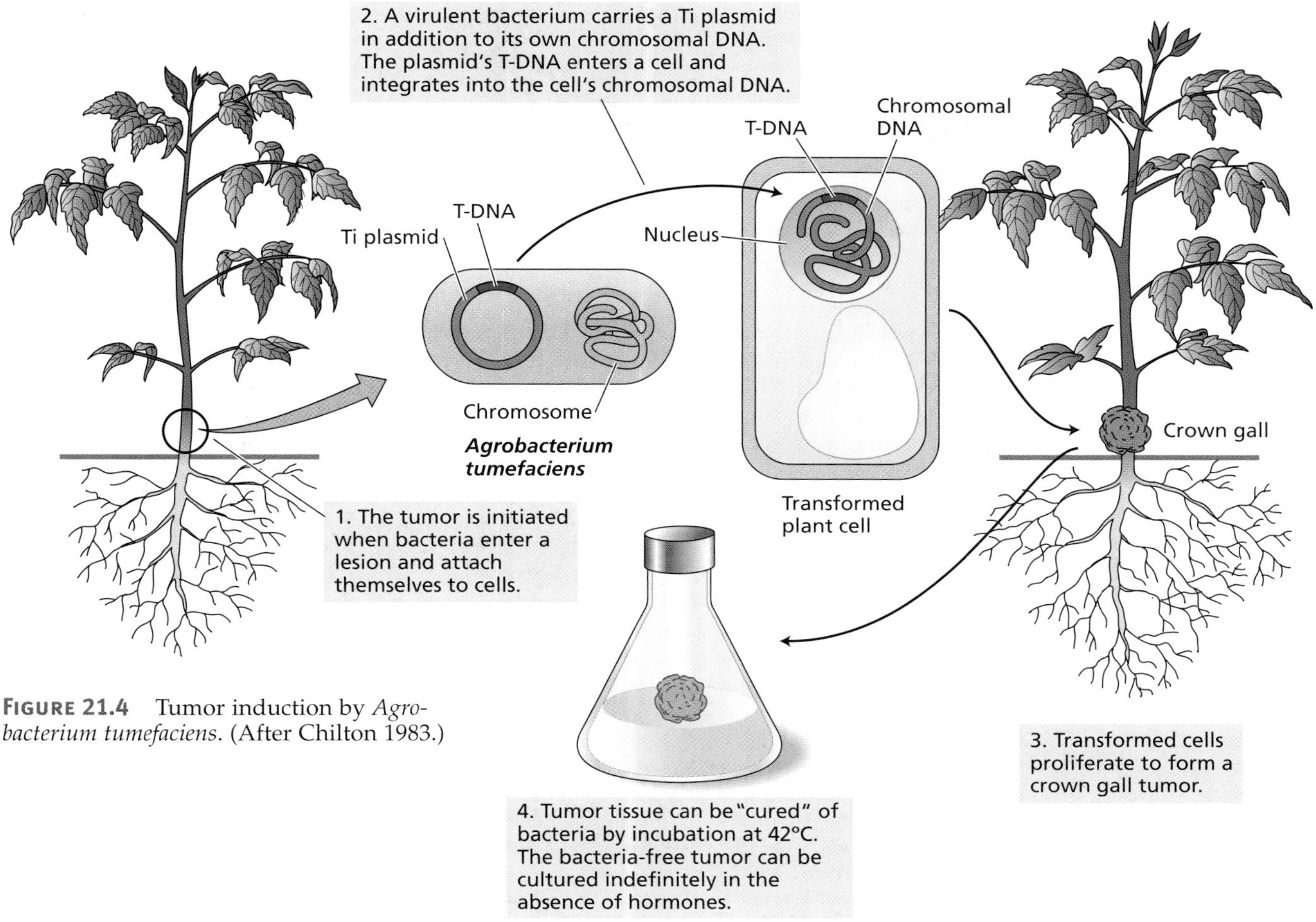
2. A virulent bacterium carries a Ti plasmid in addition to its own chromosomal DNA. The plasmid's T-DNA enters a cell and integrates into the cell's chromosomal DNA.
T-DNA
Chromosomal DNA
T-DNA
Ti plasmid
Nucleus
Chromosome
Agrobacterium tumefaciens
Crown gall
Transformed plant cell
1. The tumor is initiated when bacteria enter a lesion and attach themselves to cells.
3. Transformed cells proliferate to form a crown gall tumor.
4. Tumor tissue can be "cured" of bacteria by incubation at 42°C. The bacteria-free tumor can be cultured indefinitely in the absence of hormones.

FIGURE 21.4 Tumor induction by *Agrobacterium tumefaciens*. (After Chilton 1983.)

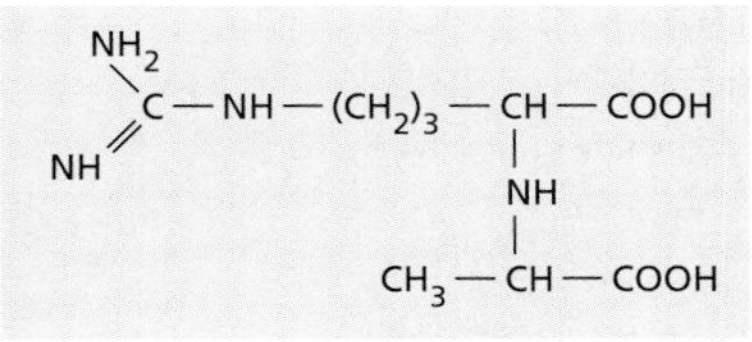

Octopine

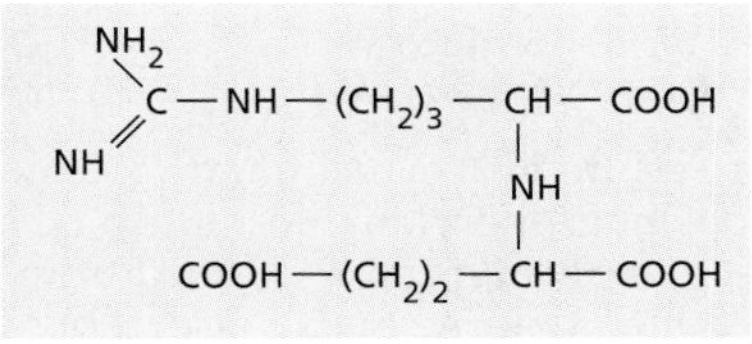

Nopaline

FIGURE 21.5 The two major opines, octopine and nopaline, are found only in crown gall tumors. The genes required for their synthesis are present in the T-DNA from *Agrobacterium tumefaciens*. The bacterium, but not the plant, can utilize the opines as a nitrogen source.

producing a zeatin-type cytokinin, rather than the iP-type cytokinin produced by the plant enzymes.

The T-DNA also contains two genes encoding enzymes that convert tryptophan to the auxin indole-3-acetic acid (IAA). This pathway of auxin biosynthesis differs from the one in nontransformed cells and involves indoleacetamide as an intermediate (see Figure 19.6). The *ipt* gene and the two auxin biosynthetic genes of T-DNA are **phyto-oncogenes**, since they can induce tumors in plants (see **Web Topic 21.5**).

Because their promoters are plant eukaryotic promoters, none of the T-DNA genes are expressed in the bacterium; rather they are transcribed after they are inserted into the plant chromosomes. Transcription of the genes leads to synthesis of the enzymes they encode, resulting in the production of zeatin, auxin, and an opine. The bacterium can utilize the opine as a carbon and nitrogen source, but cells of higher plants cannot. Thus, by transforming the plant cells, the bacterium provides itself with an expanding environment (the gall tissue) in which the host cells are directed to produce a substance (the opine) that only the bacterium can utilize for its nutrition (Bomhoff et al. 1976).

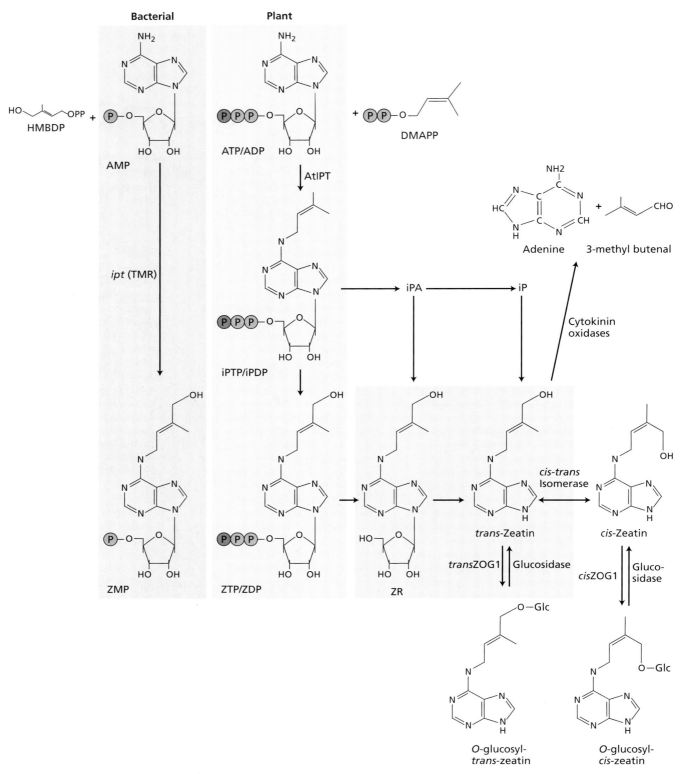

FIGURE 21.6 Biosynthetic pathway for cytokinin biosynthesis. The first committed step in cytokinin biosynthesis is the addition of the isopentenyl side chain from DMAPP (dimethylallyl diphosphate) to an adenosine moiety. The plant and bacterial IPT enzymes differ in the adenosine substrate used and the side chain donor; the plant enzyme appears to utilize both ADP and ATP and couples this to DMPP, and the bacterial enzyme utilizes AMP and couples this to HMBDP (1-hydroxy-2-methyl-2-(E)-butenyl 4-diphosphate). The products of these reactions (iPMP, iPDP, or iPTP) are converted to zeatin by a cytochrome P450 monooxygenase. The various phosphorylated forms can be interconverted, and free trans-zeatin can be formed from the riboside by enzymes of general purine metabolism. trans-Zeatin can be metabolized in various ways as shown, and these reactions are catalyzed by the indicated enzymes. The enzyme trans-zeatin O-glucosyltransferase (transZOG1) adds a glucose residue to the hydroxyl group of the side chain of zeatin, thus inactivating the hormone.

An important difference between the control of cytokinin biosynthesis in crown gall tissues and in normal tissues is that the T-DNA genes for cytokinin synthesis are expressed in all infected cells, even those in which the native plant genes for biosynthesis of the hormone are normally repressed.

IPT catalyzes the first step in cytokinin biosynthesis

The first committed step in cytokinin biosynthesis is the transfer of the isopentenyl group of dimethylallyl diphosphate to an adenosine moiety. An enzyme that catalyzes such an activity was first identified in the cellular slime mold *Dictyostelium discoideum*, and subsequently the *ipt* gene from *Agrobacterium* was found to encode such an enzyme.

As noted earlier, cytokinins are also present in the tRNAs of most cells, including plant and animal cells. The tRNA cytokinins are synthesized by modification of specific adenine residues within the fully transcribed tRNA. As with the free cytokinins, isopentenyl groups are transferred to the adenine molecules from DMAPP by an enzyme called tRNA-IPT. The genes for tRNA-IPT have been cloned from many species.

The possibility that free cytokinins are derived from tRNA has been explored extensively. Although the tRNA-bound cytokinins can act as hormonal signals for plant cells if the tRNA is degraded and fed back to the cells, it is unlikely that any significant amount of the free hormonal cytokinin in plants is derived from the turnover of tRNA.

An enzyme with IPT activity was initially identified in crude extracts of various plant tissues, but researchers were unable to purify the protein to homogeneity. Subsequently, the Arabidopsis genome was analyzed for potential *ipt*-like sequences, and nine different *IPT* genes were identified—more than are present in animal genomes, which usually contain only one or two *ipt*-like genes used in tRNA modification (Kakimoto 2001; Takei et al. 2001a).

Phylogenetic analysis revealed that one of the Arabidopsis *IPT* genes resembles bacterial tRNA-*ipt*, another resembles eukaryotic tRNA-*IPT*, and the other seven form a distinct group or clade together with other plant sequences (see **Web Topic 21.6**). The grouping of the seven Arabidopsis *IPT* genes in this unique plant clade provided a clue that these genes may encode the cytokinin biosynthetic enzyme.

The proteins encoded by these genes were expressed in *E. coli* and analyzed. It was found that, with the exception of the gene most closely related to the animal tRNA-*IPT* genes, these genes encoded proteins capable of synthesizing free cytokinins. Unlike their bacterial counterparts, however, the Arabidopsis enzymes that have been analyzed utilize adenosine triphosphate (ATP) and adenosine diphosphate (ADP) preferentially over AMP, and DMAPP rather than HMBDP (see Figure 21.6).

Cytokinins from the root are transported to the shoot via the xylem

Root apical meristems are a major site of synthesis of the free cytokinins in whole plants. The cytokinins synthesized in roots appear to move through the xylem into the shoot, along with the water and minerals taken up by the roots. This pathway of cytokinin movement has been inferred from the analysis of xylem exudate.

When the shoot is cut from a rooted plant near the soil line, the xylem sap may continue to flow from the cut stump for some time. This xylem exudate contains cytokinins. If the soil covering the roots is kept moist, the flow of xylem exudate can continue for several days. Because the cytokinin content of the exudate does not diminish, the cytokinins found in it are likely to be synthesized by the roots. In addition, environmental factors that interfere with root function, such as water stress, reduce the cytokinin content of the xylem exudate (Itai and Vaadia 1971). Conversely, resupply of nitrate to nitrogen-starved maize roots results in an elevation of the concentration of cytokinins in the xylem sap (Samuelson et al. 1992), which has been correlated to an induction of cytokinin-regulated gene expression in the shoots (Takei et al. 2001b).

Although the presence of cytokinin in the xylem is well established, grafting experiments have cast doubt on the presumed role of this root-derived cytokinin in shoot development. Tobacco plants transformed with an inducible *ipt* gene from *Agrobacterium* display increased lateral bud outgrowth and delayed senescence. To assess the role of cytokinin from the root, the tobacco root stock engineered to overproduce cytokinin was grafted to a wild-type shoot. Surprisingly, no phenotypic consequences were observed in the shoot, even though an increased concentration of cytokinin was measured in the transpiration stream (Faiss et al. 1997). Thus the excess cytokinin in the roots had no effect on the grafted shoot.

Roots are not the only parts of the plant capable of synthesizing cytokinins. Analysis of the expression of the *IPT* gene family in Arabidopsis has provided new insights into our understanding of the spatial pattern of cytokinin biosynthesis (Miyawaki et al. 2004). The seven Arabidopsis *IPT* genes are expressed in unique patterns during development in various tissues, including xylem precursor cell files in the root tip, phloem, leaf axils, ovules, immature seeds, root primordia, columella root cap cells, the upper part of young inflorescences, and fruit abscission zones. These results suggest that many different tissues are capable of synthesizing cytokinins. The expression of a subset of *IPT* genes was down-regulated by cytokinin, indicating that cytokinin exerts a negative feedback control on its own biosynthesis. A different subset of *IPT* genes was up-regulated by auxin treatment. Thus, cytokinin biosynthesis may be regulated by the interaction of the two hormones.

A signal from the shoot regulates the transport of zeatin ribosides from the root

The cytokinins in the xylem exudate are mainly in the form of zeatin ribosides. Once they reach the leaves, some of these nucleosides are converted to the free-base form or to glucosides (Noodén and Letham 1993). Cytokinin glucosides may accumulate to high levels in seeds and in leaves, and substantial amounts may be present even in senescing leaves. Although the glucosides are active as cytokinins in bioassays, often they lack hormonal activity after they form within cells, possibly because they are compartmentalized in such a way that they are unavailable. Compartmentation may explain the conflicting observations that cytokinins are transported readily by the xylem but that radioactive cytokinins applied to leaves in intact plants do not appear to move from the site of application.

Evidence from grafting experiments with mutants suggests that the transport of zeatin riboside from the root to the shoot is regulated by signals from the shoot. The *rms4* mutant of pea (*Pisum sativum* L.) is characterized by a 40-fold decrease in the concentration of zeatin riboside in the xylem sap of the roots. However, grafting a wild-type shoot onto an *rms4* mutant root increased the zeatin riboside levels in the xylem exudate to wild-type levels. Conversely, grafting an *rms4* mutant shoot onto a wild-type root lowered the concentration of zeatin riboside in the xylem exudate to mutant levels (Beveridge et al. 1997).

These results suggest that a signal from the shoot can regulate cytokinin transport from the root. Although the identity of this signal has not yet been determined, more recent studies have shown that it is probably not auxin (Beveridge 2000).

Cytokinins are rapidly metabolized by plant tissues

Free cytokinins are readily converted to their respective nucleoside and nucleotide forms. Such interconversions likely involve enzymes common to purine metabolism.

Many plant tissues contain the enzyme **cytokinin oxidase**, which cleaves the side chain from zeatin (both *cis* and *trans*), zeatin riboside, iP, and their *N*-glucosides, but not their *O*-glucoside derivatives (Figure 21.7). However, dihydrozeatin and its conjugates are resistant to cleavage.

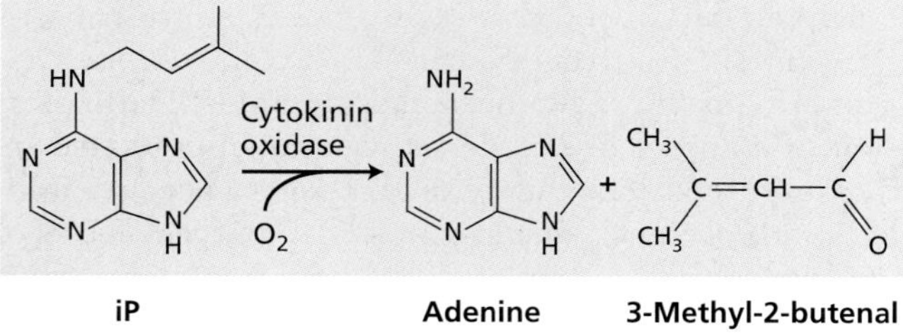

FIGURE 21.7 Cytokinin oxidase irreversibly degrades some cytokinins.

Cytokinin oxidase irreversibly inactivates cytokinins, and it could be important in regulating or limiting cytokinin effects. The activity of the enzyme is induced by high cytokinin concentrations, due at least in part to an elevation of the RNA levels for a subset of the genes.

A gene encoding cytokinin oxidase was first identified in maize (Houba-Herin et al. 1999; Morris et al. 1999). In Arabidopsis, cytokinin oxidase is encoded by a multigene family whose members show distinct patterns of expression. Interestingly, several of the genes contain putative secretory signals, suggesting that at least some of these enzymes may be extracellular.

Cytokinin levels can also be regulated by conjugation of the hormone at various positions. The nitrogens at the 3, 7, and 9 positions of the adenine ring of cytokinins can be conjugated to glucose residues. Alanine can also be conjugated to the nitrogen at the 9 position, forming lupinic acid. These modifications are generally irreversible, and such conjugated forms of cytokinin are inactive in bioassays, with the exception of the N^3-glucosides.

The hydroxyl group of the side chain of cytokinins is also the target for conjugation to glucose residues, or in some cases xylose residues, yielding *O*-glucoside and *O*-xyloside cytokinins. *O*-glucosides are resistant to cleavage by cytokinin oxidases, which may explain why these derivatives have higher biological activity in some assays than their corresponding free bases have.

Enzymes that catalyze the conjugation of either glucose or xylose to zeatin have been purified, and their respective genes have been cloned (Martin et al. 1999). These enzymes have stringent substrate specificities for the sugar donor and the cytokinin bases. Only free zeatin and dihydrozeatin bases are efficient substrates; the corresponding nucleosides are not substrates. The specificity of these enzymes suggests that the conjugation to the side chain is precisely regulated.

The conjugations at the side chain can be removed by glucosidase enzymes to yield free cytokinins, which, as discussed earlier, are the active forms. Thus, cytokinin glucosides may be a storage form, or metabolically inactive state, of these compounds. A gene encoding a glucosidase that can release cytokinins from sugar conjugates has been cloned from maize, and its expression could play an important role in the germination of maize seeds (Brzobohaty et al. 1993).

Dormant seeds often have high levels of cytokinin glucosides but very low levels of hormonally active free cytokinins. Levels of free cytokinins increase rapidly, however, as germination is initiated, and this increase in free cytokinins is accompanied by a corresponding decrease in cytokinin glucosides.

The Biological Roles of Cytokinins

Although discovered as a cell division factor, cytokinins can stimulate or inhibit a variety of physiological, metabolic, biochemical, and developmental processes when they are applied to higher plants, and it is increasingly clear that endogenous cytokinins play an important role in the regulation of these events in the intact plant.

In this section we will survey some of the diverse effects of cytokinin on plant growth and development, and discuss its role in regulating cell division. Several approaches have helped illuminate the roles that cytokinins play in growth and development. The discovery of the tumor-inducing Ti plasmid in the plant-pathogenic bacterium *A. tumefaciens* has provided plant scientists with a powerful new tool for introducing foreign genes into plants, and for studying the role of cytokinin in development. A second approach is to examine the phenotypes of mutants altered in cytokinin perception and signaling. Finally, overexpression of cytokinin oxidase enzymes can be used to decrease in vivo cytokinin levels experimentally. In addition to its role in cell proliferation, cytokinin affects many other processes, including vascular development, apical dominance, and leaf senescence.

Cytokinins regulate cell division in shoots and roots

As discussed earlier, cytokinins are generally required for cell division of plant cells in vitro. Several lines of evidence suggest that cytokinins also play key roles in the regulation of cell division in vivo.

Much of the cell division in an adult plant occurs in the meristems (see Chapter 16). Localized expression of the *ipt* gene of *Agrobacterium* in somatic sectors of tobacco leaves causes the formation of ectopic (abnormally located) meristems, indicating that elevated levels of cytokinin are sufficient to initiate cell divisions in these leaves (Estruch et al. 1991). Elevation of endogenous cytokinin levels in transgenic Arabidopsis results in overexpression of the KNOTTED homeobox transcription factor homologs *KNAT1* and *STM*—genes that are important in the regulation of meristem function (see Chapter 16) (Rupp et al. 1999). Interestingly, overexpression of *KNAT1* also appears to elevate cytokinin levels in transgenic tobacco, suggesting an interdependent relationship between *KNAT* and the level of cytokinins.

Overexpression of several of the Arabidopsis cytokinin oxidase genes in tobacco results in a reduction of endogenous cytokinin levels and a consequent strong retardation of shoot development due to a reduction in the rate of cell proliferation in the shoot apical meristem (Figures 21.8 and 21.9) (Werner et al. 2001). This finding strongly supports the notion that endogenous cytokinins regulate cell division in vivo.

FIGURE 21.8 Tobacco plants overexpressing the gene for cytokinin oxidase. The plant on the left is wild type. The two plants on the right are overexpressing two different constructs of the Arabidopsis gene for cytokinin oxidase: *AtCKX1* and *AtCKX2*. Shoot growth is strongly inhibited in the transgenic plants. (From Werner et al. 2001.)

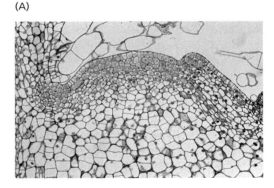

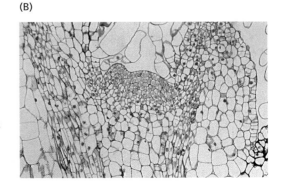

FIGURE 21.9 Cytokinin is required for normal growth of the shoot apical meristem. (A) Longitudinal section through the shoot apical meristem of a wild-type tobacco plant. (B) Longitudinal section through the shoot apical meristem of a transgenic tobacco overexpressing the gene that encodes cytokinin oxidase (*AtCKX1*). Note the reduction in the size of the apical meristem in the cytokinin-deficient plant. (From Werner et al. 2001.)

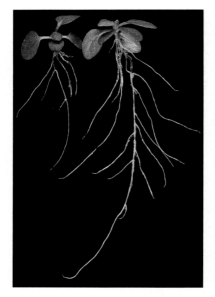

FIGURE 21.10 Cytokinin suppresses the growth of roots. The cytokinin-deficient *AtCKX1* roots (right) are larger than those of the wild-type tobacco plant (left). (From Werner et al. 2001.)

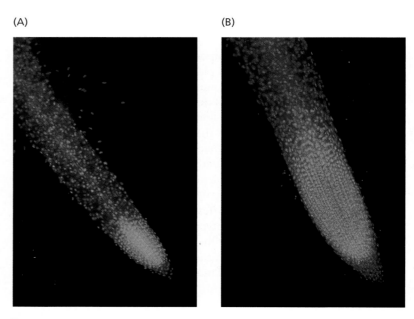

FIGURE 21.11 Cytokinin suppresses the size and cell division activity of roots. (A) Wild type. (B) *AtCKX1*. These roots were stained with the fluorescent dye 4′,6-diamidino-2-phenylindole (DAPI), which stains the nucleus. (From Werner et al. 2001.)

Surprisingly, the same overexpression of cytokinin oxidase in tobacco led to an *enhancement* of root growth (Figure 21.10), primarily by increasing the size of the root apical meristem (Figure 21.11). This result may indicate that cytokinins play opposite roles in regulating cell proliferation in root and shoot meristems.

Analyses of mutations of genes in the cytokinin signaling pathway also support a role for cytokinin in regulating meristem function. Disruption of all three cytokinin receptors in Arabidopsis results in cytokinin-insensitive plants that exhibit reduced cell division in *both* the root and shoot apical meristems (Higuchi et al. 2004; Nishimura et al. 2004). The roots of these triple-receptor mutants elongate for only a few days and then cease growth, a phenotype correlated with decreased cell division in the apical meristem. In contrast, mutants that are only partially insensitive to exogenous cytokinin (e.g., some single-receptor mutants) display an increase in root growth similar to that observed in plants overexpressing cytochrome oxidase.

How can we reconcile the observations that reduction of cytokinin levels or signaling increases root growth, yet the triple-receptor mutant shows a strong reduction in root growth? One possibility is that the dose response curve for cytokinin, like that for auxin (see Chapter 19), is bell-shaped, and that the endogenous cytokinin level in the root is supraoptimal and therefore inhibitory. In this case, a small reduction in cytokinin levels (or signaling) would increase root growth. On the other hand, if cytokinin levels or signaling fell below a critical threshold, a decrease in growth would result (Figure 21.12). By the same logic, cytokinin levels in the shoot should be optimal, since a reduction in cytokinin levels caused a reduction in size of the apical meristem.

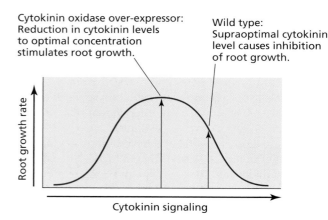

FIGURE 21.12 Hypothetical bell-shaped cytokinin response curve for root growth. The horizontal axis represents the amount of cytokinin signaling, and the vertical axis represents root growth rate. In the wild type, endogenous cytokinin levels are supraoptimal and growth is inhibited. Lowering endogenous cytokinin levels by overexpressing cytokinin oxidase genes lowers the cytokinin signaling output, resulting in increased root growth. Further reduction below the optimal amount of cytokinin signaling results in decreased root growth. (Data courtesy of F. J. Ferreira and J. J. Kieber.)

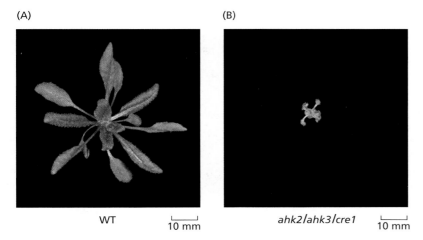

FIGURE 21.13 Comparison of the rosettes of WT Arabidopsis and the triple cytokinin receptor–knockout mutant, *ahk2/ahk3/cre1*. (From Nishimura et al. 2004.)

The shoot apical meristem of the triple-receptor mutants also displays decreased cell proliferation, leading to a stunted shoot and little or no flower production (Figure 21.13). In contrast, disruption of certain *negative regulators* of cytokinin signaling (such as type-A ARRs, discussed at the end of the chapter) results in an *increase* in cell proliferation in the inflorescence meristem of Arabidopsis and in the shoot apical meristem of maize (Figure 21.14). (Recall that disruption of a negative regulator would have a stimulatory effect on cytokinin signaling.) Interestingly, the enlargement of the apical meristems alters the phyllotaxy of the inflorescence or stem: the arrangement on the shoot of flowers in Arabidopsis and of leaves in maize (Giulini et al. 2004) (see Figure 21.14). Finally, overexpression of an activated form of the negative regulators in Arabidopsis results in a strong suppression of cell proliferation in the shoot apical meristem, presumably as a result of the constitutive inhibition of the cytokinin response pathway. Together these data are consistent with the notion that negative regulators, and by inference, cytokinins, play a key role in regulating cell proliferation and phyllotaxy in the shoot meristem.

Humans have unwittingly taken advantage of the promotive effect of cytokinin on the root apical meristem during the breeding of cultivated rice varieties. The rice varieties *japonica* and *indica* differ dramatically in their yield, with the latter generally producing many more grains in their main panicle and ultimately a higher yield. An analysis of a cross between these two varieties has lead to the identification of *q*uantitative *t*rait *l*oci (QTLs) responsible for this difference in grain number (Ashikari et al. 2005). One of the QTLs most important in increasing grain number in *indica* varieties has been linked to a decrease in the function of a cytokinin oxidase gene. As a consequence of the reduced function of this cytokinin oxidase in the *indica* varieties, cytokinin levels are higher in the inflorescence, which alters the inflorescence meristem such that it produces more reproductive organs.

Cytokinins regulate specific components of the cell cycle

Cytokinins regulate cell division by affecting the controls that govern the passage of the cell through the cell division

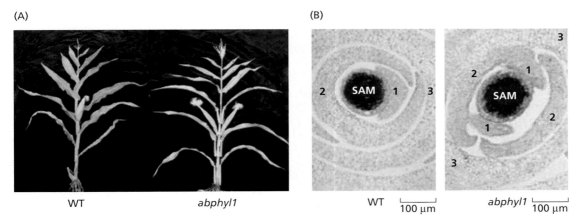

FIGURE 21.14 A type-A loss-of-function mutation in maize alters leaf phyllotaxy and the size of the shoot apical meristem. The *ABPHYL1* gene encodes a type-A response regulator. (A) Wild-type (WT) maize displays an alternate arrangement of leaves, but the type-A loss-of-function mutant *abphyl1* exhibits opposite leaf phyllotaxy. (B) The *abphyl1* mutant shows increased shoot apical meristem (SAM) size; numbers refer to individual leaves and demonstrate the difference in phyllotaxy. The switch from alternate to opposite phyllotaxy in the mutant is the result of the enlargement of the SAM, and the enlargement of the SAM is likely caused by an increase in cytokinin signaling in the apical meristem. (From Giulini et al. 2004.)

cycle. Zeatin levels peak in synchronized culture tobacco cells at the end of S phase, the G2/M phase transition, and in late G_1. Inhibition of cytokinin biosynthesis blocks cell division, and application of exogenous cytokinin allows cell division to proceed.

Cytokinins were discovered in relation to their ability to stimulate cell division in tissues supplied with an optimal level of auxin. Evidence suggests that both auxin and cytokinins participate in regulating the cell cycle and that they do so by controlling the activity of cyclin-dependent kinases. As discussed in Chapter 1, **cyclin-dependent protein kinases** (**CDKs**), in concert with their regulatory subunits, the **cyclins**, are enzymes that regulate the eukaryotic cell cycle.

The expression of the gene that encodes the major CDK, Cdc2 (*c*ell *d*ivision *c*ycle 2), is regulated by auxin (see Chapter 19). In pea root tissues, *CDC2* mRNA is induced within 10 minutes after treatment with auxin, and in tobacco pith, high levels of CDK are induced when the tissue is cultured on medium containing auxin (John et al. 1993). However, the CDK induced by auxin is enzymatically inactive, and high levels of CDK alone are not sufficient to permit cells to divide.

Cytokinin has been linked to the activation of a Cdc25-like phosphatase, whose role is to remove an inhibitory phosphate group from the Cdc2 kinase (Zhang et al. 1996). This action of cytokinin provides one potential link between cytokinin and auxin in the cell cycle: regulating the passage from G_2 to M phase.

Cytokinins also elevate the expression of the *CYCD3* gene, which encodes a *D-type cyclin* (Soni et al. 1995; Riou-Khamlichi et al. 1999). In Arabidopsis, *CYCD3* is expressed in proliferating tissues such as shoot meristems and young leaf primordia. Overexpression of *CYCD3* can bypass the cytokinin requirement for cell proliferation in culture (Figure 21.15) (Riou-Khamlichi et al. 1999). These data suggest that a major mechanism for cytokinin's ability to stimulate cell division is its increase of *CYCD3* function. This is reminiscent of animal cell cycles, in which D-type cyclins are regulated by a wide variety of growth factors and play a key role in regulating the passage through the restriction point of the cell cycle in G_1.

The auxin:cytokinin ratio regulates morphogenesis in cultured tissues

Shortly after the discovery of kinetin, it was observed that the differentiation of cultured callus tissue derived from tobacco pith segments into either roots or shoots depends on the ratio of auxin to cytokinin in the culture medium. Whereas high auxin:cytokinin ratios stimulated the formation of roots, low auxin:cytokinin ratios led to the formation of shoots. At intermediate levels the tissue grew

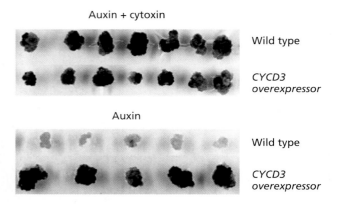

FIGURE 21.15 *CYCD3*-expressing callus cells can divide in the absence of cytokinin. Leaf explants from transgenic Arabidopsis plants expressing *CYCD3* under a cauliflower mosaic virus 35S promoter were induced to form calluses through culturing in the presence of auxin plus cytokinin or auxin alone. The wild-type control calluses required cytokinin to grow. The *CYCD3*-expressing calluses grew well on medium containing auxin alone. The photographs were taken after 29 days. (From Riou-Khamlichi et al. 1999.)

as an undifferentiated callus (Figure 21.16) (Skoog and Miller 1965).

The effect of auxin:cytokinin ratios on morphogenesis can also be seen in crown gall tumors by mutation of the T-DNA of the *Agrobacterium* Ti plasmid (Garfinkel et al. 1981). Mutating the *ipt* gene (the *tmr* locus) of the Ti plasmid

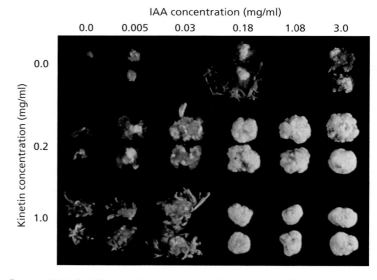

FIGURE 21.16 The regulation of growth and organ formation in cultured tobacco callus at different concentrations of auxin and kinetin. At low auxin and high kinetin concentrations (lower left), buds developed. At high auxin and low kinetin concentrations (upper right), roots developed. At intermediate or high concentrations of both hormones (middle and lower right), undifferentiated callus developed. (Courtesy of Donald Armstrong.)

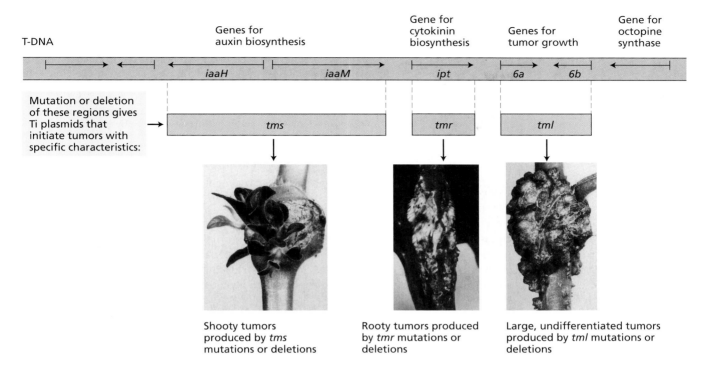

FIGURE 21.17 Map of the T-DNA from an *Agrobacterium* Ti plasmid, showing the effects of T-DNA mutations on crown gall tumor morphology. The genes *iaaH* and *iaaM* encode the two enzymes involved in auxin biosynthesis, *ipt* encodes a cytokinin biosynthesis enzyme, and *6b* encodes a transcriptional regulator that controls tumor growth. Mutations in these genes produce the phenotypes illustrated. (From Morris 1986, courtesy of R. Morris.)

blocks zeatin biosynthesis in the infected cells. The resulting high auxin:cytokinin ratio in the tumor cells causes the proliferation of roots instead of undifferentiated callus tissue. In contrast, mutating either of the genes for auxin biosynthesis (*tms* locus) lowers the auxin:cytokinin ratio and stimulates the proliferation of shoots (Figure 21.17) (Akiyoshi et al. 1983). These partially differentiated tumors are known as teratomas.

Mutations in genes at the *tml* locus result in large, undifferentiated tumors. One of the genes at this locus, *6b*, encodes a transcriptional regulator that controls tumor growth (see **Web Essay 22.1**). Tobacco cells transformed with the *6b* gene have elevated levels of flavonoids (Galis et al. 2004). The function of the *6a* gene is to facilitate the transport of nopaline and octopine.

Cytokinins modify apical dominance and promote lateral bud growth

One of the primary determinants of plant form is the degree of apical dominance (see Chapter 19). Plants with strong apical dominance, such as maize, have a single growing axis with few lateral branches. In contrast, many lateral buds initiate growth in shrubby plants. Branching patterns are normally determined by light, nutrients, and genotype. As discussed in Chapter 19, branching is also triggered by decapitation, a phenomenon that allows primary growth to continue after the loss of the terminal bud.

Physiologically, branching is regulated by a complex interplay of hormones, including auxin, cytokinin, and an as-yet unidentified root-derived signal. Auxin transported polarly from the apical bud suppresses the growth of axillary buds (see Chapters 16 and 19). In contrast, cytokinin stimulates cell division activity and outgrowth when applied directly to the axillary buds of many species, and cytokinin-overproducing mutants tend to be bushy. Recently, it was demonstrated in lateral buds that auxin inhibits the expression of a subset of *IPT* genes that encode the enzyme catalyzing the first committed step in cytokinin biosynthesis, thus providing a mechanistic link between these two hormones in regulating bud growth. This result and other data suggest that the cytokinins responsible for axillary bud growth are synthesized in the bud itself, not transported from the root.

Evidence for a root-derived signaling molecule that is involved in the suppression of branching in the shoot comes from grafting studies in mutants with highly branched phenotypes. The branching genes *MAX4* (*m*ore *ax*illary growth4) and *RMS1* (*RAMOSUS1*) in Arabidopsis and pea, respectively, are involved in the production of this root-derived signal, which appears to be a carotenoid derivative (Sorefan et al. 2003; Foo et al. 2005).

Cytokinins induce bud formation in a moss

Thus far we have restricted our discussion of plant hormones to the angiosperms. However, many plant hormones

(A) (B) (C) (D)

FIGURE 21.18 Bud formation in the moss *Funaria* begins with the formation of a protuberance at the apical ends of certain cells in the protonema filament. A–D show various stages of bud development. Once formed, the bud goes on to produce the leafy gametophyte stage of the moss. (Courtesy of K. S. Schumaker.)

are present and developmentally active in representative species throughout the plant kingdom. The moss *Funaria hygrometrica* is a well-studied example. The germination of moss spores gives rise to a filament of cells called a **protonema** (plural **protonemata**). The protonema elongates and undergoes cell divisions at the tip, and it forms branches some distance back from the tip (see **Web Essay 21.2**).

The transition from filamentous growth to leafy growth begins with the formation of a swelling or protuberance near the apical ends of specific cells (Figure 21.18). An asymmetric cell division follows, creating the **initial cell**. The initial cell then divides mitotically to produce the **bud**, the structure that gives rise to the leafy gametophyte. During normal growth, buds and branches are regularly initiated, usually beginning at the third cell from the tip of the filament.

Light, especially red light, is required for bud formation in *Funaria*. In the dark, buds fail to develop, but cytokinin added to the medium can substitute for the light requirement. Cytokinin not only stimulates normal bud development, it also increases the total number of buds (Figure 21.19). Even very low levels of cytokinin (picomolar, or 10^{-12} M) can stimulate the first step in bud formation: the swelling at the apical end of the specific protonemal cell.

Cytokinin overproduction has been implicated in genetic tumors

Many species in the genus *Nicotiana* can be crossed to generate interspecific hybrids. More than 300 such interspecific hybrids have been produced; 90% of these hybrids are normal, exhibiting phenotypic characteristics intermediate between those of both parents. The plant used for cigarette tobacco, *Nicotiana tabacum*, for example, is an interspecific hybrid. About 10% of these interspecific crosses result in progeny that tend to form spontaneous tumors called **genetic tumors** (Figure 21.20) (Smith 1988).

Genetic tumors are similar morphologically to those induced by *A. tumefaciens*, discussed at the beginning of this chapter, but genetic tumors form spontaneously in the absence of any external inducing agent. The tumors are composed of masses of rapidly proliferating cells in regions of the plant that ordinarily would contain few dividing cells. Furthermore, the cells divide without differentiating into the cell types normally associated with the tissues giving rise to the tumor.

Nicotiana hybrids that produce genetic tumors have abnormally high levels of both auxin and cytokinins. Typ-

(A)
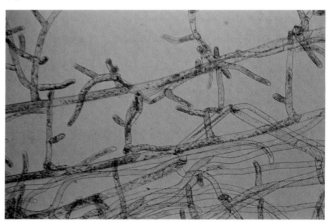

(B)
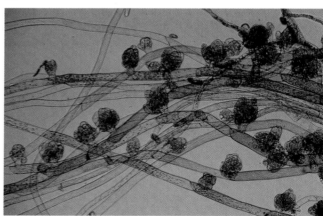

FIGURE 21.19 Cytokinin stimulates bud development in *Funaria*. (A) Control protonemal filaments. (B) Protonemal filaments treated with benzyladenine. (Courtesy of H. Kende.)

FIGURE 21.20 Expression of genetic tumors in the hybrid *Nicotiana langsdorffii* × *N. glauca*. (From Smith 1988.)

ically, the cytokinin levels in tumor-prone hybrids are five to six times higher than those found in either parent. Combinations of particular genes from each parent are responsible for the aberrant phenotype. Recent studies have shown that suppression of a receptor-like kinase, *CHRK1* (*ch*intinase-*r*elated receptor-like *k*inase*1*), in tobacco results in a mutant plant with characteristics that mimic those of the tumor-producing interspecific hybrids, including the elevated cytokinin levels (Lee et al. 2004). It is thus possible that genetic tumors are caused by the inactivation of a signaling pathway that occurs when certain genomes are mixed during hybridization.

Cytokinins delay leaf senescence

Leaves detached from the plant slowly lose chlorophyll, RNA, lipids, and protein, even if they are kept moist and provided with minerals. This programmed aging process leading to death is termed **senescence** (see Chapters 16 and 23). Leaf senescence is more rapid in the dark than in the light. Treating isolated leaves of many species with cytokinins will delay their senescence.

Although applied cytokinins do not prevent senescence completely, their effects can be dramatic, particularly when the cytokinin is sprayed directly on the intact plant. If only one leaf is treated, it remains green after other leaves of similar developmental age have yellowed and dropped off the plant. Even a small spot on a leaf will remain green if treated with a cytokinin, after the surrounding tissues on the same leaf begin to senesce. This can also be observed in the galls produced by jumping plant lice (*Pachypsylla celtidis-mamma*) on hackberry (*Celtis occidentalis* L) leaves. These galls produce green islands on an otherwise yellow, senescent leaf, which is correlated with a large increase in cytokinin content in the affected tissue.

Unlike young leaves, mature leaves produce little if any cytokinin. Mature leaves may depend on root-derived cytokinins to postpone their senescence. Senescence is initiated in soybean leaves by seed maturation—a phenomenon known as *monocarpic senescence*—and can be delayed by seed removal. Although the seedpods control the onset of senescence, they do so by controlling the delivery of root-derived cytokinins to the leaves.

The cytokinins involved in delaying senescence are primarily zeatin riboside and dihydrozeatin riboside, which may be transported into the leaves from the roots through the xylem, along with the transpiration stream (Noodén et al. 1990).

To test the role of cytokinin in regulating the onset of leaf senescence, tobacco plants were transformed with a chimeric gene in which a senescence-specific promoter was used to drive the expression of the Argrobacterial *ipt* gene (Gan and Amasino 1995). The transformed plants had wild-type levels of cytokinins and developed normally, up to the onset of leaf senescence.

As the leaves aged, however, the senescence-specific promoter was activated, triggering the expression of the *ipt* gene within leaf cells just as senescence would have been initiated. The resulting elevated cytokinin levels not only blocked senescence, but also limited further expression of the *ipt* gene, preventing cytokinin overproduction (Figure 21.21). This result suggests that cytokinins are a natural regulator of leaf senescence.

Cytokinins promote movement of nutrients

Cytokinins influence the movement of nutrients into leaves from other parts of the plant, a phenomenon known as *cytokinin-induced nutrient mobilization*. This process can be observed when nutrients (sugars, amino acids, and so on) radiolabeled with ^{14}C or ^{3}H are fed to plants after one leaf or part of a leaf is treated with a cytokinin. Subsequently the whole plant is subjected to autoradiography to reveal the pattern of movement and the sites at which the labeled nutrients accumulate.

Experiments of this nature have demonstrated that nutrients are preferentially transported to, and accumulated in, the cytokinin-treated tissues. It has been postulated that the hormone causes nutrient mobilization by creating a new source–sink relationship. As discussed in Chapter 10, nutrients translocated in the phloem move from a site of production or storage (the source) to a site of utilization (the sink). The metabolism of the treated area may be stimulated by the hormone so that nutrients move toward it. However, it is not necessary for the nutrient itself to be metabolized in

Figure 21.21 Leaf senescence is retarded in a transgenic tobacco plant containing a cytokinin biosynthesis gene, *ipt* from *Agrobacterium tumefaciens*. The *ipt* gene is expressed in response to signals that induce senescence. (From Gan and Amasino 1995, courtesy of R. Amasino.)

Plant expressing *ipt* gene remains green and photosynthetic

Age-matched control shows advanced senescence, no photosynthesis

the sink cells because even nonmetabolizable substrate analogs are mobilized by cytokinins (Figure 21.22).

Cytokinin levels are directly proportional to the level of nutrients to which plants are exposed. For example, application of nitrate to nitrogen-depleted maize or Arabidopsis seedlings results in a rapid rise in cytokinin levels in the roots, which are then mobilized to the shoots via the xylem. This increase is due, at least in part, to an induction of expression of one member of the *IPT* gene family. Cytokinin also has been implicated in the response to altered phosphate and sulfate levels.

Thus, the nutrient status of the plant regulates cytokinin levels, and in turn the relative ratio of cytokinin to auxin determines the relative growth rates of roots and shoots: High cytokinin levels promote shoot growth, and, conversely, high auxin levels promote root growth. At low nutrient levels, cytokinin levels are low, resulting in an increase in root growth, allowing the plant to more effectively acquire limiting nutrients present in the soil. In contrast, soils with optimal nutrients favor high levels of shoot growth because of increased cytokinin levels, thus maximizing photosynthetic capacity.

Cytokinins promote chloroplast development

Although seeds can germinate in the dark, the morphology of dark-grown seedlings is very different from that of

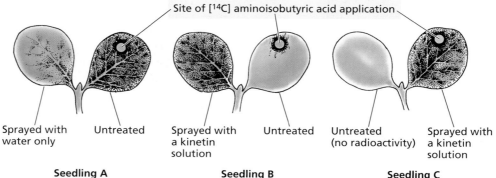

Figure 21.22 The effect of cytokinin on the movement of an amino acid in cucumber seedlings. A radioactively labeled amino acid that cannot be metabolized, such as aminoisobutyric acid, was applied as a discrete spot on the right cotyledon of each of these seedlings. (Drawn from data obtained by K. Mothes.)

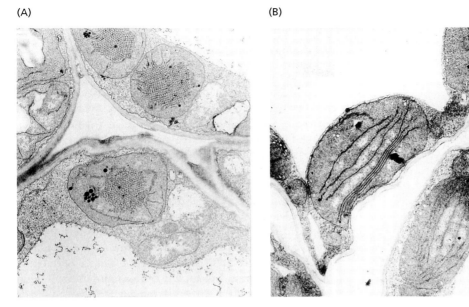

FIGURE 21.23 Cytokinin influence on the development of wild-type Arabidopsis seedlings grown in darkness. (A) Plastids develop as etioplasts in the untreated, dark grown control. (B) Cytokinin treatment resulted in thylakoid formation in the plastids of dark-grown seedlings. (From Chory et al. 1994, courtesy of J. Chory.)

light-grown seedlings (see Chapter 17): Dark-grown seedlings are said to be **etiolated**. The hypocotyl and internodes of etiolated seedlings are more elongated, cotyledons and leaves do not expand, and chloroplasts do not mature. Instead of maturing as chloroplasts, the proplastids of dark-grown seedlings develop into **etioplasts**, which do not synthesize chlorophyll or most of the enzymes and structural proteins required for the formation of the chloroplast thylakoid system and photosynthesis machinery. When seedlings germinate in the light, chloroplasts mature directly from the proplastids present in the embryo, but etioplasts also can mature into chloroplasts when etiolated seedlings are illuminated.

If etiolated leaves are treated with cytokinin before being illuminated, they form chloroplasts with more extensive grana, and chlorophyll and photosynthetic enzymes are synthesized at a greater rate upon illumination (Figure 21.23). These results suggest that cytokinins—along with other factors, such as light, nutrition, and development—regulate the synthesis of photosynthetic pigments and proteins. The ability of exogenous cytokinin to enhance de-etiolation of dark-grown seedlings is mimicked by certain mutations that lead to cytokinin overproduction. (For more on how cytokinins promote light-mediated development, see **Web Topic 21.7**.)

Cytokinins promote cell expansion in leaves and cotyledons

The promotion of cell enlargement by cytokinins is most clearly demonstrated in the cotyledons of dicots with leafy cotyledons, such as mustard, cucumber, and sunflower. The cotyledons of these species expand as a result of cell enlargement during seedling growth. Cytokinin treatment promotes additional cell expansion, with no increase in the dry weight of the treated cotyledons.

Leafy cotyledons expand to a much greater extent when the seedlings are grown in the light than in the dark, and cytokinins promote cotyledon growth in both light- and dark-grown seedlings (Figure 21.24). As with auxin-induced growth, cytokinin-stimulated expansion of radish cotyledons is associated with an increase in the mechanical extensibility of the cell walls. However, cytokinin-induced wall loosening is not accompanied by proton extrusion. Neither auxin nor gibberellin promotes cell expansion in cotyledons.

Cytokinin-regulated processes are revealed in plants that overproduce cytokinins

The *ipt* gene from the *Agrobacterium* Ti plasmid has been introduced into many species of plants, resulting in cytokinin overproduction. These transgenic plants exhibit an array of developmental abnormalities that tell us a great deal about the biological role of cytokinins.

As discussed earlier, plant tissues transformed by *Agrobacterium* carrying a wild-type Ti plasmid proliferate as tumors as a result of the overproduction of both auxin and cytokinin. And as mentioned already, if all of the other genes in the T-DNA are deleted and plant tissues are transformed with T-DNA containing only a selective antibiotic resistance marker gene and the *ipt* gene, shoots proliferate instead of callus.

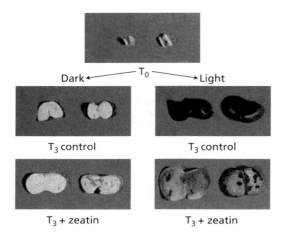

FIGURE 21.24 The effect of cytokinin on the expansion of radish cotyledons. The experiment described here shows that the effects of light and cytokinin are additive. T_0 represents germinating radish seedlings before the experiment began. The detached cotyledons were incubated for 3 days (T_3) in either darkness or light with or without 2.5 mM zeatin. In both the light and the dark, zeatin-treated cotyledons expanded more than in the control. (From Huff and Ross 1975.)

- The more extreme cytokinin-overproducing plants are stunted, with greatly shortened internodes.
- Rooting of stem cuttings is reduced, as is the root growth rate.

Some of the consequences of cytokinin overproduction could be highly beneficial for agriculture if synthesis of the hormone can be controlled. Because leaf senescence is delayed in the cytokinin-overproducing plants, it should be possible to extend their photosynthetic productivity. Indeed, when the *ipt* gene is expressed in lettuce from a senescence-inducible promoter, leaf senescence is strongly retarded (Figure 21.25), similar to the results observed in tobacco (see Figure 21.21).

In addition, cytokinin production could be linked to damage caused by predators. For example, tobacco plants transformed with an *ipt* gene under the control of the promoter from a wound-inducible protease inhibitor II gene were more resistant to insect damage. The tobacco hornworm consumed up to 70% fewer tobacco leaves in plants that expressed the *ipt* gene driven by the protease inhibitor promoter (Smigocki et al. 1993).

The shoot teratomas formed by *ipt*-transformed tissues are difficult to root, and when roots are formed, they tend to be stunted in their growth. As a result, it is difficult to obtain plants from shoots expressing the *ipt* gene under the control of its own promoter, because the promoter is a constitutive promoter and the gene is continuously expressed.

To circumvent this problem, a variety of promoters whose expression can be regulated have been used to drive the expression of the *ipt* gene in the transformed tissues. For example, several studies have employed a heat shock promoter, which is induced in response to elevated temperature, to drive inducible expression of the *ipt* gene in transgenic tobacco and Arabidopsis. In these plants, heat induction substantially increased the level of zeatin, zeatin riboside and ribotide, and N-conjugated zeatin.

These cytokinin-overproducing plants exhibit several characteristics that point to roles played by cytokinin in plant physiology and development:

- The shoot apical meristems of cytokinin-overproducing plants produce more leaves.
- The leaves have higher chlorophyll levels and are much greener.
- Adventitious shoots may form from unwounded leaf veins and petioles.
- Leaf senescence is retarded.
- Apical dominance is greatly reduced.

FIGURE 21.25 Leaf senescence is retarded in transgenic lettuce plants expressing the cytokinin biosynthesis gene *ipt* at the time of senescence. Azygous plants (upper five) lack the transgene; *SAG12-IPT* plants (lower five) utilize a "senescence-associated gene" promoter (*SAG12*) to drive the expression of *ipt* at the onset of senescence. (From McCabe et al. 2001.)

Cellular and Molecular Modes of Cytokinin Action

The diversity of the effects of cytokinin on plant growth and development is consistent with the involvement of signal transduction pathways with branches leading to specific responses. As in the case of phytochrome (see Chapter 17), cytokinin perception and signaling is mediated by a two-component system, a response pathway remarkably similar to those utilized by bacteria to sense and respond to environmental changes.

A cytokinin receptor related to bacterial two-component receptors has been identified

The first clue to the nature of the cytokinin receptor came from the discovery of the *CKI1* gene. *CKI1* was identified in a screen for genes that, when overexpressed, conferred cytokinin-independent growth on Arabidopsis cells in culture (Kakimoto, 2001). As discussed previously, plant cells generally require cytokinin to divide in culture. However, a cell line that overexpresses *CKI1* is capable of growing in culture in the absence of added cytokinin.

CKI1 encodes a protein similar in sequence to bacterial two-component sensor histidine kinases, which are ubiquitous receptors in prokaryotes (see Chapter 14 on the web site and Chapter 17). Bacterial two-component regulatory systems mediate a range of responses to environmental stimuli, such as osmoregulation and chemotaxis. Typically these systems are composed of two functional elements: a *sensor histidine kinase*, to which a signal binds, and a downstream *response regulator*, whose activity is regulated via phosphorylation by the sensor histidine kinase. The sensor histidine kinase is usually a membrane-bound protein that contains two distinct domains, called the input and histidine kinase, or "transmitter," domains (Figure 21.26).

Detection of a signal by the input domain alters the activity of the histidine kinase domain. Active sensor kinases are dimers that transphosphorylate* a conserved histidine residue. This phosphate is then transferred to a conserved aspartate residue in the receiver domain of a cognate response regulator (see Figure 21.26), and this phosphorylation alters the activity of the kinases. Most response regulators also contain *output* domains that act as transcription factors.

The phenotype resulting from *CKI1* overexpression, combined with its similarity to bacterial receptors, suggested that the CKI1 and/or similar histidine kinases are cytokinin receptors. Support for this model came from identification of the *CRE1* gene (Inoue et al. 2001).

Like *CKI1*, *CRE1* encodes a protein similar to bacterial histidine kinases. Loss-of-function *cre1* mutations were identified in a genetic screen for mutants that failed to develop shoots from undifferentiated tissue culture cells in response to cytokinin. This is essentially the opposite screen from the one just described, from which the *CKI1* gene was identified by a gain-of-function (ability to divide in the absence of cytokinin) mutation. The *cre1* mutants are also resistant to the inhibition of root elongation observed in response to cytokinin.

Convincing evidence that *CRE1* encodes a cytokinin receptor came from analysis of the expression of the protein in yeast. Yeast cells also contain a sensor histidine

---

*That is, transfer a phosphoryl group to another molecule.

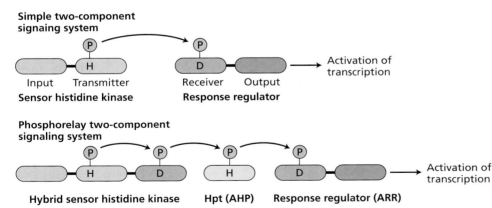

FIGURE 21.26 Simple versus phosphorelay types of two-component signaling systems. (A) In simple two-component systems, the input domain is the site where the signal is sensed. This domain regulates the activity of the histidine kinase domain, which when activated autophosphorylates a conserved His residue. The phosphate is then transferred to an Asp residue that resides within the receiver domain of a response regulator. Phosphorylation of this Asp regulates the activity of the output domain of the response regulator, which in many cases is a transcription factor. (B) In the phosphorelay-type two-component signaling system, an extra set of phosphotransfers is mediated by a histidine phosphotransfer protein (Hpt) called AHP in Arabidopsis. The Arabidopsis response regulators are called ARRs. H = histidine, D = aspartate.

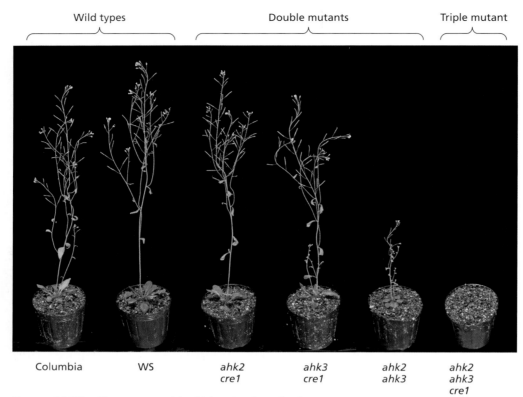

FIGURE 21.27 Phenotypes of Arabidopsis plants harboring mutations in two or all three of the cytokinin receptors (*cre1, ahk2,* and *ahk3*). The parental wild-type ecotypes (Columbia and Ws) are shown on the left. (From Nishimura et al. 2004.)

kinase, and deletion of the gene that encodes this kinase—*SLN1*—is lethal. Expression of *CRE1* in *SLN1*-deficient yeast can restore viability, *but only if cytokinins are present in the medium*. Thus the activity of CRE1 (i.e., its ability to replace SLN1) is dependent on cytokinin, which, coupled with the cytokinin-insensitive phenotype of the *cre1* mutants in Arabidopsis, unequivocally demonstrates that CRE1 is a cytokinin receptor. It remains to be determined if CKI1 is also a cytokinin receptor.

Two other genes in the Arabidopsis genome (*AHK2* and *AHK3*) are closely related to *CRE1*, suggesting that, like the ethylene receptors (see Chapter 22), cytokinin receptors are encoded by a multigene family. Indeed, it has been demonstrated that cytokinins bind to the predicted extracellular domains of AHK2 and AHK3 with high affinity, confirming that they are cytokinin receptors (Yamada et al. 2001). A triple-receptor mutant has been identified in which *CRE1*, *AHK2*, and *AHK3* are all disrupted (Higuchi et al. 2004; Nishimura et al. 2004). This triple mutant displays a variety of developmental abnormalities, including little or no floral development, greatly reduced root growth, and reduced rosette size (Figure 21.27). Nevertheless, the triple-receptor mutant is surprisingly viable, which could mean either that cytokinins are not essential for plant growth or that an alternative, second cytokinin response pathway is present in plants. Further research should help resolve this issue.

Cytokinins increase expression of the type-A response regulator genes via activation of the type-B *ARR* genes

One of the primary effects of cytokinin is to alter the expression of various genes. Among the first genes to be up-regulated in response to cytokinin are the *ARR* (*A*rabidopsis *r*esponse *r*egulator) genes. These genes are homologous to the receiver domain of bacterial two-component response regulators, the downstream target of sensor histidine kinases (see the previous section).

In Arabidopsis, response regulators are encoded by a multigene family. They fall into two basic classes: the **type-A *ARR*** genes, the products of which are made up solely of a receiver domain, and the **type-B *ARR*** genes, which encode a transcription factor domain in addition to the receiver domain (Figure 21.28 and Web Topic 21.8). The rate of transcription of the type-A genes, but not the type-B *ARR*s, is increased very rapidly in response to applied cytokinin (Figure 21.29) (D'Agostino et al. 2000). This rapid induction is specific for cytokinin and does not require new protein synthesis. Both of these features are hallmarks of primary response genes (discussed in Chapters 17 and 19).

The rapid induction of the type-A genes, coupled with their similarity to signaling elements predicted to act downstream of sensor histidine kinases, suggests that these elements act downstream of the CRE1 cytokinin receptor

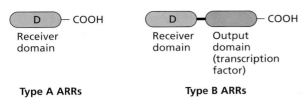

FIGURE 21.28 Comparison of the structures of the type-A and type-B ARRs. The type-A ARRs consist solely of an Asp (D)-containing receiver domain, but the type-A proteins also contain a fused output domain at the carboxy terminus.

family to mediate the primary cytokinin response. Recent analyses of single and multiple loss-of-function mutations in the type-A *ARR*s have indicated that these elements are negative regulators of cytokinin signaling with partially redundant functions (To et al., 2004).

The expression of a wide variety of other genes is altered in response to cytokinin, including the gene that encodes nitrate reductase, light-regulated genes such as *LHCB* and *SSU*, and defense-related genes such as *PR1*, as well as genes that encode an extensin (cell wall protein rich in hydroxyproline), rRNAs, cytochrome P450s, peroxidase, and various transcription factors. Cytokinin elevates the expression of these genes both by increasing the rate of transcription (as in the case of the type-A *ARR*s) and/or by a stabilization of the RNA transcript (e.g., the gene for β-expansin). The type-A gene, *ARR5*, is expressed primarily in the apical meristems of both shoots and roots, consistent with a role in regulating cell proliferation (see **Web Topic 21.9**).

The type-B *ARR*s, which contain a DNA-binding domain and transcriptional activator domain in addition to the receiver domain, have been shown to be the direct upstream activators of type-A *ARR* transcription in response to cytokinin. Increased type-B ARR function leads to an increase in the transcription of the type-A *ARR*s, and disruption of multiple type-B *ARR*s compromises the induction of the type-As by cytokinin (Hwang and Sheen, 2001; Sakai et al. 2001). Like the type-A ARRs, the type-B ARRs display partial functional redundancy, but in contrast to the type-A *ARR*s, loss-of-function mutations in the type-B *ARR*s lead to insensitivity to cytokinin. The current data suggest that phosphorylation of the receiver domain of the type-B ARRs enables them to increase the transcription of a set of genes including the type-A *ARR*s.

Histidine phosphotransferases are also involved in cytokinin signaling

From the preceding discussions we have seen that cytokinin binds to the CRE1/AHK receptors at the cell surface and initiates a phosphotransfer that ultimately leads to phosphorylation of the type-B ARRs in the nucleus. How then is the phosphate transferred from the receptors bound to the plasma membrane to the type-B ARRs in the nucleus? The answer is that another set of proteins, the AHP (*A*rabidopsis *h*istidine *p*hosphotransfer) proteins, acquire the phosphate from the activated receptors, then move into the nucleus where they transfer the phosphate group to the type-B ARRs.

In two-component systems that involve a sensor kinase with a fused receiver domain (the structure of most eukaryotic sensor histidine kinases, including those of the CRE1 family), there is an additional set of phosphotransfers that are mediated by a **histidine phosphotransfer protein** (**Hpt**). Phosphate is first transferred from ATP to a histidine within the histidine kinase domain of the sensor kinase, and then transferred to an aspartate residue on the fused receiver. From the aspartate residue, the phosphate group is then transferred to a histidine on the Hpt protein, and then finally to an aspartate on the receiver domain of the response regulator (see Figure 21.26). This phosphorylation of the receiver domain of the response regulator alters its activity. Thus, Hpt proteins are predicted to mediate the phosphotransfer between sensor kinases and response regulators.

In Arabidopsis there are five Hpt genes, called *AHP*s. The AHP proteins have been shown to physically associate with receiver domains from several Arabidopsis histidine kinases, including CRE1, and a subset of the AHPs have been demonstrated to transiently translocate from the cytoplasm to the nucleus in response to cytokinin (Figure 21.30) (Hwang and Sheen 2001). Finally, disruption of multiple *AHP* genes in Arabidopsis leads to cytokinin insensitivity. Together, these finding indicate that the

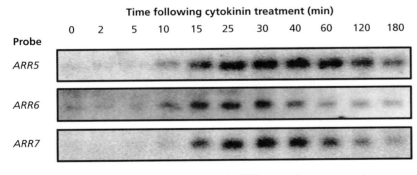

FIGURE 21.29 Induction of some type-A *ARR* genes in response to cytokinin. RNA from Arabidopsis seedlings treated for the indicated time with cytokinin was isolated and analyzed by Northern blotting. Each row shows the result of probing the Northern blot with an individual type-A gene, and each lane contains RNA derived from Arabidopsis seedlings treated for the indicated time with cytokinin. The darker the band, the higher the level of *ARR* mRNA in that sample. The *ARR5* gene is expressed preferentially in apical meristems (see **Web Topic 21.9**). (From D'Agostino et al. 2000.)

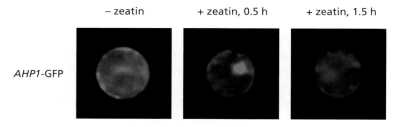

FIGURE 21.30 Cytokinin induces the transient movement of some AHP proteins into the nucleus. Arabidopsis protoplasts expressing the *AHP1* gene fused to green fluorescent protein (GFP) as a reporter were treated with zeatin and monitored for 1.5 hours. *AHP1*-GFP shows strong nuclear localization after 30 minutes, but the signal disappears by 90 minutes. (From Hwang and Sheen 2001.)

AHPs are the immediate downstream targets of the activated CRE receptors, and that these proteins transduce the cytokinin signal to the nucleus, where they phosphorylate and activate the type-B ARRs, which in turn alter the expression of various genes.

A model of cytokinin signaling is presented in Figure 21.31. Cytokinin binds to the CRE1 receptor and initiates a phosphorylation cascade that ultimately results in the phosphorylation and activation of a subset of the type-B ARR proteins.

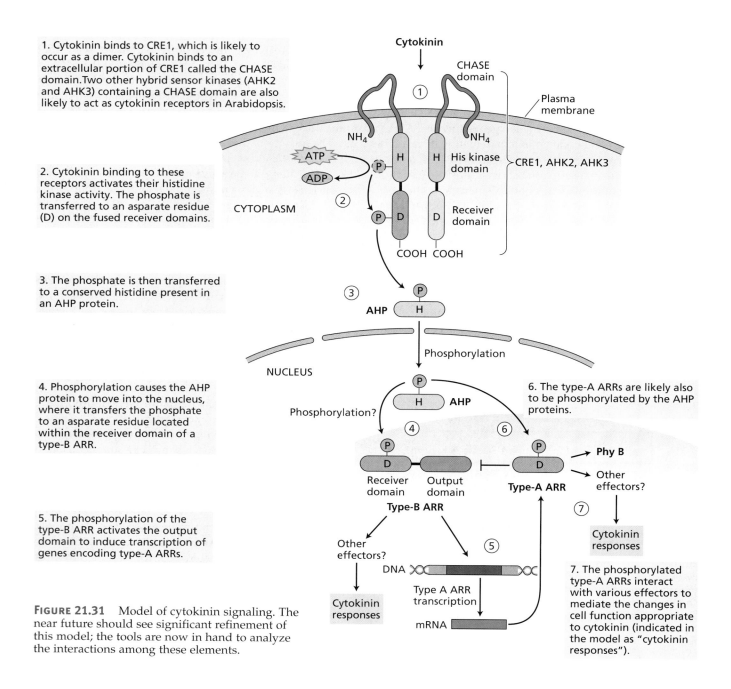

FIGURE 21.31 Model of cytokinin signaling. The near future should see significant refinement of this model; the tools are now in hand to analyze the interactions among these elements.

Activation of the type-B proteins (transcription factors) leads to the transcriptional activation of the type-A genes. The type-A ARR proteins are likely also phosphorylated in response to cytokinin, and, perhaps together with the type-B proteins, they interact with various targets to mediate the changes in cellular function, such as an activation of the cell cycle. Type-A ARRs are also able to inhibit their own expression by an unknown mechanism, providing a negative feedback loop (see Figure 21.31). Much work needs to be done to confirm and refine this model, but we are beginning to glimpse for the first time the molecular basis for cytokinin action in plants.

Summary

Mature plant cells generally do not divide in the intact plant, but they can be stimulated to divide by wounding, by infection with certain bacteria, and by plant hormones, including cytokinins. Cytokinins are N^6-substituted aminopurines that will initiate cell proliferation in many plant cells when they are cultured on a medium that also contains an auxin. The principal cytokinins of higher plants—zeatin-type cytokinins—are also present in plants as a riboside or ribotide and as glycosides. These forms are generally also active as cytokinins in bioassays through their enzymatic conversion to the free zeatin base by plant tissue.

The first committed step in cytokinin biosynthesis—the transfer of the isopentenyl group from DMAPP to the 6 nitrogen of adenosine tri- and diphosphate—is catalyzed by isopentenyl transferase (IPT). The product of this reaction is readily converted to zeatin and other cytokinins. Cytokinins are synthesized in roots, developing embryos, young leaves, fruits, and crown gall tissues. Cytokinins are also synthesized by plant-associated bacteria, fungi, insects, and nematodes.

Cytokinin oxidases degrade cytokinin irreversibly and may play a role in regulation of the levels of this hormone. Conjugation of both the side chain and the adenosine moiety to sugars (mostly glucose) also may play a role in the regulation of cytokinin levels and may target subpools of the hormone for distinct roles, such as transport. Cytokinins are also interconverted among the free base and the nucleoside and nucleotide forms.

Crown galls originate from plant tissues that have been infected with *A. tumefaciens*. The bacterium injects a specific region of its Ti plasmid called T-DNA into wounded plant cells, and the T-DNA is incorporated into the host nuclear genome. The T-DNA contains *ipt*, a gene for cytokinin biosynthesis, as well as genes for auxin biosynthesis. These phyto-oncogenes are expressed in the plant cells, leading to hormone synthesis and unregulated proliferation of the cells to form the gall.

Cytokinins are most abundant in the young, rapidly dividing cells of the shoot and root apical meristems. They do not appear to be actively transported through living plant tissues. Instead, they are transported passively into the shoot from the root through the xylem, along with water and minerals. At least in pea, however, the shoot can regulate the flow of cytokinin from the root.

Cytokinins participate in the regulation of many plant processes, including cell division, morphogenesis of shoots and roots, chloroplast maturation, cell enlargement, and senescence. Both cytokinin and auxin regulate the plant cell cycle and are needed for cell division. The roles of cytokinins have been elucidated from application of exogenous cytokinins, from the phenotypes of transgenic plants designed to overexpress cytokinins as a result of introduction of the bacterial *ipt* gene, from transgenic plants that have reduced levels of cytokinin content as a result of overexpression of cytokinin oxidase, and from the phenotypes of cytokinin signaling mutants.

In addition to cell division, the ratio of auxin to cytokinin determines the differentiation of cultured plant tissues into either roots or buds: High ratios promote roots; low ratios, buds. Cytokinins also have been implicated in the release of axillary buds from apical dominance. In the moss *Funaria*, cytokinins greatly increase the number of "buds," the structures that give rise to the leafy gametophyte stage of development.

The mechanism of action of cytokinin has been elucidated. Cytokinin receptors are transmembrane proteins that are related to bacterial two-component sensor histidine kinases. Cytokinins increase the abundance of several specific mRNAs, some of which are primary response genes similar to bacterial two-component response regulators. The mechanism of signal transduction by which CRE1 activates transcription of the type-A *ARRs* involves other homologs of two-component elements.

Web Material

Web Topics

21.1 Cultured Cells Can Acquire the Ability to Synthesize Cytokinins
The phenomenon of habituation is described, whereby callus tissues become cytokinin independent.

21.2 Structures of Some Naturally Occurring Cytokinins
The structures of various naturally occurring cytokinins are presented.

21.3 Various Methods Are Used to Detect and Identify Cytokinins
Cytokinins can be qualified using immunological and sensitive physical methods.

21.4 Cytokinins Are Also Present in Some tRNAs in Animal and Plant Cells
Modified adenosines near the 3′ end of the anticodons of some tRNAs have cytokinin activity.

21.5 The Ti Plasmid and Plant Genetic Engineering
Applications of the Ti plasmid of *Agrobacterium* in bioengineering are described.

21.6 Phylogenetic Tree of *IPT* Genes
Arabidopsis contains nine different *IPT* genes, several of which form a distinct clade with other plant sequences.

21.7 Cytokinin Can Promote Light-Mediated Development
Cytokinins can mimic the effect of the *det* mutation on chloroplast development and deetiolation.

21.8 Phylogenetic Tree of Arabidopsis Response Regulators (ARRs)

21.9 Localization of Cytokinin Signaling Elements
The *ARR5* gene is expressed preferentially in apical meristems

Web Essays

21.1 The *6b* Gene: A New Tool for the Elucidation of the Growth and/or Differentiation of Plant Cells
The *6b* gene has been shown to encode a protein that interacts with a specific nuclear protein in tobacco.

21.2 Cytokinin-Induced Form and Structure in Moss
The effects of cytokinins on the development of moss protonema are described.

Chapter References

Akiyoshi, D. E., Klee, H., Amasino, R. M., Nester, E. W., and Gordon, M. P. (1984) T-DNA of *Agrobacterium tumefaciens* encodes an enzyme of cytokinin biosynthesis. *Proc. Natl. Acad. Sci. USA* 81: 5994–5998.

Akiyoshi, D. E., Morris, R. O., Hinz, R., Mischke, B. S., Kosuge, T., Garfinkel, D. J., Gordon, M. P., and Nester, E. W. (1983) Cytokinin/auxin balance in crown gall tumors is regulated by specific loci in the T-DNA. *Proc. Natl. Acad. Sci. USA* 80: 407–411.

Akiyoshi, D. E., Regier, D. A., and Gordon, M. P. (1987) Cytokinin production by *Agrobacterium* and *Pseudomonas* spp. *J. Bacteriol.* 169: 4242–4248.

Aloni, R., Wolf, A., Feigenbaum, P., Avni, A., and Klee, H. J. (1998) The *Never ripe* mutant provides evidence that tumor-induced ethylene controls the morphogenesis of *Agrobacterium tumefaciens*-induced crown galls in tomato stems. *Plant Physiol.* 117: 841–849.

Ashikari, M., Sakakibara, H., Lin, S., Yamamoto, T., Takashi, T., Nishimura, A., Angeles, E. R., Qian, Q., Kitano, H., and Matsuoka, M. (2005) Cytokinin oxidase regulates rice grain production. *Science* 309: 741–745.

Barry, G. F., Rogers, R. G., Fraley, R. T., and Brand, L. (1984) Identification of cloned biosynthesis gene. *Proc. Natl. Acad. Sci. USA* 81: 4776–4780.

Beveridge, C. A. (2000) Long-distance signalling and a mutational analysis of branching in pea. *Plant Growth Regul.* 32: 193–203.

Beveridge, C. A., Murfet, I. C., Kerhoas, L., Sotta, B., Miginiac, E., and Rameau, C. (1997) The shoot controls zeatin riboside export from pea roots. Evidence from the branching mutant *rms4*. *Plant J.* 11: 339–345.

Bomhoff, G., Klapwijk, P. M., Kester, H. C. M., and Schilperoort, R. A. (1976) Octopine and nopaline synthesis and breakdown genetically controlled by plasmid of *Agrobacterium tumefaciens*. *Mol. Gen. Genet.* 145: 177–181.

Braun, A. C. (1958) A physiological basis for autonomous growth of the crown-gall tumor cell. *Proc. Natl. Acad. Sci. USA* 44: 344–349.

Brzobohaty, B., Moore, I., Kristoffersen, P., Bako, L., Campos, N., Schell, J., and Palme, K. (1993) Release of active cytokinin by a β-glucosidase localized to the maize root meristem. *Science* 262: 1051–1054.

Caplin, S. M., and Steward, F. C. (1948) Effect of coconut milk on the growth of the explants from carrot root. *Science* 108: 655–657.

Chilton, M.-D. (1983) A vector for introducing new genes into plants. *Sci. Am.* 248: 50–59.

Chilton, M.-D., Drummond, M. H., Merlo, D. J., Sciaky, D., Montoya, A. L., Gordon, M. P., and Nester, E. W. (1977) Stable incorporation of plasmid DNA into higher plant cells: The molecular basis of crown gall tumorigenesis. *Cell* 11: 263–271.

Chory, J., Reinecke, D., Sim, S., Washburn, T., and Brenner, M. (1994) A role for cytokinins in de-etiolation in *Arabidopsis*. Det mutants have an altered response to cytokinins. *Plant Physiol.* 104: 339–347.

D'Agostino, I. B., Deruère, J., and Kieber, J. J. (2000) Characterization of the response of the *Arabidopsis ARR* gene family to cytokinin. *Plant Physiol.* 124: 1706–1717.

Elzen, G. W. (1983) Cytokinins and insect galls. *Comp. Biochem. Physiol.* 76A(1): 17–19.

Estruch, J. J., Chriqui, D., Grossmann, K., Schell, J., and Spena, A. (1991) The plant oncogene *RolC* is responsible for the release of cytokinins from glucoside conjugates. *EMBO J.* 10: 2889–2895.

Faiss, M., Zalubìová, J., Strnad, M., and Schmülling, T. (1997) Conditional transgenic expression of the *ipt* gene indicates a function for cytokinins in paracrine signaling in whole tobacco plants. *Plant J.* 12: 410–415.

Foo, E., Bullier, E., Goussot, M., Foucher, F., Rameau, C., and Beveridge, C. A. (2005) The branching gene *RAMOSUS1* mediates interactions among two novel signals and auxin in pea. *Plant Cell* 17: 464–474.

Galis, I., Kakiuchi, Y., Simek, P., and Wabiko, H. (2004) *Agrobacterium tumefaciens AK-6b* gene modulates phenolic compound metabolism in tobacco. *Phytochemistry* 65: 169–179.

Gan, S., and Amasino, R. M. (1995) Inhibition of leaf senescence by autoregulated production of cytokinin. *Science* 270: 1986–1988.

Garfinkel, D. J., Simpson, R. B., Ream, L. W., White, F. F., Gordon, M. P., and Nester, E. W. (1981) Genetic analysis of crown gall: Fine structure map of the T-DNA by site-directed mutagenesis. *Cell* 27: 143–153.

Giulini, A., Wang, J., and Jackson, D. (2004) Control of phyllotaxy by the cytokinin-inducible response regulator homologue ABPHYL1. *Nature* 430: 1031–1034.

Hamilton, J. L., and Lowe, R. H. (1972) False broomrape: A physiological disorder caused by growth-regulator imbalance. *Plant Physiol.* 50: 303–304.

Higuchi, M., Pischke, M. S., Mahonen, A. P., Miyawaki, K., Hashimoto, Y., Seki, M., Kobayashi, M., Shinozaki, K., Kato, T., Tabata, S., et al. (2004) In planta functions of the *Arabidopsis* cytokinin receptor family. *Proc. Natl. Acad. Sci. USA* 101: 8821–8826.

Houba-Herin, N., Pethe, C., d'Alayer, J., and Laloue M. (1999) Cytokinin oxidase from *Zea mays*: Purification, cDNA cloning and expression in moss protoplasts. *Plant J.* 17: 615–626.

Huff, A. K., and Ross, C. W. (1975) Promotion of radish cotyledon enlargement and reducing sugar content by zeatin and red light. *Plant Physiol.* 56: 429–433.

Hwang, I., and Sheen, J. (2001). Two-component circuitry in *Arabidopsis* signal transduction. *Nature* 413: 383–389.

Inoue, T., Higuchi, M., Hashimoto, Y., Seki, M., Kobayashi, M., Kato, T., Tabata, S., Shinozaki, K., and Kakimoto, T. (2001) Identification of CRE1 as a cytokinin receptor from *Arabidopsis*. *Nature* 409: 1060–1063.

Itai, C., and Vaadia, Y. (1971) Cytokinin activity in water-stressed shoots. *Plant Physiol.* 47: 87–90.

John, P. C. L., Zhang, K., Don, C., Diederich, L., and Wightman, F. (1993) P34-cdc2 related proteins in control of cell cycle progression, the switch between division and differentiation in tissue development, and stimulation of division by auxin and cytokinin. *Aust. J. Plant Physiol.* 20: 503–526.

Kakimoto, T. (2001) Identification of plant cytokinin biosynthetic enzymes as dimethylallyl diphosphate: ATP/ADP isopentenyltransferases. *Plant Cell. Physiol.* 42: 677–685.

Lee, J. H., Kim, D.-M., Lim, Y. P., and Pai, H.-S. (2004) The shooty callus induced by suppression of tobacco *CHRK1* receptor-like kinase is a phenocopy of the tobacco genetic tumor. *Plant Cell Rep.* 23: 397–403.

Letham, D. S. (1973) Cytokinins from *Zea mays*. *Phytochemistry* 12: 2445–2455.

Letham, D. S. (1974) Regulators of cell division in plant tissues XX. The cytokinins of coconut milk. *Physiol. Plant.* 32: 66–70.

Martin R. C., Mok M. C., and Mok D. W. S. (1999) Isolation of a cytokinin gene, ZOG1, encoding zeatin O-glucosyltransferase from *Phaseolus lunatus*. *Proc. Natl. Acad. Sci. USA* 96: 284–289.

McCabe, M. S., Garratt, L. C., Schepers, F., Jordi, W. J., Stoopen, G. M. Davelaar, E., van Rhijn, J. H., Power, J. B., and Davey, M. R. (2001) Effects of P(SAG12)-IPT gene expression on development and senescence in transgenic lettuce. *Plant Physiol.* 127: 505–516.

Miller, C. O., Skoog, F., Von Saltza, M. H., and Strong, F. (1955) Kinetin, a cell division factor from deoxyribonucleic acid. *J. Am. Chem. Soc.* 77: 1392–1393.

Miyawaki, K., Matsumoto-Kitano, M., and Kakimoto, T. (2004) Expression of cytokinin biosynthetic isopentenyltransferase genes in *Arabidopsis*: Tissue specificity and regulation by auxin, cytokinin, and nitrate. *Plant J.* 37: 128–138.

Morris, R. O. (1986) Genes specifying auxin and cytokinin biosynthesis in phytopathogens. *Annu. Rev. Plant Physiol.* 37: 509–538.

Morris, R., Bilyeu, K., Laskey, J., and Cheikh, N. (1999) Isolation of a gene encoding a glycosylated cytokinin oxidase from maize. *Biochem. Biophys. Res. Commun.* 225: 328–333.

Nishimura, C., Ohashi, Y., Sato, S., Kato, T., Tabata, S., and Ueguchi, C. (2004) Histidine kinase homologs that act as cytokinin receptors possess overlapping functions in the regulation of shoot and root growth in *Arabidopsis*. *Plant Cell* 16: 1365–1377.

Noodén, L. D., and Letham, D. S. (1993) Cytokinin metabolism and signaling in the soybean plant. *Aust. J. Plant Physiol.* 20: 639–653.

Noodén, L. D., Singh, S., and Letham, D. S. (1990) Correlation of xylem sap cytokinin levels with monocarpic senescence in soybean. *Plant Physiol.* 93: 33–39.

Riou-Khamlichi, C., Huntley, R., Jacqmard, A., and Murray, J. A. (1999) Cytokinin activation of *Arabidopsis* cell division through a D-type cyclin. *Science* 283: 1541–1544.

Rupp, H.-M., Frank, M., Werner, T., Strnad, M., and Schmülling, T. (1999) Increased steady state mRNA levels of the STM and KNAT1 homeobox genes in cytokinin overproducing *Arabidopsis thaliana* indicate a role for cytokinins in the shoot apical meristem. *Plant J.* 18: 557–563.

Sakai, H., Honma, T., Aoyama, T., Sato, S., Kato, T., Tabata, S., and Oka, A. (2001) *Arabidopsis* ARR1 is a transcription factor for genes immediately responsive to cytokinins. *Science*. 294: 1519–1521.

Sakakibara, H., Kasahara, H., Ueda, N., Kojima, M., Takei, K., Hishiyama, S., Asami, T., Okada, K., Kamiya, Y., Yamaya, T., and Yamaguchi, S. (2005) Agrobacterium tumefaciens increases cytokinin production in plastids by modifying the biosynthetic pathway in the host plant. *Proc. Natl. Acad. Sci. USA* 102: 9972-9977

Samuelson, M. E., Eliasson, L., and Larsson, C. M. (1992) Nitrate-regulated growth and cytokinin responses in seminal roots of barley. *Plant Physiol.* 98: 309–315.

Skoog, F., and Miller, C. O. (1965) Chemical regulation of growth and organ formation in plant tissues cultured *in vitro*. In *Molecular and Cellular Aspects of Development*, E. Bell, ed., Harper and Row, New York, pp. 481–494.

Smigocki, A., Neal, J. W., Jr., McCanna, I., and Douglass, L. (1993) Cytokinin-mediated insect resistance in *Nicotiana* plants transformed with the *ipt* gene. *Plant Mol. Biol.* 23: 325–335.

Smith, H. H. (1988) The inheritance of genetic tumors in *Nicotiana* hybrids. *J. Hered.* 79: 277–284.

Soni, R., Carmichael, J. P., Shah, Z. H., and Murray, J. A. H. (1995) A family of cyclin D homologs from plants differentially controlled by growth regulators and containing the conserved retinoblastoma protein interaction motif. *Plant Cell* 7: 85–103.

Sorefan, K., Booker, J., Haurogné, K., Goussot, M., Bainbridge, K., Foo, E., Chatfield, S., Ward, S., Beveridge, C., Rameau, C., et al. (2003). MAX4 and RMS1 are orthologous dioxygenase-like genes that regulate shoot branching in *Arabidopsis* and pea. *Genes Dev.* 17: 1469–1474.

Takei, K., Sakakibara, H., and Sugiyama, T. (2001a) Identification of genes encoding adenylate isopentyltransferase, a cytokinin biosynthetic enzyme, in *Arabidopsis thaliana*. *J. Biol. Chem.* 276: 26405–26410.

Takei, K., Sakakibara, H., Taniguchi, M., and Sugiyama, T. (2001b) Nitrogen-dependent accumulation of cytokinins in roots and the translocation to leaf: Implication of cytokinin species that induces gene expression of maize response regulator. *Plant Cell Physiol.* 42: 85–93.

To, J. P., Haberer, G., Ferreira, F. J., Deruere, J., Mason, M. G., Schaller, G. E., Alonso, J. M., Ecker, J. R., and Kieber, J. J. (2004) Type-A *Arabidopsis* response regulators are partially redundant negative regulators of cytokinin signaling. Plant Cell 16:658-671.

Vogel, J. P., Woeste, K., Theologis, A., and Kieber, J. J. (1998) Recessive and dominant mutations in the ethylene biosynthetic gene AC55 of *Arabidopsis* confer cytokinin-insensitivity and ethylene overproduction respectively. *Proc. Natl. Acad. Sci. USA* 95: 4766–4771.

Werner, T., Motyka, V., Strnad, M., and Schmülling, T. (2001) Regulation of plant growth by cytokinin. *Proc. Natl. Acad. Sci. USA* 98: 10487–10492.

White, P. R. (1934) Potentially unlimited growth of excised tomato root tips in a liquid medium. *Plant Physiol.* 9: 585–600.

Yamada, H., Suzuki, T., Terada, K., Takei, K., Ishikawa, K., Miwa, K., Yamashino, T., and Mizuno, T. (2001). The *Arabidopsis* AHK4 histidine kinase is a cytokinin-binding receptor that transduces cytokinin signals across the membrane. *Plant Cell Physiol.* 42: 1017–1023.

Zhang, K., Letham, D. S., and John, P. C. L. (1996) Cytokinin controls the cell cycle at mitosis by stimulating the tyrosine dephosphorylation and activation of p34cdc2-like H1 histone kinase. *Planta* 200: 2–12.

Chapter 22

Ethylene: The Gaseous Hormone

DURING THE NINETEENTH CENTURY, when coal gas was used for street illumination, it was observed that trees in the vicinity of streetlamps defoliated more extensively than other trees. Eventually it became apparent that coal gas and air pollutants affect plant growth and development, and ethylene was identified as the active component of coal gas.

In 1901, Dimitry Neljubov, a graduate student at the Botanical Institute of St. Petersburg in Russia, observed that dark-grown pea seedlings growing in the laboratory exhibited symptoms that were later termed the *triple response*: reduced stem elongation, increased lateral growth (swelling), and abnormal, horizontal growth (see Figure 22.5A). When the plants were allowed to grow in fresh air, they regained their normal morphology and rate of growth. Neljubov identified ethylene, which was present in the laboratory air from coal gas, as the molecule causing the response.

The first indication that ethylene is a natural product of plant tissues was published by H. H. Cousins in 1910. Cousins reported that "emanations" from oranges stored in a chamber caused the premature ripening of bananas when these gases were passed through a chamber containing the fruit. However, given that oranges synthesize relatively little ethylene compared to other fruits, such as apples, it is likely that the oranges used by Cousins were infected with the fungus *Penicillium*, which produces copious amounts of ethylene. In 1934, R. Gane and others identified ethylene chemically as a natural product of plant metabolism, and because of its dramatic effects on the plant it was classified as a hormone.

For 25 years ethylene was not recognized as an important plant hormone, mainly because many physiologists believed that the effects of ethylene were due to auxin, the first plant hormone to be discovered (see Chapter 19). Auxin was thought to be the main plant hormone, and ethylene was considered to play only an insignificant and indirect physiological role. Work on ethylene was also hampered by the lack of chemical techniques for its quantification. However, after gas chromatography was introduced in ethylene research in 1959, the importance of ethylene was rediscovered and its physiological significance as a plant growth regulator was recognized (Burg and Thimann 1959).

In this chapter we will describe the discovery of the ethylene biosynthetic pathway and outline some of the important effects of ethylene on plant growth and development. At the end of the chapter we will consider how ethylene acts at the cellular and molecular levels.

Structure, Biosynthesis, and Measurement of Ethylene

Ethylene can be produced by almost all parts of higher plants, although the rate of production depends on the type of tissue and the stage of development. Ethylene production increases during leaf abscission and flower senescence, as well as during fruit ripening. Any type of wounding can induce ethylene biosynthesis, as can physiological stresses such as flooding, disease, and temperature or drought stress.

The amino acid methionine is the precursor of ethylene, and 1-aminocyclopropane-1-carboxylic acid (ACC) serves as an intermediate in the conversion of methionine to ethylene. As we will see, the complete pathway is a cycle, taking its place among the many metabolic cycles that operate in plant cells.

The properties of ethylene are deceptively simple

Ethylene is the simplest olefin (its molecular weight is 28), and it is lighter than air under physiological conditions:

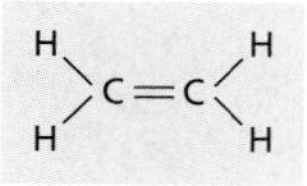

$$H_2C=CH_2$$

Ethylene

It is flammable and readily undergoes oxidation. Ethylene can be oxidized to ethylene oxide:

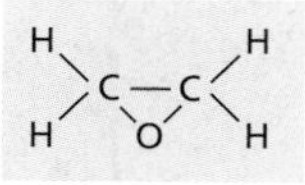

Ethylene oxide

and ethylene oxide can be hydrolyzed to ethylene glycol:

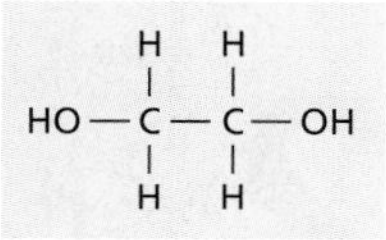

Ethylene glycol

In most plant tissues, ethylene can be completely oxidized to CO_2, in the following reaction:

Complete oxidation of ethylene

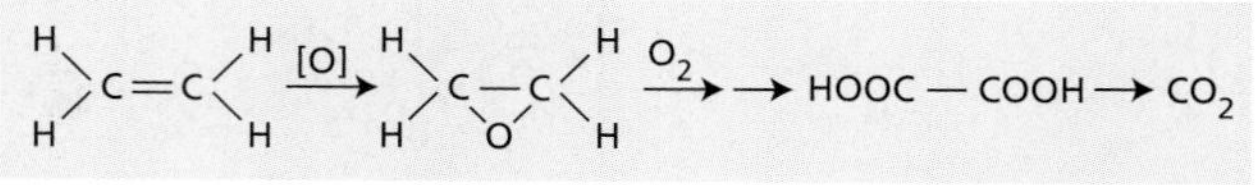

$$H_2C=CH_2 \xrightarrow{[O]} H_2C-CH_2 \xrightarrow{O_2} \rightarrow HOOC-COOH \rightarrow CO_2$$

$\qquad$ Ethylene $\qquad$ Ethylene oxide $\qquad$ Oxalic acid $\qquad$ Carbon dioxide

Ethylene is released easily from the tissue and diffuses in the gas phase through the intercellular spaces and outside the tissue. At an ethylene concentration of 1 μL L^{-1} in the gas phase at 25°C, the concentration of ethylene in water is 4.4×10^{-9} M. Ethylene concentrations are normally expressed in terms of their gas phase concentrations, which are easier to measure.

Because ethylene gas is easily lost from its tissue of origin and may affect other tissues or organs, ethylene-trapping systems are used during the storage of fruits, vegetables, and flowers. Potassium permanganate ($KMnO_4$) is an effective absorbent of ethylene and can reduce the concentration of ethylene in apple storage areas from 250 to 10 μL L^{-1}, markedly extending the storage life of the fruit.

Bacteria, fungi, and plant organs produce ethylene

Even away from cities and industrial air pollutants, the environment is seldom free of ethylene. Ethylene can be formed either photochemically (e.g., in the atmosphere and upper oceanic layers) or enzymatically by plants and microorganisms. The production of ethylene in plants is highest in senescing tissues and ripening fruits (>1.0 nL g-fresh-weight^{-1} h^{-1}), but all organs of higher plants can synthesize ethylene. Ethylene is biologically active at very low concentrations—as low as 1 part per billion (1 nL L^{-1}) (Binder et al 2004). The internal ethylene concentration in a ripe apple has been reported to be as high as 2500 μL L^{-1}.

Young developing leaves produce more ethylene than do fully expanded leaves. In kidney bean (*Phaseolus vulgaris*), young leaves produce 0.4 nL g^{-1} h^{-1}, compared with 0.04 nL g^{-1} h^{-1} for older leaves. With few exceptions, nonsenescent tissues that are wounded or mechanically perturbed will temporarily increase their ethylene production severalfold within 30 minutes. Ethylene levels later return to normal.

Gymnosperms and lower plants, including ferns, mosses, liverworts, and certain cyanobacteria, have shown the ability to produce ethylene. Ethylene production by fungi

FIGURE 22.1 Ethylene biosynthetic pathway and the Yang cycle. The amino acid methionine is the precursor of ethylene. The rate-limiting step in the pathway is the conversion of AdoMet to ACC, which is catalyzed by the enzyme ACC synthase. The last step in the pathway, the conversion of ACC to ethylene, requires oxygen and is catalyzed by the enzyme ACC oxidase. The CH_3—S group of methionine is recycled via the Yang cycle and thus conserved for continued synthesis. Besides being converted to ethylene, ACC can be conjugated to N-malonyl ACC. AOA = aminooxyacetic acid; AVG = aminoethoxyvinylglycine. (After McKeon et al. 1995.)

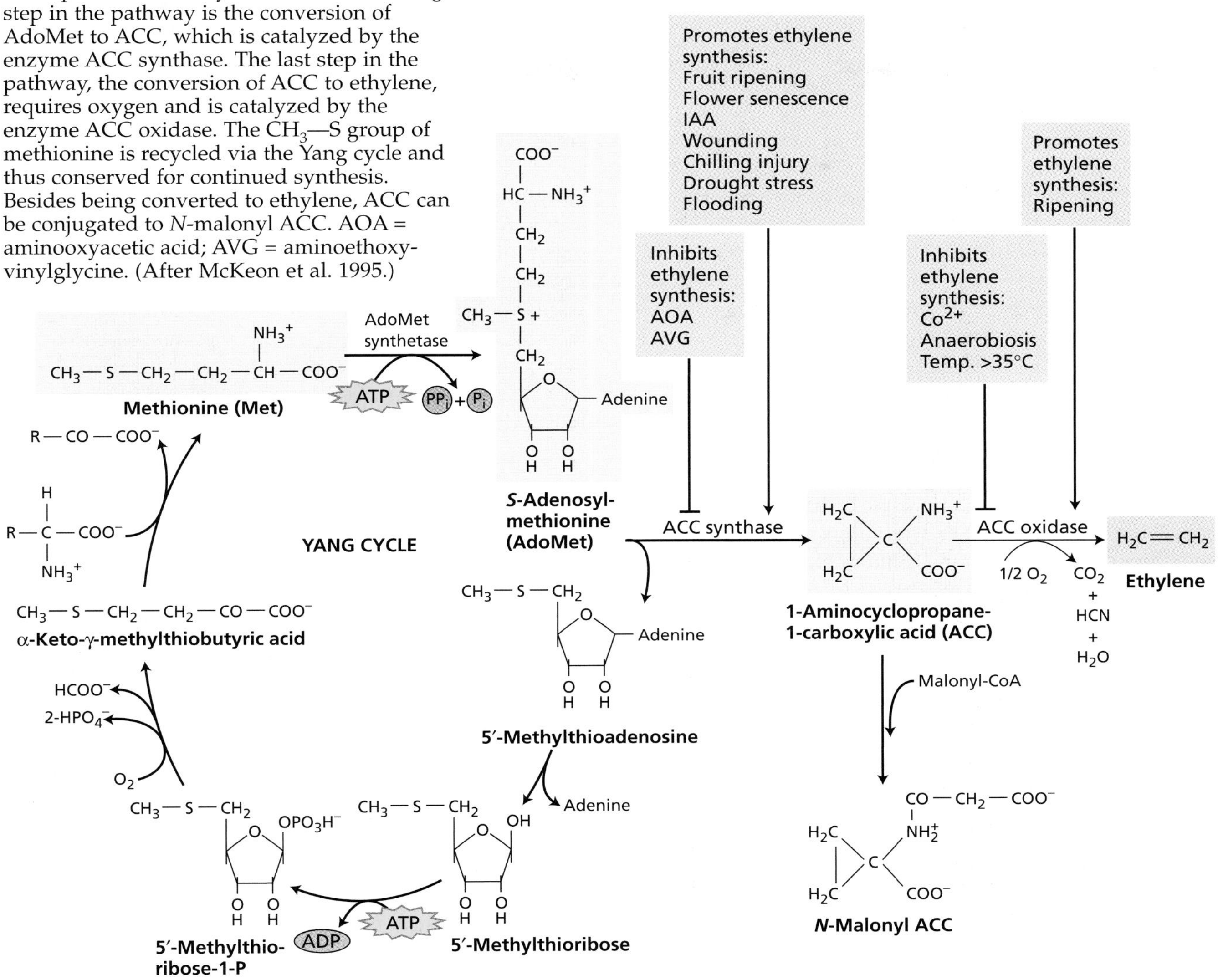

and bacteria contributes significantly to the ethylene content of soil. Certain strains of the common enteric bacterium *Escherichia coli* and of yeast (a fungus) produce large amounts of ethylene from methionine.

There is no evidence that healthy mammalian tissues produce ethylene, nor does ethylene appear to be a metabolic product of invertebrates. However, both a marine sponge and cultured mammalian cells can respond to ethylene, raising the possibility that this gaseous molecule acts as a signaling molecule in animal cells (Perovic et al. 2001).

Regulated biosynthesis determines the physiological activity of ethylene

In vivo experiments showed that plant tissues convert l-[^{14}C]methionine to [^{14}C]ethylene, and that the ethylene is derived from carbons 3 and 4 of methionine (Figure 22.1). The CH_3—S group of methionine is recycled via the Yang cycle. Without this recycling, the amount of reduced sulfur present would limit the available methionine and the synthesis of ethylene. *S*-adenosylmethionine (AdoMet), which is synthesized from methionine and adenosine triphosphate (ATP), is an intermediate in the ethylene biosynthetic pathway, and the immediate precursor of ethylene is **1-aminocyclopropane-1-carboxylic acid (ACC)** (see Figure 22.1).

The role of ACC became evident in experiments in which plants were treated with [^{14}C]methionine. Under anaerobic conditions, ethylene was not produced from the [^{14}C]methionine, and labeled ACC accumulated in the tissue. On exposure to oxygen, however, ethylene production surged. The labeled ACC was rapidly converted to ethylene in the presence of oxygen by various plant tissues, suggesting that ACC is the immediate precursor of ethylene in higher plants and that oxygen is required for the conversion.

In general, when ACC is supplied exogenously to plant tissues, ethylene production increases substantially. This observation indicates that the synthesis of ACC is usually the biosynthetic step that limits ethylene production in plant tissues.

ACC synthase (ACS), the enzyme that catalyzes the conversion of AdoMet to ACC (see Figure 22.1), has been characterized in many types of tissues of various plants. ACC synthase is an unstable, cytosolic enzyme. Its level is regulated by environmental and internal factors, such as wounding, drought stress, flooding, and auxin. Because ACC synthase is present in such low amounts in plant tissues (0.0001% of the total protein of ripe tomato) and is very unstable, it is difficult to purify the enzyme for biochemical analysis (see Web Topic 22.1).

ACS is encoded by members of a divergent multigene family that are differentially regulated by various inducers of ethylene biosynthesis. In tomato, for example, there are at least nine ACC synthase genes, different subsets of which are induced by auxin, wounding, and/or fruit ripening; the Arabidopsis genome contains eight *ACS* genes. An analysis of the purified Arabidopsis ACS proteins revealed a diversity of kinetic properties (for example, various affinities for the substrate AdoMet), suggesting that these eight isoforms might be optimized for different roles in various tissues and cell types (Yamagani et al. 2003). The deduced crystal structure of both apple and tomato ACS reveals a homodimeric protein with shared active sites, similar to aminotransferases (Capitani et al. 1999; Huai et al. 2001). (For more details, see Web Topic 22.2.)

ACC oxidase catalyzes the last step in ethylene biosynthesis: the conversion of ACC to ethylene (see Figure 22.1). In tissues that show high rates of ethylene production, such as ripening fruit, ACC oxidase activity can be the rate-limiting step in ethylene biosynthesis. The gene that encodes ACC oxidase has been cloned (see Web Topic 22.3). Like ACC synthase, ACC oxidase is encoded by a multigene family that is differentially regulated. For example, in ripening tomato fruits and senescing petunia flowers, the mRNA levels of a subset of ACC oxidase genes are highly elevated.

The deduced amino acid sequences of ACC oxidases revealed that these enzymes belong to the Fe^{2+}/ascorbate oxidase superfamily. This similarity suggested that ACC oxidase might require Fe^{2+} and ascorbate for activity—a requirement that has been confirmed by biochemical analysis of the protein. The requirement of ACC oxidase for cofactors presumably explains why the purification of this enzyme eluded researchers for so many years.

CATABOLISM Researchers have studied the catabolism of ethylene by supplying $^{14}C_2H_4$ to plant tissues and tracing the radioactive compounds produced. Carbon dioxide, ethylene oxide, ethylene glycol, and the glucose conjugate of ethylene glycol have been identified as metabolic breakdown products. However, because certain cyclic olefin compounds, such as 1,4-cyclohexadiene, have been shown to block ethylene breakdown without inhibiting ethylene action, ethylene catabolism does not appear to play a significant role in regulating the level of the hormone (Raskin and Beyer 1989).

CONJUGATION Not all the ACC found in the tissue is converted to ethylene. ACC can also be converted to a conjugated form, *N*-malonyl ACC (see Figure 22.1), which does not appear to break down and accumulates in the tissue, primarily in the vacuole. A second, minor conjugated form of ACC, 1-(γ-L-glutamylamino) cyclopropane-1-carboxylic acid (GACC), has also been identified. The conjugation of ACC may play an important role in the control of ethylene biosynthesis, in a manner analogous to the conjugation of auxin and cytokinin.

Environmental stresses and auxins promote ethylene biosynthesis

Ethylene biosynthesis is stimulated by several factors, including developmental state, environmental conditions, other plant hormones, and physical and chemical injury. Ethylene biosynthesis also varies in a circadian manner, peaking during the day and reaching a minimum at night.

FRUIT RIPENING As fruits mature, the rate of ACC and ethylene biosynthesis increases. Enzyme activities for both ACC oxidase (Figure 22.2) and ACC synthase increase, as do the mRNA levels for subsets of the genes encoding each enzyme. However, application of ACC to unripe fruits only slightly enhances ethylene production, indicating that an increase in the activity of ACC oxidase is the rate-limiting step in ripening (McKeon et al. 1995).

STRESS-INDUCED ETHYLENE PRODUCTION Ethylene biosynthesis is increased by stress conditions such as drought, flooding, chilling, exposure to ozone, or mechanical wounding. In all these cases ethylene is produced by the usual biosynthetic pathway, and the increased ethylene production has been shown to result at least in part from an increase in transcription of ACC synthase mRNA. This "stress ethylene" is involved in the onset of stress responses such as abscission, senescence, wound healing, and increased disease resistance (see Chapter 26).

AUXIN-INDUCED ETHYLENE PRODUCTION In some instances, auxins and ethylene can cause similar plant responses, such as induction of flowering in pineapple and inhibition of stem elongation. These responses might be due to the ability of auxins to promote ethylene synthesis by enhancing ACC synthase activity. These observations suggest that some responses previously attributed to auxin (indole-3-

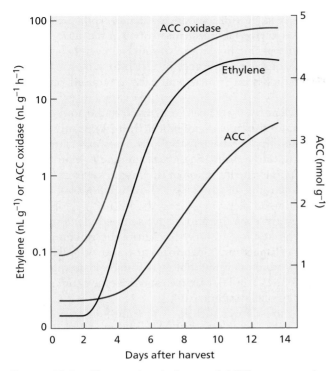

FIGURE 22.2 Changes in ethylene and ACC content and ACC oxidase activity during fruit ripening. Changes in the ACC oxidase activity and ethylene and ACC concentrations of Golden Delicious apples. The data are plotted as a function of days after harvest. Increases in ethylene and ACC concentrations and in ACC oxidase activity are closely correlated with ripening. (After Yang 1987.)

acetic acid, or IAA) are in fact mediated by the ethylene produced in response to auxin.

Inhibitors of protein synthesis block both ACC and IAA-induced ethylene synthesis, indicating that the synthesis of new ACS protein caused by auxins brings about the marked increase in ethylene production. Several ACC synthase genes have been identified whose transcription is elevated following application of exogenous IAA, suggesting that increased transcription is at least partly responsible for the increased ethylene production observed in response to auxin (Nakagawa et al. 1991; Liang et al. 1992).

Ethylene biosynthesis can be stimulated by ACC synthase stabilization

In some cases, the elevation of ethylene biosynthesis occurs through an increase in the stability of ACC synthase protein, rather than, or in addition to, an increase in transcription of ACS genes. For example, fruit ripening, pathogen attack, and cytokinin all increase ethylene biosynthesis in part by stabilizing the ACC synthase protein so that it is broken down more slowly. The carboxy-terminal domain of ACC synthase plays a key role in this regulation. This domain acts as a flag to target the protein for rapid degradation by the 26S proteosome. Phosphorylation of this domain, by either a mitogen-activated protein (MAP) kinase (activated by pathogens) or a calcium-dependent kinase blocks its ability to target the protein for rapid turnover.

Ethylene biosynthesis and action can be blocked by inhibitors

Inhibitors of hormone synthesis or action are valuable for the study of the biosynthetic pathways and physiological roles of hormones. Inhibitors are particularly helpful when it is difficult to distinguish between different hormones that have identical effects in plant tissue or when a hormone affects the synthesis or the action of another hormone.

For example, ethylene mimics high concentrations of auxins by inhibiting stem growth and causing *epinasty* (a downward curvature of leaves). Use of specific inhibitors of ethylene biosynthesis and action made it possible to discriminate between the actions of auxin and ethylene. Studies using inhibitors showed that ethylene is the primary effector of epinasty and that auxin acts indirectly by causing a substantial increase in ethylene production.

INHIBITORS OF ETHYLENE SYNTHESIS **Aminoethoxy-vinyl-glycine (AVG)** and **aminooxyacetic acid (AOA)** block the conversion of AdoMet to ACC (see Figure 22.1). AVG and AOA are known to inhibit enzymes that use the cofactor pyridoxal phosphate. The cobalt ion (Co^{2+}) is also an inhibitor of the ethylene biosynthetic pathway, blocking the conversion of ACC to ethylene by ACC oxidase, the last step in ethylene biosynthesis.

INHIBITORS OF ETHYLENE ACTION Most of the effects of ethylene can be antagonized by specific ethylene inhibitors. Silver ions (Ag^+) applied as silver nitrate ($AgNO_3$) or as silver thiosulfate $[Ag(S_2O_3)_2^{3-}]$ are potent inhibitors of ethylene action. Silver is very specific; the inhibition it causes cannot be induced by any other metal ion.

Carbon dioxide at high concentrations (in the range of 5 to 10%) also inhibits many effects of ethylene, such as the induction of fruit ripening, although CO_2 is less efficient than Ag^+. This effect of CO_2 has often been exploited in the storage of fruits, whose ripening is delayed at elevated CO_2 concentrations. The high concentrations of CO_2 required for inhibition make it unlikely that CO_2 acts as an ethylene antagonist under natural conditions. The volatile compound *trans*-cyclooctene, but not its isomer *cis*-cyclooctene, is a strong competitive inhibitor of ethylene binding (Sisler et al. 1990); *trans*-cyclooctene is thought to act by competing with ethylene for binding to the receptor. 1-methylcyclopropene (MCP) binds almost irreversibly to the ethylene receptor (Figure 22.3) (Sisler and Serek 1997), and effectively blocks multiple ethylene responses. This nearly odorless compound has been marketed under the trade name EthylBloc®, and is used in floriculture to increase the shelf life of cut flowers.

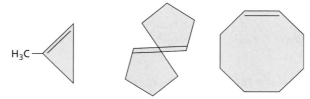

1-Methylcyclopropene (MCP)   ***trans*-Cyclooctene**   ***cis*-Cyclooctene**

FIGURE 22.3 Inhibitors that block ethylene binding to its receptor. Only the *trans* form of cyclooctene is active.

Ethylene can be measured by gas chromatography

Historically, bioassays based on the seedling triple response were used to measure ethylene levels, but they have been replaced by **gas chromatography**. As little as 5 parts per billion (ppb) (5 pL per liter)* of ethylene can be detected, and the analysis time is only 1 to 5 minutes.

Usually the ethylene produced by a plant tissue is allowed to accumulate in a sealed vial, and a sample is withdrawn with a syringe. The sample is injected into a gas chromatograph column in which the different gases are separated and detected by a flame ionization detector. Quantification of ethylene by this method is very accurate. A more sensitive, but less common method to measure ethylene uses a laser-driven photoacoustic detector that can detect as little as 50 parts per trillion (50 ppt = 0.05 pL L^{-1}) ethylene (Voesenek et al. 1997).

Developmental and Physiological Effects of Ethylene

As we have seen, ethylene was discovered in connection with its effects on seedling growth and fruit ripening. It has since been shown to regulate a wide range of responses in plants, including seed germination, cell expansion, cell differentiation, flowering, senescence, and abscission. In this section we will consider the phenotypic effects of ethylene in more detail.

Ethylene promotes the ripening of some fruits

In everyday usage, the term *fruit ripening* refers to the changes in fruit that make it ready to eat. Such changes typically include softening due to the enzymatic breakdown of the cell walls, starch hydrolysis, sugar accumulation, and the disappearance of organic acids and phenolic compounds, including tannins. From the perspective of the plant, however, fruit ripening means that the seeds are ready for dispersal.

For seeds whose dispersal depends on animal ingestion, *ripeness* and *edibility* are synonymous. Brightly colored anthocyanins and carotenoids often accumulate in the epidermis of such fruits, enhancing their visibility. However, for seeds that rely on mechanical or other means for dispersal, *fruit ripening* may mean drying followed by splitting. Because of their importance in agriculture, the vast majority of studies on fruit ripening have focused on edible fruits.

Ethylene has long been recognized as the hormone that accelerates the ripening of edible fruits. Exposure of such fruits to ethylene hastens the processes associated with ripening, and a dramatic increase in ethylene production accompanies the initiation of ripening. However, surveys of a wide range of fruits have shown that not all of them respond to ethylene.

All fruits that ripen in response to ethylene exhibit a characteristic respiratory rise before the ripening phase called a **climacteric**.* Such fruits also show a spike of ethylene production immediately before the respiratory rise (Figure 22.4). Apples, bananas, avocados, and tomatoes are examples of climacteric fruits.

In contrast, fruits such as citrus fruits and grapes do not exhibit the respiration and ethylene production rise and are called **nonclimacteric** fruits. Other examples of climacteric and nonclimacteric fruits are given in Table 22.1.

In climacteric fruits, treatment with ethylene induces the fruit to produce additional ethylene, and its action can

---

*The term climacteric can be used either as a noun, as in "most fruits exhibit a climacteric during ripening" or as an adjective, as in "a climacteric rise in respiration." The term nonclimacteric, however, is used only as an adjective, as in "nonclimacteric fruit."

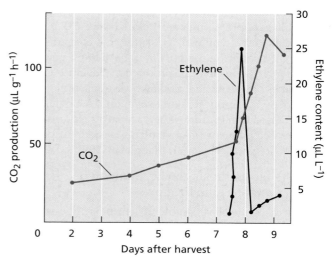

FIGURE 22.4 Ethylene production and respiration. In banana, ripening is characterized by a climacteric rise in respiration rate, as evidenced by the increased CO_2 production. A climacteric rise in ethylene production precedes the increase in CO_2 production, suggesting that ethylene is the hormone that triggers the ripening process. (After Burg and Burg 1965.)

---

*pL = picoliter = 10^{-12} L

| TABLE 22.1 Climacteric and nonclimacteric fruits ||
Climacteric	Nonclimacteric
Apple	Bell pepper
Avocado	Cherry
Banana	Citrus
Cantaloupe	Grape
Cherimoya	Pineapple
Fig	Snap bean
Mango	Strawberry
Olive	Watermelon
Peach	
Pear	
Persimmon	
Plum	
Tomato	

therefore be described as **autocatalytic**. In climacteric plants, two systems of ethylene production operate:

- **System 1**, which acts in vegetative tissue, and in which ethylene inhibits its own biosynthesis
- **System 2**, which occurs during ripening of climacteric fruit and senescence of petals in some species, and in which ethylene stimulates its own biosynthesis—that is, it is autocatalytic

The positive feedback loop for ethylene biosynthesis in System 2 is proposed to integrate ripening of the entire fruit once it has commenced. When unripe climacteric fruits are treated with ethylene, the onset of the climacteric rise is hastened. In contrast, when nonclimacteric fruits are treated with ethylene, the magnitude of the respiratory rise increases as a function of the ethylene concentration, but the treatment does not trigger production of endogenous ethylene and does not accelerate ripening. Elucidation of the role of ethylene in the ripening of climacteric fruits has resulted in many practical applications aimed at either uniform ripening or the delay of ripening.

Although the effects of exogenous ethylene on fruit ripening are straightforward and clear, establishing a causal relation between the level of endogenous ethylene and fruit ripening is more difficult. Inhibitors of ethylene biosynthesis (such as AVG) or of ethylene action (such as CO_2, MCP, or Ag^+) have been shown to delay or even prevent ripening. However, the definitive demonstration that ethylene is required for fruit ripening was provided by experiments in which ethylene biosynthesis was blocked by expression of an antisense version of either ACC synthase or ACC oxidase in transgenic tomatoes (see **Web Topic 22.3**). Elimination of ethylene biosynthesis in these transgenic tomatoes completely blocked fruit ripening, and

ripening was restored by application of exogenous ethylene (Oeller et al. 1991).

Further demonstration of the requirement for ethylene in fruit ripening came from the analysis of the *never-ripe* mutation in tomato. As the name implies, this mutation completely blocks the ripening of tomato fruit. Molecular analysis revealed that *never-ripe* was due to a mutation in an ethylene receptor that rendered it unable to bind ethylene (Lanahan et al. 1994). These experiments provided unequivocal proof of the role of ethylene in fruit ripening, and they opened the door to the manipulation of fruit ripening through biotechnology.

In tomatoes many genes that are highly regulated during ripening have been identified using tomato complementary DNA (cDNA) microarrays.* During tomato fruit ripening, the fruit softens as the result of cell wall hydrolysis, and it changes from green to red as a consequence of chlorophyll loss and the synthesis of the carotenoid pigment lycopene. At the same time, aroma and flavor components are produced.

Analysis of mRNA from fruits of wild-type and transgenic tomato plants genetically engineered to lack ethylene has revealed that gene expression during ripening is regulated by at least two independent pathways:

1. *An ethylene-dependent pathway* includes genes involved in lycopene and aroma biosynthesis, respiratory metabolism, and ACC synthase.

2. *A developmental, ethylene-independent pathway* includes genes encoding ACC oxidase and chlorophyllase.

Thus, not all of the processes associated with ripening in tomato are ethylene dependent.

Leaf epinasty results when ACC from the root is transported to the shoot

The downward curvature of leaves that occurs when the upper (adaxial) side of the petiole grows faster than the lower (abaxial) side is termed *epinasty* (Figure 22.5B). Ethylene and high concentrations of auxin induce epinasty, and it has now been established that auxin acts indirectly by inducing ethylene production. As will be discussed later in the chapter, a variety of stress conditions, such as salt stress or pathogen infection, increase ethylene production and also induce epinasty. There is no known physiological function for the response.

In tomato and other dicots, flooding (waterlogging) or anaerobic conditions around the roots enhances the synthesis of ethylene in the shoot, leading to the epinastic response. Because these environmental stresses are sensed

---

*cDNA microarrays (also known as biochips, DNA chips, or gene arrays) are small chips on which cDNAs or oligonucleotides, each representing a given gene, have been immobilized. When probed with the appropriately tagged cDNAs, DNA chips allow one to measure the expression of thousands of genes in a tissue simultaneously.

(A) Triple response

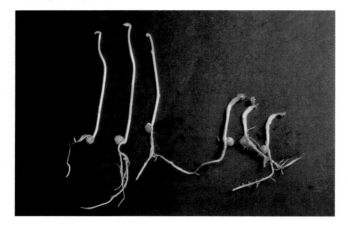

(B) Leaf epinasty

(C) Flower senescense

(D) Root hair formation

FIGURE 22.5 Some physiological effects of ethylene on plant tissue in various developmental stages. (A) Triple response of etiolated pea seedlings. Six-day-old pea seedlings were treated with 10 ppm (parts per million) ethylene (right) or left untreated (left). The treated seedlings show a radial swelling, inhibition of elongation of the epicotyl, and horizontal growth of the epicotyl (diagravitropism). (B) Epinasty, or downward bending of the tomato leaves (right), is caused by ethylene treatment. Epinasty results when the cells on the upper side of the petiole grow faster than those on the bottom. (C) Inhibition of flower senescence by inhibition of ethylene action. Carnation flowers were held in deionized water for 14 days with (left) or without (right) silver thiosulfate (STS), a potent inhibitor of ethylene action. Blocking of ethylene results in a marked inhibition of floral senescence. (D) Promotion of root hair formation by ethylene in lettuce seedlings. Two-day-old seedlings were treated with air (left) or 10 ppm ethylene (right) for 24 hours before the photo was taken. Note the profusion of root hairs on the ethylene-treated seedling. (A and B courtesy of S. Gepstein; C from Reid 1995, courtesy of M. Reid; D from Abeles et al. 1992, courtesy of F. Abeles.)

by the roots and the response is displayed by the shoots, a signal from the roots must be transported to the shoots. This signal is ACC, the immediate precursor of ethylene. ACC levels were found to be significantly higher in the xylem sap after flooding of tomato roots for 1 to 2 days (Figure 22.6) (Bradford and Yang 1980).

Because water fills the air spaces in waterlogged soil, and O_2 diffuses slowly through water, the concentration of oxygen around flooded roots decreases dramatically. The elevated production of ethylene appears to be caused by the accumulation of ACC in the roots under anaerobic conditions, since the conversion of ACC to ethylene requires oxygen (see Figure 22.1). The ACC accumulated in the anaerobic roots is then transported to shoots via the transpiration stream, where it is readily converted to ethylene.

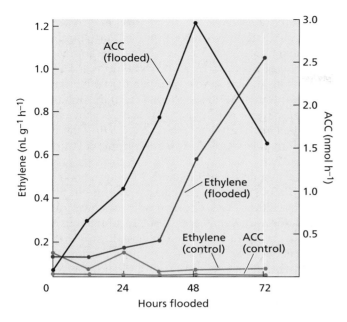

FIGURE 22.6 Changes in the amounts of ACC in the xylem sap and ethylene production in the petiole following flooding of tomato plants. ACC is synthesized in roots, but it is converted to ethylene very slowly under the anaerobic conditions of flooding. ACC is transported via the xylem to the shoot, where it is converted to ethylene. Gaseous ethylene cannot be transported, so it usually affects the tissue nearest the site of its production. In contrast, the ethylene precursor ACC is transportable and can produce ethylene far from the site of ACC synthesis. (After Bradford and Yang 1980.)

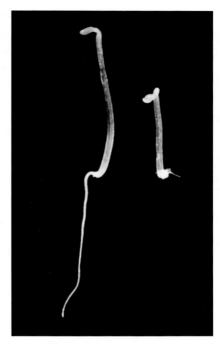

FIGURE 22.7 The triple response in Arabidopsis. Three-day-old etiolated seedlings grown in the presence (right) or absence (left) of 10 ppm ethylene. Note the shortened hypocotyl, reduced root elongation, and exaggeration of the curvature of the apical hook that results from the presence of ethylene. (Courtesy of J. Kieber.)

Ethylene induces lateral cell expansion

At concentrations above 0.1 µL L^{-1}, ethylene changes the growth pattern of seedlings by reducing the rate of elongation and increasing lateral expansion, leading to swelling of the hypocotyl or the epicotyl. These effects of ethylene are common to growing shoots of most dicots, forming part of the **triple response**. In Arabidopsis, the triple response consists of inhibition and swelling of the hypocotyl, inhibition of root elongation, and exaggeration of the curvature of the apical hook (Figure 22.7).

As discussed in Chapter 15, the directionality of plant cell expansion is determined by the orientation of the cellulose microfibrils in the cell wall. Transverse microfibrils reinforce the cell wall in the lateral direction, so that turgor pressure is channeled into cell elongation. The orientation of the microfibrils in turn is determined by the orientation of the cortical array of microtubules in the cortical (peripheral) cytoplasm. In typical elongating plant cells, the cortical microtubules are arranged transversely, giving rise to transversely arranged cellulose microfibrils.

During the seedling triple response to ethylene, the transverse pattern of microtubule alignment is disrupted, and the microtubules switch over to a longitudinal orientation. This 90° shift in microtubule orientation leads to a parallel shift in cellulose microfibril deposition. The newly deposited wall is reinforced in the longitudinal direction rather than the transverse direction, which promotes lateral expansion instead of elongation.

How do microtubules shift from one orientation to another? To study this phenomenon, pea (*Pisum sativum*) epidermal cells were injected with the microtubule protein tubulin, to which a fluorescent dye was covalently attached. The fluorescent "tag" did not interfere with the assembly of microtubules. This procedure allowed researchers to monitor the assembly of microtubules in living cells using a confocal laser scanning microscope, which can focus in many planes throughout the cell.

It was found that microtubules do not reorient from the transverse to the longitudinal direction by complete depolymerization of the transverse microtubules followed by repolymerization of a new longitudinal array of microtubules. Instead, increasing numbers of nontransversely aligned microtubules appear in particular locations (Figure 22.8). Neighboring microtubules then adopt the new alignment, so at one stage different alignments coexist before they adopt a uniformly longitudinal orientation (Yuan et al. 1994). Although the reorientations observed in this study were in response to wounding rather than induced by ethylene, it is presumed that ethylene-induced microtubule reorientation operates by a similar mechanism.

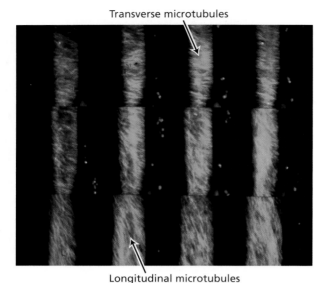

FIGURE 22.8 Reorientation of microtubules from transverse to vertical in pea stem epidermal cells in response to wounding. A living epidermal cell was microinjected with rhodamine-conjugated tubulin, which incorporates into the plant microtubules. A time series of approximately 6-minute intervals shows the cortical microtubules undergoing reorientation from net transverse to oblique/longitudinal. The reorientation seems to involve the appearance of patches of new, "discordant" microtubules in the new direction, concomitant with the disappearance of microtubules from the previous alignment. (From Yuan et al. 1994, photo courtesy of C. Lloyd.)

The hooks of dark-grown seedlings are maintained by ethylene production

Etiolated dicot seedlings are usually characterized by a pronounced hook located just behind the shoot apex (see Figure 22.7). This hook shape facilitates penetration of the seedling through the soil, protecting the tender apical meristem.

Like epinasty, hook formation and maintenance result from ethylene-induced asymmetric growth. The closed shape of the hook is a consequence of the more rapid elongation of the outer side of the stem compared with the inner side. When the hook is exposed to white light, it opens because the elongation rate of the inner side increases, equalizing the growth rates on both sides. The kinematic aspects of hook growth (i.e., maintenance of the hook shape over time) were discussed in Chapter 16.

Red light induces hook opening, and far-red light reverses the effect of red, indicating that phytochrome is the photoreceptor involved in this process (see Chapter 17). A close interaction between phytochrome and ethylene controls hook opening. As long as ethylene is produced by the hook tissue in the dark, elongation of the cells on the inner side is inhibited. Red light inhibits ethylene formation, promoting growth on the inner side, thereby causing the hook to open.

The auxin-insensitive mutation *axr1* and treatment of wild-type seedlings with NPA (*N*-1-naphthylphthalamic acid), an inhibitor of polar auxin transport, both block the formation of the apical hook in Arabidopsis. These and other results indicate a role for auxin in maintaining hook structure. The more rapid growth rate of the outer tissues relative to the inner tissues could reflect an ethylene-dependent auxin gradient, analogous to the lateral auxin gradient that develops during phototropic curvature (see Chapter 19).

A gene required for formation of the apical hook, ***HLS1*** (*HOOKLESS 1*) (so called because mutations in this gene result in seedlings lacking an apical hook), was identified in Arabidopsis (Lehman et al. 1996). In addition to regulating apical hook formation in response to ethylene, *HLS1* also functions to repress hook opening in the dark. Recently it was found that disruption of this gene severely alters the pattern of expression of auxin-responsive genes (Hai et al. 2004). When the *HLS1* was overexpressed in Arabidopsis, it caused constitutive hook formation even in the absence of ethylene.

HLS1 encodes a putative *N*-acetyltransferase that may acetylate, and thereby destabilize, one of the auxin response factors (ARFs). Recall that ARFs are a large family of transcription factors that bind to auxin-response elements in the promoters of many auxin-regulated genes (see Chapter 19). These findings indicate that hook formation and opening are regulated by the interactions of light, ethylene, and auxin, and that *HLS1* plays a role in the integration of these signals (Hai et al. 2004).

Ethylene breaks seed and bud dormancy in some species

Seeds that fail to germinate under normal conditions (water, oxygen, temperature suitable for growth) are said to be dormant (see Chapter 23). Ethylene has the ability to break dormancy and initiate germination in certain seeds, such as cereals. In addition to its effect on dormancy, ethylene increases the rate of seed germination of several species. In peanuts (*Arachis hypogaea*), ethylene production and seed germination are closely correlated. Ethylene can also break bud dormancy, and ethylene treatment is sometimes used to promote bud sprouting in potato and other tubers.

Ethylene promotes the elongation growth of submerged aquatic species

Although usually thought of as an inhibitor of stem elongation, ethylene is able to promote stem and petiole elongation in various submerged or partially submerged aquatic plants. These include the dicots *Ranunculus sceleratus*, *Nymphoides peltata*, and *Callitriche platycarpa*, and the fern *Regnellidium diphyllum*. Another agriculturally important example is deepwater rice (*Oryza sativa*), a cereal (see Chapter 20).

In these species, submergence induces rapid internode or petiole elongation, which allows the leaves or upper parts of the shoot to remain above water. Treatment with ethylene mimics the effects of submergence.

Growth is stimulated in the submerged plants because ethylene builds up in the tissues. In the absence of O_2, ethylene synthesis is diminished, but the loss of ethylene by diffusion is retarded under water. Sufficient oxygen for growth and ethylene synthesis in the underwater parts is usually provided by aerenchyma tissue.

As we saw in Chapter 20, in deepwater rice it has been shown that ethylene stimulates internode elongation by increasing the amount of, and the sensitivity to, gibberellin in the cells of the intercalary meristem. The increased sensitivity to gibberellic acid (GA) in these cells in response to ethylene is brought about by a decrease in the level of abscisic acid (ABA), a potent antagonist of GA.

Ethylene induces the formation of roots and root hairs

Ethylene is capable of inducing adventitious root formation in leaves, stems, flower stems, and even other roots. Vegetative stem cuttings from tomato and petunia make many adventitious roots in response to applied auxin, but in ethylene-insensitive mutants auxin has little or no effect, indicating that the promotive effect of auxin on adventitious rooting is mediated by ethylene (Clark et al. 1999). Ethylene also plays a role in the morphogenesis of crown gall tissue (see **Web Essay 22.1**).

Ethylene has also been shown to act as a positive regulator of root hair formation in several species (see Figure 22.5D). This relationship has been best studied in Arabidopsis, in which root hairs normally are located in the epidermal cells that overlie a junction between the underlying cortical cells (Dolan et al. 1994). In ethylene-treated roots, extra hairs form in abnormal locations in the epidermis; that is, cells not overlying a cortical cell junction differentiate into hair cells (Tanimoto et al. 1995). Seedlings grown in the presence of ethylene inhibitors (such as Ag^+), as well as ethylene-insensitive mutants, display a reduction in root hair formation. These observations suggest that ethylene acts as a positive regulator in the differentiation of root hairs.

Ethylene induces flowering in the pineapple family

Although ethylene inhibits flowering in many species, it induces flowering in pineapple and its relatives, and it is used commercially in pineapple for synchronization of fruit set. Flowering of other species, such as mango, is also initiated by ethylene. On plants that have separate male and female flowers (monoecious species), ethylene may change the sex of developing flowers (see Chapter 24). The promotion of female flower formation in cucumber is one example of this effect.

Ethylene enhances the rate of leaf senescence

As described in Chapter 16, senescence is a genetically programmed developmental process that affects all tissues of the plant. Research has provided several lines of physiological evidence that support roles for ethylene and cytokinins in the control of leaf senescence:

- Exogenous applications of ethylene or ACC (the precursor of ethylene) accelerate leaf senescence, and treatment with exogenous cytokinins delays leaf senescence (see Chapter 21).

- Enhanced ethylene production is associated with chlorophyll loss and color fading, which are characteristic features of leaf and flower senescence (see Figure 22.5C); an inverse correlation has been found between cytokinin levels in leaves and the onset of senescence.

- Inhibitors of ethylene synthesis (e.g., AVG or Co^{2+}) and action (e.g., Ag^+ or CO_2) retard leaf senescence.

Taken together, these physiological studies suggest that senescence is regulated by the balance of ethylene and cytokinin. In addition, abscisic acid has been implicated in the control of leaf senescence. The role of ABA in senescence will be discussed in Chapter 23.

SENESCENCE IN ETHYLENE MUTANTS Direct evidence for the involvement of ethylene in the regulation of leaf senescence has come from molecular genetic studies on Arabidopsis. As will be discussed later in the chapter, several mutants affecting the response to ethylene have been identified. The specific bioassay employed was the triple-response assay, in which ethylene significantly inhibits seedling hypocotyl elongation and promotes lateral expansion.

Ethylene-insensitive mutants, such as *etr1* (*e*thylene-*r*esistant *1*) and *ein2* (*e*thylene-*in*sensitive *2*), were identified by their failure to respond to ethylene (as will be described later in the chapter). The *etr1* mutant is unable to perceive the ethylene signal because of a mutation in the gene that codes for the ethylene receptor protein; the *ein2* mutant is blocked at a later step in the signal transduction pathway.

Consistent with a role for ethylene in leaf senescence, both *etr1* and *ein2* plants were found to be affected not only during the early stages of germination, but throughout the life cycle, including senescence (Zacarias and Reid 1990; Hensel et al. 1993; Grbiè and Bleecker 1995). The ethylene mutants retained their chlorophyll and other chloroplast components for a longer period of time compared to the wild type. However, because the total life spans of these mutants were increased by only 30% over that of the wild type, ethylene appears to increase the *rate* of senescence, rather than acting as a developmental switch that initiates the senescence process.

USE OF GENETIC ENGINEERING TO PROBE SENESCENCE Another very useful genetic approach that offers direct evidence for the function of specific genes is based on transgenic plants. Through genetic engineering technology, the roles of both ethylene and cytokinins in the regulation of leaf senescence have been confirmed.

One way to suppress the expression of a gene is to transform the plant with antisense DNA, which consists of the gene of interest in the reverse orientation with respect to the promoter. When the antisense gene is transcribed, the resulting antisense mRNA is complementary to the sense mRNA and will hybridize to it. Because double-stranded RNA is rapidly degraded in the cell, the effect of the antisense gene is to deplete the cell of the sense mRNA.

Transgenic plants expressing antisense versions of genes that encode enzymes involved in the ethylene biosynthetic pathway, such as ACC synthase and ACC oxidase, can synthesize ethylene only at very low levels. Consistent with a role for ethylene in senescence, such antisense mutants have been shown to exhibit delayed leaf senescence, as well as delayed fruit ripening, in tomato (see Web Topic 22.1).

Some defense responses are mediated by ethylene

Pathogen infection and disease will occur only if the interactions between host and pathogen are genetically compatible. However, ethylene production generally increases in response to pathogen attack in both compatible (i.e., pathogenic) and noncompatible (nonpathogenic) interactions.

The discovery of ethylene-insensitive mutants has allowed the role of ethylene in the response to various pathogens to be assessed. The emerging picture is that the involvement of ethylene in pathogenesis is complex and depends on the particular host–pathogen interaction. For example, blocking ethylene responsiveness does not affect the resistance responses of Arabidopsis to *Pseudomonas* bacteria or of tobacco to tobacco mosaic virus. In compatible interactions of these pathogens and hosts, however, elimination of ethylene responsiveness prevents the development of disease symptoms, even though the growth of the pathogen appears to be unaffected.

On the other hand, ethylene, in combination with jasmonic acid (see Chapter 13), is required for the activation of several plant defense genes. In addition, ethylene-insensitive tobacco and Arabidopsis mutants become susceptible to several necrotrophic (growing on dead host tissue) soil fungi that are normally not pathogenic. Thus ethylene, in combination with jasmonic acid, plays an important role in plant defense against necrotrophic pathogens. On the other hand, ethylene does not appear to play a major role in the response of plants to biotrophic (growing on living tissue) pathogens.

Ethylene regulates changes in the abscission layer that cause abscission

The shedding of leaves, fruits, flowers, and other plant organs is termed **abscission** (see Web Topic 22.5). Abscission takes place in specific layers of cells called **abscission layers**, which become morphologically and biochemically differentiated during organ development. Weakening of the cell walls at the abscission layer depends on cell wall–degrading enzymes such as cellulase and polygalacturonase (Figure 22.9).

The ability of ethylene gas to cause defoliation in birch trees is shown in Figure 22.10. The wild-type tree on the left has lost most of its leaves. The tree on the right has been transformed with a gene for the Arabidopsis ethylene receptor, ETR1, which carries the dominant *etr1-1* mutation (discussed in the next section). This tree is unable to respond to ethylene and therefore does not shed its leaves after ethylene treatment.

Ethylene appears to be the primary regulator of the abscission process, with auxin acting as a suppressor of the ethylene effect (see Chapter 19). However, supraoptimal auxin concentrations stimulate ethylene production, which has led to the use of auxin analogs as defoliants. For exam-

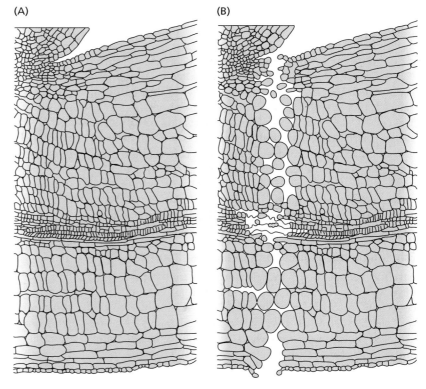

FIGURE 22.9 During the formation of the abscission layer, in this case that of jewelweed (*Impatiens*), two or three rows of cells in the abscission zone (A) undergo cell wall breakdown because of an increase in cell wall–hydrolyzing enzymes (B). The resulting protoplasts round up and increase in volume, pushing apart the xylem tracheary cells, and facilitating the separation of the leaf from the stem. (After Sexton et al. 1984.)

FIGURE 22.10 Effect of ethylene on abscission in birch (*Betula pendula*). The plant on the left is the wild type; the plant on the right was transformed with a mutated version of the Arabidopsis ethylene receptor, *etr1-1*. The expression of this gene was under the transcriptional control of its own promoter. One of the characteristics of these mutant trees is that they do not drop their leaves when fumigated for 3 days with 50 ppm ethylene. (From Vahala et al. 2003.)

ple, 2,4,5-T, the active ingredient in Agent Orange, was widely used as a defoliant during the Vietnam War. Its action is based on its ability to increase ethylene biosynthesis, thereby stimulating leaf abscission.

A model of the hormonal control of leaf abscission describes the process in three distinct sequential phases (Figure 22.11) (Reid 1995):

1. *Leaf maintenance phase.* Prior to the perception of any signal (internal or external) that initiates the abscission process, the leaf remains healthy and fully functional in the plant. A gradient of auxin from the blade to the stem maintains the abscission zone in a nonsensitive state.

2. *Shedding induction phase.* A reduction or reversal in the auxin gradient from the leaf, normally associated with leaf senescence, causes the abscission zone to become sensitive to ethylene. Treatments that enhance leaf senescence may promote abscission by interfering with auxin synthesis and/or transport in the leaf.

3. *Shedding phase.* The sensitized cells of the abscission zone respond to low concentrations of endogenous ethylene by synthesizing and secreting cellulase and other cell wall–degrading enzymes, resulting in shedding.

During the early phase of leaf maintenance, auxin from the leaf prevents abscission by maintaining the cells of the abscission zone in an ethylene-insensitive state. It has long been known that removal of the leaf blade (the site of auxin

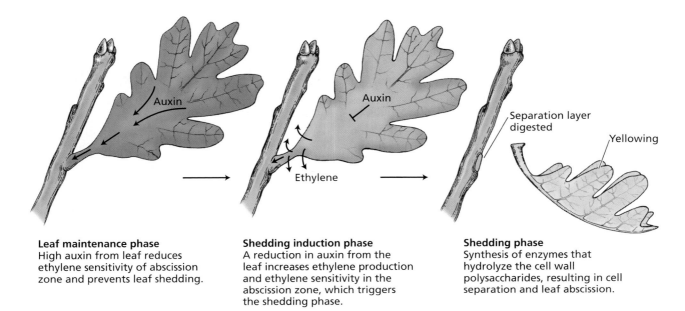

Leaf maintenance phase
High auxin from leaf reduces ethylene sensitivity of abscission zone and prevents leaf shedding.

Shedding induction phase
A reduction in auxin from the leaf increases ethylene production and ethylene sensitivity in the abscission zone, which triggers the shedding phase.

Shedding phase
Synthesis of enzymes that hydrolyze the cell wall polysaccharides, resulting in cell separation and leaf abscission.

FIGURE 22.11 Schematic view of the roles of auxin and ethylene during leaf abscission. In the shedding induction phase, the level of auxin decreases, and the level of ethylene increases. These changes in the hormonal balance increase the sensitivity of the target cells to ethylene. (After Morgan 1984.)

production) promotes petiole abscission. Application of exogenous auxin to petioles from which the leaf blade has been removed delays the abscission process. However, application of auxin to the proximal side of the abscission zone (i.e., the side closest to the stem) actually accelerates the abscission process. These results indicate that it is not the absolute amount of auxin at the abscission zone, but rather the auxin *gradient*, that controls the ethylene sensitivity of these cells.

In the shedding induction phase, the amount of auxin from the leaf decreases and the ethylene level rises. Ethylene appears to decrease the activity of auxin both by reducing its synthesis and transport and by increasing its destruction. The reduction in the concentration of free auxin increases the response of specific target cells to ethylene. The shedding phase is characterized by the induction of genes encoding specific hydrolytic enzymes of cell wall polysaccharides and proteins.

The target cells, located in the abscission zone, synthesize cellulase and other polysaccharide-degrading enzymes, and secrete them into the cell wall. The activities of these enzymes lead to cell wall loosening, cell separation, and abscission.

Ethylene has important commercial uses

Because ethylene regulates so many physiological processes in plant development, it is one of the most widely used plant hormones in agriculture. Auxins and ACC can trigger the natural biosynthesis of ethylene and in several cases are used in agricultural practice. Because of its high diffusion rate, ethylene is very difficult to apply in the field as a gas, but this limitation can be overcome if an ethylene-releasing compound is used. The most widely used such compound is Ethephon, or 2-chloroethylphosphonic acid, which was discovered in the 1960s and is known by various trade names, such as Ethrel.

Ethephon is sprayed in aqueous solution and is readily absorbed and transported within the plant. It releases ethylene slowly by a chemical reaction, allowing the hormone to exert its effects:

$$Cl-CH_2-CH_2-\overset{\overset{O}{\|}}{\underset{O^-}{P}}-OH + OH^- \longrightarrow CH_2=CH_2 + H_2PO_4^- + Cl^-$$

2-Chloroethylphosphonic acid (Ethephon) → Ethylene

Ethephon hastens fruit ripening of apple and tomato and degreening of citrus fruits, synchronizes flowering and fruit set in pineapple, and accelerates abscission of flowers and fruits. It can be used to induce fruit thinning or fruit drop in cotton, cherry, and walnut. It is also used to promote female sex expression in cucumber, to prevent self-pollination and increase yield, and to inhibit terminal growth of some plants in order to promote lateral growth and compact flowering stems.

Storage facilities developed to inhibit ethylene production and promote preservation of fruits have a controlled atmosphere of low O_2 concentration and low temperature that inhibits ethylene biosynthesis. A relatively high concentration of CO_2 (3 to 5%) prevents ethylene's action as a ripening promoter. Low pressure (vacuum) is used to remove ethylene and oxygen from the storage chambers, reducing the rate of ripening and preventing overripening.

It is estimated that from 15 to 35% of the cut flowers harvested in the U.S. are lost because of postharvest spoilage. Specific inhibitors of ethylene biosynthesis and action are useful in postharvest preservation. Silver (Ag^+) has been used extensively to increase the longevity of cut carnations and several other flowers. The potent inhibitor AVG retards fruit ripening and flower fading, but its commercial use has not yet been approved by regulatory agencies. The strong, offensive odor of *trans*-cyclooctene precludes its use in agriculture. On the other hand, the ethylene binding inhibitor 1-methylcyclopropene, which is marketed under the trade name EthylBloc®, is nearly odorless and is used for a variety of postharvest applications, most notably increasing the shelf life of cut flowers. The near future may see a variety of agriculturally important species that have been genetically modified to manipulate the biosynthesis of ethylene or its perception. The inhibition of ripening in tomato by expression of an antisense version of ACC synthase and ACC oxidase has already been mentioned. Another example of this technology is in petunia, in which ethylene biosynthesis has been blocked by transformation of an antisense version of ACC oxidase. Senescence and petal wilting of cut flowers are delayed for weeks in these transgenic plants.

Ethylene Signal Transduction Pathways

Despite the broad range of ethylene's effects on development, the primary steps in ethylene action are assumed to be similar in all cases: They all involve binding to a receptor, followed by activation of one or more signal transduction pathways (see Chapter 14 on the web site) leading to the cellular response. Ultimately, ethylene exerts its effects primarily by altering the pattern of gene expression. In recent years, remarkable progress has been made in our understanding of ethylene perception, as the result of molecular genetic studies of *Arabidopsis thaliana*.

One key to the elucidation of ethylene signaling components has been the use of the triple-response morphology of etiolated Arabidopsis seedlings to isolate mutants affected in their response to ethylene (see Figure 22.7) (Guzman and Ecker 1990). Two classes of mutants have been identified by experiments in which mutagenized Arabidopsis seeds were grown on an agar medium in the presence or absence of ethylene for 3 days in the dark:

1. Mutants that fail to respond to exogenous ethylene (ethylene-resistant or ethylene-insensitive mutants)
2. Mutants that display the response even in the absence of ethylene (constitutive mutants)

Ethylene-insensitive mutants are identified as tall seedlings extending above the lawn of short, triple-responding seedlings when grown in the presence of ethylene. Conversely, constitutive ethylene response mutants are identified as seedlings displaying the triple response in the absence of exogenous ethylene.

Ethylene receptors are related to bacterial two-component system histidine kinases

The first ethylene-insensitive mutant isolated was *etr1* (*ethylene-resistant 1*) (Figure 22.12). The *etr1* mutant was identified in a screen for mutations that block the response of Arabidopsis seedlings to ethylene. The amino acid sequence of the carboxy-terminal half of ETR1 is similar to bacterial two-component histidine kinases—receptors used by bacteria to perceive various environmental cues, such as chemosensory stimuli, phosphate availability, and osmolarity.

Bacterial two-component systems consist of a sensor histidine kinase and a response regulator, which often acts as a transcription factor (see Chapter 14 on the web site). ETR1 was the first example of a eukaryotic histidine kinase, but others have since been found in yeast, mammals, and plants. Both phytochrome (see Chapter 17) and the cytokinin receptor (see Chapter 21) also share sequence similarity to bacterial two-component histidine kinases.

The similarity to bacterial receptors and the ethylene insensitivity of the *etr1* mutants suggested that ETR1 might be an ethylene receptor. Consistent with this hypothesis, *ETR1* expression in yeast conferred the ability to bind radiolabeled ethylene with an affinity that closely parallels the dose-response curve of Arabidopsis seedlings to ethylene (see **Web Topic 22.6**).

The Arabidopsis genome encodes four additional proteins similar to ETR1 that also function as ethylene receptors: ETR2, ERS1 (*ETHYLENE-RESPONSE SENSOR 1*), ERS2, and EIN4 (Figure 22.13). Like ETR1, these receptors have been shown to bind ethylene, and missense mutations in the genes that encode these proteins, analogous to the original *etr1* mutation, prevent ethylene binding to the receptor while allowing the receptor to function normally as a regulator of the ethylene response pathway in the absence of ethylene.

All of these proteins share at least two domains:

1. The amino-terminal domain spans the membrane at least three times and contains the ethylene-binding site. Ethylene can readily access this site because of its hydrophobicity.
2. The carboxy-terminal half of the ethylene receptors contains a domain homologous to histidine kinase catalytic domains.

A subset of the ethylene receptors also have a carboxy-terminal domain that is similar to bacterial two-component receiver domains. In other two-component systems, binding of ligand regulates the activity of the histidine kinase domain, which autophosphorylates a conserved histidine residue. The phosphate is then transferred to an aspartic acid residue located within the fused receiver domain. Histidine kinase activity has been demonstrated for one of the ethylene receptors—ETR1. However, genetic studies have shown that, in contrast to bacterial two-component systems, the histidine kinase activity of ETR1 does *not* play a primary role as an ethylene receptor (Wang et al 2003). Instead, the kinase activity of ETR1 appears to have more subtle effects on ethylene signaling that may modulate the ethylene response (Qu et al. 2004; Binder et al. 2004). Several other ethylene receptors are missing critical amino acids ("Subfamily 2" in Figure 22.13), making it unlikely that they possess histidine kinase activity.

ETR1 is located on the endoplasmic reticulum, rather than on the plasma membrane as are most receptors. Such

FIGURE 22.12 Screen for the *etr1* mutant of Arabidopsis. Seedlings were grown for 3 days in the dark in ethylene. Note that all but one of the seedlings exhibit the triple response: exaggeration in curvature of the apical hook, inhibition and radial swelling of the hypocotyl, and horizontal growth. The *etr1* mutant is completely insensitive to the hormone and grows like an untreated seedling. (Photograph by K. Stepnitz of the MSU/DOE Plant Research Laboratory.)

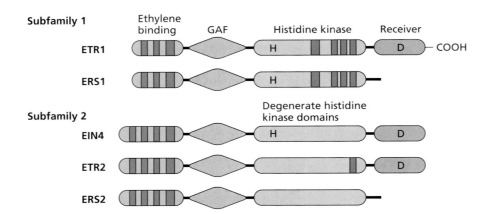

FIGURE 22.13 Schematic diagram of five ethylene receptor proteins and their functional domains. The GAF domain is a conserved cGMP-binding domain, found in a diverse group of proteins, that generally act as small molecule-binding regulatory domains. H and D are histine and aspartate residues which participate in phosphorylation. Note that EIN4, ETR2, and ERS2 have degenerate histidine kinase domains, meaning that they are missing critical, highly conserved amino acids that are required for histidine kinase catalytic activity.

an intracellular location for the ethylene receptor is consistent with the hydrophobic nature of ethylene, which enables it to pass freely through the plasma membrane into the cell. In this respect ethylene is similar to the hydrophobic signaling molecules of animals, such as steroids and the gas nitric oxide, which also bind to intracellular receptors.

High-affinity binding of ethylene to its receptor requires a copper cofactor

Even before the identification of its receptor, scientists had predicted that ethylene would bind to its receptor via a transition metal cofactor, most likely copper or zinc. This prediction was based on the high affinity of olefins, such as ethylene, for these transition metals. Recent genetic and biochemical studies have borne out these predictions.

Analysis of the ETR1 ethylene receptor expressed in yeast demonstrated that a copper ion was coordinated to the protein and that this copper was necessary for high-affinity ethylene binding (Rodriguez et al. 1999). Silver ion could substitute for copper to yield high-affinity binding, which indicates that silver blocks the action of ethylene not by interfering with ethylene binding, but by preventing the changes in the protein that normally occur when ethylene binds to the receptor.

Evidence that copper binding is required for ethylene receptor function in vivo came from identification of the *RAN1* (*RESPONSIVE-TO-ANTAGONIST 1*) gene in Arabidopsis (Hirayama et al. 1999). Strong *ran1* mutations block the formation of functional ethylene receptors (Woeste and Kieber 2000). Cloning of *RAN1* revealed that it encodes a protein similar to a yeast protein required for the transfer of a copper ion cofactor to an iron transport protein. In an analogous manner, RAN1 is likely to be involved in the addition of a copper ion cofactor necessary for the function of the ethylene receptors.

Unbound ethylene receptors are negative regulators of the response pathway

In Arabidopsis, tomato, and probably most other plant species, the ethylene receptors are encoded by multigene families. Targeted disruption (complete inactivation) of the five Arabidopsis ethylene receptors (ETR1, ETR2, ERS1, ERS2, and EIN4) has revealed that they are functionally redundant (Hua and Meyerowitz 1998). That is, disruption of any single gene encoding one of these proteins has no effect, but a plant with disruptions in multiple receptor genes exhibits a constitutive ethylene response phenotype (Figure 22.14D).

The observation that ethylene responses, such as the triple response, become constitutive when the receptors are disrupted indicates that the receptors are normally "on" (i.e., in the active state) in the *absence* of ethylene, and that the function of the receptor *minus* its ligand (ethylene), is to *shut off* the signaling pathway that leads to the response (see Figure 22.14B). Binding of ethylene "turns off" (inactivates) the receptors, thus allowing the response pathway to proceed (see Figure 22.14A).

This somewhat counterintuitive model for ethylene receptors as negative regulators of a signaling pathway is unlike the mechanism of most animal receptors, which, after binding their ligands, serve as positive regulators of their respective signal transduction pathways.

In contrast to the disrupted receptors, receptors with missense mutations at the ethylene binding site (as occurs in the original *etr1* mutant) are unable to bind ethylene, but are still active as negative regulators of the ethylene response pathway. Such missense mutations result in a plant that expresses a subset of receptors that can no longer be turned off by ethylene, and thus confer a *dominant ethylene-insensitive phenotype* (see Figure 22.14C). Even though the normal receptors can all be turned off by ethylene, the mutant receptors continue to signal the cell to suppress ethylene responses whether ethylene is present or not. In tomato, the *never-ripe* mutation, which, as the name suggests, sets fruit that fails to ripen, is such a dominant ethylene-insensitive mutation in a tomato ethylene receptor.

A serine/threonine protein kinase is also involved in ethylene signaling

The recessive *ctr1* (*constitutive triple response 1* = triple response in the absence of ethylene) mutation was identified in screens for mutations that constitutively activate

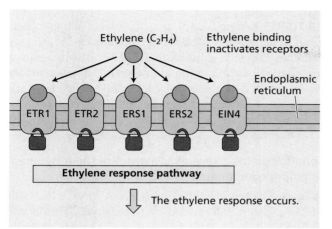
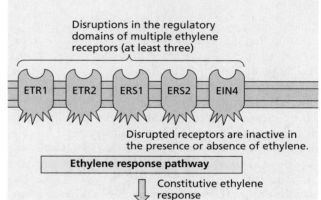
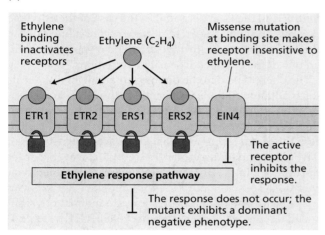

FIGURE 22.14 Model for ethylene receptor action based on the phenotype of receptor mutants. (A) In the wild type, ethylene binding inactivates the receptors, allowing the response to occur. (B) In the absence of ethylene the receptors act as negative regulators of the response pathway. (C) A missense mutation that interferes with ethylene binding to its receptor, but leaves the regulatory site active, results in a dominant negative phenotype. (D) Disruption mutations in the regulatory sites result in a constitutive ethylene response.

ethylene responses (Figure 22.15). The fact that the mutation caused an *activation* of the ethylene response suggests that the wild-type protein also acts as a *negative regulator* of the response pathway (Kieber et al. 1993), similar to the ethylene receptors.

CTR1 appears to be related to Raf, a MAPKKK (*m*itogen-*a*ctivated *p*rotein *k*inase *k*inase *k*inase) type of serine/threonine protein kinase that is involved in the transduction of various external regulatory signals and developmental signaling pathways in organisms ranging from yeast to humans (see Chapter 14 on the web site). In animal cells, the final product in the MAP kinase cascade is a phosphorylated transcription factor that regulates gene expression in the nucleus.

Various lines of evidence indicate that the CTR1 protein directly interacts with the ethylene receptors, forming part of a protein complex involved in perceiving ethylene. Genetic analysis has shown that the interaction of CTR1 with the ethylene receptors is necessary for its function, as mutations in CTR1 that block this interaction, but otherwise do not affect the protein, cause CTR1 to be inactive in the plant. The precise mechanism by which CTR1 is regulated by ETR1 and the other ethylene receptors is still not known.

EIN2 encodes a transmembrane protein

The *ein2* (*e*thylene-*in*sensitive 2) mutation blocks all ethylene responses in both seedling and adult Arabidopsis plants. The *EIN2* gene encodes a protein containing 12 membrane-spanning domains that is most similar to the N-RAMP (*n*atural *r*esistance–*a*ssociated *m*acrophage *p*rotein) family of cation transporters in animals (Alonso et al. 1999),

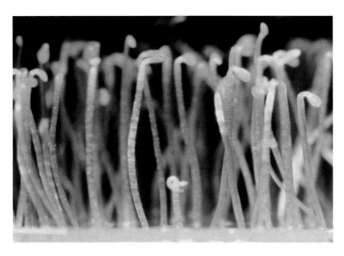

FIGURE 22.15 Screen for Arabidopsis mutants that constitutively display the triple response. Seedlings were grown for 3 days in the dark in air. A single *ctr1* mutant seedling is evident among the taller, wild-type seedlings. (Courtesy of J. Kieber.)

suggesting that it may act as a channel or pore. To date, however, researchers have failed to demonstrate a transport activity for this protein, and the intracellular location of the protein is not known.

Interestingly, mutations in the *EIN2* gene have also been identified in genetic screens for resistance to other hormones, such as jasmonic acid and ABA, which may reflect the role that ethylene plays in the response to these other hormones.

Ethylene regulates gene expression

One of the primary effects of ethylene signaling is an alteration in the expression of various target genes. Ethylene affects the mRNA transcript levels of numerous genes, including those that encode cellulase and genes related to ripening and ethylene biosynthesis. Regulatory sequences called **ethylene response elements**, or **EREs**, have been identified among the ethylene-regulated genes.

Key components mediating ethylene's effects on gene expression are the EIN3 family of transcription factors (Chao et al. 1997). There are at least four *EIN3*-like genes in Arabidopsis, and homologs have been identified in tomato and tobacco. In response to an ethylene signal, homodimers of EIN3 or closely related proteins bind to the promoters of genes that are rapidly induced by ethylene, including ***ERF1*** (*ETHYLENE RESPONSE FACTOR 1*), to activate their transcription (Solano et al. 1998).

ERF1 encodes a protein that belongs to the **ERE-binding protein** (**EREBP**) family of transcription factors, which were first identified in tobacco as proteins that bind to ERE sequences (Ohme-Takagi and Shinshi 1995). Several EREBPs are rapidly up-regulated in response to ethylene. The *EREBP* genes exist in Arabidopsis as a very large gene family, but only a few of the genes are inducible by ethylene.

As occurs in other hormone signaling pathways and in regulating the ethylene biosynthetic pathway, the regula-

tion of EIN3 protein stability plays an important role in ethylene signaling. Two redundant F-box proteins*, EBF1 and EBF2 (*EIN3-b*inding *F*-box 1 and 2), promote ubiquitination and thus targeting of EIN3 for degradation by the 26S proteosome. Ethylene inhibits this EBF1/EBF2-dependent degradation of EIN3 via an unknown mechanism, resulting in the accumulation of EIN3 and the subsequent expression of ethylene-regulated genes. Thus, ethylene acts, at least in part, by regulating the level of the EIN3 and EIN3-like proteins.

Genetic epistasis reveals the order of the ethylene signaling components

The order of action of the genes *ETR1*, *EIN2*, *EIN3*, and *CTR1* has been determined by the analysis of how the mutations interact with each other (i.e., their epistatic order). Two mutants with opposite phenotypes are crossed, and a line harboring both mutations (the double mutant) is identified in the F_2 generation. In the case of the ethylene response mutants, researchers constructed a line doubly mutant for *ctr1*, a constitutive ethylene response mutant, and one of the ethylene-insensitive mutations.

The phenotype that the double mutant displays reveals which of the mutations is epistatic to the other. For example, if an *etr1/ctr1* double mutant displays a *ctr1* mutant phenotype, the *ctr1* mutation is said to be epistatic to *etr1*. From this it can be inferred that CTR1 acts downstream of ETR1 (Avery and Wasserman 1992). In this way, the order of action of *ETR1*, *EIN2*, and *EIN3* were determined relative to *CTR1*.

The ETR1 protein has been shown to interact physically with the predicted downstream protein, CTR1, suggesting that the ethylene receptors may directly regulate the kinase activity of CTR1 (Clark et al. 1998). The model in Figure 22.16 summarizes these and other data. Genes that are similar to several of these Arabidopsis signaling genes have been found in other species (see **Web Topic 22.7**).

This model is still incomplete because other ethylene response mutations have been identified that act in this pathway. In addition, we are only beginning to understand the biochemical properties of these proteins and how they interact. Further research will be needed to provide a more complete picture of the molecular basis for the perception and transduction of the ethylene signal.

Summary

Ethylene is formed in most organs of higher plants. Senescing tissues and ripening fruits produce more ethylene than do young or mature tissues. The precursor of ethylene in vivo is the amino acid methionine, which is converted to AdoMet (*S*-adenosylmethionine), ACC (1-aminocyclopropane-1-carboxylic acid), and ethylene. The rate-limiting

---

*F-box proteins contain a sequence of about 50 amino acids that promotes protein–protein interaction.

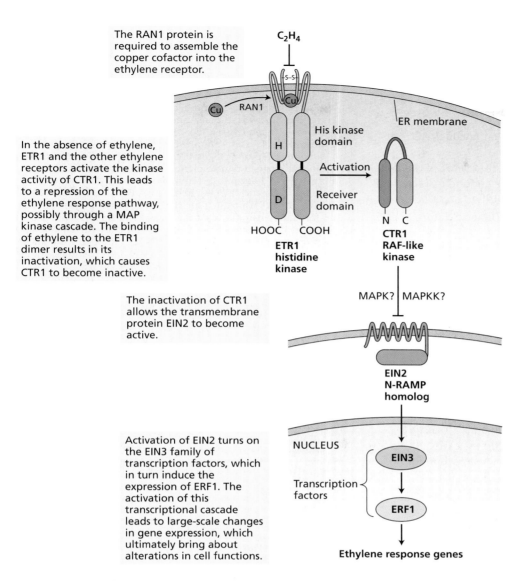

FIGURE 22.16 Model of ethylene signaling in Arabidopsis. Ethylene binds to the ETR1 receptor, which is an integral membrane protein of the endoplasmic reticulum membrane. Multiple isoforms of ethylene receptors may be present in a cell; only ETR1 is shown for simplicity. The receptor is a dimer, held together by disulfide bonds. Ethylene binds within the transmembrane domain, through a copper cofactor, which is assembled into the ethylene receptors by the RAN1 protein.

step of this pathway is the conversion of AdoMet to ACC, which is catalyzed by ACC synthase. ACC synthase is encoded by members of a multigene family that are differentially regulated in various plant tissues and in response to various inducers of ethylene biosynthesis.

Ethylene biosynthesis is triggered by various developmental processes, by auxins, and by environmental stresses. In all these cases the level of activity and of mRNA of ACC synthase increases. The physiological effects of ethylene can be blocked by biosynthesis inhibitors or by antagonists. AVG (aminoethoxy-vinylglycine) and AOA (aminooxyacetic acid) inhibit the synthesis of ethylene; carbon dioxide, silver ions, *trans*-cyclooctene, and methylcyclopropene (MCP) inhibit ethylene action. Ethylene can be detected and measured by gas chromatography.

Ethylene regulates fruit ripening and processes associated with leaf and flower senescence, leaf and fruit abscission, root hair development, seedling growth, and hook opening. Ethylene also regulates the expression of various genes, including ripening-related genes and pathogenesis-related genes.

The ethylene receptor is encoded by a family of genes that encode proteins similar to bacterial two-component histidine kinases. Ethylene binds to these receptors in a transmembrane domain through a copper cofactor. Downstream signal transduction components include CTR1, a member of the Raf family of protein kinases; and EIN2, a channel-like transmembrane protein. The pathway activates a cascade of transcription factors, including the EIN3 and EREBP families, which then modulate gene expression.

Web Material

Web Topics

22.1 Cloning of the Gene That Encodes ACC Synthase
A brief description of the cloning of the gene for ACC synthase using antibodies raised against the partially purified protein.

22.2 Demonstration of the Dimeric Nature of ACC Synthase
The dimeric nature of ACC synthase has been confirmed by an elegant study involving functional complementation.

22.3 Cloning of the Gene That Encodes ACC Oxidase
The ACC oxidase gene was cloned by a circuitous route using antisense DNA.

22.4 ACC Synthase Gene Expression and Biotechnology
A discussion of the use of the ACC synthase gene in biotechnology.

22.5 Abscission and the Dawn of Agriculture
A short essay on the domestication of modern cereals based on artificial selection for non-shattering rachises.

22.6 Ethylene Binding to ETR1 and Seedling Response to Ethylene
Ethylene binding to its receptor ETR1 was first demonstrated by expressing the gene in yeast.

22.7 Conservation of Ethylene Signaling Components in Other Plant Species
The evidence suggests that ethylene signaling is similar in all plant species.

Web Essay

22.1 Tumor-Induced Ethylene Controls Crown Gall Morphogenesis
Agrobacterium tumefaciens-induced galls produce very high ethylene concentrations, which reduce vessel diameter in the host stem adjacent to the tumor and enlarge the gall surface giving priority in water supply to the growing tumor over the host shoot.

Chapter References

Abeles, F. B., Morgan, P. W., and Saltveit, M. E., Jr. (1992) *Ethylene in Plant Biology*, 2nd ed. Academic Press, San Diego, CA.

Alonso, J. M., Hirayama, T., Roman, G., Nourizadeh, S., and Ecker, J. R. (1999) EIN2, a bifunctional transducer of ethylene and stress responses in *Arabidopsis*. *Science* 284: 2148–2152.

Avery, L., and Wasserman, S. (1992) Ordering gene function: The interpretation of epistasis in regulatory hierarchies. *Trends Genet.* 8: 312–316.

Binder, B. M., O'Malley, R. C., Wang, W., Moore, J. M., Parks, B. M., Spalding, E. P., and Bleecker, A. B. (2004) *Arabidopsis* seedling growth response and recovery to ethylene. A kinetic analysis. *Plant Physiol.* 136: 2913–2920.

Bradford, K. J., and Yang, S. F. (1980) Xylem transport of 1-aminocyclopropane-1-carboxylic acid, an ethylene precursor, in waterlogged tomato plants. *Plant Physiol.* 65: 322–326.

Burg, S. P., and Burg, E. A. (1965) Relationship between ethylene production and ripening in bananas. *Bot. Gaz.* 126: 200–204.

Burg, S. P., and Burg, K. V. (1959) The physiology of ethylene formation in apples. *Proc. Natl. Acad. Sci. USA* 45: 335–344.

Capitani, G., Hohenester, E., Feng, L., Storici, P., Kirsch, J. F., and Jansonius, J. N. (1999). Structure of 1-aminocyclopropane-1-carboxylate synthase, a key enzyme in the biosynthesis of the plant hormone ethylene. *J. Mol. Biol.* 294, 745–756.

Chao, Q., Rothenberg, M., Solano, R., Roman, G., Terzaghi, W., and Ecker, J. R. (1997) Activation of the ethylene gas response pathway in *Arabidopsis* by the nuclear protein ETHYLENE-INSENSITIVE3 and related proteins. *Cell* 89: 1133–1144.

Clark, D. G., Gubrium, E. K., Barrett, J. E., Nell, T. A., and Klee, H. J. (1999) Root formation in ethylene-insensitive plants. *Plant Phys.* 121, 53–60.

Clark, K. L., Larsen, P. B., Wang, X., and Chang, C. (1998) Association of the *Arabidopsis* CTR1 Raf-like kinase with the ETR1 and ERS ethylene receptors. *Proc. Natl. Acad. Sci. USA* 95: 5401–5406.

Dolan, L., Duckett, C. M., Grierson, C., Linstead, P., Schneider, K., Lawson, E., Dean, C., Poethig, S., and Roberts, K. (1994) Clonal relationships and cell patterning in the root epidermis of *Arabidopsis*. *Development* 120: 2465–2474.

Grbič, V., and Bleecker, A. B. (1995) Ethylene regulates the timing of leaf senescence in *Arabidopsis*. *Plant J.* 8: 595–602.

Guzman, P., and Ecker, J. R. (1990) Exploiting the triple response of *Arabidopsis* to identify ethylene-related mutants. *Plant Cell* 2: 513–523.

Hai, L., Johnson, P., Stepanova, A., Alonso, J. M., and Ecker, J. M. (2004) Convergence of signaling pathways in the control of differential cell growth in *Arabidopsis*. *Dev. Cell* 7: 193–204

Hensel, L. L., Grbič, V., Baumgarten, D. A., and Bleecker, A. B. (1993) Developmental and age-related processes that influence the longevity and senescence of photosynthetic tissues in *Arabidopsis*. *Plant Cell* 5: 553–564.

Hirayama, T., Kieber, J. J., Hirayama, N., Kogan, M., Guzman, P., Nourizadeh, S., Alonso, J. M., Dailey, W. P., Dancis, A., and Ecker, J. R. (1999) RESPONSIVE-TO-ANTAGONIST1, a Menkes/Wilson disease-related copper transporter, is required for ethylene signaling in *Arabidopsis*. *Cell* 97: 383–393.

Hua, J., and Meyerowitz, E. M. (1998) Ethylene responses are negatively regulated by a receptor gene family in *Arabidopsis thaliana*. *Cell* 94: 261–271.

Huai, Q., Xia, Y., Chen, Y., Callahan, B., Li, N., and Ke, H. (2001). Crystal structures of 1-aminocyclopropane-1-carboxylate (ACC) synthase in complex with aminoethoxyvinylglycine and pyridoxal-5'-phosphate provide new insight into catalytic mechanisms. *J. Biol. Chem.* 276, 38210–38216.

Kieber, J. J., Rothenburg, M., Roman, G., Feldmann, K. A., and Ecker, J. R. (1993) CTR1, a negative regulator of the ethylene response pathway in *Arabidopsis*, encodes a member of the Raf family of protein kinases. *Cell* 72: 427–441.

Lanahan, M., Yen, H.-C., Giovannoni, J., and Klee, H. (1994) The *Never-ripe* mutation blocks ethylene perception in tomato. *Plant Cell* 6: 427–441.

Lehman, A., Black, R., and Ecker, J. R. (1996) *Hookless1*, an ethylene response gene, is required for differential cell elongation in the *Arabidopsis* hook. *Cell* 85: 183–194.

Liang, X., Abel, S., Keller, J., Shen, N., and Theologis, A. (1992) The 1-aminocyclopropane-1-carboxylate synthase gene family of *Arabidopsis thaliana*. *Proc. Natl. Acad. Sci. USA* 89: 11046–11050.

McKeon, T. A., Fernández-Maculet, J. C., and Yang, S. F. (1995) Biosynthesis and metabolism of ethylene. In *Plant Hormones: Physiology, Biochemistry and Molecular Biology*, 2nd ed., P. J. Davies, ed., Kluwer, Dordrecht, Netherlands, pp. 118–139.

Morgan, P. W. (1984) Is ethylene the natural regulator of abscission? In *Ethylene: Biochemical, Physiological and Applied Aspects*, Y. Fuchs and E. Chalutz, eds., Martinus Nijhoff, The Hague, Netherlands, pp. 231–240.

Nakagawa, J. H., Mori, H., Yamazaki, K., and Imaseki, H. (1991) Cloning of the complementary DNA for auxin-induced 1-aminocyclopropane-1-carboxylate synthase and differential expression of the gene by auxin and wounding. *Plant Cell Physiol.* 32: 1153–1163.

Oeller, P., Min-Wong, L., Taylor, L., Pike, D., and Theologis, A. (1991) Reversible inhibition of tomato fruit senescence by antisense RNA. *Science* 254: 437–439.

Ohme-Takagi, M., and Shinshi, H. (1995) Ethylene-inducible DNA binding proteins that interact with an ethylene-responsive element. *Plant Cell* 7: 173–182.

Perovic, S., Seack, J., Gamulin, V., Müller, W. E. G., and Schröder, H. C. (2001) Modulation of intracellular calcium and proliferative activity of invertebrate and vertebrate cells by ethylene. *BMC Cell Biol.* 2: 7.

Qu, X., and Schaller, G. E. (2004) Requirement of the histidine kinase domain for signal transduction by the ethylene receptor ETR1. *Plant Physiol.* 136: 2961–2970.

Raskin, I., and Beyer, E. M., Jr. (1989) Role of ethylene metabolism in *Amaranthus retroflexus*. *Plant Physiol.* 90: 1–5.

Reid, M. S. (1995) Ethylene in plant growth, development and senescence. In *Plant Hormones: Physiology, Biochemistry and Molecular Biology*, 2nd ed., P. J. Davies, ed., Kluwer, Dordrecht, Netherlands, pp. 486–508.

Rodriguez, F. I., Esch, J. J., Hall, A. E., Binder, B. M., Schaller, E. G., and Bleecker, A. B. (1999) A copper cofactor for the ethylene receptor ETR1 from *Arabidopsis*. *Science* 283: 396–398.

Sexton, R., Burdon, J. N., Reid, J. S. G., Durbin, M. L., and Lewis, L. N. (1984) Cell wall breakdown and abscission. In *Structure, Function, and Biosynthesis of Plant Cell Walls*, W. M. Dugger and S. Bartnicki-Garcia, eds., American Society of Plant Physiologists, Rockville, MD, pp. 383–406.

Sisler, E. C., and Serek, M. (1997) Inhibitors of ethylene responses in plants at the receptor level: Recent developments. *Physiol. Plant.* 100: 577–582.

Sisler, E., Blankenship, S., and Guest, M. (1990) Competition of cyclooctenes and cyclooctadienes for ethylene binding and activity in plants. *Plant Growth Regul.* 9: 157–164.

Solano, R., Stepanova, A., Chao, Q., and Ecker, J. R. (1998) Nuclear events in ethylene signaling: A transcriptional cascade mediated by ETHYLENE-INSENSITIVE3 and ETHYLENE-RESPONSE-FACTOR1. *Genes Dev.* 12: 3703–3714.

Tanimoto, M., Roberts, K., and Dolan, L. (1995) Ethylene is a positive regulator of root hair development in *Arabidopsis thaliana*. *Plant J.* 8: 943–948.

Vahala, J., Ruonala, R., Keinänen, M., Tuominen, H., and Kangasjärvi, J. (20003) Ethylene insensitivity modulates ozone-induced cell death in birch (*Betula pendula*). *Plant Phys.* 132:185–195.

Voesenek, L. A. C. J., Banga, M., Rijnders, J. H. G. M., Visser, E. J. W., Harren, F. J. M., Brailsford, R. W., Jackson, M. B., and Blom, C. W. P. M. (1997) Laser-driven photoacoustic spectroscopy: What we can do with it in flooding research. *Ann. Bot.* 79: 57–65.

Wang, W., Hall, A. E., O'Malley, R., and Bleecker, A. B. (2003) Canonical histidine kinase activity of the transmitter domain of the ETR1 ethylene receptor from *Arabidopsis* is not required for signal transmission *Proc. Natl. Acad. Sci. USA* 100: 352–357.

Woeste, K., and Kieber, J. J. (2000) A strong loss-of-function allele of *RAN1* results in constitutive activation of ethylene responses as well as a rosette-lethal phenotype. *Plant Cell* 12: 443–455.

Yamagami, T., Tsuchisaka, A., Yamada, K., Haddon, W.F., Harden, L.A., and Theologis, A. (2003). Biochemical diversity among the 1-amino-cyclopropane-1-carboxylate synthase isozymes encoded by the *Arabidopsis* gene family. *J. Biol. Chem.* 278, 49102–49112.

Yang, S. F. (1987) The role of ethylene and ethylene synthesis in fruit ripening. In *Plant Senescence: Its Biochemistry and Physiology*, W. W. Thomson, E. A. Nothnagel, and R. C. Huffaker, eds., American Society of Plant Physiologists, Rockville, MD, pp. 156–166.

Yuan, M., Shaw, P. J., Warn, R. M., and Lloyd, C. W. (1994) Dynamic reorientation of cortical microtubules, from transverse to longitudinal, in living plant cells. *Proc. Natl. Acad. Sci. USA* 91: 6050–6053.

Zacarias, L., and Reid, M. S. (1990) Role of growth regulators in the senescence of *Arabidopsis thaliana* leaves. *Physiol. Plant.* 80: 549–554.

Chapter 23

Abscisic Acid: A Seed Maturation and Antistress Signal

THE EXTENT AND TIMING OF PLANT GROWTH are controlled by the coordinated actions of positive and negative regulators. Some of the most obvious examples of regulated nongrowth are seed and bud dormancy, adaptive features that delay growth until environmental conditions are favorable. For years, plant physiologists suspected that the phenomena of seed and bud dormancy were caused by inhibitory compounds, so they attempted to extract and isolate such compounds from a variety of plant tissues, especially dormant buds.

Early experiments used paper chromatography for the separation of plant extracts, as well as bioassays based on oat coleoptile growth. These experiments led to the identification of a group of growth-inhibiting compounds, including a substance known as *dormin* purified from sycamore leaves collected in early autumn, when the trees were entering dormancy. Upon discovery that dormin was chemically identical to a substance that promotes the abscission of cotton fruits, *abscisin II*, the compound was renamed **abscisic acid (ABA)** (see Figure 23.1) to reflect its supposed involvement in the abscission process.

As will be discussed in this chapter, ABA is now recognized as an important plant hormone that regulates growth and stomatal aperture, particularly when the plant is under environmental stress. Another important function is its regulation of seed maturation and dormancy. Ironically, ABA's effects on abscission remain controversial: In many species, ABA appears to promote senescence (i.e., the events preceding

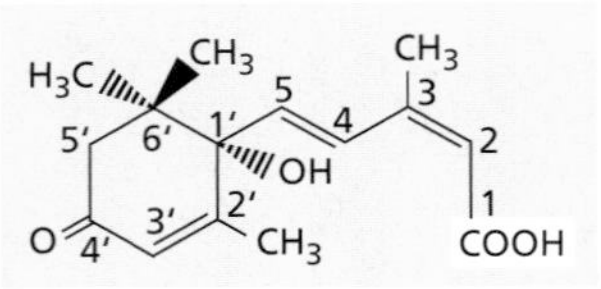

(S)-cis-ABA (naturally occurring active form)

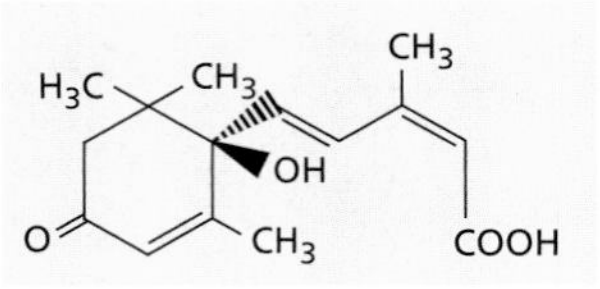

(R)-cis-ABA (inactive in stomatal closure)

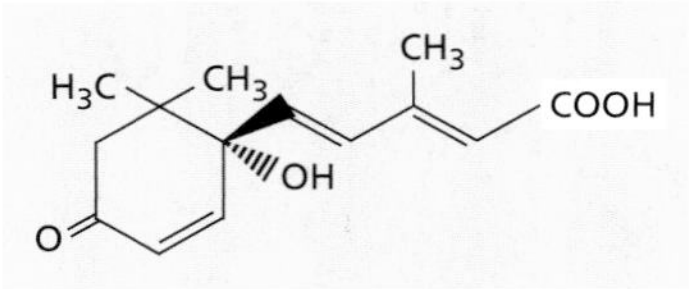

(S)-2-trans-ABA (inactive, but interconvertible with active cis form)

FIGURE 23.1 The chemical structures of the S (+, or counterclockwise array) and R (–, or clockwise array) forms of cis-ABA, and the (S)-2-trans form of ABA. The numbers in the diagram of (S)-cis-ABA identify the carbon atoms.

abscission), but not abscission itself. In retrospect, *dormin* would have been a more appropriate name for this hormone, but the name *abscisic acid* is firmly entrenched in the literature.

Occurrence, Chemical Structure, and Measurement of ABA

Abscisic acid is a ubiquitous plant hormone in vascular plants. It has been detected in mosses but appears to be absent in liverworts (see Web Topic 23.1). Several genera of fungi make ABA as a secondary metabolite (Milborrow 2001). Within the plant, ABA has been detected in every major organ or living tissue from the root cap to the apical bud. ABA is synthesized in almost all cells that contain chloroplasts or amyloplasts.

The chemical structure of ABA determines its physiological activity

ABA is a 15-carbon compound that resembles the terminal portion of some carotenoid molecules (see Figure 23.1). The orientation of the carboxyl group at carbon 2 determines the *cis* and *trans* isomers of ABA. Nearly all the naturally occurring ABA is in the *cis* form, and by convention the name *abscisic acid* refers to that isomer.

ABA also has an asymmetric carbon atom at position 1′ in the ring, resulting in the S and R (or + and –, respectively) enantiomers. The S enantiomer is the natural form; commercially available synthetic ABA is a mixture of approximately equal amounts of the S and R forms. The S enantiomer is the only one that is active in fast responses to ABA, such as stomatal closure. In long-term responses, such as seed maturation, both enantiomers are active. In contrast to the *cis* and *trans* isomers, the S and R forms cannot be interconverted in the plant tissue.

Studies of the structural requirements for biological activity of ABA have shown that almost any change in the molecule results in loss of activity (see Web Topic 23.2).

ABA is assayed by biological, physical, and chemical methods

A variety of bioassays have been used for ABA, including inhibition of coleoptile growth, germination, or gibberellic acid (GA)–induced α-amylase synthesis. Alternatively, promotion of stomatal closure and gene expression are examples of rapid inductive responses (see Web Topic 23.3).

Physical methods of detection are much more reliable than bioassays because of their specificity and suitability for quantitative analysis. The most widely used techniques are those based on gas chromatography or high-performance liquid chromatography (HPLC). Gas chromatography allows detection of as little as 10^{-13} g of ABA, but it requires several preliminary purification steps, including thin-layer chromatography. Immunoassays are also highly sensitive and specific.

Biosynthesis, Metabolism, and Transport of ABA

As with the other hormones, the response to ABA depends on its concentration within the tissue and on the sensitivity of the tissue to the hormone. The processes of biosynthesis, catabolism, compartmentation, and transport all contribute to the concentration of active hormone in the tissue at any given stage of development. The complete biosynthetic pathway of ABA was elucidated by a combination of biochemical and genetic studies. In this section we describe the biosynthetic pathway of ABA, the variability of ABA concentrations in plant tissues, the regulation of ABA levels by oxidation and conjugation, and the transport of ABA in the vascular system.

ABA is synthesized from a carotenoid intermediate

ABA biosynthesis takes place in chloroplasts and other plastids via the pathway depicted in Figure 23.2. Several ABA-deficient mutants have been identified with lesions at specific steps of the pathway. These mutants exhibit abnormal

FIGURE 23.2 ABA biosynthesis and metabolism. In higher plants, ABA is synthesized via the terpenoid pathway (see Chapter 13), as are cytokinins, brassinosteroids, and gibberellins (see Chapters 21, 24, and 20 respectively). Some ABA-deficient mutants that have been helpful in elucidating the pathway are shown at the steps at which they are blocked. The pathways for ABA catabolism include conjugation to form ABA-β-D-glucosyl ester or oxidation to form phaseic acid and then dihydrophaseic acid. NCED, 9-cis-epoxycarotenoid dioxygenase; ZEP, zeaxanthin epoxidase.

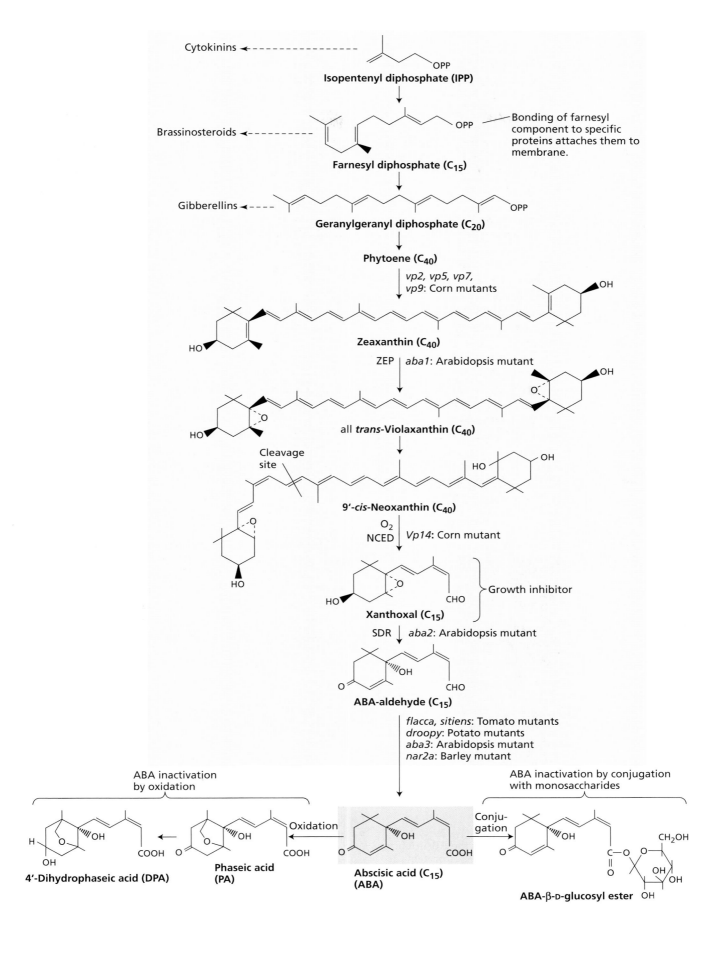

phenotypes that can be corrected by the application of exogenous ABA. For example, *flacca* (*flc*) and *sitiens* (*sit*) are "wilty mutants" of tomato in which the tendency of the leaves to wilt (due to an inability to close their stomata) can be prevented by the application of exogenous ABA. The *aba* mutants of Arabidopsis also exhibit a wilty phenotype. These and other mutants have been useful in elucidating the details of the pathway and in cloning the genes encoding ABA biosynthetic enzymes (Seo and Koshiba 2002).

The pathway begins with isopentenyl diphosphate (IPP)—the biological isoprene unit that is also a precursor of cytokinins, gibberellins, and brassinosteroids—and leads to the synthesis of the C_{40} xanthophyll (i.e., oxygenated carotenoid) **violaxanthin** (see Figure 23.2). Synthesis of violaxanthin is catalyzed by zeaxanthin epoxidase (ZEP), the enzyme encoded by the *ABA1* locus of Arabidopsis. This discovery provided conclusive evidence that ABA synthesis occurs via the "indirect" or carotenoid pathway, rather than by modification of a C_{15} isoprenoid, as in the "direct pathway" of some phytopathogenic fungi. Maize (corn; *Zea mays*) mutants (termed *viviparous* [*vp*]) that are blocked at other steps in the carotenoid pathway also have reduced levels of ABA and exhibit **vivipary**—the precocious germination of seeds in the fruit while still attached to the plant (Figure 23.3). Vivipary is a feature of many ABA-deficient seeds.

Violaxanthin is converted to the C_{40} compound **9′-*cis*-neoxanthin**, which is then cleaved to form the C_{15} compound **xanthoxal**, previously called *xanthoxin*, a neutral growth inhibitor that has physiological properties similar to those of ABA. The cleavage is catalyzed by **9-*cis*-epoxycarotenoid dioxygenase** (**NCED**), so named because it can cleave both 9-*cis*-violaxanthin and 9′-*cis*-neoxanthin.

Synthesis of NCED is rapidly induced by water stress, suggesting that the reaction it catalyzes is a key regulatory step for ABA synthesis. NCEDs are encoded by a family of genes that are differentially regulated in response to stress or developmental signals. The enzyme functions on the thylakoids, where the carotenoid substrate is located, but some family members are present in both soluble and membrane-bound forms, providing a possibility for regulating enzyme activity via effects on localization. Finally, xanthoxal is converted to ABA via oxidative steps involving the intermediate(s) **ABA-aldehyde** and/or possibly xanthoxic acid (abscisic alcohol). Conversion of xanthoxal to abscisic aldehyde is catalyzed by a short chain dehydrogenase/reductase-like (SDR) enzyme, encoded by the *ABA2* locus of Arabidopsis. The final step is catalyzed by a differentially regulated family of abscisic aldehyde oxidases (AAOs) that all require a molybdenum cofactor. Mutations in single AAO family members may have limited effects, but those mutants lacking a functional molybdenum cofactor (e.g., Arabidopsis *aba3* and tomato *flacca*) are unable to synthesize ABA.

ABA concentrations in tissues are highly variable

ABA biosynthesis and concentrations can fluctuate dramatically in specific tissues during development or in response to changing environmental conditions. In developing seeds, for example, ABA levels can increase 100-fold within a few days, reaching average concentrations in the micromolar range, and then decline to very low levels as maturation proceeds. Under conditions of water stress (i.e., dehydration stress), ABA in the leaves can increase 50-fold within 4 to 8 hours (Figure 23.4). Part of this increase is due to increased expression of biosynthetic enzymes, but the specific enzymes depend on the tissue and the signal. For example, NCED is induced in all tissues, ZEP is induced in seeds and water-stressed roots, the AAOs are differentially induced in stressed tissues, and SDR is induced by sugar but not dehydration. The mechanism by which dehydration stress is perceived by cells has not yet been identified, but might involve sensors of cellular turgor or osmosensors. Upon rewatering, the ABA level declines to normal in the same amount of time.

Biosynthesis is not the only factor that regulates ABA concentrations in the tissue. As with other plant hormones, the concentration of free ABA in the cytosol is also regulated by degradation, compartmentation, conjugation, and transport. For example, cytosolic ABA increases during water stress as a result of synthesis in the leaf, redistribution within the mesophyll cell, import from the roots, and recirculation from other leaves. The concentration of ABA declines after rewatering because of degradation and export from the leaf, as well as a decrease in the rate of synthesis.

FIGURE 23.3 Precocious germination in the ABA-deficient *vivipary 14* (*vp14*) mutant of maize. The VP14 protein catalyzes the cleavage of 9-*cis*-epoxycarotenoids to form xanthoxal, a precursor of ABA. (Courtesy of Bao Cai Tan and Don McCarty.)

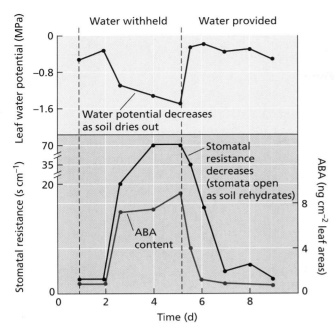

FIGURE 23.4 Changes in water potential, stomatal resistance (the inverse of stomatal conductance), and ABA content in maize in response to water stress. As the soil dried out, the water potential of the leaf decreased, and the ABA content and stomatal resistance increased. The process was reversed by rewatering. (After Beardsell and Cohen 1975.)

ABA can be inactivated by oxidation or conjugation

A major cause of the inactivation of free ABA is oxidation, yielding the unstable intermediate 6-hydroxymethyl ABA, which is rapidly converted to **phaseic acid (PA)** and **4′-dihydrophaseic acid (DPA)** (see Figure 23.2). PA is usually inactive, or it exhibits greatly reduced activity, in bioassays. However, PA can induce stomatal closure in some species, and it is as active as ABA in inhibiting gibberellic acid–induced α-amylase production in barley aleurone layers. These effects suggest that PA may be able to bind to some ABA receptors. In contrast to PA, DPA has no detectable activity in any of the bioassays tested.

Free ABA is also inactivated by covalent conjugation to another molecule, such as a monosaccharide. A common example of an ABA conjugate is **ABA-β-D-glucosyl ester (ABA-GE)**. Conjugation not only renders ABA inactive as a hormone; it also alters its polarity and cellular distribution. Whereas free ABA is localized in the cytosol, ABA-GE accumulates in vacuoles and thus could theoretically serve as a storage form of the hormone.

Esterase enzymes in plant cells could release free ABA from the conjugated form. However, there is no evidence that ABA-GE hydrolysis contributes to the rapid increase in ABA in the leaf during water stress. When plants were subjected to a series of dehydration and rewatering cycles, the ABA-GE concentration increased steadily, suggesting that the conjugated form is not broken down during water stress.

ABA is translocated in vascular tissue

ABA is transported by both the xylem and the phloem, but it is normally much more abundant in the phloem sap. When radioactive ABA is applied to a leaf, it is transported both up the stem and down toward the roots. Most of the radioactive ABA is found in the roots within 24 hours. Destruction of the phloem by a stem girdle prevents ABA accumulation in the roots, indicating that the hormone is transported in the phloem sap.

ABA synthesized in the roots can also be transported to the shoot via the xylem. Whereas the concentration of ABA in the xylem sap of well-watered sunflower plants is between 1.0 and 15.0 nM, the ABA concentration in water-stressed sunflower plants increases to as much as 3000 nM (3.0 μM) (Schurr et al. 1992). The magnitude of the stress-induced change in xylem ABA content varies widely among species, and it has been suggested that ABA also is transported in a conjugated form, then released by hydrolysis in leaves. However, the postulated hydrolases have yet to be identified.

As water stress begins, some of the ABA carried by the xylem stream may be synthesized in roots that are in direct contact with the drying soil. This transport can occur before the low water potential of the soil causes any measurable change in the water status of the leaves, leading to the suggestion that ABA is a root signal that helps reduce the transpiration rate by closing stomata in leaves (Davies and Zhang 1991). However, subsequent studies using ABA-dependent reporter gene activation to reflect localized ABA concentrations have shown that water stress–induced ABA accumulates first in shoot vascular tissue, and only later appears in roots and guard cells (Christmann et al. 2005). The lack of early ABA accumulation in roots may reflect either rapid transport of ABA or transport of a distinct long-distance signal, possibly even an ABA precursor.

Although a concentration of 3.0 μM ABA in the apoplast is sufficient to close stomata, not all of the ABA in the xylem stream reaches the guard cells. Much of the ABA in the transpiration stream is taken up and metabolized by the mesophyll cells. During the early stages of water stress, however, the pH of the xylem sap becomes more alkaline, increasing from about pH 6.3 to about pH 7.2 (Wilkinson and Davies 1997).

The major control of ABA distribution among plant cell compartments follows the "anion trap" concept: The dissociated (anion) form of this weak acid accumulates in alkaline compartments and may be redistributed according to the steepness of the pH gradients across membranes. In addition to partitioning according to the relative pH of compartments, specific uptake carriers contribute to maintaining a low apoplastic ABA concentration in unstressed plants.

Stress-induced alkalinization of the apoplast favors formation of the dissociated form of abscisic acid, ABA$^-$, which does not readily cross membranes. At the same time, dehydration also acidifies the cytosol, contributing to ABA

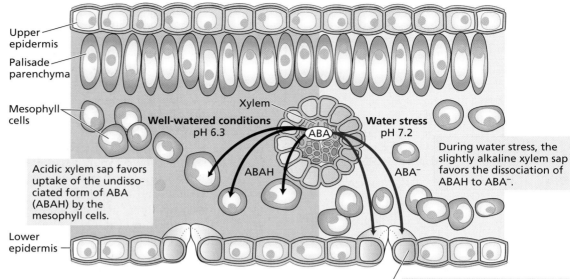

FIGURE 23.5 Redistribution of ABA in the leaf resulting from alkalinization of the xylem sap during water stress.

release from its sites of synthesis and decreasing uptake by mesophyll cells. Both of these pH changes increase the amount of ABA reaching the guard cells via the transpiration stream (Figure 23.5). Note that ABA is redistributed in the leaf in this way without any increase in the total ABA level. Therefore, the increase in xylem sap pH may function as a root signal that promotes early closure of the stomata.

Developmental and Physiological Effects of ABA

Abscisic acid plays primary regulatory roles in the initiation and maintenance of seed and bud dormancy and in the plant's response to stress, particularly water stress. In addition, ABA influences many other aspects of plant development by interacting, usually as an antagonist, with auxin, cytokinin, gibberellin, ethylene, and brassinosteroids. In this section we will explore the diverse physiological effects of ABA, beginning with its role in seed development.

ABA regulates seed maturation

Seed development can be divided into three phases of approximately equal duration:

1. During the first phase, which is characterized by cell divisions and tissue differentiation, the zygote undergoes embryogenesis and the endosperm tissue proliferates.
2. During the second phase, cell divisions cease and storage compounds accumulate.
3. In the final phase, the embryo becomes tolerant to desiccation, and the seed dehydrates, losing up to 90% of its water. As a consequence of dehydration, metabolism comes to a halt and the seed enters a **quiescent** ("resting") state. In some cases the seed becomes dormant as well. In contrast to quiescent seeds, which will germinate upon rehydration, dormant seeds require additional treatment or signals for germination to occur.

The latter two phases result in the production of viable seeds with adequate resources to support germination and the capacity to wait weeks to years before resuming growth. Typically, the ABA content of seeds is very low early in embryogenesis, reaches a maximum at about the halfway point, and then gradually falls to low levels as the seed reaches maturity. Thus there is a broad peak of ABA accumulation in the seed corresponding to mid- to late embryogenesis.

The hormonal balance of seeds is complicated by the fact that not all the tissues have the same genotype. The seed coat is derived from maternal tissues (see Web Topic 1.2); the zygote and endosperm are derived from both parents. Genetic studies with ABA-deficient mutants of Arabidopsis have shown that the zygotic genotype controls ABA synthesis in the embryo and endosperm and is essential to dormancy induction, whereas the maternal genotype controls the major, early peak of ABA accumulation and helps suppress vivipary in mid-embryogenesis (Raz et al. 2001).

ABA inhibits precocious germination and vivipary

When immature embryos are removed from their seeds and placed in culture midway through development before the onset of dormancy, they germinate precociously—that is, without passing through the normal quiescent and/or dormant stage of development. ABA added to the culture medium inhibits precocious germination. This result, in combination with the fact that the level of endogenous ABA is high during mid- to late seed development, suggests that ABA is the natural constraint that keeps developing embryos in their embryogenic state.

Further evidence for the role of ABA in preventing precocious germination has been provided by genetic studies of vivipary. The tendency toward vivipary, also known as *preharvest sprouting*, is a varietal characteristic in grain crops that is favored by wet weather. In maize, several *viviparous* mutants have been selected in which the embryos germinate directly on the cob while still attached to the plant. Several of these mutants are ABA-deficient (*vp2*, *vp5*, *vp7*, *vp9*, and *vp14*) (see Figure 23.3); one is ABA-insensitive (*vp1*). Vivipary in the ABA-deficient mutants can be partially prevented by treatment with exogenous ABA. Vivipary in maize also requires synthesis of GA early in embryogenesis as a positive signal; double mutants deficient in both GA and ABA do not exhibit vivipary (White et al. 2000).

In contrast to the maize mutants, single-gene mutants of Arabidopsis resulting in ABA deficiency or insensitivity fail to exhibit vivipary, although they are nondormant. The lack of vivipary might reflect a lack of moisture, because such seeds will germinate within the fruits under conditions of high relative humidity. However, other Arabidopsis mutants with a normal ABA response and only moderately reduced ABA levels (e.g., *fusca3*, which belongs to a class of mutants* defective in regulating the transition from embryogenesis to germination) exhibit some vivipary even at low humidities. Furthermore, double mutants combining either defects in ABA biosynthesis or ABA response with the *fusca3* mutation have a high frequency of vivipary, suggesting that redundant control mechanisms suppress vivipary in Arabidopsis (Finkelstein et al. 2002).

ABA promotes seed storage reserve accumulation and desiccation tolerance

During mid- to late embryogenesis, when seed ABA levels are highest, seeds accumulate storage compounds that will support seedling growth at germination. Another important function of ABA in the developing seed is to promote the acquisition of desiccation tolerance. As will be described in Chapter 26 (on stress physiology), desiccation can severely damage membranes and other cellular constituents. As maturing seeds begin to lose water, specific mRNAs encoding so-called **late-embryogenesis–abundant (LEA)** proteins thought to be involved in desiccation tolerance accumulate in embryos. Physiological and genetic studies have shown that ABA affects the synthesis of storage proteins and lipids, and LEAs. For example, exogenous ABA promotes accumulation of storage proteins and LEAs in cultured embryos of many species, and some ABA-deficient or -insensitive mutants have reduced accumulation of these proteins. Synthesis of some LEA proteins, or related family members, can even be induced by ABA treatment of vegetative tissues. Thus the synthesis of most LEA proteins is under ABA control (see **Web Topic 23.4**). However, synthesis of both storage proteins and LEA proteins is also reduced in other seed–developmental mutants with normal ABA levels and responses, indicating that ABA is only one of several signals controlling the expression of these genes during embryogenesis.

ABA not only regulates the accumulation of storage proteins and desiccation protectants during embryogenesis; it can also maintain the mature embryo in a dormant state until the environmental conditions are optimal for growth. Seed dormancy is an important factor in the adaptation of plants to unfavorable environments. As we will discuss in the next few sections, plants have evolved a variety of mechanisms, some of them involving ABA, that enable them to maintain their seeds in a dormant state.

The seed coat and the embryo can cause dormancy

During seed maturation, the embryo enters a quiescent phase in response to desiccation. Seed germination can be defined as the resumption of growth of the embryo of the mature seed; it depends on the same environmental conditions as vegetative growth does. Water and oxygen must be available, the temperature must be suitable, and there must be no inhibitory substances present.

In many cases a viable (living) seed will not germinate even if all the necessary environmental conditions for growth are satisfied. This phenomenon is termed **seed dormancy**. Seed dormancy introduces a temporal delay in the germination process that provides additional time for seed dispersal over greater geographic distances. It also maximizes seedling survival by preventing germination under unfavorable conditions. Two types of seed dormancy have been recognized: coat-imposed dormancy and embryo dormancy.

COAT-IMPOSED DORMANCY Dormancy imposed on the embryo by the seed coat and other enclosing tissues, such as endosperm, pericarp, or extrafloral organs, is known as **coat-imposed dormancy**. The embryos of such seeds will germinate readily in the presence of water and oxygen once the seed coat and other surrounding tissues have been either removed or damaged. There are five basic mechanisms of coat-imposed dormancy (Bewley and Black 1994):

1. *Prevention of water uptake*. This type of coat-imposed dormancy is common in plants found in arid and semi-arid regions, especially among legumes, such as clover (*Trifolium* spp.) and alfalfa (*Medicago* spp.). Waxy cuticles, suberized layers, and lignified sclereids all combine to restrict the penetration of water into the seed.

---

*Named after the Latin term for the reddish brown color of the embryos.

2. *Mechanical constraint.* The first visible sign of germination is typically the radicle (embryonic root) breaking through the seed coat. In some cases, however, the seed coat may be too rigid for the radicle to penetrate. For the seeds to germinate, the endosperm cell walls must be weakened by the production of cell wall–degrading enzymes.

3. *Interference with gas exchange.* Lowered permeability of seed coats to oxygen suggests that the seed coat inhibits germination partly by limiting the oxygen supply to the embryo.

4. *Retention of inhibitors.* The seed coat may prevent the escape of inhibitors from the seed.

5. *Inhibitor production.* Seed coats and pericarps may contain relatively high concentrations of growth inhibitors, including ABA, that can suppress germination of the embryo.

EMBRYO DORMANCY   The second type of seed dormancy is **embryo dormancy**, a dormancy that is intrinsic to the embryo and is not due to any influence of the seed coat or other surrounding tissues. In some cases, embryo dormancy can be relieved by amputation of the cotyledons. Species in which the cotyledons exert an inhibitory effect include European hazel (*Corylus avellana*) and European ash (*Fraxinus excelsior*).

A fascinating demonstration of the cotyledon's ability to inhibit growth is found in species (e.g., peach) in which the isolated dormant embryos germinate but grow extremely slowly to form a dwarf plant. If the cotyledons are removed at an early stage of development, however, the plant abruptly shifts to normal growth.

Embryo dormancy is thought to be due to the presence of inhibitors, especially ABA, as well as the absence of growth promoters, such as GA. Maintenance of dormancy in imbibed seeds requires de novo ABA biosynthesis, and the loss of embryo dormancy is often associated with a sharp decrease in the ratio of ABA to GA.

PRIMARY VERSUS SECONDARY SEED DORMANCY   Different types of seed dormancy also can be distinguished on the basis of the timing of dormancy onset rather than the cause of dormancy:

- Seeds that are released from the plant in a dormant state are said to exhibit **primary dormancy**.

- Seeds that are released from the plant in a nondormant state, but that become dormant if the conditions for germination are unfavorable, exhibit **secondary dormancy**. For example, seeds of oat (*Avena sativa*) can become dormant in the presence of temperatures higher than the maximum for germination, whereas seeds of small-flower scorpionweed (*Phacelia dubia*) become dormant at temperatures below the minimum for germination.

Environmental factors control the release from seed dormancy

Various external factors release the seed from embryo dormancy, and dormant seeds typically respond to more than one of three factors:

1. *Afterripening.* Many seeds lose their dormancy when their moisture content is reduced to a certain level by drying—a phenomenon known as **afterripening**.

2. *Chilling.* Low temperature, or **chilling**, can release seeds from dormancy. Many seeds require a period of cold (0–10°C) while in a fully hydrated (imbibed) state in order to germinate.

3. *Light.* Many seeds have a light requirement for germination, which may involve only a brief exposure, as in the case of lettuce, an intermittent treatment (e.g., succulents of the genus *Kalanchoe*), or even a specific photoperiod involving short or long days.

For further information on environmental factors affecting seed dormancy, see **Web Topic 23.5**. For a discussion of seed longevity, see **Web Topic 23.6**.

Seed dormancy is controlled by the ratio of ABA to GA

ABA mutants have been useful in demonstrating the role of ABA in seed primary dormancy. Dormancy of Arabidopsis seeds can be overcome with a period of afterripening and/or cold treatment. ABA-deficient (*aba*) mutants of Arabidopsis have been shown to be nondormant at maturity. When reciprocal crosses between *aba* and wild-type plants were carried out, the seeds exhibited dormancy only when the embryo itself produced the ABA. Neither maternal nor exogenously applied ABA was effective in inducing dormancy in an *aba* embryo.

On the other hand, maternally derived ABA constitutes the major peak present in seeds and is required for other aspects of seed development—for example, helping suppress vivipary in mid-embryogenesis. Thus the two sources of ABA function in different developmental pathways. Dormancy is also greatly reduced in seeds from the ABA-insensitive mutants *ABA-insensitive 1* (*abi1*), *abi2*, and *abi3*, even though these seeds contain higher ABA concentrations than those of the wild type throughout development, possibly reflecting feedback regulation of ABA metabolism. ABA-deficient tomato mutants seem to function in the same way, indicating that the phenomenon is probably a general one. However, other mutants with reduced dormancy, but normal ABA levels and sensitivity, point to additional regulators of dormancy (Koornneef et al. 2002). Some of these factors are being identified through genetic studies of natural variation (see **Web Topic 23.7**).

Although the role of ABA in initiating and maintaining seed dormancy is well established, other hormones contribute to the overall effect. For example, in most plants the peak of ABA production in the seed coincides with a decline in the levels of indole-3-acetic acid (IAA, also called auxin, discussed in detail in Chapter 19) and GA.

An elegant demonstration of the importance of the ratio of ABA to GA in seeds was provided by the genetic screen that led to isolation of the first ABA-deficient mutants of Arabidopsis (Koornneef et al. 1982). Seeds of a GA-deficient mutant that could not germinate in the absence of exogenous GA were mutagenized and then grown in the greenhouse. The seeds produced by these mutagenized plants were then screened for **revertants**—that is, seeds that had regained their ability to germinate.

Revertants were isolated, and they turned out to be mutants of abscisic acid synthesis. The revertants germinated because dormancy had not been induced, so subsequent synthesis of GA was no longer required to overcome it. This study elegantly illustrates the general principle that the balance of plant hormones is often more critical than are their absolute concentrations in regulating development. However, ABA and GA exert their effects on seed dormancy at different times, so their antagonistic effects on dormancy do not necessarily reflect a direct interaction.

Recent genetic screens for suppressors of ABA insensitivity during seed germination have identified additional antagonistic interactions between ABA and ethylene or brassinosteroid. In addition, many new alleles of ABA-deficient or *abi4* mutants have been identified in screens for altered sensitivity to sugar or salinity. These studies show that a complex regulatory web integrates hormonal, nutrient, and stress signaling controlling the commitment to growth of the next generation.

ABA inhibits GA-induced enzyme production

In addition to the ABA–GA antagonism affecting seed dormancy, ABA inhibits the GA-induced synthesis of hydrolytic enzymes that are essential for the breakdown of storage reserves in germinating seeds. For example, GA stimulates the aleurone layer of cereal grains to produce α-amylase and other hydrolytic enzymes that break down stored resources in the endosperm during germination (see Chapter 20). ABA inhibits this GA-dependent enzyme synthesis by inhibiting the transcription of α-amylase mRNA. ABA exerts this inhibitory effect via at least two mechanisms, one direct and one indirect:

1. VP1, a protein originally identified as an activator of ABA-induced gene expression, acts as a transcriptional repressor of some GA-regulated genes (Hoecker et al. 1995).
2. ABA represses the GA-induced expression of GAMYB, a transcription factor that mediates the GA induction of α-amylase expression (Gomez-Cadenas et al. 2001).

ABA closes stomata in response to water stress

Elucidation of the roles of ABA in freezing, salt, and water stress (see Chapter 26) led to the characterization of ABA as a stress hormone. As noted earlier, ABA concentrations in leaves can increase up to 50 times under drought conditions—the most dramatic change in concentration reported for any hormone in response to an environmental signal. Redistribution or biosynthesis of ABA is very effective in causing stomatal closure, and its accumulation in stressed leaves plays an important role in the reduction of water loss by transpiration under water stress conditions (see Figure 23.4).

Stomatal closing, resulting from ion fluxes leading to decreases in guard cell turgor pressure, can also be caused by ABA transported to the shoot. Mutants blocked in ABA synthesis exhibit permanent wilting and are called *wilty* mutants because of their inability to close their stomata. Application of exogenous ABA to such mutants causes stomatal closure and a restoration of turgor pressure. In contrast, *wilty* mutants blocked in their ability to respond to ABA are not rescued by ABA application.

ABA promotes root growth and inhibits shoot growth at low water potentials

ABA has different effects on the growth of roots and shoots, and the effects are strongly dependent on the water status of the plant. Figure 23.6 compares the growth of shoots and roots of maize seedlings grown under either abundant water conditions (high water potential [Ψ_w]) or dehydrating conditions (low Ψ_w). Two types of seedlings were used: (1) wild-type seedlings with normal ABA levels and (2) an ABA-deficient, *viviparous* mutant.

When the water supply is ample (high Ψ_w), shoot growth is greater in the wild-type plant with normal endogenous ABA levels than in the ABA-deficient mutant (see Figure 23.6A). Although the reduced shoot growth in the mutant could be due in part to excessive water loss from the leaves, the stunted shoot growth of ABA-deficient maize and tomato plants at high water potentials seems to be due to the overproduction of ethylene, which is normally inhibited by endogenous ABA. This finding suggests that endogenous ABA promotes shoot growth in well-watered plants by suppressing ethylene production. Similarly, root growth of well-watered plants is slightly greater in the wild type (normal endogenous ABA) than in the ABA-deficient mutant. Therefore, at high water potentials (when the total ABA levels are low), endogenous ABA exerts a slight positive effect on the growth of both roots and shoots

In contrast, limiting water (i.e., low water potentials), has opposite effects on root and shoot growth. Under dehydrating conditions, growth is still inhibited relative to that of either genotype when water is abundant. However, whereas shoot growth is greater in the ABA-deficient mutant than in the wild type, the growth of the roots is much higher in the wild type than in the ABA-deficient mutant (see Figure 23.6B). Endogenous ABA appears to promote root growth by inhibiting ethylene production during water stress.

To summarize, despite the traditional view of ABA as a growth inhibitor, endogenous ABA restricts shoot growth

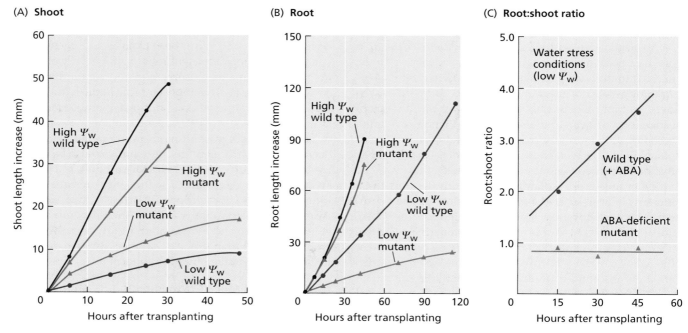

Figure 23.6 Comparison of the growth of the shoots (A) and roots (B) of normal versus ABA-deficient (*viviparous* [*vp*]) maize plants growing in vermiculite maintained either at high water potential ($\Psi w = -0.03$ MPa) or at low water potential ($\Psi w = -0.3$ MPa in A and -1.6 MPa in B). Water stress (low Ψw) depresses the growth of both shoots and roots compared to the controls. (C) Under water stress conditions (low Ψw, defined slightly differently for shoot and root), the ratio of root growth to shoot growth is much higher when ABA is present (i.e., in the wild type) than when it is absent (in the mutant). (From Saab et al. 1990.)

only under water stress conditions. Moreover, under these conditions, when ABA levels are high, endogenous ABA exerts a strong positive effect on root growth by suppressing ethylene production. The overall effect is a dramatic increase in the root:shoot ratio at low water potentials (see Figure 23.6C), which, along with the effect of ABA on stomatal closure, helps the plant cope with water stress (Sharp 2002).

An alternative response is seen under extreme stress conditions. In some species, roots may initiate many extra lateral roots whose growth is suppressed until the stress is relieved, a phenomenon known as **drought rhizogenesis** (Vartanian et al. 1994). For another example of the role of ABA in the response to dehydration, see **Web Essay 23.1**.

ABA promotes leaf senescence independently of ethylene

Abscisic acid was originally isolated as an abscission-causing factor. However, it has since become evident that ABA stimulates abscission of organs in only a few species and that the primary hormone causing abscission is ethylene. On the other hand, ABA is clearly involved in leaf senescence, and through its promotion of senescence it might indirectly increase ethylene formation and stimulate abscission. (For more discussion on the relationship between ABA and ethylene, see **Web Topic 23.8**.)

Leaf senescence has been studied extensively (the anatomical, physiological, and biochemical changes that take place during this process were described in Chapter 16). Leaf segments senesce faster in darkness than in light, and they turn yellow as a result of chlorophyll breakdown. In addition, the breakdown of proteins and nucleic acids is increased by the stimulation of several hydrolases. ABA greatly accelerates the senescence of both leaf segments and attached leaves.

ABA accumulates in dormant buds

In woody species, dormancy is an important adaptive feature in cold climates. When a tree is exposed to very low temperatures in winter, it protects its meristems with bud scales and temporarily stops bud growth. This response to low temperatures requires a sensory mechanism that detects the environmental changes (sensory signals), and a control system that transduces the sensory signals and triggers the developmental processes leading to bud dormancy.

ABA was originally suggested as the dormancy-inducing hormone because it accumulates in dormant buds and decreases after the tissue is exposed to low temperatures. However, later studies showed that the ABA content of buds does not always correlate with the degree of dormancy. As we saw in the case of seed dormancy, this apparent discrepancy could reflect interactions between ABA and other hormones as part of a process in which bud dormancy and growth are regulated by the balance between bud growth inhibitors, such as ABA, and growth-inducing substances, such as cytokinins and gibberellins.

Although much progress has been achieved in elucidating the role of ABA in seed dormancy by the use of ABA-deficient mutants, progress on the role of ABA in bud dormancy, which applies mainly to woody perennials, has lagged because of the lack of a convenient genetic system. This discrepancy illustrates the tremendous contribution that genetics and molecular biology have made to plant physiology, and underscores the need for extending such approaches to woody species.

Analyses of traits such as dormancy are complicated by the fact that they are often controlled by the combined action of several genes, resulting in a gradation of phenotypes referred to as *quantitative traits*. Recent genetic mapping studies suggest that homologs of *ABI1* may contribute to regulation of bud dormancy in poplar trees. For a description of such studies, see **Web Topic 23.7**.

ABA Signal Transduction Pathways

ABA is involved in short-term physiological effects (e.g., stomatal closure), as well as long-term developmental processes (e.g., seed maturation). Rapid physiological responses frequently involve alterations in the fluxes of ions across membranes and usually involve some gene regulation as well, whereas long-term processes inevitably involve major changes in the pattern of gene expression. Comparisons of total transcript populations have shown that ~10% of the genes in Arabidopsis or rice are regulated by ABA.

Signal transduction pathways, which amplify the primary signal generated when the hormone binds to its receptor, are required for both the short-term and the long-term effects of ABA. Genetic studies have identified more than 50 loci involved in mediating ABA responses (Finkelstein et al. 2002). Although many genes affect only a subset of ABA responses, some conserved signaling components regulate both short- and long-term responses, indicating that these responses share common signaling mechanisms. In this section we will describe some aspects of the mechanism of ABA action at the cellular and molecular levels. We will focus on mechanisms regulating stomatal aperture and gene expression, the two best-characterized ABA effects.

ABA regulates ion channels and the PM-ATPase in guard cells

As discussed in Chapter 18, stomatal closure is driven by a reduction in guard cell turgor pressure caused by a massive long-term efflux of K^+ and anions from the cell. Opening of the K^+ efflux channels requires long-term membrane depolarization, which appears to be triggered by two factors: (1) an ABA-induced transient depolarization of the plasma membrane, coupled with (2) an increase in cytosolic calcium. Both of these conditions are required to open calcium-activated slow (S-type) anion channels on the plasma membrane (see Chapter 6). ABA has been shown to activate slow anion channels in guard cells (Schroeder et al. 2001). ABA also activates another class of anion channels in guard cells, the rapid (R-type) anion channels (Raschke et al. 2003).

The prolonged opening of these slow and rapid anion channels permits large quantities of Cl^- and $malate^{2-}$ ions to escape from the cell, moving down their electrochemical gradients. (The inside of the cell is negatively charged, thus pushing Cl^- and $malate^{2-}$ out of the cell, and the outside has lower Cl^- and $malate^{2-}$ concentrations than the interior.) The outward flow of negatively charged Cl^- and $malate^{2-}$ ions generated in this way strongly depolarizes the membrane, triggering the voltage-gated K^+ efflux channels to open.

Another factor that can contribute to membrane depolarization is inhibition of the plasma membrane H^+-ATPase. ABA inhibits blue light–stimulated proton pumping by guard cell protoplasts (Figure 23.7), consistent with the model that the depolarization of the plasma membrane by ABA is partially caused by a decrease in the activity of the plasma membrane H^+-ATPase. However, ABA does not inhibit the proton pump directly.

In broad bean (*Vicia faba*), at least, the plasma membrane H^+-ATPase of the leaves is strongly inhibited by calcium. A calcium concentration of 0.3 μM blocks 50% of the activity of H^+-ATPase, and 1 μM calcium blocks the enzyme completely (Kinoshita et al. 1995). It appears that two factors contribute to ABA inhibition of the plasma membrane

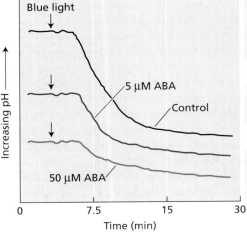

1. A pulse of blue light activates the plasma membrane H^+-ATPase, which pumps protons into the external medium and lowers the pH.

2. Addition of ABA to the medium inhibits the acidification by 40%.

3. These results demonstrate that ABA induces changes in the cell that inhibit the plasma membrane H^+-ATPase.

FIGURE 23.7 ABA inhibition of blue light–stimulated proton pumping by guard cell protoplasts. A suspension of guard cell protoplasts was incubated under red-light irradiation, and the pH of the suspension medium was monitored with a pH electrode. The starting pH was the same in all cases (the curves are displaced for ease of viewing). (After Shimazaki et al. 1986.)

proton pump: an increase in the cytosolic Ca^{2+} concentration, and alkalinization of the cytosol.

In addition to causing stomatal closure, ABA prevents light-induced stomatal opening. In this case, ABA acts by inhibiting the inward K^+ channels, which are open when the membrane is hyperpolarized by the proton pump (see Chapters 6 and 18). Inhibition of the inward K^+ channels is mediated by the ABA-induced increase in cytosolic calcium concentration. Thus, calcium and pH affect guard cell plasma membrane channels in two ways:

1. They prevent stomatal opening by inhibiting inward K^+ channels and plasma membrane proton pumps.
2. They promote stomatal closing by activating outward anion channels, thus leading to activation of K^+ efflux channels.

As a result of the large-scale (about 0.3 M) ion effluxes involved in stomatal closure, water is lost osmotically, and the surface area of the guard cell plasma membrane may contract by as much as 50%. Where does the extra membrane go? The answer seems to be that it is taken up as small vesicles by endocytosis—a process that also involves ABA-induced reorganization of the actin cytoskeleton mediated by a family of plant Rho GTPases, or "Rops" (Assmann 2004).

In the following sections, we will address mechanisms of ABA perception and signaling.

ABA may be perceived by both cell surface and intracellular receptors

Efforts to identify ABA receptors have employed both biochemical and genetic approaches. Biochemical approaches have addressed such questions as which portions of the ABA molecule are recognized by the receptor and where in the cell the receptor is located. Thus, biochemical experiments have been performed to determine the structural requirements for ABA activity, and whether the hormone must enter the cell to be effective, or whether it can act externally by binding to a receptor located on the outer surface of the plasma membrane. The results of such experiments have suggested multiple sites of interaction in the cell and multiple modes of perception, and therefore by implication, multiple receptor types, which could confound efforts to identify a receptor genetically. Consistent with this conclusion, no mutants have been identified with defects in all ABA responses. The difficulty in identifying an ABA receptor might also reflect a relatively weak affinity of the receptor for ABA, which might explain the relatively high concentrations of ABA required for some of its physiological effects.

Some experiments point to a receptor on the outer surface of the cell. For example, microinjected ABA fails to alter stomatal opening in the spiderwort *Commelina*, or to inhibit GA-induced α-amylase synthesis in barley aleurone protoplasts (Anderson et al. 1994; Gilroy and Jones 1994). Furthermore, ABA–protein conjugates that are unable to cross the membrane have been shown to activate both ion channel activity and gene expression (Schultz and Quatrano 1997; Jeannette et al. 1999). Multiple small patches of possible ABA receptors were visualized on guard cell protoplast surfaces by treating the protoplasts with biotinylated ABA, then with fluorescent avidin, a protein that binds to biotin (Yamazaki et al. 2003).

Other experiments, however, support an intracellular location for the ABA receptor:

- ABA supplied directly and continuously to the cytosol via a "patch pipette" inhibited both inward K+ channels and S-type anion channels, which are required for stomatal opening (Schwartz et al. 1994; Schroeder et al. 2001).

- Extracellular application of ABA was nearly twice as effective at inhibiting stomatal opening at pH 6.15, when it is fully protonated and readily taken up by guard cells, as it was at pH 8, when it is largely dissociated to the anionic form that does not readily cross membranes (Anderson et al. 1994).

- Microinjection of an inactive, "caged" form of ABA into guard cells of *Commelina* resulted in stomatal closure after the stomata were treated briefly with UV irradiation to activate the hormone—that is, release it from its molecular cage (Figure 23.8) (Allan et al. 1994). Control guard cells injected with a nonphotolyzable form of the caged ABA did not close after UV irradiation.

Taken together, these results indicate that extracellular perception of ABA can prevent stomatal opening and regulate gene expression, and intracellular ABA can both induce stomatal closure and inhibit channel activity required for opening. Thus, there may be both cell surface and intracellular ABA receptors.

Several ABA-binding proteins have been identified either by affinity purification using ABA derivatives or by anti-idiotypic antibodies.* Anti-idiotypic antibodies raised against anti-ABA antibodies are potentially capable of recognizing structurally similar ABA-binding sites present on ABA receptors. An anti-idiotypic antibody approach has now led to the identification of an authentic ABA receptor in Arabidopsis (Razem et al. 2006). First, the anti-idiotype antibodies were used to identify a clone from a cDNA expression library derived from ABA-treated barley aleurone tissue. The cDNA encoded a barley protein, ABAP1, which displayed saturable, reversible, high-affinity, stereospecific binding of ABA and was apparently localized on the plasma membrane.

---

*An anti-idiotypic antibody is a secondary antibody raised against the part of a primary antibody that recognizes a particular antigen. An anti-idiotype antibody thus mimics the antigen itself, in this case ABA, and potentially can interact with the hormone-binding domain on the ABA receptor.

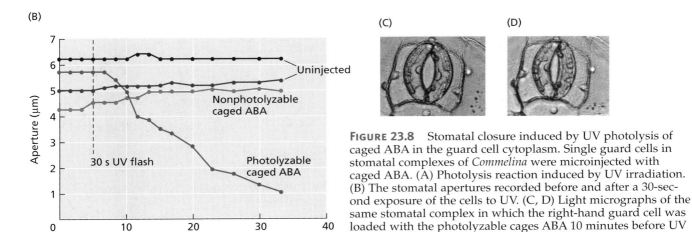

FIGURE 23.8 Stomatal closure induced by UV photolysis of caged ABA in the guard cell cytoplasm. Single guard cells in stomatal complexes of *Commelina* were microinjected with caged ABA. (A) Photolysis reaction induced by UV irradiation. (B) The stomatal apertures recorded before and after a 30-second exposure of the cells to UV. (C, D) Light micrographs of the same stomatal complex in which the right-hand guard cell was loaded with the photolyzable cages ABA 10 minutes before UV photolysis (C) and 30 minutes after photolysis (D). (A and B from Allan et al. 1994; C and D courtesy of A. Allan, from Allan et al. 1994; © American Society of Plant Biologists, reprinted with permission.)

It was therefore surprising that ABAP1 was also found to have high homology to the Arabidopsis FLOWERING TIME CONTROL PROTEIN A (FCA), known to play an important role in controlling the onset of flowering (see Chapter 26). In contrast to the presumed localization of ABAP1, FCA is an RNA-binding nuclear protein that promotes flowering in Arabidopsis by inhibiting the action of a flowering repressor, FLOWERING LOCUS C (FLC) (see Chapter 26). Nevertheless, like ABAP1, FCA was shown to bind ABA with high affinity and specificity.

In the absence of ABA, FCA forms an RNA processing complex with another protein, the polyadenylation factor FLOWERING LOCUS Y (FY). The FCA/FY complex then inhibits production of the flowering repressor FLC, and flowering is promoted (Figure 23.9) (see Chapter 25). In the presence of ABA, ABA-binding to FCA blocks the formation of the FCA/FY complex. As a result, the flowering repressor FLC builds up and flowering is inhibited.

Consequently, this mode of ABA signaling resembles that of ethylene in that ligand binding inactivates its receptor. In a further surprise, the ABA-related phenotype of FCA-minus mutants (*fca*) appears limited to a failure to delay flowering or inhibit lateral root initiation in response to ABA. These results therefore demonstrate hitherto underappreciated roles for ABA in two developmental processes,

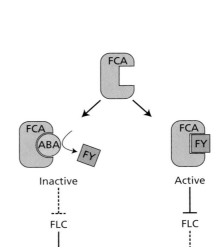

FIGURE 23.9 A model of ABA interaction with one of its receptors, FLOWERING TIME CONTROL PROTEIN A (FCA), in Arabidopsis. ABA-binding to FCA inhibits its activity as a repressor of the production of FLOWERING LOCUS C (FLC), which is itself a repressor of flowering. In the absence of ABA, FCA forms an active complex with FLOWERING LOCUS Y (FY), which represses the FLC, permitting flowering to occur. Solid arrows indicate positive regulation, solid bars indicate repression, and dotted bars indicate derepressed steps.

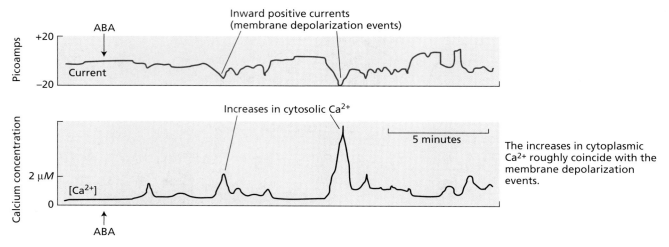

FIGURE 23.10 Simultaneous measurements of ABA-induced inward positive currents and ABA-induced increases in cytosolic Ca^{2+} concentrations in a guard cell of broad bean (*Vicia faba*). The current was measured by the patch clamp technique; calcium was measured by use of a fluorescent indicator dye. ABA was added to the system at the arrow in each case. (From Schroeder and Hagiwara 1990.)

flowering and lateral root inititation, and demonstrate that additional receptors must exist to mediate the well-characterized effects of ABA on stomatal regulation, germination, and other developmental phenomena.

ABA signaling involves both calcium-dependent and calcium-independent pathways

Some of the first changes detected after exposure of guard cells to ABA are transient membrane depolarization caused by the net influx of positive charge and transient increases in the cytosolic calcium concentration (Figure 23.10).

ABA stimulates elevations in the concentration of cytosolic Ca^{2+} by inducing both influx through plasma membrane channels and release of calcium into the cytosol from internal compartments, such as the central vacuole (Schroeder et al. 2001). Stimulation of influx occurs via NADPH oxidase–mediated production of **reactive oxygen species** (**ROS**), such as hydrogen peroxide (H_2O_2) or superoxide ($O_2^{\bullet-}$), as secondary messengers leading to plasma membrane calcium channel activation (Kwak et al. 2003).

Calcium release from intracellular stores can be induced by a variety of second messengers, including inositol 1,4,5-trisphosphate (IP_3), cyclic ADP-ribose (cADPR), and self-amplifying (calcium-induced) Ca^{2+} release. In addition, ABA stimulates the synthesis of **nitric oxide** (**NO**) by nitrate reductase in guard cells, which induces stomatal closure in a cADPR-dependent manner, indicating that NO acts upstream of cADPR in the response pathway (Desikan et al. 2002). (For background on NO signaling, see Chapter 14 on the web site.)

The combination of calcium influx and the release of calcium from internal stores raises the cytosolic calcium concentration from 50 to 350 n*M* to as high as 1100 n*M* (1.1 m*M*) (Figure 23.11) (McAinsh et al. 1990). This increase is sufficient to cause stomatal closure, as demonstrated by the following experiment.

As in the experiment described earlier, calcium was microinjected into guard cells in a caged form that could be

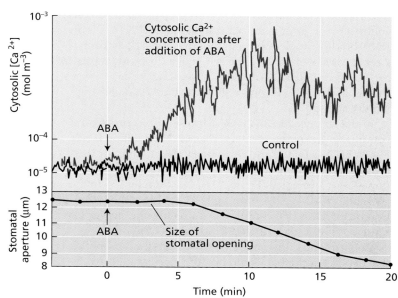

FIGURE 23.11 Time course of the ABA-induced increase in guard cell cytosolic Ca^{2+} concentration (upper panel) and ABA-induced stomatal aperture (lower panel). (From McAinsh et al. 1990.)

hydrolyzed by a pulse of UV light. This method allowed the investigators to control both the concentration of free calcium and the time of release to the cytosol. At cytosolic concentrations of 600 nM or more, release of calcium from its cage triggered stomatal closure (Gilroy et al. 1990). This level of intracellular calcium is well within the concentration range observed after ABA treatment.

In the preceding studies, intracellular free calcium was measured by the use of microinjected calcium-sensitive ratiometric fluorescent dyes*, such as fura-2 or indo-1. However, microinjections of fluorescent dyes into single plant cells are difficult and often result in cell death. Success rates of viable injections into Arabidopsis guard cells can be less than 3%. In contrast, transgenic plants expressing the gene for the calcium indicator protein **yellow cameleon** make it possible to monitor several fluorescing cells in parallel, without the need for invasive injections (Allen et al. 1999) (see **Web Topic 23.9**). Such studies have demonstrated that the cytosolic Ca^{2+} concentration oscillates with distinct periodicities, depending on the signals received (Figure 23.12).

These results support the hypothesis that an increase in cytosolic calcium, partly derived from intracellular stores, is responsible for ABA-induced stomatal closure. However, the growth hormone auxin can induce stomatal opening, and this auxin-induced stomatal opening, like ABA-induced stomatal closure, is accompanied by *increases* in cytosolic calcium. This finding suggests that the detailed characteristics of the location and periodicity of Ca^{2+} oscillations (the "Ca^{2+} signature"), rather than the overall concentration of cytosolic calcium, determine the cellular response.

In addition to increasing the cytosolic calcium concentration, ABA causes an alkalinization of the cytosol from about pH 7.67 to pH 7.94. The increase in cytosolic pH has been shown to increase the activity of the K^+ efflux channels on the plasma membrane apparently by increasing the number of channels available for activation (see Chapter 6).

Although an ABA-induced increase in cytosolic calcium concentration is a key feature of the current model for ABA-induced stomatal closure, ABA is able to induce stomatal closure even in guard cells that show no increase in cytosolic calcium (Allan et al. 1994). In other words, ABA seems to be able to act via one or more calcium-independent pathways. As previously discussed, a rise in cytosolic pH can lead to the activation of outward K^+ channels, and one effect of the *abi1* mutation is to render these K^+ channels insensitive to pH.

Redundancy in signal transduction pathways explains how guard cells are able to integrate a wide range of hormonal and environmental stimuli that affect stomatal aperture. Such redundancy in signaling pathways provides what biologists call "network robustness," which ensures that the response will occur under a variety of conditions.

---

*Ratiometric fluorescent dyes undergo a shift in their excitation and emission spectra when they bind calcium. On the basis of this property, one can determine the intracellular concentration of both forms of the dye (with and without bound calcium) by exciting them with the appropriate two wavelengths. The ratio of the two emissions provides a measure of the calcium concentration that is independent of dye concentration.

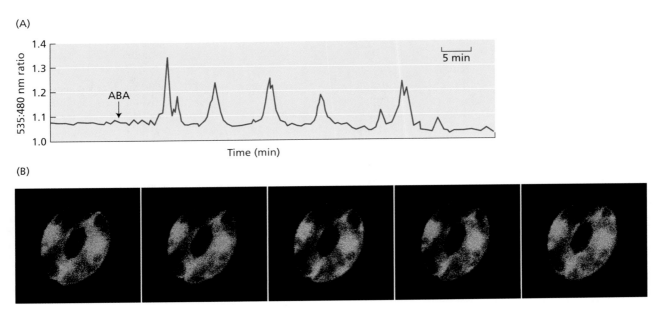

FIGURE 23.12 ABA-induced calcium oscillations in Arabidopsis guard cells expressing yellow cameleon, a calcium indicator protein dye. (A) Repetitive calcium transients elicited by ABA are indicated by increases in the ratio of fluorescence emission at 535 and 480 nm. (B) Pseudo-colored images of fluorescence in Arabidopsis guard cells; blue, green, yellow, and red represent increasing cytosolic calcium concentration. (From Schroeder et al. 2001.)

ABA-induced lipid metabolism generates second messengers

As discussed previously, much evidence supports a role for calcium both in the promotion of stomatal closing and in the inhibition of stomatal opening. According to the classic calcium-dependent signal transduction pathway of animal cells, ligand binding to a receptor on the plasma membrane leads to the activation of a heterotrimeric GTP-binding protein (G protein), composed of α, β, and γ subunits (see Chapter 14 on the web site). The activated Gα subunit then activates phospholipase C. IP_3 is released, along with diacylglycerol (DAG), when phospholipase C is activated.

Does ABA use the same pathway when it induces stomatal closure? In agreement with this model, ABA has been shown to stimulate phosphoinositide metabolism, leading to the production of IP_3 and *myo*-inositol-hexaphosphate (IP_6) in broad bean (*Vicia faba*) guard cells. To detect the effect of ABA on IP_3 production, it was necessary to include Li^+ in the incubation medium as an inhibitor of inositol phosphatase, which rapidly removes phosphate groups from IP_3. Under these conditions, a 90% ABA-induced increase in the level of IP_3 was measured within 10 seconds of hormone treatment (Lee et al. 1996). Studies in Arabidopsis using antisense DNA to block expression of an ABA-induced phospholipase C have shown that this enzyme is required for the effects of ABA on germination, growth, and gene expression (Sanchez and Chua 2001). Antisense repression of phospholipase C in tobacco also partially impaired ABA signaling in guard cells.

The Arabidopsis genome encodes single copies of the genes for the Gα and Gβ subunits, and two copies of the gene for the Gγ subunit. A role for heterotrimeric G proteins in mediating the effects of ABA on stomatal movements is supported by both pharmacological and genetic studies. For example, in *V. faba*, studies have shown that G protein activators, such as GTPγS, can inhibit the activity of the inward K^+ channels. Consistent with the inhibitor results, ABA failed to inhibit inward K^+ channels or light-induced stomatal opening in an Arabidopsis mutant with a defective Gα subunit (Wang et al. 2001). However, ABA still promoted stomatal closure in this mutant, indicating that inhibition of opening and promotion of closing take two distinct paths to the same end point—that is, closed stomata.

One Arabidopsis gene, *GCR1*, encodes a protein with significant sequence similarity to nonplant G protein–coupled receptors (GPCRs), including a predicted seven-transmembrane domain structure characteristic of GPCRs. Could the ABA receptor be the GPCR-like protein GCR1? If so, a mutant in which the *GCR1* gene was genetically knocked out should exhibit ABA insensitivity. However, the guard cells of *GCR1* knockout mutants exhibit *hypersensitivity* to ABA rather than insensitivity, indicating that G protein signaling in plants differs from that of animal cells (Pandey et al. 2004). The ligand for GCR1 is unknown, but the protein apparently functions as an inhibitor or negative regulator of Gα-mediated ABA responses rather than as an activator.

Another class of phospholipids, **sphingosine-1-phosphate (S1P)**, is produced in response to ABA. Sphingosine-1-phosphate has been shown to stimulate stomatal closure and cytosolic Ca^{2+} increases in *Commelina communis* guard cells. In addition, ABA-induced stomatal closure in this species was decreased by S1P signaling inhibitors. More recently, ABA has been shown to activate the enzyme **sphingosine kinase**, which phosphorylates sphingosine, a long-chain, unsaturated amino alcohol, to form S1P (Figure 23.13) (Coursol et al. 2003).

Where does sphingosine-1-phosphate fit into the sequence of signaling events triggered by ABA? Although S1P stimulates stomatal closure in wild-type Arabidopsis, it does not cause stomatal closure in mutants lacking Gα subunits. This suggests that sphingosine-1-phosphate acts via G proteins, and that G protein–mediated events are downstream of (after) the sphingosine-1-phosphate signal (Coursol et al. 2003).

Yet another potential second messenger mediating the ABA response is phosphatidic acid, which is released from phosphatidylcholine by the enzyme phospholipase D (PLD). Phospholipase D–mediated production of phosphatidic acid has been shown to promote ABA-induced stomatal closure and gene expression, as well as other stress responses, by multiple mechanisms (Zhang et al. 2005). The most abundant isoform of PLD is activated by ABA and binds directly to the Gα subunit of heterotrimeric G protein, modifying both activities: GTP activation of Gα increases PLD activity, whereas PLD inactivates the G protein.

The product of PLD activity, phosphatidic acid (PA), binds a variety of targets including protein phosphatases, protein kinases, and metabolic enzymes. For at least one of these, the protein phosphatase ABI1 (discussed in the following section), PA binding alters its activity and localization such that ABA signaling is de-repressed. In addition, PA may stimulate sphingosine-1-phosphate production, thereby promoting G protein-mediated ABA signaling.

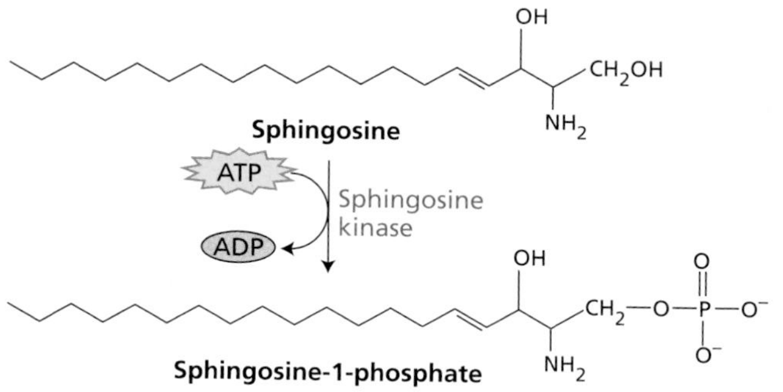

FIGURE 23.13 Sphingosine kinase catalyzes the phosphorylation of sphingosine to sphingosine-1-phosphate.

Abscisic Acid **609**

1. ABA binds to its receptors. (For clarity only the extracellular receptors are shown.)

2. ABA-binding induces the formation of reactive oxygen species (ROS) that activate Ca^{2+} channels on the plasma membrane. Phospholipase D (PLD)-mediated phosphatidic acid (PA) production contributes to ROS production.

3. The influx of calcium initiates intracellular calcium transients and promotes the further release of calcium from vacuoles.

4. ABA stimulates NO production, and NO increases cADPR levels.

5. ABA increases IP_3 levels via a signaling pathway that includes S1P, heterotrimeric G proteins, and phospholipase C and D (PLC and PLD).

6. The rise in cADPR and IP_3 activates additional calcium channels on the tonoplast, releasing more Ca^{2+} from vacuoles.

7. The rise in intracellular calcium blocks K^+_{in} channels on the plasma membrane.

8. The rise in intracellular calcium promotes the opening of Cl^-_{out} (anion) channels on the plasma membrane, causing depolarization.

9. The plasma membrane proton pump is inhibited by the ABA-induced increase in cytosolic calcium and a rise in intracellular pH, further depolarizing the membrane.

10. Membrane depolarization activates K^+_{out} channels on the plasma membrane.

11. K^+ and anions to be released across the plasma membrane are first released from vacuoles into the cytosol.

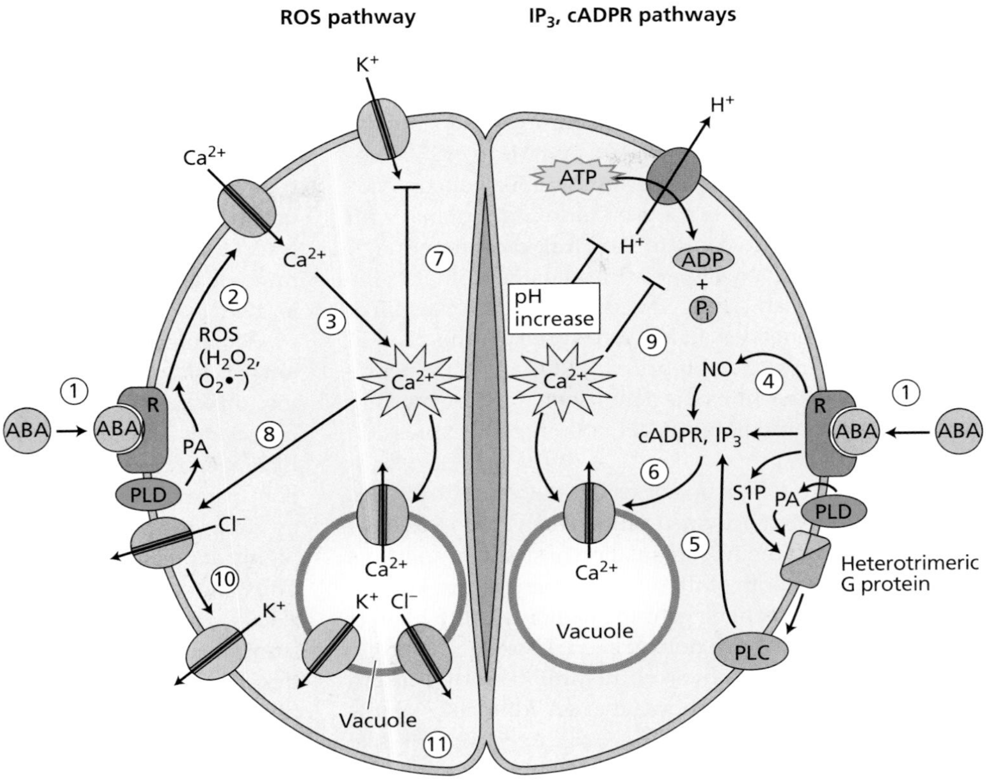

FIGURE 23.14 Simplified model for ABA signaling in stomatal guard cells. The net effect is the loss of potassium and its anion (Cl^- or malate^{2-}) from the cell. cADPR, cyclic ADP-ribose; IP_3 = inositol 1,4,5-trisphosphate; NO = nitric oxide; PA = phosphatidic acid; PLC = phospholipase C; PLD = phospholipase D; R, receptor; ROS, reactive oxygen species; S1P = sphingosine-1-phosphate.

Finally, PLD and PA also appear to mediate both the production of and the response to reactive oxygen species and phosphoinositides. Thus, PLD and phosphatidic acid are involved in multiple feedback loops and pathways that enhance ABA signaling.

A simplified general model for ABA action in stomatal guard cells is shown in Figure 23.14. For clarity, only the cell surface receptors are shown. For a more detailed model, see Mäser et al. (2002).

ABA signaling involves protein kinases and phosphatases

Nearly all biological signaling systems involve protein phosphorylation and dephosphorylation reactions at some step in the pathway. Thus we can expect that signal transduction in guard cells, with their multiple sensory inputs, involves protein kinases and phosphatases. Consistent with this expectation, artificially raising the ATP concentration inside guard cells by allowing the cytoplasm to equilibrate with the solution inside a patch pipette (see Chapter 6) strongly activates the slow (S-type) anion channels, and this activation is abolished by the inclusion of protein kinase inhibitors in the patch pipette solution (Schmidt et al. 1995). Protein kinase inhibitors also block ABA-induced stomatal closing. In contrast, lowering the concentration of ATP in the cytosol inactivates the slow anion channels. Additional experiments confirm that this inactivation is due to the presence of protein phosphatases, which remove phosphate groups that are covalently attached to proteins. In view of these results, it appears that protein phosphorylation and dephosphorylation play important roles in the ABA signal transduction pathway in guard cells.

An **ABA-activated protein kinase (AAPK)** was identified in *V. faba* guard cells that appears to be required for both ABA activation of S-type anion currents and stomatal closing (Li and Assmann 1996; Mori and Muto 1997). Subsequently, a mutant affecting a homologous gene in Arabidopsis was identified in a screen for mutant plants with leaves that appeared "cool" to an infrared camera due to excessive evaporative water loss through the stomata*; the gene was designated *OPEN STOMATA* (*OST1*) (Mustilli et al. 2002). *OST1* encodes an autophosphorylating protein kinase that acts upstream of (prior to) ROS and Ca^{2+} signaling. AAPK also modifies the activity of an RNA-binding protein that may mediate ABA effects on gene expression (Assmann 2004).

In addition to OST1, specific Ca^{2+}-dependent protein kinases and mitogen-activated protein (MAP) kinases have been implicated in the ABA regulation of stomatal aperture. Protein kinases of both of these classes may be involved in ABA regulation of seedling gene expression, although direct genetic mutant data have not yet been reported.

Pharmacological studies with inhibitors of **protein phosphatases (PPs)** have shown that several classes of serine/threonine phosphatases (PP1A, PP2A, calcineurin-like PP2Bs, and PP2C), and tyrosine phosphatases can all regulate aspects of guard cell signaling, but the effects may be positive or negative depending on numerous factors. Genetic studies permit analysis of individual protein phosphatases and have identified a specific PP2A and several PP2Cs as ABA signaling components with pleiotropic (multiple) effects on development.

The most dramatic effects on the ABA response are seen in mutants of the Arabidopsis *ABI1* and *ABI2* loci, which encode PP2Cs. The Arabidopsis *abi1-1* and *abi2-1* mutants display phenotypes consistent with a defect in ABA signaling, including reduced seed dormancy, a tendency to wilt (due to faulty regulation of stomatal aperture), and decreased expression of various ABA-inducible genes. The defects in the stomatal response include ABA insensitivity of S-type anion channels, both inward and outward K^+ channels, and actin reorganization. Although nonresponsive to ABA, the mutant stomata close when exposed to high external concentrations of Ca^{2+}, suggesting that they are defective in their ability to initiate Ca^{2+} signaling. Consistent with this finding, ABA is less effective at inducing Ca^{2+} transients in these mutants.

The Arabidopsis *abi1-1* and *abi2-1* mutations were identified in a screen for a *decreased* response to ABA, hence their designation as ABA-insensitive mutants. Based on their ABA-insensitive phenotypes, it was initially assumed that the wild-type *ABI1* genes must *promote* the ABA response. However, genetic studies showed that the original mutations were dominant: One defective copy of the gene was sufficient to disrupt the ABA response by blocking the activity of the functional gene products from the remaining wild-type allele. Subsequently, recessive mutants of *ABI1* were obtained that exhibited a simple loss of *ABI1* activity. In contrast to the dominant *ABI1* mutants, these recessive *ABI1* mutants exhibited an *increased* sensitivity to ABA (Gosti et al. 1999).

Genomic analyses have now shown that *ABI1* and *ABI2* are part of a subgroup of nine closely related genes belonging to a larger PP2C gene family in Arabidopsis (Schweighofer et al. 2004). Loss of function of many of the *ABI*-class PP2C genes results in *increased* sensitivity to ABA. In addition, overexpression of their wild-type gene products caused by reintroducing the gene into plants under control of a highly expressed promoter confers *reduced* ABA sensitivity. Thus, the wild-type function of these protein phosphatases is to *inhibit* the ABA response, presumably by dephosphorylating specific serine or threonine residues of target proteins, thereby regulating their activity. The ABI-class protein phosphatases appear to interact with many other proteins in the cell, including protein kinases, Ca^{2+}-binding proteins, and transcription factors (Himmelbach et al. 2003).

In addition, a family of calcineurin B–like proteins (CBLs) has now been identified in plants. Mammalian calcineurin phosphatases are heterodimers of a catalytic A subunit and a Ca^{2+}-binding regulatory B subunit, which may be activated by interaction with calmodulin. In Arabidopsis, different calcineurin B–like proteins have been shown to be differentially regulated by environmental stress and by ABA, and were predicted to have distinct organ and subcellular localizations (Luan et al. 2002). Although calcineurin A-like subunits have not yet been found in plants, a screen for potential targets of calcineurin B-like protein regulation identified a family of CBL-interacting protein kinases (CIPKs) that were also stress- and ABA-regulated. Genetic studies of knockout mutants for different *CBL* and *CIPK* gene family members have identified at least two loci that affect some aspects of ABA and stress signaling, *CBL9* and *CIPK3*, while others are involved in ABA-independent responses to abiotic stresses (see Chapter 26). Both CBL9 and CIPK3 repress ABA- and stress-inhibition of germination, but they have opposing effects on ABA-regulated gene expression. The mechanism by which these protein factors affect the ABA response is not yet clear, but the presence in the cell of approximately ten Ca^{2+} sensors (the CBLs) that can regulate another set of about 25 protein kinases (the CIPKs) provides plant cells with a robust network of signaling pathways for transducing the information from internal Ca^{2+} signals into physiological responses.

ABA regulates gene expression

Downstream of the early ABA signal transduction processes already discussed, ABA causes changes in gene expres-

---

*In this type of infrared screen, leaves with closed stomata are slightly warmer (~25°C) than seedlings with open stomata (~22°C) because of evapotranspiration in the latter, and the temperature differences can be detected by infrared photography.

sion. ABA has been shown to regulate the expression of numerous genes during seed maturation and under certain stress conditions, such as drought, adaptation to low temperatures, and salt tolerance (Rock 2000). Transcriptional profiling studies in Arabidopsis and rice have shown that 5 to 10% of the genome is regulated by ABA and various stresses, but only 10 to 50% of the stress-regulated genes are common to any given pair of treatments (Shinozaki et al. 2003). The ABA- and stress-induced genes are presumed to contribute to adaptive aspects of induced tolerance (see Chapter 26). They include genes encoding proteases, chaperonins, proteins similar to LEA proteins, enzymes of sugar or other compatible solute* metabolism, ion and water channel proteins, enzymes that detoxify reactive oxygen species, and regulatory proteins such as transcription factors and protein kinases.

In a few cases, stimulation of transcription by ABA has been demonstrated directly. Gene activation by ABA is mediated by transcription factors. Four main classes of regulatory sequences conferring ABA inducibility have been identified, and proteins that bind to these sequences, including members of the basic leucine zipper (bZIP), B3, MYB, and MYC families, have been characterized (see **Web Topic 23.10**).

In addition to these proteins, members of the homeodomain–leucine zipper (HD-ZIP), APETALA2 (AP2), and WRKY classes of transcription factors participate in some ABA responses. Under stress conditions, induction of gene expression may be ABA-dependent or ABA-independent, and additional transcription factors have been identified that specifically mediate responses to cold, drought, or salinity (salt stress) (see Chapter 26).

A few DNA elements have been identified that are involved in transcriptional repression by ABA. The best-characterized of these are the gibberellin response elements (GAREs) that mediate the gibberellin-inducible, ABA-repressible expression of the barley α-amylase genes (see Chapter 20).

Four transcription factors involved in ABA-inducible gene activation in maturing seeds have been identified by genetic means: VIVIPAROUS 1 (VP1) in maize and ABA-INSENSITIVE (ABI) 3, 4, and 5 in Arabidopsis. A mutation in the gene encoding any of these four transcription factors reduces the ABA responsiveness of the seeds. The maize *VP1* and Arabidopsis *ABI3* genes encode highly similar proteins, while the *ABI4* and *ABI5* genes encode members of two other transcription factor families (Finkelstein et al. 2002).

Additional members of the *ABI5* subfamily of transcription factors have been identified that are also correlated with ABA-, embryonic-, drought-, or salt stress–induced

gene expression, but loss-of-function mutations for these family members have limited effects due in part to substantial redundancy. Characterization of *vp1*, *abi3*, *abi4*, and *abi5* mutants has shown that each of these genes can either activate or repress transcription, depending on the target gene, but some of these effects may be indirect. Because the promoter of any given gene contains binding sites for a variety of regulators, it is likely that these transcription factors act in complexes made up of varying combinations of regulators whose composition is determined by the combination of available regulators and binding sites. Availability and activity of these transcription factors is controlled by developmentally and environmentally regulated gene expression, cross- and autoregulation of expression by these factors, posttranslational modification such as phosphorylation, and in some cases, stabilization against proteasomal degradation (Lopez-Molina et al. 2001; Finkelstein et al. 2002). As illustrated in Figure 23.15, many of these regulatory mechanisms require interactions among transcription factors or between transcription factors and kinases, phosphatases, or components of the degradation machinery.

Tests of anticipated interactions and searches for new interactors have made use of standard biochemical approaches (e.g., coimmunoprecipitations) and yeast two-hybrid assays (see **Web Topic 23.11**). These studies have shown that the protein ABI3/VP1 interacts physically with a variety of proteins, including ABI5 and its rice homolog (TRAB1). ABI5 also forms homodimers and heterodimers with other bZIP family members. At least one of these heterodimers actually represses ABA-inducible gene expression, illustrating another mechanism for posttranscriptional regulation of activity (Bensmihen et al. 2002). Additional interactions have been identified between specific transcription factors and kinases, phosphatases, and 14-3-3 proteins—a class of acidic proteins that facilitate protein–protein interactions involved in signaling, transport, and enzyme activity (see Chapter 14 on the web site). Recent studies have identified a class of phosphatases that regulate RNA polymerase II directly, specifically repressing transcription of ABA-inducible genes (Koiwa et al. 2002). Collectively, these studies demonstrate that many transcription factors can be present in a variety of regulatory complexes with distinct effects on ABA-induced gene expression.

Other negative regulators also influence the ABA response

As described previously, negative regulators of the ABA response (protein phosphatases) have been identified by isolation of dominant negative mutants such as *abi1* and *abi2* that result in ABA-insensitive phenotypes (analogous to the dominant negative effects of the ethylene receptor mutant *etr1*; see Chapter 22).

Other negative regulators have been identified through isolation of mutants exhibiting enhanced responses to

---

*An organic compound that can serve as a nontoxic, osmotically active solute in the cytosol; such compounds usually accumulate during water or salt stress.

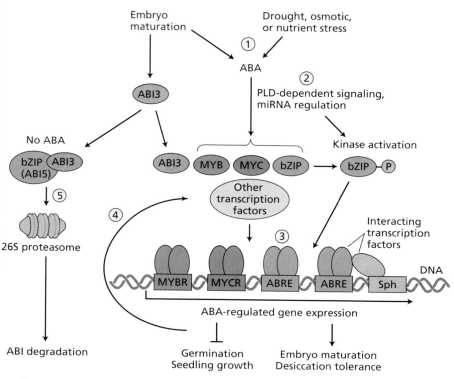

1. ABA and a variety of transcription factors are produced in response to developmental or environmental signals.

2. Phospholipase D (PLD)-dependent signaling, regulation by microRNA (miRNA), and activation of many kinases are implicated in the activation of transcription factors by ABA. The promoters of ABA-regulated genes have different combinations of recognition sites (MYBR, MYCR, etc.) that can be bound by various members of the corresponding transcription factor families. Multiple family members for each class of the transcription factors shown participate in ABA signaling.

3. In addition to forming homo- and heterodimers within families, some of these factors interact with one another and additional components of the transcription machinery. The specific combinations determine the extent to which a given gene is activated or repressed.

4. Some of the transcription factor genes are cross- and autoregulated, in some cases enhancing the ABA response by positive feedback.

5. In the absence of ABA, these ABA-Insensitive (ABI) transcription factors are degraded via the proteasome.

FIGURE 23.15 Regulatory mechanisms and transcription factors that mediate ABA-regulated gene expression.

ABA, based on either physiological criteria or reporter gene expression. Mutants showing increased sensitivity to ABA during germination include *enhanced response to ABA* (*era*) and *ABA hypersensitive* (*abh*) (Cutler et al. 1996; Hugouvieux et al. 2001). The *era* and *abh* mutants both confer ABA hypersensitivity in both stomatal closing and germination, making these mutants resistant to wilting and mildly drought-tolerant.

FARNESYLATION The *ERA1* gene encodes a subunit of the enzyme farnesyl transferase. Farnesyl transferases catalyze attachment of the isoprenoid intermediate farnesyl diphosphate (see Chapter 13) to proteins that contain a specific signal sequence of amino acids. Many proteins that have been shown to participate in signal transduction are farnesylated. Farnesylated proteins are anchored to the membrane via hydrophobic interactions between the farnesyl group and the membrane lipids (see Figure 1.6). The identification of ERA1 as part of farnesyl transferase suggests that protein(s) that normally suppress the ABA response require farnesylation for anchoring to the membrane. One such protein is the monomeric GTPase ROP10, a negative regulator of ABA signaling affecting cytoskeletal organization whose plasma membrane localization is partly dependent on ERA1-mediated farnesylation (Zheng et al. 2002).

MRNA PROCESSING AND STABILITY *ABH1* encodes an mRNA 5′ cap–binding protein that may be involved in mRNA processing of negative regulators of ABA signaling. (Recall that eukaryotic mRNAs have a "cap" consisting of methylated guanosine at the 5′ end.) Comparison of transcript accumulation in wild-type and *abh1* plants showed a small number of misexpressed genes in the mutant, including some encoding possible signaling molecules. The *SUPERSENSITIVE TO ABA AND DROUGHT 1* (*SAD1*) locus encodes a member of a small nuclear ribonucleoprotein (snRNP) complex implicated in splicing, export, and degradation of RNAs. A point mutation in this locus is sufficient to confer hypersensitivity to ABA, but the specific targets of *SAD1* regulation have not been reported. The *hyponastic leaves 1* (*hyl1*) mutation has pleiotropic effects, including ABA hypersensitivity, that can be partially explained by altered MAP kinase signaling leading to overaccumulation of the ABI5 transcription factor (Lu et al. 2002). *HYL1* encodes a double-stranded RNA–binding protein that regulates stability of specific transcripts via effects on microRNA (miRNA) production* (Vazquez et al.

---

*miRNAs are small (~21-mer) RNA molecules encoded in the genomes of plants and animals. These highly conserved RNAs regulate the expression of genes by binding to the 3′-untranslated regions of specific mRNAs and promoting their degradation.

2004), but it is not clear which targets of miRNA regulation affect the ABA response (see Figure 23.15).

ETHYLENE INSENSITIVITY *ERA3* was found to be allelic to a previously identified ethylene signaling locus, *ETHYLENE-INSENSITIVE 2 (EIN2)* (Ghassemian et al. 2000) (see Chapter 22). In addition to displaying defects in ABA and ethylene responses, mutations in this gene result in defects in the responses to auxin, jasmonic acid, and stress. This gene encodes a membrane-bound protein that appears to represent a point of "cross talk"—i.e., a common signaling intermediate—mediating the responses to *many different signals*.

IP$_3$ CATABOLISM Other screens have identified ABA signaling mutants on the basis of incorrect expression of reporter genes controlled by ABA-responsive promoters. Although the defects in some of these mutants are limited to gene expression, others affect plant growth responses. One such mutant, termed *fiery (fry)* to reflect the intensity of light emission by its ABA/stress-responsive luciferase reporter, is also hypersensitive to ABA and stress inhibition of germination and growth. The *FIERY* gene encodes an enzyme required for IP$_3$ catabolism (Xiong et al. 2001). The mutant phenotype demonstrates that the ability to attenuate, as well as induce, stress signaling is important for successful induction of stress tolerance.

Similar to the signaling mechanisms documented for other plant hormones, ABA signaling involves the coordinated action of positive and negative regulators affecting processes as diverse as transcription, RNA processing, protein phosphorylation or farnesylation, and metabolism of secondary messengers. As the signaling components are identified, and often are found to function in responses to multiple signals, the next challenge is to determine how they can lead to ABA-specific responses.

Summary

Abscisic acid plays major roles in seed and bud dormancy, as well as in responses to water stress. ABA in tissues can be measured by bioassays based on growth, germination, or stomatal closure. Gas chromatography, high-pressure liquid chromatography (HPLC), and immunoassays are the most reliable and accurate methods available for measuring ABA levels.

ABA is a 15-carbon terpenoid compound produced by cleavage of a 40-carbon carotenoid precursor that is synthesized from isopentenyl diphosphate (IPP) via the plastid terpenoid pathway. ABA is inactivated by both oxidative degradation and conjugation.

ABA is synthesized in almost all cells that contain plastids and is transported via both the xylem and the phloem. The level of ABA fluctuates dramatically in response to developmental and environmental changes. During seed maturation, ABA levels peak in mid- to late embryogenesis.

The 9-*cis*-epoxycarotenoid dioxygenases (NCEDs) are a major site of regulation, but enzymes catalyzing several steps in ABA biosynthesis are regulated by stress or developmental signals. ABA levels appear subject to feedback regulation, but the target enzymes have not yet been identified.

ABA promotes the development of desiccation tolerance in the developing embryo, the synthesis of storage proteins and lipids, and the acquisition of dormancy. Seed dormancy and germination are controlled by the ratio of ABA to gibberellic acid (GA), and ABA-deficient embryos of some species may exhibit precocious germination and vivipary. ABA is also antagonized by ethylene and brassinosteroid promotion of germination. Although less is known about the role of ABA in buds, ABA is one of the inhibitors that accumulates in dormant buds.

During water stress, the ABA level of the leaf can increase 50-fold. In addition to closing stomata, ABA increases the hydraulic conductivity of the root and increases the root:shoot ratio at low water potentials. ABA and an alkalinization of the xylem sap are thought to be two chemical signals that the root sends to the shoot as the soil dries. The increased pH of the xylem sap may allow more of the ABA of the leaf to be translocated to the stomata via the transpiration stream.

ABA exerts both short-term and long-term control over plant development. The long-term effects are mediated by ABA-induced gene expression. ABA stimulates the synthesis of many classes of proteins during seed development and water stress, including the late-embryogenesis–abundant (LEA) family, proteases and chaperonins, ion and water channels, and enzymes catalyzing compatible solute metabolism or detoxification of active oxygen species. These proteins may protect membranes and other proteins from desiccation damage, or they may aid in recovery from the deleterious effects of stress. ABA response elements and several transcription factors that bind to them have been identified. ABA also suppresses GA-induced gene expression—for example, the synthesis of GA-MYB and α-amylase by barley aleurone layers.

There is evidence for both extracellular and intracellular ABA receptors. To date, one receptor has been identified and found to encode Flowering Control Locus A (FCA), an RNA-binding protein that mediates ABA inhibition of flowering and lateral root initiation, but has no effect on germination or stomatal regulation. ABA closes stomata by causing long-term depolarization of the guard cell plasma membrane. Depolarization is believed to be caused by an increase in cytosolic Ca^{2+}, as well as alkalinization of the cytosol. The increase in cytosolic calcium is due to a combination of calcium uptake and release of calcium from internal stores. This calcium increase leads to the opening of slow anion channels, which results in membrane depolarization. Sphingosine-1-phosphate (S1P), inositol 1,4,5-trisphosphate (IP$_3$), *myo*-inositol-hexaphosphate (IP$_6$), cyclic ADP-ribose (cADPR), phosphatidic acid, and reactive oxygen species all function as secondary messengers in ABA-treated guard cells, and G proteins participate in the response. Outward K$^+$ channels open in response to membrane depolarization and to the rise in pH, bringing about massive K$^+$ efflux.

In general, ABA responses appear to be regulated by more than one signal transduction pathway, even within a single cell type. This redundancy is consistent with the ability of plant cells to respond to multiple sensory inputs. Additional redundancy has been revealed by genomic studies that have shown that the majority of ABA signaling components are members of differentially expressed families with overlapping and distinct effects. There is also genetic evidence for cross talk between ABA signaling and the signaling of all other major classes of phytohormones, as well as sugars.

Web Material

Web Topics

23.1 The Structure of Lunularic Acid from Liverworts
Although inactive in higher plants, lunularic acid appears to have a function similar to ABA in liverworts.

23.2 Structural Requirements for Biological Activity of Abscisic Acid
To be active as a hormone, ABA requires certain functional groups

23.3 The Bioassay of ABA
Several ABA-responding tissues have been used to detect and measure ABA.

23.4 Proteins Required for Desiccation Tolerance
ABA induces the synthesis of proteins that protect cells from damage due to desiccation.

23.5 Types of Seed Dormancy and the Roles of Environmental Factors
Many types of seed dormancy exist; some are affected by various environmental factors.

23.6 The Longevity of Seeds
Under certain conditions, seeds can remain dormant for hundreds of years.

23.7 Genetic Mapping of Dormancy: Quantitative Trait Locus (QTL) Scoring of Vegetative Dormancy Combined with a Candidate Gene Approach
QTL analysis is a genetic method for determining the number and chromosomal locations of genes affecting a quantitative trait affected by many unlinked genes.

23.8 ABA-Induced Senescence and Ethylene
Hormone-insensitive mutants have made it possible to distinguish the effects of ethylene from those of ABA on senescence.

23.9 Yellow Cameleon: A Noninvasive Tool for Measuring Intracellular Calcium
The yellow cameleon protein has several features that enable it to act as a reporter for calcium concentration.

23.10 Promoter Elements That Regulate ABA Induction of Gene Expression
ABA induction of gene expression is regulated by several different *cis*-acting sequences bound by distinct transcription factors

23.11 The Two-Hybrid System
The GAL4 transcription factor can be used to detect protein–protein interactions in yeast.

Web Essay

23.1 Heterophylly in Aquatic Plants
Abscisic acid induces aerial-type leaf morphology in many aquatic plants.

Chapter References

Allan, A. C., Fricker, M. D., Ward, J. L., Beale, M. H., and Trewavas, A. J. (1994) Two transduction pathways mediate rapid effects of abscisic acid in *Commelina* guard cells. *Plant Cell* 6: 1319–1328.

Allen, G. J., Kwak, J. M., Chu, S. P., Llopis, J., Tsien, R. Y., Harper, J. F., and Schroeder, J. I. (1999) Cameleon calcium indicator reports cytoplasmic calcium dynamics in *Arabidopsis* guard cells. *Plant J.* 19: 735–747.

Anderson, B. E., Ward, J. M., and Schroeder, J. I. (1994) Evidence for an extracellular reception site for abscisic acid in *Commelina* guard cells. *Plant Physiol.* 104: 1177–1183.

Assmann, S. M. (2004) Abscisic acid signal transduction in stomatal responses. In *Hormones: Biosynthesis, Signal Transduction, Action!* Davies, P. J., ed., Springer, New York.

Beardsell, M. F., and Cohen, D. (1975) Relationships between leaf water status, abscisic acid levels, and stomatal resistance in maize and sorghum. *Plant Physiol.* 56: 207–212.

Bensmihen, S., Rippa, S., Lambert, G., Jublot, D., Pautot, V., Granier, F., Giraudat, J. and Parcy, F. (2002) The homologous ABI5 and EEL transcription factors function antagonistically to fine-tune gene expression during late embryogenesis. *Plant Cell* 14: 1391–1403.

Bewley, J. D., and Black, M. (1994) *Seeds: Physiology of Development and Germination*, 2nd ed. Plenum, New York.

Christmann, A., Hoffmann, T., Teplova, I., Grill, E., and Müller, A. (2005) Generation of active pools of abscisic acid revealed by in vivo imaging of water-stressed *Arabidopsis*. *Plant Physiol.* 137: 209–219.

Coursol, S., Fan, L., Le Stunff, H., Spiegel, S., Gilroy, S., and Assmann, S. M. (2003) Sphingolipid signaling in *Arabidopsis* guard cells involves heterotrimeric G proteins. *Nature* 423: 651–654.

Cutler, S., Ghassemian, M., Bonetta, D., Cooney, S., and McCourt, P. (1996) A protein farnesyl transferase involved in abscisic acid signal transduction in *Arabidopsis*. *Science* 273: 1239–1241.

Davies, W. J., and Zhang, J. (1991) Root signals and the regulation of growth and development of plants in drying soil. *Annu. Rev. Plant Physiol. Plant Mol. Biol.* 42: 55–76.

Desikan, R., Griffiths, R., Hancock, J., and Neill, S. (2002) A new role for an old enzyme: Nitrate reductase-mediated nitric oxide generation is required for abscisic acid-induced stomatal closure in *Arabidopsis thaliana*. *Proc. Natl. Acad. Sci. USA* 99: 16314–16318.

Finkelstein, R. R, Gampala, S. S. L., and Rock, C. D. (2002) Abscisic acid signaling in seeds and seedlings. *Plant Cell* S15–S45.

Ghassemian, M., Nambara, E., Cutler, S., Kawaide, H., Kamiya, Y., and McCourt, P. (2000) Regulation of abscisic acid signaling by the ethylene response pathway in *Arabidopsis*. *Plant Cell* 12: 1117–1126.

Gilroy, S., and Jones, R. L. (1994) Perception of gibberellin and abscisic acid at the external face of the plasma membrane of barley (*Hordeum vulgare* L.) aleurone protoplasts. *Plant Physiol.* 104: 1185–1192.

Gilroy, S., Read, N. D., and Trewavas, A. J. (1990) Elevation of cytoplasmic calcium by caged calcium or caged inositol trisphosphate initiates stomatal closure. *Nature* 343: 769–771.

Gomez-Cadenas, A., Zentella, R., Walker-Simmons, M. K., and Ho, T.-H. D. (2001) Gibberellin/abscisic acid antagonism in barley aleurone cells: Site of action of the protein kinase PKABA1 in relation to gibberellin signaling molecules. *Plant Cell* 13: 667–679.

Gosti, F., Beaudoin, N., Serizet, C., Webb, A. A. R., Vartanian, N., and Giraudat, J. (1999) ABI1 protein phosphatase 2C is a negative regulator of abscisic acid signaling. *Plant Cell* 11: 1897–1909.

Himmelbach, A., Yang, Y., and Grill, E. (2003) Relay and control of abscisic acid signaling. *Curr. Opin. Plant Biol.* 6: 470–479.

Hoecker, U., Vasil, I. K., and McCarty, D. R. (1995) Integrated control of seed maturation and germination programs by activator and repressor functions of Viviparous-1 of maize. *Genes Dev.* 9: 2459–2469.

Hugouvieux, V., Kwak, J. M., and Schroeder, J. I. (2001) A mRNA cap binding protein, ABH1, modulates early abscisic acid signal transduction in *Arabidopsis*. *Cell* 106: 477–487.

Jeannette, E., Rona, J.-P., Bardat, F., Cornel, D., Sotta, B., and Miginiac, E. (1999) Induction of *RAB18* gene expression and activation of K$^+$ outward rectifying channels depend on an extracellular perception of ABA in *Arabidopsis thaliana* suspension cells. *Plant J.* 18: 13–22.

Kinoshita, T., Nishimura, M., and Shimazaki, K.-I. (1995) Cytosolic concentration of Ca^{2+} regulates the plasma membrane H$^+$-ATPase in guard cells of fava bean. *Plant Cell* 7: 1333–1342.

Koiwa, H., Barb, A. W., Xiong, L., Li, F., McCully, M. G., Lee, B. H., Sokolchik, I., Zhu, J., Gong, Z., Reddy, M. et al. (2002) C-terminal domain phosphatase-like family members (AtCPLs) differentially regulate *Arabidopsis thaliana* abiotic stress signaling, growth, and development. *Proc. Natl. Acad. Sci. USA* 99: 10893–10898.

Koornneef, M., Jorna, M. L., Brinkhorst-van der Swan, D. L. C., and Karssen, C. M. (1982) The isolation of abscisic acid (ABA) deficient mutants by selection of induced revertants in non-germinating gibberellin sensitive lines of *Arabidopsis thaliana* L. Heynh. *Theor. Appl. Genet.* 61: 385–393.

Koornneef, M., Bentsink, L., and Hilhorst, H. (2002) Seed dormancy and germination. *Curr. Opin. Plant Biol.* 5: 33–36.

Kwak, J. M., Mori, I. C., Pei, Z.-M., Leonhardt, N., Torres, M. A., Dangl, J. L., Bloom, R. E., Bodde, S., Jones, J. D. G., and Schroeder, J. I. (2003) NADPH oxidase *AtrbohD* and *AtrbohF* genes function in ROS-dependent ABA signaling in *Arabidopsis*. *EMBO J.* 22: 2623–2633.

Lee, Y., Choi, Y. B., Suh, S., Lee, J., Assmann, S. M., Joe, C. O., Kelleher, J. F., and Crain, R. C. (1996) Abscisic acid-induced phosphoinositide turnover in guard cell protoplasts of *Vicia faba*. *Plant Physiol.* 110: 987–996.

Li, J., and Assmann, S. M. (1996) An abscisic acid-activated and calcium-independent protein kinase from guard cells of fava bean. *Plant Cell* 8: 2359–2368.

Lopez-Molina, L,. Mongrand, S., and Chua, N. H. (2001) A post-germination developmental arrest checkpoint is mediated by abscisic acid and requires the ABI5 transcription factor in *Arabidopsis*. *Proc. Natl. Acad Sci. USA* 98: 4782-4787.

Lu, C., Han, M. H., Guevara-Garcia, A., and Federow, N. V. (2002) Mitogen-activated protein kinase signaling in postgermination arrest of development by abscisic acid. *Proc. Natl. Acad. Sci. USA* 99: 15812–15817.

Luan, S., Kudla, J., Rodriguez-Concepcion, M., Yalovsky, S., and Gruissem, W. (2002) Calmodulins and calcineurin B–like proteins: Calcium sensors for specific signal response coupling in plants. *Plant Cell* 14: S389–S400.

Mäser, P., Leonhardt, N., and Schroeder, J. I. (2002) The clickable guard cell: Electronically linked model of guard cell signal transduction pathways. In: *The Arabidopsis Book*. C.R. Somerville and E.M. Meyerowitz, eds. American Society of Plant Biologists, Rockville, MD. At: http://www.aspb.org/publications/arabidopsis.

McAinsh, M. R., Brownlee. C., and Hetherington, A. M. (1990) Abscisic acid-induced elevation of guard cell cytosolic Ca^{2+} precedes stomatal closure. *Nature* 343: 186–188.

Milborrow, B. V. (2001) The pathway of biosynthesis of abscisic acid in vascular plants: A review of the present state of knowledge of ABA biosynthesis. *J. Exp. Bot.* 52: 1145–1164.

Mori, I. C., and Muto, S. (1997) Abscisic acid activates a 48-kilodalton protein kinase in guard cell protoplasts. *Plant Physiol.* 113: 833–839.

Mustilli, A. C., Merlot, S., Vavasseur, A., Fenzi, F., and Giraudat, J. (2002) *Arabidopsis* OST1 protein kinase mediates the regulation of stomatal aperture by abscisic acid and acts upstream of reactive oxygen species production. *Plant Cell* 14: 3089–3099.

Pandey, S. and Assmann, S. (2004) The *Arabidopsis* putative G protein–coupled receptor GCR1 interacts with the G protein α subunit GPA1 and regulates abscisic acid signaling. *Plant Cell* 16: 1616–1632.

Raschke, K., Shabahang, M., and Wolf, R. (2003) The slow and the quick anion conductance in whole guard cells: Their voltage-dependent alternation, and the modulation of their activities by abscisic acid and CO_2. *Planta* 217: 639–650.

Raz, V., Bergervoet, J. H. W., and Koornneef, M. (2001) Sequential steps for developmental arrest in *Arabidopsis* seeds. *Development* 128: 243–252.

Razem, F., El-Kereamy, A., Abrams, S., and Hill, R. (2006) The RNA binding protein, FCA, is an abscisic acid receptor. *Nature* 439: 290–294.

Rock, C. D. (2000) Pathways to abscisic acid-regulated gene expression. *New Phytol.* 148: 357–396.

Saab, I. N., Sharp, R. E., Pritchard, J., and Voetberg, G. S. (1990) Increased endogenous abscisic acid maintains primary root growth and inhibits shoot growth of maize seedlings at low water potentials. *Plant Physiol.* 93: 1329–1336.

Sanchez, J.-P., and Chua, N.-H. (2001) *Arabidopsis* PLC1 is required for secondary responses to abscisic acid signals. *Plant Cell* 13: 1143–1154.

Schmidt, C., Schelle, I., Liao, Y.-J., and Schroeder, J. I. (1995) Strong regulation of slow anion channels and abscisic acid signaling in guard cells by phosphorylation and dephosphorylation events. *Proc. Natl. Acad. Sci. USA* 92: 9535–9539.

Schroeder, J. I., and Hagiwara, S. (1990) Repetetive increases in cytosolic Ca^{2+} of guard cells by abscisic acid activation of non-selective Ca^{2+} permeable channels. *Proc. Natl. Acad. Sci. USA* 87: 9305–9309.

Schroeder, J. I., Allen, G. J., Hugouvieux, V., Kwak, J. M., and Waner, D. (2001) Guard cell signal transduction. *Annu. Rev. Plant Phys. Plant Mol. Biol.* 52: 627–658.

Schultz, T. F., and Quatrano, R. S. (1997) Evidence for surface perception of abscisic acid by rice suspension cells as assayed by Em gene expression. *Plant Sci.* 130: 63–71.

Schurr, U., Gollan, T., and Schulze, E.-D. (1992) Stomatal response to drying soil in relation to changes in the xylem sap composition of *Helianthus annuus*. II. Stomatal sensitivity to abscisic acid imported from the xylem sap. *Plant Cell Environ.* 15: 561–567.

Schwartz, A., Wu, W.-H., Tucker, E. B., and Assmann, S. M. (1994) Inhibition of inward K^+ channels and stomatal response by abscisic acid: An intracellular locus of phytohormone action. *Proc. Natl. Acad. Sci. USA* 91: 4019–4023.

Schweighofer, A., Hirt, H., and Meskiene, I. (2004) Plant PP2C phosphatases: Emerging functions in stress signaling. *Trends Plant Sci.* 9: 236–243.

Seo, M., and Koshiba, T. (2002) Complex regulation of ABA biosynthesis in plants *Trends Plant Sci.* 7: 41–48.

Sharp, R. E. (2002) Interaction with ethylene: Changing views on the role of abscisic acid in root and shoot growth responses to water stress. *Plant Cell Environ.* 25: 211–222.

Shimazaki, K., Iino, M., and Zeiger, E. (1986) Blue light–dependent proton extrusion by guard cell protoplasts of *Vicia faba*. *Nature* 319: 324–326.

Shinozaki, K., Yamaguchi-Shinozaki, K., and Seki, M. (2003) Regulatory network of gene expression in the drought and cold stress responses. *Curr. Opin. Plant Biol.* 6: 410–417.

Vartanian, N., Marcotte, L., and Giraudat, J. (1994) Drought rhizogenesis in *Arabidopsis thaliana*. *Plant Physiol.* 104: 761–767.

Vazquez, F., Gasciolli, V., Crete, P., and Vaucheret, H. (2004) The nuclear dsRNA binding protein HYL1 is required for micro RNA accumulation and plant development, but not posttranscriptional transgene silencing. *Curr. Biol.* 14: 346–351.

Wang, X.-Q., Ullah, H., Jones, A. M., and Assmann, S. M. (2001) G protein regulation of ion channels and abscisic acid signaling in *Arabidopsis* guard cells. *Science* 292: 2070–2072.

White, C. N., Proebsting, W. M., Hedden, P., and Rivin, C. J. (2000) Gibberellins and seed development in maize. I. Evidence that gibberellin/abscisic acid balance governs germination versus maturation pathways. *Plant Physiol.* 122: 1081–1088.

Wilkinson, S., and Davies, W. J. (1997) Xylem sap pH increase: A drought signal received at the apoplastic face of the guard cell that involves the suppression of saturable abscisic acid uptake by the epidermal symplast. *Plant Physiol.* 113: 559–573.

Xiong, L., Lee, H., Ishitani, M., Zhang, C., and Zhu, J.-K. (2001) *FIERY1* encoding an inositol polyphosphate 1-phosphatase is a negative regulator of abscisic acid and stress signaling in *Arabidopsis*. *Genes Dev.* 15: 1971–1984.

Yamazaki, D., Yoshida, S., Asami, T., and Kuchitsu, K. (2003) Visualization of abscisic acid-perception sites on the plasma membrane of stomatal guard cells. *Plant J.* 35: 129–139.

Zhang, W., Yu, L., Zhang, Y., and Wang, X. (2005) Phospholipase D in the signaling networks of plant response to abscisic acid and reactive oxygen species. *Biochem. Biophys. Acta* 1736: 1–9.

Zheng, Z. L., Nafisi, M., Tam, A., Li, H. Crowell, D. N., Chary, S. N., Schroeder, J. I., Shen, A., and Yang, Z. (2002) Plasma membrane-associated ROP10 small GTPase is a specific negative regulator of abscisic acid responses in *Arabidopsis*. *Plant Cell* 14: 2787–2797.

Chapter 24 Brassinosteroids

STEROID HORMONES HAVE LONG been known in animals, but they have only recently been discovered in plants. Animal steroid hormones include the sex hormones (estrogens, androgens, and progestins) and the adrenal cortex hormones (glucocorticoids and mineralocorticoids). The brassinosteroids (BRs) are a group of steroid hormones that play pivotal roles in a wide range of developmental phenomena in plants, including cell division and cell elongation in stems and roots, photomorphogenesis, reproductive development, leaf senescence, and stress responses (Clouse and Sasse 1998).

The identification of plant steroid hormones was the result of nearly 30 years of efforts to identify novel growth-promoting substances in pollen from many different plant species (Steffens 1991). Early studies by J. W. Mitchell and colleagues showed that the greatest growth-stimulating activity was found in the organic solvent extract of pollen from the rape plant (*Brassica napus* L.). The unidentified active compounds in rape pollen were named *brassins* (Mitchell et al. 1970).

The specific growth-promoting effects of the brassins were scored in several physiological tests, including the *bean second-internode bioassay*. In this assay, brassins behaved differently than other known phytohormones, causing both cell elongation and cell division, as well as bending, swelling, and splitting of the second internode (Mandava 1988).

Based on their ability to cause dramatic changes in growth and differentiation at low concentrations, Mitchell et al. (1970) proposed that brassins constituted a new family of plant hormones. Further work demonstrated that brassins not only induced stem elongation, they also increased total biomass and seed yield. Foreseeing likely practical applications, funding became available for laboratories of the U. S. Department of Agriculture to purify and identify the active compounds in brassins.

Eventually, from 227 kg of bee-collected rape pollen investigators were able to purify 4 mg of the most bioactive brassin compound, named *brassinolide* (Grove et al. 1979). Based on X-ray analysis of the crystal structure and spectroscopic studies, the chemical structure of brassinolide was determined. The compound was shown to be a polyhydroxylated steroid similar to animal steroid hormones.

Three years later, Japanese scientists purified another phytosteroid, castasterone—thought to be the precursor of brassinolide—from chestnut galls (Yokota et al., 1982). Shortly after, the same group identified a mixture of biologically active brassinolide-like substances in the broadleaf evergreen tree *Dystilium racemosum* (Abe and Yokota 1991).

Although brassinosteroids were known to be endogenous compounds that produced dramatic growth effects in bioassays such as the bean second-internode bioassay, they were not immediately accepted as plant hormones, because their role in normal plant growth and development remained elusive for many years. It was genetic studies in Arabidopsis in the mid-1990s that finally demonstrated that brassinosteroids are authentic plant hormones that participate with other plant hormones in the regulation of numerous aspects of plant development, including shoot growth, root growth, vascular differentiation, fertility, and seed germination (Clouse and Sasse 1998).

We will begin our discussion of brassinosteroids with a brief description of the chemical structure of the BRs and the genetic experiments that led to their identification as plant hormones. We will next review the pathways for their biosynthesis, metabolism, and transport, then describe some of the important physiological processes affected by this group of hormones. Next we will examine the signaling pathways of BRs, from their receptor to their target genes. Finally, we will give a brief account of the potential applications of brassinosteroids in agriculture.

Brassinosteroid Structure, Occurrence, and Genetic Analysis

During brassinosteroid purification, two main bioassays have been used: the bean second-internode bioassay (Figure 24.1) and the rice lamina (leaf) inclination bioassay (Figure 24.2). These bioassays distinguish between biologically active BRs and their inactive intermediates or metabolites, and allow quantitation of the amount of active compound present (Figure 24.3).

The basic chemical structure of brassinosteroids was first resolved in 1979, when X-ray crystallographic analysis of the purified active substance determined that it was

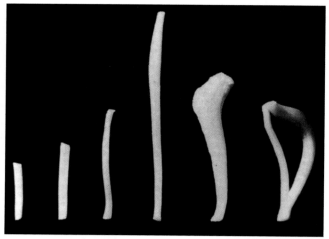

FIGURE 24.1 Bean second internode bioassay for brassinosteroids. Excised sections from the second internodes of bean plants were floated on solutions containing increasing concentrations of BRs for several days. The untreated control section is on the left. At low concentrations, BRs induce mainly elongation growth. Higher concentrations result in stem thickening, bending, and splitting. (From Mandava 1988.)

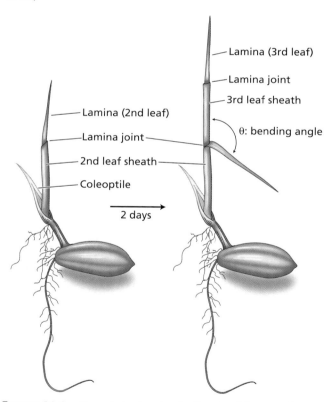

FIGURE 24.2 Dwarf rice lamina inclination bioassay for brassinosteroids. A small droplet of sample dissolved in ethanol is applied to the joint between the lamina and the leaf sheath. After incubation for 2 days in high humidity, the external angle (θ) between the lamina and the leaf sheath is measured. The angle is proportional to the amount of brassinosteroid in the sample.

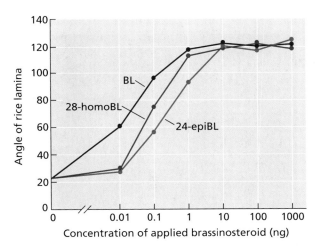

FIGURE 24.3 Dose–response curves for three active BRs in the rice lamina inclination bioassay. 24-epiBL, 24-epibrassinolide; 28-homoBL, 28-homobrassinolide; BL, brassinolide. (After Fujioka et al. 1998.)

a steroidal lactone (Figure 24.4) (Grove et al. 1979). The compound was named **brassinolide** (**BL**), and its characterization ultimately led to the chemical identification of a group of about 60 related phytosteroids called brassinosteroids (Fujioka and Yokota 2003). The direct biosynthetic precursor of BL, **castasterone** (**CS**), has weak BR activity and is therefore considered to be a BR hormone as well.

Brassinosteroids have now been identified in 27 families of seed plants (including both angiosperms and gymnosperms), one pteridophyte (*Equisetum arvense*), one bryophyte (*Marchantia polymorpha*), and one green alga (*Hydrodictyon reticulatum*). In the angiosperms they are found at low levels in pollen, anthers, seeds, leaves, stems, roots, flowers, and young vegetative tissues. Thus, brassinosteroids appear to be ubiquitous plant hormones that predate the evolution of land plants.

Knowing the molecular structure of BL allowed investigators to synthesize both naturally occurring BRs and their related analogs. By testing these compounds with either the bean second internode or rice lamina inclination bioassay, the following key requirements for BR activity were determined (Mandava 1988):

- A *cis*-vicinal glycol function at C-2 and C-3 of ring A. The absence of a single hydroxyl group at either C-2 or C-3 or any change in their configuration results in a significant loss of function.

- The seven-membered B-ring lactone. Although limited modifications of the B-ring don't eliminate activity (see, for example, the B-ring of castasterone, Figure 24.7), the activity is significantly reduced.

- The steroid side chain with hydroxyl group at C-22 and C-23. The α orientation at C-22, C-23, and C-24 confers higher activity than the corresponding β orientation.

Variations in the alkyl side chain at C-24 are tolerated, although a reduction in activity is usually detected, as in the case of **24-epibrassinolide** (**24-epiBL**) and **28-homobrassinolide** (**28-homoBL**) (see Figures 24.3 and 24.4). The different compounds can be classified as C_{27}, C_{28}, or C_{29} BRs, depending on the structure of their side chain. Since 24-epiBL can be synthesized more cheaply than brassinolide, 24-epiBL is often used in physiological experiments in preference to BL, although it is only 10% as active as brassinolide in most bioassays.

BR-deficient mutants are impaired in photomorphogenesis

Definitive proof that brassinosteroids function as plant hormones came only during the past decade from genetic analyses in Arabidopsis, which led to the isolation and description of mutants defective in BR biosynthesis and perception. The abnormal phenotypes of these mutants demonstrated that BRs were required for normal development.

The first characterized BR-deficient mutants, *det2* (*de-et*iolated 2) and *cpd* (*c*onstitutive *p*hotomorphogenesis and *d*warfism) were identified in screens for Arabidopsis seedlings that have a light-grown morphology (that is, they are de-etiolated) after growing for several days in total darkness (Figure 24.5) (Li et al. 1996; Szekeres et al. 1996). Both *det2* and *cpd* seedlings have an impaired photomorphogenic response, with short, thick hypocotyls, expanded cotyledons, young primary leaves (which are absent in

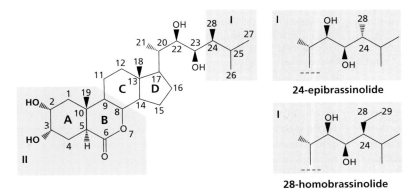

FIGURE 24.4 The structures of brassinosteroids. Brassinolide (BL) is the most widespread and active BR in plants. The structure of BL is shown with its carbons numbered and the ring types indicated by letters. Regions where variations occur are indicated by roman numerals I and II. Variations in the side chain (region I) are present in 24-epibrassinolide and 28-homobrassinolide, but do not significantly affect activity. The hydroxyls on the side chain at carbons 22 and 23 are both essential for activity.

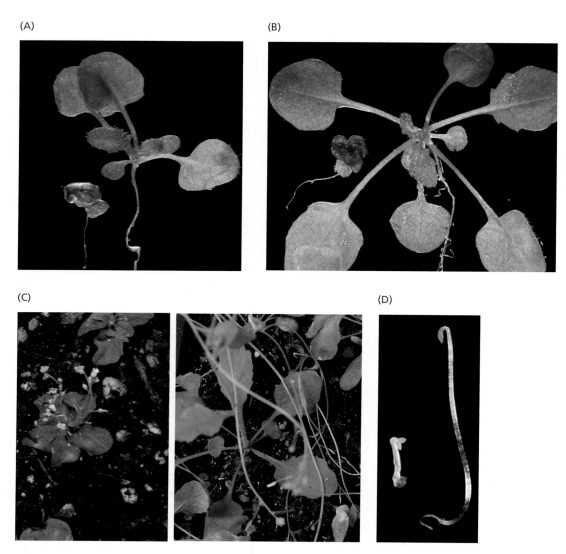

FIGURE 24.5 Phenotypes of Arabidopsis BR mutants. (A) The 3-week-old light-grown homozygous *bri1* mutant (left) is a severe dwarf compared to the heterozygous *bri1* mutant (right), which exhibits wild-type morphology. (B) The 3-week-old light-grown homozygous *cpd* mutant (left) also exhibits a dwarf phenotype; the heterozygous mutant with a wild-type phenotype is on the right. (C) The light-grown adult *det2* mutant (left) is dwarfed compared to the wild-type plant (right). (D) The dark-grown *det2* on the left has a short, thick hypocotyls and expanded cotyledons; the dark-grown wild type is on the right. (Courtesy of S. Savaldi-Goldstein.)

dark-grown seedlings), and high levels of anthocyanins—all of which in the wild type are features of light-grown seedlings but not of dark-grown seedlings. In addition, the two mutants have elevated levels of light-regulated mRNAs when grown in the dark. In contrast, dark-grown wild-type seedlings exhibit a typical etiolated phenotype (long hypocotyls, folded cotyledons, and an absence of anthocyanin pigment).

In addition to their atypical dark-grown phenotype, *det2* and *cpd* have an abnormal phenotype when grown in the light. Both mutants grow as dark-green dwarfs due to a reduction in cell size and intercellular air spaces, and they have reduced apical dominance (see Chapter 19) and male fertility. In addition, the *det2* and *cpd* mutants exhibit short roots, delayed flowering, and delayed leaf senescence even after flowering. In general, the *cpd* mutants have a more extreme phenotype than the *det2* mutants (see below).

Both mutants are impaired in brassinosteroid biosynthesis. The *DET2* locus encodes a protein with high amino acid sequence identity to that of mammalian steroid 5α-reductases (Li et al. 1996). Mammalian steroid 5α-reductases catalyze an NADPH-dependent conversion of testosterone to dihydrotestosterone—a key step in steroid metabolism that is essential for normal embryonic development of male external genitalia and the prostate. Likewise, the *CPD* gene encodes a protein homologous to

(A) *cpd*

(B) Wild type

FIGURE 24.6 BL and intermediates of the BL biosynthetic pathway (see Figure 24.7) restore normal growth to the *cpd* mutant. Wild-type and *cpd* mutant seedlings were grown for 14 days with no steroid (minus sign) or with 0.2 µM of BR intermediate compounds. Note that neither campesterol (CL) nor cathasterone (CT) have any effect on the *cpd* mutant phenotype, because these intermediates occur prior to the reaction catalyzed by CPD. In contrast, teasterone (TE), 3-dehydroteasterone (DT), typhasterol (TY), castasterone (CS), and brassinolide (BL) all rescue the phenotype because they occur after the CPD-catalyzed reaction. The wild type already contains optimal levels of these intermediates and is therefore slightly inhibited by BL and its immediate precursors. (From Szekeres et al. 1996.)

mammalian cytochrome P450 monooxygenase enzymes, including steroid hydroxylases.

The reason the *cpd* mutants have a more extreme phenotype than *det2* mutants is that *det2* mutants contain a residual amount of active BRs, while the levels of active BRs in *cpd* mutants are virtually undetectable (Szekeres et al. 1996). Treating the dwarf mutants with exogenous brassinolide restored the normal phenotype, providing further evidence that DET2 and CPD are required for both brassinosteroid production and normal photomorphogenesis (Figure 24.6).

Biosynthesis, Metabolism, and Transport of Brassinosteroids

Like gibberellin and abscisic acid, brassinosteroids are synthesized as a branch of the terpenoid pathway, starting with the polymerization of two farnesyl diphosphates to form the C_{30} triterpene **squalene** (see Chapter 13). Squalene then undergoes a series of ring closures to form the pentacyclic triterpenoid (sterol) precursor **cycloartenol**. All steroids in plants are derived from cycloartenol by a series of oxidation reactions and other modifications. Our knowledge of the BL biosynthetic pathway is the result of a combination of genetic and biochemical analyses (Fujioka and Yokota 2003).

For the biochemical studies periwinkle (*Catharanthus roseus*) cell cultures were used, as they produce BRs in relatively high amounts. Radiolabeled BR intermediates were used in feeding experiments, and their metabolic derivatives were identified by gas chromatography–mass spectroscopy analysis. Coupling this type of analysis to genetic studies with BR-deficient mutants in Arabidopsis, tomato, and other species has allowed the identification of the complete biosynthetic pathways.

Brassinolide is synthesized from campesterol

Brassinosteroids are synthesized from campesterol, sitosterol, and cholesterol. While campesterol and sitosterol are abundant in plant membranes, cholesterol is present at relatively low levels. All three sterols are metabolized to a large number of intermediates in plant cells, but only a few of these metabolites have biological activity (Clouse and Sasse 1998; Sakurai 1999).

We will illustrate the BR biosynthetic pathway starting with the sterol progenitor **campesterol**, which is ultimately derived from cycloartenol (Figure 24.7). Campesterol is first converted to campestanol in steps involving DET2. Campestanol is then converted to castasterone (CS) through either of two pathways called the *early-* and *late C-6 oxidation pathways*. Additional information about these two pathways is provided in Figure 24.7.

The two pathways merge at CS, which is then converted to BL (see Figure 24.4 and Figure 24.7). The early and the late C-6 oxidation pathways coexist and can be linked at different points in Arabidopsis, pea, and rice (Fujioka and Yokota 2003). The biological significance of having two linked pathways is currently unknown. In fact, the early C-6 oxidation branch is not detected in tomato. The presence of two linked pathways increases the complexity of BR biosynthesis and may provide an advantage under different physiological conditions, such as various types of stress.

All the mutants impaired in their ability to convert campestanol to BL have mutations in genes that encode **cytochrome p450 monooxygenases** (**CYPs**). The Arabidopsis *DWARF4* (*DWF4*) and *CPD* genes encode two such monooxygenases, CYP90B1 and CYP90A1, which hydroxylate BR intermediates at the 22 and 23 positions, respectively (Fujioka and Yokota 2003) (see Figure 24.4 and Figure 24.7).

The subcellular localization of BR biosynthesis has not yet been identified. However, since the cytochrome P450 monooxygenases involved in gibberellin biosynthesis (see Chapter 20) are located on the endoplasmic reticulum (ER), it is likely that BL biosynthesis also takes place on the ER.

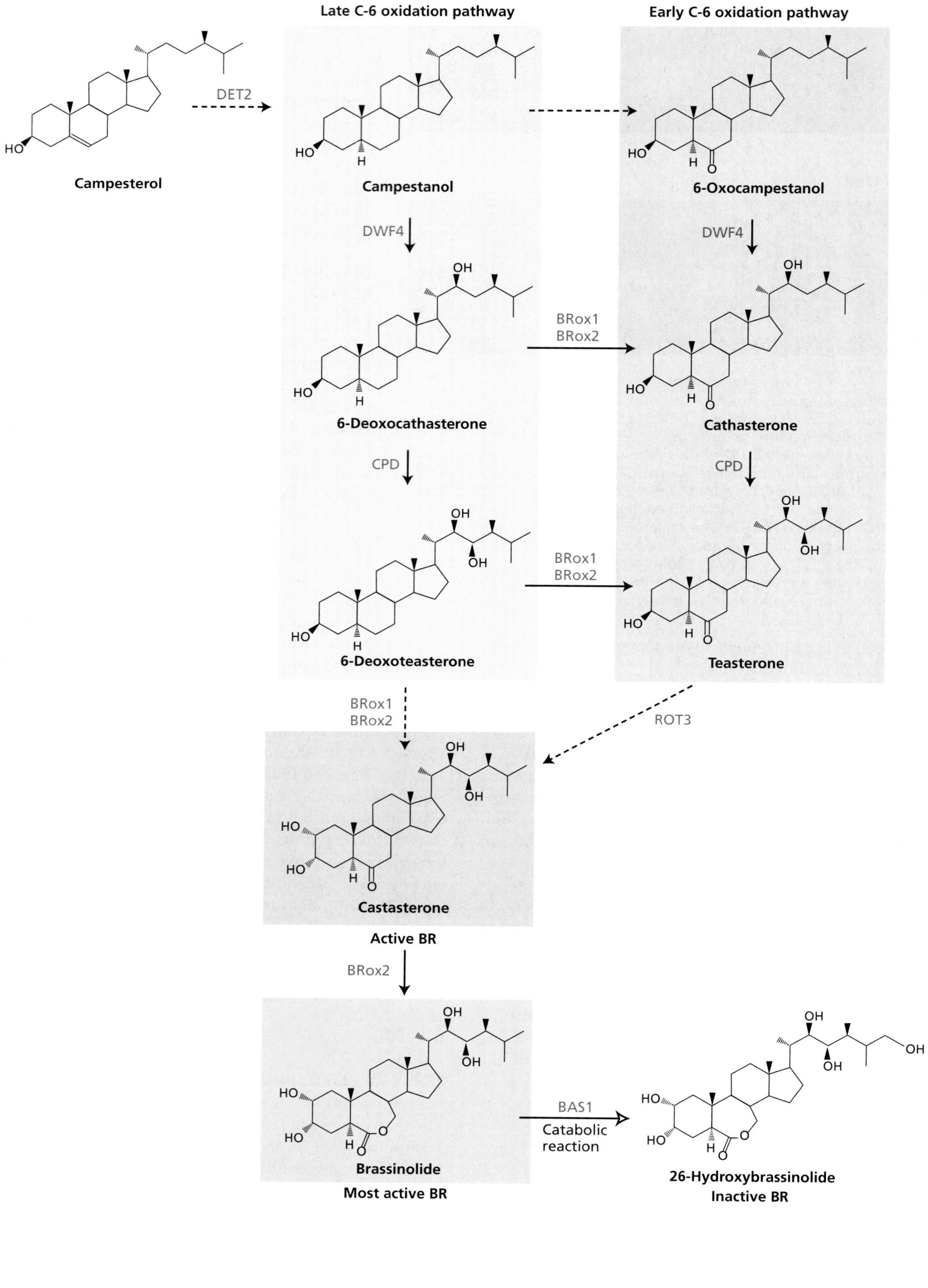

622 Chapter 24

Late C-6 oxidation pathway

Early C-6 oxidation pathway

Campesterol → (DET2) → Campestanol → 6-Oxocampestanol

Campestanol → (DWF4) → 6-Deoxocathasterone

6-Oxocampestanol → (DWF4) → Cathasterone

6-Deoxocathasterone → (BRox1, BRox2) → Cathasterone

6-Deoxocathasterone → (CPD) → 6-Deoxoteasterone

Cathasterone → (CPD) → Teasterone

6-Deoxoteasterone → (BRox1, BRox2) → Teasterone

6-Deoxoteasterone → (BRox1, BRox2) → Castasterone

Teasterone → (ROT3) → Castasterone

Castasterone — Active BR

Castasterone → (BRox2) → **Brassinolide** — Most active BR

Brassinolide → (BAS1, Catabolic reaction) → **26-Hydroxybrassinolide** — Inactive BR

FIGURE 24.7 Simplified pathways for BL biosynthesis and catabolism. The precursor for BL biosynthesis is campesterol. The sequence of biosynthetic events is represented by black arrowheads. Solid arrows indicate single reactions, dashed arrows represent multiple reactions. As shown, castasterone, the immediate precursor of BL, can be synthesized from two parallel pathways: the early and the late C-6 oxidation pathways. In the early C-6 oxidation pathway, oxidation at C-6 of the B ring occurs *before* the addition of vicinal hydroxyls at C-22 and C-23 of the side chain (refer to BL structure in Figure 24.3). In the late C-6 oxidation pathway, C-6 is oxidized *after* the introduction of hydroxyls at the side chain and C-2 of the A ring. Both the early and the late pathways may be linked at various points, creating a biosynthetic network rather than a linear pathway. The Arabidopsis enzymes that catalyze the different steps are indicated. BL catabolism is marked by an open arrowhead.

Catabolism and negative feedback contribute to BR homeostasis

The level of active BRs is also regulated by metabolic processes, which inactivate BL. Several types of reactions result in BL inactivation, including epimerization, oxidation, hydroxylation, sulphonation, and conjugation to glucose or lipids (Fujioka and Yokota 2003). Our limited knowledge in this area is based on feeding experiments in which plants are fed radiolabeled BRs, and the resulting labeled products are identified and endogenous metabolites analyzed; however, their relevance for the BR pathway in the plant is still not clear.

The isolation of the Arabidopsis *BAS1* (*phyB a*ctivation-tagged *s*uppressor1-dominant) gene encoding a cytochrome P450 monooxygenase with steroid 26-hydroxylase activity (CYP72B1) has helped to elucidate the role of at least one metabolic enzyme in controlling BL concentrations. Overexpression of *BAS1* leads to decreased BL levels and an accumulation of an inactive 26-hydroxyBL (for 26-hydroxyBL structure, see Figure 24.6), resulting in a BR-deficient dwarf phenotype (Neff et al. 1999). Thus, as in the case of other plant hormones, BL homeostasis is regulated by a balance between biosynthetic and inactivation reactions.

The levels of physiologically active BR are also regulated by negative feedback mechanisms. In other words, if an excess of the hormone accumulates, BR biosynthesis is attenuated and BR turnover is enhanced. Indeed, the mRNAs of all tested Arabidopsis BL biosynthetic genes (*DWF4, CPD, ROT3* and *BR6ox1*) are down-regulated (decreased) in response to BL application, while *BAS1* mRNA, which is involved in BR turnover, accumulates to higher levels (Figure 24.8) (Tanaka et al. 2005).

Down-regulation of BR biosynthetic genes involves a transcription factor that binds directly to a conserved promoter element found in the above biosynthetic genes, thereby repressing their expression (He et al. 2005). Accordingly, Arabidopsis mutants impaired in their ability to respond to BL accumulate high levels of the active brassinosteroids CS and BL compared to wild-type plants (Noguchi et al. 1999).

A valuable tool for the genetic, physiological, and molecular study of BRs is the specific BR biosynthesis inhibitor, **brassinazole** (**Brz**) (Figure 24.9). Brz contains a triazole ring made up of two carbon and three nitrogen atoms. Various triazole compounds can act as inhibitors of cytochrome P450 monooxygenases. The triazole Brz specifically inhibits the activity of the BL biosynthetic enzyme DWF4 (monooxygenase CYP90B1), which converts 6-oxocampestanol to cathasterone (Asami et al. 2003). Plants grown on Brz show BR-deficient phenotypes, which can be reversed by the addition of BL to their growth medium (Figure 24.10). In this experiment both Brz and BL are taken up by the root system.

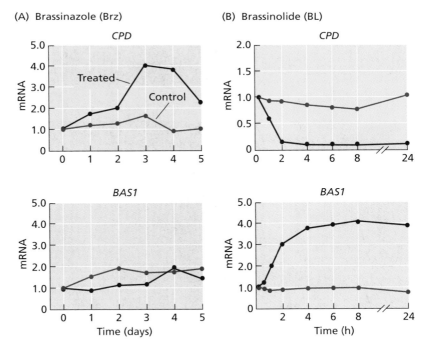

FIGURE 24.8 Brassinosteroid levels are controlled by both negative and positive feedback. The mRNA levels of CPD and BAS1 were measured in Arabidopsis seedlings treated either with (A) 5 µM brassinazole (Brz), or (B) with 5 µM Brz (to deplete the endogenous BL levels), followed by 0.1 µM BL for 2 days. Expression of the BR biosynthetic gene *CPD1* is enhanced by Brz (A) and inhibited by BL (B). *CPD1* is thus negatively regulated by BL. In contrast, the expression of the BR-degrading enzyme *BAS1* is stimulated by BL. *BAS1* is thus positively regulated by BL. (After Tanaka et al. 2005.)

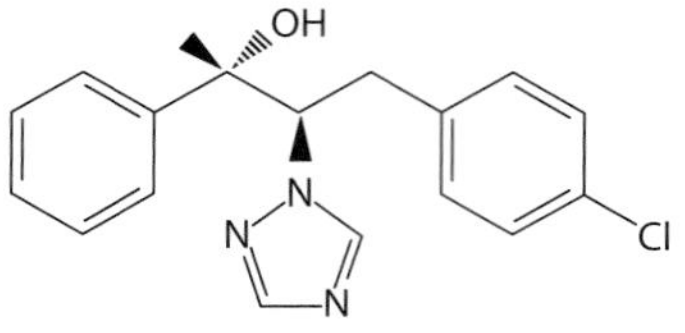

FIGURE 24.9 The structure of brassinazole [4-(4-chlorophenyl)-2-phenyl-3-(1,2,4-triazoyl)butan-2-ol], a triazole compound that inhibits brassinosteroid biosynthesis.

Numerous studies using brassinazole have yielded important information on BR homeostasis that complements the studies with BL described above. Thus, five BR biosynthetic genes (*DET2*, *DWF4*, *CPD*, *BR6ox1*, and *ROT3*) and two sterol biosynthesis genes were up-regulated in BR-depleted Arabidopsis plants grown in the presence of Brz (see Figure 24.8). Taken together, the results suggest that BR homeostasis is maintained by feedback regulation of multiple target genes (Tanaka et al. 2005).

BR homeostasis is also controlled by rate-limiting steps in the BL biosynthetic pathway. If an enzyme is rate-limiting, significant levels of its substrate should accumulate relative to its immediate product. Measurements of endogenous BRs in Arabidopsis have shown that CPD, DWF4, and BR6ox1/2 could be rate-limiting steps in BR biosynthesis and thus contribute to BR homeostasis. Indeed, overexpression of DWF4 and BR6ox2 result in increased vegetative growth of the plant (Figure 24.11) (Choe et al. 2001; Kim et al. 2005).

Brassinosteroids act locally near their sites of synthesis

An important determinant of hormone responses in general is the extent and rate of hormone transport from the site

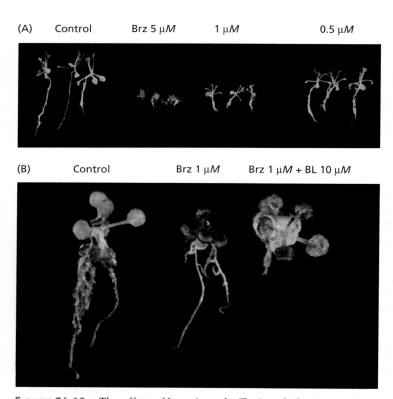

FIGURE 24.10 The effect of brassinazole (Brz) on light-grown 14-day-old Arabidopsis seedlings. (A) Control seedlings (left) and 5 μM, 1 μM, and 0.5 μM brassinazole-treated Arabidopsis seedlings (right). The Brz-treated seedlings exhibit dwarfism in a concentration-dependent manner. (B) Control light-grown 14-day-old Arabidopsis seedling (left) and seedlings treated with 1 μM brassinazole (middle) or 1 μM brassinazole plus 10 nM brassinolide (BL) (right). BRZ inhibits DWF4. The addition of BL, which occurs downstream of DWF4 catalytic reaction, rescues the inhibitory effect of Brz on growth. (From Asami et al. 2000.)

FIGURE 24.11 Overexpression of the BR biosynthetic gene, *DWF4*, in Arabidopsis results in a dramatic increase in plant size. AOD4, Arabidopsis overexpressing DWF4; Ws-2, wild type. Plants were grown for 25 days. Bar = 2 cm. (From Choe et al. 2001.)

of synthesis to the site of action. Exogenously applied 24-epibrassinolide (24-epiBL) undergoes long-distance transport from the root to the shoot. For example, when roots of cucumber, tomato, or wheat plants were treated with ^{14}C-24-epiBL, the radioactivity was readily translocated to the shoot (Schlagnhaufer and Arteca 1991; Nishikawa et al. 1994). Moreover, BR-deficient Arabidopsis mutants could be rescued when grown on agar media supplemented with BL, and the leaf petioles of wild-type plants elongated upon BL application to their root systems (Clouse and Sasse 1998).

In contrast, when ^{14}C-24-epiBL was applied to the upper surface of a young cucumber leaf it was readily taken up, but was only slowly transported out of the leaf. In all, only about 6% of the applied ^{14}C-24-epiBL was transported to the younger leaves (Nishikawa et al. 1994). These results suggest that *exogenous* BRs are readily translocated from the root to the shoot, but are poorly translocated out of leaves. Presumably, 24-epiBL taken up by the roots moves to the shoot via the xylem transpiration stream. Since the xylem stream is unidirectional, however, exogenous 24-epiBL applied to the leaf can only exit the leaf via the phloem. The lack of 24-epiBL movement out of the leaf indicates that it is poorly translocated in the phloem.

Moreover, despite the evidence for root-to-shoot transport of 24-epiBL, *endogenous* BRs do not seem to undergo root-to-shoot translocation. For example, experiments in pea and tomato indicate that reciprocal stock/scion grafting of wild-type to BR-deficient mutants does not rescue the phenotype of the latter in either the acropetal or basipetal directions (Figure 24.12) (Symons and Reid 2004; Montoya et al. 2005). These results suggest that endogenous BRs act locally, at or near their sites of synthesis.

Comparisons of the temporal and spatial distribution of BR intermediates indicates that they are present in all plant tissues, although different intermediates predominate in different organs (Shimada et al. 2003). For example, in Arabidopsis, pea, and tomato, early intermediates are more abundant in the roots, while late intermediates, such as CS, accumulate to higher levels in the shoot. Similarly, while BR biosynthetic enzymes are expressed in all tissues examined, their relative expression varies from tissue to tissue. Such variations in expression are no doubt associated with the different functions of the tissues.

In parallel with the ubiquity of the BR biosynthetic pathway, components of the BR signaling pathway (discussed at the end of the chapter) also appear to be expressed throughout the plant, especially in young growing tissues (Friedrichsen et al. 2000). Taken together, the evidence suggests that each organ synthesizes and responds to its own active BRs.

Brassinosteroids: Effects on Growth and Development

BRs were originally discovered as growth-promoting substances isolated from pollen, and their role as plant hormones was confirmed in studies of photomorphogenesis. In addition, BRs are involved in a wide range of developmental processes, such as fiber development in cotton, development of lateral roots, maintenance of apical dominance, vascular differentiation, and male sterility. Other physiological effects of BRs include their involvement in plant defense, seed germination, and leaf senescence. (For a discussion of BRs and apical hook maintenance see **Web Essay 24.1**.) Much remains to be learned about the physiological role of BRs in development. In this section we will discuss some of the better-understood BR responses, including shoot and root growth, vascular differentiation, pollen tube elongation, and seed germination.

FIGURE 24.12 Effects of reciprocal grafting between wild-type and a BR-deficient mutant of pea (*lkb*) on the phenotype of the shoot in 45-day-old plants. The grafts were made epicotyl to epicotyl using 7-day-old seedlings. The scion is labeled above the bar and the stock is below the bar. The BR-deficient dwarf shoot is not rescued by a wild-type root. Conversely, the growth of the wild-type shoot is unaffected by a BR-deficient root. Both results show an absence of long-distance BR signaling.

BRs promote both cell expansion and cell division in shoots

The growth-promoting effects of BRs are reflected in acceleration of both cell elongation and cell division. These were first characterized using the bean second-internode bioassay as discussed above (see Figure 24.1). Mandava, 1988).

The rice leaf lamina inclination bioassay (see Figure 24.2) is dependent on BR-induced cell expansion. Lamina inclination resembles the epinasty phenomenon caused by ethylene (see Chapter 22). In response to BR, the cells on the adaxial (upper) surface of the leaf near the joint region expand more than the cells on the abaxial (lower) surface, causing the vertically oriented leaf to bend outward. An increase in cell wall loosening is required for BL-induced cell expansion on the adaxial side of the leaf.

In genetic studies, the dwarf phenotype of BR mutants strongly demonstrated the requirement of BRs for normal plant growth (see Figure 24.5). Microscopic examination of leaves from BR-deficient mutants showed not only smaller cell size, but also fewer cells in the leaf blade when compared to wild-type leaves, indicating that BRs are an important class of growth hormones in shoots (Nakaya et al. 2002). Therefore it is not surprising that overexpression of the BR biosynthetic gene *DWF4* results in elevated levels of endogenous BRs, and causes an increase in plant size (see Figure 24.11) (Choe et al. 2001). Indeed, one of the prominent and early-recognized characteristics of BRs (as was first observed in the bean second-internode bioassay) is to promote both cell elongation and cell proliferation.

The stimulatory effect of BRs on growth is most pronounced in young, growing shoot tissues. The kinetics of cell expansion in response to nanomolar concentrations of BL differs from that of auxin. In soybean epicotyl sections, for example, BL begins to enhance the elongation rate after a 45-minute lag period, and reaches a maximum rate only after several hours of treatment. In contrast, auxin stimulates elongation after a 15-minute lag time and reaches a maximum rate within 45 minutes (Figure 24.13) (Zurek et al. 1994) (see Chapter 19). This suggests that BRs stimulate growth via a slower pathway involving gene transcription, whereas the rapid response to auxin may not require gene transcription. Another explanation is that the stimulation of gene expression by auxin is much greater then the gene expression induced by BL (see Chapter 19). In fact, auxin and BRs have been shown to enhance the growth of shoot tissues synergistically and in an interdependent manner, indicating that each hormone requires the presence of the other for optimal activity. A synergism between IAA and BR has also been demonstrated in the rice lamina inclination bioassay.

The process of cell expansion involves cell wall relaxation followed by the osmotic transport of water into the cell to maintain turgor pressure, and

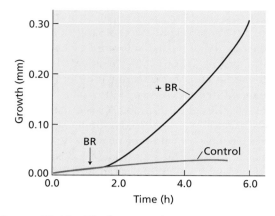

FIGURE 24.13 The kinetics of BR stimulation of soybean epicotyl elongation. A soybean epicotyl section, 1.5 cm in length, was treated with 0.1 μM BR in a sensitive growth measuring system. BR-induced growth was observed after a 45-minute lag period. Five or more hours are required to achieve the maximum steady-state growth rate. (After Zurek et al. 1994.)

cell wall synthesis to maintain wall thickness (see Chapter 15). Each of these steps is likely to be modulated by BRs. Thus, BRs are thought to affect the uptake of water through aquaporins and the activity of the vacuolar H^+-ATPase (V-ATPase). (Schumacher et al. 1999; Morillon et al. 2001). An Arabidopsis mutant, *det3*, with a mutation in one of the subunits of the V-ATPase, has a phenotype similar to that of BR-deficient mutants such as *det2*, except that *det3* mutants are less responsive to exogenous BR. BRs also enhance cell wall loosening (Figure 24.14) and induce the expression of wall-

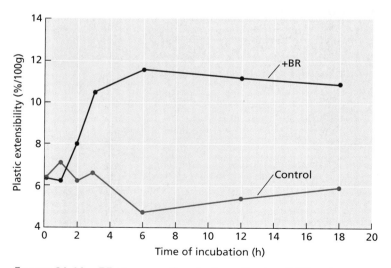

FIGURE 24.14 BRs increase the plastic wall extensibility of soybean epicotyls. Soybean epicotyl sections (1.5 cm) were incubated with or without 0.1 μM BR for the times indicated, after which their plastic extensibility was determined using an extensometer (see Chapter 15). The increase in wall plasticity in the presence of BR indicates that BR has induced cell wall loosening, which is required for cell expansion. (After Zurek et al. 1994.)

(A) Wild-type cell with transverse microtubules

(B) BR-deficient dwarf mutant cell with few nonaligned microtubules

(C) BR-deficient mutant treated with BR

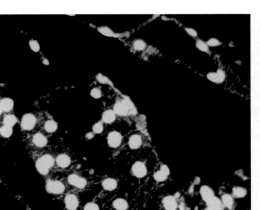
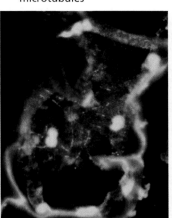
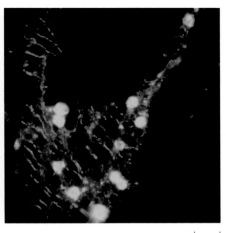

FIGURE 24.15 Effect of BR on microtubule organization in Arabidopsis seedlings. (A) Wild-type parenchyma cell, showing normal transverse microtubule arrangement. (B) BR-deficient mutant parenchyma cell with few, nonaligned microtubules. (C) BR-deficient mutant treated with BR. The normal microtubule organization has been restored. (From Catterou et al. 2001.)

modifying enzymes such as xyloglucan endo-transglucosylase/hydrolases (XTHs) and expansins (see Chapter 15).

An additional requirement for normal elongation is the control of microtubule organization. As discussed in Chapter 15, microtubule orientation helps to align cellulose microfibrils during synthesis, and transversely arranged microfibrils in the wall are required for normal cell elongation. Microscopic analyses of microtubules in a BR-deficient mutant of Arabidopsis have shown that mutant cells contain very few microtubules, and those present lack parallel organization. Treatment of the mutant with BR restored normal microtubule abundance and organization (Figure 24.15). Since BR did not increase the total amount of tubulin protein in the cell, BR must act by promoting microtubule nucleation and organization (Catterou et al. 2001).

In addition to cell elongation, BR also stimulates cell proliferation. As we saw in Chapter 21, cytokinin-induced cell division has been linked to the expression of a D-type cyclin, CYCD3. 24-epiBL has also been shown to increase *CYCD3* gene expression, and 24-epiBL can substitute for zeatin in the growth of Arabidopsis callus and cell suspension cultures (Hu et al. 2000). Thus the BRs and cytokinins appear to regulate the cell cycle via similar mechanisms.

BRs both promote and inhibit root growth

Based on the phenotypes of BR-deficient mutants, which typically exhibit reduced root growth, BRs are required for normal root elongation. However, like auxin, exogenously applied BRs may have positive or negative effects on root growth, depending on the concentration (Mussig 2005). When applied exogenously to BR-deficient mutants, BR promotes root growth at low concentrations and inhibits root growth at high concentrations. The threshold concentration for inhibition depends on the activity of the BR analog used. Thus, the threshold concentration is lower for the relatively active analog 24-epiBL than for the less active analog 24-epicastasterone.

The effects of BR on root growth are independent of both auxin and gibberellin action. An inhibitor of polar auxin transport, 2,3,5-triiodobenzoic acid (TIBA) (see Chapter 19), does not prevent BR-induced growth (Mussig 2005). When BR and auxin are applied simultaneously, both the promotive and inhibitory effects on root growth are additive. Moreover, the reduced root growth phenotype of BR-deficient mutants is not reversed by gibberellin application. Taken together, these observations indicate that BR inhibition of root growth does not involve interactions with either auxin or GA. On the other hand, high concentrations of BR, like auxin, stimulate ethylene production, so it is possible that at least some of BR's inhibitory effects on root growth are due to ethylene.

At low concentrations, BRs can also induce the formation of lateral roots (Figure 24.16) (Bao et al. 2004). In this case, however, BRs and auxin act synergistically. The current model suggests that BRs promote lateral root development partially by influencing polar auxin transport (see Chapter 19). BR treatment promotes acropetal auxin transport, which is required for the development of lateral roots, while 1-*N*-naphthylphthalamic acid (NPA), an auxin transport inhibitor, eliminates the promotive effect of BR (Bao et al. 2004). Thus BR exerts strong effects on overall root morphology, influencing both the elongation rate and the branching habit.

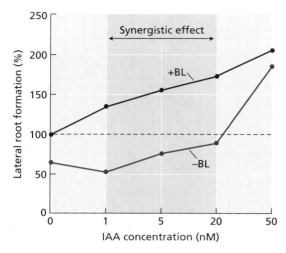

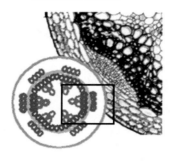 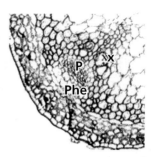

FIGURE 24.16 BL and IAA act synergistically to promote lateral root development. Arabidopsis seedlings were grown vertically on agar plates containing 0, 1, 5, 20, and 50 n*M* IAA with or without 1 n*M* BL for 8 days, and the number of lateral roots and visible lateral root primordia per centimeter of primary root was counted. The number of lateral roots per centimeter in each treatment was graphed as the percentage of the number of lateral roots per centimeter in the 1 n*M* BL, zero auxin treatment (100%, horizontal dashed line). The synergistic effect of BL and auxin occurs in the 1 to 20 n*M* IAA range. (After Bao et al. 2004.)

FIGURE 24.17 BR is required for a normal vascular development. The left panel insert shows a schematic representation of the Arabidopsis vascular system at the basal part of the inflorescence stem of a mature plant. The procambial cells (yellow) give rise to phloem tissue (red) to the outside and xylem tissue (blue) to the inside. The black box encloses a single vascular bundle. The vascular bundle of the BR-deficient *det2* mutant (right) has a lower xylem-to-phloem ratio than that of the wild type (left). P, phloem; Phe, phloem cap cells; X, xylem. (From Caño-Delgado et al. 2004.)

BRs promote xylem differentiation during vascular development

BRs play an important role in vascular development, both promoting xylem and suppressing phloem differentiation. This is evident in the impaired vasculature systems of BR mutants, which have a higher phloem to xylem ratio compared to wild type (Figure 24.17) (reviewed in Fukuda 2004). BR-deficient mutants also have a reduced number of vascular bundles with irregular spacing between the bundles. In contrast, mutants overexpressing the BR receptor protein (discussed later in the chapter) produce more xylem than the wild type.

Cell cultures of *Zinnia elegans* provide an elegant in vitro system to study the sequential stages of xylem differentiation. When single cells are mechanically isolated from young *Zinnia* leaves and cultured in liquid medium in the dark, they differentiate into tracheary elements between days 2 and 3 after culture (Figure 24.18). Measurements of BRs during xylem differentiation in this system have shown that BRs are actively synthesized in procambial-like cells and are essential for their subsequent differentiation into tracheary elements. BRs are likely to mediate the differentiation from procambium to xylem by regulating the expression of homeobox genes (see Chapter 16) that often play crucial roles in development (Fukuda 2004). Moreover, genes encoding BR receptor-like proteins (discussed later) are expressed exclusively in the vascular tissues, supporting the existence of a BR signaling pathway that is specific for the vascular system (Caño-Delgado et al. 2004).

BRs are required for the growth of pollen tubes

Pollen is a rich source of BRs, and thus it is not surprising that BRs are important for male fertility. BR has been shown to promote the growth of the pollen tube from the stigma, through the style, to the embryo sac (Mussig 2005). For exam-

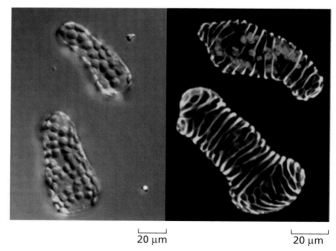

FIGURE 24.18 *Zinnia* leaf mesophyll cell before (left) and after (right) differentiation into a tracheary element. (From Fukuda 2004.)

ple, in the BR-deficient Arabidopsis mutant *cpd*, pollen tubes failed to elongate after germination on the stigma, and pollen tube elongation was shown to be partially dependent on BR application (Szekeres et al. 1996; Clouse and Sasse 1998).

Similarly, when a BR-insensitive mutant with a defective receptor gene was self-pollinated, the pollen tube failed to develop, resulting in sterile seeds. However, when the mutant was hand pollinated with wild-type pollen, fertile seeds were produced (Clouse et al. 1996). Thus, for normal pollen tube growth, both BR and the BR signaling pathway are needed.

Reduced male fertility has also been attributed to a discrepancy in the heights of the stamens versus the pistil. The Arabidopsis *dwf4* mutant is BR-deficient, and its cells fail to elongate. The stamens of the flower are also shorter than those of wild type. Because Arabidopsis is self-pollinating, the shorter filaments of *dwf4* stamens result in less pollen being deposited on the stigmatic surface. Since the pollen grains are still viable, hand pollination of the mutant flowers results in normal seed production.

BRs promote seed germination

Seeds, like pollen grains, contain very high levels of brassinosteroids, and BRs promote seed germination as well (Mussig 2005). BRs promote seed germination by interacting with other plant hormones, although the molecular basis for these interactions is not known. It is well established that GA and abscisic acid (ABA) play positive and negative roles in stimulating seed germination, respectively. BRs can enhance germination of tobacco seeds independent of GA signaling (Leubner-Metzger 2001). Moreover, BRs can rescue the delayed germination phenotype of both GA-deficient and GA-perception mutants (Figure 24.19), and BR mutants are more sensitive to the inhibition by ABA as compared to wild type (Steber and McCourt 2001). Thus, BRs can stimulate germination and are needed to overcome the inhibitory effect of ABA.

As BRs are known to stimulate cell expansion and division, it is likely that BRs facilitate germination by stimulating the growth of the embryo.

The Brassinosteroid Signaling Pathway

Although physiological studies led to the identification of brassinosteroids as plant growth regulators, it was genetic studies that led to their identification as plant hormones. Further genetic analyses of BR responses identified a plasma membrane-localized BR receptor protein and many other signaling elements in the BR signaling pathway. As a result, enormous progress has been achieved in understanding the molecular basis of BR action, although the picture is far from complete. In the final section of the chapter we will discuss the main components of the BR signal transduction pathway, focusing on results obtained with Arabidopsis.

BR-insensitive mutants identified the BR cell surface receptor

To identify components of the BR signaling pathway in Arabidopsis, genetic screens were initially carried out to isolate mutants exhibiting normal root elongation in the presence of high BL concentrations. This resulted in the isolation of a single *bri1* (*br*assinosteroid-*i*nsensitive 1) (Clouse et al. 1996). Further screens for BL-insensitive mutations all yielded additional mutant alleles of *BRI1* (e.g., Li and Chory 1997), suggesting that BRI1 is an essential component of the BR signaling pathway that is encoded by a single gene. Subsequent binding studies demonstrated that BL binds directly to BRI1 with high specificity, suggesting that BRI1 is the BR receptor (Kinoshita et al. 2005).

BL, the most active brassinosteroid, binds to the extracellular domain of the BRI1 receptor in the plasma membrane. **BRI1** is a plasma membrane-localized leucine-rich repeat (LRR)-receptor serine/threonine (S/T) kinase (Figure 24.20). The LRR-receptor kinases constitute the largest receptor class predicted in the Arabidopsis genome, with over 230 family members (see Chapter 14 on the web site).

This family has a conserved domain structure, composed of an N-terminal extracellular domain with multiple tandem (adjacent) LRR motifs, a single transmembrane domain, and a cytoplasmic kinase domain with specificity toward serine and threonine residues (see Figure 24.20). In the case of BRI1, the number of LRRs is 25. BRI1 also has a unique feature that is required for BR binding: a stretch of amino acids called the *island domain* that interrupts the LRRs between LRRs 21 and 22 (Kinoshita et al. 2005). This domain plus the flanking LRR22 compose the minimum binding site for BRs.

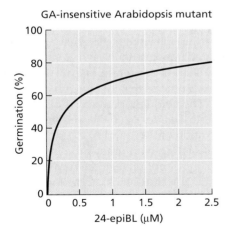

FIGURE 24.19 BR stimulates germination of Arabidopsis seeds. Seeds of a gibberellin-insensitive mutant were treated with increasing concentrations of 24-epiBL, and the percent germination was determined. The results show that the stimulation of germination by BR is independent of gibberellin. (After Steber and McCourt 2001.)

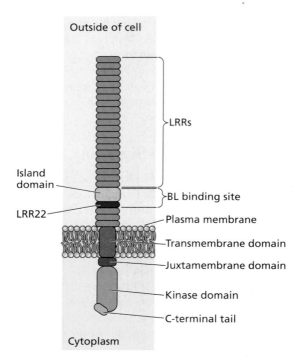

FIGURE 24.20 The domain structure of the BR receptor, BRI1. BRI1 is localized on the plasma membrane. The extracellular region consists of a stretch of leucine-rich repeat sequences (LRRs) containing an island domain that functions as the brassinolide (BL) binding site. The intracellular portion contains a juxtamembrane domain, a kinase domain, and the C-terminal tail.

Phosphorylation activates the BRI1 receptor

Analysis of a large number of BR mutant alleles indicates that both the extracellular receptor domain and the internal kinase domain are necessary for transmitting the BR signal to the rest of the cell (Friedrichsen et al. 2000; Vert et al. 2005). BL binds BRI1 via a novel steroid-binding domain of about 100 amino acids that includes the island domain and its neighboring LRR sequence (see Figure 24.20). BL binding activates the receptor, as indicated by its increased autophosphorylating activity and by its increased association with a second LRR-receptor kinase, **BRI1-associated receptor kinase 1 (BAK1)** (Figure 24.21).

In the presence of brassinolide, BRI1 becomes phosphorylated in vivo at multiple intracellular domains, including the juxtamembrane* region (JM), the C-terminal tail (CT), and the kinase itself. These phosphorylation sites play regulatory roles in receptor activity and control the interaction of BRI1 with other proteins, such as BAK1 (Wang et al. 2005a, 2005b).

As in the case of animal protein kinases, specific phosphorylation sites in the kinase domain of BRI1 are essential for its activation. In addition, the CT of BRI1 negatively

---

*The side of the plasma membrane facing the cytoplasm.

regulates the receptor. Upon ligand binding, this inhibitory effect is nullified, and BRI1 kinase activity increases (Wang et al. 2005b). However, the precise mechanism of this BL-induced activation will only become clear once a high-resolution structure for BRI1 has been determined (Wang et al. 2005b).

Receptor kinases in animal and plant cells often function as dimers in vivo. In vitro experiments have confirmed that BRI1 receptors normally function as homo-oligomers composed of identical monomers in the cell (Wang et al. 2005a).

Following BRI1's binding to and activation by its ligand, phosphorylated BRI1 forms a hetero-oligomer (i.e., composed of two different monomers) with BAK1 (see Figure 24.21). In vitro, BRI1 and BAK1 can phosphorylate each other, and like BRI1, the phosphorylation state of BAK1 is positively regulated by BL (Wang et al. 2005b). The phosphorylated BRI1/BAK1 heterodimer appears to be the activated form of the receptor that induces the BR response by inactivating a repressor called BIN2.

BIN2 is a repressor of BR-induced gene expression

The formation of the activated BRI1/BAK1 hetero-oligomer in the presence of BR initiates a signaling cascade that leads to BR-regulated gene transcription. The next known step in the signal transduction pathway involves the negative regulator **BIN2** (**b**rassinosteroid **in**sensitive-**2**) (see Figure 24.21). *BIN2* encodes a protein kinase homologous to the glycogen synthase kinase 3 of yeast and animals (Vert et al. 2005). In these organisms, the BIN2 homologs function as constitutively active S/T kinases that are involved in a wide range of signaling pathways in which they often act as repressors of gene expression.

In Arabidopsis, BIN2 is found in both the nucleus and cytosol, as well as at the plasma membrane. In the absence of BRs, BIN2 appears to act in the nucleus to constitutively phosphorylate two nuclear proteins, **BES1** (*bri1-EMS-suppressor 1*) and **BZR1** (*brassinazole-resistant 1*), at multiple regulatory sites, thus inhibiting their activity. BES1 and BZR1 are closely related transcriptional activators of BR-induced genes. They are short-lived proteins and are degraded by the 26S proteasome, a process that involves ubiquitination (see Figure 24.21). (The ubiquitin pathway is also described in Chapter 19 and in Chapter 14 on the web site).

Phosphorylation of BES1 and BZR1 by BIN2 prevents them from associating with other proteins, either themselves or other transcription factors. As a result, they are unable to bind DNA, thus blocking their activity as transcriptional regulators (Vert and Chory 2006).

In the presence of BR, the activated BRI1/BAK1 hetero-oligomer initiates a signaling cascade that blocks the activity of the BIN2 kinase by an unknown mechanism (see step 3 in Figure 24.21). This leads to the accumulation of active dephosphorylated forms of BES1 and BZR1 (Figure 24.22), in part due to the activity of the plant-specific serine/threonine phosphatase, BSU1 (*bri1 suppressor 1*), which coun-

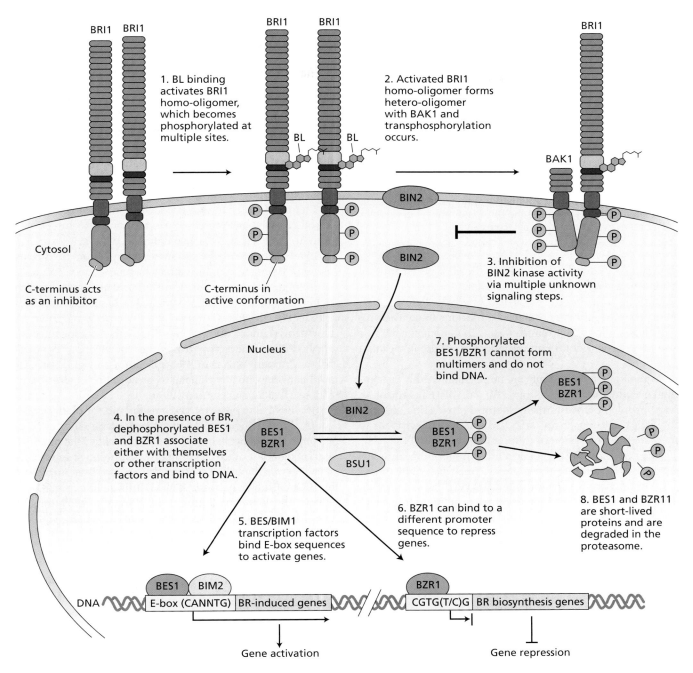

FIGURE 24.21 A model for BR signaling. Signal perception occurs at the cell surface.

teracts the effects of the BIN2 kinase. The active dephosphorylated forms of BES1 and BZR1 activate or repress BR target genes (see Figure 24.21).

BES1 and BZR1 regulate different subsets of genes

Following the sequencing of the Arabidopsis genome, techniques that enabled investigators to monitor the expression of thousands of genes simultaneously, such as *DNA microarray analysis*, became available (see **Web Topic 17.10**).

The application of these techniques to the study of BR-regulated gene expression identified hundreds of BR-induced genes, many of which are predicted to play a role in growth processes. In addition, genes were identified that are repressed by BRs. Many of these down-regulated genes are also controlled by the transcription factors BES1 and BZR1 (see Figure 24.21) (Vert et al. 2005).

The amino acid sequences of BES1 and BZR1 are 90% identical, yet they appear to regulate different subsets of

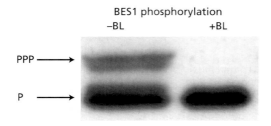

Figure 24.22 BL inhibits the phosphorylation of BES1. A Western blot was used to identify the two forms of BES1: the highly phosphorylated form (PPP) and the dephosphorylated form (P). In the Arabidopsis plants not treated with BL (–BL), much of the BES1 is in the highly phosphorylated state. In the BL-treated plants (+BL) all of the BES1 is dephosphorylated. (Antibodies generated by Y. Yin, courtesy of G. Vert.)

genes in Arabidopsis. This was predicted by their mutant phenotypes. When grown in the dark, the *bes1* and *bzr1* mutations suppress the dwarfism of weak alleles of *bri1*. This is because the *bes1* and *bzr1* mutations make the proteins less susceptible to proteolysis. In genetic terms, *bes1* and *bzr1* are semidominant mutations resulting in a *gain* of function.

In the light, *bes1* mutants are larger than wild type, similar to plants overexpressing DWF4 or BRI1. In contrast, light-grown *bzr1* mutants are semi-dwarf. Thus, despite their high sequence similarity, BZR1 and BES1 appear to mediate the expression of different target genes.

BES1 *enhances* the expression of a subset of BR-stimulated genes. To activate these genes, BES1 associates with another transcription factor called **BIM1** (*BES1-interacting Myc-like 1*). BES1/BIM1 heterodimers activate transcription by binding to a specific DNA sequence called an *E Box* that functions as a BR response element in the promoters of BR-induced genes (Yin et al. 2005).

In contrast, BZR1 acts as a repressor of BR biosynthesis. BZR1 binds directly to CGTG(T/C)G elements found in the promoter regions of various BR biosynthetic genes, turning off transcription. BZR1 thus plays a key role in the negative feedback regulation of the BR biosynthetic pathway (see above) (He et al. 2005).

Besides the distinct subsets of genes they regulate, BES1 and BZR1 are thought to regulate some genes in common. For example, both BES1 and BZR1 repress the expression of *BRI1* in the dark.

Although much progress has been made on the BR signal transduction pathway during the past decade, many important issues remain to be addressed. There are gaps in our understanding of how the BR signal is transmitted from the cell surface to the cytoplasm to regulate the expression of hundreds of genes. Are there multiple BR signaling pathways? How and to what extent do BRs interact with other hormones to control plant growth and development? Ongoing research should provide answers to many of these questions in the near future.

Prospective Uses of Brassinosteroids in Agriculture

Brassinosteroids were discovered as a class of growth-promoting hormones, and their potential applications to agriculture were immediately recognized by researchers. For the past 20 years, numerous small-scale studies have been conducted to test the ability of BRs to increase the yields of crop plants. BL has been found to increase bean crop yield (based on the weight of seeds per plant) by about 45%, and to enhance the leaf weight of different lettuce varieties by 25%. Similar increases in the yields of rice, barley, wheat, and lentils have been observed. BL also promoted potato tuber growth and increased its resistance to infections. Tomato fruit setting was also enhanced by BL.

In addition to small-scale studies, large-scale field trials using brassinosteroid derivatives have now been conducted in Japan, China, Korea, and Russia. The results of such field trials have been highly variable and appear to reflect the degree of stress under which the crop was grown. A crop grown under optimal conditions shows little effect of applied BR, while a crop grown under conditions of stress shows dramatic effects of BR application on yield. Thus BR application is most beneficial to growth under stress conditions (Ikekawa and Zhao 1991).

BRs have also proven to be a useful aid for plant propagation. Pretreatment of various woody plants, such as Norway spruce and apple trees, with BR increased the quantity and quality of rooting by cuttings. Micropropagation of cassava and pineapple by tissue culture has also been improved by BR treatment. As researchers continue to explore BR's effects on plant development, additional applications of brassinosteroids to agriculture are bound to emerge.

Summary

Brassinosteroids (BRs) are polyhydroxylated steroid hormones that regulate plant growth and development. Brassinolide (BL), the most active BR, is the endpoint of the biosynthetic pathway. Exogenous application of BL at nanomolar concentrations to young, growing shoot tissues causes dramatic cell elongation. BR-deficient Arabidopsis mutants are dark-green dwarfs and exhibit delayed senescence and reduced fertility. In the dark, the seedlings are de-etiolated, having short, thick hypocotyls and open cotyledons. Treatment of such mutants with BRs restores the wild-type phenotype.

BL is the end product of a branch of the terpenoid biosynthesis pathway, which has multiple rate-limiting steps. Brassinosteroids are synthesized from campesterol, which is derived from the plant sterol precursor, cycloartenol. Campesterol is first converted to campestanol in multiple steps involving the enzyme steroid 5α-reductase, encoded by the *DET2* gene. Campestanol is then converted to castasterone through either of two pathways, called the *early-* and *late C-6 oxidation pathways*, which are linked to each other. All the enzymes involved in the conversion of campestanol to BL are cytochrome p450 monooxygenases.

BR levels are regulated through multiple control mechanisms, including catabolism, conjugation, and negative feedback from the signaling pathway. BRs have been detected in all tissues examined. The highest level of active BRs occurs in the apical shoot, and it appears that endogenous BRs do not undergo long-distance transport.

BRs regulate growth by controlling both cell elongation and cell division, although the major effect is on cell expansion. In the shoot, BRs and auxin have synergistic effects on growth. While auxin induces a rapid growth response, the effect of BRs is much slower. In both cases, a sustained growth response is likely to be through transcriptional activation of genes required for cell expansion and division. BRs also control root growth and development, having both stimulatory and inhibitory effects on root elongation when applied at low or high concentration, respectively. The stimulatory effect is independent of auxin and gibberellins, although BRs and auxin have a synergistic effect on the formation of lateral roots. BRs are also important for vascular development. They promote xylem and suppress phloem differentiation. BR mutants have a higher phloem to xylem ratio, and fewer vascular bundles with irregular spacing between them, as compared to wild type.

The BR receptor, BRI1, is located on the plasma membrane. BRs bind directly to a 100–amino acid region located within the extracellular domain of BRI1. BRI1 is a member of a large family of plant leucine-rich repeat (LRR)-receptor serine/threonine (S/T) kinases. BL binding to BRI1 triggers the interaction between BRI1 and BAK1, a related LRR-receptor S/T kinase. In addition, BL induces the phosphorylation of both BRI1 and BAK1. BRI1 is phosphorylated at multiple sites along its intracellular domain, some of which have been shown to regulate receptor activity. The BL signal is then transmitted to the cytoplasm by an unknown mechanism where it inhibits BIN2, which is a negative regulator of the BR pathway. BIN2 is a protein kinase that interacts with and phosphorylates two nearly identical transcription factors, BES1 and BZR1, negatively regulating their activities. BSU1 dephosphorylates BES1 and BZR1 to counteract the effect of BIN2.

BRs regulate the expression of hundreds of genes. A significant portion of the up-regulated genes is predicted to play a role in growth processes. BES1 binding activity and the expression level of its target genes are enhanced synergistically by BIM1. BIM1 is another transcription factor that dimerizes with BES1 and increases its activity. Genes that are down-regulated by BR include several BR biosynthetic genes. BZR1 binds to specific elements in their promoters to repress their activity. This gene repression by BZR1 represents a negative feedback loop for the regulation of growth by BR.

Web Material

Web Essay

24.1 Brassinosteroids and the Apical Hook—An Ongoing Story in Plant Architecture
A model is proposed for the interactions between ethylene, auxin, and brassinosteroids in the formation of the hook of etiolated seedlings.

Chapter References

Abe, H., and Marumo, S. (1991). Brassinosteroids in leaves of *Distylium recemosum* Sieb. et Zucc. The beginning of brassinosteroid research in Japan. In *Brassinosteroids: Chemistry, Bioactivity, and Applications*, H. G. Cutler, T. Yokota, and G. Adam, eds., American Chemical Society, Washington, D. C., pp. 18–24.

Asami, T., Nakano, T., Nakashita, H., Sekimata, K., Shimada, Y., and Yoshida, S. (2003) The influence of chemical genetics on plant science: Shedding light on functions and mechanism of action of brassinosteroids using biosynthesis inhibitors. *J. Plant Growth Regul.* 22: 336–349.

Asami, T., Min, Y. K., Nagata, N., Yamagishi, K., Takatsuto, S., Fujioka, S., Murofushi, N., Yamaguchi, I., and Yoshida, S. (2000) Characterization of brassinazole, a triazole-type brassinosteroid biosynthesis inhibitor. *Plant Physiol.* 123: 93–100.

Bao, F., Shen, J., Brady, S. R., Muday, G. K., Asami, T., and Yang, Z. (2004) Brassinosteroids interact with auxin to promote lateral root development in *Arabidopsis*. *Plant Physiol* 134: 1624–1631.

Caño-Delgado, A., Yin, Y., Yu, C., Vafeados, D., Mora-Garcia, S., Cheng, J. C., Nam, K. H., Li, J., and Chory, J. (2004) BRL1 and BRL3 are novel brassinosteroid receptors that function in vascular differentiation in *Arabidopsis*. *Development* 131: 5341–5351.

Catterou, M., Dubois, F., Schaller, H., Aubanelle, L., Vilcot, B., Sangwan-Norreel, B. S., and Sangwan, R. S. (2001) Brassinosteroids, microtubules and cell elongation in *Arabidopsis thaliana*. II. Effects of brassinosteroids on microtubules and cell elongation in the bul1 mutant. *Planta* 212: 673–683.

Choe, S., Fujioka, S., Noguchi, T., Takatsuto, S., Yoshida, S., and Feldmann, K. A. (2001) Overexpression of DWARF4 in the brassinosteroid biosynthetic pathway results in increased vegetative growth and seed yield in *Arabidopsis*. *Plant J.* 26: 573–582.

Clouse, S. D., and Sasse, J. M. (1998) Brassinosteroids: Essential regulators of plant growth and development. *Annu. Rev. Plant Physiol. Plant Mol. Biol.* 49: 427–451.

Clouse, S.D., Langford, M., and McMorris, T. C. (1996) A brassinosteroid-insensitive mutant in *Arabidopsis thaliana* exhibits multiple defects in growth and development. *Plant Physiol* 111: 671–678.

Friedrichsen, D. M., Joazeiro, C. A., Li, J., Hunter, T., and Chory, J. (2000) Brassinosteroid-insensitive-1 is a ubiquitously expressed leucine-rich repeat receptor serine/threonine kinase. *Plant Physiol.* 123: 1247–1256.

Fujioka, S., and Yokota, T. (2003) Biosynthesis and metabolism of brassinosteroids. *Annu. Rev. Plant Biol.* 54: 137–164.

Fujioka, S. Noguchi, T., Takatsuto, S., and Yoshida, S. (1998) Activity of brassinosteroids in the dwarf rice lamina inclination bioassay. *Phytochemistry* 49: 1841–1848.

Fujioka, S., Choi, Y. H., Takatsuto, S., Yokota, T., Li, J., Chory, J., and Sakurai, A. (1996) Identification of castasterone, 6-deoxocastasterone, typhasterol and 6-deoxytyphasterol from the shoots of *Arabidopsis thaliana*. *Plant Cell Physiol.* 37, 1201-1203.

Fukuda, H. (2004) Signals that control plant vascular cell differentiation. *Nat. Rev. Mol. Cell Bio.* 5: 379–391.

Grove, M. D., Spencer, G. F., Rohwedder, W. K., Mandava, N., Worley, J. F., Warthen, J. D., Steffens, G. L., Flippenanderson, J. L., and Cook, J. C. (1979) Brassinolide, a plant growth-promoting steroid isolated from *Brassica napus* pollen. *Nature* 281: 216–217.

He, J. X., Gendron, J. M., Sun, Y., Gampala, S. S., Gendron, N., Sun, C. Q., and Wang, Z. Y. (2005) BZR1 is a transcriptional repressor with dual roles in brassinosteroid homeostasis and growth responses. *Science* 307: 1634–1638.

Hu, Y., Bao, F., and Li, J. (2000) Promotive effect of brassinosteroids on cell division involves a distinct CycD3-induction pathway in *Arabidopsis*. *Plant J.* 24: 693–701.

Ikekawa, N., and Zhao, Y. (1991) Application of 24-epibrassinolide in agriculture. In *Brassinosteroids: Chemistry, Bioactivity, and Applications*, H. G. Cutler, T. Yokota, and G. Adam, eds., American Chemical Society, Washington, D. C., pp. 280–291.

Kim, T. W., Hwang, J. Y., Kim, Y. S., Joo, S. H., Chang, S. C., Lee, J. S., Takatsuto, S., and Kim, S. K. (2005) *Arabidopsis* CYP85A2, a cytochrome P450, mediates the Baeyer-Villiger oxidation of castasterone to brassinolide in brassinosteroid biosynthesis. *Plant Cell* 17: 2397–2412.

Kinoshita, T., Cano-Delgado, A., Seto, H., Hiranuma, S., Fujioka, S., Yoshida, S., and Chory, J. (2005) Binding of brassinosteroids to the extracellular domain of plant receptor kinase BRI1. *Nature* 433: 167–171.

Leubner-Metzger, G. (2001) Brassinosteroids and gibberellins promote tobacco seed germination by distinct pathways. *Planta* 213: 758–763.

Li, J., and Chory, J. (1997) A putative leucine-rich repeat receptor kinase involved in brassinosteroid signal transduction. *Cell* 90: 929–938.

Li, J., Nagpal, P., Vitart, V., McMorris, T. C., and Chory, J. (1996) A role for brassinosteroids in light-dependent development of *Arabidopsis*. *Science* 272: 398–401.

Mandava, N. B. (1988). Plant growth-promoting brassinosteroids. *Annu. Rev. Plant Physiol. Plant Mol. Biol.y* 39: 23–52.

Mitchell, J. W., Mandava, N. B., Worley, J. F., Plimmer, J. R., and Smith, M. V. (1970) Brassins: A new family of plant hormones from rape pollen. *Nature* 225: 1065–1066.

Montoya, T., Nomura, T., Yokota, T., Farrar, K., Harrison, K., Jones, J. G., Kaneta, T., Kamiya, Y., Szekeres, M., and Bishop, G. J. (2005) Patterns of Dwarf expression and brassinosteroid accumulation in tomato reveal the importance of brassinosteroid synthesis during fruit development. *Plant J.* 42: 262–269.

Mora-Garcia, S., Vert, G., Yin, Y., Cano-Delgado, A., Cheong, H., and Chory, J. (2004) Nuclear protein phosphatases with Kelch-repeat domains modulate the response to brassinosteroids in *Arabidopsis*. *Genes Dev.* 18: 448–460.

Morillon, R., Catterou, M., Sangwan, R. S., Sangwan, B. S., and Lassalles, J. P. (2001) Brassinolide may control aquaporin activities in *Arabidopsis thaliana*. *Planta* 212: 199–204.

Mussig, C. (2005) Brassinosteroid-promoted growth. *Plant Biol. (Stuttg)* 7: 110–117.

Nakaya, M., Tsukaya, H., Murakami, N., and Kato, M. (2002) Brassinosteroids control the proliferation of leaf cells of *Arabidopsis thaliana*. *Plant Cell Physiol.* 43: 239–244.

Neff, M. M., Nguyen, S. M., Malancharuvil, E. J., Fujioka, S., Noguchi, T., Seto, H., Tsubuki, M., Honda, T., Takatsuto, S., Yoshida, S., et al. (1999) BAS1: A gene regulating brassinosteroid levels and light responsiveness in *Arabidopsis*. *Proc. Natl. Acad. Sci. U S A* 96: 15316–15323.

Nishikawa, N., Toyama, S., Shida, A., and Futatsuya, F. (1994) The uptake and the transport of ^{14}C-labeled epibrassinolide in intact seedlings of cucumber and wheat. *J. Plant Res.* 107: 125–130.

Noguchi, T., Fujioka, S., Choe, S., Takatsuto, S., Yoshida, S., Yuan, H., Feldmann, K. A., and Tax, F. E. (1999) Brassinosteroid-insensitive dwarf mutants of *Arabidopsis* accumulate brassinosteroids. *Plant Physiol.* 121: 743–752.

Sakurai, A. (1999) Biosynthesis. In *Brassinosteroids: Steroidal Plant Hormones*, A. Sakurai, T. Yokota, and S. D. Clouse, eds, Springer, Tokyo, pp. 91–112.

Schlagnhaufer, C. D., and Arteca, R. N. (1991) The uptake and metabolism of brassinosteroid by tomato (*Lycopersicon esculentum*) plants. *J. Plant Physiol.* 138: 191–194.

Schumacher, K., Vafeados, D., McCarthy, M., Sze, H., Wilkins, T., and Chory, J. (1999) The *Arabidopsis det3* mutant reveals a central role for the vacuolar H^+-ATPase in plant growth and development. *Genes Dev.* 13: 3259–3270.

Shimada, Y., Goda, H., Nakamura, A., Takatsuto, S., Fujioka, S., and Yoshida, S. (2003) Organ-specific expression of brassinosteroid-biosynthetic genes and distribution of endogenous brassinosteroids in *Arabidopsis*. *Plant Physiol.* 131: 287–297.

Steber, C. M., and McCourt, P. (2001) A role for brassinosteroids in germination in *Arabidopsis*. *Plant Physiol.* 125: 763–769.

Steffens, G. L. (1991) U.S. Department of Agriculture brassins project: 1970–1980. In *Brassinosteroids: Chemistry, Bioactivity, and Applications*, H. G. Cutler, T. Yokota, and G. Adam, eds., American Chemical Society, Washington, D.C., pp. 2–17.

Symons, G. M., and Reid, J. B. (2004) Brassinosteroids do not undergo long-distance transport in pea. Implications for the regulation of endogenous brassinosteroid levels. *Plant Physiol.* 135: 2196–2206.

Szekeres, M., Nemeth, K., Koncz-Kalman, Z., Mathur, J., Kauschmann, A., Altmann, T., Redei, G. P., Nagy, F., Schell, J., and Koncz, C. (1996) Brassinosteroids rescue the deficiency of CYP90, a cytochrome P450, controlling cell elongation and de-etiolation in *Arabidopsis*. *Cell* 85: 171–182.

Tanaka, K., Asami, T., Yoshida, S., Nakamura, Y., Matsuo, T., and Okamoto, S. (2005) Brassinosteroid homeostasis in *Arabidopsis* is ensured by feedback expressions of multiple genes involved in its metabolism. *Plant Physiol.* 138: 1117–1125.

Vert, G., and Chory, J. (2006) Downstream nuclear events in brassinosteroid signaling. *Nature* In press.

Vert, G., Nemhauser, J. L., Geldner, N., Hong, F., and Chory, J. (2005) Molecular mechanisms of steroid hormone signaling in plants. *Annu. Rev. Cell Dev. Biol.* 21: 177–201.

Wang, X., Li, X., Meisenhelder, J., Hunter, T., Yoshida, S., Asami, T., and Chory, J. (2005a) Autoregulation and homodimerization are involved in the activation of the plant steroid receptor BRI1. *Dev Cell* 8: 855–865.

Wang, X., Goshe, M. B., Soderblom, E. J., Phinney, B. S., Kuchar, J. A., Li, J., Asami, T., Yoshida, S., Huber, S. C., and Clouse, S. D. (2005b) Identification and functional analysis of in vivo phosphorylation sites of the *Arabidopsis* BRASSINOSTEROID-INSENSITIVE1 receptor kinase. *Plant Cell* 17: 1685–1703.

Wang, Z. Y., Nakano, T., Gendron, J., He, J., Chen, M., Vafeados, D., Yang, Y., Fujioka, S., Yoshida, S., Asami, T., et al. (2002) Nuclear-localized BZR1 mediates brassinosteroid-induced growth and feedback suppression of brassinosteroid biosynthesis. *Dev. Cell* 2: 505–513.

Yin, Y., Vafeados, D., Tao, Y., Yoshida, S., Asami, T., and Chory, J. (2005) A new class of transcription factors mediates brassinosteroid-regulated gene expression in *Arabidopsis*. *Cell* 120: 249–259.

Yin, Y., Wang, Z. Y., Mora-Garcia, S., Li, J., Yoshida, S., Asami, T., and Chory, J. (2002) BES1 accumulates in the nucleus in response to brassinosteroids to regulate gene expression and promote stem elongation. *Cell* 109: 181–191.

Yokota, T., Arima, M., and Takahashi, N. (1982). Castasterone, a new phytosterol with plant-hormone potency, from chestnut insect. *Tetrahedron Letters* 23: 1275–1278.

Zurek, D. M., Rayle, D. L., McMorris, T. C., and Clouse, S. D. (1994) Investigation of gene expression, growth kinetics, and wall extensibility during brassinosteroid-regulated stem elongation. *Plant Physiol.* 104: 505–513.

Chapter 25 | The Control of Flowering

MOST PEOPLE LOOK FORWARD to the spring season and the profusion of flowers it brings. Many vacationers carefully time their travels to coincide with specific blooming seasons: *Citrus* along Blossom Trail in southern California, tulips in Holland. In Washington, DC, and throughout Japan, the cherry blossoms are received with spirited ceremonies. As spring progresses into summer, summer into fall, and fall into winter, wildflowers bloom at their appointed times.

Although the strong correlation between flowering and seasons is common knowledge, the phenomenon poses fundamental questions that will be addressed in this chapter:

- How do plants keep track of the seasons of the year and the time of day?

- Which environmental signals control flowering, and how are those signals perceived?

- How are environmental signals transduced to bring about the developmental changes associated with flowering?

In Chapter 16 we discussed the role of the root and shoot apical meristems in vegetative growth and development. The transition to flowering involves major changes in the pattern of morphogenesis and cell differentiation at the shoot apical meristem. Ultimately this process leads to the production of the floral organs—sepals, petals, stamens, and carpels (see Figure 1.2A in Web Topic 1.2).

Specialized cells in the anther undergo meiosis to produce four haploid microspores that develop into pollen grains.

Similarly, a cell within the ovule divides meiotically to produce four haploid megaspores, one of which survives and undergoes three mitotic divisions to produce the cells of the embryo sac (see Figure 1.2B in **Web Topic 1.2**). The embryo sac represents the mature female gametophyte. The pollen grain, with its germinating pollen tube, is the mature male gametophyte generation. The two gametophytic structures produce the gametes (egg and sperm cells), which fuse to form the diploid zygote, the first stage of the new sporophyte generation.

Clearly, flowers represent a complex array of functionally specialized structures that differ substantially from the vegetative plant body in form and cell types. The transition to flowering therefore entails radical changes in cell fate within the shoot apical meristem. In the first part of this chapter we will discuss these changes, which are manifested as *floral development*. Recently genes have been identified that play crucial roles in the formation of the floral organs. Such studies have shed new light on the genetic control of plant reproductive development.

The events occurring in the shoot apex that specifically commit the apical meristem to produce *flowers* are collectively referred to as **floral evocation**. In the second part of this chapter we will discuss the events leading to floral evocation. The developmental signals that bring about floral evocation include endogenous factors, such as *circadian rhythms*, *phase change*, and *hormones*, and external factors, such as day length (*photoperiod*) and temperature (*vernalization*). In the case of photoperiodism, transmissible signals from the leaves, collectively referred to as the **floral stimulus**, are translocated to the shoot apical meristem. The interactions of these endogenous and external factors enable plants to synchronize their reproductive development with the environment.

Floral Meristems and Floral Organ Development

Floral meristems usually can be distinguished from vegetative meristems, even in the early stages of reproductive development, by their larger size. The transition from vegetative to reproductive development is marked by an increase in the frequency of cell divisions within the central zone of the shoot apical meristem (see Chapter 16). In the vegetative meristem, the cells of the central zone complete their division cycles slowly. As reproductive development commences, the increase in the size of the meristem is largely a result of the increased division rate of these central cells. Genetic and molecular studies have now identified a network of genes that control floral morphogenesis in Arabidopsis, snapdragon (*Antirrhinum*), and other species.

In this section we will focus on floral development in Arabidopsis, which has been studied extensively (Figure 25.1). First we will outline the basic morphological changes that occur during the transition from the vegetative to the reproductive phase. Next we will consider the arrangement of the floral organs in four whorls on the meristem, and the types of genes that govern the normal pattern of floral development. According to the widely accepted ABC model (which is described in Figure 25.6), the specific locations of floral organs in the flower are regulated by the overlapping expression of three types of floral organ identity genes.

The shoot apical meristems in Arabidopsis change with development

During the vegetative phase of growth, the Arabidopsis vegetative apical meristem, an indeterminate meristem, produces leaves with very short internodes, resulting in a basal rosette of leaves (see Figure 25.1A). When reproduc-

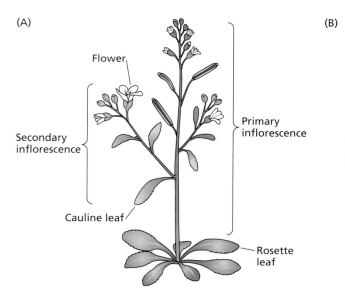

FIGURE 25.1 (A) The shoot apical meristem in *Arabidopsis thaliana* generates different organs at different stages of development. Early in development the shoot apical meristem forms a rosette of basal leaves. When the plant makes the transition to flowering, the shoot apical meristem is transformed into a primary inflorescence meristem that ultimately produces an elongated stem bearing flowers. Leaf primordia initiated prior to the floral transition become cauline leaves, and secondary inflorescences develop in the axils of the cauline leaves. (B) Photograph of an Arabidopsis plant. (Courtesy of Richard Amasino.)

FIGURE 25.2 Longitudinal sections through a vegetative (A) and a reproductive (B) shoot apical region of Arabidopsis. (Courtesy of V. Grbic and M. Nelson.)

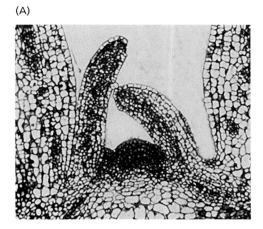
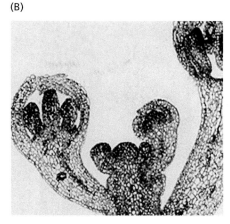

tive development is initiated, the vegetative meristem is transformed into the primary inflorescence meristem. The **primary inflorescence meristem** produces an elongated inflorescence axis bearing two types of lateral organs: cauline (or inflorescence) leaves and flowers (Figure 25.2).

The axillary buds of the cauline leaves develop into **secondary inflorescence meristems**, and their activity repeats the pattern of development of the primary inflorescence meristem, as shown in Figure 25.1A. Flowers arise from **floral meristems** that form on the flanks of the inflorescence meristem. The Arabidopsis inflorescence meristem has the potential to grow indefinitely and thus exhibits *indeterminate* growth. In contrast, flowers are formed by *determinate* growth of the floral meristem.

The four different types of floral organs are initiated as separate whorls

Floral meristems initiate four different types of floral organs: sepals, petals, stamens, and carpels (Coen and Carpenter 1993). These sets of organs are initiated in concentric rings, called **whorls**, around the flanks of the meristem (Figure 25.3). The initiation of the innermost organs, the carpels, consumes all of the meristematic cells in the apical dome, and only the floral organ primordia are present as the floral bud develops. In the wild-type Arabidopsis flower, the whorls are arranged as follows:

- The first (outermost) whorl consists of four sepals, which are green at maturity.

- The second whorl is composed of four petals, which are white at maturity.

- The third whorl contains six stamens (the male reproductive structures), two of which are shorter than the other four.

- The fourth whorl is a single complex organ, the gynoecium or pistil (the female reproductive structure), which is composed of an ovary with two fused carpels, each containing numerous ovules, and a short style capped with a stigma (Figure 25.4).

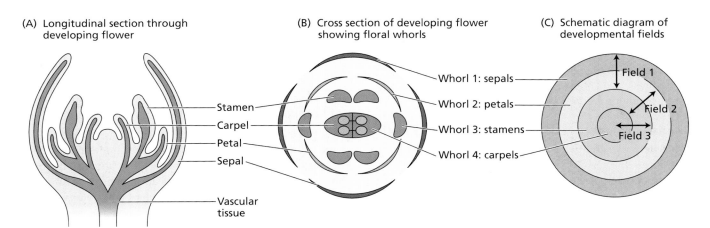

FIGURE 25.3 The floral organs are initiated sequentially by the floral meristem of Arabidopsis. (A and B) The floral organs are produced as successive whorls (concentric circles), starting with the sepals and progressing inward. (C) According to the combinatorial model, the functions of each whorl are determined by overlapping developmental fields. These fields correspond to the expression patterns of specific floral organ identity genes. (After Bewley et al. 2000.)

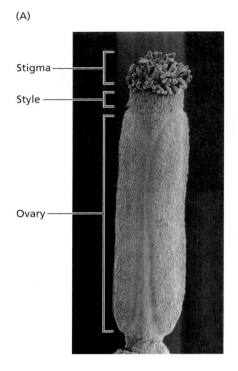
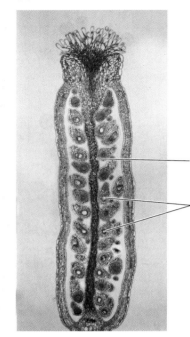

FIGURE 25.4 The Arabidopsis pistil consists of two fused carpels, each containing many ovules. (A) Scanning electron micrograph of a pistil, showing the stigma, a short style, and the ovary. (B) Longitudinal section through the pistil, showing the many ovules. (From Gasser and Robinson-Beers 1993, courtesy of C. S. Gasser, © American Society of Plant Biologists, reprinted with permission.)

Three types of genes regulate floral development

Studies of mutations have enabled identification of three classes of genes that regulate floral development: meristem identity genes, floral organ identity genes, and cadastral genes.

1. **Meristem identity genes** encode transcription factors that are necessary for the initial induction of organ identity genes. They are the positive regulators of floral organ identity in the developing floral meristem.

2. **Floral organ identity genes** directly control floral identity. The proteins encoded by these genes are transcription factors that likely control the expression of other genes whose products are involved in the formation and/or function of *floral* organs.

3. **Cadastral genes** act as spatial regulators of the floral organ identity genes by setting boundaries for their expression. (The word *cadastre* refers to a map or survey showing property boundaries for taxation purposes.)

Meristem identity genes regulate meristem function

Meristem identity genes must be active for the immature primordia formed at the flanks of the shoot or inflorescence apical meristem to become floral meristems. (Recall that an apical meristem that is forming floral meristems on its flanks is known as an inflorescence meristem.) For example, mutants of snapdragon (*Antirrhinum*) that have a defect in the meristem identity gene *FLORICAULA* (*FLO*) develop an inflorescence that does not produce flowers. Instead of causing floral meristems to form in the axils of the bracts, the loss of *flo* gene function results in the development of additional inflorescence meristems at the bract axils. Thus the wild-type *FLO* gene controls the determination step in which floral meristem identity is established.

In Arabidopsis, *SUPPRESSOR OF CONSTANS 1* (*SOC1*)*, *APETALA 1* (*AP1*), and *LEAFY* (*LFY*) are all critical genes in the genetic pathway that must be activated to establish floral meristem identity. *LFY* is the Arabidopsis version of the snapdragon *FLO* gene. *SOC1* plays a central role in floral evocation by integrating signals from several different pathways involving both environmental and internal cues (Borner et al. 2000). *SOC1* thus appears to serve as a master switch initiating floral development.

Once activated, *SOC1* triggers the expression of *LFY*, and *LFY* turns on the expression of *AP1* (Simon et al. 1996). In Arabidopsis, *LFY* and *AP1* are involved in a positive feedback loop; that is, *AP1* expression also stimulates the expression of *LFY*.

Homeotic mutations led to the identification of floral organ identity genes

The genes that determine floral organ identity were discovered as **floral homeotic mutants** (see Chapter 14 on the web site). As discussed in Chapter 14, mutations in the fruit fly, *Drosophila*, led to the identification of a set of homeotic genes encoding transcription factors that determine the locations at which specific structures develop. Such genes act as major developmental switches that activate the entire genetic program for a particular structure. The expression of homeotic genes thus gives organs their identity.

---

*Also known as *AGAMOUS-LIKE 20* (*AGL20*).

As we have seen already in this chapter, dicot flowers consist of successive whorls of organs that form as a result of the activity of floral meristems: sepals, petals, stamens, and carpels. These organs are produced when and where they are because of the orderly, patterned expression and interactions of a small group of homeotic genes that specify floral organ identity.

The floral organ identity genes were identified through homeotic mutations that alter floral organ identity so that some of the floral organs appear in the wrong places. For example, Arabidopsis plants with mutations in the *APETALA 2* (*AP2*) gene produce flowers with carpels where sepals should be, and stamens where petals normally appear.

The homeotic genes that have been identified so far encode transcription factors—proteins that control the expression of other genes. Most plant homeotic genes belong to a class of related sequences known as MADS box genes, whereas animal homeotic genes contain sequences called homeoboxes (see Chapter 14 on the web site).

Many of the genes that determine floral organ identity are MADS box genes, including the *DEFICIENS* gene of snapdragon and the *AGAMOUS* (*AG*), *PISTILLATA 1* (*PI1*), and *APETALA 3* (*AP3*) genes of Arabidopsis. The MADS box genes share a characteristic, conserved nucleotide sequence known as a *MADS box*, which encodes a protein structure known as the *MADS domain*. The MADS domain enables these transcription factors to bind to DNA that has a specific nucleotide sequence.

Not all genes containing the MADS box domain are homeotic genes. For example, *SOC1* is a MADS box gene, but it functions as a meristem identity gene.

Three types of homeotic genes control floral organ identity

Five different genes are known to specify floral organ identity in Arabidopsis: *AP1*, *AP2*, *AP3*, *PI*, and *AG* (Bowman et al. 1989; Weigel and Meyerowitz 1994). The organ identity genes initially were identified through mutations that dramatically alter the structure and thus the identity of the floral organs produced in two adjacent whorls (Figure 25.5). For example, plants with the *ap2* mutation lack sepals and petals (see Figure 25.5B). Plants bearing *ap3* or *pi* mutations produce sepals instead of petals in the second whorl, and carpels instead of stamens in the third whorl (see Figure 25.5C). Plants homozygous for the *ag* mutation lack both stamens and carpels (see Figure 25.5D).

Because mutations in these genes change floral organ identity without affecting the initiation of flowers, they are homeotic genes. These homeotic genes fall into three classes—types A, B, and C—defining three different kinds of activities (Figure 25.6):

1. Type A activity, encoded by *AP1* and *AP2*, controls organ identity in the first and second whorls. Loss of type A activity results in the formation of carpels instead of sepals in the first whorl, and of stamens instead of petals in the second whorl.

2. Type B activity, encoded by *AP3* and *PI*, controls organ determination in the second and third whorls. Loss of type B activity results in the formation of sepals instead of petals in the second whorl, and of carpels instead of stamens in the third whorl.

3. Type C activity, encoded by *AG*, controls events in the third and fourth whorls. Loss of type C activity results

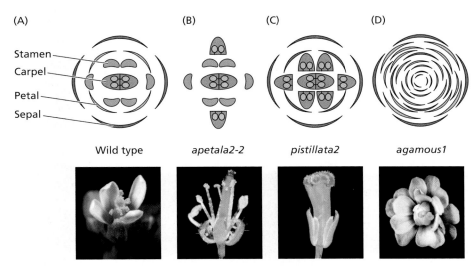

FIGURE 25.5 Mutations in the floral organ identity genes dramatically alter the structure of the flower. (A) Wild type shows normal structure in all four floral components. (B) *apetala2-2* mutants lack sepals and petals. (C) *pistillata2* mutants lack petals and stamens. (D) *agamous1* mutants lack both stamens and carpels. (From Meyerowitz et al. 2002.)

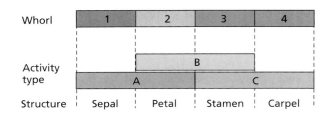

FIGURE 25.6 The ABC model for the acquisition of floral organ identity is based on the interactions of three different types of activities of floral homeotic genes: A, B, and C. In the first whorl, expression of type A (*AP2*) alone results in the formation of sepals. In the second whorl, expression of both type A (*AP1/AP2*) and type B (*AP3/PI*) results in the formation of petals. In the third whorl, the expression of type B (*AP3/PI*) and C (*AG*) causes the formation of stamens. In the fourth whorl, activity type C (*AG*) alone specifies carpels. In addition, activity type A (*AP1* and *AP2*) represses activity of C (*AG*) in whorls 1 and 2, while type C represses A in whorls 3 and 4.

in the formation of petals instead of stamens in the third whorl. Moreover, in the absence of type C activity, the fourth whorl (normally a carpel) is replaced by a *new flower*. As a result, the fourth whorl of an *ag* mutant flower is occupied by sepals. The floral meristem is no longer determinate. Flowers continue to form *within* flowers, and the pattern of organs (from outside to inside) is: sepal, petal, petal; sepal, petal, petal; etc.

The control of organ identity by type A, B, and C homeotic genes (the ABC model) is described in more detail in the next section.

The role of organ identity genes in floral development is dramatically illustrated by experiments in which two or three activities are eliminated by loss-of-function mutations (Figure 25.7). Quadruple-mutant plants (*ap1*, *ap2*, *ap3/pi*, and *ag*) produce floral meristems that develop as pseudoflowers; all the floral organs are replaced with green leaf-like structures, although these organs are produced with a whorled phyllotaxy typical of normal flowers. Evolutionary biologists, beginning with the eighteenth-century German poet and natural scientist Johann Wolfgang von Goethe (1749–1832), have speculated that floral organs are highly modified leaves, and this experiment gives direct support to these ideas.

The ABC model explains the determination of floral organ identity

In 1991 the **ABC model** was proposed to explain how homeotic genes control organ identity. The ABC model postulates that organ identity in each whorl is determined by a unique combination of the three organ identity gene activities (see Figure 25.6):

- Activity of type A alone specifies sepals.
- Activities of both A and B are required for the formation of petals.
- Activities of B and C form stamens.
- Activity of C alone specifies carpels.

The model further proposes that activities A and C mutually repress each other (see Figure 25.6); that is, both A- and C-type genes have cadastral function in addition to their function in determining organ identity.

The patterns of organ formation in wild-type flowers and most of the mutants are predicted and explained by this model (Figure 25.8). The challenge now is to understand how the expression pattern of these organ identity genes is controlled by cadastral genes; how organ identity genes, which encode transcription factors, alter the pattern of expression of other genes in the developing organ; and finally, how this altered pattern of gene expression results in the development of a specific floral organ.

FIGURE 25.7 A quadruple mutant (*ap1*, *ap2*, *ap3/pi*, *ag*) results in the production of leaf-like structures in place of floral organs. (Courtesy of John Bowman.)

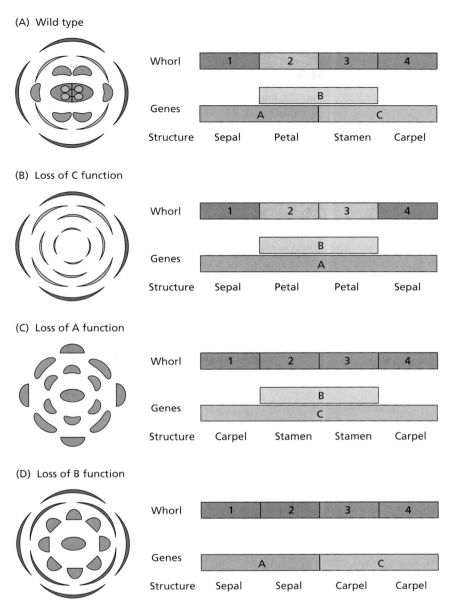

FIGURE 25.8 Interpretation of the phenotypes of floral homeotic mutants based on the ABC model. (A) All three activity types are functional in the wild type. (B) Loss of type C function results in expansion of the A function throughout the floral meristem. (C) Loss of type A function results in the spread of C function throughout the meristem. (D) Loss of type B function results in the expression of only A and C functions.

Floral Evocation: Internal and External Cues

Annual plants such as groundsel (*Senecio vulgaris*) may flower within a few weeks after germinating. Alternatively, some perennial plants, such as many forest trees, may grow for 20 or more years before they begin to produce flowers. Different species flower at widely different ages, indicating that the age, or perhaps the size, of the plant is an *internal* factor controlling the switch to reproductive development. The case in which flowering occurs strictly in response to internal developmental factors and does not depend on any particular environmental condition is referred to as *autonomous regulation*.

In contrast to plants that flower entirely through an autonomous pathway, some plants exhibit an absolute requirement for the proper environmental cues in order to flower. This condition is termed an *obligate* or *qualitative* response to an environmental cue. In other plant species, flowering is promoted by certain environmental cues but will eventually occur in the absence of such cues. This is called a *facultative* or *quantitative* response to an environmental cue. The flowering of this latter group of plants, which includes Arabidopsis, thus relies on both environmental and autonomous flowering systems.

Photoperiodism and vernalization are two of the most important mechanisms underlying seasonal responses. *Photoperiodism* is a response to the length of day or night; *vernalization* is the promotion of flowering by cold temperature. Other signals, such as total light radiation and water availability, can also be important external cues.

The evolution of both internal (autonomous) and external (environment-sensing) control systems enables plants to precisely regulate flowering at the optimal time for reproductive success. For example, in many populations of a particular species, flowering is synchronized. This synchrony favors crossbreeding and allows seeds to be produced in favorable environments, particularly with respect to water and temperature.

The Shoot Apex and Phase Changes

All multicellular organisms pass through a series of more or less defined developmental stages, each with its characteristic features. In humans, infancy, childhood, adolescence, and adulthood represent four general stages of development, with puberty as the dividing line between the nonreproductive and the reproductive phases. Higher plants likewise pass through developmental stages, but whereas in animals these changes take place throughout the entire organism, in higher plants they occur in a single, dynamic region, the **shoot apical meristem**.

Shoot apical meristems have three developmental phases

During postembryonic development, the shoot apical meristem passes through three developmental stages in sequence:

1. The juvenile phase
2. The adult vegetative phase
3. The adult reproductive phase

The transition from one phase to another is called *phase change*.

The primary distinction between the juvenile and the adult vegetative phases is that the latter has the ability to form reproductive structures: flowers in angiosperms, cones in gymnosperms. However, flowering, which represents the expression of the reproductive competence of the adult phase, often depends on specific environmental and developmental signals. Thus the absence of flowering itself is not a reliable indicator of juvenility.

The transition from juvenile to adult is frequently accompanied by changes in vegetative characteristics, such as leaf morphology, phyllotaxy (the arrangement of leaves on the stem), thorniness, rooting capacity, and leaf retention in deciduous plants (Figure 25.9; see also **Web Topic 25.1**). Such changes are most evident in woody perennials, but they are apparent in many herbaceous species as well. Unlike the abrupt transition from the adult vegetative phase to the reproductive phase, the transition from juvenile to vegetative adult is usually gradual, involving intermediate forms.

Sometimes the transition can be observed in a single leaf. A dramatic example of this is the progressive transformation of juvenile leaves of the leguminous tree, *Acacia heterophylla*, into phyllodes, a phenomenon first noted in the eighteenth century by Goethe. Whereas the juvenile pinnately compound leaves consist of rachis (stalk) and leaflets, adult phyllodes are specialized structures representing flattened petioles (Figure 25.10).

Intermediate structures also form during the transition from aquatic to aerial leaf types of aquatic plants such as common marestail (*Hippuris vulgaris*). As in the case of *A. heterophylla*, these intermediate forms possess distinct regions

FIGURE 25.9 Juvenile and adult forms of English ivy (*Hedera helix*). The juvenile form has lobed palmate leaves arranged alternately, a climbing growth habit, and no flowers. The adult form (projecting out to the right) has entire ovate leaves arranged in spirals, an upright growth habit, and flowers that develop into fruits. (Courtesy of L. Rignanese.)

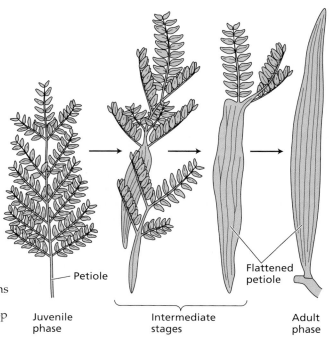

FIGURE 25.10 Leaves of *Acacia heterophylla*, showing transitions from pinnately compound leaves (juvenile phase) to phyllodes (adult phase). Note that the previous phase is retained at the top of the leaf in the intermediate forms.

The Control of Flowering

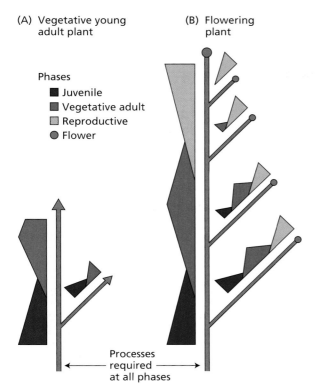

FIGURE 25.11 Schematic representation of the combinatorial model of shoot development in maize. Overlapping gradients of expression of the juvenile, vegetative adult, and reproductive phases are indicated along the length of the main axis and branches. The continuous black line represents processes that are required during all phases of development. Each of the three phases may be regulated by separated developmental programs, with intermediate phases arising when the programs overlap. (A) Vegetative young adult plant. (B) Flowering plant. (After Poethig 1990.)

TABLE 25.1 Length of juvenile period in some woody plant	
Species	**Length of juvenile period**
Rose (*Rosa* [hybrid tea])	20–30 days
Grape (*Vitis* spp.)	1 year
Apple (*Malus* spp.)	4–8 years
Citrus spp.	5–8 years
English ivy (*Hedera helix*)	5–10 years
Redwood (*Sequoia sempervirens*)	5–15 years
Sycamore maple (*Acer pseudoplatanus*)	15–20 years
English oak (*Quercus robur*)	25–30 years
European beech (*Fagus sylvatica*)	30–40 years

Source: Clark 1983.

with different developmental patterns. To account for intermediate forms during the transition from juvenile to adult in maize (corn; *Zea mays*) (see **Web Topic 25.2**), a **combinatorial model** has been proposed (Figure 25.11). According to this model, shoot development can be described as a series of independently regulated, *overlapping* programs (juvenile, adult, and reproductive) that modulate the expression of a common set of developmental processes.

In the transition from juvenile to adult leaves, the intermediate forms indicate that different regions of the same leaf can express different developmental programs. Thus the cells at the tip of the leaf remain committed to the juvenile program, while the cells at the base of the leaf become committed to the adult program. The developmental fates of the two sets of cells in the same leaf are quite different.

Juvenile tissues are produced first and are located at the base of the shoot

The sequence in time of the three developmental phases results in a spatial gradient of juvenility along the shoot axis. Because growth in height is restricted to the apical meristem, the juvenile tissues and organs, which form first, are located at the base of the shoot. In rapidly flowering herbaceous species, the juvenile phase may last only a few days, and few juvenile structures are produced. In contrast, woody species have a more prolonged juvenile phase, in some cases lasting 30 to 40 years (Table 25.1). In these cases the juvenile structures can account for a significant portion of the mature plant.

Once the meristem has switched to the adult phase, only adult vegetative structures are produced, culminating in floral evocation. The adult and reproductive phases are therefore located in the upper and peripheral regions of the shoot.

Attainment of a sufficiently large size appears to be more important than the plant's chronological age in determining the transition to the adult phase. Conditions that retard growth, such as mineral deficiencies, low light, water stress, defoliation, and low temperature tend to prolong the juvenile phase or even cause **rejuvenation** (reversion to juvenility) of adult shoots. In contrast, conditions that promote vigorous growth accelerate the transition to the adult phase. When growth is accelerated, exposure to the correct flower-inducing treatment can result in flowering.

Although plant size seems to be the most important factor, it is not always clear which specific component associated with size is critical. In some *Nicotiana* species, it appears that plants must produce a certain number of leaves to transmit a sufficient amount of the floral stimulus to the apex.

Once the adult phase has been attained it is relatively stable and is maintained during vegetative propagation or grafting. For example, in mature plants of English ivy (*Hedera helix*), cuttings taken from the basal region develop into juvenile plants, while those from the tip develop into adult plants. When scions were taken from the base of the flowering tree silver birch (*Betula verrucosa*) and grafted onto seedling rootstocks, there were no flowers on the grafts within the first 2 years. In contrast, the grafts flowered freely when scions were taken from the top of the flowering tree.

In some species, the juvenile meristem appears to be capable of flowering but does not receive sufficient floral stimulus until the plant becomes large enough. In mango (*Mangifera indica*), for example, juvenile seedlings can be induced to flower when grafted to a mature tree. In many other woody species, however, grafting to an adult flowering plant does not induce flowering.

Phase changes can be influenced by nutrients, gibberellins, and other chemical signals

The transition at the shoot apex from the juvenile to the adult phase can be affected by transmissible factors from the rest of the plant. In many plants, exposure to low-light conditions prolongs juvenility or causes reversion to juvenility. A major consequence of the low-light regime is a reduction in the supply of carbohydrates to the apex; thus carbohydrate supply, especially sucrose, may play a role in the transition between juvenility and maturity. Carbohydrate supply as a source of energy and raw material can affect the size of the apex. For example, in the florist's chrysanthemum (*Chrysanthemum morifolium*), flower primordia are not initiated until a minimum apex size has been reached.

The apex receives a variety of hormonal and other factors from the rest of the plant in addition to carbohydrates and other nutrients. Experimental evidence shows that the application of gibberellins causes reproductive structures to form in young, juvenile plants of several conifer families. The involvement of *endogenous* GAs in the control of reproduction is also indicated by the fact that other treatments that accelerate cone production in pines (e.g., root removal, water stress, and nitrogen starvation) often also result in a buildup of GAs in the plant.

On the other hand, although gibberellins promote the attainment of reproductive maturity in conifers and many herbaceous angiosperms as well, GA_3 causes rejuvenation in *Hedera* and in several other woody angiosperms. The role of gibberellins in the control of phase change is thus complex, varies among species, and probably involves interactions with other factors.

Competence and determination are two stages in floral evocation

The term *juvenility* has different meanings for herbaceous and woody species. Whereas juvenile herbaceous meristems flower readily when grafted onto flowering adult plants (see **Web Topic 25.3**), juvenile woody meristems generally do not. What is the difference between the two?

Extensive studies in tobacco (*Nicotiana tabacum*) have demonstrated that floral evocation requires the apical bud to pass through two developmental stages (Figure 25.12) (McDaniel et al. 1992). One stage is the acquisition of competence. A bud is said to be *competent* if it is able to flower when given the appropriate developmental signal.

For example, if a vegetative shoot (scion) is grafted onto a flowering stock and the scion flowers immediately, it is demonstrably capable of responding to the level of floral stimulus present in the stock and is therefore competent. Failure of the scion to flower would indicate that the shoot apical meristem has not yet attained competence. Thus the juvenile meristems of herbaceous plants are competent to flower, but those of woody species are not.

The next stage that a competent vegetative bud goes through is determination. A bud is said to be *determined* if it progresses to the next developmental stage (flowering) even after being removed from its normal context. Thus a florally determined bud will produce flowers even if it is grafted onto a vegetative plant that is not producing any floral stimulus.

In a day-neutral tobacco, for example, plants typically flower after producing about 41 leaves or nodes. In an experiment to measure the floral determination of the axillary buds, flowering tobacco plants were decapitated just above the thirty-fourth leaf (from the bottom). Released from apical dominance, the axillary bud of the thirty-fourth leaf grew

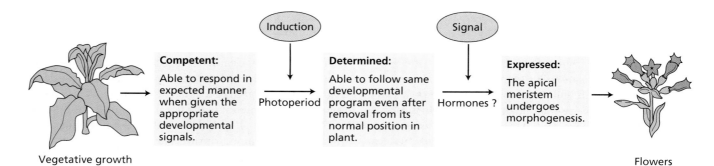

FIGURE 25.12 A simplified model for floral evocation at the shoot apex in which the cells of the vegetative meristem acquire new developmental fates. To initiate floral development, the cells of the meristem must first become competent. A competent vegetative meristem is one that can respond to a floral stimulus (induction) by becoming florally determined (committed to producing a flower). The determined state is usually expressed, but this may require an additional signal. (After McDaniel et al. 1992.)

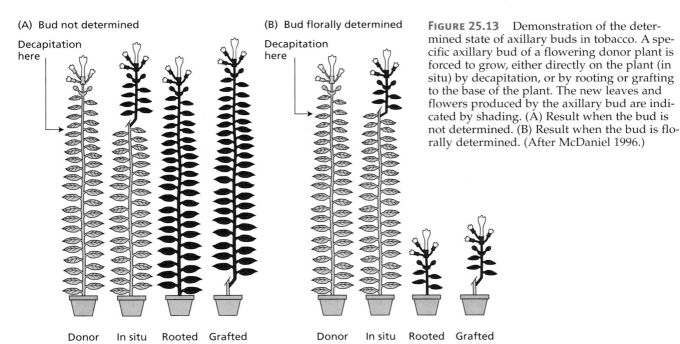

FIGURE 25.13 Demonstration of the determined state of axillary buds in tobacco. A specific axillary bud of a flowering donor plant is forced to grow, either directly on the plant (in situ) by decapitation, or by rooting or grafting to the base of the plant. The new leaves and flowers produced by the axillary bud are indicated by shading. (A) Result when the bud is not determined. (B) Result when the bud is florally determined. (After McDaniel 1996.)

out, and after producing about seven more leaves (for a total of 41), it flowered (Figure 25.13A) (McDaniel 1996). However, if the thirty-fourth bud was excised from the plant and either rooted or grafted onto a stock without leaves near the base instead of only seven leaves, it produced nearly a complete set of leaves before flowering. This result shows that the thirty-fourth bud was not yet florally determined.

In another experiment, the donor plant was decapitated above the thirty-seventh leaf. This time the thirty-seventh axillary bud flowered after producing only about four leaves *in all three situations* (see Figure 25.13B). This result demonstrates that the terminal bud became florally determined after initiating 37 leaves.

Extensive grafting of shoot tips among tobacco varieties has established that the number of nodes a meristem produces before flowering is a function of two factors: the strength of the floral stimulus from the leaves, and the competence of the meristem to respond to the signal (McDaniel et al. 1996).

In some cases the **expression** of flowering may be delayed or arrested even after the apex becomes determined, unless it receives a second developmental signal that stimulates expression (see Figure 25.12). For example, intact darnel ryegrass (*Lolium temulentum*) plants become committed to flowering after a single exposure to a long day. If the *Lolium* shoot apical meristem is excised 28 hours after the beginning of the long day and cultured in vitro, it will produce normal inflorescences in culture, but only if the hormone gibberellic acid (GA) is present in the medium. Because apices cultured from plants grown exclusively in short days never flower, even in the presence of GA, we can conclude that long days are required for determination in *Lolium*, whereas GA is required for *expression* of the determined state.

In general, once a meristem has become competent, it exhibits an increasing tendency to flower with age (leaf number). For example, in plants controlled by day length, the number of short-day or long-day cycles necessary to achieve flowering is often fewer in older plants (Figure 25.14).

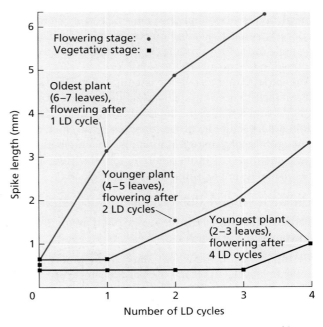

FIGURE 25.14 Effect of plant age on the number of long-day (LD) inductive cycles required for flowering in the long-day plant darnel ryegrass (*Lolium temulentum*). An inductive long-day cycle consisted of 8 hours of sunlight followed by 16 hours of low-intensity incandescent light. The older the plant is, the fewer photoinductive cycles are needed to produce flowering.

As will be discussed later in the chapter, this increasing tendency to flower with age has its physiological basis in the greater capacity of the leaves to produce a floral stimulus.

Before discussing how plants perceive day length, however, we will lay the foundation by examining how organisms measure time in general. This topic is known as **chronobiology**, or the study of **biological clocks**. The best-understood biological clock is the circadian rhythm.

Circadian Rhythms: The Clock Within

Organisms are normally subjected to daily cycles of light and darkness, and both plants and animals often exhibit rhythmic behavior in association with these changes. Examples of such rhythms include leaf and petal movements (day and night positions), stomatal opening and closing, growth and sporulation patterns in fungi (e.g., *Pilobolus* and *Neurospora*), time of day of pupal emergence (the fruit fly *Drosophila*), and activity cycles in rodents, as well as metabolic processes such as photosynthetic capacity and respiration rate.

When organisms are transferred from daily light–dark cycles to continuous darkness (or continuous dim light), many of these rhythms continue to be expressed, at least for several days. Under such uniform conditions the period of the rhythm is close to 24 hours, and consequently the term circadian rhythm is applied (see Chapter 17). Because they continue in a constant light or dark environment, these circadian rhythms cannot be direct responses to the presence or absence of light but must be based on an internal pacemaker, often called an endogenous oscillator. A molecular model for a plant endogenous oscillator was described in Chapter 17.

The endogenous oscillator is coupled to a variety of physiological processes, such as leaf movement or photosynthesis, and it maintains the rhythm. For this reason the endogenous oscillator can be considered the clock mechanism, and the physiological functions that are being regulated, such as leaf movements or photosynthesis, are sometimes referred to as the hands of the clock.

Circadian rhythms exhibit characteristic features

Circadian rhythms arise from cyclic phenomena that are defined by three parameters:

1. **Period** is the time between comparable points in the repeating cycle. Typically the period is measured as the time between consecutive maxima (peaks) or minima (troughs) (Figure 25.15A).
2. **Phase*** is any point in the cycle that is recognizable by its relationship to the rest of the cycle. The most obvious phase points are the peak and trough positions.

---

*The term *phase* in this context should not be confused with the term *phase change* in meristem development discussed earlier.

3. **Amplitude** is usually considered to be the distance between peak and trough. The amplitude of a biological rhythm can often vary while the period remains unchanged (as, for example, in Figure 25.15C).

In constant light or darkness, rhythms depart from an exact 24-hour period. The rhythms then drift in relation to solar time, either gaining or losing time depending on whether the period is shorter or longer than 24 hours. Under natural conditions, the endogenous oscillator is **entrained** (synchronized) to a true 24-hour period by environmental signals, the most important of which are the light-to-dark transition at dusk and the dark-to-light transition at dawn (see Figure 25.15B).

Such environmental signals are termed **zeitgebers** (German for "time givers"). When such signals are removed—for example, by transfer to continuous darkness—the rhythm is said to be **free-running**, and it reverts to the circadian period that is characteristic of the particular organism (see Figure 25.15B).

Although the rhythms are generated internally, they normally require an environmental signal, such as exposure to light or a change in temperature, to initiate their expression. In addition, many rhythms damp out (i.e., the amplitude decreases) when the organism is subjected to a constant environment for several cycles. When this occurs an environmental zeitgeber, such as a transfer from light to dark or a change in temperature, is required to restart the rhythm (see Figure 25.15C). Note that the clock itself does not damp out; only the coupling between the molecular clock (endogenous oscillator) and the physiological function is affected.

The circadian clock would be of no value to the organism if it could not keep accurate time under the fluctuating temperatures experienced in natural conditions. Indeed, temperature has little or no effect on the period of the free-running rhythm. The feature that enables the clock to keep time at different temperatures is called **temperature compensation**. Although all of the biochemical steps in the pathway are temperature-sensitive, their temperature responses probably cancel each other. For example, changes in the rates of synthesis of intermediates could be compensated for by parallel changes in their rates of degradation. In this way, the steady-state levels of clock regulators would remain constant at different temperatures.

Phase shifting adjusts circadian rhythms to different day–night cycles

In circadian rhythms, physiological responses are coupled to a specific time point of the endogenous oscillator so that the response occurs at a particular time of day. A single oscillator can be coupled to multiple circadian rhythms, which may even be out of phase with each other.

How do such responses remain on time when the daily durations of light and darkness change with the seasons? Investigators typically test the response of the endogenous

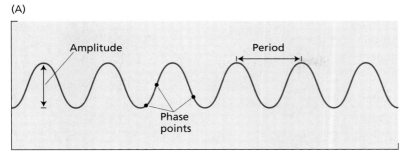

A typical circadian rhythm. The **period** is the time between comparable points in the repeating cycle; the **phase** is any point in the repeating cycle recognizable by its relationship with the rest of the cycle; the **amplitude** is the distance between peak and trough.

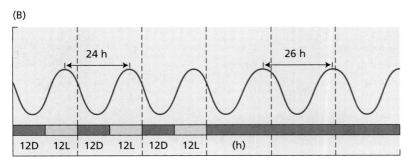

A circadian rhythm entrained to a 24 h light–dark (L–D) cycle and its reversion to the free-running period (26 h in this example) following transfer to continuous darkness.

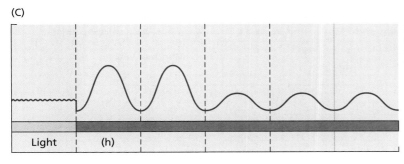

Suspension of a circadian rhythm in continuous bright light and the release or restarting of the rhythm following transfer to darkness.

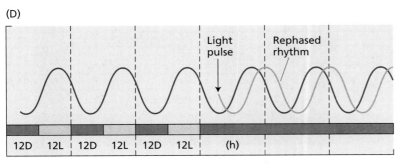

Typical phase-shifting response to a light pulse given shortly after transfer to darkness. The rhythm is rephased (delayed) without its period being changed.

FIGURE 25.15 Some characteristics of circadian rhythms.

oscillator by placing the organism in continuous darkness and examining the response to a short pulse of light (usually less than 1 hour) given at different phase points in the free-running rhythm. When an organism is entrained to a cycle of 12 hours light and 12 hours dark and then allowed to free-run in darkness, the phase of the rhythm that coincides with the light period of the previous entraining cycle is called the **subjective day**, and the phase that coincides with the dark period is called the **subjective night**.

If a light pulse is given during the first few hours of the subjective night, the rhythm is delayed; the organism interprets the light pulse as the end of the previous day (see Figure 25.15D). In contrast, a light pulse given toward the end of the subjective night advances the phase of the rhythm; now the organism interprets the light pulse as the beginning of the following day.

This is precisely the response that would be expected if the rhythm is were able to stay on local time even when the seasons change. These phase-shifting responses enable the rhythm to be entrained to approximately 24-hour cycles with different durations of light and darkness, and they demonstrate that the rhythm can adjust to different natural conditions of day length.

Phytochromes and cryptochromes entrain the clock

The molecular mechanism whereby a light signal causes phase shifting is not yet known, but studies in Arabidopsis have identified some of the key elements of the circadian oscillator and its inputs and outputs (see Chapter 17). The low levels and specific wavelengths of light that can induce phase shifting indicate that the light response must be mediated by specific photoreceptors rather than rates of photosynthesis. For example, the red-light entrainment of rhythmic nyctonastic leaf movements in *Samanea*, a semitropical leguminous tree, is a low-fluence response mediated by phytochrome (see Chapter 17).

Arabidopsis has five phytochromes, and all but one of them (phytochrome C) have been implicated in clock entrainment. Each phytochrome acts as a specific photoreceptor for red, far-red, or blue light. In addition, the CRY1 and CRY2 proteins participate in blue-light entrainment of the clock, as they do in insects and mammals (Devlin and Kay 2000). Surprisingly, CRY proteins also appear to be required for normal entrainment by red light. Since these proteins do not absorb red light, this requirement suggests that CRY1 and CRY2 may act as intermediates in phytochrome signaling during entrainment of the clock (Yanovsky and Kay 2001).

In *Drosophila*, CRY proteins interact physically with clock components and thus constitute part of the oscillator mechanism (Devlin and Kay 2000). However, this does not appear to be the case in Arabidopsis, in which *cry1/cry2* double mutants are impaired in entrainment but otherwise have normal circadian rhythms. Precisely how Arabidopsis CRY proteins interact with the endogenous oscillator mechanism to induce phase shifting remains to be elucidated (Yanovsky et al. 2001).

Photoperiodism: Monitoring Day Length

As we have seen, the circadian clock enables organisms to determine the time of day or night at which a particular molecular or biochemical event occurs. **Photoperiodism**, or the ability of an organism to detect day length, makes it possible for an event to occur at a particular time of *year*, thus allowing for a *seasonal* response. Circadian rhythms and photoperiodism have the common property of responding to cycles of light and darkness.

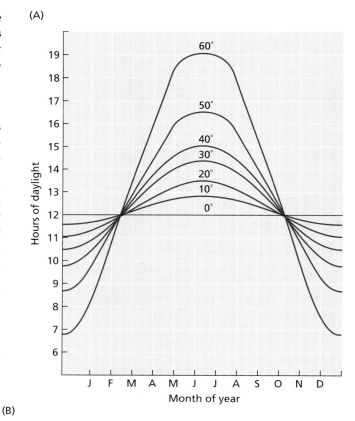

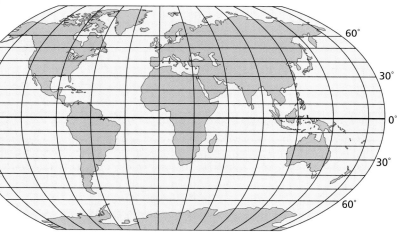

FIGURE 25.16 (A) The effect of latitude on day length at different times of the year. Day length was measured on the twentieth of each month. (B) Global map showing longitudes and latitudes.

Precisely at the equator, day length and night length are equal and constant throughout the year. As one moves away from the equator toward the poles, the days become longer in summer and shorter in winter (Figure 25.16). Plant species have evolved the ability to detect these seasonal changes in day length, and their specific photoperiodic responses are strongly influenced by the latitude from which they originated.

Photoperiodic phenomena are found in both animals and plants. In the animal kingdom, day length controls such seasonal activities as hibernation, development of summer or winter coats, and reproductive activity. Plant responses controlled by day length are numerous, including the initiation of flowering, asexual reproduction, the formation of storage organs, and the onset of dormancy.

Perhaps all plant photoperiodic responses utilize the same photoreceptors, with subsequent specific signal transduction pathways regulating different responses. Because it is clear that monitoring the passage of time is essential to all photoperiodic responses, a timekeeping mechanism must underlie both the time-of-year and the time-of-day responses. The circadian oscillator is thought to provide an endogenous time-measuring mechanism that serves as a reference point for the response to incoming light (or dark) signals from the environment. How changing photoperiods are evaluated against the circadian oscillator reference will be discussed shortly.

Plants can be classified according to their photoperiodic responses

Numerous plant species flower during the long days of summer, and for many years plant physiologists believed that the correlation between long days and flowering was a consequence of the accumulation of photosynthetic products synthesized during long days.

This hypothesis was shown to be incorrect by the work of Wightman Garner and Henry Allard, conducted in the 1920s at the U.S. Department of Agriculture laboratories in Beltsville, Maryland. They found that a mutant variety of tobacco, Maryland Mammoth, grew profusely to about 5 m in height but failed to flower in the prevailing conditions of summer (Figure 25.17). However, the plants flowered in the greenhouse during the winter under natural light conditions.

These results ultimately led Garner and Allard to test the effect of artificially providing short days by covering plants grown during the long days of summer with a light-tight tent from late in the afternoon until the following morning. These artificial short days also caused the plants to flower. This requirement for short days was difficult to reconcile with the idea that longer periods of radiation and the resulting increase in photosynthesis promote flowering in general. Garner and Allard concluded that the length of the day was the determining factor in flowering and were able to confirm this hypothesis in many different species and conditions. This work laid the foundations for the extensive subsequent research on photoperiodic responses.

The classification of plants according to their photoperiodic responses is usually based on flowering, even though many other aspects of plants' development may also be affected by day length. The two main photoperiodic response categories are short-day plants and long-day plants:

FIGURE 25.17 Maryland Mammoth mutant of tobacco (right) compared to wild-type tobacco (left). Both plants were grown during summer in the greenhouse. (University of Wisconsin graduate students used for scale.) (Courtesy of R. Amasino.)

1. **Short-day plants** (**SDPs**) flower only in short days (*qualitative* SDPs), or their flowering is accelerated by short days (*quantitative* SDPs).

2. **Long-day plants** (**LDPs**) flower only in long days (*qualitative* LDPs), or their flowering is accelerated by long days (*quantitative* LDPs).

The essential distinction between long-day and short-day plants is that flowering in LDPs is promoted only when the day length *exceeds* a certain duration, called the **critical day length**, in every 24-hour cycle, whereas promotion of flowering in SDPs requires a day length that is *less than* the critical day length. The absolute value of the critical day length varies widely among species, and only when flowering is examined for a range of day lengths can the correct photoperiodic classification be established (Figure 25.18).

Long-day plants can effectively measure the lengthening days of spring or early summer and delay flowering until the critical day length is reached. Many varieties of wheat (*Triticum aestivum*) behave in this way. SDPs often flower in the fall when the days shorten below the critical day length, as in many varieties of *Chrysanthemum morifolium*. However, day length alone is an ambiguous signal, because it cannot distinguish between spring and fall.

Plants exhibit several adaptations for avoiding the ambiguity of the day length signal. One is the presence of a juvenile phase that prevents the plant from responding to day

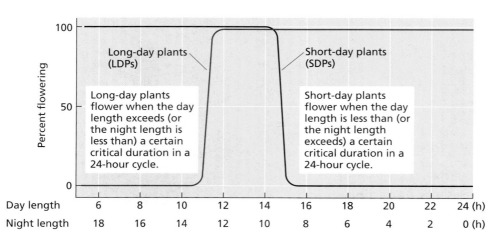

FIGURE 25.18 The photoperiodic response in long- and short-day plants. The critical duration varies between species: In this example, both the SDPs and the LDPs would flower in photoperiods between 12 and 14 h long.

length during the spring. Another mechanism for avoiding the ambiguity of day length is the coupling of a temperature requirement to a photoperiodic response. Certain plant species, such as winter wheat, do not respond to photoperiod until after a cold period (vernalization or overwintering) has occurred. (We will discuss vernalization a little later in the chapter.)

Other plants avoid seasonal ambiguity by distinguishing between *shortening* and *lengthening* days. Such "dual day length plants" fall into two categories:

1. **Long–short-day plants (LSDPs)** flower only after a sequence of long days followed by short days. LSDPs, such as *Bryophyllum*, *Kalanchoe*, and night-blooming jasmine (*Cestrum nocturnum*), flower in the late summer and fall, when the days are shortening.

2. **Short–long-day plants (SLDPs)** flower only after a sequence of short days followed by long days. SLDPs, such as white clover (*Trifolium repens*), Canterbury bells (*Campanula medium*), and echeveria (*Echeveria harmsii*), flower in the early spring in response to lengthening days.

Finally, species that flower under any photoperiodic condition are referred to as *day-neutral plants*. **Day-neutral plants (DNPs)** are insensitive to day length. Flowering in DNPs is typically under autonomous regulation—that is, internal developmental control. Some day-neutral species, such as kidney bean (*Phaseolus vulgaris*), evolved near the equator where the day length is constant throughout the year. Many desert annuals, such as desert paintbrush (*Castilleja chromosa*) and desert sand verbena (*Abronia villosa*), evolved to germinate, grow, and flower quickly whenever sufficient water is available. These are also DNPs.

The leaf is the site of perception of the photoperiodic signal

The photoperiodic stimulus in both LDPs and SDPs is perceived by the leaves. For example, treatment of a single leaf of the SDP *Xanthium* with short photoperiods is sufficient to cause the formation of flowers, even when the rest of the plant is exposed to long days. Thus, in response to photoperiod the leaf transmits a signal that regulates the transition to flowering at the shoot apex. The photoperiod-regulated processes that occur in the leaves resulting in the transmission of a floral stimulus to the shoot apex are referred to collectively as **photoperiodic induction**.

Photoperiodic induction can take place in a leaf that has been separated from the plant. For example, in the SDP *Perilla crispa*, an excised leaf exposed to short days can cause flowering when subsequently grafted to a noninduced plant maintained in long days (Zeevaart and Boyer 1987). This result indicates that photoperiodic induction depends on events that take place exclusively in the leaf.

Extensive grafting studies have shown that the induced leaf is the source of a mobile floral stimulus that is transported to the shoot apical meristem. We will describe these grafting experiments at the end of the chapter when we discuss biochemical signaling.

The floral stimulus is transported in the phloem

The leaf-derived photoperiodic floral stimulus is translocated via the phloem to the shoot apical meristem, where it promotes floral evocation. Treatments that block phloem translocation, such as girdling or localized heat-killing (see Chapter 10), block flowering by preventing the movement of the floral stimulus out of the leaf.

It is possible to measure rates of movement of the floral stimulus by removing a leaf at different times after induction, and comparing the time it takes for the signal to reach two buds located at different distances from the induced leaf. The rationale for this type of measurement is that a threshold amount of the signaling compound has reached the bud when flowering takes place, despite the removal of the leaf.

Studies using this method have shown that the rate of movement of the flowering signal is comparable to, or somewhat slower than, the rate of translocation of sugars in the phloem (see Chapter 10). For example, export of the floral stimulus from adult leaves of the SDP *Chenopodium*

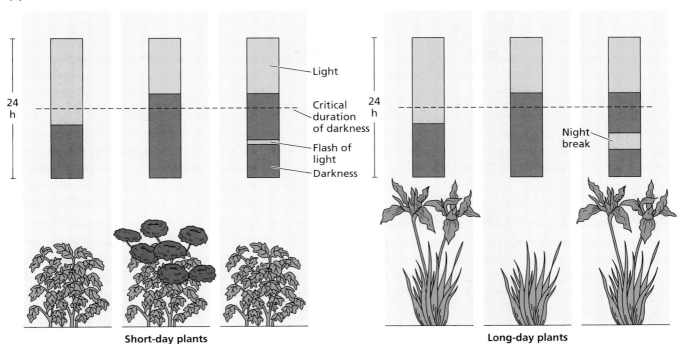

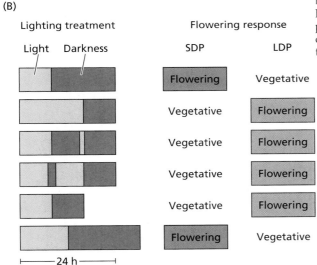

FIGURE 25.19 The photoperiodic regulation of flowering. (A) Effects on SDPs and LDPs. (B) Effects of the duration of the dark period on flowering. Treating short- and long-day plants with different photoperiods clearly shows that the critical variable is the length of the dark period.

An induced leaf positioned close to the shoot apex is more likely to cause flowering than an induced leaf at the base of a stem, which normally feeds the roots. Similarly, non-induced leaves positioned between the induced leaf and the apical bud will tend to inhibit flowering by serving as the preferred source leaves for the bud, thus preventing the floral stimulus from the more distal induced leaf from reaching its target. This inhibition also explains why a minimum amount of photosynthesis is required by the induced leaf to drive translocation.

Plants monitor day length by measuring the length of the night

Under natural conditions, day and night lengths configure a 24-hour cycle of light and darkness. In principle, a plant could perceive a critical day length by measuring the duration of either light or darkness. Much experimental work in the early studies of photoperiodism was devoted to establishing which part of the light–dark cycle is the controlling factor in flowering. Results showed that flowering of SDPs is determined primarily by the duration of darkness (Figure 25.19A). It was possible to induce flowering in

is complete within 22.5 hours from the beginning of the long night period. In the LDP *Sinapis*, movement of the floral stimulus out of the leaf is complete by as early as 16 hours after the start of the long-day treatment. These rates are consistent with a floral stimulus that moves in the phloem (Zeevaart 1976).

Because the floral stimulus is translocated along with sugars in the phloem, it is subject to source–sink relations.

SDPs with light periods longer than the critical value, provided that these were followed by sufficiently long nights (see Figure 25.19B). Similarly, SDPs did not flower when short days were followed by short nights.

More detailed experiments demonstrated that photoperiodic timekeeping in SDPs is a matter of measuring the duration of darkness. For example, flowering occurred only when the dark period exceeded 8.5 hours in cocklebur (*Xanthium strumarium*) or 10 hours in soybean (*Glycine max*). The duration of darkness was also shown to be important in LDPs (see Figure 25.19). These plants were found to flower in short days, provided that the accompanying night length was also short; however, a regime of long days followed by long nights was ineffective.

Night breaks can cancel the effect of the dark period

A feature that underscores the importance of the dark period is that it can be made ineffective by interruption with a short exposure to light, called a **night break** (see Figure 25.19A). In contrast, interrupting a long day with a brief dark period does not cancel the effect of the long day (see Figure 25.19B). Night-break treatments of only a few minutes are effective in *preventing* flowering in many SDPs, including *Xanthium* and *Pharbitis*, but much longer exposures are often required to *promote* flowering in LDPs.

In addition, the effect of a night break varies greatly according to the time when it is given. For both LDPs and SDPs, a night break was found to be most effective when given near the middle of a dark period of 16 hours (Figure 25.20).

The discovery of the night-break effect, and its time dependence, had several important consequences. It established the central role of the dark period and provided a valuable probe for studying photoperiodic timekeeping. Because only small amounts of light are needed, it became possible to study the action and identity of the photoreceptor without the interfering effects of photosynthesis and other nonphotoperiodic phenomena. This discovery has also led to the development of commercial methods for regulating the time of flowering in horticultural species, such as *Kalanchoe*, chrysanthemum, and poinsettia (*Euphorbia pulcherrima*).

The circadian clock and photoperiodic timekeeping

The decisive effect of night length on flowering indicates that measuring the passage of time in darkness is central to photoperiodic timekeeping. Most of the available evidence favors a mechanism based on a circadian rhythm (Bünning 1960). According to the **clock hypothesis**, photoperiodic timekeeping depends on an endogenous circadian oscillator of the type discussed earlier in the chapter (see also Chapter 17). The central oscillator is coupled to various physiological processes that involve gene expression, including flowering in photoperiodic species.

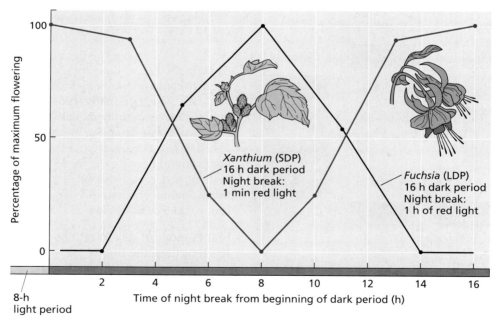

FIGURE 25.20 The time at which a night break is given determines the flowering response. When given during a long dark period, a night break promotes flowering in LDPs and inhibits flowering in SDPs. In both cases, the greatest effect on flowering occurs when the night break is given near the middle of the 16-hour dark period. The LDP *Fuchsia* was given a 1-hour exposure to red light in a 16-hour dark period. *Xanthium* was exposed to red light for 1 minute in a 16-hour dark period. (Data for *Fuchsia* from Vince-Prue 1975; data for *Xanthium* from Salisbury 1963 and Papenfuss and Salisbury 1967.)

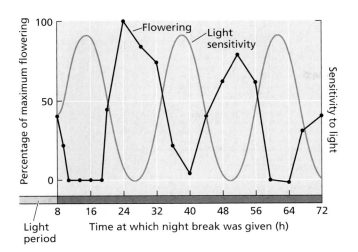

FIGURE 25.21 Rhythmic flowering in response to night breaks. In this experiment, the SDP soybean (*Glycine max*) received cycles of an 8-hour light period followed by a 64-hour dark period. A 4-hour night break was given at various times during the long inductive dark period. The flowering response, plotted as the percentage of the maximum, was then plotted for each night break given. Note that a night break given at 26 hours induced maximum flowering, while no flowering was obtained when the night break was given at 40 hours. Moreover, this experiment demonstrates that the sensitivity to the effect of the night break shows a circadian rhythm. These data support a model in which flowering in SDPs is induced only when dawn (or a night break) occurs after the completion of the light-sensitive phase. In LDPs the light break must coincide with the light-sensitive phase for flowering to occur. (Data from Coulter and Hamner 1964.)

Measurements of the effect of a night break on flowering can be used to investigate the role of circadian rhythms in photoperiodic timekeeping. For example, when soybean plants, which are SDPs, are transferred from an 8-hour light period to an extended 64-hour dark period, the flowering response to night breaks shows a circadian rhythm (Figure 25.21).

This type of experiment provides strong support for the clock hypothesis. If this SDP were simply measuring the length of night by the accumulation of a particular intermediate in the dark, any dark period greater than the critical night length should cause flowering. Yet long dark periods are not inductive for flowering if the light break is given at a time that does not properly coincide with a certain phase of the endogenous circadian oscillator. This finding demonstrates that flowering in SDPs requires both a dark period of sufficient duration and a dawn signal at an appropriate time in the circadian cycle (see Figure 25.15).

Further evidence for the role of a circadian oscillator in photoperiod measurement is the observation that the photoperiodic response can be phase-shifted by light treatments (see **Web Topic 25.4**).

The coincidence model is based on oscillating light sensitivity

How does an oscillation with a 24-hour period measure a critical duration of darkness of, say, 8 to 9 hours, as in the SDP *Xanthium*? In 1936 Erwin Bünning proposed that the control of flowering by photoperiodism is achieved by an oscillation of phases with different sensitivities to light. This proposal has evolved into the **coincidence model** (Bünning 1960), in which the circadian oscillator controls the timing of light-sensitive and light-insensitive phases.

The ability of light either to promote or to inhibit flowering depends on the phase in which the light is given. When a light signal is administered during the light-sensitive phase of the rhythm, the effect is either to *promote* flowering in LDPs or to *prevent* flowering in SDPs. As shown in Figure 25.21, the phases of sensitivity and insensitivity to light continue to oscillate in darkness. Flowering in SDPs is induced only when exposure to light from a night break or from dawn occurs after completion of the light-sensitive phase of the rhythm. If a similar experiment is performed with an LDP, flowering is induced only when the night break occurs *during* the light-sensitive phase of the rhythm. In other words, *flowering in both SDPs and LDPs is induced when the light exposure is coincident with the appropriate phase of the rhythm*. This continued oscillation of sensitive and insensitive phases in the absence of dawn and dusk light signals is characteristic of a variety of processes controlled by the circadian oscillator.

The coincidence of *CONSTANS* expression and light promotes flowering in LDPs

According to the coincidence model, plants are sensitive to light only at certain times of the day–night cycle. A key component of a regulatory pathway that promotes flowering of Arabidopsis in long days is a gene called ***CONSTANS*** (*CO*), which encodes a zinc finger transcription factor that regulates the transcription of other genes. *CO* was first identified as an Arabidopsis mutant, *co*, that was incapable of a photoperiodic flowering response. The expression of *CO* is controlled by the circadian clock, with the peak of activity occurring around dusk (Figure 25.22A). Genetic and molecular studies have shown that in Arabidopsis *CO* is a promoter of flowering, and the CO protein accumulates to levels that induce flowering only during a long day (Figure 25.22B).

As indicated in Figure 25.22B, a critical feature of the coincidence mechanism in the LDP Arabidopsis is that flowering is promoted when the *CO* gene is expressed in the leaf (the site of perception of the photoperiodic stimulus) during the light period. The increase in *CO* mRNA that occurs during short days does not lead to an increase in CO protein, because *CO* expression occurs entirely in the dark. In contrast, during long days *CO* gene expression is accompanied by a sharp increase in CO protein level, because at

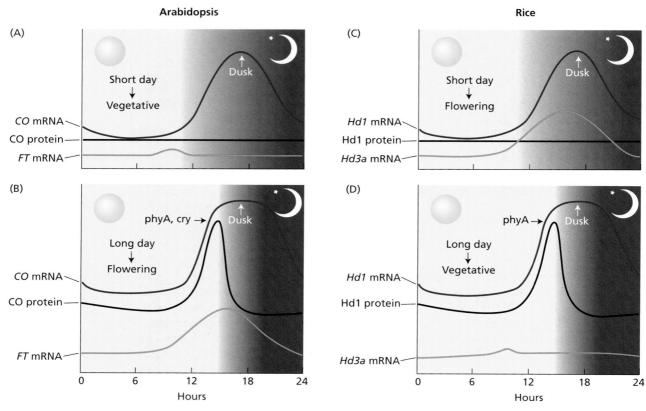

FIGURE 25.22 Molecular basis of the coincidence model in Arabidopsis (A and B) and rice (C and D). (A) In Arabidopsis under short days, there is little overlap between *CO* mRNA expression and daylight. CO protein does not accumulate to sufficient levels in the phloem to promote the expression of the transmissible floral stimulus, *FT* mRNA, and the plant remains vegetative. (B) Under long days, the peak of *CO* mRNA abundance (at hours 12 through 16) overlaps with daylight (sensed by phyA and cryptochrome), allowing CO protein to accumulate. CO activates *FT* mRNA expression in the phloem, which causes flowering when it is translocated to the apical meristem. (C) In rice under short days, the lack of coincidence between *Hd1* mRNA expression and daylight prevents the accumulation of the Hd1 protein, which acts as a repressor of the gene for the transmissible floral stimulus, *Hd3a*. In the absence of the Hd1 protein repressor, *Hd3a* mRNA is expressed and translocated to the apical meristem where it causes flowering. (D) Under long days (sensed by phytochrome), the peak of *Hd1* mRNA expression overlaps with the day, allowing the accumulation of the Hd1 repressor protein. As a result, *Hd3a* mRNA is not expressed, and the plant remains vegetative. (After Hayama and Coupland 2004.)

least some of the expression overlaps with the light period (see Figure 25.22B) (Suarez-Lopez et al. 2001). As a result, long days are inductive for Arabidopsis flowering because CO protein increases. Short days are noninductive because CO protein level does not increase in the absence of light. Thus, an important feature of the coincidence model is that there must be overlap (coincidence) between *CO* mRNA synthesis and daylight so that light can permit active CO protein to accumulate to a level that promotes flowering. The circadian oscillation of *CO* mRNA provides an explanation for the link between photoperiod perception and the circadian clock. But how does daylight bring about the accumulation of CO protein?

A clue to the function of light was provided by experiments in which *CO* was expressed from a constitutive promoter. Under these conditions, *CO* mRNA was expressed continuously, and its level remained constant throughout the day–night cycle. Nevertheless, the abundance of CO protein continued to cycle, suggesting that CO protein abundance is regulated by a posttranscriptional mechanism. The mechanism is based in part on differences in the rates of CO degradation in the light versus the dark. During the dark, CO is tagged with ubiquitin and rapidly degraded by the 26S proteasome (see Chapter 14 on the web site). Light appears to enhance the *stability* of the CO protein, allowing it to accumulate during the day. This explains why *CO* promotes flowering only when its mRNA expression coincides with the light period (Valverde et al. 2004). In the dark, CO protein does not accumulate, because it is rapidly degraded.

However, the situation is more complicated than a simple light–dark switch regulating CO turnover. The effect of light on CO stability depends on the photoreceptor involved. Different photoreceptors not only contribute to setting the phase of the circadian rhythm, but, more directly, they also affect CO protein accumulation and flowering. In the morning, phyB signaling appears to enhance CO degradation, whereas in the evening (when CO protein accumulates in long days) cryptochromes and phyA antagonize this degradation and allow the CO protein to build up (see Figures 25.22 and 25.33) (Valverde et al. 2004).

How does the CO protein stimulate flowering in long day plants? CO, a transcription factor, promotes flowering by stimulating the expression of downstream genes such as *FLOWERING LOCUS T* (*FT*) and the previously mentioned *SOC1* (also known as *AGL20*). As will be described later in the chapter, there is now evidence that *FT* mRNA is the phloem-mobile signal that stimulates flower evocation in the meristem.

The coincidence of *Heading-date 1* expression and light inhibits flowering in SDPs

Studies of flowering in the SDP rice have shown that the basic coincidence mechanism for photoperiod-sensing is conserved in rice and Arabidopsis. In the long history of rice cultivation, breeders have identified variant alleles of several genes that modify flowering behavior. The rice genes **Heading-date 1** (**Hd1**) and **Heading-date 3a** (**Hd3a**) encode proteins homologous to Arabidopsis CO and FT, respectively. In transgenic plants, overexpression of *FT* in Arabidopsis, and of *Hd3a* in rice, result in rapid flowering regardless of photoperiod, demonstrating that both *FT* and *Hd3a* are strong promoters of flowering. Moreover, expression of both the native *FT* and *Hd3a* genes is substantially elevated during inductive photoperiods—long days in Arabidopsis and short days in rice (see Figure 25.22C). In addition, rice *Hd1* and Arabidopsis *CO* exhibit similar patterns of circadian mRNA accumulation.

The difference between rice and Arabidopsis is that in the SDP rice, Hd1 acts as an *inhibitor* of *Hd3a* expression. That is, in rice the coincidence of *Hd1* expression and light signaling through phytochrome *suppresses* flowering by inhibiting the expression of *Hd3a* (see Figure 25.22D) (Izawa et al. 2003; Hayama and Coupland 2004). In contrast, CO *promotes* the expression of its downstream gene, *FT*, in the LDP Arabidopsis. Flowering in the SDP rice thus occurs only when *Hd1* is expressed exclusively in the dark. Remarkably, the different responses to photoperiod of SDPs versus LDPs are due in part to the opposite effects of this one component, CO/Hd1, of the photoperiodic sensing system.

However, it is important to note that photoperiodism is highly complex, and other regulatory mechanisms that fine-tune the responses of SDPs and LDPs to changing day length are certain to be present.

Phytochrome is the primary photoreceptor in photoperiodism

Night-break experiments are well suited for studying the nature of the photoreceptors involved in the reception of light signals during the photoperiodic response. The inhibition of flowering in SDPs by night breaks was one of the first physiological processes shown to be under the control of phytochrome (Figure 25.23).

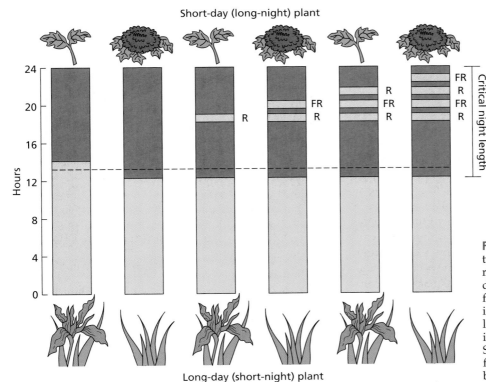

FIGURE 25.23 Phytochrome control of flowering by red (R) and far-red (FR) light. A flash of red light during the dark period induces flowering in an LDP, and the effect is reversed by a flash of far-red light. This response indicates the involvement of phytochrome. In SDPs, a flash of red light prevents flowering, and the effect is reversed by a flash of far-red light.

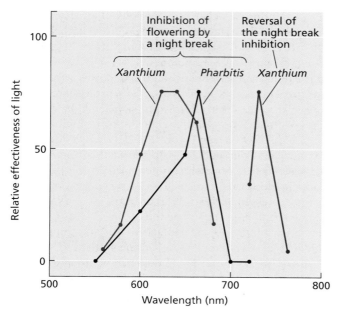

FIGURE 25.24 Action spectra for the control of flowering by night breaks implicates phytochrome. Flowering in SDPs is inhibited by a short light treatment (night break) given in an otherwise inductive period. In the SDP *Xanthium strumarium*, red-light night breaks of 620 to 640 nm are the most effective. Reversal of the red-light effect is maximal at 725 nm. In the dark-grown SDP *Pharbitis nil*, which is devoid of chlorophyll and its interference with light absorption, night breaks of 660 nm are the most effective. This 660 nm maximum coincides with the absorption maximum of phytochrome. (Data for *Xanthium* from Hendricks and Siegelman 1967; data for *Pharbitis* from Saji et al. 1983.)

In many SDPs, a night break becomes effective only when the supplied dose of light is sufficient to saturate the photoconversion of Pr (phytochrome that absorbs red light) to Pfr (phytochrome that absorbs far-red light) (see Chapter 17). A subsequent exposure to far-red light, which photoconverts the pigment back to the physiologically inactive Pr form, restores the flowering response.

Action spectra for the inhibition and restoration of the flowering response in SDPs are shown in Figure 25.24. A peak at 660 nm, the absorption maximum of Pr (see Chapter 17), is obtained when dark-grown *Pharbitis* seedlings are used to avoid interference from chlorophyll. In contrast, the spectra for *Xanthium* provide an example of the response in green plants, in which the presence of chlorophyll can cause some discrepancy between the action spectrum and the absorption spectrum of Pr. These action spectra plus the red/far-red reversibility of the night break responses confirm the role of phytochrome as the photoreceptor that is involved in photoperiod measurement in SDPs.

Another demonstration of the critical role of phytochrome in photoperiodism in SDPs comes from genetic analyses. In rice, the gene *PHOTOPERIOD SENSITIVITY 5* (*Se5*) encodes a protein similar to Arabidopsis HY1 (see Chapter 17). Se5 and HY1 are enzymes that catalyze a step in the biosynthesis of phytochrome chromophore. Mutations in Se5 cause rice to flower extremely rapidly regardless of the day length (Izawa et al. 2000).

Night break experiments with LDPs have also implicated phytochrome. Thus, in some LDPs, a night break of red light promotes flowering, and a subsequent exposure to far-red light prevents this response (see Figure 25.23).

A circadian rhythm in the promotion of flowering by far-red light has been observed in the LDPs barley (*Hordeum vulgare*), darnel ryegrass (*Lolium temulentum*), and Arabidopsis (Figure 25.25) (Deitzer 1984). The response is proportional to the irradiance and duration of far-red light and is therefore a high-irradiance response (HIR). Like other HIRs, phyA is the phytochrome that mediates the response to far-red light (see Chapter 17). Consistent with a role of phyA in the flowering of LDPs, mutations in the *PHYA* gene delay flowering in Arabidopsis (Johnson et al. 1994). However, in some LDPs the role of phytochrome is more complex than in SDPs, because a blue light photoreceptor also participates in the response.

A blue-light photoreceptor regulates flowering in some LDPs

In some LDPs, such as Arabidopsis, blue light can promote flowering, suggesting the possible participation of a blue-

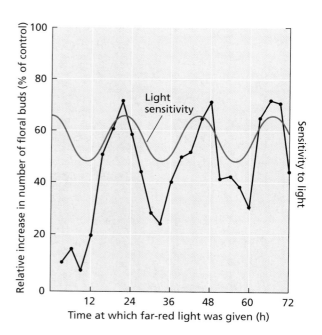

FIGURE 25.25 Effect of far-red light on floral induction in Arabidopsis. At the indicated times during a continuous 72-hour daylight period, 4 hours of far-red light was added. Data points in the graph are plotted at the centers of the 6-hour treatments. The data show a circadian rhythm of sensitivity to the far-red promotion of flowering (red line). This supports a model in which flowering in LDPs is promoted when the light treatment (in this case far-red light) coincides with the peak of light sensitivity. (After Deitzer 1984.)

light photoreceptor in the control of flowering. As discussed in Chapter 18, the cryptochromes, encoded by the *CRY1* and *CRY2* genes, are blue-light photoreceptors that control seedling growth in Arabidopsis.

As discussed in Chapter 17, the CRY protein has also been implicated in the entrainment of the circadian oscillator. The role of blue light in flowering and its relationship to circadian rhythms have been investigated by use of the luciferase reporter gene construct mentioned in **Web Topic 25.5**. In continuous white light, the cyclic luminescence has a period of 24.7 hours, but in constant darkness the period lengthens to 30 to 36 hours. Either red or blue light, given individually, shortens the period to 25 hours.

To distinguish between the effects of phytochrome and a blue-light photoreceptor, researchers transformed phytochrome-deficient *hy1* mutants, which are defective in chromophore synthesis and are therefore deficient in *all* phytochromes (see Chapter 17), with the luciferase construct to determine the effect of the mutation on the period length (Millar et al. 1995). Under continuous white light, the *hy1* plants had a period similar to that of the wild type, indicating that little or no phytochrome is required for white light to affect the period. Furthermore, under continuous red light, which would be perceived only by phyB (see Chapter 17), the period of *hy1* was significantly lengthened (i.e., it became more like constant darkness), whereas the period was not lengthened by continuous blue light. These results indicate that both phytochrome and a blue-light photoreceptor are involved in period control.

The role of blue light in regulating both circadian rhythmicity and flowering is also supported by studies with an Arabidopsis flowering-time mutant, *elf3* (*early flowering 3*) (see **Web Topics 25.5** and **25.6**). Confirmation that a blue-light photoreceptor is involved in sensing inductive photoperiods in Arabidopsis was provided by experiments demonstrating that mutations in one of the cryptochrome genes, *CRY2* (see Chapter 18), caused a delay in flowering and an inability to perceive inductive photoperiods (Guo et al. 1998). In contrast, plants carrying a gain-of-function allele of *CRY2* flower much earlier than the wild type (El-Din El-Assal 2003). In addition, the *cry1/cry2* double mutants flowered slightly later than *cry2* in long days, indicating some functional redundancy of CRY1 and CRY2 in promoting flowering time in Arabidopsis.

In addition to their role in entraining the circadian clock, it is likely that the cryptochromes, like phyA, also regulate flowering directly by stabilizing the CO protein, allowing it to accumulate under long day conditions. As noted above, the CO protein acts as a promoter of flowering in LDPs.

Vernalization: Promoting Flowering with Cold

Vernalization is the process whereby flowering is promoted by a cold treatment given to a fully hydrated seed (i.e., a seed that has imbibed water) or to a growing plant. Dry seeds do not respond to the cold treatment. Without the cold treatment, plants that require vernalization show delayed flowering or remain vegetative. In many cases these plants grow as rosettes with no elongation of the stem (Figure 25.26).

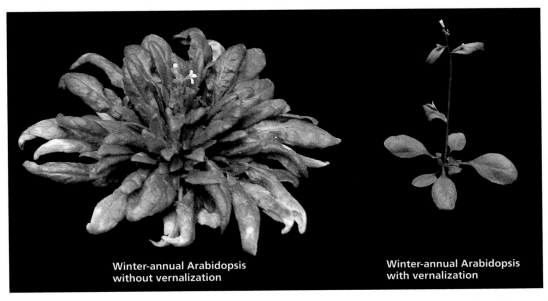

FIGURE 25.26 Vernalization induces flowering in the winter-annual types of *Arabidopsis thaliana*. The plant on the left is a winter-annual type that has not been exposed to cold. The plant on the right is a genetically identical winter-annual type that was exposed to 40 days of temperatures slightly above freezing (4°C) as a seedling. It flowered 3 weeks after the end of the cold treatment with about nine leaves on the primary stem. (Courtesy of Colleen Bizzell.)

In this section we will examine some of the characteristics of the cold requirement for flowering, including the range and duration of the inductive temperatures, the sites of perception, the relationship to photoperiodism, and a possible molecular mechanism.

Vernalization results in competence to flower at the shoot apical meristem

Plants differ considerably in the age at which they become sensitive to vernalization. Winter annuals, such as the winter forms of cereals (which are sown in the fall and flower in the following summer), respond to low temperature very early in their life cycle. They can be vernalized before germination if the seeds have imbibed water and become metabolically active. Other plants, including most biennials (which grow as rosettes during the first season after sowing and flower in the following summer), must reach a minimal size before they become sensitive to low temperature for vernalization.

The effective temperature range for vernalization is from just below freezing to about 10°C, with a broad optimum usually between about 1 and 7°C (Lang 1965). The effect of cold increases with the duration of the cold treatment until the response is saturated. The response usually requires several weeks of exposure to low temperature, but the precise duration varies widely with species and variety.

Vernalization can be lost as a result of exposure to devernalizing conditions, such as high temperature (Figure 25.27), but the longer the exposure to low temperature, the more permanent the vernalization effect.

Vernalization appears to take place primarily in the shoot apical meristem. Localized cooling causes flowering when only the stem apex is chilled, and this effect appears to be largely independent of the temperature experienced by the rest of the plant. Excised shoot tips have been successfully vernalized, and where seed vernalization is possible, fragments of embryos consisting essentially of the shoot tip are sensitive to low temperature.

In developmental terms, vernalization results in the acquisition of competence of the meristem to undergo the floral transition. Yet, as discussed earlier in the chapter, competence to flower does not guarantee that flowering will occur. A vernalization requirement is often linked with a requirement for a particular photoperiod (Lang 1965). The most common combination is a requirement for cold treatment *followed* by a requirement for long days—a combination that leads to flowering in early summer at high latitudes (see **Web Topic 25.7**). Unless devernalized, the vernalized meristem can remain competent to flower for as long as 300 days in the absence of the inductive photoperiod.

Vernalization involves epigenetic changes in gene expression

For vernalization to occur, active metabolism is required during the cold treatment. Sources of energy (sugars) and oxygen are required, and temperatures below freezing at which metabolic activity is suppressed are not effective for vernalization. Furthermore, cell division and DNA replication also appear to be required.

One model for how vernalization affects competence is that there are stable changes in the pattern of gene expression in the meristem after cold treatment. Stable changes in gene expression that do not involve alterations in the DNA sequence and which can be passed on to descendant cells through mitosis or meiosis are known as **epigenetic changes**. As such, epigenetic changes in gene expression are stable even after the signal (in this case cold) that induced the change is no longer present. Epigenetic changes of gene expression occur in many organisms, from yeast to mammals, and often require cell division and DNA replication, as is the case for vernalization.

The involvement of epigenetic regulation of a specific target gene in the vernalization process has been confirmed in the LDP Arabidopsis. In winter-annual types of Arabidopsis that require both vernalization and long days to flower, a gene that acts as a repressor of flowering has been identified: ***FLOWERING LOCUS C*** (***FLC***). *FLC* is highly expressed in nonvernalized shoot apical regions (Michaels and Amasino 2000). After vernalization, this gene is epige-

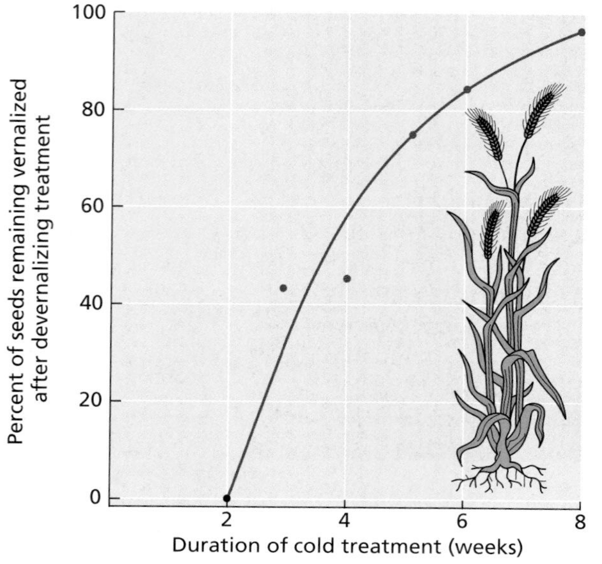

FIGURE 25.27 The duration of exposure to low temperature increases the stability of the vernalization effect. The longer that winter rye (*Secale cereale*) is exposed to a cold treatment, the greater the number of plants that remain vernalized when the cold treatment is followed by a devernalizing treatment. In this experiment, seeds of rye that had imbibed water were exposed to 5°C for different lengths of time, then immediately given a devernalizing treatment of 3 days at 35°C. (Data from Purvis and Gregory 1952.)

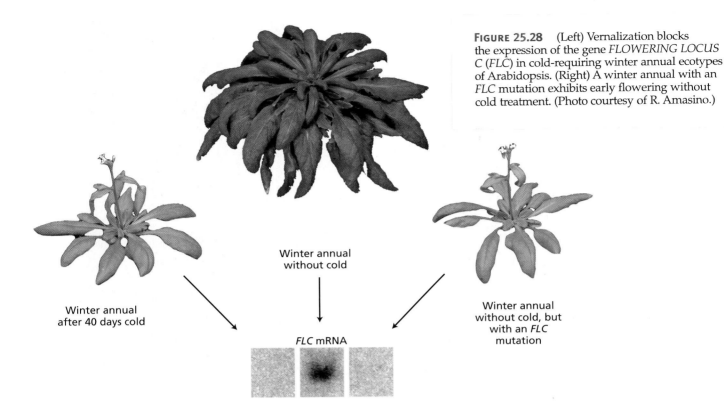

FIGURE 25.28 (Left) Vernalization blocks the expression of the gene *FLOWERING LOCUS C* (*FLC*) in cold-requiring winter annual ecotypes of Arabidopsis. (Right) A winter annual with an *FLC* mutation exhibits early flowering without cold treatment. (Photo courtesy of R. Amasino.)

netically switched off for the remainder of the plant's life cycle, permitting flowering in response to long days to occur (Figure 25.28). In the next generation, however, the gene is switched on again, restoring the requirement for cold. Thus in Arabidopsis, the state of expression of the *FLC* gene represents a major determinant of meristem competence (Michaels and Amasino 2000).

The epigenetic regulation of *FLC* involves stable changes in chromatin structure resulting from **chromatin remodeling**. As discussed in Chapter 1, in eukaryotes nuclear DNA can exist in a transcriptionally inactive, compact form called heterochromatin or a more dispersed, transcriptionally active form called euchromatin. Heterochromatin and euchromatin are characterized by distinct sets of covalent modifications of certain histones in the nucleosome (short stretch of DNA wrapped around a histone core—see Chapter 1), and these modifications are thought to favor the formation of either heterochromatin or euchromatin. The set of covalent modifications that favors a particular chromatin structure is often called the **histone code** (Jenuwein and Allis 2001). Vernalization causes the chromatin of the *FLOWERING LOCUS C* (*FLC*) gene, a flowering repressor, to lose histone modifications characteristic of euchromatin and to acquire modifications, such as the methylation of specific lysine residues, characteristic of heterochromatin (Bastow et al. 2004; Sung and Amasino 2004). The cold-induced conversion of *FLC* from euchromatin to heterochromatin effectively silences the gene.

A variety of vernalization mechanisms may have evolved

Many vernalization-requiring plants germinate in the fall, taking advantage of the cool and moist conditions optimal for their growth. The vernalization requirement of such plants ensures that flowering does not occur until spring has actually arrived. For the vernalization requirement to work as intended, plants must not only sense cold exposure but also have a mechanism to measure the duration of cold exposure. For example, if a plant is exposed to a short period of cold early in the fall, followed by a return of warm temperatures later that fall, it is important for the plant not to perceive the brief exposure to cold as winter and the following warm weather as spring. Accordingly, vernalization occurs only after an exposure to a duration of cold sufficient to indicate that a complete winter season has passed. A similar system of measuring the duration of cold before buds can be released from dormancy operates in many perennials that grow in temperate climates. The mechanism that plants have evolved to measure the duration of cold is not known, but in Arabidopsis there are genes that are induced only after exposure to a long period of cold, and these genes are critical to the vernalization process (Amasino 2004; Sung and Amasino 2004).

The vernalization pathway may not be conserved among all flowering plants. As discussed above, *FLC* is the flowering repressor that is responsible for the vernalization requirement in Arabidopsis. *FLC* encodes a MADS box protein that is related to regulatory proteins discussed earlier in the chapter, such as DEFICIENS and AGAMOUS, that are involved in floral development. In cereals, a gene encoding a different type of protein, a zinc finger-containing protein called VRN2 (*vernalization 2*), acts as the flowering repressor that creates a vernalization requirement (Yan et al. 2004).

It is possible that the major groups of flowering plants evolved in warm climates and therefore did not evolve a mechanism to measure the duration of winter. When these groups of plants adapted to temperate climates, vernalization and bud dormancy responses may have evolved independently in each group (Amasino 2004).

Biochemical Signaling Involved in Flowering

In the preceding sections we examined the influence of environmental conditions (such as temperature and day length) versus that of autonomous factors (such as age) on flowering. Although floral evocation occurs at the apical meristems of the shoots, some of the events that result in floral evocation are triggered by biochemical signals arriving at the apex from other parts of the plant, especially from the leaves. Mutants have been isolated that are deficient in the floral stimulus (see **Web Topic 25.5**).

In this section we will consider the nature of the biochemical signals arriving from the leaves and other parts of the plant in response to photoperiodic stimuli. Such signals may serve either as activators or as inhibitors of flowering. After years of investigation, no single substance has been identified as the universal floral stimulus, although certain hormones, such as gibberellins and ethylene, can induce flowering in some species. Hence, most current models of the floral stimulus are based on multiple factors.

Grafting studies have provided evidence for a transmissible floral stimulus

The production in photoperiodically induced leaves of a biochemical signal that is transported to a distant target tissue (the shoot apex) where it stimulates a response (flowering) satisfies an important criterion for a hormonal effect. In the 1930s, Mikhail Chailakhyan, working in Russia, postulated the existence of a universal flowering hormone, which he named **florigen**.

The evidence in support of florigen comes mainly from early grafting experiments in which noninduced receptor plants were stimulated to flower by being joined by grafting to a leaf or shoot from photoperiodically induced donor plants. For example, in the SDP *Perilla crispa*, a member of the mint family, grafting a leaf from a plant grown under inductive short days onto a plant grown under noninductive long days causes the latter to flower (Figure 25.29). Moreover, the floral stimulus seems to be the same

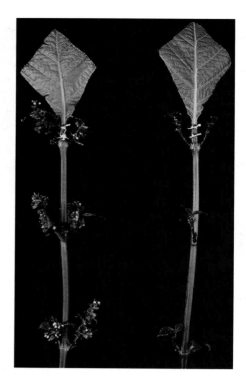

FIGURE 25.29 Demonstration by grafting of a leaf-generated floral stimulus in the SDP *Perilla*. (Left) Grafting an induced leaf from a plant grown under short days onto a noninduced shoot causes the axillary shoots to produce flowers. The donor leaf has been trimmed to facilitate grafting, and the upper leaves have been removed from the stock to promote phloem translocation from the scion to the receptor shoots. (Right) Grafting a noninduced leaf from a plant grown under LDs results in the formation of vegetative branches only. (Courtesy of J. A. D. Zeevaart.)

in plants with different photoperiodic requirements. Thus, grafting an induced shoot from the LDP *Nicotiana sylvestris*, grown under long days, onto the SDP Maryland Mammoth tobacco, caused the latter to flower under noninductive (long day) conditions.

The leaves of DNPs have also been shown to produce a graft-transmissible floral stimulus (Table 25.2). For example, grafting a single leaf of a day-neutral variety of soybean, Agate, onto the short-day variety, Biloxi, caused flowering in Biloxi even when the latter was maintained in noninductive long days. Similarly, a shoot from a day-neutral variety of tobacco (*Nicotiana tabacum*, cv. Trapezond) grafted onto the LDP *Nicotiana sylvestris* induced the latter to flower under noninductive short days.

In a few cases, flowering has been induced by grafts between different genera. The SDP *Xanthium strumarium* flowered under long-day conditions when shoots of flowering *Calendula officinalis* were grafted onto a vegetative *Xanthium* stock. Similarly, grafting a shoot from the LDP *Petunia hybrida* onto a stock of the cold-requiring biennial henbane (*Hyoscyamus niger*) caused the latter to flower

TABLE 25.2
Transmissible factors regulate flowering

Donor plants maintained under flower-inducing conditions	Photoperiod type[a,b]	Vegetative receptor plant induced to flower	Photoperiod type[a,b]
Helianthus annus	DNP in LD	*H. tuberosus*	SDP in LD
Nicotiana tabacum Delcrest	DNP in SD	*N. sylvestris*	LDP in SD
Nicotiana sylvestris	LDP in LD	*N. tabacum* Maryland Mammoth	SDP in LD
Nicotiana tabacum Maryland Mammoth	SDP in SD	*N. sylvestris*	LDP in SD

Note: The successful transfer of a flowering induction signal by grafting between plants of different photoperiodic response groups shows the existence of a transmissible floral hormone that is effective.

[a]LDPs = Long-day plants; SDPs = Short-day plants; DNPs = Day-neutral plants.
[b]LD, long days; SD, short days.

FIGURE 25.30 Successful transfer of the floral stimulus between different genera: The scion (right branch) is the LDP *Petunia hybrida*, and the stock is nonvernalized henbane (*Hyoscyamus niger*). The graft combination was maintained under LDs. (Courtesy of J. A. D. Zeevaart.)

under long days, even though it was nonvernalized (Figure 25.30).

In *Perilla* (see Figure 25.29), the movement of the floral stimulus from a donor leaf to the stock across the graft union correlated closely with the translocation of ^{14}C-labeled assimilates from the donor, and this movement was dependent on the establishment of vascular continuity across the graft union (Zeevaart 1976). These results confirmed earlier girdling studies showing that the floral stimulus is translocated along with photoassimilates in the phloem.

Indirect induction implies that the floral stimulus is self-propagating

In at least three cases—*Xanthium* (SDP), *Bryophyllum* (SLDP), and *Silene* (LDP)—the induced state appears to be self-propagating (Zeevaart 1976). That is, young leaves that develop on the receptor plant after it has been induced to flower by a donor leaf can themselves be used as donor leaves in subsequent grafting experiments, even though these leaves have never been subjected to an inductive photoperiod. This phenomenon is called *indirect induction*.

It is characteristic of indirect induction that the strength of the floral stimulus from the donor leaf remains constant even after serial grafting of new donors to several plants has taken place (Figure 25.31A). This suggests that the induced state is in some way propagated throughout the plant. Although this feature of the floral stimulus has sometimes been described as virus-like, it is unlikely that the floral stimulus can replicate itself like a virus. Rather, the floral stimulus is likely to be a molecule that induces its own production in a positive feedback loop. In cocklebur (*Xanthium*), removal of all buds from the shoot blocks indirect induction, indicating that meristematic tissue, or perhaps auxin, is required for propagation of the induced state.

On the other hand, indirect induction does not occur in the SDP *Perilla*. In *Perilla*, only the leaf actually given an

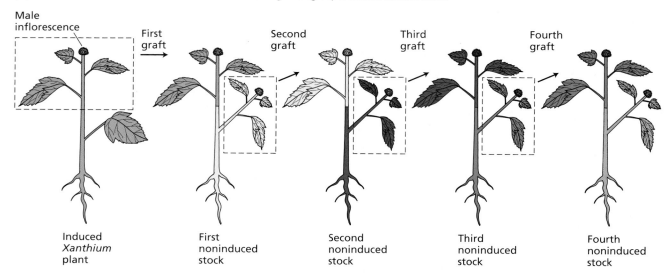

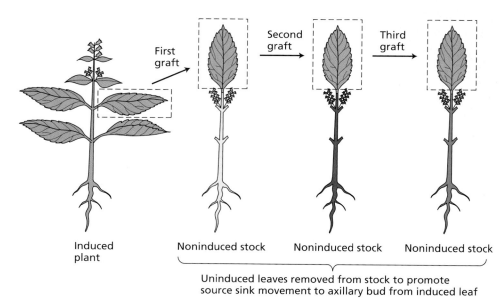

FIGURE 25.31 Different types of leaf induction in *Xanthium* and *Perilla*. (A) *Xanthium* exhibits indirect induction. Noninduced leaves from a plant induced to flower are capable of inducing other plants to flower even though they have never received an inductive photoperiod. This suggests that the floral stimulus is self-propagating. (B) In *Perilla*, only the leaf given the inductive photoperiod is capable of serving as a donor for the floral stimulus. In *Perilla* as well as *Xanthium*, one leaf can continue to induce flowering in successive grafting experiments. (Lang 1965.)

inductive photoperiod is capable of transmitting the floral stimulus in a grafting experiment (see Figure 25.31B). Thus the floral stimulus of *Perilla* is not self-propagating as it is in *Xanthium*, *Bryophyllum*, and *Silene*. Either the mechanism for a positive feedback loop is absent in *Perilla* leaves, or translocation of the floral stimulus is restricted to the meristem so that it never enters the leaves.

Unlike *Xanthium*, which requires the presence of a bud for stable induction, *Perilla* leaves can be stably induced even when detached from the plant. Once induced, *Perilla* leaves cannot be uninduced, and the same leaf can continue to serve as a donor of the floral stimulus in successive grafting experiments without any reduction in potency (Zeevaart 1976).

Evidence for antiflorigen has been found in some LDPs

Grafting studies have implicated transmissible inhibitors in flowering regulation as well. Such inhibitors have been called **antiflorigen**, but (like florigen) antiflorigen may consist of multiple compounds. For example, grafting an uninduced leafy shoot from the LDP *Nicotiana sylvestris* onto the day-neutral tobacco cultivar Trapezond suppressed flowering in the day-neutral plant under short days but not

The Control of Flowering   **663**

FIGURE 25.32 Graft transmission of an inhibitor of flowering. Noninduced rosettes from the LDP *Nicotiana sylvestris* were grafted onto the day-neutral tobacco (*N. tabacum*, cv. Trapezond). Flowering of the day-neutral plant was suppressed under short days (left branch of plant on right), but not under long days (left branch of plant on left). Arrows indicate graft unions. (From Lang et al. 1977.)

long-days (Figure 25.32). On the other hand, when an uninduced donor from the SDP Maryland Mammoth was grafted onto Trapezond, it had no effect on flowering in either short-days or long-days. This and similar results suggest that the leaves of LDPs, but not SDPs, produce flowering inhibitors under noninductive conditions.

Similar studies in peas have led to the identification of several genetic loci that regulate steps in the biosynthetic pathways of both floral activators and floral inhibitors (see **Web Topic 25.6**).

Florigen may be a macromolecule

Over the years, there have been many unsuccessful attempts to isolate and characterize both florigen and antiflorigen. Based on the structures of the known plant hormones, nearly all of these early studies focused on small molecules. Typically, extracts from induced and uninduced leaf tissue were prepared and tested for their ability either to promote or inhibit flowering. In one notable case, an investigator laboriously collected and tested phloem sap obtained from hundreds of aphids stationed on induced plants (see Chapter 10). Although positive results were occasionally reported, none of them has stood the test of time. The frustrating lack of progress using this approach led to the speculation that florigen may be a macromolecule.

As discussed in Chapter 16, macromolecular traffic of mRNAs and proteins between cells via plasmodesmata plays essential roles in normal meristem development and function. Particles as large as viruses can move from cell to cell through plasmodesmata, and throughout the plant via the phloem. Phloem translocation of messenger RNAs and small double-stranded RNAs, called *short interfering RNA* (siRNA) and *microRNA* (miRNA), has been implicated in the long-distance spread of viral resistance as well as other developmental signals throughout plants (Hamilton and Baulcombe 1999; Yoo et al. 2004) (see also Chapter 14 on the web site and Chapter 16). It is therefore possible that florigen is an RNA or protein molecule that is translocated from the leaf to the apical meristem via the phloem in induced plants (Corbesier and Coupland 2005).

As will be discussed below, a specific mRNA has been identified that fits many of the known characteristics of florigen.

FLOWERING LOCUS T is a candidate for the photoperiodic floral stimulus

According to the coincidence model, flowering in LDPs such as Arabidopsis occurs when the *CONSTANS* gene is expressed during the light period. *CO* gene expression appears to be highest in the phloem of leaves and stems (An et al. 2004). The downstream target gene of *CO*, ***FLOWERING LOCUS T*** (*FT*), is also specifically expressed in the phloem.

Consistent with a phloem localization of CO, *co* mutants with a defective photoperiodic response could be rescued by expressing *CO* specifically in the phloem of the minor veins of mature leaves using a promoter construct specific for companion cells (An et al. 2004; Ayre and Turgeon 2004). In contrast, expressing *CO* in the apical meristems of *co* mutants did not restore the photoperiodic response. Thus, *CO* seems to act specifically in the phloem of leaves to stimulate flowering in response to long days. In addition, flowering could be induced in the *co* mutant by grafting transgenic shoots expressing CO in the phloem of their leaves onto the mutant. This observation suggests that CO expression gives rise to a graft-transmissible floral stimulus that can cause flowering at the apical meristem (Ayre and Turgeon 2004).

It is believed that some of the effects of *CO* on flowering are mediated by the *FT* gene. In Arabidopsis, *CO* expression during long days results in an increase in *FT* mRNA abundance in the phloem. However, unlike *CO*, *FT* stimulates flowering when expressed in either the phloem or the apical meristem (An et al. 2004; Corbesier and Coupland 2005). Biochemically, FT is related to a small group of regulatory proteins called RAF kinase inhibitors. In mammals, RAF kinases help to initiate mitogen-activated protein (MAP) kinase cascades involved in gene regulation (see Chapter 14 on the web site for a description of RAF kinases).

Could *FT* mRNA or protein be the long-sought phloem-mobile floral stimulus ("florigen") involved in photoperiodic signaling to the apical meristem? Investigators have now obtained evidence that *FT* mRNA can move from the leaves to the apical meristem (Huang et al. 2005). Transgenic Arabidopsis plants were constructed expressing *FT* under the control of a heat shock promoter. This made it possible to activate *FT* expression in a single leaf of the transgenic plant. Heat-treating a single rosette leaf caused a transient rise in *FT* expression in the leaf, and induced floral bud formation 2 weeks later. After 5 hours of heat treatment, an increase in *FT* mRNA was also detected in the apical meristem. Because the leader sequences of the transgenic and endogenous *FT* mRNAs were different, it was possible to show that the increased *FT* mRNA in the apical meristem was due to *FT* mRNA translocated from the heat-treated leaf. These results suggest that *FT* mRNA is the phloem-mobile floral stimulus in Arabidopsis (Huang et al. 2005).

It has now been established that the FT protein interacts with another protein called FD, a basic leucine zipper (bZIP) transcription factor that is expressed in the meristem (Abe et al. 2005; Wigge et al. 2005). According to the current model, *FT* mRNA moves via the phloem from the leaves to the meristem under inductive photoperiods. Once in the meristem, the *FT* mRNA is translated into FT protein, and the FT protein forms a complex with FD. The complex of FT and FD then activates floral identity genes such as *APETALA 1*. It is also possible that FT protein can be translocated from leaves to the meristem, although this has not been demonstrated (Figure 25.33).

In Arabidopsis, the translocation of some leaf-derived *FT* mRNA to the meristem appears to activate expression of the *FT* gene in the meristem, creating a positive feedback loop of *FT* expression (Huang et al. 2005). This positive *FT* feedback loop may be the reason that once Arabidopsis initiates flowering it never reverts to a nonflowering state. A positive feedback loop of *FT* expression could also account for the phenomenon of indirect induction discussed earlier, and if such feedback loop occurs in leaves it could account for the permanence of the induced state in *Perilla* leaves as well.

Gibberellins and ethylene can induce flowering in some plants

Among the naturally occurring growth hormones, gibberellins (GAs) (see Chapter 20) can have a strong influence on flowering (see **Web Topic 25.8**). Exogenous gibberellin can evoke flowering when applied either to rosette LDPs like Arabidopsis, or to dual–day-length plants such as *Bryophyllum*, when grown under short days (Lang 1965; Zeevaart 1985).

Gibberellin appears to promote flowering in Arabidopsis by activating expression of the *LEAFY* gene (Blazquez and Weigel 2000). The activation of *LFY* by GA is mediated by the transcription factor GAMYB, which is negatively regulated by DELLA proteins (see Chapter 20). In addition, GAMYB levels are also modulated by a microRNA that promotes the degradation of the GAMYB transcript (Achard et al. 2004) (see Chapter 20).

Exogenously applied GAs can also evoke flowering in a few SDPs in noninductive conditions and in cold-requiring plants that have not been vernalized. As previously discussed, cone formation can also be promoted in juvenile plants of several gymnosperm families by addition of GAs. Thus, in some plants exogenous GAs can bypass the endogenous trigger of age in autonomous flowering, as well as the primary environmental signals of day length and temperature.

As discussed in Chapter 20, plants contain many GA-like compounds. Most of these compounds are either precursors to, or inactive metabolites of, the active forms of GA. In some situations different GAs have markedly different effects on flowering and stem elongation, such as in the long-day plant darnel ryegrass (*Lolium temulentum*) (see **Web Topic 25.9**).

These observations suggest that the regulation of flowering may be associated with specific GAs, and they may serve as a leaf-derived floral stimulus in certain species (see **Web Essay 25.1**). Indeed, a certain level of GA is likely to be required for flowering in many species, but other pathways to flowering are necessary as well. For example, a mutation in GA biosynthesis renders the quantitative LDP Arabidopsis unable to flower in noninductive short days but has little effect on flowering in long days, demonstrating that endogenous GA is required for flowering in specific situations (Wilson et al. 1992).

Considerable attention has been given to the effects of day length on GA metabolism in the plant (see Chapter 20). For example, in the LDP spinach (*Spinacia oleracea*), the levels of gibberellins are relatively low in short days, and the plants maintain a rosette form. After the plants are transferred to long days, the levels of all the gibberellins of the 13-hydroxylated pathway ($GA_{53} \rightarrow GA_{44} \rightarrow GA_{19} \rightarrow GA_{20} \rightarrow GA_1$; see Chapter 20) increase. However, the fivefold increase in the physiologically active gibberellin, GA_1, is what causes the marked stem elongation that accompanies flowering.

In addition to GAs, other growth hormones can either inhibit or promote flowering. One commercially important example is the striking promotion of flowering in pineapple (*Ananas comosus*) by ethylene and ethylene-releasing compounds—a response that appears to be restricted to members of the pineapple family (Bromeliaceae). As discussed in Chapter 23, abscisic acid (ABA) can delay flowering in Arabidopsis by inactivating its receptor, FCA, which functions as a repressor of FLC. GA and ABA thus appear to have antagonistic effects on flowering, as they do on many other aspects of development.

The transition to flowering involves multiple factors and pathways

It is clear that the transition to flowering involves a complex system of interacting factors that include, among others, carbohydrates, gibberellins, cytokinins, and, in the bromeliads, ethylene (see **Web Topic 25.10**). Leaf-generated transmissible signals are required for determination of the shoot apex in both autonomously regulated and photoperiodic species. Determining whether these transmissible signals consist of single or multiple components is a major challenge for the future.

Genetic studies have established that there are four distinct developmental pathways that control flowering in the LDP Arabidopsis (see Figure 25.33):

1. The *photoperiodic pathway* begins in the leaf and involves phytochromes and cryptochromes. (Note that PHYA and PHYB have contrasting effects on flowering; see **Web Topic 25.11**.) In LDPs under long-day conditions, the interaction of these photoreceptors with the circadian clock initiates a pathway that results in the expression of *CO* in the phloem companion cells of the leaf.

CO activates the expression of its downstream target gene, *FT*, in the phloem. *FT* mRNA is an important component of the phloem-mobile signal (formerly referred to as "florigen") that stimulates flowering in the apical meristem. As shown in the enlargement of the meristem, *FT* mRNA is translated into protein in the apex, and the FT protein forms a complex with the transcription factor, FD. The FD/FT complex then

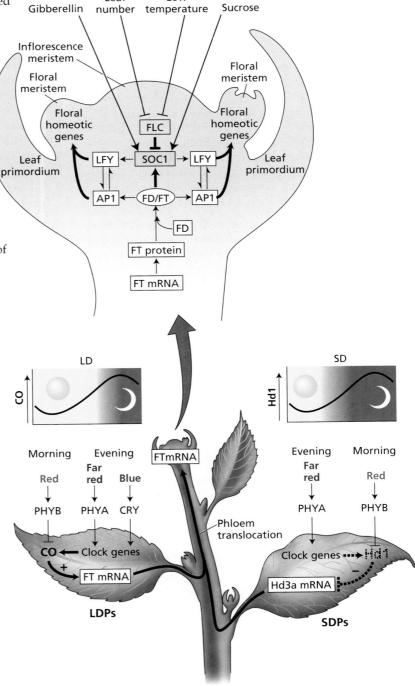

FIGURE 25.33 Multiple developmental pathways for flowering in Arabidopsis: photoperiodism, the autonomous (leaf number) and vernalization (low temperature) pathways, the energy (sucrose) pathway, and the gibberellin pathway. The photoperiodic pathway is located in the leaves and involves the production of a transmissible floral stimulus, believed to be *FT* mRNA. In LDPs such as Arabidopsis, *FT* mRNA is produced in the phloem in response to CO protein accumulation under long days. It is then translocated via sieve tubes to the apical meristem. In SDPs such as rice, the transmissible floral stimulus *Hd3a* mRNA is expressed when the repressor protein, Hd1, fails to accumulate under short days. In the meristem, *FT* or *Hd3a* mRNA are translated to proteins. In Arabidopsis, FT interacts with another protein, FD. The FT/FD complex then activates the *AP1* and *SOC1* genes, which trigger *LFY* gene expression. *LFY* and *AP1* then trigger the expression of the floral homeotic genes. The autonomous (leaf number) and vernalization (low temperature) pathways act in the apical meristem to negatively regulate *FLC*, a negative regulator of *SOC1*. The sucrose and gibberellin pathways, also localized to the meristem, promote *SOC1* expression. (After Blázquez 2005.)

activates downstream target genes such as *SOC1*, *AP1*, and *LFY*, which turn on floral homeotic genes on the flanks of the inflorescence meristem.

In rice, a SDP, the *CO* homolog *Heading-date 1* (*Hd1*) acts as an inhibitor of flowering. During inductive short-day conditions, however, Hd1 protein is not produced. The absence of Hd1 stimulates the expression of the *Hd3a* gene in the phloem companion cells. *Hd3a* mRNA is then translocated via the sieve tubes to the apical meristem, where it is believed to stimulate flowering by a pathway similar to that in Arabidopsis.

2. In the *autonomous and vernalization pathways*, flowering occurs either in response to internal signals—the production of a fixed number of leaves—or to low temperatures. In the autonomous pathway of Arabidopsis, all of the genes associated with the pathway are expressed in the meristem. The autonomous pathway acts by reducing the expression of the flowering repressor gene *FLOWERING LOCUS C* (*FLC*), an inhibitor of *SOC1* expression (Michaels and Amasino 2000). Vernalization also represses *FLC*, but perhaps by a different mechanism (an epigenetic switch). Because the *FLC* gene is a common target, the autonomous and vernalization pathways are grouped together.

3. The *carbohydrate*, or *sucrose*, *pathway* reflects the metabolic state of the plant. Sucrose stimulates flowering in Arabidopsis by increasing *LFY* expression, although the exact genetic pathway is unknown.

4. The *gibberellin pathway* is required for early flowering and for flowering under noninductive short days. The gibberellin pathway involves GAMYB as an intermediary, which promotes the *LFY*; GA may also interact with *SOC1* by a separate pathway.

All four pathways converge by increasing the expression of the key floral meristem identity gene *SOC1*. The role of *SOC1*, a MADS box–containing transcription factor, is to integrate the signals coming from all four pathways into a unitary output. Obviously the strongest output signal occurs when all four pathways are activated.

Figure 25.34 shows the level of *SOC1* gene expression in the shoot apical meristem of an Arabidopsis plant after shifting from noninductive short days (8-hour day length) to inductive long days (16-hour day length). Note that an increase in *SOC1* expression can be detected as early as 18 hours after the start of the long-day treatment (Borner et al. 2000). Thus it takes only 10 hours beyond an 8-hour short day for the meristem to begin responding to the floral stimulus from the leaves. This timing is consistent with previous measurements of the rates of export of the floral stimulus from induced leaves (discussed earlier in the chapter).

Although many pathways feed into *SOC1*, there must be some redundancy in the system, because flowering is only delayed, but not completely blocked, in *soc1* mutants. Thus, one or two other genes must be able to take over the role of *SOC1* when it is mutated.

Once turned on by *SOC1*, *LFY* activates the floral homeotic genes—*APETALA1* (*AP1*), *AP3*, *PISTILLATA* (*PI*), and *AGAMOUS* (*AG*)—that are required for floral organ development. *AP2* is expressed in both vegetative and floral meristems and is therefore not affected by *LFY*. However, as discussed earlier in the chapter, *AP2* exerts a negative effect on *AG* expression (see Figure 25.6).

Besides serving as a floral homeotic gene, *AP1* functions as a meristem identity gene in Arabidopsis because it is involved in a positive feedback loop with *LFY*. Consequently, once the transition to flowering has reached this stage, flowering is irreversible.

Short days to long days at time 0

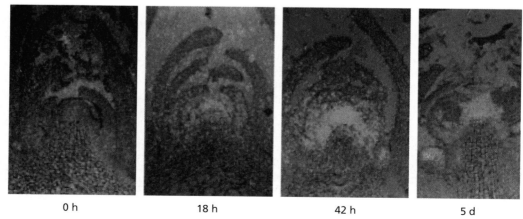

0 h   18 h   42 h   5 d

FIGURE 25.34 Increase in expression of *SOC1* (indicated by red staining) during floral evocation in the shoot apical meristem of Arabidopsis. The times after shifting the plants from SDs to LDs are indicated. (From Borner et al. 2000.)

The existence of multiple flowering pathways provides angiosperms with maximum reproductive flexibility, enabling them to produce seeds under a wide variety of conditions. Redundancy within the pathways ensures that reproduction, the most crucial of all physiological functions, will be relatively insensitive to mutations and evolutionarily robust.

The details of the pathways undoubtedly vary among different species. In maize, for example, at least one of the genes involved in the autonomous pathway is expressed in leaves (see **Web Topic 25.12**). Nevertheless, the presence of multiple flowering pathways is probably universal among angiosperms.

Summary

Flower formation occurs at the shoot apical meristem and is a complex morphological event. The rosette plant Arabidopsis has been an important model for studies on floral development. The four floral organs (sepals, petals, stamens, and carpels) are initiated as successive whorls. Three classes of genes regulate floral development. The first class contains positive regulators of the floral meristem identity. *APETALA1* (*AP1*) and *LEAFY* (*LFY*) are the most important Arabidopsis floral meristem identity genes.

Meristem identity genes are positive regulators of another class of genes that determine floral organ identity. There are five known floral organ identity genes in Arabidopsis: *AP1*, *AP2*, *AP3*, *PISTILLATA* (*PI*), and *AGAMOUS* (*AG*). Cadastral genes make up the third group. Cadastral genes act as spatial regulators of the floral organ identity genes by setting boundaries for their expression.

The genes that control floral organ identity are homeotic. Most homeotic genes in plants contain the MADS box. Mutations in these genes alter the identity of the floral organs produced in two adjacent whorls. The ABC model seeks to explain how the floral homeotic genes control organ identity through the unique combinations of their products. Type A genes control organ identity in the first and second whorls. Type B activity controls organ determination in the second and third whorls. The third and fourth whorls are controlled by type C activity.

The ability to flower (i.e., to make the transition from juvenility to maturity) is attained when the plant has reached a certain age or size. In some plants, the transition to flowering then occurs independently of the environment (autonomously). Other plants require exposure to appropriate environmental conditions. The most common environmental inputs for flowering are day length and temperature.

The response to day length—photoperiodism—promotes flowering at a particular time of year, and several different categories of responses are known. The photoperiodic signal is perceived by the leaf. Exposure to low temperature—vernalization—is required for flowering in some plants, and this requirement is often coupled with a day length requirement. Vernalization occurs at the shoot apical meristem. Photoperiodism and vernalization interact in several ways.

Daily rhythms—circadian rhythms—can locate an event at a particular time of day. Timekeeping in these rhythms is based on an endogenous circadian oscillator. Keeping the rhythm on local time depends on the phase response of the rhythm to environmental signals. The most important signals are dawn and dusk.

Short-day plants (SDPs) flower when a critical duration of darkness is exceeded. Long-day plants (LDPs) flower when the length of the dark period is less than a critical value. Light given at certain times in a dark period that is longer than the critical value—a night break—prevents the effect of the dark period. Light also acts on the circadian oscillator to entrain the photoperiodic rhythm, an effect that is important for timekeeping in the dark. The photoperiodic mechanism shows some variation in short-day and long-day responses, but both appear to involve phytochrome and a circadian oscillator.

When photoperiod-responsive plants are induced to flower by exposure to appropriate day lengths, leaves send a chemical signal to the apex to bring about flowering. This transmissible signal is able to cause flowering in plants of different photoperiodic response groups. In noninductive day lengths, a transmissible inhibitor of flowering may be produced by the leaves of LDPs.

Although physiological experiments, especially grafting, indicate the existence of a transmissible floral stimulus and, in some cases, flowering inhibitors, the chemical identity of these factors is not known. Plant growth hormones, especially the gibberellins, can modify flowering in many plants.

The mechanism of photoperiodic time sensing in both LDPs and SDPs is based on the coincidence model. Flowering in LDPs occurs when the expression of the gene *CONSTANS* (*CO*) coincides with the light period. *CO*, located in the phloem, then activates its downstream target gene, *FLOWERING LOCUS T* (*FT*), which leads to the activation of a floral stimulus. The floral stimulus ("florigen") is then translocated via the phloem to the apical meristem, where it causes flowering. The FT protein, itself, or its mRNA, are candidates for the mobile floral stimulus. In SDPs, CO during the light period acts as an inhibitor of flowering, so flowering occurs when CO is synthesized exclusively in the dark.

The transition to flowering is regulated by multiple signals and multiple pathways. In Arabidopsis, flowering is controlled by four pathways: the photoperiodic, autonomous/vernalization, sucrose, and GA pathways. All of these pathways converge to regulate the meristem identity genes *SUPPRESSOR OF CONSTANS 1* (*SOC1*) and *LFY*. *SOC1* and *LFY*, in turn, regulate the floral homeotic genes to produce the floral organs. The existence of multiple pathways for flowering provides angiosperms with the flexibility to reproduce under a variety of environmental conditions, thus increasing their evolutionary fitness.

Web Material

Web Topics

25.1 Contrasting the Characteristics of Juvenile and Adult Phases of English Ivy (*Hedera helix*) and Maize (*Zea mays*)
A table of juvenile vs. adult morphological characteristics is presented.

25.2 Regulation of Juvenility by the *TEOPOD* (*TP*) Genes in Maize
The genetic control of juvenility in maize is discussed.

25.3 Flowering of Juvenile Meristems Grafted to Adult Plants
The competence of juvenile meristems to flower can be tested in grafting experiments.

25.4 Characteristics of the Phase-Shifting Response in Circadian Rhythms
Petal movements in *Kalanchoe* have been used to study circadian rhythms.

25.5 Support for the Role of Blue-Light Regulation of Circadian Rhythms
ELF3 plays a role in mediating the effects of blue light on flowering time.

25.6 Genes That Control Flowering Time
A discussion of genes that control different aspects of flowering time is presented.

25.7 Regulation of Flowering in *Canterbury bell* by Both Photoperiod and Vernalization
Short days acting on the leaf can substitute for vernalization at the shoot apex in Canterbury bell.

25.8 Examples of Floral Induction by Gibberellins in Plants with Different Environmental Requirements for Flowering
A table of the effects of gibberellins on plants with different photoperiodic requirements is presented.

25.9 The Different Effects of Two Different Gibberellins on Flowering (Spike Length) and Elongation (Stem Length)
GA_1 and GA_{32} have different effects on flowering in *Lolium*.

25.10 The Influence of Cytokinins and Polyamines on Flowering
Other growth regulators in addition to gibberellins may participate in the flowering response.

25.11 The Contrasting Effects of Phytochromes A and B on Flowering
PhyA and phyB affect flowering in Arabidopsis and other species.

25.12 A Gene That Regulates the Floral Stimulus in Maize
The *INDETERMINATE 1* gene of maize regulates the transition to flowering and is expressed in young leaves.

Web Essay

25.1 The Role of Gibberellins in Floral Evocation of the Grass *Lolium temulentum*
Evidence that GA functions as a leaf-derived floral stimulus in *Lolium* is presented.

Chapter References

Abe, M., Kobayashi, Y., Yamamoto, S., Daimon, Y., Yamaguchi, A., Ikeda, Y., Ichinoki, H., Notaguchi, M., Goto, K., and Araki, T. (2005) FD, a bZIP protein mediating signals from the floral pathway integrator FT at the shoot apex. *Science* 309: 1052–1056.

Achard, P., Herr, A., Baulcombe, D. C., and Harberd, N. P. (2004) Modulation of floral development by a gibberellin-regulated microRNA. *Development* 131: 3357–3365.

Amasino, R. M. (2004) Vernalization, competence, and the epigenetic memory of winter. *Plant Cell* 16: 2553–2559.

An, H., Roussot, C., Suárez-López, P., Corbesier, L., Vincent, C., Piñeiro, M., Hepworth, S., Mouradov, A., Justin, S., Turnbull, C., and Coupland, G. (2004) CONSTANS acts in the phloem to regulate a systemic signal that induces photoperiodic flowering of *Arabidopsis*. *Development* 131: 3615–3626.

Aukerman, M. J., and Sakai, H. (2003) Regulation of flowering time and floral organ identity by a microRNA and its *APETALA2*-like target genes. *Plant Cell* 15: 2730–2741.

Ayre, B. G., and Turgeon, R. (2004) Graft transmission of a floral stimulant derived from CONSTANS. *Plant Physiol.* 135: 2271–2278.

Bastow, R., Mylne, J. S., Lister, C., Lippman, Z., Martienssen, R. A., and Dean, C. (2004) Vernalization requires epigenetic silencing of FLC by histone methylation. *Nature* 427: 164–167.

Bewley, J. D., Hempel, F. D., McCormick, S., and Zambryski, P. (2000) Reproductive Development. In: *Biochemistry and Molecular Biology of Plants*, B. B. Buchanan, W. Gruissem, and R. L. Jones, eds., American Society of Plant Biologists, Rockville, MD.

Blázquez. M. A. (2005) The right time and place for making flowers. *Science* 309: 1024–1025.

Blazquez, M. A., and Weigel, D. (2000) Integration of floral inductive signals in *Arabidopsis*. *Nature* 404: 889–892.

Borner, R., Kampmann, G., Chandler, J., Gleissner, R., Wisman, E., Apel, K., and Melzer, S. (2000) A *MADS* domain gene involved in the transition to flowering in *Arabidopsis*. *Plant J.* 24: 591–599.

Bowman, J. L., Smyth, D. R., and Meyerowitz, E. M. (1989) Genes directing flower development in *Arabidopsis*. *Plant Cell* 1: 37–52.

Bünning, E. (1960) Biological clocks. *Cold Spring Harbor Symp. Quant. Biol.* 15: 1–9.

Clark, J. R. (1983) Age-related changes in trees. *J. Arboriculture* 9: 201–205.

Coen, E. S., and Carpenter, R. (1993) The metamorphosis of flowers. *Plant Cell* 5: 1175–1181.

Corbesier, L. and Coupland, G. (2005) Photoperiodic flowering in *Arabidopsis*: Integrating genetic and physiological approaches to characterization of the floral stimulus. *Plant Cell Environ.* 28: 54–66.

Coulter, M. W., and Hamner, K. C. (1964) Photoperiodic flowering response of Biloxi soybean in 72 hour cycles. *Plant Physiol.* 39: 848–856.

Deitzer, G. (1984) Photoperiodic induction in long-day plants. In *Light and the Flowering Process*, D. Vince-Prue, B. Thomas, and K. E. Cockshull eds., Academic Press, New York, pp. 51–63.

Devlin, P. F., and Kay, S. A. (2000) Cryptochromes are required for phytochrome signaling to the circadian clock but not for rhythmicity. *Plant Cell* 12: 2499–2509.

El-Din El-Assal, S., Alonso-Blanco, C., Peeters, A. J., Wagemaker, C., Weller, J. L., and Koornneef, M. (2003) The role of cryptochrome 2 in flowering in *Arabidopsis*. *Plant Physiol.* 133: 1504–1516.

Gasser, C. S., and Robinson-Beers, K. (1993) Pistil development. *Plant Cell* 5: 1231–1239.

Guo, H., Yang, H., Mockler, T. C., and Lin, C. (1998) Regulation of flowering time by *Arabidopsis* photoreceptors. *Science* 279: 1360–1363.

Hamilton, A. J., and Baulcombe, D. C. (1999) A species of small antisense RNA in posttranscriptional gene silencing in plants. *Science* 286: 950–952.

Hayama, R., and Coupland, G. (2004) The molecular basis of diversity in the photoperiodic flowering responses of *Arabidopsis* and rice. *Plant Physiol.* 135: 677–684.

Hendricks, S. B., and Siegelman, H. W. (1967) Phytochrome and photoperiodism in plants. *Comp. Biochem.* 27: 211–235.

Huang, T., Böhlenius, H., Eriksson, S., Parcy, F., and Nilsson, O. (2005) The mRNA of the *Arabidopsis* gene FT moves from leaf to shoot apex and induces flowering. *Science* 309: 1694–1696.

Izawa, T., Oikawa, T., Tokutomi, S., Okuno, K., and Shimamoto, K. (2000) Phytochromes confer the photoperiodic control of flowering in rice (a short-day plant). *Plant J.* 22: 391–399.

Izawa T., Takahashi Y., and Yano M. (2003) Comparative biology comes into bloom: Genomic and genetic comparison of flowering pathways in rice and *Arabidopsis*. *Curr. Opin. Plant Biol.* 6: 113–120

Jenuwein, T., and Allis, C. D. (2001) Translating the histone code. *Science* 293: 1074–1080.

Johnson E., Bradley M., Harberd, N., and Whitelam, G. C. (1994) Photoresponces of light grown phyA mutants of *Arabidopsis*. *Plant Physiol.* 105: 141–149.

Lang, A. (1965) Physiology of flower initiation. In *Encyclopedia of Plant Physiology* (Old Series, Vol. 15), W. Ruhland, ed., Springer, Berlin, pp. 1380–1535.

Lang, A., Chailakhyan, M. K., and Frolova, I. A. (1977) Promotion and inhibition of flower formation in a day-neutral plant in grafts with a short-day plant and a long-day plant. *Proc. Natl. Acad. Sci. USA* 74: 2412–2416.

McDaniel, C. N. (1996) Developmental physiology of floral initiation in *Nicotiana tabacum* L. *J. Exp. Bot.* 47: 465–475.

McDaniel, C. N., Hartnett, L. K., and Sangrey, K. A. (1996) Regulation of node number in day-neutral *Nicotiana tabacum*: A factor in plant size. *Plant J.* 9: 56–61.

McDaniel, C. N., Singer, S. R., and Smith, S. M. E. (1992) Developmental states associated with the floral transition. *Dev. Biol.* 153: 59–69.

Meyerowitz, E. M. (2002) Plants compared to animals: The broadest comparative study of development. *Science* 295: 1482–1485.

Michaels, S. D., and Amasino, R. M. 2000. Memories of winter: Vernalization and the competence to flower. *Plant Cell Environ.* 23: 1145–1154.

Millar, A. J., Carre, I. A., Strayer, C. A., Chua, N.-H., and Kay, S. A. (1995) Circadian clock mutants in *Arabidopsis* identified by luciferase imaging. *Science* 267: 1161–1163.

Papenfuss, H. D., and Salisbury, F. B. (1967) Aspects of clock resetting in flowering of *Xanthium*. *Plant Physiol.* 42: 1562–1568.

Poethig, R. S. (1990) Phase change and the regulation of shoot morphogenesis in plants. *Science* 250: 923–930.

Purvis, O. N., and Gregory, F. G. (1952) Studies in vernalization of cereals. XII. The reversibility by high temperature of the vernalized condition in Petkus winter rye. *Ann. Bot.* 1: 569–592.

Reid, J. B., Murfet, I. C., Singer, S. R., Weller, J. L., and Taylor, S.A. (1996) Physiological genetics of flowering in *Pisum*. *Sem. Cell Dev. Biol.* 7: 455–463.

Saji, H., Vince-Prue, D., and Furuya, M. (1983) Studies on the photoreceptors for the promotion and inhibition of flowering in dark-grown seedlings of *Pharbitis nil* Choisy. *Plant Cell Physiol.* 67: 1183–1189.

Salisbury, F. B. (1963) Biological timing and hormone synthesis in flowering of *Xanthium*. *Planta* 49: 518–524.

Simon, R., Igeno, M. I., and Coupland, G. (1996) Activation of floral meristem identity genes in *Arabidopsis*. *Nature* 384: 59–62.

Suarez-Lopez, P., Wheatley, K., Robson, F., Onouchi, H., Valverde, F., and Coupland, G. (2001) CONSTANS mediates between the circadian clock and the control of flowering in *Arabidopsis*. *Nature* 410: 1116–1120.

Sung, S., and Amasino, R. M. (2004) Vernalization in *Arabidopsis thaliana* is mediated by the PHD finger protein VIN3. *Nature* 427: 159–164.

Valverde, F., Mouradov, A., Soppe, W., Ravenscroft, D., Samach, A., and Coupland, G. (2004) Photoreceptor regulation of CONSTANS protein in photoperiodic flowering. *Science* 303: 1003–1006.

Vince-Prue, D. (1975) *Photoperiodism in Plants*. McGraw-Hill, London.

Weigel, D., and Meyerowitz, E. M. (1994) The ABCs of floral homeotic genes. *Cell* 78: 203–209.

Wigge, P. A., Kim, M. C., Jaeger, K. E., Busch, W., Schmid, M., Lohmann, J. U., and Weigel, D. (2005) Integration of spatial and temporal information during floral induction in *Arabidopsis*. *Science* 309: 1056–1059.

Wilson, R. A., Heckman, J. W., and Sommerville, C. R. (1992) Gibberellin is required for flowering in *Arabidopsis thaliana* under short days. *Plant Physiol.* 100: 403–408.

Yan, L., Loukoianov, A., Blechl, A., Tranquilli, G., Ramakrishna, W., San Miguel, P., Bennetzen, J. L., Echenique, V., and Dubcovsky, J. (2004) The wheat *VRN2* gene is a flowering repressor down-regulated by vernalization. *Science* 303: 1640–1644.

Yanovsky, M. J., and Kay, S. A. (2001) Signaling networks in the plant circadian rhythm. *Curr. Opin. Plant Biol.* 4: 429–435.

Yanovsky, M. J., Mazzella, M. A., Whitelam, G. C., and Casal, J. J. (2001) Resetting the circadian clock by phytochromes and cryptochromes in *Arabidopsis*. *J. Biol. Rhythms* 16: 523–530.

Yoo, B. C., Kragler, F., Varkonyi-Gasic, E., Haywood, V., Archer-Evans, S., Lee, Y. M., Lough T. J., and Lucas, W. J. (2004) A systemic small RNA signaling system in plants. *Plant Cell* 16: 1979–2000.

Zeevaart, J. A. D. (1976) Physiology of flower formation. *Annu. Rev. Plant Physiol.* 27: 321–348.

Zeevaart, J. A. D. (1985) Bryophyllum. In *Handbook of Flowering*, Vol. II, A. H. Halevy, ed., CRC Press, Boca Raton, FL, pp. 89–100.

Zeevaart, J. A. D. (1986) Perilla. In *Handbook of Flowering*, Vol. 5, A. H. Halevy, ed., CRC Press, Boca Raton, FL, pp. 239–252.

Zeevaart, J. A. D., and Boyer, G. L. (1987) Photoperiodic induction and the floral stimulus in *Perilla*. In *Manipulation of Flowering*, J. G. Atherton, ed., Butterworths, London, pp. 269–277.

Chapter 26 | Stress Physiology

IN BOTH NATURAL AND AGRICULTURAL CONDITIONS, plants are frequently exposed to environmental stresses. Some environmental factors, such as air temperature, can become stressful in just a few minutes; others, such as soil water content, may take days to weeks, and factors such as soil mineral deficiencies can take months to become stressful. It has been estimated that because of stress resulting from suboptimal climatic and soil conditions (abiotic factors), the yield of field-grown crops in the United States is only 22% of the genetic potential yield (Boyer 1982).

In addition, stress plays a major role in determining how soil and climate limit the distribution of plant species. Thus, understanding the physiological processes that underlie stress injury and the adaptation and acclimation mechanisms of plants to environmental stress is of immense importance to both agriculture and the environment.

The concept of plant stress is often used imprecisely, and stress terminology can be confusing, so it is useful to start our discussion with some definitions. **Stress** is usually defined as an external factor that exerts a disadvantageous influence on the plant. This chapter will concern itself with environmental or abiotic factors that produce stress in plants, although biotic factors such as weeds, pathogens, and insect predation can also produce stress. In most cases, stress is measured in relation to plant survival, crop yield, growth (biomass accumulation), or the primary assimilation processes (CO_2 and mineral uptake), which are related to overall growth.

The concept of stress is intimately associated with that of **stress tolerance**, which is the plant's fitness to cope with an unfavorable environment. In the literature the term *stress resistance* is often used interchangeably with *stress tolerance*, although the latter term is preferred. Note that an environment that is stressful for one plant may not be stressful for another. For example, pea (*Pisum sativum*) and soybean (*Glycine max*) grow best at about 20°C and 30°C, respectively. As temperature increases, the pea shows signs of heat stress much sooner than the soybean. Thus the soybean has greater heat stress tolerance.

If tolerance increases as a result of exposure to prior stress, the plant is said to be **acclimated** (or hardened). Acclimation can be distinguished from **adaptation**, which usually refers to a *genetically* determined level of resistance acquired by a process of selection over many generations. Unfortunately, the term *adaptation* is sometimes used in the literature to indicate acclimation. And to add to the complexity, we will see later that gene expression plays an important role in acclimation.

Adaptation and acclimation to environmental stresses result from integrated events occurring at all levels of organization, from the anatomical and morphological level to the cellular, biochemical, and molecular level. For example, the wilting of leaves in response to water deficit reduces both water loss from the leaf and exposure to incident light, thereby reducing heat stress on leaves.

Cellular responses to stress include changes in the cell cycle and cell division, changes in the endomembrane system and vacuolization of cells, and changes in cell wall architecture, all leading to enhanced stress tolerance of cells. At the biochemical level, plants alter metabolism in various ways to accommodate environmental stresses, including producing osmoregulatory compounds such as proline and glycine betaine. The molecular events linking the perception of a stress signal with the genomic responses leading to tolerance have been intensively investigated in recent years.

In this chapter we will examine these principles, and the ways in which plants adapt and acclimate to water deficit, salinity, chilling and freezing, heat, and oxygen deficiency in the root biosphere. Air pollution, an important source of plant stress, is discussed in **Web Essay 26.1**. Although it is convenient to examine each of these stress factors separately, most are interrelated, and a common set of cellular, biochemical, and molecular responses accompanies many of the individual acclimation and adaptation processes.

For example, water deficit is often associated with salinity in the root biosphere and with heat stress in the leaves (resulting from decreased evaporative cooling due to low transpiration), and freezing leads to reductions in water activity and osmotic stress. We will also see that plants often display cross-tolerance—that is, tolerance to one stress induced by acclimation to another. This behavior implies that mechanisms of resistance to several stresses share many common features.

Water Deficit and Drought Tolerance

In this section we will examine some drought resistance mechanisms, which are divided into several types. First we can distinguish between **desiccation postponement** (the ability to maintain tissue hydration) and **desiccation tolerance** (the ability to function while dehydrated), which are sometimes referred to as drought tolerance at high and low water potentials, respectively. The older literature often uses the term *drought avoidance* (instead of *drought tolerance*), but this term is a misnomer because drought is a meteorological condition that is tolerated by all plants that survive it and avoided by none. A third category, **drought escape**, comprises plants that complete their life cycles during the wet season, before the onset of drought. These are the only true "drought avoiders."

Among the desiccation postponers are water savers and water spenders. *Water savers* use water conservatively, preserving some in the soil for use late in their life cycle; *water spenders* aggressively consume water, often using prodigious quantities. The mesquite tree (*Prosopis* sp.) is an example of a water spender. This deeply rooted species has ravaged semiarid rangelands in the southwestern United States, and because of its prodigious water use, it has prevented the reestablishment of grasses that have agronomic value.

Drought resistance strategies can vary

The water-limited productivity of plants (Table 26.1) depends on climate and soil conditions that influence the total amount of water available and on the water-use efficiency of the plant (see Chapter 4). A plant that is capable of acquiring more water or that has higher water-use efficiency will resist drought better. Some plants possess adaptations, such as the C_4 and CAM modes of photosynthesis,

TABLE 26.1
Yields of corn and soybean crops in the United States

	Crop yield (percentage of 10-year average)		
Year	Corn	Soybean	
1979	104	106	
1980	87	88	Severe drought
1981	104	100	
1982	108	104	
1983	77	87	Severe drought
1984	101	93	
1985	112	113	
1986	113	110	
1987	114	111	
1988	80	89	Severe drought

Source: U.S. Department of Agriculture 1989.

that allow them to exploit more arid environments. In addition, plants possess acclimation mechanisms that are activated in response to water deficit or osmotic stress.

Water deficit can be defined as any water content of a tissue or cell that is below the highest water content exhibited at the most hydrated state. When water deficit develops slowly enough to allow changes in developmental processes, water stress has several effects on growth, one of which is a limitation in leaf expansion. Leaf area is important because photosynthesis is usually proportional to it. However, rapid leaf expansion can adversely affect water availability.

If precipitation occurs only during winter and spring, and summers are dry, accelerated early growth can lead to large leaf areas, rapid water depletion, and too little residual soil moisture for the plant to complete its life cycle. In this situation, only plants that have some water available for reproduction late in the season or that complete the life cycle quickly, before the onset of drought (exhibiting drought escape), will produce seeds for the next generation. Either strategy will allow some reproductive success.

The situation is different if summer rainfall is significant but erratic. In this case, a plant with large leaf area, or one capable of developing large leaf area very quickly, is better suited to take advantage of occasional wet summers. One acclimation strategy in these conditions is a capacity for both vegetative growth and flowering over an extended period. Such plants are said to be *indeterminate* in their growth habit, in contrast to *determinate* plants, which develop preset numbers of leaves and flower over only very short periods.

In the discussions that follow, we will examine several acclimation strategies, including inhibited leaf expansion, leaf abscission, enhanced root growth, and stomatal closure.

Decreased leaf area is an early response to water deficit

Typically, as the water content of the plant decreases, cells shrink and the turgor pressure against cell walls relaxes (see Chapter 3). This decrease in cell volume resulting from lower turgor pressure subsequently concentrates solutes in cells. The plasma membrane becomes thicker and more compressed because it covers a smaller area than before. Because turgor reduction is the earliest significant biophysical effect of water stress, turgor-dependent activities such as leaf expansion and root elongation are the most sensitive to water deficits (Figure 26.1).

Cell expansion is a turgor-driven process and is extremely sensitive to water deficit. Cell expansion is described by the relationship:

$$GR = m(\Psi_p - Y) \tag{26.1}$$

where GR is growth rate, Ψ_p is turgor, Y is the yield threshold (the pressure below which the cell wall resists plastic, or nonreversible, deformation), and m is the wall extensibility (the responsiveness of the wall to pressure).

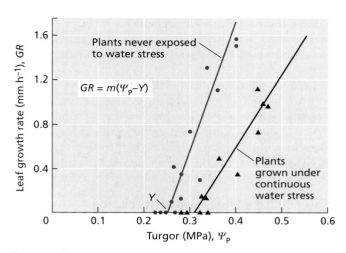

FIGURE 26.1 Dependence of leaf expansion on leaf turgor. Sunflower (*Helianthus annuus*) plants were grown either with ample water or with limited soil water to produce mild water stress. After rewatering, plants of both treatment groups were stressed by the withholding of water, and leaf growth rates (GR) and turgor (Ψ_p) were periodically measured. Both decreased extensibility (m) and increased threshold turgor for growth (Y) limit the leaf's capacity to grow after exposure to stress. (After Matthews et al. 1984.)

This equation shows that a decrease in turgor causes a decrease in growth rate. Note also that besides showing that growth slows down when stress reduces Ψ_p, Equation 26.1 shows that Ψ_p need decrease only to the value of Y, not to zero, to eliminate expansion. In normal conditions, Y is usually only 0.1 to 0.2 megapascals (MPa) less than Ψ_p, so small decreases in water content and turgor can slow down or fully stop growth.

Water stress not only decreases turgor, but also decreases m and increases Y. Wall extensibility (m) is normally greatest when the cell wall solution is slightly acidic. In part, stress decreases m because cell wall pH typically rises during stress. The effects of stress on Y are not well understood, but presumably they involve complex structural changes of the cell wall (see Chapter 15) that may not be readily reversed after relief of stress. Water-deficient plants tend to become rehydrated at night, and as a result substantial leaf growth occurs at that time. Nonetheless, because of changes in m and Y, the growth rate is still lower than that of unstressed plants having the same turgor (see Figure 26.1).

Because leaf expansion depends mostly on cell expansion, the principles that underlie the two processes are similar. Inhibition of cell expansion results in a slowing of leaf expansion early in the development of water deficits. The smaller leaf area transpires less water, effectively conserving a limited water supply in the soil over a longer period. *Reduction in leaf area can thus be considered a first line of defense against drought.*

In indeterminate plants, water stress limits not only leaf size, but also leaf number, because it decreases both the number and the growth rate of branches. Stem growth has been studied less than leaf expansion, but stem growth is probably affected by the same forces that limit leaf growth during stress.

Keep in mind, too, that cell and leaf expansion also depend on biochemical and molecular factors beyond those that control water flux. Much evidence supports the view that plants change their growth rates in response to stress by coordinately controlling many other important processes such as cell wall and membrane biosynthesis, cell division, and protein synthesis (Burssens et al. 2000).

Water deficit stimulates leaf abscission

The total leaf area of a plant (number of leaves × surface area of each leaf) does not remain constant after all the leaves have matured. If plants become water stressed after a substantial leaf area has developed, leaves will senesce and eventually fall off (Figure 26.2). Such a leaf area adjustment is an important long-term change that improves the plant's fitness in a water-limited environment. Indeed, many drought-deciduous, desert plants drop all their leaves during a drought and sprout new ones after a rain. This cycle can occur two or more times in a single season. Abscission during water stress results largely from enhanced synthesis of and responsiveness to the endogenous plant hormone ethylene (see Chapter 22).

Water deficit enhances root growth

Mild water deficits also affect the development of the root system. Root-to-shoot biomass ratio appears to be governed by a functional balance between water uptake by the root and photosynthesis by the shoot (see Figure 23.6). Simply stated, *a shoot will grow until it is so large that water uptake by the roots becomes limiting to further growth*; conversely, *roots will grow until their demand for photosynthate from the shoot equals the supply*. This functional balance is shifted if the water supply decreases.

As discussed already, leaf expansion is affected very early when water uptake is curtailed, but photosynthetic activity is much less affected. Inhibition of leaf expansion reduces the consumption of carbon and energy, and a greater proportion of the plant's assimilates can be distributed to the root system, where they can support further root growth. At the same time, the root apices in dry soil lose turgor.

All these factors lead to a preferential root growth into the soil zones that remain moist. As water deficits progress, the upper layers of the soil usually dry first. Thus, plants commonly show a mainly shallow root system when all soil layers are wetted, and a loss of shallow roots and proliferation of deep roots as water in top layers of the soil is depleted. Deeper root growth into wet soil can be considered a second line of defense against drought.

Enhanced root growth into moist soil zones during stress requires allocation of assimilates to the growing root tips. During water deficit, assimilates are directed to the fruits and away from the roots (see Chapter 10). For this reason the enhanced water uptake resulting from root growth is less pronounced in reproductive plants than in vegetative plants. Competition for assimilates between roots and fruits is one explanation for the fact that plants are generally more sensitive to water stress during reproduction.

Abscisic acid induces stomatal closure during water deficit

The preceding sections focused on changes in plant development during slow, long-term dehydration. When the onset of stress is more rapid or the plant has reached its full leaf area before initiation of stress, other responses protect the plant against immediate desiccation. Under these conditions, stomatal closure reduces evaporation from the existing leaf area. Thus, stomatal closure can be considered a third line of defense against drought.

Uptake and loss of water in guard cells changes their turgor and modulates stomatal opening and closing (see Chapters 4 and 18). Because guard cells are located in the leaf epidermis, they can lose turgor as a result of a direct loss of water by evaporation to the atmosphere. **Hydropassive stomatal closure** results from such evaporative water loss. This closing mechanism is likely to operate in air of low humidity, when direct water loss from the guard cells is too rapid to be balanced by water movement into the guard cells from adjacent epidermal cells.

A second mechanism, referred to as **hydroactive stomatal closure**, occurs when the whole leaf and/or the roots are dehydrated, and it depends on metabolic processes in the guard cells. A reduction in the solute content of the guard cells (increase in osmotic potential) results in water loss and decreased turgor, causing the stomata to close;

FIGURE 26.2 The leaves of young cotton (*Gossypium hirsutum*) plants abscise in response to water stress. The plants at left were watered throughout the experiment; those in the middle and at right were subjected to moderate stress and severe stress, respectively, before being watered again. Only a tuft of leaves at the top of the stem is left on the severely stressed plants. (Courtesy of B. L. McMichael.)

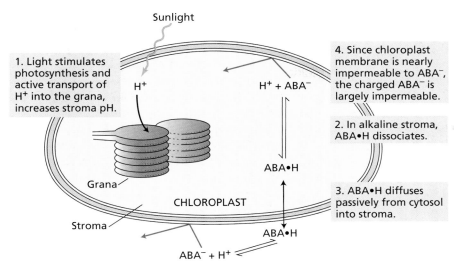

FIGURE 26.3 Accumulation of ABA by chloroplasts in the light. Light stimulates proton uptake into the grana, making the stroma more alkaline. The increased alkalinity causes the weak acid ABA•H to dissociate into H⁺ and the ABA⁻ anion. The concentration of ABA•H in the stroma is lowered below the concentration in the cytosol, and the concentration difference drives the passive diffusion of ABA•H across the chloroplast membrane. At the same time, the concentration of ABA⁻ in the stroma increases, but the chloroplast membrane is almost impermeable to the anion (red arrows), which thus remains trapped. This process continues until the ABA•H concentrations in the stroma and the cytosol are equal. But as long as the stroma remains more alkaline, the total ABA concentration (ABA•H + ABA⁻) in the stroma greatly exceeds the concentration in the cytosol.

thus the hydraulic mechanism of hydroactive closure is a reversal of the mechanism of stomatal opening. However, the control of hydroactive closure differs in subtle but important ways from stomatal opening.

Solute loss from guard cells can be triggered by a decrease in the water content of the leaf, and abscisic acid (ABA) (see Chapter 23) plays an important role in this process. Abscisic acid is synthesized continuously at a low rate in mesophyll cells and tends to accumulate in the chloroplasts. When the mesophyll becomes mildly dehydrated, two things happen:

1. Some of the ABA stored in the chloroplasts is released to the apoplast (the cell wall space) of the mesophyll cell (Hartung et al. 1998). The redistribution of ABA depends on pH gradients within the leaf, on the weak-acid properties of the ABA molecule, and on the permeability properties of cell membranes (Figure 26.3). The redistribution of ABA makes it possible for the transpiration stream to carry some of the ABA to the guard cells.

2. ABA is synthesized at a higher rate, and more ABA accumulates in the leaf apoplast. The higher ABA concentrations resulting from the higher rates of ABA synthesis appear to enhance or prolong the initial closing effect of the stored ABA. The mechanism of ABA-induced stomatal closure is discussed in Chapter 23.

Stomatal responses to leaf dehydration can vary widely both within and across species. The stomata of some dehydration-postponing species, such as cowpea (*Vigna unguiculata*) and cassava (*Manihot esculenta*), are unusually responsive to decreasing water availability, and stomatal conductance and transpiration decrease so much that leaf water potential (Ψ_w; see Chapters 3 and 4) may remain nearly constant during drought.

Chemical signals from the root system may affect the stomatal responses to water stress (Davies et al. 2002). Stomatal conductance is often much more closely related to soil water status than to leaf water status, and the only plant part that can be directly affected by soil water status is the root system. In fact, dehydrating only part of the root system may cause stomatal closure even if the well-watered portion of the root system still delivers ample water to the shoots.

When corn (maize; *Zea mays*) plants were grown with roots trained into two separate pots and water was withheld from only one of the pots, the stomata closed partially as guard cell turgor was reduced similar to the response in the dehydration postponers already described. These results show that stomata can respond to conditions sensed in the roots. Besides ABA (Sauter et al. 2001), other signals, such as pH and inorganic ion redistribution, appear to play a role in long-distance signaling between the roots and the shoots (Davies et al. 2002).

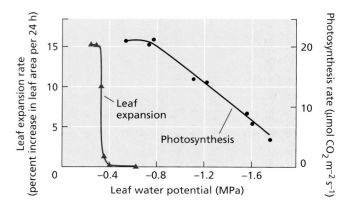

FIGURE 26.4 Effects of water stress on photosynthesis and leaf expansion of sunflower. This species is typical of many plants in which leaf expansion is very sensitive to water stress, and it is completely inhibited under mild stress levels that hardly affect photosynthetic rates. (After Boyer 1970.)

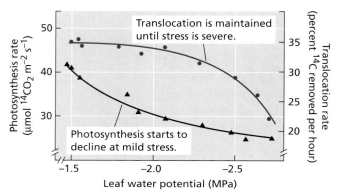

FIGURE 26.5 Relative effects of water stress on photosynthesis and translocation in sorghum (*Sorghum bicolor*). Plants were exposed to $^{14}CO_2$ for a short time interval. The radioactivity fixed in the leaf was taken as a measure of photosynthesis, and the loss of radioactivity after removal of the $^{14}CO_2$ source was taken as a measure of the rate of assimilate translocation. Photosynthesis was affected by mild stress, whereas translocation was unaffected until stress became severe. (After Sung and Krieg 1979.)

More information on the effects of stomatal conductance on productivity of pima cotton can be found in **Web Topic 26.1**.

Water deficit limits photosynthesis

The photosynthetic rate of the leaf (expressed per unit leaf area) is typically not as responsive to mild water stress as is leaf expansion (Figure 26.4), because photosynthesis is much less sensitive to a decrease in turgor than is leaf expansion. However, mild water stress does usually affect both leaf photosynthesis and stomatal conductance. As stomata close during early stages of water stress, water-use efficiency (see Chapter 4) may increase (i.e., more CO_2 may be taken up per unit of water transpired) because stomatal closure inhibits transpiration more than it decreases intercellular CO_2 concentrations.

As stress becomes severe, however, the dehydration of mesophyll cells inhibits photosynthesis, mesophyll metabolism is impaired, and water-use efficiency usually decreases. Results from many studies have shown that the relative effect of water stress on stomatal conductance is significantly larger than on photosynthesis. The response of photosynthesis and stomatal conductance to water stress can be partitioned by exposure of stressed leaves to air containing high concentrations of CO_2. Any effect of the stress on stomatal conductance is eliminated by the high CO_2 supply, and differences between photosynthetic rates of stressed and unstressed plants can be directly attributed to damage from the water stress to photosynthesis.

Does water stress directly affect translocation? Water stress decreases both photosynthesis and the consumption of assimilates in the expanding leaves. As a consequence, water stress indirectly decreases the amount of photosynthate exported from leaves. Because phloem transport depends on turgor (see Chapter 10), decreased water potential in the phloem during stress may inhibit the movement of assimilates. However, experiments have shown that translocation is unaffected until late in the stress period, when other processes, such as photosynthesis, have already been strongly inhibited (Figure 26.5).

This relative insensitivity of translocation to stress allows plants to mobilize and use reserves where they are needed (e.g., in seed growth), even when stress is extremely severe. The ability to continue translocating assimilates is a key factor in almost all aspects of plant resistance to drought.

Osmotic adjustment of cells helps maintain water balance

As the soil dries, its matric potential (see **Web Topic 3.5**) becomes more negative. Plants can continue to absorb water only as long as their water potential (Ψ_w) is lower (more negative) than that of the soil water. Osmotic adjustment, or accumulation of solutes by cells, is a process by which cellular water potential can be decreased without an accompanying decrease in turgor or decrease in cell volume. Recall Equation 3.6 from Chapter 3: $\Psi_w = \Psi_s + \Psi_p$. The change in cell water potential results simply from changes in solute potential (Ψ_s), the osmotic component of Ψ_w.

Osmotic adjustment is a net increase in solute concentration per cell that is independent of the volume changes that result from loss of water. The decrease in Ψ_s is typically limited to about 0.2 to 0.8 MPa, except in plants adapted to extremely dry conditions. Most of the adjustment can usually be accounted for by increases in concentration of a variety of common solutes, including sugars, organic acids, amino acids, and inorganic ions (especially K^+).

Cytosolic enzymes of plant cells can be severely inhibited by high concentrations of ions. The accumulation of ions during osmotic adjustment appears to be restricted to the vacuoles, where the ions are kept out of contact with enzymes in the cytosol or subcellular organelles. Because of this compartmentalization of ions, other solutes must accumulate in the cytoplasm to maintain water potential equilibrium within the cell.

These other solutes, called **compatible solutes** (or compatible osmolytes), are organic compounds that do not interfere with enzyme functions. Commonly accumulated compatible solutes include the amino acid proline, sugar alcohols (e.g., sorbitol and mannitol), and a quaternary amine called glycine betaine. Synthesis of compatible solutes helps plants adjust to increased salinity in the rooting zone, as discussed later in this chapter.

Osmotic adjustment develops slowly in response to tissue dehydration. Over a time course of several days, other changes (such as growth and photosynthesis) are also taking place. Thus, it can be argued that osmotic adjustment is not an independent and direct response to water deficit, but a result of another factor, such as decreased growth rate. Nonetheless, leaves that are capable of osmotic adjustment clearly can maintain turgor at lower water potentials than nonadjusted leaves. Maintaining turgor enables the continuation of cell elongation and facilitates higher stomatal conductance at lower water potentials. This suggests that osmotic adjustment is an acclimation that enhances dehydration tolerance.

How much extra water can be acquired by the plant because of osmotic adjustment in the leaf cells? Most of the extractable soil water is held in spaces (filled with water and air) from which it is readily removed by roots (see Chapter 4). As the soil dries, this water is used first, leaving behind the small amount of water that is held more tightly in small pores.

Osmotic adjustment enables the plant to extract more of this tightly held water, but the increase in total available water is small. Thus, the cost of osmotic adjustment in the leaf is offset by rapidly diminishing returns in terms of water availability to the plant, as can be seen by a comparison of the water relations of adjusting and nonadjusting species (Figure 26.6). These results show that osmotic adjustment promotes dehydration tolerance but does not substantially affect productivity (McCree and Richardson 1987).

Osmotic adjustment also occurs in roots, although compared to leaves the process is less well understood. As with leaves, these changes may increase water extraction from the previously explored soil only slightly. However, osmotic adjustment can occur in the root meristems, enhancing turgor and maintaining root growth. This is an important component of the changes in root growth patterns as water is depleted from the soil.

Does osmotic adjustment increase plant productivity? Researchers have engineered the accumulation of osmoprotective solutes by conventional plant breeding, by physiological methods (inducing adjustment with controlled

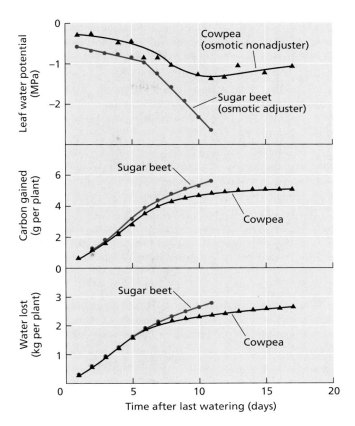

FIGURE 26.6 Water loss and carbon gain by sugar beet (*Beta vulgaris*), an osmotically adjusting species, and cowpea (*Vigna unguiculata*), a nonadjusting species that conserves water during stress by stomatal closure. Plants were grown in pots and subjected to water stress. On any given day after the last watering, the sugar beet leaves maintained a lower water potential than the cowpea leaves, but photosynthesis and transpiration during stress were only slightly greater in the sugar beet. The major difference between the two plants was the leaf water potential. These results show that osmotic adjustment promotes dehydration tolerance but does not have a major effect on productivity. (After McCree and Richardson 1987.)

water deficits), and through the use of transgenic plants expressing genes for solute synthesis and accumulation. However, the engineered plants grow more slowly, and they are only slightly more tolerant to osmotic stresses. Thus, the use of osmotic adjustment to improve agricultural performance appears to be only a component of what is required to obtain stress tolerance.

Water deficit increases resistance to water flow

When a soil dries, its resistance to the flow of liquid-phase water increases very sharply, particularly near the *permanent wilting point*. Recall from Chapter 4 that at the permanent wilting point (usually about −1.5 MPa), plants cannot regain turgor pressure even if all transpiration stops (for more details on the relationship between soil hydraulic conductivity and soil water potential, see Figure 4.2.A in **Web Topic 4.2**). Because of the very large soil resistance to water flow, water delivery to the roots at the permanent wilting point is too slow to allow the overnight rehydration of plants that have wilted during the day.

Rehydration is further hindered by the resistance within the plant, which has been found to be larger than the resistance within the soil over a wide range of water deficits (Blizzard and Boyer 1980). Several factors may contribute to the increased plant resistance to water flow during drying. As plant cells lose water, they shrink. When roots shrink, the root surface can move away from the soil particles that hold the water, and the delicate root hairs may be damaged. In addition, as root extension slows during soil drying, the outer layer of the root cortex (the hypodermis) often becomes more extensively covered with suberin, a water-impermeable lipid (see Figure 4.3), increasing the resistance to water flow.

Another important factor that increases resistance to water flow is *cavitation*, or the breakage of water columns under tension within the xylem. As we saw in Chapter 4, transpiration from leaves "pulls" water through the plant by creating a tension on the water column. The cohesive forces that are required to support large tensions are present only in very narrow columns in which the water adheres to the walls.

Cavitation begins in most plants at moderate water potentials (–1 to –2 MPa), and the largest vessels cavitate first. For example, in trees such as oak (*Quercus*), the large-diameter vessels that are laid down in the spring function as a low-resistance pathway early in the growing season, when ample water is available. As the soil dries out during the summer, these large vessels cease functioning, leaving the small-diameter vessels produced during the stress period to carry the transpiration stream. This shift may have long-lasting consequences: Even if water becomes available, the original low-resistance pathway may not function efficiently.

Water deficit increases leaf wax deposition

A common developmental response to water stress is the production of a thicker cuticle that reduces water loss from the epidermis (cuticular transpiration). Although waxes are deposited in response to water deficit both on the surface and within the cuticle inner layer, the inner layer may be more important in controlling the rate of water loss in ways that are more complex than by just increasing the amount of wax present (Jenks et al. 2002).

A thicker cuticle also decreases CO_2 permeability, but leaf photosynthesis remains unaffected because the epidermal cells underneath the cuticle are nonphotosynthetic. Cuticular transpiration, however, accounts for only 5 to 10% of the total leaf transpiration, so it becomes significant only if stress is extremely severe or if the cuticle has been damaged (e.g., by wind-driven sand).

Water deficit alters energy dissipation from leaves

Recall from Chapter 9 that evaporative heat loss lowers leaf temperature. This cooling effect can be remarkable: In Death Valley, California—one of the hottest places in the world—leaf temperatures of plants with access to ample water were measured to be 8°C below air temperatures. In warm, dry climates, an experienced farmer can decide whether plants need water simply by touching the leaves, because a rapidly transpiring leaf is distinctly cool to the touch. When water stress limits transpiration, the leaf heats up unless another process offsets the lack of cooling. Because of these effects of transpiration on leaf temperature, water stress and heat stress are closely interrelated (see the discussion of heat stress later in this chapter).

Maintaining a leaf temperature that is much lower than the air temperature requires evaporation of vast quantities of water. This is why adaptations that cool leaves by means other than evaporation (e.g., changes in leaf size and leaf orientation) are very effective in conserving water. When transpiration decreases and leaf temperature becomes warmer than the air temperature, some of the extra energy in the leaf is dissipated as sensible heat loss (see Chapter 9). Many arid-zone plants have very small leaves, which minimize the resistance of the boundary layer (the thin film of still air at the surface of the leaf) to the transfer of heat from the leaf to the air (see Figure 9.16).

Because of their low boundary layer resistance, small leaves tend to remain close to air temperature even when transpiration is greatly slowed. In contrast, large leaves have higher boundary layer resistance and dissipate less thermal energy (per unit leaf area) by direct transfer of heat to the air.

In larger leaves, leaf movement can provide additional protection against heating during water stress. Leaves that orient themselves away from the sun are called *paraheliotropic*; leaves that gain energy by orienting themselves normal (perpendicular) to the sunlight are referred to as *diaheliotropic* (see Chapter 9). Figure 26.7 shows the strong effect of water stress on leaf position in soybean. Other factors that can alter the interception of radiation include wilting, which changes the angle of the leaf, and leaf rolling in grasses, which minimizes the profile of tissue exposed to the sun.

Absorption of energy can also be decreased by hairs on the leaf surface or by layers of reflective wax outside the cuticle. Leaves of some plants have a gray-white appearance because densely packed hairs reflect a large amount of light. This hairiness, or **pubescence**, keeps leaves cooler by reflecting radiation, but it also reflects the visible wavelengths that are active in photosynthesis and thus it decreases carbon assimilation. Because of this problem, attempts to breed pubescence into crops to improve their water-use efficiency have been generally unsuccessful.

CAM plants are adapted to water stress

Some plants have the ability to induce crassulacean acid metabolism (CAM), an adaptation in which stomata open

(A) Well-watered

(B) Mild water stress

(C) Severe water stress

FIGURE 26.7 Leaf movements in soybean in response to osmotic stress. Orientation of leaflets of field-grown soybean (*Glycine max*) plants in the normal, unstressed position (A); during mild water stress (B); and during severe water stress (C). The large leaf movements induced by mild stress are quite different from wilting, which occurs during severe stress. Note that during mild stress (B), the terminal leaflet has been raised, whereas the two lateral leaflets have been lowered; each is almost vertical. (Courtesy of D. M. Oosterhuis.)

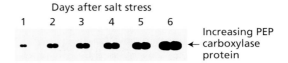

FIGURE 26.8 Phosphoenolpyruvate (PEP) carboxylase levels change during the shift from C_3 metabolism to CAM. Increases in the content of PEP carboxylase in ice plant (*Mesembryanthemum crystallinum*) during the salt-induced shift from C_3 metabolism to CAM. Salt stress was induced by the addition of 500 mM NaCl to the irrigation water. The PEP carboxylase protein was revealed in the gels by the use of antibodies and a stain. (After Bohnert et al. 1989.)

at night and close during the day (see Chapters 8 and 9). The leaf-to-air vapor pressure difference that drives transpiration is much reduced at night, when both leaf and air are cool. As a result, the water-use efficiencies of CAM plants are among the highest measured. A CAM plant may gain 1 g of dry matter for only 125 g of water used—a ratio that is three to five times greater than the ratio for a typical C_3 plant.

CAM is very prevalent in succulent plants such as cacti. Some succulent species display facultative CAM, switching to CAM when subjected to water deficits or saline conditions (see Chapter 8). This switch in metabolism is a remarkable adaptation to stress, involving accumulation of the enzymes phosphoenolpyruvate (PEP) carboxylase (Figure 26.8), pyruvate–orthophosphate dikinase, and NADP–malic enzyme, among others.

As discussed in Chapters 8 and 9, CAM metabolism involves many structural, physiological, and biochemical features, including changes in carboxylation and decarboxylation patterns, transport of large quantities of malate into and out of the vacuoles, and reversal of the periodicity of stomatal movements. Thus, CAM induction is a remarkable adaptation to water deficit that occurs at many levels of organization.

Osmotic stress changes gene expression

As noted earlier, the accumulation of compatible solutes in response to osmotic stress requires the activation of the metabolic pathways that biosynthesize these solutes. Several genes coding for enzymes associated with osmotic adjustment are turned on (up-regulated) by osmotic stress and/or salinity, and cold stress. These genes encode enzymes such as the following (Buchanan et al. 2000):

- Δ'^1-Pyrroline-5-carboxylate synthase, a key enzyme in the proline biosynthetic pathway
- Betaine aldehyde dehydrogenase, an enzyme involved in glycine betaine accumulation
- *myo*-Inositol 6-*O*-methyltransferase, a rate-limiting enzyme in the accumulation of the cyclic sugar alcohol called pinitol

Several other genes that encode well-known enzymes are induced by osmotic stress. The expression of glyceraldehyde-3-phosphate dehydrogenase increases during osmotic stress, perhaps to allow an increase of carbon flow into organic solutes for osmotic adjustment. Enzymes involved in lignin biosynthesis are also controlled by osmotic stress.

Reduction in the activities of key enzymes also takes place. The accumulation of the sugar alcohol mannitol in response to osmotic stress appears not to be brought about by the up-regulation of genes producing enzymes involved in mannitol biosynthesis, but rather by the down-regulation of genes associated with sucrose production and mannitol degradation. In this way mannitol accumulation is enhanced during episodes of osmotic stress.

Other genes regulated by osmotic stress encode proteins associated with membrane transport, including ATPases (Niu et al. 1995) and the water channel proteins, *aquaporins* (see Chapters 3 and 4) (Maggio and Joly 1995). Several protease genes are also induced by stress, and these enzymes may degrade (remove and recycle) other proteins that are denatured by stress episodes. The protein *ubiquitin* tags proteins that are targeted for proteolytic degradation. Synthesis of the mRNA for ubiquitin increases in Arabidopsis upon desiccation stress. In addition, some *heat shock proteins* are osmotically induced and may protect or renature proteins inactivated by desiccation.

The sensitivity of cell expansion to osmotic stress (see Figure 26.1) has stimulated studies of various genes that encode proteins involved in the structural composition and integrity of cell walls. Genes coding for enzymes such as *S*-adenosylmethionine synthase and peroxidases, which may be involved in lignin biosynthesis, have been shown to be controlled by stress.

A large group of genes that are regulated by osmotic stress was discovered by examination of naturally desiccating embryos during seed maturation. These genes code for so-called **LEA proteins** (named for *l*ate *e*mbryogenesis *a*bundant), and they are suspected to play a role in cellular membrane protection. Although the function of LEA proteins is not well understood (Table 26.2), they accumulate in vegetative tissues during episodes of osmotic stress, and are typically hydrophilic and strongly bind water. Their protective role might be associated with an ability to retain water and to prevent crystallization of important cellular proteins and other molecules during desiccation. They might also contribute to membrane stabilization.

More recently, microarray techniques have been used to examine the expression of whole genomes of some plants in response to stress. Such studies have revealed that large numbers of genes display changes in expression after plants are exposed to stress. Stress-controlled genes reflect up to 10% of the total number of rice genes examined (Kawasaki et al. 2001).

Osmotic stress typically leads to the accumulation of ABA (see Chapter 23), so it is not surprising that products of ABA-responsive genes accumulate during osmotic stresses. Studies of ABA-insensitive and ABA-deficient mutants have shown that numerous genes that are induced by osmotic stress are in fact induced by the ABA accumulated during the stress episode. However, not all genes that are up-regulated by osmotic stresses are ABA regulated. As discussed in the next section, other mechanisms for regulating gene expression of osmotic stress–regulated genes have been uncovered.

ABA-dependent and ABA-independent signaling pathways regulate stress tolerance

Gene transcription is controlled through the interaction of regulatory proteins (transcription factors) with specific regulatory sequences in the promoters of the genes they regulate (Chapter 14 on the web site discusses these processes in detail). Different genes that are induced by the same signal (desiccation or salinity, for example) are controlled by a signaling pathway leading to the activation of these specific transcription factors.

Studies of the promoters of several stress-induced genes have led to the identification of specific regulatory sequences for genes involved in different stresses. For example, the *RD29* gene contains DNA sequences that can be activated by osmotic stress, by cold, and by ABA (Yamaguchi-Shinozaki and Shinozaki 1994; Stockinger et al. 1997).

The promoters of ABA-regulated genes contain a six-nucleotide sequence element referred to as the **ABA response element** (**ABRE**), which probably binds transcription factors involved in ABA-regulated gene activation (see Chapter 23). The promoters of some of these genes, which are regulated by osmotic stress in an ABA-dependent manner, also contain an alternative nine-nucleotide regulatory sequence element, the **dehydration response element** (**DRE**) which is recognized by an alternative set of proteins regulating transcription. Thus, the genes that are regulated by osmotic stresses appear to be regulated either by signal transduction pathways mediated by the action of ABA (**ABA-dependent genes**), or by an **ABA-independent**, osmotic stress–responsive signal transduction pathway.

There are at least two signaling pathways that have been implicated in the regulation of gene expression in an ABA-independent manner (Figure 26.9). *Trans*-acting transcription factors (called DREB1 and DREB2) that bind to the DRE elements discussed above are apparently activated by an ABA-independent signaling cascade while other ABA-independent, osmotic stress–responsive genes appear to be directly controlled by the so-called mitogen-activated protein (MAP) kinase signaling cascade of protein kinases (discussed in detail in Chapter 14 on the web site). Other changes in gene expression appear to be mediated via other mechanisms not involving DREBs.

This complexity and "cross talk" found in signaling cascades, exemplified here by both ABA-dependent and ABA-

TABLE 26.2
The five groups of late embryogenesis abundant (LEA) proteins found in plants

Group (family name)[a]	Protein(s) in the group	Structural characteristics and motifs	Functional information/ proposed function
Group 1 (D-19 family)	Cotton D-19 Wheat Em (early methionine-labeled protein) Sunflower Ha ds10 Barley B19	Conformation is predominantly random coil with some predicted short α helices Charged amino acids and glycine are abundant	Contains more water of hydration than typical globular proteins Overexpression confers water deficit tolerance on yeast cells
Group 2 (D-11 family) (also referred to as dehydrins)	Maize DHN1, M3, RAB17 Cotton D-11 Arabidopsis pRABAT1, ERD10, ERD14 *Craterostigma* pcC 27-04, pcC 6-19 Tomato pLE4, TAS14 Barley B8, B9, B17 Rice pRAB16A Carrot pcEP40	Variable structure includes α helix–forming lysine-rich regions The consensus sequence for group 2 dehydrins is EKKGIMDKIKELPG The number of times this consensus repeats per protein varies Often contains a poly(serine) region Often contains regions of variable length rich in polar residues and either Gly or Ala., and Pro	Often localized to the cytoplasm or nucleus More acidic members of the family are associated with the plasma membrane May act to stabilize macromolecules at low water potential
Group 3 (D-7 family)	Barley HVA1 (ABA-induced) Cotton D-7 Wheat pMA2005, pMA1949 *Craterostigma* pcC3-06	Eleven amino-acid consensus sequence motif TAQAAKEKAXE is repeated in the protein Contains apparent amphipathic α helices Dimeric protein	Transgenic plants expressing HVA1 demonstrate enhanced water deficit stress tolerance D-7 is an abundant protein in cotton embryos (estimated concentration 0.25 mM) Each putative dimer of D-7 may bind as many as ten inorganic phosphates and their counterions
Group 4 (D-95 family)	Soybean D-95 *Craterostigma* pcC27-45	Slightly hydrophobic N-terminal region is predicted to form amphipathic α helices	In tomato, a gene encoding a similar protein is expressed in response to nematode feeding
Group 5 (D-113 family)	Tomato LE25 Sunflower Hads11 Cotton D-113	Family members share sequence homology at the conserved N terminus N-terminal region is predicted to form α helices C-terminal domain is predicted to be a random coil of variable length and sequence Ala, Gly, and Thr are abundant in the sequence	Binds to membranes and/or proteins to maintain structure during stress Possibly functions in ion sequestration to protect cytosolic metabolism When LE25 is expressed in yeast, it confers salt and freezing tolerance D-113 is abundant in cottonseeds (up to 0.3 mM)

[a]The protein family names are derived from the cotton seed proteins that are most similar to the family.

Source: After Bray et al. 2000.

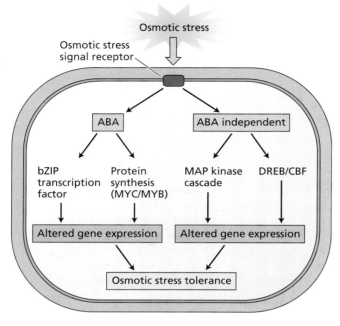

FIGURE 26.9 Signal transduction pathways for osmotic stress in plant cells. Osmotic stress is perceived by an as-yet unknown receptor in the plasma membrane that activates ABA-independent and an ABA-dependent signal transduction pathways. Protein synthesis participates in one of the ABA-dependent pathways involving MYC/MYB. The bZIP ABA-dependent pathway involves recognition of ABA-responsive elements in gene promoters. Two ABA-independent pathways, one involving the MAP kinase signaling cascade and the other involving DREB/CBF-related transcription factors have also been demonstrated. (After Shinozaki and Yamaguchi-Shinozaki 2000.)

Plant	Heat-killing temperature (°C)	Time of exposure
Nicotiana rustica (wild tobacco)	49–51	10 min
Cucurbita pepo (squash)	49–51	10 min
Zea mays (corn)	49–51	10 min
Brassica napus (rape)	49–51	10 min
Citrus aurantium (sour orange)	50.5	15–30 min
Opuntia (cactus)	>65	—
Sempervivum arachnoideum (succulent)	57–61	—
Potato leaves	42.5	1 hour
Pine and spruce seedlings	54–55	5 min
Medicago seeds (alfalfa)	120	30 min
Grape (ripe fruit)	63	—
Tomato fruit	45	—
Red pine pollen	70	1 hour
Various mosses		
Hydrated	42–51	—
Dehydrated	85–110	—

TABLE 26.3 Heat-killing temperatures for plants

Source: After Table 11.2 in Levitt 1980.

independent pathways, is typical of eukaryotic signaling. Such complexity reflects the wealth of interaction between gene expression and the physiological processes mediating adaptations to osmotic stress.

Heat Stress and Heat Shock

Most tissues of higher plants are unable to survive extended exposure to temperatures above 45°C. Nongrowing cells or dehydrated tissues (e.g., seeds and pollen) can survive much higher temperatures than hydrated, vegetative, growing cells (Table 26.3). Actively growing tissues rarely survive temperatures above 45°C, but some dry seeds can endure 120°C, and pollen grains of some species can endure 70°C. In general, only certain single-celled eukaryotes are known to be able to complete their life cycle at temperatures above 50°C, and only certain prokaryotes and some Archea can divide and grow above 60°C.

Periodic brief exposure to sublethal heat stresses often induces tolerance to otherwise lethal temperatures, a phenomenon referred to as **induced** or **acquired thermotolerance**. The mechanisms mediating induced thermotolerance will be discussed later in the chapter. As mentioned earlier, drought and temperature stress are interrelated; shoots of most C_3 and C_4 plants with access to abundant water supply are maintained below 45°C by evaporative cooling even at elevated ambient temperatures. However, if environmental conditions reduce evaporative cooling, either because of low water availability or high atmospheric relative humidity, tissue temperatures can increase. Emerging seedlings in moist soil may constitute an exception to this general rule. These seedlings may be exposed to greater heat stress than those in drier soils because wet, bare soil is typically darker and absorbs more solar radiation than drier soil. Additionally, not all plants are exposed to stressful temperatures in the same way, at the same rate, or for the same duration. Consequently, plants have evolved a complex set of responses to stressful high temperatures that are regulated by several possibly interrelated signaling pathways.

High leaf temperature and minimal evaporative cooling lead to heat stress

Many succulent higher plants that have CAM, such as *Opuntia* and *Sempervivum*, are adapted to high temperatures and can tolerate tissue temperatures of 60 to 65°C under conditions of intense solar radiation in summer (see Table 26.3). Because CAM plants keep their stomata closed during the day, they cannot cool by transpiration. Instead, they dissipate the heat from incident solar radiation by re-emission of long-wave (infrared) radiation and loss of heat by conduction and convection (see Chapter 9).

In contrast, typical, nonirrigated C_3 and C_4 plants rely on transpirational cooling to lower leaf temperature. In these plants, leaf temperature can readily rise 4 to 5°C above ambient air temperature in bright sunlight near midday, when soil water deficit causes partial stomatal closure or when high relative humidity reduces the potential for evaporative cooling. Increases in leaf temperature during the day can be pronounced in plants from arid and semiarid regions experiencing drought and high irradiance from sunshine. Heat stress independent of drought is also a potential danger in tropical regions and in greenhouses, where high humidity and poor air circulation can decrease the rate of leaf cooling. A moderate degree of heat stress slows growth of the whole plant. Some irrigated crops, such as cotton, use transpirational cooling to dissipate heat. In irrigated cotton, enhanced transpirational cooling is associated with higher agronomic yields (see **Web Topic 26.1**).

At high temperatures, photosynthesis is inhibited before respiration

Both photosynthesis and respiration are inhibited at high temperatures, but as temperature increases, photosynthetic rates drop before respiratory rates (Figure 26.10A and B). The temperature at which the amount of CO_2 fixed by photosynthesis equals the amount of CO_2 released by respiration in a given time interval is called the **temperature compensation point**.

At temperatures above the temperature compensation point, photosynthesis cannot replace the carbon used as a substrate for respiration. As a result, carbohydrate reserves decline, and fruits and vegetables lose sweetness. This imbalance between photosynthesis and respiration is one of the main reasons for the deleterious effects of high temperatures.

In the same plant the temperature compensation point is usually lower for shade leaves than for sun leaves that are exposed to light (and heat). Enhanced respiration rates relative to photosynthesis at high temperatures are more detrimental in C_3 plants than in C_4 or CAM plants, because the rates of both dark respiration and photorespiration are increased in C_3 plants at higher temperatures (see Chapter 8).

Plants adapted to cool temperatures acclimate poorly to high temperatures

The extent to which plants that are genetically adapted to a given temperature range can acclimate to a contrasting temperature range is illustrated by a comparison of the responses of two C_4 species: frosted orache (*Atriplex sabulosa*, family Chenopodiaceae) and Arizona honeysweet (*Tidestromia oblongifolia*, family Amaranthaceae).

A. sabulosa is native to the cool climate of coastal northern California, and *T. oblongifolia* is native to the very hot climate of Death Valley, California, where it grows in a tem-

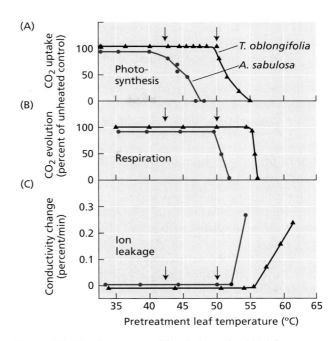

FIGURE 26.10 Response of frosted orache (*Atriplex sabulosa*) and Arizona honeysweet (*Tidestromia oblongifolia*) to heat stress. Photosynthesis (A) and respiration (B) were measured on attached leaves, and ion leakage (C) was measured in leaf slices submerged in water. At the beginning of the experiment, control rates were measured at a noninjurious 30°C. Attached leaves were then exposed to the indicated temperatures for 15 minutes and returned to the initial control conditions before the rates were recorded. Arrows indicate the temperature thresholds for inhibition of photosynthesis in each of the two species. Photosynthesis, respiration, and membrane permeability were all more sensitive to heat damage in *A. sabulosa* than in *T. oblongifolia*. In both species, however, photosynthesis was more sensitive to heat stress than were either of the other two processes, and photosynthesis was completely inhibited at temperatures that were noninjurious to respiration. (After Björkman et al. 1980.)

perature regime that is lethal for most plant species. When these species were grown in a controlled environment and their growth rates were recorded as a function of temperature, *T. oblongifolia* barely grew at 16°C, while *A. sabulosa* was at 75% of its maximum growth rate. In contrast, the growth rate of *A. sabulosa* began to decline between 25 and 30°C, and growth ceased at 45°C, the temperature at which *T. oblongifolia* growth showed a maximum (see Figure 26.10A) (Björkman et al. 1980). Clearly, neither species could acclimate to the temperature range of the other.

This same phenomenon is observable with crop plants. Crop species and varieties that are adapted to cool climates typically are not highly productive in high heat environments, and crops that are adapted to grow in high heat environments are typically not as productive in lower temperature environments. The exact cellular mechanisms that generate this behavior are presently not well understood, but it is likely that many of the mechanisms of thermotolerance discussed in this chapter play a critical role in this behavior. However, the specific mechanisms involved vary from one species or variety to another and for each specific environment. Thus, the ecological adaptation of plants to temperature is a complex interaction of a number of cellular mechanisms of stress tolerance and not simply explained by any one cellular mechanism alone.

Temperature affects membrane stability

The stability of various cellular membranes is important during high-temperature stress, just as it is during chilling and freezing. Excessive fluidity of membrane lipids at high temperatures is correlated with loss of physiological function. In oleander (*Nerium oleander*), acclimation to high temperatures is associated with a greater degree of saturation of fatty acids in membrane lipids, which makes the membranes less fluid (Raison et al. 1982). Mutants of *Arabidopsis thaliana* with reduced amounts of omega-3 fatty acid desaturases show increased thermotolerance of photosynthesis, presumably because the degree of saturation of chloroplast lipids is increased.

At high temperatures there is a decrease in the strength of hydrogen bonds and electrostatic interactions between polar groups of proteins within the aqueous phase of the membrane. High temperatures thus modify membrane composition and structure and can cause leakage of ions (see Figure 26.10C). Membrane disruption also causes the inhibition of processes such as photosynthesis and respiration that depend on the activity of membrane-associated electron carriers and enzymes.

Photosynthesis is especially sensitive to high temperature (see Chapter 9). In their study of *Atriplex* and *Tidestromia*, O. Björkman and colleagues (1980) found that electron transport in photosystem II was more sensitive to high temperature in the cold-adapted *A. sabulosa* than in the heat-adapted *T. oblongifolia*. The enzymes ribulose-1,5-bisphosphate carboxylase, NADP–glyceraldehyde-3-phosphate dehydrogenase, and phosphoenolpyruvate carboxylase were less stable at high temperatures in *A. sabulosa* compared to *T. oblongifolia*. However, the temperatures at which these enzymes began to denature and lose activity were distinctly higher than the temperatures at which photosynthesis began to decline. These results suggest that early stages of heat injury to photosynthesis are more directly related to changes in membrane properties and to uncoupling of the energy transfer mechanisms in chloroplasts than to a general denaturation of proteins.

Several adaptations protect leaves against excessive heating

In environments with intense solar radiation and high temperatures, plants avoid excessive heating of their leaves by decreasing their absorption of solar radiation. This adaptation is important in warm, sunny environments in which a transpiring leaf is near its upper limit of temperature tolerance. In these conditions, any further warming arising from decreased evaporation of water or increased energy absorption can damage the leaf.

Both drought resistance and heat resistance depend on the same adaptations: reflective leaf hairs and leaf waxes; leaf rolling and vertical leaf orientation; and growth of small, highly dissected leaves to minimize the boundary layer thickness and thus maximize convective and conductive heat loss (see Chapters 4 and 9). Some desert shrubs—for example, white brittlebush (*Encelia farinosa*, family Compositae)—have dimorphic leaves to avoid excessive heating: Green, nearly hairless leaves found in the winter are replaced by white, pubescent leaves in the summer.

At higher temperatures, plants produce protective proteins

In response to sudden, 5 to 10°C rises in temperature, plants produce a unique set of proteins referred to as **heat shock proteins** (**HSPs**). Most HSPs function to help cells withstand heat stress by acting as molecular chaperones. Heat stress causes many cell proteins that function as enzymes or structural components to become unfolded or misfolded, thereby leading to loss of proper enzyme structure and activity. Such misfolded proteins often aggregate and precipitate, creating serious problems within the cell. HSPs act as molecular chaperones and serve to attain a proper folding of misfolded, aggregated proteins and to prevent misfolding of proteins. This facilitates proper cell functioning at suddenly elevated, stressful temperatures.

Heat shock proteins were discovered in the fruit fly (*Drosophila melanogaster*) and appear to be ubiquitous, having since been identified in other animals, and in humans, as well as in plants, fungi, and microorganisms. For example, when soybean seedlings are suddenly shifted from 25 to 40°C (just below the lethal temperature), synthesis of the set of mRNAs and proteins commonly found in the cell is suppressed, while transcription and translation of a set of 30 to 50 other proteins (HSPs) is enhanced. New mRNA transcripts for HSPs can be detected 3 to 5 minutes after heat shock (Sachs and Ho 1986).

Although plant HSPs were first identified in response to sudden changes in temperature (25 to 40°C) that rarely occur in nature, HSPs are also induced by more gradual rises in temperature that are representative of the natural environment, and they occur in plants under field conditions. Some HSPs are found in normal, unstressed cells, and some essential cellular proteins are homologous to HSPs but do not increase in response to thermal stress (Vierling 1991).

Plants and most other organisms make HSPs of different sizes in response to temperature increases (Table 26.4). The molecular masses of the HSPs range from 15 to 104 kilodaltons (kDa), and they can be grouped into five classes based on size. Different HSPs are localized to the nucle-

TABLE 26.4
The five classes of heat shock proteins found in plants

HSP class	Size (kDa)	Examples (Arabidopsis / prokaryotic)	Cellular location
HSP100	100–114	AtHSP101 / ClpB, ClpA/C	Cytosol, mitochondria, chloroplasts
HSP90	80–94	AtHSP90 / HtpG	Cytosol, endoplasmic reticulum
HSP70	69–71	AtHSP70 / DnaK	Cytosol/nucleus, mitochondria, chloroplasts
HSP60	57–60	AtTCP-1 / GroEL, GroES	Mitochondria, chloroplasts
smHSP	15–30	Various AtHSP22, AtHSP20, AtHSP18.2, AtHSP17.6 / IBPA/B	Cytosol, mitochondria, chloroplasts, endoplasmic reticulum

Source: After Boston et al. 1996.

us, mitochondria, chloroplasts, endoplasmic reticulum, and cytosol. Members of the HSP60, HSP70, HSP90, and HSP100 groups act as molecular chaperones, involving ATP-dependent stabilization and folding of proteins, and the assembly of oligomeric proteins. Some HSPs assist in polypeptide transport across membranes into cellular compartments. HSP90s are associated with hormone receptors in animal cells and may be required for their activation, but there is no comparable information for plants.

Low-molecular-weight (15–30 kDa) HSPs are more abundant in higher plants than in other organisms. Whereas plants contain five to six classes of low-molecular-weight HSPs, other eukaryotes show only one class (Buchanan et al. 2000). The different classes of 15- to 30-kDa HSPs (smHSPs) in plants are distributed in the cytosol, chloroplasts, ER and mitochondria. These small HSPs may bind to aggregating proteins and keep them accessible for refolding by HSP70 and possibly HSP100 chaperones.

Cells that have been induced to synthesize HSPs show improved thermal tolerance and can tolerate exposure to temperatures that are otherwise lethal. Some of the HSPs are not unique to high-temperature stress. They are also induced by widely different environmental stresses or conditions, including water deficit, ABA treatment, wounding, low temperature, and salinity. Thus, cells previously exposed to one stress may gain cross-protection against another stress. Such is the case with tomato fruits, in which heat shock (48 hours at 38°C) has been observed to promote HSP accumulation and to protect cells for 21 days from chilling at 2°C.

A transcription factor mediates HSP accumulation

All cells seem to contain molecular chaperones that are constitutively expressed and function like HSPs. These chaperones are called **heat shock cognate proteins**. However, when cells are subjected to a stressful but nonlethal heat episode, the synthesis of HSPs dramatically increases while the continuing translation of other proteins is dramatically lowered or ceases. This heat shock response appears to be mediated by one or more signal transduction pathways, one of which involves a specific transcription factor (**heat shock factor**, or **HSF**) that acts on the transcription of HSP mRNAs.

In the absence of heat stress, HSF exists in the cytosol as monomers bound to HSP70 and/or HSP90 that are incapable of binding to DNA and directing transcription (Figure 26.11). Stress causes HSF monomers to associate into trimers that are then able to bind to specific sequence elements in DNA referred to as **heat shock elements** (**HSEs**). Once bound to the HSE, the trimeric HSF is phosphorylated and promotes the transcription of HSP mRNAs. HSP70 subsequently binds to HSF, leading to the dissociation of the HSF/HSE complex, and the HSF is recycled to the monomeric HSF form. Thus, by the action of HSF, HSPs accumulate until they become abundant enough to bind to HSF, leading to the cessation of HSP mRNA production. Control of the process involves competition between unfolded proteins in the cell and HSFs for HSP70 molecules inside the cell.

HSPs mediate tolerance to high temperatures

Conditions that induce heat stress tolerance in plants closely match those that induce the accumulation of HSPs, but that correlation alone does not prove that HSPs play an essential role in acclimation to heat stress. More conclusive experiments show that expression of an activated HSF induces constitutive synthesis of HSPs and increases the thermotolerance of Arabidopsis. Studies with Arabidopsis plants containing an antisense DNA sequence that reduces HSP70 synthesis showed that the high-temperature extreme at which the plants could survive was reduced by 2°C compared with controls, although the mutant plants grew normally at optimum temperatures (Lee and Schoeffl 1996). Other studies with both Arabidopsis mutants (Hong and Vierling 2000) and transgenic plants (Queitsch et al. 2000) demonstrate that at least HSP101 is a critical component of both induced and constitutive thermotolerance in plants. Presumably, failure to synthesize the entire range of HSPs that are usually induced in the plant would lead to a much more dramatic loss of thermotolerance. However, the number, nature, and complexity of the HSP and HSF systems in plants have made it difficult to test this hypothesis.

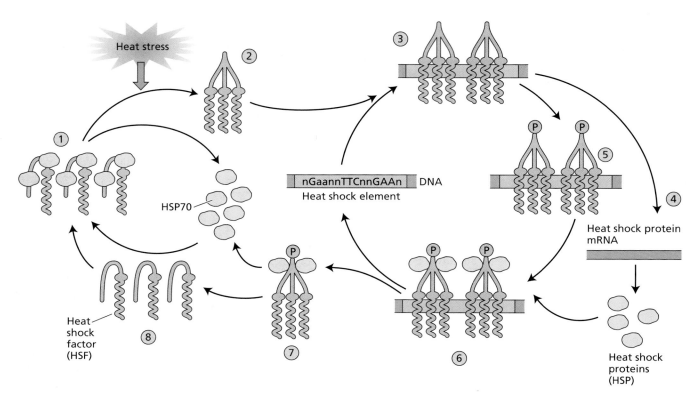

FIGURE 26.11 The heat shock factor cycle activates the synthesis of heat shock protein mRNAs. In nonstressed cells, HSF normally exists in a monomeric state (1) associated with HSP70 proteins. Upon the onset of an episode of heat stress, HSP70 dissociates from HSF, which subsequently trimerizes (2). Active trimers bind to heat shock elements in the promoter of heat shock protein genes (3), and activate the transcription of HSP mRNAs leading to the translation of HSPs, among which are HSP70 (4). The HSF trimers associated with the HSE are phosphorylated (5), facilitating the binding of HSP70 to the phosphorylated trimers (6). The HSP70 trimer complex (7) dissociates from the HSE and disassembles and dephosphorylates into HSF monomers (8), which subsequently bind HSP to re-form the resting HSP70/HSF complex. (After Bray et al. 2000.)

Several signaling pathways mediate thermotolerance responses

It is clear that a number of additional signaling pathways that likely mediate thermotolerance through mechanisms other than HSP production also exist. For example, both ABA and salicylic acid (SA) treatment can lead to better survival under heat stress. Several ABA- and SA-signaling mutants of Arabidopsis are more sensitive to heat stress and demonstrate an inability to acquire thermotolerance, but show no alterations in HSP101 and small HSPs (Larkindale et al. 2005). This suggests that ABA and SA increase thermotolerance through an HSP-independent pathway. Ethylene biosynthesis is increased during heat stress. Ethylene pretreatment induces a low level of thermotolerance in some plants, and mutants in the ethylene response pathway demonstrate increased sensitivity to heat. However, there is no evidence that HSP production is influenced by ethylene, and the mechanism(s) by which ethylene mediates thermotolerance are unknown at this time.

Heat stress leads to oxidative damage of plant tissues, presumably through the production of reactive oxygen species. It has only recently been recognized that H_2O_2 can act as a signaling molecule increasing plant thermotolerance and leading to the production of HSPs.

Calcium signaling appears to play a role in thermotolerance, but it is not well understood. Pharmacological experiments using calcium chloride and various calcium inhibitors have demonstrated that treatments expected to increase cytosolic calcium increase thermotolerance, while treatments expected to reduce cytosolic calcium reduce tolerance to high temperatures. Oscillations of cytosolic calcium levels have been implicated as a signaling mechanism in guard cells (see Chapter 23) and may be involved in thermotolerance signaling, since calcium is intimately involved in the production of γ-aminobutyric acid (GABA), a nonprotein amino acid that accumulates in response to heat stress (Figure 26.12). **Web Topic 26.2** summarizes recent studies that focus on the role GABA plays in the integration of metabolic responses to stress.

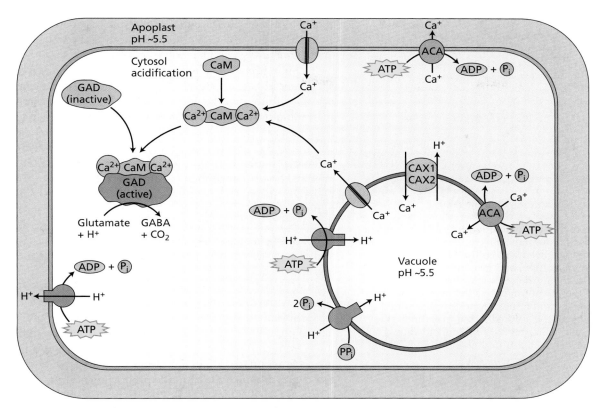

FIGURE 26.12 Regulation of GABA accumulation during heat stress. Heat stress causes a reduction in cytosolic pH from the normal, slightly alkaline value, probably by inhibiting ATPases and pyrophosphatases that pump protons across the plasma membrane or into the vacuole. Additionally, heat stress alters calcium homeostasis inside the cell by affecting the influx of calcium into the cytosol through either plasma membrane or vacuolar calcium channels, or by action on efflux ATPases or proton cotransporters. This increase in cytosolic calcium leads to the activation of calmodulin (CaM), which binds to glutamate decarboxylase (GAD) and converts it from the inactive to the active form. Glutamate is then converted to GABA, and protons are consumed in the process, mediating an increase in cytosolic pH. CAX1 and CAX2 are transport proteins; ACA, Ca^{2+}-ATPase.

Chilling and Freezing

Chilling temperatures are too low for normal growth but not low enough for ice to form. Typically, tropical and subtropical species are susceptible to chilling injury. Among crops, corn, *Phaseolus* bean, rice, tomato, cucumber, sweet potato, and cotton are chilling sensitive. *Passiflora*, *Coleus*, and *Gloxinia* are examples of susceptible ornamentals.

When plants growing at relatively warm temperatures (25 to 35°C) are cooled to 10 to 15°C, **chilling injury** occurs: Growth is slowed, discoloration or lesions appear on leaves, and the foliage looks soggy, as if soaked in water for a long time. If roots are chilled, the plants may wilt.

Species that are generally sensitive to chilling can show appreciable variation in their response to chilling temperatures. Genetic adaptation to the colder temperatures associated with high altitude can improve chilling resistance (Figure 26.13). In addition, resistance often increases if plants are first hardened (acclimated) by exposure to cool, but noninjurious, temperatures. Chilling damage thus can be minimized if exposure is slow and gradual. Sudden exposure to temperatures near 0°C, called *cold shock*, greatly increases the chances of injury. **Freezing injury**, on the other hand, occurs at temperatures below the freezing point of water. Full induction of tolerance to freezing, as with chilling, requires a period of acclimation at cold temperatures.

In the discussion that follows we will examine how chilling injury alters membrane properties, how ice crystals damage cells and tissues, and how ABA, gene expression, and protein synthesis mediate acclimation to freezing.

Membrane properties change in response to chilling injury

Leaves from plants injured by chilling show inhibition of photosynthesis, slower carbohydrate translocation, lower respiration rates, inhibition of protein synthesis, and increased degradation of existing proteins. All of these responses appear to depend on a common primary mechanism involving loss of membrane function during chilling.

For instance, solutes leak from the leaves of chilling-sensitive conch apple (*Passiflora maliformis*) floated on water at

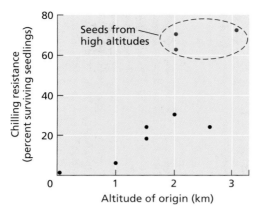

FIGURE 26.13 Survival at low temperature of seedlings of different populations of tomato collected from different altitudes in South America. Seed was collected from wild tomato (*Lycopersicon hirsutum*) and grown in the same greenhouse at 18 to 25°C. All seedlings were then chilled for 7 days at 0°C and then kept for 7 days in a warm growth room, after which the number of survivors was counted. Seedlings from some of the seed collected from high altitudes showed greater resistance to chilling (cold shock) than those from seed collected from lower altitudes. (After Patterson et al. 1978.)

0°C, but not from those of chilling-resistant passionflower (*Passiflora caerulea*). Loss of solutes to the water reflects damage to the plasma membrane and possibly also to the tonoplast. In turn, inhibition of photosynthesis and of respiration reflects injury to chloroplast and mitochondrial membranes.

Why are membranes affected by chilling? Plant membranes consist of a lipid bilayer interspersed with proteins and sterols (see Chapters 1 and 11). The physical properties of the lipids greatly influence the activities of the integral membrane proteins, including H$^+$-ATPases, carriers, and channel-forming proteins that regulate the transport of ions and other solutes (see Chapter 6), as well as the transport of enzymes on which metabolism depends.

In chilling-sensitive plants, the lipids in the bilayer have a high percentage of saturated fatty acid chains, and membranes with this composition tend to solidify into a semicrystalline state at a temperature well above 0°C. Keep in mind that saturated fatty acids that have no double bonds, and lipids containing *trans*-monounsaturated fatty acids solidify at higher temperatures than do membranes composed of lipids that contain unsaturated fatty acids.

As the membranes become less fluid, their protein components can no longer function normally. The result is inhibition of H$^+$-ATPase activity, of solute transport into and out of cells, of energy transduction (see Chapters 7 and 11), and of enzyme-dependent metabolism. In addition, chilling-sensitive leaves exposed to high photon fluxes and chilling temperatures are photoinhibited (see Chapter 7), causing acute damage to the photosynthetic machinery.

Membrane lipids from chilling-resistant plants often have a greater proportion of unsaturated fatty acids than those from chilling-sensitive plants (Table 26.5), and during acclimation to cool temperatures the activity of desaturase enzymes increases and the proportion of unsaturated lipids rises (Williams et al. 1988; Palta et al. 1993). This modification lowers the temperature at which the membrane lipids begin a gradual phase change from fluid to semicrystalline and allows membranes to remain fluid at lower temperatures. Thus, desaturation of fatty acids provides some protection against damage from chilling.

The importance of membrane lipids to tolerance of low temperatures has been demonstrated by work with mutant and transgenic plants in which the activity of particular enzymes led to a specific change in membrane lipid composition independent of acclimation to low temperature. For example, Arabidopsis was transformed with a gene

TABLE 26.5
Fatty acid composition of mitochondria isolated from chilling-resistant and chilling-sensitive species

	Percent weight of total fatty acid content					
	Chilling-resistant species			Chilling-sensitive species		
Major fatty acids[a]	Cauliflower bud	Turnip root	Pea shoot	Bean shoot	Sweet potato	Maize shoot
Palmitic (16:0)	21.3	19.0	17.8	24.0	24.9	28.3
Stearic (18:0)	1.9	1.1	2.9	2.2	2.6	1.6
Oleic (18:0)	7.0	12.2	3.1	3.8	0.6	4.6
Linoleic (18:2)	16.1	20.6	61.9	43.6	50.8	54.6
Linolenic (18:3)	49.4	44.9	13.2	24.3	10.6	6.8
Ratio of unsaturated to saturated fatty acids	3.2	3.9	3.8	2.8	1.7	2.1

[a]Shown in parentheses are the number of carbon atoms in the fatty acid chain and the number of double bonds.
Source: After Lyons et al. 1964.

from *Escherichia coli* that raised the proportion of high-melting-point (saturated) membrane lipids. This gene greatly increased the chilling sensitivity of the transformed plants.

Similarly, the *fab1* mutants of Arabidopsis have increased levels of saturated fatty acids, particularly 16:0 (see Table 26.5, and Tables 11.3 and 11.4). During a period of 3 to 4 weeks at chilling temperatures, photosynthesis and growth were gradually inhibited, and exposure to chilling temperatures eventually destroyed the chloroplasts of this mutant. At nonchilling temperatures, the mutant grew as well as wild-type controls did (Wu et al. 1997). (For additional transformation examples, see **Web Topic 26.3**.)

Ice crystal formation and protoplast dehydration kill cells

The ability to tolerate freezing temperatures under natural conditions varies greatly among tissues. Seeds, other partly dehydrated tissues, and fungal spores can be kept indefinitely at temperatures near absolute zero (0 K, or −273°C), indicating that these very low temperatures are not intrinsically harmful.

Fully hydrated, vegetative cells can also retain viability if they are cooled very quickly to avoid the formation of large, slow-growing ice crystals that would puncture and destroy subcellular structures. Ice crystals that form during very rapid freezing are too small to cause mechanical damage. Conversely, rapid warming of frozen tissue is required to prevent the growth of small ice crystals into crystals of a damaging size, or to prevent loss of water vapor by sublimation, both of which take place at intermediate temperatures (−100 to −10°C).

Under natural conditions, cooling of intact, multicellular plant organs is never fast enough to limit crystal formation in fully hydrated cells to only small, harmless ice crystals. Ice usually forms first within the intercellular spaces, and in the xylem vessels, along which the ice can quickly propagate. This ice formation is not lethal to hardy plants, and the tissue recovers fully if warmed. However, when plants are exposed to freezing temperatures for an extended period, the growth of extracellular ice crystals results in the movement of liquid water from the protoplast to the extracellular ice, causing excessive dehydration (for a detailed description of this process, see **Web Topic 26.4**).

During rapid freezing, the protoplast, including the vacuole, supercools; that is, the cellular water remains liquid even at temperatures several degrees below its theoretical freezing point. Several hundred molecules are needed for an ice crystal to begin forming. The process by which these hundreds of water molecules start to form a stable ice crystal is called **ice nucleation**, and it strongly depends on the properties of the involved surfaces. Some large polysaccharides and proteins facilitate ice crystal formation and are called ice nucleators.

Some ice nucleation proteins made by bacteria appear to facilitate ice nucleation by aligning water molecules along repeated amino acid domains within the protein. In plant cells, ice crystals begin to grow from endogenous ice nucleators, and the resulting, relatively large intracellular ice crystals cause extensive damage to the cell and are usually lethal.

Limitation of ice formation contributes to freezing tolerance

Several specialized plant proteins may help limit the growth of ice crystals by a noncolligative mechanism—that is, an effect that does not depend on the lowering of the freezing point of water by the presence of solutes. These *antifreeze proteins* are induced by cold temperatures, and they bind to the surfaces of ice crystals to prevent or slow further crystal growth.

In rye (*Secale cereale*) leaves, antifreeze proteins are localized in the epidermal cells and cells surrounding the intercellular spaces, where they can inhibit the growth of extracellular ice. Plants and animals may use similar mechanisms to limit ice crystals: A cold-inducible gene identified in Arabidopsis has DNA homology to a gene that encodes the antifreeze protein in fishes such as winter flounder. Antifreeze proteins are discussed in more detail later in the chapter.

Sugars and some of the cold-induced proteins are suspected to have cryoprotective (*cryo-* = "cold") effects; they stabilize proteins and membranes during dehydration induced by low temperature. For example, in winter wheat, as sucrose concentrations increase, the level of freezing tolerance increases. Sucrose predominates among the soluble sugars associated with freezing tolerance that function in a colligative fashion, but in some species raffinose, fructans, sorbitol, or mannitol serve the same function.

During cold acclimation of winter cereals, soluble sugars accumulate in the cell walls, where they may help restrict the growth of ice. A cryoprotective glycoprotein has been isolated from leaves of cold-acclimated cabbage (*Brassica oleracea*). In vitro, the protein protects thylakoids isolated from nonacclimated spinach (*Spinacia oleracea*) against damage from freezing and thawing.

Some woody plants can acclimate to very low temperatures

When in a dormant state, some woody plants are extremely resistant to low temperatures. Resistance is determined in part by previous acclimation to cold, but genetics plays an important role in determining the degree of tolerance to low temperatures. Native species of *Prunus* (cherry, plum, and other pit fruits) from northern cooler climates in North America are hardier after acclimation than those from milder climates. When the species were tested together in the laboratory, those with a northern geographic distribution showed a greater ability to avoid intracellular ice formation, underscoring distinct genetic differences (Burke and Stushnoff 1979).

Under natural conditions, woody species acclimate to cold in two distinct stages (Weiser 1970):

1. In the first stage, hardening is induced in the early autumn by exposure to short days and nonfreezing chilling temperatures, which combine to stop growth. A diffusible factor that promotes acclimation (probably ABA) moves in the phloem from leaves to overwintering stems and may be responsible for the changes. During this period, woody species also withdraw water from the xylem vessels, thereby preventing the stem from splitting in response to the expansion of water during later freezing. Cells in this first stage of acclimation can survive temperatures well below 0°C, but they are not fully hardened.

2. In the second stage, direct exposure to freezing is the stimulus; no known translocatable factor can confer the hardening resulting from exposure to freezing. When fully hardened, the cells can tolerate exposure to temperatures of –50 to –100°C.

In many species of the hardwood forests of southeastern Canada and the eastern United States, acclimation to freezing involves the suppression of ice crystal formation at temperatures far below the theoretical freezing point (see **Web Topic 26.3** for details). This *deep supercooling* is seen in species such as oak, elm, maple, beech, ash, walnut, hickory, rose, rhododendron, apple, pear, peach, and plum (Burke and Stushnoff 1979). Deep supercooling also takes place in the stem and leaf tissue of tree species such as Engelmann spruce (*Picea engelmannii*) and subalpine fir (*Abies lasiocarpa*) growing in the Rocky Mountains of Colorado.

Resistance to freezing is quickly weakened once growth has resumed in the spring (Becwar et al. 1981). Stem tissues of subalpine fir, which undergo deep supercooling and remain viable to below –35°C in May, lose their ability to suppress ice formation in June and can be killed at –10°C.

Cells can supercool only to about –40°C, at which temperature ice forms spontaneously. Spontaneous ice formation sets the *low-temperature limit* at which many alpine and subarctic species that undergo deep supercooling can survive. It also explains why the altitude of the timberline in mountain ranges is at or near the –40°C minimum isotherm.

The cell protoplast suppresses ice nucleation when undergoing deep supercooling. In addition, the cell wall acts as a barrier both to the growth of ice from the intercellular spaces into the wall, and to the loss of liquid water from the protoplast to the extracellular ice, which is driven by a steep vapor pressure gradient (Wisniewski and Arora 1993).

Many flower buds (e.g., grape, blueberry, peach, azalea, and flowering dogwood) survive the winter by deep supercooling, and serious economic losses, particularly of peach, can result from the decline in freezing tolerance of the flower buds in the spring. The cells then no longer supercool, and ice crystals that form extracellularly in the bud scales draw water from the apical meristem, killing the floral apex by dehydration.

The floral buds of apple and pear, the vegetative buds of all temperate fruit trees, and the living cells in their bark do not supercool, but they resist dehydration during extracellular ice formation. Resistance to cellular dehydration is highly developed in woody species that are subject to average annual temperature minima below –40°C, particularly species found in northern Canada, Alaska, northern Europe, and Asia.

Ice formation starts at –3 to –5°C in the intercellular spaces, where the crystals continue to grow, fed by the gradual withdrawal of water from the protoplast, which remains unfrozen. Resistance to freezing temperatures depends on the capacity of the extracellular spaces to accommodate the volume of growing ice crystals and on the ability of the protoplast to withstand dehydration.

This restriction of ice crystal formation to extracellular spaces, accompanied by gradual protoplast dehydration, may explain why some woody species that are resistant to freezing are also resistant to water deficit during the growing season. Species of willow (*Salix*), white birch (*Betula papyrifera*), quaking aspen (*Populus tremuloides*), pin cherry (*Prunus pennsylvanica*), chokecherry (*Prunus virginiana*), and lodgepole pine (*Pinus contorta*) tolerate very low temperatures by limiting the formation of ice crystals to the extracellular spaces. However, acquisition of resistance depends on slow cooling and gradual extracellular ice formation and protoplast dehydration. Sudden exposure to very cold temperatures before full acclimation causes intracellular freezing and cell death.

Some bacteria living on leaf surfaces increase frost damage

When leaves are cooled to temperatures in the –3 to –5°C range, the formation of ice crystals on the surface (frost) is accelerated by certain bacteria that naturally inhabit the leaf surface, such as *Pseudomonas syringae* and *Erwinia herbicola*, which act as ice nucleators. When artificially inoculated with cultures of these bacteria, leaves of frost-sensitive species freeze at warmer temperatures than leaves that are bacteria free (Lindow et al. 1982). The surface ice quickly spreads to the intercellular spaces within the leaf, leading to cellular dehydration.

Bacterial strains can be genetically modified so that they lose their ice-nucleating characteristics. Such strains have been used commercially in foliar sprays of valuable frost-sensitive crops such as strawberry to compete with native bacterial strains and thus minimize the number of potential ice nucleation points.

Acclimation to freezing involves ABA and protein synthesis

In seedlings of alfalfa (*Medicago sativa* L.), tolerance to freezing at –10°C is greatly improved by previous exposure to cold (4°C) or by treatment with exogenous ABA without exposure to cold. These treatments cause changes in the pattern of newly synthesized proteins that can be resolved on two-dimensional gels. Some of the changes are unique to the particular treatment (cold or ABA), but some of the newly synthesized proteins induced by cold appear to be the same as those induced by ABA (see Chapter 23) or by mild water deficit.

Protein synthesis is necessary for the development of freezing tolerance, and several distinct proteins accumulate during acclimation to cold, as a result of changes in gene expression (Guy 1999). Isolation of the genes for these proteins reveals that several of the proteins that are induced by low temperature share homology with the RAB/LEA/DHN (RESPONSIVE TO ABA, LATE EMBRYO ABUNDANT, and DEHYDRIN, respectively) protein family. As described earlier in the section on gene regulation by osmotic stress, these proteins accumulate in tissues exposed to different stresses, such as osmotic stress. These proteins appear to have both osmoprotective and cryoprotective functions, although the exact mechanisms of cryo- and osmoprotection are not well understood.

ABA appears to have a role in inducing freezing tolerance. Winter wheat, rye, spinach, and *Arabidopsis thaliana* are all cold-tolerant species that increase their freezing tolerance when they are hardened by water shortages, a condition that increases endogenous ABA in the leaves. Plants develop freezing tolerance at nonacclimating temperatures when treated with exogenous ABA. Many of the genes or proteins expressed at low temperatures or under water deficit are also inducible by ABA under nonacclimating conditions. All these findings support a role of ABA in tolerance to freezing.

Mutants of Arabidopsis that are insensitive to ABA (*abi1*) or are ABA deficient (*aba1*) are unable to undergo low-temperature acclimation to freezing. Only in *aba1*, however, does exposure to ABA restore the ability to develop freezing tolerance (Mantyla et al. 1995). On the other hand, not all the genes induced by low temperature are ABA dependent, and it is not yet clear whether expression of ABA-induced genes is critical for the full development of freezing tolerance. For instance, research on the tolerance of rye crowns to freezing has found that the lethal temperature for 50% of the crowns (LT_{50}) is −2 to −5°C for controls grown at 25°C, −8°C for ABA-treated crowns, and −28°C after acclimation at 2°C.

Clearly, exogenous ABA cannot confer the same freezing acclimation that exposure to low temperatures does. Cell cultures of bromegrass (*Bromus inermis*) show a more dramatic induction of freezing tolerance when treated with ABA: Whereas controls grown at 25°C could survive to −9°C, 7 days of exposure to ABA improved the freezing tolerance to −40°C (Gusta et al. 1996).

Typically, a minimum of several days of exposure to cool temperatures is required for freezing resistance to be induced fully. Potato requires 15 days of exposure to cold. On the other hand, when rewarmed, plants lose their freezing tolerance rapidly, and they can become susceptible to freezing once again in 24 hours.

The need for cool temperatures to induce acclimation to chilling or freezing, and the rapid loss of acclimation upon warming, explain the susceptibility of plants in the southern United States (and similar climatic zones with highly variable winters) to extremes of temperature in the winter months, when air temperature can drop from greater than 20°C to below 0°C in a few hours.

Numerous genes are induced during cold acclimation

Expression of certain genes and synthesis of specific proteins are common to both heat and cold stress, but some aspects of cold-inducible gene expression differ from that produced by heat stress (Thomashow 2001). Whereas during cold episodes the synthesis of "housekeeping" proteins (proteins made in the absence of stress) is not substantially down-regulated, during heat stress housekeeping-protein synthesis is essentially shut down.

The synthesis of several heat shock proteins that can act as molecular chaperones is up-regulated under cold stress in the same way that it is during heat stress. This suggests that protein destabilization accompanies both heat and cold stress, and that mechanisms for stabilizing protein structure during both heat and cold episodes are important for survival.

Another important class of proteins whose expression is up-regulated by cold stress is the **antifreeze proteins**. Antifreeze proteins were first discovered in fishes that live in water under the polar ice caps. As discussed earlier, these proteins have the ability to inhibit ice crystal growth in a noncolligative manner, thus preventing freeze damage at intermediate freezing temperatures. Antifreeze proteins confer to aqueous solutions the property of *thermal hysteresis* (transition from liquid to solid is promoted at a lower temperature than is transition from solid to liquid), and thus they are sometimes referred to as **thermal hysteresis proteins** (THPs).

Several types of cold-induced, antifreeze proteins have been discovered in cold-acclimated winter-hardy monocots. When the specific genes coding for these proteins were cloned and sequenced, it was found that all antifreeze proteins belong to a class of proteins such as endochitinases and endoglucanases, which are induced upon infection of different pathogens. These proteins, called **pathogenesis-related (PR) proteins**, are thought to protect plants against pathogens. It thus appears that at least in monocots, the dual role of these proteins as antifreeze and pathogenesis-related proteins might protect plant cells against both cold stress and pathogen attack.

Another group of proteins found to be associated with osmotic stress (see the discussion earlier in this chapter) are also up-regulated during cold stress. This group includes proteins involved in the synthesis of *osmolytes*, proteins for membrane stabilization, and the LEA proteins. Because the formation of extracellular ice crystals generates significant osmotic stresses inside cells, coping with freezing stress also requires the means to cope with osmotic stress.

A transcription factor regulates cold-induced gene expression

More than 100 genes are up-regulated by cold stress. Because cold stress is clearly related to ABA responses and to osmotic stress, not all the genes up-regulated by cold stress necessarily need to be associated with cold tolerance, but many of them likely are. Many cold stress–induced genes are activated by transcriptional activators called **C-repeat binding factors** (**CBF1**, **CBF2**, **CBF3**; also called DREB1b, DREB1c, and DREB1a, respectively) (Shinozaki and Yamaguchi-Shinozaki 2000).

The CBF proteins are a series of three cold-induced transcription factors that bind to **CRT/DRE elements** (C-repeat/dehydration-responsive, ABA-independent sequence elements) in gene promoter sequences, which were discussed earlier in the chapter. CBF proteins are involved in the transcriptional response of the CBF regulon, a collection of numerous cold- and osmotic stress–regulated genes whose expression is regulated by the CBF proteins (Fowler et al. 2005). The CBF regulon contains numerous proteins involved in stress tolerance and metabolism of cryoprotective compounds, as well as additional transcriptional activators and signal transduction proteins likely leading to the induction of additional proteins not directly containing the CRT/DRE elements. CBF1/DREB1b is unique in that it is specifically induced by cold stress and not by osmotic or salinity stress, whereas the DRE-binding elements of the DREB2 type (discussed earlier in the section on osmotic stresses) are induced only by osmotic and salinity stresses and not by cold.

The cold-induced expression of all three CBF proteins appears to be controlled by a set of transcription factors called ICE (*i*nducer of *C*BF *e*xpression). ICE transcription factors do not appear to be induced by cold, but rather are made in plants at warm temperatures, and ICE proteins are posttranscriptionally activated by a cold-responsive signal transduction pathway, permitting increased expression of CBF proteins, but the precise signaling pathway(s) of cold perception and signaling leading to the activation of ICE have not been elucidated. The promoters of all three CBF genes contain two critical sequence elements required for cold-induced expression of CBF proteins, but only one ICE transcription factor (ICE1) regulating the expression of CBF3 has been elucidated to date. ICE1 is a MYC-like transcription factor, and there are numerous candidates for the transcription factors recognizing similar sequence elements that should lead to the elucidation of the regulation of the other CBF proteins.

Transgenic plants constitutively expressing CBF1 have more cold–up-regulated gene transcripts than wild-type plants have, suggesting that numerous cold–up-regulated proteins that may be involved in cold acclimation are produced in the absence of cold in these CBF1 transgenic plants. In addition, CBF1 transgenic plants are more cold tolerant than control plants.

Salinity Stress

Under natural conditions, terrestrial higher plants encounter high concentrations of salts close to the seashore and in estuaries where seawater and freshwater mix or replace each other with the tides. Far inland, natural salt seepage from geologic marine deposits can wash into adjoining areas, rendering them unusable for agriculture. Evaporation and transpiration remove pure water (as vapor) from the soil solution, and this water loss concentrates salts. Water droplets from oceans are dispersed over land and evaporate, adding to increased soil salinity. However, man is a major cause of soil salinization. Intensive agriculture and improper water management practices have caused (since the beginning of cultivation) and continue to cause substantial salinization of crop lands. When irrigation water contains a high concentration of solutes and when there is no opportunity to flush out accumulated salts to a drainage system, salts can quickly reach levels that are injurious to salt-sensitive species, in particular crops. Salinity is a major threat to sustainable irrigation required to meet the food demands of increasing human population growth (Flowers 2004).

In this section, we discuss how plant function is affected by water and soil salinity, and we examine the processes that assist plants in avoiding salinity stress.

Salt accumulation in irrigated soils impairs plant function

In discussing the effects of salts in the soil, we distinguish between high concentrations of Na^+, referred to as **sodicity**, and high concentrations of total salts, referred to as **salinity**. The two are often related, but in some areas Ca^{2+}, Mg^{2+}, and SO_4^{2-}, as well as NaCl, can contribute substantially to salinity. The high Na^+ concentration of a sodic soil not only injures plants directly but also degrades the soil structure, decreasing porosity and water permeability. A sodic clay soil known as caliche is so hard and impermeable that dynamite is sometimes required to dig through it! Throughout the remainder of the chapter, the collective term salinity is used to describe salt accumulation.

In the field, the salinity of soil water or irrigation water is typically determined by its electrical conductivity, which is correlates with the osmotic potential. Pure water is a very poor conductor of electric current; the conductivity of a water sample is due to the ions dissolved in it. Higher salt concentration in water increases electrical conductivity and

TABLE 26.6
Properties of seawater and of good quality irrigation water

Property	Seawater	Irrigation water
Concentration of ions (mM)		
Na^+	457	<2.0
K^+	9.7	<1.0
Ca^{2+}	10	0.5–2.5
Mg^{2+}	56	0.25–1.0
Cl^-	536	<2.0
SO_4^{2-}	28	0.25–2.5
HCO_3^-	2.3	<1.5
Osmotic potential (MPa)	–2.4	–0.039
Total dissolved salts (mg L^{-1} or ppm)	32,000	500

lowers the osmotic potential (higher osmotic pressure) (Table 26.6).

The quality of irrigation water in semiarid and arid regions is often poor. In the United States the salt content of the headwaters of the Colorado River is only 50 mg L^{-1}, but about 2000 km downstream, in southern California, the salt content of the same river reaches about 900 mg L^{-1}, enough to preclude growth of some salt-sensitive crops, such as maize. Water from some wells used for irrigation in Texas may contain as much as 2000 to 3000 mg salt L^{-1}. An annual application of irrigation water totaling 1 m from such wells would add 20 to 30 tons of salts per hectare (8–12 tons per acre) to the soil. These levels of salts are damaging to all but the most tolerant crops.

Plants show great diversity for salt tolerance

Salinity can impair plant function, growth, and developmental processes. In the extreme, it can reduce survival. Plants can be divided into two broad groups based on their responses to high concentrations of salts. **Halophytes** are native to saline soils and complete their life cycles in that environment. **Glycophytes** (literally "sweet plants"), or nonhalophytes, are not able to tolerate salts to the same degree as halophytes. Among crops, maize, onion, rice, citrus, pecan, lettuce, and bean are highly sensitive to salt; cotton and barley are moderately tolerant; and sugar beet and date palms are more tolerant (Greenway and Munns 1980). Some highly tolerant halophytes, such as *Suaeda maritima* (a salt marsh plant) and *Atriplex nummularia* (a saltbush) are called euhalophytes; these plants exhibit growth stimulation at Cl^- concentrations that are lethal to some sensitive species (Figure 26.14) Although it is convenient to group species as either halophytic or glycophytic, both groups exhibit substantial genetic variation for salt tolerance. Even salt-sensitive crops and their relatives can be relatively salt tolerant.

Salt stress causes multiple injury effects

OSMOTIC EFFECTS Dissolved solutes in the root zone cause a low (more negative) osmotic potential that reduces the soil water potential (see Chapter 4). The general water balance of plants is thus affected, because root hydraulic conductance and transport into cells of leaves requires devel-

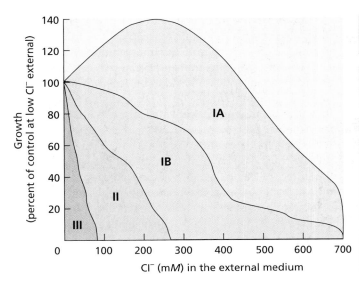

Group IA (halophytes) includes sea blite (*Suaeda maritima*) and salt bush (*Atriplex nummularia*). These species show growth stimulation with Cl^- levels below 400 m*M*.

Group IB (halophytes) includes Townsend's cordgrass (*Spartina x townsendii*) and sugar beet (*Beta vulgaris*). These plants tolerate salt, but their growth is retarded.

Group II (halophytes and nonhalophytes) includes salt-tolerant halophytic grasses that lack salt glands, such as red fescue (*Festuca rubra* subsp. *littoralis*) and *Puccinellia peisonis*, and nonhalophytes, such as cotton (*Gossypium* spp.) and barley (*Hordeum vulgare*). All are inhibited by high salt concentrations. Within this group, tomato (*Lycopersicon esculentum*) is intermediate, and common bean (*Phaseolus vulgaris*) and soybean (*Glycine max*) are sensitive.

The species in **Group III (very salt-sensitive nonhalophytes)** are severely inhibited or killed by low salt concentrations. Included are many fruit trees, such as citrus, avocado, and stone fruits.

FIGURE 26.14 The growth of different species subjected to salinity relative to that of unsalinized controls. The curves dividing the regions are based on data for different species. Plants were grown for 1 to 6 months. (After Greenway and Munns 1980.)

opment of even lower water potentials to maintain a "downhill" gradient of water potential that facilitates water movement from the soil into leaves (see Chapter 4). This osmotic effect of dissolved solutes is similar to that of a soil water deficit (as discussed earlier in this chapter), and initial plant responses to excessive levels of soil salinity are the same as described earlier for water deficit.

A major difference between the low-water-potential environments caused by salinity versus soil desiccation is the total amount of water available. During soil desiccation, a finite amount of water can be obtained from the soil profile by the plant, causing ever-decreasing soil water potentials. In most saline environments, a large (essentially unlimited) amount of water at a constant, low water potential is available. Of particular importance here is the fact that plants can adjust osmotically when growing in saline soils. Such adjustment prevents loss of turgor (which would slow cell growth; see Figure 26.1) while generating a lower water potential that allows plants to access water in the soil solution for growth. However, these plants often continue to grow more slowly after this adjustment for an unknown reason that curiously is not related to insufficient turgor (Bressan et al. 1990). This response is similar to that of plants after osmotic adjustment to water deficit (Figure 26.1).

ION TOXICITY In addition to effects of low water potential stress, specific ion **toxicity effects** also occur when injurious concentrations of ions—particularly Na^+, Cl^-, or SO_4^{2-}—accumulate in cells. Under nonsaline conditions, the cytosol of higher-plant cells contains about 100 mM K^+ and less than 10 mM Na^+, an ionic environment in which enzymes are operationally functional. An abnormally high ratio of Na^+ to K^+ and high concentrations of total ions inactivate enzymes and inhibit protein synthesis. Enzymes isolated from halophytes are just as sensitive to the presence of NaCl as are enzymes from glycophytes. Hence halophytes do not possess salt-tolerant metabolism. Instead, other mechanisms are necessary to minimize salt injury and facilitate metabolic function, as is discussed in a following section.

At a high concentration, Na^+ can displace Ca^{2+} from the plasma membrane, resulting in a change in plasma membrane permeability that can be detected as leakage of K^+ from the cells (Cramer et al. 1985). Na^+ negatively disturbs ion homeostasis, affecting plant nutrient status in a variety of ways, such as by inhibiting acquisition of the essential element K^+ both by competition for sites on transport proteins and through intracellular processes yet to be fully deciphered. Photosynthesis is inhibited when high concentrations of Na^+ and/or Cl^- accumulate in chloroplasts. Since photosynthetic electron transport appears relatively insensitive to salts, either carbon metabolism or photophosphorylation may be affected.

SECONDARY EFFECTS The above are examples of primary deleterious effects of salinity on plants; however, secondary events also inhibit plant function. These arise from disruption of cell membrane integrity and metabolism, production of toxic molecules such as reactive oxygen species, and death of cells. It is not known how these primary and secondary effects of salt stress perturb cell division and expansion, but these processes are substantially modulated at sublethal salt concentrations.

Plants use multiple strategies to reduce salt stress

Reduced water availability is the initial stress perceived by shoots during a salt incursion into the environment of roots, and responses are similar to those of plants under water deficit, as discussed previously. Plants minimize salt injury by reducing salt exposure of meristems, particularly in the shoot, and from leaves that are actively growing and photosynthesizing.

Recall from Chapter 4 that the Casparian strip in root endodermal cells imposes a restriction to the movements of ions into the xylem. To bypass the Casparian strip barrier, ions must move from the apoplastic to the symplastic (transmembrane) transport pathway in order to cross cell membranes and move through the endodermis. This transition of transport pathways allows plants to restrict movement of ions to the xylem through active transport processes (see Chapter 6), partially excluding harmful ions from the transpirational stream, as explained in more detail below.

Na^+ enters roots passively (by moving down an electrochemical-potential gradient; see Chapter 6), so root cells must use energy to extrude Na^+ actively back to the apoplast (Niu et al. 1995). In contrast, Cl^- entry into the cytosol of cells is restricted by a negative plasma membrane potential. Ions that accumulate in the xylem are transported to the shoot in water that is moved by transpiration. However, as water (xylem sap) moves from roots to shoots, ion absorption from the transpirational stream into adjacent cells along the transpirational pathway can occur, and this can lower the salt load to the leaves. Transpiration, albeit detrimental in the context of toxic ion transport, is essential for movement of water and essential nutrients (see Chapter 4).

Some halophytes, such as salt cedar (*Tamarix* sp.) and salt bush (*Atriplex* sp.) have evolved specialized salt glands on the surface of leaves that allow these plants to accommodate the salt load. Ions are transported to these glands, where the salt crystallizes and is no longer harmful to plants. These glands are unique structures with a highly specialized function in salt tolerance. In general, halophytes accumulate more ions in shoot cells than glycophytes.

As discussed earlier in relation to water deficit, plant cells can adjust their water potential (Ψ_w) in response to osmotic stress by lowering their solute potential (Ψ_s). Two intracellular processes contribute to the decrease in Ψ_s: the accumulation of potentially toxic ions in the vacuole and the synthesis of compatible solutes in the cytosol. This osmotic adjustment allows the plant to withstand hyperosmotic ion stress, and it is necessary for growth and development in saline environments. Interestingly, this process is conserved in both glycophytes and halophytes, but is more effectively implemented by salt-tolerant plants.

Compatible organic solutes include glycine betaine, proline, sorbitol, mannitol, pinitol, and sucrose. K^+ is also a major compatible inorganic solute. Some of these solutes have an osmoprotectant function in that they protect plants from toxic byproducts produced as secondary effects during osmotic and/or ion stress. Specific plant families tend to use one or two of these compounds in preference to others. The amount of carbon used for the synthesis of these organic solutes can be rather large (about 10% of the plant weight).

Besides making adjustments in water potential, plants adjusting to salinity stress undergo the other osmotic stress–related acclimations described earlier for water deficit. For example, plants subjected to salt stress can reduce leaf area and/or drop leaves via leaf abscission. In addition, many of the changes in gene expression associated with osmotic stress are similarly associated with salinity stress. Genes that code for proteins that function in cellular protection and/or injury recovery are especially involved (see osmotic, heat, and cold stress sections above). Keep in mind, however, that in addition to acclimation to a low-water-potential environment, plants experiencing salinity stress must cope with the toxicity of the associated high ion concentrations, and thus, salinity stress–specific genes are also involved.

Ion exclusion and compartmentalization reduce salinity stress

Abundant evidence makes it clear that both halophytes and glycophytes alike accumulate ions intracellularly during salt adaptation and use these for osmotic adjustment necessary for cell expansion (Hasegawa et al. 2000). Since high ion concentrations are toxic to the cellular metabolism of all plants, both halophytes and glycophytes compartmentalize cytotoxic ions into the vacuole or actively pump them outside the cell to the apoplast. Cell expansion of differentiated cells occurs primarily through an increase in volume of the vacuole, making it an active sink during growth that augments the flux of ions caused by the transpirational demands on plants.

The prevailing hypothesis is that plants sense both excess ions and hyperosmolarity, although the determinants and processes are still being determined. Relaying of the stress signature occurs through signal transduction pathways (discussed in several chapters, including Chapter 14 on the web site) that coordinate the plant's responses necessary for acclimation, including regulation of ion transport proteins that facilitate the delicate balancing act of osmotic adjustment and ion compartmentalization. Lipid membranes function as barriers to the movement of ions and solutes (see Chapter 1), and transport proteins facilitate ion compartmentalization.

More than 40 years ago, Epstein and colleagues discovered how plant cells cope with excess external Na^+. As discussed above and in Chapter 6, the electrochemical potential can facilitate Na^+ accumulation into the cytosol 100- to 1000-fold above the external concentrations. Na^+ cannot move through the plasma membrane lipid bilayer, but the ion is transported through both low- and high-affinity transport systems, many of which are necessary for K^+ acquisition. External Ca^{2+} at physiological levels (i.e., at millimolar concentrations) facilitates K^+ and minimizes Na^+ uptake through mechanisms that reduce uptake and facilitate efflux to the apoplast (discussed below). Low-affinity (high capacity) flux of Na^+ occurs through nonselective or voltage-insensitive cation–cyclic nucleotide–gated channels that transport cytotoxic cations (passively) into cells, when such channels are open, Na^+ diffuses into the cytosol where it can be substantially concentrated. Physiological Ca^{2+} causes closure of these channels restricting Na^+ uptake (Rus et al. 2001). Additionally, plant cells possess Na^+-specific transport systems, one of which (HKT) is implicated in intracellular uptake of Na^+ that is Ca^{2+} independent (Rus et al. 2001).

As discussed in Chapter 6, a P-type ATPase in the plasma membrane provides the driving force (H^+ electrochemical potential) for secondary active transport of ions (Figure 26.15) and is required for cellular extrusion of excess ions associated with plant responses to salinity stress. Na^+ efflux across the plasma membrane occurs via the SOS1 $Na^+–H^+$ antiporter (see **Web Topic 26.5**). SOS1 activation occurs in response to high NaCl and is mediated through the Ca^{2+}-signaling SOS pathway (Figure 26.16) (see **Web Topic 26.5**). Ca^{2+} activates the pathway through binding to SOS3, which in turn activates a serine/threonine kinase, SOS2. SOS2 phosphorylates SOS1, which activates its $Na^+–H^+$ antiporter function. By this mechanism, plants have the capacity to control net Na^+ flux across the plasma membrane through the regulation of influx and efflux. Na^+ efflux is the most crucial process for salt tolerance in plants. Furthermore, overexpression of SOS1 in transgenic plants enhances salt tolerance.

Two H^+ pumps in the tonoplast generate the electrochemical gradient for secondary active and passive transport of ions into the vacuole: a V-type H^+-ATPase and a H^+-pyrophosphatase (see Chapter 6). Cation–H^+ transporters such as the $H^+–Na^+$ antiporter AtNHX (see Figure 26.15) are responsible for influx of Na^+ into the vacuole. It is unclear whether Cl^- influx to the vacuole is active or passive. Transgenic Arabidopsis and tomato plants overexpressing the gene that encodes AtNHX1 exhibit enhanced salt tolerance (Apse et al. 1999; Quintero et al. 2000). (See **Web Topic 26.6** for details on molecular studies of Na^+ compartmentalization.)

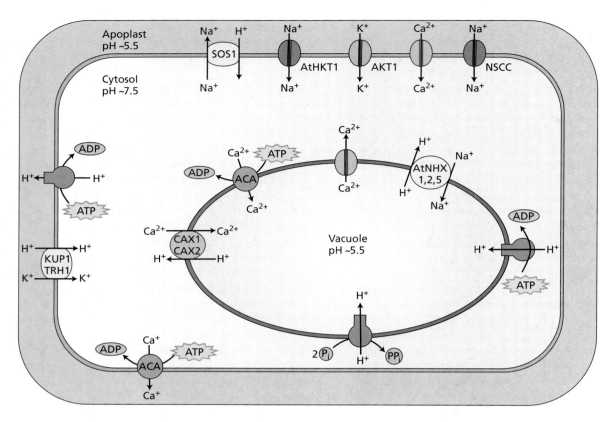

FIGURE 26.15 Membrane transport proteins mediating sodium, potassium, and calcium transport during salinity stress. SOS1, plasma membrane Na^+–H^+ antiporter; ACA, plasma/tonoplast membrane Ca^{2+}-ATPase; KUP1/TRH1, high-affinity K^+–H^+ cotransporter; AtHKT1, sodium influx transporter; AKT1, inwardly rectifying K^+ channel; NSCC, nonselective cation channel; CAX1 and 2, Ca^{2+}–H^+ antiporter; AtNHX1, 2, and 5, endomembrane Na^+–H^+ antiporter. Also indicated in the figure are proteins that have been implicated in ion homeostasis, but whose molecular identity either is not presently known or is unconfirmed in plants. These include plasma membrane and tonoplast calcium channel proteins and vacuolar proton-pumping ATPases and pyrophosphatases. The membrane potential difference across the plasma membrane is typically 120 to 200 mV, negative inside (cytosol); across the tonoplast 0 to 20 mV, positive inside (vacuole).

Salt-sensitive plants generally tolerate moderate salinity mainly because of root mechanisms that reduce movements of potentially harmful ions to the shoots. In contrast, halophytes generally have a greater capacity to accumulate ions in the shoot. As mentioned, halophytes have a greater capacity for vacuolar compartmentalization of ions in leaf cells. Furthermore, it is likely that halophytes have a greater ability to restrict net intracellular Na^+ uptake in leaf cells. Consequently, these plants are likely to cope with greater salt flux from the transpirational stream.

Restricting root xylem load apparently is less necessary for halophytes than for glycophytes. However, it is clear that salt tolerance of both halophytes and glycophytes, regardless of the degree, is dependent on the ion transport processes that control net ion uptake across the plasma membrane and compartmentalization into the vacuole. (**Web Topic 26.7** provides details about results of transgenic plant experiments to improve salt tolerance of glycophytes.)

Plant adaptations to toxic trace elements

Some metals can be highly toxic. Plants have evolved two basic strategies to tolerate the presence of elevated concentrations of various toxic trace elements in the environment, including As, Cd, Cu, Ni, Zn, and Se: i.e., *exclusion*, by which the concentration of these elements is maintained below a toxic threshold value in the plant, and *internal tolerance*, through which various biochemical adaptations allow the plant to tolerate elevated concentrations of these elements (see **Web Essay 26.2**). Interestingly, the extreme of this strategy is the *hyperaccumulation* of certain trace elements. These plants can accumulate foliar concentrations of various trace elements, including As, Cd, Ni, Zn, and Se to more than 1% (10,000 µg g^{-1}) of their shoot dry weight. Hyperaccumulation has been identified in over 400 plant taxa, and it occurs even when the plants are growing on soils with low concentrations of the hyperaccumulated element. Hyperaccumulation is an active process and appears to protect the plants against pathogens and insect herbi-

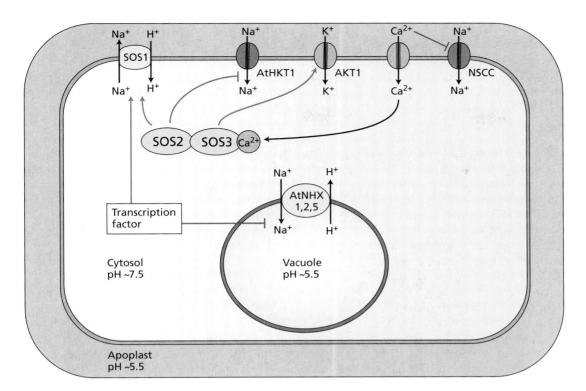

Figure 26.16 The regulation of ion homeostasis by the SOS signal transduction pathway, salinity stress, and calcium levels. Red arrows indicate positive regulation of the affected transport protein, while blue arrows indicate negative regulation. Proteins shown in yellow are activated by salinity stress. SOS1, plasma membrane Na^+–H^+ antiporter; SOS2, serine/threonine kinase; SOS3, Ca^{2+}-binding protein; HKT1, sodium influx transporter; AKT1, inwardly rectifying K^+ channel; NSCC, nonselective cation channel; NHX1, 2, and 5, endomembrane Na^+–H^+ antiporter; shown in orange is an undetermined calcium channel protein. Salinity stress activates a calcium channel, leading to an increase in cytosolic calcium that activates the SOS cascade through SOS3. The SOS cascade must negatively regulate HKT1, which in turn secondarily regulates AKT1. At the same time, the SOS cascade increases the activity of SOS1 and AKT1. Working through an as-yet undefined transcription factor, the SOS cascade increases transcription of SOS1 while decreasing transcription of the gene or genes encoding NHX. At low calcium concentrations, NSCC can also function as an alternative sodium influx system, but this transporter is inhibited at high calcium levels. The membrane potential difference across the plasma membrane is typically 120 to 200 mV, negative inside (cytosol), that of the tonoplast is 0 to 20 mV, positive inside (vacuole).

vores. To achieve hyperaccumulation, these plants not only contain a remarkable internal hypertolerance mechanism that allows them to resist the cytotoxic burden of the accumulated trace elements, but also a powerful scavenging mechanism for the efficient uptake of these potentially toxic elements from the soil. Hyperaccumulation involves alterations at the transcriptional level that allow overproduction of the ion transporters involved in uptake of these elements from the soil, including ZIP family members, P-type ATP-dependent metal transporters, and Nramp ion transporters (Assunção et al. 2001; Becher et al. 2004). Long-distance transport from roots to shoots is also a critical process for hyperaccumulation of metals in shoot tissues. Both the iron chelator nicotianamine and the free amino acid histidine have been implicated in chelation of metals during this transport process (Ingle et al. 2005). Metal hyperaccumulators also require hypertolerance mechanisms to resist the potentially acute cytotoxic effects of the hyperaccumulated elements. Cd, Ni, and Zn are compartmentalized in the vacuole of various hyperaccumulator species, where they are bound by organic acids (Salt et al. 1999). Constitutive overexpression of zinc transporters has been implicated in at least 1 case of such a compartmentalization process (Assunção et al. 2001). Hyperaccumulation of various trace elements has the potential to cause extensive oxidative damage to plant tissues. To protect against such oxidative damage, Ni hyperaccumulators in the *Thlaspi* genus overaccumulate the antioxidant glutathione. Such overaccumulation is driven by constitutive activation of the sulfur assimilation pathway, through allosteric activation of serine acetyltransferase (Freeman et al. 2004). Glutathione accumulation in *Thlaspi* hyperaccumulators appears related to the constitutive overaccumulation of salicylic acid, suggesting a direct interaction between metal hyperaccumulation and plant–pathogen interactions, given that salicylic acid is a key signaling molecule involved in plant–pathogen responses. Both genome-wide transcriptional profiling and physiological studies show that hyperaccumulation is driven by changes in multiple genes and biochemical processes. Genetic studies reveal that hypertolerance and hyperaccumulation are independent genetic traits.

Oxygen Deficiency

Roots usually obtain sufficient oxygen (O_2) for aerobic respiration (see Chapter 11) directly from the gaseous space in the soil. Gas-filled pores in well-drained, well-structured soil readily permit the diffusion of gaseous O_2 to depths of several meters. Consequently, the O_2 concentration deep in the soil is similar to that in humid air. However, poorly drained soil can become flooded or waterlogged when rain or irrigation is excessive. Water then fills the pores and blocks the diffusion of O_2 in the gaseous phase. Dissolved oxygen diffuses so slowly in stagnant water that only a few centimeters of soil near the surface remain oxygenated.

When temperatures are low and plants are dormant, oxygen depletion is very slow and the consequences are relatively harmless. However, when temperatures are higher (greater than 20°C), oxygen consumption by plant roots, soil fauna, and microorganisms can totally deplete the oxygen from the bulk of the soil water in as little as 24 hours.

Flooding-sensitive plants are severely damaged by 24 hours of anoxia. The growth and survival of many plant species are greatly depressed under such conditions, severely reducing crop yields. For example, 24 hours of flooding can halve flooding-sensitive garden pea (*Pisum sativum*) yields. Other plants, particularly many crop plants and other species not adapted to grow in continually wet conditions are affected by flooding in a milder way, and are considered flooding-tolerant plants. **Flooding-tolerant plants** can withstand anoxia (lack of oxygen) temporarily, but not for periods of more than a few days.

In contrast, specialized natural vegetation found in wetlands, and crops such as rice, are well adapted to resist oxygen deficiency in the root environment. Wetland plants can resist anoxia, allowing them to grow and survive for extended periods (up to months) with their root systems in **anoxic** (anaerobic) conditions. Most of these plants have special adaptations that permit oxygen available in nearby environments to reach the tissues held in anoxic conditions. Nearly all plants require oxygen when they are engaging in rapid metabolic activity, and plants can be classified according to the time they can withstand anoxic conditions in their root environment without suffering substantial damage.

Anaerobic microorganisms are active in water-saturated soils

When soil is depleted of molecular O_2, the function of soil microbes becomes significant for plant life, because anaerobic microorganisms (anaerobes) produce toxic compounds. Anaerobes derive their energy from the reduction of nitrate (NO_3^-) to nitrite (NO_2^-) or to nitrous oxide (N_2O) and molecular nitrogen (N_2). These gases (N_2O and N_2) are lost to the atmosphere in a process called denitrification. As conditions become more reducing, anaerobes reduce Fe^{3+} to Fe^{2+}, and because of its greater solubility, Fe^{2+} can rise to toxic concentrations when some soils are anaerobic for many weeks. Other anaerobes may reduce sulfate (SO_4^{2-}) to hydrogen sulfide (H_2S), which is a respiratory poison.

When anaerobes have an abundant supply of organic substrates, bacterial metabolites such as acetic acid and butyric acid are released into the soil water, and these acids along with reduced sulfur compounds account for the unpleasant odor of waterlogged soil. All of these substances made by microorganisms under anaerobic conditions are toxic to plants at high concentrations.

Roots are damaged in anoxic environments

Even before O_2 is completely depleted and the root environment becomes anoxic, root respiration rate and metabolism are affected in these **hypoxic** (partly deficient in oxygen) conditions. The **critical oxygen pressure (COP)** is the oxygen pressure at which the respiration rate is first slowed by O_2 deficiency. The COP for a maize root tip growing in a well-stirred nutrient solution at 25°C is about 0.20 atmosphere (20 kilopascals [KPa], or 20% O_2 by volume), almost the concentration in ambient air. At this oxygen partial pressure (for a discussion of partial pressures, see **Web Topic 9.4**), the rate of diffusion of dissolved O_2 from the solution into the tissue and from cell to cell barely keeps pace with the rate of O_2 utilization. However, a root tip is metabolically very active, with respiration and ATP turnover rates comparable to those of mammalian tissues.

In older zones of the root, where cells are mature and fully vacuolated and the respiration rate is lower, the COP is often in the range of 0.1 to 0.05 atmosphere. When O_2 concentrations are below the COP, the center of the root becomes hypoxic or anoxic.

The COP decreases when respiration slows down at cooler temperatures; its level also depends on how bulky the organ is and how tightly the cells are packed. Large, bulky fruits are able to remain fully aerobic because of the large intercellular spaces that readily allow gaseous diffusion. For single cells, an O_2 partial pressure as low as 0.01 atmosphere (1% O_2 in the gaseous phase) can be adequate because diffusion over short distances ensures an adequate O_2 supply to mitochondria. A very low partial pressure of O_2 at the mitochondrion is sufficient to maintain oxidative phosphorylation.

The K_m value (Michaelis–Menten constant; see Chapter 2 on the web site) for cytochrome oxidase is 0.1 to 1.0 μM dissolved O_2. This is a tiny fraction of the concentration of dissolved O_2 at equilibrium with air (277 μM at 20°C). The large difference between the COP values for an organ or tissue and the O_2 requirements of mitochondria is explained by the slow diffusion of dissolved O_2 in aqueous media.

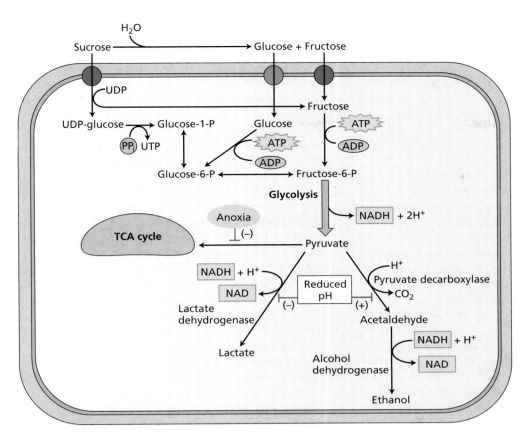

FIGURE 26.17 During episodes of anoxia, pyruvate produced by glycolysis is initially fermented to lactate. Proton production by glycolysis and other metabolic pathways, and decreased proton translocation out of the cytosol across the plasma membrane (out of cell) and tonoplast (into vacuole) lead to a lowering of cytosolic pH. At lower pH, lactate dehydrogenase activity is inhibited, and pyruvate decarboxylase is activated. This leads to an increase in the fermentation to ethanol and a decrease in the fermentation to lactate at lower pH. The pathway of ethanol fermentation consumes more protons than does the pathway of lactate fermentation. This increases the cytosolic pH and enhances the ability of the plant to survive the episode of anoxia.

In the absence of O_2, electron transport and oxidative phosphorylation in mitochondria cease, the tricarboxylic acid (TCA) cycle cannot operate, and ATP can be produced only by fermentation. Thus, when the supply of O_2 is insufficient for aerobic respiration, roots first begin to ferment pyruvate (formed in glycolysis; see Chapter 11) to lactate, through the action of lactate dehydrogenase (LDH) (Figure 26.17). In the root tips of maize, lactate fermentation is transient because lowered intracellular pH quickly leads to a switch from lactate fermentation to ethanol fermentation. The shift occurs because of the different pH optima of the cytosolic enzymes involved.

At acidic pH, LDH is inhibited and pyruvate decarboxylase is activated. The net yield of ATP in fermentation is only 2 moles of ATP per mole of hexose sugar respired (compared with 36 moles of ATP per mole of hexose respired in aerobic respiration). Thus, injury to root metabolism by O_2 deficiency originates in part from a lack of ATP to drive essential metabolic processes (Drew 1997).

Nuclear magnetic resonance (NMR) spectroscopy was used to measure the intracellular pH of living maize root tips under nondestructive conditions (Roberts et al. 1992). In healthy cells, the vacuolar contents are more acidic (pH 5.8) than the cytoplasm (pH 7.4). However, under conditions of extreme O_2 deficiency, protons gradually leak from the vacuole into the cytoplasm, adding to the acidity generated in the initial burst of lactic acid fermentation. These changes in pH (called *cytosolic acidosis*) are associated with the onset of cell death.

Apparently, active transport of H^+ into the vacuole by tonoplast ATPases is slowed by lack of ATP, and without ATPase activity the normal pH gradient between cytosol and vacuole cannot be maintained. Cytosolic acidosis irreversibly disrupts metabolism in the cytoplasm of higher-plant cells, as it does in anoxic cells of animals. Cytosolic acidosis is the primary cause of damage, and the timing and degree to which acidosis can be limited are the primary factors distinguishing flooding-sensitive from flooding-tolerant species.

Damaged O_2-deficient roots injure shoots

Anoxic or hypoxic roots lack sufficient energy to support physiological processes on which the shoots depend. Experiments have shown that the failure of the roots of wheat or barley to absorb nutrient ions and transport them to the xylem (and from there to the shoot) quickly leads to a shortage of ions within developing and expanding tissues. Older leaves senesce prematurely because of reallocation of phloem-mobile elements (N, P, K) to younger leaves. The decreased permeability of roots to water often leads to a decrease in leaf water potential and wilting, although this decrease is temporary if stomata close, preventing further water loss by transpiration.

Hypoxia accelerates production of the ethylene precursor **ACC (1-aminocyclopropane-1-carboxylic acid)** in roots (see Chapter 22). In tomato, ACC travels via the xylem sap to the shoot, where, in contact with oxygen, it is converted by ACC oxidase to ethylene. The upper (adaxial) surfaces of the leaf petioles of tomato and sunflower have ethylene-responsive cells that expand more rapidly when ethylene concentrations are high. This expansion results in epinasty, the downward growth of the leaves such that they appear to droop. Unlike wilting, epinasty does not involve loss of turgor.

In some species (e.g., pea and tomato), flooding induces stomatal closure apparently without detectable changes in leaf water potential. Oxygen shortage in roots, like water deficit or high concentrations of salts, can stimulate abscisic acid production mostly in lower, older, wilted leaves on the plant. The subsequent movement of this ABA to younger, more photosynthetically active, turgid leaves leads to stomatal closure (Zhang and Zhang 1994).

Submerged organs can acquire O_2 through specialized structures

In contrast to flooding-sensitive and flooding-tolerant species, wetland vegetation is well adapted to grow for extended periods in water-saturated soil. Even when shoots are partly submerged, they grow vigorously and show no signs of stress. In some wetland species, such as the water lily (*Nymphoides peltata*), submergence traps endogenous ethylene, and the hormone stimulates cell elongation of the petiole, extending it quickly to the water surface so that the leaf is able to reach the air. Internodes of deep-water (floating) rice respond similarly to trapped ethylene, so the leaves extend above the water surface despite increases in water depth. In the case of pondweed (*Potamogeton pectinatus*), an aquatic monocot, stem elongation is insensitive to ethylene; instead elongation is promoted even under anaerobic conditions by acidification of the surrounding water caused by the accumulation of respiratory CO_2.

In most wetland plants, and in many plants that acclimate well to wet conditions, the stem and roots develop longitudinally interconnected, gas-filled channels that provide a low-resistance pathway for movement of oxygen and other gases. The gases (air) enter through stomata, or through lenticels on woody stems and roots, and travel by molecular diffusion, or by convection driven by small pressure gradients.

In many wetland plants, exemplified by rice, cells are separated by prominent, gas-filled spaces that form a tissue called **aerenchyma**, that develop in the roots independently of environmental stimuli. In a few nonwetland plants, however, including both monocots and dicots, oxygen deficiency induces the formation of aerenchyma in the stem base and newly developing roots (Figure 26.18).

In the root tip of maize, hypoxia stimulates the activity of ACC synthase and ACC oxidase, thus causing ACC and ethylene to be produced faster. The ethylene leads to the death and disintegration of cells in the root cortex. The spaces these cells formerly occupied provide the gas-filled voids that facilitate movement of O_2.

Ethylene-signaled cell death is highly selective; cells not destined to die in the root are unaffected. A rise in cytosolic Ca^{2+} concentration is thought to be part of the ethylene signal transduction pathway leading to cell death. Chemicals that elevate cytosolic Ca^{2+} concentration promote cell death under noninducing conditions; conversely, chemicals that lower cytosolic Ca^{2+} concentration block cell death in hypoxic roots that would normally form aerenchyma. Ethylene-dependent cell death in response to hypoxia is an example of *programmed cell death*, which was discussed in Chapter 16 (Drew et al. 2000).

Some plants (or parts of them) can tolerate exposure to strictly anaerobic conditions for an extended period (weeks or months) before developing aerenchyma. These include the embryo and coleoptile of rice and of rice grass (*Echinochloa crus-galli* var. *oryzicola*), and rhizomes (underground horizontal stems) of giant bulrush (*Schoenoplectus lacustris*), salt marsh bulrush (*Scirpus maritimus*), and narrow-leafed cattail (*Typha sp.*). These rhizomes can survive for several months and expand their leaves in an anaerobic atmosphere.

In nature, rhizomes overwinter in anaerobic mud at the edges of lakes. In spring, once the leaves have expanded above the mud or water surface, O_2 diffuses down through the aerenchyma into the rhizome. Metabolism then switches from an anaerobic (fermentative) to an aerobic mode, and roots begin to grow using the available oxygen. Likewise, during germination of paddy (wetland) rice and of rice grass, the coleoptile breaks the water surface and becomes a diffusion pathway (a "snorkel") for O_2 to the rest of the plant. (Although rice is a wetland species, its roots are as intolerant of anoxia as maize roots are.)

As the root extends into oxygen-deficient soil, the continuous formation of aerenchyma just behind the tip allows oxygen movement within the root to supply the apical zone. In roots of rice and other typical wetland plants, structural barriers composed of suberized and lignified cells prevent O_2 diffusion outward to the soil. The O_2 thus retained supplies the apical meristem and allows growth to proceed 50 cm or more into anaerobic soil.

In contrast, roots of nonwetland species such as maize leak O_2, failing to conserve it to the same extent as aerenchyma-containing wetland species. Thus, in the root apex of these plants, internal O_2 becomes insufficient for

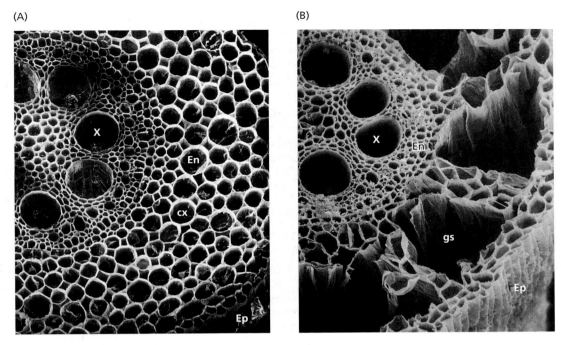

FIGURE 26.18 Scanning electron micrographs of transverse sections through roots of maize, showing changes in structure with oxygen supply. (150×) (A) Control root, supplied with air, with intact cortical cells. (B) Oxygen-deficient root growing in a nonaerated nutrient solution. Note the prominent gas-filled spaces (gs) in the cortex (cx), formed by degeneration of cells. The stele (all cells interior to the endodermis, En) and the epidermis (Ep) remain intact. X, xylem. (Courtesy of J. L. Basq and M. C. Drew.)

aerobic respiration, and this lack of O_2 severely limits the depth to which such roots can extend into anaerobic soil.

Most plant tissues cannot tolerate anaerobic conditions

Most tissues of higher plants cannot survive prolonged anaerobic conditions. For example, root tips of maize suddenly deprived of O_2 remain viable for only 20 to 24 hours. Under anoxia some ATP is generated slowly by fermentation, but the energy status of cells gradually declines because of cytosolic acidosis. The precise combination of biochemical characteristics that allow some cells to tolerate anoxia for long periods is not fully understood. Root tips of maize and other cereals show a modest degree of acclimation—if they first are made hypoxic they can survive anoxia for up to 4 days.

Acclimation to anaerobic conditions is associated with expression of the genes that encode anaerobic stress proteins (see the next section). After acclimation, the ability to carry out ethanolic fermentation under anoxia (thereby producing enough ATP to keep some metabolism going) is improved, and this improvement is accompanied by an ability to transport lactate out of the cytosol to the external medium, thus minimizing cytosolic acidosis (Drew 1997).

The ability of organs of wetland plants to tolerate chronic anoxia may depend on strategies similar to those just described, but they are clearly employed to greater effect: Critical features appear to be control of cytosolic pH, continued generation of ATP by glycolysis and fermentation, and sufficient storage of fuel for anaerobic respiration over extended periods. The synthesis of alanine, succinate, and γ-aminobutyric acid by the GABA shunt pathway during anoxia has been suggested to consume protons and further minimize cytosolic acidosis. Evidence for this has been found in anoxia-tolerant shoots of rice and rice grass, but not in anoxia-sensitive shoots of wheat and barley.

Organs of species that alternate between anaerobic and aerobic metabolism need to deal with the consequences of the entry of O_2 following anoxia. Highly reactive oxygen species are generated during aerobic metabolism, and they are normally detoxified by cellular defense mechanisms that involve superoxide dismutase (SOD). This enzyme converts superoxide radicals to hydrogen peroxide, which is then converted to water by peroxidase.

In anoxia-tolerant rhizomes of yellow flag (*Iris pseudacorus*), SOD activity increases 13-fold during 28 days of anoxia. This increase is not observed in rhizomes of other *Iris* species that are not anoxia tolerant. In the tolerant species, SOD may be available to cope with the influx of O_2 that occurs when the leaves emerge into the air from water or mud, so it may assist in resisting postanoxic stress.

Synthesis of anaerobic stress proteins leads to acclimation to O_2 deficit

When maize roots are made anoxic, protein synthesis ceases except for the continued production of about 20 polypeptides (Sachs and Ho 1986). In *A. thaliana*, anaerobic stress-induced proteins are a diverse set of proteins including enzymes of carbohydrate metabolism in glycolytic and fermentation pathways, lipid metabolism, ethylene biosynthesis, auxin-mediated processes, reactive oxygen scavenging, and calcium and reactive oxygen signaling (see Bailey-Serres and Chang [2005] for references).

Homologs of the animal and fungal genes involved in directly sensing reduced oxygen levels under hypoxic or anoxic conditions are not found in plants, and thus plants either possess a unique direct oxygen deficit sensing mechanism or they utilize only indirect mechanisms. These indirect sensing mechanisms include decreases in cytosolic pH, reduction in ATP levels, increases in cytosolic calcium levels, and increases in reactive oxygen species following reduced oxygen levels. One of the earliest and best-studied events that occur following lowering of O_2 levels is an elevation of the intracellular Ca^{2+}. Evidence suggests that this calcium signal is involved in the signal transduction of anoxia. Within minutes of the onset of anoxia, a rise in cytosolic Ca^{2+} concentration acts as a signal leading to increases in mRNA levels of alcohol dehydrogenase (ADH) and sucrose synthase in maize cells in culture.

Chemicals that block a rise in intracellular Ca^{2+} concentration also prevent the induction by anoxia of ADH and sucrose synthase gene expression, and they greatly enhance the sensitivity of maize seedlings to anoxia (Sachs et al. 1996). Further research is needed to resolve these mechanisms and to explain how intracellular Ca^{2+} concentration signals both the early survival of cells under anoxia and the induction of cell death and aerenchyma formation during prolonged hypoxia. Different patterns of calcium oscillations could signal for different outputs, similar to the situation in guard cells (see Chapter 23).

The accumulation of mRNAs of anaerobic stress genes results from changes in the rate of transcription of these genes. Analysis of common sequence elements in the promoters of the ADH genes of maize and Arabidopsis and of the other anaerobic stress genes has led to the identification of a GT-motif anaerobic stress sequence element that is bound by a MYB-class transcription factor (AtMYB2). Other candidates for anaerobic stress–induced transcription factors also exist (Bailey-Serres and Chang 2005), although exactly how oxygen deficiency is sensed, and how the signal is transduced through elevations in cytosolic Ca^{2+} to cause alterations in the transcription of specific genes, are not yet understood. One component of this pathway appears to be ROP, a monomeric Rho-like GTPase of plants. In the active state, ROP has a bound GTP, whereas in the inactive state ROP contains GDP. The conversion of ROP–GTP to inactive ROP–GDP is modulated by a ROP–GAP regulatory protein. ROP–GDP primarily modulates carbohydrate conservation and antioxidant defenses when adequate oxygen is present, while ROP–GTP favors activation of ethanolic fermentation and developmental adaptations to low oxygen availability.

There is also strong evidence for some type of translational control of anaerobic stress genes. The efficiency with which mRNAs for nonanaerobic stress–regulated genes are translated following hypoxic stress is dramatically lower than that of stress-regulated genes such as that encoding ADH.

Summary

Stress is usually defined as an external factor that exerts a disadvantageous influence on the plant. Under both natural and agricultural conditions, plants are exposed to unfavorable environmental conditions that result in some degree of stress. Water deficit, heat stress and heat shock, chilling and freezing, salinity, and oxygen deficiency are major stress factors restricting plant growth such that biomass or agronomic yields at the end of the season express only a fraction of the plant's genetic potential.

The capacity of plants to cope with unfavorable environments is known as stress resistance. Plant adaptations that confer stress resistance, such as CAM metabolism, are genetically determined. Acclimation improves resistance as a result of prior exposure of a plant to stress.

Drought resistance mechanisms vary with climate and soil conditions. Indeterminate growth patterns such as that of sorghum and soybean allow these species to take advantage of late-occurring rains; plants with a determinate growth pattern, such as that of corn, lack this form of resistance to water stress. Inhibition of leaf expansion is one of the earliest responses to water stress, occurring when decreases in turgor ensuing from water deficit reduce or eliminate the driving force for cell and leaf expansion. Additional stress resistance mechanisms in response to water stress include leaf abscission, root extension into deeper, wetter soil, and stomatal closure.

Stress caused by water deficit leads to the expression of sets of genes involved in acclimation and adaptation to the stress. These genes mediate many cellular and whole-plant responses. The sensing and activation of signal transduction cascades mediating these changes in gene expression involve both an ABA-dependent pathway and an ABA-independent pathway.

Heat stress and heat shock are caused by high temperatures. Some CAM species can tolerate temperatures of 60 to 65°C, but most leaves are damaged above 45°C. The temperature of actively transpiring leaves is usually lower than that of the ambient air, but water deficit curtails transpiration and causes overheating and heat stress. Heat stress inhibits photosynthesis and impairs membrane function and protein stability.

Adaptations that confer heat resistance include responses that decrease light absorption by the leaves, such as rolling of leaves and decreases in leaf size that minimize boundary layer resistance and increase conductive heat loss. High temperatures trigger the expression of heat shock proteins, which act as molecular chaperones to promote stabilization and correct folding of cell proteins, and biochemical responses that lead to pH and metabolic homeostasis.

Chilling and freezing stress ensue from low temperatures. Chilling injury occurs at temperatures that are too low for normal growth but are above freezing, and it is typical of species of tropical or subtropical origin exposed to temperate climates. Chilling injuries include slow growth, leaf lesions, and wilting. The primary cause of most chilling injuries is the loss of membrane properties ensuing from changes in membrane fluidity. Membrane lipids of chilling-resistant plants often have a greater proportion of unsaturated fatty acids than those of chilling-sensitive plants.

Freezing injury is associated primarily with damage caused by ice crystals formed within cells and organs. Freezing-resistant species have mechanisms that limit the growth of ice crystals to extracellular spaces. Mechanisms that confer the resistance to freezing that is typical of woody plants include dehydration and supercooling.

Cold stress reduces water activity and leads to osmotic stress within the cells. This osmotic stress effect leads to the activation of osmotic stress–related signaling pathways and the accumulation of proteins involved in cold acclimation. Other cold-specific, nonosmotic stress–related genes are also activated. Transgenic plants overexpressing cold stress–activated signaling components demonstrate increased cold tolerance.

Salinity stress results from salt accumulation in the soil. Some halophyte species are highly tolerant to salt, but salinity depresses growth and photosynthesis in sensitive species. Salt injury ensues from a decrease in the water potential of the soil that makes soil water less available and from toxicity of specific ions accumulated at injurious concentrations. Plants avoid salt injury by exclusion of excess ions from leaves or by compartmentation of ions in vacuoles. Some of the molecular determinants of Na^+ exclusion and vacuolar partitioning have been determined, and a signaling pathway, the SOS pathway, regulating the expression of these genes involved in ion homeostasis has been established.

Oxygen deficiency is typical of flooded or waterlogged soils. Oxygen deficiency depresses growth and survival of many species. In contrast, plants of marshes and swamps, and crops such as rice, are well adapted to resist oxygen deficiency in the root environment. Most tissues of higher plants cannot survive anaerobically, but some tissues, such as the embryo and coleoptiles of rice, can survive for weeks under anoxic conditions. The metabolic pathways for resisting anoxic damage and their regulation have been uncovered.

Web Material

Web Topics

26.1 Stomatal Conductance and Yields of Irrigated Crops
Stomatal conductance predicts yields of irrigated crops grown in hot environments.

26.2 The GABA Shunt Pathway Integrates Stress Metabolic Responses
GABA accumulates in plant cells upon exposure to many stress stimuli and appears to function as a signaling molecule.

26.3 Membrane Lipids and Low Temperatures
Lipid enzymes from mutant and transgenic plants mimic the effects of low-temperature acclimation.

26.4 Ice Formation in Higher-Plant Cells
Heat is released when ice forms in intercellular spaces.

26.5 Ca^{2+} Signaling and Activation of the Salt Overly Sensitive (SOS) Signal Pathway
Three genetically linked loci control ion homeostasis and salt tolerance.

26.6 Na^+ Transport across the Plasma Membrane and Vacuolar Compartmentation
SOS1 is a Na^+–H^+ antiporter that controls Na^+ fluxes across the plasma membrane.

26.7 Gene Transfer and Stress Tolerance
Transgenic plants are valuable tools for studying stress tolerance.

Web Essays

26.1 The Effect of Air Pollution on Plants
Polluting gases inhibit stomatal conductance, photosynthesis, and growth.

26.2 An Extreme Plant Lifestyle: Metal Hyperaccumulation
Plants can overaccumulate highly toxic metals.

Chapter References

Apse, M. P., Aharon, G. S., Snedden, W. A., and Blumwald, E. (1999) Salt tolerance conferred by over expression of vacuolar Na^+/H^+ antiport in Arabidopsis. *Science* 285: 1256–1258.

Assunção, A. G. L., Da Costa Martins, P., De Folter, S., Vooijs, R., Schat, H., and Aarts, M. G. M. (2001) Elevated expression of metal transporter genes in three accessions of the metal hyperaccumulator *Thlaspi caerulescens*. *Plant Cell Environ.* 24: 217–226.

Bailey-Serres, J., and Chang, R. (2005) Sensing and signaling in response to oxygen deprivation in plants and other organisms. *Ann. Bot.* 96: 507–518.

Becher, M., Talke, I. N., Krall, L., and Krämer, U. (2004) Cross-species microarray transcript profiling reveals high constitutive expression of metal homeostasis genes in shoots of the zinc hyperaccumulator *Arabidopsis halleri*. *Plant J.* 37: 251–268.

Becwar, M. R., Rajashekar, C., Bristow, K. J. H., and Burke, M. J. (1981) Deep undercooling of tissue water and winter hardiness limitations in timberline flora. *Plant Physiol.* 68: 111–114.

Björkman, O., Badger, M. R., and Armond, P. A. (1980) Response and adaptation of photosynthesis to high temperatures. In *Adaptation of Plants to Water and High Temperatures Stress*, N. C. Turner and P. J. Kramer, eds., Wiley, New York, pp. 233–249.

Blizzard, W. E., and Boyer, J. S. (1980) Comparative resistance of the soil and the plant to water transport. *Plant Physiol.* 66: 809–814.

Bohnert, H. J., Ostrem, J. A., and Schmitt, J. M. (1989) Changes in gene expression elicited by salt stress in Mesembryanthemum crystallinum. In *Environmental Stress in Plants*, J. H. Cherry, ed., Springer, Berlin, pp. 159–171.

Boston, R. S., Viitanen, P. V., and Vierling, E. (1996) Molecular chaperones and protein folding in plants. *Plant Mol. Biol.* 32: 191–222.

Boyer, J. S. (1970) Leaf enlargement and metabolic rates in corn, soybean, and sunflower at various leaf water potentials. *Plant Physiol.* 46: 233–235.

Boyer, J. S. (1982) Plant productivity and environment. *Science* 218: 443–448.

Bray, E. A., Bailey-Serres J., and Weretilnyk, E. (2000) Responses to abiotic stresses. In *Biochemistry and Molecular Biology of Plants*, B. Buchanan, W. Gruissem, and R. Jones, eds., American Society of Plant Physiologists, Rockville, MD, pp. 1158–1203.

Bressan, R. A., Nelson, D. E., Iraki, N. M., LaRosa, P. C., Singh, N. K., Hasegawa, P. M., and Carpita, N. C. (1990) Reduced cell expansion and changes in cell wall of plant cells adapted to NaCl. In *Environmental Injury to Plants*, F. Katterman, ed., Academic Press, New York, pp. 137–171.

Buchanan, B. B., Gruissem, W., and Jones, R. eds. (2000) *Biochemistry and Molecular Biology of Plants*. American Society of Plant Physiologists, Rockville, MD.

Burke, M. J., and Stushnoff, C. (1979) Frost hardiness: A discussion of possible molecular causes of injury with particular reference to deep supercooling of water. In *Stress Physiology in Crop Plants*, H. Mussell and R. C. Staples, eds., Wiley, New York, pp. 197–225.

Burssens, S., Himanen, K., van de Cotte, B., Beeckman, T., Van Montagu, M., Inze, D., and Verbruggen, N. (2000) Expression of cell cycle regulatory genes and morphological alterations in response to salt stress in *Arabidopsis thaliana*. *Planta* 211: 632–640.

Cramer, G. R., Läuchli, A., and Polito, V. S. (1985) Displacement of Ca^{2+} by Na^+ from the plasmalemma of root cells. A primary response to salt stress? *Plant Physiol.* 79: 207–211.

Davies, W. J., Wilkinson, S., and Loveys, B. (2002) Stomatal control by chemical signaling and the exploitation of this mechanism to increase water-use efficiency in agriculture. *New Phytol.* 153: 449–460.

Drew, M. C. (1997) Oxygen deficiency and root metabolism: Injury and acclimation under hypoxia and anoxia. *Annu. Rev. Plant Physiol. Plant Mol. Biol.* 48: 223–250.

Drew, M. C., He, C. J., and Morgan P. W. (2000) Programmed cell death and aerenchyma formation in roots. *Trends Plant Sci.* 5: 123–127.

Epstein, E. (1956) Mineral Nutriton of Plants: Mechanisms of Ion Uptake and Transport. *Annual Review of Plant Physiology* 7: 1–24.

Flowers, T. J. (2004) Improving crop salt tolerance. *J. Exp. Bot.* 55: 639–648.

Fowler, S., Cook, D., and Thomashow, M. F. (2005) The CBF cold-response pathway. In *Plant Abiotic Stress*, M. A. Jenks and P. M. Hasegawa, eds., Blackwell Publishing Ltd, Oxford, England, pp. 70–99.

Freeman, J. L., Persans, M. W., Nieman, K., Albrecht, C., Peer, W., Pickering, I. J., and Salt, D. E. (2004) Increased glutathione biosynthesis plays a role in nickel tolerance in Thlaspi nickel hyperaccumulators. *Plant Cell* 16: 2176–2191.

Greenway, H., and Munns, R. (1980) Mechanisms of salt tolerance in nonhalophytes. *Annu. Rev. Plant Physiol. Plant Mol. Biol.* 31: 149–190.

Gusta, L. V., Wilen, R. W., and Fu, P. (1996) Low-temperature stress tolerance: The role of abscisic acid, sugars, and heat-stable proteins. *Hort. Sci.* 31: 39–46.

Guy, C. L. (1999) Molecular responses of plants to cold shock and cold acclimation. *J. Mol. Microbiol. Biotechnol.* 1: 231–242.

Hartung, W., Wilkinson, S., and Davies, W. J. (1998) Factors that regulate abscisic acid concentrations at the primary site of action at the guard cell. *J. Exp. Bot.* 49: 361–367.

Hasegawa, P. M., Bressan, R. A., Zhu, J. K., and Bohnert, H. J. (2000) Plant cellular and molecular responses to high salinity. *Annu. Rev. Plant Physiol. Plant Mol. Biol.* 51: 463–499.

Hong, S. W., and Vierling, E. (2000) Mutants of *Arabidopsis thaliana* defective in the acquisition of tolerance to high temperature stress. *Proc. Natl. Acad. Sci. USA* 97: 4392–4397.

Ingle, R. A., Mugford, S. T., Rees, J. D, Campbell, M. M., and Smith, J. A. (2005) Constitutively high expression of the histidine biosynthetic pathway contributes to nickel tolerance in hyperaccumulator plants. *Plant Cell* 17: 2089–2106.

Jenks, M. A., Eigenbrode, S., and Lemeiux, B. (2002) Cuticular waxes of *Arabidopsis*. In *The Arabidopsis Book*, C. Somerville and E. Meyerowitz, eds., American Society of Plant Physiologists, Rockville, MD. doi/10.1199/tab.0009, http://www.aspb.org/publications/arabidopsis/

Kawasaki, S., Brochert, C., Deyholos, M., Wang, H., Brazille, S., Kawai, K., Galbraith, D. W., and Bohnert, H. J. (2001) Gene expression profiles during the initial phase of salt stress in rice. *Plant Cell* 13: 889–906.

Larkindale, J., Hall, J. D., Knight, M. R., and Vierling, E. (2005) Heat stress phenotypes of Arabidopsis mutants implicate multiple signaling pathways in the acquisition of thermotolerance. *Plant Physiol.* 138: 882–897.

Lee, J. H., and Schoeffl, F. (1996) An Hsp70 antisense gene affects the expression of HSP70/HSC70, the regulation of HSF, and the acquisition of thermotolerance in transgenic *Arabidopsis thaliana*. *Mol. Gen. Genet.* 252: 11–19.

Levitt, J. (1980) *Responses of Plants to Environmental Stresses*, Vol. 1, 2nd ed. Academic Press, New York.

Lindow, S. E., Arny, D. C., and Upper, C. D. (1982) Bacterial ice nucleation: A factor in frost injury to plants. *Plant Physiol.* 70: 1084–1089.

Lyons, J. M., Wheaton, T. A., and Pratt, H. K. (1964) Relationship between the physical nature of mitochondrial membranes and chilling sensitivity in plants. *Plant Physiol.* 39: 262–268.

Maggio, A., and Joly, R. J. (1995) Effects of mercuric chloride on the hydraulic conductivity of tomato root systems: Evidence for a channel-mediated water pathway. *Plant Physiol.* 109: 331–335.

Mantyla, E., Lang, V., and Palva, E. T. (1995) Role of abscisic acid in drought-induced freezing tolerance, cold acclimation, and accumulation of LTI78 and RAB18 proteins in *Arabidopsis thaliana*. *Plant Physiol.* 107: 141–148.

Matthews, M. A., Van Volkenburgh, E., and Boyer, J. S. (1984) Acclimation of leaf growth to low water potentials in sunflower. *Plant Cell Environ.* 7: 199–206.

McCree, K. J., and Richardson, S. G. (1987) Stomatal closure vs. osmotic adjustment: A comparison of stress responses. *Crop Sci.* 27: 539–543.

Niu, X., Bressan, R. A., Hasegawa, P. M., and Pardo, J. M. (1995) Ion homeostasis in NaCl stress environments. *Plant Physiol.* 109: 735–742.

Palta, J. P., Whitaker, B. D., and Weiss, L. S. (1993) Plasma membrane lipids associated with genetic variability in freezing tolerance and cold acclimation of Solanum species. *Plant Physiol.* 103: 793–803.

Patterson, B. D., Paull, R., and Smillie, R. M. (1978) Chilling resistance in *Lycopersicon hirsutum* Humb. & Bonpl., a wild tomato with a wide altitudinal distribution. *Aust. J. Plant Physiol.* 5: 609–617.

Queitsch, C., Hong, S. W., Vierling, E., and Lindquist, S. (2000) Heat shock protein 101 plays a crucial role in thermotolerance in *Arabidopsis*. *Plant Cell* 12: 479–492.

Quintero, F. J., Blatt, M. R., and Pardo, J. M. (2000) Functional conservation between yeast and plant endosomal Na^+/H^+ antiporters. *FEBS Lett.* 471: 224–228.

Raison, J. K., Pike, C. S., and Berry, J. A. (1982) Growth temperature-induced alterations in the thermotropic properties of *Nerium oleander* membrane lipids. *Plant Physiol.* 70: 215–218.

Roberts, J. K. M., Hooks, M. A., Miaullis, A. P., Edwards, S., and Webster, C. (1992) Contribution of malate and amino acid metabolism to cytoplasmic pH regulation in hypoxic maize root tips studied using nuclear magnetic resonance spectroscopy. *Plant Physiol.* 98: 480–487.

Rus, A., Yokoi, S., Sharkhuu, A., Reddy, M., Lee, B. H., Matsumoto, T. K., Koiwa, H., Zhu, J. K., Bressan, R.A., and Hasegawa, P. M. (2001) AtHKT1 is a salt tolerance determinant that controls Na^+ entry into plant roots. *Proc Natl Acad Sci USA* 98: 14150–14155.

Sachs, M. M., and Ho, D. T. H. (1986) Alteration of gene expression during environmental stress in plants. *Annu. Rev. Plant Physiol. Plant Mol. Biol.* 37: 363–376.

Sachs, M. M., Subbaiah, C. G., and Saab, I. N. (1996) Anaerobic gene expression and flooding tolerance in maize. *J. Exp. Bot.* 47: 1–15.

Salt, D. E., Prince, R. C., Baker, A. J. M., Raskin, I., and Pickering, I. J. (1999) Zinc ligands in the metal hyperaccumulator *Thlaspi caerulescens* as determined using X-ray absorption spectroscopy. *Environ. Sci. Tech.* 33: 713–717.

Sauter, A., Davies W. J., and Hartung W. (2001) The long distance abscisic acid signal in the droughted plant: The fate of the hormone on its way from the root to the shoot. *J. Exp. Bot.* 52: 1–7.

Shi, H., Ishitani, M., Kim, C., and Zhu, J. K. (2000) The *Arabidopsis thaliana* salt tolerance gene SOS1 encodes a putative Na^+/H^+ antiporter. *Proc. Natl. Acad. Sci. USA* 97: 6896–6901.

Shinozaki, K., and Yamaguchi-Shinozaki, K. (2000) Molecular responses to dehydration and low temperature: Differences and cross-talk between two stress signaling pathways. *Curr. Opin. Plant Biol.* 3: 217–223.

Stockinger, E. J., Gilmour, S. J., and Thomashow, M. F. (1997) *Arabidopsis thaliana* CBF1 encodes an AP2 domain-containing transcriptional activator that binds to the C-repeat-DRE, a cis-acting DNA regulatory element that stimulates transcription in response to low temperature and water deficit. *Proc. Natl. Acad. Sci. USA* 94: 1035–1040.

Sung, F. J. M., and Krieg, D. R. (1979) Relative sensitivity of photosynthetic assimilation and translocation of ^{14}carbon to water stress. *Plant Physiol.* 64: 852–856.

Thomashow, M. (2001) So what's new in the field of plant cold acclimation? Lots! *Plant Physiol.* 125: 89–93.

U. S. Department of Agriculture (1989) *Agricultural Statistics*, U. S. Government Printing Office, Washington DC.

Vierling, E. (1991) The roles of heat shock proteins in plants. *Annu. Rev. Plant Physiol. Plant Mol. Biol.* 42: 579–620.

Weiser, C. J. (1970) Cold resistance and injury in woody plants. *Science* 169: 1269–1278.

Williams, J. P., Khan, M. U., Mitchell, K., and Johnson, G. (1988) The effect of temperature on the level and biosynthesis of unsaturated fatty acids in diacylglycerols of *Brassica napus* leaves. *Plant Physiol.* 87: 904–910.

Wisniewski, M., and Arora, R. (1993) Adaptation and response of fruit trees to freezing temperatures. In *Cytology, Histology and Histochemistry of Fruit Tree Diseases*, A. Biggs, ed., CRC Press, Boca Raton, FL, pp. 299–320.

Wu, J., Lightner, J., Warwick, N., and Browse, J. (1997) Low-temperature damage and subsequent recovery of *fab1* mutant *Arabidopsis* exposed to 2°C. *Plant Physiol.* 113: 347–356.

Yamaguchi-Shinozaki, K., and Shinozaki, K. (1994) A novel cis-acting element in an Arabidopsis gene is involved in responsiveness to drought, low-temperature, or high-salt stress. *Plant Cell* 6:251–264.

Zhang, J., and Zhang, X. (1994) Can early wilting of old leaves account for much of the ABA accumulation in flooded pea plants? *J. Exp. Bot.* 45: 1335–1342.

Glossary

ABA-response element (ABRE) A six-nucleotide sequence element found in the promoters of genes regulated by abscisic acid (ABA), a plant hormone.

ABC model Proposal for the way in which floral homeotic genes control organ formation in flowers. According to the model, organ identity in each whorl is determined by a unique combination of the three organ identity gene activities.

ABC transporters See ATP-binding cassette transporters.

Abiotic Referring to what is nonliving. In the context of plant stress refers to stresses caused by environmental factors.

ABP1 **gene** Gene that encodes the ABP1 protein.

ABP1 protein A glyscosylated *a*uxin-*b*inding *p*rotein found in the endoplasmic reticulum (ER).

Abscission The shedding of leaves, flowers, and fruits from a living plant. The process whereby specific cells in the leaf petiole (stalk) differentiate to form an abscission layer, allowing a dying/dead organ to separate from the plant. *See* abscission layer.

Abscission layer Located within the abscission zone, a distinct layer of cells with weakened cell walls that permit abscission, usually of a leaf or fruit.

Abscission zone A region that contains the abscission layer and is located near the base of the petiole of leaves.

Absorption spectrum A graphic representation of the amount of light energy absorbed by a substance plotted against the wavelength of the light.

ACC The immediate precursor of ethylene; 1-*a*mino*c*yclopropane-1-*c*arboxylic acid.

ACC oxidase Catalyzes the conversion of ACC to ethylene, the last step in ethylene biosynthesis.

ACC synthase The enzyme that catalyzes the synthesis of ACC from *S*-adenosylmethionine.

Acclimation (hardening) The increase in plant stress tolerance due to exposure to prior stress. May involve gene expression. Contrast with adaptation.

Acetogenic bacteria Obligatory anaerobes that use CO_2 and H_2 as electron acceptor and donor, respectively, in anaerobic respiration for the production of acetyl-CoA. The latter compound serves as building blocks for biosynthetic processes while the excess is excreted as acetate.

Acid growth A characteristic of growing cell walls in which they extend more rapidly at acidic pH than at neutral pH.

Acid-growth hypothesis Stress relaxation and expansion of primary cell walls that is induced by protons transported across the plasma membrane.

Acropetal From the base to the tip of an organ, such as a stem, root, or leaf.

Action spectrum A graphic representation of the magnitude of a biological response to light as a function of wavelength.

Active transport The use of energy to move a solute across a membrane against a concentration gradient, a potential gradient, or both (electrochemical potential). Uphill transport.

Acyl Refers to the fatty acid residue linked to another compound, often glycerol (e.g., triacylglycerol).

Acyl carrier protein (ACP) A low-molecular-weight, acidic protein to which growing acyl chains are covalently bonded on fatty acid synthetase.

Acyl-ACP A fatty acid chain bonded to the acyl carrier protein.

Adaptation (to stress) An inherited level of stress resistance acquired by a process of selection over many generations. Contrast with acclimation.

Adaxial Refers to the upper surface of a leaf.

Adenosine-5′-phosphosulfate (APS) A short-lived, activated form of sulfate, formed by the reaction between sulfate and ATP. Product of first reaction in the several pathways from sulfate to the amino acid cysteine. *See* 3′-phosphoadenosine-5′-phosphosulfate and thiosulfonate.

AdoMet *S*-adenosylmethionine.

ADP/ATP transporter A protein that catalyzes an antiport exchange of ADP and ATP across the inner mitochondrial membrane.

ADP/O ratio The ratio of consumed ADP to $\frac{1}{2} O_2$ in oxidative phosphorylation. Provides the number of ATP synthesized per two electrons transferred to oxygen.

ADPG pyrophosphorylase Catalyzes ATP + α-D-glucose 1-phosphate → pyrophosphate + ADP-glucose.

Adventitious roots Roots that arise from structures other than roots, such as stems or leaves.

Aerenchyma Anatomical feature of roots found in hypoxic conditions, showing large, gas-filled intercellular spaces in the root cortex.

Agar block A semi-solid cube of agar used to donate or receive auxin when in contact with plant tissue. Used in studies of polar auxin transport.

Agravitropic mutants Mutants that do not respond to gravity. Useful in understanding the mechanism of gravitropism.

***AHPs* genes** Arabidopsis *H*istidine *P*hosphotransfer genes involved in cytokinin signal propagation from the plasma membrane receptor to the nucleus.

Alcohol dehydrogenase (ADH) The enzyme that catalyzes the conversion of acetaldehyde to ethanol.

Alcoholic fermentation Metabolism of pyruvate from glycolysis produces ethanol and CO_2, while reoxidizing NADH to NAD^+. Allows the TCA cycle to function in the absence of oxygen.

Aldose A sugar with a terminal aldehyde.

Aleurone cells Cells of the aleurone layer enclosed in thick primary cell walls and containing large numbers of protein-storing organelles called protein bodies.

Aleurone layer Layer of aleurone cells surrounding and distinct from starchy endosperm of cereal grains.

***alf* mutants** A series of Arabidopsis mutants (*a*berrant *l*ateral root *f*ormation mutants) in which the role of auxin in the initiation of lateral roots is altered.

Alkaloids A large family of nitrogen-containing secondary metabolites found in many vascular plants. Defend against predators, especially mammals. Includes toxins such as strychnine and atropine, and medicinals such as morphine, codeine, atropine, and ephedrine. Others are stimulants and sedatives (e.g., cocaine, nicotine, and caffeine).

Allelopathy Release by plants of substances into the environment that have harmful effects on neighboring plants.

Allocation The regulated diversion of photosynthate into storage, utilization, and/or transport.

α-amylase Catalyzes the endohydrolysis of α-D-1,4 glucosidic linkages in polysaccharides containing three or more 1,4-α-linked D-glucose units. The term α indicates the initial anomeric configuration of the free sugar group released and not the configuration of the linkage hydrolysed. In the aleurone layer of cereal grains, α-amylase is synthesized de novo in response to gibberellin.

α-(1,4)-glucan branching enzyme Catalyzes the transfer of a segment of an α-D-1,4-linked glucan chain to a primary hydroxyl group in a similar glucan chain.

α-glucosidase Catalyzes the hydrolysis of terminal, nonreducing 1,4-linked α-D-glucose residues with the release of α-D-glucose.

Alternative oxidase A plant- and fungal-specific enzyme in the mitochondrial electron transport chain, which reduces oxygen to water.

Alternative pathway The pathway comprising the oxidation of ubiquinol and the reduction of oxygen by the alternative oxidase. Also known as cyanide-resistant pathway.

Ambiphotoperiodic Describes plants that flower in long days or short days but not at intermediate day lengths.

Amide A nitrogen-containing compound formed by the reaction of an amine and a carboxylic acid to form a $-CONR_2$ group.

Amide exporters Legumes from temperate region that convert toxic ammonia to amides, such as the amino acids asparagine or glutamine for transport to the shoot via the xylem. Contrast with ureide exporters.

Amine A nitrogen-containing compound derived from ammonia by replacing hydrogens with carbon-containing groups.

Aminotransferases Enzymes that carry out transaminations.

Amphipathic In a molecule, the quality of having both hydrophilic and hydrophobic regions.

Amplitude In a biological rhythm, the distance between peak and trough; it can often vary while the period remains unchanged.

Amylopectin The constituent of starch having a polymeric, branched structure in which an α-D-1,6 bond occurs every 20–30 glucose units linked via α-D-1,4 bonds.

Amylose The constituent of starch in which α-D-1,4 glucosidic bonds form linear chains of 200–2000 glucose units.

Anaplerotic Quality of a reaction to provide a reactant that is limiting in another reaction or pathway. For example, the PEP carboxylase reaction replenishes oxaloacetate to the citric acid cycle that has been used for biosynthesis.

Anastomoses Vascular interconnections that provide a pathway between source and sink tissues that are not directly connected.

Ancymidol (A-Rest) A commercial inhibitor of gibberellin biosynthesis. Like paclobutrazol, ancymidol blocks the oxidation reaction between kaurene and kaurenoic acid on the endoplasmic reticulum.

Angiosperm Flowering plants, distinguished from gymnosperms by the presence of a carpel that encloses the seeds.

Anisotropic Characteristic of materials showing different mechanical properties in different directions. Usually due to a bias in the orientation or linkage of constituent molecules. Contrast with isotropic.

Anomer Epimers that differ only in the configuration at the carbonyl carbon (C=O) of aldoses or ketoses.

Anoxic Refers to the absence of oxygen. Contrast with hypoxic.

Anoxygenic organism Photosynthetic organism that does not produce molecular oxygen.

Antenna complex A group of pigment molecules that cooperate to absorb light energy and transfer it to a reaction center complex.

Antenna pigments Chlorophylls and carotenoids are the pigments found in the antenna complex.

Anterograde transport The movement of membrane cisternae or vesicles in the forward direction.

Anther The apical structure of stamen in which pollen forms and from which it is released.

Anthocyanidins Pigmented flavonoids derived from anthocyanin by removal of attached sugars.

Anthocyanins Pigmented glycosylated flavonoids responsible for most of the red, pink, purple, and blue colors in plants.

Anticlinal Pertaining to the orientation of the cell plate at right angles to the longitudinal axis during cytokinesis. Transverse.

Antiflorigen Hypothetical hormone produced by uninduced leaves and translocated to the shoot apical meristem. Proposed to inhibit the formation of flowers in certain long-day plants under noninductive conditions.

Antifreeze proteins Proteins that confer to aqueous solutions the property of thermal hysteresis. When induced by cold temperatures, these plant proteins bind to the surfaces of ice crystals to prevent or slow further crystal growth, thereby limiting or preventing freeze damage. Some antifreeze proteins may be identical to pathogenesis-related proteins.

Antimycin a A specific inhibitor of complex III of the electron transport chain of the mitochondrion.

Antiport A type of secondary active transport in which the passive (downhill) movement of protons or other ions drives the active (uphill) transport of a solute in the opposite direction.

Antiporter A protein involved in antiport.

Antisense DNA DNA of a gene whose transcription produces antisense mRNA, which hybridizes with sense mRNA and inhibits its translation.

A number An A number is assigned to a newly characterized, naturally-occurring gibberellin for which the structure has been fully determined. Numbers are assigned gibberellins in the order of their discovery.

AOA Aminooxyacetic acid, an inhibitor of ethylene biosynthesis, blocking the conversion of *AdoMet* to ACC.

***AP1* gene** In Arabidopsis, *APETALA1* (*AP1*) is a gene involved in establishing floral meristem identity.

***ap2* mutants** In Arabidopsis, mutations in the *APETALA2* (*AP2*) homeotic gene produce flowers with carpels where sepals should be and stamens where petals normally appear.

Apical Relating to the apex or tip. Distinguishing one end of an axis. Contrast with *basal*.

Apical cell In ferns and other primitive vascular plants, the single initial or stem cell of roots and shoots that gives rise to all the other cells of the organ. In angiosperm embryogenesis, the smaller, cytoplasm-rich cell formed by the first division of the zygote.

Apical dominance In most higher plants, the growing apical bud's inhibition of the growth of lateral buds (axillary buds).

Apoplast The mostly continuous system of cell walls, intercellular air spaces, and xylem vessels in a plant.

Apoplastic pathway The route by which water and water-soluble solutes move exclusively through the cell walls without crossing any membranes.

Apoprotein The polypeptide component which may be bound to a chromophore or other cofactor or prosthetic group to form an active protein, the holoprotein.

Apoptosis Programmed cell death showing characteristic morphological and biochemical changes, including fragmentation of nuclear DNA between the nucleosomes. Occurs in some senescing plant tissues, differentiating xylem tracheary elements, and in the hypersensitive response against pathogens.

APS See adenosine-5′-phosphosulfate.

Aquaporins Integral membrane proteins that form water-selective channels across the membrane. Such channels facilitate water movement across the membrane.

Arabidopsis Response Regulator (ARR) Arabidopsis genes that are similar to bacterial two-component signaling proteins called response regulators. There are two classes: Type-A ARRs, whose transcription is up-regulated by cytokinin; type-B ARR, whose expression is not affected by cytokinin.

Arabinogalactan proteins (AGPs) Water soluble, highly glycosylated cell-wall proteins that may function in cell adhesion and signaling during cell differentiation.

Arbuscules Branched structures of mycorrhizal fungi that form within penetrated cells; the sites of nutrient transfer between the fungus and the host plant.

ARR See Arabidopsis Response Regulator.

Asparagine synthetase (AS) Enzyme transferring nitrogen as an amino group from glutamine to aspartate, forming asparagine.

Aspartate aminotransferase (AspAT) An aminotransferase that transfers the amino group from glutamate to the carboxyl atom of oxaloacetate to form aspartate.

Assimilation See nutrient assimilation.

Assimilatory power The combined energy available in NADPH and ATP that can be used to drive the photosynthetic fixation of atmospheric CO_2 into organic molecules.

ATI See auxin transport inhibitor.

***AtNHX1* gene** In Arabidopsis, gene encoding a Na^+/H^+ antiporter that is active in vacuolar compartmentalization of Na^+.

ATP-binding cassette transporters (ABC transporters) A group of active transport proteins, of distinctive structure, energized by ATP hydrolysis and involved in moving organic molecules across a membrane.

ATP synthase (ATPase or CF_o–CF_1) The enzyme that synthesizes ATP from ADP and phosphate (P). Consists of two parts: a hydrophobic membrane-bound portion (CF_o) and a portion that sticks out into the stroma (CF_1).

Autocatalysis The capacity of the Calvin cycle to use most of the fixed carbon (triose phosphate) to build up the concentrations of intermediates to adequately function at steady state. As a result, the concentration of the acceptor (ribulose-1,5-bisphosphate) doubles every five revolutions of the cycle.

Autocatalytic An action or reaction that promotes the action or reaction. For example, ethylene production by fruit is stimulated by ethylene.

Autocatalytic Relating to a process that occurs spontaneously when purified components of a system are mixed in vitro, requiring no additional proteins or cofactors.

Autoinhibitory domain A domain at the C-terminal end of the plasma membrane H^+-ATPase that inhibits the activity of the enzyme. Regulated by phosphorylation and by agents such as fusicoccin.

Autoradiography A technique used to determine the location of a radioactive isotope that has been supplied to living cells, using photographic film (overall localization) or photographic emulsion (precise cellular localization).

Autotrophic Cells or organisms that synthesize cellular components from structural elements found in the most oxidized states (for example, carbohydrates from CO_2). As the reduction is usually endothermic, the energy is supplied either by sunlight (photoautotrophs) or by chemical energy (chemoautotrophs).

Aux/IAA **genes** Family of primary response genes encoding short-lived transcription factors that function as repressors or activators of the expression of late auxin-inducible genes.

Auxin A compound with biological activities similar to, but not necessarily identical with, those of IAA. Induces cell elongation in isolated coleoptile or stem sections, cell division in callus tissues in the presence of a cytokinin, lateral root formation at the cut surfaces of stems, parthenocarpic fruit growth, and ethylene formation.

Auxin response domains (AuxRDs) Auxin responsive promoter sequences composed of multiple *AuxREs*.

Auxin response elements (AuxREs) DNA promoter sequences that modulate gene expression when bound by auxin-responsive transcription factors.

Auxin transport inhibitor (ATI) Any compound that prevents the polar transport of auxin, usually by preventing auxin exit (efflux) from cells. *See* NPA, TIBA, and phytotropins.

Auxin transport The energy-requiring, polar movement of IAA from the shoot apex to the root tip, and subsequent redistribution of auxin from the root tip to the basal portion of the root. Consists of protein-mediated auxin efflux from cells driven by the membrane potential, ΔE, and auxin uptake into cells driven by the total proton motive force ($\Delta E + \Delta pH$).

Auxin-Induced Proton Extrusion Increased rate of proton extrusion by both activation of preexisting plasma membrane H^+-ATPases and synthesis of new H^+-ATPases.

AVG Aminoethoxy-vinylglycine, an inhibitor of ethylene biosynthesis, blocking the conversion of AdoMet to ACC.

avr **genes** Pathogen avirulence genes that encode for specific elicitors of plant defense responses.

Axil The angled juncture between the upper side of a leaf and the stem to which it is attached. Usual site for insertion of an axillary bud.

Axillary buds Secondary meristems that are formed in the axils of leaves. If they are also vegetative meristems, they will have a structure and developmental potential similar to that of the vegetative apical meristem. Axillary buds can also form flowers, as in inflorescences. *See* lateral bud.

Axis (plant axis) The hypothetical central line of a body around which parts are arranged. *See* linear axis, radial axis.

axr1 **mutant** Arabidopsis auxin-resistant mutant deficient in many auxin response deficiencies, including gravitropism.

Bacteriochlorophyll Light-absorbing pigments active in photosynthesis in anoxygenic photosynthetic organisms.

Bacteroids Nitrogen-fixing organelles that develop from endosymbiotic bacteria upon a signal from the host plant.

BAK1 A second LRR-receptor kinase which associates with the BR receptor, BRI1, forming an active complex.

Bark Collective term for all the tissues outside the cambium of a woody stem or root, and composed of phloem and periderm.

Basal Relating to the base. Distinguishing one end of an axis. Contrast with apical.

Basal cell In embryogenesis, the larger, vacuolated cell formed by the first division of the zygote. Gives rise to the suspensor.

Basipetal From the growing tip of a shoot or root toward the base (junction of the root and shoot).

Basipetal transport Transport away from the apical meristems, in both root and shoot.

β-amylase Catalyzes the hydrolysis of β-D-1,4 glucosidic linkages in polysaccharides so as to remove successive maltose units from the nonreducing ends of the chains. The term β denotes the initial anomeric configuration of the free sugar group released and not the configuration of the linkage hydrolysed.

β-amylolysis The stepwise hydrolysis of alternate glucosidic bonds in starch-type polysaccharides with the liberation of maltose.

Beta-(β-) oxidation Oxidation of fatty acids into fatty acyl-CoA, and the sequential breakdown of the fatty acids into a number of pairs of two molecules of acetyl-CoA. NADH and $FADH_2$ are also produced.

Bidirectional transport Simultaneous transport in two directions in a single sieve element.

Biennial Plant that requires two growing seasons to flower and produce seed.

Bilayer *See* lipid bilayer.

BIN2 A repressor of brassinosteroid-induced gene expression.

Bioassay Quantitation of a known or suspected biologically active substance by measuring the effect the substance has on living material.

Biological nitrogen fixation Nitrogen fixation carried out by bacteria or blue-green algae (cyanobacteria); about 90% of all nitrogen fixation on Earth.

Biotic Referring to what is living.

Bleached The loss of chlorophyll's characteristic absorbance due to its conversion into another structural state, often by oxidation.

Blue-light responses Responses of plant cells and organs to blue light (400 to 500 nm). Includes phototropism, chloroplast movement within cells, sun tracking by leaves, inhibition of hypocotyl elongation, stimulation of chlorophyll and carotenoid synthesis, activation of gene expression, and stomatal movements.

Bolting Premature elongation of the stem of a rosette plant, usually associated with flowering.

Boundary layer resistance (r_b) The resistance to the diffusion of water vapor due to the layer of unstirred air next to the leaf surface. A component of diffusional resistance.

Boundary layer A thin film of still air at the leaf surface. Its resistance to water vapor diffusion is proportional to its thickness.

Bowen ratio The ratio of sensible heat loss to evaporative heat loss, the two most important processes in the regulation of leaf temperature.

Bract Small leaf-like structure with underdeveloped blade.

Brassinolide (BL) A plant steroidal hormone with growth-promoting activity, first isolated from *Brassica napus* pollen. One of a group of plant steroidal hormones with similar activities called Brassinosteroids.

Brassinazole (Brz) A triazole inhibitor of cytochrome p450 monooxygenases which acts as a specific inhibitor of brassinolide biosynthesis.

BRI1 The plasma membrane receptor protein for brassinosteroids. BRI1 belongs to the general class of leucine-rich repeat (LRR)-receptor serine/threonine (S/T) kinases.

Bulk flow (or mass flow) The concerted movement of molecules en masse, most often in response to a pressure gradient.

Bundle sheath One or more layers of closely packed cells surrounding the small veins of leaves and the primary vascular bundles of stems.

Bundle sheath cells Chloroplast-containing cells found in the leaves of C_4 plants but not in C_3 plants.

Bünning hypothesis Hypothesis that photoperiodic control of flowering is achieved by a coincidence of the light or dark phase of an inductive photoperiod with a particular phase of oscillation of the circadian rhythm that has a different sensitivity to light.

C_3 metabolism *See* Calvin cycle.

C_3 plants Plants in which the first stable product of photosynthetic CO_2 fixation is a three-carbon compound (i.e., 3-phosphoglycerate). *See* Calvin cycle.

C_4 cycle The photosynthetic carbon metabolism of certain plants in which the initial fixation of CO_2 and its subsequent reduction take place in different cells, the mesophyll and bundle sheath cells respectively. The initial carboxylation is catalyzed by phosphoenylpyruvate carboxylase, (not by rubisco as in C_3 plants), producing a four-carbon compound (oxaloacetate), which is immediately converted to malate or aspartate.

C_4 metabolism *See* C_4 cycle.

C_4 plants Plants in which the first stable product of CO_2 assimilation in mesophyll cells is a four-carbon compound that is immediately transported to bundle sheath cells and decarboxylated. The CO_2 released enters the Calvin cycle. *See* C_4 cycle.

C_{19}-gibberellins Gibberellins that have only 19 carbons because they have lost carbon-20 by metabolism.

C_{20}-gibberellins Gibberellins that possess the full diterpenoid complement of 20 carbon atoms, and which are precursors of C_{19}-GAs.

Cadastral **genes** Genes that act as spatial regulators of the floral organ identity genes by setting boundaries for their expression.

Callose A β-1,3-glucan synthesized in the plasma membrane and deposited between the plasma membrane and the cell wall. Synthesized by sieve elements in response to damage, stress, or as part of a normal developmental process.

Callus tissue The product of the disorganized growth of undifferentiated plant cells in tissue culture.

Calmodulin A conserved calcium-binding protein found in all eukaryotes that regulates many calcium-driven, metabolic reactions.

Calvin cycle The biochemical pathway for the reduction of CO_2 to carbohydrate. The cycle involves three phases: the carboxylation of ribulose-1,5-bisphosphate with atmospheric CO_2, catalyzed by rubisco, the reduction of the formed 3-phosphoglycerate to trioses phosphate by 3-phosphoglycerate kinase and NADP-glyceraldehyde-3-phosphate dehydrogenase, and the regeneration of ribulose-1,5-bisphospate through the concerted action of ten enzymatic reactions.

CAM plants Plants that fix CO_2 during the night into a four-carbon compound (malate) that, after storage in the vacuole, is transported out of the vacuole and decarboxylated during the day. The CO_2 released is assimilated by the Calvin cycle in the chloroplast stroma.

Cambium (vascular cambium) Layer of meristematic cells between the xylem and phloem that produces cells of these tissues and results in the lateral (secondary) growth of the stem or root.

Campesterol The sterol progenitor of brassinosteroids, ultimately derived from cycloartenol.

Canavanine Toxic, nonprotein amino acid that is a close analog of the protein amino acid arginine.

Capillarity The movement of water for small distances up a glass capillary tube or within the cell wall, due to water's cohesion, adhesion, and surface tension.

Carbon dioxide compensation point The CO_2 concentration (partial pressure of CO_2 in the intercellular air space) at which CO_2 fixed by photosynthesis and CO_2 produced by respiration equal each other.

Carboxylation The reaction in which a carboxylase catalyzes the formation of a carbon–carbon bond between CO_2 and the carbon atom of an organic molecule.

Cardenolides Steroidal glycosides that taste bitter and are extremely toxic to higher animals through their action on Na^+K^+-activated ATPases. Extracted from foxglove (*Digitalis*) for treatment of human heart disorders.

Carotenoids Linear polyenes arranged as a planar zigzag chain, with the repeating conjugated double-bond system —CH=CH—CCH$_3$=CH—. These orange pigments serve both as antenna pigments and photoprotective agents.

Carpels The structure that contains the ovules of flowering plants, consisting of ovary, style, and stigma. The carpel develops into the fruit and the ovule develops into the seed. *See also* ovule, sepals, petals, and stamens.

Carrier-mediated transport Active or passive transport of a solute accomplished by a carrier.

Carriers Membrane-transport proteins that bind to a solute, undergo conformational change, and release the solute on the other side of the membrane.

Cascade A succession of interactions so that each interaction derives from or acts upon the product of the preceding. A series of interactions that involves an amplification of the initial action.

Casparian strip A band in the cell walls of the endodermis that is impregnated with the waxlike, hydrophobic substance suberin. Prevents water and solutes from entering the xylem by moving between the endodermal cells.

Cation exchange The replacement of mineral cations adsorbed to the surface of soil particles by other cations.

Cavitation The collapse of tension in a column of water resulting from the indefinite expansion of a tiny gas bubble.

CCCP (Carbonyl cyanide m-chlorophenylhydrazone) An ionophore that makes the plasma membrane highly permeable to protons, thus abolishing the proton gradient.

Cdc2 protein The major cyclin-dependent protein kinase, *c*ell *di*vision *c*ycle 2. Biosynthesis stimulated by auxin.

CDK *See* cyclin-dependent protein kinase.

Cell number A convenient means to measure the growth of unicellular organisms. In multicellular plants, increasing numbers of cells are usually associated with cell expansion, particularly in meristems.

Cell plate Wall-like structure that separates newly divided cells. Formed by the phragmoplast and later becomes the cell wall.

Cell wall matrix Plant cell wall material consisting of hemicelluloses and pectins plus a small amount of structural protein.

Cellobiose A 1→4-linked β-D-glucose disaccharide that makes up cellulose.

Cellulose Linear chains of 1→4-linked β-D-glucose. The repeating unit is cellobiose.

Cellulose microfibril Thin, ribbon-like structure of indeterminate length and variable width composed of 1→4-linked β-D-glucan chains tightly packed in crystalline arrays alternating with less organized amorphous regions. Provides structural integrity to the cell walls of plants and determines the directionality of cell expansion.

Cellulose synthase Enzyme that catalyzes the synthesis of individual 1→4-linked β-D-glucans that make up the cellulose microfibril.

Central zone A central cluster of relatively large, highly vacuolate, slow-dividing cells in shoot apical meristems, comparable to the quiescent center of root meristems.

Cereal grains Seeds of grasses consisting of the diploid embryo, the triploid endosperm, and the fused seed coat–fruit wall.

CF_0F_1 ATPase A multi-protein complex associated with the thylakoid membrane that couples the passage of protons across the membrane to the synthesis of ATP from ADP and phosphate. Similar to F_0F_1ATP synthase in oxidative phosphorylation but much less sensitive to oligomycin.

Channels Transmembrane proteins that function as selective pores for passive transport of ions or water across the membrane.

Chelates Substances such as EGTA that form a complex with divalent cations eliminating their biological activity.

Chelator A carbon compound that can form a noncovalent complex with certain cations facilitating their uptake (e.g., malic acid, citric acid).

Chemical potential The free energy associated with a substance that is available to perform work.

Chemical potential of water *See* water potential.

Chemical-potential gradient A change in the free energy per mole of a substance, measured over a given distance. A substance moving spontaneously does so down its chemical potential gradient.

Chemiosmotic mechanism The mechanism whereby the electrochemical gradient of protons established across the thylakoid membrane by the electron transport process between photosystems II and I is used to drive energy requiring cellular processes such as ATP synthesis. It also operates in mitochondrial respiration and at the cell plasma membrane.

Chilling injury Changes that occur when plants growing at 25 to 35°C are cooled to 10 to 15°C. Includes slowed growth, leaf discoloration, and/or lesions. Contrast with freezing injury.

Chilling-sensitive plants Plants that experience a sharp reduction in growth rate at temperatures between 0 and 12°C.

2-chloroethylphosphonic acid *See* ethephon.

Chlorophyll A group of light absorbing green pigments active in photosynthesis.

Chlorophyll *a/b* antenna proteins *See* light harvesting complex proteins.

Chlorophyllase An enzyme that removes the phytol from chlorophyll as part of the chlorophyll breakdown process.

Chlorophyte Unicellular photosynthetic eukaryote whose chloroplasts contain chlorophyll *a* and *b* (green algae).

Chloroplast The organelle that is the site of photosynthesis in eukaryotic photosynthetic organisms.

Chlorosis The yellowing of older, lower plant leaves characteristic of prolonged nitrogen deficiency.

Cholodny–Went model Early mechanism proposed for tropisms involving stimulation of the bending of the plant axis by lateral transport of auxin in response to a stimulus, such as light, gravity, or touch. The original model has been supported and expanded by recent experimental evidence.

Chromophore A light-absorbing pigment molecule that is usually bound to a protein (an apoprotein).

Chronic photoinhibition Photoinhibition of photosynthetic activity in which both quantum efficiency and the maximum rate of photosynthesis are decreased. Occurs under excess light.

Cinnamic acid A phenylpropanoid derived from the amino acid phenylalanine that is a key intermediate in the biosynthesis of many phenolic compounds.

Circadian rhythm A biological activity that shows a cycle of high-activity and low-activity independent of external stimuli, with a regular periodicity of about 24 hours (L. *circa diem*: about a day).

Cisternal maturation/progression model Model for Golgi membrane development in which the Golgi stack is not fixed, but is a dynamic structure in which cisternae progress through *cis*, *medial*, and *trans* faces, carrying their cargo with them.

Citric acid cycle (Krebs cycle, tricarboxylic acid cyle) A cycle of reactions localized in the mitochondrial matrix that catalyze the oxidation of pyruvate to CO_2, ATP, and NADH are generated in the oxidation process.

***CKI1* gene** Gene whose overexpression confers cytokinin-independent growth on Arabidopsis cells in culture. Encodes a protein similar to bacterial histidine kinases functioning in signal transduction.

Climacteric Marked rise in respiration at the onset of ripening that occurs in all fruits that ripen in response to ethylene, and in the senescence process of detached leaves and flowers.

Clock hypothesis Currently accepted hypothesis of how plants measure night length. Proposes that photoperiodic timekeeping depends on the endogenous oscillator of circadian rhythms. *See* hourglass hypothesis.

CoA *See* coenzyme A.

Coconut milk The liquid endosperm of coconut seeds that contains cytokinins and other nutritional factors. Stimulates the growth of normal stem tissues when added to liquid culture media.

Coenzyme A A coenzyme with an —SH group that serves as an acyl group carrier for many enzymatic reactions.

Cohesion–tension theory A model for sap ascent in the xylem up the stem of the plant, stating that evaporation of water from the leaves at the top of the stem causes a tension (negative hydrostatic pressure) that pulls water up the long water columns in the xylem.

Coincidence model A model for flowering in photoperiodic plants in which the circadian oscillator controls the timing of light-sensitive and light-insensitive phases during the twenty-four hour cycle.

Colchicine A drug that destroys microtubules and blocks cell division.

Coleoptile A modified ensheathing leaf that covers and protects the young primary leaves of a grass seedling as it grows through the soil. Unilateral light perception, especially blue light, by the tip results in asymmetric growth and bending due to unequal auxin distribution in the lighted and shaded sides.

Collenchyma A specialized parenchyma with irregularly thickened, pectin-rich, primary cell walls that function in support in growing parts of a stem or leaf.

Colligative properties Properties of solutions that depend on the number of dissolved particles and not on their chemical characteristics.

Columella root cap The central region of the root cap that contains the statocytes—cells containing large, dense amyloplasts that function in gravity perception during root gravitropism.

Columella stem cells Root cap stem cells that divide to generate a sector of the root cap, the columella.

Combinatorial model Proposal that during the transition from juvenile to adult shoot in maize, a series of independently regulated, overlapping programs (juvenile, adult, and reproductive) modulate the expression of a common set of developmental processes.

Companion cell In angiosperms, a metabolically active cell that is connected to its sieve element by large, branched plasmodesmata and takes over many of the metabolic activities of the sieve element. In source leaves, it functions in the transport of photosynthate into the sieve elements.

Compatible solutes (compatible osmolytes) Organic compounds that are accumulated in the cytosol during osmotic adjustment. Compatible solutes do not inhibit cytosolic enzymes as do high concentrations of ions. Examples of compatible solutes include proline, sorbitol, mannitol, glycine betaine.

Competence The capacity of a particular cell or group of cells to respond in the expected manner when given the appropriate developmental signal.

Complex I A protein complex in the mitochondrial electron transport chain that oxidizes NADH and reduces ubiquinone.

Complex II A protein complex in the mitochondrial electron transport chain that oxidizes succinate and reduces ubiquinone.

Complex III A protein complex in the mitochondrial electron transport chain that oxidizes reduced ubiquinone (ubiquinol) and reduces cytochrome *c*.

Complex IV A protein complex in the mitochondrial electron transport chain that oxidizes reduced cytochrome *c* and reduces O_2 to H_2O.

Complex V *See* F_oF_1-ATP synthase.

Compound (or mixed) fertilizers Contain two or more mineral nutrients; numbers such as 10–14–10 refer to the effective percentages of nitrogen, phosphorus, and potassium.

Condensed tannins Tannins that are polymers of flavonoid units. Require use of strong acid for hydrolysis.

Constitutive Constantly present or expressed, whether there is demand or not. Refers to the ongoing synthesis of a particular protein. Contrast with inducible.

Constitutive ethylene response mutants Mutants that show the ethylene triple response in the absence of exogenous ethylene. *See ctr* mutant.

COP *See* critical oxygen pressure.

COP1 COP1 is an E3 ubiquitin ligase that has been shown to interact with several proteins involved in the phytochrome response.

Cork cambium A layer of lateral meristem that develops within mature cells of the cortex and the secondary phloem. Produces the secondary protective layer, the periderm.

Corpus The internal cytohistological zones of the shoot apical meristem: central zone, peripheral zone, rib meristem.

Cortical cytoplasm The outer region or layer of cytoplasm adjacent to the plasma membrane.

Cortical–endodermal stem cells A ring of stem cells that surround the quiescent center and generate the cortical and endodermal layers in roots.

Cotransport The simultaneous transport of two solutes by the same carrier. Usually one solute is moving down its chemical-potential gradient, while the other is moving against its chemical-potential gradient. *See* symport and antiport.

Cotyledons The one or more seed leaves contained in the seed of seed plants. In some seeds they are storage organs supporting early non-photosynthetic growth of the seedling. In other seeds, they absorb and transmit to the seedling resources stored in the endosperm. *See also* monocot and dicot.

Coumarins A group of phenylpropanoid compounds including the phototoxic furanocoumarins, and other substances responsible for the odor of fresh hay.

Coupled reactions *See* coupling.

Coupling A process by which a chemical reaction releasing free energy is linked to a reaction requiring free energy.

Crassulacean acid metabolism (CAM) A biochemical process for concentrating CO_2 at the carboxylation site of rubisco. Found in the family Crassulaceae (*Crassula, Kalanchoe, Sedum*) and numerous other families of angiosperms. In CAM, CO_2 uptake and fixation take place at night, and decarboxylation and reduction of the internally released CO_2 occur during the day.

***CRE1* gene** Arabidopsis gene that encodes a cytokinin receptor protein, similar to bacterial two-component histidine kinases.

Cristae Folds in the inner mitochondrial membrane that project into the mitochondrial matrix. The enzymes of the electron transport chain and of oxidative phosphorylation are localized in the cristae.

Cytochrome *c* A peripheral, mobile component of the mitochondrial electron transport chain that oxidizes complex III and reduces complex IV.

Critical concentration (of a nutrient) The minimum tissue content of a mineral nutrient that is correlated with maximal growth or yield.

Critical day length The minimum length of the day required for flowering of a long-day plant; the maximum length of day that will allow short-day plants to flower. However, studies have shown that it is the length of the night, not the length of the day, that is important. *See* critical night length.

Critical night length The night length that must be exceeded for flowering of short-day plants, or for inhibition of flowering in long-day plants.

Critical oxygen pressure (COP) The oxygen pressure at which the respiration rate is first decreased by O_2 deficiency.

Crown gall A tumor-forming plant disease resulting from wound infection of the stem or trunk by the soil-dwelling bacterium *Agrobacterium tumefaciens*. A tumor resulting from the disease.

***CRY1* gene** The gene encoding cryptochrome, a flavoprotein implicated in many blue light responses that has homology with photolyase. Formerly *HY4*, *see also hy4* mutant.

***cry1* mutant** Arabidopsis mutant that lacks the blue light-stimulated inhibition of hypocotyl elongation.

Cryptochrome Flavoprotein involved in the inhibition of stem elongation, possibly as a blue-light receptor or as a signaling intermediate.

***cry^s* mutant** *See la* mutant.

Crystalline Pertaining to solids having a highly ordered and repetitive geometric form.

***CTR1* gene** Encodes for a negative regulator of ethylene responses.

***ctr1* mutant** In Arabidopsis, a recessive mutation causing the constitutive expression of the ethylene triple responses (*c*onstitutive *t*riple *r*esponse 1 = triple response in the absence of ethylene).

CTR1 protein Regulator of the ethylene triple response; resembles serine/threonine protein kinases involved in signal transduction.

Curvature test A bioassay for auxin using the curvature of the *Avena* coleoptile in response to asymmetrically applied auxin in an agar block.

Cutan A lipid polymer made up of long-chain hydrocarbons that is a constituent of the cuticle. *See* cutin.

Cuticle A multilayered structure that coats the outer cell walls of the epidermis and restricts the passage of water and gases into and out of the plant. Includes cutin, cutan, and waxes.

Cutin A rigid three-dimensional polymer of hydroxyl-bearing fatty acids that are attached to each other by ester linkages. The principal constituent of the cuticle.

Cyanide An inhibitor of complex IV and other heme-containing enzymes.

Cyanide-resistant pathway *See* alternative pathway.

Cyanogenic glycosides Nonalkaloid, nitrogenous protective compounds that break down to give off the poisonous gas hydrogen cyanide when the plant is crushed.

Cyclic electron flow In photosystem I, flow of electrons from the electron acceptors through the cytochrome b_6f complex and back to *P700*, coupled to proton pumping into the lumen. This electron flow energizes ATP synthesis but does not oxidize water or reduce $NADP^+$.

Cyclins Regulatory proteins associated with CDKs that play a crucial role in regulating the cell cycle.

Cyclin-dependent protein kinase (CDK) Protein kinases that regulate the transitions from G_1 to S, and from G_2 to mitosis, during the cell cycle.

Cycloartenol The direct biochemical precursor of all plant steroid compounds, formed from squalene by a series of ring closures.

Cycloheximide Inhibits eukaryotic protein synthesis on 80S ribosomes, but does not block protein synthesis in prokaryotes, mitochondria, or chloroplasts.

Cytochalasin B A drug that destroys actin filaments.

Cytochrome b_6f complex A large multi-subunit protein containing two *b*-type hemes, one *c*-type heme (cytochrome *f*), and a Rieske iron–sulfur protein. A nonmobile protein distributed equally between the grana and the stroma regions of the membranes.

Cytochrome p450 monooxygenases (CYPs) A generic term for a large number of related, but distinct, mixed-function oxidative enzymes localized on the endoplasmic reticulum. CYPs participate in a variety of oxidative processes, including steps in the biosynthesis of gibberellins and brassinosteroids.

Cytohistological zones Regions of the shoot apical meristem showing differences in morphological appearance and rates of mitosis.

Cytokinesis In plant cells, following nuclear division, the separation of daughter nuclei by the formation of new cell wall.

Cytokinin oxidase Enzyme that inactivates cytokinins by removing the iosprene moiety from adenine or its derivatives.

Cytokinin synthase Plant enzyme that transfers the isopentenyl group from isopentenyl diphosphate to AMP to form isopentenyl adenine ribotide, the first unique intermediate in the synthesis of cytokinin. An isopentenyl transferase.

Cytokinins Compounds with many developmental effects on plants, including leaf senescence, nutrient mobilization, apical dominance, the formation and activity of shoot apical meristems, floral development, the breaking of bud dormancy, and seed germination. Cytokinins mediate many light-regulated processes, including chloroplast development and metabolism, and expansion of leaves and cotyledons. May exist as free and bound forms. Operationally defined as compounds with biological activities similar to those of *trans*-zeatin.

Cytoplasm The cellular matter enclosed by the plasma membrane exclusive of the nucleus, that contains the cytosol, ribosomes, and the cytoskeleton which, in eukaryotes, surrounds intracellular and membrane-limited organelles (chloroplasts, mitochondria, endoplasmic reticulum, etc.).

Cytoplasmic male sterility (*cms*) A plant trait caused by mutations in mtDNA in which viable pollen is not formed.

Cytosol The colloidal-aqueous phase of the cytoplasm containing dissolved solutes but excluding supramolecular structures, such as ribosomes and components of the cytoskeleton.

d8 mutant In maize, phenotypically dwarf mutant that is insensitive to added GA.

Day-neutral plant A plant whose flowering is not regulated by day length.

DCMU (Dichlorophenyldimethylurea, diuron) An herbicide that blocks photosynthetic electron flow by displacing Q_B from the photosystem II reaction center complex. It also causes a partial inhibition of light-stimulated stomatal opening.

de novo synthesis Synthesis and/or assembly from simple molecular species.

Deciduous trees Trees that shed their leaves seasonally.

Decussate phyllotaxy Phyllotactic pattern in which the two opposite leaves at one node are at right angles to the two leaves at the next node.

De-differentiation Process by which cells lose their differentiated characteristics, reinitiate cell division, and may, under appropriate conditions, regenerate whole plants.

De-embryonate To remove the embryo from a cereal seed, leaving only the starchy endosperm, aleurone layer, and fused seed coat-fruit wall. De-embryonated half-seeds are dependent on exogenous GA to produce α-amylase.

De-etiolation Rapid developmental changes associated with loss of the etiolated form due to the action of light. *See* photomorphogenesis.

Deficiency zone Concentrations of a mineral nutrient in plant tissue below the critical concentration that reduces plant growth.

Dehydration-response element (DRE) A nine-nucleotide regulatory sequence element found in the promoters of genes regulated by abscisic acid (ABA), a plant hormone.

DELLA domain N-terminal domain consisting of 17 amino acids in GRAS transcriptional regulators that is necessary for their proteolysis.

Denitrification Process by which soil anaerobic microbes convert nitrate (NO_3^-) or nitrite (NO_2^-) to the gases, nitrous oxide (N_2O) and molecular nitrogen (N_2), which are then lost to the atmosphere.

Depolarization Refers to a decrease in the usually negative membrane potential difference across the plasma membrane of plant cells. May be caused by the activation of anion channels and loss of anions, such as chloride from the cell interior, which is negative with respect to the outside.

Desiccation postponement A plant's ability to maintain tissue hydration when environmental water is limited.

Desiccation tolerance (drought avoidance) Plant's ability to function while dehydrated.

Determinate growth Inability to grow beyond the mature, reproductive stage due to the loss of meristematic activity. For example, plants that form flowers after the development of a preset numbers of leaves. Flowering times are generally short and all meristems of the plant are used up in the production of flowers.

Determination Developmental state in which the plant follows a developmental program even after it is removed from its normal physical and environmental context.

Dextranase Catalyzes the endohydrolysis of α-D-1,6 glucosidic linkages in dextran.

Diaheliotropism Leaf movements that maximize light interception by solar tracking and minimize overexposure to light.

Dicot One of the two classes of flowering plants characterized by two seed leaves (cotyledons) in the embryo. Contrast with monocot.

Differentiation Process by which a cell acquires metabolic, structural, and functional properties that are distinct from those of its progenitor cell. In plants, differentiation is frequently reversible, when excised differentiated cells are placed in tissue culture.

Diffuse growth A type of cell growth in plants in which expansion occurs more or less uniformly over the entire surface. Contrast with tip growth.

Diffusion The movement of substances due to random thermal agitation from regions of high free energy (high concentration) to regions of low free energy (low concentration).

Diffusion coefficient (D_s) The proportionality constant that measures how easily a specific substance *s* moves through a particular medium. The diffusion coefficient is a characteristic of the substance and depends on the medium.

Diffusion potential The potential (voltage) difference that develops across a semipermeable membrane as a result of the differential permeability of solutes with opposite charges (for example K^+ and Cl^-).

Diffusional resistance Restriction posed by the boundary layer and the stomata to the free diffusion of gases from and into the leaf.

Dioecious Refers to plants in which male and female flowers are found on different individuals (e.g., spinach [*Spinacia*] and hemp [*Cannabis sativa*]). Contrast with monoecious.

Dioxygenases Group of oxygenase enzymes that incorporate the two oxygen atoms from O_2 into one or two-carbon compounds.

Direct-inhibition model Hypothesis for *apical dominance* stating that the growth of lateral buds is directly inhibited by the relatively high auxin concentration in the stem produced by the shoot apex.

Disproportionating enzyme Catalyzes the transfer of a segment of a α-D-1,4 glucan to a new position in an acceptor, which may be glucose or a 1,4-linked α-D-glucan.

Diterpenes Terpenes having twenty carbons, four five-carbon isoprene units.

Dormancy A living condition in which growth does not occur under conditions that are normally favorable to growth.

Drought avoidance *See* desiccation tolerance.

Drought escape Capacity of a plant to grow and complete its life cycle during the wet season, before the onset of drought.

Drought resistance Plant's capacity to limit and control the consequences of water deficit. Mechanisms include desiccation postponement and desiccation tolerance.

Dry weight Weight of desiccated (dried) tissue. Often used to measure growth. Avoids variation in water content when fresh weight measurements are made.

DTT (dithiothreitol) An inhibitor of the enzyme that converts violaxanthin to zeaxanthin. It inhibits blue light-stimulated stomatal opening. A reducing agent that reduces S—S bonds to –SH groups.

Dynamic photoinhibition Photoinhibition of photosynthesis in which quantum efficiency decreases but the maximum photosynthetic rate remains unchanged. Occurs in moderate, not high, excess light.

Early genes *See* primary response genes.

Ectotrophic mycorrhizal fungi A dense sheath of fungal mycelium around the roots and extending into the surrounding soil; some hyphae may penetrate between, but not into, the root cortical cells.

EGTA (ethylene glycol-bis[β-aminoethyl ether]-N,N,N′,N′-tetraacetic acid) A compound that chelates calcium ions, preventing their uptake by cells and inhibiting both root gravitropism and the asymmetric distribution of auxin in response to gravity. *See* plasmalemma central control model.

Eicosanoids A group of substances involved in the mammalian inflammatory response and similar to jasmonic acid found in plants.

***ein* mutants** In Arabidopsis, mutants blocked in their ethylene responses (*ethylene-in*sensitive).

***EIN2* gene** In Arabidopsis, ethylene response gene encoding a protein that may act as a membrane channel protein. *See ein* mutants.

***EIN3* gene** In Arabidopsis, ethylene response gene encoding a transcription factor. *See ein* mutants.

Electrochemical potential The chemical potential of an electrically charged solute.

Electrogenic transport Active ion transport involving the net movement of charge across a membrane.

Electron spin resonance (ESR) A magnetic resonance technique that detects unpaired electrons in molecules. Instrumental measurements that identify intermediate electron carriers in the photosynthetic electron transport system.

Electron transport chain (in the mitochondrion) A series of protein complexes in the inner mitochondrial membrane linked by the mobile electron carriers ubiquinone and cytochrome *c*, that catalyze the transfer of electrons from NADH to O_2. In the process a large amount of free energy is released. Some of that energy is conserved as an electrochemical proton gradient.

Electronegative Having a negative electric charge.

Electroneutral transport Active ion transport that involves no net movement of charge across a membrane.

***elf3* mutant** In Arabidopsis, a flowering-time mutant (*early flowering 3*).

Elicitors Specific pathogen molecules or cell wall fragments that bind to plant proteins and thereby signal for plant defense against a pathogen. *See avr* genes.

ELISA *E*nzyme-*l*inked *i*mmuno*s*orbent *a*ssay. A very sensitive detection method that uses radioisotopes or chemiluminescent compounds linked to antibodies to detect compounds (like IAA) or proteins in tissue extracts.

Elongation Growth of the plant axis or cell primarily in the longitudinal direction.

Elongation zone (of a root) The region of rapid and extensive root cell elongation showing few, if any, cell divisions.

Embryo Immature plant formed after sexual or asexual reproduction. Found in seed (of seed plants) and consists of an embryonic axis bearing a terminal bud (plumule), root (radicle), and one or more seed leaves (cotyledons).

Embryo sac In flowering plants, large oval cell (arising from the megaspore) in the ovule that develops into the female gametophyte and in which fertilization of the egg and development of the embryo take place.

Embryogenesis The development of a zygote into a multicellular embryo. In plants, the processes of cell division and differentiation that take place in the ovule and immature seed and establishes the basic developmental patterns of the adult plant: the radial pattern of tissues; the apical–basal axis; and the primary meristems.

Embryonic axis The hypothetical central line of an embryo around which lateral organs are arranged.

***emf* mutant** In Arabidopsis, a mutant of *EMBRYONIC FLOWER* (*EMF*) that produces flowers soon after germination; the shoot apical meristem initiates no vegetative structures, but immediately initiates floral organs during germination.

Endodermis A specialized layer of cells with a Casparian strip surrounding the vascular tissue in roots and some stems.

Endogenous oscillator An internal, molecular pacemaker that maintains circadian rhythms in a constant light or dark environment.

Endogenous rhythm A rhythm that persists in the absence of external controlling factors such as light.

Endogenous Relating to the interior of a living system. Concerning what is found within or originates within such a system.

Endosymbiosis A theory that explains the evolutionary origin of the chloroplast and mitochondrion through formation of a symbiotic relationship between a prokaryotic cell and a simple nonphotosynthetic eukaryotic cell, followed by extensive gene transfer to the nucleus.

Endosymbiotic origin A theory in which the engulfment of a free-living ancestral prokaryotic cell by a eukaryotic host evolved into a cell with semi-autonomous chloroplasts and mitochondria.

Energy transfer In the light reactions of photosynthesis, the direct transfer of energy from an excited molecule, such as carotene, to another molecule, such as chlorophyll. Energy transfer can also take place between chemically identical molecules, such as chlorophyll-to-chlorophyll transfer.

Enhancement effect The synergistic (higher) effect of red and far-red light on the rate of photosynthesis, as compared with the sum of the rates when the two different wavelengths are delivered separately.

Entrainment The synchronizing effects of external controlling factors, such as light and darkness, on the period of biological rhythms.

Envelope The double-membrane system surrounding the chloroplast or the nucleus. The outer membrane of the nuclear envelope is continuous with the endoplasmic reticulum.

24-Epibrassinolide (24-epiBL) A brassinosteroid hormone differing from brassinolide (BL) in the alkyl side chain at C-24, and possessing slightly lower activity than BL.

Epicotyl The region of the seedling stem above the cotyledons.

Epidermis (epidermal cells) The outermost layer of plant cells, typically one cell thick.

Epimers A pair of isomers that differ from each other only in their configuration at a single asymmetric center.

Epinasty A downward curvature of leaves due to asymmetric growth of the petiole. A response to ethylene production during flooding.

Equilibrium For a particular solute, a state or condition in which there is no gradient of (electro)chemical potential, and therefore no net passive transport of the solute.

ERE See ethylene response element.

ERE-binding proteins (EREBPs) Proteins that bind to *ERE* sequences.

Escape from photoreversibility The loss of photoreversibility by far-red light of phytochrome-mediated red light–induced events after a short period of time.

ESR See electron spin resonance.

Essential element A chemical element that is part of a molecule that is an intrinsic component of the structure or metabolism of a plant. When the element is in limited supply, a plant suffers abnormal growth, development, or production.

Essential oils Mixtures of volatile terpenes and other secondary metabolites that give characteristic odors to some plants, e.g., peppermint, lemon, basil, and sage.

Ethephon An ethylene-releasing compound, 2-chloroethylphosphonic acid, that makes practical the application of the plant hormone ethylene gas under field conditions. Trade name, Ethrel.

Ethylene Ethylene gas ($CH_2{=}CH_2$) functions as a plant hormone that is synthesized from the amino acid methionine via ACC. Has important effects on plant growth and development, including stimulating or inhibiting elongation of stems, roots, depending on conditions and species; enhances fruit development; suppresses flowering in most species; increases abscission of flowers and fruits. Increases RNA transcription of numerous genes. See also ethylene triple response.

Ethylene response element Key regulatory sequence found in genes regulated by ethylene.

Ethylene triple response Common responses to ethylene of growing etiolated seedlings of most dicots and of coleoptiles and mesocotyls of grass seedlings (e.g., wheat and oats). Reduced rate of elongation, increased lateral expansion, and swelling in the region below the hook.

Ethylene-insensitive mutants Mutants that do not respond to ethylene. In Arabidopsis, tall seedlings protruding above short seedlings showing ethylene triple response when grown in the presence of ethylene.

Etiolated seedlings Dark-grown seedlings in which the hypocotyl and stem are more elongated, cotyledons and leaves do not expand, and chloroplasts do not mature.

Etiolation The form and growth of seedlings grown in darkness. A pale, unusually tall and slender appearance, dramatically different from the stockier, green appearance of seedlings grown in the light.

Etioplast Photosynthetically inactive form of chloroplast found in etiolated seedlings. Does not synthesize chlorophyll or most of the enzymes and structural proteins required for the formation of thylakoids and operation of photosynthesis. Contains an elaborate system of interconnected membrane tubules called the prolamellar body.

etr1 **mutant** In Arabidopsis, a dominant mutation that blocks the response to ethylene, (*et*hylene-*r*esistant 1).

Evaporative heat loss Loss of heat due to the cooling resulting from the evaporation of water.

Exodermis (or hypodermis) In mature roots, an outer layer of cells of the cortex that is relatively impermeable to water.

Exogenous Relating to or coming from outside a living system. Originating outside such a system.

Expansins Class of wall-loosening proteins that accelerate wall stress relaxation and cell expansion, typically with an optimum at acidic pH. Appears to mediate acid growth.

Expansion Cell and tissue growth due to increase in cell size, not cell number.

Export The movement of photosynthate in sieve elements away from the source tissue.

Facilitated diffusion Passive transport across a membrane using a carrier.

FAD See flavin adenine dinucleotide.

Fermentative metabolism Fermentative metabolism. The metabolism of pyruvate in the absence of oxygen, leading to the oxidation of the NADH generated in glycolysis to NAD^+. See also alcoholic fermentation and lactic acid fermentation.

Ferredoxin (Fd) A small, water-soluble, iron–sulfur protein involved in electron transport in photosystem I.

Ferredoxin–thioredoxin system Three chloroplast proteins (ferredoxin, ferredoxin-thioredoxin reductase, thioredoxin). The concerted action of the three proteins uses reducing power from the photosynthetic electron transport system to reduce protein disulfide bonds by a cascade of thiol/disulfide exchanges. As a result, light controls the activity of several enzymes of the Calvin cycle.

Fiber An elongated, tapered sclerenchyma cell that provides support in vascular plants.

Fibrous root system The complex root system of monocots, which lack a main root axis; all roots have about the same diameter.

Fick's first law The rate of diffusion is directly proportional to the concentration gradient, defined as the difference in concentration of a substance between two points separated by the distance Δx.

Field capacity The water content of a soil after it has been saturated with water and excess water has been allowed to drain away. The moisture-holding capacity of soils.

Fixed carbon See photosynthate.

Flavin adenine dinucleotide (FAD) A riboflavin-containing cofactor that undergoes a reversible one or two electron reduction to produce FADH or $FADH_2$.

Flavin hypothesis Discredited suggestion that riboflavin activated by blue light could participate in the in vivo photodestruction of an auxin. Although blue light can reduce a flavin such as riboflavin in vitro, and the reduced flavin can in turn reduce cytochrome c, no in vivo role for these photoreactions has been shown.

Flavin mononucleotide (FMN) A riboflavin-containing cofactor that undergoes a reversible one or two electron reduction to produce FMNH or $FMNH_2$.

Flavones Group of ultraviolet light-absorbing, protective flavonoids that may also attract pollinating insects to flowers. Secreted along with flavonoids into the soil by legume roots, they mediate interaction with nitrogen-fixing symbionts.

Flavonoids A large group of plant phenolics with the basic carbon structure of two aromatic rings connected by three carbons. Includes the anthocyanins, the flavones, the flavonols, and the isoflavones. Functions in plant pigmentation, protection against UV irradiation, and defense against herbivores and pathogens.

Flavonols Group of ultraviolet light-absorbing, protective flavonoids that may also attract pollinating insects to flowers. Secreted into the soil by legume roots, they mediate interactions with nitrogen-fixing symbionts.

Flavoprotein ferredoxin-NADP reductase (Fp) In photosystem I, a membrane-associated flavoprotein that reduces $NADP^+$, using electrons from ferredoxin. This reaction completes the noncyclic electron transport that begins with the oxidation of water.

Flooding-sensitive Refers to plants that are severely damaged by 24 hours of anoxia due to flooding of roots.

Flooding-tolerant Refers to plants that can withstand anoxia temporarily, but not for more than a few days.

Floral evocation The events occurring in the shoot apex that specifically commit the apical meristem to produce flowers.

Floral homeotic genes Key regulatory genes that determine the positions and identities of floral organs in flowers.

Floral meristem Forms floral (reproductive) organs: sepals, petals, stamens, and carpels. May form directly from vegetative meristems or indirectly via an inflorescence meristem.

Floral organs Angiosperm organs involved directly or indirectly in sexual reproduction; sepals, petals, stamens, and carpels.

Floral primordia A primordium is the earliest recognizable stage when a group of cells begins to form a structure. In the developing Arabidopsis flower, four discrete whorls of floral primordia are formed from a floral meristem, and these primordia give rise to sepals, petals, stamens, ovules, style, and stigma.

Floral reversion The conversion of a floral meristem to a vegetative or inflorescence meristem, causing a shoot or inflorescence to grow directly out of the developing flower.

Floral stimulus In photoperiodism, signals that are translocated from the leaves to the shoot apical meristem. See florigen.

Florigen The hypothetical, universal flowering hormone synthesized by leaves and translocated to the shoot apical meristem via the phloem. So far, it has not been isolated or characterized.

Flower Specialized reproductive shoot structure of angiosperms. Consists of nonreproductive organs (sepals and petals) and reproductive organs (stamens and carpels).

Fluence The number of photons absorbed per unit surface area.

Fluence rate A unit for the measurement of light falling on a spherical sensor from many directions expressed as watts per square meter (W m^{-2}) or moles of photons per square meter per second (mol m^{-2} s^{-1}). See irradiance.

Fluorescence Following light absorption, the emission of light at a slightly longer wavelength (lower energy) than the wavelength of the absorbed light.

Flux density (J_s) The rate of transport of a substance s across a unit area per unit time. J_s may have units of moles per square meter per second [mol m^{-2} s^{-1}].

FMN See flavin mononucleotide.

F_oF_1-ATP synthase A multi-protein complex associated with the inner mitochondrial membrane that couples the passage of protons across the membrane to the synthesis of ATP from ADP and phosphate. The subscript 'o' in F_o refers to the binding of the inhibitor oligomycin. Similar to CF_0–CF_1 ATPsynthase in photophosphorylation.

Foliar application The application and absorption of some mineral nutrients to leaves as sprays.

Formative divisions Cell divisions that form new cell files in an axis, as in a root apical meristem, usually longitudinally oriented.

Free energy A thermodynamic term representing a capacity to perform work.

Free-running Designation of the biological rhythm that is characteristic for a particular organism when environmental signals (zeitgebers) are removed, as in total darkness.

Freezing injury Injury that occurs when plants are cooled below the freezing point of water. Contrast with chilling injury.

Frequency (V) A unit of measurement that characterizes waves, in particular light energy. The number of wave crests that pass an observer in a given time.

Fresh weight Weight of living tissue.

Fruit In angiosperms, one or more mature ovaries containing seeds and sometimes adjacent attached parts.

Fruit set The start of fruit growth following pollination.

Furanocoumarins Group of coumarins with attached furan rings whose toxicity may result from exposure to light.

Fusicoccin A fungal toxin that induces acidification of plant cell walls by activating an H^+-ATPase in the plasma membrane. Fusicoccin stimulates rapid acid growth in stem and coleoptile sections. It also stimulates stomatal opening by stimulating proton pumping at the guard cell plasma membrane.

Fusiform stem cells Elongated, vacuolate stem cells of the vascular cambium that divide longitudinally and whose derivatives form the conducting cells of the secondary xylem and phloem.

G protein GTP-binding protein involved in signal transduction.

GA *See* gibberellins.

GA_1 *See* gibberellin A_1.

GA_3 *See* gibberellin A_3.

GA_4 *See* gibberellin A_4.

GA_{12}-aldehyde The end-product of the second stage of gibberellin biosynthesis and the precursor for all other gibberellins.

***GAI* gene** In Arabidopsis, encodes a transcription factor that represses GA responses in the absence of a GA signal.

***GA1* mutants** In Arabidopsis, dwarf mutants deficient in biologically active gibberellins.

Gall An unorganized mass of tumorlike plant tissue resulting from uncontrolled cell division.

Gas chromatography (GC) Method that separates components of a mixture according to their affinity for inert column material versus their tendency to volatilize and move through the column on a current of inert gas.

Gate A structural domain of the channel protein that opens or closes the channel in response to external signals such as voltage changes, hormone binding, or light.

G-box A specific sequence of DNA nucleotides that bind G-box type *cis*-acting transcription factors, leading to the transcriptional activation of genes.

GDH *See* glutamate dehydrogenase.

Gel A highly-hydrated, dispersed network of long polymers, typically with elastic properties intermediate between that of a liquid and that of a solid.

Gene activators Proteins that, either alone or in concert with other proteins, increase gene expression. *See* transcription factor.

Genetic tumors Spontaneous tumors produced by certain genotypes. Form in about ten percent of interspecific crosses of the genus *Nicotina* due to overproduction of cytokinin.

Geranylgeranyl diphosphate (GGPP) A precursor in the synthesis of gibberellins.

Germinaton The beginning or resumption of growth by a spore, seed, or bud.

GGPP *See* geranylgeranyl diphosphate.

***GH3* genes** In soybean and Arabidopsis, family of primary response genes that are stimulated by auxin within 5 minutes of treatment.

Gibberellic acid Identical to gibberellin A_3.

Gibberellin A_1 (GA_1) A chemically distinct form of gibberellin. The primary active GA in stem growth for most species.

Gibberellin A_3 (GA_3) The main gibberellin found in fungal cultures; commonly available and used in the management of fruit crops, the malting of barley, and the extension of sugarcane (to increase sugar yield). Occurs rarely in plants.

Gibberellin A_4 (GA_4) Thought to be the primary active gibberellin for stem growth in Arabidopsis.

Gibberellin-aldehyde *See* GA_{12}-aldehyde.

Gibberellin glycosides Inactive or stored forms of gibberellin in which the hormone is covalently linked to a sugar (usually glucose).

Gibberellins (GAs) A large group of chemically related plant hormones synthesized by a branch of the terpenoid pathway and associated with the promotion of stem growth (especially in dwarf and rosette plants), seed germination, and many other functions. *See* gibberellic acid, gibberellin-aldehyde, Gibberellin A_1.

***GID1* gene** *Gibberellin insensitive dwarf 1* encodes a soluble gibberellin receptor in rice.

***GID2* gene** *Gibberellin insensitive dwarf 2* encodes an F-box protein in rice that targets SLR repressor protein for proteolytic degradation.

Girdling Removal of a ring of bark from a woody stem that severs the vascular system.

Globular stage embryo The first stage of embryogenesis. A radially symmetrical, but not developmentally uniform, sphere of cells produced by initially synchronous cell divisions of the zygote. *See* heart stage, torpedo stage.

Gluconeogenesis The synthesis of carbohydrates through the reversal of glycolysis.

Glucosinolates (mustard oil glycosides) Plant glycosides that break down to release volatile substances that defend against herbivores and are responsible for the odor and flavor of broccoli, cabbage, and other cruciferous vegetables.

Glutamate dehydrogenase (GDH) Catalyzes a reversible reaction that synthesizes or deaminates glutamate as part of the nitrogen assimilation process.

Glutamate synthase Enzyme that transfers the amide group of glutamine to 2-oxoglutarate, yielding two molecules of glutamate. Also known as glutamine:2-oxoglutarate aminotransferase (GOGAT).

Glutamine synthetase Enzyme that converts ammonium into the amino acid glutamine.

Glutathione The form in which assimilated sulfur is exported from leaves to sites of protein synthesis (shoot and root apices, and fruits). Signal molecule that coordinates root absorption of sulfate and shoot assimilation of sulfate.

Glyceroglycolipids Glycerolipids in which sugars form the polar head group. Glyceroglycolipids are the most abundant glycerolipid in chloroplast membranes.

Glycerolipids Polar lipids that form the lipid bilayer of cellular membranes.

Glycerophospholipids (phospholipids) Glycerolipids in which the polar head group contains phosphate. Most important glycerolipids in nonphotosynthetic tissue.

Glycine oxidation The part of the photorespiratory carbon cycle in which glycine is converted into serine in the mitochondrial matrix, producing NADH, CO_2, and NH_4^+.

Glycolysis A series of reactions in which glucose is partly oxidized to produce two molecules of pyruvate. A small amount of ATP and NADH is produced.

Glycophytes Plants that are not able to resist salts to the same degree as halophytes. Show growth inhibition, leaf discoloration, and loss of dry weight at soil salt concentrations above a threshold. Contrast with halophytes.

Glycoside Compound containing an attached sugar or sugars.

Glycosidic bond A bond between the C-1 of glucose and the oxygen atom of another hexose moeity sugar can be linked to each other by O-glycosidic bonds to form oligo- and polysaccharides.

Glyoxylate cycle The sequence of reactions that convert two molecules of acetyl-CoA to succinate in the glyoxysome.

Glyoxysome An organelle found in the oil-rich storage tissues of seeds in which fatty acids are oxidized. A type of microbody.

***GNOM* gene** Arabidopsis gene for the development of roots and cotyledons. Homozygous *GNOM* mutant produces seedlings lacking both roots and cotyledons.

GOGAT *See* glutamate synthase.

Goldman diffusion potential The diffusion potential calculated from the Goldman equation.

Goldman equation An equation that predicts the diffusion potential across a membrane, as a function of the concentrations and permeabilities of all ions (e.g., K^+, Na^+ and Cl^-) that permeate the membrane.

Gossypol An aromatic 30-carbon sesquiterpene dimer that defends against insects, fungi, and bacterial pathogens in cotton.

Grana lamellae Stacked thylakoid membranes within the chloroplast. Each stack is called a granum, while the exposed membranes in which stacking is absent are known as stroma lamellae.

GRAS domain The C-terminal domain in GRAS proteins that functions as a transcriptional repressor.

GRAS proteins A family of transcriptional regulators named after the first three to be characterized (GAI, RGA and SCR).

Gravitropism Plant growth in response to gravity, enabling roots to grow downward into the soil and shoots to grow upward.

Gravity potential, Ψ_g A component of water potential determined by the effect of gravity on the free energy of water.

Greenhouse effect The warming of Earth's climate, caused by the trapping of long-wavelength radiation by CO_2 and other gases in the atmosphere. Term derived from the heating of a greenhouse resulting from the penetration of long-wavelength radiation through the glass roof, the conversion of the long-wave radition to heat, and the blocking of the heat by the glass roof.

Green Revolution The introduction of high-yielding dwarf varieties of wheat and rice into Latin America and Southeast Asia in the 1960s to keep abreast of human population growth.

Green-leaf volatiles A mixture of lipid-derived 6-carbon aldehydes, alcohols, and esters released by plants in response to mechanical damage.

Ground meristem In the plant meristems, cells that will give rise to the cortical and pith tissues and, in the root and hypocotyl, will produce the endodermis.

***GSA* gene** Gene of unicellular alga *Chlamydomonas reinhardtii* whose expression is mediated solely by a blue light-sensing system. Encodes the enzyme glutamate-1-semialdehyde aminotransferase, a key enzyme in the chlorophyll biosynthesis pathway.

Guard cells A pair of specialized epidermal cells which surround the stomatal pore and regulate its opening and closing.

GUS β-Glucuronidase from bacteria used as a reporter when fused to the promoter of a gene of interest in transgenic plants. Activity shows as blue staining or fluorescence depending on the substrate used to assay its activity.

Guttation An exudation of liquid from the leaves due to root pressure.

Gymnosperm An early type of seed plant. Distinguished from angiosperm by having seeds borne unprotected in cones.

H^+-PPase (proton pumping pyrophosphatase) An electrogenic pump that moves protons into the lumen of the vacuole and the Golgi cisternae, energized by the hydrolysis of pyrophosphate.

Halophytes Plants that are native to saline soils and complete their life cycles in that environment. Contrast with glycophytes.

Hartig net A fungal network of hyphae that surround but do not penetrate the cortical cells of roots.

Heart stage embryo The second stage of embryogenesis. A bilaterally symmetrical structure produced by rapid cell divisions in two regions on either side of the future shoot apex. *See* globular stage, torpedo stage.

Heat shock cognate proteins Molecular chaperon proteins that are constitutively expressed and function like HSPs.

Heat shock proteins (HSPs) A specific set of proteins that are induced by a rapid rise in temperature, and by other factors that lead to protein denaturation. Most act as molecular chaperones.

Heat shock response The increased synthesis of HSPs and the reduced synthesis of other proteins following a stressful but nonlethal heat episode.

Heliotropism Movements of leaves toward or away from the sun.

Hemicelluloses Heterogeneous group of polysaccharides that bind to the surface of cellulose, linking cellulose microfibrils together into a network. Typically solubilized by strong alkali solutions.

Herb A plant with no enduring above-ground parts, as distinct from trees and shrubs.

Herbaceous *See* herb.

Herbivores Plant-feeding animals, including many species of insects and mammals.

Hermaphroditic flowers Containing both carpels and stamens. *See* perfect flowers.

Heterocyclic ring A ring structure that contains both carbon and non-carbon (nitrogen or oxygen) atoms.

Heterocysts Specialized, thick-walled structures formed by cyanobacteria within plant cells to create an anaerobic environment for nitrogen fixation.

Heterotrimeric G protein A membrane-bound GTP-binding protein composed of three subunits, α, β, and γ; mediates the signal transduction pathways of G-protein-linked membrane receptors.

Heterotrophic cells Organisms that rely upon reduced forms of carbon for the synthesis of structural elements and the supply of metabolic energy.

Hexose monophosphate shunt See pentose phosphate pathway.

HIR (high irradiance response) Phytochrome responses whose magnitude is proportional to the irradiance (rather than fluence). HIRs saturate at fluences 100 times higher than LFRs and are not photoreversible. HIRs do not obey the law of reciprocity.

Histogenic layers The tissue-producing regions of the shoot apical meristem: tunica layers and corpus.

Hoagland solution A nutrient solution for plant growth, originally formulated by Dennis R. Hoagland.

Holoprotein An apoprotein bound to a smaller, nonprotein molecule such as a chromophore.

Homeobox A sequence coding for the homeodomain, a 60-amino acid domain in transcription factors that binds to specific regions of DNA.

Homeotic genes First discovered in *Drosophila*, homeotic genes regulate the locations of body parts in flies. These transcription factors are typically characterized by homeodomains.

Homeotic mutations In *Drosophila*, homeotic mutations in homeobox genes cause body segments and other structures to form at inappropriate locations. In plants, mutations with similar phenotypic effect on flower development, but do not involve homeobox genes.

28-Homobrassinolide (28-homoBL) A brassinosteroid hormone differing from brassinolide (BL) in the alkyl side chain at C-24, and possessing slightly lower activity than BL.

Hormone Organic molecule often (but not always) synthesized in one location of the organism and transported to another, where it produces dramatic effects on growth or development at vanishingly low concentrations. *See* plant hormone.

Hourglass hypothesis One hypothesis of how plants measure night length. Proposes that time is measured by a unidirectional series of biochemical reactions that start at the beginning of the dark period. *See* clock hypothesis.

Hpt protein In Arabidopsis, *h*istidine *p*hospho*t*ransfer protein.

HSF A specific transcription factor that acts during the transcription of heat shock proteins.

***hy* mutants** Arabidopsis mutants in which white light does not inhibit hypocotyl growth as it does in wild-type. Some *hy* mutants cannot synthesize phytochrome.

***hy4* mutant** See *cry1* mutant.

Hydathodes Specialized pores associated with vein endings at the leaf margin from which xylem sap may exude when there is positive hydrostatic pressure in the xylem. Also a site of auxin synthesis in immature leaves of Arabidopsis.

Hydroactive stomatal closure Closing of stomata in response to a closing signal which includes water stress. Depends on metabolic processes that reduce guard cell solute content resulting in water loss. Reversal of the mechanism of stomatal opening.

Hydrogen bond A weak chemical bond formed between a hydrogen atom and an oxygen or nitrogen atom.

Hydrolyzable tannins Tannins that are polymers of phenolic acids, especially gallic acid, and simple sugars. May be hydrolyzed by dilute acid.

Hydropassive closure Closing of stomata due to loss of water directly from guard cells and the consequent loss of cell turgor. Operates at low external humidity, when direct water loss is not rapidly compensated.

Hydrophilic Ability of an atom or a molecule to attract water molecules. For example, substances that can engage in hydrogen bonding are hydrophilic.

Hydrophobic Substances, molecules, or functional groups that repel water molecules

Hydroponics A technique for growing plants with their roots immersed in nutrient solution without soil.

Hydroquinone (QH$_2$) A fully reduced form of quinone.

Hydrostatic pressure Pressure generated by compression of water into a confined space. Measured in units called pascals (Pa) or, more conveniently, megapascals (MPa).

Hydroxyproline-rich proteins (HRGPs) A class of wall structural proteins, rich in hydroxyproline, with possible roles in protection against pathogens and desiccation.

Hyperpolarization An increase in the normally negative inside, electrical potential (mV) that exists across the plasma membrane of a cell.

Hypersensitive response A common plant defense following microbial infection, in which cells immediately surrounding the infection site die rapidly, depriving the pathogen of nutrients and preventing its spread.

Hyphae (singular hypha) The small, tubular filaments of fungi.

Hypocotyl hook An inverted "J" formed when the apical end of the dicot hypocotyl bends back on itself as dicot seedlings emerge from the seed coat. Protects the shoot apex from damage during growth through the soil.

Hypocotyl The region of the seedling stem below the cotyledons and above the root.

Hypophysis In seed plant embryogenesis, the apical-most progeny of the basal cell which contributes to the embryo and will form part of the root apical meristem.

Hypoxic Refers to low concentrations (pressures) of oxygen. Contrast with anoxic.

IAA *See* indole-3-acetic acid.

IAM *See* indole-3-acetamide.

IAN *See* indole-3-acetonitrile.

Ice nucleation Process by which water molecules start to form a stable ice crystal. The surface properties of some polysaccharides and proteins facilitate ice crystal formation.

Immunocytochemistry The use of specific antibodies attached to identifying molecules in order to indicate the presence and perhaps location of molecules with antigenic properties.

Imperfect flowers Flowers that lack either the male (stamens) or female (pistils) structures. Unisexual flowers.

Import The movement of photosynthate in sieve elements into sink organs.

In vitro Biological experiments performed in "the test tube" isolated from a whole organism. Contrast with in vivo.

In vivo Within the intact organism. Contrast with in vitro.

Indeterminate growth Capacity for both vegetative growth and flowering over an extended period of time. Growth is not genetically limited, and will continue as long as environmental conditions and resources permit.

Indole A precursor of IAA biosynthesis by a tryptophan-independent pathway.

Indole-3-acetamide (IAM) A biosynthetic intermediate for the synthesis of IAA in various pathogenic bacteria, such as *Pseudomonas savastanoi* and *Agrobacterium tumefaciens*.

Indole-3-acetic acid (IAA) The most common naturally occurring auxin. When protonated, IAA is lipophilic and diffuses across lipid bilayer membranes, but when dissociated, the negatively charged IAA cannot cross membranes unaided. Free IAA is biologically active; IAA covalently bound to other molecules is inactive.

Indole-3-acetonitrile (IAN) An intermediate in one of three biosynthetic pathways for the synthesis of IAA in plants. IAN converted to IAA by action of nitrilase.

Indole-3-glycerol phosphate A precursor of IAA biosynthesis by a tryptophan-independent pathway.

Indole-3-pyruvic acid (IPA) An intermediate in one of three biosynthetic pathways for the synthesis of IAA in plants. Formed by deamination of the amino acid tryptophan.

Induced thermotolerance Tolerance to lethal, high temperatures accomplished by periodic brief exposure to sublethal heat stress.

Inducible The capacity for increased synthesis of a particular protein or proteins in response to a particular external signal, such as a hormone.

Induction period The period of time (time lag) elapsed between the perception of a signal and the activation of the response. In the Calvin cycle, the time elapsed between the onset of illumination and the full activation of the cycle.

Infection thread An internal tubular extension of the plasma membrane of root hairs through which rhizobia enter root cortical cells.

Inflorescence meristem Produces cauline leaves and inflorescence meristems in the axils of the leaves, as well as bracts and floral meristems in the axils of the bracts, and does not directly produce floral organs.

Inner membrane The inner of a mitochondrion's or chloroplast's two membranes.

Integral membrane proteins Proteins embedded in, and often crossing, the lipid bilayer of biological membranes.

Integument The outer tissue layers surrounding the nucellus of an ovule; develops into the seed coat.

Intercalary meristem Meristem located near the base, rather than the tip, of a stem or leaf, as in grasses.

Intercellular air space resistance The resistance or hindrance that slows down the diffusion of CO_2 inside leaf, from the substomatal cavity to the walls of the mesophyll cells.

Intermediary cell A type of companion cell with numerous plasmodesmatal connections to surrounding cells, particularly to the bundle sheath cells.

Intermediate-day plant A plant that flowers only between narrow day length limits (for example between 12 and 14 hours).

Intermembrane space The fluid-filled space between the two mitochondrial or chloroplast membranes.

Internode Portion of a stem between nodes.

Invertase An enzyme that catalyzes the hydrolysis of sucrose to glucose and fructose.

Ionophore A molecule that allows ions to cross lipid bilayers. There are two classes: carriers and channels. Carriers, like valinomycin, form cage-like structures around specific ions, diffusing freely through the hydrophobic regions of the bilayer. Channels, like gramicidin, form continuous aqueous pores through the bilayer, allowing ions to diffuse through.

IPA *See* indole-3-pyruvic acid.

ipt **gene** The T-DNA gene involved in cytokinin biosynthesis; encodes cytokinin synthase, an isopentenyl transferase enzyme. Also called the *tmr* locus.

Irradiance The amount of energy that falls on a flat sensor of known area per unit time. Expressed as watts per square meter (W m^{-2}). Note, time (seconds) is contained within the term watt: 1 W = 1 joule (J) s^{-1}, or as moles of quanta per square meter per second (mol m^{-2} s^{-1}), also referred to as fluence rate.

isoamylase Catalyzes the hydrolysis of α-D-1,6 glucosidic branch linkages in glycogen, amylopectin, and their β-limit dextrins.

Isoflavonoids (isoflavones) Group of flavonoids with antimicrobial activity.

Isoprene (2-methyl-1,3-butadiene) A gaseous, branched five-carbon molecule emitted by many plants. Isoprene emission protects leaves against damage from high temperatures. Provides the structure of the basic five-carbon unit in terpene formation.

Isotropic Characteristic of materials showing uniform structural and mechanical properties in all directions. Contrast with anisotropic.

Isozymes Proteins that are structurally similar, but not identical, and share the same catalytic activity. Isozymes might be regulated by different mechanisms.

Jasmonate *See* Jasmonic acid.

Jasmonic acid A plant signaling molecule derived from linolenic acid (18:3) found in membrane lipids. Activates plant defenses against insects, fungal pathogens, and regulates plant growth including the development of anthers and pollen. Activates the expression of genes involved in response to various biotic and abiotic stresses, such as proteinase inhibitors.

Ketose Sugar with a terminal ketone group.

Kinases Enzymes that have the capacity to transfer phosphate groups from ATP to other molecules. *See* protein kinase.

Kinematics The concepts and numerical methods applicable to the motion of fluid particles and the shape changes that the fluids undergo. Useful in analyzing meristematic growth.

Kinetin Substance originally isolated from autoclaved herring sperm DNA that greatly promotes cell division. Not a naturally occurring cytokinin. Chemically, a derivative of adenine (or aminopurine), 6-furfurylaminopurine. *See* zeatin.

Kranz anatomy (G: *kranz*: wreath or halo.) The wreathlike arrangment of mesophyll cells around a layer of large bundle-sheath cells. The two concentric layers of photosynthetic tissue surrounds the vascular bundle. This anatomical feature is typical of leaves of C_4 plants.

Krebs cycle *See* citric acid cycle.

la **mutant** Occurs in peas, when the cry^s mutation is present. Results in an ultra tall constitutive response mutant caused by the loss of function of a negative regulator. *See also* cry^s mutant.

Lactate dehydrogenase (LDH) The enzyme that catalyzes the reversible conversion of pyruvate to lactate using the coenzyme NAD.

Lactic acid fermentation Reaction in which pyruvate from glycolysis is reduced to lactate using NADH and thus regenerating NAD^+.

Late genes *See* secondary response genes.

Latent heat of vaporization The energy needed to separate molecules from the liquid phase and move them into the gas phase at constant temperature.

Lateral bud Undeveloped shoot consisting of an axillary meristem, a short stem and immature leaves, often covered with bud scales and located above the point of attachment of a leaf to the stem.

Lateral meristems Secondary meristems found in mature woody stems and roots as cylinders of meristematic cells. Their activity increases stem or root circumference. *See* vascular cambium, cork cambium.

Lateral roots (branch roots) Arise from the pericycle in mature regions of the root through establishment of secondary meristems that grow out through the cortex and epidermis, establishing a new growth axis.

Latex A complex, often milky solution exuded from cut surfaces of some plants that represents the cytoplasm of laticifers and may contain defensive substances.

Laticifers In many plants, elongated, often interconnected phloem cells that contain rubber, latex, and other secondary metabolites.

Law of reciprocity The reciprocal relationship between fluence rate (mol $m^{-2}\,s^{-1}$) and duration of light exposure characteristic of many photochemical reactions as well as some developmental responses of plants to light. Total fluence depends on two factors: the fluence rate and the irradiation time. A brief light exposure can be effective with bright light; conversely, dim light requires a long exposure time. Also referred to as Bunsen–Roscoe Law.

Le **gene** Dominant allele for tall stems in peas, first studied by Mendel; codes for an enzyme, 3β-hydroxylase, that hydroxylates GA_{20} to produce GA_1.

le **mutant** In peas, a recessive mutant allele that causes a dwarf phenotype by interfering with the synthesis of active gibberellin. Mendel's dwarf gene.

LEA proteins Proteins that stabilize cell membranes during desiccation. Encoded by a group of genes that are regulated by osmotic stress first characterized in desiccating embryos during seed maturation.

Leaf laminae The blade of a leaf.

Leaf primordia Region of the shoot apical meristem that will form a leaf during the normal course of development.

Lectins Defensive plant proteins that bind to carbohydrates; or carbohydrate-containing proteins inhibiting their digestion by a herbivore.

Leghemoglobin An oxygen-binding heme protein found in the cytoplasm of infected nodule cells that facilitates the diffusion of oxygen to the respiring symbiotic bacteria.

Legume A member of the family Leguminosae often associated with rhizobia. Legumes include the pea (*Pisum*), clover (*Trifolium*), broad bean (*Vicia*), lentil (*Lens*), soybean (*Glycine*), kidney bean (*Phaseolus*), peanut (*Arachis*), and southern pea (*Vigna*).

LFR (low fluence response) Phytochrome responses whose magnitude is proportional to low fluence (1.0 to 1000 µmol m^{-2}). These include the classic red/far-red photoreversible responses.

***LFY* gene** In Arabidopsis, *LEAFY* (*LFY*) is a gene involved in establishing floral meristem identity.

***LHCB* gene family** Genes encoding the chlorophyll *a/b*–binding proteins (also called CAB proteins) of photosystem II. Regulated at the transcriptional level by both circadian rhythms and phytochrome.

Light channeling In photosynthetic cells, the propagation of some of the incident light through the central vacuole of the palisade cells and through the air spaces between the cells.

Light compensation point The amount of light reaching a photosynthesizing leaf at which photosynthetic CO_2 uptake exactly balances respiratory CO_2 release.

Light harvesting complex proteins (LHC proteins) Chlorophyll-containing proteins associated with one or the other of the two photosystems in eukaryotic organisms. Also known as chlorophyll *a/b* antenna proteins.

Light scattering The randomization of the direction of photon movement within plant tissues due to the reflecting and refracting of light from the many air–water interfaces. Greatly increases the probability of photon absorption within a leaf.

Light A form of radiant energy with properties of both particles and waves.

Lignin Highly branched phenolic polymer with a complex structure made up of phenylpropanoid alcohols that may be associated with celluloses and proteins. Deposited in secondary walls, it adds strength allowing upward growth and permitting conduction through the xylem under negative pressure. Lignin has significant defensive functions.

Limit dextrinase Catalyzes the hydrolysis of α-D-1,6 glucosidic linkages in α- and β-limit dextrins of amylopectin and glycogen, and in amylopectin and pullulan.

Limiting factor In any physiological process, the factor whose operation limits the rate at which the entire process can operate. For example, leaf photosynthesis can be limited by light or by ambient CO_2 concentrations.

Limonoids Antiherbivore triterpenoids (30 carbons) which give citrus fruits their bitter taste.

Linear axis In most plants, the body pattern in which the root and the shoot are at opposite ends. *See* radial axis.

Lipid bilayer The core of cellular membranes formed by two layers of phospholipid molecules facing each other through their nonpolar tails.

Liquid phase resistance The resistance or hindrance that slows down the diffusion of CO_2 inside a leaf, from the walls of the mesophyll cells to the carboxylation sites in the chloroplast.

Lipid rafts Detergent-resistant aggregates of lipids and proteins that may represent transient microdomains of tightly packed fatty acid chains enriched in sphingolipids and sterols. Lipid rafts may play a role in membrane signaling.

Lodging The bending of cereal stalks to the ground because of the weight of moisture collecting on the ripened heads. Makes mechanical harvesting ineffective.

Long-day plant (LDP) A plant that flowers only in long days (qualitative LDP) or whose flowering is accelerated by long days (quantitative LDP).

Lumen The cavity or space within a tube or sac, especially the space inside thylakoid membranes.

LUX **genes** Bacterial genes for luciferase that catalyze a light emitting reaction. Used as a reporter gene in transgenic plants in order to visibly indicate the activity of another gene sharing the same promoter.

Lysophosphatidylcholine A product of PLA_2 phospholipid degradation that activates protein kinases; present on plant membranes in vitro and may be a second messenger in the auxin signal transduction pathway.

Macronutrient Minerals obtained from the soil and present in plant tissues at concentrations usually greater than 30 µmol g^{-1} dry matter. Nitrogen, potassium, calcium, magnesium, phosphorus, sulfur, silicon.

Magnesium dechelatase The enzyme that removes magnesium from chlorophyll as part of the chlorophyll breakdown process.

Malting The first step in the brewing process. Germination of barley seeds at temperatures that maximize the production of hydrolytic enzymes.

Malto-oligosaccharides (maltos**e, malto***tri***ose, malto***tetra***ose, malto***penta***ose, malto***hexa***ose)** The series of linear oligosaccharides composed of two, three, four, five and six, respectively, units of glucose all linked via a α-D-1,4 bond.

Maltose phosphorylase Catalyzes the phosphorolysis of **maltose** yielding D-glucose and β-D-glucose 1-phosphate.

Mass spectrometry (MS) A method that identifies chemical compounds by their molecular mass to charge (m/z) ratios and fragmentation patterns. MS can detect as little as 10^{-12} g (1 picogram, or pg) of IAA in plant extracts.

Mass transfer rate The quantity of material passing through a given cross section of phloem or sieve elements per unit time.

Material Pertaining to matter, the amount and kind of matter under consideration.

Matric potential, Ψ_m (or matric pressure) The sum osmotic potential (Ψ_s) + hydrostatic pressuer (Ψ_p). Useful in situations (dry soils, seeds, and cell walls) where the separate measurement of Ψ_s and Ψ_p is difficult or impossible.

Matrix The colloidal-aqueous phase contained within the inner membrane of a mitochondrion.

Maturation zone The region of the root that has completed its differentiation and shows root hairs for the absorption of water and solutes; and competent vascular tissue.

MCP A strong competitive inhibitor of ethylene binding 1-methylcyclopropene.

Megapascals (MPa) 10^6 Pa.

Membrane depolarization *See* depolarization.

Membrane permeability The extent to which a membrane permits or restricts the movement of a substance.

Meristem identity genes Genes necessary for the initial induction of the floral organs.

Meristematic zone A region at the tip of the root containing the meristem that generates the body of the root. Located just above the root cap.

Mesophyll Leaf tissue found between the upper and lower epidermal layers, consisting of palisade parenchyma and spongy mesophyll.

Mesophyll cells Photosynthetic cells found in the mesophyll of leaves of both C_4 and C_3 plants.

Methanogenic bacteria Obligate anaerobes that use CO_2 and H_2 as electron acceptor and donor, respectively, in anaerobic respiration for the production of methane.

Methemoglobinemia A disease of humans and livestock caused by the consumption of plant material high in nitrate. The liver reduces nitrate to nitrite, which combines with hemoglobin blocking its ability to bind oxygen.

Michaelis–Menten constant (K_m) A constant in the Michaelis-Menten equation for enzyme kinetics. The constant reflects the binding affinity of the substrate for the enzyme or a solute for its carrier and corresponds to the concentration of substrate that gives ½ the maximum velocity that the enzyme can catalyze.

Microaerobic conditions Reduced oxygen conditions maintained by some aerobic nitrogen-fixing bacteria such as *Azotobacter* through high rates of cellular respiration

Microfilament A component of the cell cytoskeleton made of actin; it is involved in organelle motility within cells.

Micronutrient Minerals obtained from the soil and present in plant tissues at concentrations usually less than 3 µmol g^{-1} dry matter. Chloride, iron, boron, manganese, sodium, zinc, copper, nickel, molybdenum.

MicroRNA A noncoding RNA molecule, 20–25 nucleotides long, which regulates the expression of other genes.

Microtubule Component of the cell cytoskeleton made of tubulin, a component of the mitotic spindle, and a player in the orientation of cellulose microfibrils in the cell wall.

Middle lamella A thin layer of pectin-rich material at the junction where the primary walls of neighboring cells come into contact. Originates as the cell plate during cell division.

Mineral nutrients Inorganic ions absorbed from soil. *See also* macronutrients, micronutrients.

Mineralization Process of breaking down organic compounds by soil microorganisms that releases mineral nutrients in forms that can be assimilated by plants.

Mixed-function oxidases *See* monooxygenases.

Moiety A part of a larger molecule or structure.

Molecular motors Large protein complexes associated with microtubules that move flagella, vesicles, chromosomes, and the cellulose synthase complex.

Monocarpic senescence Senescence of the entire plant after a single reproductive cycle, initiated by fruit and seed development. Can be prevented by continual removal of flowers.

Monocot One of the two classes of flowering plants characterized by a single seed leaf (cotyledon) in the embryo. Contrast with dicot.

Monoecious Refers to plants in which male and female flowers are found on the same individuals, e.g., cucumber (*Cucumis sativus*) and maize (*Zea mays*). Contrast with dioecious.

Monooxygenases Group of oxygenase enzymes that incorporate only one of the oxygen atoms from O_2 into a carbon compound; the other oxygen atom is reduced to water using NADH or NADPH as an electron donor.

***MONOPTEROS* gene** Arabidopsis gene whose mutations produce seedlings lacking both a hypocotyl and a root, but have cotyledons and a shoot apical meristem.

Monoterpenes Terpenes having ten carbons, two five-carbon isoprene units.

mtDNA Mitochondrial DNA.

Multidrug resistance/P-glycoproteins (MDR/PGPs) MDR/PGP proteins are integral membrane proteins that function as ATP-dependent hydrophobic anion carriers that facilitate cellular auxin efflux during polar transport.

Mutualism A symbiotic relationship in which both organisms benefit.

Mycelium The mass of hyphae that forms the body of a fungus.

Mycorrhizae (singular mycorrhiza, from the Greek words for "fungus" and "root"). The symbiotic (mututalistic) association of certain fungi and plant roots. Facilitate the uptake of mineral nutrients by roots.

***na* mutant** In peas, a recessive mutant allele causing an extreme dwarf phenotype by completely blocking gibberellin biosynthesis—no GA_{12}-aldehyde is produced.

Natural products *See* secondary metabolites.

Necrosis A type of cell death.

Necrotic lesions Regions within an organ in which cells die while surrounding regions remain alive.

Necrotic spots Small spots of dead leaf tissue. For example, a characteristic of phosphorus deficiency.

Nectar Sugary fluid produced in specialized glands of some flowers to attract insects as pollinators. Sometimes contains amino acids and other nitrogenous substance.

Nectar guides On a flower, symmetric patterns of stripes, spots, or concentric circles formed by flavonols and other compounds to attract and orient pollinating insects.

Neoplasm *See* tumor.

Nernst equation An equation that predicts the electrical potential at which a charged ion will be in equilibrium across a membrane, as a function of the relative concentrations of the ion on each side.

Nernst potential The electrical potential described by the Nernst equation.

Night break An interruption of the dark period with a short exposure to light that makes the entire dark period ineffective.

NIT gene One of the genes (*NIT1* through *NIT4*) encoding nitrilase enzymes that convert IAN to IAA.

Nitrate reductase Enzyme that reduces nitrate (NO_3^-) to nitrite (NO_2^-). Catalyzes the first step by which nitrate absorbed by roots is assimilated into organic form.

Nitrite reductase The enzyme that reduces nitrite (NO_2^-) to ammonium (NH_4^+).

Nitrogen fixation The natural or industrial processes by which atmospheric nitrogen N_2 is converted to ammonia (NH_3) or nitrate (NO_3^-).

Nitrogenase The enzyme complex that catalyzes the reduction of N_2 to $2\ NH_3$. The reaction is coupled to the reduction of two protons to H_2.

***nod* genes** *See* nodulation genes.

***Nod* genes** *See* nodulin genes.

Node Position on the stem where leaves are attached.

Nodulation genes (*nod*) Rhizobial genes, the products of which participate in nodule formation.

Nodule primordium Rhizobia-infected root cortical cells that dedifferentiate and start dividing.

Nodules Specialized organs of a plant host containing symbiotic nitrogen-fixing prokaryotes.

Nodulin genes (*Nod*) Plant genes specific to nodule formation.

Nonclimacteric fruits Fruits such as citrus fruits and grapes that do not exhibit the increased cellular respiration and ethylene production rise seen in climacteric fruits.

Nonphotochemical quenching The quenching of chlorophyll fluorescence by processes other than photochemistry—the converting of excess excitation into heat.

Nonreducing end The terminal glucose moiety of the starch molecule whose C-1 is β-linked to the C-4 of the preceding moiety.

Nonreducing sugar A sugar in which the aldehyde or ketone group is reduced to an alcohol or combined with a similar group on another sugar, e.g., sucrose. Less reactive than a reducing sugar.

Northern-blot analysis A method for detecting and quantifying specific RNAs (after electrophoresis and transfer of the RNA onto nitrocellulose paper or a nylon membrane filter) by hybridization with complementary strands of RNA or DNA.

NPA (1-N-naphthylphthalamic acid) A noncompetitive inhibitor of auxin anion efflux carriers that blocks the polar transport of auxin.

***nph1* mutant** Arabidopsis mutant (*non*phototropic *h*ypocotyl) whose blue light–dependent phototropism of the hypocotyl is blocked. Also defective in a blue light–stimulated phosphorylation. The mutant *nph1* is genetically independent of the *hy4* (*cry1*) mutant. Recently renamed *phot 1*, for *phototropin 1*.

***npq1* mutant** An Arabidopsis mutant (*n*on*p*hotochemical *q*uenching) defective in the enzyme that converts violaxanthin into zeaxanthin. Lacks a specific stomatal response to blue light.

Nuclear division Mitosis or meiosis. Distinguished from cytokinesis.

Nuclear localization signal A specific amino acid sequence required for a protein to gain entry into the nucleus.

Nuclear pore complex (NPC) An elaborate structure, 120 nm wide, composed of more than a hundred different nucleoporin proteins arranged octagonally. The NPC forms a large, protein-lined pore in the nuclear membrane.

Nutrient assimilation The incorporation of mineral nutrients into carbon compounds such as pigments, enzyme cofactors, lipids, nucleic acids, or amino acids.

Nutrient depletion zone The region surrounding the root surface showing diminished nutrient concentrations due to slow diffusion.

Nutrient solution A solution containing only inorganic salts that supports the growth of plants in sunlight without soil or organic matter.

Nyctinasty Sleep movements of leaves. Leaves extend horizontally to face the light during the day and fold together vertically at night.

Oleosomes Organelles bounded by an unusual single layer phospholipid membrane that store triacylglycerols.

Oligogalacturonans Pectin fragments (10 to 13 residues) resulting from plant cell wall degradation that elicit multiple defense responses. May also function during the normal control of cell growth and differentiation.

Oligomycin Specific inhibitor of the F_oF_1 ATP synthase.

Oligosaccharins Fragments resulting from plant cell wall degradation that affect plant development or defenses.

Opines Nitrogen-containing compounds found only in crown gall tumors. Genes required for their synthesis found only in Ti plasmid T-DNA of *Agrobacterium tumefaciens*, which can utilize the opines, not the plant. The bacterium can utilize the opine as a nitrogen source, but cells of higher plants cannot.

Ordinary companion cell A type of companion cell with relatively few plasmodesmata connecting it to any of the surrounding cells other than its associated sieve element.

Organic fertilizers Contain nutrient elements derived from natural sources without any synthetic additions.

Organ identity genes Genes that directly control floral development. Genes for transcription factors that likely control the expression of other genes whose products are involved in the formation and/or function of floral organs.

***orp* mutant** In maize, a mutation that inactivates the enzyme for the final step in tryptophan biosynthesis. Prevents conversion of tryptophan to IAA. Named for *orange pericarp* mutant of maize.

Orthostichy A set of leaves inserted in the stem directly above or below one another.

Osmolality A unit of concentration expressed as moles of total dissolved solutes per liter of water [mol L^{-1}].

Osmoregulation Control of the osmotic potential of a cell or organism.

Osmosis The movement of water across a selectively permeable membrane toward the region of more negative water potential, Ψ_w (lower concentration of water).

Osmotic adjustment The ability of the cell to accumulate compatible solutes and lower water potential during periods of osmotic stress.

Osmotic pressure (π) The negative of the osmotic potential, (Ψ_s). Because Ψ_s has negative values, π has positive values

Osmotic solutes The major osmotically active solutes in guard cells. Include potassium, chloride, malate, and sucrose.

Outer membrane The outer membrane of the mitochondrion or chloroplast.

Over-expression The engineering of a specific gene to cause its greater activity than in the nonengineered form.

Ovule In seed plants, the structure that contains the embryo sac and develops into the seed after fertilization of the egg contained within it.

OxIAA (oxindole-3-acetic acid) End product of oxidative degradation of IAA.

Oxidation The chemical reaction whereby electrons or hydrogen atoms are removed from a substance.

Oxidation number Number used to keep track of the redistribution of electrons during chemical reactions. The sum of the oxidation numbers of all atoms in a neutral compound is zero. A useful method to balance redox equations.

Oxidative pentose phosphate pathway See pentose phosphate pathway.

Oxidative phosphorylation Transfer of electrons to oxygen in the mitochondrial electron transport chain that is coupled to ATP synthesis from ADP and phosphate by the ATP synthase.

Oxigenic organism Photosynthetic organisms that produce molecular oxygen as an end product of photosynthesis.

Oxygenases Group of enzymes that add oxygen from O_2 directly to organic compounds. See dioxygenases, monooxygenases.

Oxygenation The oxygenase-catalyzed reaction in which the carbon of an organic molecule binds the oxygen atoms of the O_2 molecule.

P protein A set of proteins abundant in the sieve elements of most angiosperms but absent from gymnosperms. Occurs in tubular, fibrillar, granular, and crystalline forms depending on the species and maturity of the cell. Seals off damaged sieve elements by plugging up the sieve plate pores. May play other roles, such as defense against phloem-feeding insects. Formerly called "slime."

P-protein bodies Discrete spheroidal, spindle-shaped, or twisted/coiled structures of P-proteins present in the cytosol of immature sieve tube elements. Generally disperse into tubular or fibrillar forms during cell maturation.

P680 The chlorophyll of the photosystem II reaction center that absorbs maximally at 680 nm in its neutral state. The P stands for pigment.

P700 The chlorophyll of the photosystem I reaction center that absorbs maximally at 700 nm in its neutral state. The P stands for pigment.

P870 The reaction center bacteriochlorophyll from purple photosynthetic bacteria that absorbs maximally at 870 nm in its neutral state. The P stands for pigment.

Paclobutrazol (aka: Bonzi) A commercial inhibitor of gibberellin biosynthesis. Paclobutrazol inhibits P450 mono-oxygenases, blocking the synthesis GA_{12}-aldehyde on the endoplasmic reticulum.

Palisade cells Below the leaf upper epidermis, the top one to three layers of pillar-shaped photosynthetic cells.

PAPS See 3'-phosphoadenosine-5'-phosphosulfate.

PAR See photosynthetically active radiation.

Paraheliotropism Movement of leaves away from incident sunlight.

Paraquat An herbicide that blocks photosynthetic electron flow by accepting electrons from photosystem I.

Parenchyma Metabolically active plant tissue consisting of thin-walled cells, with air-filled spaces at the cell corners.

Parthenocarpy The production of fruits without fertilization, consequently fruits lacking mature functioning seeds. Occurs naturally in banana and pineapple.

Particle rosettes (terminal complexes) Large, ordered protein complexes that are embedded in the plasma membrane and contain cellulose synthase.

Partitioning The differential distribution of photosynthate to multiple sinks within the plant.

Pascal (Pa) SI unit of pressure: 1 Pa = 1 kg m^{-1} s^{-2}, or 1 Pa = 1 J m^{-3}.

Passive transport Diffusion across a membrane. The spontaneous movement of a solute across a membrane in the direction of a gradient of (electro)chemical potential (from higher to lower potential). Downhill transport.

Patch-clamp Electrophysiological method used to study ion pumps and single ion channels.

Pathogenesis-related proteins (PR proteins) Group of proteins produced by pathogen attack; includes hydrolytic enzymes that attack the cell wall of the pathogen, particularly fungi, e.g., glucanases, chitinases.

Pathogens Microorganisms capable of infecting other organisms and causing disease.

P-ATPase An ion pump in which the enzyme protein is phosphorylated by ATP during the catalytic cycle, e.g., the plasma membrane H$^+$-ATPase.

PCD See programmed cell death.

PCMBS (p-chloromercuribenzenesulfonic acid) A reagent that inhibits the transport of sucrose across plasma membranes but does not permeate the cell membrane.

Pectins A heterogeneous group of complex cell wall polysaccharides that form a gel in which the cellulose–hemicellulose network is embedded. Typically contain acidic sugars such as galacturonic acid, and neutral sugars such as rhamnose, galactose, and arabinose. Often include calcium as a structural component, allowing extractions from the wall with chelators or dilute acids.

Pentose phosphate pathway (hexose monophosphate shunt) A cytosolic pathway that oxidizes glucose and produces NADPH and a number of sugar phosphates.

Perfect flowers Flowers with both male (stamens) and female (pistils) structures present. Hermaphroditic flowers.

Perforation plate The perforated end walls of vessel elements.

Periclinal division Cell divisions parallel to the surface or longitudinal axis.

Periderm Tissue produced by the cork cambium that contributes to the outer bark of stems and roots during secondary growth of woody plants, replacing the epidermis. Also forms over wounds and abscission layers after the shedding of plant parts.

Period In cyclic (rhythmic) phenomena, the time between comparable points in the repeating cycle, such as peaks or troughs.

Peripheral zone A doughnut-shaped region surrounding the central zone in shoot apical meristems consisting of small, actively dividing cells with inconspicuous vacuoles. Leaf primordia are formed in the peripheral zone.

Permanent wilting point Water content of the soil (or soil water potential) from which plants cannot regain turgor and therefore remain wilted even if all water loss through transpiration stops.

Peroxisome Organelle in which organic substrates are oxidized by O_2. These reactions generate H_2O_2 that is broken down to water by the peroxysomal enzyme catalase.

Petals Conspicuous, often brightly colored flower structures surrounded at their base by sepals. *See also* sepals stamens, and carpels.

Petiole The leaf stalk that joins the leaf blade to the stem.

Pfr The far-red light–absorbing form of phytochrome converted from Pr by the action of red light. The blue-green colored Pfr is converted back to Pr by far-red light. Pfr is the physiologically active form of phytochrome.

Phase In cyclic (rhythmic) phenomena, any point in the cycle recognizable by its relationship to the rest of the cycle, for example, the maximum and minimum positions.

Phase change The phenomenon in which the fates of the meristematic cells become altered in ways that cause them to produce new types of structures.

Phenolics Plant secondary metabolites that contain a hydroxyl group on an aromatic ring, a phenol. Many plant phenolic compounds function as defenses against insect herbivores and pathogens, UV protectants, pollinator attractants and allelopathic agents.

Phenylalanine ammonia lyase (PAL) Catalyzes the conversion of phenylalanine to form cinnamic acid with loss of an ammonia molecule. An important regulatory step between primary and secondary metabolism.

Phenylpropanoids Phenolic derivatives of cinnamic acid containing a benzene ring and a three-carbon side chain.

Pheophytin A chlorophyll in which the central magnesium atom has been replaced by two hydrogen atoms.

Phloem The tissue that transports the products of photosynthesis from mature leaves to areas of growth and storage, including the roots.

Phloem loading The movement of photosynthetic products from the mesophyll chloroplasts to the sieve elements of mature leaves. Includes short-distance transport steps and sieve-element loading. *See also* phloem unloading.

Phloem unloading The movement of photosynthate from the sieve elements to the sink cells that store or metabolize them. Includes sieve-element unloading and short-distance transport. *See also* phloem loading.

3′-phosphoadenosine-5′-phosphosulfate (PAPS) The first stable intermediate in the bacterial and fungal reduction of sulfate (SO_4^{2-}) to sulfite (SO_3^{2-}) and then to sulfide (S^{2-}). Probably formed by reaction of adenosine-5′-phosphosulfate (APS) with ATP. A few plants have more specialized day length requirements or no requirement.

Phosphate Orthophosphate; inorganic phosphate, HPO_4^{2-}, P_i.

Phosphate translocator (phosphate/triose phosphate translocator) An integral membrane protein that catalyzes the reversible transport of inorganic phosphate in exchange for trioses phosphates (or 3-phosphoglycerate) across the chloroplast envelope. Under illumination, this protein facilitates the increase of trioses phosphates in the cytoplasm while concurrently replenishing inorganic phosphate in the chloroplast for further photosynthetic CO_2 fixation.

Phosphorolysis The cleavage of a glycosidic linkage by the addition of a phosphoryl group to one of the products.

Photoassimilation The coupling of nutrient assimilation to photosynthetic electron transport.

Photochemistry Very rapid chemical reactions in which light energy absorbed by a molecule causes a chemical reaction to occur.

Photoinhibition The inhibition of photosynthesis by excess light.

Photolyase A blue light–activated enzyme that repairs pyrimidine dimers in DNA that have been damaged by ultraviolet radiation. Contains an FAD and a pterin.

Photomorphogenesis The influence and specific roles of light on plant development. In the seedling, light-induced changes in gene expression to support above-ground growth in the light rather than below-ground growth in the dark.

Photon A discrete physical unit of radiant energy.

Photon irradiance Measurement of light energy expressed in moles per square meter per second (mol m^{-2} s^{-1}), where 1 mol of light = 6.02 × 10^{23} photons, Avogadro's number.

Photoperiod The amount of time per day that a plant is exposed to light or darkness. May control various aspects of sexual or vegetative reproductive development, including flowering and tuberization.

Photoperiodic induction The photoperiod-regulated processes that occur in leaves resulting in the transmission of a floral stimulus to the shoot apex.

Photoperiodism A biological response to the length and timing of day and night, making it possible for an event to occur at a particular time of year.

Photophosphorylation The formation of ATP from ADP and inorganic phosphate (P_i) using light energy stored in the proton gradient across the thylakoid membrane.

Photoprotection A carotenoid-based system for dissipating excess energy absorbed by chlorophyll in order to avoid forming singlet oxygen and damaging pigments. Involves quenching.

Photorespiration Uptake of atmospheric O_2 with a concomitant release of CO_2 by illuminated leaves. Molecular oxygen serves as substrate for rubisco and the formed 2-phosphoglycolate enters the photorespiratory carbon oxidation cycle. The activity of the cycle recovers some of the carbon found in 2-phosphoglycolate, but some is lost to the atmosphere.

Photorespiratory carbon oxidation (PCO) cycle *See* photorespiration.

Photoreversibility Relating to phytochrome, the interconversion of the Pr and Pfr forms.

Photostationary state Relating to phytochrome under natural light conditions, the equilibrium of 97% Pr and 3% Pfr.

Photosynthate The carbon-containing products of photosynthesis.

Photosynthesis The conversion of light energy to chemical energy by photosynthetic pigments using water and CO_2, and producing carbohydrates.

Photosynthetic electron transport Electron flow from light-excited chlorophyll and the oxidation of water, through PSII and PSI, to the final electron acceptor NADP$^+$.

Photosynthetic photon flux density (PPFD) Photosynthetically active radiation expressed on a quantum basis, quanta (mol m^{-2} s^{-1}).

Photosynthetically active radiation (PAR) Light of wavelengths between 400 to 700 nm range that corresponds to the wave band absorbed by photosynthetic pigments.

Photosystem A functional unit in the chloroplast that harvests light energy to power electron transfer and to generate a proton motive force used to synthesize ATP.

Photosystem I (PSI) A system of photoreactions that absorbs maximally far-red light (700 nm), oxidizes plastocyanin and reduces ferredoxin.

Photosystem II (PSII) A system of photoreactions that absorbs maximally red light (680 nm), oxidizes water and reduces plastoquinone. Operates very poorly under far-red light.

Phototropin Protein encoded by *PHOT1* that has been postulated to be a blue-light receptor in phototropism, chloroplast movements, and stomatal opening.

Phototropism The alteration of plant growth patterns in response to the direction of incident radiation.

Phragmoplast An assembly of microtubules, membranes, and vesicles that forms during late anaphase or early telophase and precedes fusion of vesicles to form cell plate.

***PHY* genes** The genes for the apoproteins of phytochrome; phytochrome gene family members. In Arabidopsis these are *PHYA*, *PHYB*, *PHYC*, *PHYD*, and *PHYE*.

Phyllotaxy The arrangement of leaves on the stem.

Phytoalexins Chemically diverse group of secondary metabolites with strong antimicrobial activity that are synthesized following infection and accumulate at the site of infection.

Phytochrome A plant growth-regulating photoreceptor protein that absorbs primarily red light and far-red light, but also absorbs blue light.

The holoprotein that contains the chromophore phytochromobilin.

Phytochromobilin The linear tetrapyrrole chromophore of phytochrome.

Phytoecdysones Group of plant steroids that are toxic to insects because of chemical similarity to the insect molting hormone.

Phytoferritin An iron–protein complex which stores surplus iron in plant cells.

Phytoglycogen A minor constituent of starch having a branched structure in which an α-D-1,6 bond occurs every 10–15 glucose units linked via α-D-1,4 bonds.

Phytohormone *See* plant hormone.

Phytomere A developmental unit consisting of one or more leaves, the node to which the leaves are attached, the internode below the node, and one or more axillary buds.

Phytosiderophores A special class of iron chelators produced by grasses and made of amino acids that are not found in proteins.

Phytol The long hydrocarbon chain found on the chlorophyll molecule that anchors it to the thylakoid membrane.

P_i *See* phosphate.

PIF3 A basic helix-loop-helix transcription factor that interacts with both phytA and phyB.

pin1-1 mutant An Arabidopsis mutant with flowers similar to those of plants treated with auxin transport inhibitors. The inflorescence is impaired in polar auxin transport. Useful in investigating the role of polar auxin transport in floral development.

PIN1 Membrane protein that forms part of the complex that transports auxin out of the basal ends of conducting cells during polar auxin transport.

Pit membrane The porous layer between pit pairs, consisting of two thinned primary walls and a middle lamella.

Pit pairs The adjacent pits of adjoined tracheid cells. A low-resistance path for water movement between tracheids.

Pit Microscopic regions where the secondary wall of tracheids is absent and the primary wall is thin and porous, facilitating sap movement between one tracheid and the adjacent one.

Plant growth substance *See* plant hormone.

Plant hormones Substances that influence plant growth and development at low concentrations. Major classes are abscisic acid, auxin, brassinosteroid, cytokinin, ethylene, and gibberellin.

Plasma membrane H^+-ATPase A P-ATPase that transports H^+ across the plasma membrane.

Plasmalemma *See* plasma membrane.

Plasmalemma central control model (PCC model) Proposed mechanism for gravitropism in which calcium channels in the plasma membrane open in response to gravity-induced changes in the distribution of forces exerted by the protoplast, the cytoskeleton, or the cell wall. Contrast with starch–statolith hypothesis.

Plasmid Circular pieces of extrachromosomal bacterial DNA that are not essential for the life of the bacterium. Plasmids frequently contain genes that enhance the ability of the bacterium to survive in special environments.

Plasmodesmata (singular plasmodesma) Microscopic membrane-lined channel connecting adjacent cells through the cell wall and filled with cytoplasm and a central rod derived from the ER called the desmotubule. Allows the movement of molecules from cell to cell through the symplast. The pore size can apparently be regulated by globular proteins lining the channel inner surface and the desmotubule to allow particles as large as viruses to pass through.

Plasmolysis Shrinking of the protoplasm of a cell placed in an hypertonic (low osmotic potential) solution, away from the cell wall, due to loss of water.

Plastocyanin A small (10.5 kDa), water-soluble, copper-containing protein that transfers electrons between the cytochrome $b_6 f$ complex and P700. This protein is found in the lumenal space.

Plastohydroquinone (PQH_2) The fully reduced form of plastoquinone.

Plastoquinone A small, nonpolar molecule that diffuses readily in the nonpolar core of the thylakoid membrane and is capable of undergoing reduction to plastohydroquinone. A mobile electron carrier connecting PSII and PSI. Chemically and functionally, similar to ubiquinone in the mitochondrial electron transport chain.

Pleiotropic Referring to a gene having more than one (perhaps many) phenotypic effects.

Plumule *See* embryo.

Pneumatophores Plant structures that protrude out of the water and provide a gaseous pathway for oxygen diffusion to the roots growing in water or in water-saturated soils.

Pollen Small structures (microspores) produced by anthers of seed plants. Contain haploid male nuclei that will fertilize egg in ovule.

Pollination Transfer of pollen from anther to stigma.

Polymer-trapping model A model that explains the specific accumulation of sugars in the sieve elements of symplastically loading species.

Polysaccharide A polymer made of many sugar residues.

Polyterpenes ($[C_5]_n$) High-molecular-weight terpenes containing 1500 to 15,000 isoprene units; e.g., rubber.

Polyterpenoids Terpenes having more than fifty carbons.

Positional information The concept that cell position, cell relationships, and associations rather than cell lineage is the important determinant of cell differentiation and tissue formation.

PPFD *See* photosynthetic photon flux density.

PP_i *See* pyrophosphate.

Pr Red light–absorbing phytochrome form. This is the form in which phytochrome is assembled. The blue-colored Pr is converted by red light to the far-red light–absorbing form, Pfr.

Preprophase band A circular array of microtubules and microfilaments formed in the cortical cytoplasm just prior to cell division that encircles the nucleus and predicts the plane of cytokinesis following mitosis.

Pressure potential (Ψ_p) The hydrostatic pressure of a solution in excess of ambient atmospheric pressure.

Pressure-flow model A widely accepted model of phloem translocation in angiosperms. It states that transport in the sieve elements is driven by a pressure gradient between source and sink. The pressure gradient is osmotically generated and results from the loading at the source and unloading at the sink.

Prevacuolar compartment (PVC) A membrane compartment equivalent to the late endosome in animal cells where sorting occurs before cargo is delivered to the lytic vacuole.

Primary active transport The direct coupling of a metabolic energy source such as ATP hydrolysis, oxidation-reduction reaction, or light absorption to active transport by a carrier protein.

Primary meristem Meristems that are formed during embryogenesis and are found at the tip of the root and the shoot. *See* secondary meristem.

Primary phloem Phloem derived from the procambium during growth and development of a vascular plant.

Primary response genes ("early genes") Genes whose expression is necessary for plant morphogenesis and that are expressed rapidly following exposure to a light signal. Often regulated by phytochrome-linked activation of transcription factors. Genes whose expression does not require protein synthesis. *See* secondary response genes.

Primary root Root generated directly by growth of the embryonic root or radicle.

Primary walls The first-formed, unspecialized cell walls with similarities in molecular architecture in diverse types of growing plant cells. About 85% polysaccharide and 10% protein by dry weight.

Primary xylem Xylem derived from the procambium during growth and development of a vascular plant.

Primordia Localized region of the shoot apical meristem characterized by higher cell division, leading to the formation or growth of cells with identifiable future, such as leaf.

pro **mutant** In tomato, an ultra-tall constitutive response mutant resulting from the loss of function of a negative regulator.

Procambium Primary meristematic cells that differentiate into xylem, phloem, and cambium.

Programmed cell death (PCD) Process whereby individual cells activate an intrinsic senescence program accompanied by a distinct set of morphological and biochemical changes, called apoptosis.

Prohexadione (BX-112) Inhibitor of GA biosynthesis. Inhibits gibberellin 3β-hydroxylase that converts inactive GA_{20} to growth-active GA_1.

Promeristem That part of the shoot or root apical meristem containing the stem cells and their immediate derivatives which have not yet begun to differentiate.

Proplastid Type of immature, undeveloped plastid found in meristematic tissue that can convert to various specialized plastid types, such as chloroplasts, amyloplasts, and chromoplasts, during development.

Prosthetic group A metal ion or small carbon compound (other than an amino acid) that is covalently bound to a protein and is essential for its function.

14-3-3 Proteins Phosphoserine-binding proteins that regulate the activities of a wide array of targets via direct protein–protein interactions. The 14-3-3 proteins were first identified as abundant brain proteins and named after their elution position on ion exchange chromatography and mobility in starch gel electrophoresis.

Protein bodies Protein storage organelles enclosed by a single unit membrane; found mainly in seed tissues.

Protein kinases Enzymes that have the capacity to transfer phosphate groups from ATP to specific amino acid, such as histidine, serine, threonine, or tyrosine, located either within themselves or on other proteins. Play important roles in enzyme regulation, gene expression, and signal transduction.

Protein phosphatases Enzymes that remove regulatory phosphate groups from proteins. Have important roles in enzyme regulation, gene expression, and signal transduction.

Protoderm In the plant embryo, the surface layer one cell thick that covers both halves of the embryo and will generate the epidermis.

Protomeristem An embryonic structure that will become the root or shoot meristem upon germination.

Proton motive force (PMF, Δp) Gradient of electrochemical potential for H^+ across a membrane, expressed in units of electrical potential.

Protonema (plural protonemata) An algal-like filament of moss cells generated by germination of moss spores.

Protoplasm Classically, the living substance of all cells. The plasma membrane and all the active organelles, substances, and processes required for life and contained within the plasma membrane.

Protoplast The living contents of a cell enclosed by the plasma membrane: the cytosol, organelles, and nucleus. What remains after removal of the cell wall.

Pteridine Nitrogen-containing compound composed of two six-membered rings. Component of riboflavin and parent compound of pterins.

Pterin A carbon compound to which molybdenum is complexed, forming a prosthetic group in the nitrate reductases of higher plants. Also a blue light-absorbing chromophore in the DNA-repair enzyme, photolyase. Chemically derived from pteridines.

Pullulanase Catalyzes the hydrolysis of α-D-1,6 glucosidic linkages in pullulan (a linear polymer of α-1,6-linked maltotriose units) and in amylopectin and glycogen, and the α- and β-limit dextrins of amylopectin and glycogen.

Pulvinus (plural pulvini) A turgor-driven organ found at the junction between the blade and the petiole of the leaf that provides a mechanical force for leaf movements.

Pumps Membrane proteins that carry out primary active transport across a biological membrane. Most pumps transport ions, such as H^+ or Ca^{2+}.

Pyrethroids Monoterpene esters with high insect toxicity. Both natural and synthetic forms are popular ingredients in commercial insecticides.

Pyridoxal phosphate (vitamin B6) A cofactor required by all transamination reactions.

Pyrophosphate Two phosphate groups linked by a phosphate ester bond, PP_i.

Q cycle A mechanism of electron and proton flow in which plastohydroquinone is oxidized, with one of the two electrons passed along a linear electron transport chain toward photosystem I, while the other electron goes through a cyclic process that increases the number of protons pumped across the membrane.

Quantum (plural quanta) A discrete packet of energy contained in a photon.

Quantum efficiency Photosynthetic yield per photon flux of absorbed light.

Quantum yield The ratio of the yield of a particular product of a photochemical process to the total number of quanta absorbed.

Quenching The process by which energy stored in light excited chlorophylls is rapidly dissipated mainly by excitation transfer or photochemistry.

Quercetin An endogenous flavonoid compound that may inhibit and regulate auxin transport by blocking auxin efflux.

Quiescent center Central region of root meristem where cells divide more slowly than surrounding cells, or do not divide at all.

Quinone A small, nonpolar molecule that diffuses readily in the nonpolar core of the membrane and is capable of undergoing reduction to hydroquinone.

R genes Resistance genes that function in plant defense against fungi, bacteria, and nematodes in some cases by encoding protein receptors that bind to specific pathogen molecules, elicitors.

Radial axis The pattern of concentric tissues extending from the outside of a root or stem into its center. *See* linear axis.

Radicle The embryonic root. Usually the first organ to emerge on germination.

Ray stem cells Small cells whose derivatives include the radially oriented files of parenchyma cells known as rays.

***rcn1* mutant** Arabidopsis (roots curl in NPA) mutant that shows high sensitivity to NPA, inhibition of hypocotyl elongation, and auxin efflux. *RCN1* gene is closely related to the regulatory subunit of protein phosphatase 2A, a serine/threonine phosphatase.

Reaction center complex A group of electron transfer proteins that receive energy from the antenna complex and convert it to chemical energy using oxidation-reduction reactions.

Reactive oxygen species Toxic forms of oxygen, including the superoxide anion (O_2^-), hydrogen peroxide (H_2O_2), and the hydroxyl radical (•OH). Generated by cells in the vicinity of infection and preceding the hypersensitive response.

Redox reactions Chemical reactions involving simultaneous oxidation and reduction of molecular species.

Reducing sugar A sugar with an aldehyde or ketone group available for oxidation, e.g, glucose, fructose.

Reduction The chemical process by which electrons or hydrogen atoms are added to a substance.

Reductive pentose phosphate (RPP) cycle *See* Calvin cycle.

Rejuvenation The reversion of adult shoots to juvenile shoots. May be promoted by hormones, mineral deficiencies, low light, water stress, defoliation, and low temperature.

Reporter gene A gene whose expression conspicuously reveals the activity of another gene. Gene engineered to share the same promoter as another gene.

Repressors Proteins that either alone or in concert with other proteins repress expression of a gene. *See* transcription factor.

Resistivity (specific electrical resistance) A measure indicating how strongly a material opposes the flow of electric current. A low resistivity indicates a material that readily allows the movement of electrons. The SI unit for electrical resistivity is the ohm meter.

Resonance transfer The nonradiative, molecule-to-molecule transfer of energy from one excited molecule to another, such as the transfer on energy from an antenna complex to the reaction center.

Respiration The complete oxidation of carbon compounds to CO_2 and H_2O, using oxygen as the final electron acceptor. Energy is released and conserved as ATP.

Respiratory control ratio The ratio of oxygen-consumption rates of isolated mitochondria in the presence and absence of ADP. It is used as a measure of the quality of a mitochondrial preparation.

Respiratory quotient The ratio of CO_2 evolution to O_2 consumption.

Retrograde vesicular transport Maintains the spatial distribution of enzymes and other functional proteins within the Golgi stack by acting as a countercurrent to cisternal progression.

***RGA* gene** This gene encodes the repressor of ga1 protein, one of five DELLA-domain GRAS proteins in Arabidopsis that function as repressors of gibberellin response.

Rheological properties Referring to the flow properties of a material. *See* viscoelastic properties.

Rhizobia Collective term for the genera of soil bacteria that form symbiotic (mutualistic) relationships with members of the plant family Leguminosae.

Rhizosphere The immediate microenvironment surrounding the root.

Rib meristem zone Meristematic cells beneath the central zone that give rise to the internal tissues of the stem in shoot apical meristems.

Riboflavin A vitamin that is part of FAD and FMN.

Rieske iron–sulfur protein A protein in which two iron atoms are bridged by two sulfur atoms, with two histidine and two cysteine ligands.

Root cap Cells at the root apex that cover and protect the meristematic cells from mechanical injury as the root moves through the soil. Site for the perception of gravity and signaling for the gravitropic response.

Root cap–epidermal stem cells Generate the epidermis of the root cap by anticlinal cell divisions and generate the lateral root cap by periclinal divisions followed by anticlinal divisions.

Root cap stem cells Meristematic cells that give rise to the root cap.

Root hairs Microscopic extensions of root epidermal cells that greatly increase the surface area of the root, thus providing greater capacity for absorption of soil ions and, to a lesser extent, soil water.

Root pressure A positive hydrostatic pressure in the xylem of roots.

Rotenone Specific inhibitor of complex I.

R-type channels A type of gated channel for anions that opens or closes very rapidly in response a voltage change.

Rubisco The acronym for the chloroplast enzyme *ribu*lose *bis*phosphate *c*arboxylase/*o*xygenase. In a carboxylase reaction, rubisco uses atmospheric CO_2 and ribulose-1,5-bisphosphate to form two molecules of 3-phosphoglycerate. It also functions as an oxygenase that incorporates O_2 to ribulose-1,5-bisphosphate to yield one molecule of 3-phosphoglycerate and another of 2-2-phosphoglycolate. The competition between CO_2 and O_2 for ribulose-1,5-bisphosphate limits net CO_2 fixation.

Rubisco activase An enzyme that facilitates the dissociation of sugar bisphosphates-rubisco complexes and, in so doing, activates rubisco.

RUBs A family of small, ubiquitin-related proteins. Proteins linked to RUB are usually activated rather than degraded.

SAG genes Senescence-associated genes whose expression is induced during senescence.

Salicyl hydroxamic acid Specific inhibitor of the alternative oxidase.

Salicylic acid A benzoic acid derivative believed to be an endogenous signal for *SAR*.

Salinity Refers to high concentrations of total salts in the soil. Contrast with sodicity.

Salinization The accumulation of mineral ions, particularly sodium chloride and sodium sulfate, in soil often due to irrigation.

Salt stress The adverse effects of excess minerals on plants.

Salt-tolerant plants Plants that can survive or even thrive in high-salt soils. *See also* halophytes.

Sap Fluid content of the xylem, sieve elements of the phloem and the cell vacuole.

Saponins Toxic glycosides of steroids and triterpenes with detergent properties. They may interfere with sterol uptake from the digestive system or disrupt cell membranes.

SAR *See* systemic acquired resistance.

Saturation Refers to a condition under which an increase in a stimulus does not elicit a further increase in a response. A maximum state; not capable of further increase, movement, or inclusion.

SAUR **genes** In soybean, a group of primary response genes stimulated by auxin within 2 to 5 minutes of treatment.

Sclereid A type of nonelongated sclerenchyma cell commonly found in hard structures such as seed coats.

Sclerenchyma Plant tissue composed of cells, often dead at maturity, with thick, lignified secondary cell walls. It functions in support of nongrowing regions of the plant.

SCR **gene** Arabidopsis *SCARECROW* gene that controls tissue organization and cell differentiation in the embryo, hypocotyl, primary roots, and secondary roots.

scr **mutant** Arabidopsis mutant in which hypocotyl and inflorescence are agravitropic and lack both endodermis and starch sheath.

Scutellum The single cotyledon of the grass embryo, specialized for nutrient absorption from the endosperm.

Second messenger Intracellular molecule (e.g., cyclic AMP, cyclic GMP, calcium, IP_3, or diacylglycerol) whose production has been elicited by a systemic hormone (the primary messenger) binding to a receptor (often on the plasma membrane). Diffuses intracellularly to the target enzymes or intracellular receptor to produce and amplify the response.

Secondary active transport Active transport that uses energy stored in the proton motive force or other ion gradient, and operates by symport or antiport.

Secondary meristems Meristems that are formed after seed germination and include axillary meristems and lateral meristems. Their activity may be suppressed by active primary meristem.

Secondary metabolites (secondary products) Plant compounds that have no direct role in plant growth and development, but function as defenses against herbivores and microbial infection by microbial pathogens, attractants for pollinators and seed-dispersing animals, and as agents of plant–plant competition.

Secondary products (natural products) *See* secondary metabolites.

Secondary response genes ("late genes") Genes whose expression requires protein synthesis and follows that of primary response genes.

Secondary transport Active transport driven by the proton gradient established by the proton pump.

Secondary wall Cell wall synthesized by nongrowing cells. Often multilayered and containing lignin, it differs in composition and structure from the primary wall. Forms during cell differentiation after cell expansion ceases.

Seed Develops from the ovule after fertilization of the egg, consisting of protective layers enclosing embryo of seed plants. May contain nutritive endosperm tissue separate from the embryo.

Seed coat The outer layer of the seed, derived from the integument of the ovule.

Seed plants Plants in which the embryo is protected and nourished within a seed. The gymnosperms and angiosperms.

Selectively permeability (of a membrane) Membrane property that allows diffusion of some molecules across the membrane to a different extent than other molecules.

Selectivity filter The domain of a channel protein that determines its specificity of transport.

Self-assembly Tendency for large molecules under appropriate conditions to aggregate spontaneously into organized structures.

Senescence An active, genetically controlled, developmental process in which cellular structures and macromolecules are broken down and translocated away from the senescing organ (typically leaves) to actively growing regions that serve as nutrient sinks. Initiated by environmental cues, regulated by hormones.

Sense RNA RNA capable of translation into a functional protein. *See* antisense RNA.

Sensible heat loss Loss of heat from leaf surfaces to the air circulating around the leaf, under conditions in which leaf surface temperature is higher than that of the air.

Sepals Green, leaf-like structures that form the outermost part of a flower. In bud, they enclose and protect other flower parts. *See also* petals, stamens, and carpels.

Sesquiterpene lactones Bitter, antiherbivore, 15-carbon terpenes found in members of the composite family, such as sunflower and sagebrush.

Sesquiterpenes Terpenes having fifteen carbons, three five-carbon isoprene units.

Sex determination The process whereby unisexual flowers are produced by the early selective abortion of either the stamen or the pistil primordia. Genetically regulated, but also influenced by photoperiod and nutritional status. Mediated by GA.

Shade avoidance response A response to shading; includes lengthening of the stem.

Shikimic acid pathway Reactions that convert simple carbohydrate precursors to the aromatic amino acids—phenylalanine, tyrosine, and tryptophan.

Shoot apex Consists of the shoot apical meristem plus the most recently formed leaf primordia (organs derived from the apical meristem).

Shoot apical meristem Meristem at the tip of a shoot. Consists of tunica layers, central zone, peripheral zone, rib meristem.

Short-day plant (SDP) A plant that flowers only in short days (qualitative SDP) or whose flowering is accelerated by short days (quantitative SDP).

Short-distance transport Transport over a distance of only two or three cell diameters. Involved in phloem loading, when sugars move from the mesophyll to the vicinity of the smallest veins of the source leaf, and in phloem unloading, when sugars move from the veins to the sink cells.

Sieve cells The relatively unspecialized sieve elements of gymnosperms. Contrast with sieve tube elements.

Sieve effect The penetration of photosynthetically active light through several layers of cells due to the gaps between chloroplasts permitting the passage of light.

Sieve element loading The movement of sugars into the sieve elements and companion cells of source leaves, where they become more concentrated than in the mesophyll.

Sieve element unloading The process by which imported sugars leave the sieve elements of sinks.

Sieve element–companion cell complex A functional unit consisting of a sieve element and its companion cell.

Sieve elements Cells of the phloem that conduct sugars and other organic materials throughout the plant. Refers to both sieve tube elements (angiosperms) and sieve cells (gymnosperms).

Sieve plates Sieve areas found in angiosperm sieve-tube elements; they have larger pores (sieve-plate pores) than other sieve areas and are generally found in end walls of sieve tube elements.

Sieve tube elements The highly differentiated sieve elements typical of the angiosperms. Contrast with sieve cells.

Sieve tube Tube formed by the joining together of individual sieve tube elements at their end walls.

Signal transduction A sequence of processes by which an extracellular signal (typically light, a hormone or neurotransmitter) interacts with a receptor at the cell surface, causing a change in the level of a second messenger and ultimately a change in cell functioning.

Signal transduction cascade See cascade, signal transduction.

Singlet oxygen ($^1O_2^*$) An extremely reactive and damaging form of oxygen formed by reaction of excited chlorophyll with molecular oxygen. Causes damage to cellular components especially lipids.

Sink Any organ that imports photosynthate, including nonphotosynthetic organs and organs that do not produce enough photosynthetic products to support their own growth or storage needs, e.g., roots, tubers, developing fruits, and immature leaves. Contrast with source.

Sink activity The rate of uptake of photosynthate per unit weight of sink tissue.

Sink size The total weight of the sink.

Sink strength The ability of a sink organ to mobilize assimilates toward itself. Depends on two factors: sink size and sink activity.

Slender mutants Plants that are phenotypically very tall, due either to constitutive gibberellin response (as in *slr1* and *sln1* mutants of rice and barley respectively) or to elevated levels of bioactive gibberellins (as in the *sln* mutant of pea).

***sln* mutant** In peas, a mutant having abnormally high levels of GA_{20} in the seed due to the impairment of a hydroxylation step in GA deactivation. In barley, an ultra-tall mutant resulting from a recessive allele causing negation of a negative signal transduction factor.

***SLY* gene** *Sleepy* gene encodes an F-box protein in Arabidopsis that targets GAI and RGA for proteolytic degradation.

SnRK1 The plant protein SnRK1 is homologous to the product of the *SNF1* (sucrose *n*on-*f*ermenting-*1*) gene that was identified genetically in screenings of budding yeast mutants. These variants fail to express the invertase gene, *SUC2*, in response to glucose deprivation. SnRK1 phosphorylates and, in so doing, inactivates *in vitro* sucrose 6F-phosphate phosphatase, 3-hydroxy-3-methylglutaryl-Coenzyme A reductase (linked to sterol/isoprenoid synthesis), and nitrate reductase (associated to nitrogen assimilation).

Sodicity Refers to the high concentration of Na^+ in the soil. Contrast with salinity.

Soil hydraulic conductivity A measure of the ease with which water moves through the soil.

Solar tracking The movement of leaf blades throughout the day so that the planar surface of the blade remains perpendicular to the sun's rays.

Solute potential (or osmotic potential) Ψ_s The effect of dissolved solutes on water potential.

Source Any exporting organ that is capable of producing photosynthetic products in excess of its own needs, e.g., a mature leaf or a storage organ. Contrast with sink.

Spatial Pertaining to space, the form and distribution of matter in space.

Specific heat Ratio of the heat capacity of a substance to the heat capacity of a reference substance, usually water. Heat capacity is the amount of heat needed to change the temperature of a unit mass by 1° C. The heat capacity of water is 1 calorie per gram per degree Celsius.

Spectrophotometer Instrument that measures the amount of light absorbed at different wavelengths by a substance.

Spectrophotometry The technique used to measure the absorption of light by a sample.

Spiral phyllotaxy One leaf is produced at each node, with each subsequent leaf produced at a 137° angle to the previously formed leaf.

Spongy mesophyll Mesophyll cells of very irregular shape located below the palisade cells and surrounded by large air spaces.

spy **mutant** In Arabidopsis, an ultra-tall constitutive response mutant resulting in the loss of function of a negative regulator.

Squalene A triterpene ($C_{30}H_{50}$) which serves as the starting point for the synthesis of the whole family of steroids in both plants and animals.

Stamen Male reproductive organ of the flower that produces pollen. Consists of stalk (filament) and an anther. *See also* sepals, petals, and carpels.

Starch phosphorylase Catalyzes the phosphorolysis of (α-D-1,4 glucosyl)n yielding (α-D-1,4 glucosyl)n-1 and α-D-glucose 1-phosphate.

Starch sheath A layer of cells that surrounds the vascular tissues of the shoot and coleoptile and is continuous with the root endodermis. Required for gravitropism in Arabidopsis shoots.

Starch–statolith hypothesis Proposed mechanism for gravitropism involving sedimentation of statoliths in statocytes.

Starch synthase Catalyzes the reaction: ADP-glucose + (α-D-1,4 glucosyl)n → ADP + (α-D-1,4 glucosyl)n+1.

Statocytes Specialized gravity-sensing plant cells that contain statoliths.

Statoliths Cellular inclusions such as amyloplasts that act as gravity sensors by having a high density relative to the cytosol and sedimenting to the bottom of the cell.

Steady state The condition in which influx and efflux of a given solute between two compartments separated by a membrane are equal. In the presence of active transport, steady state will differ from equilibrium.

Stele stem cells Root stem cells immediately behind the quiescent center that give rise to the pericycle and vascular tissue.

Stem cells Uncommitted, slowly dividing initial cells that produce all the cells in the meristem and thereby all the cells in the entire plant body (stem and root). Upon cell division, one daughter cell remains a stem cell, while the other is committed to a developmental pathway.

Stigma The receptive surface for pollen atop the style. *See* carpel.

STM **gene** Arabidopsis *SHOOT-MERISTEMLESS* gene that suppresses cell differentiation, ensuring that shoot meristem cells remain undifferentiated. Function required for formation of shoot protomeristem.

Stoichiometry The quantitative relationship between the amounts of reactants consumed and the products produced during a chemical reaction.

Stoma (plural stomata) A microscopic pore in the leaf epidermis surrounded by a pair of guard cells and in some species, also including subsidiary cells. Stomata regulate the gas exchange (water and CO_2) of leaves by controlling the dimension of stomatal pore.

Stomata *See* stoma.

Stomatal apertures The opening in the leaf epidermis through which gases enter and leave the interior spaces of the leaf. Changes in stomatal apertures are controlled by guard cells.

Stomatal complex The guard cells, subsidiary cells, and stomatal pore, which together regulate leaf transpiration.

Stomatal conductance A measurement of the flux of water and carbon dioxide through the stomata, in and out of the leaf. The inverse of stomatal resistance.

Stomatal movements Opening and closing of the stomata due to tugor changes in the guard cells.

Stomatal pore An opening through the epidermis to the leaf interior. Surrounded by a pair of guard cells.

Stomatal resistance (r_s) A measurement of the limitation to the free diffusion of gases from and into the leaf posed by the stomatal pores. The inverse of stomatal conductance.

Straight fertilizers Chemical fertilizers that contain inorganic salts of only one of the three macronutrients: nitrogen, phosphorus, and potassium (e.g., superphosphate, ammonium nitrate, or muriate of potash, a source of potassium).

Straight-growth test A bioassay based on the ability of auxin to stimulate the elongation of *Avena* coleoptile sections.

Stratification In some plants, a cold temperature requirement for seed germination. The term is derived from the former practice of breaking dormancy by allowing seeds to overwinter in small mounds of alternating layers of seeds and soil.

Stress Disadvantageous influences exerted on a plant by external abiotic or biotic factor(s), such as infection, or heat, water, and anoxia. Measured in relation to plant survival, crop yield, biomass accumulation, or CO_2 uptake.

Stress relaxation Selective loosening of bonds between primary cell wall polymers, allowing the polymers to slip by each other, simultaneously increasing the wall surface area and reducing the physical stress in the wall.

Stress resistance *See* stress tolerance.

Stress tolerance (stress resistance) A plant's ability to cope with an unfavorable environment.

Stroma lamellae Unstacked thylakoid membranes within the chloroplast.

Stroma reactions The carbon fixation and reduction reactions of photosynthesis that take place in the stroma of the chloroplast.

Style A stalk-like extension of the carpel. *See* carpel.

S-type channels A type of gated channel for anions that remains open for the duration of the stimulus.

Suberin A wax-like lipid polymer similar to cutin that acts as a barrier to water and solute movement through the Casparian strips of the endodermis, the outer cell walls of underground organs, cork cells of the periderm, and sites of leaf abscission or wounding.

Subjective day When an organism is placed in total darkness, the phase of the rhythm that coincides with the light period of a preceding light/dark cycle. *See* subjective night.

Subjective night When an organism is placed in total darkness, the phase of the rhythm that coincides with the dark period of a preceding light/dark cycle. *See* subjective day.

Subsidiary cells Specialized epidermal cells that flank the guard cells and work with the guard cells in the control of stomatal apertures.

Substomatal cavity The air space within the leaf to which the stomatal pore opens and which is bounded by mesophyll cells.

Subunit A polypeptide that is part of a protein complex.

SUC2 protein A sucrose-H$^+$ symporter found in the plasma membrane of companion cells.

Sucrose-H$^+$ symporter An active carrier that couples the energy dissipated by a proton, diffusing back into the cell to the uptake of a sucrose molecule. Mediates the transport of sucrose from the apoplast into the sieve element-companion cell complex in source leaves of apoplastically loading species.

Sunflecks Patches of sunlight that pass through openings in a forest canopy to the forest floor. Major source of incident radiation for plants growing under the forest canopy.

Superoxide (O_2^-) A reduced form of oxygen that is damaging to biological membranes.

Superoxide dismutase (SOD) This enzyme converts superoxide radicals to hydrogen peroxide.

Surface tension A force exerted by water molecules at the air–water interface, resulting from the cohesion and adhesion properties of water molecules. This force minimizes the surface area of the air–water interface.

Suspensor In seed plant embryogenesis, the structure that develops from the basal cell following the first division of the zygote. Supports, but is not part of, the embryo that develops from the apical cell and hypophysis.

SUT1 protein One of several sucrose-H$^+$ symporters found in the plasma membrane of sieve elements.

Symbionts Either one of the two organisms associated in a symbiotic relationship that may or may not be mutually beneficial. *See* mutualistic.

Symbiosis The close association of two organisms in a relationship that may or may not be mutually beneficial. Often applied to beneficial (mutualistic) relationships. *See* mutualism.

Symplast The continuous system of cell protoplasts interconnected by plasmodesmata.

Symplast pathway The route by which xylem and phloem sap travel from one cell to the next via the plasmodesmata.

Symport A type of secondary active transport in which two substances are moved in the same direction across a membrane.

Symporter A protein carrier involved in symport.

Syntaxins Proteins that integrate into membranes, permitting the membranes to fuse.

Synthetic auxin A substance with auxin activity that is not produced by plants, for example, 2,4-D and dicamba. Often used as herbicides because they are not degraded by the plant as quickly as IAA.

Systemic acquired resistance (SAR) The increased resistance throughout a plant to a range of pathogens following the infection of a pathogen at one site.

Systemin A polypeptide plant hormone that signals production of some defenses not continuously present in plants; initiates the biosynthesis of jasmonic acid.

TAM *See* tryptamine.

Tannins Plant phenolic polymers that often bind proteins and function as defenses against microorganisms, insects, and many mammals, such as cattle, deer, and apes.

Taproot The main single root axis from which lateral roots develop.

T-DNA A small portion of the *Ti* plasmid that is incorporated into the nuclear DNA of the host plant cell carries genes necessary for the biosynthesis of *trans*-zeatin and auxin, and opines. Because their promoters are plant eukaryotic promoters, none of these T-DNA genes are expressed in the bacterium, but are transcribed after they are inserted into the plant chromosomes.

Temperature compensation point The temperature at which the amount of CO_2 fixed by photosynthesis equals the amount of CO_2 released by respiration in a given time.

Tensile strength The ability to resist a pulling force. Water has a high tensile strength.

Tension Often used to refer to the partial pressure of a gas.

Teratomas Tumors that contain partially developed structures. Mutated *tmr* locus of T-DNA produced teratomas with abnormal proliferation of roots.

Terpenes (terpenoids or isoprenoids) A large class of plant compounds formed from five-carbon isoprene units, many of which are secondary metabolites with anti-herbivore activity.

Terpenoids (isoprenoids) Class of plant lipids that includes carotenoids and sterols.

Tetraterpenes Terpenes having forty carbons, eight five-carbon isoprene units. The latter compound serves as building blocks for anabolic processes while the excess is excreted as acetate.

Thermal hysteresis Phenomenon in which the liquid to solid transition is promoted at a lower temperature than the solid to liquid transition.

Thermal hysteresis proteins *See* antifreeze proteins.

Thigmotropism Plant growth in response to touch, enabling roots to grow around rocks and shoots of climbing plants to wrap around structures for support.

Thiosulfide (R–S$^-$) An intermediate in the reduction of sulfate. Formed from thiosulfonate. Reacts with O-acetylserine to form cysteine.

Thiosulfonate (R–SO$_3^-$) An enzyme-bound intermediate in the reduction of sulfate. Formed from APS. *See* thiosulfide.

Threshold The magnitude of a stimulus that needs to be exceeded to elicit a response.

Thylakoid reactions The chemical reactions of photosynthesis that occur in the specialized internal membranes of the chloroplast (called thylakoids). Include photosynthetic electron transport and ATP synthesis.

Thylakoids The specialized, internal, chlorophyll containing membranes of the chloroplast where light absorption and the chemical reactions of photosynthesis take place.

Ti plasmid A large tumor-inducing plasmid found in virulent strains of *Agrobacterium tumefaciens*.

TIBA (2,3,5-triiodobenzoic acid) A competitive inhibitor of polar auxin transport.

Tip growth Localized growth at the tip of a plant cell caused by localized secretion of new wall polymers. Occurs in pollen tubes, root hairs, some sclerenchyma fibers, and cotton fibers, as well as moss protonema and fungal hyphae.

TIR1/AFB proteins A family of F box soluble proteins, subunits of the ubiquitin E3 ligase complex, which include the principal auxin receptors.

***tir3* mutant** Arabidopsis (*transport inhibitor response 3*) mutant that exhibits reduced polar auxin transport and NPA binding. *tir3* contains a mutation in the BIG calossin-like protein, which is also defective in the *doc1* mutant.

Tissue culture Growth of isolated plant cells, tissues, or organs in the laboratory on an artificial medium consisting of various essential minerals, vitamins, hormones, and a carbon source.

***tmr* locus** Mutations at this locus of T-DNA produced teratomas with abnormal proliferation of roots. "Rooty" mutations. *See ipt* gene.

***tms* loci** Mutations in these sites of T-DNA resulted in an abnormal proliferation of shoots. "Shooty" mutations. *See tmr* locus.

Torpedo stage embryo The third stage of embryogenesis. The structure produced by elongation of the axis of the heart stage embryo and further development of the cotyledons. *See* globular stage.

Torus A central thickening found in the pit membranes of tracheids of most gymnosperms.

Totipotency In differentiated cells, retention of full genetic capacity for the development into a complete plant.

Toxicity effects Injuries caused by high concentrations of ions (particularly Na^+, Cl^-, or SO_4^{2-}) that accumulate in cells at low water potentials.

Tracheary elements Water transporting cells of the xylem.

Tracheids Spindle-shaped, water-conducting cells with tapered ends and pitted walls without perforations found in the xylem of both angiosperms and gymnosperms.

Transamination Reversible reactions catalyzed by transaminases in which nitrogen as an amino group is transferred from an α-amino acid to an α-keto acid.

Transcellular current A positive current, due largely to calcium ions, that flows into the cell on one side and loops back through the external medium. Establishes polarity prior to morphogenesis in *Fucus* zygotes, root hairs and germinating pollen grains.

Transcription factors Proteins that interact with DNA promoter elements to modulate gene expression (RNA transcription).

***trans*-cyclooctene** A strong competitive inhibitor of ethylene binding.

Transfer cells A type of companion cell similar to an ordinary companion cell, but with finger-like projections of the cell wall that greatly increase the surface area of the plasma membrane and increase the capacity for solute transport across the membrane from the apoplast.

Transgenic plants A plant expressing a foreign gene introduced by genetic engineering techniques.

Transit peptide An N-terminal amino acid sequence that facilitates the passage of a precursor protein through both the outer and the inner membranes of an organelle such as the chloroplast. The transit peptide is then clipped off.

Translocation The movement of photosynthate from sources to sinks in the phloem.

Translocator A membrane-based protein that functions as a carrier in the transport of one or more substances across the membrane.

Translocons Protein-lined channels in the rough endoplasmic reticulum that form associations with SRP receptors and enable proteins synthesized on ribosomes to enter the ER lumen.

Transmembrane pathway The route followed by the xylem sap that sequentially passes from cell to cell, crossing the cell plasma membrane both on entering and exiting the cell. Transport across the tonoplast may also be involved.

Transpiration The evaporation of water from the surface of leaves and stems.

Transpiration ratio The ratio of water loss to photosynthetic carbon gain. Measures the effectiveness of plants in moderating water loss while allowing sufficient CO_2 uptake for photosynthesis.

Transport Molecular or ionic movement from one location to another; may involve crossing a diffusion barrier such as one or more membranes.

***trans*-zeatin** The principal free cytokinin; chemically similar to kinetin. Exogenously applied in presence of auxin, induces cell division in callus cells and promotes bud or root formation from callus cultures. Endogenous *trans*-zeatin delays senescence of leaves and promotes expansion of dicot cotyledons. *See* zeatin.

Triacylglycerols (triglycerides) Three fatty acyl groups in ester linkage to three hydroxyl groups of glycerol. Fats and oils.

Tricarboxylic acid cycle *See* citric acid cycle.

Triterpenes Terpenes having thirty carbons, six five-carbon isoprene units.

Tropism Oriented plant growth in response to a perceived directional stimulus from light, gravity, or touch.

Tryptamine (TAM) An intermediate in one of three biosynthetic pathways for the synthesis of IAA in plants. Formed by decarboxylation of the amino acid tryptophan.

Tumor A mass of rapidly dividing, undifferentiated, and disorganized cells.

Tunica layers The outer cell layers of the shoot apical meristem. Outermost tunica layer generates the shoot epidermis.

Turgor pressure (or hydrostatic pressure) Force per unit area in a liquid. In a plant cell, the turgor pressure pushes the plasma membrane against the rigid cell wall and provides a force for cell expansion.

Turgor The firmness of a cell resulting from its hydrostatic or turgor pressure.

Ubiquinone A mobile electron carrier of the mitochondrial electron transport chain. Chemically and functionally similar to plastoquinone in the photosynthetic electron transport chain.

Ubiquitin A small polypeptide that is covalently attached to proteins by the enzyme ubiquitin ligase using energy from ATP, and that serves as a recognition site for a large proteolytic complex, the proteasome.

Ubiquitination The tagging of a protein with the small protein ubiquitin, thus targeting the protein for destruction by the proteasome.

Uncoupler A chemical compound that increases the proton permeability of membranes and thus uncouples the formation of the proton gradient from ATP synthesis.

Uncoupling protein A protein that increases the proton permeability of the inner mitochondrial membrane and thereby decreases energy conservation.

Uncoupling A process by which coupled reactions are separated in such a way that the free energy released by one reaction is not available to drive the other reaction.

Unisexual *See* imperfect flowers.

Ureide exporters Legumes of tropical origin that convert toxic ammonia to ureides such as allantoin, allantoic acid, and citrulline for transport to the shoot via the xylem. Contrast with amide exporters.

Vacuolar H⁺-ATPase A large, multi-subunit enzyme complex, related to F_oF_1-ATPases, present in endomembranes (tonoplast, Golgi). Acidifies the vacuole and provides the proton motive force for the secondary transport of a variety of solutes into the lumen.

van't Hoff equation Relates the solute potential, Ψ_s, to the solute concentration.

Vanadate Inhibitor of the H⁺-ATPase. The H⁺-ATPase is phosphorylated as part of the catalytic cycle that hydrolyze ATP. Because of this phosphorylation step, the plasma membrane ATPases are strongly inhibited by orthovanadate (HVO_4^{2-}), a phosphate (HPO_4^{2-}) analog that competes with phosphate from ATP for the aspartic acid phosphorylation site on the enzyme.

Vascular bundles Strands of primary phloem and xylem, separated by the vascular cambium and often surrounded by a bundle sheath, found in shoots, but continuous with the vascular cylinder of the root.

Vascular cambium A lateral meristem consisting of fusiform and ray stem cells, giving rise to secondary xylem and phloem elements, as well as ray parenchyma.

Vascular tissue Plant tissues specialized for the transport of water (xylem) and photosynthetic products (phloem).

V-ATPase A vacuolar H⁺-ATPase.

Veins The fine branches and intricate network of leaf vascular bundles.

Vernalization In some species, the cold temperature requirement for flowering. The term is derived from the word for "spring."

Vesicular-arbuscular mycorrhizal fungi A nondense mycelium within the root itself and extending into the soil; some hyphae penetrate individual cells of the root cortex as well as extending through the regions between cells.

Vesicular shuttle model Model for Golgi membrane development in which the *cis*, *medial*, and *trans* cisternae are stable structures. Cargo movement is via vesicles.

Vessel elements Nonliving water-conducting cells with perforated end walls found only in angiosperms and a small group of gymnosperms.

Viscoelastic properties Properties that are intermediate between those of a solid and those of a liquid and combine viscous and elastic behavior.

Vitamin B6 *See* pyridoxal phosphate.

VLFR (very low fluence response) Phytochrome responses whose magnitude is proportional to very low fluence (1 to 100 nmol m⁻²).

Wall creep In isolated primary cell walls, the time-dependent irreversible extension of cell wall polymers due to their slippage relative to one another.

Wall extensibility (m) During primary cell wall expansion, the coefficient that relates growth rate to the turgor pressure that is in excess of the yield threshold.

Water deficit Any water content of a plant cell or tissue below its water content when fully hydrated.

Water potential, Ψ_w Water potential is a measure of the free energy associated with water per unit volume (J m⁻³). These units are equivalent to pressure units such as pascals. Ψ_w is a function of the solute potential, the pressure potential, and the potential due to gravity: $\Psi_w = \Psi_s + \Psi_p + \Psi_g$. The term Ψ_g is often ignored because it is negligible for heights under five meters.

Water potential difference, $\Delta\Psi_w$ The water potential difference across the plasma membrane (expressed in megapascals).

Wavelength (λ) A unit of measurement for characterization of light energy. The distance between successive wave crests. In the visible spectrum, corresponds to a color.

Waxes Complex mixtures of extremely hydrophobic lipids synthesized and secreted by epidermal cells and that make up the protective cuticle that reduces water loss from exposed plant tissues. Waxes are mostly alkanes and alcohols of 25- to 35-carbon atoms.

Wetlands Land that is saturated with water on a regular basis.

Whorled phyllotaxy Several leaves or floral organs of a given type are attached to the stem at the same node.

Wilting Plant loss of rigidity, leading to a flaccid state, due to turgor pressure falling to zero.

Wound callose Callose deposited in the sieve pores of damaged sieve elements that seals them off from surrounding intact tissue. As sieve elements recover, the callose disappears from pores.

Xanthophyll cycle The interconversion of violaxanthin and zeaxanthin via the intermediate antheraxanthin. Zeaxanthin accumulates under conditions of excess energy.

Xanthophylls Carotenoids involved in nonphotochemical quenching. The xanthophyll zeaxanthin is associated with the quenched state of photosystem II, and violaxanthin is associated with the unquenched state.

Xylem The vascular tissue that transports water and ions from the root to the other parts of the plant.

Xylem loading The process whereby ions exit the symplast and enter the conducting cells of the xylem.

Xyloglucan A hemicellulose with a backbone of 1→4-linked β-D-glucose residues and short side chains that contain xylose, galactose and sometimes fucose. It is the most abundant hemicellulose in the primary walls of most plants (in grasses it is present, but less abundant).

Xyloglucan endotransglucosylase/hydrolases (XHTs) A large family of enzymes, which includes xyloglucan endotransglucosylase (XET), having the ability to cut the backbone of a xyloglucan in the cell wall and join one end of the cut chain with the free end of an acceptor xyloglucan.

Yield threshold (Y) The minimum value for turgor pressure at which measurable extension of the cell wall begins.

Yielding Long-term irreversible stretching that is characteristic of growing (expanding) cell walls. Nearly lacking in nonexpanding walls.

Z ("zigzag") scheme Arrangement of the reaction center and antenna complexes of photosystem II and photosystem I, and the electron transport chain that links them by their midpoint redox potential. The resulting alignment is shaped like a letter "Z."

Zeatin A naturally occurring cytokinin that stimulates mature plant cells to divide when added to a culture medium along with an auxin. Chemically, *trans*-6-(4-hydroxy-3-methylbut-2-enylamino) purine. *See trans-zeatin.*

Zeatin ribosides Zeatin with a ribose attached to the amino purine moiety. The main cytokinin in the xylem exudate.

Zeaxanthin A carotenoid implicated as a blue-light photoreceptor. A component of the xanthophyll cycle of chloroplasts, which protects from excess excitation energy.

Zeitgebers Environmental signals such as light-to-dark or dark-to-light transitions that synchronize the endogenous oscillator to a 24-hour periodicity.

Zygote The single celled product of the union of an egg and a sperm.

$\Psi_s = -RTc_s$ Where R is the gas constant, T is the absolute temperature (in degrees Kelvin, or K), and c_s is the solute concentration of the solution, expressed as osmolality.

Author Index

Abe, H., 618
Abe, M., 392, 664
Abeles, F., 578
Aber, J. D., 74
Abrahams, J. P., 269
Achard, P., 535, 664
Adams, P., 181
Adams, W. W., 152, 206, 207
Aerts, R. J., 321
Ahmad, I., 86
Ahmad, M., 455, 456
Aida, M., 403
Akiyoshi, D. E., 547, 548, 557
Alabadi, D., 433
Alberts, B., 11
Albertson, R., 26
Alborn, H. T., 334
Allan, A., 605
Allan, A. C., 604
Allard, H., 649
Allen, G. J., 607
Allen, J. F., 136, 137, 153, 156
Allis, D. C., 659
Aloni, R., 58, 228, 471, 499, 544
Alonzo, J. M., 587
Amasino, R., 559, 560, 636, 649, 658, 659, 660, 666
Amodeo, G., 450
Amor, Y., 355, 356
An, H., 663
Andel, F., 423
Anderson, B. E., 604
Anderson, J. M., 203
Aoki, K., 250
Ap Rees, T., 262
Appel, H. M., 329
Apse, M. P., 695
Arango, M., 113
Arimura, G.-I., 338
Arioli, T., 354
Armstrong, D., 556
Arnold, W., 131, 132
Arnon, D. I., 74, 148
Arora, R., 690
Asada, K., 151, 153
Asami, T., 623, 624
Asher, C. J., 77
Ashikari, M., 555
Aspe, M. P., 112
Assmann, S. M., 105, 452, 604, 610
Assunção, A. G. L., 697

Atkin, O. K., 277
Aukerman, M. J., 440
Avers, C. J., 129
Avery, L., 588
Ayre, B. G., 663

Babcock, G. T., 142
Badger, M. R., 173
Bailar, J. C., Jr., 77, 78
Bailey-Serres, J., 702
Baker, D. A., 228
Baker, N. R., 206
Balagué, C., 111
Baldwin, I. T., 338
Balmer, Y., 166, 167
Bange, G. G. J., 66
Bao Cai Tan, 596
Bao, F., 627, 628
Bar-Yosef, B., 88
Barber, J., 143
Barbier-Brygoo, H., 112
Barkla, B. J., 115
Barnola, J. M., 212
Baroja-Fernandez, E., 183
Barry, G. F., 548
Bartel, B., 473, 475
Barton, M. K., 396, 397
Baskin, T., 367
Basq, J. L., 701
Bassham, J. A., 160
Bastow, R., 659
Bauer, D., 431
Baulcombe, D. C., 663
Beale, S. I., 154, 449
Beardsell, M. F., 597
Becher, M., 697
Becker, T. W., 311
Becker, W. M., 135
Becwar, M. R., 690
Ben-Shem, A., 136, 137, 139, 146, 147
Benfey, P. N., 30, 393, 409
Benkova, E., 490
Bennet, M., 483
Bennett, A. B., 372
Bensmihen, S., 611
Benson, A, 160
Berenbaum, M. R., 324
Bergmann, L., 305
Berleth, T., 388
Berry, J. A., 210, 214
Berry, W. L., 86

Bethke, P., 17, 527, 528, 531
Beveridge, C. A., 552
Bewley, J. D., 599, 637, 639
Beyer, E. M., Jr., 574
Bibikova, T. N., 370
Bienfait, H. F., 308
Bilger, W., 152
Binder, B. M., 585
Bisseling, T., 301
Bizzell, C., 657
Björkman, O., 205, 210, 683, 684
Björn, L. O., 199, 200, 202
Black, M., 599
Blackman, F. F., 198
Blakely, S. D., 257
Blakeslee, J. J., 482, 483, 490
Blankenship, R. E., 125, 140, 141, 156
Blázquez, M. A., 664
Bleecker, A. B., 581
Blilou, I., 389, 402, 480
Blizzard, W. E., 678
Bloom, A. J., 74, 76, 77, 86, 89, 289, 292, 293, 310
Blumwald, E., 86, 112
Bohnert, H. J., 679
Boldt, R., 172
Boller, T., 340
Bomhoff, G., 549
Bonner, J., 180
Bonner, W., 180
Booker, J., 497, 498
Borlaug, N., 524
Borner, R., 666
Borthwick, H. A., 418
Bortolotti, C., 248
Boston, R. S., 685
Bouma, D., 82
Bowes, G., 212
Bowling, D. J. F., 118
Bowman, J., 398, 399, 406, 408, 639, 640
Boyer, G. L., 650
Boyer, J. S., 56, 671, 676, 678
Boyer, P. D., 149
Boysen-Jensen, P., 469
Bradford, K. J., 578, 579
Bradley, K. J., 339
Brady, N. C., 85
Brand, M. D., 268
Braun, A. C., 545
Bray, E. A., 681, 686

Brentwood, B., 226
Bret-Harte, M. S., 88
Brett, C. T., 373
Brettel, K., 146
Briarty, L. G., 382
Briggs, W. R., 421, 456, 457, 489
Brisson, L. F., 372
Broadley, M. R., 80
Brown, D. E., 483
Brown, R. M., Jr., 354
Browse, J. A., 282
Brudvig, G. W., 143
Bruick, R. K., 154
Brundrett, M. C., 90, 91
Brutnell, T., 439
Brzobohaty, B., 552
Buchanan, B., 167
Buchanan, B., 8, 22, 103, 147, 166, 303, 679, 685
Bucher, M., 112
Buddle, R. J. A., 276
Bünning, E., 652, 653
Bunsen, R. W., 422
Burg, E. A., 576
Burg, S. P., 572, 576
Burke, M. J., 689, 690
Burnell, J. N., 180
Burnstein, N., 67
Burris, R. H., 298, 299, 302
Burssens, S., 674
Butler, L, G., 329
Butler, W. L., 419

Calvin, M., 160
Campbell, G. S., 209
Campbell, W. H., 292
Caño-Delgado, A., 628
Capitani, G., 574
Caplin, S. M., 545
Carlson, R. W., 300
Carpenter, R., 637
Carpita, N. C., 353, 358, 360, 361, 362
Cashmore, A. R., 455
Catterou, M., 627
Cavalier-Smith, T., 154
Cerda-Olmedo, E., 447
Cerling, T. E., 217
Chailakhyan, M., 660
Chandler, P. M., 522, 523, 524
Chang, C., 379
Chang, R., 702

Chao, Q., 588
Chen, J. G., 504
Chen, M., 425
Chen, X., 154
Chia, T., 186
Chiang, H. H., 518
Chilton, M. -D., 548
Chitnis, P. R., 146
Choe, S., 624, 626
Choi, D., 537
Chollet, R., 174
Chory, J., 440, 561, 629, 630
Christie, J. M., 456, 457
Christmann, A., 597
Chua, N. -H., 608
Cilia, M. L., 249
Clack, T., 428
Clark, A. M., 225
Clark, D. G., 581
Clark, J. R., 643
Clark, K. L., 588
Clark, S. E., 401
Clarke, S. M., 81
Clarkson, D. T., 88, 90
Cleland, R. E., 370, 485, 486
Cleland, W. W., 167
Clough, S. J., 111
Clouse, S. D., 617, 618, 621, 625, 629
Coen, E. S., 637
Cohen, D., 597
Cohen, J. D., 474
Colmer, T. D., 89
Conner, D. J., 73, 74
Cooper, A., 77
Corbesier, L., 663
Corder, R., 328
Cosgrove, D. J., 352, 368, 370, 371, 372, 446, 449
Coulter, M. W., 653
Coupland, G., 655, 663
Coursol, S., 608
Cousins, H. H., 571
Craig, S., 175
Cramer, G. R., 694
Cramer, W. A., 144
Crawford, M. N., 292
Critchley, J. H., 186
Cronshaw, J., 226
Croteau, R., 321
Curry, G. M., 446
Cushman, J. C., 180, 181
Cutler, S., 612

D'Agostino, J. B., 564, 565
Dahlqvist, A., 283
Daie, J., 238
Dainty, J., 48
Davidson, S. E., 522
Davies, P., 515, 520
Davies, W. J., 597, 675
Davin, L. B., 325
Davis, J. F., 83
Davis, S. D., 63
De Boer, A. H., 119
de Vries, G. E., 300, 303
Deisenhofer, J., 137
Deitzer, G., 656
Delledonne, M., 338
Delwiche, C. F., 156
Demming-Adams, B., 152, 206
Deng, X. W., 430
Denison, R. F., 300
Denning, D. P., 9

Dennis, D. T., 257, 260
Desikan, R., 606
Dever, L. V., 176
Devlin, P. F., 648
Dharmasiri, N., 501
Dhugga, K. S., 354
Di Laurenzio, L., 393
Dill, A., 524, 525
Dinant, S., 225
Ding, B., 29, 250
Dion, R. O. D., 303
Dittmer, H. J., 86
Dixon, R., 301
Dixon, R. A., 338
Doblin, M. S., 354, 356
Dodd, A. N., 435
Doering-Saad, C., 229
Dolan, L., 404, 581
Dong, Z., 299
Douce, R., 262, 271
Downton, J. S., 214
Drew, M. C., 699, 700, 701
Drincovich, M. F., 180
Driouich, A., 14
Dunlop, J., 118
Durnford, D. G., 137, 139

Eaton-Rye, J. J., 81
Eberhart, W., 298, 322
Ecker, J. R., 584
Eckhardt, U., 154
Edwards, D. G., 77
Edwards, G. E., 171, 176, 178
Ehleringer, J. R., 201, 203, 205, 211, 215, 218
Ehlers, K., 225
Eigenbrode, S. D., 317
Eisner, T., 327
El-Din El Assal, S., 657
Elzen, G. W., 548
Emerson, R., 131, 132, 133
Emery, J. F., 409
Engel, R. E., 81
Engelberth, J., 338
Engelmann, T. W., 130
Epstein, E., 73, 74, 76, 77, 78, 80, 695
Eriksson, M. E., 521
Esau, K., 57, 378
Eshed, Y., 398, 399, 406, 408
Estruch, J. J., 553
Etzler, M. E., 301
Evans, H. J., 75
Evans, M. L., 494, 495
Evert, R., 30, 224, 236

Fabian, T., 537
Faiss, M., 551
Fan, L. M., 455
FAOSTAT, 291
Farquhar, G. D., 198, 217
Fasano, J. M., 493
Felton, G. W., 329
Fenn, M. E., 74
Ferguson, S. J., 268
Ferreira, F. J., 554
Ferreira, K. N., 144
Fiers, M., 405
Finazzi, G., 153
Finkelstein, A., 43
Finkelstein, R. R., 599, 603, 611
Firml, J., 389, 406
Firn, R. D., 446, 447
Fisher, D. B., 235

Flint, L. H., 418
Flowers, T. J., 692
Flügge, U. I., 186, 245
Föhse, D., 89
Fondy, B. R., 237
Foo, E., 557
Forde, B. J., 292
Forsberg, J., 136, 137, 153
Forseth, I., 203
Foster, C. E., 395
Fowlers, S., 692
Foyer, C. H., 276
Franklin, K. A., 428
Frasch, W., 149
Frechilla, S., 458, 461
Frederick, S. E., 18
Freeman, J. L., 697
Frensch, J., 57
Fricker, M., 207, 208
Friedman, M. H., 44
Friedrichsen, D. M., 625, 630
Friml, J., 490, 496
Fry, S. C., 362, 363
Fu, X., 512
Fu, Y., 366
Fujioka, S., 619, 621, 623
Fukshansky, L., 201
Fukuda, H., 628

Galis, I., 557
Gallagher, K. L., 30
Galston, A., 500
Gälweiler, L., 479
Gan, S., 559, 560
Gane, R., 571
Gardiner, J. C., 366
Garfinkel, D. J., 556
Garner, W., 649
Gaspar, Y., 361, 362
Gasser, C. S., 638
Gates, P., 222
Gaupels, F., 341
Gaymard, F., 119
Geiger, D. R., 210, 234, 237, 245
Geisler, M., 475, 476, 482, 483, 495
Geldner, N., 389, 483
Gepstein, S., 578
Gericke, W. F., 76
Gershenzon, J., 321
Geurts, R., 301
Ghassemian, M., 613
Gibson, A. C., 216
Gifford, E. M., 395
Gilbert, S. F., 482
Gilroy, S., 531, 604, 607
Giordano, M., 173
Giulini, A., 555
Givan, C. V., 260
Gocal, G. F. W., 532, 534
Goldstein, G., 48
Golecki, G., 248
Gomez, G., 249
Gomez-Cadenas, A., 601
Gonzales-Meler, M. A., 278
Goodchild, D. J., 175
Gorton, H. L., 207
Gosti, F., 610
Gottwald, J. R., 112
Gowan, E., 239
Graham, R. D., 308
Grbic, J., 637
Grbiè, V., 581
Green, B. R., 137, 139

Greenway, H., 693
Gregory, F. G., 658
Gresshoff, P. M., 301
Grossman, A. R., 137, 139
Grove, M. D., 618, 619
Grusak, M. A., 237
Gruszecki, W. I., 462
Gubler, F., 532, 535
Guerinot, M. L., 112, 308
Guilfoyle, T., 490
Gunning, B. E. S., 13, 14, 20, 21, 62, 263, 316, 356
Guo, H., 657
Gusta, L. V., 691
Guy, C. L., 691
Guzman, P., 584

Haberlandt, G., 528, 544
Hacke, U. G., 63
Haecker, A., 389
Haehnel, W., 82
Hagiwara, S., 606
Hai, L., 580
Hain, R., 339
Haldrup, A., 153
Halford, N. G., 247
Hall, A., 434
Hall, N. G., 228
Hamann, T., 389
Hamilton, A. J., 663
Hamilton, J. L., 548
Hamner, K. C., 653
Han, T., 214
Harada, J., 383
Harberd, N. P., 512
Hardham, A., 367
Harling, H., 81
Harmer, S. L., 435
Harter, B. L., 300
Hartman, H., 156
Hartmann, H. T., 477
Hartmann, K. M., 423
Hartmann, T., 330, 331
Hartung, W., 675
Harvey, G. W., 204
Harwood, J. L., 22
Hasegawa, P. M., 695
Hasenstein, K. H., 495
Hatch, M. D., 173, 174, 176, 180
Hatfield, R., 325
Haupt, W., 207, 448
Hausler, R. E., 190
Havau, M., 207
Hayama, R., 655
Hayashi, H., 229
Hayashi, T., 357
Haywood, V., 249, 395
He, J. Z., 623, 632
Hedden, P., 515, 518, 521, 524
Heidstra, R., 300
Helariutta, Y., 393, 394
Heldt, H. W., 168, 186
Hell, R., 290, 304
Hendricks, S. B., 656
Hensel, L. L., 581
Hepler, P. K., 67, 79, 450
Herrmann, K. M., 323
Heytler, P. G., 303
Hibberd, J. M., 178
Higgins, T. J. V., 532
Higinbotham, N., 100, 101, 186
Higuchi, M., 554, 564
Hikosaka, K., 200
Hill, A. E., 112

Hill, R., 132
Himmelbach, A., 610
Hirayama, T., 586
Hirsch, R. E., 111
Hirschi, K. D., 112
Hisamatsu, T., 519
Ho, D. T. H., 684, 702
Hoagland, D. R., 77
Hoagson, C. W., 142
Hoecker, U., 601
Hoefnagel, M. H. N., 256, 276
Holbrook, N. M., 64
Holt, N. E., 152
Hong, S. W., 685
Horie, T., 112
Horton, P., 152
Horwitz, B. A., 446
Hoson, T., 371
Houba-Herin, N., 552
Howe, G. A., 336
Howell, S. H., 520
Hsiao, T. C., 50
Hsieh, H. L., 502
Hu, Y., 627
Hua, J., 586
Huai, Q., 574
Huala, E., 447
Huang, A. H. C., 22
Huang, J., 276
Huang, T., 664
Huber, J. L., 192
Huber, S. C., 190, 192
Huff, A. K., 562
Hugouvieu, V., 612
Hwang, I., 565, 566

Iino, M., 453, 462, 489
Ikeda, A., 522, 523
Ikekawa, N., 632
Im, C., 449
Imlau, A., 248
Inderjit, 325
Ingenhousz, J., 130
Ingle, R. A., 697
Ingram, G. D., 400
Ingram, T. J., 520
Inoue, T., 392, 562
Ishii, T., 358
Itai, C., 551
Itoh, H., 524
Itoh, J., 384
Izawa, T., 655, 656

Jackson, D., 249
Jackson, G. E., 63
Jacobs, M., 482, 487
Jacobson, J. V., 531
Jagendorf, A., 148, 149
Jarvis, P. G., 47, 67, 206
Javot, H., 112, 118
Jeannette, E., 604
Jeffree, C. E., 316
Jenik, P. D., 396, 397
Jenks, M. A., 678
Jenuwein, T., 659
Jeuffroy, M. -H., 246
Jin, H., 323
John, M., 373
John, P. C. L., 556
Johnson, E., 656
Johnson, P. W., 299
Johnson, R., 336
Johnstone, M., 81
Joly, R. J., 680

Jones, H. D., 529
Jones, O. T. G., 308
Jones, R. L., 17, 604
Jongsma, M. A., 336
Jordan, P., 146
Joy, K. W., 228
Jürgens, G., 388, 389, 390

Kahn, D., 301
Kaiser, W. M., 293
Kakimoto, T., 551, 562
Kamiya, Y., 519
Kaneko, M., 519, 529, 534
Kaplan, A., 173
Kapulnik, Y., 300
Karban, R., 338
Karpilov, Y., 173
Karplus, P. A., 147
Kashirad, A., 88
Kawasaki, S., 680
Kay, S. A., 648
Keeling, C. D., 212
Keiber, J. J., 554
Kende, H., 536, 537, 558
Kepinski, S., 501
Kessel, G. S. -R., 60
Kester, D. E., 477
Kidner, C. A., 409
Kieber, J. J., 586, 587, 588
Kim, H. Y., 437
Kim, I., 395
Kim, T. W., 624
Kimura, S., 354, 356
King, R. W., 533
Kinoshita, T., 450, 459, 461, 603, 629
Kirkby, E. A., 74, 75, 84, 89
Kiss, J., 492
Kleinhofs, A., 292, 293
Kluge, C., 114, 116
Knoblauch, M., 225, 234
Knop, W., 76, 77
Koch, K. E., 247
Koller, D., 202
Koller, H. R., 246
Kombrink, E., 339
Kondo, T., 326
Koornneef, M., 427, 517, 522, 523, 600, 601
Kortschack, H. P., 173
Koshiba, T., 596
Kotani, H., 153
Kozaki, A., 171
Krab, K., 273
Kramer, P. J., 56
Krause, G. H., 152
Krebs, H. A., 262
Krieg, D. R., 676
Krömer, S., 274
Kruger, N. J., 257
Kruse, M., 325
Ku, S. B., 171
Kühlbrandt, W., 139
Kuhn, C., 238
Kumar, P. P., 249
Kurisu, G., 144, 145
Kuzma, M. M., 299
Kwak, J. M., 606

Lagarias, J. C., 424
Laloi, M., 273
Lam, H. -M., 294, 296
Lamb, C., 338, 412
Lambers, H., 275

Lanahan, M., 577
Lang, A., 658, 662, 663, 664
Lariguet, P., 431
Larkindale, J., 686
Larsen, Calvin, 321
Lau , T., 387, 392, 397
Layzell, D. B., 289
Lea, P. J., 294
Lee, D. J., 533
Lee, J. H., 559, 685
Lee, Y., 117, 608
Leegood, R. C., 168
Lehman, A., 580
Leng, Q., 103
Leon, P., 272
Lester, D. R., 520
Letarjet, R., 200
Letham, D. S., 545, 552
Leubner-Metzger, G., 629
Leustek, T., 305
Leverenz, J. W., 206
Levings, C. S., III, 276
Levitt, J., 682
Lewis, N. G., 325
Lev-Yadun, S., 476
Li, J., 116, 327, 610, 620, 629
Li, L., 424
Li, X. P., 152
Liang, X., 575
Libertini, E., 371
Lichtenthaler, H. K., 318
Liepman, A. H., 354
Lin, C., 456
Lin, W., 107, 108
Lindow, S. E., 690
Lipson, E. D., 447
Liu, C., 386
Liu, K. H., 110
Liu, Z. F., 139
Ljung, K., 471, 494
Lloyd, C., 580
Lloyd, F. E., 453
Lloyd, J. R., 183
Lobreau, D., 308
Lodish, H., 10
Logemann, E., 323
Long, S. P., 152, 209
Loo, D. D. F., 47
Loomis, R. S., 73, 74
Lopez-Molina, L., 611
Lorimer, G., 166, 168, 171
Lott, J. N. A., 222
Lott, N. A., 321
Lowe, R. H., 548
Lu, C., 612
Lu, P., 392
Luan, S., 610
Lucas, R. E., 83
Lucas, W. J., 29, 117
Ludwig, M., 190
Ludwig, R. A., 300, 303
Lunau, K., 327
Lund, J. E., 192
Lüttge, U., 186
Luu, D.-T., 112
Lyons, J. M., 688

Maathuis, F. J. M., 119
Macek, T., 74
MacMillan, J., 509, 511
Maggio, A., 680
Mahonen, A. P., 392, 393
Maier, R. M., 165

Malamy, J. E., 409
Maldonaldo, A. M., 341
Maloof, J. N., 440
Mandava, N. B., 617, 618, 619, 626
Mandoli, D. F., 421
Mansfield, D., 69
Mansfield, S. G., 382
Mansfield, T. A., 67
Mantyla, E., 691
Mao, J., 459, 460
Marga, F., 351, 365
Margulis, L., 20
Marienfeld, J., 272
Maroco, J. P., 180
Marschner, H., 75, 81, 91, 294, 300, 303, 306
Martienssen, R. A., 409
Martin, C., 323
Martin, D. N., 520
Martin, R. C., 552
Martinoia, E., 110
Martre, P., 57
Mäser, P., 609
Mathews, S., 426, 428
Matsuoka, M., 529
Matsushita, T., 424
Matters, G. L., 449
Matthews, M. A., 673
Maurel, C., 112, 118
Mauseth, J. D., 90
Mayer, K. F., 398
Mayer, U., 388
Mayfield, S. P., 154
McAinsh, M. R., 606
McCabe, M. S., 562
McCabe, T. C., 272
McCann, M., 351, 355, 358, 360, 361, 362
McCarty, D., 596
McClung, C. R., 433
McClure, B. A., 490
McConn, M., 335
McCourt, P., 629
McCree, K. J., 200, 677
McDaniel, C. N., 644, 645
McEnvoy, J. P., 143
McGinnis, K. M., 524
McIntosh, C. A., 265, 272, 273
McIntyre, D., 48, 435
McKeon, T. A., 573, 574
McKown, A., 175
McMichael, B. L., 674
Meckes, A., 5
Medville, R., 241
Meidner, H., 69
Melis, A., 152, 203
Memelink, J., 335
Mendel, R. R., 292
Mengel, K., 74, 75, 84, 89
Meyerowitz, E. M., 586, 639
Michaels, S. D., 658, 659, 666
Michel, H., 137
Mierzwa, R. J., 236
Mikkelsen, R., 184
Milanowska, J., 462
Milborow, B. V., 594
Millar, A. A., 535
Millar, A. H., 266, 272
Millar, A. J., 433, 657
Miller, C. O., 545, 556
Mitchell, J. M., Jr., 38
Mitchell, J. W., 617

Author Index

Mitchell, P., 148, 268
Miyawaki, K., 551
Möller, I. M., 256, 267, 273
Monte, E., 428, 431
Montoya, T., 625
Moore, A. L., 268
Mora-Garcia, S., 166, 167
Mordue, A. J., 321
Morgan, P. W., 582
Mori, I. C., 610
Morillon, R., 626
Morris, R., 552, 557
Mothes, K., 560
Muissig, C., 627, 628, 629
Müller, M., 116
Multani, D. S., 481
Münch, E., 231
Munns, R., 693
Murillo, I., 250
Murphy, A. S., 476, 482, 483
Mustilli, A. C., 610
Muto, S., 610
Mylona, P., 301

Nagatani, A., 427
Nakagawa, J. H., 575
Nakajima, K., 393, 394
Nakaya, M., 626
Nakazawa, M., 502
Napier, J. A., 279
Nebenführ, A., 14
Neff, M. M., 165, 623
Neftel, A., 212
Neinhuis, S., 317
Nelson, M., 637
Nelson, N., 136, 137, 147
Neuffer, M. G., 522
Neuhaus, H. E., 183
Newcomb, E. H., 18, 19, 26
Nicholls, D. G., 268
Nielsen, T. H., 190
Niittyla, T., 186
Nishikawa, N., 625
Nishimura, C., 554, 555, 564
Nishio, J. N., 214
Niu, X., 680, 694
Niyogi, K. K., 458
Nobel, P. S., 43, 44, 59, 65, 96, 216
Noctor, G., 276
Noguchi, T., 623
Noh, B., 482
Nolan, B. T., 74
Nonhebel, H. M., 472
Noodén, L., 411, 552, 559
Normanly, J. P., 475
Nürnberger, T., 340
Nye, P. H., 88, 91

O'Neill, D., 537
O'Neill, M. A., 353, 358
Oaks, A., 293, 296
Oeller, P., 577
Ogawa, M., 518
Ogawa, T., 173
Ogren, W. L., 166, 168, 308
Oh, S.-H., 81
Ohlrogge, J. B., 282
Ohme-Takagi, M., 588
Okamura, M. Y., 144
Oliver, D. J., 265, 273
Olszewski, N., 523
Oosterhuis, D. M., 679
Oparka, K. J., 29, 248

Opperta, T., 336
Ort, D. R., 206
Osmond, C. B., 207, 208
Otegui, M. S., 17
Otte, O., 372
Ottenschlager, I., 495, 496
Outlaw, W. H., Jr., 453, 455
Ozga, J. A., 537

Paál, A., 469
Palevitz, B. A., 26, 67
Pallas, V., 249
Palme, K., 479, 483, 490
Palmer, J. D., 156
Palmgren, M. G., 114, 115
Palta, J. P., 688
Pandey, S., 608
Pantoja, O., 115
Papadopoulou, K., 338
Papenfuss, H. D., 652
Pare, P. W., 335
Park, E., 431
Park, W. W., 535
Park, Y. W., 372
Parks, B. M., 421, 457
Parson, W. W., 137, 139
Pate, J. S., 289, 294
Patel, Samir, 9
Patterson, B. D., 688
Paul, M., 190, 247
Paulson, H., 139
Pauly, M., 357
Pear, J. R., 354
Pearcy, R. W., 202
Peer, W. A., 483, 484
Peiter, E., 111
Peng, J. R., 522, 523, 524
Peng, L., 355
Pennell, R. I., 412
Perovic, S., 573
Persus-Barbeoch, L., 105
Peuke, A. D., 231
Peumans, W. J., 336
Pfündel, E., 152
Pharis, R. P., 513
Phillips, A. L., 515, 518, 521
Phillips, D. A., 300
Phinney, B., 510, 521
Plaxton, W. C., 260
Poethig, R. S., 399, 400, 643
Portis, A. R., 166
Poulton, J. E., 332
Preisig, O., 300
Preiss, J., 245
Price, G. D., 173
Priestley, J., 130
Prince, R. C., 140, 141
Pullerits, T., 137
Purton, S., 165
Purvis, O. N., 658

Qu, X., 585
Quail, P. H., 420, 426, 438
Quatrano, R. S., 604
Queitsch, C., 685
Quick, P., 178
Quiñones, M. A., 447
Quintero, F. J., 695

Rachmilevitch, S., 172
Raison, J. K., 684
Randall, D. D., 276
Ranson-Hodgkins, W. D., 239
Raschke, K., 603

Raskin, I., 272, 574
Rasmusson, A. G., 256, 267
Rausch, C., 112
Raveh, E., 450
Raven, J. A., 77
Ray, P. M., 487
Rayle, D. L., 370
Raz, V., 598
Razem. R., 604
Rea, P. A., 116
Rees, D. A., 307
Reid, J. B., 518, 520, 521, 625
Reid, M. S., 578, 581, 583
Reinfelder, J. R., 179
Reinhardt, D., 406, 498
Reis, V. M., 299
Rennenberg, H., 305
Renwick, J. A. A., 333
Rexach, M., 9
Rhoades, M. W., 535
Richardson, S. G., 677
Richter, T., 201
Rignanesse, L., 642
Riou-Khamlichi, C., 556
Ritchie, S., 531
Robards, A. W., 29
Roberts, A. G., 29, 242
Roberts, J. K. M., 699
Roberts, M. R., 114
Robinson, J. M., 310
Robinson-Beers, K., 30, 638
Robson, P. R., 439
Rock, C. D., 611
Rodriguez, F. I., 586
Rodriguez, R. L., 247
Roelfsema, R. G., 453, 455
Rojo, E., 401
Roland, J. C., 352
Rolfe, B. G., 301
Roscoe, H. E., 422
Rose, A., 24
Rose, J. K. C., 362, 372
Rosenthal, G. A., 333
Ross, C. W., 562
Ross, J. J., 520, 537
Ross, S. D., 513
Rothwell, G. W., 476
Rovira, A. D., 90
Rupert, C. S., 200
Rupp, H. -M., 553
Rus, A., 695
Russell, R. S., 88
Ryals, J. A., 341
Ryan, C. A., 336
Ryden, P., 358
Ryu, J. S., 431

Saab, I. N., 602
Sabatini, S., 403
Sachs, M. M., 260, 684, 702
Sack, F. D., 68
Sage, R. F., 178, 210
Saito, K., 305
Saji, H., 656
Sakai, H., 565
Sakakibara, H., 548
Sakamoto, T., 521, 535, 536
Sakurai, A., 621
Salerno, G. L., 192
Salisbury, F. B., 652
Salleo, S., 64
Salnikov, V. V., 355
Salome, P. A., 433
Salt, D. E., 697

Salvucci, M. E., 166
Samuelson, M. E., 551
Sanchez, J. -P., 608
Sandberg, S. L., 324
Sanders, D., 79, 112
Santa Cruz, S., 248
Sasaki, A., 524
Sasaki, Y., 281
Sasse, J. M., 617, 618, 621, 625, 629
Sauter, A., 675
Sauter, M., 536, 537
Savaldi-Goldstein, S., 620
Sawers, R. J., 438
Schäffner, A. R., 49
Scheel, D., 340
Scheer, J. M., 336
Scheible, W. R., 357
Scheres, B., 385
Scheuerlein, R., 207
Schiefelbein, J. W., 405
Schilmiller, A. L., 336
Schlefelbein, J., 405
Schlesinger, W. H., 290, 291
Schmidt, C., 609
Schmidt, R. C., 472
Schnell, D. J., 154
Schoeffl, F., 685
Schoof, H., 402
Schopfer, P., 370
Schroeder, J. I., 112, 283, 453, 603, 604, 606, 607
Schultz, C. J., 361
Schultz, T. F., 604
Schulz, A., 224, 225, 238, 246, 247
Schulze, W., 110
Schumacher, K., 626
Schumaker, K. S., 558
Schurr, U., 597
Schwacke, R., 102
Schwartz, A., 450, 604
Schweighofer, A., 610
Scott, G. K., 548
Seton, R., 582
Seagull, R. W., 367
Seelert, H., 150
Senger, H., 445
Sentenac, H., 110, 111
Seo, H. S., 433
Seo, M., 596
Serek, M., 575
Serrano, E. E., 452
Servaites, J. C., 210
Settleman, J., 366
Sharkey, T. D., 171, 198, 210
Sharma, R., 425
Sharma, V. K., 401
Sharp, R. E., 89, 602
Sharpe, P. J. H., 68
Sharrock, R. A., 426, 428
Shaw, S., 494
Sheen, J., 565, 566
Shelp, B. J., 80
Shen, Y., 431
Shi, H., 112
Shimada, Y., 625
Shimazaki, K., 450, 452, 461, 603
Shimmen, T., 27
Shinomura, T., 428
Shinozaki, K., 611, 680, 682
Shinshi, H., 588
Shropshire, W. Jr., 422
Shulaev, V., 341

Siebke, K., 177
Siedow, J. N., 276
Siefritz, F., 57
Siegel, L. M., 293
Siegelman, H. W., 656
Sieren, D., 321
Sievers, A., 492
Sievers, R. E., 77, 78
Silk, W. K., 88
Silverstone, A. I., 518, 519, 522, 523, 524, 525
Simon, R., 638
Singh, D. P., 514
Sisler, E. C., 575
Sitbon, F. E. A., 475
Sivasankar, S., 293, 296
Skoog, F., 496, 497, 545, 556
Slack, C. R., 173, 174, 176
Small, J., 116
Smart, D. R., 293
Smigocki, A., 562
Smith, A. M., 183
Smith, F. A., 77
Smith, H., 201, 202, 437
Smith, H. H., 559
Smith, R. C., 363
Smith, S. E., 90, 91
Sokkermans, T. J. W., 300
Solano, R., 588
Solliday, J., 350
Somers, D. E., 434
Somssich, I. E., 339
Sondergaard, T. E., 112, 114
Sorefan, K., 557
Sorger, G. J., 75
Soriano, I., 322
Soupene, E., 173
Sovonick, S. A., 234
Spalding, E. P., 421, 449, 456
Spanswick, R. M., 101
Spence, I., 322
Sperry, J. S., 63, 64
Sponsel, V., 517
Srivastava, A., 450, 458
Stacey, G., 110
Stadler, R., 248
Staehelin, L. A., 14
Staswick, P. E., 502
Steber, C. M., 629
Steer, M., 356
Steer, M. M., 62
Steer, M. W., 13, 14, 20, 21, 263, 316
Steeves, T. A., 378, 398, 399
Steffens, G. L., 617
Stegink, S. J., 269
Stein, W. D., 43
Stepnitz, K., 585
Steudle, E., 48, 57, 62
Steward, F. C., 545
Stewart, G. R., 86
Stintzi, A., 283
Stitt, M., 190
Stock, D., 150
Stockinger, E. J., 680
Stoner, J. D., 74
Stout, P. R., 74
Stracke, R., 532
Strayer, C., 433
Stroebel, D., 144
Sturcova, A., 354
Stushnoff, C., 689, 690
Suarez-Lopez, P., 654
Sugimoto, K., 368

Sun, T. -P., 519, 525
Sundström, V., 137
Sung, F. J. M., 676
Sung, S., 659
Susse , I. M., 378, 398, 399
Swafford, J., 135
Swain, S. M., 514
Swartz, T. E., 456
Swarup, R., 479, 483
Symons, G. M., 625
Szekeres, M., 621, 629
Szymanski, D., 366

Tabata, S., 153
Taiz, L., 411
Takamiya, K. -I., 154
Takeba, G., 171
Takeda, T., 371
Takei, K., 551
Talbott, L. D., 453, 455, 459, 460, 461, 462
Tallman, G., 455, 458
Tanaka, K., 623, 624
Tang, G. -Q., 246
Tanimoto, M., 581
Taylor, A. R., 89
Tazawa, M., 108
Tepperman, J. M., 430
Terasaka, K., 482
Terashima, I., 200, 203
Thimann, K. V., 446, 496, 497, 572
Thomas, B. R., 247
Thomashow, M., 691
Thompson, J. E., 362
Thompson, M. V., 235
Thompson, P., 276
Thorne, J. H., 246
Tilney-Basset, R. A. E., 406
Timmers, A. C. J., 301
Tinker, P. B., 88, 91
Tlalka, M., 207, 208
To, J. P., 565
Törnroth-Horsefield, S., 58
Torres-Ruiz, R. A., 390
Tournaire-Rou , C., 58
Trapp, S., 321
Trass, J., 391
Trebst, A., 135
Trelease, R. N., 285
Tsay, Y. F., 110
Tsunoyama, Y., 448
Tucker, G. A., 276
Turgeon, R., 226, 239, 241, 242, 243, 663
Tyerman, S. D., 49, 113
Tyree, M. T., 47, 63

Ueguchi-Tanaka, M., 526
Ullrich, C. I., 471
Ulmasov, T., 471

Vaadia, Y., 551
Valverde, F., 655
van Bel, A. J. E., 234, 239, 240, 249, 341
Van Damme, E. J. M., 336
Van den Berg, C., 405
Van den Fits, L., 335
Van der Mark, F., 308
van der Veen, J. H., 517
van Grondelle, R., 137
van Niel, C. B., 130
van Rhijn, P., 301

van Wijk, K. J., 153
Vande Broek, A., 289, 299
Vanderleyden, J., 289, 299
Vanlerberghe, G. C., 272
Vartanian, N., 602
Vaughn, M. W., 238, 239
Vazquez, F., 612–613
Vedel, F., 277
Vercesi, A. E., 273
Vermerris, W., 325
Vernou, T., 498
Vert, G., 630, 631, 632
Very, A. A., 110, 111
Vidal, J., 174
Vierling, E., 684, 685
Vierstra, R. D., 420, 433
Vietor, R. J., 354
Vince-Prue, D., 652
Vincken, J. P., 360
Vogelmann, T. C., 198, 199, 200, 201, 202, 214, 448
Volkmann, D., 492
Volkov, V., 119
Voesenek, L. A. C., 576
von Caemmerer, S., 177
von Mohl, H., 453

Waaland, Robert, 2
Wada, M., 457
Wagner, A. M., 273
Wagner, J. R., 425
Waites, R., 406
Waldron, K., 373
Walker, D., 176
Wallace, A., 86
Walz, C., 229
Wang, H., 430
Wang, W., 585
Wang, X., 283, 630
Wang, X. -Q., 608
Warembourg, F. R., 246
Warmbrodt, R. D., 223
Warner, R. L., 292
Wasserman, S., 588
Wayne, R. O., 79
Weathers, P. J., 77
Weaver, J. E., 87
Weaver, L. M., 323
Webb, D., 175, 364
Webb, J. A., 243
Weber, A. P., 112, 186
Weig, A., 49
Weigel, D., 639, 664
Weijers, D., 389
Weis, E., 152
Weiser, C. J., 690
Welch, R. M., 308
Weller, J. L., 429
Went, F. W., 469
Wergin, W. P., 19
Werner, T., 553, 554
West, M. A. L., 383
Wheeler, C. T., 301, 303
White, C. N., 599
White, P. J., 80
White, P. R., 544
Whitehouse, D. G., 268
Whittaker, R. H., 38
Whorf, T. P., 212
Wigge, P. A., 664
Wilco, H. E., 89
Wilkerson, J. Q., 293
Wilkins, M. B., 418, 419, 470, 491, 494, 497

Wilkinson, S., 597
Willats, W., 360
Willemsen, B., 385, 388
Williams, J. P., 688
Wilson, R. A., 664
Wilson, R. H., 357
Wipf, D., 110
Wisniewski, M., 690
Wittstock, U., 333
Wittwer, S., 510, 514
Woeste, K., 586
Wolf, S., 29
Wollman, F. -A., 154
Wolosiuk, R. A., 165, 166
Wolter, F. P., 283
Wood, B. W., 82
Woodward, W., 473, 475
Wray, J. L., 294
Wright, A. D., 474
Wu, D., 440
Wyse, R. E., 238

Xiong, J., 156
Xiong, L., 613
Xu, W., 537
Xu, Y. -L., 518

Yachandra, V. K., 143
Yaley, J. R., 512
Yamada, H., 547
Yamada, K., 153
Yamagani, T., 574
Yamaguchi, R., 427
Yamaguchi, S., 518, 519
Yamaguchi-Shinozaki, K., 680, 682
Yamamoto, H. Y., 462
Yamazaki, D., 604
Yan, L., 660
Yang, S. F., 575, 578, 579
Yang, T., 537
Yanovsky, M. J., 648
Yasuda, R., 150
Yee, K. M., 383
Yi, Y., 308
Yin, Y., 632
Yokota, E., 27
Yokota, T., 618, 619, 621, 623
Yoo, B. -C., 249, 663
York, W. S., 353, 373
Yu, H., 249
Yu, T. S., 184
Yuan, M., 579, 580

Zacarias, L., 581
Zeevaart, J. A. D., 533, 650, 651, 660, 661, 662, 664
Zeiger, E., 450, 453, 455, 458, 462
Zhang, H. X., 112
Zhang, J., 700
Zhang, K., 556
Zhang, L. -Y., 241
Zhang, N., 166
Zhang, W., 597, 608
Zhang, X., 700
Zhao, Y., 632
Zheng, Z. L., 612
Ziegler, H., 67, 84
Zimmermann, M. H., 60, 231
Zobel, R. W., 77
Zrenner, R., 190
Zurek, D. M., 626

Subject Index

The letter **f** or **t** following a page number indicates that the entry is derived from a **figure** or a **table**, respectively.

AAPK (abscisic acid-activated protein kinase), 610
ABA. *see* abscisic acid
ABA-dependent genes, 680
ABA hypersensitive (abh) mutants, 612
ABA-independent genes, 680
ABA-insensitive 1 (abi1) mutants, 600
ABA-INSENSITIVE (ABI) transcription factors, 611
aba mutants, 600
ABA response element (ABRE), 680
ABAP1 protein, 604–605
ABC model, 640, 640f, 641f
ABI loci, 610
ABL1 gene, 603
ABP1 (auxin-binding protein 1) function, 503–504
ABPHYL1 gene, 555f
Abronia villosa (desert sand verbena), 650
abscisic acid (ABA), 593–616
  accumulation of, 675f, 680
  assays for, 594
  in axillary buds, 498
  caged, 605f
  deficiency of, 494
  desiccation tolerance and, 599
  enantiomers of, 594, 594f
  flowering and, 664
  freezing tolerance and, 691
  gene expression and, 610–611, 612f
  gibberellic acid and, 531f
  inactivation of, 597
  leaf senescence and, 602
  lipid metabolism induced by, 608–609
  localization in water stress, 598f
  model for signaling, 609f
  root growth and, 602f
  seed dormancy and, 600
  seed storage reserve and, 599
  shoot growth and, 602f
  signal transduction pathways, 603–613
  structure, 319
  synthesis, 594–596, 595f, 675
  translocation, 597–598
  transport, 247
  water stress and, 674–676
abscisic acid-activated protein kinase (AAPK), 610
abscisic acid aldehyde, 595f, 596

abscisic acid-β-D-glucosyl ester (ABA-GE), 597
abscisic aldehyde oxidases (AAOs), 596
abscisin II. *see* abscisic acid (ABA)
abscission
  definition, 582
  effect of auxin on, 500
  ethylene and, 602
  leaf, 674f
  phases of, 583
  senescence and, 410
abscission layer, 500, 582f, 582–584
abscission zone, 500
absorption spectra, 127
Acacia heterophylla, 642, 642f
ACC (1-aminocyclopropane-1-carboxylic acid), 700
accessory pigments, 128
acclimation
  to cold, 689–690
  definition, 672
  to freezing, 690–691
  gene induction and, 691–692
  to high temperature, 683–684
  optimal temperatures and, 211
  osmotic adjustment, 677
  by plants, 203
Acer pseudoplatanus (sycamore maple), 643t
acetoacetyl-acyl carrier protein (acyl-ACP), 281, 281f
Acetobacter diazotrophicus, 299
acetyl-CoA, 318, 320f
acetyl-CoA carboxylase, 388
O-acetylserine (OAS), 304f, 305
acid growth hypothesis, 486–487, 487f
acid-induced growth, 370–371, 370f
acid invertase, 246
acid rain, 304–305
acquired thermotolerance, 682
acropetal transport, 476
actin-binding proteins (ABPs), 24
actin filaments, 14, 207
actinorhizal plants, 297, 299
actins, 24, 29
action spectra
  of blue-light responses, 446, 446f
  photosynthesis, 130, 130f, 131f
active transport
  chemical potential and, 97f

  description, 96, 97f
  energy for, 101, 105–106
  Nernst equation and, 100–101
  primary, 105
  secondary, 105, 106, 106f, 107f
acyl carrier protein (ACP), 281
adaptation
  definition, 672
  to light levels, 203
  to light quality, 437
  optimal temperatures and, 211
  reciprocal evolutionary changes, 339
adaxial cell layers, 398
adenosine-5′-phosphosulfate (APS), 304f
adenosine diphosphate. *see* ADP
adenosine triphosphate. *see* ATP
S-adenosylmethionine (AdoMet), 573
adequate zones, 82f, 83
adhesion, definition, 39
ADP (adenosine diphosphate), 269f
ADP-glucose pyrophosphorylase, 183, 184f
adult phases, 513
aerenchyma, 277, 700
aerobic respiration, 253–254, 271, 271t
aeroponic growth system, 76f, 77
afterripening, 600
AGAMOUS (AG) gene, 639, 666
agamous1 (ag1) mutation, 639, 639f
AGAMOUS protein, 660
agave (*Agave* spp.), 180
Agent Orange, 583
Agrobacterium rhizogenes, 548
Agrobacterium tumefaciens, 471, 544, 544f, 547, 548
AHP (Arabidopsis histidine phosphotransfer) proteins, 565–567, 566f
air seeding, 63
Albizia, 436f
alcohol dehydrogenase (ADH), 259, 702
aldolase, 186
aleurone layer
  α-amylase synthesis, 528f, 530f
  barley, 528f
  description, 527
  gibberellin receptors in, 529
  protoplasts, 531f
  provasculature, 518
alfalfa (*Medicago* spp.), 87f, 447, 599, 690–691
alkaloids, 329–332, 330t

allantoic acid, 230f, 303, 303f
allantoin, 230f, 303, 303f
Allard, Henry, 649
Allium cepa (onion), 451f
allocation, 244–247
allopathy, 324–325
Alonsoa warscewiczii (heartleaf maskflower), 226f
alternate pattern, leaves, 406, 406f
alternative oxidase, 272–273
aluminum (Al), 75
Amaranthus edulis, 176
amide transport, 303–304
amino acids
    ammonium conversion to, 294–296
    biosynthesis, 296, 297f
    movement, 560f
    in phloem sap, 228, 228t
    transporters, 110
1-aminocyclopropane-1-carboxylic acid (ACC), 573
1-aminocyclopropane-1-carboxylic acid oxidase, 574, 575f
1-aminocyclopropane-1-carboxylic acid synthase (ACS), 574, 575
aminoethyoxy-vinyl-glycine (AVG), 575
5-aminolevulinic acid (ALA), 155f
aminooxyacetic acid (AOA), 575
ammonia, 289
ammonium ion
    assimilation, 294–296
    formation of, 289
    metabolism of, 295f
    root uptake of, 88, 89
    toxic effects, 291–292, 292f
amphiphathic, definition, 6
amplitude, of rhythms, 646
α-amylase, 186, 514, 527, 530f, 532
α-amylase inhibitors, 336
α-amylase mRNA, 531–532
amylases, production of, 528f
amylopectin, 183
amyloplasts, 18, 20, 21, 491
amylose, 183
Anabaena, 297, 299f
anaerobic conditions, 698, 701
anaerobic metabolism, 277
anaerobic stress proteins, 702
Ananas comosus (pineapple), 180, 664
anaphase, 26f
anaplerotic reactions, 265
anastomoses, vascular, 227
anchored proteins, 6, 8f
angiosperms
    description, 2
    phloem, 221
    shoot apexes in, 399
    sieve elements, 4f, 224t, 235
    vessel elements, 59
anion-selective channels, 119
anion transporters, 112
"anion trap" concept, 597
anions, uptake of, 100
anisotropic arrangement, 364–365
anoxia, 699f
anoxic conditions, 698
anoxygenic organisms, 137
antenna complexes, 130–132, 137–139, 138–139
antheraxanthin, 152, 206–207
anthers, 277, 522, 635
anthocyanidin, 326, 326f, 327t
anthocyanins, 326–327, 326f, 456f
anti-idiotypic antibodies, 604n

anticlinal cell division, 398
antiflorigen, 662–663
antifreeze proteins, 689, 691
antiport, 106
antiporters, 106, 109f
Antirrhinum (snapdragon), 407, 636, 638
ap2 mutation, 639f
apetala2-2 (ap2-2) mutation, 639f
APETALA (AP) genes, 638, 639, 664, 666
apetala2 (ap2) mutation, 639
apical dominance, 409–410, 496–498, 557
apical hook, 417
apical meristems, 2, 3f, 636f
apical–basal axis, 384
apical–basal patterning, 387–388
apoplasts
    acidification of, 490
    composition of, 401
    ion transport, 117–118
    loading, 240f
    pH, 436
    phloem loading, 236–237, 240f
    phloem unloading, 241–242
    potassium availability, 239
    solute movement, 117
    sucrose concentration, 238
    sugar movement, 237
    water movement and, 56, 57f
    water potential, 50
apoproteins, 423
apoptosis, 412. *see also* programmed cell death
apple (*Malus* spp.), 643t
aquaporins
    brassinosteroids and, 626
    functions, 49, 49f, 113
    regulation of, 58, 680
    spinach plasma membrane, 58f
    water transport and, 58f
Arabidopsis
    ABA-deficient mutants, 494
    aba mutants, 596
    abi mutants, 610
    AP2 mutants, 639
    ASYMMETRIC LEAVES 1 (AS1) gene, 407
    AtMYB33 gene, 532, 534
    AUX1 function, 477
    auxin synthesis, 471
    auxin transport, 480f
    BL-deficient mutants, 619, 620f
    blue-light responses, 447, 447f
    cation channels, 111f
    clock genes, 433–434, 434
    coincidence model, 654f
    CONSTANS (CO) gene, 653
    control of embryogenesis, 388–412
    cop1 mutants, 460
    cortical microtubules, 367f
    dwarf mutants, 521
    elf3 (early flowering 3) mutant, 657
    embryogenesis, 380f, 382, 383f
    ethylene signaling, 589f
    etr1 mutant, 585f
    evocation of flowering, 664
    fab1 mutants, 689
    FLC expression, 659, 659f
    floral organ initiation, 637f
    flower initiation, 406
    flowering, 440, 653–654, 656, 665f
    GA activity, 521
    GA-deficient, 517f
    GCR1 gene, 608
    gene regulation in, 603
    gibberellin expression, 518, 519f

    gibberellin mutants, 523f
    gibberellin overexpression, 514
    GNOM genes, 388f
    gravity perception in, 492f
    hormone experiments, 618
    jasmonic acid in, 335
    lateral root formation, 410f
    leaf initiation, 406
    LFR action spectrum, 422, 422f
    metal uptake, 113
    mutant screening, 588f
    nph1 mutant, 460f
    ovule, 380f
    pavement cells, 366f
    phot genes, 456
    PHY genes, 426
    phytochrome encoding in, 426
    pin mutants, 386f, 478, 479f, 484
    pinoid mutant, 483
    pistel, 638f
    quiescent center, 404
    rcn1 (root curling in NPA1) mutant, 483
    resistance to fungal pathogens, 339
    response to *Pseudomonas*, 582
    RGA-GFP localization, 524, 525f
    root development, 393f
    R2R3-MYB factors, 532
    scarecrow (scr) mutant, 491
    seed dormancy in, 600
    seedling growth, 657
    shoot apical meristems in, 398
    shoot apical region, 637f
    SIG genes, 448
    stem elongation, 457f
    systemic acquired resistance in, 341
    thermotolerance, 684
    triple response in, 579f, 584
    UV protection, 327
    vernalization, 659–660, 659f
    VLFR in, 428
    Wassilewskija (Ws) accession, 440
Arabidopsis thaliana
    anerobic stress proteins, 702
    apical meristem development, 636f
    coiled-coil proteins in genome of, 24
    companion cells, 237
    floral organs, 379f
    genome, 153
    mature plant, 379f
    nuclear genome size, 8
    PHY gene function, 429f
    postembryonic development, 381f
    starch degradation, 185, 185f
    vernalization, 657f
ARABIDOPSIS THALIANA MERISTEM LAYER 1 (ATML1) gene, 392, 392f
5-arabinan, 359f
arabinogalactan, type I, 359f
arabinogalactan proteins (AGPs), 361, 362f
Araceae (arum family), 272
Arachis hypogaea (peanut), 580
arbuscules, mycorrhizal, 90
arginine, 333, 333f, 334
arid areas, 693–694
Arizona honeysweet (*Tidestromia oblongifolia*), 214f, 683, 683f, 684
Arnold, William, 131–132
Arnon, Daniel, 148
ARRs (Arabidopsis response regulators), 564–565, 565f
ascorbate peroxidase, 153
ash (*Fraxinus excelsior*), 222f
ASK1 protein, 501
asparagine synthetase (AS), 296

aspartate, 173–174, 295f
aspartate aminotransferase (Asp-AT), 296
assimilatory quotient, 310f
Astragalus, 75
ASYMMETRIC LEAVES 1 (AS1) gene, 407, 409f
atmospheric nitrogen deposition, 74
AtMYB2, 702
AtMYB33 gene, 532, 534, 535
AtMYB65 gene, 532, 534
AtNHX, 695, 697f
AtNHX1 gene, 112
ATP (adenosine triphosphate), 305
   aerobic respiration, 271
   mitochondrial synthesis of, 268–269
   synthesis of, 125, 148–151, 149f
ATP-binding cassette (ABC) proteins, 110
ATP synthase, 101, 149
   chloroplast, 19f
   mitochondrial synthesis of, 268–269
   proton gradients and, 18
   structure, 150
ATPase, 149
Atriplex sabulosa (frosted orache), 683, 683f, 684
Atriplex sp. (salt bush), 176, 694
atropine, 331
autocatalytic, 424, 576–577
autonomous regulation, 641
autophosphorylation, 425, 426f
AUX1, 477, 479f, 481f
AUX/IAA proteins, 503f
auxin. *see* indole-3-acetic acid (IAA)
auxin 1-naphthylene acetic acid (1-NAA), 478
auxin-binding protein 1 (ABP1), 487
auxin efflux inhibitors (AEIs), 482
auxin receptors, 501f, 503f
auxin response elements (AuxREs), 502
auxin response factors (ARFs), 388, 502, 580
auxin signaling F-box proteins (AFBs), 502
Avena sativa (oat). *see* oat *(Avena sativa)*
Avicennia, 277
avr (avirulence) genes, 340
axial polarity, 384
axillary buds, 2
AXR1 protein, 497
Azadirachta indica (neem tree), 321
azadirachtin, 321, 322f
azetidine-2-carboxylic acid, 333, 333f
Azobacter, 298–299
Azolla (water fern), 297, 298, 367f
Azorhizobium, 297

bacteria
   electron flow, 150f
   ethylene production, 572–573
   ice crystal formation and, 690
   nitrogen fixation, 291
bacteriochlorophylls, 128, 129f
bacteroids, 301
bafilomycin, 114
balsam fir *(Abies balsamea)*, 548f
Balsamorhiza sagittata (spring sunflower), 321f
banana family (Musaceae), 472
barley *(Hordeum vulgare)*
   aleurone protoplasts, 531f
   calcium absorption, 88
   flowering, 656
   GA response, 521
   malting, 514
   structure of grain, 528f
BAS1 gene, 623

Baskin, T. I., 366
Bassham, J. A., 160
bean *(Phaseolus vulgaris)*, 233
bean second-internode bioassay, 617, 618, 618f
beans *(Phaseoulus* spp.), 217
bending, blue-light responses, 446–447
Benson, A, 160
benzyladenine (BA), 546, 547
Bermuda grass *(Cynodon dactylon)*, 18f
BES1 *(bri1*-EMS-suppressor 1), 630–631
Beta maritima (wild beet), 227
Beta vulgaris (sugar beet), 227, 236f, 677f
β-oxidation pathway, 335
Betula papyrifera (white birch), 690
Betula pendula (birch), 583f
Betula verrucosa (silver birch), 643–644
Bienertia cyclopotera, 174f, 176, 178, 178f
bilin lyase domain (BLD), 424, 425f
bilin pigments, 129f
biliverdin, 424, 425f
BIM1 (BES-interacting Myc-like 1), 632
BIN2 (brassinosteroid insensitive-2), 630–631
biological clocks, 646
Bipolaris maydis, 276
birch *(Betula pendula)*, 583f
black-eyed Susan *(Rudbeckia* sp.), 327f
black necrosis, 80
Blackman, F. F., 198
Bloeothece, 299
blue light
   chloroplast movement and, 27, 207
   photoreceptors, 423, 656–657
   phototropin stimulation by, 489
   phytochrome absorption of, 420
   solar tracking and, 202
blue-light responses, 445–465
   chloroplast movement, 457
   elongation, 448
   gene expression, 448–449
   guard cell osmoregulation, 453
   kinetics, 452–453
   sensory transduction cascade, 457f
   stomatal, 449–451, 450f, 451f, 462f
BODENLOS (BDL) gene, 389
body plans, 3f
bolting, definition, 533
bordered pits, 4f
Borlaug, Norman, 524
boron, 75t, 80
Borszczowia aralocaspica, 174f, 176, 178, 178f
boundary layer resistance, 65, 213, 213f
Boussingault, Jean-Baptiste-Joseph-Dieudonné, 76
Bowen ratio, 209–210
BR2 (Brahytic 2) gene, 481f
brace roots, 87
Bradyrhizobium, 297
branching genes, 557
branching signals, 497–498
Brassica juncea, 386f
Brassica napus (canola, rape), 209, 333
Brassica oleracea (cabbage), 317f, 511f, 689
Brassicaceae (mustard family), 89, 472
brassinazole (Brz), 623, 624f, 625f
brassinolide (BL)
   bioassays, 618
   byosinthetic pathway, 621–623, 621f, 622f
   catabolism, 622–624
   homeostasis, 622–624, 623f
   IAA and, 628f
   identification, 619
   receptor for, 336

brassinosteroids, 617–634
   in agriculture, 632
   bioassays, 619f
   cell division and, 27
   development and, 625–629
   function, 387
   growth and, 625–629
   root growth and, 627–628
   signaling pathway, 629–632, 631f
   structure, 319
BRI1-associated receptor kinase (BAK1), 630
bri mutants, 620f, 629
BRI1 receptor, 629, 630, 630f
bristlecone pine trees, 377–378
broad bean *(Vicia faba)*, 234f, 237
   guard cell calcium ions, 606f
   guard cell protoplasts, 452f
   guard cells, 608
   mitochondrion, 263f
   proplastid, 21f
   proton pump regulation, 603–604
   stomatal opening, 450, 450f
broad-leaved plantain *(Plantago major)*, 237
broad-leaved weeds, 501
bromegrass *(Bromus inermis)*, 691
Bromus inermis (bromegrass), 691
Bryophyllum, 180, 650, 661–662, 664
Buchanan, Bob, 166
buds
   apical dominance and, 497
   competence in, 644
   deep supercooling, 690
   determination, 644, 645f
   dormancy, 580, 602–603
   on moss, 557, 558, 558f
Bulbochaete, 13f
bulk flow, 42, 55
bundle sheath cells, 174, 176–178, 222, 226f, 240f
bundle sheath parenchyma, 3f
Bunsen, R. W., 422
BZR1 (brassinazole-resistant 1), 630–631

C-14 reductase, 388
C_2 oxidative photosynthetic cycle, 168–172, 170t
C_3 plants
   atmospheric carbon dioxide and, 215f, 215–216
   carbon isotope ratios in, 217–218
   latitude and, 211f
   PEP carboxylase and, 679f
   photosynthetic response, 204f, 205f
   temperature effects, 210, 210f
C_4 plants
   atmospheric carbon dioxide and, 215f, 215–216
   carbon cycle, 173–180, 174f, 175t
   carbon dioxide concentration, 178–179
   carbon isotope ratios in, 217–218
   drought resistance, 672–673
   in hot, dry climates, 180
   latitude and, 211f
   leaves with Kranz anatomy, 177f
   light regulation, 178–180
   sodium requirement, 81
   stages, 176
   temperature effects, 210, 210f
   temperature range, 683–684
   transpiration, 69
C-repeat binding factors (CBFs), 692
cabbage *(Brassica oleracea)*, 317f, 511f, 689
cacti (Cactaceae)
   CAM, 180, 679

CAM idling in, 216
water relations, 48, 48f
cadastral genes, 638
caffeic acid, 325f
caffeine, 330f
calcineurin B-like proteins (CBLs), 610
calcium
    absorption by barley, 88
    aleurone protoplasts, 531, 531f
    biochemical function, 75t
    deficiency, 79, 80–81
    GABA production and, 686, 687f
    pectin network binding by, 360f
    transport of, 100
calcium ions
    abscisic acid signaling and, 607
    efflux, 111
    in gravitropic response, 493, 494f
    guard cell levels, 606f
    oxidative burst and, 340
    oxygen deficits and, 702
    sodium uptake and, 695
    transport, 696f
    vacuolar storage, 112
calcium pectate, 307f
Calendula officinalis, 660
Callitriche platycarpa, 580
callose production, 225
callus tissues, 545, 556f
calmodulin, 80
Calvin, M., 160
Calvin cycle, 133, 160–165
    atmospheric carbon dioxide and, 215
    component regeneration, 164
    energy use, 164–165
    ferredoxin–thioredoxin system and, 166–168, 167f, 167t
    location, 310
    reactions of, 161t
    regulation of, 165–168
    stages, 160–161, 161f, 162f
Campanula medium (Canterbury bells), 650
campesterol, 621, 622f
canary grass (*Phalaris canariensis*), 468
Canavalia ensiformis (jack bean), 334
canavanine, 333, 333f, 334
Cannabis sativa (hemp), 513
canola (*Brassica napus*, rape), 333
canopies, sunlight, 202f
Canterbury bells (*Campanula medium*), 650
capillarity, 39–40, 40f
capsidiol, 340f
carbohydrates
    conversion to pyruvate, 256–257
    flowering and, 666
    glycolysis, 256–257
    from storage lipids, 283
    synthesis of, 125
carbon
    fixation reactions, 126
    isotope ratios, 216–218, 217f
    metabolism, 319f
    mobilization, 182f
carbon dioxide
    atmospheric, 212f, 212–213
    concentrating mechanisms, 172
    diffusion of, 212–213, 213f
    fixation, 168–172, 213–214
    inhibition of ethylene action, 575
    photosynthesis and, 210f, 211–216
    plant uptake of, 38
    pumps, 173
    respiration rates and, 277–278
    rubisco and, 166f
    transpiration ratio and, 69
carbon dioxide compensation point, 214–215
carbonic anhydrase, 173
carbonyl cyanide *m*-chlorophenylhydrazone (CCCP), 452
cardenolides, 322
β-carotene, 129f
carotenoids
    description, 128
    energy transfer, 139
    function, 326
    quenching by, 151, 152
    structure, 129f
carpels, initiation, 637, 637f
carrier proteins
    function, 101, 102f, 105
    saturation kinetics, 107f
carrot plants (*Daucus carota*), 246
Casparian strip, 57, 57f, 88, 118, 694
cassava (*Manihot esculenta*), 332, 675
castasterone (CS), 618, 619, 621
Castilleja chromosa (desert paintbrush), 650
castor bean (*Ricinus communis*), 228t, 257
Catharanthus roseus (Madagascar periwinkle), 335, 621
cations
    assimilation, 306–308
    carriers, 112
    channels, 110–111, 111f
    exchange, 85, 85f
    transport, 114f
cavitation, 41, 63–64, 678
CBL-interacting protein kinases (CIPKs), 610
CCA1 gene, 433–434
Cdc2 (cell division cycle 2) gene, 556
Cecropia obtusifolia (trumpet tree), 438
celery, 324
cell cycle kinases, 537
cell cycles, 27–29, 27f, 555–556
cell division, 535–536, 536f, 626–627
cell fates, 385f
cell plates, 26f, 362
cell walls
    acid-induced growth, 370f
    cessation of expansion, 372
    degradation, 372–373
    elasticity, 626f
    elongation, 487f
    expansion, 364, 371f
    extensibility, 369, 486, 673
    functions, 350
    of guard cells, 67–68
    ice crystal formation and, 690
    lignan in, 325
    matrices, 353
    noncellulose polysaccharides, 15
    pores, 224
    response to pathogens, 339
    rigid, 2
    stress relaxation, 368
    structural components, 353t, 361–362, 361t
    structure, 350–353
    types of, 2, 3f
cellobiose, 353, 354f
cells
    definition, 1
    division of, 544–545
    elongation, 368–372, 484–488, 535–536
    lateral expansion, 579
    patterns of expansion, 364–368
    structure, 5–23, 5f
    transport processes, 109f
    turgor pressure, 368
    water entry, 45
    water exit, 45–47
    yielding properties, 368
cellulose
    microfibrils, 68, 69f, 353–357, 355f
    in primary cell walls, 352f
    synthesis, 355, 356f
    tensile strength, 354
cellulose synthases, 354, 366
Celtis occidentalis L. (hackberry), 559
central vacuoles, 17
central zone, 398, 399f
centromeric regions, 25
cereal grains, 527
CesA (cellulose synthase A) gene, 354
Cestrum nocturnum (night-blooming jasmine), 650
CF_0-CF_1, 149
Chailakhyan, Michail, 660
chalcone synthase, 448
chalcones, 324f
channel proteins, 101, 102f, 103–104, 109f
Chara (green alga), 26, 492
chelators, 77, 78f
chemical potential, 43, 96–97, 97f, 148
chemiosmotic hypothesis, 268, 269
chemiosmotic mechanism, 148
chemiosmotic model, 477, 478f
Chemopodium album (pigweed), 420–421
Chenopodiaceae, 89, 176
Chenopodium, 650–651
chilling, definition, 600
chilling injury, 687–692
Chlamydomonas reinhardtii, 449f
chlorates, 110
Chlorella pyrenoidosa, 131
chloride ion channels, 603
chloride ions, 340, 694
chlorine (Cl), 75t, 81
4-chloroindole-3-acetic acid (4-Cl-IAA), 470f
chlorophyll *a*, 138
    absorption spectrum, 129f
    abundance, 128
    energy transfer, 139
    structure, 129f, 155f, 307f
chlorophyll *a/b* antenna proteins, 139
CHLOROPHYLL a/b BINDING (CAB) gene. *see LHCB* gene
chlorophyll *b*, 138
    absorption spectrum, 129f
    abundance, 128
    energy transfer, 139
    in shade leaves, 203
    structure, 129f
chlorophyll *c*, 128
chlorophyll *d*, 128
chlorophyllide *a*, 155f
chlorophylls
    biosynthesis, 154, 155f
    bleaching, 142
    breakdown, 154
    color, 127
    description, 126
    energy capture, 140–142
    energy disposition by, 127–128
    light absorption, 128f, 200
    light emission, 128f
    oxygen production, 131f, 132
    reaction centers, 141f
chloroplasts
    abscisic acid accumulation, 675f
    ATP synthesis, 148–151, 149f

bundle sheath cells, 176–178
C₂ oxidative photosynthetic cycle, 172f
carbon reactions, 160f
composition, 18
conversion of, 21
deterioration of, 411
development, 560–561
distribution, 208f
electron flow, 150f
energy conversion in, 17–20
envelopes, 135
function, 20
genome, 20, 153
glycerolipid synthesis, 282f
guard cell, 453
light reactions, 160f
membrane composition, 6, 279, 279t
movement, 27, 207, 430, 457
nutrient assimilation, 310
photorespiratory cycle, 169f
photosynthesis in, 134–135
plant cell, 5f
protein import, 154
proton transport, 148–151
reproduction, 153–154
starch synthesis, 183, 184t
chlorosis, 79, 80, 81, 249
chokecherry *(Prunus virginiana)*, 690
choline, 7f
Cholodny, Nicolai, 488
Cholodny–Went model, 488–23, 490
CHRK1 (chintinase-related receptor-like kinase 1) gene, 559
chromatids, 11f
chromatin, 5f, 10, 11f
chromophores, 423
chromoplasts, 18, 20f
chromosomes, 10, 11f, 20
chronic photoinhibition, 208–209
chronobiology, 646
chrysanthemum *(Chrysanthemum morifolium)*, 321, 644, 649
cinnabar moth *(Tyria jacobeae)*, 331
cinnamic acid, 323
CIRCADIAN CLOCK-ASSOCIATED 1 (CCA1) gene, 433, 434, 434f
circadian resonance, 435
circadian rhythms
    characteristics, 646
    definition, 433
    fitness and, 435
    gene expression and, 434–435
    nyctinasty, 436f
    phase shifting, 646–647
    phytochromes and, 433–435
cis double bonds, 6
cisternae, 11, 14, 14f
cisternal maturation/progression model, 15
citrate, structure, 264f
citric acid cycle, 262–265, 699, 699f
    description, 255
    glycolysis precursors and, 275f
    intermediates, 273
    pyruvate oxidation, 263–265
    reactions, 264f
citrulline, 230f, 303, 303f
Citrus spp., 392, 643t
CKII gene, 563
cladodes, 216
Class III homeodomain–leucine–zipper (HD-Zip) gene family, 398, 409
clathrin, 15
clathrin-coated vesicles, 16f
CLAVATA (CLV) genes, 400–401

CLAVATA (CLV) receptor kinase signaling cascade, 401f
clays, 54, 54t, 85
climacteric, definition, 576
climacteric plants, 576–577, 577f
clock hypothesis, 652–653
clonal analysis, 399
clover *(Trifolium* spp.), 222f, 350f, 599
CO (CONSTANS) gene, 653, 663
co (constans) mutant, 653
CO (CONSTANS) protein, 654
co-evolution, 339
coarse sand, 54t, 85
coat-imposed dormancy, 599–600
coat proteins, 15, 16f
cobalamin (vitamin B₁₂), 75–76
cobalt, 75, 575
cocaine, structure, 330f
Cochliobolus heterostrophus. see *Bipolaris maydis*
cocklebur *(Xanthium strumarium)*, 294, 652
codeine, 331
cohesion, definition, 39
cohesion–tension theory, 61–62
coiled-coils, 23, 23f
coincidence model, 653, 654f
cold acclimation, 691–692
cold temperatures, 657–660
coleoptile stage, 384
coleoptiles
    blue-light responses, 447f
    definition, 446
    description, 468
    elongation, 470f
    gravity sensing by, 491
    growth of, 484–485
    phototropism, 489f
    response to auxin, 486f
coleus *(Coleus blumei)*, 13f, 239, 241
collenchyma, 4f, 351
color, light absorption and, 127
columella, 384, 405, 491, 493f
combinatorial model, 643
Commelina, 326, 451f, 604, 605f
companion cells, 4f
    A. thaliana, 237
    function, 225–227
    ordinary, 226
    osmotic potential, 237
    pressure-flow model, 232f
    sieve elements and, 222
    soluble proteins, 9t
compatible solutes, 677
competence, 644, 658
competion for sunlight, 201–202
compound middle lamella, 5f
concentration, 43–45
conch apple *(Passiflora maliformis)*, 687
condensed tannins, 328, 328f
condensing enzyme, 281
conifers. see also specific conifers
    conebud formation, 513f
    monoterpenes in, 321
    tracheary elements, 60f
coniferyl alcohol, 325
coniine, 331
Conium maculatum (poison hemlock), 331
CONSTANS (CO) gene, 653, 663
constans (co) mutant, 653
constitutive defense responses, 334, 339
CONSTITUTIVE PHOTOMORPHOGENE-SIS (COP) genes, 432–433
contact angles, 39
coordination bonds, 306

coordination complexes, 307f
cop1 mutants, 460
COP9 signalosome (CSN), 432–433, 432f
copper, 75t, 82
copper cofactors, 586
cork cambrium, 5
cork cells, peridermal, 317
corn. see maize (corn: *Zea mays*)
Cornus florida (dogwood), 277
cortex, root, 88
cortical endoplasmic reticulum, 11
cortical tissues, 393–394
Corylus avellana (European hazel), 600
Corynebacterium fascians, 547
cotyledons, 561, 562f, 600
p-coumaric acid, 323, 324f
coumarins, 323
p-coumaroyl-CoA, 324f
coumaryl alcohol, 325
Cousins, H. H., 571
cowpea *(Vigna unguiculata)*, 675, 677f
CPD1 gene, 623f
cpd mutants, 619, 620f
CRABS CLAW (CRC) gene, 407–408, 408f
crassulacean acid metabolism (CAM) plants, 180–192, 181f, 216–218
    blue-light responses, 458
    carbon dioxide uptake, 180–181
    drought resistance, 672–673
    in hot, dry climates, 180
    sodium requirement, 81
    stomata, 682–683
    water stress and, 678–679
creep, cell wall slippage and, 370
creosote bush *(Larrea divaricata)*, 214f
cress *(Lepidium sativum)*, 7f
cristae, function, 17
critical concentrations, 82f, 83
critical day length, 649
critical oxygen pressures (COPs), 698
crown gall bacteria, 544f
crown galls, 544, 548–551, 549f
CRT/DRT elements, 692
cry1/cry2 double mutants, 657
CRY proteins, 423, 440, 648, 657
cryoprotective effects, 689
cryptochrome 1 (CRY1), 455, 456f, 555–456
cryptochrome 2 (CRY2), 456
cryptochromes
    circadian clocks and, 434
    entrainment by, 648
    homologs, 456
    stem elongation and, 455–456
crystalline ribbons, 353
CslA (cellulose synthase-like) superfamily, 354
ctr1 (constitutive triple response 1) mutation, 586
cucumber *(Cucumis sativus)*, 248, 351f, 449f, 499f, 513
Cucumis melo (melon), 239
Cucurbita maxima (pumpkin; winter squash), 224f, 248, 515
Cucurbita pepo (summer squash), 239, 243f
Cucurbitaceae (squash family), 472, 521
CUP-SHAPED COTYLEDON gene, 396, 397f
Curbita pepo (zucchini), 477
current–voltage relationships, 104f
cuticle, 3f, 212, 315, 316f, 678
cutin, 316, 316f, 317
cutinase, 317
cuttings, rooting of, 632
cyanide, 101f
cyanobacteria

chlorophylls in, 128
cytochrome b_6f, 146f
genome, 153
heterocysts, 298, 299f
nitrogen fixation, 291, 298
reaction centers, 144f
cycads, 398
CYCD3 gene, 556, 556f
cyclic adenosine diphosphate-ribose (cADPR), 609f
cyclic electron flow, 147
cyclic nucleotide-gated cation channels, 111
cyclin-dependent protein kinases (CDKs), 27–28, 27f, 537, 556
cyclins, 27, 27f, 556
cycloartenol, 621
1-*cis*-cyclootene, 576f
1-*trans*-cyclootene, 575, 576f
Cycocel, 514
Cynodon dactylon (Bermuda grass), 18f
Cyperaceae (sedges), 176
cysteine, 304f, 305
cytochimeras, 399f
cytochrome *a*, 266
cytochrome a_3, 266
cytochrome bc_1 complex, 266
cytochrome b_6f, 137, 139, 146f
cytochrome b_6f complex, 144
cytochrome *c*, 266
cytochrome *c* oxidase, 266
cytochrome *f*, 144, 145
cytochrome P450, 309f
cytochrome p40 monooxygenases (CYPs), 621
cytokinesis, 25–26, 26f, 362
cytokinin-induced nutrient mobilization, 559
cytokinin oxidase, 552f, 553, 553f
CYTOKININ RECEPTOR 1 gene. *see* *WOODEN LEG (WOL)* gene
cytokinin receptors, 564f
cytokinins, 247
biological roles, 553–562
biosynthetic pathway, 550f, 551
cell cycles and, 555–556
cell division and, 27, 543–569
chloroplast development, 560–561
degradation, 552f
forms of, 547
function, 392–393
histidine kinases and, 585
hormonally active form, 547
leaf senescence and, 559, 581
metabolism, 552
modes of action, 563–567
overproduction, 561–562
root growth and, 554–555, 554f
shoot growth and, 555f
signaling, 566f
synthesis, 548–551
transport, 551
cytoplasmic male sterility (cms), 276
cytoplasmic sleeves, 29
cytoplasmic streaming, 26–27
cytoskeleton, 23–27
cytosolic acidosis, 699

2,4-D, 500
D1 protein, 135f, 152–153
D2 protein, 448
dahlia (*Dahlia pinnata*), 227
dandelion (*Taraxacum* sp.), 201
dark reactions. *see* photosynthesis, carbon reactions

darnel ryegrass (*Lolium temulentum*), 645, 645f, 656, 664
Darwin, Charles, 468, 488
Darwin, Francis, 468, 488
Daucus carota (carrot), 246
day length, 648f, 651–652
day-neutral plants (DNPs), 533, 650
dayflower (*Commelina communis*), 326, 451f
day–night cycles, 646–647, 647f
decussate pattern, 406, 406f
DE-ETIOLATED (DET) genes, 432–433
de novo synthesis, 153
de Saussure, Nicholas-Théodore, 76
Death Valley, California, 683
deep supercooling, 690
defense responses, 582
deficiency zones, 82, 82f
DEFICIENS gene, 639
DEFICIENS protein, 660
dehydration response element (DRE), 680
Deinococcus radiodurans, 424, 425f
DELLA domain, 523f, 523–527, 525f, 532–533
Delphinium (larkspur), 330
1-deoxy-D-xylulose-5-phosphate, 319, 320f
deoxy sugars, 354f
dermal tissues, 3f, 4f
desert paintbrush (*Castilleja chromosa*), 650
desert sand verbena (*Abronia villosa*), 650
desiccation postponement, 672
desiccation tolerance, 599, 672
desmotubules, 29, 117, 117f
det2 mutants, 619, 620f
determinate plants, 673
determination
in buds, 644, 645f
of cells, 385
development
analysis of, 381–382
brassinosteroids and, 625–629
intercellular protein movement, 395f
light control of, 417–443
reproductive, 381
root apical meristem, 402–405
shoot apical meristem (SAM), 396–402
sporophytic, 378–381
vegetative phase of, 377, 381
dhurrin, 332
diacylglycerol (DAG), 282
diaheliotropic solar tracking, 203
diatoms, 179
dicamba (2-methoxy-3,6-dichlorobenzoic acid), 471f, 500
2,4-dichlorophenoxyacetic acid (2,4-D), 471f
dichlorophenyldimethylurea (DCMU), 147, 148f, 450
dicots
action of auxins, 485
body plan, 3f
embryogenesis, 382, 383f
mycorrhizal fungi associated with, 89
phytochrome genes, 429f
root systems, 87
dicotyledons, 238t, 357
diethylenetriaminepentaacetic acid (DPTA, pentetic acid), 77, 78f
difference in water vapor concentration, 65
differential growth, 489
differentiation, 382
diffuse growth, 364, 364f
diffusion
of carbon dioxide, 212–213
definition, 41
distance relationships, 41–42

thermal agitation and, 42f
water flow across membranes, 49f
diffusion coefficients, 41
diffusion potential, 98, 98f
diffusion resistance, 65
Digitalis (foxglove), 322
dihydroflavonols, 324f
4'-dihydrophaseic acid (DPA), 597
dihydroxyacetone phosphate, 187
dihydrozeatin (DZ), 546, 553
dihydrozeatin riboside, 559
dimethylallyl diphosphate (DMAPP), 548, 550f, 551
dimethylallyl diphosphate (DPP), 319, 320f
dioxygenases, 309, 517
N,N'-diphenylurea, 546
DIR1 (defective in induced resistance 1) gene, 341
distal elongation zones (DEZ), 495
diterpenes, 318, 320f
Dittmer, H. J., 86
diuron, 147
DNA
chloroplast, 153
microarray analysis, 430
packing, 11f
photolyases, 455
dogwood (*Cornus florida*), 277
dolichol diphosphate, 14
dolichols, 319
donor–receiver agar block method, 476, 476f
dormancy
breaking of, 580
buds, 580, 602–603
definition, 382
embryo, 599–600
seeds, 580, 599–600
dormin. *see* abscisic acid (ABA)
dorsal motor cells, 436, 436f
DR5 gene, 503
DR5 reporters, 387f
Drosophila, 433, 648, 684
drought avoidance. *see* drought tolerance
drought escape, 672
drought resistance, 672–673
drought rhizogenesis, 602
drought tolerance, 65
duckweed (*Lemna*), 1, 208f
dwarf phenotypes, 521–523
DWF4 enzyme, 623
DWF4 gene, 624f
dynamic photoinhibition, 208–209
dyneins, 27
Dystilium racemosum, 618

E1 enzyme, 497
EARLY FLOWERING 3 (ELF3) genes, 434
early genes, 502–503
ebb and flow growth system, 76f, 77
α-ecdysone, 322f
Echeveria (*Echeveria harmsii*), 650
Echionochloa curs-galli var. *oryzicola* (rice grass), 700
ecosystem productivity, 38f
ectotrophic mycorrhizal fungi, 90
EGases (endo-(1→4)β-D-glucanases), 371
ein2 (ethylene-insensitive 1) mutant, 581, 587–588
electric potential, 96
electrochemical potential, 97
electrogenic pumps, 101, 114f
electrogenic transport, 105
electromagnetic spectrum, 126f
electron spin resonance (ESR), 142

Subject Index

electron transport, 139–148, 171, 265–274, 267f
electronegativity, 38
electroneutral transport, 105
electrostatic bonds, 306, 307f
elf3 (early flowering 3) mutant, 657
elicitors, 334, 340
elongation
   cells, 368–372, 484–488, 535–536
   coleoptiles, 470f
   distal elongation zones, 495
   epicotyls, 626f
   hypocotyls, 423f
   leaf sheath, 515f
   rates, 448
   stems, 455–456, 456f, 457f, 519–521
elongation centers, 88
elongation growth, 580–581
elongation phase, 24
elongation zone, 404, 404f
emboli, xylem, 63
embryo dormancy, 599–600
embryo lethality, 387
embryo sacs, 636. *see also* gametophytes
embryogenesis
   Arabidopsis, 380f
   cell fates within, 385f
   description, 378–381
   dicot, 383f
   early patterning in, 390f
   gene function, 388
   monocot, 383f
   polarity and, 382–396
   position-dependent signaling, 384–386
   radial patterning in, 391f
   root formation, 403f
embryos, gibberellin synthesis, 528f, 528–529
Emerson, Robert, 131–132, 133
Encelia farinosa (white brittlebush), 684
end wall perforations, 4f
endergonic reactions, 303
endocytic recycling, 15
endocytosis, 604
endoderm, 3f, 57f, 88, 113, 390, 393–394, 405
endogenous factors, 433
endogenous oscillators, 646
endoplasmic reticulum (ER)
   components, 7f
   description, 11
   glycerolipid synthesis, 282f
   membrane composition, 279t
   triacylglycerol synthesis and, 279
endosperm, 353, 379, 527
endosymbiosis, 154
endothelin-1, 328
energetics, nutrient assimilation, 310–311
energy
   dissipation from leaves, 678
   efficiency, 132
   sites of conversion, 17–20
   storage of, 278
   transfer, 128, 131f
Engelmann, T. W., 130
English ivy *(Hedera helix)*, 400f, 513, 642f, 643t
English oak *(Quercus robur)*, 643t
enhanced response to ABA (era) mutants, 612
enhancement effect, 133, 133f
entrainment
   blue-light, 648
   cryptochromes and, 648
   of endogenous oscillators, 646
   phytochromes and, 648
   of rhythms, 433

envelopes, chloroplast, 135
environment
   circadian rhythms and, 646–647, 647f
   flowering and, 641
   signals, 445
   stresses, 574
24-epibrassinolide (24-epiBL), 619
epicotyl elongation, 626f
epidermal cells
   aquaporins in, 113
   cellulose microfibrils, 69f
   pavement cells, 365
   walls of, 351, 352f
epidermis, 3f, 57f, 392f, 405
epigenetic changes, 658
epigenetic regulation, 10
epinasty, 575, 577
9-*cis*-epoxy-carotenoid dioxygenase (NCED), 595f, 596
Equisetaceae (scouring rushes), 80
Equsetum arvense, 619
ERE-binding protein (EREBP), 588
ERS (ETHYLENE-RESPONSE SENSORs), 585
Erwinia herbicola, 690
escape from photoreversibility, 421
Escherichia coli, 455, 573
essential elements, 74, 74t
essential oils, 321
esterases, 597
ethane methyl sulfonate (EMS), 427
Ethephon, 584
Ethrel, 584
EthylBloc®, 575, 584
ethylene, 571–592
   abscission and, 602
   ACC conversion to, 700
   biosynthesis, 537
   catabolism of, 574
   commercial uses, 584
   conjuction, 574
   defense responses and, 582
   effects of, 578f, 581
   flowering and, 664–665
   fruit development and, 500
   fruit ripening and, 575f, 576–577
   induction by auxin, 485
   inhibitors of, 575–576, 576f
   insensitivity to, 613
   leaf abscission and, 583f
   leaf senescence and, 581–582
   measurement of, 576
   model of signaling by, 589f
   oxidation of, 572, 572f
   physiological activity, 573–574
   production, 572, 601
   properties, 572
   regulation of gene expression, 588
   signal transduction, 584–589
   structure, 572f
   triple response to, 571, 579, 579f
ethylene glycol, 572f
ETHYLENE-INSENSITIVE 2 (EIN2) locus, 613
ethylene oxide, 572f
ethylene receptors, 586, 586f
ethylene response elements (EREs), 588
ethylenediaminetetraacetic acid (EDTA), 77
etiolation, 417, 418f, 561
etioplasts, 20–21, 561
etr1 (ethylene-resistant 1) mutant, 581, 585, 585f
Eucalyptus regnans (mountain ash), 61
euchromatin, 10

euhalophytes, 693
eukaryotes, 5, 110
eukaryotic pathway, 282
European ash *(Fraxinus excelsior)*, 600
European beech *(Fagus sylvatica)*, 643t
European hazel *(Corylus avellana)*, 600
evaporative cooling, 682–683
evening element (AAAATATCT), 435
evolution, 154, 156, 339
exodermis, 56
expansins, 370–371, 537
export, sucrose, 236
extensin, 309
extracellular spaces, 117. *see also* apoplasts

F-box proteins, 400, 501, 524
F-boxes, function, 434
Fabaceae (legume family), 472
facilitated diffusion, 105
FACKEL gene, 388
facultative responses, 641
Fagus sylvatica (European beech), 643t
far-red light
   absorption of, 203
   chloroplast movement and, 27
   effect on floral induction, 656f
   end of day, 435
   morphogenesis and, 418
   photoreception, 655f
   photosystem absorption of, 133
   phyA-mediated responses, 427–428
Faraday's constant, 96
farnesyl diphosphate (FPP), 319, 320f, 595f, 612
farnesylation, 612
FASS gene, 390, 391f, 392
fate mapping, 399
fats, energy storage, 278
fatty acid-amino acid conjugates, 334
fatty acid synthase, 281
fatty acids
   β-oxidation, 285
   common, 279t
   in cutin, 316f
   in membranes, 6
   metabolism of, 284f
   mitochondrial, 688t
   modification, 282
   in suberin, 316f
   synthesis, 281f
   in waxes, 316f
fava bean *(Vicia faba)*, 214f
FCA (FLOWERING TIME CONTROL PROTEIN A), 605, 605f
FD protein, 664
fermentation, 256, 258f
   efficiency of, 259
   glycolysis and, 256–260
   process of, 699, 699f
fermentative metabolism, 259f
ferns, vessel elements, 59
ferredoxin, 147, 293
ferredoxin–NADP reductase (FNR), 147
ferredoxin–thioredoxin system, 166–168, 167f, 167t
ferrochelatase reaction, 308, 308f
fertilization, 378–379. *see also* syngamy
fertilizers
   crop yields and, 83–84
   leaching, 74
   organic, 84
   primary mineral elements, 73
ferulic acid, 325f
Fe–S centers, 147

FeS$_A$, 147
FeS$_B$, 147
FeS$_X$, 147
F$_0$F$_1$-ATP synthase, 267, 268–269
fiberglass, 365
Fick, Adolf, 41
Fick's law, 41, 43f, 96
field capacity, 54
FIERY gene, 613
FIMBRIATA gene, 400
fine sand, 54t, 85
firs, monoterpenes in, 321
Fisher, D. B., 233
fitness, circadian rhythms and, 435
flacca (flc) mutants, 596
Flaveria australasica, 175f, 177
flavin adenine dinucleotide (FAD), 264–265, 264f, 292, 455
flavin mononucleotide (FMN), 266, 456, 457f
flavinoids, 326, 326f, 327–328
flavones, 324f
flavonols, 324f, 482f
flax *(Linum usitatissimum)*, 3f
FLC (FLOWERING LOCUS C), 605, 605f
flooding-sensitive plants, 698
flooding-tolerant plants, 698
floral buds, 498
floral development, 638
floral evocation, 636, 644–646, 644f
floral homeotic mutants, 638
floral induction, 656f
floral initiation, 513
floral meristems, 637
floral organ identity, 639–640, 640
floral organ identity genes, 638–639
floral stimuli, 636, 661–662, 663–664
FLORICAULA (FLO) gene, 638
florigen, 533, 660
flowering
 action spectra for, 656, 656f
 control of, 635–669
 developmental pathways, 665f
 effect of night breaks on, 652f, 653f
 ethylene and, 664–665
 expression of, 645–646
 GAMYB and, 534–535
 gibberellins and, 533–535, 664–665
 inhibitors of, 662–663, 663f
 photoperiodic regulation of, 651f
 transition to, 665–666
 transmissible factors regulating, 661t
 vernalization and, 657–660
FLOWERING LOCUS C (FLC), 605, 605f
FLOWERING LOCUS C (FLC) gene, 658–659, 659f
FLOWERING LOCUS T (FT) gene, 249, 663–664, 664, 665
FLOWERING LOCUS Y (FY), 605, 605f
flowering plants. *see* angiosperms
FLOWERING TIME CONTROL PROTEIN A (FCA), 605, 605f, 664–665
flowers, 276, 411f, 690
fluence
 definition, 421
 phototropism and, 489
 total, 422
fluence rate. *see* irradiance
fluid-mosaic model, 6
fluorescence, 128
fluorescence resonance energy transfer (FRET), 138
flux density, 41
FLY gene, 534
foliar application, 84

forward genetic screens, 387
fossil fuels, 304
14-3-3 inhibitor proteins, 293
14-3-3 proteins, 114
foxglove *(Digitalis)*, 322
Fraxinus excelsior (ash), 222f, 600
free energy, 43
free-running rhythms, 646
freezing, 687–692
frequency, definition, 126
Fritillaria assyriaca, 8
frosted orache *(Atriplex sabulosa)*, 683, 683f, 684
fructans, 183
fructose, 230f
fructose-1,6-bisphosphate, 186, 187, 190, 260
fructose-2,6-bisphosphate, 190
fructose-1,6-bisphosphate phosphate, 165, 187
fructose-6-phosphate, 190, 260
fruit
 climacteric, 577f
 development, 500
 nonclimacteric, 577f
 ripening, 276, 575f, 576–577
 seedless, 514
 set, 500, 514
fruit fly *(Drosophila melanogaster)*, 684
FT (FLOWERING LOCUS T) gene, 663–664
FT protein, 664
Fuchsia, 653f
fumarate, 264f
Funaria hygrometrica, 558, 558f
fungal symbionts, 331–332
fungi, 572–573, 594
furan rings, 325f
furanocoumarins, 324
FUSCA gene, 432–433
fusicoccin, 114, 487
FY (FLOWERING LOCUS Y), 605, 605f

G-actin, 23
G protein, 608
GA 2-oxidase, 518
GA 3-oxidase, 518
GA 20-oxidase, 517–518
GA-3 oxidase genes, 528–529
GA-20 oxidase genes, 528–529
GA5 gene, 518
GA-insensitive dwarf1 (gid2) mutant, 524
GA insensitive (gai-1) mutant, 521
GA-MYB transcription factor, 601
GA response elements (GAREs), 532–533
galactose, 240f
gallic acid, 328, 328f
galls, 544–544
gametophytes, 378
γ-aminobutyric acid (GABA), 686, 687f, 701
gamyb mutants, 534f
GAMYB proteins, 532f, 532–533, 534–535, 664
Gane, R., 571
GA2ox gene, 537
GA3ox gene, 537
Garner, Wightman, 649
gas bubbles, 40f
gas chromatography, 576
gas exchange, 600
gas-liquid interfaces, 40f
gates, in channel proteins, 103
GCR1 gene, 608
Geiger, D. R., 233
gene expression, steps in, 12f
genetic engineering, 521, 581–582

genetic tumors, 558–559, 559f
genistein, 482f
β-gentiobiose, 332f
geranyl diphosphate (GPP), 319, 320f
geranylgeranyl diphosphate (GGPP), 319, 320f, 515, 516f, 595f
germination
 definition, 382
 far-red light and, 418
 gibberellin and, 514
 IAA-*myo*-inositol translocation, 474
 phytochrome and, 439
 precocious, 596f, 598–599
 seed size and, 438–439
 structures during, 528f
GH3 early genes, 502
giant bulrush *(Schoenoplectus lacustris)*, 700
giant sequoia *(Sequoiadendron giganteum)*, 513f
Giaquinta, R. T., 233
Gibberella fujikuroi, 510
GIBBERELLIN-INSENSITIVE (GA1) genes, 393
GIBBERELLIN-INSENSITIVE (GAI) protein, 523, 523f
gibberellin receptors, 526–527, 529
gibberellin response elements (GAREs), 611
gibberellins (GAs)
 in *Arabidopsis*, 519f
 auxin and, 537–538
 bioactivity, 519–521
 biosynthesis, 509, 514–515, 518–519
 cell cycle kinases and, 537
 cell elongation and, 368, 535–536
 commercial uses, 514
 discovery, 510
 effects of, 512–514
 ent-gibberellane skeleton, 511–512
 flowering and, 645–646, 664–665, 666
 identification, 509
 induction of α-amylase, 530f
 metabolism of, 518
 negative regulators, 522–523
 nomenclature, 511
 purification, 510–511
 receptor binding, 526f
 seed dormancy and, 600–601
 shoot apex transitions and, 644
 signaling during growth, 527f, 529–531
 stem elongation and, 519–521
 stem growth and, 535–538
 structure, 319
 synthesis in embryos, 528–529
 transport, 247
GID1 gene, 526
gid1 mutation, 526
GIGANTEA (GI) protein, 434
globular stage, 382, 384, 391f
glucanases, 372
glucan–water dikinase, 183
glucomannans, 353, 357
gluconeogenesis, 257
glucose, 230f, 254
glucose-1-phosphate, 186
glucose-6-phosphate, 192
glucosinolates, 332–333, 333f
β-glucuronidase (GUS), 518
glucuronoarabinozylan, 353, 357, 358f
glutamate, 295f
glutamate-1-semialdehyde aminotransferase (GSA), 449
glutamate dehydrogenase (GHD), 296
glutamic acid, 155f, 230f
glutamine, 230f, 295f, 334f

glutamine synthase (glutamine-2-oxo-glutarate aminotransferase, GOGAT), 294, 295f
glutamine synthetase (GS), 294
glutathione-S-transferases (GTSs), 503
glycans, 353
N-linked glycans, 14
glyceolin I, 340f
glyceraldehyde-3-phosphate, 187, 318, 320f
glyceraldehyde-3-phosphate dehydrogenase, 257, 680
glyceroglycolipids, 279, 280f
glycerol, 7f, 278f
glycerolipids, 278f, 279t, 282–283, 282f
glycerophospholipids, 279
glycine, 170–171, 275
Glycine max. see soybean
glycolate, 168, 170
glycolysis
  alternative reactions, 257–259
  control of, 260
  description, 254
  pathways, 275f
  phosphate incorporation and, 306
  process of, 256–262, 258f–259f
  pyruvate production, 699f
glycophytes, 693, 695
N-linked glycoproteins, 14
glycosides, 332, 332f, 546
glycosylglycerides, 6
glycosylphosphatidylinositol (GPI)-anchored, 6
glyoxylate cycle, 21, 285
glyoxysomes, 21, 283, 284f
glyphosate (Roundup®), 323
Gnetales, 59
GNOM (GN) genes, 388f, 389
GNOM protein, 484
Goethe, Johan Wolfgang von, 640, 642
Goldman diffusion potential, 101
Goldman equation, 100–101
Golgi apparatus
  description, 14
  gene expression and, 12f
  matrix polymer synthesis in, 357
  micrograph, 14f
  protein processing, 15
Golgi bodies, 5f, 12f, 352f
Gramineae (Poaceae), 176
grana, 18, 19f
grana lamellae, 135, 137
grape (*Vitis* spp.), 514, 643t
GRAS transcription factors, 393
grass family (Poaceae), 472
  coleoptiles, 353
  fungal symbionts, 331
  siderophores, 307
  symbiotic relationships, 299
gravel, 85
gravistimulation, 492–494, 496f
gravitropism, 88, 488, 490, 491f, 495f
gravity, 44–45, 492–494
green fluorescent protein (GFP), 248, 248f, 400, 518
green-leaf volatiles, 336
green-light responses, 461–462, 462f
greenhouse effect, 212
ground tissues, 3f, 4f
groundsel (*Senecio*), 330, 641
growth
  auxin-induced, 484–488
  blue-light responses, 446–447, 447f
  brassinosteroids and, 625–629
  differential, 489

excess minerals in soil and, 86
GA signaling model, 527f
indeterminant, 377
phototropism and, 488–490, 489f
water potential and, 50
growth rate equation, 486
growth zones, 468
GSA gene, 449, 449f
guanine nucleotide exchange factor (GEF), 388, 389
guanosine triphosphate (GTP), 24–25
guard cells, 3f
  abscisic acid signaling in, 609f
  blue-light photoreception, 457–461
  calcium ion levels, 606f
  cell walls, 67–68
  cellulose microfibrils, 69f
  dicot, 69f
  effect of sucrose, 453–455
  function, 67
  osmoregulation, 453, 454f
  PM-ATPase in, 603–604
  protoplasts, 450, 452f, 603f
  zeaxanthin content, 458, 458f
Gunnera, 297, 299
Gunning, B. E. S., 25
GURKE gene, 388
GUS (β-glucuronidase) enzyme, 471
GUS (β-glucuronidase) reporter gene, 385f, 471, 471f, 528
guttation, 58, 58f
guttation fluid, 471
gymnosperms
  description, 2
  ethylene production, 572
  mycorrhizal fungi associated with, 89
  phloem, 221–221
  sieve cells, 4f
  sieve tube elements, 224t
  translocation, 235
  vessel elements, 59
gynoecium initiation, 637
gypsum, 85

H^+-ATPase, 105, 113–114, 452f, 487, 626
H^+-pyrophosphatase, 105, 116
Haber–Bosch process, 291
Haberlandt, Gottlieb, 528, 544
habitats, shady, 201–202
hackberry (*Celtis occidentalis* L.), 559
hadathodews, 58
Haemophilus influenzae, 102
halophytes
  definition, 693
  description, 86
  ion accumulation, 695–696
  salt glands, 694
  water potential, 50
Hartig net, 90, 90f
Hatch, M. D., 173, 176
Hd1 (Heading-date 1) gene, 654f, 655, 666
Hd1 (heading-date 1) mRNA, 654f
Hd3a (Heading-date 3a) gene, 654f, 655
head groups, phospholipid, 6
heart stage, 382, 383f, 395f, 397f
heartleaf maskflower (*Alonsoa warscewiczii*), 226f
heat
  dissipation of, 209, 209–210
  emission by chlorophyll, 128
heat shock, 684–687
heat shock cognate proteins, 685
heat shock factor (HSF), 685, 686f
heat shock promoters, 562

heat shock proteins (HSPs)
  accumulation, 685
  cycle, 686f
  induction of, 680
  production of, 684–685, 685f
  temperature tolerance and, 685
heat stress, 682–683
heavy metals, 74, 86
Hedden, Peter, 511
Hedera spp., 400f, 513, 642f, 643t, 644
Helianthus annuus (sunflower), 661t, 673f, 676f
heliotropism, 203
hemes, 144–145, 292
hemicelluloses
  in cell walls, 353
  cellulose binding by, 357
  modification, 372
  in primary cell walls, 352f, 353t
  structures, 358f
hemp (*Cannabis sativa*), 513
henbane (*Hyoscyamus niger*), 660–661, 661f
herbicides, function of, 147–148
herbivores, 318
heterochromatin, 10
heterocyclic rings, 329
heterocysts, 299f
hexose monophosphate shunt. see oxidative pentose phosphate pathway
hexose phosphates, 186–191, 187f, 190–191
hexoses, 256, 354f
high-irradiance responses (HIRs), 421–422, 421f, 422–423, 422f, 423, 656
Hill, Robert, 132
Hippuris vulgaris (marestail), 642
histidine kinases, 425, 563
histidine phosphotransfer protein (Hpt), 565–567
histidine phosphotransferases, 565–567
histidines, 697
histogenesis, 382
histone code, 659
histones, 10
HLS1 (HOOKLESS 1) gene, 580
HLT family, 112
HmT-toxin, 276
Hoagland, Dennis R., 77
Hoagland solution, 77, 78t
Hoffler diagrams, 47f
holoproteins, 423
homeostasis, ions, 697f
homeotic genes, 640, 640f, 641f, 666
28-homobrassinolide (28-homoBL), 619
homogalacturonan, 357, 359f
Hooke, Robert, 1
hooks, dark-grown seedlings, 580
Hordeum vulgare. see barley (*Hordeum vulgare*)
hormone-insensitive lipases, 526
hormones, 467
hy1 mutants, 657
HY1 protein, 656
HY4 protein. see cryptochrome 1 (CRY1)
hydathodes, 471
hydraulic conductivity, 48, 65f
hydroactive stomatal closure, 674
Hydrodictyon reticulatum, 619
hydrogen bonds, 38
hydrogen cyanide, 332
hydrogen ions, 340
hydrogen peroxide, 338, 686
hydrolysis, 285
hydrolyzable tannins, 328, 328f
hydropassive stomatal closure, 674

hydroponic growth system, 76–77, 76f
hydrostatic pressure, 40, 41f, 44, 54f
hydroxyl radicals, 338
hydroxyproline, 309
hydroxyproline-rich glycoprotein (HRGP), 361, 361f, 361t
Hyoscyamus niger (henbane), 660–661, 661f
hyperaccumulation, 696–698
hyperacidification, 115
hyperosmolarity, 695
hypersensitive responses, 338, 412
hyphae, mycorrhizal, 89
hypocotyls, elongation, 423f
hypodermis, 56
hyponastic leaves 1 (hyl1) mutants, 612
hypophyses, 384
hypoxia, 698. see also oxygen, deficiency

ice crystal formation, 689
ice nucleation, 689
ice plant (*Mesembryanthemum crystallinum*), 458, 679f
ice transcription factor 1 (ICE1), 692
illite clay, 85, 85t
immature leaves, 398
α-importin, 10, 10f
β-importin, 10, 10f
indeterminant growth, 377
indeterminate plants, 673
indirect induction, 661
indole, 330t
indole-3-acetamide (IAM) pathway, 472, 473f
indole-3-acetic acid (IAA), 467–507
  abscisic acid production and, 600–601
  apical shoot meristem development and, 396, 397f, 398
  biosynthesis, 471–472, 473f
  brassinolide and, 628f
  cell elongation and, 368, 484–488
  coleoptile elongation, 470f
  conjugation, 474
  degradation, 475, 475f
  developmental effects, 496–501
  dose-response curve, 486f
  early experiments, 469f
  efflux, 478–480
  in embryonic development, 386f
  ethylene biosynthesis and, 574–575
  fruit development and, 500
  function, 470–471
  GA function and, 537–538
  gene activation and, 488f
  gibberellin content and, 538f
  H⁺-ATPase stabilization and, 488f
  identification, 468
  influx, 476, 477–478
  leaf abscission and, 583f
  leaf initiation and, 406
  localization, 475–476
  as morphogen, 386–387
  phototropism and, 488–490, 489f
  polar transport, 484
  protein trafficking and, 488f
  proton extrusion and, 488f
  receptor binding, 503f
  root apical meristem development, 402, 403f
  root cap inhibition, 494
  in root caps, 494–496
  root development and, 498–499
  signal transduction pathways, 501–504
  sites of accumulation, 471f
  structure, 470f

  synthesis, 471
  transport, 476–484, 478f, 480f, 481f, 483–484
  transport inhibitors, 482–484, 482f, 483f
  vascular differentiation and, 499–500
indole-3-acetonitrile (IAN) pathway, 472, 473f
indole-3-butyric acid (IBA), 470f, 474
indole-3-glycerol phosphate, 472–474
indole-3-pyruvate (IPA) pathway, 472
induced defense responses, 334, 339
induced thermotolerance, 682
inductive effects, 394–395
infection thread, 301, 302f
infections, 338, 339f
Ingenhousz, Jan, 130
initial cells, 558
inner membranes, 262, 267f
inner plastid membranes, 112
inorganic soils, 85
inorganic solutes, 228–229, 228t
inositol trisphosphate (InsP$_3$), 111
insect herbivores, 334
insects, saliva, 334–335
integral proteins, 6
intercalcary meristems, 396
intercellular air space resistance, 213, 213f
intercisternal elements, 14
intermediary cells, 226f, 227, 239
intermediate filaments (IFs), 23, 23f
intermembrane spaces, 262
internodal growth, 81–82
internodes, 3f
inwardly rectifying K⁺ channels, 103
ion channels. see also specific ion channels
  abscisic acid regulation of, 603–604
  anion-selective, 119
  in membranes, 6
ion fluxes, 430, 436f
ions
  concentration in cytosol, 100f
  concentration in roots, 100t
  homeostasis, 697f
  toxicity, 694
  transport across membranes, 98–101
  transport in roots, 116–119
IP$_3$ catabolism, 613
Ipomoea acuminata (morning glory), 411f
IPT gene, 560
ipt gene, 559, 560f, 561–562
Iris pseudacorus (yellow flag), 701
iron
  absorption, 308f
  availability, 698
  biochemical function, 75t
  chelation, 307
  chelation of, 78f
  deficiency, 81, 112
  ferrochelatase reaction, 308f
  in nutrient solutions, 77
  root uptake of, 88
irradiance, 199, 200, 421
island domains, 629
isocitrate, 264f
isoflavones, 324f
isoflavonoids, 328
isopentenyl adenine (iP), 546, 547
isopentenyl diphosphate (IPP)
  in abscisic acid synthesis, 595f, 596
  in terpene biosynthesis, 318, 319, 320f
isopentenyl transferase (IPT), 548, 550f, 551
isoprene units, 318, 320f
isoprenoid pathway, 6
isoprenoids. see terpenes

isoquinoline, 330t
isotropic arrangement, 364–365

jack bean (*Canavalia ensiformis*), 334
Jagendorf, André, 148
jasmonate, 283
jasmonic acid (JA), 335–336, 337f
jewelweed (*Impatiens*), 582f
jumping plant lice (*Pachypsylla celtidismamma*), 559
juvenile phases, 513
juvenile vegetative stage, 384

kaempherol, 483
Kalanchoe, 600, 650
Kamiya, Yuri, 511
KANADI genes, 408–33, 408f
kaolinite clay, 85, 85t
Karlsson, P. E., 451
Karpilov, Y., 173
Keeling, C. David, 212
keratins, 23
key resources, 38
kidney bean (*Phaseolus vulgaris*)
  apical dominance in, 496
  ethylene production, 572
  evolution, 650
  seedlings, 418f
kinesins, 27
kinetin, 545, 545f
kinetochores, 25
KNAT1 gene, 553
Knoblauch, M., 233
Knop, Wilhelm, 76
KNOTTED 1-LIKE HOMEOBOX (KNOX) genes, 407, 535
KNOX homeodomain, 535
Koornneef, M., 427
Kortschack, H. P., 173
Kranz anatomy, 174, 175f, 176, 177f
Krebs, Hans, A., 262
Krebs cycle. see citric acid cycle
KUP/HAK/KT family, 112

L1 cell lines, 399, 399f
L2 cell lines, 399, 399f
L3 cell lines, 399, 399f
lactic dehydrogenase (LDH), 699
Laminum sp., 316f
larkspur (*Delphinium*), 330
Larrea divaricata (creosote bush), 214f
LATE ELONGATED HYPOCYTYLS (LHY) gene, 433–434, 434f
LATE ELONGATED HYPOCYTYLS (LHY) protein, 433
late-embryogenesis-abundant (LEA) proteins, 599
late genes, 502, 503
latent heat of vaporization, 39
lateral buds, 557
lateral root cap, 405
lateral roots, 3f
laticifers, 222
latitude, 211f, 648f
lauric acid, 279t
Lavatera (Malvaceae), 202
LE gene, 518, 519
LEA (late embryogenesis abundant) proteins, 680, 681f
leaf primordia, 3f, 398, 406
leaf stomatal resistance, 65
LEAFY (*LEY*) gene, 638, 664
leaves. see also phyllotactic patterns
  abscisic acid localization, 598f

abscission, 500, 583f, 674, 674f, 695
adaptations, 684
anatomy, 198f
angle of, 202f, 202–203
area of, 673–674, 695
arrangements, 406f
cell expansion, 561
day length perception in, 533
development, 407f
dissipation of energy, 206–207, 678
ethylene production, 572
expansion, 676f
formation sites, 406f
heat dissipation, 209f, 209–210
immature, 398
induction, 662f
with Kranz anatomy, 177f
lamina, 408f
light absorptoin, 200–201
movement of, 202f, 202–203, 207–208
movements, 648, 679f
nitrogen assimilation, 310f
nutrient absorption by, 84
optical properties, 201f
photoperception, 650
photosynthetic responses, 203–209
planar form, 407
primordia, 398, 498f
senescence, 276, 410, 411f, 559, 560f, 562f, 581–582, 602
senescent, 17, 154
sidedness, 409
sleep movements, 435–437
source-to-sink transition, 242–243, 243f
sulfate assimilation, 305–306
thigmonastic movements, 435
transformation, 642, 642f
turgor, 673–674, 673f
water movement and, 62f, 64–69
water pathway through, 65f
water stress and, 676f
wax deposition, 678
legmenoglobin, 300
legume family (Fabaceae), 299, 472. *see also specific legumes*
Leguminosae, 297
Lemna (duckweed), 1, 208f
lemons, 115
lens effects, 448
Lepidium sativum (cress), 7f
lettuce, 419f, 423f
leucoplasts, 18
LHCB gene family, 433, 434–435
light. *see also* sunlight
absorption patterns, 213–214, 214f
adaptation to, 437
characteristics, 126–127
circadian clocks and, 434
day length and latitude, 648f
description, 126f
development and, 417–443
ecological parameters, 437t
entrainment by, 648
expression of flowering and, 645–646, 645f
gradients, 448
leaf anatomy and, 198f
leaf movement and, 436
parameters of, 199
phototropism and, 488–490
quantification, 199f
seed dormancy and, 600
sensing of, 448
sensors, 199f
signaling by, 445

solar spectrum, 127f
units of measurement, 199–203
light-absorbing antenna systems, 137–139
light channeling, 201
light compensation points, 204
light-harvesting antennas, 130–132, 203
light-harvesting complex II (LHCII), 139, 139f, 153
light reactions, 125–158
light-response curves, 204, 204f, 208, 208f
light scattering, 201
light–dark cycles, 651–652
lightening, 291
lignification, 325
lignins
    biosynthesis, 680
    description, 2, 325, 363
    function, 351
    phenolic subunits, 363f
lime, 83, 84
limiting factors, 198
limonene, 321, 321f
limonoids, 321
lineage-dependent mechanisms, 385
linoleic acid, 6, 279t, 334
linolenic acid, 279t, 334, 334f, 335f
N-linolenoyl-glutamine, 334f
Linum usitatissimum (flax), 3f
lipases, 22–23, 285
lipid bilayers, 6–8, 102f, 279
lipid bodies. *see* oil bodies
lipid metabolism, 278–285, 608–609
lipid rafts, 6, 8
lipids, 283–285
lipoxygenases, 309f
liquid phase resistance, 213, 213f
Liriodendron tulipifera (tulip tree poplar), 277
Lloyd, F. E., 453
lodgepole pine *(Pinus contorta)*, 690
lodging, definition, 514
Lolium temulentum (ryegrass), 512, 533, 645, 645f, 656, 664
long-day plants (LDPs)
    critical dark periods and, 651f
    description, 533, 649
    flowering, 533–535
    photoperiodic response, 650f
long-distance transport, 236
long–short-day plants (LSDPs), 650
Lorimer, George, 166
lotus *(Lotus japonicus)*, 317
LOV domains, 456
low-fluence responses (LFRs), 422, 422f
lupine *(Lupinus, Fabaceae)*, 202, 330
lupinic acid formation, 553
Lupinus albus (white lupine), 294
Lupinus (lupines), 202, 330
Lupinus succulentus, 202f
lycopene, 20f
Lycopersicon esculentum (tomato), 20f, 77, 238
Lycopersicon hirsutum (tomato), 688f
lytic vacuoles, 16, 17

MacMillan, Jake, 511
macronutrients, 74
Madagascar periwinkle *(Catharanthus roseus)*, 335
MADS box genes, 639
magnesium, 75t, 81
Magnifera indica (mango), 644
maize (corn; *Zea mays*)
    anther ear1 mutant, 522
    blue-light transmission, 448
    br2 mutants, 481f

carbon isotope ratios, 217
cms mutant, 276
dehydration and, 675
dwarf mutants, 521
effect of gibberellins, 510f
effect of green-leaf volatiles on, 339
GA-deficient mutants, 522f
hypoxia, 700
invertase activity, 246
orange pericarp (orp) mutant, 472, 474
plant density, 439f
precocious germination, 596f
roots, 701f
secondary metabolites, 335
seedlings, 418f
shoot development, 643f
U.S. yields, 672t
viviparous (vp) mutants, 596, 602f
vivipary in, 599
yields, 38f
malate, 173–174, 264f, 284f, 307f
male fertility, 534–535
malic acid, 307f
malonic acid pathway, 323f
maltose, 186
Malus spp. (apple), 643t
Malvaceae, 202
manganese, 75t, 81, 142
mango *(Magnifera indica)*, 644
mangrove swamps, 277
Manihot esculenta (cassava), 332, 675
mannitol, 229–230, 230f
mannose, 230f
maple *(Acer)*, 26f
Marchantia polymorpha, 619
marestail *(Hippuris vulgaris)*, 642
Maryland Mammoth mutant, 649, 649f, 660, 661t
mass flow, 42
mass transfer rate, 231
maternal inheritance, 154
maturation stage, 384
maturation zone, 88, 404, 404f
mature stage, 382
Mauna Loa, Hawaii, 212, 212f
MAX4 (more axillary growth 4) gene, 557
maximum quantum yield, 204–205
MDR/PGP proteins (PGPs), 480, 480f
Medicago spp. (alfalfa), 447, 599, 690–691
medicarpin, 340f
megapascals, 40
melon *(Cucumis melo)*, 239
membrane potential, 98–100, 99f, 101f, 430
membranes
    anchored proteins in, 8f
    during chilling, 687
    composition of, 6–8, 263
    effect of chilling, 688
    glycerolipid components, 279t
    ion transport, 98–101, 696f
    lipids in, 280f, 283
    permeability, 98, 102f
    stability, 684
    structure, 7f
    transport, 101–108, 680
    water flow across, 45–47, 49f
    water-selective channels, 49, 49f
Mendel, Gregor, 519
menthol, 321f
meristem identity genes, 638
meristematic zone, 87, 88f, 404, 404f
meristemoids, 396
meristems. *see also* root apical meristem; shoot apical meristem

description, 2–5
IAA synthesis in, 471
necrosis of, 81
Mesembryanthemum crystallinum (ice plant), 458, 679f
mesophyll cells
    beet leaf, 236f
    C_4 cycle and, 174
    chloroplasts, 126
    dehydration, 676
    differentiated, 628f
mesophyll resistance, 213
mesquite (*Prosopis* sp.), 86, 672
messenger RNA (mRNA), 11, 249, 612, 653
metals
    accumulation of, 696–698
    chelation, 697
    transport, 112–113
metaphase, 11f, 26f
methemoglobinemia, 292
methionine, 305–306
2-methoxy-3,6-dichlorobenzoic acid (dicamba), 471f
2-C-methyl-D-erythritol 4-phosphate (MEP), 319
N-methyl pyrrolinium structure, 331f
methyl salicylate, 341f
1-methylcyclopropene (MCP), 575, 576f
methylerythritol phosphate (MEP) pathway, 318, 320f
mevalonic acid pathway, 318, 320f
MEX1 locus, 186
microbes, anaerobic, 698
microbodies, 21, 23f
microfibrils, 364–366, 365f
microfilaments, 23, 24–25
micronutrients, 74
microRNA (miRNA), 249, 409, 535, 653
microtubule-associated proteins (MAPs), 24
microtubule organizing centers (MTOCs), 24
microtubules
    arrangements, 25f
    assembly, 24–25
    description, 23
    function, 25–26
    microfibril orientation and, 366
    organization, 627, 627f
    orientation, 367f, 579, 580f
middle lamellae, 4f, 5f, 351
Mimosa pudica, 435f
Mimulus cardinalis (scarlet monkey flower), 226f
mineral nutrients, 289–313
mineral nutrition, 73–93
    classification of nutrients, 75t
    deficiencies, 77–82
    element mobility, 79
    foliar application, 84
    mycorrhizae facilitated, 89–91
    root absorption of, 88–89
mineralization, 84
miR159a, 535
Mitchell, J. W., 617
Mitchell, Peter, 148, 268
mitochondria
    ADP:O ratios, 268t
    electron flow, 150f
    electron transport, 267f
    energy conversion in, 17–20
    fatty acid composition, 688t
    function, 17, 20
    genome size, 20
    matrix, 17

membrane composition, 279t
photorespiratory cycle, 169f, 170
plant cell, 5f
pollen development and, 276–277
reproduction, 153–154
structure, 18f, 262–263, 263f
succinate conversion and, 285
transmembrane transport, 270f
mitosis, 25–26, 26f, 535–536, 544
mitotic spindles, 25
molybdenum, 75t, 82
molybdenum cofactors, 596
monocarpic senescence, 410–411, 411f, 559
monocots
    embryogenesis, 382, 383f, 384
    mycorrhizal fungi associated with, 89
    phytochrome genes, 429f
    root development, 87
monogalactosyldiacylglycerol, 7f
monooxygenases, 309–310
MONOPTEROS (*MP*) genes, 388–389, 388f, 403f
monoterpenes, 318, 320f, 321
monovinyl protochlorophyllide *a*, 155f
montmorillonite clay, 85, 85t
morning glory (*Ipomoea acuminata*), 411f
morphine, 330f, 331
morphogenesis, 382
morphogens, 386–387
mosses
    abscisic acid in, 594
    bud formation, 557–558, 558f
motor proteins, 27
Mougeotia, 430
mountain ash (*Eucalyptus regnans*), 61
Mu8 transpondon, 481f
mucigel, 87
mucopolysaccharides (slime), 403–404
multinet growth hypothesis, 365
Münch, Ernst, 231
mung bean (*Vigna radiata*), 269f
Mus musculus (mouse), 433
Musaceae (banana family), 472
mustard family (Brassicaceae), 472
mustard oil glycosides, 332–333
mutants
    screening for, 588f
    seedling defective, 388
mutations. *see specifc* mutations
MYB genes, 434f
MYB regulation, 535
mycelium, 89
mycorrhizae, 89–91
mycorrhizal fungi
    ectotrophic, 90, 90f
    nutrient transfer to roots, 90
    vesicular-arbuscular, 90, 90f
myosins, 27, 29
myristic acid, 6, 279t
myrosinase, 332

NAD-malic enzyme, 178
NADH dehydrogenase, 266
NADP-glyceraldehyde-3-phosphate dehydrogenase, 165
NADP (nicotinamide adenine dinucleotide phosphate), 125, 132–133, 140, 146–147, 256, 331f
    production, 260, 261f
    redox reactions and, 255–256
natural products. *see* secondary metabolites
necrosis, 410
necrotic lesions, 412
necrotic spots, 80

nectar guides, 327
neem tree (*Azadirachta indica*), 321
negative feedback regulation, 518
negative regulators, 522
Neljubov, Dimitry, 571
Neospora crassa, 433
9′-*cis*-neoxanthin, 595f, 596
Nepenthes alata (pitcher plant), 110
Nerium oleander (oleander), 684
Nernst equation, 99, 100–101
Nernst potential, 99
nickel, 75t, 82, 697
Nicotiana attenuata (tobacco), 336
Nicotiana hybrids, 558–559, 559f
Nicotiana sylvestris, 660, 661t
Nicotiana tabacum. *see* tobacco
NICOTIANA TABACUM HOMEOBOX 15 (*NTH15*) gene, 535, 536f
nicotinamide adenine dinucleotide dehydrogenase, 273
nicotinamide adenine dinucleotide phosphate. *see* NADP
nicotine, 330f, 331f
nicotinic acid, 331f
night-blooming jasmine (*Cestrum nocturnum*), 650
night breaks, 652, 652f, 656f
NIT (nitrilase) genes, 472
Nitella (green alga), 26, 365
nitrate reductase, 292, 292f, 293–294, 293f
nitrates
    assimilation, 172, 289, 292–294
    in drinking water, 74
    transport, 110
    in xylem sap, 294f
nitric oxide (NO), 338, 607, 609f
nitric oxide synthase (NOS), 338
nitrilase enzymes, 472
nitrites, 293–294, 293f
nitrogen
    assimilation, 310f
    atmospheric deposition, 74
    biochemical function, 74–75, 75t
    biogeochemical cycle, 290–291, 290t
    cycles, 291f
    deficiency of, 79
    environmental, 290–292
    fertilizers, 291
    in Hoagland solution, 77
    in phloem sap, 228
nitrogen fixation, 289
    anaerobic conditions for, 297–299
    biological, 296–304
    description, 291
    energetics of, 301, 303
    nodule formation, 301
    organisms involved in, 298t
    process of, 291
nitrogenase enzyme complex, 301–303, 303f, 312t
nitrogenase enzymes, 297
nitrous oxide (N_2O), 293
1-NOA (1-naphthoxyacetic acid) structure, 482f
Nobel, P. S., 232
nodal roots, 87
NodD factors, 300–301, 300f
nodulation (*nod*) genes, 300
nodulin (*Nod*) genes, 300
non-Mendelian inheritance, 154
nonclimacteric plants, 576, 577f
noncovalent bonds, 306
NONPHOTOTROPIC HYPOCOTYL-4 (*NPH4*) gene, 402, 403f

NONPHOTOTROPIC HYPOCOTYL-4 (NPH4) protein, 396
nonprotein amino acids, 333–334
nopaline, 549f
Nostoc, 297
NPA (*N*-1-naphthylphthalamic acid), 482f, 580
nph1 (nonphototropic hypocotyl) mutant, 456, 458–459, 459–461, 460f
Nramp ion transporters, 697
nuclear envelopes, 5f, 8, 9f
nuclear genome, 8
nuclear laminins, 23–24
nuclear localization sequences (NLSs), 424, 425
nuclear localization signals, 9
nuclear pore complex (NPC), 9, 9f, 10f
nuclear pores, 8, 9f
nucleation phase, 24
nuclei, 4f
　characterization, 8–11
　genome, 153
　plant cell, 5f
　protein import into, 10f
nucleoids, 20
nucleolar organizers, 10
nucleoli, 5f, 9f, 10
nucleoporin proteins, 9
nucleoside diphosphate kinase 2 (NKPK2), 431
nucleosomes, 10, 11f
nutrient depletion zones, 89, 89f
nutrients
　assimilation, 289–313, 310–311
　availability in soil, 83f
　basic groups, 74–75, 75t
　film growth system, 76f, 77
　pH of soil and, 86
　solution composition, 76
nutrition
　deficiencies, 83–84
　mineral, 73–93
nyctinasty, 435, 435f
Nymphoides pelata (water lily), 580, 700

oat (*Avena sativa*)
　auxin-induced growth, 485f
　coleoptiles, 468
　gravitropism in, 490, 491f
　mesophil cell, 21f
　response to auxin, 486, 486f
　seed dormancy, 600
obligate responses, 641
octadecanoid pathway, 335
octant embryos, 382
octopine, 549f
oil bodies, 21–23, 278, 284f
oils, 278
oilseed rape (*Brassica napus*), 209
oleander (*Nerium oleander*), 684
oleic acid, 6, 279t
oleosins, 22, 22f, 278
oleosomes. *see* oil bodies
oligogalacturonans, 373
O-linked oligosaccharides, 15
oligosaccharins, 373, 373f
omega-3 fatty acid desaturases, 684
onion (*Allium cepa*), 351f, 451f
OPEN STOMATA (OST1) GENE, 610
opines, 548
opposite patterns, 406f
Opuntia ficus-indica, 48, 216f
orange pericarp (orp) mutant, 472, 474
orchids (Orchidaceae), 180

ordinary companion cells, 226
organic fertilizers, 84
organogenesis. *see also specific* organs
　definition, 382
　lateral, 409f
　vegetative, 405–410
ornithine, 331f
orthostichy, 227
orthovanadate, 113
Oryza sativa. see rice (*Oryza sativa*)
oryzalin, 366, 367f
oscillators, 433–435, 434f
OsEXP4 gene, 537, 537f
osmolality, 44
osmolytes, 692
osmosis, 42–43
osmotic adjustment, 676–677
osmotic potential
　aquaporins and, 113
　definition, 44
　temperature and, 44f
　water potential and, 54
　xylem, 58
osmotic stress, 679–680, 679f, 682f
outer membranes, 262
outwardly rectifying K$^+$ channels, 103
ovules, 380f, 636
oxaloacetate (OAA), 257, 264f
oxidases, 309, 363
oxidative bursts, 338, 372–373
oxidative pentose phosphate pathway, 260–262
oxidative phosphorylation, 255, 265–266
2-oxoglutarate, 264f, 276, 295f
oxygen
　acquisition, 700–701
　assimilation, 308–310
　deficiency of, 698–702
　deficit sensing, 702
　diffusion, 57–58, 700
　electronegativity, 38
　generation of, 125
　production, 132
　respiration rates and, 277
oxygenase reactions, 309f
oxygenases, 308

P680, 142
P700, 142, 145, 146
P870, 142
P-glycoproteins, 480–482, 481f
P-protein bodies, 225
P-proteins
　function, 225
　location, 233
　in phloem sap, 229
　synthesis, 248
P-type ATP-dependent metal transporters, 697
Pachypsylla celtidis-mamma (jumping plant lice), 559
palisade cells, 200–201
palisade parenchyma, 3f
palmitic acid, 6, 279t
Paphiopedium, 459–460
paraquat, 147, 148f
parenchyma, 4f
parenchymal cells, 350–351
parthenocarpy, 500, 514
partitioning, 244–247
PAS domain, 424
pascals, 40
Passiflora maliformis (conch apple), 687
passionflower (*Passiflora caerulea*), 688

passive transport, 96, 97f, 100–101, 105
Pasteur, Louis, 259
Pasteur effect, 259
patch clamp electrophysiology, 103
pathogenesis-related (PR) proteins, 339, 691
pathogenesis-related proteins (PRms), 250
pathogens, 318, 338–341
pea (*Pisum sativum*), 579, 580f
　auxin transport, 483
　chloroplast, 134f
　dwarf, 510
　flooding sensitivity, 698
　growing temperature, 672
　nana mutants, 520, 522
　rms4 mutant, 553
　seed abortion, 521f
　sieve elements, 226f
peanut (*Arachis hypogaea*), 580
pear (*Pyrus communis*), 514
pectin esterases, 361
pectinases, 412
pectins
　in cell walls, 353
　de-esterification, 372
　functions, 357–361
　modification, 372
　in primary cell walls, 352f, 353t
　structure, 360f
Penicillium, 571
pentose phosphate pathway, 254–255, 260–262, 261f, 275f
pentoses, 354f
peppermint oil, 321
peptide transporters, 111
perforation plates, 59
pericarps, 600
periclinal cell division, 406
periclinal chimeras, 399–400, 400f
periclinal divisions, 389–390
pericycle, 2, 3f, 390
periderm, 315, 317
Perilla crispa
　floral stimulus, 661
　graft experiments, 660, 660f
　leaf induction, 662f
　photoperception in, 650
perinuclear endoplasmic reticulum, 11
perinuclear spaces, 8
period, of rhythms, 433, 646
peripheral proteins, 6
peripheral zone, 398, 399f
periwinkle (*Catharanthus roseus*), 621
periwinkle (*Vinca rosea*), 548
permanent wilting points, 55, 677–678
permeability, selective, 42–43
peroxidases, 363
peroxisomes
　function, 21
　mesophyll cell, 22f
　photorespiratory cycle, 169f, 170, 170t
　plant cell, 5f
petals, initiation, 637, 637f
Petunia hybrida, 660, 661f
Pfr phytochrome, 419
　cellular partioning of, 425
　conversion to Pr, 448
　function, 420
　shade avoidance response and, 437
　structure, 423f
pH, 83f, 84, 86, 699
PHABULOSA (PHB) gene, 409
Phacelia dubia (small-flower scorpionweed), 600

Phalaris canariensis (canary grass), 468
PHANTASTICA (PHAN) gene, 407
Pharbitis nil, 656, 656f
phase, of rhythms, 646, 646n, 647f
Phaseolus coccineus (runner beans), 511
Phaseolus vulgaris (kidney bean), 233, 418f, 496, 650
phasic acid (PA), 597
PHAVOLUTA (PHV) gene, 409
phenolic compounds
    activation by light, 323–324
    biosynthesis, 322–323, 323f, 324f
    release into soil, 324–325
phenolic groups, 372
phenylalanine, 322–323, 324f
phenylalanine amonia lyase (PAL), 323
phenylpropanoid lactones, 323
phenylpropanoids, 323
pheophytin, 143
Phleum pratense (timothy grass), 19f
phloem, 3f, 4f
    development of, 393f
    function, 221
    loading, 235–241, 236f
    mass transfer rate, 231
    pressure-flow model, 231–235, 232f
    pressure gradients, 233–235
    root, 88
    sap, 228–229, 228t, 229t, 230f
    secondary, 223f
    signal molecules, 247–250
    source-to-sink patterns, 228f
    sugar transport, 183
    translocated materials, 228–230, 228t
    unloading, 241–242
    vascular differentiation and, 499–500, 499f
    velocity, 231
    walls of, 351
phosphate ions
    assimilation, 306
    root uptake of, 88, 89
    translocators, 112
    transporter protein, 269
phosphate–H^+ symporters, 112
phosphatidic acid, 282, 609
phosphatidylcholine, 7f
phosphatidylinositol-4,5-bisphosphate (PIP_2), 283
3′-phosphoadenosine-5′-phosphosulfate (PAPS), 304f, 305
phosphoenolpyruvate carboxylase (PEPCase), 174, 174f, 178–179, 181
phosphoenolpyruvate (PEP), 257, 260, 265f
phosphoenolpyruvate (PEP) carboxylase, 679f
phospholipase C, 608, 609f
phospholipase D, 608, 609, 609f, 612f
phospholipid bilayers, 6–8
phospholipids, 6, 101
phosphorelay systems, 563f
phosphorus, 75t, 77, 80
phot1 gene, 456
phot2 gene, 456
phot1 mutants, 459
photoassimilation, 310
photochemical reactions, 291
photochemistry, 128
photodamage, 151–153, 151f
photoinduction, 645, 645f
photoinhibition, 152–153, 208–209
photolyases, 455
photolysis, 605f
photomorphogenesis
    brassinolide and, 619–621
    definition, 430, 446
    transition to, 417
photon irradiance, 199
photons, 126, 132
PHOTOPERIOD SENSITIVITY 5 (Se5) gene, 656
photoperiodic induction, 650
photoperiodism
    coincidence model, 653
    definition, 641, 648
    flowering and, 533
    night breaks and, 652
    plant classification by, 649–650
    site of perception, 650
    timekeeping, 652
photophosphorylation, 148
photoprotection, 151–153
photoreceptors, 455–462, 457
photorespiration cycle, 168–172, 169f, 171–172
photoreversibility, 419–420, 421
Photorhizobium, 297
photostationary states, 420
photosynthesis, 197–220
    action spectra, 130, 130f
    apparatus, 134–137
    carbon dioxide and, 206, 211–216, 212–213
    carbon reactions, 159–195
    chemical reaction, 130, 132
    chloride accumulation, 694
    energy for, 128, 132
    light reactions, 125–158
    quantum yield, 132, 204–205
    respiration during, 275–276
    responses of leaves, 203–209
    sodium accumulation and, 694
    steps in, 198
    temperature and, 209–211, 683
    transpiration and, 66–67
    water stress and, 676, 676f
    Z scheme, 134f
photosynthetic photon flux density (PPFD), 200
photosynthetically active radiation (PAR), 200
photosystem I (PSI), 133, 134f
    far-red light absorption, 203
    NADP reduction, 140, 146–147
    protection of, 153
    structure, 147f
    Z scheme, 140f
photosystem II (PSII), 133, 134f
    plastoquinones, 145f
    reaction center, 137, 142, 143f, 152–153
    temperature sensitivity, 684
    water oxidation, 139
    Z scheme, 140f
photosystems
    description, 126
    electron transfer, 146
    separation of, 137
phototropins, 456–457, 489
phototropism
    action spectra, 446f
    auxin and, 488–490
    blue-light responses, 447f
    definition, 445, 446, 488
    description, 468
phragmoplasts, 25–26, 362
PHY domain, 424
PHY genes
    Arabidopsis, 426
    effect of, 656
    evolution, 428
    interactions, 428
phycoerythrobilin, 129f
Phycomyces, 448
phylloquinones, 147
phyllotactic patterns, 406, 406f, 407–409
phyllotaxy, 498, 555f
phyto-oncogenes, 549
phytoalexins, 339–340
phytochrome interacting factors (PIFs), 430–431
phytochrome kinase substrate 1 (PKS1), 431
phytochromes
    action spectra, 656f
    autophosphorylation, 425, 426f
    chloroplast movement and, 207
    circadian rhythms and, 433–435
    classes of, 425–426
    deficiencies, 428–429, 429f
    development and, 417–443
    ecological functions, 435–440
    ecotypic responses, 440
    encoding of, 425–426
    entrainment by, 648
    flowering and, 665
    function, 418
    functional domains, 424–425
    gene expression induced by, 432–433
    genetic analysis of function, 427–429
    histidine kinases and, 585
    hook opening and, 580
    modulation of, 440
    phosphorylation status, 431f
    photoreceptor function, 655–656, 655f
    photoreversible responses, 419–420, 419t
    phyA mutants, 427–428, 434, 438f
    phyB mutants, 434, 438f
    properties of, 418–420
    red light activation of, 424f
    responses induced by, 420–423
    signaling pathways, 430–433
    stomatal responses, 459–461
phytochromobilin, 423, 423f
phytoecdysones, 321–322
phytoene, 595f
phytoferritin, 308
phytohormones, 301
Phytophthora, 373
Picea glauca (white spruce), 513f
Picea sitchensis (Sitka spruce), 47f, 206f
PIF-like proteins (PILs), 431
pigments, 128
pigweed (*Chemopodium album*), 420–421
pin cherry (*Prunus pennsylvanica*), 690
PIN genes, 389
pin mutants, 386f, 478, 479f, 484, 498f
PIN proteins, 480f, 481f
    auxin movement dependent on, 387f
    auxin redirection and, 495–496
    cycling of, 484, 484f
    function, 478, 479f
    leaf primordia and, 406f
pineapple (*Ananas comosus*), 180, 664
pineapple family, 581
pines, monoterpenes in, 321
PINOID gene, 483
pinoid mutant, 483
Pinus contorta (lodgepole pine), 690
Pinus resinosa (conifer), 225f
Piper auritum (Veracruz pepper), 438
piperidine, 330t
pistels, components, 638f
PISTILLATA (PI) genes, 639, 666

pistillata2 (pi2) mutation, 639, 639f
Pisum sativum. see pea
pit membranes, 59
pit pairs, 2, 59
pitcher plant *(Nepenthes alata)*, 110
pith, 223f
pits, 59
Planck's constant, 127
Planck's law, 127
plant tissue analysis, 82–83, 82f
Plantago major (broad-leaved plantain), 237
plants
  body plan, 3f
  design elements, 2
  structure, 2–5, 3f
  transport processes and, 41–51
  unifying principles, 1–2
plasma membrane H$^+$ATPase, 105, 113–114, 115f
plasma membranes, 3f
  cellulose microfibril synthesis at, 353–357
  components, 7f
  function, 6
  plant cell, 5f
  in primary cell walls, 352f
  transport processes, 109f
plasmalemma. *see* plasma membranes
plasmodesmata
  function, 117, 117f, 351
  internal structure, 29
  macromolecular traffic, 29–30
  role in signaling, 249–250
  SHR protein movement and, 395
  structure, 30f
  sucrose diffusion, 240f
  types, 29
plastids
  abscisic acid synthesis in, 594–596
  chloroplasts and, 18
  development of, 561f
  gravity sensors, 490–492
  membranes of, 6
  plasticity, 20–21
plastocyanin (PC), 137, 146
plastohydroquinone (PQH$_2$), 139, 144–145, 146f
plastoquinone (PQ), 137, 144, 145f, 146
PLETHORA (PLT) genes, 402–403, 403f
pluripotent state, 396
pneumatophores, 277
Poa sp., 175f
Poaceae (grass family), 383f, 472
Podocarpus, 364f
Poiseuille, Jean-Léonard-Marie, 42
Poiseuille's equation, 42, 59, 61
poison hemlock *(Conium maculatum)*, 331
polar molecules, 38
polar transport
  of auxin, 484
  chemiosmotic model, 477–480, 478f
  description, 476
  energy for, 476–477
  model, 481f
polarity
  adaxial–abaxial, 408f
  axial, 384
  developmental, 379
  embryogenesis and, 384
  origins of, 382–396
pollen, 276–277, 513–514, 635
pollen tube growth, 27, 362, 513–514, 628–629
polygalacturonic acid, 306, 307f, 357
polymer creep, 370, 371

polymer-trapping model, 239, 240f
Polypodium vulgare (fern), 321–322
polyterpenes, 318
pondweed *(Potamogeton pectinatus)*, 700
Populus tremuloides (quaking aspen), 521, 690
porphobilinogen (PBG), 155f
position-dependent signaling, 385
positive feedback regulation, 518
positive regulators, 522
Potamogeton pectinatus (pondweed), 700
potassium ion channels
  cytosol pH and, 607
  effect of abscisic acid on, 609f
  inwardly rectifying, 103, 103f
  outwardly rectifying, 103
  phytochrome regulation and, 437
  regulation, 603
  Shaker-type, 103f, 110–111
potassium ions
  accumulation, 100
  biochemical function, 75t
  deficiency, 80
  diffusion potential, 98f
  electrostatic bonds and, 306
  guard cell osmoregulation and, 453
  R gene activation and, 340
  root uptake of, 88, 89
  transport, 696f
potassium malate, 307f
potassium permanganate, 572
potato *(Solanum tuberosum)*, 217, 237–238, 245, 691
The Power of Movement in Plants (Darwin and Darwin), 468
PP1 protein, 225, 248
PP2 protein, 225, 248
Pr phytochrome, 419, 423f, 425, 437, 448
precipitation, *38*, 673
preharvest sprouting. *see* vivipary
preprophase bands (PPBs), 25
pressure
  gradients, 231
  units of, 40, 41f
  water potential and, 43–44
pressure-flow model, 231–235, 232f
pressure potential, 44
prevacuolar components (PVCs), 15, 16f
Priestley, Joseph, 130
primary active transport, 105
primary cell walls, 4f
  assembly, 362–363
  composition, 351–353
  description, 2, 351
  fluorescence-labeled, 360f
  illustration, 3f
  Nomarski optic view, 351f
  plant cell, 5f
  structural components, 352f
primary dormancy, 600
primary growth, 2
primary inflorescence meristem, 637
primary plasmodesmata, 29
primary response genes, 430, 502
primary root axes, 87
Primula kewensis, 13f
PRms protein, 248
productivity, water and, 38
proembryos, 384
programmed cell death (PCD), 277, 412, 700. *see also* apoptosis
prokaryotic pathway, 282
prolamellar bodies, 21
proline, 333f

proline-rich protein (PRP), 361, 361t
prolyl hydroxylase, 309f
prometaphase, 26f
prophase, 26f
prophase spindles, 25
Prosopis (mesquite), 86
prosystemin, 336
Proteaceae, 89
proteasomes, 27–28
protection mechanisms, 318–344
protein bodies, 17, 527
protein kinases, 425, 431–432, 680
protein phosphatases (PPs), 425, 431–432, 610
protein-storing vacuoles (PSVs), 16
proteins
  folding, 135f
  import into nuclei, 10f
  lipid bilayers and, 6, 8
  in phospholipid membranes, 6, 8
  processing, 15
  as signaling molecules, 248–249
  structural, 353
  synthesis, 10, 11
protists, 128
protochlorophyllide, 21
protoderm, 390, 392
PROTODERMAL FACTOR 2 (PDF2) GENE, 392, 392f
protofilaments, 23, 24
proton motive force (PMF), 106, 148–149
proton pumps
  blue-light activation, 451–452
  blue-light stimulated, 603f
  H$^+$-pyrophosphatase as, 116
  plasma membrane, 481f
  regulation, 603–604
  tonoplast, 695
protons
  extrusion, 100, 487
  transport, 101
protoplasts, 689
protoporphyrin IX, 155f
provacuoles, 17
Prunus pennsylvanica (pin cherry), 690
Prunus sp., 689
Prunus virginiana (chokecherry), 690
psbD gene, 448
Pseudomonas, 582
Pseudomonas syringae, 690
psoralen, 325f
pterins, 292, 292f, 455
pulvini, 203, 436
pump proteins, 101, 102f, 105, 109f
pumpkin *(Cucurbita maxima)*, 248, 515
pyrethroids, 321
pyrophosphatase, 184f
pyrrolidine, 330t
pyrrolizidine, 330t
pyrrolizidine alkaloids, 331f
Pyrus communis (pear), 514
pyruvate, 256–257, 263–265, 320f
pyruvate decarboxylase, 259
pyruvate dehydrogenase, 273, 273f, 276

Q cycle, 144–145
quaking aspen *(Populus tremuloides)*, 521, 690
qualitative responses, 641
quanta, definition, 126–126, 199
quantitative responses, 641
quantitative trait loci (QTLs), 555
quantitative traits, 603
quantum efficiency, 132

quantum yield, 132
quenching, 151
quercetin, 482f
Quercus robur (English oak), 643t
quiescent center (QC), 88, 384, 403, 404
quinolizidine, 330t
quinone, 145f

R genes, 339f, 340
radial axis, 384
radial patterning, 389–391
radiometric fluorescent dyes, 607n
RAF kinase inhibitors, 663–664
raffinose, 229–230, 240f
Ran protein, 9, 10f
Ranunculus sceleratus, 580
rape (*Brassica napus*, canola), 333, 617
rcn1 (root curling in NPA1) mutant, 483
RD29 gene, 680
reaction center complexes, 130–132, 138f
reaction centers, 137, 144f
reactive oxygen species (ROS), 338, 607, 609f
red drop effect, 133, 133f
red light
    bud formation and, 558
    chloroplast movement and, 27
    effects on morphogenesis, 418
    far-red light and, 435
    photoreception, 655f
    photosystem absorption of, 133, 134f
Reduced height (Rht) mutations, 524
reductive pentose phosphate cycle. *see* Calvin cycle
redwood (*Sequoia sempervirens*), 61, 643t
Regnellidium diphyllum, 580
rehydration, 678
rejuvenation, 643, 644
REPRESSOR OF ga1-3 (RGA), 523, 523f
REPRESSOR OF GA1-3 (RGA) gene, 393
reproductive development, 381
reservatrol, 339
respiration
    bioenergetics, 255f
    critical oxygen pressures, 698
    ethylene production and, 576f
    overview, 253–256, 254f
    photosynthetic yield and, 274–276
    rates, 211, 276
    regulation of, 274f
response regulator proteins, 425
retrograde vesicular transport, 15
revertants, 601
rhamnogalacturonan I (RG I), 357–358, 359f
rhamnogalacturonan II (RG II), 357–358, 359f
rheological properties, 368
Rheus protein, 172
rhizobia, 297, 299t, 302f
Rhizobium, 297
Rhizobium japonicum, 298f
Rhizobium leguminosarum bv. *viciae*, 300
Rhizobium meliloti, 300
rhizomes, 700
Rhizophora, 277
rhizosphere, 86
Rho GTPases, 604, 612
Rhus semialata (sumac), 328f
rib zone, 398, 399f
ribonucleoproteins (RNPs), 249
ribosides, 546, 552, 559
ribosomes, 5f, 10, 12f
ribotides, 546
ribulose-1,5-bisphosphate
    carboxylation of, 161–163, 162f

intermediates, 168
oxygenation, 168
regeneration, 163–164, 163f
ribulose-1,5-bisphosphate carboxylase/oxygenase. *see* rubisco
ribulose-5-phosphate kinase, 165
rice grass (*Echionochloa curs-galli* var. *oryzicola*), 700
rice lamina inclination bioassay, 618, 618f, 619f, 626
rice (*Oryza sativa*)
    aerenchyma, 700
    carbon isotope ratios, 217
    coincidence model, 654f
    elongation growth, 580
    embryogenesis, 382, 383f, 384
    fungal pathogen, 510
    genome, 153
    germinating seeds, 529f
    leaf sheath elongation, 515f
    PHOTOPERIOD SENSITIVITY 5 (Se5) gene, 656
    PHY gene function, 429f
    slender rice 1 mutant, 522
    SLR1-GFP localization, 524
    uptake of ammonium, 89
    yields, 555
Ricinus communis (castor bean), 228t
RICs (*ROP*-interacting CRIB motif-containing proteins), 365–366, 366f
Rieske iron–sulfur protein, 144, 145
rishitin, 340f
RMS1 (IRAMOSUS1) gene, 557
rms4 mutant, 553
RNAs. *see also* messenger RNA (mRNA); microRNA; transfer RNA (tRNA)
    messenger, 11
    pathogenic, 249
    in phloem sap, 229
    as signaling molecules, 249
    transfer (tRNA), 20
root apical meristem (RAM)
    axial polarity and, 384
    cellular organization, 403–405
    development, 402–405
    function, 396
    initials, 404–405
    lateral root initials, 409
    quiescent center, 384
root caps, 3f
    auxin distribution, 494–496
    endocytic recycling in, 15–16
    function, 87, 403
    Golgi apparatus, 14f
    initials, 405
root hairs, 3f, 55f, 56, 88, 581
root-knot nematodes, 548
root pressure, 58
root systems, 86–87, 87–88
root tips, 3f, 7f, 496f, 698
rooting, of cuttings, 632
roots
    adventitious, 477f, 498–499
    anaerobic, 57–58
    anoxic environments and, 698–699
    apical region, 88f
    brassinosteroids and, 627–628
    cation assimilation and, 306–308
    cellular organization, 404f
    cross-section, 3f
    cytokinin and growth of, 554–555, 554f
    development, 393f, 553–555
    distal elongation zones (DEZ), 495
    effect of abscisic acid, 601–602

    ethylene and, 581
    extension, 674
    flooding of, 577–578, 579f
    gravitropic responses, 88
    gravitropic signal transduction, 493, 494f
    gravity sensing by, 491–492
    growth, 484–485, 602f, 674
    initials, 404–405
    ion transport in, 116–119
    iron absorption, 308f
    lateral, 409–410, 410f, 498–499
    maize, 701f
    mycorrhizal fungus-infected, 90f
    nitrate assimilation, 294
    nodule formation, 301, 302f
    oxygen absorption, 698
    oxygen-deficient, 699–700
    shrinkage, 678
    thigmotropism, 493–494
    tissue organization in, 118f
    water absorption, 56–58
    water uptake, 56f
    zeatin riboside transport, 552
    zones, 88, 88f
ROP (*Rho*-related) GTPases, 365, 366f, 702
Roscoe, H. E., 422
rose family (Rosaceae), 472
rose (*Rosa*), 643t
rough endoplasmic reticulum (RER)
    description, 11
    micrograph, 13f
    plant cell, 5f
    protein secretion and, 11–14, 12f, 13f
R2R3-MYB factors, 532
rubisco (ribulose-1,5-bisphosphate carboxylase/oxygenase)
    antisense suppression, 177
    function, 20, 161–163, 163f
    light-dark modulation, 165, 166, 166f
    in sun leaves, 203
Rudbeckia sp. (black-eyed Susan), 327f
runner beans (*Phaseolus coccineus*), 511
rye (*Secale cereale*), 689
ryegrass (*Lolium temulentum*), 512, 533

S states, 142–143
Saccharum officinarum (sugarcane), 175f
salicylhydroxamic acid (SHAM), 272
salicylic acid, 325f
salinity, 86, 692, 693f
salinity stress, 86, 611, 692–698, 694–696
Salix (willow), 690
salt bush (*Atriplex* sp.), 694
salt cedar (*Tamarix* sp.), 694
salt glands, 694
Salt Overly Sensitive (SOS1) antiporter, 112
salt stress. *see* salinity stress
salt-tolerant plants, 86
Samanea, 437, 648
sands, 54t, 85
sap, 598f
sap ascent, 61–62
saponins, 322, 338
saturated fatty acids, 6
Sauromatum guttatum (voodoo lily), 272
SAURs (small auxin up-regulated RNAs), 490, 502
SCARECROW (SCR) gene, 393, 394–395, 394f, 403f
scarecrow (scr) mutant, 491
SCARECROW (SCR) protein, 523
scarlet monkey flower (*Mimulus cardinalis*), 226f

SCF protein, 501–502
schelerenchyma cells, 4f
Schlagnhaufer, C. D., 625
schlereids, 4f
Schoenoplectus lacustris (giant bulrush), 700
scions, 644
sclereids, 222, 364f
scopolamine, 331
SCR^TIR1, 501–502
Se5 protein, 656
seasonal responses, 648, 648f
seawater, 693t
Secale cereale (rye), 658f, 689
secondary active transport, 105, 105f, 106, 107f
secondary cell walls, 2, 3f, 4f, 351, 363
secondary dormancy, 600
secondary growth, 2, 4
secondary inflorescence meristems, 637
secondary metabolites, 315–344
secondary phloem, 223f
secondary plasmodesmata, 29
secondary products. see secondary metabolites
secondary response genes, 430, 502
secondary xylem, 223f
secretory vesicles, 352f
sedoheptulose-1,7-bisphosphate phosphatase, 165
seed coats, 599–600, 600
seed plant categories, 2
seedless fruits, 514
seedling defective mutants, 388
seedlings, 579
  dark-grown, 423f
  etiolated, 417
  hooks, 580
  screening for mutants, 588f
  temperature sensitivity, 688f
seeds
  abortion, 521f
  auxin bound in, 474
  desiccation tolerance, 599
  development, 514
  dispersal, 576
  dormancy, 580, 599–600
  germination, 283, 284f, 419f, 438–439, 629
  hormonal balance in, 598
  maturation, 594, 598, 611
  phases of development, 598
  storage reserve, 599
  temperature tolerance, 689
selective permeability, 42–43
selenium, 75
Senecio (groundsel), 330, 641
senescence
  abscisic acid and, 593–594
  changes involved in, 411–412
  definition, 410, 559
  flowers, 411f
  genetic engineering studies, 581–582
  leaf, 559, 562f, 602
  programmed cell death and, 412
  respiration during, 276
  types of, 410–411
senescence-associated genes (SAGs), 411–412
senescence down-regulated genes (SDGs), 411
sensor proteins, 425
sepals, initiation, 637, 637f
Sequoia sempervirens (redwood), 61, 643t
Sequoiadendron giganteum (giant sequoia), 513f

serine, 304f
serine acetyltransferase, 305
serine/threonine kinases, 425, 586–587
serine/threonine phosphatases, 610
sesquiterpenes, 318, 320f, 339, 340f
sex determination, 513
shade avoidance response, 437–438, 438f
shade habitats, 203
shade plants, 204
shells of hydration, 39
shikimic acid pathway, 322–323, 323f
shoot apical meristem (SAM)
  axial polarity and, 384
  cellular organization, 398–400, 399f
  competence, 658
  cytokinins and, 555, 555f
  development, 396–402
  embryonic, 398
  formation, 396, 397f, 398
  gibberellins and, 535
  growth of, 553f
  intercellular protein movement and, 400
  phase changes, 641–646, 644
shoot apices, 398
SHOOT MERISTEMLESS (STM) gene, 396, 409f
shoot tips, 3f
shoots
  axes of development, 407f
  damage to, 699–700
  development, 553–555
  effect of abscisic acid, 601–602
  effect of brassinosteroids, 626–627
  gravity sensing by, 491
  growth of, 602
  lateral organs of, 409–410
  nitrate assimilation, 294
  seedling development, 417
  water deficit and, 674
short-day plants (SDPs), 533, 649, 650f, 651f, 655
short-distance transport, 235, 241–242, 242f
short-interfering RNAs (siRNAs), 249, 653
short–long-day plants (SLDPs), 650
SHORTROOT (SHR) gene, 393, 394–395, 394f, 403f
siderophores, 307
sieve cells, 222, 237
sieve effect, 201
sieve element–companion cell complex, 235
sieve elements
  angiosperms, 235
  characteristics, 224t
  damaged, 224–225
  description, 222
  loading, 238–239, 238f
  mature, 223–224, 223f
  open sieve plate ports, 224f
  pressure-flow model, 232f
  soluble proteins, 9t
  translocation and, 234f
sieve plates, 4f, 224, 224f, 233
sieve tube elements, 222
sieve tubes, 224
SIG genes, 448
signal peptide sequence, 14
signal recognition particle receptors, 14
signal recognition particles (SRPs), 14
signaling molecules, 247–250, 248–249, 283
Silene, 661–662
silica clays, 85t
silicone, 75t, 80
silt, 54t, 85
silver birch (*Betula verrucosa*), 643–644

silver ions, 575, 584
simple pits, 2, 3f, 4f
Sinapis, 651
Sinapis alba (white mustard), 422
sinapyl alcohol, 325
sinks
  activity, 246
  availability, 246
  photosynthate export and, 244–245
  size, 246
  sources and, 227, 387
Sinorhizobium, 297
siol hydraulic conductivity, 55
sitiens (sit) mutants, 596
Sitka spruce (*Picea sitchensis*), 47f, 206f
size exclusion limits (SELs), 29, 249–250
Skoog, Folke, 496, 497, 545
skotomorphogenesis, 417
Slack, C. R., 173, 176
sleepy 1 (sly1) mutant, 524
slender genes, 522
slender mutants, 523
SLENDER (SLN) gene, 518
SLN1 repressor gene, 524f
SLY protein, 525f
small-flower scorpionweed (*Phacelia dubia*), 600
smooth endoplasmic reticulum (SER), 5f, 11, 12f, 13f, 224
snapdragon (*Antirrhinum*), 636, 638
sodicity, 692
sodium ions
  biochemical function, 75t
  deficiency of, 81
  extrusion, 694
  pumping of, 100
  transport, 696f
  uptake, 694
soil
  acidification, 306
  analysis, 82
  bulk flow of water, 55
  cation exchange capacity, 85
  description, 84–91
  inorganic, 85, 85f
  lime addition to, 83
  pH of, 83f, 84
  physical characteristics, 54t
  salt accumulation, 692–693
  water in, 54–55
  water-saturated, 700
  water supply, 673–674
  waterlogged, 698
Solanaceae (tobacco family), 340f, 472, 579
Solanum tuberosum (potato), 237–238
solar energy, 2, 197, 200f, 201–202. see also light; sunlight
solar spectrum, 127f
solar tracking, 202–203
solute potential, 44
solutes
  chemical potential, 96
  salt stress and, 695
  transport of, 95–121
  transport properties, 108f
solution culture systems, 76f
sorbitol translocation, 229–230
sorghum (*Sorghum bicolor*), 217, 332, 429f, 676f
SOS1 pathway, 695, 697f
source-to-sink ratio, 246
sources, 227, 387
source–sink translocations, 476
southern corn leaf blight, 276

soybean (*Glycine max*)
　Agate variety, 660
　Biloxi variety, 660
　effect of night breaks on, 653f
　growing temperature, 672
　leaf movements, 679f
　monocarpic senescence, 411f
　response to auxin, 486, 486f
　root nodules, 298f
　SAUR accumulation, 490
　source-to-sink pathways, 227
　timekeeping, 652
　U.S. yields, 672t
specific heat, 39
spectrophotometers, 127f
spherosomes. see oil bodies
sphingolipids, 6, 279
sphingosine-1-phosphate (S1P), 608, 608f
sphingosine kinase, 608f
Spinacea oleracea (spinach), 214f, 512, 513, 664, 689
spinach plasma membrane aquaporin (SoPIP2;1), 58f
spinach (*Spinacea oleracea*), 214f, 512, 513, 664, 689
spindle midzones, 25
SPINDLY (SPY) protein, 527
spiral patterns, 406, 406f
spongy mesophyll, 3f, 201
sporeangiophores, 448
sporophytic development, 378–381
spring sunflower (*Balsamorhiza sagittata*), 321f
sprouts, germination, 417
squalene, 621
squash (*Cucurbita pepo*), 239
squash family (Cucurbitaceae), 472
stachyose, 229–230
stamens, initiation, 637, 637f
Stanleya, 75
starch branching enzyme, 183, 184f
starch synthase, 183, 184f
starches
　biosynthesis, 183
　degradation, 183–186, 185f, 256, 527–528
　production, 182–180
　synthesis, 256
　synthesis in chloroplasts, 184t
　transitory, 182
starch–statolith hypothesis, 491
statocytes, 491, 492f
statoliths, 491
steady-state phase, 24
stearic acid, 279t
Steer, M. W., 25
stelar outwardly rectifying K+ channel (SKOR), 119
stele, 88
stem cells, 396, 400–402, 402f, 403f
stems
　auxin action on, 485
　cross-section, 3f
　elongation, 455–456, 456f, 457f, 520f
　growth, 484–485, 512, 519–521, 535–538
sterol-glucosides, 355, 357f
sterols, 6, 279, 319
STM genes, 553
stoma, CAM plants, 180
stomata, 3f
　abscisic acid signaling, 609f
　blue-green reversibility, 461f
　blue-light responses, 450f, 451f, 458–461
　closure, 594, 674–676
　daily changes, 455f

effect of abscisic acid, 601
green-light responses, 461–462
leaf transpiration and, 66–67
micrographs, 67f
opening, 68–69, 449–451
regulation of, 604, 605f
resistance, 213, 213f
sensory transduction cascade, 459f
stomatal complex, 68
stress physiology, 671–705
stress proteins, 502
stress relaxation, 368–369, 370f
stress resistance, 672
stress tolerance, 672, 680–682
stroma, 18, 19f, 135
stroma lamellae, 18, 19f, 135, 137
stromules, 21
structural proteins, 353
strychnine, 331
suberin, 57, 316f, 317
subjective day, 647
subjective night, 647
subsidiary cells, 68
substrate-level phosphorylation, 257
SUC2 symporter, 237, 248f
succinate, 264f, 284f
succinate dehydrogenase, 264, 266
succulent plants, 679, 682–683
sucrose
　apoplastic pathway, 237
　biosynthetic reactions, 188t–189t
　chemical potential, 96–97
　diffusion across membranes, 96–97
　flowering and, 666
　guard cells and, 453–455
　lipid conversion to, 283–285, 284f
　production, 182–180
　short-distance transport, 235
　synthesis, 240f
　synthesis regulation, 191–192, 191f, 192f
　translocation, 229–230
　transporters, 238t
sucrose-6^F phosphate synthase, 192
sucrose synthase, 256, 355, 702
sucrose–H+ symporters, 237–238
sugar beet (*Beta vulgaris*), 87f, 217, 227, 236f, 677f
sugar-nucleotide polysaccharide glycosyl-transferases, 355
sugarcane (*Saccharum officinarum*), 175f, 217, 299, 514
sugars
　carbon isotopes in, 217
　metabolism, 241
　storage, 241
　translocation, 222–223, 229–230
sulfate ions, 304–305, 698
sulfides, 304f
sulfite reductase, 305
sulfites, 304f
5-sulfoglutathione, 304f
sulfur, 74–75, 75t, 79–80, 305–306, 305f
sulfur dioxide, 305
sulfuric acid, 304–305
sumac (*Rhus semialata*), 328f
summer squash (*Cucurbita pepo*), 239, 243f
sun plants, 204, 205f
sunflecks, 201–202
sunflower (*Helianthus annuus*), 207f, 661t, 673f, 676f
sunlight, 202f, 209–210. see also light; solar energy
sunny habitats, 203
supercooling, 690

superoxide dismutase (SOD), 153, 701
superoxide ions, 153, 338
SUPERSENSITIVE TO ABA AND DROUGHT 1 (SAD1) locus, 612
SUPPRESSOR OF CONSTANS 1 (SOC1) gene, 638, 639, 665f, 666, 666f
surface tension, 39, 40f
surfaces, wettable, 40f
suspensors, 384
sycamore maple (*Acer pseudoplatanus*), 643t
symbiosis, 84, 297, 300
symplastic loading, 240f
symplasts, 29, 56, 117–118, 236–237, 240F
symport, 106, 107
symporters, 106, 109f
Synechoccus, 433
Synechocystis, 153
syngamy, 378–379. see also fertilization
systemic acquired resistance (SAR), 338, 339f, 340–341, 341f
systemin, 336, 337f

T-DNA, 548–549, 549f, 556, 557f
Takahashi, Nobutaka, 511
Talbott, L. D., 453
Tamarix sp. (salt cedar), 694
tannins, 328–329, 328f, 329f
tap roots, 87, 87f
tartaric acid, 307f
telophase, 26f
temperature
　chilling, 687–689
　heat-killing, 682t
　leaf adaptations, 684
　membrane stability and, 684
　photosynthesis and, 683
　photosynthetic responses to, 209–211
　respiration rates and, 277
　seedling sensitivity to, 688f
　stress and, 672
　water vapor and, 66t
temperature compensation, 433, 646, 683
tensile strength, 40
terpenes
　biosynthesis, 318–319, 319f, 320f
　formation, 318–319
　pathways, 515
　roles, 319–321
terpenoid pathways, 514–515
tetraterpenes, 318, 319, 320f
Thalassiosira pseudonana, 179
Thalassiosira weissflogii, 179
thermal agitation, 41, 42f
thermal hysteresis proteins (THPS), 691
Thermopsis montana, 198f
Thermosynechococcus elongatus, 144f
thermotolerance, 682, 686
thidiazuron, 546
thigmonastic leaf movements, 435f
thigmotropism, 488, 494
Thimann, Kenneth V., 496, 497
thioglucosidase, 332
Thlasip, 697
Thompson's seedless grapes, 514f
"three finger pattern," 446, 450
thylakoid reactions, 126, 133
thylakoids
　description, 18, 134
　membrane proteins, 135–137
　membranes, 136f
　stacking, 153
Ti plasmid, 400, 471, 548–549, 549f, 553
TIBA (2,3,5-trilodobenzoic acid), 482f, 496
Tidestromia oblongifolia (Arizona hon-

eysweet), 214f, 683, 683f, 684
TIME FOR COFFEE (TIC) gene, 434
timothy grass *(Phleum pratense)*, 19f
tip growth, 364, 364f
TIR1/AFB, 501
TIR1 (transport inhibitor response 1) protein, 501, 502
tissue systems, 5
tml mutations, 557
tobacco family (Solanaceae), 472
tobacco mosaic virus, 582
tobacco *(Nicotiana attenuata)*, 336
tobacco *(Nicotiana tabacum)*, 241
    bud determination states, 645f
    cv. Trapeszond, 660, 662, 663f
    floral evocation in, 644
    guard cells, 69f
    ipt gene, 553, 553f
    leaf veins, 243f
    Maryland Mammoth mutant, 649, 649f
    NTH15 overexpression, 535, 536f
    regulation of flowering, 661t
    root cap cell, 14f
TOC (TIMING of CAB [LCHB]) genes, 433–434, 434, 434f
toc (timing of CAB [LHCB]) mutants, 433
tomato *(Lycopersicon esculentum)*, 238
    chromoplast, 20f
    crown gall, 544f
    effect of insect feeding on, 336
    flacca (flc) mutants, 596
    flower initiation, 406
    heat shock, 685
    hydroponics, 77
    leaf initiation, 406
    never-ripe mutant, 577
    sitiens (sit) mutants, 596
tomato *(Lycopersicon hirsutum)*, 688f
TONNEAU 2 (TON2) gene. *see* FASS gene
tonoplasts, 5f, 109f, 114–116
torpedo stage, 382, 383f, 395f
torus, 59
toxic zones, 82f, 83
toxicity effects, 694
trace elements, 696–698
tracheary elements, 59, 60f, 61f
tracheids, 4f, 59, 60f, 325
trafficking mechanisms, 249–250
trans-cinnamic acid, 323, 324f, 386f
trans-cisternae, 14f
trans-Golgi network (TGN), 14, 14f
transcription, 11, 12f
transfer cells, 226–227
transfer RNA (tRNA), 20, 547
transit peptides, 154
translation, 11, 12f
translocation, 222–227, 227–228, 234f
translocons, 14
transmembrane pathways, 56
transpiration, 38, 66–67, 682–683
transpiration flux, 66f
transpiration ratio, 69
transport. *see also* active transport; passive transport
    definition, 95
    electrogenic, 105
    electroneutral, 105
    kinetic analyses, 107–109
    long-distance, 236
    membranes processes, 101–108
    of nitrates, 110
    of protons, 101
    of solutes, 95–121
    water and, 41–51

transport proteins, 101, 102f
transporters, 108, 269–271
trees, water transport, 63
triacylglycerols, 278–279, 278f, 284f
Trifolium repens (white clover), 650
Trifolium spp. (clover), 222f, 350f, 599
Triodia irritans, 175f
triose phosphates, 186–191, 188t–189t
triple response, 571, 579, 579f
triploid endosperm, 379
triterpenes, 318, 319, 320f
Triticum aestivuum (wheat)
    critical day length in, 649
    gibberellin action on, 512
    nitrogen application, 84
    root systems, 87f
tropane, 330t
trumpet tree *(Cecropia obtusifolia)*, 438
tryptamine (TAM) pathway, 472, 473f
tryptophan, 471, 472f
tryptophan synthase β gene, 474
tubulin, 23
γ-tubulin, 24
γ-tubulin ring complexes (γ-TuRCs), 24
tulip tree poplar *(Liriodendron tulipifera)*, 277
turgor, leaf, 673f
turgor pressure, 247
    aquaporins and, 113
    cellular, 368
    changes in, 47–48
    definition, 37
    guard cells, 68–69
    Hoffler diagrams, 47f
    hydrostatic pressure and, 44
    leaf movement and, 436
    stress relaxation and, 370f
    water potential, 50
Tween 80, 84
two-pore domain channels (TPCs), 111
2,4,5-T, 583
type A *ARR* genes, 564–565, 565f
type B *ARR* genes, 564–565, 565f
Tyria jacobeae (cinnabar moth), 331
tyrosine phosphatases, 610

ubiquinone, 266, 273
ubiquitin, 28, 680
ubiquitin 3 ligases, 432, 433
ubiquitination, 28, 432, 433, 654
UDP-glucose, 191–192
UDP-glucose pyrophosphorylase, 256
ultraviolet (UV) irradiation, 327–328, 605f
Umbelliferae family, 324
umbelliferone, 325f
unsaturated fatty acids, 6
UNUSUAL FLORAL ORGANS (UFO) gene, 400
ureides, 230f, 303–304
URF13 protein, 276
uridine diphosphate D-glucose (UDP-glucose), 355
uronic acids, 354f

vaculolar H$^+$ATPase (V-ATPase), 105, 114–116, 116f
vacuoles, 5f, 17, 112
van Bel, A. J. E., 233
van Niel, C. B., 130
vanillin, 325f
van't Hoff equation, 44, 44f
vaporization, 39
variegation, 406
vascular cambium, 223f

vascular cambria, 3f, 5
vascular elements, 4f, 628f. *see also* phloem; xylem
vascular initials, 405
vascular tissues, 3f, 182–183, 182f, 390
vegetative development, 381
vegetative organogenesis, 405–410
ventral motor cells, 436, 436f
Veracruz pepper *(Piper auritum)*, 438
verbascose, 229–230
vernalization
    Arabidopsis thaliana, 657f
    competence and, 658
    definition, 641
    description, 657–660
    gene expression and, 658–659
    mechanisms of, 659–660
    pathways, 666
VERNALIZATION (VRN) protein, 660
very low-fluence responses (VLFRs), 421–422, 421f, 428
vesicles, 15–16, 16f, 90
vesicular-arbuscular mycorrhizal fungi, 90, 90f
vesicular shuttle model, 15
vessel elements, 59, 60f
vessels, description, 59. *see also* phloem; xylem
Vicia faba. see broad bean
Vigna radiata (mung bean), 269f
Vigna unguiculata (cowpea), 675, 677f
Vinca rosea (periwinkle), 548
violaxanthin, 152, 206–207
trans-violaxanthin, 595f, 596
viroids, pathogenic, 249
viscoelastic properties, 368
vitamin B$_{12}$ (cobalamin), 75–76
vitamin K$_1$, 147
Vitis spp. (grape), 514, 643t
VIVIPAROUS 1 (VP1) transcription factors, 611
viviparous (vp) mutants, 596, 596f, 602f
vivipary, 596, 598–599
volicitin, 334f, 335
von Mohl, H., 453
von Sachs, Julius, 76, 467
voodoo lily *(Sauromatum guttatum)*, 272
VP1 protein, 601
VP14 protein, 596f
VRN (vernalization) protein, 660

wall extensibility, 369
water
    absorption by roots, 56–58
    adhesive properties, 39–40
    in cell walls, 353
    cohesive properties, 39–40
    ecosystem productivity and, 38f
    evaporation via leaves, 62f
    fertilizer leaching into, 74
    flow, 54f
    hydrogen bonding, 38–39
    for irrigation, 693t
    lift to treetops, 61
    loss of, 65–66
    molecular structure, 39f
    in plant cells, 37–52
    polarity of, 38–39
    root systems and, 87f
    stagnant, 698
    status of plants, 49–50
    storage in cacti, 48
    tensile strength, 40–41
    transport processes and, 41–51

uptake pathways, 57f
uptake rates, 56f
xylem transport, 59–64
water balance, 53–72, 676–677
water deficits
    definition, 673
    leaf abscission and, 674
    root extension and, 674
    water flow and, 677–678
water flow, 677–678
water lily (*Nymphoides pelata*), 700
water potential
    ABA and changes in, 597f
    cellular, 369f
    contributing factors, 43–45
    definition, 43
    environmental, 694
    growth and, 602f
    Hoffler diagrams, 47f
    plant function and, 46f
    of plants, 50f
    rate of transport and, 49f
    root growth and, 601–602
    shoot growth and, 601–602
    in soil, 54–55
    water flow and, 54f
    xylem, 58
water saturation, 277
water savers, 672
water spenders, 672
water stress
    abscisic acid and, 674–676
    abscisic acid in response to, 597–598, 597f
    abscisic acid localization in, 598f
    leaf area and, 673–674
    photosynthesis and, 676f
water use efficiency, 69
water vapor, 54f, 65, 65t, 66t
wavelength, 126
waxes
    composition, 316–317, 316f
    deposition, 678
    function, 317
    surface deposits, 317f
Welvischia mirabilis, 4f
Went, Frits, 468, 488
wheat (*Triticum aestivuum*)
    carbon isotope ratios, 217
    critical day length in, 649
    dwarfing, 521, 521f
    gibberellin action on, 512
    nitrogen application, 84
    root systems, 87f

whiptail disease, 82
white birch (*Betula papyrifera*), 690
white brittlebush (*Encelia farinosa*), 684
white clover (*Trifolium repens*), 650
white lupine (*Lupinus albus*), 294
white mustard (*Sinapis alba*), 422
white spruce (*Picea glauca*), 513f
whorls
    ABC model of homeotic genes, 640, 640f
    floral organ initation and, 637
    leaf patterns, 406f
wild beet (*Beta maritima*), 227
willow (*Salix*), 690
wilt
    plant cells, 45
    turgor pressure and, 51
wilting points, 677–678
wilty mutants, 601
winter rye (*Secale cereale*), 658f
winter squash (*Cucurbita maxima*), 224f, 248, 515
witches' broom, 547–548, 548f
wood density, 63
WOODEN LEG (WOL) gene, 392, 393f
woody plants, 643t, 689–690
wound callose, 225
wound healing, 544
WOX genes, 389
Wright, J. P., 233
WUSCHEL (WUS) gene, 398, 401f, 402, 405

Xanthium (cocklebur)
    floral stimulus, 661–662
    flowering, 656, 656f
    graft experiments, 660
    leaf induction, 662f
    nitrate metabolism, 294
    photoperception in, 650
    timekeeping, 652, 652f
xanthophyll cycle, 203, 206–207, 207f, 462
xanthophylls, 152
xanthoxal, 595f, 596
Xenopus oocytes, 108
XTH5 enzyme, 518
XTH family, 371
xtricarboxylic acid (TCA) cycle. *see* citric acid cycle
xylem, 3f, 4f
    cavitation, 63–64
    cell walls, 351
    cytokinin transport, 551
    differentiation, 628
    ion transport in roots, 116–119

    lignan in, 325
    loading, 118–119
    osmotic potential, 58
    parenchyma, 113, 118–119
    root, 88
    secondary, 223f
    tracheary elements, 59
    transport in trees, 63
    vascular differentiation and, 499–500, 499f
    vessel elements, 232f
    walls of, 351
    water potential, 58
    water transport, 59–64, 61–62
xyloglucan, 353, 357, 358f
xyloglucan endotransglucosylase/hydrolases (XETs), 362, 363f
Xylorhiza, 75

YABBI gene family, 407–409, 407
Yang cycle, 573, 573f
yeast transport mutants, 108
yeasts, 27
yellow chameleon protein, 607, 607f
yellow flag (*Iris pseudacorus*), 701
yield thresholds, 369
yielding of cells, 368
Y_z, electron carrier, 143

Z scheme, 139–140, 140f
Zea mays. *see* maize
zeatin, 545–546, 552, 556
zeatin isomerase, 546
zeatin riboside, 559
zeaxanthin, 206–207
    in abscisic acid synthesis, 595f
    absorption spectrum, 458, 458f
    function, 457–461
    quenching by, 152
zeaxanthin epoxidase (ZEP), 595f, 596
zebra plant (*Zebrina pendula*), 66f
zinc, 75t, 81–82
Zinnia, 367f
Zinnia elegans, 628, 628f
ZIP family proteins, 112–113, 697
zonation, 398
zucchini (*Curbita pepo*), 477
zygotes, 378–379
zygotic stage, 382, 384